Concepts of Calculus with Applications

Martha Goshaw

Seminole Community College

Boston San Francisco New York
London Toronto Sydney Tokyo Singapore Madrid
Mexico City Munich Paris Cape Town Hong Kong Montreal

Publisher: Greg Tobin
Editor in Chief: Deirdre Lynch
Senior Acquisitions Editor: William Hoffman
Senior Project Editor: Rachel S. Reeve
Assistant Editor: Susan Whalen
Senior Managing Editor: Karen Wernholm
Senior Production Supervisor: Peggy McMahon
Design Supervision and Cover Design: Barbara Atkinson
Photo Researcher: Beth Anderson
Digital Assets Manager: Marianne Groth
Media Producer: Christine Stavrou
Software Development: Bob Carroll and Ted Hartman
Marketing Coordinator: Caroline Celano
Senior Author Support/Technology Specialist: Joe Vetere
Senior Prepress Supervisor: Caroline Fell
Rights and Permissions Advisor: Dana Weightman
Senior Manufacturing Buyer: Carol Melville
Production Coordination and Composition: Progressive Information Technologies
Text Design and Art Editing: The Davis Group, Inc.
Illustrations: Scientific Illustrators

Cover photo: © David Doubilet, Undersea Images, Inc.

For permission to use copyrighted material, grateful acknowledgment has been made to the copyright holders on page C-1 in the back of the book, which is hereby made part of this copyright page.

Library of Congress Cataloging-in-Publication Data

Goshaw, Martha T.
Concepts of calculus / Martha T. Goshaw. – 1st ed.
p. cm.
Includes index.
ISBN 0-321-32078-6
1. Fractional calculus. 2. Calculus. I. Title.

QA314.G67 2007
515–dc22 2006051567

ISBN-13: 978-0-321-32078-0 ISBN-10: 0-321-32078-6

1 2 3 4 5 6 7 8 9 10—VHP—11 10 09 08 07

To my husband Dave, for his continuing love and support of all my endeavors

In memory of Betty B. Taylor —this one's for you, Mom!

Contents

Unit 0 Function Review

Unit 1 Limits and Derivatives

Unit 2 Applications of the Derivative

Preface

Throughout my teaching career, numerous students have suggested that I should write my own book. While obviously pleased at the compliment, it occurred to me that perhaps I should try to organize my copious lecture notes and various teaching strategies into a written manuscript. I love mathematics, but I love teaching even more. And as I wrote this book, I tried to capture those twin passions in my writing. The book is written in much the same way that I teach my courses. I put my care for both the mathematics and the student into this book. I know where students will stumble, no matter how marvelously the material is presented, and I know how I try to help them. This book is my voice and my experience, which is offered for the benefit and encouragement of the student. It incorporates my insistence on realistic applications of mathematics, my love for using technology to demonstrate or enhance a topic, and my desire to present the material using words that students can understand without sacrificing the mathematics.

This textbook is written to be read. I know most students begin reading a math book by going to the exercises assigned as homework. And I know they work backwards through the exercises to the examples. But this book, while both precise and mathematically correct, is more than a road map to solve exercises that baffle students. This book motivates students through understanding ("What is this question really asking me to do?"), through interpreting ("What does this answer really mean now that I have it?"), and through emphasis of understanding the correct application of rules rather than rote memorization of rules ("Which rule should I use, and how does it work?"). This book guides students to think.

All calculus teachers know that it isn't the calculus that causes problems for students; it's the algebra! My text addresses this specific issue by introducing the algebra skills necessary for real understanding at the beginning of each topic. I have also addressed the cumulative nature of the course by inserting calculus warm-up exercises where appropriate. Both of these features foster student confidence by providing help and guidance at critical points in their mastery of the material.

As you scan the Contents of *Concepts of Calculus with Applications*, you will notice that the book is divided into units and topics, rather than chapters and sections. Why did I choose a topical format for this book?

1. A topical format is how teachers teach a course. We don't teach sections and chapters. We teach topics. Often with chapter/section layouts, material may be spread over two or three sections so that each section stays the same length. Rather, it was my intent to include everything related to one concept within that topic.

2. Applied calculus is cumulative in nature. The sequential numbering emphasizes that each topic builds on the previous ones. In the long run, students will benefit from this progression through the topics because they are continually reinforcing what was previously learned.
3. The audience for applied calculus is composed of students who learn better if the mathematics is presented slowly and in small segments. Topics provide smaller bites of the material, allowing time to digest that material before moving on.
4. By arranging the material in topics, there are added opportunities to reinforce understanding and build confidence. I have included Warm-up Exercises at the beginning of each topic that show students exactly what prerequisite skills are needed in that topic. The Check Your Understanding Exercises within each topic allow students to quickly assess mastery of a concept before continuing to the next concept.

The topics are grouped into five units in a manner somewhat similar to the way tests might be given. The five units cover a *review of functions* (what students are supposed to already know but probably don't remember), *an introduction to the derivative* (what is a derivative, what does it mean, and how do we find it), *applications of the derivative* (graph behavior and optimization of functions), *an introduction to integrals* (what is an integral, how do we find it, and what does it mean), and *a brief discussion of multi-variable calculus* (which some but not all courses in applied calculus have time to cover).

A variety of real-life applications taken from fields such as business, life sciences, economics, and physics appear throughout this book. Many data sets are also provided for those who wish to use regression to find models for the data. The applications give this book the applied flavor that makes the course it supports one of the most useful and interesting that a student will take. Students who ask "What's this stuff good for?" will have their answer.

Key Features

There are several features of this textbook that distinguish it from other books.

- **Topical Format.** As noted before, this textbook is organized into units and topics rather than the traditional chapter/section format. Each topic concludes with a Topic Review, which presents an overview of the major concepts of that topic. Each unit concludes with a Unit Review and Unit Test, plus a Unit Project that applies the skills learned in that unit.
- **Warm-up Exercises.** Each topic begins with a quick review of the algebra skills and/or calculus skills needed for that particular topic. These exercises show students exactly what prerequisite knowledge is expected of them. Answers to the exercises appear at the end of the exercise set. A more complete solution with accompanying algebra review is provided in the Student Study Guide.
- **Check Your Understanding Exercises.** Within each topic, after a concept is introduced, several exercises are provided to allow students to

practice on that concept before moving on to the next concept. Answers to these exercises are provided at the end of that topic. A more complete solution is provided in the Student Study Guide.

- **Calculator Corner.** Throughout the text, where appropriate, Calculator Corners provide additional insight into using the graphing calculator to enhance or further demonstrate a topic. Placing this feature in boxes does not interrupt the flow of the text for those instructors who do not use calculators, and it allows students to see how technology can enhance their learning of the material. The calculator of choice is the TI-84 Plus Silver Edition.
- **Mathematics Corner.** This feature delves more deeply into the mathematics of a concept. Placing the theoretical material in a separate location caters to varying levels of student interest and ability and allows instructors the option of whether or not to present the material.
- **Warning Boxes.** Warning boxes are inserted at those places where errors are commonly made in calculation or in the application of a rule. Pointing out potential danger zones early will help students avoid mistakes later.
- **Tips.** Tips are provided, where appropriate, to guide students in the application of a rule or to give a shorter or more efficient method of performing a calculation. These tips are intended to help students improve on mastering the material. It's like having the professor beside you as you learn.

Strengths of Text

In applied calculus, the concepts are as important as the computation. This book focuses not just on applying a rule but also on understanding the result. It is as important to know what the answer means as how to get to the answer.

Technology is also an integral part of the learning of mathematics. While technology should never be used to replace learning a concept, it can be useful in extending that concept to real problems that don't have such nice answers. Technology can be used to analyze real data and to make informed decisions about that data. A topic on modeling is provided for those instructors who wish to include it.

Another major strength of this text is the exercise sets. Each topic concludes with a set of exercises, which include a number of exercises using real data. The exercises graduate in difficulty and include a variety of applications. At the end of the exercises are several special exercise sets. *Sharpen the Tools* exercises are slightly more challenging and may require more algebra and reasoning on the part of the student. *Calculator Connection* exercises require the use of a graphing calculator for solving. These exercises help students see that not all equations can be solved algebraically. *Communicate* gives students the opportunity to verbalize mathematics. The best way to learn something is to explain it to someone else! The ability to communicate your thoughts is vital in the work force and in society.

Print Supplements

Student's Study Guide and Solutions Manual

10-digit ISBN 0-321-33473-6
13-digit ISBN 978-0-321-33473-2

The *Study Guide* portion, written by Martha Goshaw, gives students tips on how to effectively use this text to study and provides worked out annotated solutions to the Warm-up Exercises and Check Your Understanding Exercises. The remainder of the manual contains completely worked-out solutions with step-by-step annotations for most of the odd-numbered exercises in the exercise sets in the text and all end-of-chapter material.

Online Graphing Calculator Manual

The *Online Graphing Calculator Manual* by Ian Walters of D'Youville College contains keystroke-level instruction for the Texas Instruments TI-83/TI-84 Plus® models. Actual examples and exercises from the text *Concepts of Calculus* demonstrate how to use a graphing calculator. The order of topics in the *Online Graphing Calculator Manual* mirrors that of the text, providing a just-in-time mode of instruction. This manual is available online at http://www.aw-bc.com and as part of MyMathLab.

Instructor's Solutions Manual

10-digit ISBN 0-321-33474-4
13-digit ISBN 978-0-321-33474-9

The *Instructor's Solutions Manual* contains worked-out solutions to all exercises in the exercise sets.

Printed Test Bank

10-digit ISBN 0-321-33471-X
13-digit ISBN 978-0-321-33471-8

The *Printed Test Bank* contains alternate test forms for each Topic test. These can be easily copied and handed out to students.

Media Supplements

Video Lectures on CD with Optional Captioning

10-digit ISBN 0-321-33472-8
13-digit ISBN 978-0-321-33472-5

An engaging team of mathematics instructors present coverage of most Topics from the text. The lecturers' presentations include examples and exercises from the text and support an approach that emphasizes visualization and problem solving. The video lectures for this text are available on CD-ROM, making it easy and convenient for students to watch the videos from a computer at home or on campus. The complete digitized video set, affordable and portable for students, is ideal for distance learning or supplemental instruction.

TestGen®

10-digit ISBN 0-321-33469-8
13-digit ISBN 978-0-321-33469-5

TestGen® enables instructors to build, edit, print, and administer tests using a computerized bank of questions developed to cover all the objectives of the text. TestGen® is algorithmically based, allowing instructors to create multiple but equivalent versions of the same question or test with the click of a button. Instructors can also modify test bank questions or add new questions. Tests can be printed or administered online. The software is available on a dual-platform Windows/Macintosh CD-ROM.

MathXL® http://mathxl.com

MathXL® is a powerful online homework, tutorial, and assessment system. With MathXL, instructors can create, edit, and assign online homework and tests using algorithmically generated exercises correlated at the objective level to the textbook. They can also create and assign their own online exercises and import TestGen® tests for added flexibility. All student work is tracked in MathXL's online gradebook. Students can take chapter tests in MathXL and receive personalized study plans based on their test results. The study plan diagnoses weaknesses and links students directly to tutorial exercises for the objectives they need to study and retest. Students can also access supplemental animations and video clips directly from selected exercises.

MyMathLab® http://mymathlab.com

A complete course online, MyMathLab® offers all the features found in MathXL®, plus the complete textbook online, additional course management features and communication tools, and access to the Addison Wesley Math Tutor Center. Powered by CourseCompass™ (Pearson Education's online teaching and learning environment) and MathXL® (our online homework, tutorial, and assessment system), MyMathLab™ gives you the tools you need to deliver all or a portion of your course online, whether students are in a lab setting or working from home.

Acknowledgments

My gratitude extends to a number of people.

I am extremely grateful to my husband, who has endured countless hours watching me struggle over some facet of this book. His never-ending love and support are always the charge that my batteries need.

Over the last thirty years of teaching I have had countless students tell me at the end of a course that I should write my own book. And so, to all of you out there who will never know I actually did it, thanks for planting the idea. And to Carol Britz, my Pearson Addison-Wesley sales representative, thanks for setting the wheels in motion.

My mother also gets a huge acknowledgement. She was my mentor, my tutor, my shoulder to cry on, and my biggest cheerleader. Though she has been gone for several years, her spirit still guides me.

Without the dedicated team at Pearson Addison-Wesley, there would, of course, be no book and I am extremely grateful to the team that has patiently guided me through this process. Carter Fenton recruited me and provided invaluable encouragement and assistance throughout the early stages of the book's development. Bill Hoffman (Senior Acquisitions Editor), Rachel Reeve (Senior Project Editor), Peggy McMahon (Senior Production Supervisor), Susan Whalen (Assistant Editor), and Caroline Celano (Marketing Assistant) were always there to answer my stupid questions and provide support when I needed it, which was quite often. Thanks also go to Phyllis Hubbard and Becky Anderson (Marketing Managers), Barbara Atkinson (Senior Designer, cover design), and Geri Davis (interior design) for their hard work in designing and promoting the project.

I am deeply indebted to Ian Walters, John Samons, and Angela Spalsbury, the accuracy checkers. Their comments were invaluable and have resulted in what I hope is a textbook as error-free as possible. Thanks also to Paul Lorczak, for his help in reviewing the text for developmental edits and consistency and to Mike Elia, the developmental editor for the first draft. And I would be remiss if I did not extend my heartfelt thanks to the Focus Group who reviewed the preliminary manuscript: Denise Brown, Nancy Eschen, Fred Feldon, Charles Odion, Jane Tanner, and Debbie Woods.

The following reviewers provided many helpful comments and suggestions for the first edition of this text.

Carolyn Autrey, *University of West Georgia*
Camie Bates, *University of Denver*
Viola Lee Bean, *Boise State University*
Susan Caldiero, *Cosumnes River College*
Calandra M. Davis, *Georgia Perimeter College*
Nancy Eschen, *Florida Community College at Jacksonville*
Fred Feldon, *Coastline Community College*
Elaine Fitt, *Bucks County Community College*
David French, *Tidewater Community College*
Sharda Gudehithlu, *Wright College*
JoBeth Horney, *South Plains College*
Steve Howard, *Rose State College*
Jean-Marie Magnier, *Springfield Tech Community College*
Carla Monticelli, *Camden County Community College*
John C. Nardo, *Oglethorpe University*
Irene Palacios, *Grossmont College*
Ariel Ramirez, *Robert Morris College*
Ed Slaminka, *Auburn University*
Angela Spalsbury, *Youngstown State University*
William Summons, *Southwest Tennessee Community College*
Amy Tankersley, *Pellissippi State Technical Community College*
Dr. James Verner, *Simon Fraser University*
Gary Walls, *West Texas A&M University*
Jennifer Walsh, *Daytona Beach Community College*

To the Student

You are about to embark on the study of a fascinating subject, and it is my fervent hope that you will learn to appreciate calculus for how it enables you to solve problems and analyze situations. Calculus is a unique mathematics course because of its cumulative nature. Each concept you learn builds on the previous ones, so you will constantly be reinforcing what you learn. Don't let yourself get behind. Attend class, take good notes, do all the homework, and ask questions when you don't understand something. Learn the rules, but don't get bogged down in them. Always remember that each rule is an important tool that allows you to apply calculus to some specific situation.

Mathematics is a language. Calculus has its own set of vocabulary, so learn those words and terms and be a good calculus communicator. Calculus has its own set of symbols that help you say what you want to do in a more concise fashion. Recognize them and use them.

Remember that the concepts are just as important as the calculations. Yes, it's important to know how to get the answer, but it's just as important to know what that answer means. You'll see me ask you to interpret your answers in every exercise set in this book, but I hope you'll soon learn to appreciate why I do so. It is vital that you know what your answer means once you get it and that you be able to communicate that meaning clearly and concisely.

Mathematics is important because it is useful. Throughout this book you will see hundreds of examples of how calculus is used to analyze various types of data. These examples and exercises expand upon ways for you to apply your knowledge to problems in business, economics, physics, or social science. That's the most important tool you can carry away from this course.

0

Function Review

Starbucks Coffee Shops began in Seattle in 1971. In the mid-1980s, a major effort was undertaken to increase the number of Starbucks locations.

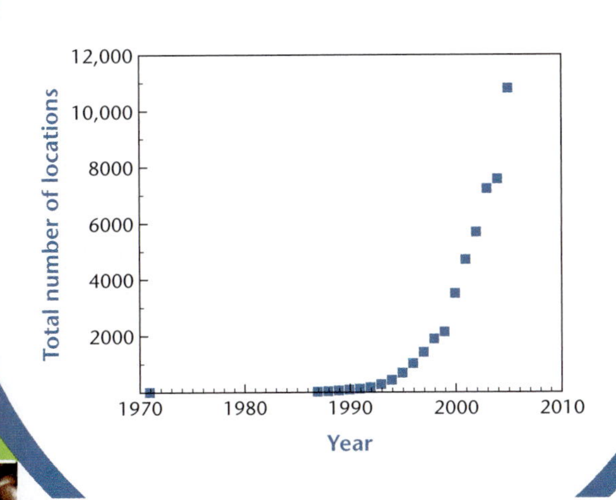

Year	Number of Locations	Year	Number of Locations	Year	Number of Locations
1971	1	1993	272	2000	3,501
1987	17	1994	425	2001	4,709
1988	33	1995	676	2002	5,688
1989	55	1996	1,015	2003	7,225
1990	84	1997	1,412	2004	7,569
1991	116	1998	1,886	2005	10,801
1992	165	1999	2,135		

Data from: www.starbucks.com/aboutus/timeline.asp.

Since 2000, the company has expanded to include international markets. The table and graph above show the growth of the total number of Starbucks locations by the end of each year since the firm was started in 1971. (From 1971 until 1987, there was only the one Seattle location.)

Data such as those shown here are readily available to consumers today. Much of the data can be represented in an algebraic form and displayed graphically. An informed consumer must be able to read tables, draw and interpret graphs, and create models of the data, from which it is possible to infer meaningful predictions about the future of the firm or industry the data describe.

This unit discusses functions, the basic algebraic models and their graphs, and regression techniques for determining algebraic models of data. Because this book assumes you have already successfully completed a college algebra course, the focus of this unit is to *review* the topics from that prerequisite course, not teach the concepts from scratch.

TOPIC 1

Linear and Absolute Value Functions

The concept of function is central to the study of calculus. Consider the following examples.

- Suppose a worker is paid an hourly wage. The amount of money the worker earns each week depends on, or is a function of, the number of hours worked that week.
- A driver travels down an interstate highway at a constant speed of 65 mph. The distance traveled depends on, or is a function of, the number of hours driven.
- Colleges usually charge tuition based on a fixed amount per credit hour. The total tuition paid by a student depends on, or is a function of, the total number of credit hours the student is taking that term.

Each of these examples demonstrates a *linear function*, which is the focus of Topic 1. Before starting your review of linear functions, however, you should complete the Topic 1 Warm-up Exercises to be sure you have the necessary algebra skills.

IMPORTANT TOOLS IN TOPIC 1

- *The basic concept of function*
- *Domain and range of functions*
- *Function notation and evaluating functions*
- *Difference quotients*
- *Linear functions and their graphs*
- *Slope and y-intercept of linear functions*
- *Absolute value functions and their graphs*
- *Transformations to draw graphs*
- *Break-even points and equilibrium points*
- *Straight-line depreciation*

TOPIC 1 WARM-UP EXERCISES

Be sure you can successfully complete the following exercises before starting Topic 1.

1. Given $y = 2x - 5$,

a. Solve for y if $x = 7$.

b. Solve for x if $y = 7$.

c. Graph the equation.

2. Given $3x - 2y = 6$,

a. Find the x-intercept.

b. Find the y-intercept.

c. Graph the equation.

3. State the degree of

a. x^2 **b.** x^4

c. 2 **d.** $3x - 2$

4. Match the type of equation to the actual equation.

linear equation	**a.** $2x^2 - 5 = 7$
quadratic equation	**b.** $2x - 5 = 3$
cubic equation	**c.** $\dfrac{3}{x - 2} = 8$
radical equation	**d.** $x^3 + 1 = 9$
rational equation	**e.** $\sqrt{x} + 5 = 10$

5. Find the slope and y-intercept of each of the following functions.

a. $y = 2x - 3$

b. $2x + 5y = 4$

6. Write the equation of each of the following lines in slope–intercept form.

a. with slope $\frac{1}{2}$ and y-intercept $(0, -4)$

b. passing through $(2, -3)$ and $(5, 2)$

7. Match the line shown in each graph to its equation.

a. $y = 2x - 3$ **b.** $y = -2x - 3$

c. $y = -2x + 3$ **d.** $y = 2x + 3$

8. Solve for y in terms of x.

a. $5x - 3y = 10$ **b.** $4x + 7y = 8$

9. Evaluate each of the following.

a. $|5|$ **b.** $|-3|$

c. $|0|$ **d.** $-|5|$

10. Solve for x.

a. $x - 3 = 0$ **b.** $|x| + 3 = 7$

11. Given $f(x) = 2x^2 - 5x + 6$, evaluate each of the following functions.

a. $f(3)$ **b.** $f(a)$ **c.** $f(a + h)$

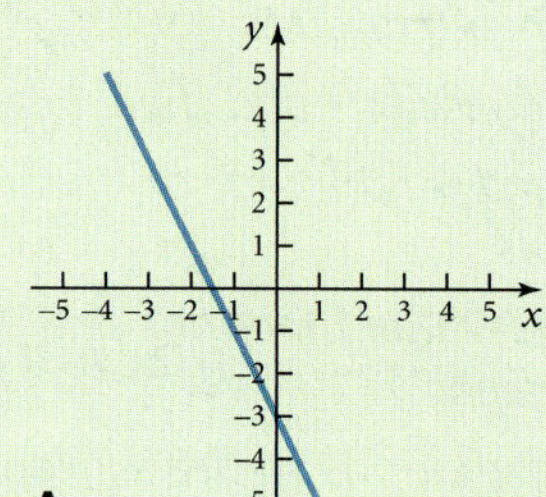
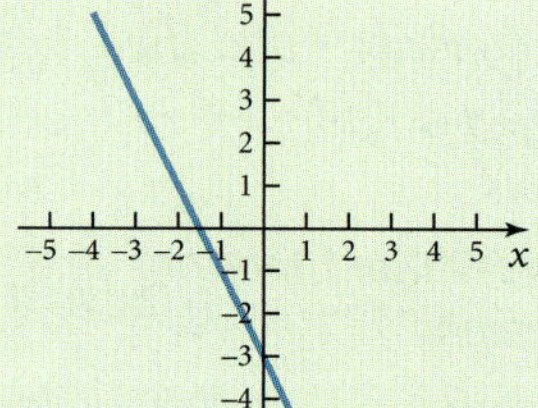

A.

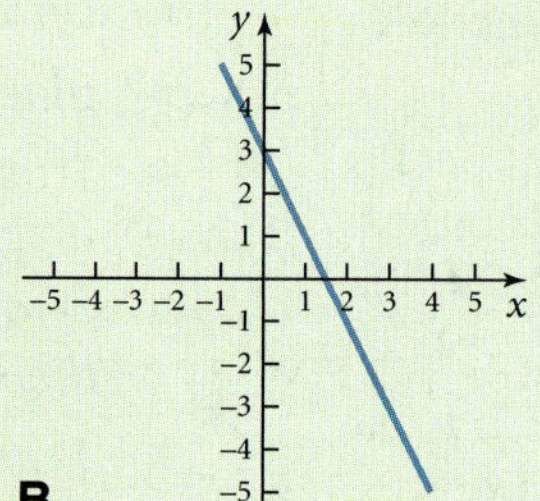
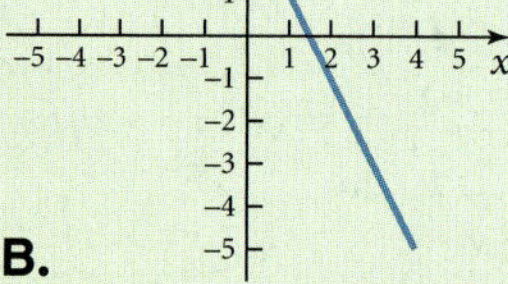

B.

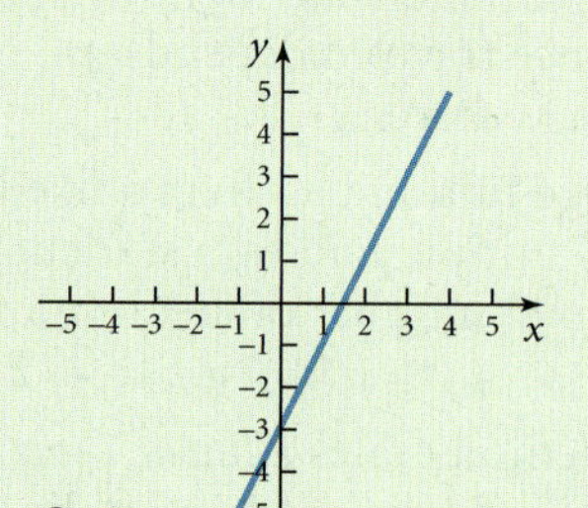

C.

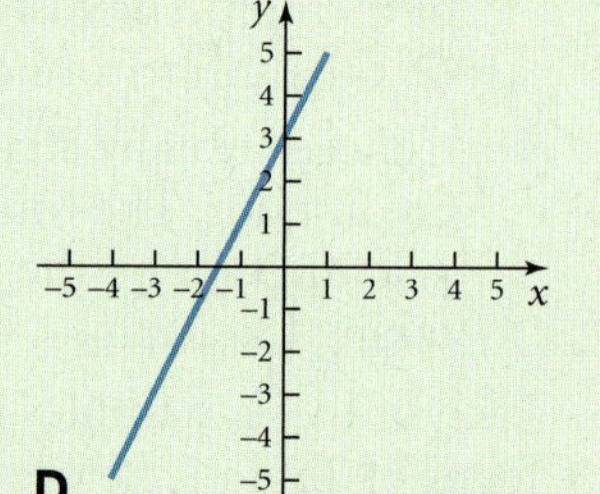

D.

Answers to Warm-up Exercises

1. a. 9 **b.** 6

c.

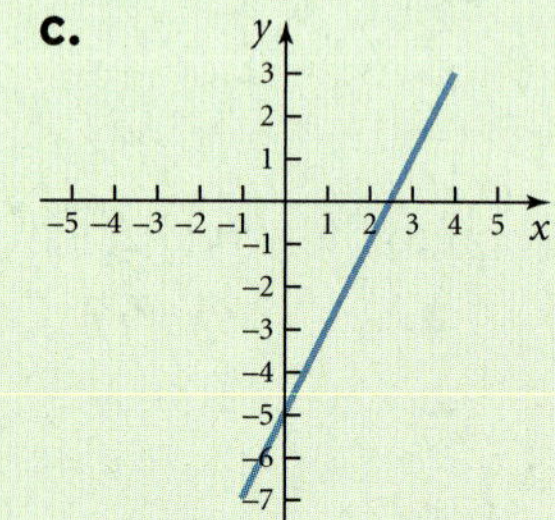

2. a. $(2, 0)$ **b.** $(0, -3)$

c.

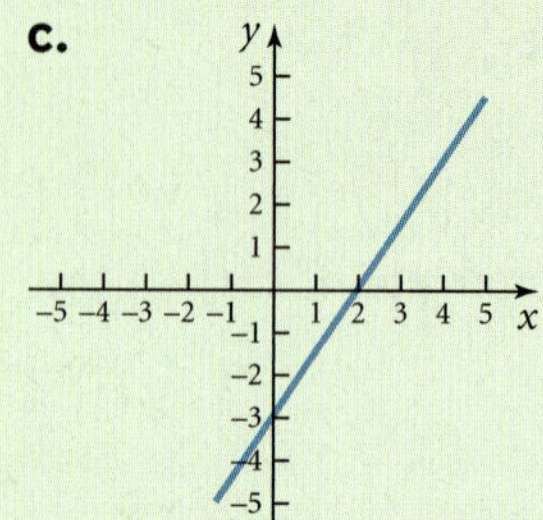

3. a. 2 **b.** 4 **c.** 0 **d.** 1

4. a. quadratic **b.** linear

c. rational **d.** cubic

e. radical

5. a. slope $= 2$, $(0, -3)$

b. slope $= -\frac{2}{5}$, $(0, \frac{4}{5})$

6. a. $y = \frac{1}{2}x - 4$ **b.** $y = \frac{5}{3}x - \frac{19}{3}$

7. a. C **b.** A **c.** B **d.** D

8. a. $y = \frac{5}{3}x - \frac{10}{3}$ **b.** $y = -\frac{4}{7}x + \frac{8}{7}$

9. a. 5 **b.** 3 **c.** 0 **d.** -5

10. a. $x = 3$

b. $4, -4$

11. a. $f(3) = 9$

b. $f(a) = 2a^2 - 5a + 6$

c. $f(a + h) = 2(a + h)^2 - 5(a + h) + 6$ or $2a^2 + 4ah + h^2 - 5a - 5h + 6$

Functions

Calculus is built on the concept of the function. A review of some basic functions provides a more solid foundation for learning what calculus is all about.

Definition: A **function** is a correspondence between two sets that assigns to each member of the first set (the **domain**) one and only one element from the second set (the **range**).

Suppose a worker is paid an hourly wage. The amount of money the worker earns each week depends on the number of hours worked that week. In other words, the amount of money earned each week is a *function* of the number of hours worked. This simple example illustrates three concepts:

- In this function, the amount of money earned each week depends on the number of hours worked that week.
- The domain is the set of all possible numbers of hours worked that week.
- The range is the set of all possible amounts of money earned during that week.

Based on what you have just read, consider the following correspondences between sets. Which are functions? Why or why not?

Example 1:

Set 1: the students enrolled in your course
Set 2: each student's Social Security number

The domain is the set of all students in the class (set 1). The range is the set of social security numbers of each student (set 2). This example represents a function because each student in the class (each member of set 1) has only one Social Security number (set 2). The function here is the correspondence between a student and his or her unique Social Security number. ■

Example 2:

Set A: the students enrolled in your class
Set B: the number of credit hours each student is taking this term

The domain is set of all students in the class (set A). The range is the set of number of credit hours (set B). This example represents a function because each student can take only one particular number of credit hours this term. Suppose two students from set A are both taking 12 credit hours this term. The definition of function is still fulfilled because each individual student from set A is still paired with only one number of credit hours from set B. ■

Example 3:

Set x: the students enrolled in your class
Set y: the names of the professors teaching the classes each student is taking this term

The domain is the set of all students in the class (set x). The range is the set of names of each of the student's professors this term (set y). Because each student could be paired with more than one professor, this example does not represent a function. It is quite probable that at least one student will have more than one professor this term. ■

A function may be described in several ways:

- Using a set of ordered pairs containing each member of the domain and its corresponding range value
- Stating the equation relating the domain and range
- Drawing the graphical representation of the relation between the domain and range on a coordinate plane

Functions as Sets of Ordered Pairs

In algebra, you learned that an **ordered pair** looks like

$$(x, y)$$

where x is an element of the domain and y is an element of the range. Thus, the order of the two terms is important, which is why (x, y) is called an *ordered* pair. The pair (2, 5) is not the same as the pair (5, 2).

A **relation** is *any* set of ordered pairs (x, y). Examples of relations are $\{(2,3),(5,4),(3,-2),(6,7)\}$ and $\{(2,3),(5,4),(2,-2),(6,7)\}$. To determine if a relation describes a function, check to see that no first term (or domain element or x value) is paired with more than one second term (or range element or y value). If a first term is repeated in a subsequent ordered pair with a different second term, the relation is not a function.

Example 4: $\{(2,3),(5,4),(3,-2),(6,7)\}$ *describes a function.*

The domain is {2, 3, 5, 6}, and the range is $\{-2,3,4,7\}$. Because each domain element is paired with only one range element, this set of ordered pairs is a function. ■

Example 5: $\{(2,3),(5,4),(2,-2),(6,7)\}$ *does not describe a function.*

The domain is {2, 5, 6}, and the range is $\{-2,3,4,7\}$. Because the domain element 2 is paired with both 3 and -2 in the range, this set of ordered pairs does not describe a function. ■

Functions as Equations

Many functions are described using an equation with two variables to show the relationship between the domain and the range. The domain of the function, usually represented by x in the equation, is referred to as the **independent variable.** The range, usually represented by y in the equation, is the **dependent variable** because the range value depends on the domain value. The domain of the function consists of all possible values for x. We assume the domain is the set of all real numbers unless the equation contains a term that would restrict the choice of x. For instance, if the equation contains a $\sqrt{x}$ term, we could not choose a negative x value, because $\sqrt{x}$

is only defined for $x \geq 0$. If the equation contained a rational term such as $\frac{1}{x}$, we could not choose 0 for x because fractions are undefined if the denominator has a value of 0. We could, however, choose any other positive or negative value for x.

To determine if an equation of two variables represents a function,

1. Choose an arbitrary acceptable value for the independent variable.
2. Substitute that value for the independent variable in the equation.
3. Solve the equation for all possible values of the dependent variable.
4. If there is more than one possible value for the dependent variable, the equation does not represent a function.
5. If there is only one possible value for the dependent variable, repeat steps 1 through 3 with a different value of the independent variable.

If only one possible solution is obtained for the dependent variable for *any* value of the independent variable, the equation represents a function. If more than one possible solution is obtained for the dependent variable for at least one value of the independent variable, the equation does not represent a function.

Example 6: Identify the acceptable values for x and determine whether or not the equation represents a function. State the domain of each function.

a. $y = 2x - 3$ **b.** $y = 4 - x^2$

c. $x + \sqrt{y} = 0$ **d.** $x = 4 - y^2$

Solution:

a. For $y = 2x - 3$, any real number is an acceptable value for x because any number can be multiplied by 2. Choosing $x = 5$ yields $y = 2(5) - 3 = 7$. You should see that for any real number, multiplying by 2 and subtracting 3 will always yield one answer. Thus, because one value of x always produces one value for y, this equation describes a function. The domain of the function is the set of all real numbers.

b. For $y = 4 - x^2$, any real number is an acceptable value for x because any number can be squared. Choosing $x = 5$ yields $y = 4 - (5)^2 = -21$. You should see that for any real number, squaring and subtracting from 4 will always yield one answer. Because one value of x always produces one value for y, this equation describes a function. The domain of the function is the set of all real numbers.

c. If $x + \sqrt{y} = 0$, then $x = -\sqrt{y}$. Because of the square root, we know that y cannot have a negative value. For zero or any positive value of y, the value of $\sqrt{y}$ is always zero or some positive number, meaning that $-\sqrt{y}$ is always zero or some negative value. Thus x can be replaced by any nonpositive numbers (nonpositive means either negative or zero). Choosing $x = -9$ yields $-9 = -\sqrt{y}$, so $y = 81$. You should see that any other nonnegative value for x will also produce one value for y. Because one value of x produces one value for y, this equation describes a function. The domain is all $x \leq 0$.

d. For $x = 4 - y^2$, the value of y^2 is always zero or some positive number, so the value of $4 - y^2$ is 4 or some number less than 4. Thus, the acceptable values for x are $x \leq 4$. Choosing $x = -5$ yields $-5 = 4 - y^2$, so $y^2 = 9$, which means that $y = 3$ or -3. You should see that any other x value less than 4 also produces two values for y. Because at least one value of x produces more than y value, this equation does *not* describe a function. ■

Tip: Two types of equations that do not represent functions are equations that have either a y^2 term or a $|y|$ term. For some number $k > 0$,

1. Solving $y^2 = k$ will yield both $y = \sqrt{k}$ and $y = -\sqrt{k}$ as solutions.
2. Solving $|y| = k$ will yield both $y = k$ and $y = -k$ as solutions.

Because two solutions are obtained for the dependent variable, the function definition is violated.

Function Notation

A special notation denotes an equation that represents a function. Recall from Example 6a that $y = 2x - 3$ represented a function. We can replace y by $f(x)$ and write $f(x) = 2x - 3$ to indicate that the given equation is a function. The function $f(x) = 2x - 3$ is read "f of x equals $2x$ minus 3." The notation $f(x)$ means the same thing as y and tells us that the equation to follow describes a function. The notation $f(5)$, read "f of five," means "evaluate the given function for $x = 5$." Thus, $f(5)$ indicates that 5 is the replacement value for the independent variable. For the above function,

$$f(5) = 2(5) - 3 = 7$$

Thus, $f(5)$ is simply another way of denoting the y value of a function when $x = 5$.

Warning! ***The notation f(5) does not imply multiplication!*** Rather, it means to evaluate $f(x)$ by replacing every x in the definition of the function by 5.

Calculator Corner 1.1

Use your calculator to evaluate $f(x) = -2x^3 + 5x^2 - 4x + 3$ if $x = -2$, $x = 1$ and $x = 3.6$.

Example 7: Given $f(x) = 2x^2 - 5x + 1$, evaluate each of the following functions.

a. $f(3)$ **b.** $f(-1)$ **c.** $f(a)$ **d.** $f(a + h)$

Solution: To evaluate $f(x)$ for a specific value, replace the independent variable in the function by the number or expression within parentheses. For $f(x) = 2x^2 - 5x + 1$,

a. $f(3) = 2(3)^2 - 5(3) + 1 = 4$

b. $f(-1) = 2(-1)^2 - 5(-1) + 1 = 8$

c. $f(a) = 2a^2 - 5a + 1$

d. $f(a + h) = 2(a + h)^2 - 5(a + h) + 1$
$= 2(a^2 + 2ah + h^2) - 5(a + h) + 1$
$= 2a^2 + 4ah + 2h^2 - 5a - 5h + 1$

■

Difference Quotients

An expression used extensively in calculus is the **difference quotient**, which has the form $\frac{f(a+h)-f(a)}{h}$.

Simplifying a difference quotient is easiest using this four-step process:

Step 1. Find $f(a)$.
Step 2. Find $f(a+h)$.
Step 3. Find $f(a+h)-f(a)$.
Step 4. Evaluate $\frac{f(a+h)-f(a)}{h}$.

Example 8: Given $f(x) = 5x - 2$, simplify the difference quotient.

Solution: For $f(x) = 5x - 2$, we know that

1. $f(a) = 5a - 2$
2. $f(a+h) = 5(a+h) - 2$

Then

3. $f(a+h) - f(a) = [5(a+h) - 2] - (5a - 2)$
$= 5a + 5h - 2 - 5a + 2$ — Distribute
$= 5h$ — Combine like terms

4. Substituting $f(a+h) - f(a) = 5h$ into the difference quotient and simplifying yields

$$\frac{f(a+h)-f(a)}{h} = \frac{5h}{h} = 5$$

■

Example 9: Given $f(x) = x^2 - 2x + 3$, simplify the difference quotient.

Solution: For $f(x) = x^2 - 2x + 3$, we know that

1. $f(a) = a^2 - 2a + 3$
2. $f(a+h) = (a+h)^2 - 2(a+h) + 3$

Then

3. $f(a+h) - f(a) = [(a+h)^2 - 2(a+h) + 3] - (a^2 - 2a + 3)$
$= a^2 + 2ah + h^2 - 2a - 2h + 3 - a^2 + 2a - 3$ — Expand and distribute
$= 2ah + h^2 - 2h$ — Combine like terms

4. Substituting $f(a+h) - f(a) = 2ah + h^2 - 2h$ into the difference quotient and simplifying yields

$$\frac{f(a+h)-f(a)}{h} = \frac{2ah + h^2 - 2h}{h}$$

$$= \frac{h(2a + h - 2)}{h}$$ — Factor h from numerator

$$= 2a + h - 2$$ — Divide h from numerator and denominator

■

Difference quotients are a very important tool in the development of the derivative. Some facility in evaluating and simplifying difference quotients will lead to a better understanding of the derivative.

Functions as Graphs

Any equation in two variables can be represented graphically by some curve in the xy coordinate plane. How do you know by looking at the graphical representation of an equation that the equation is a function? You can answer that question with the Vertical Line Test.

> **Vertical Line Test**
> If every vertical line drawn through any point on the graph of a relation intersects the graph only once, the equation represented by that graph represents a function.

Example 10: Which of the following graphs (Figure 1.1) represent a function?

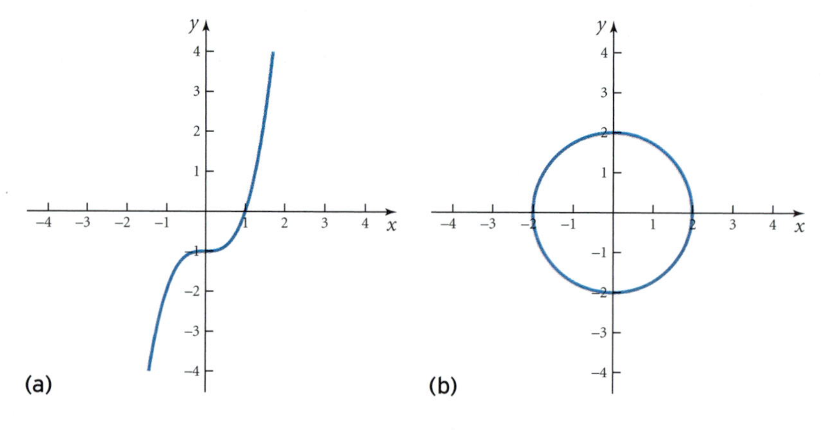

Figure 1.1

Solution: The curve in graph a describes a function, because any vertical line would intersect the graph at only one point. In other words, no x value is paired with more than one y value. (See Figure 1.2.)

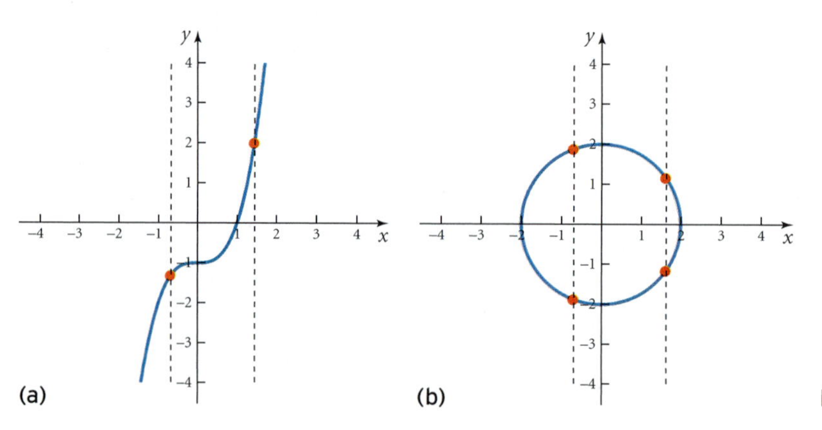

Figure 1.2

The curve in graph b does not represent a function, because there is at least one vertical line that intersects the graph in more than one point. In other words, there is at least one x value that is paired with more than one y value. ■

Calculator Corner 1.2

Use your graphing calculator to draw the graphs of the following functions. Use the standard window of $[-10, 10]$ for both x and y.

1. $f(x) = x^3$
2. $f(x) = 3x - 2$
3. $f(x) = 4 - x^2$

Check Your Understanding 1.1

For Questions 1 through 6, state whether a function is being described. Explain your answer. For the functions, state the domain.

1. $\{(2, 3), (5, -2), (7, 3), (4, -2)\}$

2. $y = 3x + 6$

3.

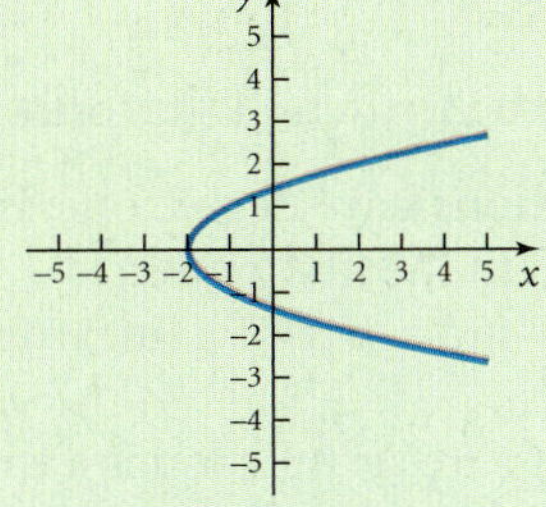

4. $(2, 3), (4, -1), (5, 0), (4, -6), (3, 2)$

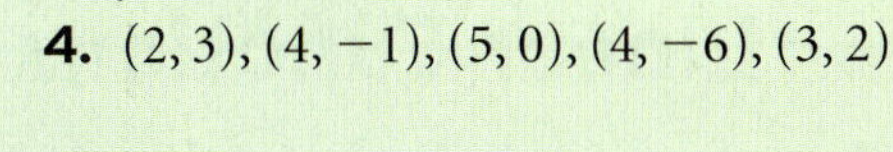

5. $x^2 + y^2 = 4$

6.

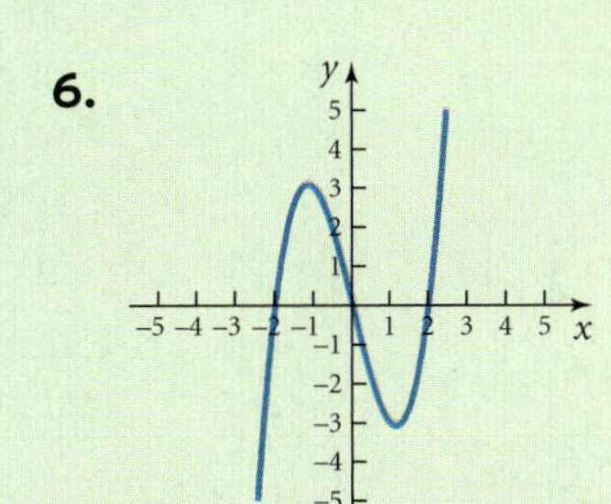

7. Evaluate the following functions at the indicated value.

a. $h(t) = -3t^2 + 8t + 5$; evaluate $h(3)$.

b. $C(p) = 120p + 5000$; evaluate C(75).

c. $k(a) = \dfrac{120 - 5a}{a^2 + 3}$; evaluate k(8).

8. Given $f(x) = x^2 + 2x + 1$, simplify the difference quotient.

The Basic Algebraic Functions

Functions may be classified as algebraic or nonalgebraic.

- **Algebraic functions** include the polynomial functions (linear, quadratic, cubic, and higher), radical functions, and rational functions. Algebraic functions are constructed using the algebraic operations of addition, subtraction, multiplication, division, powers, and roots with polynomials.
- **Nonalgebraic functions**, sometimes referred to as **transcendental functions,** include the exponential and logarithmic functions and the trigonometric functions. In this book, we consider only exponential and logarithmic functions.

The basic algebraic functions with which you should be familiar are the linear, quadratic, cubic, square root, and rational functions. You should be able to

- Relate each function with its particular graph.
- Identify the domains of each function.
- Sketch the graph of each function without using a calculator.

These functions and indicated skills should already be part of your Algebra Tool Kit. Nevertheless, a brief review of the basic algebraic functions is provided here to sharpen those tools. Exponential and logarithmic functions will be discussed in Topic 12.

Linear Functions

Linear functions can be written in the form $f(x) = mx + b$.

- **Linear functions** have degree one, meaning that the independent variable is raised to the first power.
- The domain and range of a linear function are always the set of real numbers.
- The graph of a linear function is a straight line. Recall that m is the **slope** of the line and $(0, b)$ is the ***y*-intercept** of the line, which is the point where the line passes through the vertical axis. The slope represents the steepness of the line, measured by the ratio of the change in y to the change in x.

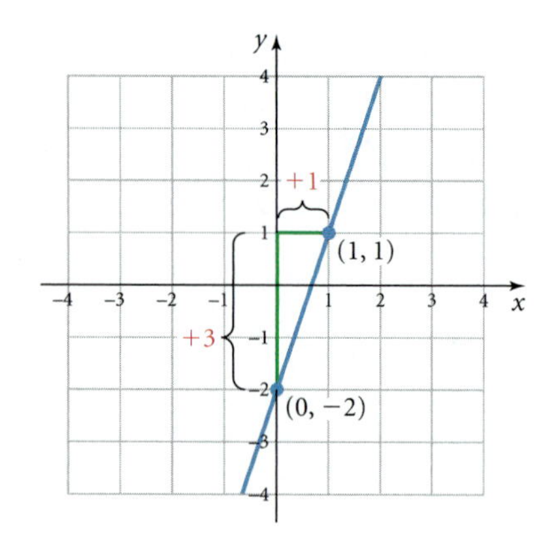

Figure 1.3

Example 11: State the slope and y-intercept and sketch the graph of $f(x) = 3x - 2$.

$$f(x) = mx + b$$
$$\downarrow \quad \downarrow$$
$$f(x) = 3x - 2$$

Remember that $3x - 2 = 3x + (-2)$

This function is expressed in the slope–intercept form $y = mx + b$, so the slope is 3 and the y-intercept is $(0, -2)$. The y-intercept is a point on the graph, so it is expressed as an ordered pair.

To draw the graph of the line, first plot the y-intercept. Then use the slope to find a second point by counting from the y-intercept up 3 units and right 1 unit (Figure 1.3). ■

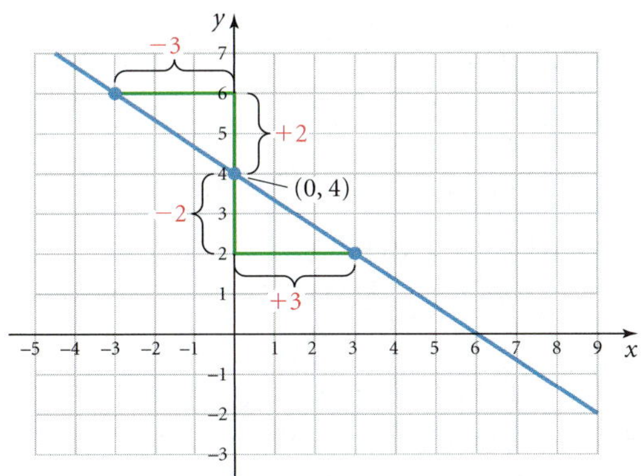

Figure 1.4

Example 12: State the slope and y-intercept and sketch the graph of $2x + 3y = 12$.

Solution: This equation is not expressed in slope–intercept form, so we must first put it in that form. Solving for y gives . The slope is $m = -\frac{2}{3}$, and the y-intercept is $(0, 4)$. To draw the graph of the line, first plot the y-intercept and then use the slope to find a second point by counting from the y-intercept down 2 units and right 3 units (or up 2 units and left 3 units). (See Figure 1.4.) ■

Calculator Corner 1.3

The graph of the equation $y = mx + b$ is a straight line with slope m. For any equation of this form,

1. The line rises if $m > 0$ and falls if $m < 0$, as we move from left to right. To verify, graph $y_1 = 2x$ and $y_2 = -2x$.
2. The larger the m value, the steeper the line will be. To verify, graph $y_1 = x$, $y_2 = 3x$, and $y_3 = \frac{1}{2}x$.

Hint: Don't forget to enclose the $\frac{1}{2}$ in parentheses when you enter it in your calculator.

3. How many x-intercepts may the graph of a linear function have?
   Note: The graph of a linear function will always have one x-intercept. The only exception is a linear function of the form $f(x) = k$, which is a horizontal line with slope zero and no x-intercept.

MATHEMATICS CORNER 1.1

Graphing Linear Equations Using Both Intercepts

Linear functions of the form $ax + by = c$ can also be graphed by determining the x-intercept and the y-intercept. For example, you were asked to graph $2x + 3y = 12$ in Example 12. The x-intercept is found by letting $y = c$, yielding $(6, 0)$. The y-intercept is found by letting $x = c$, yielding $(0, 4)$. Plot the two intercepts and draw the line between them, which gives the same graph as that obtained in Example 12.

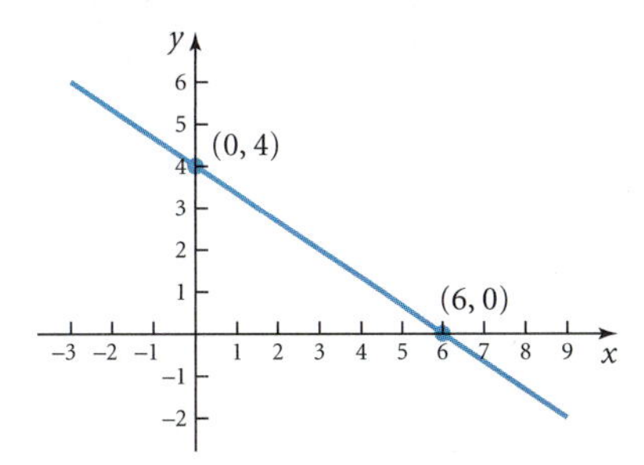

You can then use the two intercepts to find the slope of the line.

$$m = \frac{y_2 - y_1}{x_2 - x_1} = \frac{4 - 0}{0 - 6} = \frac{4}{-6} = -\frac{2}{3}$$

Example 13: The relationship between your height and your shoe size is approximately linear. For a group of female calculus students, suppose the relationship between their height and shoe size is given by $s = 0.48h - 23.8$, where s is women's shoe size and h is height in inches.

a. State a reasonable domain for the function.

b. Estimate the shoe size of a woman who is 67 inches tall.

c. Estimate the height of a woman who wears a size 7 shoe.

Solution:

a. The domain is the height of the women. Because women are usually between 4 feet 10 inches and 6 feet 2 inches, a reasonable domain might be heights between 58 inches and 74 inches: $58 \leq h \leq 74$.

b. If a woman is 67 inches tall, her shoe size is $s(67) = 0.48(67) - 23.8 = 8.36$. A woman who is 67 inches tall probably wears a size $8\frac{1}{2}$ shoe.

c. To estimate the height of a woman wearing a size 7 shoe, or $s = 7$, we solve $7 = 0.48h - 23.87$ for h and get $h = 64.16$. A woman who wears a size 7 shoe would be approximately 64 inches tall, or 5 feet 4 inches tall. ■

Absolute Value Functions

For any real number n, the **absolute value** of that number, denoted $|n|$, is defined as *the distance from* n *to zero on the number line.* Absolute value measures only the distance from zero, not the direction. Because distance cannot be negative, the absolute value of a number will always be a nonnegative number.

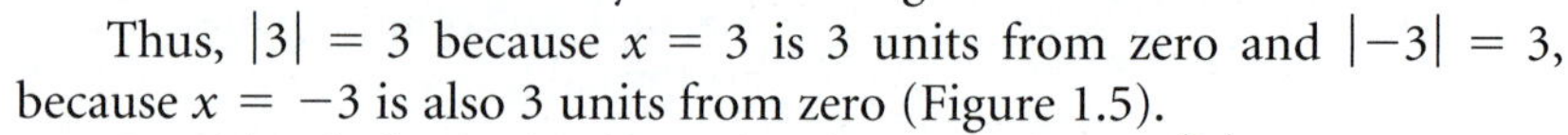

Thus, $|3| = 3$ because $x = 3$ is 3 units from zero and $|-3| = 3$, because $x = -3$ is also 3 units from zero (Figure 1.5).

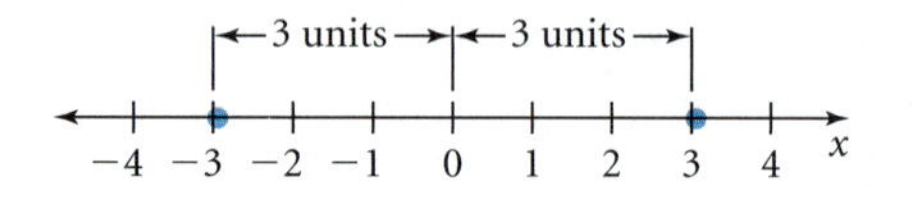

Figure 1.5

Consider the basic absolute value function, $f(x) = |x|$. The domain is any real number. The range must be nonnegative because $|x| \geq 0$ for all values of x. What does the graph look like? (See Figure 1.6.)

x	$f(x)$
-3	$\lvert -3 \rvert = 3$
-2	$\lvert -2 \rvert = 2$
-1	$\lvert -1 \rvert = 1$
0	$\lvert 0 \rvert = 0$
1	$\lvert 1 \rvert = 1$
2	$\lvert 2 \rvert = 2$
3	$\lvert 3 \rvert = 3$

(a)

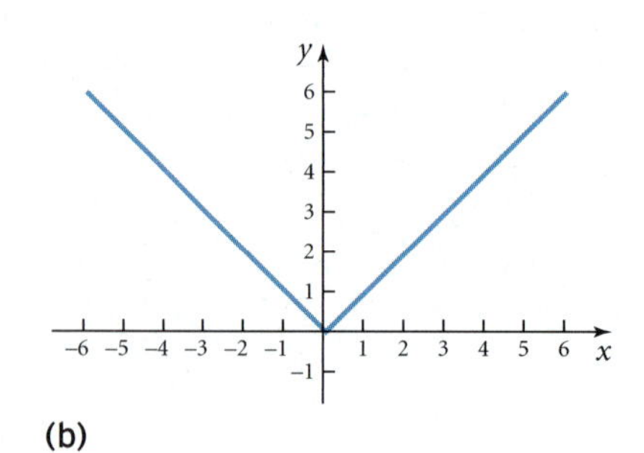

(b)

Figure 1.6

Look carefully at the graph and you will see that the graph of $f(x) = |x|$ is essentially the graph of $f(x) = -x$ for $x \leq 0$ and $f(x) = x$ for $x \geq 0$. The graph has a corner point at $(0, 0)$ where the two lines meet. The domain is all real numbers. The range is $y \geq 0$ because absolute value is never a negative number.

Transformations

The basic graph of any function can be moved to a different location in the coordinate plane by using **transformations**. The three basic transformations are

- Horizontal translations
- Vertical translations
- Reflections

The ability to recognize and understand these three basic transformations provides a quick tool for sketching and analyzing graphs of functions.

Horizontal Translation
Given the graph of $f(x)$,

$f(x - h)$ shifts the graph of $f(x)$ h units to the right.
$f(x + h)$ shifts the graph of $f(x)$ h units to the left.

Example 14: The graphs of $f(x) = |x|$, $f(x) = |x + 3|$, and $f(x) = |x - 3|$ are shown in Figure 1.7. The graph of $f(x) = |x + 3|$ is the graph of $f(x) = |x|$ shifted to the left 3 units, so that the corner point is at $(-3, 0)$. The graph of $f(x) = |x - 3|$ is the graph of $f(x) = |x|$ shifted to the right 3 units so that the corner point is at $(3, 0)$. The basic shape of all three graphs is exactly the same, but the corner points have been moved due to the translation. All three functions have $y \geq 0$ as the range.

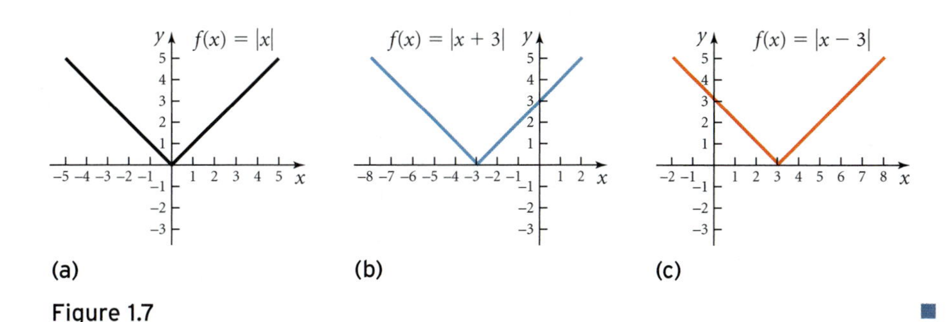

Figure 1.7

Vertical Translation
Given the graph of $f(x)$,

$f(x) + k$ shifts the graph of $f(x)$ up k units.
$f(x) - k$ shifts the graph of $f(x)$ down k units.

Example 15: The graphs of $f(x) = |x|$, $f(x) = |x| + 3$, and $f(x) = |x| - 3$ are shown in Figure 1.8. The graph of $f(x) = |x| + 3$ is the graph of $f(x) = |x|$ shifted up 3 units, moving the corner point to $(0, 3)$ and making the range $y \geq 3$. The graph of $f(x) = |x| - 3$ is the graph of $f(x) = |x|$ shifted down 3 units, moving the corner point to $(0, -3)$ and making the range $y \geq -3$. The shape of the three graphs is exactly the same; only the corner points are different. The three graphs also have different ranges.

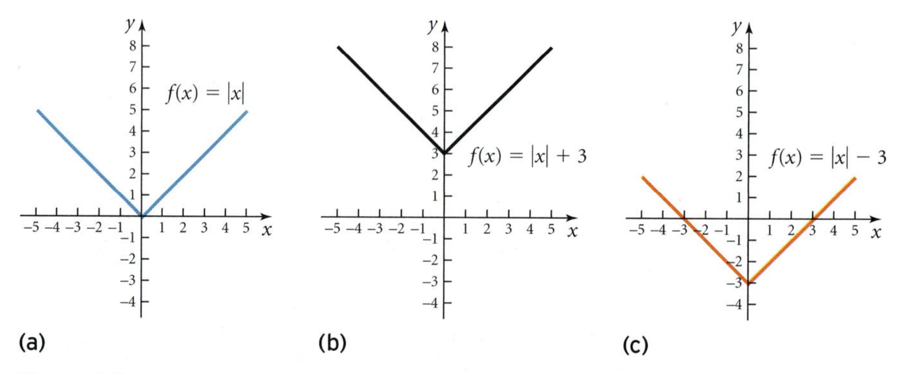

Figure 1.8

Reflection

Given the graph of $f(x)$,

$-f(x)$ reflects the graph of $f(x)$ about the x-axis.

$f(-x)$ reflects the graph of $f(x)$ about the y-axis.

Example 16: The graphs of $f(x) = |x|$ and $f(x) = -|x|$ are shown in Figure 1.9. The graph of $f(x) = -|x|$ is the graph of $f(x) = |x|$ inverted, or reflected, about the x-axis. The location of the corner point is unchanged, but the range changes to $y \leq 0$.

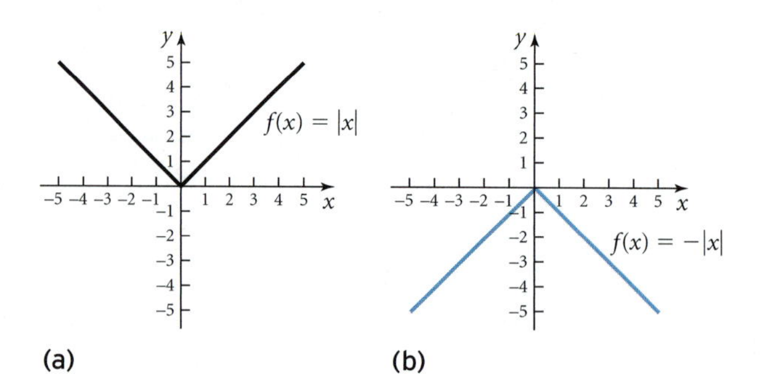

Figure 1.9

Example 17: Sketch the graph of each of the following functions. State the domain and range of each.

a. $f(x) = |x| - 4$ **b.** $f(x) = |x - 4|$ **c.** $f(x) = 4 - |x|$

Solution:

a. The graph of $f(x) = |x| - 4$ is the graph of $f(x) = |x|$ shifted down 4 units. The corner point is $(0, -4)$ The domain is all real numbers; the range is $y \geq -4$. See Figure 1.10.

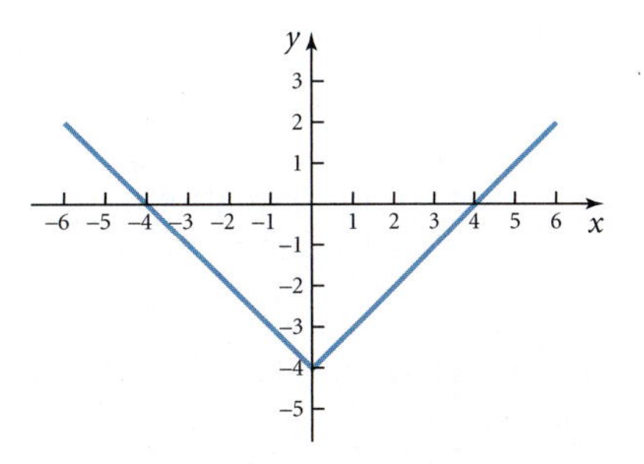

Figure 1.10

b. The graph of $f(x) = |x - 4|$ is the graph of $f(x) = |x|$ shifted 4 units to the right. The corner point is (4, 0). The domain is all real numbers; the range is $y \geq 0$. See Figure 1.11.

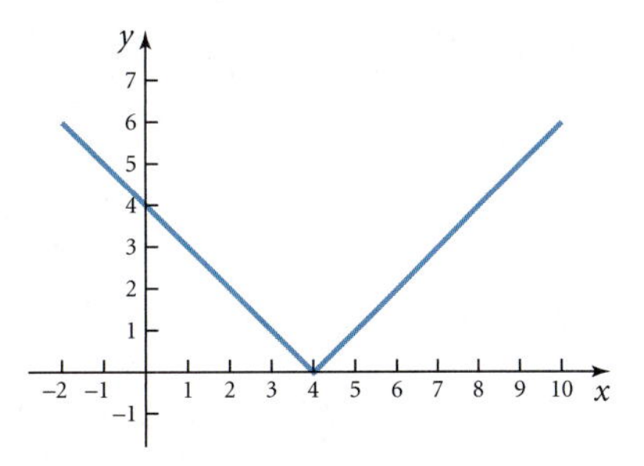

Figure 1.11

c. The graph of $f(x) = 4 - |x|$ is the graph of $f(x) = |x|$ reflected about the x-axis and shifted up 4 units. The corner point is (0, 4). The domain is all real numbers; the range is $y \leq 4$. See Figure 1.12.

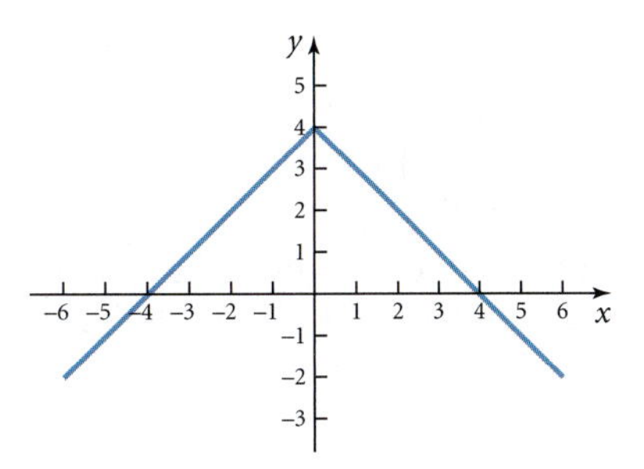

Figure 1.12 ■

Business Applications

Several business applications can be analyzed using linear functions. We conclude Topic 1 with a discussion of straight-line depreciation, break-even points, and supply and demand.

Straight-line depreciation estimates the value of an asset, such as a piece of equipment, as it loses value, or depreciates, with use over time. Given its original price P, its expected lifetime L, and its scrap value C (its value at the end of its expected lifetime), we can model the value V of the asset at any time t during its lifetime by

$$\textit{value} = \textit{price} - \left(\frac{\textit{price} - \textit{scrap value}}{\textit{lifetime}}\right) \cdot t \quad \text{or} \quad V = P - \left(\frac{P - C}{L}\right) \cdot t$$

The slope of this equation represents the rate of depreciation, and the y-intercept is the original value of the asset.

Example 18: A small business buys a piece of equipment for \$75,000 and estimates its lifespan to be 20 years, after which it will be worth \$6000.

a. Find a formula for the value of the equipment after t years, $V(t)$, and state a reasonable domain.

b. What is the value of the equipment after 5 years? After 10 years?

c. When will the equipment be worth \$20,000?

d. What is the rate of depreciation?

Solution:

a. With $P = \$75{,}000$, $L = 20$, and $C = \$6000$, we have

$$V(t) = \$75{,}000 - \left(\frac{\$75{,}000 - \$6000}{20}\right) \cdot t \rightarrow V(t) = \$75{,}000 - \$3450t$$

The lifetime of the equipment is 20 years, so the domain is $0 \leq t \leq 20$.

b. After 5 years, the equipment has a value of $V(5) = \$57{,}750$. After 10 years, the equipment has a value of $V(10) = \$40{,}500$.

c. To determine when the equipment is worth \$20,000, let $V = 20{,}000$ and solve for t.

$$20{,}000 = 75{,}000 - 3450t \rightarrow -55{,}000 = -3450t \rightarrow t = 15.94$$

It will take nearly 16 years for the equipment to depreciate to a value of \$20,000.

d. The rate of depreciation is the slope, $-\$3450$, meaning that the equipment depreciates by \$3450 per year. ■

Calculator Corner 1.4

Graph $y_1 = 75{,}000 - 3450x$ using a window of $[0, 20]$ for x with $xScl = 5$ and a window of $[0, 80{,}000]$ for y with $yScl = 10{,}000$. Use the TRACE feature to find the approximate point where $y = 20{,}000$.

Example 19: *Break-Even Point.* The Good Buy Company manufactures DVD players. Their marketing department finds that if the company manufactures and sells x DVD players per day, its costs are $C(x) = 150x + 10{,}000$ and its revenues are $R(x) = 400x$. The break-even point gives the number of DVD players for which costs and revenues are the same amount. Find the break-even point for Good Buy's DVD players.

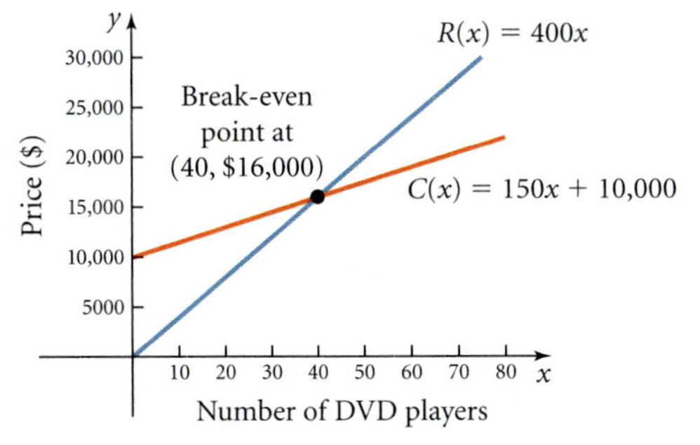

Figure 1.13

Solution: Setting costs equal to revenue, $C(x) = R(x)$.

$$150x + 10{,}000 = 400x$$
$$10{,}000 = 250x$$
$$x = 40 \text{ units}$$

Good Buy will break even if it manufactures and sells 40 DVD players. You should see that $C(40) = R(40) = \$16{,}000$. (See Figure 1.13.)

If Good Buy manufactures and sells fewer than 40 DVD players, costs will exceed revenue and the company will suffer a loss. To make any money, the company should manufacture and sell more than 40 DVD players. ■

Calculator Corner 1.5

The break-even point can be found using the Intersect (ISECT) feature of the TI-83/84. First graph $y_1 = 150x + 10{,}000$ and $y_2 = 400x$ using a window of $[0, 50]$ for x and $[0, 25{,}000]$ for y. Then enter **2nd TRACE** (Calc) 5: intersect. Use the left and right arrows to trace to a point close to the intersection point. Press **ENTER** three times to see the coordinates of the break-even point.

In economics, the **supply function** $S(p)$ refers to the quantity of units sellers are willing to supply (or sell) at price p. The **demand function** $D(p)$ refers to the quantity of units demanded by consumers at price p. Common sense tells us that a low price means a high demand by buyers but less willingness by suppliers to sell (low supply). The **equilibrium point** is the price for which the quantity producers supply equals the quantity demanded by buyers. In other words, the equilibrium point is the price at which both sellers and consumers are satisfied.

Warning! The break-even point and the equilibrium point are not equivalent.

Example 20: *Supply and Demand.* The supply and demand functions for oranges (both in thousands of bushels) are given by $S(p) = 1500p - 15{,}000$ and $D(p) = 17{,}500 - 1000p$, where p is the price per bushel. Determine the equilibrium point and explain what is represented by that point.

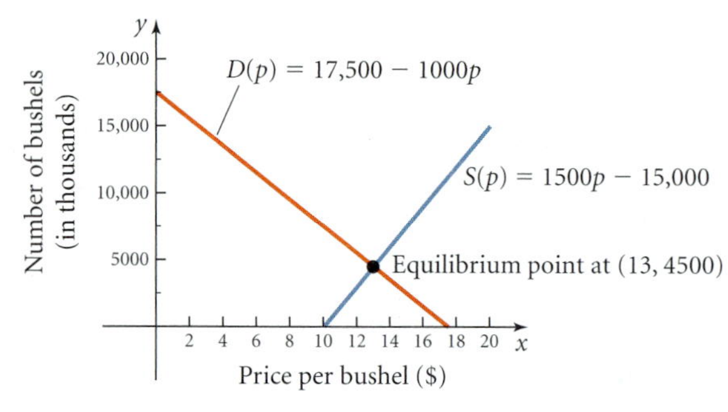

Figure 1.14

Solution: The equilibrium point is the price for which $S(p) = D(p)$.

$$1500p - 15{,}000 = 17{,}500 - 1000p$$
$$2500p = 32{,}500$$
$$p = 13$$

Equilibrium results if the price is \$13 per bushel.
If $p = \$13$, then $S(13) = D(13) = 4500$ thousand, or 4,500,000 bushels. (See Figure 1.14.) ■

Calculator Corner 1.6

Graph $S(p)$ and $D(p)$ from Example 20 using a suitable window. Use the Intersect (ISECT) feature of your calculator to determine the equilibrium point.

Check Your Understanding 1.2

State the slope and y-intercept and draw the graph of each of the following functions.

1. $f(x) = 2x + 3$ **2.** $f(x) = -\frac{2}{3}x + 1$ **3.** $4x - 3y = 12$

Sketch the graph of the following functions. State the domain and range for each.

4. $f(x) = |x| + 3$ **5.** $f(x) = |x + 3|$ **6.** $f(x) = |x - 2| - 1$

7. $f(x) = 2 - |x + 1|$

8. A college buys a copying machine for \$60,000 and estimates that it will last for 5 years, when it will be worth \$15,000.

  a. Find a formula for the value of the machine after t years.

  b. What is its value after 2 years?

9. Tracey's Gym has weekly costs of $C(x) = 25x + 1800$ and weekly revenue of $R(x) = 40x$, where x is the number of members that week. Find the break-even point.

10. The supply and demand functions for a case of bottled water are $S(p) = 200p - 450$ and $D(p) = 1200 - 150p$, where p is the price of the case. Determine the equilibrium point and interpret the answer.

Check Your Understanding Answers

Check Your Understanding 1.1

1. yes; domain $= \{2, 4, 5, 7\}$

2. yes; domain is all real numbers

3. no

4. no

5. no

6. yes; domain is all real numbers

7. a. 2 **b.** 14,000 **c.** $\frac{80}{67} \approx 1.194$

8. $2a + h + 2$

Check Your Understanding 1.2

1. $m = 2$, $(0, 3)$

2. $m = -\frac{2}{3}$, $(0,1)$

3. $m = \frac{4}{3}$, $(0,-4)$

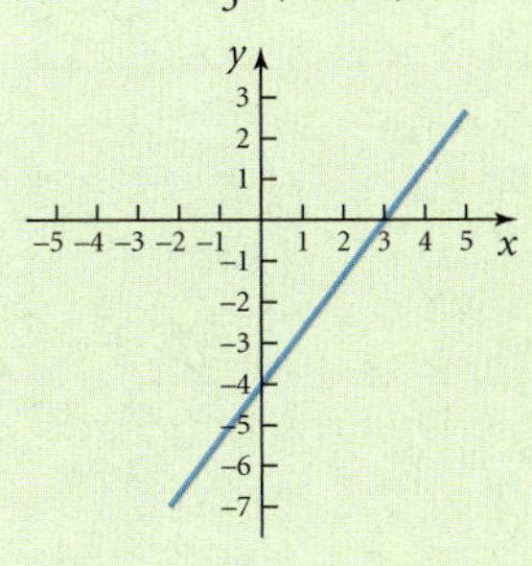

4. Domain: reals
range: $y \geq 3$

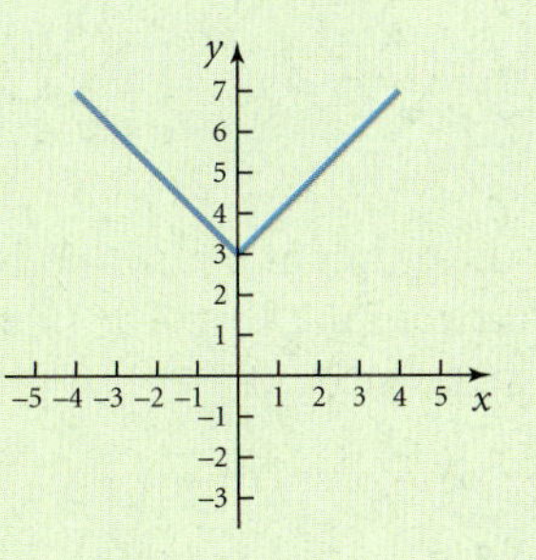

5. Domain: reals
range: $y \geq 0$

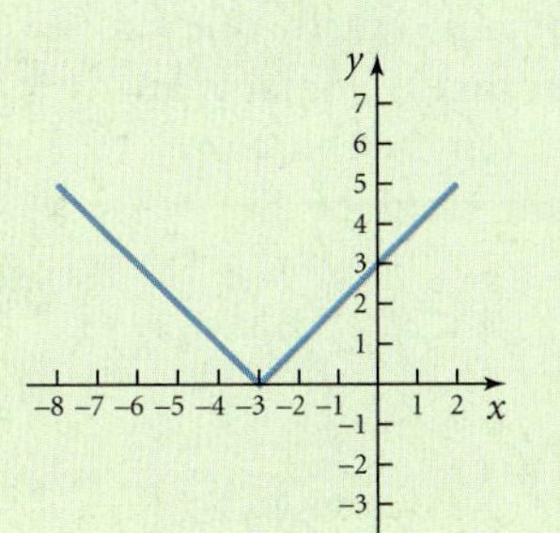

6. Domain: reals
range: $y \geq -1$

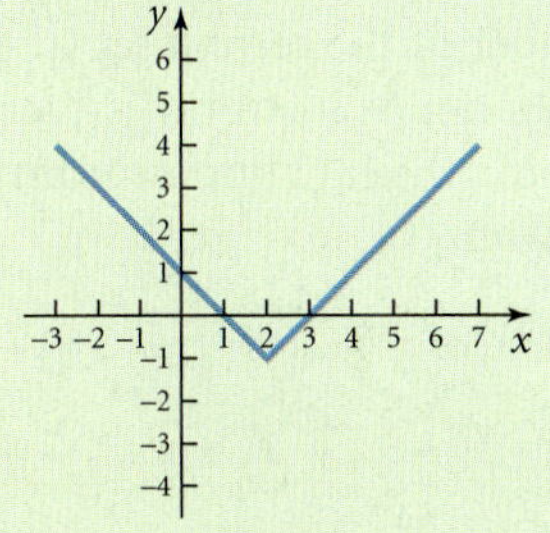

7. Domain: reals
range: $y \leq 2$

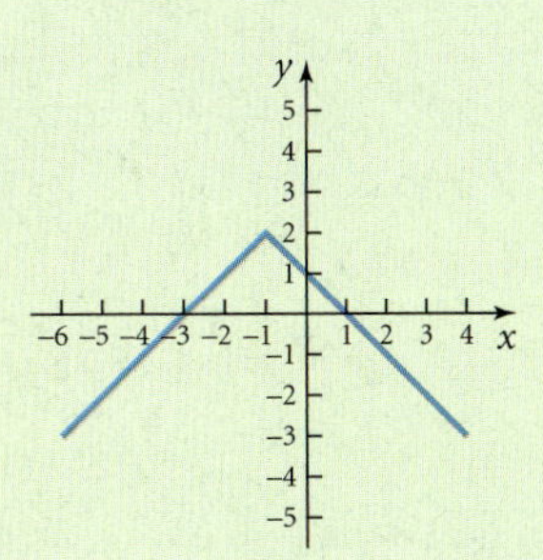

8. a. $v(t) = 60{,}000 - 9000t$
b. \$42,000

9. 120 members

10. $p =$ \$4.71 per case

Topic 1 Review

This topic reviewed the basic concepts of function, domain, and range. Linear and absolute value functions were also reviewed.

CONCEPT	EXAMPLE
A **function** is a correspondence that assigns to each member of the first set (the **domain**) exactly one member of the second set (the **range**). Domains and ranges can be identified either from the equation defining the function or from the graph of the function.	$\{(2, 3), (5, -3), (7, -2), (8, 1)\}$ is a function. The domain is the set $\{2, 5, 7, 8\}$, and the range is the set $\{-3, -2, 1, 3\}$. $f(x) = x^3 - 1$ is a function because any value of x produces only one value for y. The domain and range are the set of all real numbers. 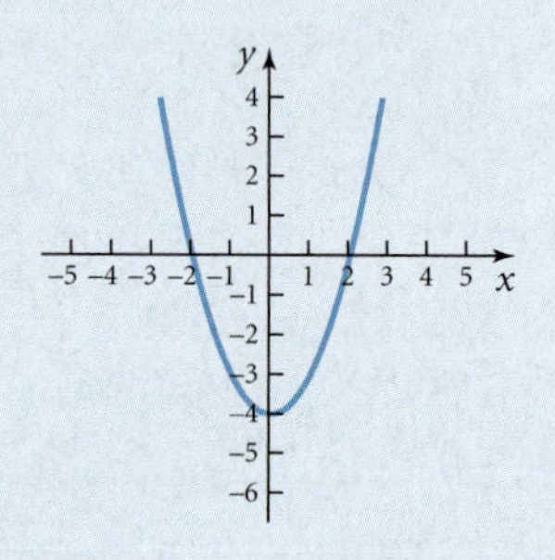 is the graph of a function because is satisfies the **Vertical Line Test**. The domain is all real numbers. The range is all $y \geq -4$
Linear functions are of the form $f(x) = mx + b$. The graph of a linear function is a straight line. The **slope** of the line is m, and the ***y*-intercept** is $(0, b)$. The slope of the line is a measure of its steepness. The y-intercept is the point where the line passes through the y-axis. Equations of the form $y = mx + b$ are said to be in slope–intercept form.	$f(x) = 5 - 2x$ is a linear function in slope–intercept form. The slope is -2, and the y-intercept is $(0, 5)$. The graph is a line. 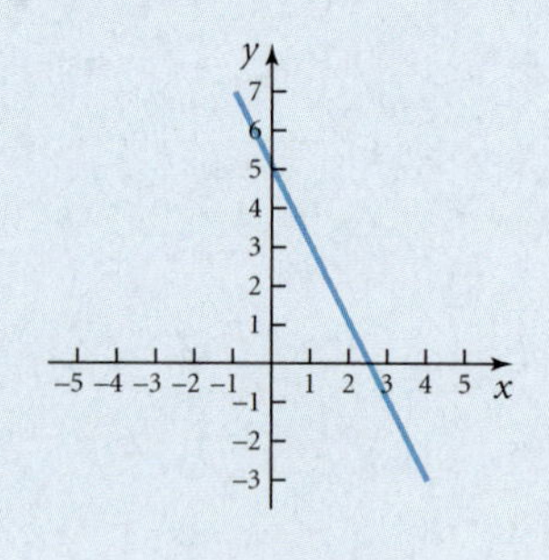

(continued)

For the function $f(x)$, the notation $f(a)$ indicates that the given function is to be evaluated by replacing the independent variable, x, with the value a.	For $f(x) = -3x^2 + 7x - 2$, we have $f(-1) = -3(-1)^2 + 7(-1) - 2 = -12$.
The **difference quotient** is the expression $\dfrac{f(a + h) - f(a)}{h}$.	For $f(x) = -3x^2 + 7x - 2$, the difference quotient is $\dfrac{[-3(a+h)^2 + 7(a+h) - 2] - (-3a^2 + 7a - 2)}{h}$, which simplifies to $-6a - 3h + 7$.
The domain of an **absolute value function** is all real numbers. The range is restricted because absolute value measures distance from zero on the number line and therefore is nonnegative.	$f(x) = \|x - 3\|$ is an absolute value function. The domain is all real numbers. The range is all nonnegative real numbers.
The graph of the basic absolute value function $f(x) = \|x\|$ is a V-shape with a **corner point** at (0, 0).	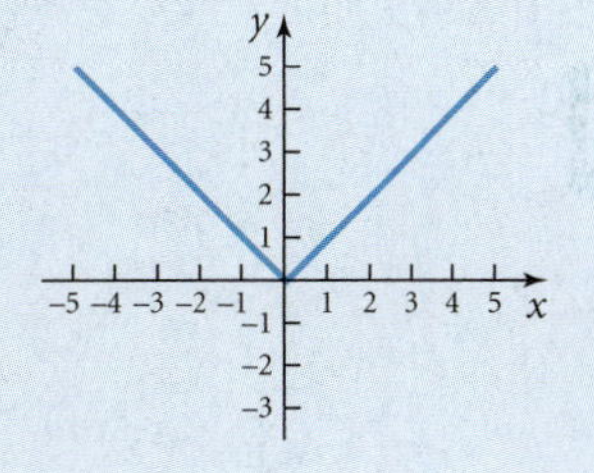
Three **transformations** that change the location, but not the shape, of a graph are the following: • **Horizontal translations.** $f(x - h)$ shifts the graph of $f(x)$ h units to the right and $f(x + h)$ shifts the graph of $f(x)$ h units to the left. • **Vertical translations.** $f(x) + k$ shifts the graph of $f(x)$ up k units if $k > 0$or down k units if $k < 0$. • **Reflections.** $-f(x)$ reflects, or inverts, the graph of $f(x)$ about the x-axis. $f(-x)$ reflects the graph of $f(x)$ about the y-axis.	The graph of $f(x) = \|x - 3\|$ is the graph of $f(x) = \|x\|$ shifted 3 units to the right. 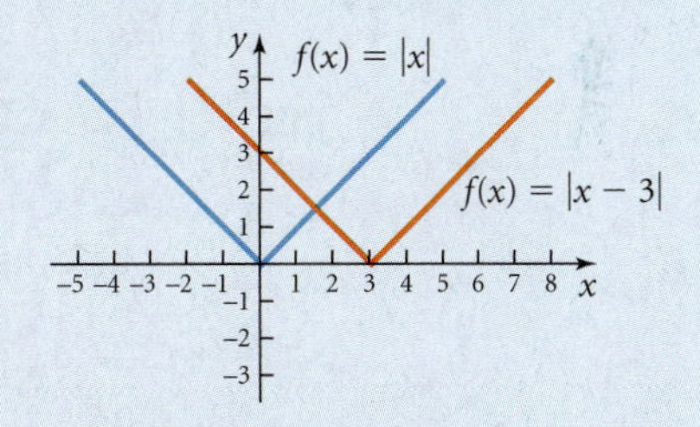 The graph of $f(x) = \|x\| - 3$ is the graph of $f(x) = \|x\|$ shifted down 3 units. 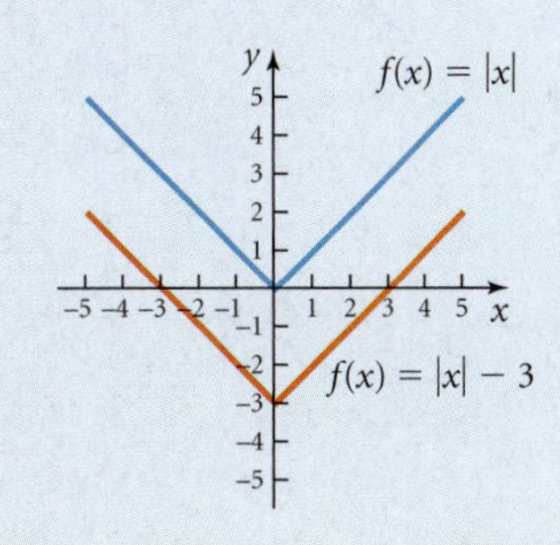

(continued)

The graph of $f(x) = -|x|$ is the graph of $f(x) = |x|$ inverted about the x-axis.

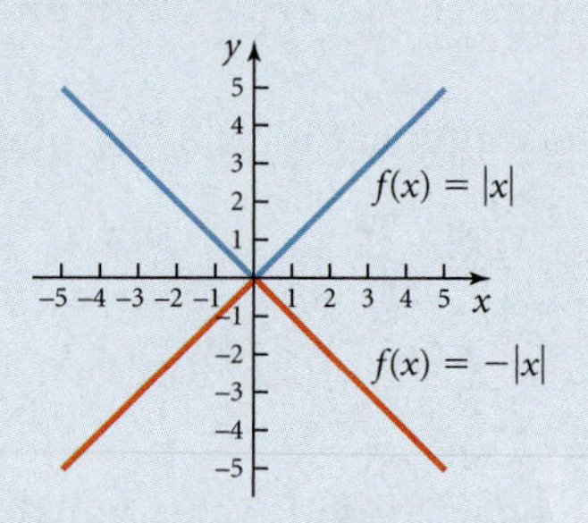

Linear functions occur in many business applications, including **straight-line depreciation, break-even points** of cost and revenue functions, and **equilibrium points** of supply and demand functions.

Depreciation: A machine purchased for \$50,000 is expected to last 10 years, at which time it will be worth \$15,000. The value of the machine after t years is given by

$$\begin{aligned} V(t) &= 50{,}000 - \left(\frac{50{,}000 - 15{,}000}{10}\right)t \\ &= 50{,}000 - 3500t. \end{aligned}$$

The rate of depreciation is \$3500 per year.

Break-even point: The cost of producing x Super G motorcycles is given by $C(x) = \$4800x + 71{,}100$. The revenue generated from selling x Super G motorcycles is given by $R(x) = \$28{,}500x$. The break-even point occurs if $C(x) = R(x)$, or if $x = 3$. If three Super G cycles are produced and sold, the plant will break even because costs and revenues are both \$85,500.

Equilibrium point: The supply function and demand function for Super G motorcycles are given by $S(p) = 1800p - 30{,}000$ and $D(p) = 72{,}000 - 1200p$, where p is the price of the cycle in thousands of dollars. Equilibrium is achieved if $S(p) = D(p)$, or if $p = 34$. Supply and demand are equal if the price is \$34,000.

NEW TOOLS IN THE TOOL KIT

- Understanding functions as a correspondence between two sets
- Stating the domain and range of a function
- Identifying the slope and y-intercept of a linear function
- Sketching graphs of linear functions, without a graphing calculator, using the slope and y-intercept
- Identifying domain and range of absolute value functions
- Sketching the graph of absolute value functions without a graphing calculator and identifying the corner point
- Using horizontal translations, vertical translations, and reflections to describe and graph functionsm
- Evaluating difference quotients

Topic 1 Exercises

1

State whether each of the following represent a function.

1. Set M: residents of New York

Set N: home address of the resident

2. Set 1: textbooks in the bookstore

Set 2: author of the textbook

3. $\{(4, 1), (5, 7), (9, 2), (5, -3), (10, 1)\}$

4. $x^2 + y^2 = 25$

5. $y = x^2 - 4x - 5$

6. $\{(4, 1), (5, 7), (9, 2), (10, 7)\}$

7. $\{(1.1, 2), (1.3, 4), (1.7, 5), (1.9, 2)\}$

8.

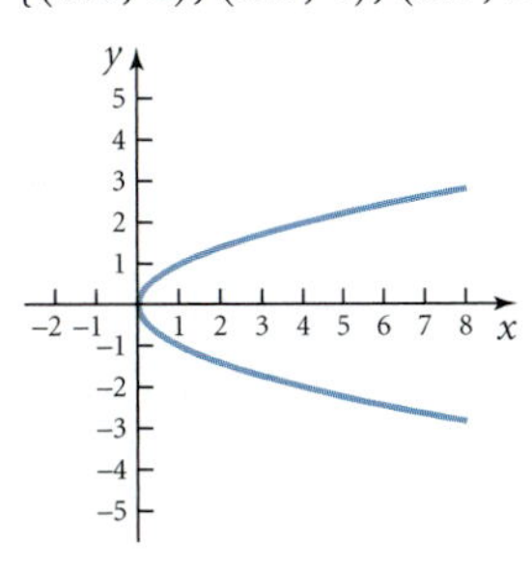

9. $x + |y| = 4$

10. $y = x^3 - 4$

11.

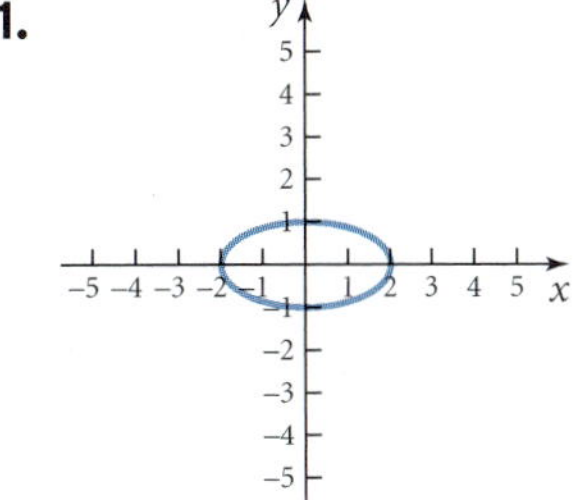

12. $\{(1.1, 2), (1.3, 4), (1.7, 5), (1.1, 9)\}$

13.

14.

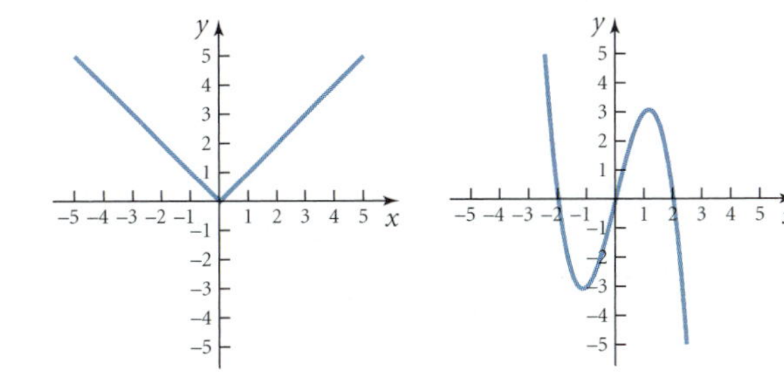

Evaluate each of the following functions at the indicated value.

15. $f(x) = 3x - 5; f(2)$

16. $f(x) = 7 - 2x; f(-1)$

17. $f(x) = x^2 - 3x + 4; f(1)$

18. $f(x) = -2x^2 + x - 5; f(-3)$

19. $f(x) = -2x^2 + x - 5; f(a)$

20. $f(x) = x^3 + 2x; f(b)$

21. $f(x) = -2x^2 + x - 5; f(a + h)$

22. $f(x) = x^2 - 3x + 4; f(a + h)$

23. $f(x) = x^2 - 3x; f(a + h) - f(a)$

24. $f(x) = 4 - x^2; f(a + h) - f(a)$

Simplify the difference quotient for each function using the four-step process.

25. $f(x) = 3x - 5$

26. $f(x) = 10$

27. $f(x) = x^2 + 4$

28. $f(x) = 3 - x^2$

29. $f(x) = 2x^2 - 3x + 5$

30. $f(x) = 6 - 2x + x^2$

31. $f(x) = x^3$

32. $f(x) = x^4$

For Exercises 33 through 40, match each equation to its graph.

33. $y = 3x - 2$

34. $y = 3x + 2$

35. $y = -3x + 2$

36. $y = -3x - 2$

37. $y = \frac{1}{3}x + 2$

38. $y = -\frac{1}{2}x + 2$

39. $y = -\frac{1}{2}x - 2$

40. $y = \frac{1}{3}x - 2$

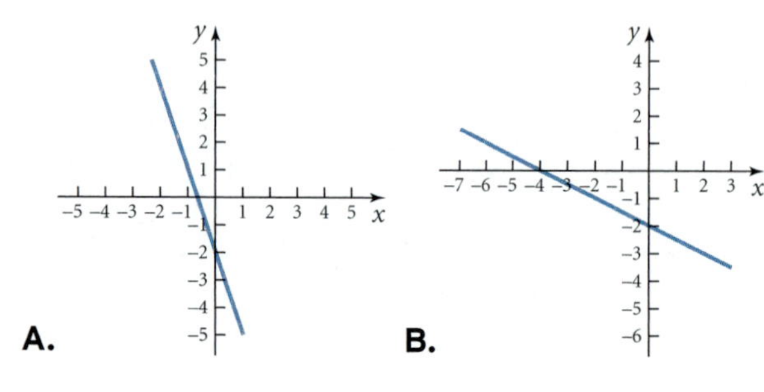

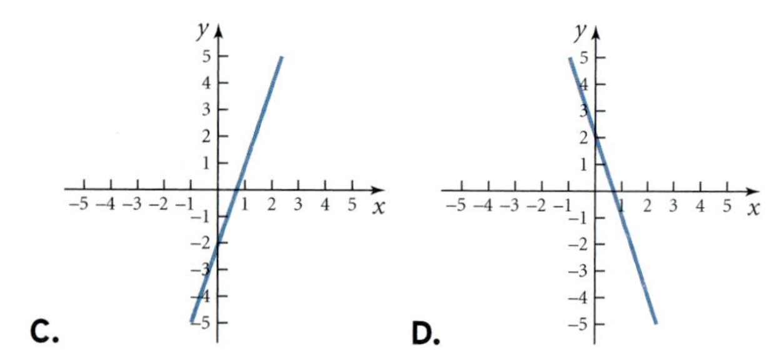

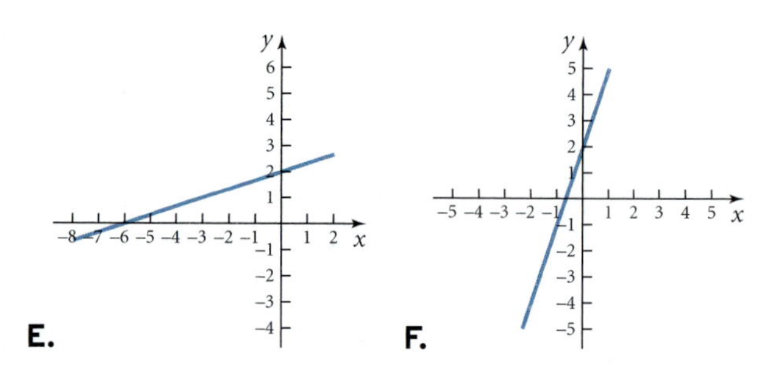

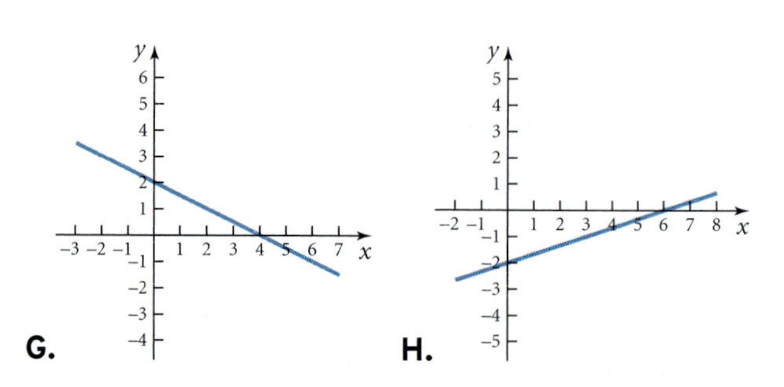

Determine the slope and y-intercept of each of the following linear equations. Draw the graph without using a graphing calculator.

41. $y = 3x - 4$

42. $y = 2 - 5x$

43. $2x - 3y = 12$

44. $5x + 2y = 20$

45. $x = 3$

46. $y = 3$

47. $y = 6$

48. $x = 6$

In Exercises 49 through 54, match each equation to its graph.

49. $y = |x + 3|$

50. $y = 2 - |x|$

51. $y = |x| + 3$

52. $y = |x - 2|$

53. $y = 2 - |x + 1|$

54. $y = |x + 3| - 2$

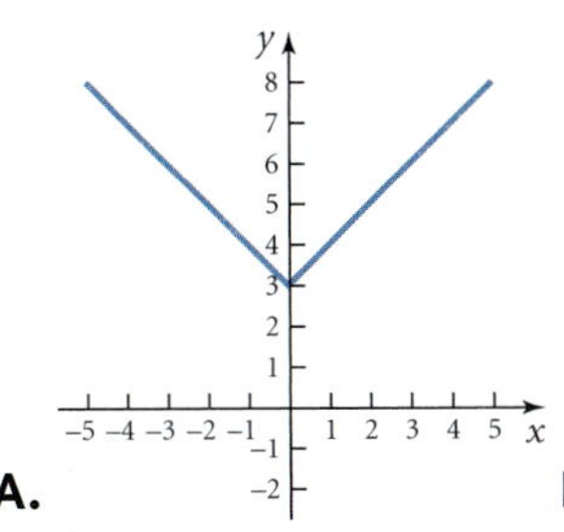

A.

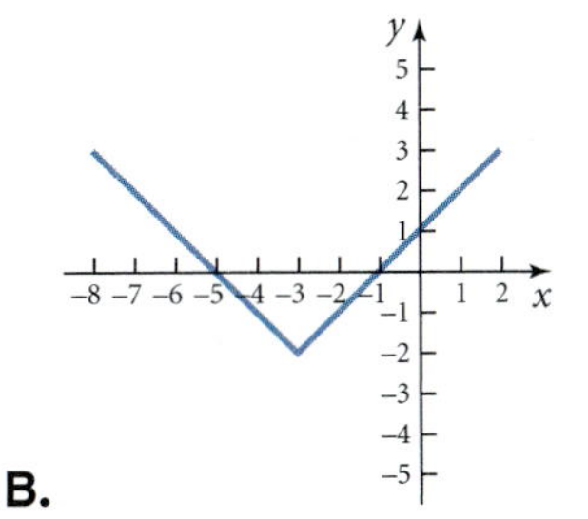

B.

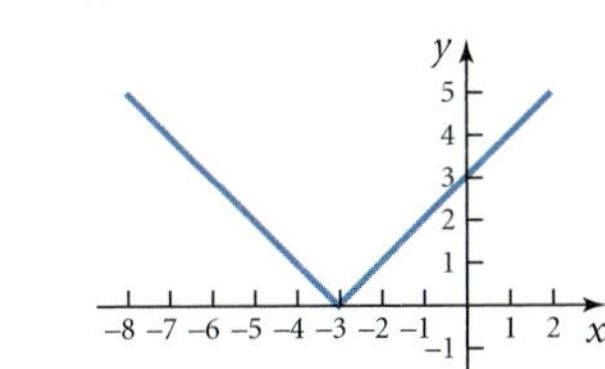

C.

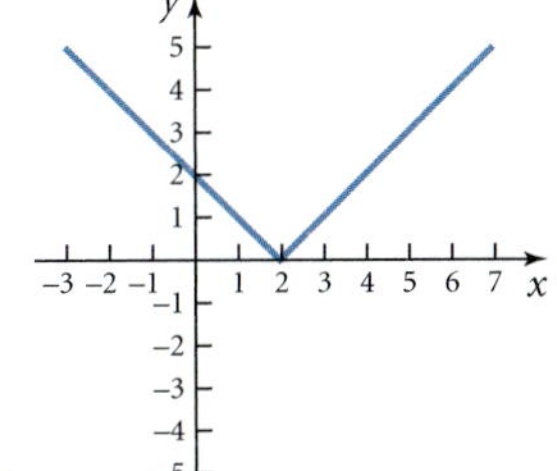

D.

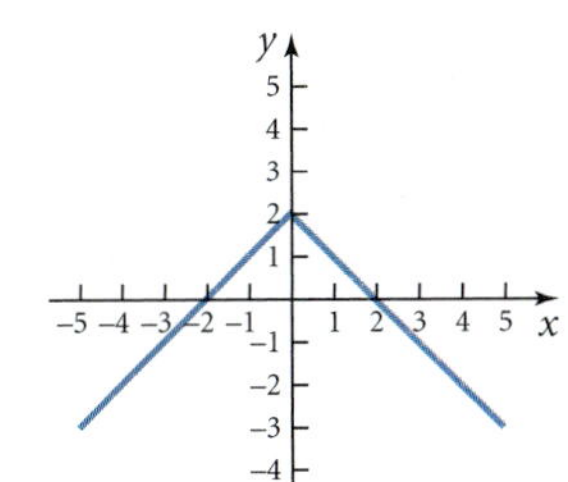

E.

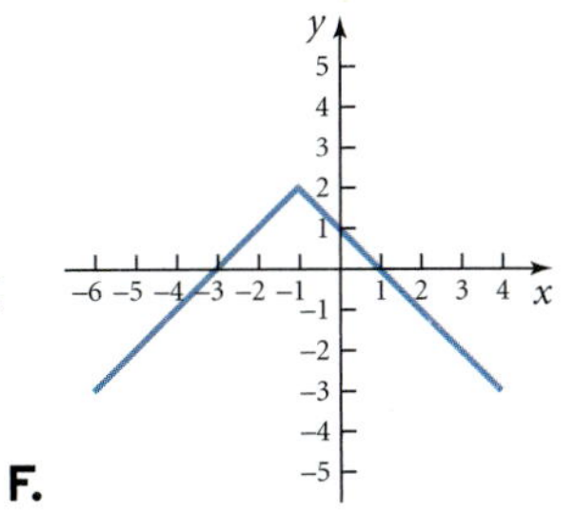

F.

Draw the graph of each of the following functions without using your graphing calculator. State the domain and range of each.

55. $f(x) = |x - 3|$

56. $f(x) = |x + 5|$

57. $f(x) = |x + 2| - 3$

58. $f(x) = |x - 4| + 1$

59. $f(x) = 3 - |x - 1|$

60. $f(x) = 5 - |x + 3|$

61. A company rents copiers to small businesses. The charge is \$125 per month plus \$0.02 per copy.

- **a.** Write the linear function that describes the company's total rental charges per month.
- **b.** Determine the charge to a business that makes 5600 copies per month.

62. A limousine rental agency charges a flat fee of \$45 plus \$0.23 per mile.

- **a.** Write the linear function that describes the agency's total rental charges.
- **b.** Determine the charge for a business that used the limousine for a 362-mile trip.

63. The median age of women at their first marriage has risen slowly since 1980 and can be modeled by $y = 22.1 + 0.17x$, where x is the number of years after 1980.

- **a.** What was the median age of a first-time bride in 2002?
- **b.** Graph the equation, using a window of $[0, 25]$ for x and $[0, 30]$ for y.
- **c.** Predict the year during which the median age for first-time brides will be 28.

64. The median age of men at their first marriage has risen slowly since 1980 and can be modeled by $y = 24.8 + 0.13x$, where x is the number of years after 1980.

- **a.** What was the median age of a first-time groom in 2002?
- **b.** Graph the equation, using a window of $[0, 25]$ for x and $[0, 40]$ for y.
- **c.** Predict the year during which the median age for first-time grooms will be 30.

65. Life expectancy in the United States has increased steadily since 1960. For men, life expectancy can be modeled by $y = 0.203x + 65.92$, where x is the birth year measured in years after 1960. For instance, a man born in 1960 ($x = 0$) had a life expectancy of 65.92 years.

- **a.** Predict the life expectancy of a man born in 1990.
- **b.** What is the slope of the equation? Interpret the slope with regard to birth year and life expectancy.

66. Life expectancy in the United States has increased steadily since 1960. For women, life expectancy can be modeled by $y = 0.171x + 73.32$, where x is the birth year measured in years after 1960. For instance, a woman born in 1960 $x = 0$ had a life expectancy of 73.32 years.

- **a.** Predict the life expectancy of a woman born in 1990.
- **b.** What is the slope of the equation? Interpret the slope with regard to birth year and life expectancy.

67. A cleaning service charges a flat fee of $50 plus $10 per hour. Write a function describing their total charges for any number of hours, h. If Malcolm uses the service for 6 hours, what is his charge?

68. A company rents copiers to small businesses. The charge is $125 per month plus $0.025 per copy. Write a function describing their total fees for any number of copies, x. Barry's Accounting Services used a copier this month to print 25,300 copies. What was Barry's charge for the use of the copier?

69. The Lifetime Printing Company buys a new printing press for $950,000. The press is expected to last 25 years, at which time it will be worth $75,000. Write a function that describes the value of the machine at any time during its lifetime. Be sure to state the domain of the function. What will the press be worth after 10 years? When will its value be $125,000?

70. A fence builder buys a posthole digger for $85,000. The digger is expected to last 15 years, after which its value will be $10,000. Write a function that describes the value of the equipment at any time during its lifetime. Be sure to state the domain of the function. What is the worth of the equipment after 5 years? When will the equipment be worth $50,000?

71. Digital Concepts installs home theater systems. If the firm installs x systems each month, its monthly costs are $C(x) = 25{,}000 + 540x$ and its monthly revenue is $R(x) = 1400x + 7800$. Determine the break-even point and interpret what it means for this firm.

72. Fifi's Dog Groomers has weekly costs of $C(x) = 800 + 12x$ and weekly revenue of $R(x) = 30x + 440$, where x is the number of dogs groomed that week. Determine the break-even point and interpret what it means for the firm.

73. Market supply and demand, in hundreds, for cell phones is given by $S(x) = -20 + x$ and $D(x) = 59.2 - 0.76x$, where x is the monthly cost of the phone. Find the equilibrium value and interpret its meaning.

74. Market supply and demand, in thousands, for bottled water is given by $S(x) = 4x - 1.5$ and $D(x) = 10.5 - 2x$, where x is the price per bottle. Find the equilibrium value and its meaning.

CALCULATOR CONNECTIONS

Your graphing calculator can be used to estimate solutions to equations by determining the x-intercepts of a graph.

75. **a.** Graph $f(x) = |x| - 4$. What are the x-intercepts of this graph?

b. Solve $|x| - 4 = 0$.

c. What do you notice about your answers to parts a and b?

Calculator Corner 1.7

Graph $y_1 = |x| - 4$. To access the absolute value sign, press **MATH**, scroll right to highlight **NUM**, and then press **1:abs**. Enter the quantity within the absolute value sign in parentheses.

76. **a.** Graph $f(x) = 3 - |x + 5|$. What are the x-intercepts of this graph?

b. Solve $3 - |x + 5| = 0$.

c. What do you notice about your answers to parts a and b?

In Exercises 75 and 76, you discovered that the x-intercepts of the graph of a function $f(x)$ are equivalent to the solutions of the equation $f(x) = 0$. Use this relationship to complete Exercises 77 and 78.

77. **a.** Graph $f(x) = 3 - |x^2 - 4x|$.

b. Estimate the solutions of $3 - |x^2 - 4x| = 0$ to three decimal places.

78. **a.** Graph $f(x) = |3 - 2x| - |x^2 - 7|$.

b. Estimate the solutions of $|3 - 2x| - |x^2 - 7| = 0$ to three decimal places.

Calculator Corner 1.8

To estimate the x-intercepts of a graph from the graph, press **2nd TRACE** (CALC). To set the left bound, trace to a point immediately to the left of the x-intercept and press **ENTER**. To set the right bound, continue tracing to a point immediately to the right of the x-intercept and press **ENTER ENTER**. The calculator will then show the location of the x-intercept.

PROJECT 1.1
Linear Functions

Dave works at a local grocery store and is paid an hourly rate. His pay stubs for the past eight weeks show his take-home pay for each week, which reflects the deductions taken out of his paycheck each week.

Hours Worked	Take-home Pay
38.48	$276.15
34.42	$245.89
38.95	$279.64
36	$257.68
34.5	$246.48
34	$242.77
34.48	$246.34
35.58	$254.54

1. Plot the points, using hours worked on the horizontal axis and take-home pay on the vertical axis.
2. Write a linear equation relating the number of hours worked to the take-home pay. Select any two points and find the equation of the line between them.
3. How much take-home pay will Dave receive in a week when he worked 38 hours?
4. If Dave's take-home pay last week was $245.31, how many hours did he work?
5. Draw the graph of the linear equation you found in Question 2.
6. Suppose Dave gets a raise in his hourly rate. How will the graph change?
7. Suppose the amount deducted from Dave's paycheck each week is reduced. How will the graph change?
8. How many hours must Dave work each week to cover the amount being deducted from his paycheck?

TOPIC

2 Nonlinear Functions

IMPORTANT TOOLS IN TOPIC 2

- *Quadratic functions*
- *Vertex, domain, and range of quadratic functions*
- *Graphs of quadratic functions*
- *Cubic functions*
- *Domain, range, and point of inflection*
- *Graphs of cubic functions*
- *Square-root functions*
- *Domain and range of square-root functions*
- *Graphs of square-root functions*
- *Maximum profit*

Not all the algebraic functions used in calculus are linear functions. For example, the proportion of accidents that occur each year in the United States is extremely high for younger drivers and for older drivers, but declines to a low point for middle-aged drivers.

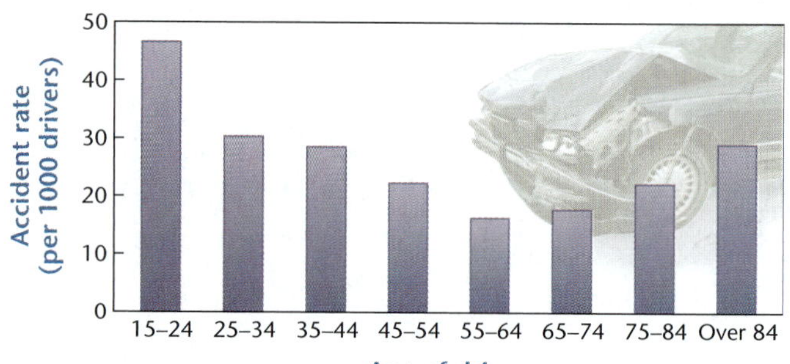

Data from: National Highway Traffic Safety Administration

The curve that approximates the shape of these data shows another type of algebraic function called a quadratic function. In this topic, you learn about quadratic and other types of nonlinear functions. Before starting your review of nonlinear functions, you should complete the Topic 2 Warm-up Exercises to review the prerequisite algebra skills needed for this topic.

TOPIC 2 WARM-UP EXERCISES

Be sure you can successfully complete the following exercises before starting Topic 2.

Graph each of the following functions without using a graphing calculator. State the domain and range for each.

1. $f(x) = 3x - 5$ **2.** $3x - 5y = 10$

3. $f(x) = |x - 2| - 3$

Solve each of the following equations for x.

4. $x - 3 = 0$ **5.** $2x + 5 = 7$

6. $x^2 - 4 = 0$ **7.** $x^2 - 4x = 0$

8. $x^2 + 4 = 0$ **9.** $x^2 - 3x + 1 = 0$

10. $x^3 - 4x = 0$ **11.** $\sqrt{x} - 3 = 0$

12. $\sqrt{4 - 2x} + 3 = 0$ **13.** $x + 3 \geq 0$

14. $5 - 2x \geq 0$

Answers to Warm-up Exercises

1.

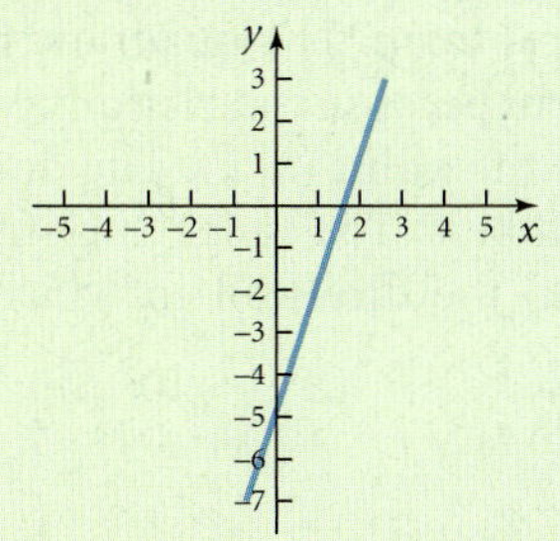

D: all real numbers
R: all real numbers

2.

D: all real numbers
R: all real numbers

3.

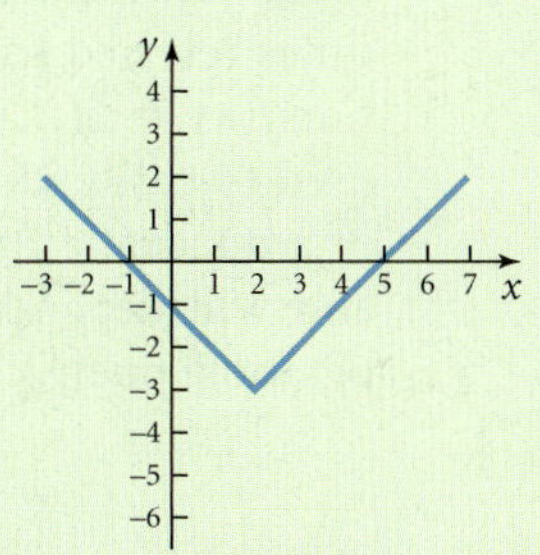

D: all real numbers
R: $y \geq -3$

4. $x = 3$ **5.** $x = 1$

6. $x = 2, -2$ **7.** $x = 0, 4$

8. no real solution **9.** $x = \dfrac{3 \pm \sqrt{5}}{2}$

10. $x = 0, 2, -2$ **11.** $x = 9$

12. no solution; $x = -\frac{5}{2}$ is an extraneous solution because $x = -\frac{5}{2}$ does not satisfy the original equation when substituted

13. $x \geq -3$ **14.** $x \leq \frac{5}{2}$

Having just reviewed the fundamentals of linear functions and some basic applications, we now review some nonlinear functions and several related applications.

Quadratic Functions

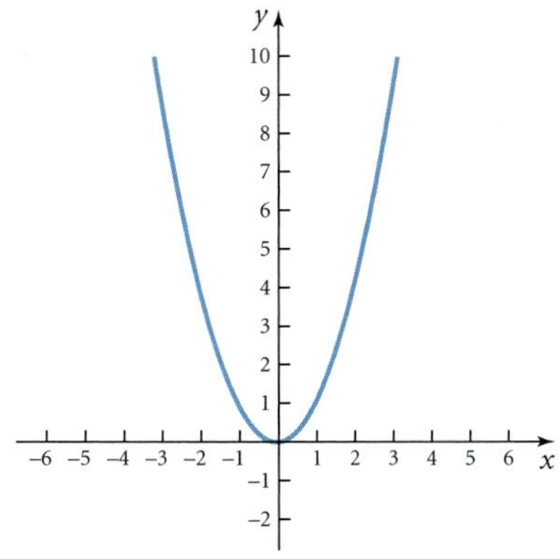

Figure 2.1

Quadratic functions are of the form $f(x) = ax^2 + bx + c$ or $f(x) = a(x - h)^2 + k$.

- A **quadratic function** has degree two, meaning that the independent variable is raised to the second power.
- The domain of a quadratic function is the set of all real numbers. The range is restricted.
- The graph of the basic quadratic function $f(x) = x^2$ is a **parabola** with **vertex** at the origin; the parabola opens upward from the vertex. (See Figure 2.1.) The vertex is the **turning point** of the graph, or the point where the direction of the graph changes.

Calculator Corner 2.1

Graph $y_1 = x^2$, $y_2 = x^2 - 2$, $y_3 = (x - 2)^2$, and $y_4 = 2 - x^2$. How are these four graphs related?

The quadratic function $f(x) = ax^2 + bx + c$ is said to be in **polynomial form** because the function is expressed as a polynomial. Polynomial form is sometimes referred to as the **standard form** or the **general form**. The quadratic function $f(x) = a(x - h)^2 + k$ is said to be in **vertex form** because translations can be used to easily identify the vertex, which is always at the point (h, k). The domain of a quadratic function is always the set of real numbers. The range is restricted and will depend on the location of the vertex and the direction in which the parabola opens.

Graphs of Quadratic Functions

Given $f(x) = a(x - h)^2 + k$, the graph is a parabola with vertex at (h, k). The domain is the set of real numbers.

If $a > 0$, the parabola opens upward and the range is $y \geq k$.
If $a < 0$, the parabola opens downward and the range is $y \leq k$.

You should be able to identify the vertex and sketch the graph of quadratic functions in vertex form *without* using your graphing calculator!

Example 1: For each of the following functions, state the domain, range, and vertex and sketch the graph without using your graphing calculator.

a. $f(x) = x^2 - 4$

b. $f(x) = (x - 4)^2$

c. $f(x) = 4 - x^2$

d. $f(x) = (x + 1)^2 - 4$

Solution:

a. $f(x) = x^2 - 4$. The domain is all real numbers; the vertex is $(0, -4)$. Because $a = 1 > 0$, the parabola opens upward and the range is $y \geq -4$. (See Figure 2.2.)

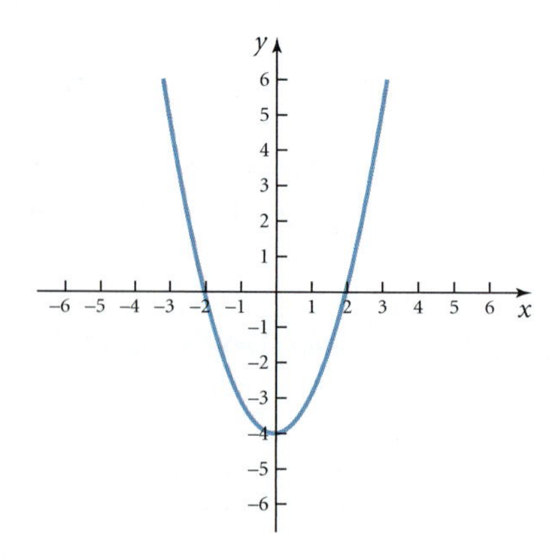

Figure 2.2

b. $f(x) = (x - 4)^2$. The domain is all real numbers; the vertex is $(4, 0)$. Because $a = 1 > 0$, the parabola opens upward and the range is $y \geq 0$. (See Figure 2.3.)

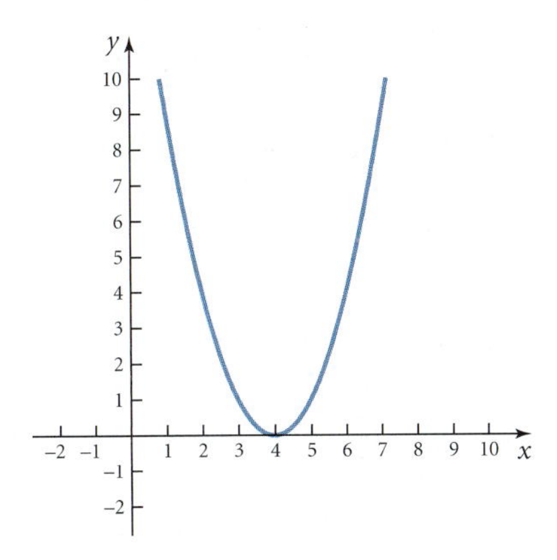

Figure 2.3

c. $f(x) = 4 - x^2$. The domain is all real numbers; the vertex is $(0, 4)$. Because $a = -1 < 0$, the parabola opens downward and the range is $y \leq 4$. (See Figure 2.4.)

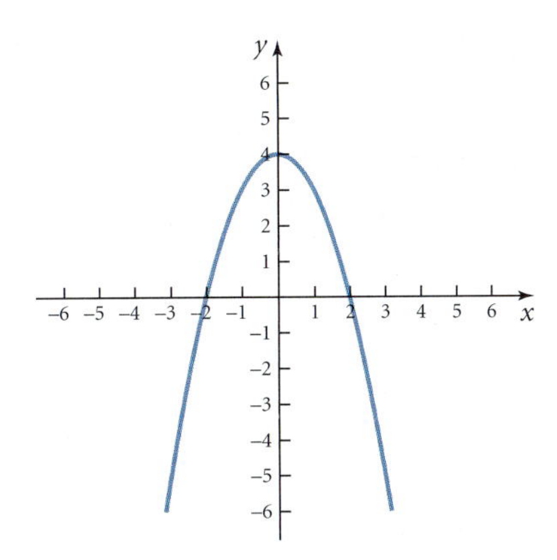

Figure 2.4

d. $f(x) = (x + 1)^2 - 4$. The domain is all real numbers; the vertex is $(-1, -4)$. Because $a = 1 > 0$, the parabola opens upward and the range is $y \geq -4$. (See Figure 2.5.)

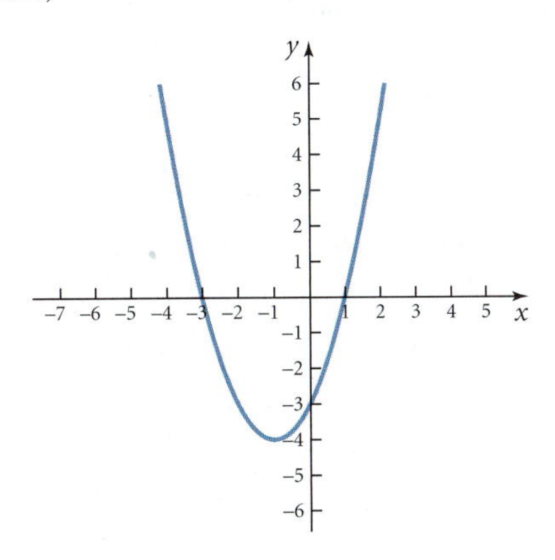

Figure 2.5 ■

Calculator Corner 2.2

How many x-intercepts does the graph of a quadratic function have? How many turning points does it have? Graph the following functions to help you decide.

$$y_1 = (x - 3)^2,\ y_2 = x^2 + 3, \text{ and } y_3 = x^2 - 3$$

(The graph of a quadratic function need not have any x-intercepts. The graph may have one x-intercept or at most two x-intercepts. The graph of a quadratic function will always have one turning point, the vertex.)

The vertex of a parabola is an important point because it gives the location of the maximum or minimum point of the graph. In later topics, we see that turning points are quite important. If the quadratic function is given in vertex form, the vertex can be found quickly and easily. The vertex of a quadratic function in polynomial form can be found with the following substitution formula.

Vertex of a Quadratic Function

If $f(x) = ax^2 + bx + c$, the vertex (h, k) is given by

$$h = -\frac{b}{2a} \quad \text{and} \quad k = f(h)$$

(For the derivation of this formula, see Mathematics Corner 2.2.)

Example 2: Use the formula above to find the vertex of each of the following functions.

a. $f(x) = x^2 - 8x + 5$ **b.** $f(x) = 2x^2 - 8x + 5$

Solution:

a. For $f(x) = x^2 - 8x + 5$, $a = 1$, $b = -8$, and $c = 5$. Substituting these values in the formula yields

$$h = -\frac{b}{2a} = -\frac{-8}{2(1)} = 4$$
$$k = f(4) = 4^2 - 8(4) + 5 = -11$$

The vertex is at $(4, -11)$.

b. For $f(x) = 2x^2 - 8x + 5$, $a = 2$, $b = -8$, and $c = 5$. Substituting these values in the formula yields

$$h = -\frac{b}{2a} = -\frac{-8}{2(2)} = 2$$
$$k = f(2) = 2(2)^2 - 8(2) + 5 = -3$$

The vertex is at $(2, -3)$. See Mathematics Corner 2.1 for an alternate derivation of the vertices of these two functions. ■

MATHEMATICS CORNER 2.1

Completing the Square

Quadratic functions given in polynomial form $f(x) = ax^2 + bx + c$ can be converted to vertex form $f(x) = a(x - h)^2 + k$ by a process called **completing the square.** From algebra, we know that $x^2 - 2hx + h^2 = (x - h)^2$. The expression $x^2 - 2hx + h^2$ is called a **perfect square trinomial.** Given the expression $x^2 + bx$, you can create a perfect square trinomial by completing the square as follows:

1. Take half of the linear coefficient (the coefficient of the term raised to the first power), which is $\frac{1}{2}(b) = \frac{b}{2}$.
2. Square the number obtained in the first step, giving $\left(\frac{b}{2}\right)^2 = \frac{b^2}{4}$.
3. Add the value from step 2 to the original expression, giving $x^2 + bx + \frac{b^2}{4}$.

The resulting expression, $x^2 + bx + \frac{b^2}{4}$, is then a perfect square trinomial, which will factor as $\left(x + \frac{b}{2}\right)^2$. The factor $x + \frac{b}{2}$ gives the horizontal translation of the vertex. For example, completing the square on $x^2 - 8x$ would yield $x^2 - 8x + 16$, which factors as $(x - 4)^2$.

To determine the vertex of $f(x) = x^2 - 8x + 5$, we convert to vertex form as follows:

$f(x) = (x^2 - 8x + 16) + 5 - 16$ Complete the square and then "undo" the operation

$f(x) = (x - 4)^2 - 11$ Factor and simplify

Thus, we see that the vertex is at $(4,-11)$.

If the coefficient of the quadratic term is not 1, the first step is to factor that value from the quadratic and linear terms before completing the square. To find the vertex of $f(x) = 2x^2 - 8x + 5$:

$f(x) = 2x^2 - 8x + 5$

$\rightarrow f(x) = 2(x^2 - 4x) + 5$ Factor 2 from the first two terms

$\rightarrow f(x) = 2(x^2 - 4x + 4) + 5 - 4 \cdot 2$ Complete the square and "undo"

$\rightarrow f(x) = 2(x - 2)^2 - 3$ Factor and simplify

Thus, the vertex is at $(2, -3)$.

Example 3: Sales of a newly released movie on DVD grow according to $S(t) = 1.4t + 0.1t^2$, where S is sales in millions of dollars and t is the number of weeks after the release of the DVD.

- **a.** State a reasonable domain for the function.
- **b.** Estimate the sales after four weeks.
- **c.** Estimate when sales will hit \$10 million.

Solution:

a. Because the domain represents time after the DVD's release and time cannot be negative, a reasonable domain would be $t \geq 0$.

b. After four weeks, the sales of the DVD will be $S(4) = 1.4(4) + 0.1(4)^2 = 7.2$. After four weeks, sales of the DVD are about \$7.2 million.

c. To reach \$10 million in sales, $S = 10$. Thus, $10 = 1.4t + 0.1t^2$ or $0.1t^2 + 1.4t - 10 = 0$. This quadratic equation does not factor, so the Quadratic Formula is used to solve for t:

$$10 = 1.4t + 0.1t^2 \rightarrow 0.1t^2 + 1.4t - 10 = 0$$

By the Quadratic Formula,

$$t = \frac{-1.4 \pm \sqrt{1.4^2 - 4(0.1)(-10)}}{2(0.1)}$$
$$= \frac{-1.4 \pm \sqrt{5.96}}{0.2}$$
$$\approx \frac{-1.4 \pm 2.44}{0.2} = \frac{1.04}{0.2} \text{ or } \frac{-3.84}{0.2}$$
$$\approx 5.2 \text{ or } -19.2$$

Quadratic Formula
If $ax^2 + bx + c = 0$, then
$$x = \frac{-b \pm \sqrt{b^2 - 4ac}}{2a}.$$
It is assumed that $a \neq 0$.

Because $t \geq 0$, the only acceptable answer is $t \approx 5.2$. Thus, sales will hit $10 million dollars during the fifth week after the DVD's initial release. ■

MATHEMATICS CORNER 2.2

Derivation of the Quadratic Formula by Completing the Square

By completing the square of $f(x) = ax^2 + bx + c$, we can obtain a formula for locating the vertex of a quadratic given in polynomial form.

Given $f(x) = ax^2 + bx + c$, complete the square as follows:

$$f(x) = ax^2 + bx + c$$

$$f(x) = a\left(x^2 + \frac{b}{a}x\right) + c$$ Factor a from quadratic and linear terms

$$f(x) = a\left(x^2 + \frac{b}{a}x + \frac{b^2}{4a^2}\right) + c - a\cdot\frac{b^2}{4a^2}$$ Complete the square and "undo"

$$f(x) = a\left(x + \frac{b}{2a}\right)^2 + c - \frac{b^2}{4a}$$ Factor the perfect square trinomial

You see that the x-coordinate of the vertex is $h = -\frac{b}{2a}$. You know that the y-coordinate is $k = f(h)$, which, from completing the square, is $k = c - \frac{b^2}{4a}$.

Calculator Corner 2.3

The TI-84 can be used to solve the equation $0.1t^2 + 1.4t - 10 = 0$. Under **APPS**, choose PolySimult and then PolyRoot Finder. Enter the degree of the equation (2) and then each of the coefficients. Press **SOLVE** to see the decimal approximation of the solutions.

Graph $y_1 = 0.1x^2 + 1.4x - 10$. Verify that the graph crosses the x-axis at the values obtained algebraically. Note that the calculator uses the variables x and y for writing equations, so the independent variable t in the original equation is replaced by x when entering in the calculator.

Check Your Understanding 2.1

Sketch the graph of each of the following functions without using a graphing calculator. State the vertex, domain, and range of each.

1. $f(x) = x^2 - 4$ **2.** $f(x) = (x - 4)^2$

3. $f(x) = 4 - x^2$ **4.** $f(x) = -(x + 2)^2$

5. $f(x) = (x - 1)^2 - 3$ **6.** $f(x) = 4 - (x + 3)^2$

Use your graphing calculator to draw the graph of each of the following functions. State the vertex, domain, and range of each.

7. $f(x) = x^2 - 3x - 4$ **8.** $f(x) = 8 - 2x - x^2$

9. $f(x) = 2x^2 + 5x + 3$

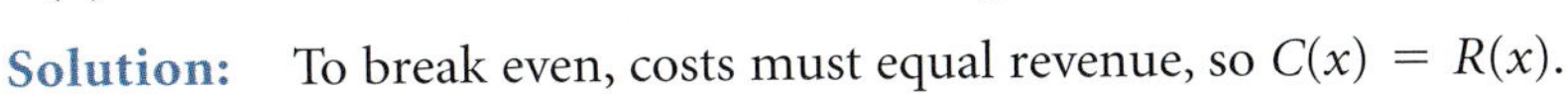

Quadratic Functions in Business Applications

Example 4: *Break-Even Point.* The Good Buy Company manufactures DVD players. Their marketing department finds that if the company manufactures x DVD players per day, its costs are $C(x) = 150x + 10{,}000$ and its revenues are $R(x) = -2x^2 + 600x$. Find the break-even point for Good Buy's DVD players.

Solution: To break even, costs must equal revenue, so $C(x) = R(x)$.

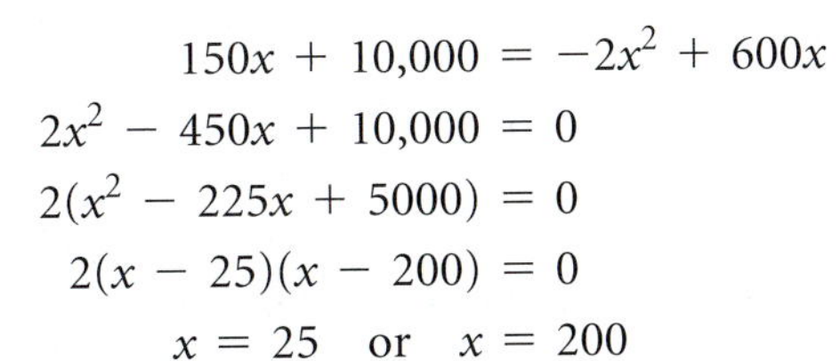

$$150x + 10{,}000 = -2x^2 + 600x$$
$$2x^2 - 450x + 10{,}000 = 0$$
$$2(x^2 - 225x + 5000) = 0$$
$$2(x - 25)(x - 200) = 0$$
$$x = 25 \quad \text{or} \quad x = 200$$

Good Buy breaks even if either 25 DVD players or 200 DVD players are manufactured and sold per day.

Calculator Corner 2.4

The break-even point can be found using the Intersect (ISECT) feature of your calculator. First graph $y_1 = 150x + 10{,}000$ and $y_2 = -2x^2 + 600x$ using a suitable window. Then enter **2nd TRACE** (Calc) 5: to determine the break-even point.

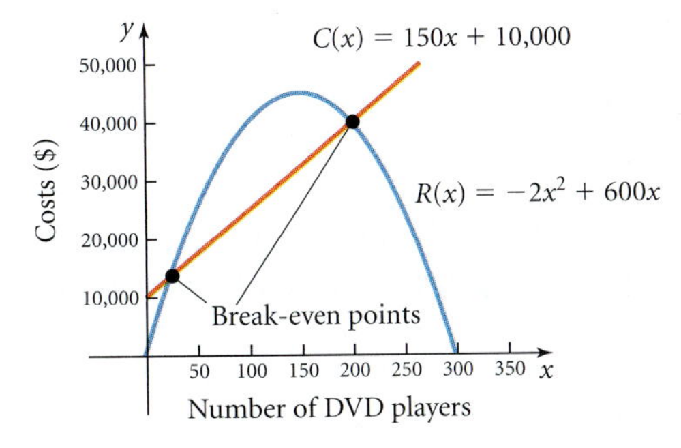

Figure 2.6

To see why there are two break-even points, look closely at the graph (Figure 2.6) for the cost and revenue functions. From the graph, you can see that if fewer than 25 or more than 200 DVD players are manufactured and sold per day, costs will exceed revenue and the firm will not make any money. To make money, the firm should manufacture and sell between 25 and 200 DVD players per day.

How many DVD players should the firm manufacture and sell to make the most amount of money possible? ■

Example 5: *Maximum Profit.* **Profit** occurs when revenues exceed costs.

Definition: **profit = revenue − cost**

For Good Buy's DVD players in Example 4, how many DVD players should be manufactured and sold per day to maximize profit?

Solution: The profit function is given by $P(x) = R(x) - C(x)$, so

$$P(x) = (-2x^2 + 600x) - (150x + 10{,}000)$$
$$P(x) = -2x^2 + 450x - 10{,}000$$

The profit function is a quadratic function whose graph is a parabola opening downward. The maximum profit will occur at the vertex of the graph.

Algebraically, the vertex is (h, k), where $h = -\frac{b}{2a}$ and $k = P(h)$.

$$h = -\frac{450}{2(-2)} = 112.5 \quad \text{and} \quad k = P(112.5) = 15{,}312.5$$

Thus, maximum profit (of \$15,312.50) occurs if Good Buy manufactures and sells 112.5 DVD players per day, which actually means either 112 or 113 DVDs per day. ■

Calculator Corner 2.5

To find the maximum of the function in Example 5 graphically, we will use the fmax feature of the graph menu. First, graph $y_1 = -2x^2 + 450x - 10{,}000$ using a suitable window. Then enter **2nd TRACE** (Calc) 4: to find the maximum value of the graph.

Cubic Functions

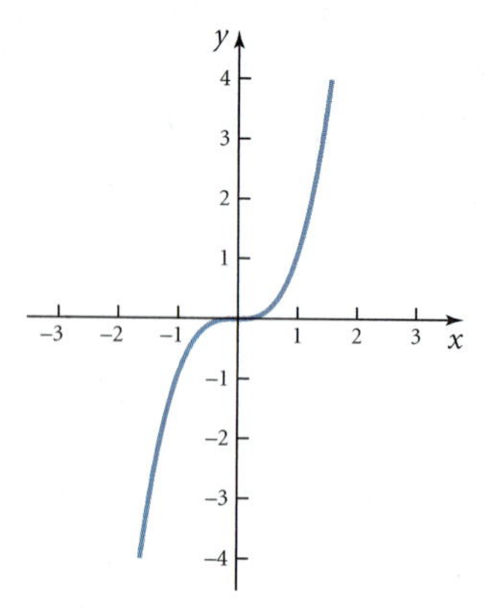

Figure 2.7

Cubic functions are of the form $f(x) = ax^3 + bx^2 + cx + d$ or $f(x) = a(x - h)^3 + k$.

- A **cubic function** has degree three, meaning that the highest power of the independent variable is the third power.
- The domain and range of cubic functions are the set of real numbers.
- The graph of a cubic function is an S-shape, which may have two turning points.
- The basic cubic function is $f(x) = x^3$. The domain and range are the set of all real numbers. The graph (Figure 2.7) is an elongated S-shape passing through the origin.
- For the graph of the basic cubic function $f(x) = x^3$, you should see that the origin is not a turning point, or vertex, because the graph does not turn or change direction at that point. The shaping of the graph does

change at the origin, however. When $x < 0$, the graph opens downward, but when $x > 0$, the graph opens upward. A point for which the shaping of a function changes is called a **point of inflection**. Points of inflection are discussed in much more detail in Topic 15.

Transformations can also be applied to cubic functions to transform their location in the plane. The point of inflection of $f(x) = (x - h)^3 + k$ is (h, k).

Example 6: Determine the point of inflection and sketch the graph for each function. Use your graphing calculator *only* when a translation is not being used.

a. $f(x) = x^3 - 1$
b. $f(x) = (x - 1)^3$
c. $f(x) = 3x - x^3$

Solution:

a. The graph of $f(x) = x^3 - 1$ is the graph of $f(x) = x^3$ shifted down 1 unit. The point of inflection is at $(0, -1)$. The graph (Figure 2.8) is an increasing S-shape passing through $(0, -1)$.

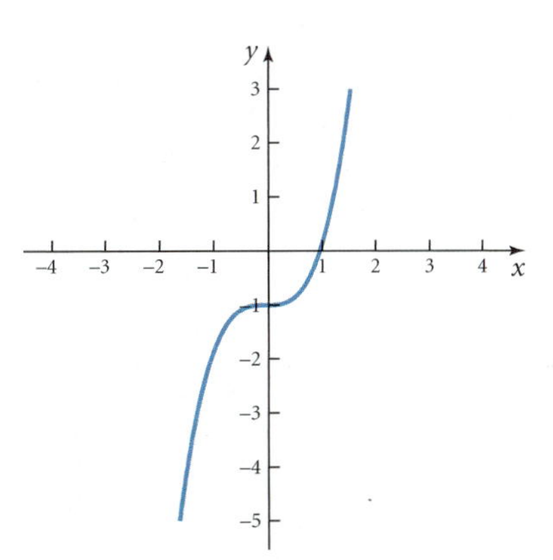

Figure 2.8

b. The graph of $f(x) = (x - 1)^3$ is the graph of $f(x) = x^3$ shifted to the right 1 unit. The point of inflection is at $(1, 0)$. The graph (Figure 2.9) is an increasing S-shape passing through $(1, 0)$.

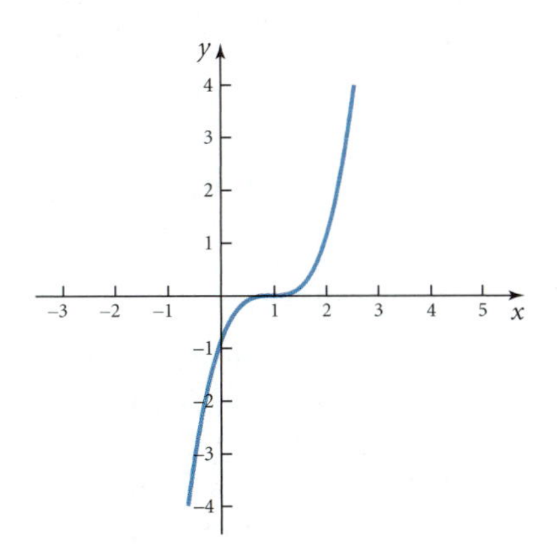

Figure 2.9

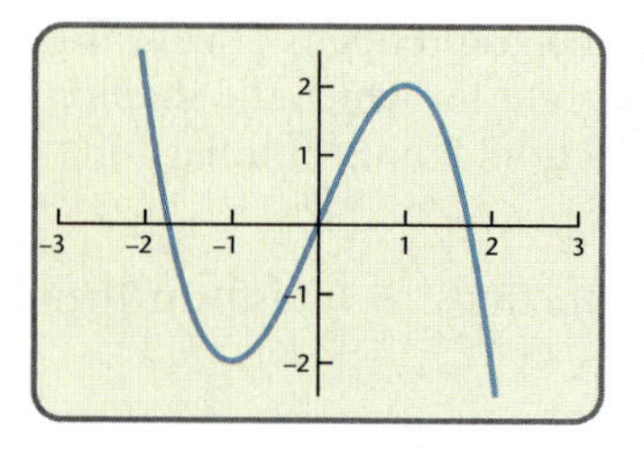

Figure 2.10

c. The graph of $f(x) = 3x - x^3$ will be inverted because of the $-x^3$. The linear term, $3x$, however, makes this function a little more difficult to analyze simply by looking at it. Use your graphing calculator to see what the graph of this function looks like. (See Figure 2.10.) You should see an S-shape with three x-intercepts and with turning points at $x = 1$ and $x = -1$. You should also see an inflection point at $x = 0$. ■

Tip: For polynomial functions, the domain is always the set of real numbers. If the polynomial has an odd degree, the range is also the set of real numbers. If the polynomial has an even degree, the range is restricted and is most easily found by examining the graph of the function.

Calculator Corner 2.6

How many x-intercepts does the graph of a cubic function have? How many turning points does it have? To help you decide, graph these three cubic functions: $y_1 = x^3$, $y_2 = x^3 - 3x$, and $y_3 = x^3 - 3x^2$.

(You should find that the graph of a cubic function has at least one and at most three x-intercepts. The graph of a cubic function has either no turning point or two turning points.)

Check Your Understanding 2.2

Sketch the graph of each of the following functions without using your graphing calculator. Identify the point of inflection of each.

1. $f(x) = x^3 - 3$
2. $f(x) = 2 - x^3$
3. $f(x) = (x + 3)^3 - 12$

Use your calculator to draw the graph of each of the following functions. Determine the point of inflection.

4. $f(x) = x^3 - 4x$
5. $f(x) = x^3 - 3x^2$

Square-Root Functions

A square-root function is of the form $f(x) = \sqrt{p(x)}$, where the expression $p(x)$ is called the **radicand**.

- The domain of a **square-root function** must be restricted because even roots of negative numbers are not defined. The radicand must be nonnegative.
- The range of a square-root function is also restricted because evaluating a square root always yields a nonnegative number.

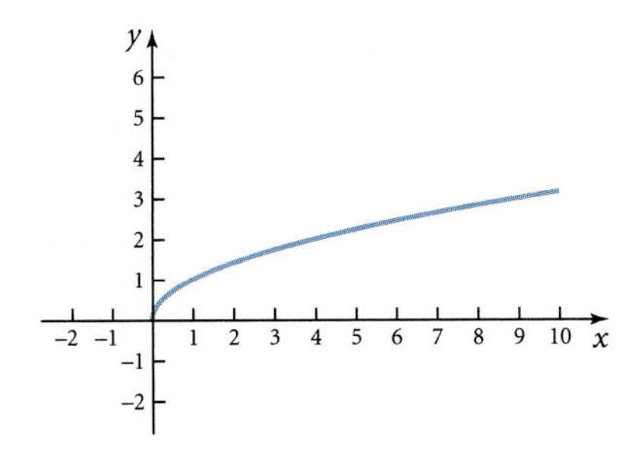

Figure 2.11

- The graph of the basic square-root function $f(x) = \sqrt{x}$ is shown in Figure 2.11. The domain of $f(x) = \sqrt{x}$ is all $x \geq 0$, and the range is all $y \geq 0$. The end point of the graph is $(0, 0)$.

Transformations can also be applied to square-root functions. In the following examples, we sketch several square-root graphs without using a graphing calculator and state the domain and range of each.

Example 7: $f(x) = \sqrt{x} + 4$

Solution: The domain is all $x \geq 0$. The graph is the graph of $f(x) = \sqrt{x}$ shifted 4 units upward, so the range is all $y \geq 4$ and the end point of the graph is $(0, 4)$. (See Figure 2.12.)

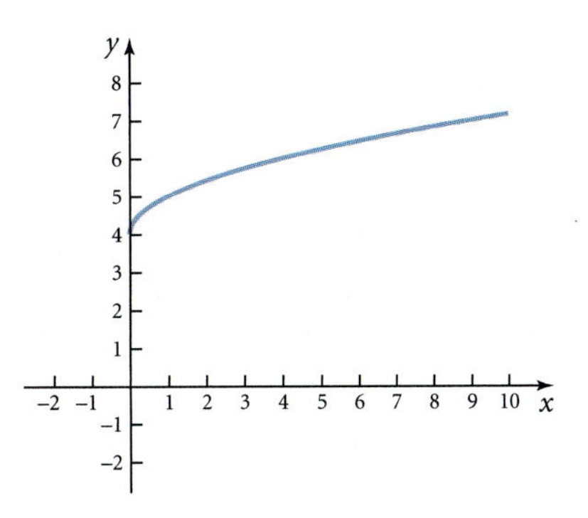

Figure 2.12 ■

Example 8: $f(x) = \sqrt{x + 4}$

Solution: The domain is all values of x for which $x + 4 \geq 0$ or $x \geq -4$. The range is all nonnegative numbers. The end point of the graph is $(-4, 0)$. The graph is the graph of $f(x) = \sqrt{x}$ shifted 4 units to the left. (See Figure 2.13.)

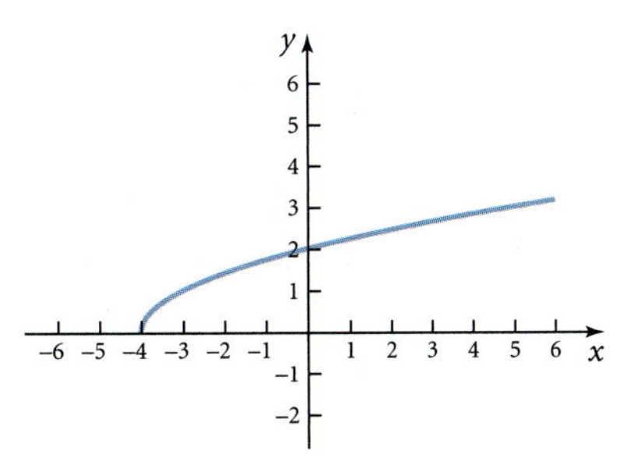

Figure 2.13 ■

(Calculator Corner 2.7)

Use your graphing calculator to graph the square-root function $f(x) = \sqrt{x + 4}$. When you enter the function, be sure to enclose $x + 4$ in parentheses because the entire expression is under the radical sign.

Example 9: $f(x) = 4 - \sqrt{x - 2}$

The domain is all values of x for which $x - 2 \geq 0$ or $x \geq 2$. The graph is the graph of $f(x) = \sqrt{x}$ reflected about the x-axis because of the negative sign in front of the radical and shifted up 4 units and to the right 2 units. The range is all $y \leq 4$. The end point of the graph is (2, 4). (See Figure 2.14.)

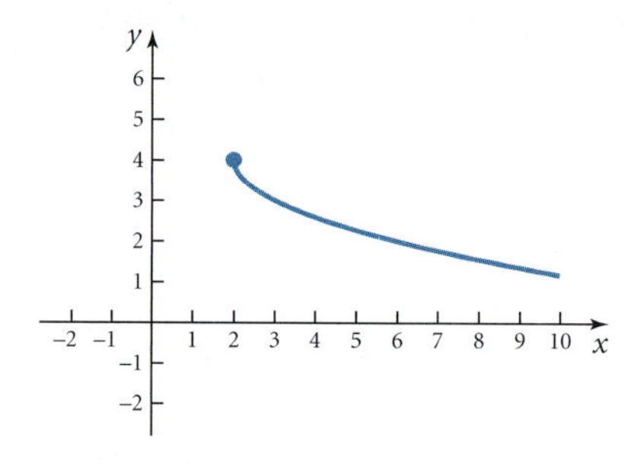

Figure 2.14

Example 10: $f(x) = \sqrt{-x}$

The domain is all values of x for which $-x \geq 0$ or $x \leq 0$. The range is all non-negative numbers. The $-x$ under the radical sign reflects the graph about the y-axis. The graph of $f(x) = \sqrt{-x}$ is the graph of $f(x) = \sqrt{x}$ reflected about the y-axis. The end point of the graph is still at (0, 0). (See Figure 2.15.)

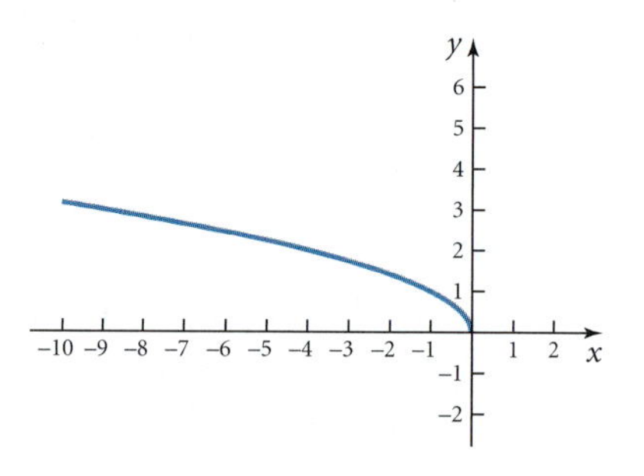

Figure 2.15

Example 11: Weekly operating costs for an airline are given by $C(x) = \sqrt{4 + 1.2x}$, where x is the number of passengers in thousands and C is the cost in millions of dollars.

a. State a reasonable domain for the function.

b. Estimate the weekly costs when there are 20,000 passengers in that week.

c. Estimate the number of passengers in a week with costs of $3 million.

Solution:

a. Because x represents the number of passengers, x cannot be a negative number. The domain is $x \geq 0$.

b. If there are 20,000 passengers, then $x = 20$ and the weekly costs would be $C(20) = \sqrt{4 + 1.2(20)} = 5.29$. If there are 20,000 passengers in a week, then the weekly costs would be about $5.29 million.

c. If weekly costs are $3 million, then $C = 3$ and $3 = \sqrt{4 + 1.2x}$, so $9 = 4 + 1.2x$, which means that $x = 4.167$. Thus, if weekly costs were $3 million, there were about 4167 passengers that week.

MATHEMATICS CORNER 2.3

Visualizing Domain and Range from the Graph of a Function

The domain and range of a function can be determined from the graph of the function. To determine the domain, imagine the shadow on the x-axis if a light were shining directly from above. For example, consider the graph of $y = \sqrt{x - 1} + 2$.

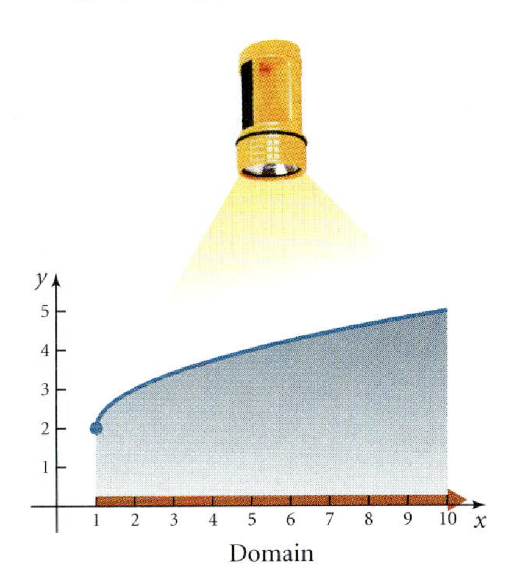

The shadow along the x-axis would run from $x = 1$ out to the right; thus, the domain is $x \geq 1$.

To determine the range, imagine the shadow on the y-axis if a light were shining directly from the side. Again, consider the graph of $y = \sqrt{x - 1} + 2$.

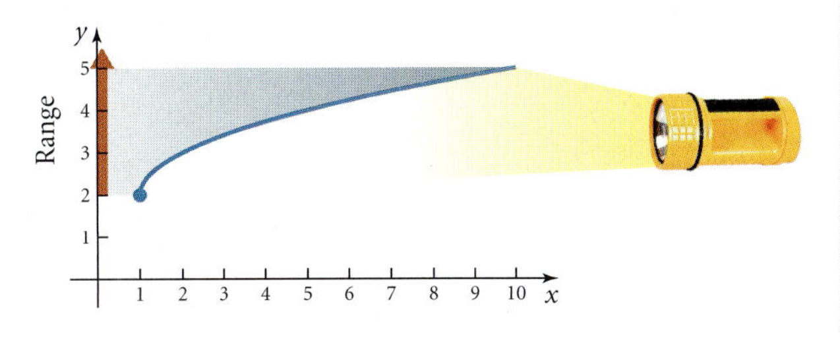

The shadow along the y-axis would run from $y = 2$ upward; thus, the range is $y \geq 2$.

Calculator Corner 2.8

To evaluate $C(20)$ for Example 11, be sure to enter the entire radicand in parentheses. Enter $\sqrt{(4 + 1.2*20)}$. The TI-83/84 will automatically enter the left parenthesis for you.

Check Your Understanding 2.3

Sketch the graph of each of the following square-root functions without using a graphing calculator. State the domain and range of each.

1. $f(x) = \sqrt{x + 2}$
2. $f(x) = 2 - \sqrt{x}$
3. $f(x) = \sqrt{x - 4} - 2$
4. $f(x) = \sqrt{4 - x}$

Check Your Understanding Answers

Check Your Understanding 2.1

1. vertex at $(0, -4)$
domain: all real numbers
range: $y \geq -4$

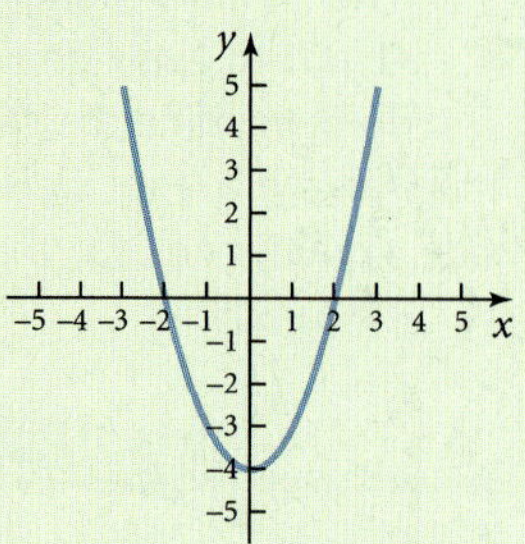

2. vertex at $(4, 0)$
domain: all real numbers
range: $y \geq 0$

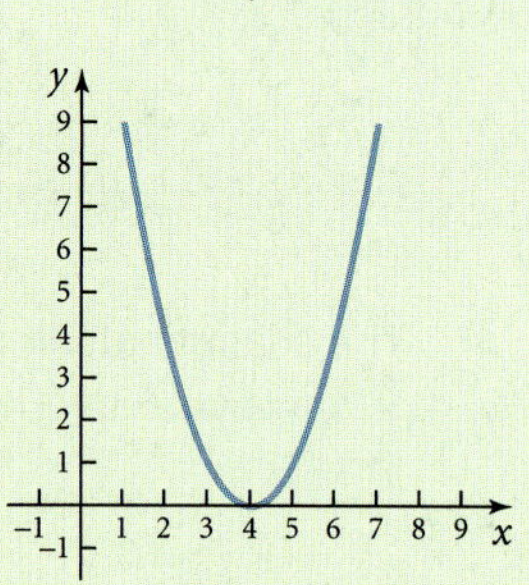

3. vertex at $(0, 4)$
domain: all real numbers
range: $y \leq 4$

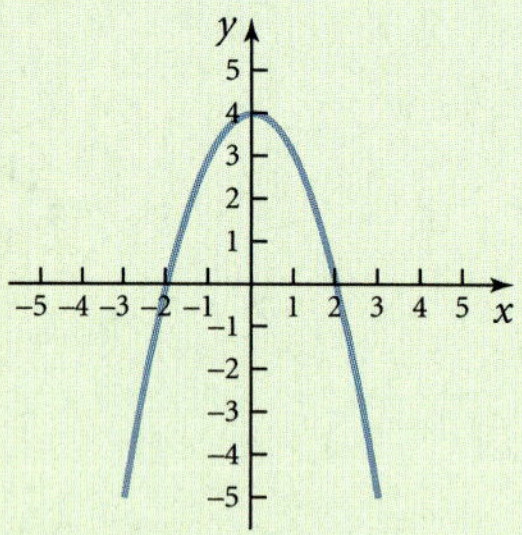

4. vertex at $(-2, 0)$
domain: all real numbers
range: $y \leq 0$

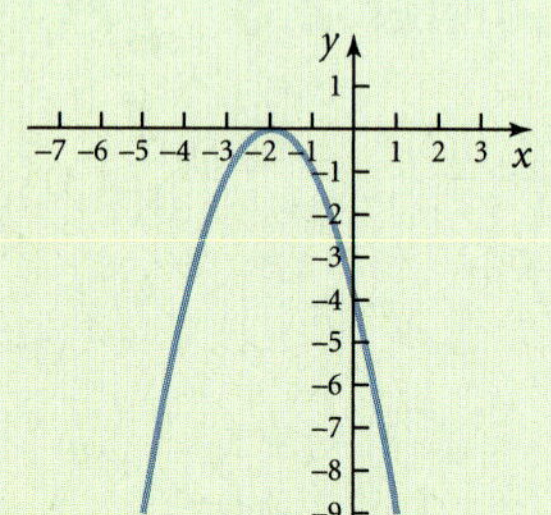

5. vertex at $(1, -3)$
domain: all real numbers
range: $y \geq -3$

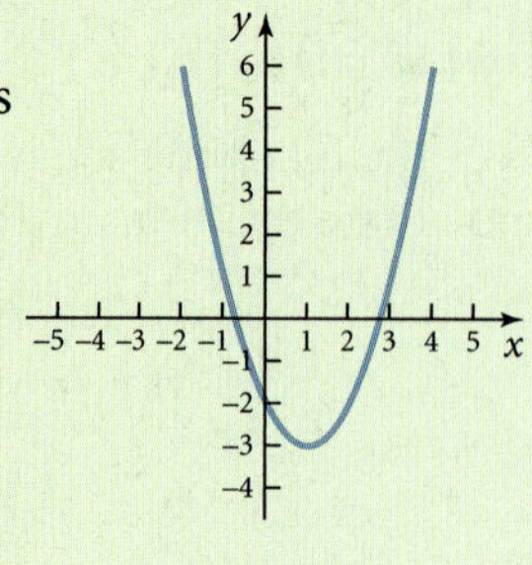

6. vertex at $(-3, 4)$
domain: all real numbers
range: $y \leq 4$

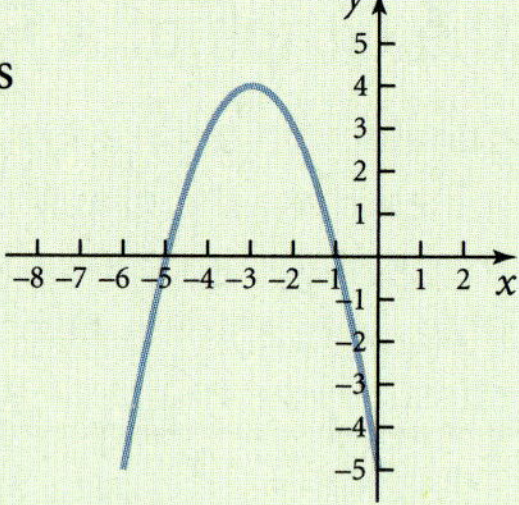

7. vertex at $(1.5, -6.25)$
domain: reals
range: $y \geq -6.25$

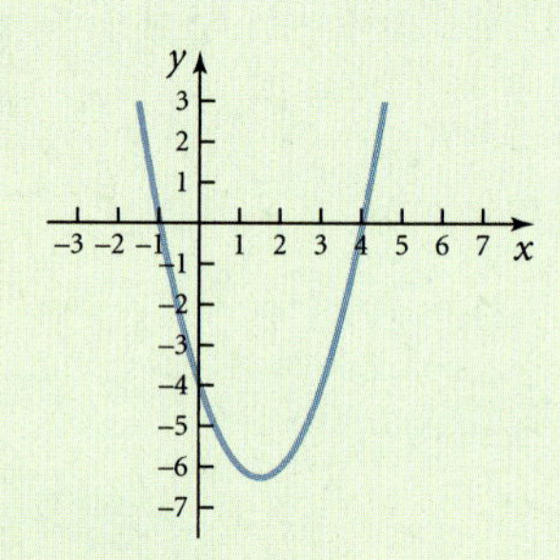

8. vertex at $(-1, 9)$
domain: reals
range: $y \leq 9$

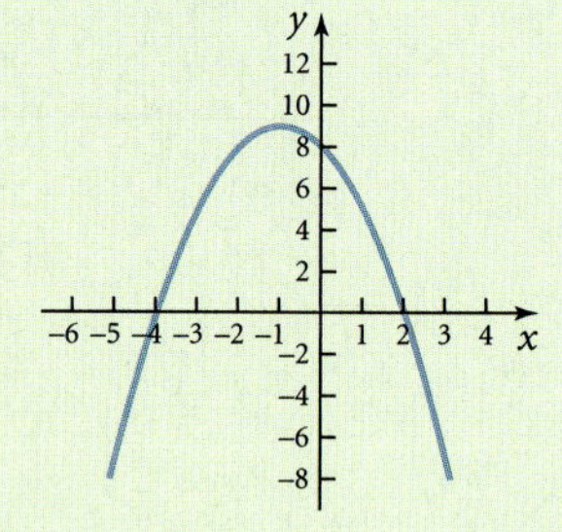

9. vertex at$(-1.25, -0.125)$
domain: reals
range $y \geq -0.125$

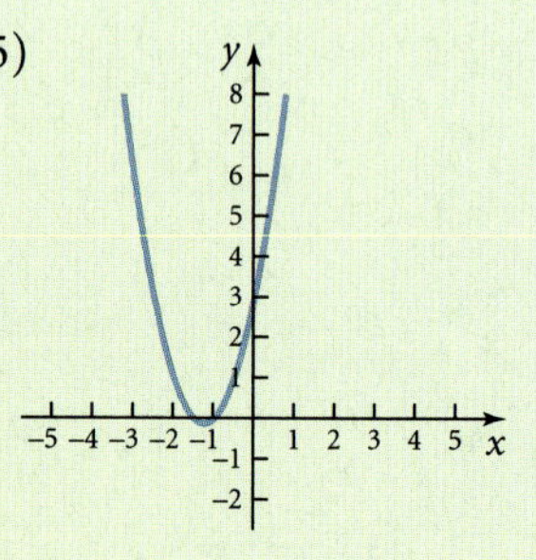

Check Your Understanding 2.2

1. point of inflection at $(0, -3)$

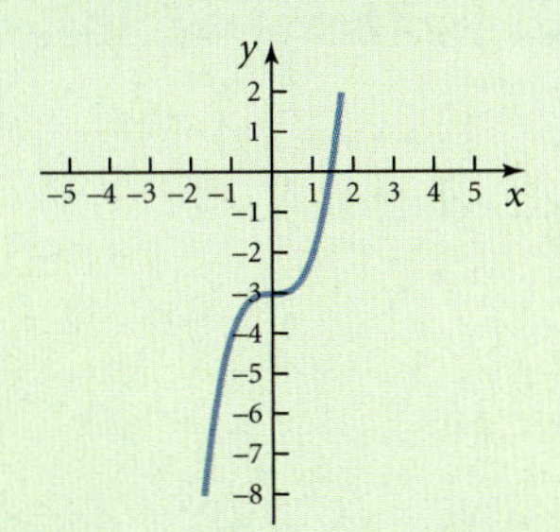

2. point of inflection at $(0, 2)$

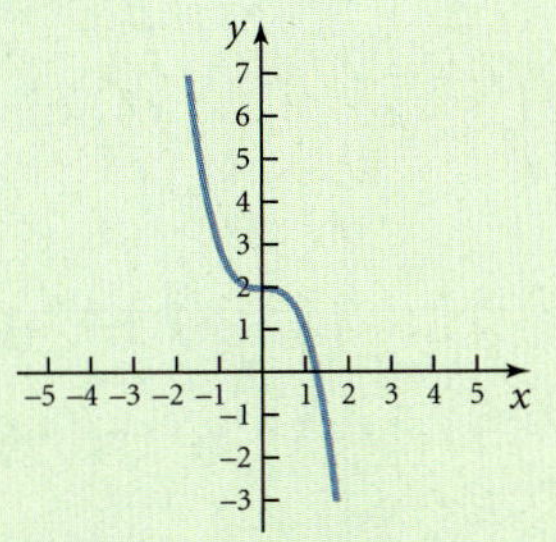

3. point of inflection at $(-3, -1)$

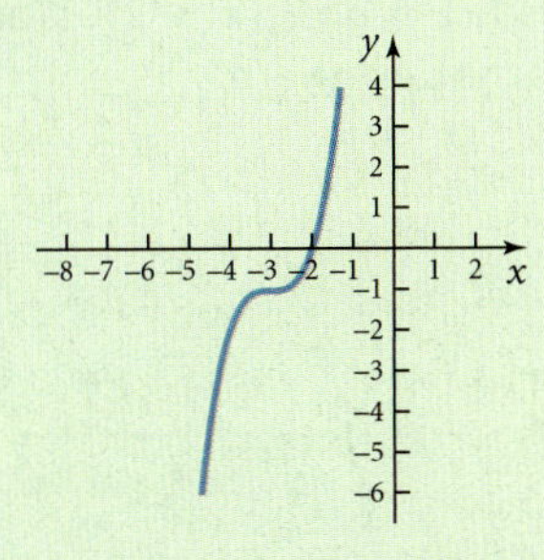

4. point of inflection at $(0, 0)$

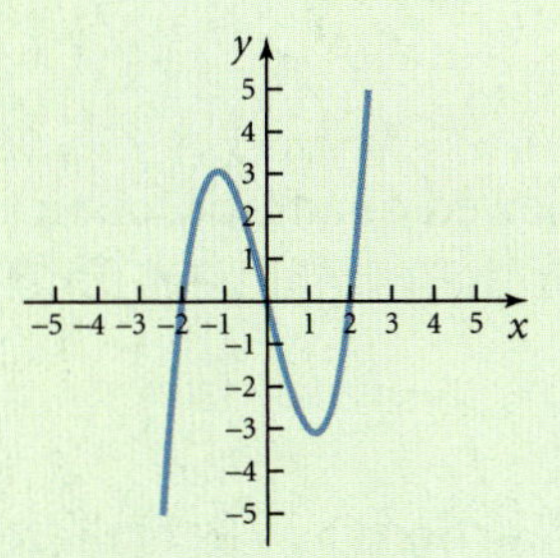

5. point of inflection at $(1, -2)$

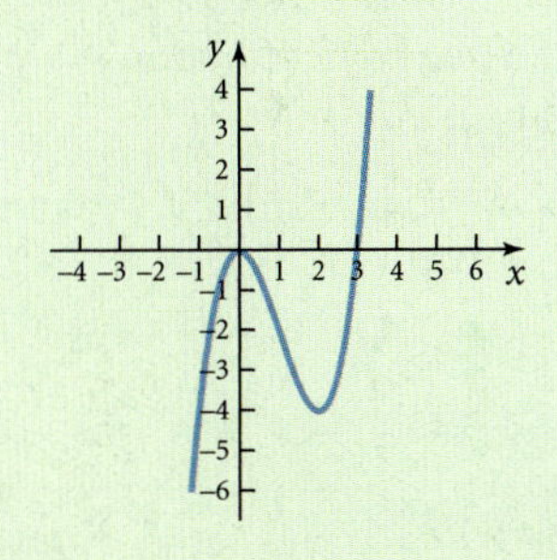

Check Your Understanding 2.3

1. domain: $x \geq -2$
range: $y \geq 0$

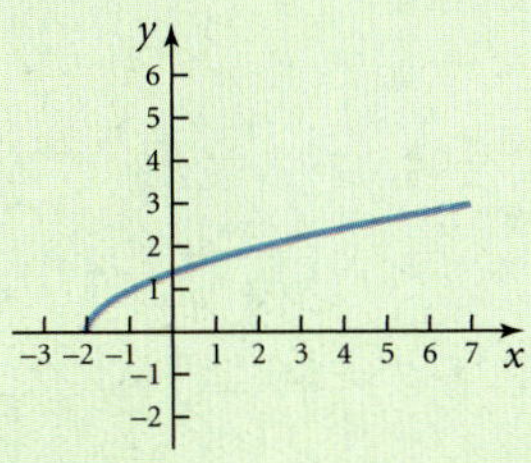

2. domain: $x \geq 0$
range: $y \leq 2$

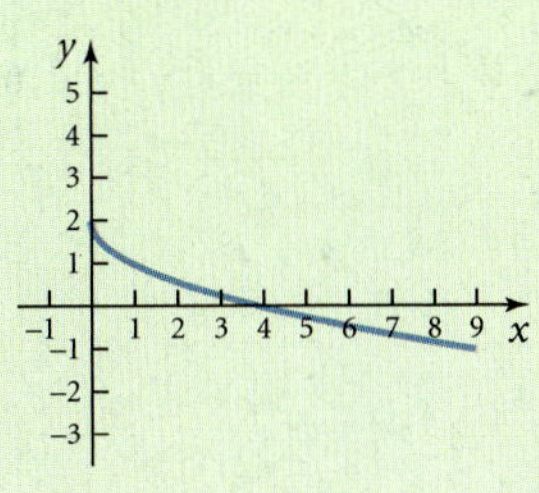

3. domain: $x \geq 4$
range: $y \geq -2$

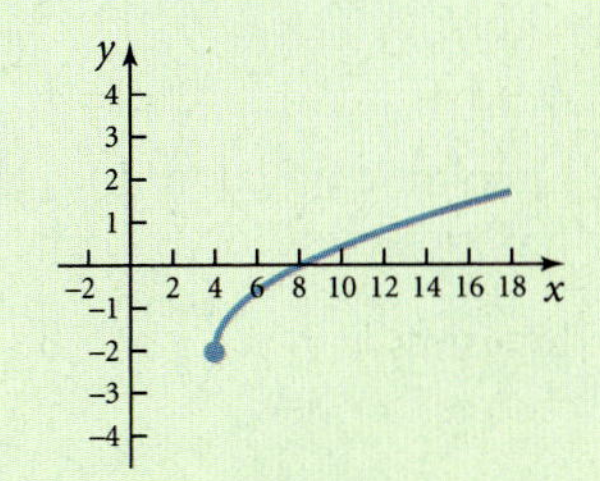

4. domain: $x \leq 4$
range: $y \geq 0$

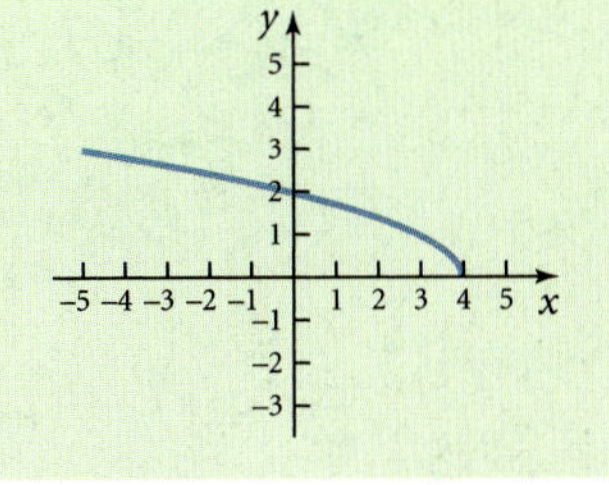

Topic 2 Review

This topic reviewed quadratic, cubic, and square-root functions and related applications of these nonlinear functions.

CONCEPT	EXAMPLE
Quadratic functions may be of the form $f(x) = a(x - h)^2 + k$. The graph of a quadratic function is a **parabola**, with **vertex** at (h, k). The parabola opens upward if $a > 0$ and opens downward if $a < 0$.	The vertex of $f(x) = (x - 2)^2 - 4$ is at $(2, -4)$. The graph opens upward. 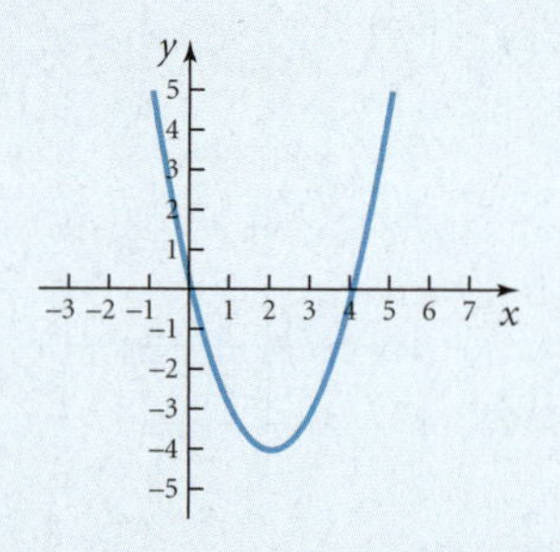 The vertex of $f(x) = 4 - (x + 2)^2$ is at $(-2, 4)$. The graph opens downward. 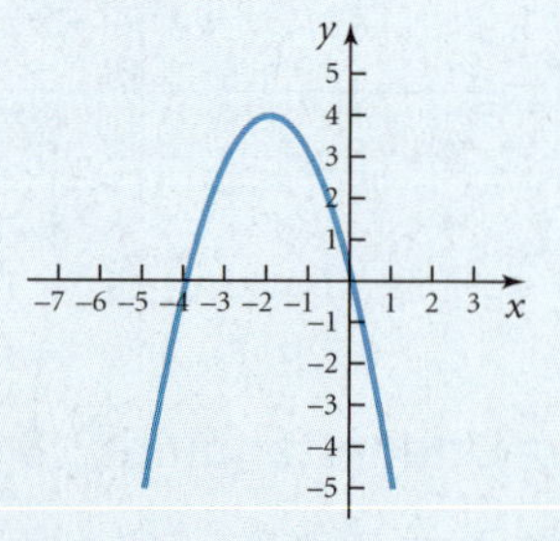
Quadratic functions may also be of the form $f(x) = ax^2 + bx + c$. The vertex may then be found by using $h = -\frac{b}{2a}$ and $k = f(h)$.	The vertex of $f(x) = 2x^2 - 8x + 3$ is at $h = -\frac{-8}{2(2)} = 2$ and $k = f(2) = -5$. 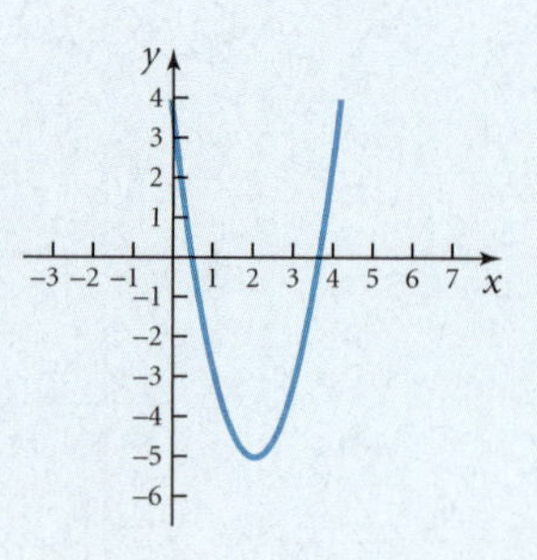

Cubic functions are of the form $f(x) = ax^3 + bx^2 + cx + d$. The graph is an S-shape and has a point of inflection. Cubic functions of the form $f(x) = (x - h)^3 + k$ have a point of inflection at (h, k) and can be graphed using transformations.

$f(x) = (x - 1)^3 - 4$ has a point of inflection at $(1, -4)$.

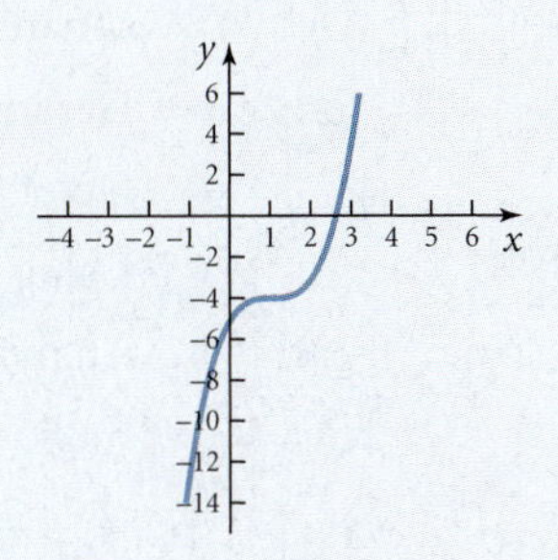

The domain of a **square-root function** is restricted to those values of x for which the radicand is nonnegative. The range is also restricted because evaluating a square-root always yields a nonnegative number.

The domain of $f(x) = \sqrt{x - 3} + 2$ is all $x \geq 3$. The range is all $y \geq 2$.

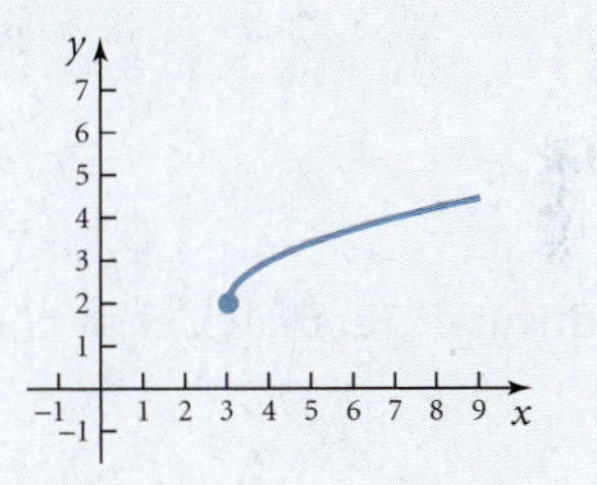

In business applications, the **break-even point** is the point for which costs equal revenue. **Profit** is defined as *revenue minus costs*. If profit is given by a quadratic function, the maximum profit occurs at the vertex of the graph.

The cost of manufacturing x canoes is $C(x) = 4{,}000 + 250x$, and the revenue generated from selling x canoes is $R(x) = 850x - x^2$. The break-even point occurs if $C(x) = R(x)$, or $4{,}000 + 250x = 850x - x^2$, which gives $x \approx 6.74, 593.3$. The company breaks even (costs equal revenue) if either 7 canoes or 593 canoes are produced and sold. The profit function is

$$\begin{aligned} P(x) &= (850x - x^2) - (4000 + 250x) \\ &= -x^2 + 600x - 4000. \end{aligned}$$

Maximum profit occurs at $x = 300$. If 300 canoes are produced and sold, the company will receive a maximum profit of \$86,000.

NEW TOOLS IN THE TOOL KIT

- Sketching graphs of quadratic functions without a graphing calculator, using transformations
- Sketching graphs of cubic functions without a graphing calculator, using transformations
- Sketching graphs of square-root functions without a graphing calculator, using transformations

- Drawing graphs of quadratic functions, cubic functions, and square-root functions with a graphing calculator
- Identifying the domain and range of quadratic functions, cubic functions, and square-root functions
- Determining the vertex of a quadratic function
- Determining the point of inflection of a cubic function
- Using cost, revenue, and profit functions to determine maximum profit
- Analyzing business applications using quadratic functions

Topic 2 Exercises

2

In Exercises 1 through 12, match each equation to its graph.

1. $y = x^2 - 3$

2. $y = 3 - x^2$

3. $y = (x - 3)^2$

4. $y = (x - 2)^2 - 1$

5. $y = x^3 + 2$

6. $y = (x + 2)^3$

7. $y = 4 - x^3$

8. $y = 2 - (x - 1)^3$

9. $y = -\sqrt{x}$

10. $y = \sqrt{x - 2}$

11. $y = \sqrt{x} - 2$

12. $y = \sqrt{-x}$

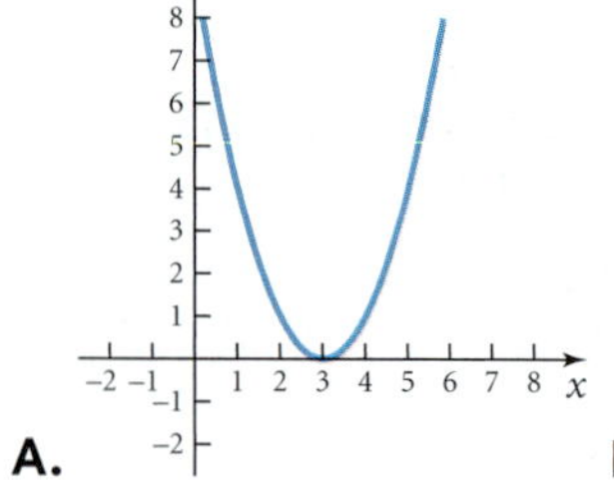

A.

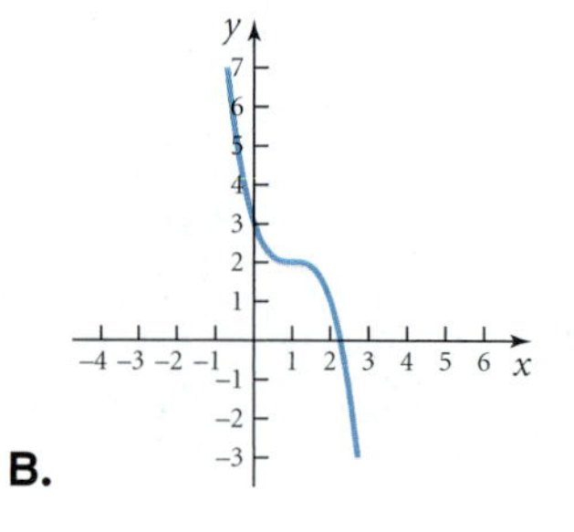

B.

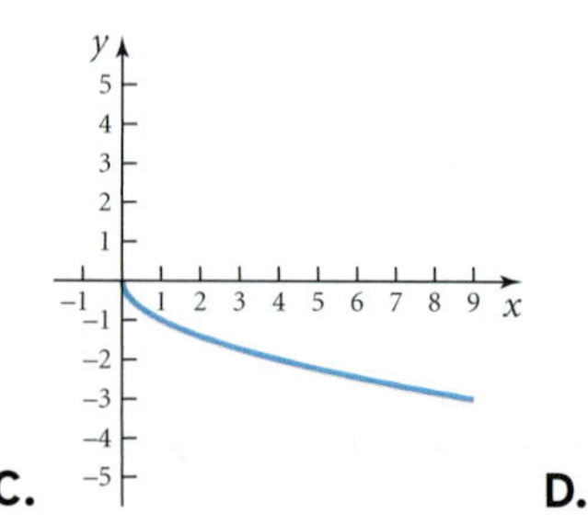

C.

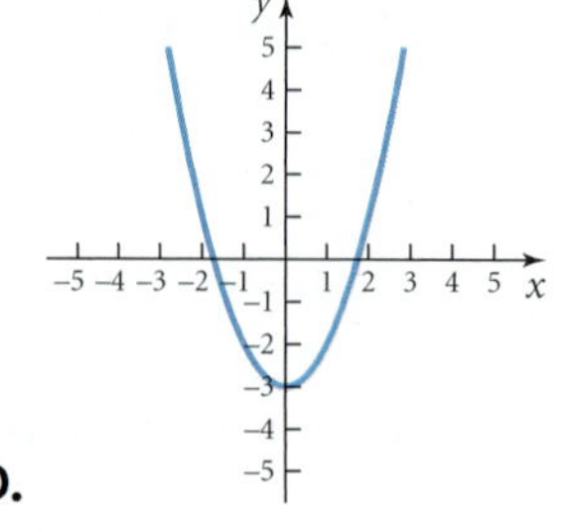

D.

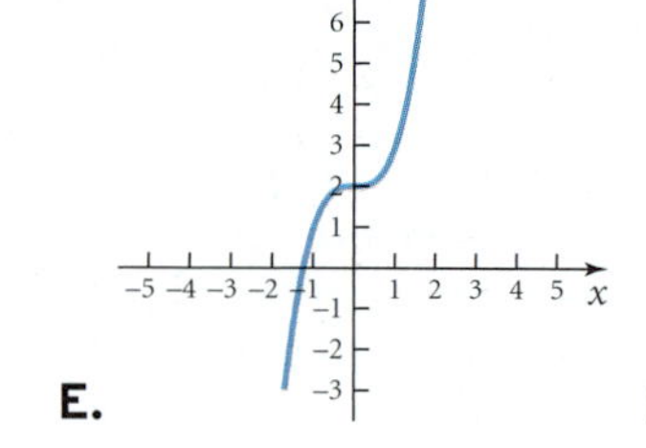

E.

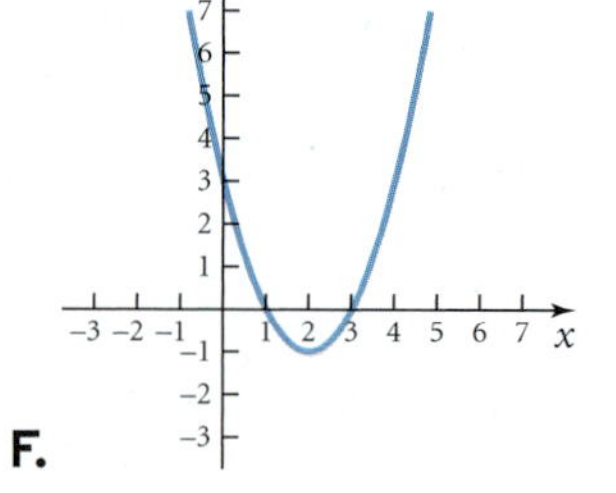

F.

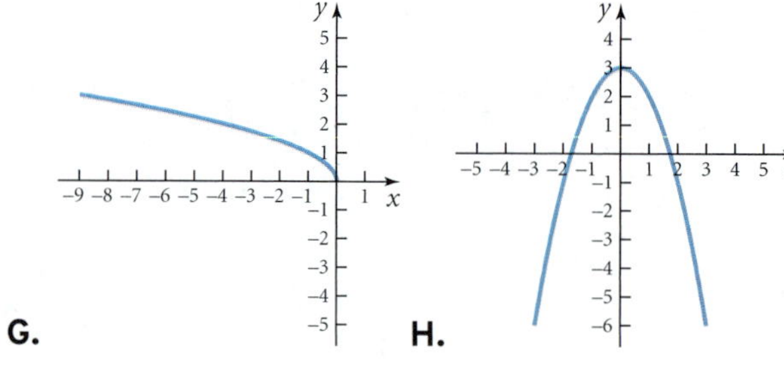

G. **H.**

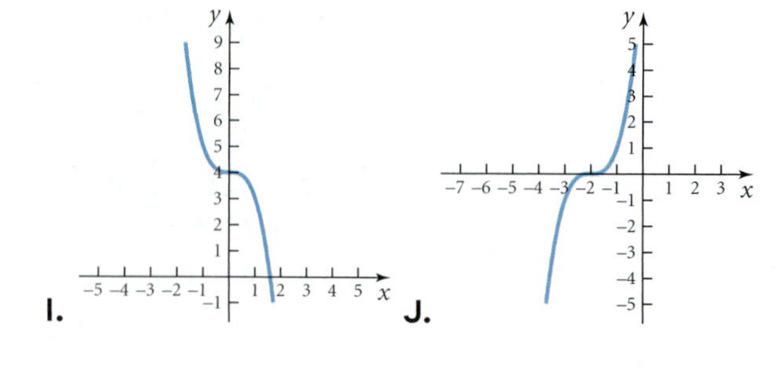

I. **J.**

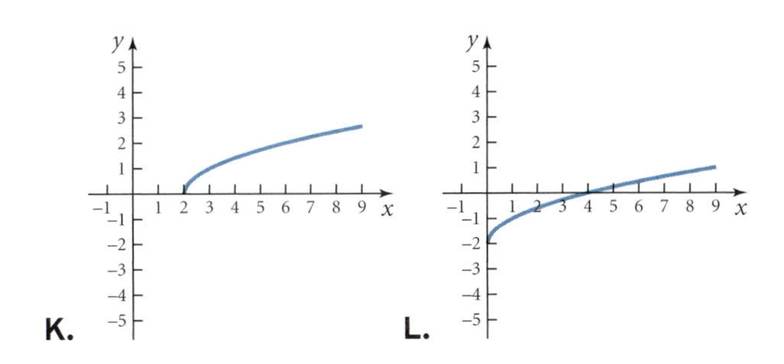

Determine the vertex of each of the following quadratic functions. Sketch the graph without using a graphing calculator. State the domain and range of each.

13. $y = x^2 - 2$ **14.** $y = x^2 + 5$

15. $y = 3 - x^2$ **16.** $y = -2 - x^2$

17. $y = (x + 4)^2 + 3$ **18.** $y = (x - 2)^2 + 1$

19. $y = 5 - (x - 1)^2$ **20.** $y = 3 - (x + 2)^2$

Determine the point of inflection of each of the following cubic functions. Draw the graph of each without using a graphing calculator.

21. $y = x^3 - 2$ **22.** $y = x^3 + 1$

23. $y = (x - 2)^3$ **24.** $y = (x + 1)^3$

25. $y = (x - 1)^3 - 2$ **26.** $y = (x + 2)^3 + 1$

Use a graphing calculator to graph each of the following functions. Determine the vertex of the quadratic functions and the point of inflection of the cubic functions. State the domain and range of each function.

27. $y = x^2 - 4x$ **28.** $y = x^2 + 6x$

29. $y = x^2 + 6x + 4$ **30.** $y = x^2 - 8x - 2$

31. $y = x^3 + 1$ **32.** $y = 2 - x^3$

33. $y = 2 - (x - 1)^3$ **34.** $y = (x + 2)^3 - 4$

Draw the graph of each of the following square-root functions without using a graphing calculator. State the domain and range of each function.

35. $y = \sqrt{x} - 3$ **36.** $y = \sqrt{x} + 4$

37. $y = \sqrt{x} + 2$ **38.** $y = \sqrt{x} - 5$

39. $y = 4 - \sqrt{x}$ **40.** $y = 3 - \sqrt{x}$

41. $y = \sqrt{x - 5}$ **42.** $y = \sqrt{x + 2}$

43. $y = 1 + \sqrt{x - 3}$ **44.** $y = 2 + \sqrt{x + 1}$

45. $y = 2 - \sqrt{x + 3}$ **46.** $y = 3 - \sqrt{x + 1}$

47. $y = 3 + \sqrt{2 - x}$ **48.** $y = 2 - \sqrt{3 - x}$

49. The number of new AIDS cases reported each year in the United States can be modeled by $A(x) = -680x^2 + 19{,}307x - 74{,}310$, where x is the number of years after 1980.

a. Graph the function, using a window of [0, 25] for x and [0, 75,000] for y. Set xScl at 5 and yScl at 5000.

b. Find the vertex of the equation and interpret its meaning.

50. The annual unemployment rates in the United States can be modeled by $E(x) = 0.005x^2 - 0.2x + 7.5$, where x is the number of years after 1980.

a. Graph the function, using a window of [0, 60] for x and [0, 10] for y. Set xScl at 5 and yScl at 1.

b. Find the vertex of the equation and interpret its meaning.

51. Annual hotel industry profit since 2000 can be modeled by $P(x) = 1.1x^2 - 5.97x + 22.1$, where x is the number of years after 2000 and P is profit in billions of dollars.

a. Estimate the profit in 2007.

b. About what year were profits at their minimum level? What was that profit?

52. The height, in feet, of a baseball t seconds after being hit is given by $h(t) = -5.8t^2 + 46.5t - 2.7$.

a. Estimate the height of the ball after three seconds.

b. What is the ball's maximum height?

53. The Consumer Price Index (CPI) tracks prices of consumer goods in the United States. For instance, the CPI was 13.9 in 1940 and 168.8 in 2000, which means that an item costing $13.90 in 1940 cost $168.80 in 2000. The CPI can be modeled by $P(x) = 0.00047x^3 - 0.042x^2 + 1.25x + 2.77$, where x is the number of years after 1900. Estimate the CPI for 1950, 1980, and 2005 and interpret the meaning of your answers.

54. The percent of young adult men (age 18 – 24) in the United States who smoke cigarettes can be modeled by $M(x) = 0.0013x^3 - 0.2713x^2 + 17.96x - 315$, where x is the number of years after 1900. Estimate the percent of young adult men who smoked in 1975, 1996, and 2005 and interpret the meaning of your answers.

55. The number of consumer bankruptcy filings, in thousands, is given by $B(x) = 6.9x^3 - 190.4x^2 + 1744x - 3931$, where x is the number of years after 1995.

a. Approximately how many consumer bankruptcy filings were there in 2003?

b. Graph the function and estimate when there will be two million filings.

56. The population of West Virginia, in thousands of people, since 1900 can be modeled by $P(x) = 0.00225x^3 - 0.554x^2 + 41.8x + 907$.

a. Estimate the population in 2005.

b. Graph the function and estimate when the population will be two million.

57. A video store finds that the revenue function for a new DVD is $D(p) = \sqrt{200 - p^2}$, where p is the price in dollars of the DVD and D is the number of units demanded, in thousands, at that price. What is the demand if the price is $10?

58. A certain chemotherapy dosage depends on the patient's height and weight and is given by $s = \frac{\sqrt{hw}}{15}$, where height, h, is in inches; weight, w, is in pounds; and s is in milligrams.

a. For a man who is 70 inches tall and weighs 220 pounds, what is the dosage?

b. If the man's weight drops to 190 pounds, what is the dosage?

59. A manufacturer of outboard motors determines its weekly cost to be $C(x) = 40{,}000 + 200x$ and its weekly revenue to be $R(x) = -2x^2 + 800x$, where x is the number of motors manufactured each week. Graph the cost and revenue functions. Determine the break-even point and interpret its meaning.

60. A-to-Z Electronics installs home security systems. The firm's monthly costs are $C(x) = 42{,}000 + 120x$, and its monthly revenues are $R(x) = -4x^2 + 1160x$, where x is the number of systems installed that month. Graph the cost and revenue functions. Determine the break-even point and interpret its meaning.

61. Refer to the outboard motor manufacturer information in Exercise 59. How many motors must be manufactured each week for maximum weekly profit? What is the maximum profit that week?

62. Refer to the A-to-Z Electronics information in Exercise 60. How many systems should be installed each month for maximum monthly profit? What is the maximum profit for that month?

CALCULATOR CONNECTIONS

Use your graphing calculator as needed to answer the following questions.

63. **a.** Factor $x^3 - 4x$.

b. Solve $x^3 - 4x = 0$.

c. Graph $y = x^3 - 4x$ and determine the x-intercepts.

d. Explain how your answers to parts a, b, and c are related.

64. **a.** Factor $x^4 - 3x^2 - 4$.

b. Solve $x^4 - 3x^2 - 4 = 0$.

c. Graph $y = x^4 - 3x^2 - 4$ and determine the x-intercepts.

d. Explain how your answers to parts a, b, and c are related.

65. **a.** Graph $f(x) = 0.5x^2 - 2\sqrt{x} + 1$.

b. Estimate the solutions of $0.5x^2 - 2\sqrt{x} + 1 = 0$ to three decimal places.

66. **a.** Graph $f(x) = \sqrt{x^3 + 4x^2} - 2$.

b. Estimate the solutions of $\sqrt{x^3 + 4x^2} - 2 = 0$ to three decimal places.

In Exercises 67 and 68, use a graphing calculator to graph the function. From the graph, determine the domain and the range of the function.

67. $f(x) = \sqrt{x^2 - 4x}$

68. $f(x) = \sqrt{x^3 - 4x}$

69. Market supply and market demand, in hundreds, for CD players are given by $S(x) = 200 + 15x$ and $D(x) = -1.5x^2 + 100x$, where x is the cost of a CD player. Find the equilibrium point and interpret its meaning.

70. Market supply and market demand, in thousands, for calculators are given by $S(x) = 50 + 5x$ and $D(x) = \sqrt{400 + 2500x}$, where x is the cost of the calculator. Find the equilibrium point and interpret its meaning.

PROJECT 2.1

Quadratic Functions

According to the American Red Cross, the percent of Americans in various age groups who donate blood is given in the following graph.

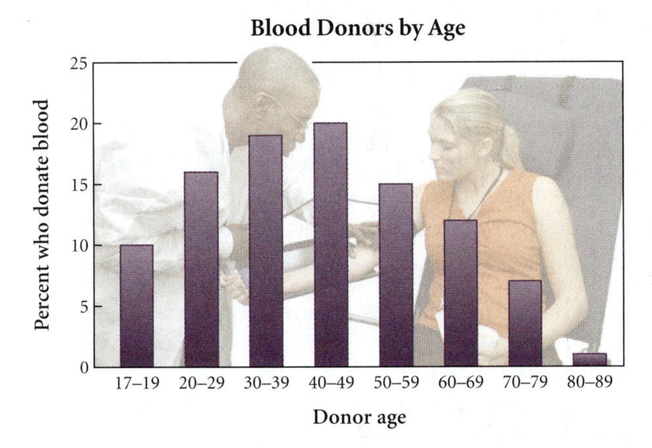

Data from: The American Red Cross

1. Why is the graph of the data considered quadratic? What is the domain of the data? What is the range of the data?
2. Write a quadratic function relating the percent of each age group who donate blood. Use the midpoint of each age group as the x-coordinate ($x = 18, 25, 35, 45, 55, 65, 75$, and 85). Select three data points and substitute them into one of the general forms of a quadratic equation, $f(x) = ax^2 + bx + c$ or $f(x) = a(x - h)^2 + k$. You will then have a system of three equations in three variables to solve.
3. As donor age increases, what is the effect on blood donorship?
4. Use the graph above to locate the maximum donorship. Now locate the vertex of the quadratic function from Question 2. How closely does your equation reflect the maximum donorship age?

PROJECT 2.2

Cubic Functions

In 2000, the U.S. Bureau of the Census reported the percentage of the U.S. population that is foreign born.

Year	% Foreign born	Year	% Foreign born
1900	13.6	1960	5.4
1910	14.7	1970	4.8
1920	13.2	1980	6.2
1930	11.6	1990	7.9
1940	8.8	1998	9.3
1950	6.9	2000	11.1
		2005	12.1

1. Plot the data points. Why do the data seem to follow a cubic pattern?
2. As time increases, how does the percentage of the U.S. population that is foreign born change? Where are the turning points, and what is their meaning? Where are the points of inflection?
3. Write a cubic equation that describes the data. Select four data points and substitute them into the general form of a cubic equation, $f(x) = ax^3 + bx^2 + cx + d$. You will have a system of four equations with four unknowns (a, b, c, and d) to solve.

GROUP PROJECTS

These projects provide the opportunity to explore in more detail the behavior of the graphs of quadratic functions (degree two), cubic functions (degree three), and quartic functions (degree four).

GROUP PROJECT 2.1

Quadratic Functions

The graph of the quadratic function $y = ax^2 + bx + c$ is a parabola with one turning point called the vertex. Let's explore the roles of the coefficients a, b, and c on the graph.

1. Graph $y_1 = x^2, y_2 = 2x^2, y_3 = \left(\frac{1}{2}\right)x^2$, and $y_4 = -2x^2$. You will see that the vertex of each graph is the origin. Describe in your own words how the different values of the quadratic coefficient, a, affect the graph.

2. Graph $y_1 = x^2, y_2 = x^2 + 3$, and $y_3 = x^2 - 4$. Do the graphs of these three functions have the same vertex? Describe in your own words how the values of the constant, c, affect the graph.
3. Now let's explore the effect of the linear coefficient, b. Graph $y_1 = x^2, y_2 = x^2 - 2x, y_3 = x^2 - 4x$, $y_4 = x^2 + 2x, y_5 = x^2 + 4x$, $y_6 = x^2 - 6x$, and $y_7 = x^2 + 6x$. How does the linear coefficient, b, change the graph? Make a plot of the vertices of each of the graphs. What do you see?
4. How many x-intercepts might a parabola have? Give specific examples to support your answer.
5. How many vertices might a parabola have?
6. Write a general description for the graph of a quadratic function.

GROUP PROJECT 2.2

Cubic Functions

The general form of a cubic function is $f(x) = ax^3 + bx^2 + cx + d$. The graph of a cubic function is an S-shaped figure. Let's explore the roles of the coefficients a, b, c, and d on the graph.

1. What is the role of the cubic coefficient, a? Graph $y_1 = x^3, y_2 = 2x^3, y_3 = \left(\frac{1}{3}\right)x^3, y_4 = -2x^3$, $y_5 = x^3 - 4x$, and $y_6 = -x^3 + 4x$. Based on your graphs, how does the value of a affect the graph?
2. What is the role of the constant, d? Graph $y_1 = x^3 - 4x, y_2 = x^3 - 4x + 3$, and $y_3 = x^3 - 4x - 2$. Based on your graphs, how does the value of d affect the graph?
3. What is the role of the quadratic coefficient, b, and the linear coefficient, c? Graph $y_1 = x^3$, $y_2 = x^3 - 4x^2, y_3 = x^3 - 4x, y_4 = x^3 + 2x$, $y_5 = x^3 + 0.1x$, and $y_6 = -x^3 - 4x^2 + 2x - 3$. Based on your graphs, how do the various values of b and c affect the graph?
4. How many x-intercepts might the graph of a cubic function have? Give specific examples to support your answer.
5. How many turning points might the graph of a cubic function have? Give specific examples to support your answer.
6. Write a general description for the graph of a cubic function.

GROUP PROJECT 2.3

Quartic Functions

The general form of a quartic equation is $f(x) = ax^4 + bx^3 + cx^2 + dx + e$. The graph may be a U-shape similar to a parabola or a W-shape. Let's explore the roles of the coefficients a, b, c, d, and e on the graph.

1. What is the role of the quartic coefficient, a, on the graph? Graph $y_1 = x^4, y_2 = 2x^4, y_3 = \left(\frac{1}{3}\right)x^4$, $y_4 = -x^4$, and $y_5 = -3x^4$. Based on your graphs, how does the value of a affect the graph?
2. What is the role of the constant, e, on the graph? Graph $y_1 = x^4, y_2 = x^4 + 5$, and $y_3 = x^4 - 2$. Based on your graphs, how does the value of e affect the graph?
3. What is the role of the quadratic coefficient, c, on the graph? Graph $y_1 = x^4 - 2, y_2 = x^4 + 4x^2 - 2$, and $y_3 = x^4 - 3x^2 - 2$. Based on your graphs, how does the value of c affect the graph?
4. What are the roles of the cubic coefficient, b, and the linear coefficient, d, on the graph? Graph $y_1 = x^4 - 2x^3$, $y_2 = x^4 - 2x^3 + 3x$, $y_3 = x^4 - 4x$, $y_4 = x^4 + 4x^3$, and $y_5 = x^4 + 4x$. Based on your graphs, how do the various values of b and d affect the graph?
5. How many x-intercepts might the graph of a quartic function have? Give specific examples to support your answers.
6. How many turning points might the graph of a quartic function have? Give specific examples to support your answer.
7. Write a general description for the graph of a quartic function.

TOPIC 3

Rational and Piecewise Defined Functions

Many of the functions used in calculus are not simple linear, quadratic, or cubic functions. For instance, working Americans must pay federal income taxes each year. These taxes are based on a person's income, but are graduated so that those with higher incomes pay a higher percentage in taxes. The function describing the tax due is a series of linear functions, one for each income bracket. Functions of this sort are called piecewise defined functions.

Suppose a refrigerator is purchased for $2000 with an additional service contract costing $150 per year. The average cost of the refrigerator per year is given by $C(x) = \frac{2000 + 150x}{x} = \frac{2000}{x} + 150$. Here the variable in the denominator creates another type of algebraic function called a rational function.

Both rational functions and piecewise defined functions are reviewed in this topic. Before beginning the review, be sure to complete the Warm-up Exercises to sharpen the necessary algebra skills.

IMPORTANT TOOLS IN TOPIC 3

- *Rational functions*
- *Horizontal and vertical asymptotes*
- *Graphs of rational functions*
- *Piecewise defined functions*
- *Step functions*

TOPIC 3 WARM-UP EXERCISES

Be sure you can successfully complete the following exercises before starting Topic 3.

1. Solve each of the following equations for x.

a. $x + 3 = 0$ **b.** $5 - 3x = 0$

c. $x^2 - 4 = 0$ **d.** $x^2 - 4x = 0$

e. $x^2 - 4x + 3 = 0$ **f.** $x^3 - 9x = 0$

2. State the domain of each of the following functions.

a. $f(x) = x^2 - 3$ **b.** $f(x) = \sqrt{x - 3}$

c. $f(x) = \frac{1}{x}$

3. Given $f(x) = \dfrac{2x^3 - 5x + 1}{x - 4}$, evaluate each of the following.

a. $f(1)$ **b.** $f(-2)$

c. $f(0)$ **d.** $f(4)$

4. Graph each of the following equations. State the slope and y-intercept of the linear equations, the vertex of the quadratic equations, and the point of inflection of the cubic equations.

a. $y = 2x - 5$ **b.** $3x + 2y = 6$

c. $y = x^2 - 3$ **d.** $y = (x - 3)^2$

e. $y = 3 - x^3$

5. Simplify $\dfrac{3x - 3}{x^2 - 1}$.

6. Divide $\dfrac{3x^2}{x - 4}$.

Answers to Warm-up Exercises

1. a. $x = -3$ **b.** $x = \frac{5}{3}$

c. $x = 2, -2$ **d.** $x = 0, 4$

e. $x = 1, 3$ **f.** $x = 0, 3, -3$

2. a. all real numbers

b. $x \geq 3$

c. $x \neq 0$

3. a. $\frac{2}{3}$ **b.** $\frac{5}{6}$

c. $-\frac{1}{4}$ **d.** undefined

4. a. slope 2; y-intercept $(0, -5)$

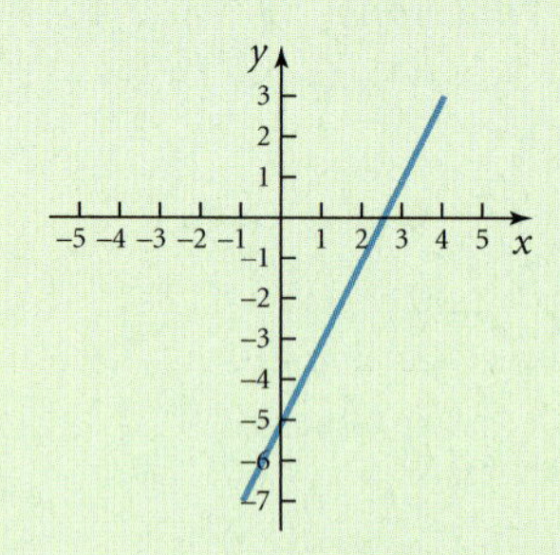

b. slope $-\frac{3}{2}$; y-intercept $(0, 3)$

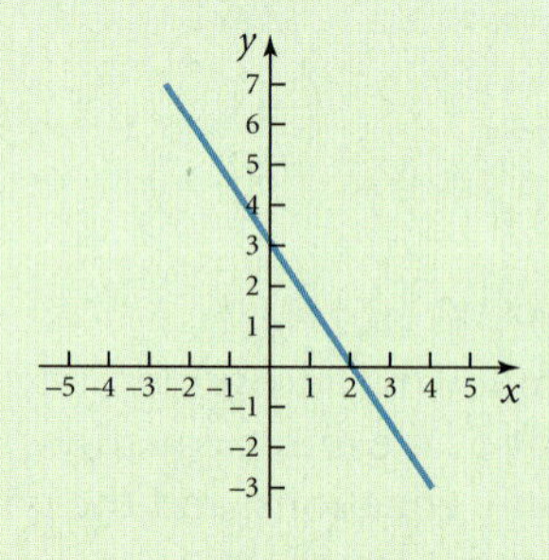

c. Vertex $(0, -3)$

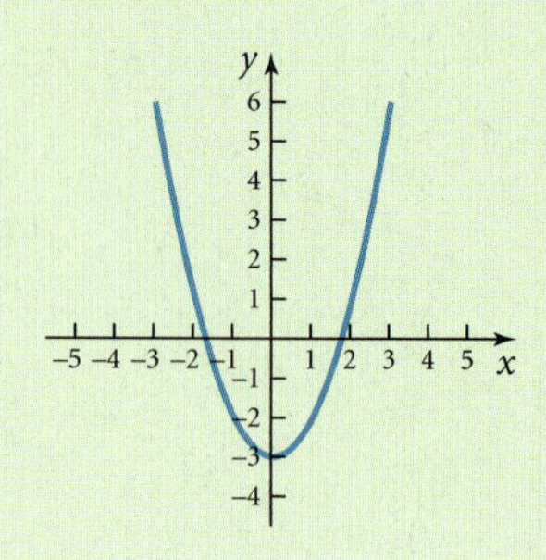

d. Vertex $(3, 0)$

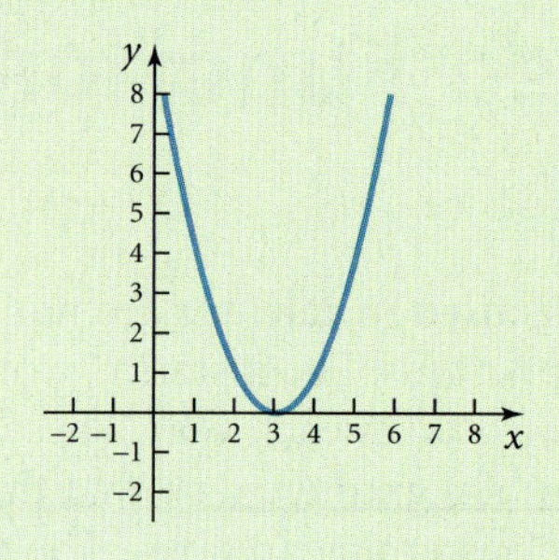

e. point of inflection $(0, 3)$

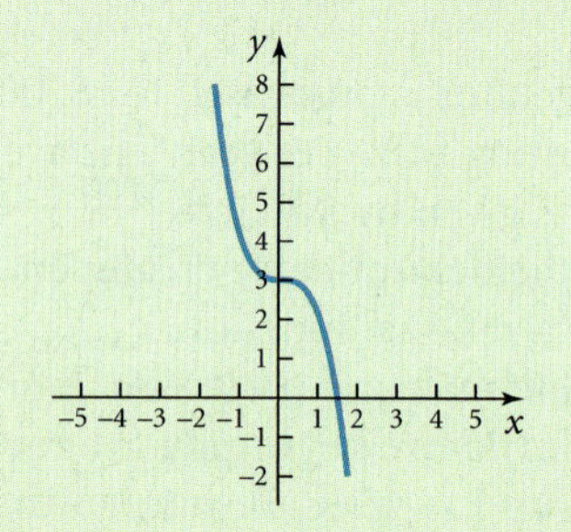

5. $\dfrac{3}{x+1}$

6. $\dfrac{3x^2}{x-4} = 3x + 12 + \dfrac{48}{x-4}$

Rational Functions

Recall that a **polynomial** is an algebraic expression whose terms are all of the form ax^n, where a is a real number and n is a nonnegative integer. A **rational function** is a ratio of two polynomial functions, written in the form $f(x) = \frac{p(x)}{q(x)}$.

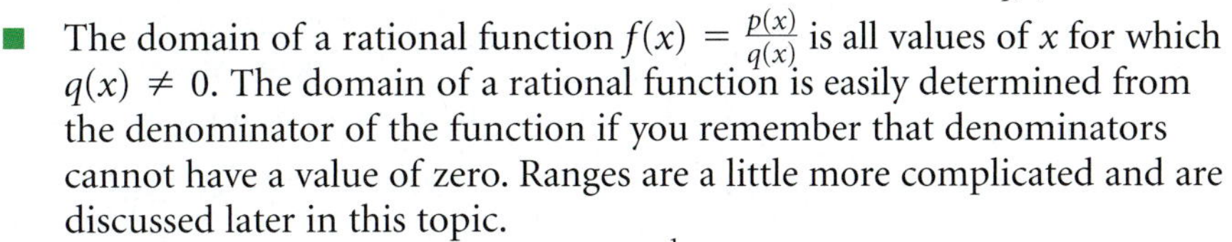

- The domain of a rational function $f(x) = \frac{p(x)}{q(x)}$ is all values of x for which $q(x) \neq 0$. The domain of a rational function is easily determined from the denominator of the function if you remember that denominators cannot have a value of zero. Ranges are a little more complicated and are discussed later in this topic.
- The basic rational function is $f(x) = \frac{1}{x}$, whose domain is all $x \neq 0$ and whose range is all $y \neq 0$.
- The graph of the basic rational function $f(x) = \frac{1}{x}$, shown here, is a **hyperbola.**

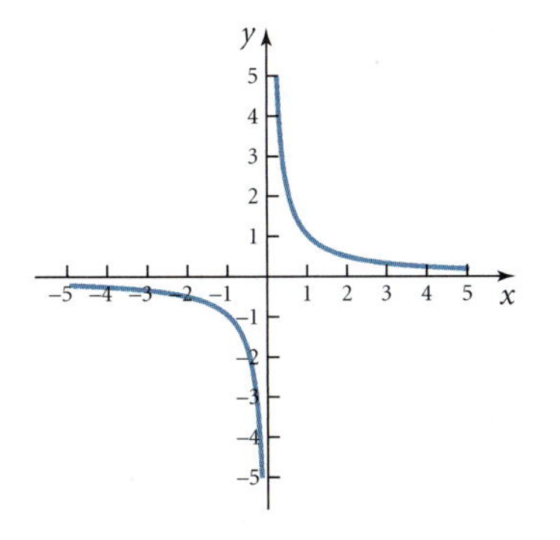

Figure 3.1

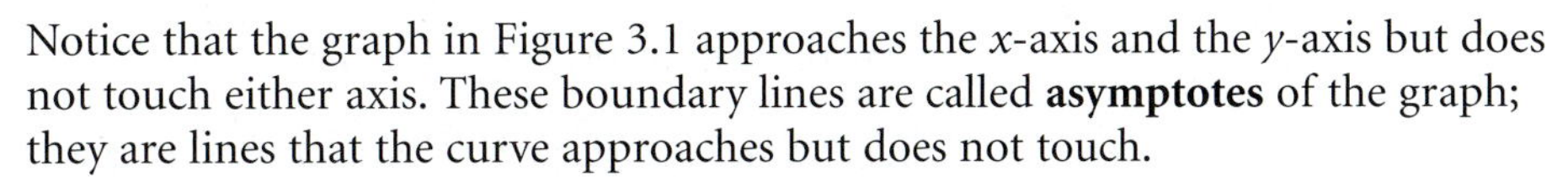

Notice that the graph in Figure 3.1 approaches the x-axis and the y-axis but does not touch either axis. These boundary lines are called **asymptotes** of the graph; they are lines that the curve approaches but does not touch.

- The graph of a rational function with a linear denominator and a constant or linear numerator is always a hyperbola. If we know the asymptotes of a rational function and a point or two that satisfy the function, we can sketch a good approximation of the graph of the function.

MATHEMATICS CORNER  **3.1** 

Denominators of rational functions cannot have a value of zero. To see why that is true, consider two cases.

1. The fraction $\frac{k}{0}$, with k as some nonzero constant, is **undefined** because any number multiplied by zero is zero, not the constant.
2. The fraction $\frac{0}{0}$ is **indeterminate**. There is no unique answer because any number multiplied by 0 is 0.

Vertical Asymptotes

How do we determine the asymptotes of a rational function? First, we discuss vertical asymptotes.

Definition: Let $y = \frac{p(x)}{q(x)}$ be a rational function in simplified form. If $q(a) = 0$, then $x = a$ is a **vertical asymptote** of the graph of $y = \frac{p(x)}{q(x)}$.

A rational expression is in simplified form if the numerator and denominator have no common factors. For example, $f(x) = \frac{x}{x - 3}$ is in simplified form, but $f(x) = \frac{2x - 6}{x^2 - 9}$ is not in simplified form because both numerator and denominator contain the factor $x - 3$.

Example 1: State the domain and the vertical asymptote(s) of each of the following.

a. $f(x) = \dfrac{2}{x + 3}$ **b.** $f(x) = \dfrac{2x}{x + 3}$ **c.** $f(x) = \dfrac{2x^2}{x - 1}$

Solution: The domain is the set of all values of x for which the denominator is not equal to zero. Set the denominator equal to zero and solve for x to determine the location of the vertical asymptote.

a. The domain of $f(x) = \frac{2}{x+3}$ is all $x \neq -3$; the vertical asymptote is $x = -3$.

> **Warning!** ***Asymptotes are lines and must be stated as equations.*** It is incorrect to say that "the asymptote is -3."

b. The domain of $f(x) = \frac{2x}{x+3}$ is all $x \neq -3$; the vertical asymptote is $x = -3$.

c. The domain of $f(x) = \frac{2x^2}{x-1}$ is all $x \neq 1$; the vertical asymptote is $x = 1$. ■

Not all rational functions have exactly one vertical asymptote. It is also possible that there may be more than one vertical asymptote or that there is no vertical asymptote.

Example 2: State the domain and the vertical asymptote(s) of each of the following functions.

a. $f(x) = \dfrac{2x}{x^2 - 1}$ **b.** $f(x) = \dfrac{2x^2}{x^2 + 1}$

Solution:

a. The domain of $f(x) = \frac{2x}{x^2-1}$ is all $x \neq 1, -1$; the vertical asymptotes are $x = 1$ and $x = -1$.

b. The domain of $f(x) = \frac{2x^2}{x^2+1}$ is all real numbers because there is no value of x for which $x^2 + 1 = 0$. There is no vertical asymptote. ■

The definition of a vertical asymptote required that the rational function be in simplified form, meaning that the numerator and denominator have no common factors. To see why this requirement is necessary, look at Example 3.

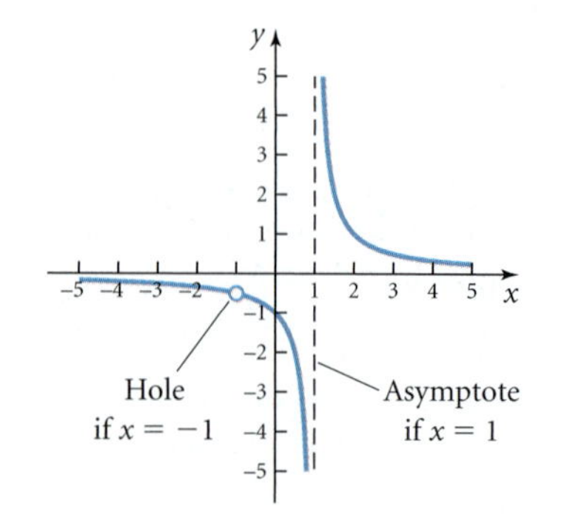

Figure 3.2

Example 3: State the domain and the vertical asymptote of $f(x) = \dfrac{x + 1}{x^2 - 1}$.

Solution: The domain of $f(x) = \frac{x+1}{x^2-1}$ is all $x \neq 1, -1$. Does that mean that there are two vertical asymptotes?

By factoring, we see that the numerator and denominator have a common factor. Factoring and simplifying gives $\frac{x+1}{x^2-1} = \frac{x+1}{(x+1)(x-1)} = \frac{1}{x-1}$. The simplified form of this function is $f(x) = \frac{1}{x-1}$, which has a vertical asymptote at $x = 1$. Why is $x = -1$ not a vertical asymptote?

When we factored the original function and simplified it by dividing out the common factor, we eliminated $x = -1$ as an asymptote because the factor $x + 1$ no longer appeared in the denominator. Graphically, there is a "hole" in the graph at $x = -1$ because the function is not defined at that point. (See Figure 3.2.) ■

Horizontal Asymptotes

Horizontal asymptotes are a little more difficult to determine than vertical asymptotes. The horizontal asymptote of the graph of a rational function is *the value* y *approaches when* |x| *gets quite large* (remember this terminology when we get to limits in Topic 5.) To explore horizontal asymptotes in more detail, we will revisit the functions from Examples 1 and 2.

Example 4: Find the horizontal asymptote of $f(x) = \frac{2}{x + 3}$.

If we let $|x|$ get very large, we see that the value of $f(x)$ decreases and approaches the value of zero.

x	$f(x)$	x	$f(x)$
10	2/13 or 0.1538	−10	−2/7 or −0.2857
100	2/103 or 0.0194	−100	−2/97 or −0.0206
1,000	2/1003 or 0.001994	−1,000	−2/997 or −0.002006
100,000	0.00002	−100,000	−0.00002
1,000,000	0.000002	−1,000,000	−0.000002

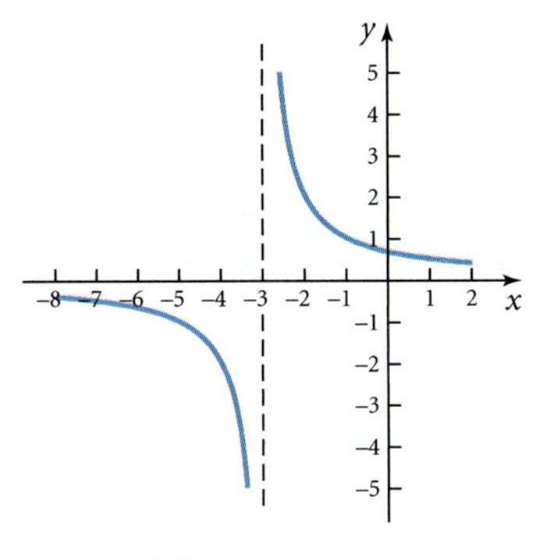

Figure 3.3

Graphically, we see that as $|x|$ gets very large, the curve approaches the x-axis ($y = 0$). (See Figure 3.3.)

Thus, the horizontal asymptote is $y = 0$. ■

Calculator Corner 3.1

Graph $y = \frac{2}{x + 3}$. Remember to enclose the entire denominator, $x + 3$, in parentheses. To eliminate the vertical line appearing along the vertical asymptote, try one of the following methods:

1. Zoom decimal. After entering the equation, press **ZOOM 4:** ZDecimal **ENTER**.
2. Dot mode. After entering the equation, press **MODE**. Scroll down to line 5 and highlight DOT.

Then press **ENTER 2nd MODE** (QUIT). Then press **GRAPH**. To examine the value of the function as x gets large, do the following.

1. Press **TRACE**. Hold the right arrow down and trace along the graph for larger values of x. What is happening to the y values?
2. Press **2nd WINDOW** (TBLSET). Start at 1 with ΔTbl $= 10$. Then press **2nd GRAPH** (TABLE). Scroll down the table. As x increases, what is happening to y?

Example 5: Find the horizontal asymptote of $f(x) = \dfrac{2x}{x^2 - 1}$.

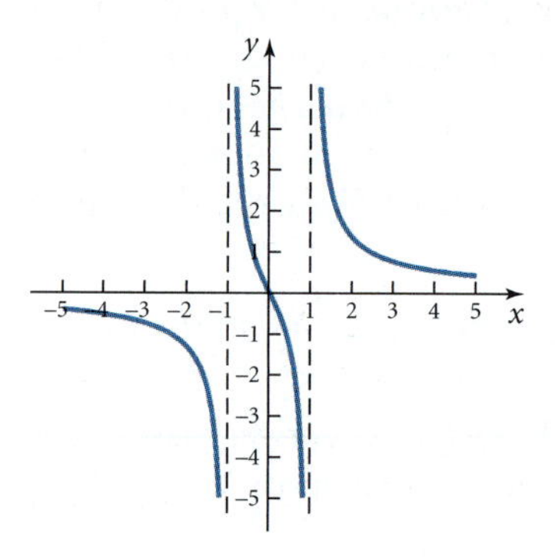

Figure 3.4

By examining the graph of the function (Figure 3.4), we see that as $|x|$ gets large, the curve approaches the x-axis. Thus, the horizontal asymptote is $y = 0$. ■

Calculator Corner 3.2

Complete the table for $f(x) = \dfrac{2x}{x^2 - 1}$.

x	$f(x)$
10	
1,000	
100,000	
1,000,000	
−10	
−1,000	
−100,000	
−1,000,000	

Based on the table, what value does y seem to be getting close to as $|x|$ gets larger? Now draw the graph of the function. Use **TABLE** on your calculator to see that the y value is approaching 0 as $|x|$ gets large. Press **2nd WINDOW** (TblSet). Set TblStart $= 1{,}000$ and ΔTbl $= 10{,}000$. Then press **2nd GRAPH** and scroll through the table of values to see that y approaches 0 as $|x|$ gets large. *Note:* The y values will be given in scientific notation. 1.5E-8 means $1.5(10)^{-8}$ or 0.000000015, which is virtually zero.

In both Examples 4 and 5, you should see that *the denominator has a higher degree than the numerator and the horizontal asymptote is* y $=$ *0* (the x-axis).

Example 6: Find the horizontal asymptote of each of following functions.

a. $f(x) = \dfrac{2x}{x + 3}$ **b.** $f(x) = \dfrac{2x^2}{x^2 - 1}$

Solution:

a. Numerically, as $|x|$ gets very large, the value of $f(x) = \frac{2x}{x + 3}$ approaches $\frac{2x}{x}$ or 2. When $|x|$ is large, the "+3" contributes very little to the value of the denominator.

Calculator Corner 3.3

Complete the table for $f(x) = \dfrac{2x}{x + 3}$.

x	$f(x)$
10	1.53846
100	1.94175
10,000	1.9994
1,000,000	1.999994
−10	2.8571
−100	2.0619
−10,000	2.0006
−1,000,000	2.000006

Based on the table, what value does y seem to be approaching? Now draw the graph of $f(x) = \frac{2x}{x + 3}$. Use the **TABLE** feature of your calculator to see that the y value is approaching 2 as $|x|$ gets large.

Graphically, we see that as $|x|$ gets very large, the curve approaches the horizontal line $y = 2$. (See Figure 3.5.)

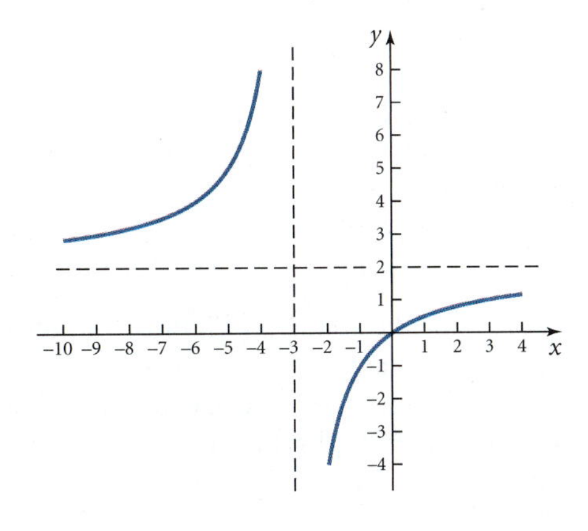

Figure 3.5

b. Numerically, as x gets very large, the value of $f(x) = \frac{2x^2}{x^2 - 1}$ approaches $\frac{2x^2}{x^2}$ or 2. When $|x|$ is large, the "-1" contributes very little to the value of the denominator.

Calculator Corner 3.4

Complete the following table for $f(x) = \frac{2x^2}{x^2 - 1}$. As $|x|$ gets large, what value does y seem to be approaching?

x	$f(x)$
10	
100	
1,000	
10,000	
−10	
−100	
−1,000	
−10,000	

Graphically, we see that as $|x|$ gets large, the curve approaches $y = 2$. Thus, the horizontal asymptote is $y = 2$. (See Figure 3.6.)

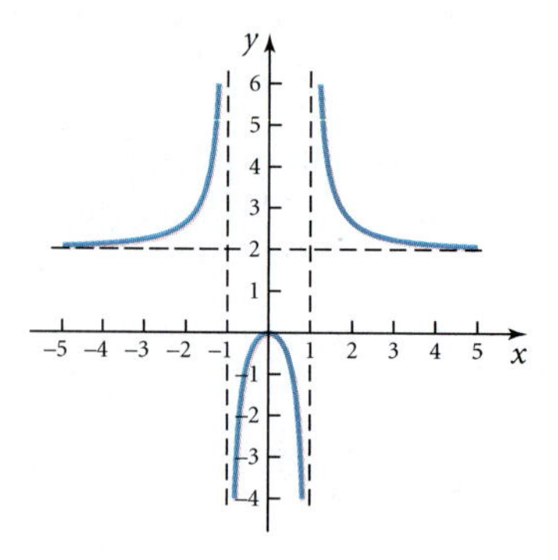

Figure 3.6

In both parts of Example 6, be sure to see that *the numerator and denominator have the same degree, so as |x| gets large, the ratio of the leading coefficients determines the horizontal asymptote.* (The leading coefficient is the coefficient of the term with the highest power.)

Example 7: Determine the horizontal asymptote of $f(x) = \dfrac{2x^2}{x - 1}$.

Solution: As $|x|$ gets larger, the value of y continues to grow without bound.

Calculator Corner 3.5

How does $f(x) = \frac{2x^2}{x - 1}$ behave as $|x|$ gets large? Complete the table and see what happens.

x	$f(x)$
10	
1,000	
10,000	
100,000	
−10	
−1,000	
−10,000	
−100,00	

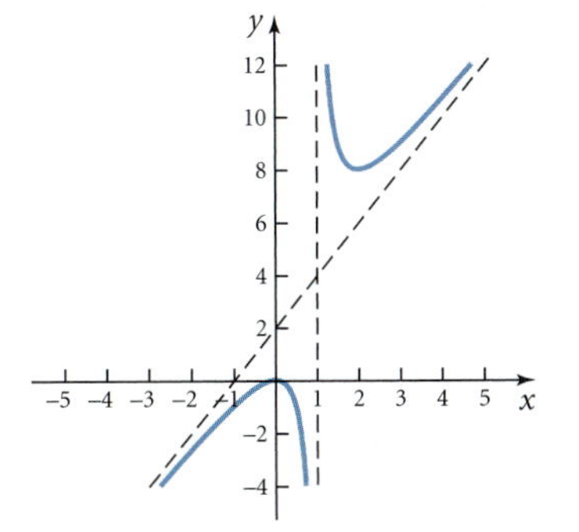

Figure 3.7

By examining the graph (Figure 3.7), you can see that as $|x|$ gets large, the value of y approaches infinity. Thus, there is no horizontal asymptote. In this example, you should see that *the degree of the numerator is one higher than the degree of the denominator and there is no horizontal asymptote.* As $|x|$ gets large, the value of the numerator grows larger than the value of the denominator, so the value of the fraction continues to get larger and approaches infinity. ■

Calculator Corner 3.6

1. Use your graphing calculator to graph $y_1 = \frac{2x^2}{x - 1}$. Use a window of $[-6.3, 6.3]$ for x and $[-12.4, 12.4]$ for y.
2. Algebraically reduce $\frac{2x^2}{x - 1}$ by dividing the polynomials. You should obtain $2x + 2 + \frac{2}{x - 1}$.
3. Now graph $y_2 = 2x + 2$. You should notice that this line describes the behavior of the rational function as x gets large or small. The line $y_2 = 2x + 2$ is referred to as a **slant**, or **oblique**, **asymptote.** Slant asymptotes occur when there is no horizontal asymptote and when the numerator has a degree one higher than the degree of the denominator.

Let's summarize the results.

Asymptotes of Rational Functions

Given $f(x) = \frac{p(x)}{q(x)} = \frac{ax^n + \cdots + c}{bx^m + \cdots + d}$, where $f(x)$ is in simplified form and $q(x) \neq 0$:

The **vertical asymptote**(s) is (are) $x = k$, where $q(k) = 0$.
The **horizontal asymptote** is determined by examining the degrees of $p(x)$ and $q(x)$:

If $n < m$, the horizontal asymptote is $y = 0$.
If $n = m$, the horizontal asymptote is $y = \frac{a}{b}$.
If $n > m$, there is no horizontal asymptote.

Example 8: State the horizontal asymptote of each of following functions.

a. $f(x) = \frac{3}{x - 2}$ **b.** $f(x) = \frac{-3x}{2x + 7}$

c. $f(x) = \frac{3x^2}{x^2 - 16}$ **d.** $f(x) = \frac{x^2}{x + 5}$

Solution:

a. $f(x) = \frac{3}{x - 2}$. The denominator has the higher degree, so the horizontal asymptote is $y = 0$.

b. $f(x) = \frac{-3x}{2x + 7}$. The numerator and denominator have the same degree, so the horizontal asymptote is $y = -\frac{3}{2}$.

c. $f(x) = \frac{3x^2}{x^2 - 16}$. The numerator and denominator have the same degree, so the horizontal asymptote is $y = \frac{3}{1}$ or $y = 3$.

d. $f(x) = \frac{x^2}{x + 5}$. The numerator has the higher degree, so there is no horizontal asymptote. ■

Example 9: A woman on a weight-loss program weighs herself every week. Her weight at any time is given by $W(t) = \frac{790 + 132t}{t + 4.4}$, where W is her weight in pounds and t is the number of weeks after she started the program.

a. State a reasonable domain for the function.

b. What was her initial weight?

c. What was her weight loss after four weeks on the program?

d. Her goal weight is 145 pounds. Approximately how long will it take her to reach her goal weight?

Solution:

a. The domain is time, so a logical domain is $t \geq 0$.

b. Her initial weight is given by $W(0) = \frac{790 + 132(0)}{0 + 4.4} = \frac{790}{4.4} = 179.55$. Her initial weight was approximately 179.6 pounds.

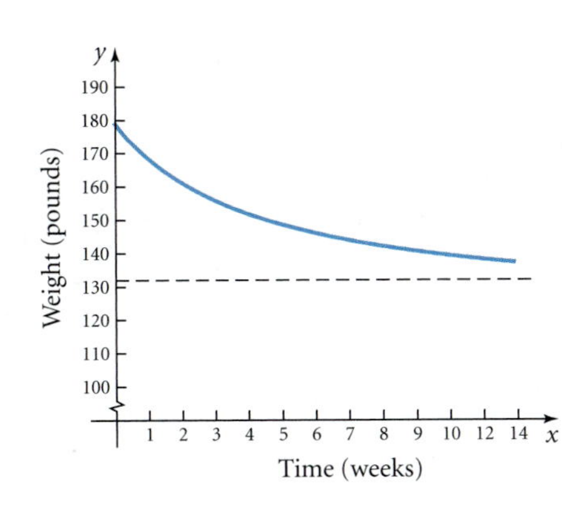

Figure 3.8

c. After four weeks, her weight would be $W(4) = \frac{790 + 132(4)}{4 + 4.4} = 156.9$ pounds. She weighed 179.6 pounds at the beginning, so her weight loss is $179.6 - 156.9 = 22.7$ pounds.

d. Her goal is 145 pounds, so $W = 145$. Thus, $145 = \frac{790 + 132t}{t + 4.4}$.

$\rightarrow 145(t + 4.4) = 790 + 132t$ **Multiply both sides by $t + 4.4$**

$\rightarrow 145t + 638 = 790 + 132t$ **Distribute**

$\rightarrow 13t = 152$ **Isolate variable**

$\rightarrow t \approx 11.7$ **Solve for t**

She should reach her goal weight of 145 pounds in 11.7 weeks.

Look at the graph of her weight (Figure 3.8). Do you see that her weight will stabilize after a long time? Is it possible for her to ever get to a weight of 110 pounds? The horizontal asymptote of this function is $y \approx 132$, so realistically she will never get below that weight. ■

Example 10: A refrigerator is purchased for $2000 with an additional service contract costing $150 per year. The total cost of the refrigerator for x years is $C(x) = 2{,}000 + 150x$. What is the average annual cost of the refrigerator during the first five years?

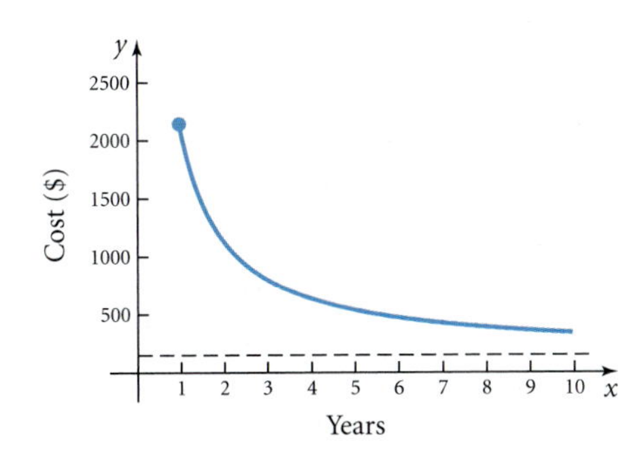

Figure 3.9

Solution: The average cost of the refrigerator per year for the x years is

$$AveC(x) = \frac{2000 + 150x}{x} = \frac{2000}{x} + 150$$

For the first five years, the average annual cost is

$$AveC(5) = \frac{2000}{5} + 150 = 400 + 150 = \$550$$

See Figure 3.9. ■

Check Your Understanding 3.1

For each of the following rational functions, state the horizontal and vertical asymptotes., if they exist.

1. $f(x) = \frac{4}{x + 5}$ **2.** $f(x) = \frac{4x}{x - 2}$ **3.** $f(x) = \frac{x}{x^2 - 9}$

4. $f(x) = \frac{2x^2}{x^2 - 9}$ **5.** $f(x) = \frac{x^3}{x^2 + 9}$ **6.** $f(x) = \frac{3}{x^2 + 1}$

7-12. Sketch the graphs of the functions in Exercises 1 through 6. Use your graphing calculator only when necessary. State the domain and range of each function.

Piecewise Defined Functions

Every year, workers in the United States pay a federal income tax. How is that income tax determined? Obviously, the amount of tax you pay depends on your income, but different levels of earnings are taxed at different rates.

A veterinarian is prescribing a medication for a dog. The amount of medication to administer depends on the weight of the dog. There is one dosage for small dogs, another dosage for medium-sized dogs, and a bigger dosage for large dogs. Dosage is a function of the weight of the dog, but different weight levels are used to determine that dosage.

Both of these examples demonstrate a type of function called a piecewise defined function. A **piecewise defined function** is one that has a different formula for different intervals in the domain.

Example 11: Consider the following piecewise defined function.

$$f(x) = \begin{cases} x^2 & \text{if } x \leq -1 \\ 2x - 3 & \text{if } -1 < x < 2 \\ 7 & \text{if } x \geq 2 \end{cases}$$

Find:

a. $f(-2)$ **b.** $f(0)$ **c.** $f(2)$

Solution: To evaluate a piecewise defined function for a given value of x, simply determine which "piece" of the function the x value belongs to and substitute that value into the formula for $f(x)$.

a. $x = -2$ belongs to the first domain interval because $-2 < -1$, so $f(-2) = (-2)^2 = 4$.

b. $x = 0$ belongs to the second domain interval because $-1 < 0 < 2$, so $f(0) = 2(0) - 3 = -3$.

c. $x = 2$ belongs to the third domain interval so $f(2) = 7$.

To graph a piecewise defined function, consider each piece as a separate function with a specified domain. For the function in Example 11, we first graph $y_1 = x^2$ on the domain $x \leq -1$, so only that part of the graph is drawn. Next, graph the $y_2 = 2x - 3$ on the domain $-1 < x < 2$, showing only that part of the graph. Finally, graph $y_3 = 7$. The domain for this piece is only those values of x for which $x \geq 2$, so only that part of the graph is drawn. (See Figure 3.10.) ■

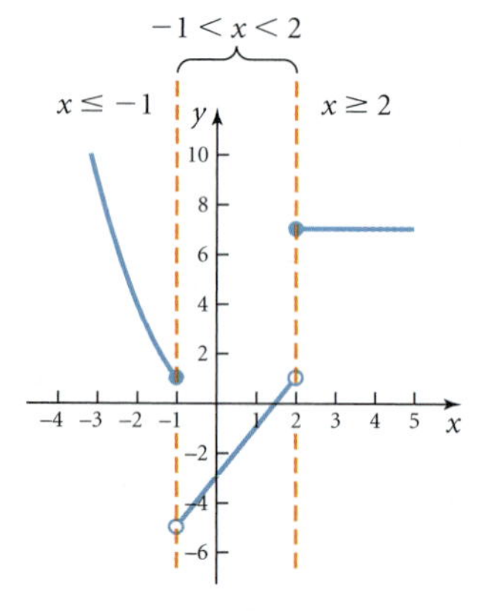

Figure 3.10

Calculator Corner 3.7

To graph the piecewise defined function in Example 11 on your calculator, enter the following:

$$y_1 = (x^2)/(x \leq -1)$$
$$y_2 = (2x - 3)/(x > -1)(x < 2)$$
$$y_3 = (7)/(x \geq 2)$$

The inequality symbols are located under **2nd MATH** (Test) on the TI-84.

Example 12: The 2005 U.S. Federal Income Tax Tables for taxpayers using the Single Filing Status shows various tax rates, depending on the taxable income.

$$T(d) = \begin{cases} 0.10d & \text{if } d \leq \$7300 \\ \$730 + 0.15(d - \$7300) & \text{if } \$7300 < d \leq \$29{,}700 \\ \$4{,}090 + 0.25(d - \$29{,}700) & \text{if } \$29{,}700 < d \leq \$71{,}950 \\ \$14{,}625.50 + 0.28(d - \$71{,}950) & \text{if } \$71{,}950 < d \leq \$150{,}150 \\ \$36{,}548.50 + 0.33(d - \$150{,}150) & \text{if } \$150{,}150 < d \leq \$326{,}450 \\ \$94{,}727.50 + 0.35(d - \$326{,}450) & \text{if } d > \$326{,}450 \end{cases}$$

where T is income tax due and d is taxable income.

Find the tax for each of the following situations.

a. A single filer whose taxable income is $15,600
b. A single filer whose taxable income is $35,400
c. A single filer whose taxable income is $150,150

Solution:

a. $\$7300 < \$15{,}600 < \$29{,}700$, so

$$\begin{aligned} T(\$15{,}600) &= \$730 + 0.15(\$15{,}600 - \$7300) \\ &= \$1975 \end{aligned}$$

The income tax due is $1975.

b. $\$29{,}700 < \$35{,}400 < \$71{,}950$, so

$$\begin{aligned} T(\$35{,}400) &= \$4090 + 0.25(\$35{,}400 - \$29{,}700) \\ &= \$5515 \end{aligned}$$

The income tax due is $5515.

c. $\$71{,}950 < \$150{,}150 \leq \$150{,}150$, so

$$\begin{aligned} T(\$150{,}150) &= \$14{,}625.50 + 0.28(\$150{,}150 - \$71{,}950) \\ &= \$36{,}521.50 \end{aligned}$$

The income tax due is $36,521.50. ■

Step Functions

Another type of function commonly seen in calculus is the **step function**. This function is similar to a piecewise defined function, but the formulas that define each piece of the function are constants.

Example 13:

$$f(x) = \begin{cases} 3 & \text{if } x \le -2 \\ -2 & \text{if } -2 < x \le 1 \\ 1 & \text{if } x > 1 \end{cases}$$

Find each of the following function values.

a. $f(0)$ **b.** $f(6)$ **c.** $f(-2)$

Solution: The graph is shown to the right (Figure 3.11).

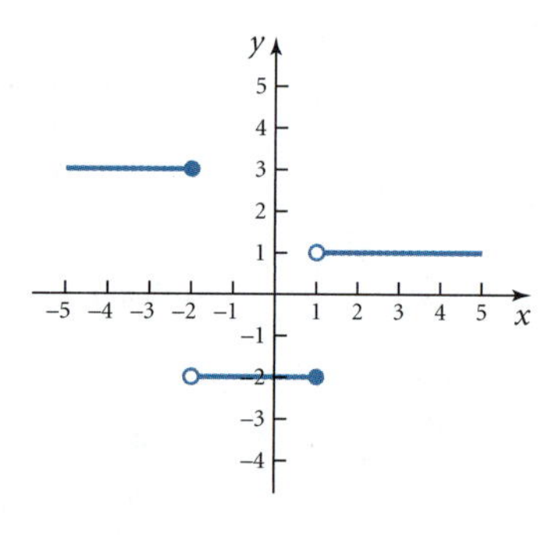

Figure 3.11

a) Because $-2 < 0 < 1$, use the second piece of the function and $f(0) = -2$.
b) Because $6 > 1$, use the third piece of the function and $f(6) = 1$.
c) Because $x = -2$, use the first piece of the function and $f(-2) = 3$. ■

Example 14: Dosages for baby aspirin tablets depend on the baby's weight.

Weight (in pounds)	Dosage (number of tablets)
Less than 12 pounds	2 tablets
12 pounds to 18 pounds	4 tablets
More than 18 pounds	6 tablets

How many tablets would you give a baby who weighs 10 pounds? 14 pounds? 20 pounds?

Solution: The table above represents a function. Weight is the domain, and the dosage is the range. The function describing this table is

$$T(w) = \begin{cases} 2 & \text{if } w < 12 \\ 4 & \text{if } 12 \le w \le 18 \\ 6 & \text{if } w > 18 \end{cases}$$

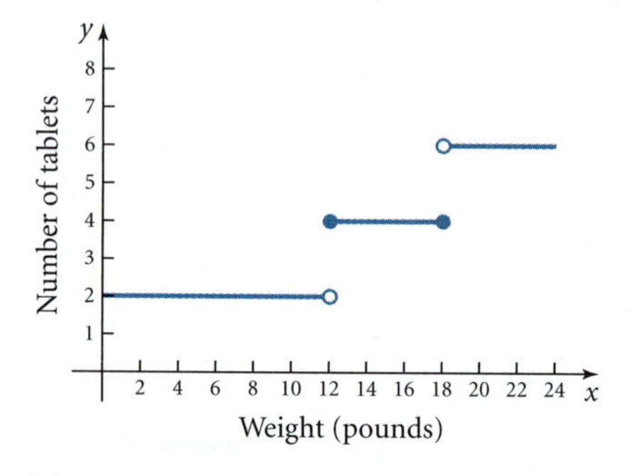

Figure 3.12

A baby weighing 10 pounds would receive two tablets because $10 < 12$.

A baby weighing 14 pounds would receive four tablets because $12 < 14 < 18$.

A baby weighing 20 pounds would receive six tablets because $20 > 18$.

The graph of this function is shown in Figure 3.12. Notice that the graph resembles "steps." ■

MATHEMATICS CORNER 3.2

One special type of step function is the **greatest integer function**, denoted $f(x) = [\![x]\!]$. The greatest integer function is defined as *the greatest integer equal to or less than x*. For example, $[\![5]\!] = 5$, $[\![6.47]\!] = 6$, $[\![\pi]\!] = 3$, and $[\![\sqrt{14}]\!] = 3$. The graph of the greatest integer function is a step function as shown below.

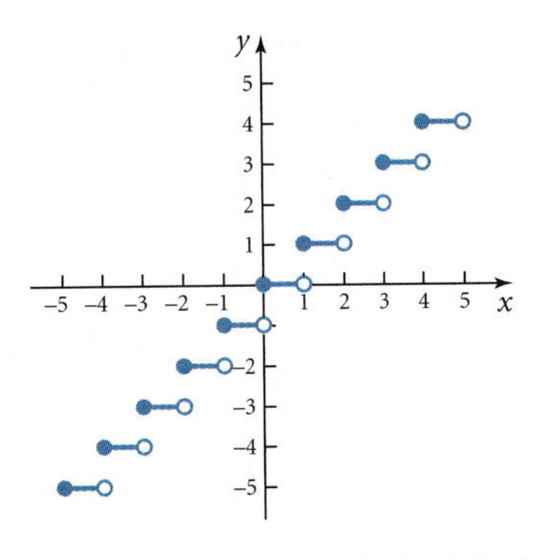

Check Your Understanding 3.2

1. Given

$$f(x) = \begin{cases} x - 2 & \text{if } x < -2 \\ x^2 + 1 & \text{if } -2 \leq x < 4 \\ 3 - x & \text{if } x \geq 4 \end{cases}$$

a. Evaluate $f(-6)$.
b. Evaluate $f(-2)$.
c. Evaluate $f(1)$.
d. Evaluate $f(4)$.
e. Draw the graph of the function.

2. Given

$$f(x) = \begin{cases} -1 & \text{if } x \leq -2 \\ 1 & \text{if } -2 < x \leq 2 \\ 3 & \text{if } x > 2 \end{cases}$$

a. Draw the graph of $f(x)$.
b. Evaluate $f(1.5)$, $f(2.333)$, and $f(-2.1)$.

Check Your Understanding Answers

Check Your Understanding 3.1

1. $x = -5, y = 0$

2. $x = 2, y = 4$

3. $x = 3, x = -3, y = 0$

4. $x = 3, x = -3, y = 2$

5. no horizontal or vertical

6. $y = 0$, no vertical

7. domain: $x \neq -5$

range: $y \neq 0$

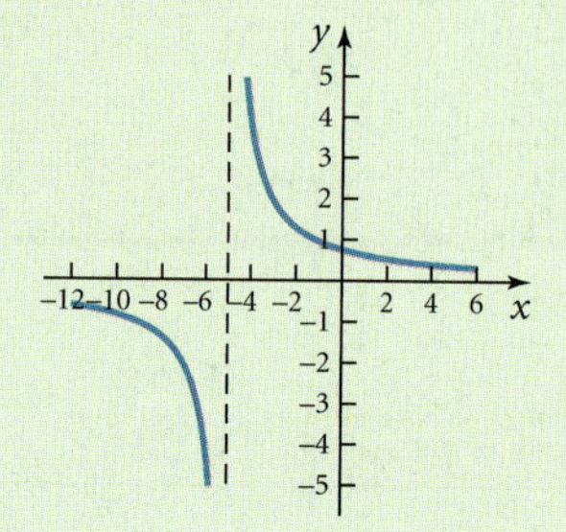

8. domain: $x \neq 2$

range: $y \neq 4$

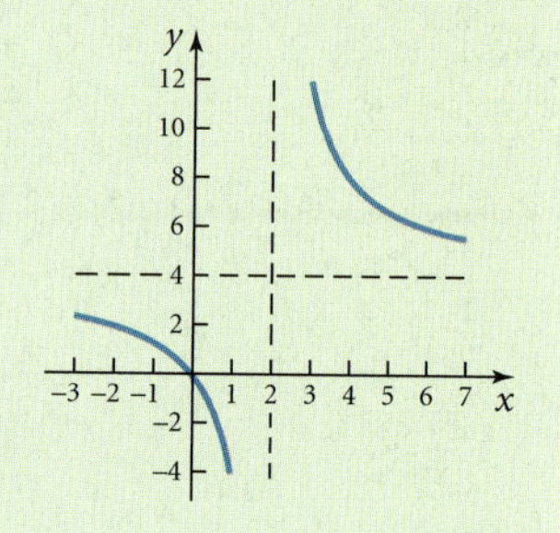

9. domain: $x \neq -3,3$

range: all real numbers

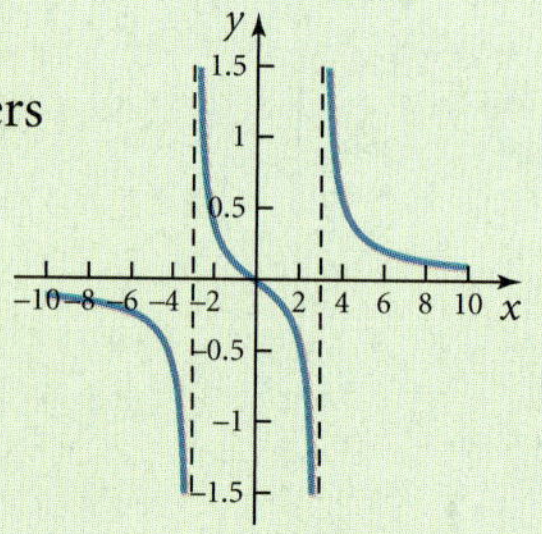

10. domain: $x \neq -3,3$

range: $y \leq 0, y > 2$

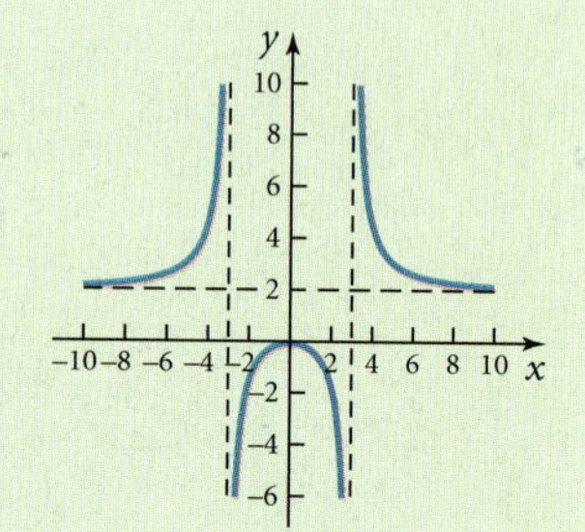

11. domain: all real numbers

range: all real numbers

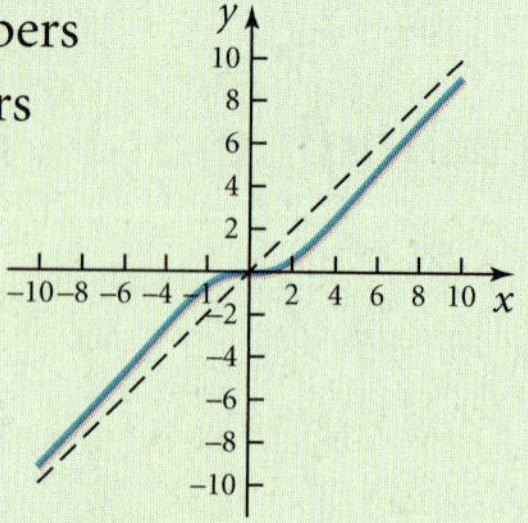

12. domain: all real numbers

range: $0 < y \leq 3$

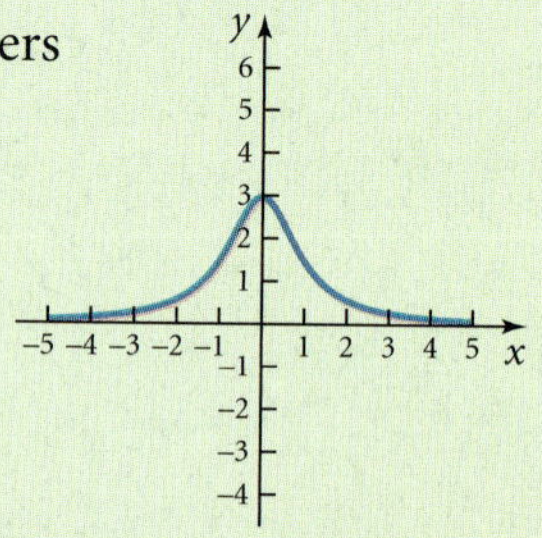

Check Your Understanding 3.2

1. a. $f(-6) = -8$ **b.** $f(-2) = 5$

c. $f(1) = 2$ **d.** $f(4) = -1$

e.

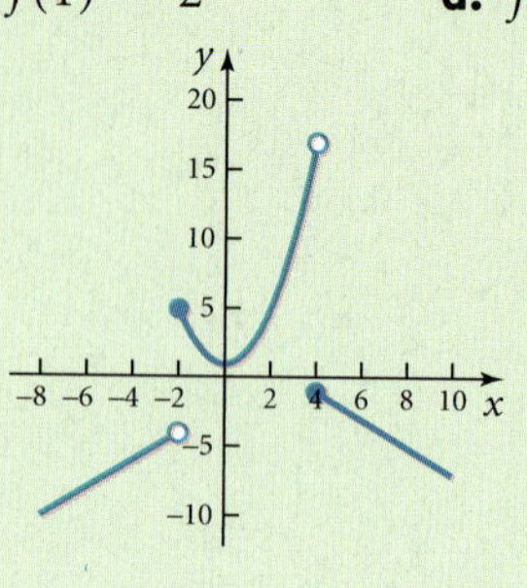

2. a.

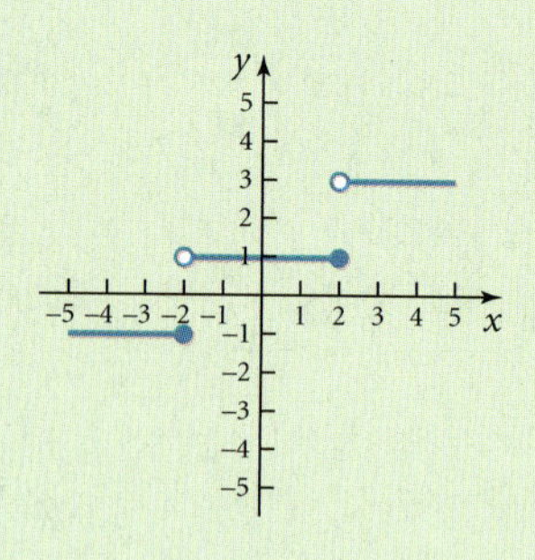

b. $f(1.5) = 1, f(2.333) = 3, f(-2.1) = -1$

Topic 3 Review

3

This topic reviewed rational functions and piecewise defined functions.

CONCEPT	EXAMPLE
A **rational function** is a ratio of two polynomial functions, $f(x) = \frac{p(x)}{q(x)}$.	$f(x) = \frac{6}{x-3}$ and $f(x) = \frac{3x}{x^2-4}$ are rational functions.
The **domain** of a rational function $f(x) = \frac{p(x)}{q(x)}$ is the set of all values of x for which $q(x) \neq 0$.	The domain of $f(x) = \frac{6}{x-3}$ is all $x \neq 3$. The domain of $f(x) = \frac{3x}{x^2-4}$ is $x \neq -2, 2$. The domain of $f(x) = \frac{3x}{x^2+4}$ is all real numbers.
The graph of a rational function may be a **hyperbola**.	The graph of $f(x) = \frac{6}{x-3}$ is a hyperbola. 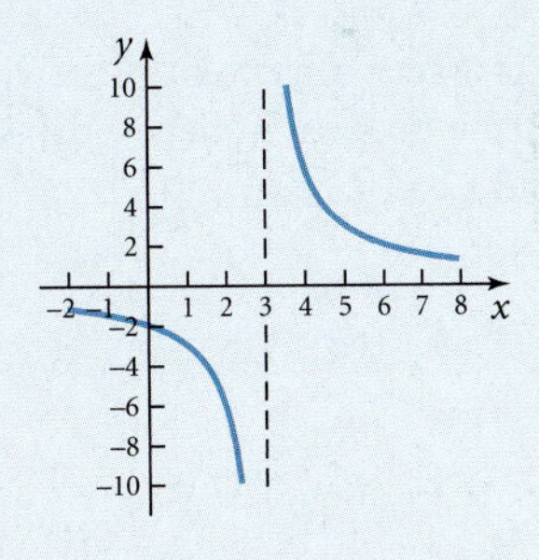 The graph of $f(x) = \frac{3x}{x^2-4}$ is 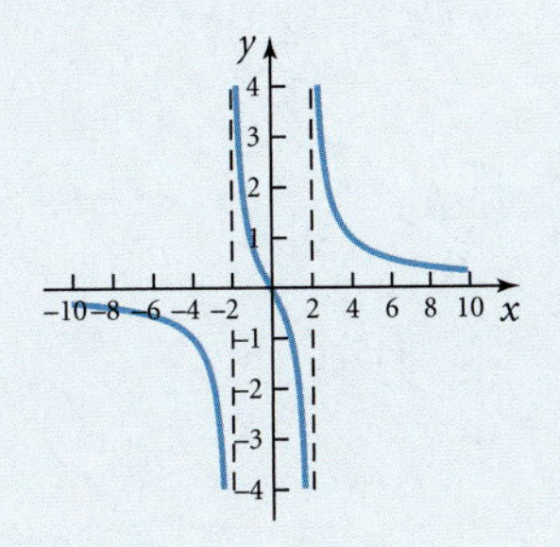

(continued)

The graph of a rational function may have one or more **vertical asymptotes**, occurring at the values for which $f(x)$ is undefined. To determine the vertical asymptotes, set the denominator equal to zero and solve for x. If $q(a) = 0$, then $x = a$ is a vertical asymptote.	The vertical asymptote of $f(x) = \frac{6}{x - 3}$ is $x = 3$. The vertical asymptotes of $f(x) = \frac{3x}{x^2 - 4}$ are $x = -2$ and $x = 2$.
The **horizontal asymptote** is determined by the degrees of the numerator and the denominator. • If the degree of the numerator is less than the degree of the denominator, the horizontal asymptote is $y = 0$. • If the degree of the numerator and the degree of the denominator are the same, the horizontal asymptote is the ratio of the leading coefficients. • If the degree of the numerator is higher than the degree of the denominator, there is no horizontal asymptote.	The horizontal asymptote of $f(x) = \frac{6}{x - 3}$ is $y = 0$. The horizontal asymptote of $f(x) = \frac{3x^2}{x^2 - 4}$ is $y = 3$. $f(x) = \frac{3x^3}{x^2 - 4}$ has no horizontal asymptote.
Piecewise defined functions are defined using different formulas for specific intervals in the domain.	$f(x) = \begin{cases} 3x - 2 & \text{if } x < -2 \\ 1 - x^2 & \text{if } -2 \le x < 3 \\ \frac{6}{x} & \text{if } x \ge 3 \end{cases}$ is a piecewise defined function.
A **step function** is a special type of piecewise defined function in which each piece of the function is defined as a constant.	$f(x) = \begin{cases} -3 & \text{if } x < -2 \\ 1 & \text{if } -2 \le x < 3 \\ 6 & \text{if } x \ge 3 \end{cases}$ is a step function.

NEW TOOLS IN THE TOOL KIT

- Determining the domain and range of a rational function
- Determining the horizontal and vertical asymptotes of a rational function
- Sketching the graph of a rational function by identifying the asymptotes
- Graphing a rational function using a graphing calculator
- Evaluating piecewise defined functions
- Graphing piecewise defined functions
- Evaluating step functions
- Graphing step functions

Topic 3 Exercises

3

State the vertical and horizontal asymptotes of each of the following functions.

1. $y = \dfrac{3}{x - 2}$ **2.** $y = \dfrac{-4}{x + 3}$

3. $y = \dfrac{-2x}{3x + 7}$ **4.** $y = \dfrac{3x}{2x - 5}$

5. $y = \dfrac{5x^2}{x^2 - 16}$ **6.** $y = \dfrac{-2x^2}{x^2 - 9}$

7. $y = \dfrac{-3x}{x^2 - 4}$ **8.** $y = \dfrac{3x}{x^2 + 4}$

9. $y = \dfrac{x^2}{x^2 + 2}$ **10.** $y = \dfrac{x^3}{x^2 - 36}$

11. $y = \dfrac{3x}{x^2 + 1}$ **12.** $y = \dfrac{-2x}{x^2 - 4}$

Draw the graph of each of the following functions without using a graphing calculator. State the domain, range, and asymptotes of each function.

13. $y = \dfrac{2}{x - 3}$ **14.** $y = \dfrac{-4}{x + 2}$

15. $y = \dfrac{-3x}{x + 1}$ **16.** $y = \dfrac{2x}{x - 3}$

Draw the graph of each of the following functions using a graphing calculator. State the domain, range, and asymptotes of each function.

17. $y = \dfrac{2x}{x^2 - 9}$ **18.** $y = \dfrac{-x}{x^2 - 4}$

19. $y = \dfrac{x^2}{x^2 - 9}$ **20.** $y = \dfrac{-x^2}{x^2 - 4}$

21. $y = \dfrac{3}{x^2 + 9}$ **22.** $y = \dfrac{2x^2}{x^2 + 1}$

Evaluate each function for the given values of x.

23. $f(x) = \begin{cases} x - 3 & \text{if } x < 1 \\ x^2 & \text{if } x \geq 1 \end{cases}$

a. -3 **b.** 1 **c.** 4

24. $f(x) = \begin{cases} 5 - x & \text{if } x \leq 4 \\ \sqrt{x} & \text{if } x > 4 \end{cases}$

a. -5 **b.** 1 **c.** 9

25. $f(x) = \begin{cases} 3 & \text{if } x \le -1 \\ 2x + 5 & \text{if } -1 < x \le 2 \\ 3 - x & \text{if } x > 2 \end{cases}$

a. -3 **b.** 1 **c.** 2 **d.** 4

26. $f(x) = \begin{cases} x^3 & \text{if } x < 0 \\ 1 & \text{if } 0 \le x < 2 \\ x - 1 & \text{if } x \ge 2 \end{cases}$

a. -2 **b.** 0 **c.** 2 **d.** 4

27. $f(x) = \begin{cases} x^2 - 4 & \text{if } x < 0 \\ 2 & \text{if } x = 0 \\ x^2 - 4 & \text{if } x > 0 \end{cases}$

a. -3 **b.** 0 **c.** 3

28. $f(x) = \begin{cases} x - 3 & \text{if } x < 1 \\ 5 & \text{if } x = 1 \\ 3 - x & \text{if } x > 1 \end{cases}$

a. 0 **b.** 1 **c.** 2

29. $f(x) = \begin{cases} 2x + 3 & \text{if } x < 1 \\ 5 & \text{if } x = 1 \\ \sqrt{x - 1} & \text{if } x > 1 \end{cases}$

a. 0 **b.** 1 **c.** 5

30. $f(x) = \begin{cases} 1 - x^2 & \text{if } x < 0 \\ 1 & \text{if } x = 0 \\ x^3 + 2 & \text{if } x > 0 \end{cases}$

a. -3 **b.** 0 **c.** 2

31. $f(x) = \begin{cases} -3 & \text{if } x < -1 \\ 4 & \text{if } -1 \le x < 2 \\ 1 & \text{if } x \ge 2 \end{cases}$

a. -1 **b.** 1.99999 **c.** 2.1

32. $f(x) = \begin{cases} -2 & \text{if } x \le -2 \\ 1 & \text{if } -2 < x < 2 \\ 3 & \text{if } x \ge 2 \end{cases}$

a. -2 **b.** 1.999999 **c.** 2

33-42. Draw the graphs of each of the functions in Exercises 23 through 32.

43. A cab fare in Las Vegas is \$2.20 for the first mile and \$1.50 for each additional mile or fraction thereof. (There are also charges for sitting at red lights and for additional passengers that are not considered here.) Data from: Las Vegas Taxi Cab Authority.

a. What is the fare for a 4-mile trip?
b. What is the fare for a 6.4-mile trip?
c. Write the piecewise defined function that describes cab fares in Las Vegas. Assume a domain of 6 miles.
d. Graph the function for a 6-mile domain.

44. A cab fare in New York City is \$2.00 for the first mile plus \$.30 for each additional 1/5 mile. Data from: www.ny.com/transportation/taxis

a. What is the fare for a 4-mile trip?
b. What is the fare for a 2.4-mile trip?
c. There is an additional charge of \$0.20 for each minute of idle time. Suppose a passenger rode for 4.7 miles with 10.2 idle minutes. What would that fare be?

45. The 2005 U.S. Federal Income Tax Tables for Married Filing Jointly filing status shows the following tax rates.

$$T(d) = \begin{cases} 0.10d & \text{if } d \le \$14{,}600 \\ \$1460 + 0.15(d - \$14{,}600) & \text{if } \$14{,}600 < d \le \$59{,}400 \\ \$8180 + 0.25(d - \$59{,}400) & \text{if } \$59{,}400 < d \le 119{,}950 \\ \$23{,}317.50 + 0.28(d - \$119{,}950) & \text{if } \$119{,}950 < d \le \$182{,}800 \\ \$40{,}915.50 + 0.33(d - \$182{,}800) & \text{if } \$182{,}800 < d \le \$326{,}450 \\ \$88{,}320 + 0.35(d - \$326{,}450) & \text{if } d > \$326{,}450 \end{cases}$$

where T is income tax due and d is taxable income.

Find the income tax due for a married couple filing jointly for each of the following combined taxable incomes.

a. \$15,800 **b.** \$68,400 **c.** \$202,600

46. The 2005 U.S. Federal Income Tax Tables for Single filing status shows the following tax rates.

$$T(d) = \begin{cases} 0.10d & \text{if } d \le \$7300 \\ \$730 + 0.15(d - \$7300) & \text{if } \$7300 < d \le \$29{,}700 \\ \$4090 + 0.25(d - \$29{,}700) & \text{if } \$29{,}700 < d \le \$71{,}950 \\ \$14{,}652.50 + 0.28(d - \$71{,}950) & \text{if } \$71{,}950 < d \le \$150{,}150 \\ \$36{,}548.50 + 0.33(d - \$150{,}150) & \text{if } \$150{,}150 < d \le \$326{,}450 \\ \$94{,}727.50 + 0.35(d - \$326{,}450) & \text{if } d > \$326{,}450 \end{cases}$$

where T is income tax due and d is taxable income.

Find the income tax due for a single filer for each of the following taxable incomes.

a. \$25,800 **b.** \$56,800 **c.** \$92,600

47. Annual hotel industry profit is estimated by $P(x) = -0.6x^2 + 5.97x + 22.1$, where P is the profit in billions of dollars and x is the number of years after 2000.

a. Find the average annual profit function for x years after 2000.

b. Find the average profit for 2010.

c. Graph the average annual profit function on the interval $[1, 15]$.

48. The number of new AIDS cases reported each year in the United States is modeled by $A(x) = -680x^2 + 19{,}307x - 74{,}310$, where x is the number of years after 1990.

a. Find the average number of cases per year for x years after 1990.

b. Find the average number of new cases for 2008.

c. Graph the average number of new AIDS cases on the interval $[1, 20]$.

49. Bernie's Mowing Service purchased a new riding mower for \$5600 plus a maintenance agreement costing \$100 for the first year and an additional \$20 per year for each year after that. The total cost of the mower and maintenance agreement for x years is $C(x) = 5600 + 70x + 30x^2$.

a. Find the average cost of the mower per year for x years.

b. What is the average cost of the mower after 10 years? After 30 years?

c. Graph the average cost function over the interval $[1, 40]$.

d. When should Bernie replace the mower?

50. A college purchased a high-speed copier for \$35,000. The maintenance contract cost \$2500 for the first year with an increase of \$500 per year after that. The total cost of the copier and maintenance contract is $C(x) = 35{,}000 + 2000x + 500x^2$.

a. Find the average cost of the copier per year for x years.

b. What is the average cost of the copier after 5 years? After 15 years?

c. Graph the average cost function on the interval $[1, 20]$.

d. When should the copier be replaced?

51. A builder is fencing a rectangular 4000-square-foot storage space. The perimeter of the storage space is given by $P(x) = 2x + 2(\frac{4000}{x})$, where x is the length of the space.

a. What is the domain of the perimeter function?

b. Graph the perimeter function on the interval $[1, 250]$.

c. Use the graph to determine the length of side x that gives the smallest perimeter.

52. A farmer is enclosing a rectangular grazing space beside a stream. The space has an area of 4000 square feet, with the side beside the stream left open. The amount of fencing needed is given by $F(x) = x + 2(\frac{4000}{x})$, where x is the length of the side parallel to the stream.

a. What is the domain of the fence function?

b. Graph the fence function on the interval $[1, 250]$.

c. Use the graph to determine the length of side x that gives the smallest amount of fencing.

53. Postage rates for parcel airlift are given in the following table.

Weight Not More Than ...	Fee
2 pounds	$0.45
3 pounds	$0.85
4 pounds	$1.25
30 pounds	$1.70

a. Determine the postage for a parcel weighing 1.5 pounds.

b. Determine the postage for a parcel weighing 4 pounds.

c. Determine the postage for a parcel weighing 20 pounds.

d. Write a function describing postage rates for parcel airlift based on weight.

54. Postage rates for express mail are given in the following table.

Weight	Rate
Up to 8 ounces	$13.65
Over 8 ounces up to 2 pounds	$17.85
Up to 3 pounds	$21.05
Up to 4 pounds	$24.20
Up to 5 pounds	$27.30
Up to 6 pounds	$30.40

a. Determine the postage for an express mail package weighing 5 ounces.

b. Determine the postage for an express mail package weighing 2 pounds.

c. Determine the postage for an express mail package weighing 3.8 pounds.

d. Write a function that describes express mail rates based on weight.

SHARPEN THE TOOLS

55. According to the 2005 U.S. Federal Income Tax Rate Schedules, the tax due for taxpayers using the Head of Household filing status is given in the following table.

If the Amount of Line 40 Is Over ...	But Not Over ...	Enter on Line 41	Of the Amount Over ...
$0	$10,450	10%	$0
$10,450	$39,800	$1045 + 15%	$10,450
$39,800	$102,800	$5447.50 + 25%	$39,800
$102,800	$166,450	$21,197.50 + 28%	$102,800
$166,450	$326,450	$39,019.50 + 33%	$166,450
$326,450		$91,819.50 + 35%	$326,450

a. Write a piecewise defined function that describes the tax due for this filing status.

b. Determine the tax due for a taxpayer whose taxable income is $20,000.

c. Determine the tax due for a taxpayer whose taxable income is $50,200.

56. According to the 2005 U.S. Federal Income Tax Rate Schedules, the tax due for filers using the Married Filing Separately status is given in the following table.

If the Amount of Line 40 Is Over . . .	But Not Over . . .	Enter on Line 41	Of the Amount Over . . .
$0	$7300	10%	$0
$7300	$29,700	$730 + 15%	$7300
$29,700	$59,975	$4090 + 25%	$29,700
$59,975	$91,400	$11,658.75 + 28%	$59,975
$91,400	$163,225	$20,457.75 + 33%	$91,400
$163,225		$44,160 + 35%	$163,225

a. Write a piecewise defined function that describes the tax due for this filing status.

b. Determine the tax due for a taxpayer whose taxable income is $20,000.

c. Determine the tax due for a taxpayer whose taxable income is $50,200.

TOPIC

4 Regression and Modeling

IMPORTANT TOOLS IN TOPIC 4

- *Scatter plots of paired data*
- *Regression equations to model paired data*
- *Predictions using regression models*

A recent newspaper article on a county's school enrollment showed a projected enrollment for the next school year of 617,450 students. A company's latest annual financial report indicated projected earnings for next year of \$6.4 million. The Federal Aviation Administration predicted in January 2004 that the number of airline passengers will increase in coming years.

AIR TRAVEL REBOUNDING

The FAA projects that the number of people flying will continue to increase in the next several years.

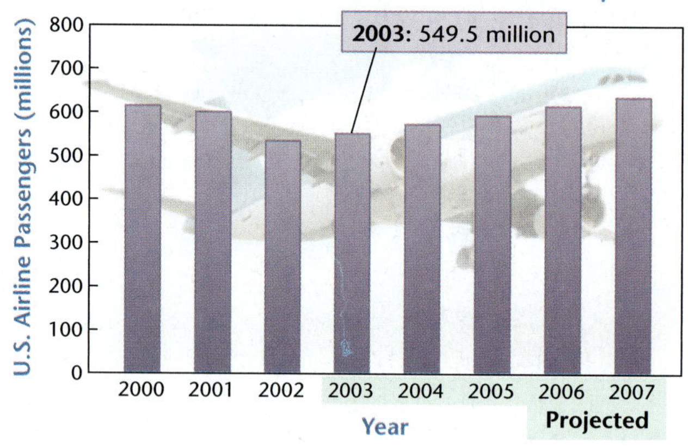

Source: Federal Aviation Administration

How are school systems, companies, and airlines able to make predictions about future time periods? These projections are made using current data and a mathematical process called regression, which is the focus of Topic 4.

Before beginning your study of regression, be sure to complete the Warm-up Exercises to sharpen the necessary algebra skills.

TOPIC 4 WARM-UP EXERCISES

Be sure you can successfully complete the following exercises before starting Topic 4.

For Exercises 1 through 4, match the graph to the appropriate type of equation.

1. linear
2. quadratic
3. cubic
4. quartic (fourth power)

A.
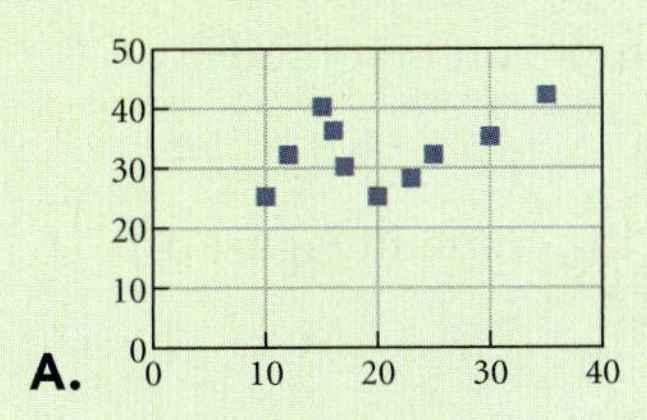

B.

C.
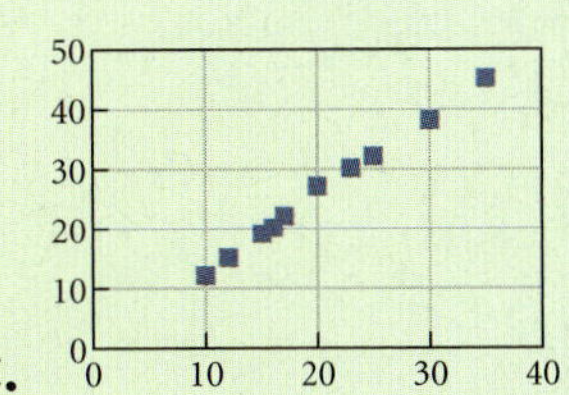

D.

5. Given $f(x) = 2x^3 - 5x^2 + 7x - 3$, evaluate each of the following functions.

a. $f(0)$ **b.** $f(2)$ **c.** $f(-5)$ **d.** $f(9)$

Algebra Warm-up Answers

1. C
2. B
3. A
4. D
5. a. -3 **b.** 7
c. -413 **d.** 1113

Regression is a powerful mathematical tool that transforms real data into a function that models the data. With the availability of the graphing calculator and Excel, a variety of regression functions are now quite easily accessible. To create a model of a set of data using regression, follow these steps:

- Make a plot of the data points. This graph is called a **scatter plot**.
- Look at the scatter plot and identify any trends or patterns in the data. Do the data appear to be linear, quadratic, or cubic? The type of function most closely resembled by the scatter plot is used in the next step.
- Enter the data into the calculator or Excel and select the function that best models the data.

Scatter Plots

The first steps in regression are to draw the scatter plot and identify the type of equation that best models the data.

Example 1: The caloric content and the sodium content (in milligrams) for 10 types of beef hot dogs are given in the table below.

Calories, x	186	181	176	149	184	190	158	139	175	148
Sodium, y	495	477	425	322	482	587	370	322	479	375

Plot the data points and determine which type of function best models the data.

Solution: The scatter plot of the data looks like the graph in Figure 4.1.

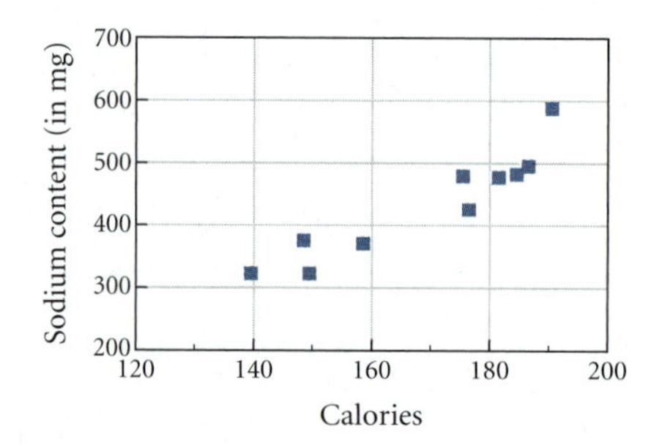

Figure 4.1

The graph shows that as x increases in value, y also increases at a fairly constant rate. Thus, a linear model would be the best model of this data. ■

Example 2: Unemployment claims for Volusia County, Florida, for recent years are given below.

Year	Number of Claims
1992	1502
1993	1268
1994	1250
1995	1052
1996	1110
1997	1004
1998	959
1999	727
2000	725
2001	1126
2002	1292
2003	1286
2004	1887
2005	1209

Source: Daytona Beach News-Journal.

Plot the data points and determine which type of function best models the data.

Solution: The scatter plot of the data looks like the graph in Figure 4.2.

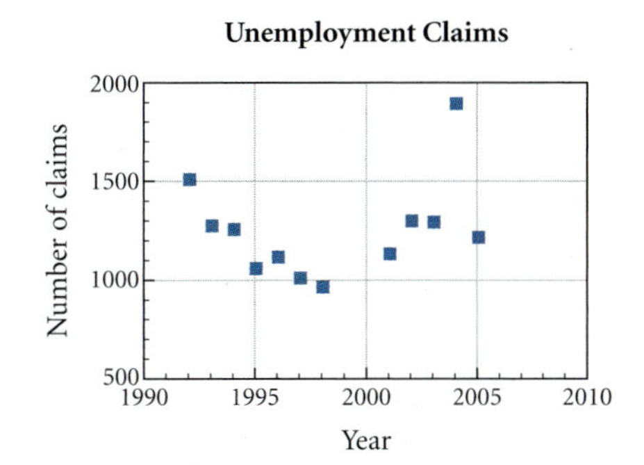

Figure 4.2

From the graph, it appears that the number of claims decreased until about 1999 and then increased again. (The spike in 2004 was caused by three major hurricanes that swept through the area that summer.) Thus, a quadratic model is the best function for the data. ■

Example 3: The percentage of the U.S. population that is foreign born is given in the table below.

Year	Percent
1900	13.6
1910	14.7
1920	13.2
1930	11.6
1940	8.8
1950	6.9
1960	5.4
1970	4.7
1980	6.2
1990	8.0
2000	10.4
2005	12.1

Data from: World Almanac.

Plot the data points and determine which type of function best models the data.

Solution: The scatter plot of the data looks like the graph in Figure 4.3.

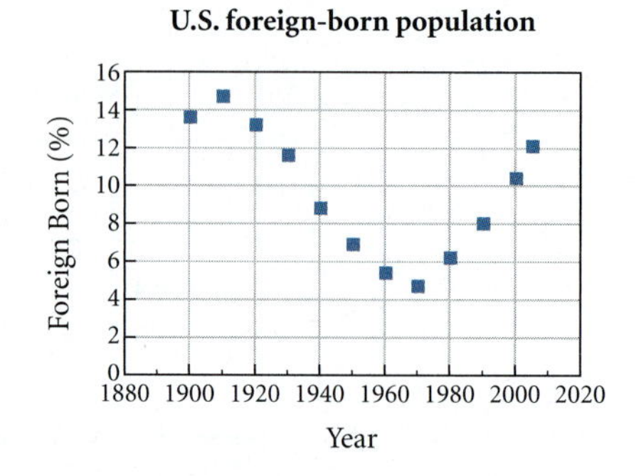

Figure 4.3

The graph clearly shows a cubic pattern. ■

Example 4: The federal hourly minimum wage has increased over the years according to the following table.

Year	Hourly Minimum Wage
1938	\$0.25
1939	\$0.30
1945	\$0.40
1950	\$0.75
1956	\$1.00
1961	\$1.15
1963	\$1.25
1967	\$1.40
1968	\$1.60
1974	\$2.00
1975	\$2.10
1976	\$2.30
1978	\$2.65
1979	\$2.90
1980	\$3.10
1981	\$3.35
1990	\$3.80
1991	\$4.25
1996	\$4.75
1997	\$5.15

Data from: U.S. Department of Labor

Plot the data points and determine which type of function best models the data.

Solution: The scatter plot of the data looks like the graph in Figure 4.4.

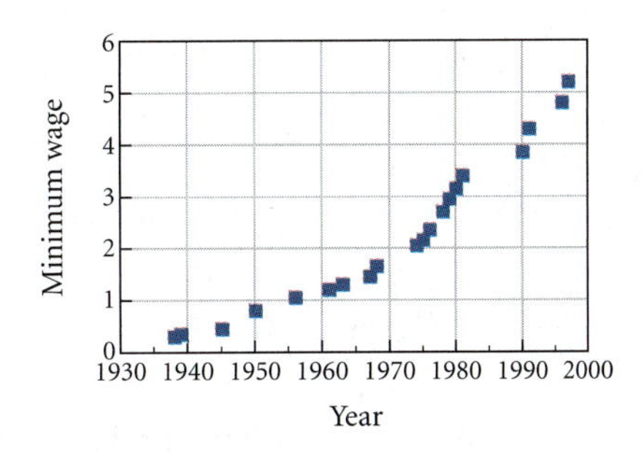

Figure 4.4

The graph shows an upward trend in the data, but the increase is not always at a constant rate. The data might be modeled by a linear function or by a power function. A power function is x raised to some real number power. ■

Regression Equations

After the type of pattern has been established, the regression model can then be found. To determine the function that best models the data, use the regression features of the graphing calculator or the Chart Wizard in Excel. The chart below explains how to perform regression using two of the more popular calculators on the market and Excel.

	TI-83/84	Excel
Enter Data	**STAT** → **EDIT** → enter the x data in list 1 and the y data in list 2.	Enter x data in column A and y data in column B.
Regression	**STAT** → **CALC** → choose the number of the appropriate regression model → **2nd 1** (L1), **2nd 2** (L2) → **ENTER**. The calculator gives the form of the equation and the values of the coefficients.	From the Chart Wizard, select the Scatter option and draw the plot. Then right click on any point and choose "Add Trendline." Select the appropriate regression type. Then click on Options and check the box for "Display equation on chart." Then click OK.

There are several types of regression models from which to choose:

- LinReg linear $y = ax + b$
- QuadReg quadratic $y = ax^2 + bx + c$
- CubicReg cubic $y = ax^3 + bx^2 + cx + d$
- QuartReg quartic $y = ax^4 + bx^3 + cx^2 + dx + e$
- LnReg logarithmic $y = a + b\ln x$
- ExpReg exponential $y = ab^x$
- PwrReg power $y = ax^b$

The calculator provides the values of the coefficients to be substituted into the equation. Excel displays the actual function. (Exponential and logarithmic models are discussed in Topic 12.) Excel and the calculator also can be made to display the **r^2 value**, which is a good measure of how well the equation fits the data. The closer the r^2 value is to 1, the better the function fits the data. Using this value can be a big help in determining which of several models is the best.

Calculator Corner 4.1

To set your calculator to display the r^2 value when doing regression, press **2nd 0** (Catalog). Scroll down and highlight Diagnostic On, then press **ENTER** twice. Each time you determine a regression equation, the r^2 value will automatically be displayed.

To have Excel display the r^2 value when doing regression, draw the scatter plot. Right click on any dot and choose "Add Trendline." Select the appropriate regression type and then click on "Options." Check the box that says "Display r^2 value on chart." Then click **OK**.

Example 5: Find the regression model for the hot dog data in Example 1.

Solution: Enter the calorie data as x in List 1 (TI-83/84) and the sodium data as y in List 2. The scatter plot (see Example 1) showed that a linear model was best, so choose the LinR regression model. (See Figure 4.5.)

$y = 4.3469x - 299.48;\ r^2 = 0.8709$

Figure 4.5

The linear regression model for the hot dog data is $y = 4.3469x - 299.48$, where y is sodium content in milligrams and x is the number of calories. The r^2 value of 0.8709 indicates that this model is a good one for the data. ■

4.1

The r^2 Coefficient

The r^2 value gives the percent of variation in the y variable that is caused by variations in the x variable. For example, in Example 5, an r^2 value of 0.8709 means that about 87% of the variation in sodium content for these hot dogs is explained by the number of calories, leaving only 13% to be explained by other variables.

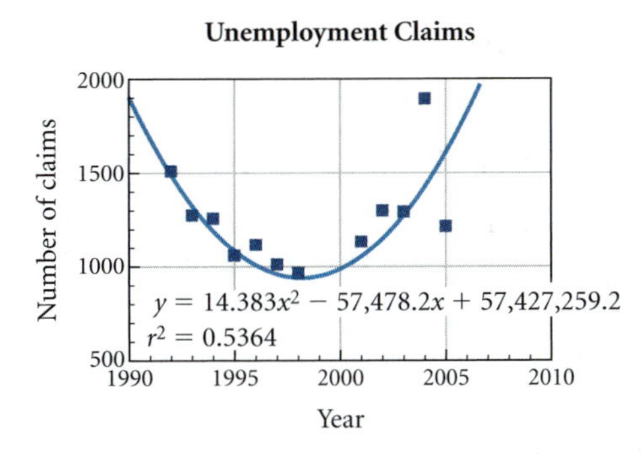

Figure 4.6

Example 6: Find the regression model for the unemployment claims data in Example 2.

Solution: Entering the data exactly as it is given yields the regression function $y = 14.383x^2 - 57,478.2x + 57,427,259.2$. (See Figure 4.6.) When the independent variable is time, it is usually better to let $x = 0$, rather than the exact year itself, be the starting point. Entering x as the number of years after 1990 gives the equation $y = 14.383x^2 - 235.68x + 1901.6$, which has the advantage of somewhat smaller coefficients. Be sure to indicate the domain used when stating the regression equation. The r^2 value of 0.5364 indicates that this model is a fairly good one for the data. ■

Example 7: Find the regression model for the data from Example 3 representing the percent of the U.S. population that is foreign born.

Solution: Entering the data exactly as given yields $y = 0.00005x^3 - 0.3039x^2 + 588.67x - 379,902$, as displayed in Figure 4.7.

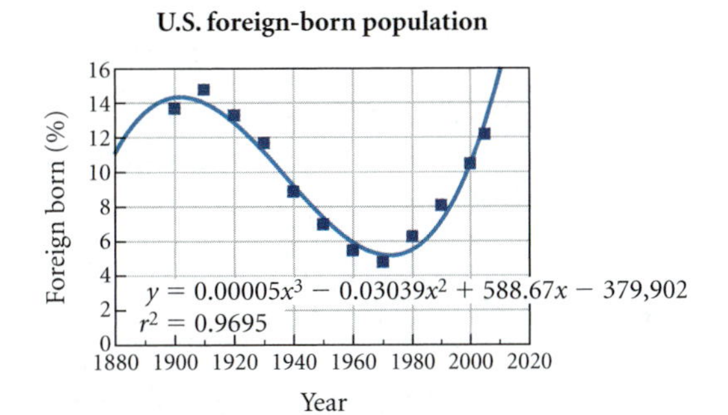

Figure 4.7

As before, if x were the number of years after 1900, the regression equation would be given by $y = 0.00005x^3 - 0.00587x^2 + 0.0246x + 14.27$. The r^2 value is 0.9695, which shows a very good fit for the data. ■

MATHEMATICS CORNER 4.2

Scientific Notation

The coefficient 5E-05 in Example 7 is written in scientific notation and means 5×10^{-5}. Move the decimal point after the 5 five places to the left to obtain the number 0.00005. A positive exponent would require moving the decimal point to the right.

Example 8: Find the regression model for the minimum-wage data given in Example 4.

Solution: With x as the number of years after 1900, there are two possible models:

Linear model	$y = 0.0801x - 3.4$	$r^2 = 0.9213$
Power model	$y = 0.0000026x^{3.17}$	$r^2 = 0.9907$

Often, more than one regression model may seem to describe the data. To decide which model is better, compare r^2 values and see which value is closer to 1. You can also check each model at several points in the domain and see which value is closer to the actual value. In Example 4, check the two models for the years 1950, 1975, and 1990.

$1950 \rightarrow x = 50$	Using the linear model, $y = \$0.61$. Using the power model, $y \approx \$0.63$, which is closer to the actual value of \$0.75.
$1975 \rightarrow x = 75$	Using the linear model, $y = \$2.61$. With the power model, $y \approx \$2.29$. Again the power model is closer to the actual wage of \$2.10.
$1990 \rightarrow x = 90$	Using the linear model, $y = \$3.81$. The power model gives $y \approx \$4.07$. This time, the linear model is closer to the actual wage of \$3.80.

By considering projected future increases in the minimum wage, it seems that the power model is the better predictor of the data. The r^2 values verify this conclusion. ■

Using Models to Make Predictions

After the regression model has been established, it can be used to make projections or predictions about the data.

Example 9: Estimate the sodium content of a 180-calorie beef hot dog.

Solution: From Example 5, the regression model is $y = 4.3469x - 299.48$, where y is the sodium content in milligrams and x is the number of calories. So, $y(180) = 4.3469(180) - 299.48$, or 482.962. We can estimate that a beef hot dog with 180 calories has about 483 mg of sodium. ■

Example 10: Estimate the number of unemployment claims in Volusia County, Florida, in 2006.

Solution: From Example 6, the regression model is $y = 14.383x^2 - 235.68x + 1901.6$, where x is the number of years after 1900. The year 2006 is 16 years after 1990, so $y(16) = 14.383(16)^2 - 235.68(16) + 1901.6 = 1812.768$. Thus, we can estimate that in 2006, there were approximately 1813 unemployment claims in Volusia County. ■

Example 11: Estimate the percentage of U.S. residents who were foreign born in 1975 and in 1998. Predict the percentage of U.S. residents who will be foreign born in 2020.

Solution: From Example 7, the regression model is $y = 0.00005x^3 - 0.00587x^2 + 0.0246x + 14.27$, where x is the number of years after 1900.

For 1975, which is 75 years after 1900, we have $y(75) = 0.00005(75)^3 - 0.00587(75)^2 + 0.0246(75) + 14.27 \approx 4.19$. We estimate that approximately 4.19% of U.S. residents in 1975 were foreign born.

For 1998, which is 98 years after 1900, we have $y(98) = 0.00005(98)^3 - 0.00587(98)^2 + 0.0246(98) + 14.27 \approx 7.36$. We estimate that approximately 7.36% of U.S. residents in 1998 were foreign born.

For 2020, which is 120 years after 1900, we have $y(120) = 0.00005(120)^3 - 0.00587(120)^2 + 0.0246(120) + 14.27 \approx 19.1$. We predict that approximately 19.1% of U.S. residents in 2020 will be foreign born.

The answers for 1975 and 1998 seem logical, based on the graph and common sense. What about 2020, however? Does it seem logical to predict such a high percentage of foreign-born residents? It is probably not logical, which points out a possible error when using regression models to make predictions. If the prediction is being made for a value far outside the domain used to determine the model, the prediction may not be accurate. ■

Topic 4 Review

4

This topic introduced the concept of regression to model a set of paired data.

<table>
<tr><th>CONCEPT</th><th>EXAMPLE</th></tr>
<tr><td>The graph of a paired data set is called a scatter plot.</td><td>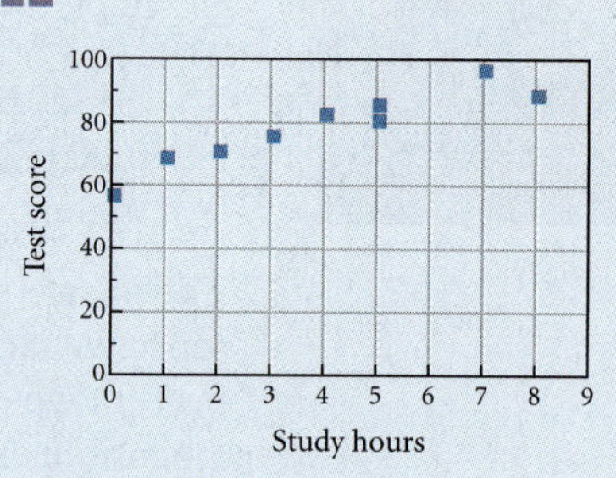
</td></tr>
<tr><td>Regression is a mathematical tool that creates a function to model the data displayed in the scatter plot.</td><td>The data appears to be linear. As study hours increase, test scores seem to increase at a fairly constant rate.</td></tr>
<tr><td>From the scatter plot, the best regression model (linear, quadratic, cubic, or power) can be determined. The graphing calculator will give the equation of the model. The r^2 value helps determine if the model is a good fit for the data. The closer the r^2 value is to 1, the more closely the model fits the data.</td><td>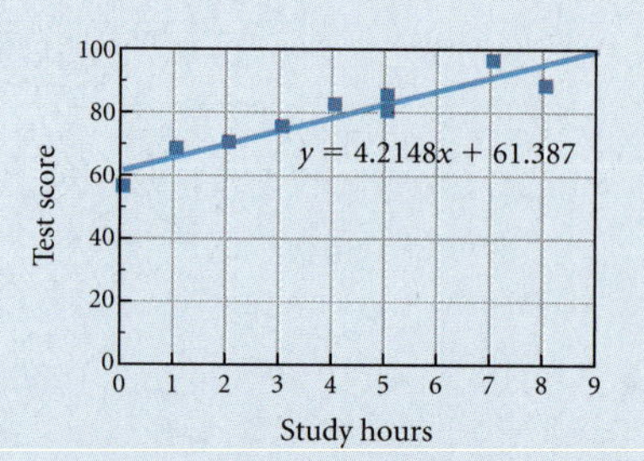
</td></tr>
<tr><td></td><td>The r^2 value of 0.8791 shows that the equation is a good model for the data.</td></tr>
<tr><td>The regression model can be used to estimate or make predictions about the data.</td><td>A student who studied for 6.5 hours can expect a test score of $y = 4.2148(6.5) + 61.387 = 88.7832$, or about 89.</td></tr>
</table>

NEW TOOLS IN THE TOOL KIT

- Drawing scatter plots of data
- Examining the scatter plot to determine which type of function best describes the data
- Determining the regression models using a calculator or Excel
- Using regression equations to make predictions about the data

Topic 4 Exercises

4

Match each scatter plot to the best type of regression model.

1.

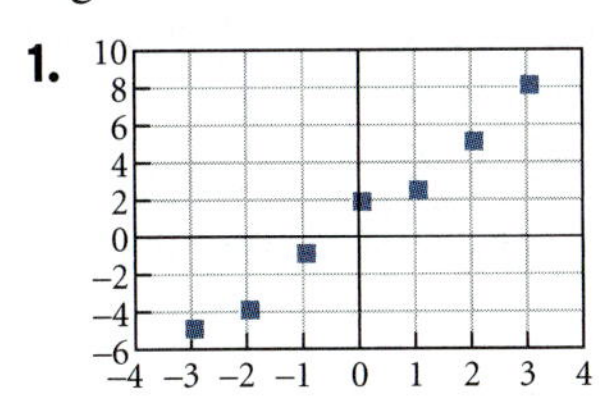

2.

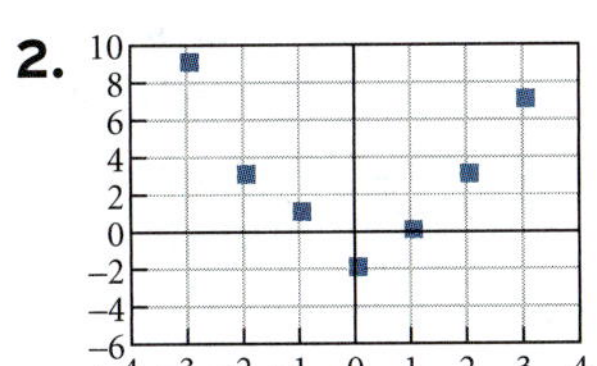

3.

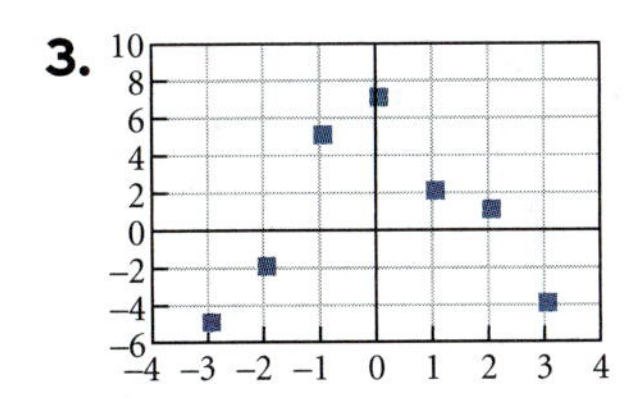

4.

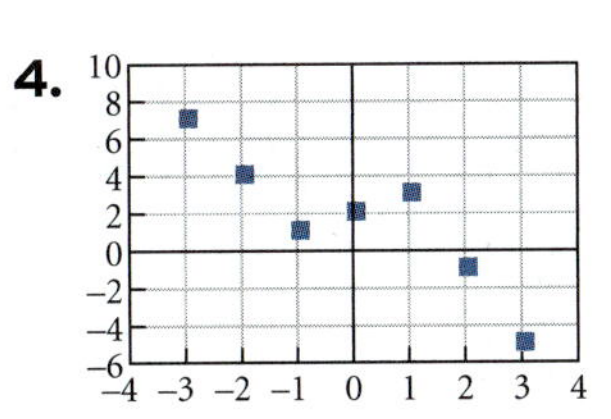

5.

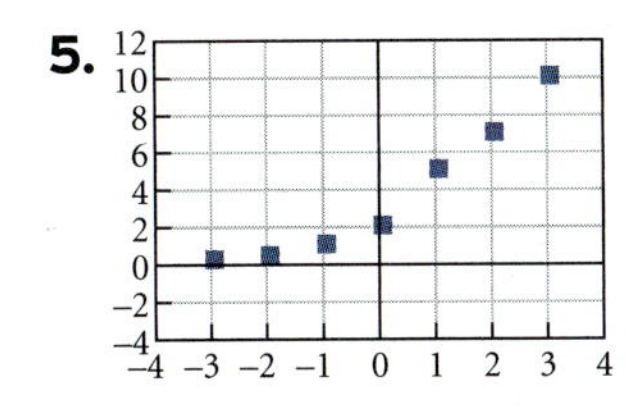

6.

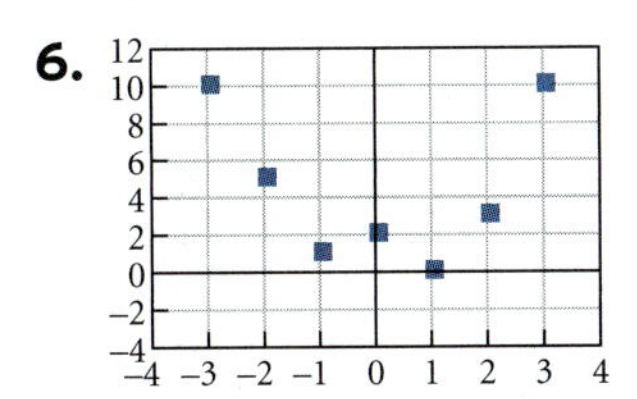

7.

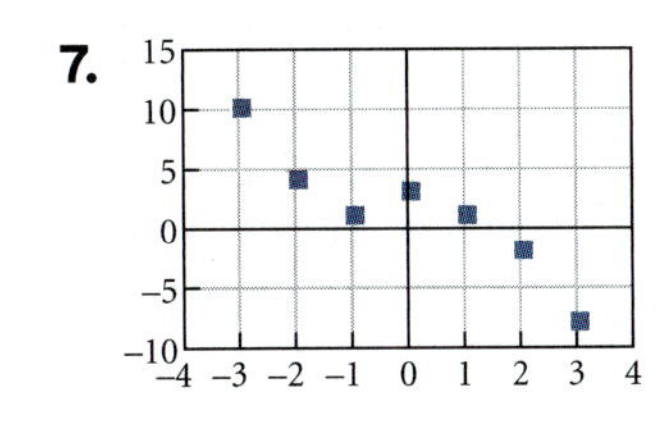

8.

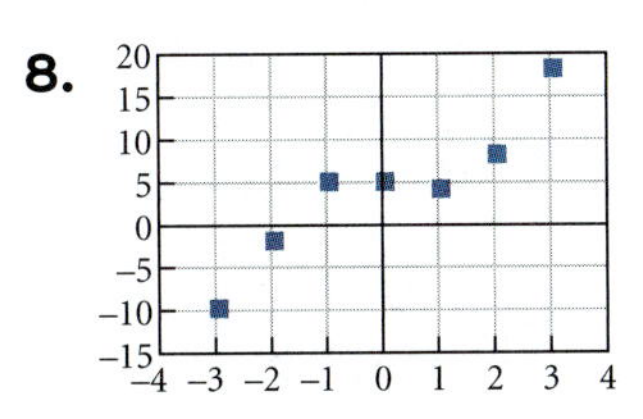

A. linear **B.** quadratic
C. cubic **D.** power

9. The 2003 *World Almanac* reports that the percent of men aged 65 or older in the labor force has steadily declined since 1890. Data are given in the following table.

Year	Percent
1890	68.3
1900	63.1
1920	55.6
1930	54.0
1940	41.8
1950	41.4
1960	30.5
1970	24.8
1980	19.3
1990	17.6
2000	18.6

a. Draw a scatter plot of the data.

b. Find a linear regression function for the data. Let $x = 0$ represent 1890.

c. Use the model from part b to estimate the percent of men aged 65 or older in the work force in 1960. How close is the estimate to the actual percent?

d. Predict the percent of men aged 65 or older in the labor force in 2008.

10. According to the *Orlando Sentinel*, ticket prices at Disney World (without tax) have risen steadily.

Year	Ticket Price	Year	Ticket Price
1988	$28.00	1997	$39.75
1989	$29.00	1998	$42.00
1990	$31.00	1999	$44.00
1991	$33.00	2000	$46.00
1992	$34.00	2001	$48.00
1993	$35.00	2002	$50.00
1994	$36.00	2003	$52.00
1995	$37.00	2004	$54.75
1996	$38.50	2005	$59.75

a. Draw a scatter plot of the data.

b. Find a linear regression function for the data. Let $x = 0$ represent 1985.

c. Use the model from part b to estimate ticket prices in 2002. How close is the estimate to the actual price?

d. Predict the ticket price in 2008.

11. According to 2002 statistics from the Florida Highway Patrol, the fatal crash rate per 1000 licensed drivers is given in the following table.

Driver Age	Midpoint Age, x	Crash Rate per 1000 Drivers
15–24	20	46.5
25–34	30	30.4
35–44	40	28.5
45–54	50	22.3
55–64	60	17.8
65–74	70	16.3
75–84	80	22.3
85 and older	90	29.2

a. Draw a scatter plot of the data. Use the midpoint age of each group for x.

b. Find a quadratic regression function for the data. Use the midpoint value, x, of each age group.

c. Use the model from part b to estimate the crash rate for 40-year-old drivers. How close is the estimate to the actual crash rate?

12. An analysis done by the Bureau of Health Professions shows that the demand for registered nurses in Florida will exceed the supply by 2020.

Year	Number of Registered Nurses Available
2000	112,735
2005	120,285
2010	126,075
2015	126,257
2020	123,904

a. Draw a scatter plot of the data.

b. Find a quadratic regression function that models the data. Let $x = 0$ represent the year 2000.

c. Use the model from part b to estimate the number of registered nurses available in 2007.

13. Daily high temperatures for New York City for selected days in November are given in the following table.

Day of Month	Temperature (°F)
1	77
5	58
8	45
16	51
19	64
24	58
30	51

a. Draw a scatter plot of the data.

b. Find a cubic regression function that models that data.

c. Use the model in part b to estimate the temperature on November 16. How close is the estimate to the actual temperature of 51°F?

d. Use the model to predict the temperature on November 12.

e. Use the model to predict the temperature on December 3.

14. Annual median sales prices for existing homes sold in a two-county area are given in the following table.

Year	Median Sales Price
1995	$71,100
1996	$75,200
1997	$75,500
1998	$79,000
1999	$84,500
2000	$88,300
2001	$96,770
2002	$109,768
2003	$131,300
2004	$157,800
2005	$187,550

Source: Florida Association of Realtors.

a. Draw a scatter plot of the data.

b. Find a cubic regression function that models the data. Let x represent the number of years after 1990.

c. Use the model in part b to estimate the median sales price in 2001. How close is the prediction to the actual price?

d. Use the model to predict the median sales price in 2008.

15. The *Statistical Abstract of the United States* shows the following trends in the percent of public school classrooms with Internet access.

Year	Years Since 1990	Percent of Classrooms
1994	4	3
1995	5	8
1996	6	14
1997	7	27
1998	8	51
1999	9	64
2000	10	77
2002	12	92

a. Draw a scatter plot of the data.

b. Find a power regression function that models the data. Let x be the number of years after 1990.

c. Use the model from part b to estimate the percent of classrooms that had Internet access in 1997. How close is the estimate to the actual percent?

16. The growth of Starbucks locations since its beginnings in 1971 is shown in the following table.

Year	Number of Locations
1971	1
1987	17
1988	33
1989	55
1990	84
1991	116
1992	165
1993	272
1994	425
1995	676
1996	1015
1997	1412
1998	1886
1999	2135
2000	3501
2001	4709
2002	5688
2003	7225
2004	7569

a. Draw a scatter plot of the data.

b. Find a power regression function that models the data. Let x be the number of years after 1970.

c. Use the model in part b to estimate the number of Starbucks locations in 1999. How close is the estimate to the actual number?

d. Use the model to predict the number of locations in 2010. Does the prediction seem reasonable?

17. The percent of 3- to 5-year-olds in preschool has increased over the years, as shown in the following table.

Year	Percent in Preschool
1970	37.5
1975	48.6
1980	52.5
1985	54.6
1990	59.4
1992	55.5
1993	55.1
1994	61
1995	61.8
2000	64
2001	63.9
2002	65.4

Source: U.S. Bureau of the Census.

a. Draw a scatter plot of the data.

b. Find at least two regression functions for the data. Be sure to state what x represents in the model.

c. For the models found in part b, estimate the percent of 3- to 5-year-olds in preschool in 1975 and in 1993. Which model seems to be the best for these data? Why?

d. Use the best model as determined in part b to predict what percent of 3- to 5-year-olds will be in preschool in 2009.

18. Health care spending in the United States has risen steadily since 1970, as shown in the following table.

Year	Per Capita Spending on Health Care
1970	$362
1980	$1005
1988	$2012
1993	$3250
1997	$3981
2000	$4670
2001	$4995
2002	$5440

Source: Health Affairs.

a. Draw a scatter plot of the data.

b. Find at least two regression functions for the data. Be sure to state what x represents in the model.

c. For the models found in part b, estimate the per capita spending on health care in 1988 and 2001. Which model seems to be the best for the data? Why?

d. Use the best model as determined in part b to predict per capita spending on health care in 2008.

19. Monthly premiums for Medicare have more than doubled since 1990, as shown in the following table.

Year	Monthly Premium
1990	$28.60
1991	$29.90
1992	$31.80
1993	$36.60
1994	$41.10
1995	$46.10
1996	$42.50
1997	$43.80
1998	$43.80
1999	$45.50
2000	$45.50
2001	$50.00
2002	$54.00
2003	$54.70
2004	$66.60
2005	$78.20
2006	$88.50

Source: Centers for Medicare and Medicaid Services.

a. Draw a scatter plot of the data.

b. Find the best regression model for the data. Why did you choose this type of model? Be sure to state the domain of your model.

c. Use your model to predict the cost of premiums in 2009.

20. Nationally, fewer than 5% of healthy Americans donate blood. A breakdown of blood donors by age is given on the next page.

Age of Donor	Percentage of That Age Group Who Donate Blood
18	10
25	16
35	19
45	20
55	15
65	12
75	7
85	1

a. Draw a scatter plot of the data.

b. Find the best regression model for the data. Why did you choose this type of model? Be sure to state the domain.

c. Use your model to estimate the percentage of 20-year-olds who donate blood. How good do you think your estimate is?

d. Use your model to estimate the percentage of 50-year-olds who donate blood. How good do you think your estimate is?

CALCULATOR CONNECTIONS

21. The graph of the shortage of registered nurses as presented by the Bureau of Health Professions is as follows.

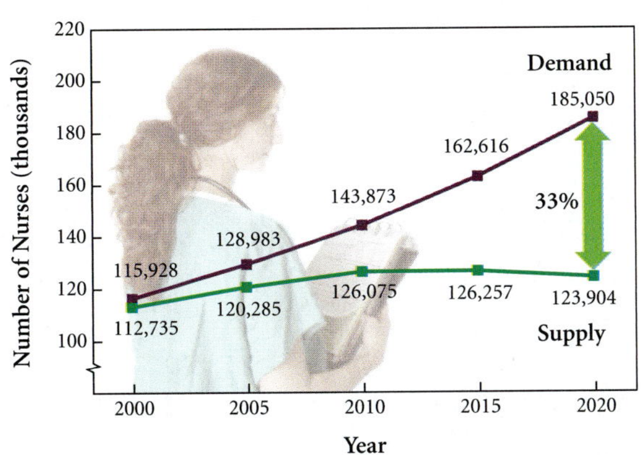

Source: "FHS Study on Nurse Staffing Issues in Florida," October 2002. "Projected Supply, Demand and Shortages of Registered Nurses: 2000–2020, " HHS, HRS Bureau of Health Professionals, National Center for Health Workforce Analysis, July 2002.

a. Determine a linear regression function that models the demand for registered nurses.

b. Determine a quadratic regression function that models the supply of registered nurses. (See Exercise 12.)

c. Based on your models, predict the shortage of registered nurses in 2008 and in 2015.

22. A publishing company is preparing to publish a new textbook. The costs of producing x units of the book, in thousands of dollars, and the revenue generated, in thousands of dollars, are given in the following chart.

Number of Units (thousands)	Cost (thousand $)	Revenue (thousand $)
1	200	50
2	270	100
4	370	200
6	420	360
8	500	680
10	600	1120

a. Determine the power regression model for cost and revenue.

b. Graph the cost and revenue equations. Determine the break-even point and interpret your answer.

c. Estimate the profit if 8000 textbooks are sold.

UNIT 0 Review

Unit 0 reviewed the concepts from algebra necessary for a solid understanding of the calculus.

TOPIC 1	■ The basic concepts of function, domain, and range ■ Function notation and evaluating functions ■ Difference quotients ■ Linear functions and their graphs ■ Slope and y-intercept of linear functions ■ Absolute value functions and their graphs ■ Using transformations to draw a graph on the coordinate plane ■ Business applications
TOPIC 2	■ Quadratic functions and their graphs ■ Determining the vertex of a quadratic function ■ Cubic functions and their graphs ■ Determining the point of inflection of a cubic function ■ Applications of quadratic and cubic functions ■ Square-root functions and their graphs ■ Graphing using transformations
TOPIC 3	■ Rational functions and their graphs ■ Determining the vertical and horizontal asymptotes of a rational function ■ Piecewise defined functions
TOPIC 4	■ Scatter plots ■ Determining the best model for a scatter plot using regression techniques ■ Making predictions about a data set using the regression model

Having completed Unit 0, you should now be able to

1. Examine ordered pairs, graphs, and equations to determine if they describe a function or not.
2. State the domain and range of functions expressed as ordered pairs, graphs, or equations.
3. Evaluate functions expressed in function notation.
4. Evaluate difference quotients.
5. Determine the slope and the y-intercept of a linear function.
6. Determine the vertex of a quadratic function.
7. Determine the point of inflection of a cubic function.
8. Determine the end point of a square-root function.
9. Determine the horizontal and vertical asymptotes of a rational function.
10. Determine the corner point of an absolute value function.
11. Sketch the graphs of linear, quadratic, cubic, radical, rational, and absolute value functions without using a graphing calculator.
12. State the domain and range of a function.
13. Use the graphing calculator to draw graphs of functions, when necessary, using a suitable window for x and y.
14. Evaluate and graph piecewise defined functions.
15. Solve application problems involving algebraic functions.
16. Draw a scatter plot of a set of paired data.
17. Use the graphing calculator or Excel to determine a regression model for a set of paired data.
18. Use the regression model to make predictions or estimations about a set of paired data.

UNIT 0 TEST

State whether each of the following in Exercises 1 through 6 represent a function. For each function, state the domain and the range.

1. $\{(3,5),(4,-1),(3,7),(9,0)\}$
2. $y = x^3 - 4x$
3. 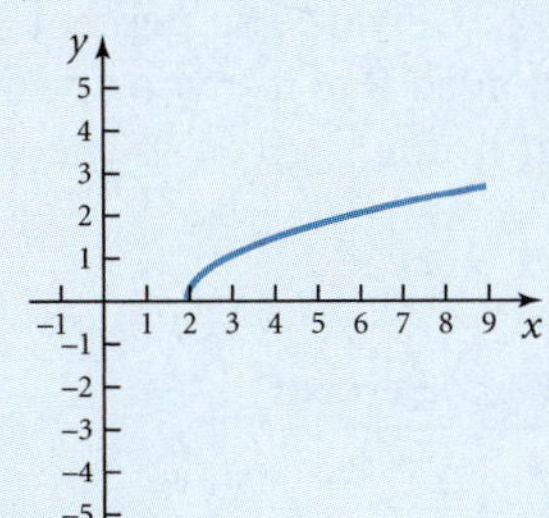
4. 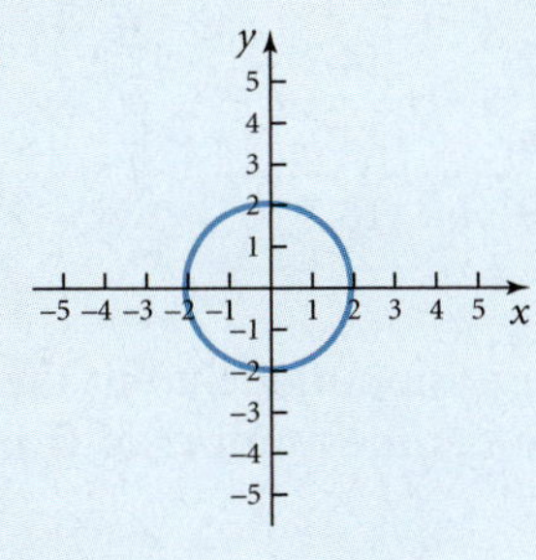
5. $x = y^2 + 1$
6. $\{(3,5),(6,5),(-1,4)\}$
7. Given $f(x) = -2x^2 + 3x - 4$, evaluate each of the following functions.
   a. $f(3)$ b. $f(-1)$
   c. $f(a)$ d. $f(a+h)$
8. Determine the slope and y-intercept of each of the following functions.
   a. $y = 5 - 3x$
   b. $5x - 2y = 10$

Without using a graphing calculator, sketch the graph of each of the following functions. State the domain and range of each function. Give the vertex, point of inflection, corner point, end point, or asymptotes of each function as applicable.

9. $f(x) = x^2 - 3$
10. $f(x) = 4 - (x+1)^2$
11. $f(x) = 2 - x^3$
12. $f(x) = (x+1)^3 - 4$

13. $f(x) = \sqrt{x - 2}$

14. $f(x) = \dfrac{4}{x - 3}$

15. $f(x) = |x + 4| - 3$

For Exercises 16 and 17, use a graphing calculator to draw the graph of the function. State the domain and range of each function.

16. $y = 2 - 3x + x^2$

17. $y = x^3 - x^2 + 3x$

18. Given $f(x) = \begin{cases} x - 3 & \text{if } x < -2 \\ 4 & \text{if } -2 \le x < 1 \\ \sqrt{x} & \text{if } x \ge 1 \end{cases}$

a. Evaluate $f(0)$.

b. Evaluate $f(1)$.

c. Evaluate $f(-2)$.

d. Evaluate $f(-4)$.

e. Graph the function.

State the horizontal and vertical asymptotes of each rational function in Exercises 19 through 22.

19. $f(x) = \dfrac{4}{x - 3}$

20. $f(x) = \dfrac{6x}{x - 4}$

21. $f(x) = \dfrac{3x^2}{x^2 + 1}$

22. $f(x) = \dfrac{x^3}{x^2 - 1}$

23. Starting salaries for workers at a utility company are based on x years of experience, according to the function.

$$S(x) = \begin{cases} \$28{,}600 & \text{if } x \le 3 \\ \$35{,}900 & \text{if } 3 < x < 8 \\ \$46{,}200 & \text{if } x \ge 8 \end{cases}$$

If Steve has five years of experience when he is hired, what is his starting salary?

24. An appliance repair service charges a $65 service call fee plus $30 per hour (or any fractional part of an hour).

a. Write a function that describes the total charge to a customer for a service call.

b. Calculate the fee for a service call lasting 1.4 hours.

25. Williams Loader Service bought a backhoe for $325,000. It is expected to last for 20 years, at which time it will be worth $50,000.

a. Write a function describing the value of the machine at any time during its lifetime. Be sure to state the domain.

b. Find the value of the backhoe after seven years.

c. If the backhoe is replaced when its value is $100,000, how long can the owner expect to have the machine?

26. Moodz Hair Salon finds that its weekly costs are $C(x) = 200 + 7x$ and its weekly revenues are $R(x) = -x^2 + 40x$, where x is the number of customers per week. Determine the break-even point and interpret its meaning.

27. For the Moodz Hair Salon information in Exercise 26, determine the number of weekly customers required for maximum profit. What is the maximum weekly profit?

28. Market supply and market demand, in thousands of units, for digital cameras are given by $S(p) = 0.1p - 10$ and $D(p) = -0.2p + 130$ where p is the cost per camera. Find the equilibrium point and interpret its meaning.

29. According to the U.S. Census, the percent of U.S. schoolchildren aged 5 to 17 who speak another language at home is given in the following table.

Year	Percent
1979	8.5
1989	12.6
1992	14.2
1995	14.1
1999	16.7

a. Find a linear regression model to fit the data. Let x represent the number of years after 1970.

b. Use the model from part a to estimate the percent of U.S. schoolchildren in 1994 who spoke another language at home.

c. Use the model from part a to predict the percent of U.S. schoolchildren in 2010 who speak another language at home.

30. According to the U.S. Census, the number of bachelor's degrees in mathematics conferred from 1971 to 2002 is given in the following table.

Year	Bachelor's Degrees Conferred
1971	24,937
1980	11,872
1990	15,176
1995	13,723
2002	12,395

a. Find a quadratic regression model that fits the data. Let x represent the number of years after 1970.

b. Use the model from part a to estimate the number of mathematics degrees conferred in 1985.

c. Use the model to predict the number of mathematics degrees to be conferred in 2009.

UNIT 0 PROJECT

For this project, you will gather data on a topic that interests you and apply regression techniques to determine the best model for the data.

1. Select a topic that interests you and gather data on that topic. Data can be found in magazines, in newspapers, on the Internet, and even in your other textbooks. Be sure you have from 5 to 15 data points.
2. Plot the data points on a scatter plot. Use regression techniques to determine the equation of at least two possible models for your data. Which function do you think best models the data and why? State the domain of your model and indicate what $x = 0$ represents.
3. Draw the graph of your function, superimposed over the data points.
4. Choose any x value from your data list and evaluate your function for that value. How close did your model estimate the actual value?
5. Choose an x value not included in your data list and evaluate your function for that value. How good do you think the prediction is and why?

Limits and Derivatives

UNIT 1

Here's the seasonal average attendance at the Orlando Magic home basketball games.

SEASON	AVERAGE ATTENDANCE
1995-1996	17,248
1996-1997	17,199
1997-1998	17,113
1998-1999	16,444
1999-2000	14,059
2000-2001	14,757
2001-2002	15,149
2002-2003	14,545
2003-2004	14,352
2004-2005	14,507

(Source: Orlando Sentinel.)

- What trends or patterns do you notice in the attendance data?
- Could you find a function that describes that trend or pattern?
- Could you predict the average home attendance for the 2007–2008 season?
- At what rate did the average home attendance increase or decrease between any two seasons?

You will soon be able to determine the answer to all these questions. You already know that a graph of the data shows the trends or patterns and that algebra provides tools for determining functions that model or describe the data. You will soon see that calculus provides a more thorough analysis of the behavior of the function describing the data.

What Is Calculus?

The introduction of calculus in the 1600s stimulated rapid growth in mathematics as well as in other areas, such as navigation, astronomy, and physics. What exactly does calculus do? Calculus deals with the behavior of functions. Calculus may be defined as the *mathematical tool used to analyze changes in variable quantities.* A variable quantity may change at a constant rate, such as the earnings of a worker paid by the hour, or it may change at a nonconstant rate, such as the rate at which planets orbit around the sun, the rate at which sales of a new product change over time, or the rate at which a population of bacteria grows. Calculus provides a powerful mathematical tool to examine the rates of change for these variable quantities, which are described by functions.

The word *calculus* comes from the same Latin word, which means "a stone used in counting." People used to use stones to solve mathematical problems. Sometimes you may see "the calculus" used to distinguish it from other methods of computation.

Baron Gottfried Wilhelm Leibniz (1646-1716)

The basic ideas of calculus were known and explored by Archimedes (287–212 BC), but they were not formalized until 2000 years later, in the 17th century. The two founders of calculus were a German mathematician, Baron Gottfried Wilhelm Leibniz (1646–1716), and an English mathematician and physicist, Sir Isaac Newton (1642–1727). If you delve into the history of mathematics, you will find that both men discovered the same ideas at about the same time. Newton apparently developed the basic ideas of calculus and shared them with his circle of friends, but he was reluctant to publish them. Leibniz independently developed the same basic ideas of calculus about 10 years after Newton, but he immediately published them and had them distributed by his supporters. The story is that one of Newton's colleagues heard one of Leibniz's ideas and was astonished, because he knew that Newton had done the same work years earlier! Thus, both men are usually credited as being the founders of calculus.

Sir Isaac Newton (1642-1727)

During the Renaissance, there were many great advances in mathematics. The exploration of new lands and continents called for better mathematics for navigation. The growth of business demanded better mathematics for banking and finance. The beginnings of calculus were developed to answer two basic questions that were arising in science at the time: how do we find the slope of a curve, and how do we find the area of nongeometric figures?

Finding the Slope of a Curve

Before calculus, mathematicians knew how to find the slope of a line. As scientific discoveries progressed, however, scientists soon realized that the planets, for instance, do not move in straight line paths, but in curved

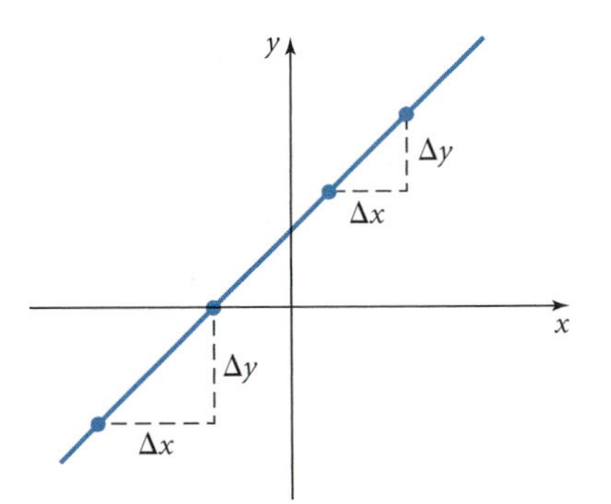

Figure 1

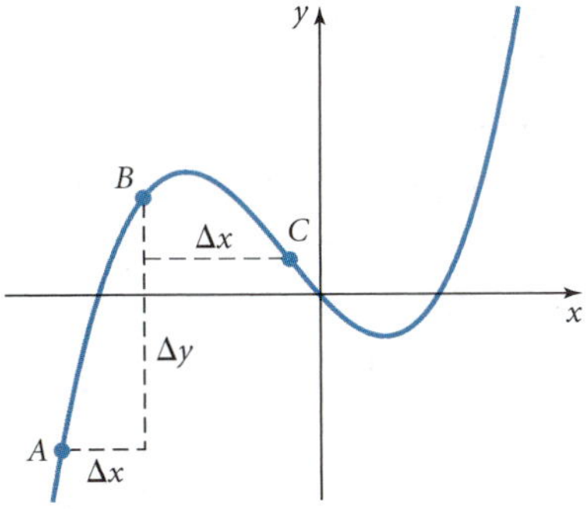

Figure 2

orbital paths. Because Earth is curved, not flat, navigational calculations could not be done as linear arithmetic.

For the line in Figure 1, the slope, or the rate of change of y with respect to x, or $\frac{\Delta y}{\Delta x}$, between any two points on the line is always the same. (*Note:* $\frac{\Delta y}{\Delta x}$ is read "delta y over delta x". The capital Greek letter delta is used in calculus to symbolize a small change in some quantity.) For the curve in Figure 2, however, the rate of change of y with respect to x is not always the same. The slope of the line between points A and B is positive, but the slope of the line between points B and C is negative.

For scientific advances to continue, scientists needed to be able to determine rates of change for nonlinear quantities, so they turned to the mathematicians for help.

Finding the Area of "Nongeometric" Figures

"Geometric" figures are rectangles, triangles, circles, trapezoids, and other such figures for which area formulas exist. What about areas of regions such as that displayed in Figure 3?

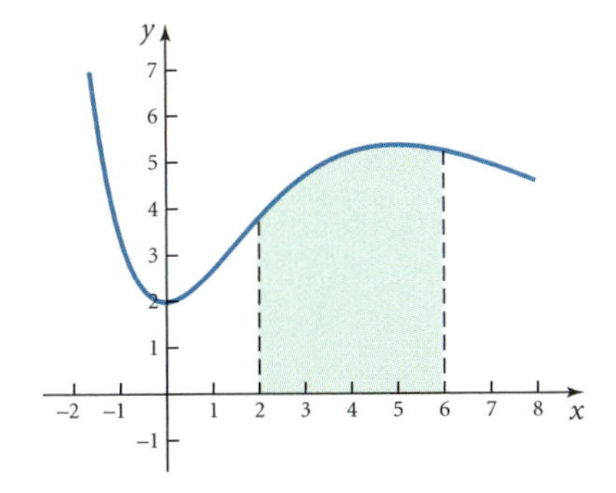

Figure 3

There is obviously an area enclosed within the bounded region. Because one of the boundaries is not a straight line, this region is not one of the recognizable geometric figures for which an area formula exists. We will refer to such figures as "nongeometric" figures, meaning figures whose area cannot be found using existing area formulas. Nevertheless, there must be a way to find the area of such nongeometric regions.

Of course, there are answers to these questions, or there would be no need for this book! To answer these questions, we discuss the three fundamental concepts in calculus:

- Limit
- Derivative
- Integral

By the time you have completed your study of calculus, you will see how these three concepts are all related to one another and how they are applied in practical situations.

Calculus is a mathematical tool, and a very powerful one at that. As we progress through this book, you will be adding new tools to your mathematical Tool Kit. Any carpenter or mechanic begins with a tool box containing just the basic tools and then adds more tools to the tool box as the need arises. Your mathematical Tool Kit now contains the basic tools needed to begin a study of calculus: an understanding of functions and domains, the ability to evaluate functions and solve basic equations, and a working knowledge of the basic functions and their graphs. These tools have served you well up to this point, but the time has come to add more tools to the Tool Kit.

TOPIC

Introduction to Limits

5

Warm-Up Exercises

Page 101

The warm up exercises allow students to brush up on the algebra and calculus skills they'll need to fully comprehend the subsequent material.

T
i
a
a
ti
i
s

T
w

IMPORTANT TOOLS IN TOPIC 5

- *Limits of functions*
- *Left-hand and right-hand limits*
- *Limits of rational functions*
- *Limits of difference quotients*
- *Limits of piecewise defined functions*
- *Limits at infinity*

TOPIC 5 WARM-UP EXERCISES

Be sure you can successfully complete the following exercises before starting Topic 5.

1. Factor each of the following expressions, if possible.

a. $3x - 12$ **b.** $3x^2 - 12$ **c.** $x^2 - 9$

d. $x^2 + 9$ **e.** $x^3 - 4x$ **f.** $x^2 - 4x + 3$

g. $2x^2 - 5x + 3$

2. If $f(x) = 2x^2 - 5x + 3$, evaluate each of the following functions.

a. $f(0)$ **b.** $f(2)$ **c.** $f(-3)$

d. $f(a)$ **e.** $f(a + h)$

3. State the domain of each of the following functions.

a. $f(x) = 3x - 2$

b. $f(x) = \sqrt{x + 3}$

c. $f(x) = \dfrac{6}{x^2 - 4}$

d. $f(x) = 2x^2 + 5x - 3$

e. $f(x) = x^3 - 1$

f. $f(x) = \dfrac{3x}{x^3 - 4x}$

4. State the horizontal and vertical asymptotes of the following functions, without drawing the graph.

a. $f(x) = \dfrac{3}{x + 2}$

b. $f(x) = \dfrac{6x}{x + 2}$

c. $f(x) = \dfrac{2x}{x^2 - 9}$

d. $f(x) = \dfrac{2x^3}{x^2 - 9}$

5. Use your knowledge of translations to describe how the graph of each of the following functions is related to the basic graph of $f(x) = x^2$. You should be able to do so without drawing the graphs.

a. $f(x) = x^2 - 4$

b. $f(x) = (x - 4)^2$

c. $f(x) = 4 - x^2$

d. $f(x) = (x + 3)^2 + 2$

Answers to Warm-up Exercises

1. a. $3(x-4)$ b. $3(x+2)(x-2)$
   c. $(x+3)(x-3)$ d. cannot be factored
   e. $x(x+2)(x-2)$ f. $(x-3)(x-1)$
   g. $(2x-3)(x-1)$
2. a. 3 b. 1
   c. 36 d. $2a^2-5a+3$
   e. $2a^2+4ah+2h^2-5a-5h+3$
3. a. reals b. $x \geq -3$
   c. $x \neq 2, -2$ d. reals
   e. reals f. $x \neq 0, 2, -2$
4. a. $y=0, x=-2$ b. $y=6, x=-2$
   c. $y=0, x=3, x=-3$
   d. no horizontal, $x=3, x=-3$
5. a. The graph is shifted 4 units down, so the vertex is at $(0,-4)$.
   b. The graph is shifted 4 units to the right, so the vertex is at $(4,0)$.
   c. The graph is shifted 4 units up and inverted, so the vertex is at $(0,4)$ and the graph opens downward.
   d. The graph is shifted 3 units left and 2 units up, so the vertex is $(-3,2)$.

Unit/Topic Organization

Page 102

The organization of this text breaks difficult concepts down into bite-sized chunks allowing students to completely understand a topic before they move on to a new concept.

...ppose Dave is ... I will pay for ...ing value that ... end up pay-... ...sider a tennis ... ball bounces ... from a height ...er the second, ... will the ball ...eight of 0 feet, ... 0 feet. These

Sequence...

... a brief discussion of limits of sequences.

Definition: A **sequence** is a function whose domain is the set of positive integers and whose range is the set of real numbers. In other words, a sequence is a list of numbers. The domain is the set of term numbers $\{1, 2, 3, 4, \ldots\}$, and the range is the set of values of each term.

The terms of a sequence usually follow some type of pattern. Consider the following special types of sequences.

- **Arithmetic sequence**: Each term is the previous term plus some constant difference.
- **Geometric sequence**: Each term is a constant multiple of the previous term.
- **Constant sequence**: There is no change from one term to the next.

This list is not exhaustive because there are many other types of sequences. Consider the next few examples.

Example 1: The following sequences are arithmetic sequences.

$2, 5, 8, 11, \ldots$	The constant difference is 3.
$3, 1, -1, -3, \ldots$	The constant difference is -2.

Be sure to see the dots (...) at the end of the sequence. These dots mean that the pattern established by the first few terms of the sequence continues indefinitely throughout all remaining terms of the sequence. The domain of each sequence is $\{1, 2, 3, \ldots\}$, indicating the term number. The range of each sequence is the list of terms in the sequence. ■

Example 2: The following sequences are geometric sequences.

$1, 3, 9, 27, \ldots$	The constant multiple is 3.
$16, -8, 4, -2, \ldots$	The constant multiple is $-\frac{1}{2}$.

■

The sequences Examples 1 and 2 are **infinite sequences** because there are an infinite number of terms in the sequence. In calculus, we are interested in how the terms of the sequence behave after a large number of terms. Do the terms of the sequence seem to be growing without bound, or are they approaching some specific value? This behavior, referred to as a *limit*, is an important concept in calculus.

Definition: The **limit** of a sequence is the unique real number that the terms of the sequence are approaching.

From the definition of the limit of a sequence, you need to understand the following points.

1. The terms of the sequence may or may not actually reach the limiting number, but they will eventually be extremely close to that limit.
2. The limit must be unique; that is, there must be only one value for the limit.
3. If the terms do not approach a unique real number, we say that the limit *does not exist*.
4. If the terms grow without bound, the limit is ∞ (or $-\infty$).

To determine the limit of a sequence, figure out the next few terms and determine what number, if any, the terms of the sequence seem to be approaching.

Example 3: The sequence $16, -8, 4, -2, \ldots$ has a limit of 0.

This sequence is a geometric sequence. The constant multiple is $-\frac{1}{2}$, so the next few terms are $1, -\frac{1}{2}, \frac{1}{4}$, and $-\frac{1}{8}$. Because the magnitude of each term is half as large as that of the previous term, the terms of the sequence approach a value of 0. ■

Example 4: The sequence 2, 5, 8, 11, ... has a limit of ∞.

This sequence is an arithmetic sequence. The constant being added is 3, so the next few terms are 14, 17, 20, and 23. By continually adding 3, the terms continue to grow infinitely large. ■

Example 5: The sequence 5.1, 5.01, 5.001, 5.0001, ... has the limit 5.

The next few terms of the sequence are 5.00001, 5.000001, and 5.0000001. The decimal part of the number is being reduced by a factor of 10 each time, so gradually the decimal part of the number approaches 0, leaving the number 5 as the limit. ■

Example 6: The sequence 2, 2, 2, 2, ... has the limit 2.

The next few terms would each be 2. The 1000th term would still be 2. Thus, the limit is 2 because this sequence is a constant sequence that has no other value. ■

Limits can also be viewed graphically. The graph of the sequence {2, 5, 8, 11, ...} is shown in Figure 5.1.

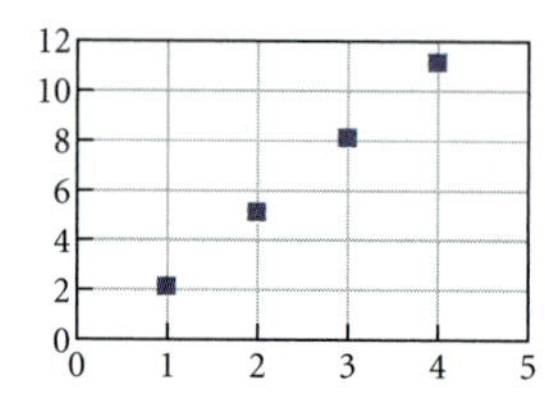

Figure 5.1

From the graph, you can see that the terms are growing without bound and are approaching infinity.

The graph of the sequence {16, −8, 4, −2, ...} is shown in Figure 5.2.

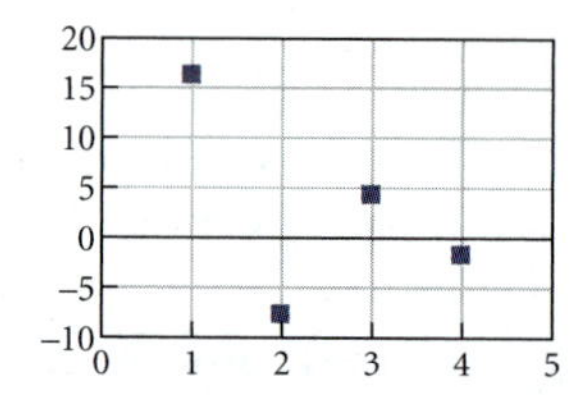

Figure 5.2

The graph shows that the terms are alternating between positive and negative values, but are gradually approaching the x-axis and a value of 0.

The graph of the sequence {5.1, 5.01, 5.001, ...} is shown in Figure 5.3.

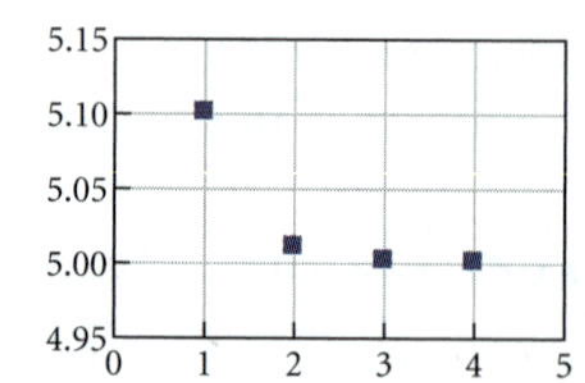

Figure 5.3

The graph shows that the terms of the sequence are gradually decreasing and approaching a value of 5.

Example 7: For each of the following sequences, determine the limit, if it exists.

a. 125, 25, 5, ...
b. 1, 4, 16, ...
c. 5, 5, 5, ...
d. 4.9, 4.99, 4.999, ...
e. 4.49, 4.499, 4.4999, ...
f. 0.3, 0.33, 0.333, ...
g. 1, 2, 1, 2, 1, ...

Solution:

a. This sequence is a geometric sequence with constant multiple of $\frac{1}{5}$. The terms are growing successively smaller and approach, but never reach, 0. Thus, the limit of the sequence is 0.

b. This sequence is a geometric sequence with a constant multiple of 4. The terms are increasing without bound, so the limit is ∞.

c. This sequence is a constant sequence, so the limit is 5.

d. The next three terms of the sequence are 4.9999, 4.99999, and 4.999999. In each term, we are adding one more 9 to the decimal part of the number. Imagine what the 100th term of the sequence would be. The limit is 5.

e. Here the decimal part of the number is always 4 followed by a series of 9s. The next three terms are 4.49999, 4.499999, and 4.4999999. Again, imagine what the 100th term of the sequence would look like. The limit is 4.5.

f. In this sequence, each term simply adds another 3 to the decimal part of the number. The limit is $\frac{1}{3}$. To envision this process, simply divide 1 by 3 and look at the result. The terms of this sequence are simply the values of $\frac{1}{3}$ rounded to various decimal places.

g. The terms of this sequence alternate between 1 and 2. If you go infinitely far in the sequence, the term will be either 1 or 2, depending on whether it is an odd-numbered term or an even-numbered term. Recall from the definition of limit that a limit is a *unique* real number. Because this sequence does not approach a unique number, this sequence has no limit. ■

Warning! ***When there is more than one possible value for the limit of a sequence, the limit of that sequence does not exist.***

Limits of Functions

Functions can be considered a type of sequence. The ordered pairs that satisfy the function could conceivably be listed with the x values in numerical order, creating an **infinite sequence**. The **limit of a function** refers to how the y values of the function behave as x approaches a certain value.

There are three ways to examine limits of functions:

- Numerically
- Graphically
- Algebraically

Consider the function $f(x) = 3x - 2$. What is the limit of this function as x approaches some specific value, say 2? In other words, how does the function—the

y value—behave as x approaches 2? Symbolically, we write "$x \to 2$" to denote "x approaching 2."

Numerically, construct a table to analyze the behavior of $f(x)$ as x gets close to 2 from the left, that is, using values less than 2.

x	$f(x)$
1.9	$f(1.9) = 3(1.9) - 2 = 3.7$
1.99	$f(1.99) = 3(1.99) - 2 = 3.97$
1.999	$f(1.999) = 3(1.999) - 2 = 3.997$
1.9999	$f(1.9999) = 3(1.9999) - 2 = 3.9997$
$\downarrow$	$\downarrow$
2	?

In the x column, begin with $x = 1.9$, a number close to but smaller than 2, and evaluate the function for $x = 1.9$. Then choose numbers successively closer to 2 and evaluate the function for those values. As $x \to 2$, what value is y approaching? Looking at the y values of 3.7, 3.97, 3.997, and 3.9997, we see a list of numbers approaching 4. This limit is called a **left-hand limit** because the values chosen for x were less than 2, which means approaching 2 from the left on a horizontal number line.

Definition: The **left-hand limit** of $f(x)$ is the value that $f(x)$ approaches as x approaches a from the left using only values of x that are smaller than a. Symbolically, the left-hand limit is written as

$$\lim_{x \to a^-} f(x)$$

Now construct a table to analyze the behavior of $f(x)$ as x gets close to 2 from the right, using values greater than 2.

x	$f(x)$
2.1	$f(2.1) = 3(2.1) - 2 = 4.3$
2.01	$f(2.01) = 3(2.01) - 2 = 4.03$
2.001	$f(2.001) = 3(2.001) - 2 = 4.003$
2.0001	$f(2.0001) = 3(2.0001) - 2 = 4.0003$
$\downarrow$	$\downarrow$
2	?

In the x column, choose 2.1, a number close to but greater than 2, and evaluate the function for $x = 2.1$. Then choose numbers successively closer to 2 than 2.1 and evaluate the function for those values. As $x \to 2$, what value is y approaching? Looking at the y values of 4.3, 4.03, 4.003, and 4.0003, we see a list of numbers approaching 4. This limit is called a **right-hand limit** because the values chosen for x were greater than 2, meaning that they approach 2 from the right on a horizontal number line.

Definition: The **right-hand limit** of $f(x)$ is the value that $f(x)$ approaches as x approaches a from the right using only values of x that are larger than a. Symbolically, the right-hand limit is written as

$$\lim_{x \to a^+} f(x)$$

Calculator Corner 5.1

Graph $y_1 = 3x - 2$. Then use the TABLE feature of the calculator to explore the limit as $x \to 2$. To examine the left-hand limit, press **2nd WINDOW** (TblSet). Set TblStart = 1.9 and ΔTbl = .001. Press **2nd GRAPH** (Table) and scroll down the table to see what value y is approaching as $x \to 2$. To examine the right-hand limit, press **2nd WINDOW** (TblSet). Set TblStart = 2.1 and ΔTbl = $-.001$. Press **2nd GRAPH** (Table) and scroll down the table to see what value y is approaching as $x \to 2$.

For $f(x) = 3x - 2$ as x approaches 2, the left-hand limit and the right-hand limit are the same. It would appear that the limit of $f(x) = 3x - 2$ is 4 as $x \to 2$ from either the left or the right. This limit is written symbolically as $\lim_{x \to 2}(3x - 2) = 4$. *For the limit of a function to exist, it is necessary that the left-hand limit and the right-hand limit be the same.*

Definition: $\lim_{x \to a} f(x)$ exists if and only if $\lim_{x \to a^-} f(x) = \lim_{x \to a^+} f(x)$.

1. If $\lim_{x \to a^-} f(x) = \lim_{x \to a^+} f(x) = L$, then $\lim_{x \to a} f(x) = L$.
2. If $\lim_{x \to a^-} f(x) = L$ and $\lim_{x \to a^+} f(x) = M$, where $L \neq M$, then $\lim_{x \to a} f(x)$ does not exist.

Graphically, consider the graph of the function $f(x) = 3x - 2$ in Figure 5.4.

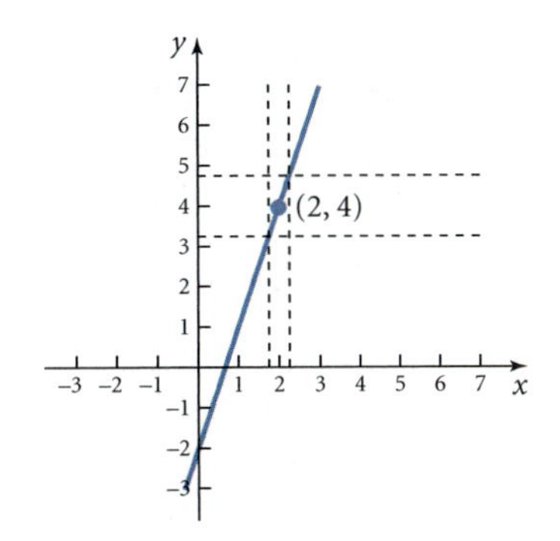

Figure 5.4

Choose a small interval around $x = 2$ and allow that interval to "close in" on $x = 2$. Notice that the corresponding y values are also closing in and creating a "box" around the point (2, 4). By continuing to close in around $x = 2$, the box continues to shrink until only the point $(2, f(2))$ is included. Thus, the limit is the y value of the point on the graph that corresponds to $x = 2$. As in the numerical analysis of this limit, we see that $\lim_{x \to 2}(3x - 2) = 4$.

MATHEMATICS CORNER 5.1

Epsilon-Delta Definition of Limit

We have examined $\lim_{x \to a}(3x - 2) = 4$ numerically and graphically. Let's expand on the graphical "box" discussion.

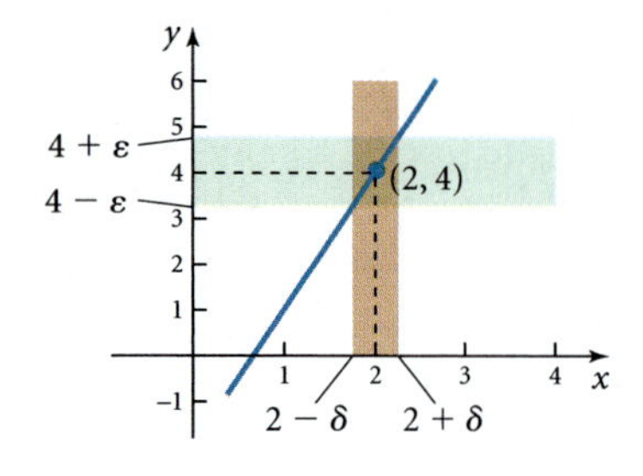

Select a small interval around $x = 2$. The boundaries of this interval are $(2 - \delta, 2 + \delta)$, where δ is a very small value approaching 0 (δ is the lowercase Greek letter "delta"). There is also a small interval around the limiting value of $L = f(2) = 4$. The boundaries of this interval are $(4 - \epsilon, 4 + \epsilon)$, where ϵ is also a very small value approaching 0 (ϵ is the lowercase Greek letter "epsilon"). If $\lim_{x \to 2}(3x - 2)$ exists, as $\delta \to 0$, $\epsilon \to 0$ also. In other words, we can make $f(x)$ closer and closer to 4 by choosing x sufficiently close to 2.

Generalizing to $\lim_{x \to a} f(x)$, we have the following property, called the "*epsilon-delta* definition of limit."

> $\lim_{x \to a} f(x) = L$ means that for all $\epsilon > 0$, there exists a $\delta > 0$ such that if $0 < |x - a| < \delta$, then $|f(x) - L| < \epsilon$.

In other words, if the difference between x and a is sufficiently small, the difference between $f(x)$ and the limit is also infinitesimally small.

Algebraically, why does the ordered pair (2, 4) represent the coordinates of the point on the graph? In other words, what is the value of the function $f(x) = 3x - 2$ when $x = 2$? If $x = 2$, the y value is $y = f(2) = 4$. For this polynomial function, evaluating $f(2)$ to obtain 4 is the quickest way to the limit of this function. For algebraic functions, it may be easiest to first try to evaluate $f(a)$ when trying to determine the limit of the function as $x \to a$.

Substitution Property

To evaluate the limit of an algebraic function, first try to evaluate $f(a)$. If $\lim_{x \to a} f(x) = f(a) = L$, where L is a unique real number, then L is the value of the limit.

If the function is a polynomial or a radical, the Substitution Property alone should be sufficient to evaluate the limit. The Substitution Property is also sufficient to evaluate limits of rational functions at points within the domain of the function.

Warning! ***The Substitution Property is not valid for all functions.***

1. The Substitution Property fails if $f(a)$ is indeterminate $\left(\frac{0}{0}\right)$ or if $f(a)$ is undefined $\left(\frac{k}{0}\right)$. Remember that division by zero is not possible. Rational functions are discussed in more detail later in this topic. See Mathematics Corner 3.1 for a more detailed discussion of indeterminate and undefined.
2. The Substitution Property also fails if $f(x)$ is a piecewise defined function with a break point at $x = a$. At a break point, the left-hand limit and the right-hand limit are not equal.

Warning Boxes

Page 109

These boxes alert students to common errors made during calculation or the application of a rule.

ate $\lim\limits_{x\to -2}(4x+5)$.

$4(-2) + 5 = -3$. ■

ate $\lim\limits_{x\to 2}\sqrt{2x+5}$.

$2) + 5} = \sqrt{9} = 3$.

dered because functions

■

ich is a constant function.

■

$\frac{4}{5-2(-3)} = \frac{4}{11}$. ■

valuating limits.

Properties of Limits

Given functions $f(x)$ and $g(x)$ such that $\lim\limits_{x\to a} f(x)$ and $\lim\limits_{x\to a} g(x)$ both exist:

1. $\lim\limits_{x\to a} c = c$
2. $\lim\limits_{x\to a}[f(x) + g(x)] = \lim\limits_{x\to a} f(x) + \lim\limits_{x\to a} g(x)$
3. $\lim\limits_{x\to a}[f(x) - g(x)] = \lim\limits_{x\to a} f(x) - \lim\limits_{x\to a} g(x)$
4. $\lim\limits_{x\to a}[f(x)\cdot g(x)] = \lim\limits_{x\to a} f(x)\cdot \lim\limits_{x\to a} g(x)$
5. $\lim\limits_{x\to a}\frac{f(x)}{g(x)} = \frac{\lim\limits_{x\to a} f(x)}{\lim\limits_{x\to a} g(x)}$ if $g(x) \neq 0$ and $\lim\limits_{x\to a} g(x) \neq 0$.

Property 1 states that the limit of a constant is that constant. See Example 10.

Property 2 states that the limit of the sum of two (or more) functions is the sum of the limits of the terms of the function. Example 8 could have been evaluated as

$$\begin{aligned}\lim_{x\to -2}(4x+5) &= \lim_{x\to -2}(4x) + \lim_{x\to -2} 5 && \text{Property 2}\\ &= 4(-2) + 5 && \text{Substitution Property and Property 1}\\ &= -3\end{aligned}$$

Property 3 states that the limit of the difference of two (or more) functions is the difference of the limits of each of the terms of the function. Consider, for instance, $\lim_{x\to 3}(x^2 - 5x - 2)$. Using Property 3, we may break this limit into three separate limits:

$$\lim_{x\to 3}(x^2 - 5x - 2) = \lim_{x\to 3} x^2 - \lim_{x\to 3}(5x) - \lim_{x\to 3}(2) \qquad \textbf{Property 3}$$
$$= 3^2 - 5(3) - 2 \qquad \textbf{Substitution Property and}$$
$$= -8 \qquad \textbf{Property 1}$$

Property 4 states that the limit of the product of two functions is the product of the limits of the individual factors. Consider, for instance, $\lim_{x\to 2}[(x^2 - 2)(7\sqrt{x} - 1)]$. Using Property 4, we could rewrite this limit as

$$\lim_{x\to 2}[(x^2 - 2)(7\sqrt{x} - 1)] = \lim_{x\to 2}(x^2 - 2) \cdot \lim_{x\to 2}(7\sqrt{x} - 1) \qquad \textbf{Property 4}$$
$$= (2)(7) \qquad \textbf{Substitution Property}$$
$$= 14$$

Property 5 states that the limit of the quotient of two functions is the quotient of the limits of the function in the numerator and the function in the denominator. For example,

$$\lim_{x\to 1}\left(\frac{x + 3}{\sqrt{x + 3}}\right) = \frac{\lim_{x\to 1}(x + 3)}{\lim_{x\to 1}\sqrt{x + 3}} \quad \text{assuming } x > -3 \qquad \textbf{Property 5}$$
$$= \frac{4}{2} \qquad \textbf{Substitution Property}$$
$$= 2$$

Check Your Understanding 5.1

Evaluate each of the following limits using the Substitution Property.

1. $\lim_{x\to 10} 17$

2. $\lim_{x\to 2}(7x - 4)$

3. $\lim_{x\to -1}(2x^3 - 5x^2 + 4)$

4. $\lim_{x\to 0}\frac{4x - 3}{x^2 + 5}$

5. $\lim_{x\to -2}\sqrt{5 - 2x}$

More on Limits of Rational Functions

Because rational functions are either undefined or indeterminate at those values of x for which the denominator has a value of zero, care must be taken when evaluating limits as x approaches those values.

Example 12: Evaluate $\lim_{x\to 3}\frac{1}{x - 3}$, if it exists.

Solution: $f(x) = \frac{1}{x - 3}$ has a vertical asymptote at $x = 3$, so direct substitution yields $\frac{1}{0}$, which is undefined. Examining one-sided limits shows that $\lim_{x\to 3^-}\frac{1}{x - 3} = -\infty$ and $\lim_{x\to 3^+}\frac{1}{x - 3} = \infty$. Thus, $\lim_{x\to 3}\frac{1}{x - 3}$ does not exist. ■

MATHEMATICS CORNER 5.2

Nonexistence of $\lim_{x \to 3} \frac{1}{x-3}$

In Example 12, we saw that $\lim_{x \to 3} \frac{1}{x-3}$ did not exist. Let's explore this limit using the epsilon–delta definition.

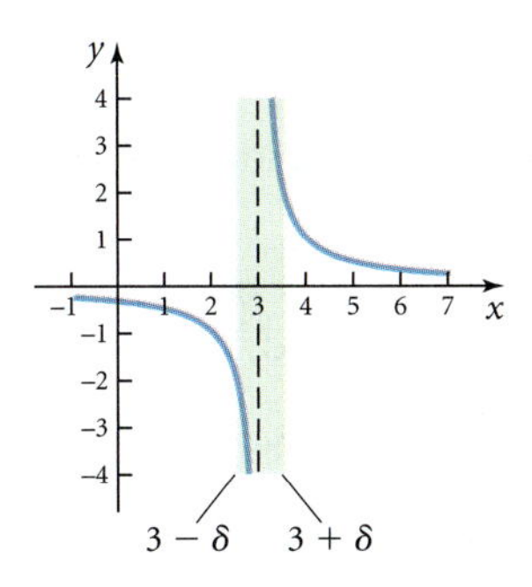

Choose a small δ interval around $x = 3$. For the limit to exist, we must be able to choose a small ϵ interval around the limit. In the interval $(3 - \delta, 3 + \delta)$, however, the value of $f(x)$ is not approaching a unique value because $f(x) > 0$ for $x > 3$ and $f(x) < 0$ for $x < 3$. Thus, no matter how small δ gets, there is no unique ϵ interval around $f(x)$. Therefore, $\lim_{x \to 3} \frac{1}{x-3}$ does not exist.

Example 13: Evaluate $\lim_{x \to 3} \frac{x-3}{x^2-9}$, if it exists.

Solution: The domain of this function is $x \neq 3, -3$. Evaluating $f(3)$ yields $\frac{0}{0}$, which is an indeterminate result. Does that mean there is no limit? Absolutely not! Consider the following table of values for this function.

x	$f(x)$	x	$f(x)$
2.9	0.16949153	3.1	0.16393443
2.99	0.16694491	3.01	0.16638935
2.999	0.16669445	3.001	0.16663889
$\to 3$	$\to ?$	$\to 3$	$\to ?$

You should see from the table that both the left-hand limit and the right-hand limit appear to be approaching a value near 0.1666667.

To find the value of the limit, some algebraic simplification must be done.

$$\lim_{x \to 3} \frac{x-3}{x^2-9} = \lim_{x \to 3} \frac{x-3}{(x+3)(x-3)} \qquad \text{Factor } x^2 - 9$$

$$= \lim_{x \to 3} \frac{1}{x+3} \qquad \text{Cancel common factors}$$

$$= \frac{1}{6} \qquad \text{Evaluate the limit}$$

So, $\lim_{x \to 3} \frac{x-3}{x^2-9} = \frac{1}{6}$.

Warning! ***Beware of limits for which direct substitution yields the indeterminate form $\frac{0}{0}$! When that happens, it is always necessary to simplify the function algebraically to determine the value of the limit.*** The most common methods of algebraic simplification are

1. Factoring and simplifying
2. Rationalizing the numerator
3. Simplifying a complex fraction

To better see what is happening here, look more closely at the function. The domain of $f(x) = \dfrac{x-3}{x^2-9}$ is all values of x except 3 and -3. When factored and simplified, the equivalent function obtained is $f(x) = \dfrac{1}{x+3}$, which is undefined only at $x = -3$. The graph of the original function in Figure 5.5 provides a clearer understanding of what is happening.

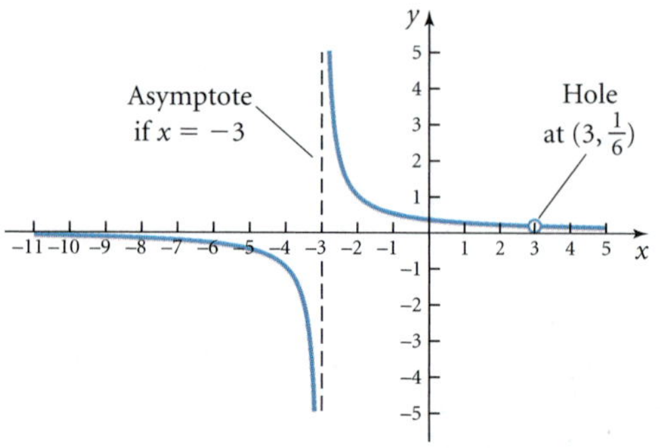

The graph of $f(x) = \dfrac{x-3}{x^2-9}$ has an asymptote at $x = -3$, which was excluded from the domain of both the original function and its simplified form, and a "hole" at $x = 3$, because the original function is not defined there. The limit as $x \to 3$ exists and is equal to the y value at the "hole", or $\frac{1}{6}$. The limit as $x \to -3$ does not exists because the function is undefined if $x = -3$. ■

Figure 5.5

Calculator Corner 5.2

1. Graph $y_1 = \dfrac{x-3}{x^2-9}$ using Zoom Decimal. Remember to enclose both $x - 3$ and $x^2 - 9$ in parentheses. Trace along the graph to the point where $x = 3$, the hole in the graph. You should see that there is no y value given. Look at the y values for $x = 2.9$ and $x = 3.1$. You should see the limiting value of 0.166667, or $\frac{1}{6}$.
2. Graph $y_1 = \dfrac{x-3}{x^2-9}$ using Zoom Standard. Remember to enclose both $x - 3$ and $x^2 - 9$ in parentheses. You can also use **TABLE** to see the y value of a hole in the graph. Press **2nd WINDOW** and set TblStart = 2.9 with ΔTbl = .005. Then press **2nd GRAPH** and scroll down the table of values. As the x values get closer to 3 from either direction, what are the y values getting close to? Be sure to see that when $x = 3$ (the hole in the graph), there is no y value given.

Example 14: Evaluate $\lim\limits_{x \to -2} \dfrac{2x + 4}{x^2 - 4}$, if it exists.

Solution: Because $f(-2)$ gives $\frac{0}{0}$, we must simplify the function by factoring.

$$\lim_{x \to -2} \frac{2x + 4}{x^2 - 4} = \lim_{x \to -2} \frac{2(x + 2)}{(x + 2)(x - 2)} \quad \text{Factor}$$

$$= \lim_{x \to -2} \frac{2}{x - 2} \quad \text{Cancel common factors}$$

$$= \frac{2}{-4} = -\frac{1}{2} \quad \text{Evaluate the limit}$$

Thus, $\lim\limits_{x \to -2} \dfrac{2x + 4}{x^2 - 4} = -\frac{1}{2}$. ■

Calculator Corner 5.3

Graph $y_1 = \dfrac{2x + 4}{x^2 - 4}$ using Zoom Decimal. Trace along the graph to the point where $x = -2$ and verify that the limit is $-\frac{1}{2}$. Then graph $y_1 = \dfrac{2x + 4}{x^2 - 4}$ using Zoom Standard. Use the table to verify that the limit as x approaches -2 is $-\frac{1}{2}$.

Example 15: Evaluate $\lim\limits_{x \to 3} \dfrac{x + 3}{x^2 - 9}$, if it exists.

Solution: Here, $f(3)$ is undefined because the denominator has a value of zero and the numerator has a value of 6. The left-hand limit approaches $-\infty$, and the right-hand limit approaches ∞. The limit does not exist, because $\lim\limits_{x \to 3^-} f(x) \neq \lim\limits_{x \to 3^+} f(x)$.

Simplifying this function as in Example 14 gives $\lim\limits_{x \to 3} \dfrac{x + 3}{x^2 - 9} = \lim\limits_{x \to 3} \dfrac{1}{x - 3}$ which is still undefined and yields no limit.

Consider this function graphically in Figure 5.6.

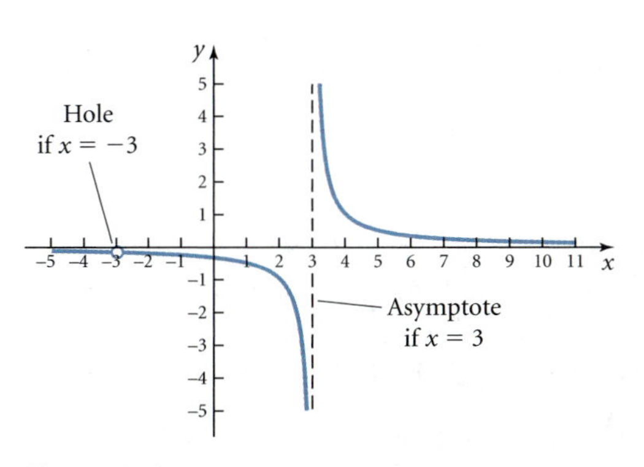

Figure 5.6

The original function $f(x) = \dfrac{x + 3}{x^2 - 9}$ is undefined at $x = 3$ because $f(3)$ has a denominator of zero and is indeterminate at $x = -3$ because $f(-3)$ has its numerator and denominator both equal to zero. The simplified form of this function, $f(x) = \dfrac{1}{x - 3}$, is undefined at $x = 3$ only. The graph of the original function has an asymptote at $x = 3$ and a "hole" at $x = -3$. As $x \to 3$, the graph approaches the asymptote, where the left-hand limit is $-\infty$ and the right-hand limit is ∞. Because $\lim\limits_{x \to 3^-} f(x) \neq \lim\limits_{x \to 3^+} f(x)$, the right-hand limit does not equal the left-hand limit and $\lim\limits_{x \to 3} f(x)$ does not exist. ■

Calculator Corner 5.4

Graph $y_1 = \dfrac{x+3}{x^2-9}$ using Zoom Decimal. Trace along the graph to the point where $x = 3$. You should notice that there is no y value given. Compare the y values for $x = 2.9$ and $x = 3.1$. Do you see that the limit does not exist?

Graph $y_1 = \dfrac{x+3}{x^2-9}$ using Zoom Standard. Set TblStart $= 2.9$ with ΔTbl $= .005$. Then press **2nd GRAPH** and scroll down the table of values to the point where $x = 3$. You should notice that there is no y value given. Compare the y values for $x = 2.995$ and $x = 3.005$. Do you see that the limit does not exist?

So far, two possibilities have been presented: functions for which the limit approaches a unique real number and functions for which the limit does not exist.
...its for rational functions.

Calculator Corner

Page 114

These boxes provide instruction on using the graphing calculator to enhance or further demonstrate a topic.

...ts.

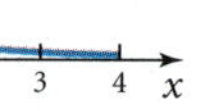

Figure 5.7

...see that the left-hand limit and right-
...efinition required the function to
...not necessarily unique. Because the
left-hand and right-hand limits both "do the same thing," however, we say that

$\lim\limits_{x \to 0} \dfrac{1}{x^2} = \infty.$ ■

Here is a summary of the results obtained in evaluating limits of functions.

Existence of Limits

1. If $\lim\limits_{x \to a^-} f(x) = L$ and $\lim\limits_{x \to a^+} f(x) = M$, where $L \neq M$, then $\lim\limits_{x \to a} f(x)$ does not exist.
2. If $\lim\limits_{x \to a^-} f(x) = -\infty$ and $\lim\limits_{x \to a^+} f(x) = \infty$, then $\lim\limits_{x \to a} f(x)$ does not exist.
3. If $\lim\limits_{x \to a^-} f(x) = \infty$ and $\lim\limits_{x \to a^+} f(x) = \infty$, then $\lim\limits_{x \to a} f(x) = \infty$.
4. If $\lim\limits_{x \to a^-} f(x) = -\infty$ and $\lim\limits_{x \to a^+} f(x) = -\infty$, then $\lim\limits_{x \to a} f(x) = -\infty$.

Evaluating Limits

To evaluate $\lim_{x \to a} f(x)$, first try to evaluate $f(a)$.

- If $f(x)$ is polynomial, then $f(a)$ is a real number L and $\lim_{x \to a} f(x) = f(a) = L$.
- If $f(a)$ is indeterminate $\left(\frac{0}{0}\right)$, simplify $f(x)$ algebraically and evaluate $f(a)$ to find the limit.
- If $f(a)$ is undefined ($\frac{k}{0}$ for any nonzero constant k), the limit equals ∞ or $-\infty$ or does not exist.

Check Your Understanding 5.2

Evaluate each limit, if it exists.

1. $\lim_{x \to 5} \frac{3}{x - 2}$
2. $\lim_{x \to 2} \frac{3}{x - 2}$
3. $\lim_{x \to 2} \frac{3x - 6}{x^2 - 4}$
4. $\lim_{x \to -2} \frac{3x - 6}{x^2 - 4}$
5. $\lim_{x \to 3} \frac{2x - 6}{x^2 - 4x + 3}$
6. $\lim_{x \to -1} \frac{4x + 4}{x^3 - x}$
7. $\lim_{x \to 0} \frac{3}{x^2}$

Limits of Difference Quotients

We now work with a special expression called a difference quotient.

Definition: Given $f(x)$, the **difference quotient** is the expression $\frac{f(a + h) - f(a)}{h}$, where $x = a$ is a value in the domain of $f(x)$.

To evaluate a difference quotient of a function $f(x)$, follow these steps.

Step 1: Evaluate $f(a)$.

Step 2: Evaluate $f(a + h)$.

Step 3: Evaluate $f(a + h) - f(a)$.

Step 4: Evaluate $\frac{f(a + h) - f(a)}{h}$.

Example 17: Given $f(x) = x^2 - 3$, evaluate the difference quotient.

Solution:

Step 1: $f(a) = a^2 - 3$

Step 2:
$$\begin{aligned} f(a + h) &= (a + h)^2 - 3 \\ &= a^2 + 2ah + h^2 - 3 \end{aligned}$$

Step 3:
$$\begin{aligned} f(a + h) - f(a) &= (a^2 + 2ah + h^2 - 3) - (a^2 - 3) \\ &= a^2 + 2ah + h^2 - 3 - a^2 + 3 \\ &= 2ah + h^2 \end{aligned}$$

Step 4:
$$\begin{aligned} \frac{f(a + h) - f(a)}{h} &= \frac{2ah + h^2}{h} \\ &= \frac{h(2a + h)}{h} && \text{Factor } h \text{ from numerator} \\ &= 2a + h && \text{Cancel the common factor, } h \end{aligned}$$

■

In subsequent topics, it will be necessary to evaluate the limit of a difference quotient.

Example 18: For the function $f(x)$ in Example 17, find $\lim_{h \to 0} \frac{f(a + h) - f(a)}{h}$.

Solution: Using the results of Example 17, we obtain

$$\lim_{h \to 0} \frac{f(a + h) - f(a)}{h} = \lim_{h \to 0} (2a + h) = 2a + 0 = 2a$$

■

Example 19: Find $\lim_{h \to 0} \frac{f(a + h) - f(a)}{h}$ if $f(x) = 4x - x^2$.

Solution:

Step 1: $f(a) = 4a - a^2$

Step 2:
$$\begin{aligned} f(a + h) &= 4(a + h) - (a + h)^2 \\ &= 4a + 4h - a^2 - 2ah - h^2 \end{aligned}$$

Step 3:
$$\begin{aligned} &f(a + h) - f(a) \\ &\quad = (4a + 4h - a^2 - 2ah - h^2) - (4a - a^2) \\ &\quad = 4a + 4h - a^2 - 2ah - h^2 - 4a + a^2 \\ &\quad = 4h - 2ah - h^2 \end{aligned}$$

Step 4:
$$\begin{aligned} \lim_{h \to 0} \frac{f(a + h) - f(a)}{h} &= \lim_{h \to 0} \frac{4h - 2ah - h^2}{h} \\ &= \lim_{h \to 0} \frac{h(4 - 2a - h)}{h} && \text{Factor } h \text{ from numerator} \\ &= \lim_{h \to 0} (4 - 2a - h) && \text{Cancel common factor, } h \end{aligned}$$

$$= 4 - 2a - 0$$ Evaluate the limit

$$= 4 - 2a$$ ■

Example 20: Find $\lim_{h \to 0} \frac{f(a + h) - f(a)}{h}$ if $f(x) = \frac{4}{x - 3}$.

Solution:

Step 1: $f(a) = \frac{4}{a - 3}$

Step 2: $f(a + h) = \frac{4}{a + h - 3}$

Step 3: $f(a + h) - f(a)$

$$= \frac{4}{a + h - 3} - \frac{4}{a - 3}$$

$$= \frac{4(a - 3)}{(a + h - 3)(a - 3)} - \frac{4(a + h - 3)}{(a - 3)(a + h - 3)}$$

Rewrite both terms with the least common denominator $(a + h - 3)(a - 3)$

$$= \frac{4a - 12}{(a + h - 3)(a - 3)} - \frac{4a + 4h - 12}{(a + h - 3)(a - 3)}$$ Simplify numerators

$$= \frac{4a - 12 - (4a + 4h - 12)}{(a + h - 3)(a - 3)}$$ Write as one fraction

$$= \frac{4a - 12 - 4a - 4h + 12}{(a + h - 3)(a - 3)}$$ Distribute

$$= \frac{-4h}{(a + h - 3)(a - 3)}$$ Simplify numerator

We now substitute this expression into the difference quotient.

Step 4: $\lim_{h \to 0} \frac{f(a + h) - f(a)}{h} = \lim_{h \to 0} \frac{-4h}{h(a + h - 3)(a - 3)}$

Cancel the factor h from the numerator and the denominator:

$$\lim_{h \to 0} \frac{-4}{(a + h - 3)(a - 3)}$$

Evaluate the limit:

$$\frac{-4}{(a - 3)^2}$$ ■

MATHEMATICS CORNER 5.3

Evaluating Limits of Difference Quotients Using Complex Fractions

The limit in Example 20 can also be evaluated by substituting f(a) and f(a + h) directly into the difference quotient and simplifying the resulting complex fraction.

$$\lim_{h\to 0}\frac{f(a+h)-f(a)}{h}=\lim_{h\to 0}\frac{\dfrac{4}{a+h-3}-\dfrac{4}{a-3}}{h}$$

The Substitution Property yields the indeterminate form $\frac{0}{0}$, so we must simplify the complex fraction by multiplying the numerator and the denominator by the lowest common denominator (LCD), which is $(a + h - 3)(a - 3)$.

$$\lim_{h\to 0}\frac{\dfrac{4}{a+h-3}-\dfrac{4}{a-3}}{h}\cdot\frac{(a+h-3)(a-3)}{(a+h-3)(a-3)}$$

$$=\lim_{h\to 0}\frac{4(a-3)-4(a+h-3)}{h(a+h-3)(a-3)}$$

Expand the numerator algebraically and combine like terms:

$$\lim_{h\to 0}\frac{4a-12-4a-4h+12}{h(a+h-3)(a-3)}$$

$$=\lim_{h\to 0}\frac{-4h}{h(a+h-3)(a-3)}$$

Cancel the common factor, h:

$$\lim_{h\to 0}\frac{-4}{(a+h-3)(a-3)}$$

Evaluate the limit:

$$\frac{-4}{(a+0-3)(a-3)}=\frac{-4}{(a-3)^2}$$

Limits of Piecewise Defined Functions

Limits can also be determined for piecewise defined functions. If the x value being approached is not a break point in the domain of the function, simply evaluate the function at that value. If the x value being approached is a break point in the domain of the function, left- and right-hand limits must be evaluated.

Example 21: Given

$$f(x)=\begin{cases}x+2 & \text{if } x<1\\ x^2-3 & \text{if } x\geq 1\end{cases}$$

evaluate each limit, if it exists.

a. $\lim_{x\to 0} f(x)$ **b.** $\lim_{x\to 3} f(x)$ **c.** $\lim_{x\to 1^-} f(x)$ **d.** $\lim_{x\to 1^+} f(x)$ **e.** $\lim_{x\to 1} f(x)$

Solution:

a. x is approaching 0, which belongs to the domain of the first part of the function. So, $\lim_{x\to 0} f(x) = f(0) = 0 + 2 = 2$.

b. x is approaching 3, which belongs to the domain of the second part of the function. So, $\lim_{x\to 3} f(x) = f(3) = 3^2 - 3 = 6$.

c. x is approaching 1 from the left, which means that $x < 1$, which is the domain of the first piece of the function, $f(x) = x + 2$. Thus, $\lim_{x \to 1^-} f(x) = 1 + 2 = 3$.

d. x is approaching 1 from the right, which means that $x > 1$, which is in the domain of the second piece of the function, $f(x) = x^2 - 3$. Thus, $\lim_{x \to 1^+} f(x) = 1^2 - 3 = -2$.

e. x is approaching 1, which is the break point of the domains of the two parts of the function. Thus, the left-hand limit would be comprised of values from the domain of the first part of the function, $f(x) = x + 2$ for $x < 1$, and the right-hand limit would be comprised of values from the domain of the second part of the function, $f(x) = x^2 - 3$ for $x \geq 1$. Thus, $\lim_{x \to 1^-} f(x) = 1 + 2 = 3$ and $\lim_{x \to 1^+} f(x) = 1^2 - 3 = -2$. Because the left-hand limit and the right-hand limit are not equal, $\lim_{x \to 1} f(x)$ does not exist. ■

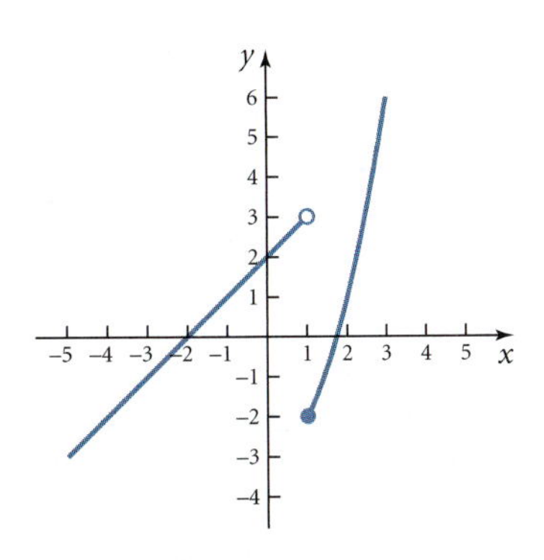

Figure 5.8

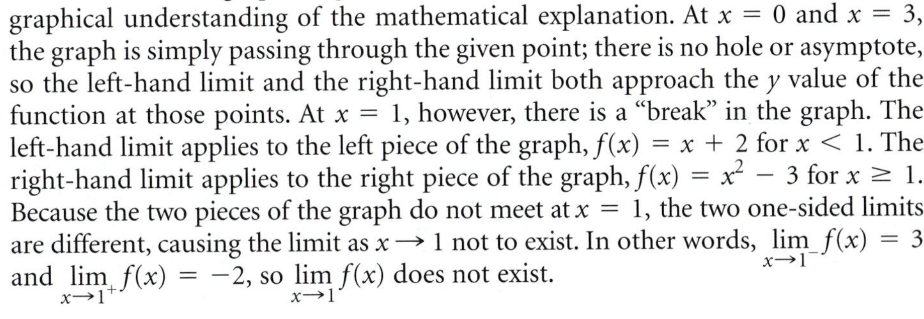
Consider the graph (Figure 5.8) of this piecewise defined function for a graphical understanding of the mathematical explanation. At $x = 0$ and $x = 3$, the graph is simply passing through the given point; there is no hole or asymptote, so the left-hand limit and the right-hand limit both approach the y value of the function at those points. At $x = 1$, however, there is a "break" in the graph. The left-hand limit applies to the left piece of the graph, $f(x) = x + 2$ for $x < 1$. The right-hand limit applies to the right piece of the graph, $f(x) = x^2 - 3$ for $x \geq 1$. Because the two pieces of the graph do not meet at $x = 1$, the two one-sided limits are different, causing the limit as $x \to 1$ not to exist. In other words, $\lim_{x \to 1^-} f(x) = 3$ and $\lim_{x \to 1^+} f(x) = -2$, so $\lim_{x \to 1} f(x)$ does not exist.

Example 22: Tax tables for the 2005 U.S. federal income tax show that the income tax due for those filing as head of household is calculated as follows:

$$T(d) = \begin{cases} 0.10d & \text{if } d \leq \$10{,}450 \\ \$1045 + 0.15(d - \$10{,}450) & \text{if } \$10{,}450 < d \leq \$39{,}800 \\ \$5{,}447.50 + 0.25(d - \$39{,}800) & \text{if } \$39{,}800 < d \leq \$102{,}800 \\ \$21{,}197.50 + 0.28(d - \$102{,}800) & \text{if } \$102{,}800 < d \leq \$166{,}450 \\ \$39{,}019.50 + 0.33(d - \$166{,}450) & \text{if } \$166{,}450 < d \leq \$326{,}450 \\ \$91{,}819.50 + 0.35(d - \$326{,}450) & \text{if } d > \$326{,}450 \end{cases}$$

where $T(d)$ is income tax due and d is taxable income.

a. Find the income tax due for a person filing as head of household whose taxable income is \$85,420.

b. Determine $\lim_{d \to \$39{,}800} T(d)$ and interpret your answer.

Solution:

a. Here, $d = \$85{,}420$, so evaluate $T(\$85{,}420)$. Since $\$39{,}800 < \$85{,}420 \leq \$102{,}800$, the third piece of the function is used for evaluation:

$$\begin{aligned} T(\$85{,}420) &= \$5447.50 + 0.25(\$85{,}420 - \$39{,}800) \\ &= \$16{,}852.50 \end{aligned}$$

A person filing as head of household with taxable income of \$85,420 would have owed \$16,852.50 in income tax for 2005.

b. Because $d \to \$39{,}800$, which is the break point of the second and third pieces of the function, the left- and right-hand limits must be determined to evaluate the limit:

$$\begin{aligned}\lim_{d \to \$39{,}800^-} T(d) &= \$1045 + 0.15(\$39{,}800 - \$10{,}450) \\ &= \$5447.50 \\ \lim_{d \to \$37{,}450^+} T(d) &= \$5447.50 + 0.25(\$39{,}800 - \$39{,}800) \\ &= 5447.50\end{aligned}$$

Thus, $\lim_{d \to \$39{,}800} T(d) = \$5447.50$, which means that a person whose income approaches \$39,800 will pay income tax of \$5447.50.

It should seem logical that a person whose income approaches a specific value should pay the same tax, regardless of whether the income increases or decreases to reach that amount. ■

Evaluating Limits Graphically

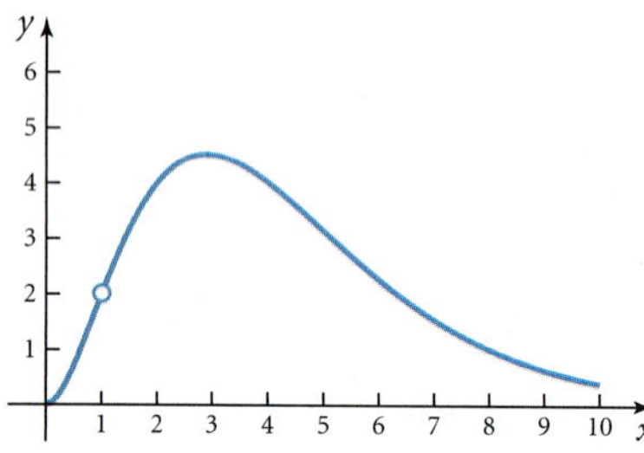

Figure 5.9

It is very important to be able to determine limits of functions graphically as well as algebraically using the function itself. For Figure 5.9, evaluate $\lim_{x \to 1} f(x)$, if it exists.

Look at the graph and consider both the left-hand and right-hand limits. The left-hand limit would involve x values that are less than 1; the right-hand limit involves x values that are greater than 1. In both cases, the limit is 2. Thus, $\lim_{x \to 1} f(x) = 2$, which can be true even though there is a hole in the graph!

Tip: Remember that ***the limit is the number that* y *approaches, not necessarily the actual* y *value of the function at that point.***

For Figure 5.10, evaluate $\lim_{x \to 2} f(x)$, if it exists.

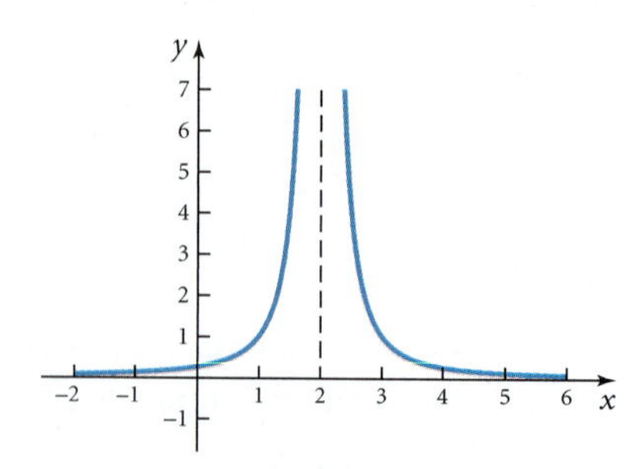

Figure 5.10

There is an asymptote at $x = 2$, so consider the left-hand and right-hand limits. As $x \to 2^-$, the value of $f(x)$ increases without bound, so $\lim_{x \to 2^-} f(x) = \infty$. As $x \to 2^+$, the values of $f(x)$ also increase without bound, so $\lim_{x \to 2^+} f(x) = \infty$. Because both one-sided limits approach infinity, $\lim_{x \to 2} f(x) = \infty$.

For Figure 5.11, evaluate $\lim_{x \to 1} f(x)$, if it exists.

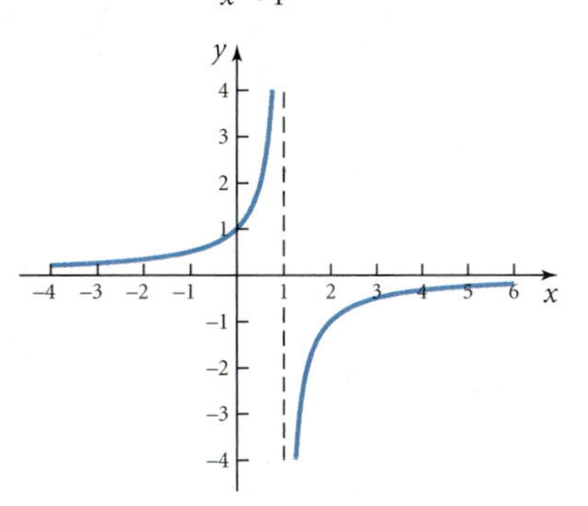

Figure 5.11

There is an asymptote at $x = 1$, so determine the left-hand and right-hand limits. As $x \to 1^-$, the value of $f(x)$ increases without bound, so $\lim_{x \to 1^-} f(x) = \infty$. As $x \to 1^+$, the value of $f(x)$ deceases without bound, so $\lim_{x \to 1^+} f(x) = -\infty$. Because the left-hand and right-hand limits are different, $\lim_{x \to 1} f(x)$ does not exist.

For Figure 5.12, evaluate $\lim_{x \to 1} f(x)$, if it exists.

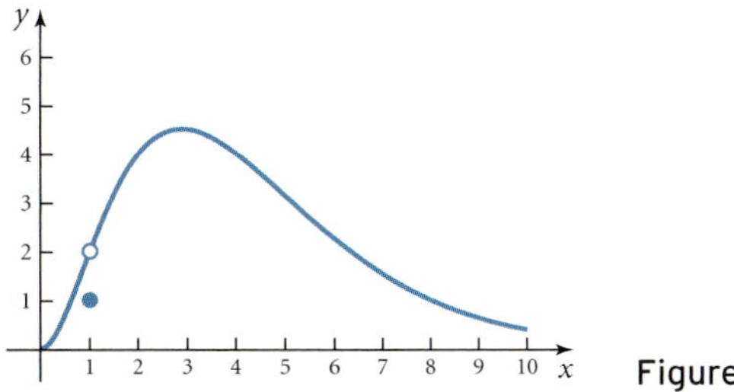

Figure 5.12

This graph is similar to that in Figure 5.9 except that a point has been added to define $f(1)$. The left-hand limit, using values of x less than 1, still approaches the hole where $y = 2$. The right-hand limit, using values of x greater than 1, also approaches $y = 2$. Thus, $\lim_{x \to 1} f(x) = 2$ as in Figure 5.9. Defining $f(1)$ by adding the point to the graph did not change the limit.

For Figure 5.13, evaluate $\lim_{x \to 1} f(x)$, if it exists.

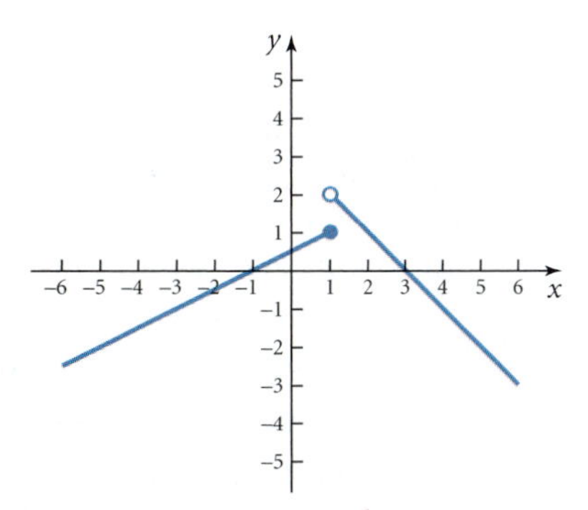

Figure 5.13

The left-hand limit corresponds to the left piece of the function, so $\lim_{x \to 1^-} f(x) = 1$. The right-hand limit corresponds to the right piece of the function, so $\lim_{x \to 1^+} f(x) = 2$. Because the left-hand and right-hand limits are different, $\lim_{x \to 1} f(x)$ does not exist, which should be apparent from the break in the graph at $x = 1$.

Check Your Understanding 5.3

1. Evaluate $\lim\limits_{h\to 0}\dfrac{f(a+h)-f(a)}{h}$ for each of the following functions.

a. $f(x) = 2x - 5$ **b.** $f(x) = 2x^2 - 5$ **c.** $f(x) = \dfrac{2}{x-5}$

2. For

$$f(x) = \begin{cases} 3x - 6 & \text{if } x < 0 \\ x^2 & \text{if } 0 \le x < 2 \\ \sqrt{x} & \text{if } x \ge 2 \end{cases}$$

evaluate each limit, if it exists.

a. $\lim\limits_{x\to -2} f(x)$ **b.** $\lim\limits_{x\to 0} f(x)$ **c.** $\lim\limits_{x\to 1} f(x)$

d. $\lim f(x)$ **e.** $\lim\limits_{x\to 5} f(x)$

Check Your Understanding Exercises

Page 122

These in-text exercises follow new material, encouraging students to confirm their comprehension of a concept.

...own here.

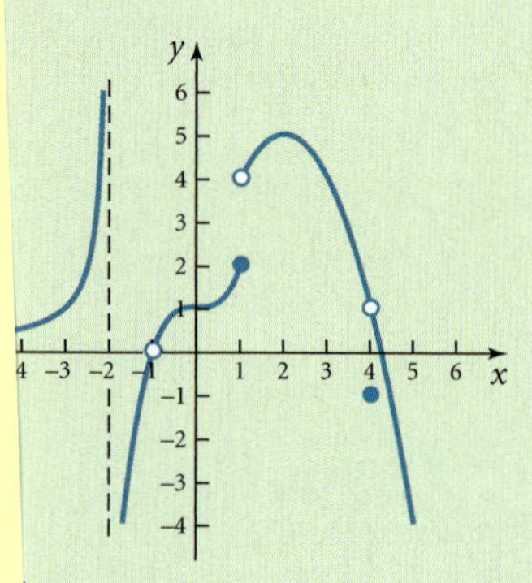

...ts.

b. $\lim\limits_{x\to -1} f(x)$ **c.** $\lim\limits_{x\to 0} f(x)$

e. $\lim\limits_{x\to 4} f(x)$

Evaluating Limits at Infinity

Suppose we are interested in how a function behaves as $|x|$ becomes very large rather than as x approaches some particular point within its domain. For instance, how are sales of a new product performing after a long time? A question of this sort implies that we need to determine what range value is being approached as the domain values increase without bound. This type of limit is called a **limit at infinity** and is denoted as $\lim\limits_{x\to\infty} f(x)$ or $\lim\limits_{x\to -\infty} f(x)$.

For nonconstant polynomial functions $P(x)$, $\lim\limits_{x\to\infty} P(x) = \infty$ (or $-\infty$). This equality should be obvious because, as x increases (or decreases) in value, so do any positive integer powers of x.

For instance,

$$\lim_{x\to\infty} x^2 = \infty \quad \text{and} \quad \lim_{x\to\infty} (4 - x^2) = -\infty$$

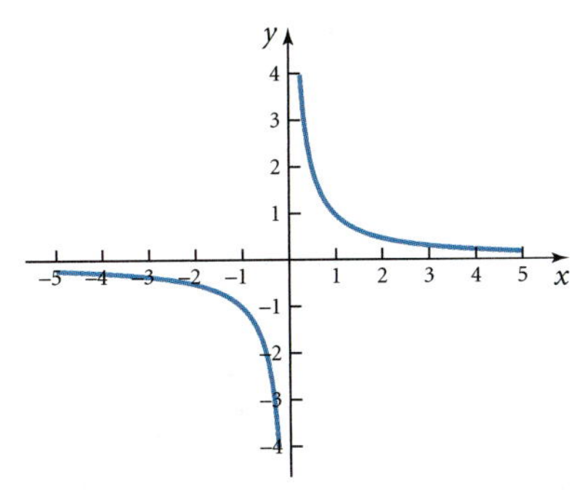

Figure 5.14

For rational functions $f(x) = \frac{p(x)}{q(x)}$, where $q(x) \neq 0$, evaluating $\lim_{x\to\infty} f(x)$ is exactly the same as finding the horizontal asymptotes. To verify, refer to the definition of horizontal asymptote in Topic 3.

Consider the basic rational function $f(x) = \frac{1}{x}$. The domain is all values of $x \neq 0$, and the graph is a hyperbola with asymptotes along the vertical axis (where $x = 0$) and the horizontal axis (where $y = 0$). The graph of $f(x) = \frac{1}{x}$ is shown in Figure 5.14.

Numerically, evaluate $f(x)$ for successively larger values of x to see how y behaves:

$$f(100) = \frac{1}{100} = 0.01$$
$$f(1000) = \frac{1}{1000} = 0.001$$
$$f(1{,}000{,}000) = \frac{1}{1{,}000{,}000} = 0.000001$$

As x grows larger, the y values are diminishing and approaching 0. So, $\lim_{x\to\infty}\left(\frac{1}{x}\right) = 0$.

Calculator Corner 5.5

Graph $y_1 = \frac{1}{x}$. Press **2nd WINDOW** (TblSet) and set TblStart = 10 with ΔTbl = 10. Scroll down the table of values to see that $y \to 0$ as $x \to \infty$. Press **2nd WINDOW** (TblSet) and set TblStart = −10 with ΔTbl = −10. Scroll down the table of values to see that $y \to 0$ as $x \to -\infty$.

Graphically, you should see the curve decreasing and approaching the x-axis (where $y = 0$) as x increases in value.

Algebraically, what is the horizontal asymptote of $f(x) = \frac{1}{x}$? The denominator has a degree one larger than the numerator, so the horizontal asymptote is $y = 0$. Thus, $\lim_{x\to\infty} \frac{1}{x} = 0$, which is the same as the horizontal asymptote of the function $(y = 0)$.

Is $\lim_{x\to\infty} \frac{p(x)}{q(x)}$ always zero? How does horizontal shifting or vertical shifting affect the limit of the basic rational function? The following examples address these questions.

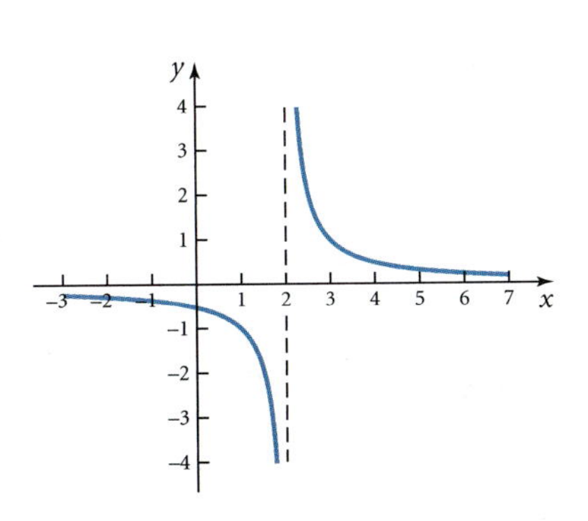

Figure 5.15

Example 23: State the asymptotes, draw the graph, and evaluate $\lim_{x\to\infty} f(x)$ for $f(x) = \frac{1}{x-2}$.

Solution: The denominator has a value of zero if $x = 2$, so the vertical asymptote is $x = 2$. The denominator has a higher degree than the numerator, so the horizontal asymptote is $y = 0$. The graph of this function is the graph of $f(x) = \frac{1}{x}$ shifted to the right 2 units. (See Figure 5.15.)

From the graph, you should see that as the x values increase, the y values decrease and approach 0. The limit is still approaching 0. Thus, $\lim\limits_{x\to\infty} \dfrac{1}{x-2} = 0.$ ■

Tip: **Horizontal shifts do not affect the horizontal asymptote of $f(x) = \dfrac{p(x)}{q(x)}$, so the value of $\lim\limits_{x\to\infty} f(x)$ is still 0.**

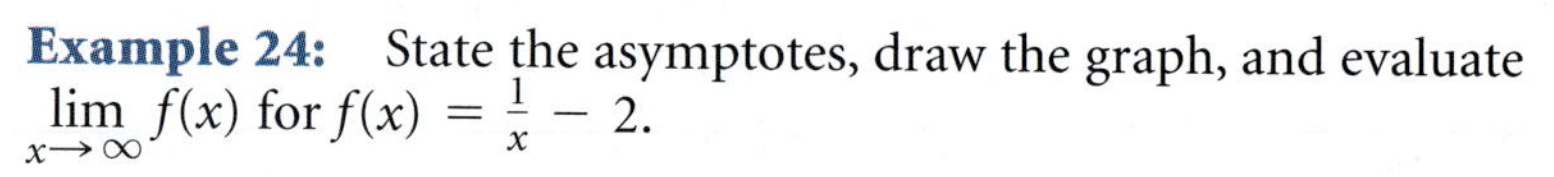

Example 24: State the asymptotes, draw the graph, and evaluate $\lim\limits_{x\to\infty} f(x)$ for $f(x) = \frac{1}{x} - 2$.

Solution: The graph of this function is the graph of $f(x) = \frac{1}{x}$ shifted down 2 units, so the asymptotes are $x = 0$ and $y = -2$. (See Figure 5.16.)

Thus, $\lim\limits_{x\to\infty} \left(\frac{1}{x} - 2\right) = \lim\limits_{x\to\infty} \frac{1}{x} - \lim\limits_{x\to\infty} 2 = 0 - 2 = -2.$ ■

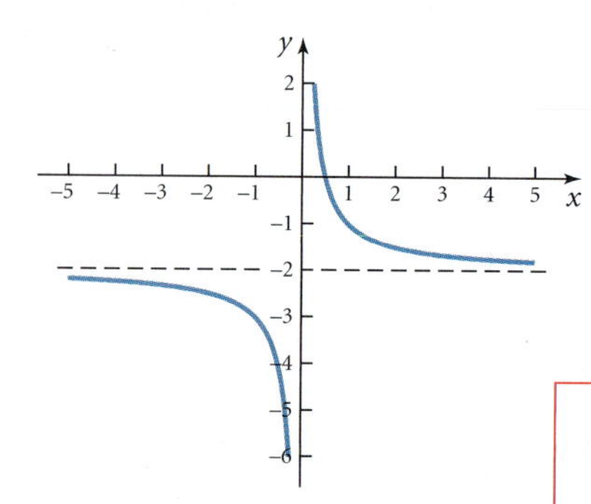

Figure 5.16

Tip: **Vertical shifts *do* affect the limit at infinity.** In fact, $\lim\limits_{x\to\infty} \dfrac{p(x)}{q(x)}$ determines the value of the horizontal asymptote of the function. If $\lim\limits_{x\to\infty} \dfrac{p(x)}{q(x)} = m$, the horizontal asymptote of the function is $y = m$.

Example 25: For each function, state the asymptotes, draw the graph, and evaluate $\lim\limits_{x\to\infty} f(x)$.

a. $f(x) = \dfrac{3x}{x-2}$ **b.** $f(x) = \dfrac{3x}{x^2-4}$

c. $f(x) = \dfrac{3x^2}{x^2-4}$ **d.** $f(x) = \dfrac{3x^2}{x-2}$

Solution:

a. The asymptotes are $x = 2$ and $y = 3$, so $\lim\limits_{x\to\infty} \dfrac{3x}{x-2} = 3$. (See Figure 5.17.)

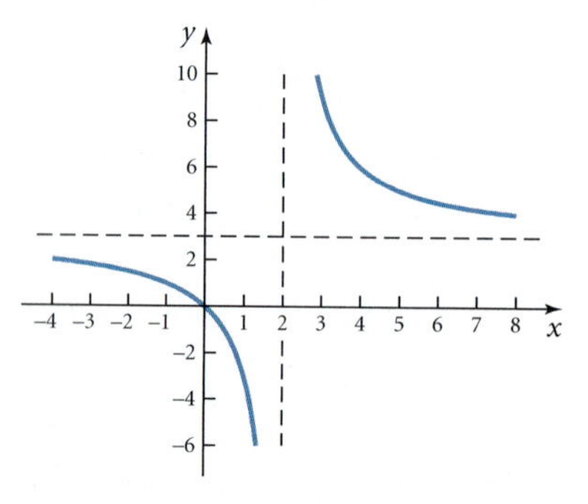

Figure 5.17

b. The vertical asymptotes are $x = 2$ and $x = -2$ because $x^2 - 4 = 0$ if $x = 2$ or $x = -2$. The denominator has the higher degree, so the horizontal asymptote is $y = 0$. Thus, $\lim\limits_{x\to\infty} \dfrac{3x}{x^2-4} = 0$. (See Figure 5.18.)

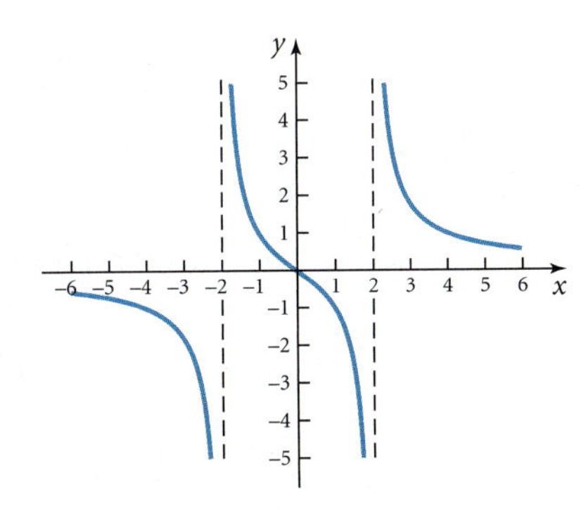

Figure 5.18

c. The vertical asymptotes are $x = 2$ and $x = -2$ because $x^2 - 4 = 0$ if $x = 2$ or $x = -2$. The numerator and denominator have the same degree, so divide the leading coefficients to get the horizontal asymptote of $y = 3$. Thus, $\lim_{x \to \infty} \dfrac{3x^2}{x^2 - 4} = 3$. (See Figure 5.19.)

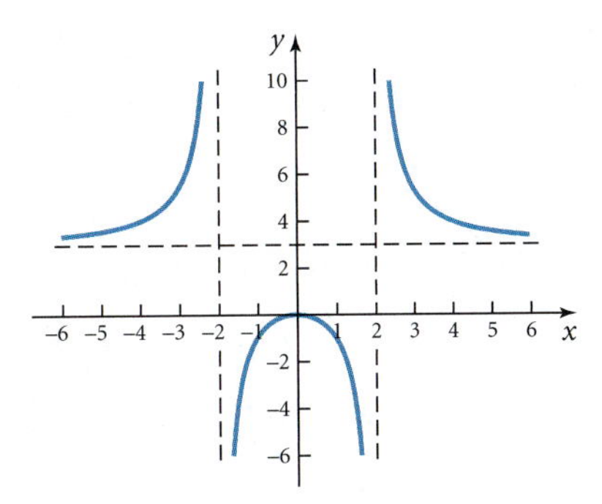

Figure 5.19

d. The vertical asymptote is at $x = 2$. The numerator is one degree higher than the denominator, so there is no horizontal asymptote. Dividing and simplifying the function gives $f(x) = 3x + 6 + \dfrac{12}{x - 2}$ (Figure 5.20). Thus,

$$\begin{aligned}\lim_{x \to \infty}\left(3x + 6 + \frac{12}{x - 2}\right) &= \lim_{x \to \infty}(3x + 6) + \lim_{x \to \infty}\frac{12}{x - 2} \\ &= \lim_{x \to \infty}(3x + 6) + 0 \\ &= \infty + 0 \\ &= \infty\end{aligned}$$

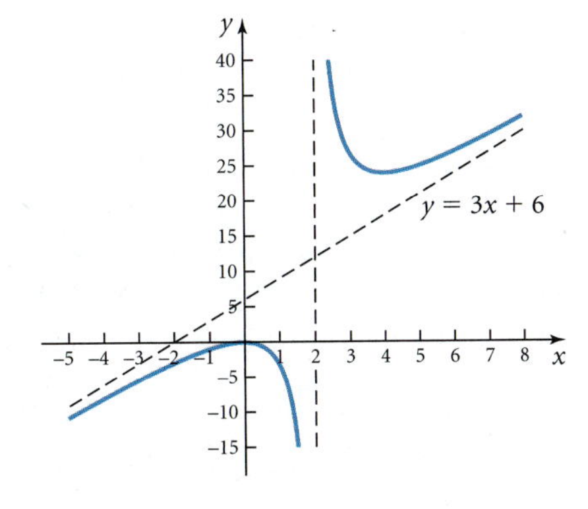

Figure 5.20

The line $y = 3x + 6$, which is the nonrational term of $f(x) = 3x + 6 + \dfrac{12}{x - 2}$, actually serves as an asymptote for this function. This asymptote is referred to as a **slant**, or **oblique, asymptote**. ■

The results of these examples can be summarized as follows.

Limits of Rational Functions at Infinity

Given a rational function $\dfrac{p(x)}{q(x)}$, where $q(x) \neq 0$ and $\dfrac{p(x)}{q(x)} = \dfrac{ax^n + \cdots + c}{bx^m + \cdots + d}$,

$$\lim_{x \to \infty} \frac{p(x)}{q(x)} = 0 \quad \text{if the degree of } p(x) < \text{the degree of } q(x)$$

$$= \frac{a}{b} \quad \text{(ratio of leading coefficients) if the degree of } p(x) = \text{the degree of } q(x)$$

$$= \infty \quad \text{if the degree of } p(x) > \text{the degree of } q(x)$$

Example 26: The concentration of a drug in the bloodstream over time is as follows.

T (hours)	*C* (mg/cm^3)
0	500
1	620
2	225
3	175
4	125
5	100

The concentration at any time can be approximated by $C(t) = \dfrac{1750}{3t + 2}$, where t is the number of hours after injection and $C(t)$ is the concentration in mg/cm^3.

a. Find the concentration of the drug in the bloodstream after two hours and after five hours. How well does the function model the actual data?

b. Find the concentration of the drug in the bloodstream after a long time.

Solution:

a. The concentration after two hours is given by $C(2) = 218.75$ or approximately 219 mg, which is close to the actual concentration of 225 mg. The concentration after 5 hours is given by $C(5) = 102.9$ or approximately 103 mg, which is close to the actual concentration of 100 mg.

b. To find the concentration "after a long time," we must determine the limit as $t \to \infty$. So, $\displaystyle\lim_{t \to \infty} \frac{1750}{3t + 2} = 0$, which implies that after a long time, there is no longer any drug remaining in the bloodstream. ■

MATHEMATICS CORNER 5.4

Another Method for Evaluating $\lim_{x \to \infty} \frac{p(x)}{q(x)}$

There is another method for evaluating $\lim_{x \to \infty} \frac{p(x)}{q(x)}$. The first step is to divide each term in the fraction by the highest power of x that appears in the denominator. Then use $\lim_{x \to \infty} \frac{1}{x} = 0$ and $\lim_{x \to \infty} \frac{k}{x^n} = 0$ for k, a constant, to evaluate the limit and thereby determine the horizontal asymptote.

Example 5.4.1:

$$\lim_{x \to \infty} \frac{3x}{x-2} = \lim_{x \to \infty} \frac{\frac{3x}{x}}{\frac{x}{x} - \frac{2}{x}} \quad \text{Divide each term in the fraction by } x$$

$$= \lim_{x \to \infty} \frac{3}{1 - \frac{2}{x}} \quad \text{Simplify the fractions}$$

$$= \frac{3}{1-0} \quad \text{Evaluate the limit, with } \lim_{x \to \infty} \frac{2}{x} = 0$$

$$= 3$$

Thus, $\lim_{x \to \infty} \frac{3x}{x-2} = 3$ and the horizontal asymptote of $y = \frac{3x}{x-2}$ is $y = 3$. This answer agrees with the result obtained in Example 25a.

Example 5.4.2:

$$\lim_{x \to \infty} \frac{3x}{x^2 - 4} = \lim_{x \to \infty} \frac{\frac{3x}{x^2}}{\frac{x^2}{x^2} - \frac{4}{x^2}} \quad \text{Divide each term in the fraction by } x^2$$

$$= \lim_{x \to \infty} \frac{\frac{3}{x}}{1 - \frac{4}{x^2}} \quad \text{Simplify the fractions}$$

$$= \frac{0}{1-0} \quad \text{Evaluate the limit, with } \lim_{x \to \infty} \frac{3}{x} = 0 \text{ and } \lim_{x \to \infty} \frac{4}{x^2} = 0$$

$$= 0$$

Thus, $\lim_{x \to \infty} \frac{3x}{x^2 - 4} = 0$ and the horizontal asymptote of $y = \frac{3x}{x^2 - 4}$ is $y = 0$, which agrees with the result obtained in Example 25b.

Example 5.4.3:

$$\lim_{x \to \infty} \frac{3x^2}{x-2} = \lim_{x \to \infty} \frac{\frac{3x^2}{x}}{\frac{x}{x} - \frac{2}{x}} \quad \text{Divide each term in the fraction by } x$$

$$= \lim_{x \to \infty} \frac{3x}{1 - \frac{2}{x}} \quad \text{Simplify the fractions}$$

$$= \infty \quad \text{Evaluate the limit, with } \lim_{x \to \infty} \frac{2}{x} = 0 \text{ and } \lim_{x \to \infty} 3x = \infty$$

Thus, there is no horizontal asymptote, which agrees with the result obtained in Example 25c.

Check Your Understanding 5.4

Evaluate each limit, if it exists.

1. $\lim_{x \to \infty} \frac{6}{x-2}$
2. $\lim_{x \to \infty} \frac{6x}{x+2}$
3. $\lim_{x \to \infty} \frac{3x^2}{x^2-1}$
4. $\lim_{x \to \infty} \frac{x^2}{x^3+1}$
5. $\lim_{x \to \infty} \frac{x^2}{x^2+2}$
6. $\lim_{x \to \infty} \frac{x^3}{x^2+1}$

Check Your Understanding Answers

Check Your Understanding 5.1

1. 17
2. 10
3. -3
4. $-\frac{3}{5}$
5. 3

Check Your Understanding 5.2

1. 1
2. does not exist
3. $\frac{3}{4}$
4. does not exist
5. 1
6. 2
7. ∞

Check Your Understanding 5.3

1. **a.** 2 **b.** $4a$ **c.** $\dfrac{-2}{(a-5)^2}$
2. **a.** -12 **b.** does not exist **c.** 1 **d.** does not exist **e.** $\sqrt{5}$
3. **a.** does not exist **b.** 0 **c.** 1 **d.** does not exist **e.** 1

Check Your Understanding 5.4

1. 0
2. 6
3. 3
4. 0
5. 1
6. ∞

Topic 5 Review

5

This topic introduced the **limit**, which is the foundation upon which the derivative and integral will be developed. The limit was explored numerically, graphically, and algebraically.

CONCEPT	EXAMPLE
The **limit of a function** refers to the value that y is approaching as x gets close to some number. Limits of functions are defined only if the **left-hand limit** and the **right-hand limit** are the same. You should be able to determine limits from numerical tables, from graphs, or algebraically by evaluating the function at the specified value. The left-hand limit is denoted symbolically as $\lim_{x \to a^-} f(x)$. The right-hand limit is denoted symbolically as $\lim_{x \to a^+} f(x)$.	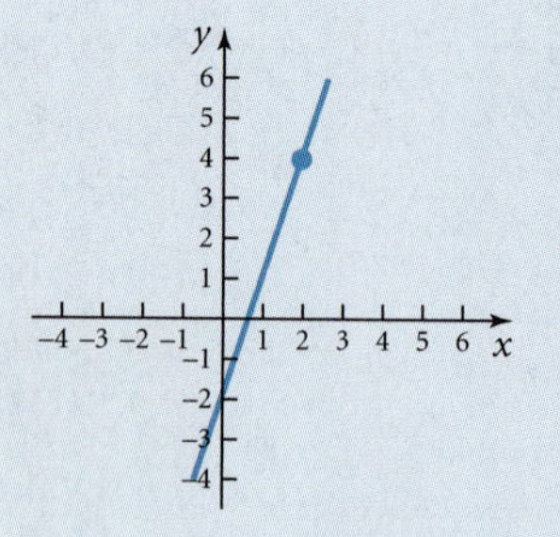 $\lim_{x \to 2} (3x - 2) = 4$

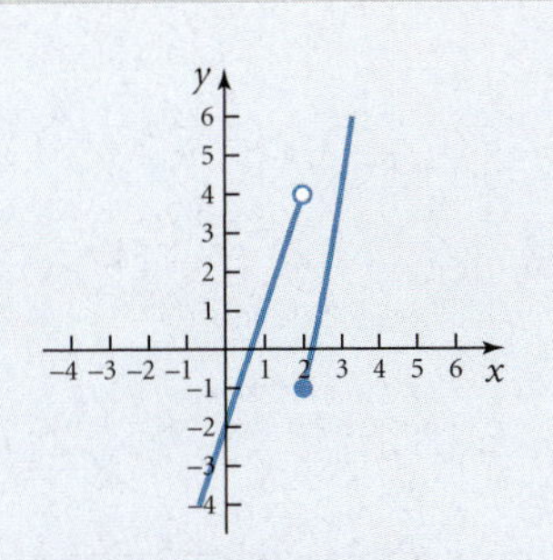

$\lim_{x \to 2^-} f(x) = 4$ and $\lim_{x \to 2^+} f(x) = -1$, so $\lim_{x \to 2} f(x)$ does not exist.

For algebraic functions, $\lim_{x \to a} f(x) = f(a) = L$, if L is a unique real number.

For $f(x) = x^2 - \sqrt{2x + 3}$,
$\lim_{x \to 3} f(x) = 3^2 - \sqrt{2(3) + 3} = 9 - \sqrt{9} = 6.$

For rational functions $\dfrac{p(x)}{q(x)}$, with $q(x) \neq 0$,

$$\lim_{x \to a} \frac{p(x)}{q(x)} = \frac{p(a)}{q(a)}.$$

- If $\dfrac{p(a)}{q(a)}$ yields a real number, that number is the limit.
- If $\dfrac{p(a)}{q(a)}$ is indeterminate, the expression must be simplified algebraically and reevaluated to determine the limit.
- If $\dfrac{p(a)}{q(a)}$ is undefined, the limit does not exist. The limit is approaching an asymptote, so $\lim_{x \to a} \dfrac{p(x)}{q(x)} = \infty$, or $\lim_{x \to a} \dfrac{p(x)}{q(x)} = -\infty$, or $\lim_{x \to a^-} \dfrac{p(x)}{q(x)} = \infty$ and $\lim_{x \to a^+} \dfrac{p(x)}{q(x)} = -\infty$,

meaning that the limit does not exist.

- $\lim_{x \to 4} \dfrac{5x - 7}{2x + 3} = \dfrac{13}{11}$
- $\lim_{x \to 4} \dfrac{5x - 20}{x^2 - 16} = \lim_{x \to 4} \dfrac{5(x - 4)}{(x - 4)(x + 4)}$
  $= \lim_{x \to 4} \dfrac{5}{x + 4} = \dfrac{5}{8}$
- $\lim_{x \to 4} \dfrac{5}{x - 4}$ does not exist
- $\lim_{x \to 0} \dfrac{4}{x^2} = \infty$
- $\lim_{x \to 0} \dfrac{-4}{x^2} = -\infty$

The **difference quotient** is defined as $\dfrac{f(a + h) - f(a)}{h}$.
It is used extensively later in the development of the derivative. You should be able to evaluate a difference quotient for specified functions and determine the $\lim_{h \to 0}$ for the difference quotient.

Given $f(x) = x^2 - 3x + 1$, evaluate

$$\lim_{h \to 0} \frac{f(a + h) - f(a)}{h}.$$

$$\lim_{h \to 0} \frac{f(a + h) - f(a)}{h} =$$

(continued)

	$\lim_{h\to 0}\dfrac{[(a+h)^2-3(a+h)+1]-(a^2-3a+1)}{h}$ $=\lim_{h\to 0}\dfrac{2ah+h^2-3h}{h}$ $=\lim_{h\to 0}(2a+h-3)=2a-3$
To evaluate **limits of piecewise defined functions,** evaluate $f(a)$ using the appropriate piece of the domain of the function. If the value that x is approaching is one of the break points of the function, evaluate the left- and right-hand limits at the point.	Given $f(x)=\begin{cases}5-3x & \text{if } x<2\\ x^2+1 & \text{if } x\geq 2\end{cases}$ • $\lim_{x\to -1} f(x)=5-3(-1)=8$ • $\lim_{x\to 2} f(x)$ does not exist because $\lim_{x\to 2^-} f(x)=-1$ and $\lim_{x\to 2^+} f(x)=5$
Limits can also be determined as x approaches infinity. Limits of rational functions as $x\to\infty$ are the same as the horizontal asymptote of the function.	$\lim_{x\to\infty}\dfrac{5}{x-3}=0$, which means that $y=0$ is the horizontal asymptote of $f(x)=\dfrac{5}{x-3}$.
For any nonzero constant k, $\lim_{x\to\infty}\dfrac{k}{x^n}=0$, where n is a positive number.	$\lim_{x\to\infty}\dfrac{3}{x^2}=0$

NEW TOOLS IN THE TOOL KIT

- Evaluating limits numerically by examining a table of values
- Evaluating limits graphically by examining the behavior of the graph near a specified point
- Evaluating limits symbolically by evaluating the function with the Substitution Property
- Evaluating difference quotients
- Evaluating limits of difference quotients
- Evaluating limits at infinity
- Evaluating limits of piecewise defined functions

Topic 5 Exercises

Topic Review Exercises

Page 131

Exercise sets conclude each topic and range in level—from routine to more advanced, applied problems.

Sharpen the tools exercises challenge students to practice different concepts within a single exercise.

Calculator Connection exercises encourage students to use their graphing calculator when solving problems.

Communicate problems require short, verbal answers.

For each of the following limits, determine the indi cated limit, if it exists.

1. $\lim_{x \to 3^+} f(x) = ?$

x	$f(x)$
3.1	1.2
3.01	1.02
3.001	1.002
3.0001	1.0002
↓	↓
3	?

2. $\lim_{x \to 2^+} f(x) = ?$

x	$f(x)$
2.1	2.3
2.01	2.03
2.001	2.003
2.0001	2.0003
↓	↓
2	?

3. $\lim_{x \to 4^-} f(x) = ?$

x	$f(x)$
3.9	−3.7
3.99	−3.97
3.999	−3.997
3.9999	−3.9997
↓	↓
4	?

5. $\lim_{x \to 2} f(x) = ?$

x	$f(x)$	x	$f(x)$
1.9	−30	2.1	30
1.99	−300	2.01	300
1.999	−3000	2.001	3000
↓	↓	↓	↓
2	?	2	?

6. $\lim_{x \to 3} f(x) = ?$

x	$f(x)$	x	$f(x)$
2.9	−20	3.1	20
2.99	−200	3.01	200
2.999	−2000	3.001	2000
↓	↓	↓	↓
3	?	3	?

7. $\lim_{x \to 1} f(x) = ?$

x	$f(x)$	x	$f(x)$
0.9	0.81	1.1	4.1
0.99	0.9801	1.01	4.01
0.999	0.998001	1.001	4.001
↓	↓	↓	↓
1	?	1	?

8. $\lim_{x \to 2} f(x) = ?$

x	$f(x)$	x	$f(x)$
1.9	−36.1	2.1	44.1
1.99	−396.01	2.01	404.01
1.999	−3996.001	2.001	4004.001
↓	↓	↓	↓
2	?	2	?

Use the graph of the function to determine the indicated limit, if it exists.

9. $\lim_{x \to 3} f(x) = ?$

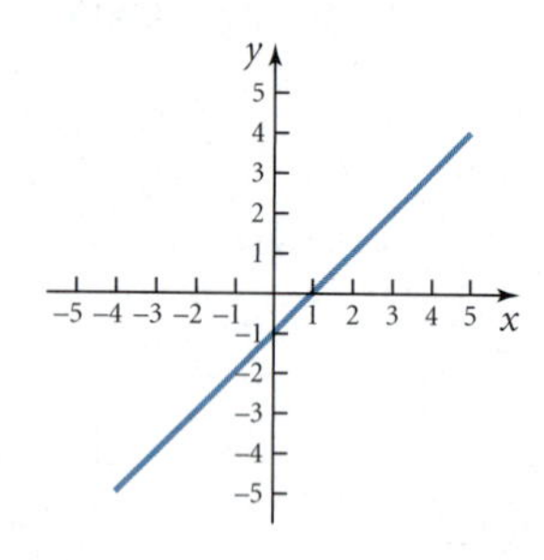

10. $\lim_{x \to 2} f(x) = ?$

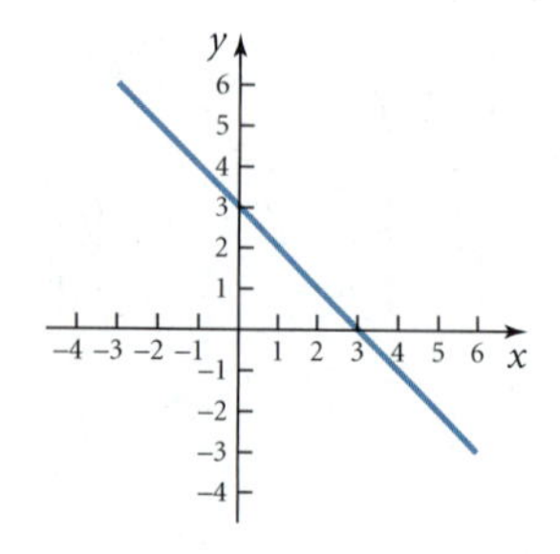

11. $\lim_{x \to 3} f(x) = ?$

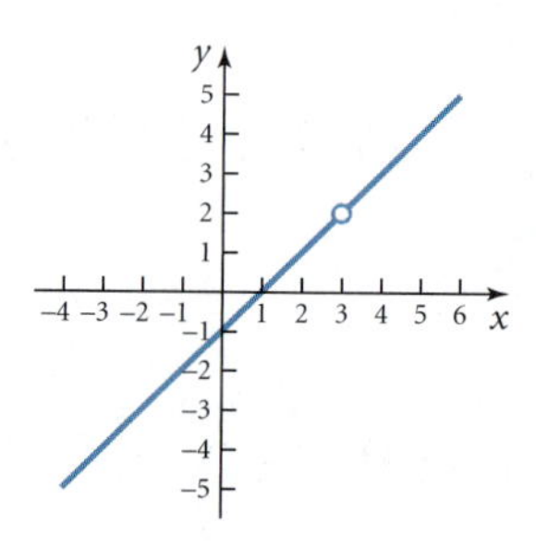

12. $\lim_{x \to 2} f(x) = ?$

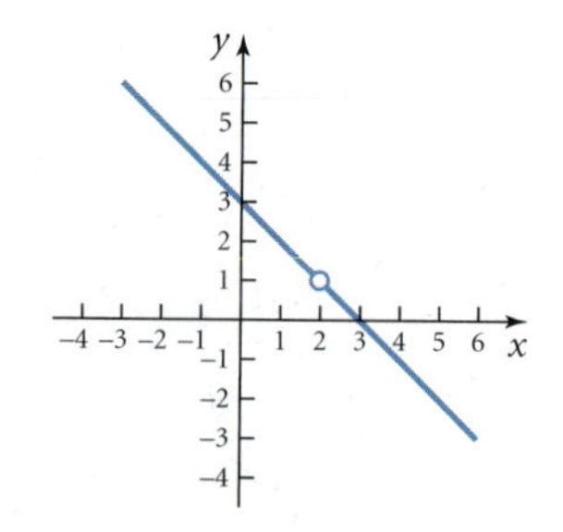

13. $\lim_{x \to 3} f(x) = ?$

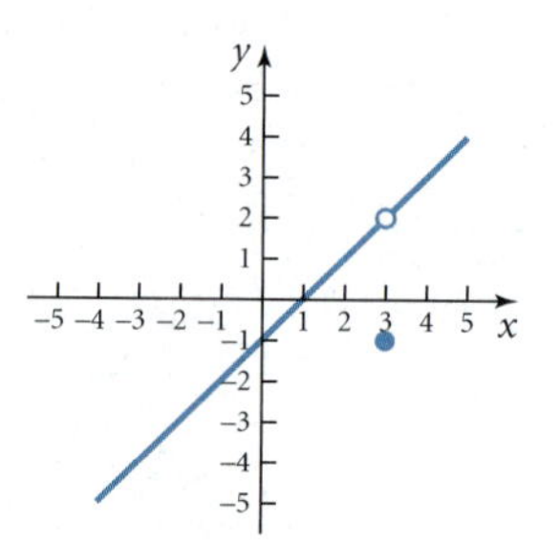

14. $\lim_{x \to 2} f(x) = ?$

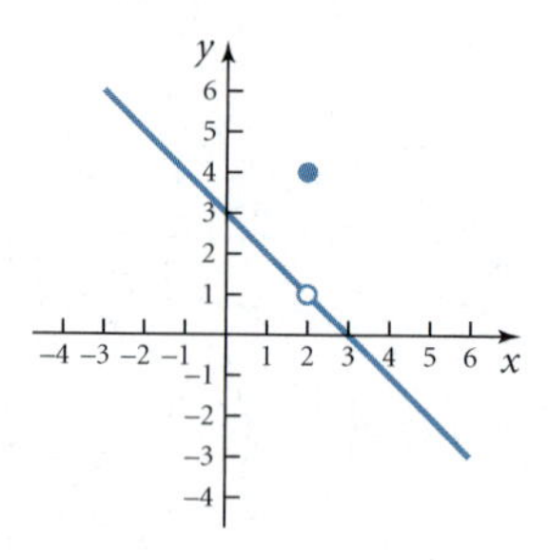

15. $\lim_{x \to 3} f(x) = ?$

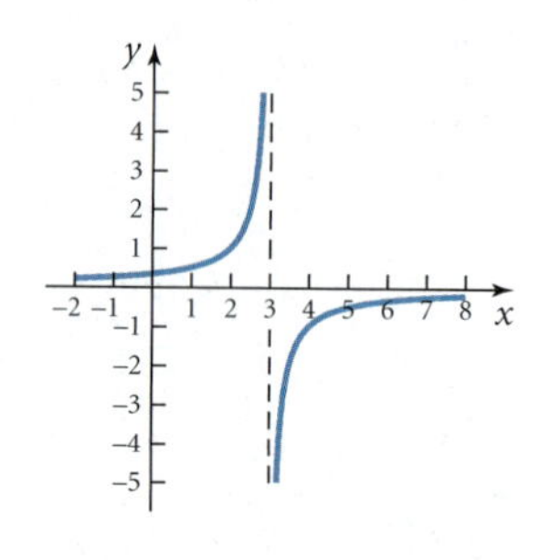

16. $\lim_{x \to 2} f(x) = ?$

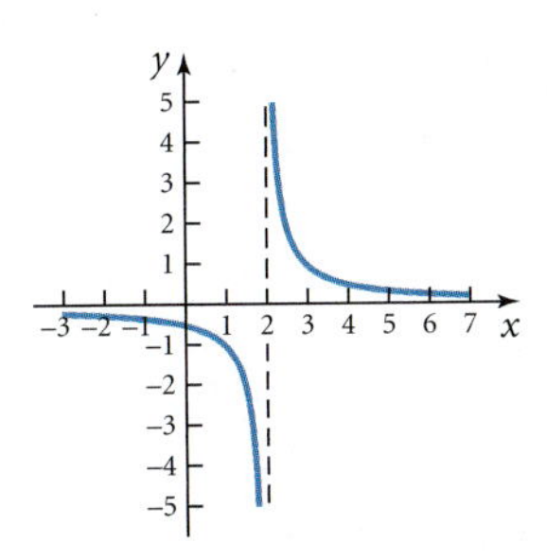

17. $\lim_{x \to 3} f(x) = ?$

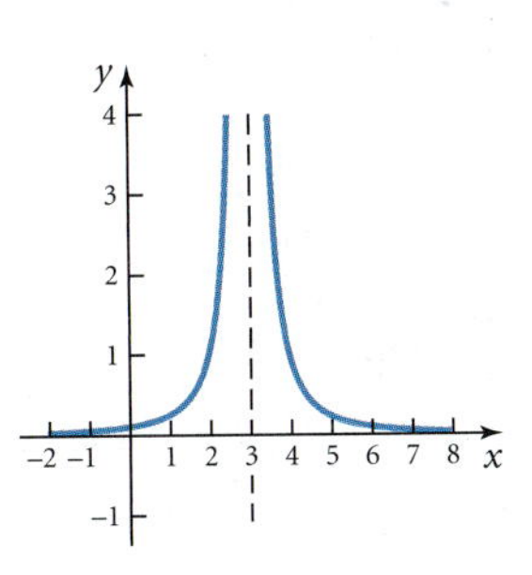

18. $\lim_{x \to 2} f(x) = ?$

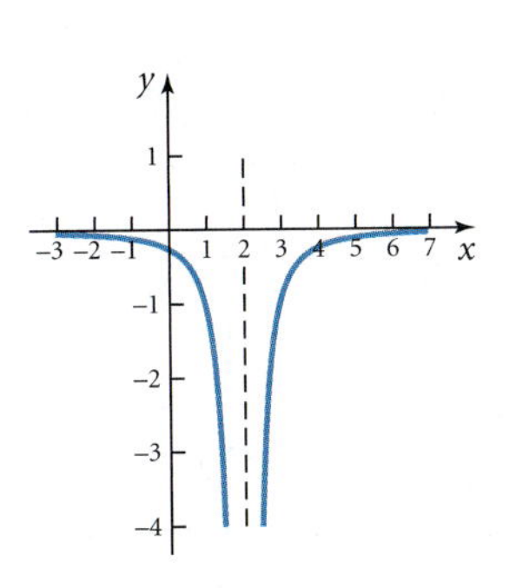

19. **a.** $\lim_{x \to 3^-} f(x) = ?$

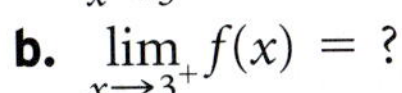

b. $\lim_{x \to 3^+} f(x) = ?$

c. $\lim_{x \to 3} f(x) = ?$

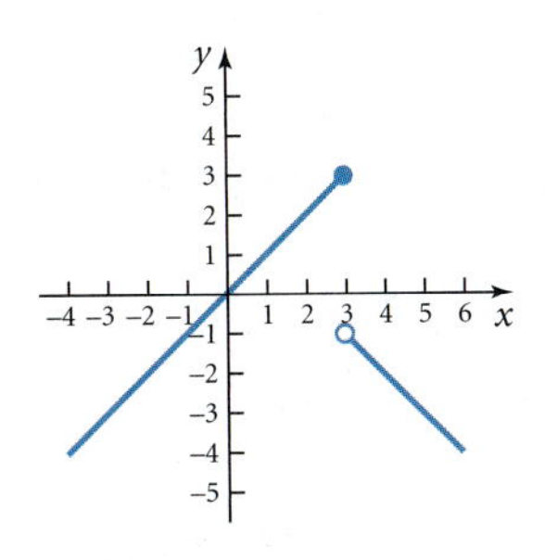

20. **a.** $\lim_{x \to 2^-} f(x) = ?$

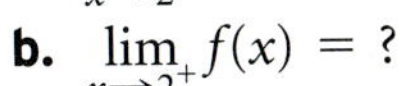

b. $\lim_{x \to 2^+} f(x) = ?$

c. $\lim_{x \to 2} f(x) = ?$

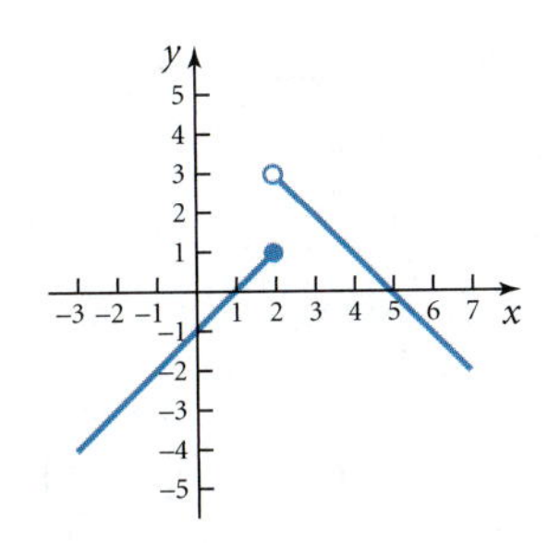

21. **a.** $\lim_{x \to 1^-} f(x) = ?$

b. $\lim_{x \to 1^+} f(x) = ?$

c. $\lim_{x \to 1} f(x) = ?$

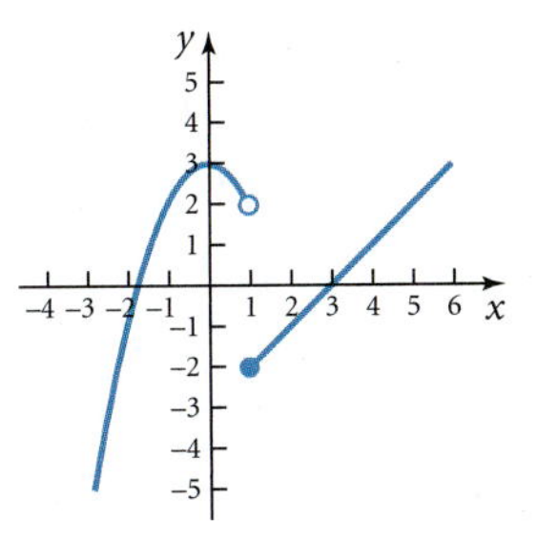

22. **a.** $\lim_{x \to 1^-} f(x) = ?$

b. $\lim_{x \to 1^+} f(x) = ?$

c. $\lim_{x \to 1} f(x) = ?$

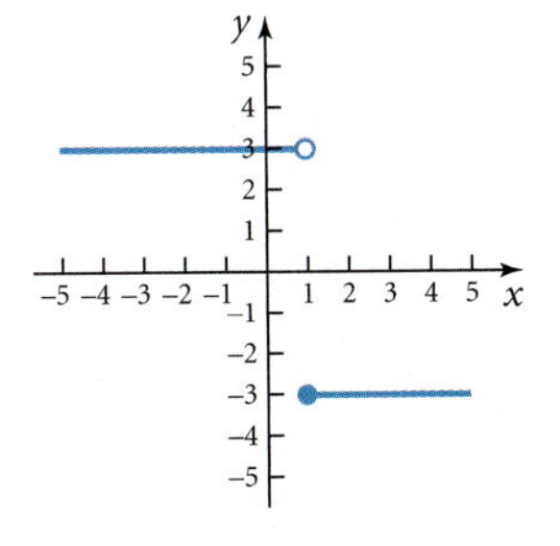

23. For the following graph, determine $\lim_{x \to a} f(x)$, if it exists, at the indicated values of a.

a. $\lim_{x \to -4} f(x)$ **b.** $\lim_{x \to -2} f(x)$

c. $\lim_{x \to -1} f(x)$ **d.** $\lim_{x \to 2} f(x)$

e. $\lim_{x \to 3} f(x)$ **f.** $\lim_{x \to 5} f(x)$

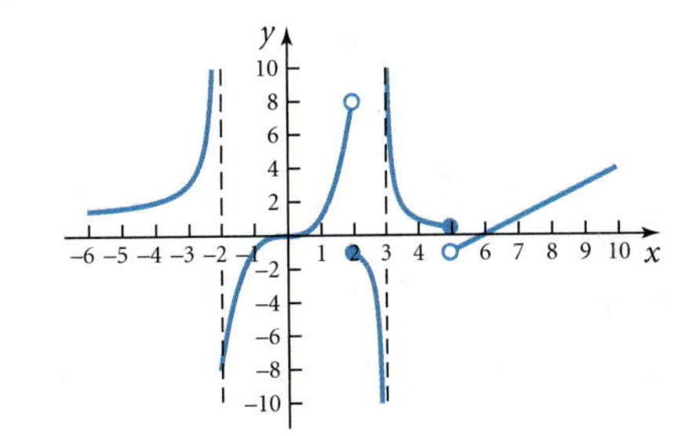

24. For the following graph, determine $\lim_{x \to a} f(x)$, if it exists, at the indicated values of a.

a. $\lim_{x \to -3} f(x)$ **b.** $\lim_{x \to -1} f(x)$

c. $\lim_{x \to 0} f(x)$ **d.** $\lim_{x \to 1} f(x)$

e. $\lim_{x \to 3} f(x)$ **f.** $\lim_{x \to 4} f(x)$

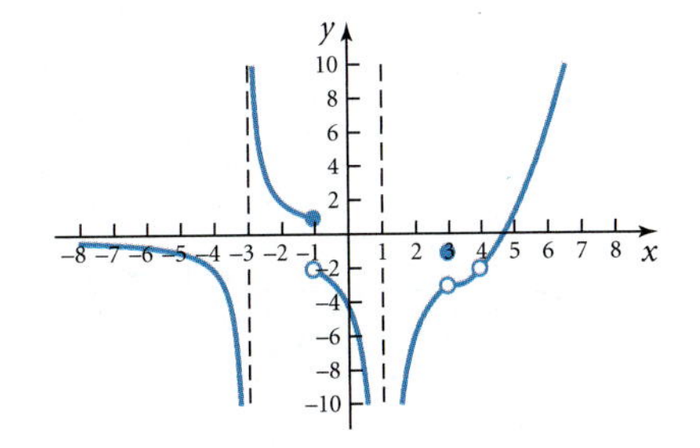

25. If the graph of $f(x)$ has a "hole" at $x = a$, with $f(a)$ not defined, which of the following statements must be true?

a. $\lim_{x \to a} f(x)$ does not exist.

b. $f(a)$ is undefined.

c. Both statements A and B are true.

d. Neither statement A nor B is true.

26. If the graph of $f(x)$ has a jump at $x = a$, which of the following statements must be true?

a. $\lim_{x \to a} f(x)$ does not exist.

b. $f(a)$ does not exist.

c. Both statements A and B are true.

d. Neither statement A nor B is true.

27. Use the following graph to decide if each of the following statements are true or false.

a. $\lim_{x \to 1} f(x) = 1$ **b.** $\lim_{x \to 0^-} f(x) = -2$

c. $\lim_{x \to 0} f(x) = 0$ **d.** $f(-3) = 3$

e. $\lim_{x \to -3} f(x) = 3$ **f.** $\lim_{x \to 3} f(x) = 4$

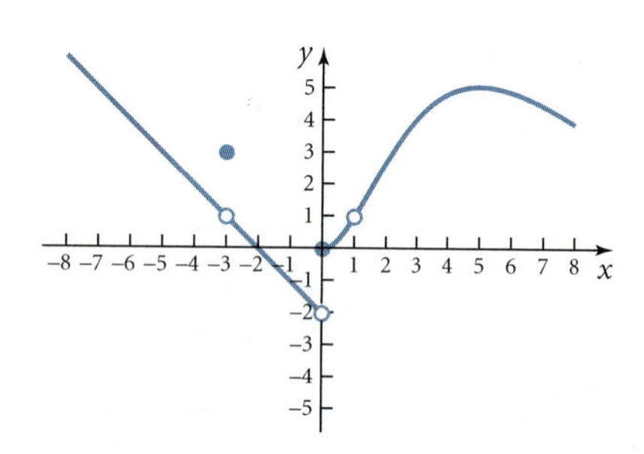

28. Use the following graph to decide if each of the following statements is true or false.

a. $\lim_{x \to 0} f(x) = -2$ **b.** $\lim_{x \to 0^+} f(x) = -2$

c. $\lim_{x \to 2} f(x) = 0$ **d.** $\lim_{x \to 3} f(x) = 1$

e. $f(3) = 4$ **f.** $\lim_{x \to 3^-} f(x) = 3$

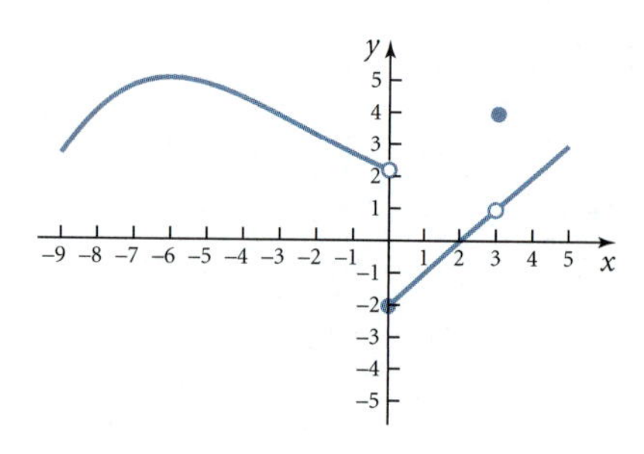

In Exercises 29 through 50, determine $\lim_{x \to a} f(x)$, if it exists.

29. $\lim_{x \to 3} (3x - 2)$

30. $\lim_{x \to 2} (x^2 - 5)$

31. $\lim_{x \to 7} 42$

32. $\lim_{x \to 11} 54$

33. $\lim_{x \to -3} \frac{2x}{x + 4}$

34. $\lim_{y \to -2} \frac{3y}{y + 5}$

35. $\lim_{d \to 2} \sqrt{2d + 5}$

36. $\lim_{a \to 5} \sqrt[3]{5a + 2}$

37. $\lim_{t \to 1} (2t^3 - 5t + 4)$

38. $\lim_{x \to 2} \frac{-2x}{x - 3}$

39. $\lim_{x \to 1} \frac{3x}{x^2 - 1}$

40. $\lim_{x \to 3} \frac{4x + 12}{x^2 - 9}$

41. $\lim_{x \to 1} \frac{3x - 3}{x^2 - 1}$

42. $\lim_{x \to 3} \frac{4x - 12}{x^2 - 9}$

43. $\lim_{x \to -3} \frac{x^2 + 3x}{x^2 - 9}$

44. $\lim_{x \to 2} \frac{3x - 6}{x^2 - 4}$

45. $\lim_{x \to 4} \frac{3x + 12}{x^2 - 16}$

46. $\lim_{x \to 2} \frac{x^2 + 4}{x^2 - 2x}$

47. $\lim_{x \to 4} \frac{x^3}{x^2 - 16}$

48. $\lim_{x \to 5} \frac{-x^3}{x^2 - 25}$

49. $\lim_{h \to 0} \frac{4a^2h - 3ah + h^2}{h}$

50. $\lim_{h \to 0} \frac{-6a^2h - 12ah - h^2}{h}$

51. Graph

$$f(x) = \begin{cases} 2x - 3 & \text{if } x < 1 \\ x^2 & \text{if } x \geq 1 \end{cases}$$

Use the graph of $f(x)$ to evaluate each limit, if it exists.

a. $\lim_{x \to 0} f(x)$ **b.** $\lim_{x \to 3} f(x)$ **c.** $\lim_{x \to 1} f(x)$

52. Graph

$$f(x) = \begin{cases} 1 - x^2 & \text{if } x \leq -2 \\ 3x + 1 & \text{if } x > -2 \end{cases}$$

Use the graph of $f(x)$ to evaluate each limit, if it exists.

a. $\lim_{x \to -5} f(x)$ **b.** $\lim_{x \to 0} f(x)$ **c.** $\lim_{x \to -2} f(x)$

53. Given

$$f(x) = \begin{cases} x + 5 & \text{if } x < -1 \\ -x^2 - 3 & \text{if } -1 \leq x < 2 \\ \sqrt{x - 1} & \text{if } x \geq 2 \end{cases}$$

find

a. $\lim_{x\to 0} f(x)$ **b.** $\lim_{x\to -1} f(x)$ **c.** $\lim_{x\to 2} f(x)$

54. Given

$$f(x) = \begin{cases} \frac{x^2}{2} & \text{if } x \le -3 \\ -x^2 - 3 & \text{if } -3 < x \le 2 \\ \sqrt{x+2} & \text{if } x > 2 \end{cases}$$

find

a. $\lim_{x\to 0} f(x)$ **b.** $\lim_{x\to -3} f(x)$ **c.** $\lim_{x\to 2} f(x)$

For the functions in Exercises 55 through 68,

a. Find the difference quotient $\frac{f(a+h)-f(a)}{h}$.

b. Determine $\lim_{h\to 0} \frac{f(a+h)-f(a)}{h}$.

55. $f(x) = 5x - 7$

56. $f(x) = -46$

57. $f(x) = x^2 + 2$

58. $f(x) = 3 - 2x$

59. $f(x) = 12$

60. $f(x) = 3 - x^2$

61. $f(x) = -2x^2 + 7x - 1$

62. $f(x) = 3x^2 - 2x + 4$

63. $f(x) = \frac{2}{x+1}$

64. $f(x) = \frac{4}{2-x}$

65. $f(x) = x^3$

66. $f(x) = x^4$

67. $f(x) = \sqrt{x}$

68. $f(x) = \frac{1}{\sqrt{x}}$

Evaluate each limit, if it exists.

69. $\lim_{x\to\infty} (2x - 3)$

70. $\lim_{x\to\infty} (6 - x)$

71. $\lim_{x\to-\infty} (5x + 2)$

72. $\lim_{x\to-\infty} (4 - 3x)$

73. $\lim_{x\to\infty} (3x^5 - 7x + 11)$

74. $\lim_{x\to\infty} (-3x^4 - 2x^2 + 5)$

75. $\lim_{x\to 3} \sqrt{16 - 4x}$

76. $\lim_{x\to -2} \sqrt{x^2 + 5}$

77. $\lim_{x\to\infty} \frac{4}{x}$

78. $\lim_{x\to-\infty} \frac{2}{x}$

79. $\lim_{x\to-\infty} \frac{-3}{x - 5}$

80. $\lim_{x\to-\infty} \frac{2}{x+1}$

81. $\lim_{x\to\infty} \frac{5x}{x+6}$

82. $\lim_{x\to\infty} \frac{3x}{2 - 6x}$

83. $\lim_{x\to\infty} \frac{5x^2}{x+6}$

84. $\lim_{x\to\infty} \frac{3x^3}{x^2 - 4}$

85. $\lim_{x\to\infty} \frac{3x}{x^2 + 5}$

86. $\lim_{x\to\infty} \frac{3x}{x^2 - 4}$

87. $\lim_{x\to\infty} \frac{-2x^3}{3x^3 - 6}$

88. $\lim_{x\to\infty} \frac{3x^3}{x^3 + 1}$

89. The annual salary for professors at a certain college is based on years of experience. The salaries are modeled by

$$S(t) = \begin{cases} \$30{,}000 + \$1200t & \text{if } t \le 3 \\ \$35{,}000 + \$1500t & \text{if } 3 < t \le 10 \\ \$51{,}000 + \$2000t & \text{if } t > 10 \end{cases}$$

where S is the annual salary and t is the number of years of experience.

a. What is the salary of a professor with six years of experience?

b. How many years of experience does a professor have if she makes an annual salary of $32,400?

c. Evaluate $\lim_{t\to 3} S(t)$, if it exists, and interpret your answer.

90. The 2005 U.S. federal income tax table for single filers shows various tax rates, depending on taxable income:

$$T(d) = \begin{cases} 0.10d & \text{if } d \le \$7300 \\ \$730 + 0.15(d - \$7300) & \text{if } \$7300 < d \le \$29{,}700 \\ \$4090 + 0.25(d - \$29{,}700) & \text{if } \$29{,}700 < d \le \$71{,}950 \\ \$14{,}652.50 + 0.28(d - \$71{,}950) & \text{if } \$71{,}950 < d \le \$150{,}150 \\ \$36{,}548.50 + 0.33(d - \$150{,}150) & \text{if } \$150{,}150 < d \le \$326{,}450 \\ \$94{,}727.50 + 0.35(d - \$326{,}450) & \text{if } d > \$326{,}450 \end{cases}$$

Data from: The United States Internal Revenue Service.

where T is income tax due and d is taxable income.

a. Find the tax due for a single person whose taxable income is $30,562.

b. Estimate the taxable income of a single person whose tax was $6725.

c. Evaluate $\lim_{d\to \$71{,}950} T(d)$, if it exists, and interpret your answer.

91. A woman on a weight-loss program weighs herself each week during a 10-week period with the following results.

Week	Weight (lbs)
0	180
1	172
2	164
3	159
4	155
5	153
6	152
7	150
8	148
9	146
10	145

Her weight at any time can be approximated by the rational function $W(t) = \dfrac{790 + 132t}{t + 4.4}$, where W is her weight in pounds and t is the number of weeks after starting the program.

a. Evaluate $W(5)$ and $W(9)$ using the model. How closely does the model predict the actual weight?

b. Find $\lim\limits_{t \to \infty} W(t)$, if it exists, and interpret your answer.

92. The value of a car depreciates over time. The value of a new car purchased for \$32,500 during the next five years has the following values.

Year	Value
0	\$32,500
1	\$26,000
2	\$20,800
3	\$16,640
4	\$13,300
5	\$10,500

The value of the car at any time during the first five years can be modeled by $V(t) = \dfrac{0.75t + 97.5}{t + 3}$, where V is the value of the car in thousands of dollars and t is the number of years after purchase.

a. Evaluate $V(2)$ and $V(5)$ using the model. How closely does the model predict the actual value of the car?

b. Find $\lim\limits_{t \to \infty} V(t)$, if it exists, and interpret your answer.

93. Cab fares in Las Vegas are determined by a flat rate of \$2.20 for the first mile plus \$1.50 for each additional mile (or fractional part of a mile). In the graph of this function shown here, the horizontal axis represents miles and the vertical axis represents dollars.

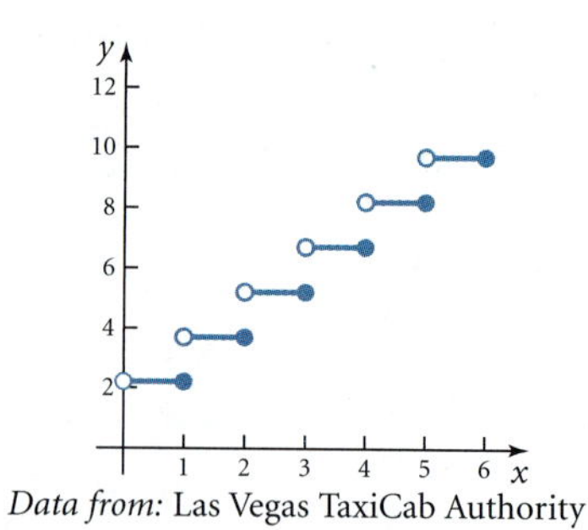

Data from: Las Vegas TaxiCab Authority

Evaluate each of the following functions or limits.

a. $C(3.5)$ **b.** $\lim\limits_{x \to 3.5} C(x)$

c. $C(5)$ **d.** $\lim\limits_{x \to 5} C(x)$

94. Rates for parcel airlift, according to the U.S. Postal Service, are as follows.

Weight (x) Not More Than . . .	Fee
2 pounds	\$0.45
3 pounds	\$0.85
4 pounds	\$1.25
30 pounds	\$1.70

Evaluate each of the following functions or limits.

a. $C(2.5)$ **b.** $\lim\limits_{x \to 2.5} C(x)$

c. $C(4)$ **d.** $\lim\limits_{x \to 4} C(x)$

SHARPEN THE TOOLS

Evaluate each limit, if it exists, graphically and algebraically.

95. $\lim\limits_{x \to \infty} \dfrac{5x^3}{x + 6}$

96. $\lim\limits_{x \to \infty} \dfrac{-2x}{\sqrt{x^2 - 3}}$

97. $\lim\limits_{x \to \infty} \dfrac{\sqrt{4x^2 + 2}}{3x + 1}$

98. $\lim\limits_{x \to \infty} \dfrac{2x^4 - 3x^2 + 5}{3x^4 + 2x + 5}$

99. The 2005 tax rate schedule for taxpayers using the married filing jointly status is given here. Calculate the income tax due according to the table.

If Taxable Income Is More Than ...	But Not More Than ...	The Tax Is ...
$0	$14,600	10% of the amount over $0
$14,600	$59,400	$1460 plus 15% of the amount over $14,600
$59,400	$119,950	$8180 plus 25% of the amount over $59,400
$119,950	$182,800	$23,317.50 plus 28% of the amount over $119,950
$182,800	$326,450	$40,915.50 plus 33% of the amount over $182,800
$326,450	No limit	$88,320 plus 35% of the amount over $326,450

Data from: The United States Internal Revenue Service.

a. Write a piecewise defined function that describes the amount of tax due.

b. Find the tax due for a married couple with combined income of $92,340.

c. Evaluate $\lim_{d \to \$65,000} T(d)$, if it exists.

d. Evaluate $\lim_{d \to \$59,400} T(d)$, if it exists.

100. The 2005 U.S. tax rate schedule for taxpayers using the married filing separately status is given here. Calculate the income tax due according to the table.

If Taxable Income Is More Than ...	But Not More Than ...	The Tax Is ...
$0	$7300	10% of the amount over $0
$7300	$29,700	$730 plus 15% of the amount over $7300
$29,700	$59,975	$4090 plus 25% of the amount over $29,700
$59,975	$91,400	$11,658.75 plus 28% of the amount over $59,975
$91,400	$163,225	$20,457.75 plus 33% of the amount over $91,400
$163,225	No limit	$44,160 plus 35% of the amount over $163,225

Data from: The United States Internal Revenue Service.

a. Write a piecewise defined function that describes the amount of tax due.

b. Find the tax due for a taxpayer using the married filing separately status with an income of $52,340.

c. Evaluate $\lim_{d \to \$65,000} T(d)$, if it exists.

d. Evaluate $\lim_{d \to \$59,975} T(d)$, if it exists.

TOPIC

6 Continuity

IMPORTANT TOOLS IN TOPIC 6

- *Continuous functions*
- *Continuity at a point*
- *Continuity of piecewise defined functions*

Think for a minute about what the word ***continuous*** implies. Imagine a line that extends indefinitely or water that flows nonstop from the faucet. The concept of a continuous function plays a big role in calculus. In this topic you will learn what a continuous function is and how continuity is related to limits.

Before beginning the study of continuous functions, be sure to complete the Warm-up Exercises to brush up on the algebra and calculus tools you will need in Topic 6.

TOPIC 6 WARM-UP EXERCISES

Be sure you can successfully complete the following exercises before starting Topic 6.

For Exercises 1 through 4,

$$f(x) = \begin{cases} 3x + 5 & \text{if } x < -2 \\ x^2 & \text{if } -2 \le x \le 1 \\ \sqrt{x} & \text{if } x > 1 \end{cases}$$

1. Evaluate the following functions.

a. $f(-4)$ **b.** $f(0)$ **c.** $f(4)$

d. $f(-2)$ **e.** $f(1)$

2. Evaluate the following limits.

a. $\lim_{x \to 1^-} f(x)$

b. $\lim_{x \to 1^+} f(x)$

c. $\lim_{x \to 1} f(x)$

3. Evaluate the following limits.

a. $\lim_{x \to -2^-} f(x)$

b. $\lim_{x \to -2^+} f(x)$

c. $\lim_{x \to -2} f(x)$

4. Evaluate $\lim_{x \to a} f(x)$, if it exists, for the following values.

a. $a = -4$ **b.** $a = 0$ **c.** $a = 4$

d. $a = -2$ **e.** $a = 1$

5. State the domain of each of the following functions.

a. $f(x) = 7x - 11$

b. $f(x) = \frac{1}{2}x^2 - 4x + 6$

c. $f(x) = \sqrt{x + 2}$

d. $f(x) = \dfrac{6}{x - 7}$

Given $f(x)$ as shown:

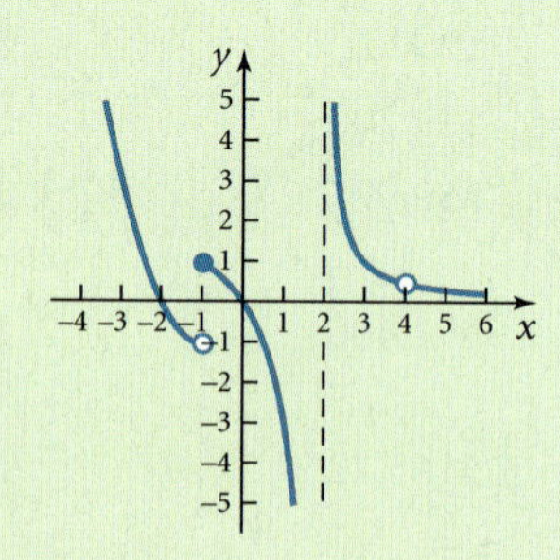

6. Evaluate the following functions.

a. $f(-1)$ **b.** $f(0)$

c. $f(2)$ **d.** $f(4)$

7. Evaluate $\lim_{x \to a} f(x)$ for the following values.

a. $a = 0$ **b.** $a = 4$

c. $a = 2$ **d.** $a = -1$

Answers to Warm-up Exercises

1. a. -7 **b.** 0 **c.** 2
d. 4 **e.** 1

2. a. 1 **b.** 1 **c.** 1

3. a. -1 **b.** 4 **c.** does not exist

4. a. -7
b. 0
c. 2
d. does not exist
e. 1

5. a. reals **b.** reals
c. $x \geq -2$ **d.** $x \neq 7$

6. a. 1 **b.** 0
c. undefined **d.** undefined

7. a. 0
b. 1
c. does not exist
d. does not exist

We begin the discussion of continuity by exploring what it means for a function to be continuous over its domain and at a specific point in its domain.

Continuity

Intuitively, we might say that a function $f(x)$ is **continuous** if its graph can be drawn without lifting the pencil. Furthermore, a function is **continuous at a point** in its domain if the graph has no holes, asymptotes, or jumps at that point.

There is nothing wrong with this intuitive understanding, but in calculus a more mathematically precise definition of continuity needs to be developed. The following examples explore the necessary requirements for continuity of functions at a point.

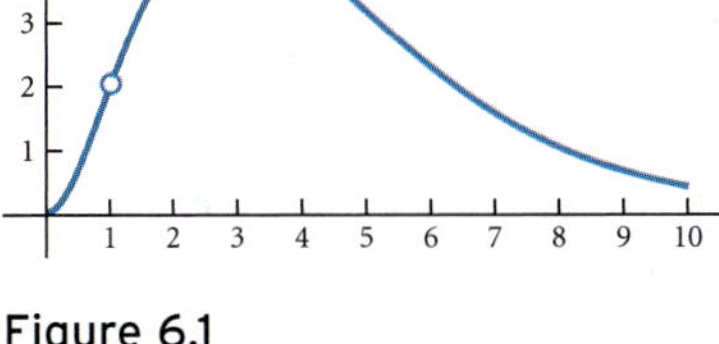

Figure 6.1

Example 1: Is the graph in Figure 6.1 continuous throughout its domain?

Solution: The graph in Figure 6.1 is not continuous throughout its domain because the hole at $x = 1$ would require lifting the pencil when drawing the graph. So, we say that $f(x)$ is **discontinuous** at $x = 1$. The hole in the graph indicates that $f(x)$ is not defined for that value of x. This situation is referred to as a **removable discontinuity.**

This example clearly demonstrates the first requirement for continuity of a function at a point: *the function must be defined* at that point. Stated symbolically, if we are given a particular domain value c, then $f(c)$ must exist. Graphically, there must be a dot on the graph at the point where $x = c$. As the next example shows, this requirement is necessary but not sufficient for continuity. ■

Example 2: Is the graph in Figure 6.2 continuous throughout its domain?

Solution: The graph in Figure 6.2 is not continuous throughout its domain because of the break in the graph at $x = 1$, so again $f(x)$ is discontinuous at $x = 1$. As in Example 1, the first requirement for continuity at a point is that the function be defined at the point. In this example, $f(x)$ is

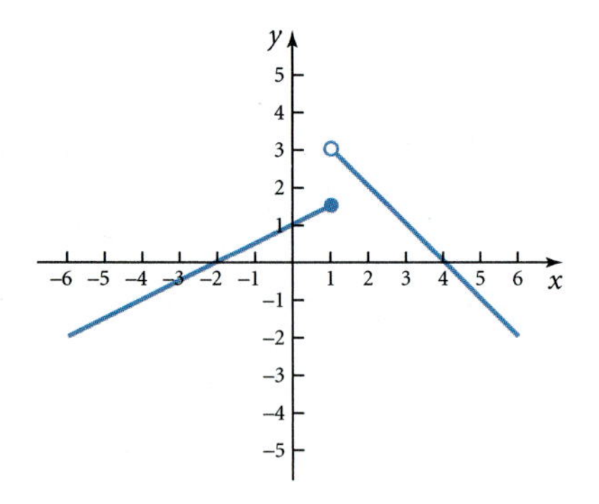

Figure 6.2

defined because $f(1) = 1$ (the dot on the graph). Nevertheless, the function is discontinuous at $x = 1$. Why? The discontinuity occurs because of the break in the graph at $x = 1$. Examining limits shows that $\lim_{x \to 1^-} f(x) = 1$, but $\lim_{x \to 1^+} f(x) = 3$, so $\lim_{x \to 1} f(x)$ does not exist. This situation is referred to as a **jump discontinuity**. ■

Example 2 shows the second requirement for continuity of a function at a point: *the limit of the function must exist* as x approaches that point. To determine if these two requirements are sufficient for continuity, consider one more example.

Example 3: Is the graph in Figure 6.3 continuous throughout its domain?

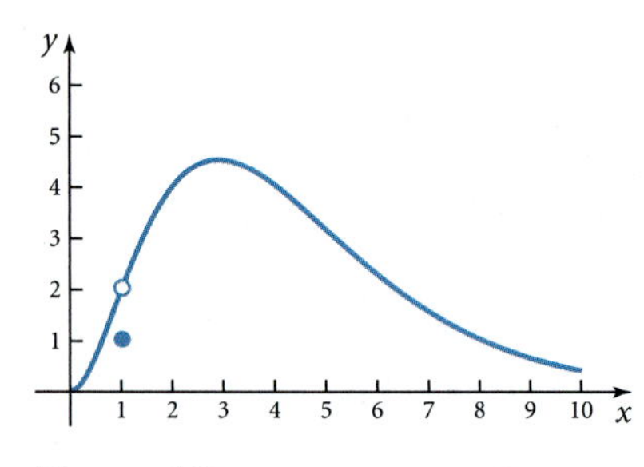

Figure 6.3

Solution: The graph in Figure 6.3 is not continuous throughout its domain. Because of the hole at $x = 1$ and the additional point at $(1, 1)$, there is still a need to lift the pencil when drawing the graph.

Because of the dot on the graph, $f(1) = 1$, so the function is defined, which satisfies the first requirement for continuity. Also $\lim_{x \to 1^-} f(x) = \lim_{x \to 1^+} f(x) = 2$, so $\lim_{x \to 1} f(x) = 2$. The function has a limit as $x \to 1$, which satisfies the second requirement for continuity. The graph is still discontinuous at $x = 1$, however. Why? The discontinuity occurs because the dot is separated from the rest of the graph; in other words, the value of the limit as $x \to 1$ and the value of function when $x = 1$, $f(1)$, are not the same. ■

Example 3 shows the third necessary requirement for continuity at a point: *the value of the function at the point and the value of the limit as x approaches that point must be the same.*

Continuity at a Point

A function $f(x)$ is **continuous at a point** $(a, f(a))$ if the following three conditions are met:

1. $f(a)$ is defined.
2. $\lim_{x \to a} f(x)$ exists.
3. $f(a) = \lim_{x \to a} f(x)$.

Tip:

Requirement 1 for continuity at a point—that the function be defined—eliminates any holes or asymptotes that cause discontinuity.

Requirement 2 for continuity at a point—that the limit exist—eliminates any breaks causing discontinuity of piecewise defined functions.

Requirement 3 for continuity at a point eliminates any other jumps or stray dots in the definition of the function.

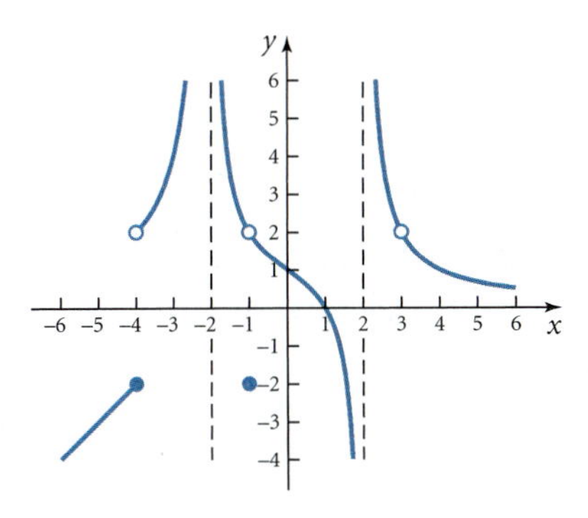

Figure 6.4

Example 4: For the graph given in Figure 6.4, state the points of discontinuity and identify which requirement for continuity is violated.

Solution: The discontinuities are at $x = -4, -2, -1, 2$, and 3.

At $x = -4$, the *limit does not exist* because $\lim_{x\to-4^-} f(x) = -2$ and $\lim_{x\to-4^+} f(x) = 1$.

At $x = -2$, the *function is not defined* because of the asymptote, so $f(-2)$ is undefined and $\lim_{x\to-2} f(x)$does not exist.

At $x = -1$, *the value of the function does not equal the value of the limit.* At $x = -1, f(-1) = -2$, but $\lim_{x\to-1} f(x) = 1$.

At $x = 2$, the *function is not defined* and there is no limit because of the asymptote.

At $x = 3$, the *function is not defined* because of the hole, so $f(3)$ is undefined. ■

Check Your Understanding 6.1

Each of the following four graphs is discontinuous at $x = 2$. State the requirement that is violated.

1.

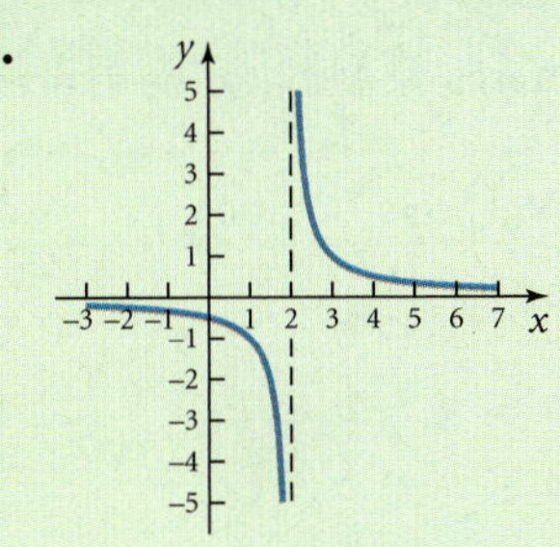

2.

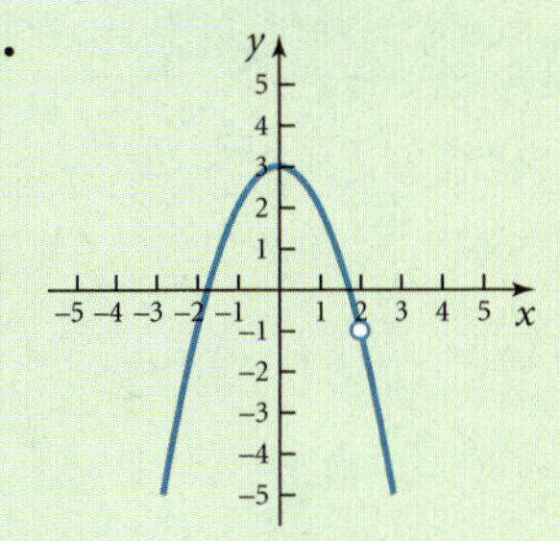

3.

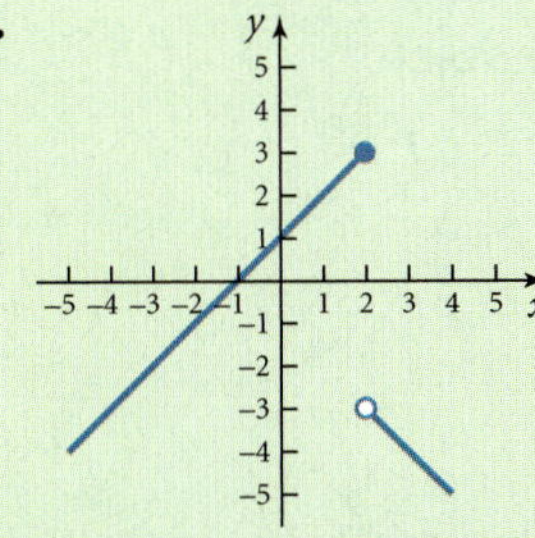

4.

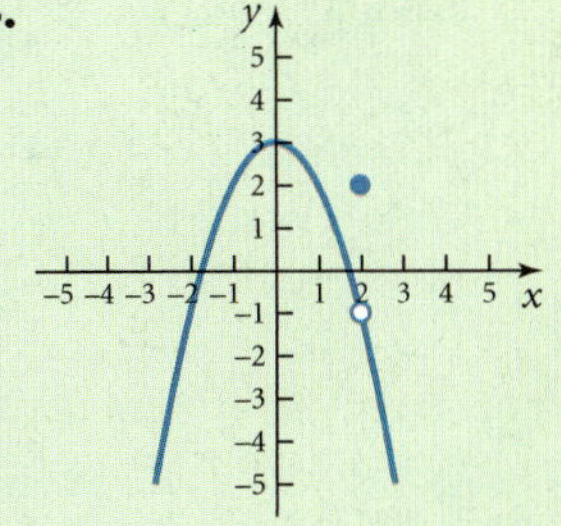

5. For the following graph, state the points of discontinuity and the requirement that is violated.

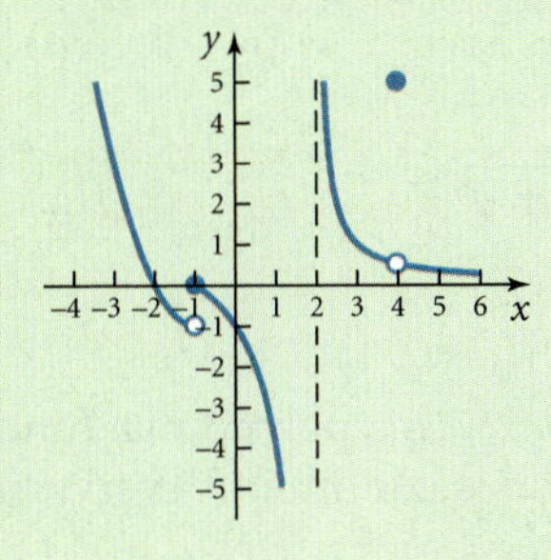

Finding Intervals of Continuity for Algebraic Functions

In the definition of continuity at a point, one requirement was that the function be defined at that point. The set of all points for which a function is defined is the domain of the function. For algebraic functions, the interval(s) of continuity can be determined by identifying the domain.

- Polynomial functions are continuous everywhere.
- Square-root functions are continuous for all values of x for which the radicand is nonnegative. (Nonnegative means positive or zero. The radicand is the expression under the radical sign.)
- Rational functions are continuous for all values of x for which the denominator is nonzero.

Interval Notation

Intervals of real numbers can also be expressed in interval notation.

Algebraic Notation	Interval Notation
$a < x < b$	(a, b)
$a \leq x \leq b$	$[a, b]$
$a < x \leq b$	$(a, b]$
$a \leq x < b$	$[a, b)$
$x > a$	(a, ∞)
$x \geq a$	$[a, \infty)$
$x < b$	$(-\infty, b)$
$x \leq b$	$(-\infty, b]$

Example 5: For what interval(s) is each of the following functions continuous?

a. $f(x) = x^4 + 5x - 7$

b. $f(d) = \dfrac{3d}{d^2 - 4}$

c. $P(t) = \sqrt{t + 5}$

d. $f(x) = \dfrac{6}{\sqrt{x + 5}}$

Solution:

a. $f(x) = x^4 + 5x - 7$ is a polynomial function, so the domain is all real numbers. The function is continuous everywhere for $(-\infty, \infty)$.

b. $f(d) = \dfrac{3d}{d^2 - 4}$ is a rational function, so the domain is all real numbers except 2 and -2. The function is continuous for $d \neq 2, -2$, or $(-\infty, -2) \cup (-2, 2) \cup (2, \infty)$.

c. $P(t) = \sqrt{t + 5}$ is a square-root function, so the domain is all real numbers such that $t + 5 \geq 0$. The function is continuous for $t \geq -5$ or $[-5, \infty)$.

d. $f(x) = \dfrac{6}{\sqrt{x + 5}}$ is a fraction, so the denominator cannot be zero. The denominator is also a square-root function (see part c). The domain of the function is all values of x such that $x + 5 > 0$ ($x + 5$ cannot equal zero because it is in the denominator). Thus, the function is continuous for $x > -5$ or $(-5, \infty)$. ■

MATHEMATICS CORNER 6.2

Set Notation with Intervals

In set notation, the symbol $\cup$ means the **union** of two sets. In Example 5b, the points of discontinuity are at $d = 2$ and $d = -2$. The function is continuous for all other values: $d < -2$ *or* $-2 < d < 2$ *or* $d > 2$. The word *or* is replaced by the union symbol, $\cup$. In interval notation, we write $(-\infty, -2) \cup (-2, 2) \cup (2, \infty)$.

Check Your Understanding 6.2

State the interval(s) for which each of the following functions are continuous.

1. $g(x) = 7 - 3x$

2. $f(x) = -2x^2 + 8$

3. $h(v) = \sqrt{v - 2}$

4. $f(x) = x^3 - 2x + 5$

5. $f(x) = \dfrac{x}{x^2 - 9}$

Continuity of Piecewise Defined Functions

The second requirement for continuity of a function at a point is that the limit of the function exist as x approaches that point. For piecewise defined functions, care must be taken when evaluating limits as $x \rightarrow a$. If $x = a$ is not a break point of the function, then $\lim_{x \to a} f(x) = f(a)$. If $x = a$ is a break point of the domain of the function, the left-hand and right-hand limits must be examined.

Example 6: Determine whether $f(x)$ is continuous at the given value of a.

$$f(x) = \begin{cases} 2x - 3 & \text{if } x < 1 \\ x^2 - 1 & \text{if } x \geq 1 \end{cases}$$

for

a. $a = 0$ **b.** $a = 1$

Solution: Remember the requirements for continuity at a point:

1. $f(a)$ must be defined.
2. $\lim_{x \to a} f(x)$ must exist.
3. $f(a) = \lim_{x \to a} f(x)$.

a. If $a = 0$, then $f(0) = -3$ because $0 < 1$. The function is defined at $a = 0$. Because $a = 0$ is not a break point of the domain, then $\lim_{x \to 0} f(x) = f(0) = -3$. Thus, $f(0) = -3$ and $\lim_{x \to 0} f(x) = -3$, so $f(0) = \lim_{x \to 0} f(x)$ and the function is continuous at $a = 0$.

b. If $a = 1$, then $f(1) = 1^2 - 1 = 0$. The function is defined at $a = 1$. Here, $a = 1$ is a break point, so $\lim_{x \to 1^-} f(x) = -1$ and $\lim_{x \to 1^+} f(x) = 0$. Thus, $\lim_{x \to 1} f(x)$ does not exist because the left-hand and right-hand limits are different, $f(x)$ is not continuous at $a = 1$ because the limit does not exist. ■

Example 7: Determine whether $f(x)$ is continuous at the given value of a.

$$f(x) = \begin{cases} \dfrac{3}{x} & \text{if } x < -2 \\ 4 - x & \text{if } -2 \leq x < 3 \\ x^2 - 8 & \text{if } x \geq 3 \end{cases}$$

for

a. $a = -2$ **b.** $a = 0$ **c.** $a = 3$

Solution:

a. If $a = -2$, then $f(-2) = 4 - (-2) = 6$, so the function is defined. Because $a = -2$ is a break point of the domain, $\lim_{x \to -2^-} f(x) = -\frac{3}{2}$ and $\lim_{x \to -2^+} f(x) = 6$. Thus, $\lim_{x \to -2} f(x)$ does not exist and $f(x)$ is not continuous at $a = -2$.

b. If $a = 0$, then $f(0) = 4 - 0 = 4$, so the function is defined. Because $-2 < 0 < 3$, $\lim_{x \to 0} f(x) = f(0) = 4$ and $\lim_{x \to 0} f(x)$ exists. Because $f(0) = \lim_{x \to 0} f(x)$, $f(x)$ is continuous at $a = 0$.

c. If $a = 3$, then $f(3) = 3^2 - 8 = 1$, so the function is defined. Because $a = 3$ is a break point of the domain, $\lim_{x \to 3^-} f(x) = 1$ and $\lim_{x \to 3^+} f(x) = 1$. Thus, $\lim_{x \to 3} f(x)$ exists. Because $f(3) = \lim_{x \to 3} f(x)$, $f(x)$ is continuous at $a = 3$. ■

Example 8: Cab fares in Las Vegas are \$2.20 for the first mile and \$1.50 for each additional mile. (There are also charges for sitting at red lights and for additional passengers that are not considered here.) Graph the function for the first 6 miles and identify the points of discontinuity.

Solution: The graph in Figure 6.5 is a piecewise step function, and the points of discontinuity are $x = 1, 2, 3, 4$, and 5.

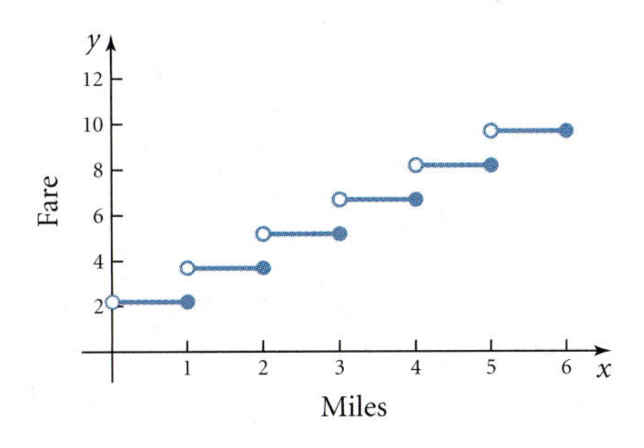

Figure 6.5

Any fractional part of a mile is charged as 1 whole mile, so the changes in fares come at the integer values of x. ■

Check Your Understanding 6.3

Determine whether or not each of the following functions is continuous at the indicated value. If not, explain what requirement is violated.

1. $f(x) = \begin{cases} \dfrac{3}{x-2} & \text{if } x < 2 \\ x^2 + 1 & \text{if } x \geq 2 \end{cases}$

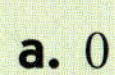

a. 0

b. 2

c. 5

2. $f(x) = \begin{cases} x^2 & \text{if } x < -1 \\ 2x + 3 & \text{if } -1 \leq x \leq 2 \\ \sqrt{x+2} & \text{if } x > 2 \end{cases}$

a. 2

b. 0

c. -1

3. $f(x) = \begin{cases} x^2 - 1 & \text{if } x < 0 \\ 4 & \text{if } x = 0 \\ x^2 + 1 & \text{if } x > 0 \end{cases}$

a. 2 **b.** -2 **c.** 0

4.

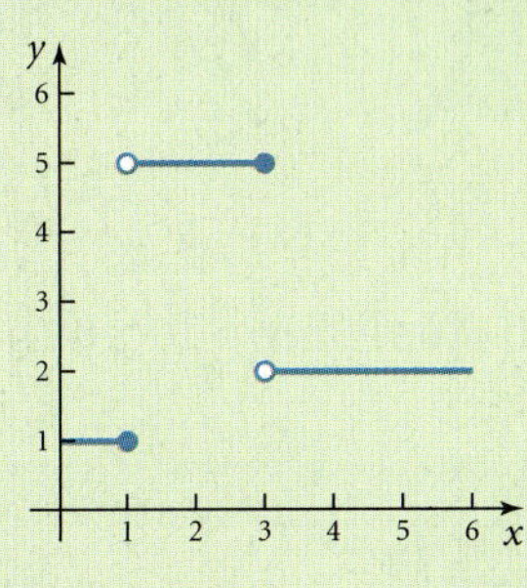

a. 1

b. 2

c. 3

Check Your Understanding Answers

Check Your Understanding 6.1

1. $f(2)$ is undefined, $\lim_{x \to 2} f(x)$ does not exist
2. $f(2)$ is undefined
3. $\lim_{x \to 2} f(x)$ does not exist
4. $f(2) \neq \lim_{x \to 2} f(x)$
5. discontinuous at $x = -1$ because $\lim_{x \to -1} f(x)$ does not exist
   discontinuous at $x = 2$ because $f(2)$ is undefined
   discontinuous at $x = 4$ because $f(4) \neq \lim_{x \to 4} f(x)$

Check Your Understanding 6.2

1. all real numbers $(-\infty, \infty)$
2. all real numbers $(-\infty, \infty)$
3. $v \geq 2$ or $[2, \infty)$
4. all real numbers $(-\infty, \infty)$
5. $x \neq 3, -3$ or $(-\infty, -3) \cup (-3, 3) \cup (3, \infty)$

Check Your Understanding 6.3

1. **a.** continuous
   b. discontinuous; $\lim_{x \to 2} f(x)$ does not exist
   c. continuous
2. **a.** discontinuous; $\lim_{x \to 2} f(x)$ does not exist
   b. continuous **c.** continuous
3. **a.** continuous **b.** continuous
   c. discontinuous; $\lim_{x \to 0} f(x)$ does not exist
4. **a.** discontinuous; $\lim_{x \to 1} f(x)$ does not exist
   b. continuous
   c. discontinuous; $\lim_{x \to 3} f(x)$ does not exist

Topic 6 Review

6

This topic introduced the concept of **continuity**.

CONCEPT	EXAMPLE
Intuitively, a curve is **continuous** if its graph can be drawn without lifting the pencil. A curve is **continuous at a point** if there are no jumps, holes, or asymptotes at that point. A point for which the graph of a function is not continuous is called a **point of discontinuity**.	$f(x) = x^3 - 4x$ is continuous throughout its domain because its graph has no jumps, holes, or asymptotes. 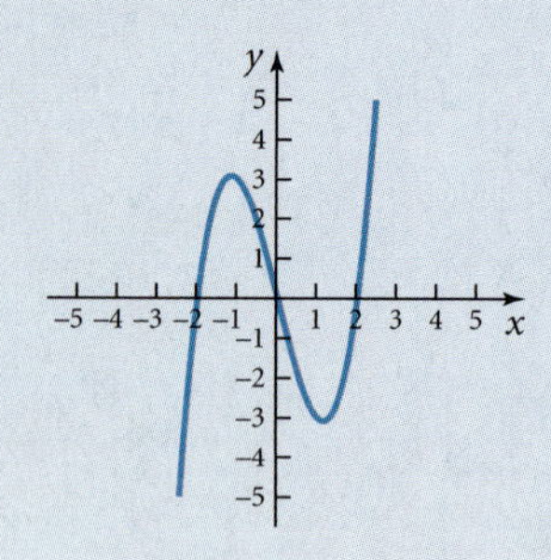

$f(x) = \dfrac{x}{x^2 - 4}$ is not continuous at $x = -2$ or $x = 2$ because of the vertical asymptotes on its graph. The function is discontinuous at $x = -2$ and $x = 2$.

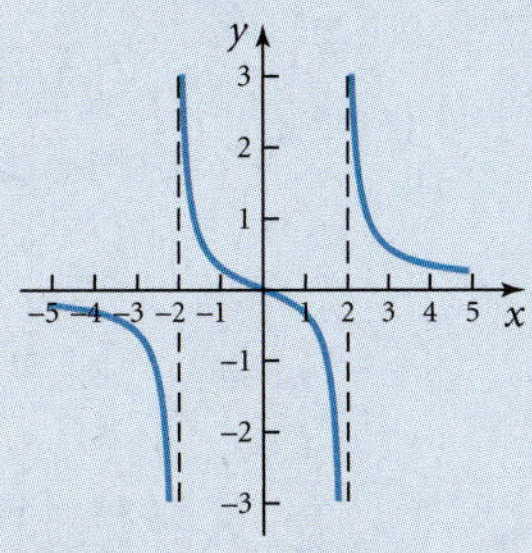

$$f(x) = \begin{cases} 2x - 3 & \text{if } x < 2 \\ \sqrt{x + 2} & \text{if } x \geq 2 \end{cases}$$

is not continuous at $x = 2$ because of the jump in the graph. The function is discontinuous at $x = 2$.

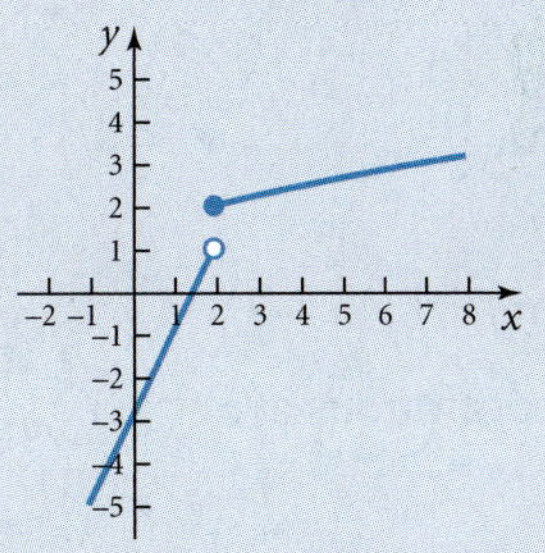

To prove continuity of $f(x)$ at a point $(a, f(a))$, three requirements must be met.

1. $f(a)$ must be defined.
2. $\lim_{x \to a} f(x)$ must exist.
3. $f(a) = \lim_{x \to a} f(x)$.

For

$$f(x) = \begin{cases} 2x - 3 & \text{if } x < 2 \\ \sqrt{x + 2} & \text{if } x \geq 2 \end{cases}$$

the function is discontinuous at $x = 2$ because $\lim_{x \to 2} f(x)$ does not exist.

For $f(x) = \dfrac{x}{x - 2}$, the function is discontinuous at $x = 2$ because $f(2)$ is undefined.

For algebraic functions, determining intervals of continuity is the same as determining the domain of the function.

$f(x) = \sqrt{2x - 5}$ is continuous for $x \geq \frac{5}{2}$, which is also the domain of the function.

For piecewise defined functions, determining continuity at a point requires validating that the three requirements of continuity are satisfied. This step may involve examining the left- and right-hand limits at break points in the domain of the function.

Is

$$f(x) = \begin{cases} x^2 & \text{if } x < 0 \\ 4 & \text{if } x = 0 \\ x^2 & \text{if } x > 0 \end{cases}$$

continuous at $x = 0$?

1. $f(0) = 4$, so the function is defined.
2. $\lim_{x \to 0^-} f(x) = \lim_{x \to 0^+} f(x) = 0$, so the limit exists.
3. $f(0) \neq \lim_{x \to 0} f(x)$, so the function is discontinuous at $x = 0$.

NEW TOOLS IN THE TOOL KIT

- Continuity of a function throughout its domain
- Continuity at a point
- Determining continuity from a graph
- Determining continuity from a function
- Jump continuity and removable discontinuity

Topic 6 Exercises

6

1. For

$$f(x) = \begin{cases} x - 5 & \text{if } x < 2 \\ 3 - x & \text{if } x \geq 2 \end{cases}$$

a. Find $f(2)$.

b. Find $\lim_{x \to 2} f(x)$.

c. Is $f(x)$ continuous at $x = 2$?

2. For

$$f(x) = \begin{cases} 2x + 5 & \text{if } x \leq 1 \\ 4 + 3x & \text{if } x > 1 \end{cases}$$

a. Find $f(1)$.

b. Find $\lim_{x \to 1} f(x)$.

c. Is $f(x)$ continuous at $x = 1$?

3. For

$$f(x) = \begin{cases} x^2 + 3 & \text{if } x < 1 \\ \frac{4}{x} & \text{if } x \geq 1 \end{cases}$$

a. Find $f(1)$.

b. Find $\lim_{x \to 1} f(x)$.

c. Is $f(x)$ continuous at $x = 1$?

4. For

$$f(x) = \begin{cases} 2 - x^2 & \text{if } x < -3 \\ \sqrt{1 - x} & \text{if } x \geq -3 \end{cases}$$

a. Find $f(-3)$.

b. Find $\lim_{x \to -3} f(x)$.

c. Is $f(x)$ continuous at $x = -3$?

5. For

$$f(x) = \begin{cases} 3x + 2 & \text{if } x \leq -2 \\ -x^2 & \text{if } -2 < x < 2 \\ \frac{2}{x} & \text{if } x \geq 2 \end{cases}$$

a. Find $f(-1)$.

b. Find $f(2)$.

c. Find $\lim_{x \to -1} f(x)$.

d. Find $\lim_{x \to 2} f(x)$.

e. Is $f(x)$ continuous at $x = -1$?

f. Is $f(x)$ continuous at $x = 2$?

6. For

$$f(x) = \begin{cases} 5 - 3x & \text{if } x < -1 \\ -\frac{8}{x} & \text{if } -1 \leq x < 2 \\ 2 - x^2 & \text{if } x \geq 2 \end{cases}$$

a. Find $f(-1)$.

b. Find $f(2)$.

c. Find $\lim_{x \to -1} f(x)$.

d. Find $\lim_{x \to 2} f(x)$.

e. Is $f(x)$ continuous at $x = -1$?

f. Is $f(x)$ continuous at $x = 2$?

7. For

$$f(x) = \begin{cases} x - 4 & \text{if } x < 1 \\ 6 & \text{if } x = 1 \\ 4 - x & \text{if } x > 1 \end{cases}$$

a. Find $f(1)$.

b. Find $\lim_{x\to 1} f(x)$.

c. Is $f(x)$ continuous at $x = 1$?

8. For

$$f(x) = \begin{cases} \frac{2}{x} & \text{if } x < 0 \\ -4 & \text{if } x = 0 \\ \frac{x}{2} & \text{if } x > 0 \end{cases}$$

a. Find $f(0)$.

b. Find $\lim_{x\to 0} f(x)$.

c. Is $f(x)$ continuous at $x = 0$?

9. For

$$f(x) = \begin{cases} x^2 & \text{if } x \neq 3 \\ 4 & \text{if } x = 3 \end{cases}$$

a. Find $f(3)$.

b. Find $\lim_{x\to 3} f(x)$.

c. Is $f(x)$ continuous at $x = 3$?

10. For

$$f(x) = \begin{cases} 8 - 3x & \text{if } x < 4 \\ 10 & \text{if } x = 4 \\ \dfrac{4}{x - 5} & \text{if } x > 4 \end{cases}$$

a. Find $f(4)$.

b. Find $\lim_{x\to 4} f(x)$.

c. Is $f(x)$ continuous at $x = 4$?

For each of the graphs in Exercises 11 through 20, state the point(s) of discontinuity and the requirement that is violated.

11.

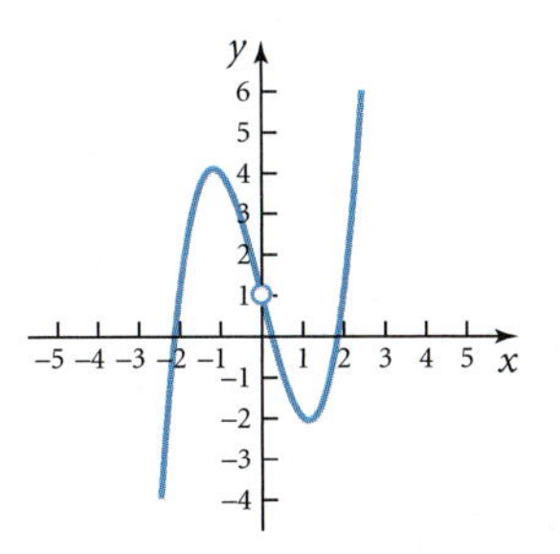

12.

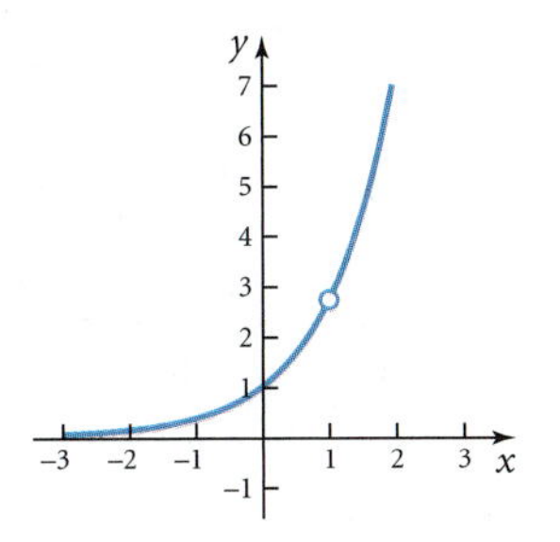

13.

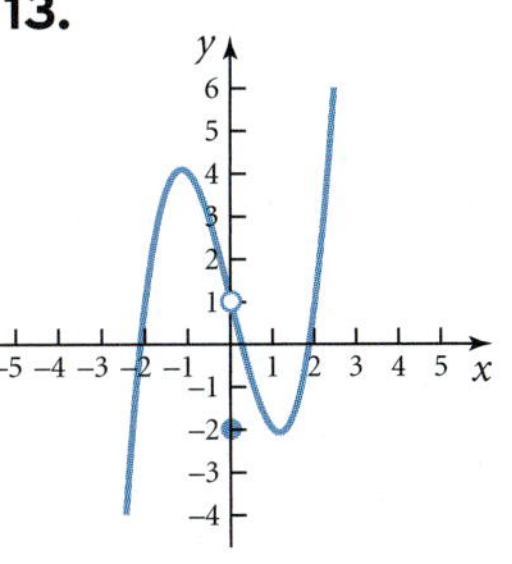

14.

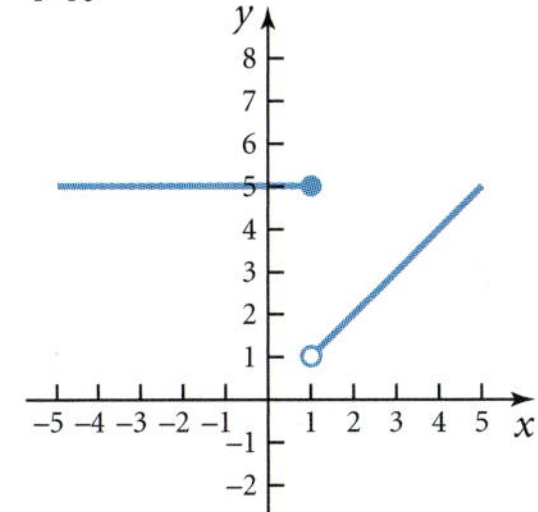

15.

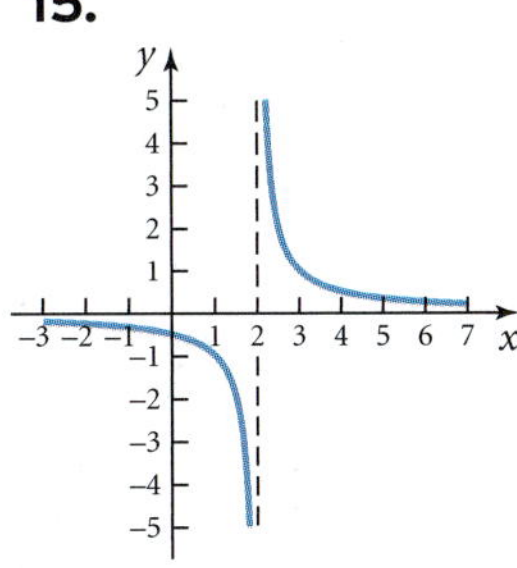

16.

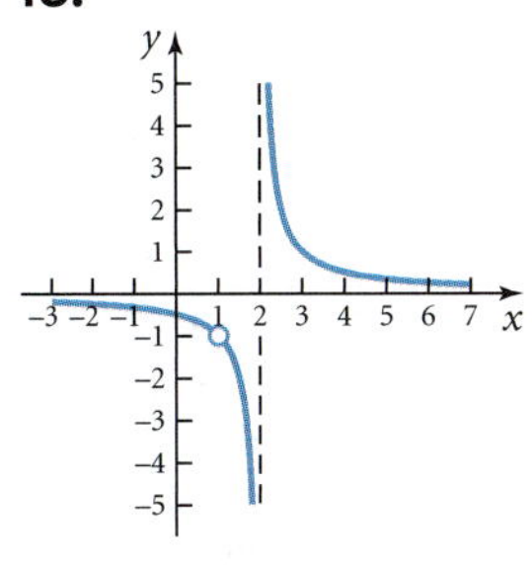

17.

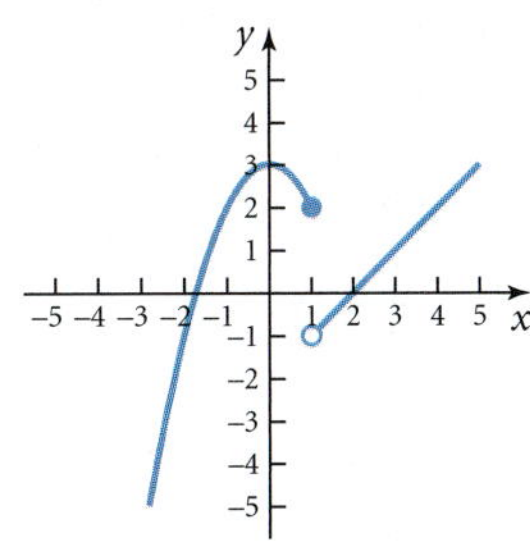

18.

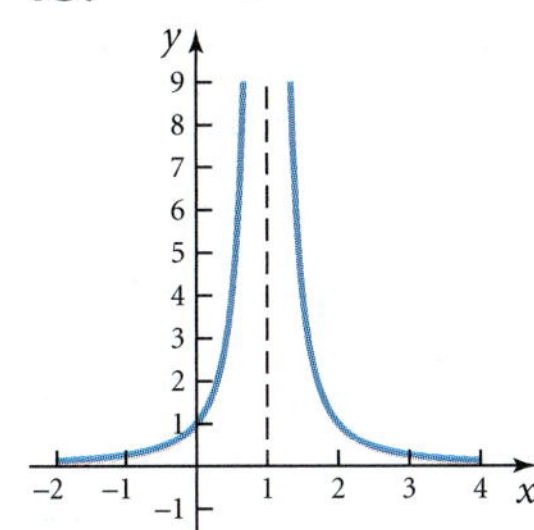

19.

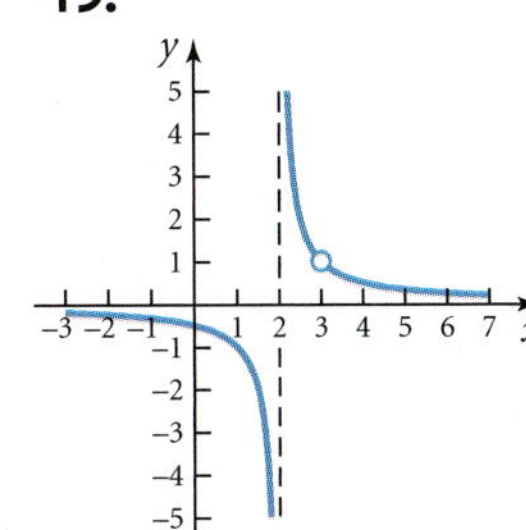

20.

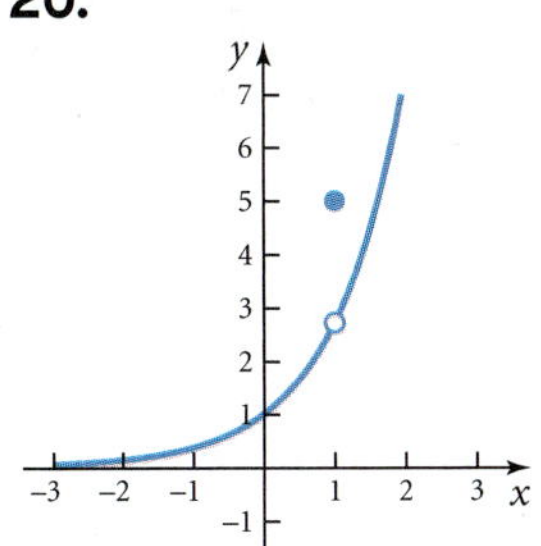

In Exercises 21 and 22, use the graph to determine if the statements are true or false.

21.

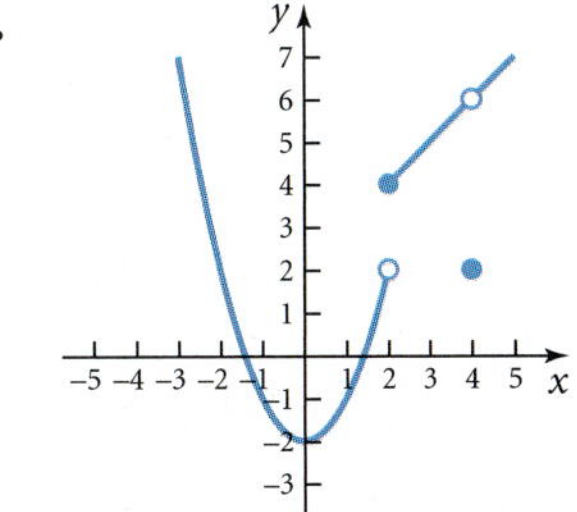

a. $\lim_{x\to4} f(x) = 2$
b. $\lim_{x\to2^-} f(x) = 2$
c. $\lim_{x\to2^+} f(x) = 4$
d. $\lim_{x\to2} f(x) = 4$
e. $f(4) = 2$
f. $\lim_{x\to0} f(x) = -2$
g. $f(2) = 4$
h. $f(x)$is continuous at $x = 2$
i. $\lim_{x\to0} f(x) = f(0)$
j. $f(x)$ is continuous at $x = 4$
k. $f(x)$ is continuous at $x = 0$

22.

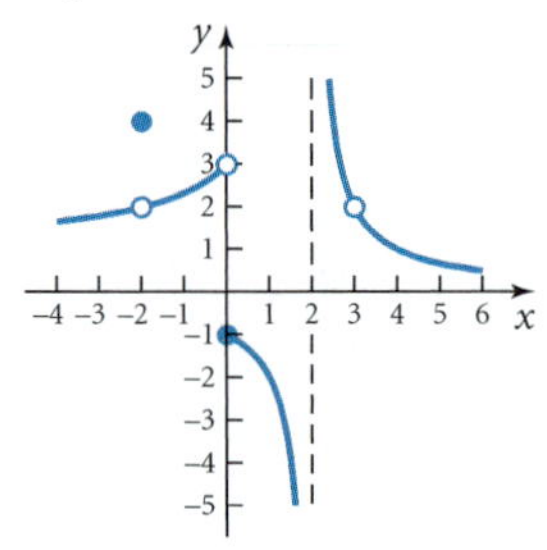

a. $\lim_{x\to3} f(x) = 2$
b. $\lim_{x\to0} f(x) = -1$
c. $\lim_{x\to0^+} f(x) = -1$
d. $\lim_{x\to-2} f(x) = 2$
e. $f(-2) = 4$
f. $f(0) = 3$
g. $\lim_{x\to0^-} f(x) = 3$
h. $\lim_{x\to-2} f(x) = f(-2)$
i. $f(x)$ is continuous at $x = 1$
j. $f(x)$ is continuous at $x = 3$
k. $f(x)$ is continuous at $x = -2$

In Exercises 23 through 30, use the graph of $f(x)$ shown below and the indicated value of a.

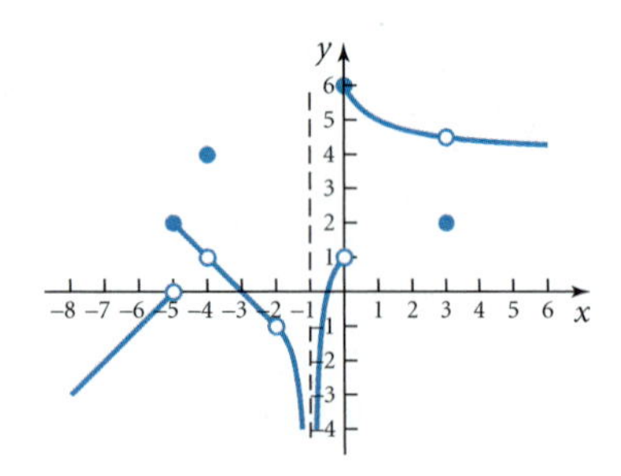

a. Find $\lim_{x\to a} f(x)$.
b. Find $f(a)$.
c. Is $f(x)$ continuous at $x = a$? If not, state the requirement violated.

23. $a = -4$
24. $a = -2$
25. $a = -1$
26. $a = 1$
27. $a = 0$
28. $a = 3$
29. $a = -3$
30. $a = -5$

For each of the following functions, state the interval(s) for which the function is continuous.

31. $f(x) = 2x^2 - 5x + 7$
32. $f(x) = \sqrt{3x + 12}$
33. $f(x) = 12x - 5$
34. $f(x) = 4x + 9$
35. $g(x) = \sqrt{4 - 8x}$
36. $h(t) = t^3 + 5t - 4$
37. $f(a) = \dfrac{6a}{a + 3}$
38. $T(c) = \dfrac{-2}{3c - 7}$
39. $f(x) = \dfrac{5x}{x^2 - 4}$
40. $f(x) = \dfrac{3}{x^2 - 9x}$
41. $f(x) = \dfrac{8x}{\sqrt{x - 2}}$
42. $f(x) = \dfrac{4x + 6}{\sqrt{7 - x}}$
43. $m(t) = \dfrac{3}{t^2 + 1}$
44. $P(h) = \dfrac{-4h}{h^3 - 8}$
45. $f(z) = \sqrt{z^2 - 16}$
46. $Q(a) = \sqrt{a^3 - 25a}$
47. $f(x) = \sqrt[3]{x + 5}$
48. $g(x) = \sqrt[4]{x + 5}$

In Exercises 49 through 52, select the letter of the correct answer.

49. $f(x)$ is discontinuous at $x = 2$ because

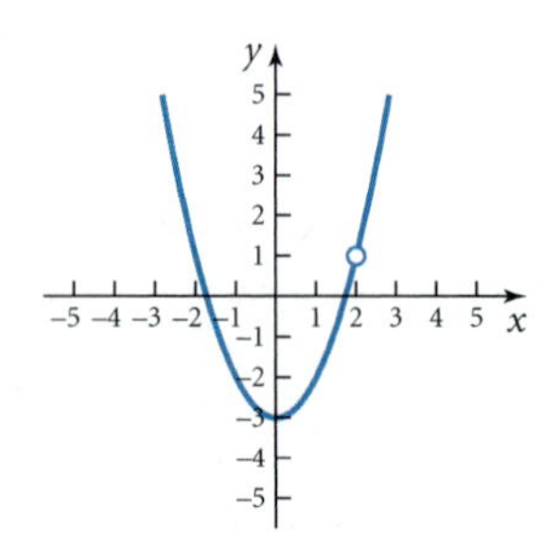

a. $f(2)$ is undefined.

b. $\lim_{x \to 2} f(x)$ does not exist.

c. $f(2) \neq \lim_{x \to 2} f(x)$.

d. None of these answers is correct.

50. $f(x)$ is discontinuous at $x = -1$ because

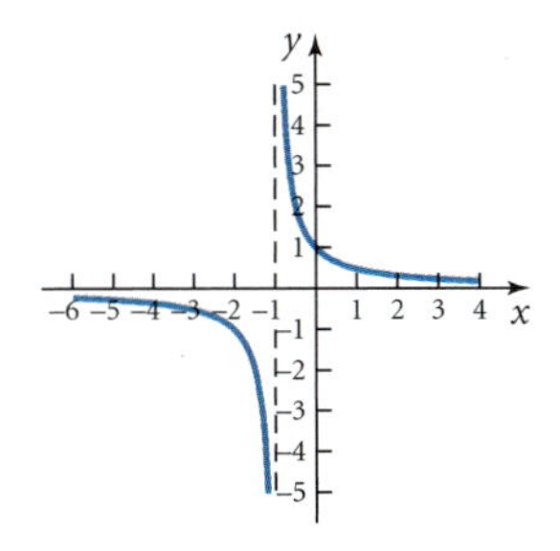

a. $f(-1)$ is undefined.

b. $\lim_{x \to -1} f(x)$ does not exist.

c. $f(-1) \neq \lim_{x \to -1} f(x)$.

d. None of these answers is correct.

51. $f(x)$ is discontinuous at $x = 2$ because

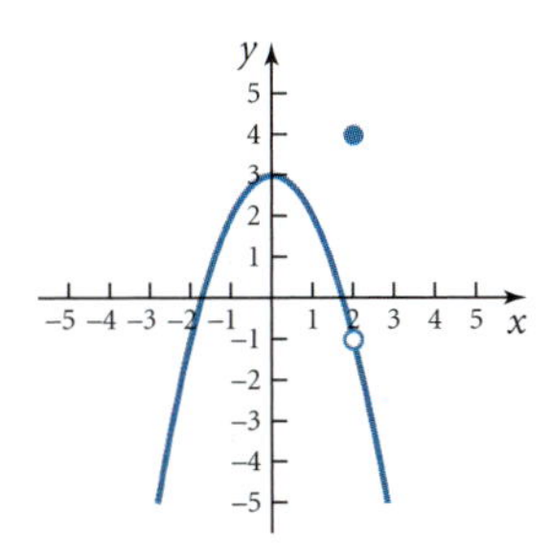

a. $f(2)$ is undefined.

b. $\lim_{x \to 2} f(x)$ does not exist.

c. $f(2) \neq \lim_{x \to 2} f(x)$.

d. None of these answers is correct.

52. $f(x)$ is discontinuous at $x = 3$ because

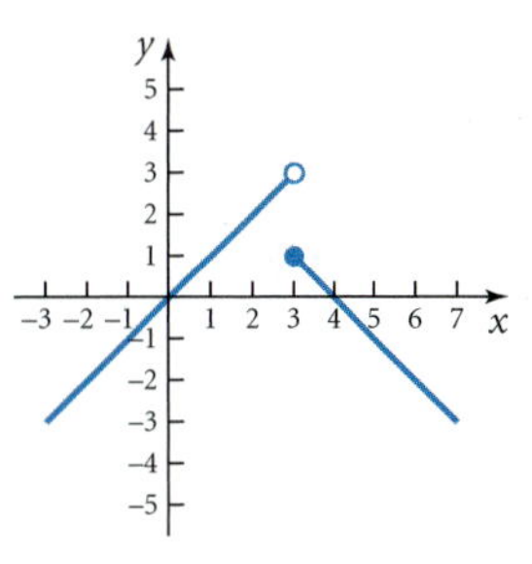

a. $f(3)$ is undefined.

b. $\lim_{x \to 3} f(x)$ does not exist.

c. $f(3) \neq \lim_{x \to 3} f(x)$.

d. None of these answers is correct.

For Exercises 53 and 54, use the following graph to select the letter of the correct answer.

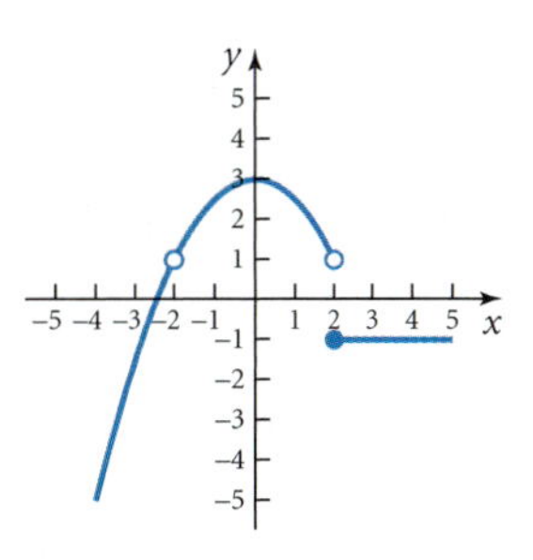

53. $f(x)$ is discontinuous at $x = 2$ because

a. $f(2)$ is undefined.

b. $\lim_{x \to 2} f(x)$ does not exist.

c. $f(2) \neq \lim_{x \to 2} f(x)$.

54. $f(x)$ is discontinuous at $x = -2$ because

a. $f(-2)$ is undefined.

b. $\lim_{x \to -2} f(x)$ does not exist.

c. $f(-2) \neq \lim_{x \to -2} f(x)$.

55. The number of students enrolled in a calculus class at the start of a given week in a term is shown in the following table.

Week	1	2	3	4	5	6	7	8	9	10	11	12	13	14	15
# Students	36	42	40	40	40	36	36	35	35	35	32	32	32	32	32

a. Draw a graph of the function.

b. Identify the points of discontinuity. What event might be occurring that would cause the discontinuity?

56. Package fees for parcel airlift are given in the following table.

Weight Not More Than ...	Fee
2 pounds	\$0.45
3 pounds	\$0.85
4 pounds	\$1.25
30 pounds	\$1.70

a. Draw a graph of the function.

b. Identify the points of discontinuity of the function.

COMMUNICATE

57. Briefly explain how continuity and limits are related.

58. Explain how you would find the points of discontinuity of a graph.

SHARPEN THE TOOLS

For what interval(s) are each of the following functions continuous?

59. $f(x) = x^{2.3} - 5x^{1.8} + 3.64$

60. $f(x) = \sqrt{x^3 - 4x}$

61. $f(x) = \dfrac{x + 3}{\sqrt{x^2 - 9x}}$

62. $f(x) = \dfrac{3x^{1.9}}{x^{2.4} - 4x^{1.2} + 3}$

TOPIC 7

Rates of Change and Slope

A car company recently announced that its sales had increased 26% during the past quarter. Retail stores stated that holiday sales had decreased 7% this year. The National Endowment for the Arts reported that the reading rate of young people is decreasing. The idea of **rates of change** is central to the development of calculus and is the focus of this topic.

Before beginning the study of rates of change, though, be sure to complete the Topic 7 Warm-up Exercises to brush up on the algebra and calculus skills you will need.

IMPORTANT TOOLS IN TOPIC 7

- *Average rate of change*
- *Instantaneous rate of change*
- *Slope of a curve at a point*
- *Slope of a curve at any point*
- *Velocity of an object in motion*

TOPIC 7 WARM-UP EXERCISES

Be sure you can successfully complete the following exercises before starting Topic 7.

Algebra Warm-up Exercises

1. Given $f(x) = x^2 - 3x + 1$, find the following values.

a. $f(3)$ **b.** $f(-2)$ **c.** $f(a)$

d. $f(a + h)$ **e.** $\dfrac{f(a + h) - f(a)}{h}$

2. Give the conjugate of $5 + \sqrt{3}$.

3. Multiply $\left(\sqrt{x + 2} + \sqrt{x}\right)\left(\sqrt{x + 2} - \sqrt{x}\right)$.

4. Rationalize the denominator: $\dfrac{6}{5 + \sqrt{3}}$.

5. Find the slope of the line between the points $(1, -1)$ and $(3, 4)$.

6. Find the equation of the line between $(1, -1)$ and $(3, 4)$.

Calculus Warm-up Exercises

Evaluate each limit, if it exists.

1. $\lim_{h\to 0} (3h - 5)$ **2.** $\lim_{h\to 0} \dfrac{3ah - h^2}{h}$

Given $f(x) = -2x^2 + 6x$, find the following values.

3. $f(a)$ **4.** $f(a + h)$

5. $\dfrac{f(a + h) - f(a)}{h}$ **6.** $\lim_{h\to 0} \dfrac{f(a + h) - f(a)}{h}$

Answers to Warm-up Exercises

Algebra Warm-up Exercises

1. a. 1 **b.** 11 **c.** $a^2 - 3a + 1$

d. $a^2 + 2ah + h^2 - 3a - 3h + 1$

e. $2a + h - 3$

2. $5 - \sqrt{3}$ **3.** 2 **4.** $\dfrac{15 - 3\sqrt{3}}{11}$

5. $\dfrac{5}{2}$ **6.** $5x - 2y - 7 = 0$ or $y = \dfrac{5}{2}x - \dfrac{7}{2}$

Calculus Warm-up Exercises

1. -5 **2.** $3a$ **3.** $-2a^2 + 6a$

4. $-2a^2 - 4ah - 2h^2 + 6a + 6h$

5. $-4a - 2h + 6$ **6.** $-4a + 6$

alculus is the study of rates of change for variable quantities. The next major tool in calculus, the **derivative**, is used to measure rates of change of functions and determine slopes of curves. Before defining the derivative, we first discuss rates of change.

Average Rates of Change

The **average rate of change** of a quantity over an interval is given by the following ratio.

$$\textbf{Average rate of change} = \frac{\text{amount of change in the quantity over the interval}}{\text{amount of change in the interval}}$$

Consider the Orlando Magic average home attendance data presented in the introduction to Unit 1. What was the average rate of change of home attendance between the 1999–2000 season and the 2001–2002 season?

$$\begin{aligned}\text{average rate of change} &= \frac{\text{change in attendance}}{\text{change in time}}\\ &= \frac{15{,}149 - 14{,}059}{2001 - 1999}\\ &= \frac{1090 \text{ people}}{2 \text{ years}}\\ &= 545 \text{ people per year}\end{aligned}$$

Season	Average Attendance
1995–1996	17,248
1996–1997	17,199
1997–1998	17,113
1998–1999	16,444
1999–2000	14,059
2000–2001	14,757
2001–2002	15,149
2002–2003	14,545
2003–2004	14,352
2004–2005	14,507

Source: Orlando Sentinel.

Because the average rate of change is **positive**, the average home attendance **increased** at a rate of 545 people per year between 1999 and 2001.

What was the average rate of change of home attendance between the 1995–1996 season and the 1998–1999 season?

$$\begin{aligned}\text{average rate of change} &= \frac{\text{change in attendance}}{\text{change in time}}\\ &= \frac{16{,}444 - 17{,}248}{1998 - 1995}\\ &= \frac{-804 \text{ people}}{3 \text{ years}}\\ &= -268 \text{ people per year}\end{aligned}$$

Because the average rate of change is **negative**, the average home attendance **decreased** at a rate of 268 people per year between the 1995 and the 1998 seasons.

Plot the data to see graphically the rates of change just calculated. Using the first year of the season interval for plotting purposes, draw the lines between the points representing the two time frames, as shown in Figure 7.1.

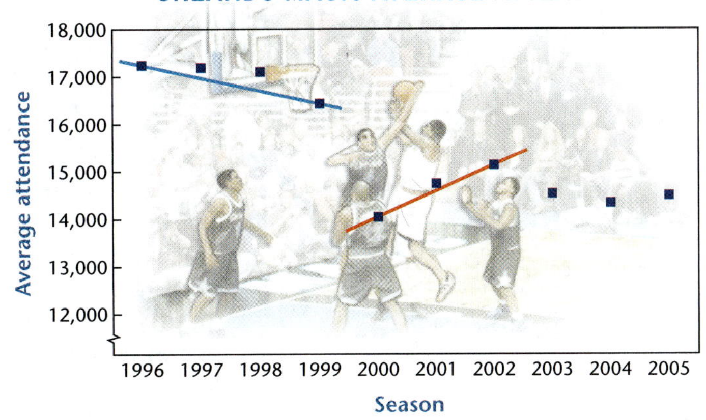

Figure 7.1

The red line is the line between the two points representing the 1999 and 2001 seasons. This line rises from left to right, indicating a *positive* slope. The average rate of change during this time period showed an *increase* in attendance. The blue line is the line between the two points representing the 1995 and 1998 seasons. This line falls from left to right, indicating a *negative* slope. The average rate of change during this time period showed a *decrease* in attendance.

The average rate of change in each instance is the slope of the line between the two indicated points. These lines are called **secant lines** because they pass through two points of the graph. The average rate of change corresponds to the slope of the secant line between the two points.

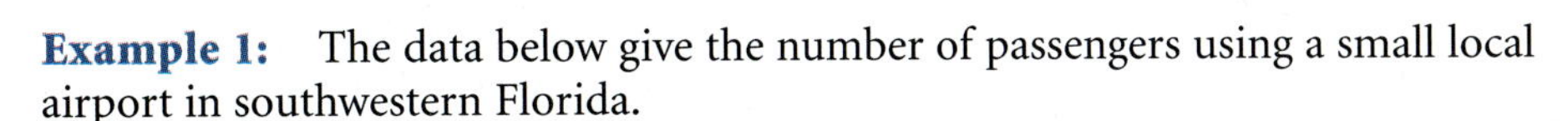

Example 1: The data below give the number of passengers using a small local airport in southwestern Florida.

a. What is the rate of change in number of passengers between 1995 and 1997?

b. What is the rate of change in number of passengers between 2000 and 2003?

Year	Number of Passengers
1994	8,386
1995	27,017
1996	46,028
1997	64,241
1998	62,033
1999	48,994
2000	57,396
2001	39,716
2002	40,156
2003	45,115

Source: Fort Myers News-Press.

Solution:

a.
$$\text{average rate of change} = \frac{\text{change in number of passengers}}{\text{change in time}}$$
$$= \frac{64{,}241 - 27{,}017}{1997 - 1995}$$
$$= \frac{37{,}224 \text{ passengers}}{2 \text{ years}}$$
$$= 18{,}612 \text{ passengers per year}$$

The number of passengers increased by 18,612 passengers per year between 1995 and 1997.

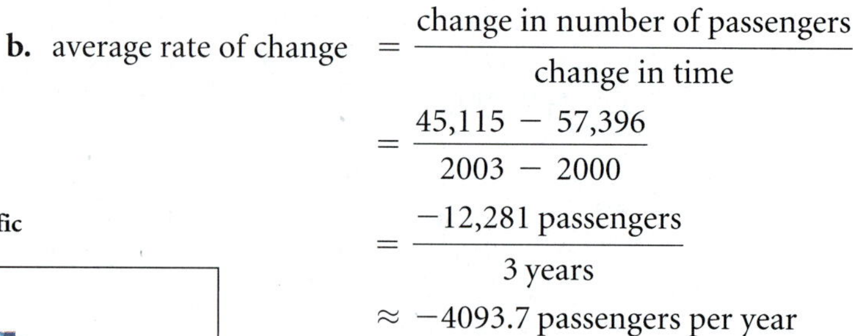

b.
$$\text{average rate of change} = \frac{\text{change in number of passengers}}{\text{change in time}} = \frac{45{,}115 - 57{,}396}{2003 - 2000} = \frac{-12{,}281 \text{ passengers}}{3 \text{ years}} \approx -4093.7 \text{ passengers per year}$$

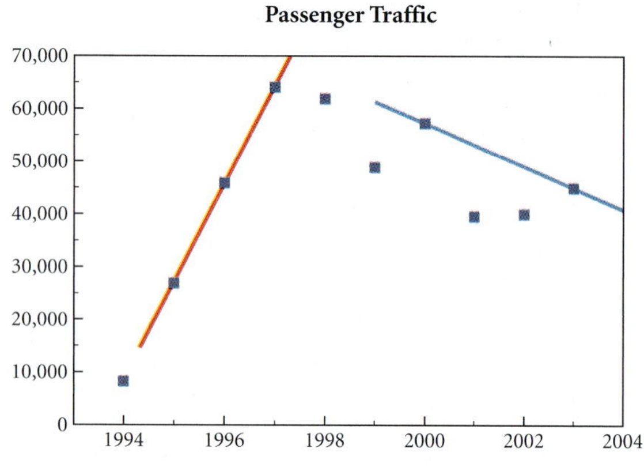

Figure 7.2

The number of passengers decreased by approximately 4094 passengers per year between 2000 and 2003.

Plotting the data shows the rates of change just calculated. (See Figure 7.2.) The red line is the line between the two points representing 1995 and 1997. The line has a positive slope, meaning that the number of passengers increased, on average, between these two years. The blue line is the line between the two points representing 2000 and 2003. This line has a negative slope, showing that the number of passengers decreased, on average, between these two years. ■

Suppose the quantity being analyzed is described by a function, $f(x)$. Then the average rate of change of the function over the interval $[a, b]$ is given by the following ratio.

The **average rate of change** of a function $f(x)$ between points $(a, f(a))$ and $(b, f(b))$ is given by $\dfrac{f(b) - f(a)}{b - a}$.

Tip: Do you see that the ratio used to calculate average rate of change is the same as the one used to calculate slope? The numerator, $f(b) - f(a)$, is the difference in two y values, and the denominator, $b - a$, is the difference in two x values.

Example 2: According to the 2003 *World Almanac*, the number of new cases of AIDS reported each year in the United States can be modeled by $N(t) = -680t^2 + 19{,}307t - 74{,}310$, where t is the number of years after 1980.

a. Find the average rate of change in the number of new AIDS cases reported between 1992 and 1995.

b. Find the average rate of change in the number of new cases reported between 1998 and 2000.

Solution:

a. The year 1992 is 12 years after 1980, and 1995 is 15 years after 1980, so the average rate of change between those years is

$$\text{average rate of change} = \frac{N(15) - N(12)}{15 - 12} = \frac{62{,}295 - 59{,}454}{15 - 12}$$

$$= \frac{2841 \text{ cases}}{3 \text{ years}}$$
$$= 947 \text{ cases per year}$$

The number of new AIDS cases reported in the United States between 1992 and 1995 increased by an average of 947 cases per year.

b. The year 1998 is 18 years after 1980, and 2000 is 20 years after 1980, so the average rate of change between those years is

$$\text{average rate of change} = \frac{N(20) - N(18)}{20 - 18}$$
$$= \frac{39{,}830 - 52{,}896}{20 - 18}$$
$$= \frac{-13{,}066 \text{ cases}}{2 \text{ years}}$$
$$= -6533 \text{ cases per year}$$

The number of new AIDS cases reported in the United States between 1998 and 2000 decreased by an average of 6533 cases per year.

The graph in Figure 7.3 shows the function and the two lines representing the average rates of change calculated for this example. The year 1980 is represented on the horizontal axis as zero. ■

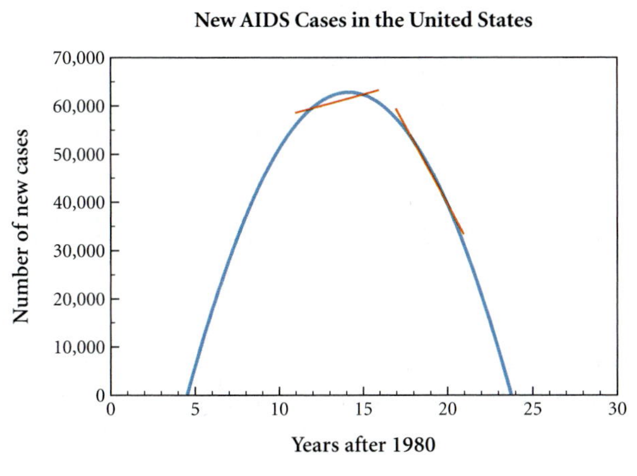

Figure 7.3

Data from: Center for Disease Control

Projectile Motion

Many of the functions studied in calculus—such as the path of a baseball after being hit into the air, a rocket after being launched, or an object after being thrown or dropped—involve **projectile motion**. The path of such objects in motion is parabolic, and the object's height above the ground after t seconds is given by $h(t) = at^2 + vt + c$. The quadratic coefficient, a, is negative and accounts for the pull of gravity causing the object to return to the ground. The linear coefficient, v, is the rate at which the object was put into motion and is referred to as the **initial velocity**. The constant term, c, represents the **initial height** of the object.

Example 3: The height of a rock t seconds after being thrown upward into the air from a platform 50 feet above the ground is given by the function $h(t) = -16t^2 + 40t + 50$, where $h(t)$ is the height above the ground in feet. Find the average rate of change in the height of the rock between

a. 0.5 second and 1 second **b.** 2 seconds and 3 seconds

Solution:

a. $$\text{average rate of change in height} = \frac{\text{change in height}}{\text{change in time}}$$
$$= \frac{h(1) - h(0.5)}{1 - 0.5}$$

$$= \frac{74 - 66}{0.5}$$

$$= \frac{8}{0.5}$$

$$= 16 \text{ feet per second}$$

The height of the rock between 0.5 second and 1 second after release increased by 16 feet per second, on average. In other words, the rock's height above the ground was increasing, or rising, at an average rate of 16 feet per second.

b. average rate of change in height $= \dfrac{\text{change in height}}{\text{change in time}}$

$$= \frac{h(3) - h(2)}{3 - 2}$$

$$= \frac{26 - 66}{1}$$

$$= -40 \text{ feet per second}$$

The height of the rock between 2 seconds and 3 seconds after release decreased by 40 feet per second, on average. In other words, the rock's height above the ground was decreasing, or falling, at an average rate of 40 feet per second. ■

Instantaneous Rate of Change

Now let's change the question slightly. Suppose we want to know the rate at which the rock in Example 3 is moving at the exact instant 2 seconds after its release. Average rate of change is calculated as *change in height/change in time*, but this new question does not provide two specific points with which to determine the actual change in height and the actual change in time. The point for which the rate of change is desired is the point $(2, f(2))$. (See Figure 7.4.) Create a second point close to this point and designate it as $(2 + h, f(2 + h))$. Then $f(2)$ represents the height of the rock at 2 seconds and $f(2 + h)$ represents the height of the rock at some time very close to 2 seconds. The parameter h represents the difference between the two points, or the time elapsed.

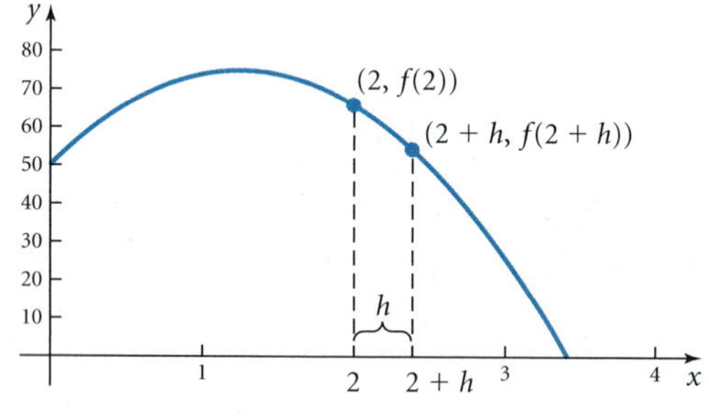

Figure 7.4

Given $f(t) = -16t^2 + 40t + 50$ representing the height of the rock at any time t, complete the following table for times that approach $t = 2$ seconds.

$t + h$	Average Rate of Change
2.01	$\dfrac{f(2.01) - f(2)}{2.01 - 2} = \dfrac{65.7584 - 66}{0.01} = -24.16$
2.001	$\dfrac{f(2.001) - f(2)}{2.001 - 2} = \dfrac{65.975984 - 66}{0.001} = -24.016$
2.0001	$\dfrac{f(2.0001) - f(2)}{2.0001 - 2} = \dfrac{65.99759984 - 66}{0.0001} = -24.0016$

From the table, can you guess the instantaneous rate of change at $t = 2$? It appears that the closer t gets to two seconds, the average rate of change is approaching -24 feet per second. Let's see how to find the instantaneous rate of change algebraically.

The height of the rock at any time was given by

$$f(t) = -16t^2 + 40t + 50$$

Thus,

$$\begin{aligned} f(2) &= -16(2)^2 + 40(2) + 50 \quad \text{and} \\ f(2+h) &= -16(2+h)^2 + 40(2+h) + 50 \end{aligned}$$

Then the average rate of change is given as

$$\begin{aligned} \text{average rate of change} &= \frac{\text{change in height}}{\text{change in time}} \\ &= \frac{f(2+h) - f(2)}{(2+h) - 2} \\ &= \frac{[-16(2+h)^2 + 40(2+h) + 50] - [-16(2)^2 + 40(2) + 50]}{h} \\ &= \frac{[-64 - 64h - 16h^2 + 80 + 40h + 50] - [-64 + 80 + 50]}{h} \\ &= \frac{(-16h^2 - 24h + 66) - 66}{h} \\ &= \frac{-16h^2 - 24h}{h} \\ &= \frac{h(-16h - 24)}{h} \\ &= -16h - 24 \text{ feet per second} \end{aligned}$$

To determine the rate of change at the exact instant 2 seconds after release, we need to know how the average rate of change will behave as the point where $x = 2 + h$ gets closer and closer to the point where $x = 2$. If the two points are getting closer together, the distance between them, denoted by h, is getting smaller and smaller and approaching 0. In other words, *to find the rate of change at the instant when $x = 2$, we need to find the limit of the average rate of change as $h \to 0$*. Thus, the rate of change at the exact instant 2 seconds after release is given by

$$\lim_{h \to 0} (-16h - 24) = -24 \text{ feet per second}$$

This rate is called the **instantaneous rate of change**. Generalizing this result gives a more formal definition.

Definition: The **instantaneous rate of change** of $f(x)$ when $x = a$ is given by

$$\lim_{h \to 0} \frac{f(a+h) - f(a)}{h}$$

For an object in motion, the instantaneous rate of change of the object's position is called the **velocity** of the object. Positive velocity indicates the rate at which an object rises; negative velocity indicates the rate at which an object falls. The **speed** of an object represents only how fast it is moving, not its direction, and is simply $|velocity|$.

Example 4: What is the instantaneous rate of change—the velocity—of the rock in Example 3 one second after its release?

Solution: The height of the rock is given by $f(t) = -16t^2 + 40t + 50$. In this case, $a = 1$ second, so the instantaneous velocity of the rock after 1 second is found by

$$\begin{aligned}
\text{instantaneous rate of change} &= \lim_{h\to 0}\frac{f(1+h)-f(1)}{h} \\
&= \lim_{h\to 0}\frac{[-16(1+h)^2+40(1+h)+50]-[-16(1)^2+40(1)+50]}{(1+h)-1} \\
&= \lim_{h\to 0}\frac{(-16h^2+8h+74)-74}{h} \\
&= \lim_{h\to 0}\frac{-16h^2+8h}{h} \\
&= \lim_{h\to 0}\frac{h(-16h+8)}{h} \\
&= \lim_{h\to 0}(-16h+8) \\
&= 8 \text{ feet per second}
\end{aligned}$$

After one second, the rock is rising at a rate of 8 feet per second. ■

Example 5: According to the 2003 *World Almanac*, annual unemployment rates since 1980 can be modeled by the function $E(t) = 0.005t^2 - 0.2t + 7.5$, where $E(t)$ is the percentage of the workforce that is unemployed and t is the number of years after 1980.

a. What was the unemployment rate in 1990?

b. Find the instantaneous rate of change in the employment rate in 1990 and interpret the answer.

c. Use the answers from parts a and b to predict the unemployment rate in 1991. Check the estimate using the model.

Solution:

a. The year 1990 is 10 years after 1980, so the unemployment rate in 1990 is $E(10) = 6$ or 6%.

b. The instantaneous rate of change in 1990 is given by

$$\begin{aligned}
\lim_{h\to 0}\frac{E(10+h)-E(10)}{(10+h)-10} &= \lim_{h\to 0}\frac{[0.005(10+h)^2-0.2(10+h)+7.5]-[0.005(10)^2-0.2(10)+7.5]}{h} \\
&= \lim_{h\to 0}\frac{(0.5+0.1h+0.005h^2-2-0.2h+7.5)-(0.5-2+7.5)}{h}
\end{aligned}$$

$$= \lim_{h \to 0} \frac{(0.005h^2 - 0.1h + 6) - 6}{h}$$

$$= \lim_{h \to 0} \frac{0.005h^2 - 0.1h}{h}$$

$$= \lim_{h \to 0} (0.005h - 0.1)$$

$$= -0.1$$

In 1990, the unemployment rate was dropping by an average of 0.1% per year.

c. In 1990, the unemployment rate was 6% (part a) and the rate was dropping by 0.1% per year (part b). The unemployment rate in 1991 is the unemployment rate in 1990 plus the predicted change. In 1991 the unemployment rate can be estimated as

$$6 - 0.1 = 5.9\%$$

Using the model, the unemployment rate in 1991 is $E(11) = 5.905\%$. ■

MATHEMATICS CORNER 7.1

Actual Change in y versus Predicted Change in y

Why did the predicted value of the unemployment rate for 1991 (5.9%) derived in Example 5 using the instantaneous rate of change not equal the actual value of the function $E(11) = 5.905\%$? The answer has to do with the linear property of the prediction and the nonlinear property of the function.

The **actual change** in the unemployment rate from 1990 to 1991 is based on the actual function and is given by $\Delta y = E(11) - E(10) = 5.905 - 6 = -0.095\%$.

The **predicted change** in the unemployment rate from 1990 to 1991 is given by the instantaneous rate of change in 1990, calculated by

$$dy = \lim_{h \to 0} \frac{E(10 + h) - E(10)}{h} = -0.1.$$

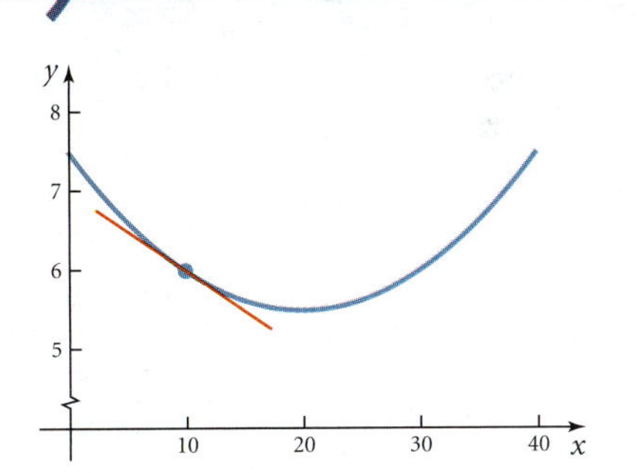

The graph above shows the graph of the unemployment rate data along with the graph of the tangent line at the point representing 1990. The actual change in the unemployment rate is the difference in the actual points on the graph. The predicated change is based on the points on the tangent line, not on the actual graph.

Check Your Understanding 7.1

1. The temperatures during one day in May in a midwestern city were as follows:

Time of Day	Temperature (°F)
7 a.m.	49
8 a.m.	58
9 a.m.	66
10 a.m.	72
11 a.m.	76
12 noon	79
1 p.m.	80
2 p.m.	81
3 p.m.	78
4 p.m.	74
5 p.m.	69
6 p.m.	63

Find the average rate of change in temperature between the following times of day.

a. 7 a.m. and 9 a.m.

b. 2 p.m. and 6 p.m.

2. A rock thrown in the air follows the path $h(t) = -16t^2 + 32t + 10$, where $h(t)$ is height above the ground in feet and t is the number of seconds after the rock was thrown. Find the average rate of change in the height of the rock between the following times.

a. 0.5 second and 1 second

b. 1 second and 2 seconds

3. For the rock in Question 2, find the instantaneous rate of change in the height of the rock after two seconds. Interpret your answer.

4. Use the unemployment rate model in Example 5 to answer the following questions.

a. What was the unemployment rate in 2001?

b. Find the instantaneous rate of change in the unemployment rate in 2001.

c. Use the answers from parts a and b to predict the unemployment rate in 2002. Use the model to check the answer.

Slopes of Curves

One question that led to the development of the calculus was how to find the slope of a curve. The answer is very closely related to rates of change.

> **Definition:** The **slope of a curve** at a point on the curve is the slope of the tangent line to the curve at that point.

The word *tangent* is derived from the Latin word *tangere*, meaning "touching." In geometry, a **tangent line to a circle**, as shown in Figure 7.5, is a line that touches the circle only once.

For more complicated curves, however, this definition does not suffice. For example, in Figure 7.6, the line at point A is a tangent line even though its extension intersects the curve again in a second point. The line at point B is not a tangent line even though it intersects the curve only once. A **tangent line to a function at a point** can touch the curve only once in an interval around that point. A tangent line does not usually pass through the curve at that point.

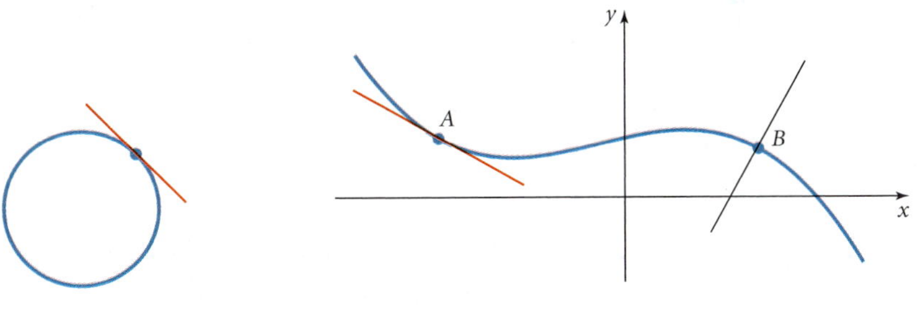

Figure 7.5 Figure 7.6

Tangent Lines That Pass Through a Curve

At times, a tangent line to a curve at a point will pass through the curve at that point. Consider, for instance, the curve shown here, which has a vertical tangent line at $x = 2$.

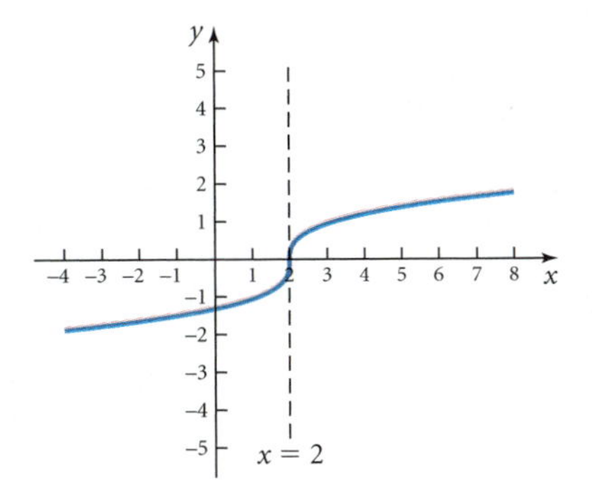

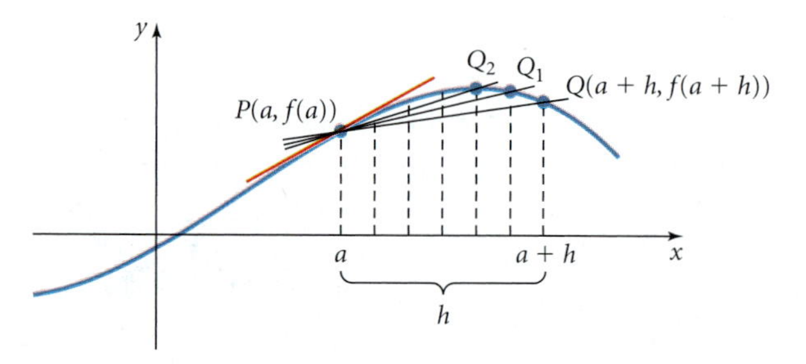

Figure 7.7

With that background, let's return to slopes of curves. Consider the function shown in Figure 7.7. The slope of $f(x)$ at the point $P(a, f(a))$ is the slope of the tangent line to $f(x)$ at P, which is shown in red on the graph. Determining the slope of a line requires two points, so locate another point on the curve that is close to $P(a, f(a))$. Call this point $Q(a + h, f(a + h))$, where h is the horizontal distance between the x values of the two points.

Next find the slope of the secant line between points P and Q. The slope of this line is

$$m = \frac{y_2 - y_1}{x_2 - x_1} \quad \text{or} \quad m = \frac{f(a + h) - f(a)}{(a + h) - a}$$

$$= \frac{f(a + h) - f(a)}{h}$$

This expression should look quite familiar because it is the same expression we used to find the average rate of change of a function at a specific point.

The slope of a curve, however, requires the slope of the tangent line, not the secant line. Move point Q closer to point P and call this new point Q_1. Draw the secant line PQ_1. The slope of the secant line is now a little closer to the slope of the tangent line. Move point Q_1 closer to point P and call this new point Q_2. The slope of the secant line PQ_2 is even closer to the slope of the tangent line. You should also see that as point Q moves closer and closer to point P, the distance between their x values, denoted by h, gets smaller and smaller, approaching a value of 0. So, *to find the slope of the tangent line, we need to find the limit of the slope of the secant line as* $h \to 0$.

Definition: The **slope of the curve $f(x)$ at the point $(a, f(a))$** is given by

$$m = \lim_{h \to 0} \frac{f(a + h) - f(a)}{h}$$

This definition is exactly the same as the definitions of instantaneous rate of change and velocity.

Example 6: Find the slope of the parabola $f(x) = x^2$ at the point (2, 4).

Solution: In this example, $f(x) = x^2$ and $a = 2$, so $f(a) = f(2)$ and $f(a + h) = f(2 + h)$. Thus,

$$m = \lim_{h \to 0} \frac{f(a + h) - f(a)}{(a + h) - a} = \lim_{h \to 0} \frac{f(2 + h) - f(2)}{(2 + h) - 2}$$

$$= \lim_{h \to 0} \frac{(2 + h)^2 - 2^2}{h}$$

$$= \lim_{h \to 0} \frac{(4 + 4h + h^2) - 4}{h}$$

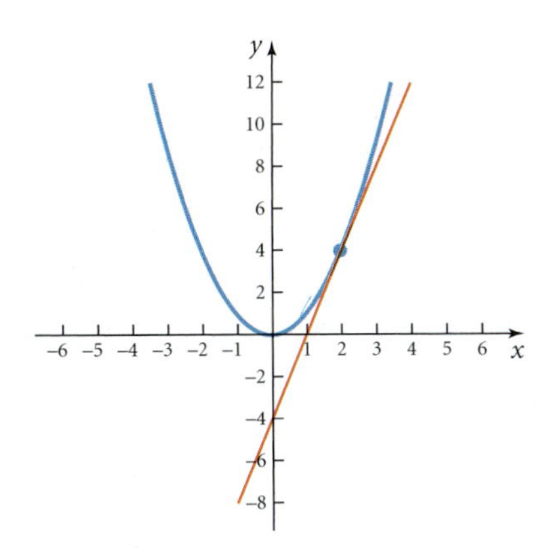

Figure 7.8

$$= \lim_{h\to 0} \frac{4h + h^2}{h}$$

$$= \lim_{h\to 0} \frac{h(4 + h)}{h}$$

$$= \lim_{h\to 0} (4 + h)$$

$$= 4$$

Thus, the slope of the parabola $f(x) = x^2$ at the point $(2, 4)$ is 4.

The slope of the graph $f(x) = x^2$ at $(2, 4)$ is the slope of the tangent line to the graph at the point. Figure 7.8 shows both the function and its tangent line. A slope of 4 seems quite reasonable for this line. ■

Example 7: Find the slope of the parabola $f(x) = x^2$ at the point where $x = -1$.

Solution: Changing the point of tangency on the curve changes the value of a in the definition. Reevaluate the definition using $a = -1$:

$$m = \lim_{h\to 0} \frac{f(-1 + h) - f(-1)}{h}$$

$$= \lim_{h\to 0} \frac{(-1 + h)^2 - (-1)^2}{h}$$

$$= \lim_{h\to 0} \frac{-2h + h^2}{h}$$

$$= \lim_{h\to 0} (-2 + h)$$

$$= -2$$

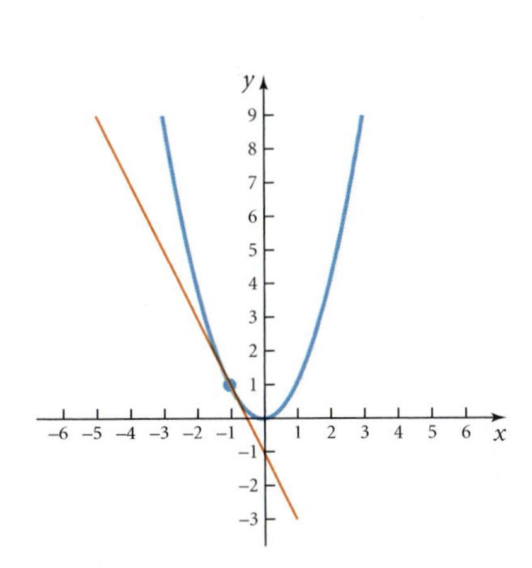

Figure 7.9

The slope of the parabola $f(x) = x^2$ at the point where $x = -1$ is -2.

The slope of the graph $f(x) = x^2$ at $(-1, 1)$ is the slope of the tangent line to the graph at the point. Figure 7.9 shows both the function and its tangent line. A slope of -2 seems quite reasonable for this line. ■

Example 8: Find the equation of the tangent line to $f(x) = x^2 - 3x$ at the point where $x = 2$.

Solution: If we know the slope of a line and a point on the line, the equation of the line is given by the point–slope form $y - y_1 = m(x - x_1)$, where m is the slope and (x_1, y_1) is a point on the line.

The slope of the tangent line is the slope of the curve and is found by

$$m = \lim_{h\to 0} \frac{f(2 + h) - f(2)}{h}$$

$$= \lim_{h\to 0} \frac{[(2 + h)^2 - 3(2 + h)] - [2^2 - 3(2)]}{h}$$

$$= \lim_{h\to 0} \frac{(4 + 4h + h^2 - 6 - 3h) - (-2)}{h}$$

$$= \lim_{h \to 0} \frac{h^2 + h}{h}$$

$$= \lim_{h \to 0} (h + 1)$$

$$= 1$$

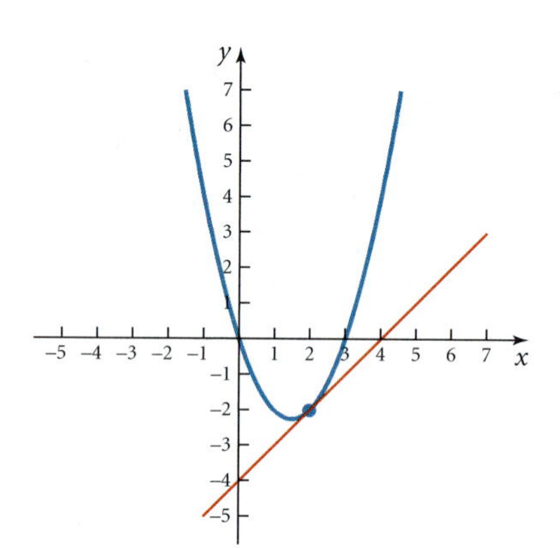

Figure 7.10

At $x = 2$, the slope of the tangent line is 1.

To find the equation of the tangent line, we must also know the y-coordinate of the point where $x = 2$:

$$y = f(2) = -2$$

So the point of tangency is $(2, -2)$.

The equation of the tangent line is given by

$$y - y_1 = m(x - x_1)$$

where $m = 1$ and $(x_1, y_1) = (2, -2)$, so

$$y - (-2) = 1(x - 2) \text{ or } y = x - 4.$$

Figure 7.10 verifies that the line is tangent to the curve at $x = 2$. ■

Tip: To find the equation of the tangent line to a function $f(x)$ at the point where $x = a$, use the point–slope form for the equation of a line, $y - y_1 = m(x - x_1)$, with the following substitutions:

x_1 is the given $x = a$, $y_1 = f(x_1)$, and $m = \lim_{h \to 0} \frac{f(x_1 + h) - f(x_1)}{h}$.

Check Your Understanding 7.2

1. Find the slope of $f(x) = 2x^2 - 3$ at the indicated points.

a. $(1, f(1))$ **b.** $(0, f(0))$ **c.** $(-2, f(-2))$

2. Find the equation of the tangent line to $f(x) = 2x^2 - 3$ at the point where

a. $x = 1$ **b.** $x = 0$ **c.** $x = -2$

Slopes of Curves at Any Point

The slope of a curve at a specific point is the slope of the tangent line to the curve at that point. The slope of the tangent line is the instantaneous rate of change of the function at that point. The only drawback to the definition so far is that changing the point of tangency requires reworking the entire limit problem. It would be easier if there were a way to find slopes of curves at different points and only work the limit part of the problem once. Fortunately, there is such a way! The slope of a curve definition involves the point $(a, f(a))$. Because a represents any x value in the domain of the function, we can replace the a in the definition by x. Then we have a generic definition that works no matter where the point of tangency is located.

Definition: The **slope of a curve $f(x)$ at any point** x is given by

$$m = \lim_{h \to 0} \frac{f(x+h) - f(x)}{h}.$$

Example 9: Find the slope of the parabola $f(x) = x^2$ at any point.

$$\begin{aligned} m &= \lim_{h \to 0} \frac{f(x+h) - f(x)}{h} \\ &= \lim_{h \to 0} \frac{(x+h)^2 - x^2}{h} \\ &= \lim_{h \to 0} \frac{(x^2 + 2xh + h^2) - x^2}{h} \\ &= \lim_{h \to 0} \frac{2xh + h^2}{h} \\ &= \lim_{h \to 0} (2x + h) \\ &= 2x \end{aligned}$$

The slope of the parabola $f(x) = x^2$ at *any* point is $m = 2x$. The slope at the point where $x = 2$ is $m = 2(2) = 4$ (see Example 6). The slope at the point where $x = -1$ is $m = 2(-1) = -2$ (see Example 7). ■

Example 10: What is the slope of $f(x) = x^2$ at the point where $x = 3$? Where $x = -4$?

Solution: The slope of $f(x) = x^2$ at any point is $m = 2x$.
At $x = 3$, $m = 2(3) = 6$. At $x = -4$, $m = 2(-4) = -8$. ■

Example 11: Find the slope of the curve $f(x) = 1/x$ at any point.

Solution:

$$\begin{aligned} m &= \lim_{h \to 0} \frac{f(x+h) - f(x)}{h} \\ &= \lim_{h \to 0} \frac{\dfrac{1}{x+h} - \dfrac{1}{x}}{h} \end{aligned}$$

Direct substitution yields the indeterminate form 0/0, so algebraic simplification is needed. Because there is no factoring to perform, we will multiply the numerator and denominator by the LCD—the lowest common denominator—of the two fractions in the numerator, which is $x(x + h)$.

$$m = \lim_{h \to 0} \frac{\dfrac{1}{x+h} - \dfrac{1}{x}}{h} \cdot \frac{x(x+h)}{x(x+h)}$$

Multiply numerator and denominator by $x(x + h)$

$$= \lim_{h \to 0} \frac{x - (x+h)}{hx(x+h)}$$

Distribute in numerator and denominator

$$= \lim_{h \to 0} \frac{-h}{hx(x + h)}$$ Combine terms in numerator

$$= \lim_{h \to 0} \frac{-1}{x(x + h)}$$ Divide h from numerator and denominator

$$= -\frac{1}{x^2}$$ Evaluate the limit

What is the slope at $x = 1$? At $x = 2$?
At $x = 1$, $m = -1/1^2 = -1$. At $x = -2$, $m = -1/(-2)^2 = -\frac{1}{4}$. ■

Example 12: Find the slope of the curve $f(x) = \sqrt{x}$ at any point.

Solution:
$$m = \lim_{h \to 0} \frac{f(x + h) - f(x)}{h}$$
$$= \lim_{h \to 0} \frac{\sqrt{x + h} - \sqrt{x}}{h}$$

Direct substitution yields the indeterminate form $0/0$, so algebraic simplification is needed.

Because there is no factoring to perform, we multiply the numerator and denominator by the conjugate of the numerator, $\sqrt{x + h} + \sqrt{x}$.

$$m = \lim_{h \to 0} \frac{\sqrt{x + h} - \sqrt{x}}{h} \cdot \frac{\sqrt{x + h} + \sqrt{x}}{\sqrt{x + h} + \sqrt{x}}$$ Multiply numerator and denominator by conjugate

$$= \lim_{h \to 0} \frac{(x + h) - x}{h(\sqrt{x + h} + \sqrt{x})}$$ Distribute in numerator and denominator

$$= \lim_{h \to 0} \frac{h}{h(\sqrt{x + h} + \sqrt{x})}$$ Combine terms in numerator

$$= \lim_{h \to 0} \frac{1}{\sqrt{x + h} + \sqrt{x}}$$ Divide h from numerator and denominator

$$= \frac{1}{2\sqrt{x}}$$ Evaluate the limit ■

Earlier, we defined the instantaneous rate of change of a function $f(x)$ at the point $(a, f(a))$ as $\lim_{h \to 0} \frac{f(a + h) - f(a)}{h}$. The instantaneous rate of change of $f(x)$ at any point can be found as $\lim_{h \to 0} \frac{f(x + h) - f(x)}{h}$.

MATHEMATICS CORNER 7.3

Conjugates and Rationalizing the Denominator

The **conjugate** of a binomial expression is another binomial expression with the same two terms but opposite operation. For example, the conjugate of $2 + \sqrt{x}$ is $2 - \sqrt{x}$. When a binomial containing a radical term is multiplied by its conjugate, the resulting product will be a rational number containing no radical. For example,

$$\left(2 + \sqrt{x}\right)\left(2 - \sqrt{x}\right) = 4 - 2\sqrt{x} + 2\sqrt{x} - x = 4 - x$$

Conjugates were used in your previous algebra course to eliminate radicals appearing in the denominator of a rational expression. The process was called **rationalizing the denominator**. For example,

$$\frac{6}{2 + \sqrt{x}} = \frac{6}{2 + \sqrt{x}} \cdot \frac{2 - \sqrt{x}}{2 - \sqrt{x}} = \frac{6\left(2 - \sqrt{x}\right)}{4 - x}$$

Example 13: The average home game attendance for the Orlando Magic basketball team (see the Unit 1 Introduction) can be approximated by $A(t) = 45t^2 - 865t + 18{,}552$, where $t = 1$ is the 1995–1996 season.

a. Determine the instantaneous rate of change in home attendance for any season.

b. What was the instantaneous rate of change in home attendance during the 1999–2000 season (let $t = 5$) and during the 2005–2006 season (let $t = 11$)?

Solution:

a. The instantaneous rate of change in home game attendance in any season is given by

$$\begin{aligned}
\lim_{t\to 0}\frac{A(t+h) - A(t)}{h} &= \lim_{h\to 0}\frac{[45(t+h)^2 - 865(t+h) + 18{,}552] - (45t^2 - 865t + 18{,}552)}{h} \\
&= \lim_{h\to 0}\frac{[45t^2 + 90th + 45h^2 - 865t - 865h + 18{,}552] - 45t^2 + 865t - 18{,}552}{h} \\
&= \lim_{h\to 0}\frac{90th + 45h^2 - 865h}{h} \\
&= \lim_{h\to 0}(90t + 45h - 865) \\
&= 90t - 865
\end{aligned}$$

where $t = 1$ is the 1995–1996 season.

b. The instantaneous rate of change in home attendance at any time is given by $IRC(t) = 90t - 865$.

For the 1999–2000 season, let $t = 5$. So, $90(5) - 865 = -415$. During the 1999–2000 season, average home attendance dropped by 415 attendees per year.

For the 2005–2006 season, let $t = 11$. So, $90(11) - 865 = 125$. During the 2005–2006 season, average home attendance increased by 125 attendees per year. ■

Check Your Understanding 7.3

1. Find the slope of $f(x) = 4 - x^2$ at any point.
2. Use your answer to Question 1 to find the slope of $f(x) = 4 - x^2$ at $x = 1$ and $x = -3$.
3. Find the equation of the tangent line to $f(x) = 4 - x^2$, where $x = 1$.
4. According to the 2003 *World Almanac*, annual unemployment rates since 1980 can be modeled by $E(t) = 0.005t^2 - 0.2t + 7.5$, where t is the number of years after 1980.
   a. Find the instantaneous rate of change in the unemployment rate in 1985 and in 1992. Interpret your answers.
   b. Use your answers from part a to predict the unemployment rate in 1986 and in 1993.

Check Your Understanding Answers

Check Your Understanding 7.1

1. a. 8.5°F per hour. Between 7 a.m. and 9 a.m., the temperature increased, on average, by 8.5°F per hour.
b. -4.5°F per hour. Between 2 p.m. and 6 p.m., the temperature decreased, on average, by 4.5°F per hour.

2. a. 8 feet per second. Between 0.5 second and 1 second after release, the rock's height increased by 8 feet per second, on average.
b. -16 feet per second. Between 1 second and 2 seconds after release, the rock's height decreased by 16 feet per second, on average.

3. -32 feet per second. Two seconds after release, the rock is falling at a rate of 32 feet per second.

4. a. 5.505%
b. 0.01. The unemployment rate is increasing by 0.01% per year in 2001.
c. $E(22) = 5.505 + 0.01 = 5.515\%$ using parts a and b. The model says $E(22) = 5.52\%$

Check Your Understanding 7.2

1. a. 4 **b.** 0 **c.** -8

2. a. $y = 4x - 5$ **b.** $y = -3$
c. $y = -8x - 11$

Check Your Understanding 7.3

1. $-2x$ **2.** $-2, 6$

3. $y = -2x + 5$

4. a. For 1985, -0.15, so the unemployment rate is dropping by 0.15% per year in 1985. For 1992, -0.08, so the unemployment rate is dropping by 0.08% per year in 1992.
b. For 1986, the predicted unemployment rate is

$$E(5) + (-0.15) = 6.625 - 0.15 = 6.475\%$$

For 1993, the predicted unemployment rate is

$$E(12) + (-0.08) = 5.82 = 0.08 = 5.74\%$$

Topic 7 Review

7

This topic introduced the concepts of **rate of change** and **slope of a curve**. These concepts are central to the development of the derivative.

CONCEPT	EXAMPLE
Average rate of change is the rate of change in a quantity between two points. It is calculated as a ratio comparing the change in the quantity between the two points with the change in its domain between the two points. Determining an average rate of change is the same as finding the slope of the secant line between two points. $\text{Average rate of change} = \dfrac{f(b) - f(a)}{b - a}$ *or* $\dfrac{f(x + h) - f(x)}{h}$	The per capita cost of health care over the past years can be estimated by $C(x) = 4x^2 + 23x + 357$, where x is the number of years after 1970. The average rate of change in per capita health care costs between 2000 and 2008 is given by $\dfrac{C(38) - C(30)}{38 - 30} = \dfrac{7007 - 4647}{8} = 295$ which means that per capita health care costs rose an average of \$295 dollars per year between 2000 and 2008. The graph of the per capita health care cost function is shown below. The line between the points where $x = 30$ and $x = 38$ has slope of 295. 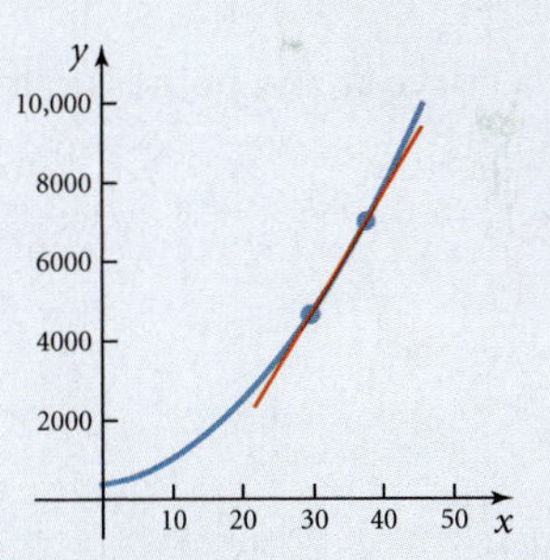
Instantaneous rate of change is the rate of change in a quantity at the specific point. It is calculated as $\lim\limits_{h \to 0} \dfrac{f(a + h) - f(a)}{h}$. The value of this limit also represents the slope of the tangent line to the graph of $f(x)$ at the point $x = a$.	The per capita cost of health care over the past years can be estimated by $C(x) = 4x^2 + 23x + 357$, where x is the number of years after 1970. The instantaneous rate of change in per capita health care costs in 2008 is given by $\lim\limits_{h \to 0} \dfrac{[4(38 + h)^2 + 23(38 + h) + 357] - [4(38)^2 + 23(38) + 357]}{h}$ $= \lim\limits_{h \to 0} (4h + 327) = 327$ In 2008, per capita health care costs rose by \$327.

(continued)

	The tangent line to the graph of the cost function at $x = 8$ has a slope of 327. 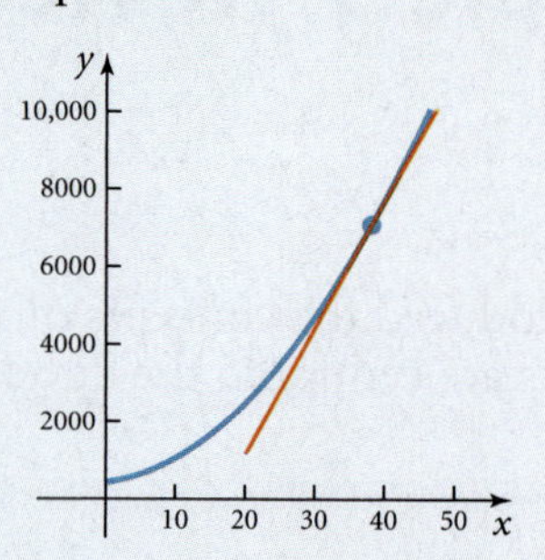
The slope of a curve at a specific point on the curve is the slope of the tangent line at that point. This slope is calculated using $\lim\limits_{h \to 0} \dfrac{f(a + h) - f(a)}{h}$.	The slope of $f(x) = x^2 - 3x + 4$ at $x = 1$ is given by $\lim\limits_{h \to 0} \dfrac{f(1 + h) - f(1)}{h} = \lim\limits_{h \to 0} \dfrac{h^2 - h}{h} = -1$ 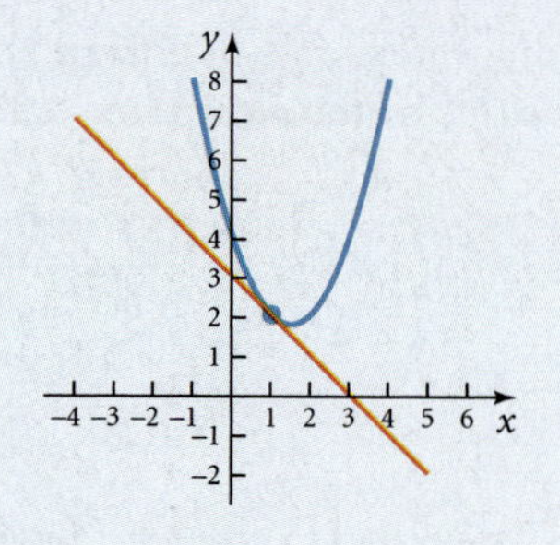
The slope of a curve at any point is calculated using $\lim\limits_{h \to 0} \dfrac{f(x + h) - f(x)}{h}$.	The slope of $f(x) = x^2 - 3x + 4$ at any point is given by $\lim\limits_{h \to 0} \dfrac{f(x + h) - f(x)}{h}$ $= \lim\limits_{h \to 0} \dfrac{h^2 + 2xh - 3h}{h} = 2x - 3$ At the point where $x = 1$, the slope is $2(1) - 3 = -1$.
The **velocity** of the object in motion at any time is the instantaneous rate of change of its position over time.	The height in feet above the ground of a ball is given by $h(t) = -16t^2 + 20t + 30$, where t is seconds after the release of the ball. The velocity of the ball after two seconds is given by $\lim\limits_{h \to 0} \dfrac{f(2 + h) - f(2)}{h}$ $= \lim\limits_{h \to 0} \dfrac{-16h^2 - 44h}{h} = -44$ After two seconds, the ball is falling to the ground at a rate of 44 feet per second.

NEW TOOLS IN THE TOOL KIT

- Determining the average rate of change of a quantity between two points
- Determining the instantaneous rate of change of a quantity at a point
- Finding the slope of a curve at a point
- Finding the slope of a curve at any point
- Finding the velocity of an object in motion

Topic 7 Exercises

7

1. The following table shows the average price of a new car for selected years.

Year	Price
1930	$610
1940	$850
1950	$1,511
1960	$2,610
1970	$3,979
1980	$7,201
1990	$16,012

Find the average rate of change in the average price of a new car between the following years. Interpret your answers.

a. 1930 and 1940

b. 1950 and 1980

c. 1970 and 1990

2. The average price of a gallon of gasoline for selected years is given in the following table.

Year	Price of a Gallon of Gasoline
1925	$0.12
1938	$0.10
1948	$0.16
1955	$0.23
1965	$0.30
1973	$0.47
1977	$0.65
1980	$1.15
1985	$1.09
1990	$1.34
2002	$1.57

Find the average rate of change in the price of a gallon of gasoline between the following years. Be sure to interpret your answers.

a. 1925 and 1938

b. 1948 and 1973

c. 1980 and 1990

3. The average annual income in the United States for selected years is given in the following table.

Year	Average Annual Income
1920	$2,160
1930	$1,973
1940	$1,725
1950	$3,216
1960	$3,199
1970	$9,357
1980	$17,173
1990	$28,906
2000	$56,644
2004	$60,070

Data from: U.S. Bureau of the Census

Find the average rate of change in average annual income between the following years. Interpret your answers.

a. 1930 and 1940

b. 1950 and 1970

c. 1990 and 2004

4. The following table shows the number of accidents per 1000 licensed drivers for various age groups of drivers.

Age of Driver	Number of Accidents per 1000 Licensed Drivers
16	294
17	198
18	176
19	152
25	92
35	76
45	53
50	38

Source: National Highway Traffic Safety Administration.

Find the average rate of change of number of accidents by drivers between the following ages.

a. 16 and 17

b. 19 and 25

c. 45 and 50

5. The average cost of a new home for selected years is given in the following table.

Year	Average Cost of New Home
1920	$6,296
1930	$7,146
1940	$3,925
1950	$8,450
1960	$12,675
1970	$23,400
1980	$68,714
1990	$123,000

Data from: U.S. Bureau of the Census

Find the average rate of change in the average cost of a new home between the following years. Interpret your answer.

a. 1930 and 1940

b. 1950 and 1970

c. 1980 and 1990

6. Annual median home prices for a county are given in the following table.

Year	Median Home Price
1995	$71,100
1996	$75,200
1997	$75,500
1998	$79,000
1999	$84,500
2000	$88,300
2001	$96,770
2002	$109,768
2003	$125,942

Source: Daytona Beach News-Journal, August 2003.

Find the average rate of change in median home prices between the following years.

a. 1996 and 1997

b. 1998 and 2000

c. 2000 and 2003

7. The following graph shows union membership as a percentage of the labor force.

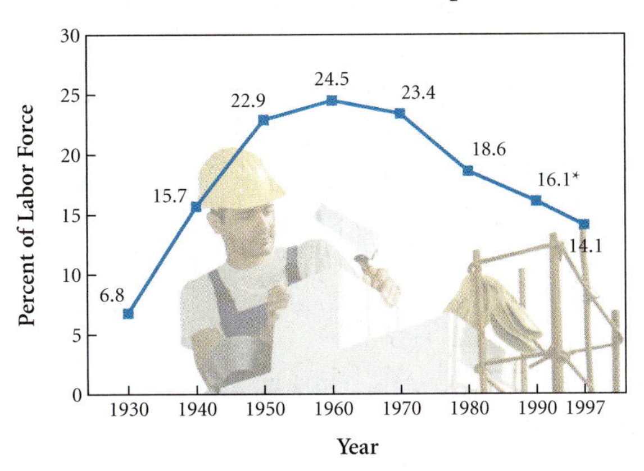

* After 1980, the numbers reflect union membership among all wage and salary employees. Before 1980, the numbers refelect all union members in the work force, whether employed or unemployed.

Source: The New Democrat, March/April 1998, pp. 18–20.

Find the average rate of change in union membership between the following years.

a. 1930 and 1950

b. 1960 and 1980

c. 1980 and 1997

8. The graph gives the size of the White House staff (full-time employees) for selected administrations and years.

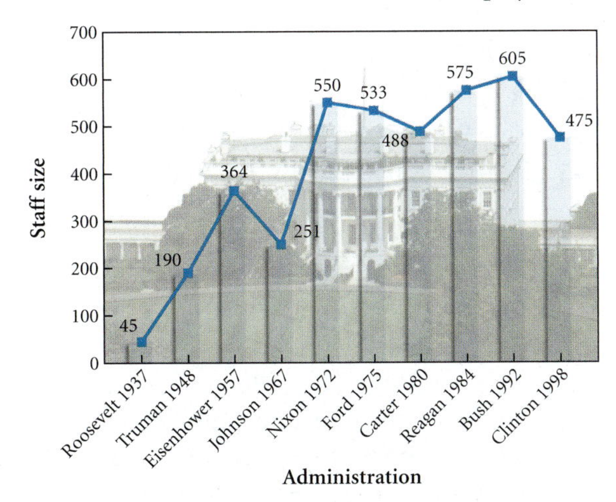

Source: Benjamin Ginsberg; Theodore J. Lowi, and Margaret Weir. *We the People: An Introduction to American Politics.* W.W. Norton, 4th ed., 2001; p. 547.

Find the average rate of change in staff size between the following years.

a. 1937 and 1957

b. 1967 and 1975

c. 1984 and 1998

9. The height of a projectile launched into the air is given by $h(t) = -16t^2 + 80t + 100$, where h is the height of the projectile in feet above the ground and t is the number of seconds after its launch. In parts a through c, find the average rate of change in the height of the projectile during the following time intervals. Interpret your answers.

a. 1 second and 2 seconds

b. 2 seconds and 2.5 seconds

c. 3 seconds and 4 seconds

d. When will the projectile strike the ground?

10. A ball is thrown upward into the air. Its height in feet above the ground is given by $h(t) = -16t^2 + 64t + 10$, where t is the number of seconds after the ball was thrown. In parts a through c, find the average rate of change in the height of the ball during the following time intervals. Interpret your answers.

a. 0.5 second and 1 second

b. 1.5 seconds and 2 seconds

c. 2 seconds and 3 seconds

d. When will the ball hit the ground?

11. Using the projectile in Exercise 9, find the instantaneous rate of change in the height of the projectile at the following times. Interpret your answers.

a. 1 second

b. 4 seconds

c. 2.5 seconds

12. Using the ball in Exercise 10, find the instantaneous rate of change in the height of the ball at the following times. Interpret your answers.

a. 1 second

b. 3.5 seconds

c. 2 seconds

13. Union membership as a percentage of the labor force (see Exercise 7) can be modeled by $M(x) = 0.0003x^3 - 0.066x^2 + 4.64x - 81$, where M is membership as a percentage of the labor force and x is the number of years after 1900.

a. Use the model to estimate union membership in 1950.

b. Find the instantaneous rate of change of union membership in 1950 and interpret your answer.

c. Find the instantaneous rate of change of union membership in 1980 and interpret your answer.

14. The number of full-time White House staff (see Exercise 8) can be modeled by $N(t) = -0.004t^3 + 0.703t^2 - 26t + 277$, where N is the number of full-time staff members and t is the number of years after 1900.

a. Use the model to estimate the number of full-time staff members in 1960.

b. Find the instantaneous rate of change in number of staff in 1960 and interpret your answer.

c. Find the instantaneous rate of change in number of staff in 1995 and interpret your answer.

15. According to the National Highway Traffic Safety Administration the number of motorcyclists killed on highways each year can be modeled by $N(t) = 20t^2 - 571t + 6550$, where N is the number of motorcyclists killed and t is the number of years after 1980.

a. Find the instantaneous rate of change in the number of motorcyclists killed in 1992 and interpret your answer.

b. Use the model to estimate the number of motorcyclists killed in 2002.

c. Find the instantaneous rate of change in number of motorcyclists killed in 2002 and interpret your answer.

d. Use your answers to parts b and c to predict the number killed in 2003.

16. The amount that workers contribute monthly for health insurance premiums can be modeled by $A(t) = 0.07t^3 - 3.4t^2 + 53.7t - 240$, where A is the monthly amount contributed and t is the number of years after 1980.

a. Find the instantaneous rate of change in monthly contribution in 1995 and interpret your answer.

b. Use the model to estimate the monthly contribution of workers in 2002.

c. Find the instantaneous rate of change in monthly contribution in 2002 and interpret your answer.

d. Predict the contribution in 2003 using your answers to parts b and c.

In Exercises 17 through 20, find the slope of the curve at the two indicated points.

17. $f(x) = 3 - 5x$ at

a. $(2, f(2))$ **b.** $(-2, f(-2))$

18. $f(x) = 3x + 7$ at

a. $(-2, f(-2))$ **b.** $(1, f(1))$

19. $f(x) = x^2 - 5x + 1$ at

a. $(3, f(3))$ **b.** $(-2, f(-2))$

20. $f(x) = x^2 + 3x$ at

a. $(-2, f(-2))$ **b.** $(3, f(3))$

Find the equation of the tangent line to each of the following functions at the indicated point. Confirm by drawing a graph of the function showing the tangent line.

21. $f(x) = 4x - x^2$ at $x = 1$

22. $f(x) = 2x^2 + 3x - 4$ at $x = 1$

In Exercises 23 through 36, find the function giving the slope of the curve at any point.

23. $f(x) = 3x - 4$ **24.** $f(x) = 2x + 5$

25. $f(x) = x^2 + 3$ **26.** $f(x) = 4 - x^2$

27. $f(x) = 2x^2 + 5x - 3$

28. $f(x) = 3x^2 - 2x - 4$

29. $f(x) = 3x - x^2$ **30.** $f(x) = -6x + 2x^2$

31. $f(x) = x^3$ **32.** $f(x) = x^4$

33. $f(x) = \sqrt{x + 1}$ **34.** $f(x) = \sqrt{x - 3}$

35. $f(x) = \dfrac{2}{x + 1}$ **36.** $f(x) = \dfrac{3}{x - 2}$

CALCULATOR CONNECTION

37. Use the data on average yearly income in Exercise 3 to find a quadratic function that models the data. See Topic 4 for a review of regression.

a. What is an appropriate domain for the model?

b. Use the model to estimate the yearly income in 1960, 1980, and 2000. How close is the model to the actual value?

c. Use the model to find the average rate of change in yearly income between 1990 and 2004. How closely does the model estimate the actual rate of change found in Exercise 3?

d. Based on your results in parts b and c, do you think that the model is a good one?

38. Use the data on accidents in Exercise 4 to find a quadratic function that models the data. See Topic 4 for a review of regression.

a. What is an appropriate domain for the model?

b. Use the model to estimate the number of accidents for 18-year-old drivers, 35-year-old drivers, and 50-year-old drivers. How close is the model to the actual value?

c. Use the model to find the average rate of change in the number of accidents between 16-year-old drivers and 17-year-old drivers, between 19-year-old drivers and 25-year-old drivers, and between 45-year-old drivers and 50-year-old drivers. How close did the model estimate the actual rate of change found in Exercise 4?

d. Based on the results in parts b and c, do you think the model is a good one?

39. Using the model you created in Exercise 37, determine the instantaneous rate of change in yearly income in 1960, 1980, and 2000. Interpret your answers.

40. Using the model you created in Exercise 38, determine the instantaneous rate of change in number of accidents for 17-year-old drivers, 25-year-old drivers, and 45-year-old drivers. Interpret your answers.

41. The following table gives unemployment rates in the United States for selected years since 1970.

Year	Percent of the Workforce That Is Unemployed
1970	4.9
1975	8.5
1980	7.1
1985	7.2
1990	5.6
1995	5.6
2000	4.0
2003	5.9
2005	5.1

Data from: U.S. Bureau of Labor Statistics

a. Find a cubic function that models the data. See Topic 4 for a review of regression.

b. Find the instantaneous rate of change in unemployment rate for any given year.

c. Find the instantaneous rate of change in unemployment in 1980 and 2000 and interpret your answers.

d. Use the results of part c to predict the unemployment rate in 1981 and 2001.

GROUP CALCULATOR CONNECTION PROJECTS

These projects will help you visualize why the slope of a curve is given by the slope of its tangent line. Calculator directions are for the TI-83/84.

Project 1

1. Graph $y = 0.5x^3 - 1.5x^2 + 1$ using Zoom Decimal with GRID ON. To access Zoom Decimal, press **ZOOM** 4. To turn the grid on, press **2nd ZOOM**, highlight **GRID ON**, and press **ENTER**.

2. Draw the tangent line to the graph at the point where $x = 0.5$. Press **TRACE** and trace to the point on the graph where $x = 0.5$. Then press **2nd PRGM** (DRAW), select option 5: Tangent, and press **ENTER**. The tangent line should now appear on the graph.

3. Use the grid dots to estimate the slope of the tangent line. Find two dots that the line goes through and calculate the change in y and the change in x. Then estimate the slope.

4. Now find $\frac{dy}{dx}$. Press **2nd TRACE** (CALC) and select option 6: dy/dx. Trace to the point where $x = 0.5$ and press **ENTER**.

5. How close was your dot-counting estimate to the actual slope given by the calculator? Your estimate should be close to the actual value shown on the graph.

Project 2

1. Graph $y = 0.5x^3 - 1.5x^2 + 1$ using Zoom Decimal with GRID OFF. To access Zoom Decimal, press **ZOOM** 4 . To turn the grid off, press **2nd ZOOM**, highlight **GRID OFF**, and press **ENTER**.

2. Zoom in on the curve at the point where $x = 0.5$. Press ZOOM and select option 2: Zoom In. Trace to the point on the graph where $x = 0.5$ and press **ENTER**. The graph will then redraw to focus only on the portion of the graph around $x = 0.5$.
3. Zoom in on the curve again at the point where $x = 0.5$ by pressing **ENTER**. (If your cursor is not on the curve, use the arrows to adjust its placement.) Do you see the curve straightening out?
4. Zoom in once more by pressing **ENTER** one more time. By now, the curve should resemble a line. This property is referred to as *local linearity*. Because the curve behaves like a line for a small neighborhood around the point, the slope of the tangent line can be used as the slope of the curve at that point.

TOPIC

8

Introduction to the Derivative

With the tools of limits and rates of change now in the Tool Kit, we can move on to the next level of the house of calculus, the *derivative*. The derivative is the name for the mathematical process of evaluating limits to find an instantaneous rate of change or the slope of a curve at any point. In this topic, we focus on a more efficient way of evaluating those limits. Before starting, however, be sure to complete the Topic 8 Warm-up Exercises to sharpen the algebra and calculus tools you will need.

IMPORTANT TOOLS IN TOPIC 8

- *Definition of the derivative*
- *Finding derivatives*
- *Evaluating derivatives at specific points*
- *Nonexistence of derivatives*
- *Meaning of derivatives*

TOPIC 8 WARM-UP EXERCISES

Be sure you can successfully complete the following exercises before starting Topic 8.

Algebra Warm-up Exercises

1. Given $f(x) = x^2 + x - 2$, evaluate the following functions.

a. $f(3)$

b. $f(a)$

c. $f(a + h)$

2. Find the slope of each linear equation.

a. $y = 4x$ **b.** $y = 2x - 3$

c. $y = 4 - 7x$ **d.** $y = 4$

Calculus Warm-up Exercises

1. Evaluate the following limits.

a. $\lim_{h \to 0} (2x + 3h)$ **b.** $\lim_{h \to 0} (2x - 3 + 4h)$

c. $\lim_{h \to 0} 16$

2. Evaluate $\lim_{h \to 0} \dfrac{f(x + h) - f(x)}{h}$ for the following functions.

a. $f(x) = 2x - 3$ **b.** $f(x) = x^2 + x - 3$

3. Find the instantaneous rate of change of $f(x) = x^2 + x - 3$ at the point where $x = 1$.

4. Find the slope of the curve $f(x) = x^2 + x - 3$ at any point.

Answers to Warm-up Exercises

Algebra Warm-up Exercises

1. a. $f(3) = 10$ **b.** $a^2 + a - 2$

c. $a^2 + 2ah + h^2 + a + h - 2$

2. a. 4 **b.** 2 **c.** -7 **d.** 0

Calculus Warm-up Exercises

1. a. $2x$ **b.** $2x - 3$ **c.** 16

2. a. 2 **b.** $2x + 1$

3. 3 **4.** $m = 2x + 1$

he concepts of limit and rate of change provide the foundation upon which the house of calculus is built. With these tools in the Tool Kit, we can now develop the next major tool of calculus, the derivative.

The Derivative

In Topic 7, we defined the **slope of a curve at the point** $(a, f(a))$ as the slope of the tangent line to the curve at that point. The **slope of the tangent line** at $(a, f(a))$ is determined by evaluating $\lim_{h \to 0} \frac{f(a + h) - f(a)}{h}$. This same formula also gave the **instantaneous rate of change of $f(x)$ at the point** $(a, f(a))$.

We also defined the **slope of a curve $f(x)$ at any point** as $\lim_{h \to 0} \frac{f(x + h) - f(x)}{h}$, which is the same formula used for finding the instantaneous rate of change of $f(x)$ for any x. These two concepts, slope of a curve and instantaneous rate of change, lead to the definition of the **derivative**.

Definition: The **derivative** of $f(x)$ is found by evaluating $\lim_{h \to 0} \frac{f(x + h) - f(x)}{h}$, if the limit exists. If the derivative exists, then $f(x)$ is said to be **differentiable**. The process of finding a derivative is called **differentiation**.

Symbolically, the derivative of $f(x)$ can be written in several ways:

$$f'(x) \qquad y' \qquad \frac{dy}{dx} \qquad D_x(f) \qquad \frac{d}{dx}f$$

Example 1: Given $f(x) = x^2$, find $f'(x)$.

Solution:

$$\begin{aligned} f'(x) = \lim_{h \to 0} \frac{f(x + h) - f(x)}{h} &= \lim_{h \to 0} \frac{(x + h)^2 - x^2}{h} \\ &= \lim_{h \to 0} \frac{x^2 + 2xh + h^2 - x^2}{h} \\ &= \lim_{h \to 0} \frac{2xh + h^2}{h} \\ &= \lim_{h \to 0} (2x + h) \\ &= 2x \end{aligned}$$

So $f'(x) = 2x$. Remember that simply means that the slope of $f(x) = x^2$ at any point on the curve is given by $m = 2x$, or the instantaneous rate of change of the function at any point is given by $2x$. ■

Example 2: Given $y = 2x + 5$, find y'.

Solution:

$$y' = \lim_{h \to 0} \frac{[2(x+h)+5] - (2x+5)}{h} = \lim_{h \to 0} \frac{2x + 2h + 5 - 2x - 5}{h}$$
$$= \lim_{h \to 0} \frac{2h}{h}$$
$$= \lim_{h \to 0} 2$$
$$= 2$$

So $y' = 2$. This means the slope of $y = 2x + 5$ is 2, which you should already know because $y = 2x + 5$ is a linear equation. ■

Example 3: Find $\frac{d}{dx}(3 - 5x)$.

Solution: $f(x) = 3 - 5x$ is a linear function, so its slope is -5. Thus, $\frac{d}{dx}(3 - 5x) = -5$. Using the formula, we obtain

$$\frac{d}{dx}(3 - 5x) = \lim_{h \to 0} \frac{[3 - 5(x+h)] - (3 - 5x)}{h}$$
$$= \lim_{h \to 0} \frac{3 - 5x - 5h - 3 + 5x}{h}$$
$$= \lim_{h \to 0} \frac{-5h}{h}$$
$$= \lim_{h \to 0} (-5)$$
$$= -5$$

■

> **Warning!** When finding derivatives of functions, beware of getting too bogged down in the calculations themselves. Always remember that a derivative is representing the instantaneous rate of change of a function, the slope of a curve, or the velocity of an object in motion.

Example 4: Find $D_x(x^2 - 3)$.

Solution:

$$D_x(x^2 - 3) = \lim_{h \to 0} \frac{[(x+h)^2 - 3] - (x^2 - 3)}{h}$$
$$= \lim_{h \to 0} \frac{(x^2 + 2xh + h^2 - 3) - x^2 + 3}{h}$$
$$= \lim_{h \to 0} \frac{2xh + h^2}{h}$$
$$= \lim_{h \to 0} (2x + h)$$
$$= 2x$$

So $D_x(x^2 - 3) = 2x$. ■

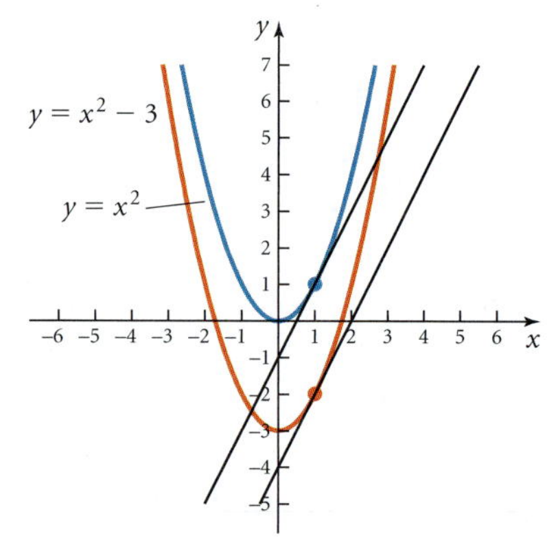

Figure 8.1

Be sure to see from Examples 1 and 4 that the derivatives of x^2 and $x^2 - 3$ are both the same: $2x$. This result should seem logical. The graph of $f(x) = x^2 - 3$ is the graph of $f(x) = x^2$ shifted down 3 units, so the slopes should be the same. For example, at the point where $x = 1$, the slope of both graphs should be $m = 2(1) = 2$. As you can see from Figure 8.1, the two tangent lines are parallel because their slopes are the same.

Example 5: Find $f'(x)$ for $f(x) = 8$.

Solution:

$$f'(x) = \lim_{h \to 0} \frac{f(x+h) - f(x)}{h}$$

$$= \lim_{h \to 0} \frac{8 - 8}{h}$$

$$= \lim_{h \to 0} \frac{0}{h}$$

$$= \lim_{h \to 0} 0$$

$$= 0$$

This result should seem logical. The graph of $f(x) = 8$ is a horizontal line, and we know that horizontal lines have zero slope. The function $f(x) = 8$ is a constant function, so there is no vertical change from one point on the function to the next. ■

Example 6: Find $f'(x)$ for each of the following functions.

a. $f(x) = x^2 - 4x + 3$ **b.** $f(x) = x^3$

c. $f(x) = \sqrt{x}$ **d.** $f(x) = \dfrac{1}{x}$

Solution:

a. $f(x) = x^2 - 4x + 3$

$$f'(x) = \lim_{h \to 0} \frac{[(x+h)^2 - 4(x+h) + 3] - (x^2 - 4x + 3)}{h}$$

The Substitution Property yields $\frac{0}{0}$, so algebraic simplification is needed.

$$= \lim_{h \to 0} \frac{x^2 + 2xh + h^2 - 4x - 4h + 3 - x^2 + 4x - 3}{h}$$ Expand

$$= \lim_{h \to 0} \frac{2xh + h^2 - 4h}{h}$$ Combine terms

$$= \lim_{h \to 0} (2x + h - 4)$$ Factor h and reduce

$$= 2x - 4$$ Evaluate limit

Thus, $\dfrac{d}{dx}(x^2 - 4x + 3) = 2x - 4$.

This result should seem logical. Separate the function as $f(x) = (x^2) + (-4x + 3)$. It was shown in Example 1 that the derivative of x^2 is $2x$. Because $-4x + 3$ is linear, its derivative is its slope, -4. Thus, it should seem logical that the derivative would be $2x + (-4)$ or $2x - 4$.

b. $f(x) = x^3$

$$f'(x) = \lim_{h \to 0} \frac{(x + h)^3 - x^3}{h}$$

The Substitution Property yields $\frac{0}{0}$, so algebraic simplification is needed.

$$f(x) = \lim_{h \to 0} \frac{x^3 + 3x^2h + 3xh^2 + h^3 - x^3}{h}$$

Note: $(x + h)^3 = (x + h)(x + h)(x + h) = (x + h)(x^2 + 2xh + h^2)$

$$= \lim_{h \to 0} \frac{3x^2h + 3xh^2 + h^3}{h}$$ Combine terms

$$= \lim_{h \to 0} (3x^2 + 3xh + h^2)$$ Factor h and reduce

$$= 3x^2$$ Evaluate limit

Thus, $\frac{d}{dx}(x^3) = 3x^2$.

c. $f(x) = \sqrt{x}$

$$f'(x) = \lim_{h \to 0} \frac{\sqrt{x + h} - \sqrt{x}}{h}$$

The Substitution Property gives the indeterminate form $\frac{0}{0}$, so algebraic simplification is needed.

$$f(x) = \lim_{h \to 0} \frac{\sqrt{x + h} - \sqrt{x}}{h} \cdot \frac{\sqrt{x + h} + \sqrt{x}}{\sqrt{x + h} + \sqrt{x}}$$ Rationalize the numerator

$$= \lim_{h \to 0} \frac{x + h - x}{h(\sqrt{x + h} + \sqrt{x})}$$ Multiply the numerators and denominators

$$= \lim_{h \to 0} \frac{h}{h(\sqrt{x + h} + \sqrt{x})}$$ Combine like terms

$$= \lim_{h \to 0} \frac{1}{\sqrt{x + h} + \sqrt{x}}$$ Cancel the common factor, h

$$= \frac{1}{2\sqrt{x}}$$

Thus, $\frac{d}{dx}\left(\sqrt{x}\right) = \frac{1}{2\sqrt{x}}$.

d. $f(x) = \frac{1}{x}$

$$f'(x) = \lim_{h \to 0} \frac{\frac{1}{x + h} - \frac{1}{x}}{h}$$

The Substitution Property gives the indeterminate form $\frac{0}{0}$, so algebraic simplification is needed.

$$f(x) = \lim_{h\to 0} \frac{\frac{1}{x+h} - \frac{1}{x}}{h} \cdot \frac{x(x+h)}{x(x+h)}$$ Complex fraction, so multiply by LCD

$$= \lim_{h\to 0} \frac{x - (x+h)}{hx(x+h)}$$ Multiply the numerators and denominators

$$= \lim_{h\to 0} \frac{-h}{hx(x+h)}$$ Combine like terms

$$= \lim_{h\to 0} \frac{-1}{x(x+h)}$$ Cancel the common factor, h

$$= \frac{-1}{x^2} \text{ or } -\frac{1}{x^2}$$

Thus, $\frac{d}{dx}\left(\frac{1}{x}\right) = -\frac{1}{x^2}$. ■

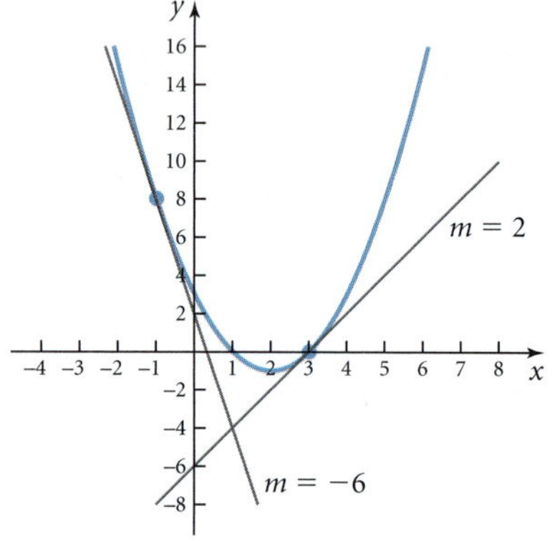
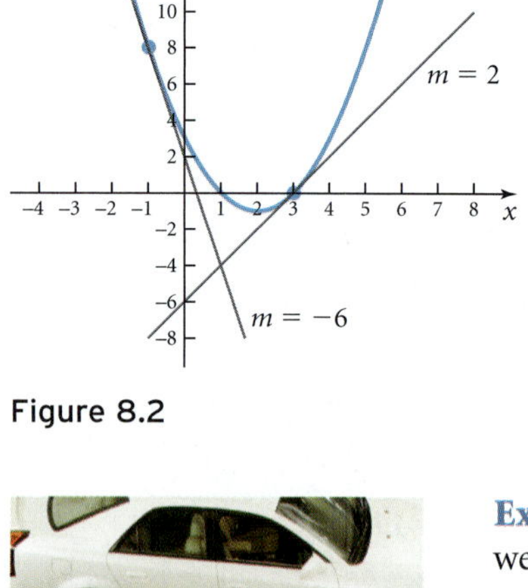

Figure 8.2

Example 7: Given $f(x) = x^2 - 4x + 3$, find $f'(-1)$ and $f'(3)$ and interpret your answers.

Solution: From Example 6a, we know that the derivative is $f'(x) = 2x - 4$.

Thus,

$$f'(-1) = 2(-1) - 4 = -6 \quad \text{and} \quad f'(3) = 2(3) - 4 = 2$$

The slope of $f(x) = x^2 - 4x + 3$ at the point where $x = -1$ is -6, and the slope of the curve at the point where $x = 3$ is 2. Figure 8.2 provides verification. ■

Example 8: A car dealer finds that the number of cars sold on day x of a weeklong ad campaign is $S(x) = 10x - x^2$, where $0 \le x \le 7$.

a. Find $S'(x)$ and interpret your answer.

b. Find $S(3)$ and interpret your answer.

c. Find the rate at which sales are changing on the third day.

d. Estimate the number of cars sold on the fourth day using $S(3)$ and $S'(3)$.

Solution:

a. $$S'(x) = \lim_{h\to 0} \frac{[10(x+h) - (x+h)^2] - (10x - x^2)}{h}$$

$$= \lim_{h\to 0} \frac{10x + 10h - x^2 - 2xh - h^2 - 10x + x^2}{h}$$

$$= \lim_{h\to 0} \frac{10h - 2xh - h^2}{h}$$

$$= \lim_{h\to 0} (10 - 2x - h)$$

$$= 10 - 2x$$

The rate at which sales are changing at any time is $10 - 2x$ cars per day.

b. $S(3)$ does not require the derivative; rather, it just asks for an evaluation of the original function. Thus, $S(3) = 10(3) - 3^2 = 21$. On the third day, 21 cars were sold.

c. The *rate of change* on day 3 requires a derivative (why?).

$$S'(3) = 10 - 2(3) = 4$$

On the third day, the sales were increasing (because $4 > 0$) at a rate of 4 cars per day.

d. There were 21 cars sold on day 3 (see part b). Sales are increasing by 4 cars per day (see part c), so we could predict about $21 + 4$ or 25 cars to be sold on day 4. ■

Check Your Understanding 8.1

1. Find $f'(x)$ for each of the following functions.

a. $f(x) = 7x - 2$ **b.** $f(x) = 2x^2 - 4x + 1$ **c.** $f(x) = 2$

2. For $f(x) = 2x^2 - 4x + 1$, evaluate $f'(2)$ and $f'(3)$ and interpret your answers.

Nonexistence of Derivatives

At times, the derivative of a function does not exist.

1. **If $f(x)$ is discontinuous at a point, the derivative does not exist at that point.**

The function $f(x) = \dfrac{1}{x}$ shown in Figure 8.3 is discontinuous at $x = 0$ because $\lim_{x \to 0} 1/x$ does not exist. Because $f(0)$ is undefined, it is not possible to draw a tangent line at $x = 0$, which means there is no derivative at that point.

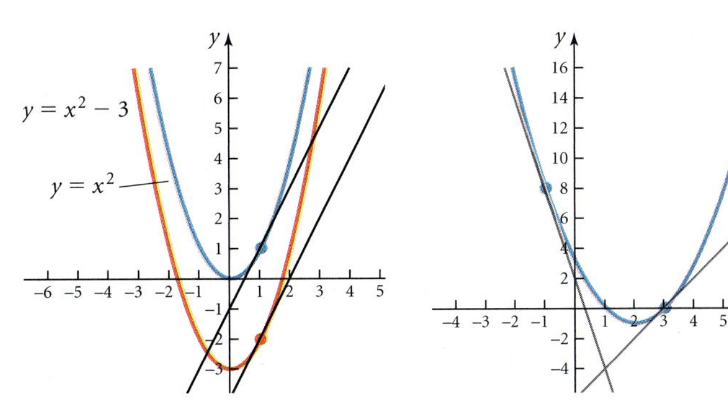

Figure 8.3

Figure 8.4

Consider the piecewise defined function shown in Figure 8.4. The function represented in Figure 8.4 is discontinuous at $x = 1$. Because

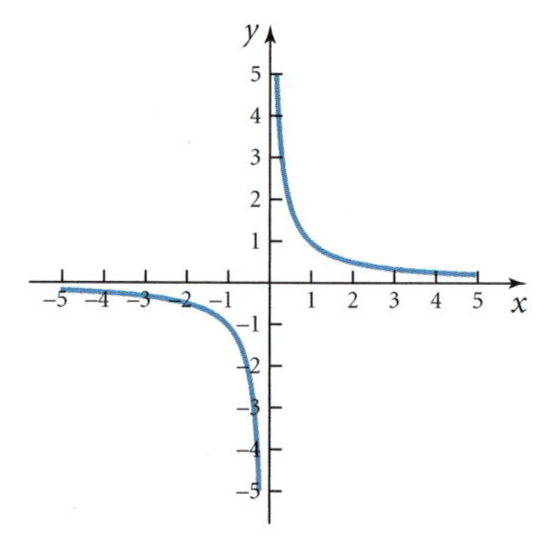

Figure 8.5

$\lim_{x \to 1^-} f(x) = 1$ and $\lim_{x \to 1^+} f(x) = 3$, there is no unique tangent line and $f'(1)$ does not exist.

2. **If the graph of $f(x)$ has a "sharp corner" at some point, the derivative does not exist at that point.**

   For the graph in Figure 8.5, there is a sharp corner at $x = 1$. For $x < 1$, the slope of the tangent line is -1. For $x > 1$, however, the slope of the tangent line is not -1 but some positive value. Thus, there is no unique tangent line at the point, so there is no unique slope and no derivative.

3. **If the graph of $f(x)$ has a vertical tangent line at some point, the derivative does not exist at that point.**

   Remember that vertical tangent lines have no slope, so the derivative is undefined.

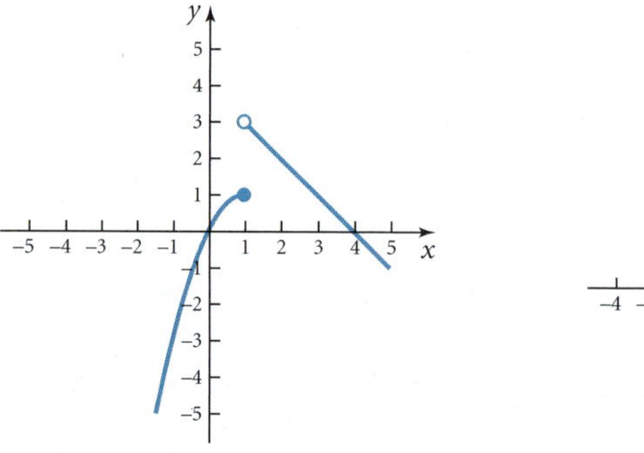

Figure 8.6

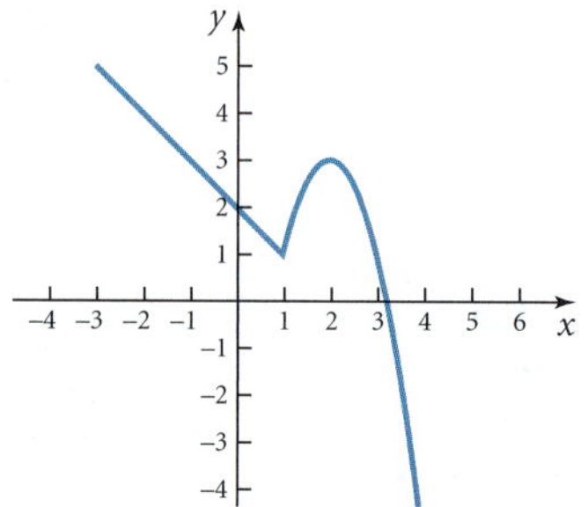

Figure 8.7

In both Figures 8.6 and 8.7, the graph has a vertical tangent line at $x = 1$, so the function has no derivative at that point.

For a function to be differentiable at a point, the function must be continuous at that point and must have a unique tangent line with a defined slope at that point.

Here is a summary of the instances for which a derivative does not exist.

Nonexistence of Derivatives

If $f'(x)$ does not exist at some point, then $f(x)$ is said to be **nondifferentiable** at that point.

Given a function $f(x)$, $f(x)$ is nondifferentiable at $x = a$ if

- The graph of $f(x)$ is discontinuous at $x = a$.
- The graph of $f(x)$ has a sharp corner at $x = a$.
- The graph of $f(x)$ has a vertical tangent line at $x = a$.

Check Your Understanding Answers

Check Your Understanding 8.1

1. a. $f'(x) = 7$
   b. $f'(x) = 4x - 4$
   c. $f'(x) = 0$

2. $f'(2) = 4$, which means the slope of the curve at $x = 2$ is 4. $f'(3) = 8$, which means the slope of the curve at $x = 3$ is 8.

Topic 8 Review

8

This topic introduced the derivative, which is one of the major mathematical tools of calculus.

CONCEPT	EXAMPLE
The **derivative** of a function gives the **instantaneous rate of change** of the function or the **slope** of the graph of the function.	If $h(t) = -16t^2 + 80t + 45$ is the height of an object above the ground t seconds after being thrown upward, then $h'(t) = -32t + 80$ is the *instantaneous rate of change* in the height of the object with respect to time. The derivative gives the rate at which the object is rising or falling at a particular time. If $f(x) = x^2 - 3$, then $f'(x) = 2x$ gives the *slope* of the graph of $f(x) = x^2 - 3$ at any point.
Derivatives can be written symbolically as $f'(x), y', \frac{dy}{dx}, D_x(f)$, or $\frac{d}{dx} f(x)$.	To find the derivative of $y = f(x) = x^2 - 3$, we can write $f'(x), y', \frac{dy}{dx}, D_x(x^2 - 3)$, or $\frac{d}{dx}(x^2 - 3)$.
The **derivative of a function at the point $(a, f(a))$** is found by evaluating $$f'(a) = \lim_{h \to 0} \frac{f(a+h) - f(a)}{h}$$ if the limit exists.	The derivative of $f(x) = x^2 - 3$ at the point $(2, f(2))$ is found by evaluating $$f'(2) = \lim_{h \to 0} \frac{[(2+h)^2 - 3] - (2^2 - 3)}{h}$$

(continued)

The **derivative of a function at any point** is found by evaluating $$f'(x) = \lim_{h \to 0} \frac{f(x+h) - f(x)}{h}$$ if the limit exists.	The derivative of $f(x) = x^2 - 3$ at any point is found by evaluating $$f'(x) = \lim_{h \to 0} \frac{[(x+h)^2 - 3] - (x^2 - 3)}{h}$$
Derivatives of functions do not exist at the following points: • where $f(x)$ is discontinuous • where $f(x)$ has a sharp corner • where $f(x)$ has a vertical tangent line	$f(x) = \frac{1}{x-2}$ is nondifferentiable at $x = 2$ because $f(x)$ is discontinuous at $x = 2$. The graph of $f(x) = \frac{1}{x-2}$ has no tangent line at $x = 2$.
If a function does not have a derivative at some point, it is said to be **nondifferentiable** at that point.	$$f(x) = \begin{cases} x + 3 & \text{if } x \le 4 \\ \sqrt{x} - 1 & \text{if } x > 4 \end{cases}$$ is nondifferentiable at $x = 4$ because the graph has a sharp corner at $x = 4$. The graph of $f(x)$ is continuous at $x = 4$, but there is no unique tangent line at that point. 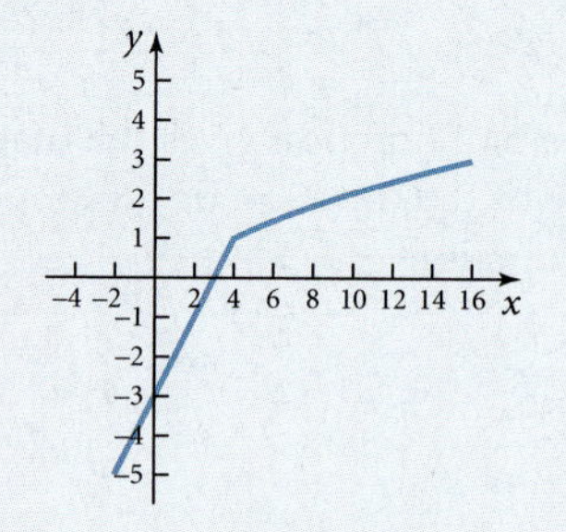

NEW TOOLS IN THE TOOL KIT:

- Derivatives of functions at a specific point
- Derivatives of functions at any point
- Nonexistence of derivatives

Topic 8 Exercises

8

Find $f'(x)$ for each of the following functions.

1. $f(x) = 3x - 7$

2. $f(x) = -6$

3. $f(x) = x^2 + 4$

4. $f(x) = 3x^2 - 2$

5. $f(x) = 11$

6. $f(x) = 6 - 2x$

7. $f(x) = 3 - 2x^2$

8. $f(x) = x^3 - x$

Find $\dfrac{dy}{dx}$ for each of the following functions.

9. $f(x) = 2x^2 - 5x + 3$

10. $f(x) = -3x^2 + 6x$

11. $f(x) = 4x - x^3$

12. $f(x) = 2x - 3x^3$

Find each of the following derivatives.

13. $D_x(11x - 5)$

14. $D_x(17 - 3x)$

15. $D_x(x^2 + 6x - 2)$

16. $D_x(2x - 5x^2)$

17. $D_x\dfrac{3}{x}$

18. $D_x\dfrac{4}{x - 2}$

For each of the following functions, find $f'(x)$ for the indicated value of x.

19. $f(x) = x^2 + 4x - 3$ for $x = 2$

20. $f(x) = 5x - 2x^2$ for $x = -1$

21. $f(x) = 6x + 4$ for $x = -3$

22. $f(x) = 9 - 5x$ for $x = 2$

23. $f(x) = \sqrt{x - 1}$ for $x = 5$

24. $f(x) = \sqrt{x + 3}$ for $x = 1$

25. $f(x) = \dfrac{-2}{x - 3}$ for $x = 4$

26. $f(x) = \dfrac{3}{x + 5}$ for $x = -2$

For each graph, determine the value(s) of x for which $f(x)$ is nondifferentiable.

27.

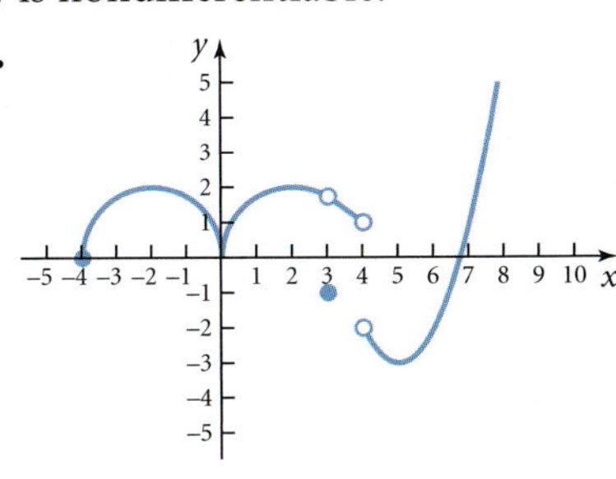

28.

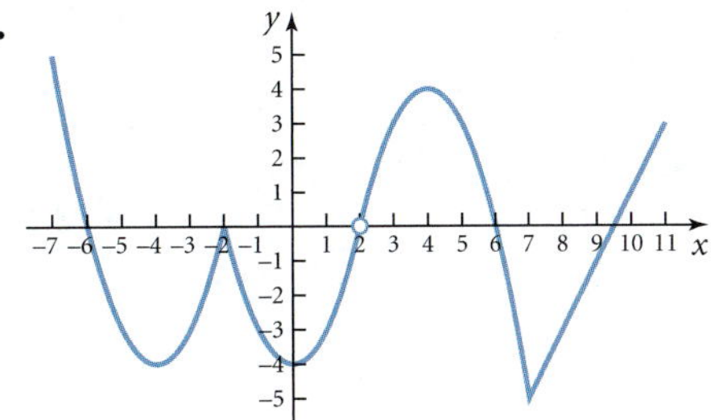

For each graph, determine if $f'(x)$ is positive, negative, zero, or undefined at the indicated points.

29.

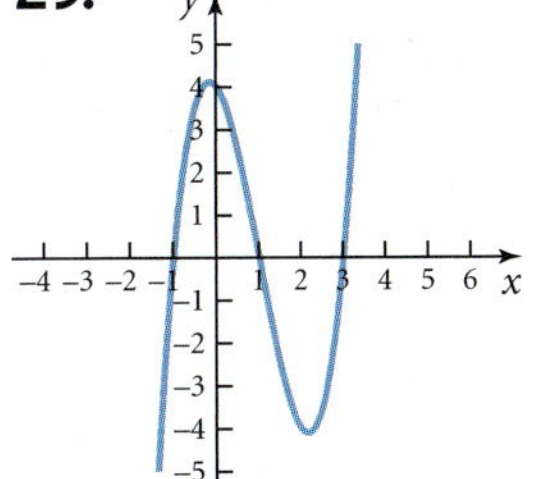

a. $x = -1$

b. $x = 1$

c. $x = 2$

30.

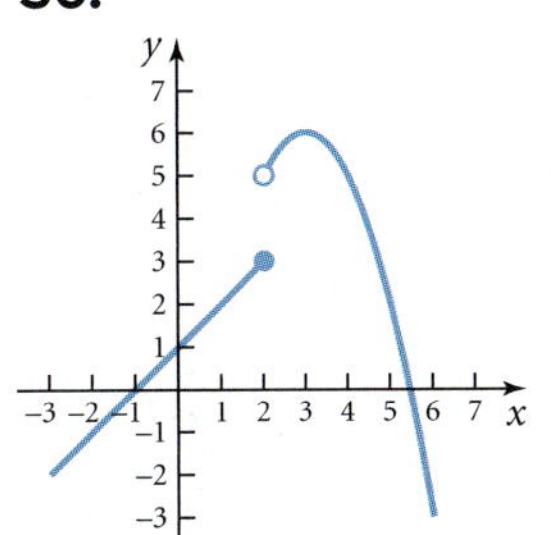

a. $x = 0$

b. $x = 2$

c. $x = 4$

31. The price of a gallon of gasoline between 1980 and 2004 can be modeled by $P(t) = 0.00043t^2 + 0.0117t + 1.114$, where P is the price in dollars and t is the number of years after 1980.

a. Find $P'(t)$.

b. Find $P'(15)$ and interpret your answer.

32. According to the National Highway Traffic Safety Administration, the fatality rate on the nation's highways is a function of the driver's age and can be modeled by $F(t) = 0.0038t^2 - 0.35t + 8.3$, where F is the fatality rate per 100 million vehicle miles traveled and t is the driver's age.

a. Find $F'(t)$.

b. Find $F'(25)$ and interpret your answer.

33. Union membership as a percentage of the total labor force can be modeled by $M(x) = 0.0003x^3 - 0.066x^2 + 4.64x - 81$ where M is membership as a percentage of the labor force and x is the number of years after 1900.

a. Find $M(85)$ and interpret your answer.

b. Find $M'(85)$ and interpret your answer.

c. Use your answers from parts a and b to estimate the union membership percentage in 1986.

34. The number of full-time White House staff members can be modeled by $N(t) = -0.004t^3 + 0.7t^2 - 26t + 277$, where N is the number of full-time staff members and t is the number of years after 1900.

a. Find $N(75)$ and interpret your answer.

b. Find $N'(75)$ and interpret your answer.

c. Use your answers from parts a and b to estimate the number of full-time staff members in 1976.

35. A ball is thrown into the air, and its height above the ground, in feet, t seconds after release is given by $h(t) = -16t^2 + 96t + 30$.

a. Find $h'(t)$ and interpret your answer.

b. Find $h'(1)$ and $h'(2.5)$ and interpret your answers.

c. At what time will the derivative equal 0? Interpret in the context of the problem.

36. A rocket is launched into the air and its height above the ground, in feet, after t seconds is given by $h(t) = -16t^2 + 480t$.

a. Find $h'(t)$ and interpret your answer.

b. Find $h'(10)$ and $h'(25)$ and interpret your answers.

c. At what time will the derivative equal 0? Interpret in the context of the problem.

37. For $f(x)$,

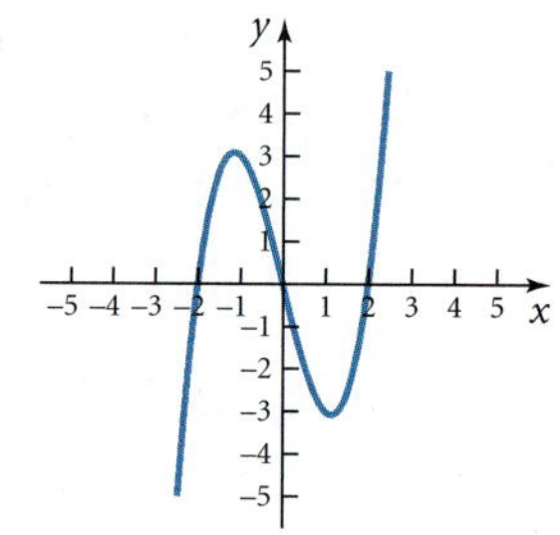

a. For what x interval(s) is $f'(x)$ positive?

b. For what x interval(s) is $f'(x)$ negative?

c. For what x value(s) is $f'(x)$ zero?

d. Which has a value closest to -2, $f'\left(\frac{1}{2}\right)$ or $f'(3)$?

38. For $f(x)$,

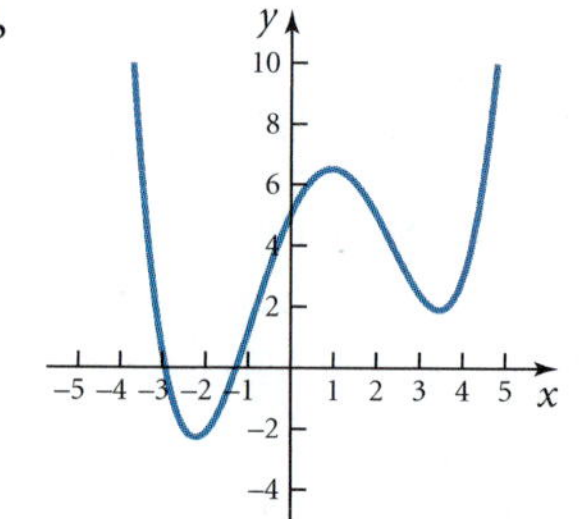

a. For what x interval(s) is $f'(x)$ positive?

b. For what x interval(s) is $f'(x)$ negative?

c. For what x value(s) is $f'(x)$ zero?

d. Which has a value closest to 2, $f'(-1)$ or $f'(2)$?

CALCULATOR CONNECTION

For each of the following functions:

a. Draw a graph of the function.

b. Find $f'(x)$.

c. Evaluate $f'(1)$ and $f'(-2)$.

d. Draw the tangent line to $f(x)$at $x = 1$ and $x = -2$.

39. $f(x) = x^3 - 4x$

40. $f(x) = x^4$

41. $f(x) = \frac{1}{x^2}$

42. $f(x) = \sqrt[4]{x}$ (evaluate at $x = 1$ only)

43. The percentage of the U.S. population living in poverty is given in the following table.

Year	Percent Living in Poverty
1992	14.8
1993	15.1
1994	14.5
1995	13.0
1996	13.1
1997	12.8
1998	12.5
1999	11.8
2000	11.3
2001	11.7
2002	12.3
2003	12.5
2004	12.7

Source: Census Bureau and *New York Times*.

a. Draw a scatter plot of the data, letting x be the number of years after 1990.

b. Find a quadratic function that models the data. Be sure to state the domain of the model. See Topic 4 for a discussion of regression.

c. Use your model to estimate the percentage living in poverty in 1998 and 2004. How good is the model?

d. Find the derivative, $f'(x)$, of your model.

e. Evaluate the derivative for 1998 and 2004. Interpret your answers.

f. At what rate was the percentage living in poverty changing in 1999?

g. Predict the percentage living in poverty in 2005 using $f(14)$ and $f'(14)$.

44. Average monthly unemployment claims for a certain area are given in the following table.

Year	Number of Claims
1992	1502
1993	1268
1994	1250
1995	1052
1996	1110
1997	1004
1998	959
1999	727
2000	725
2001	1126
2002	1292
2003	1286
2004	1887
2005	1431

Source: Daytona Beach News-Journal.

a. Draw a scatter plot of the data, letting x be the number of years after 1990.

b. Find a quadratic function that models the data. Be sure to state the domain of the model.

c. Use your model to estimate the number of claims in 1998 and in 2004. How good is the model?

d. Find the derivative, $f'(x)$, of your model.

e. Evaluate the derivative for 1998 and 2004. Interpret your answers.

f. At what rate was the number of claims changing in 2005?

g. Predict the number of claims in 2006 using $f(15)$ and $f'(15)$.

TOPIC

9 Derivatives of Algebraic Functions

IMPORTANT TOOLS IN TOPIC 9

- *Power Rule*
- *Derivative of a constant function*
- *Derivative of a linear function*
- *Sum and Difference Properties*
- *Constant Multiple Property*
- *Marginal analysis*

One major question leading to the development of calculus was how to find the slope of a curve. You now know that the answer to that question is the derivative! You also know that the derivative is calculated by evaluating the limit of the difference quotient:

$$f'(x) = \lim_{h \to 0} \frac{f(x + h) - f(x)}{h}$$

In this topic, you learn some rules that will make the evaluation of this limit much simpler to perform.

Before starting the study of these derivative rules, be sure to complete the Topic 9 Warm-up Exercises to be sure you have the algebra and calculus tools you will need.

TOPIC 9 WARM-UP EXERCISES

Be sure you can successfully complete the following exercises before starting Topic 9.

Algebra Warm-up Exercises

Convert each radical to exponential notation.

1. $\sqrt{x}$ **2.** $\sqrt[5]{x}$ **3.** $\sqrt[3]{x^2}$

Convert to radical form.

4. $x^{1/3}$ **5.** $x^{-1/2}$ **6.** $x^{3/5}$

Convert to exponential notation.

7. $\frac{1}{x}$ **8.** $-\frac{3}{x^5}$

Rewrite using only positive exponents.

9. x^{-6} **10.** $-2x^{-3}$

11. Given functions $f(x) = 4 - x^2$ and $g(x) = x + 2$,

a. Draw the graphs on one set of axes.

b. Determine the coordinates of the points of intersection.

c. For what values of x will $f(x) > g(x)$?

Calculus Warm-up Exercises

Find the derivative of each function.

1. $f(x) = 3x - 5$ **2.** $f(x) = 31$

3. $f(x) = x^2 + 2$ **4.** $f(x) = 2x^2 - 4x$

5. $f(x) = x^3 - 1$ **6.** $f(x) = \sqrt{x}$

7. $f(x) = \frac{3}{x}$

Answers to Warm-up Exercises

Algebra Warm-up Exercises

1. $x^{1/2}$ **2.** $x^{1/5}$ **3.** $x^{2/3}$ **4.** $\sqrt[3]{x}$ **5.** $\frac{1}{\sqrt{x}}$

6. $\sqrt[5]{x^3}$ **7.** x^{-1} **8.** $-3x^{-5}$ **9.** $\frac{1}{x^6}$ **10.** $-\frac{2}{x^3}$

11. a.

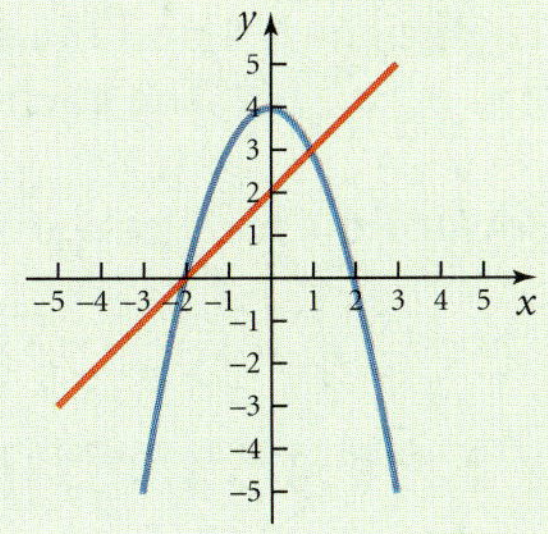

b. $x = -2, x = 1$ **c.** $-2 < x < 1$

Calculus Warm-up Exercises

1. 3 **2.** 0

3. $2x$ **4.** $4x - 4$

5. $3x^2$ **6.** $\dfrac{1}{2\sqrt{x}}$

7. $-\dfrac{3}{x^2}$

You now know that the derivative is calculated by evaluating the limit of the difference quotient:

$$f'(x) = \lim_{h \to 0} \frac{f(x+h) - f(x)}{h}$$

There is absolutely nothing wrong with continuing to use this definition because it will work for any differentiable function. You likely have become painfully aware, however, that the algebra required to simplify the difference quotient is often tedious and difficult. Imagine, for instance, how much fun it would be to use the difference quotient to find the derivative of $f(x) = x^{10}$!

Examine some of the derivatives we have calculated so far. Can you see a pattern that might simplify the work?

$f(x)$	$f'(x)$
k (constant function)	0
$mx + b$ (linear function)	m
x^2	$2x$
x^3	$3x^2$

We know that

- The derivative of a constant function is zero because the graph of a constant function is a horizontal line, which has slope of zero.
- The derivative of a linear function is its slope, the m value in the equation.
- The derivatives of the other functions in the table were derived using the limit formula and difference quotient.

Looking at these four functions and derivatives above, do you see a pattern? Could you determine the derivative of x^4 or x^{10}?

- You should first observe that *the power of the derivative is always one less than the power of the function*. Thus, the derivative of $f(x) = x^4$ would have exponent 3 and the derivative of $f(x) = x^{10}$ would have exponent 9.
- You should also observe that *the coefficients of the derivatives are the exponents of the original function.*

What would the derivative of $f(x) = x^4$ be? The coefficient is 4 and the power is 3, so $f'(x) = 4x^3$. What would the derivative of $f(x) = x^{10}$ be? The coefficient is 10 and the power is 9, so $f'(x) = 10x^9$.

We can state this pattern as a rule called the **Power Rule for Derivatives.**

Power Rule for Derivatives
If $f(x) = x^n$, then $f'(x) = D_x(x^n) = nx^{n-1}$, where n is any real number.

The Power Rule is valid for any real value of n, as Examples 1 and 2 show.

Example 1: Verify that $D_x\sqrt{x} = \dfrac{1}{2\sqrt{x}}$. (See Example 6c in Topic 8.)

Solution: First rewrite $\sqrt{x}$ using exponents:

$$\sqrt{x} = x^{1/2}$$

Then use the Power Rule:

$$\begin{aligned} D_x x^{1/2} &= \tfrac{1}{2}x^{1/2-1} \\ &= \tfrac{1}{2}x^{-1/2} \end{aligned}$$

Rewrite the negative exponent:

$$D_x x^{1/2} = \frac{1}{2} \cdot \frac{1}{x^{1/2}}$$

Rewrite as a radical and multiply:

$$D_x x^{1/2} = \frac{1}{2\sqrt{x}}$$ ■

Example 2: Verify that $D_x\dfrac{1}{x} = -\dfrac{1}{x^2}$. (See Example 6d in Topic 8.)

Solution: First rewrite $\dfrac{1}{x}$ using exponents:

$$\frac{1}{x} = x^{-1}$$

Use the Power Rule:

$$D_x x^{-1} = -1x^{-2}$$

Rewrite with a positive exponent:

$$D_x x^{-1} = -\frac{1}{x^2}$$ ■

Example 3: Find $f'(x)$ for each of the following functions using the Power Rule.

a. x^{17} **b.** x^{-4} **c.** $x^{2.3}$ **d.** $x^{1/3}$

Solution:

a. $17x^{16}$

b. $-4x^{-4-1} = -4x^{-5}$

c. $2.3x^{2.3-1} = 2.3x^{1.3}$

d. $\frac{1}{3}x^{1/3-1} = \frac{1}{3}x^{-2/3}$ or $\dfrac{1}{3} \cdot \dfrac{1}{x^{2/3}}$ or $\dfrac{1}{3\sqrt[3]{x^2}}$ ■

Check Your Understanding 9.1

Find the derivative of each of the following functions using the Power Rule.

1. $f(x) = 11x - 5$ **2.** $f(x) = x^{11}$ **3.** $f(x) = 11$

4. $f(x) = \sqrt[4]{x}$ **5.** $f(x) = \dfrac{1}{x^{11}}$

You now have three properties of derivatives in your Tool Kit: the Power Rule and its two corollaries. (A corollary is a result that follows logically from another statement.)

Properties of Derivatives, Part I

Power Rule for Derivatives: $D_x x^n = nx^{n-1}$

Corollaries: $D_x k = 0$ (where k is a constant)

$D_x(mx + b) = m$

Warning! Be sure to see that **the limit of a constant is the constant,** but **the derivative of a constant is zero.**

The following additional properties of derivatives will make differentiation even easier.

Properties of Derivatives, Part II

Given functions $f(x)$ and $g(x)$,

Sum Rule: $D_x(f(x) + g(x)) = f'(x) + g'(x)$

Difference Rule: $D_x(f(x) - g(x)) = f'(x) - g'(x)$

Constant Multiple Rule: $D_x(c \cdot f(x)) = c \cdot f'(x)$

MATHEMATICS CORNER 9.1

Power Rule Corollaries

The Power Rule states that $D_x x^n = nx^{n-1}$. We also know that any constant, k, can be written as kx^0. We can now verify the two corollaries to the Power Rule.

$$D_x(k) = D_x(kx^0) = k \cdot 0x^{0-1} = 0$$

$$D_x(mx + b) = D_x(mx) + D_x(b) = mx^0 + 0 = m$$

Example 4:

$$\begin{aligned} D_x(x^2 + 3x - 6) &= D_x x^2 + D_x(3x) + D_x(-6) \\ &= 2x + 3 + 0 \\ &= 2x + 3 \end{aligned}$$

This result should seem logical because we already know that $D_x x^2 = 2x$ and $D_x(3x - 6) = 3$. ■

Example 5:

$$\begin{aligned} D_x(2x^3) &= 2 \cdot D_x x^3 \\ &= 2(3x^2) \\ &= 6x^2 \end{aligned}$$

In practice, you will apply the Constant Multiple Property without actually writing each step. Here, for instance, we simply multiplied the exponent, 3, by the coefficient, 2, to obtain the coefficient for the derivative. The exponent of the derivative is one less than the exponent of the function, as per the Power Rule. ■

Example 6: Find the derivative of each of the following functions.

a. $4x^{-3}$ **b.** $2\sqrt{x}$ **c.** $\dfrac{6}{x^3}$

d. $2x^5 - 7x^3 + 4x^2 - 11$ **e.** $2.1x^2 - 0.025x^3$ **f.** $3x^{2.3} - 7x^{0.2}$

g. $3x^{2/3} + 5x^{1/3}$ **h.** $\dfrac{5}{2x^3} + \dfrac{5}{2}x^3$

Solution:

a.
$$\begin{aligned} D_x(4x^{-3}) &= 4 \cdot D_x x^{-3} && \text{Constant Multiple Property} \\ &= 4(-3x^{-3-1}) && \text{Power Rule} \\ &= -12x^{-4} \end{aligned}$$

b.
$$\begin{aligned} D_x(2\sqrt{x}) &= 2 \cdot D_x\sqrt{x} && \text{Constant Multiple Property} \\ &= 2 \cdot D_x x^{1/2} && \text{Rewrite } \sqrt{x} \text{ using exponents} \\ &= 2\left(\frac{1}{2} \cdot x^{1/2-1}\right) && \text{Power Rule} \\ &= x^{-1/2} \quad \text{or} \quad \frac{1}{\sqrt{x}} \end{aligned}$$

c.
$$\begin{aligned} D_x\frac{6}{x^3} &= D_x(6x^{-3}) && \text{Rewrite the rational expression} \\ &= 6(-3x^{-4}) && \text{Constant Multiple Property and Power Rule} \\ &= -18x^{-4} \quad \text{or} \quad -\frac{18}{x^4} && \text{Rewrite the negative exponent} \end{aligned}$$

d.
$$\begin{aligned} D_x(2x^5 - 7x^3 + 4x^2 - 11) &= D_x(2x^5) - D_x(7x^3) + D_x(4x^2) - D_x 11 && \text{Sum and Difference Properties} \\ &= 10x^4 - 21x^2 + 8x && \text{Constant Multiple Property and Power Rule} \end{aligned}$$

Warning! In this text, the general rule of thumb when finding derivatives will be to *write the derivative in the same form as the original function.* There are two reasons for doing so:

1. ***Determining Points of Discontinuity.*** The points of discontinuity of $x^{-1/2}$ and $-18x^{-4}$ may not be obvious. The points of discontinuity of $\frac{1}{\sqrt{x}}$ and $-\frac{18}{x^4}$ are more easily determined, knowing that denominators of fractions cannot be zero and that no negative numbers may appear under a square-root sign.
2. ***Solving Equations.*** In later topics, it will be necessary to set the derivative equal to zero and solve for x. Seeing $x^{-1/2} = 0$ or $-18x^{-4} = 0$ may seem to imply that $x = 0$ is the solution. Seeing $\frac{1}{\sqrt{x}} = 0$ or $-\frac{18}{x^4} = 0$ should make it a little more apparent that $x \neq 0$ and that no value of x solves the equation.

e. $D_x(2.1x^2 - 0.025x^3) = 2.1(2x) - 0.025(3x^2)$ Constant Multiple Property and Power Rule

$$= 4.2x - 0.075x^2$$

f. $D_x(3x^{2.3} - 7x^{0.2}) = 3(2.3x^{2.3-1}) - 7(0.2x^{0.2-1})$ Constant Multiple Property and Power Rule

$$= 6.9x^{1.3} - 1.4x^{-0.8}$$

g. $D_x(3x^{2/3} + 5x^{1/3}) = 3\left(\tfrac{2}{3}x^{2/3-1}\right) + 5\left(\tfrac{1}{3}x^{1/3-1}\right)$

$$= 2x^{-1/3} + \tfrac{5}{3}x^{-2/3}$$

h. $D_x\left(\frac{5}{2x^3} + \frac{5}{2}x^3\right) = D_x\left(\frac{5}{2x^3}\right) + D_x\left(\frac{5}{2}x^3\right)$

In the first term, x^3 is part of the denominator, but in the second term it is not. Rewrite the first term using negative exponents and then apply the Power Rule.

$$D_x\left(\frac{5}{2x^3} + \frac{5}{2}x^3\right) = D_x\left(\frac{5}{2}x^{-3}\right) + D_x\left(\frac{5}{2}x^3\right)$$

The coefficient 2 in the denominator did not move; it stayed in the denominator because the exponent applies only to the x. Now apply the Power Rule:

$$D_x\left(\frac{5}{2x^3} + \frac{5}{2}x^3\right) = \frac{5}{2}(-3x^{-4}) + \frac{5}{2}(3x^2)$$

$$= -\frac{15}{2}x^{-4} + \frac{15}{2}x^2$$

Now rewrite the first term as a rational expression, which is the way the original function was given.

$$D_x\left(\frac{5}{2x^3} + \frac{5}{2}x^3\right) = -\frac{15}{2x^4} + \frac{15}{2}x^2$$ ■

> **Warning!** It is important that you not get bogged down in the formulas and rules for differentiation. Always keep in mind that *finding the derivative of a function means calculating its instantaneous rate of change or determining the slope of the curve at any point.*

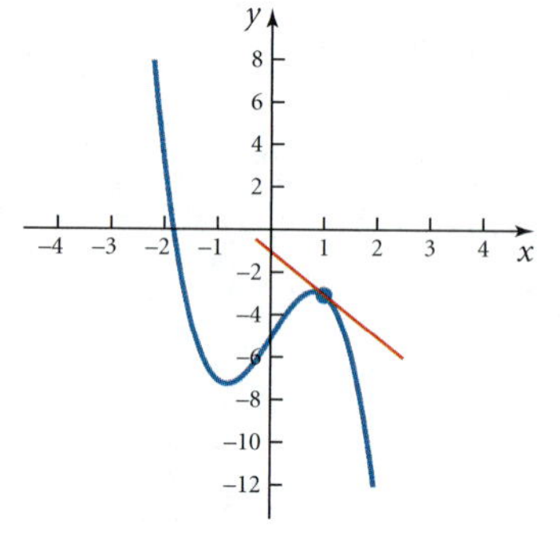

Figure 9.1

Example 7: Given $f(x) = -2x^3 + 4x - 5$, find the slope of the curve at the point where $x = 1$. Sketch a graph of the function and the tangent line to verify your result.

Solution:

$$f'(x) = -6x^2 + 4$$

The slope of the curve when $x = 1$ is the value of the derivative at $x = 1$. Evaluating gives $f'(1) = -6(1)^2 + 4 = -2$. So the slope of the curve at the point where $x = 1$ is -2. Figure 9.1 shows the function and the tangent line where $x = 1$. ■

Calculator Corner 9.1

Numerical derivatives can be evaluated using the TI-83/84. To evaluate a numerical derivative, press **MATH**, select option 8: nDeriv, and press **ENTER**. You will now see nDeriv(on the screen. Next enter the function, followed by comma, x, comma, and the value for which you want to evaluate the derivative. Press **ENTER**. The calculator gives you the value of the derivative at that point.

For instance, to evaluate $f(x) = -2x^3 + 4x - 5$ when $x = 1$, you would enter nDeriv($-2x^3 + 4x - 5, x, 1$) on the TI-83/84. The calculator returns a value of -2.

Example 8: Union membership as a percentage of the total labor force can be modeled by $M(x) = 0.0002x^3 - 0.049x^2 + 3.75x - 68$, where M is the membership as a percentage of the labor force and x is the number of years after 1900. Find the rate at which membership is changing in 1980.

Solution: The key words in this problem are *rate* and *changing*. Rate of change means derivative, so

$$M'(x) = 0.0006x^2 - 0.098x + 3.75$$

We are seeking the rate of change for 1980, which is 80 years after 1900, so we calculate

$$M'(80) = 0.0006(80)^2 - 0.098(80) + 3.75$$
$$M'(80) = -0.25$$

Union membership as a percentage of the labor force was decreasing by 0.25% per year in 1980. ■

Example 9: A ball is thrown into the air from a 20-foot-high platform. Its height t seconds after being thrown is given by $h(t) = -16t^2 + 64t + 20$, where h is feet above the ground. Evaluate $h'(1)$, $h'(2)$, and $h'(3.5)$ and interpret your answers.

Solution:

$h(t) = -16t^2 + 64t + 20$ so $h'(t) = -32t + 64$

$h'(1) = 32$ feet per second. — 1 second after its release, the velocity of the ball was 32 feet per second, meaning the ball was rising at a rate of 32 feet per second at that time.

$h'(2) = 0$ feet per second. — 2 seconds after its release, the velocity of the ball was 0 feet per second, meaning the ball is not rising or falling at this point. This point must be the peak, or highest point, in the path of the ball.

$h'(3.5) = -48$ feet per second. — 3.5 seconds after its release, the velocity of the ball is -48 feet per second, which means the ball is dropping at a rate of 48 feet per second. ■

Check Your Understanding 9.2

In Exercises 1 through 4, find the derivative of each function.

1. $f(x) = 2x^4 - 3x^2 + 7x - 9$

2. $f(x) = \dfrac{3}{x^5}$

3. $f(x) = 4\sqrt[3]{x}$

4. $f(x) = -3x^{-2} + 5x^{1.8}$

5. Given $f(x) = 2x^4 - 3x^2 + 7x - 9$, find the slope of the curve at $x = -1$.

Marginal Analysis

In business and economics, **productivity** is defined as the ratio of output units to input units, such as labor or materials. **Marginal productivity** is the additional output resulting from adding one more unit of input. To economists, the concept of "marginal" is key. The word *marginal* means "change" and always refers to the change in one quantity when a related quantity changes by 1 unit. For instance, consider from an economist's viewpoint the marginal benefits of reading the next section of this lesson. The marginal benefit here would be the change in benefits (a better understanding of the material or an improved test score) you receive from an increase in study time (perhaps an additional 20 minutes to read the next section). When you decide whether to continue reading, you weigh the marginal benefits of more study (the added understanding or improved test score) with the marginal costs of more study (the things you would have to give up to continue reading). If the marginal benefits outweigh the marginal costs, you should continue reading the section. Many decisions in business and industry follow the same path.

To make intelligent decisions, management usually looks at the effect that changes in production will have on cost and revenue.

Costs may be *fixed* or *variable*. **Fixed costs** are those that are not dependent on the number of units produced, such as management salaries, property taxes, maintenance, or security. **Variable costs** are dependent on the number of items produced and include such items as cost of materials and labor. **Total costs** are the sum of the fixed costs and the variable costs. Let x be the number of units produced, k the cost per unit, and F the fixed costs. Then the total costs are given by $C(x) = kx + F$, which is a linear cost model. You should realize that $C(x)$ is always a positive value.

Revenue is the amount received from the sale of the units produced. Let p be the selling price per unit and x be the number of units produced. Then revenue = (the number of units produced) $\times$ (the selling price per unit), or $R(x) = px$, which is a linear revenue model if price stays constant. Again you should see that $R(x)$ is always a positive value.

Profit is defined as *revenue minus cost*. Thus, $P(x) = R(x) - C(x)$. Profit may be positive, negative, or zero. $P(x) > 0$ denotes a *gain*; $P(x) < 0$ denotes a *loss*.

Suppose a retail chain is trying to decide whether or not to open a new store in a particular area. From a financial standpoint, management wants to open the store if the new store brings in more money for the chain. Before making the decision, however, management must also consider the additional costs and expenses of opening the new store. Management has to determine if the additional revenues outweigh the additional costs. These additional costs and additional revenues are called the marginal cost and marginal revenue. **Marginal cost** is the additional cost of adding one more input unit; **marginal revenue** is the additional revenue generated from the addition of one more input unit. *Marginal* always implies a *change*, so the derivative will be used in approximating marginal cost or marginal revenue.

Let x be the number of stores the chain currently operates. If a new store is added, there will be $x + 1$ stores. The marginal cost of adding the additional store is $C(x + 1) - C(x)$. The rate at which costs are changing at that point can be found by $\dfrac{C(x + 1) - C(x)}{(x + 1) - x}$, which is the average rate of change of costs. If the graph of the cost function is not changing too quickly at this point, the average rate of change is quite close to the instantaneous rate of change. Thus, **marginal cost** is approximately equal to $C'(x)$.

Similarly, the marginal revenue generated from adding one more store is $R(x + 1) - R(x)$. The rate at which revenue is changing at that point is found by $\dfrac{R(x + 1) - R(x)}{(x + 1) - x}$. Just as we considered for costs, the average rate of change of revenue at this point is quite close to the instantaneous rate of change. Thus, **marginal revenue** is approximately equal to $R'(x)$.

Profit is revenue minus cost, so **marginal profit** is defined as marginal revenue minus marginal cost.

Marginal Cost, Revenue, and Profit

If x is the number of units produced over some time interval, then

Total cost $= C(x)$ | Total revenue $= R(x)$

Marginal cost $= C'(x)$ | **Marginal revenue** $= R'(x)$

Total profit: $P(x) = R(x) - C(x)$

Marginal profit: $P'(x) = R'(x) - C'(x)$

Thus, for a function $f(x)$, the derivative, $f'(x)$, may represent

- Instantaneous rate of change of $f(x)$
- Velocity of an object whose path is modeled by $f(x)$
- Slope of the graph of $f(x)$ at any point
- Marginal cost, marginal revenue, marginal profit

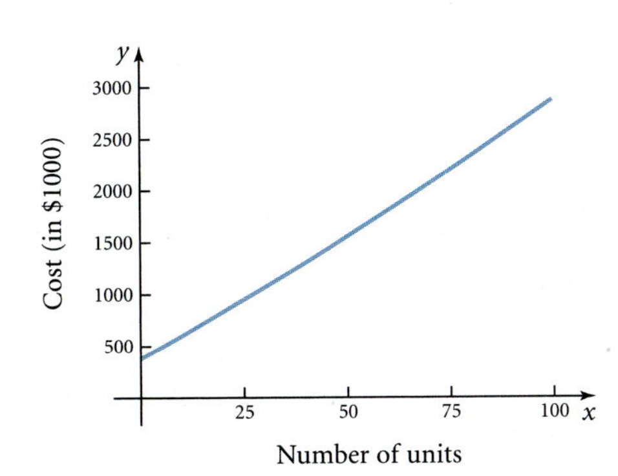

Figure 9.2

Example 10: The annual cost of operating a large plant (in thousands of dollars) is given by $C(x) = 0.03x^2 + 22x + 377$, where x is the number of units produced annually in thousands. Find $C'(10)$ and $C'(25)$ and interpret your answers.

Solution: Figure 9.2 shows that costs increase as more units are produced. The function is quadratic, so costs rise more rapidly as more units are produced.

$C'(x) = 0.06x + 22$

$C'(10) = 22.6$. When 10,000 units are produced, the cost of producing one thousand more units is \$22,600.

$C'(25) = 23.5$. When 25,000 units are produced, the cost of producing one thousand more units is \$23,500. ■

Example 11: The cost and revenue functions for production of table saws are $C(x) = 7200 + 6x$ and $R(x) = 20x - 0.0033x^2$, where x is number of saws produced.

a. Find the marginal cost.
b. Find the marginal revenue.
c. Evaluate $R'(1500)$ and $R'(4500)$ and interpret your answers.
d. Graph $C(x)$ and $R(x)$ on the same axes. Determine the break-even point(s) and the regions of gain and loss.
e. Find the profit function, $P(x)$.
f. Find the marginal profit.
g. Evaluate $P'(1500)$ and $P'(4500)$ and interpret your answers.

Solution:

a. Marginal cost = $C'(x) = 6$, because $C(x)$ is a linear function. The additional cost of producing one more saw is \$6.

b. Marginal revenue = $R'(x) = R'(x) = 20 - 0.0066x$

c. $R'(1500) = 10.1$. When 1500 saws are produced, the additional revenue expected by producing one more saw is \$10.10.

$R'(4500) = -9.70$. When 4500 saws are produced, the expected revenue for producing one more saw decreases by \$9.70.

d.

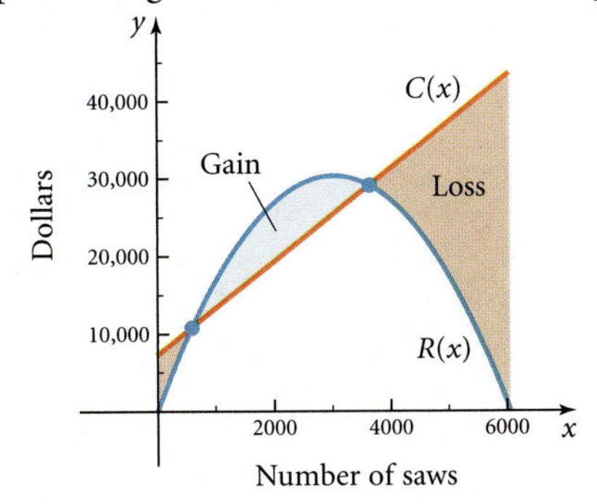

The break-even points are the points where $C(x) = R(x)$.

$$\begin{aligned} C(x) &= R(x) \\ 7200 + 6x &= 20x - 0.0033x^2 \\ 0.0033x^2 - 14x + 7200 &= 0 \end{aligned}$$

Solving with the Quadratic Formula gives $x_1 \approx 598.8$ and $x_2 \approx 3643.6$. Thus, the break-even points are approximately (599, \$10,794) and (3644, \$29,064). You should notice that when 599 saws are produced, $C(599) = \$10{,}794$ and $R(599) = \$10{,}796$, so the cost and revenue are not exactly the same because of rounding. Similarly, when 3644 saws are produced, $C(3644) = \$29{,}064$ and $R(3644) \approx \$29{,}060$, so again the cost and revenue are not exactly the same because of rounding.

When costs exceed revenue, or $C(x) > R(x)$, there is a loss. The company shows a loss if fewer than 599 saws or more than 3644 saws are produced.

When revenue exceeds cost, or when $C(x) < R(x)$, there is a gain, which is certainly what the company wants. The company shows a gain when producing between 599 and 3644 saws.

e. $$\begin{aligned} P(x) &= R(x) - C(x) \\ &= (20x - 0.0033x^2) - (7200 + 6x) \\ &= -0.0033x^2 + 14x - 7200 \end{aligned}$$

f. Marginal profit $= P'(x) = -0.0066x + 14$

g. $P'(1500) = 4.1$. When 1500 saws are produced, the company's profit is expected to increase by \$4.10 with the production of one more saw.

$P'(4500) = -15.7$. When 4500 saws are produced, the company's profit is expected to decrease by \$15.70 with the production of one more saw. ■

Calculator Corner 9.2

Graph $y_1 = 7200 + 6x$ and $y_2 = 20x - 0.0033x^2$ using a window of $[0, 6000]$ for x and $[0, 40{,}000]$ for y. To find the points of intersection, press **2nd TRACE** (CALC) and select option 5:intersect. Trace to a point on the curve close to the intersection point and press **ENTER** three times. The calculator will give the coordinates of the point of intersection. Repeat the process to find the other point.

Check Your Understanding Answers

Check Your Understanding 9.1

1. 11

2. $11x^{10}$

3. 0

4. $\frac{1}{4}x^{-3/4}$ or $\dfrac{1}{4\sqrt[4]{x^3}}$

5. $-11x^{-12}$ or $-\dfrac{11}{x^{12}}$

Check Your Understanding 9.2

1. $8x^3 - 6x + 7$

2. $-\dfrac{15}{x^6}$

3. $\dfrac{4}{3\sqrt[3]{x^2}}$

4. $6x^{-3} + 9x^{0.8}$

5. 5

Topic 9 Review

9

This topic introduced several basic rules for differentiation.

CONCEPT	EXAMPLE
The **Power Rule** for differentiation of algebraic (polynomial, rational, or radical) functions states that $D_x x^n = nx^{n-1}$	$D_x x^{12} = 12x^{11}$ $D_x x^{-1.2} = -1.2x^{-2.2}$ $D_x \sqrt[6]{x} = D_x x^{1/6} = \frac{1}{6}x^{-5/6} = \dfrac{1}{6\sqrt[6]{x^5}}$
The **derivative of a constant function** is 0: $D_x k = 0$	$D_x 93 = 0$
The **derivative of a linear function** is its slope: $D_x(mx + b) = m$	$D_x(5x + 7) = 5$ $D_x(7 - 4x) = -4$
Finding derivatives is made even simpler with these additional properties. **Sum Property** $D_x[f(x) + g(x)] = f'(x) + g'(x)$ **Difference Property** $D_x[f(x) - g(x)] = f'(x) - g'(x)$ **Constant Multiple Property** $D_x[c \cdot f(x)] = c \cdot f'(x)$	$D_x(x^3 - x^2 + 5) = 3x^2 - 2x$ $D_x(7x^4) = 7 \cdot D_x(x^4) = 28x^3$
Marginal analysis in economics refers to the change in production for one additional unit of input. If x is the number of units produced, **Marginal cost** is $C'(x)$. **Marginal revenue** is $R'(x)$. **Marginal profit** is $P'(x) = R'(x) - C'(x)$.	The cost of producing x units of a product is given by $C(x) = 60 + 3x$, and the revenue generated from selling x units of the product is given by $R(x) = x^2 + 4x + 18$. When 4 units are produced and sold, Marginal cost $= C'(4) = 3$ Marginal revenue $= R'(4) = 12$ Marginal profit $= 9$

NEW TOOLS IN THE TOOL KIT

- Power Rule for finding the derivative of x^n
- Sum Property for finding derivatives of sums of functions
- Difference Property for finding derivatives of differences of functions
- Constant Multiple Property for finding derivatives of cx^n
- Marginal cost
- Marginal revenue
- Marginal profit

Topic 9 Exercises

9

Find $f'(x)$ for each of the following functions.

1. x^{12}

2. x^{27}

3. x^{-4}

4. $x^{5/4}$

5. $\sqrt[5]{x}$

6. x^{-7}

7. $x^{2/3}$

8. $\sqrt[4]{x^3}$

9. $x^{1.7}$

10. $x^{0.73}$

11. $\dfrac{2}{x^3}$

12. $\dfrac{x^6}{3}$

13. $\dfrac{x^3}{2}$

14. $\dfrac{3}{x^6}$

15. $\dfrac{2}{3x^4}$

16. $-\dfrac{3}{2x^5}$

17. $4 - 3x$

18. $5x - 2$

19. $3x^4 - 2x^3 + 7x + 11$

20. $-2x^5 + 7x^4 - 3x^2 + 9$

21. $x^3 - 4x + \dfrac{6}{x}$

22. $5x^{2/3} - 4x^{-3}$

23. $2x^{3/4} + 5x^{-1}$

24. $2x^5 + 3x - \dfrac{2}{x^3}$

25. $4\sqrt{x} - 3x^{-1.3}$

26. $4.2x^{-0.8} - 2\sqrt[3]{x}$

27. $\dfrac{2x^3 - x^2}{4x}$

28. $\dfrac{x^4 + 3x^2}{x^3}$

29. $5x^4(x^2 - 7)$

30. $2x^{1/3}(6x^{2/3} - 4x^{-1/3})$

31. $17\sqrt[3]{x} + \dfrac{5}{x^4} - 3x^2 + 2$

32. $7\sqrt[4]{x} - \dfrac{4}{x^3} + 2x^3 - 5$

33. $(5x - 3)^2$

34. $(2 - 7x)^2$

Evaluate the derivative of the given function at the indicated value.

35. $f(x) = x^2 - 3$; find $f'(3)$

36. $f(x) = 2x - x^2$; find $f'(-1)$

37. $f(x) = x^3$; find $f'\left(-\frac{1}{2}\right)$

38. $f(x) = x^4$; find $f'\left(-\frac{1}{2}\right)$

39. $f(x) = \sqrt{x}$; find $f'(4)$

40. $f(x) = \sqrt[3]{x}$; find $f'(-8)$

41. $f(x) = x^2 + \dfrac{2}{x} - 3$; find $f'(1)$

42. $f(x) = 3x - \dfrac{4}{x^2}$; find $f'(-1)$

43. Find the slope and equation of the tangent line to $f(x) = x^3 + 2x^2 + 5x - 3$ at the following points.

a. $x = -1$ **b.** $x = 2$

44. Find the slope and equation of the tangent line to $f(x) = \sqrt[3]{x}$ at the following points.

a. $x = 8$ **b.** $x = -1$

45. A ball is thrown into the air from a 6-foot-high platform. Its height, in feet, above the ground after t seconds is given by $h(t) = -16t^2 + 50t + 60$.

a. Find the velocity of the ball after 1 second.

b. Find the velocity of the ball after 3 seconds.

c. When is the velocity of the ball 0?

46. A projectile is launched into the air. Its height, in feet, above the ground after t seconds is given by $h(t) = -16t^2 + 84t + 100$.

a. Find the velocity of the projectile after 2 seconds.

b. Find the velocity of the projectile after 4.5 seconds.

c. When is the velocity of the projectile 0?

47. Sales of a new CD, in thousand dollars, t months after its release is given by $S(t) = 400\sqrt{t}$.

a. Find the rate at which sales are changing after 4 months.

b. Find the rate at which sales are changing after 10 months.

48. Union membership as a percentage of the labor force can be modeled by $M(x) = 0.0003x^3 - 0.066x^2 + 4.64x - 81$, where x is the number of years after 1900 and M is membership as a percentage of the labor force.

a. Find the rate at which membership is changing in 1960.

b. Find the rate at which membership is changing in 1985.

49. The population of a colony of bacteria grows according to the model $P(t) = 450t^{1.87}$, where P is the number of bacteria and t is in days.

a. Evaluate $P(4)$ and interpret your answer.

b. Evaluate $P'(4)$and interpret your answer.

50. The number of full-time White House staff members can be modeled by $N(t) = -0.004t^3 + 0.7t^2 - 26t + 277$, where N is the number of full-time staff members and t is the number of years after 1900.

a. Evaluate $N(80)$ and interpret your answer.

b. Evaluate $N'(80)$ and interpret your answer.

51. A newspaper has a circulation of 500,000 readers. An ad that runs for x days will be seen by $A(x) = 500{,}000 - (250{,}000/x)$ people. How fast is the number of potential customers growing when the ad has run for one week?

52. The intensity of light is inversely proportional to the square of the distance from the source and is given by $I(d) = 0.65/d^2$, where and I is the intensity in foot-candles and d is the distance in feet from the source. How fast is the intensity of a light changing at a point 10 feet from the source?

53. Given $C(x) = 40 + 2x$ and $R(x) = 0.01x^2 + 2.3x + 10$, find the following functions.

a. $P(x)$

b. $C(50)$, $R(50)$, and $P(50)$. Interpret your answers.

c. $C'(x)$, $R'(x)$, and $P'(x)$

d. $C'(50)$, $R'(50)$, and $P'(50)$. Interpret your answers.

54. Given $R(x) = 45x - 0.4x^2$ and $C(x) = 200 + 5x$, find the following.

a. $P(x)$

b. $C(20)$, $R(20)$, and $P(20)$. Interpret your answers.

c. $C'(x)$, $R'(x)$, and $P'(x)$

d. $C'(20)$, $R'(20)$, and $P'(20)$. Interpret your answers.

55. A company that manufactures cell phones finds that the cost of producing x phones is given by $C(x) = 250 + 40\sqrt{x}$.

a. Find the marginal cost function.

b. Find the marginal cost when 100 phones are produced. Interpret your answer.

56. The cell phone manufacturing company in Exercise 55 finds that the profit from producing x phones is given by $P(x) = 0.2x + 0.25x^2$ dollars.

a. Find the marginal profit function.

b. Find the marginal profit when 100 phones are produced. Interpret your answer.

MATHEMATICS CORNER  

Measures of Central Tendency

There are three ways to describe a "typical member" of a set of data. In statistics, they are referred to as **measures of central tendency**.

1. The **mean** is the average of the data. The mean is calculated by adding all the data values and then dividing by the number of entries.
2. The **median** is the exact middle of the data, if the data are arranged in order. Half the entries will fall above this number, and half the entries will fall below this number.
3. The **mode** is the most frequently occurring number in the data set. Modes may or may not be representative of the entire set. There may be more than one mode for a given set of data.

CALCULATOR CONNECTION

57. The annual median sales prices for existing houses sold by real estate agents in Volusia and Flagler counties are given below.

Year	Median Sales Price
1995	$71,100
1996	$75,200
1997	$75,500
1998	$79,000
1999	$84,500
2000	$88,300
2001	$96,770
2002	$109,768
2003	$131,300
2004	$142,000
2005	$183,800
2006	$224,600

Source: Daytona Beach News-Journal.

a. Find the cubic regression function, $f(x)$, that best models the data. Let x be the number of years after 1990.

b. Sketch the graph of $f(x)$.

c. How fast was the median sales price changing in 2002?

d. Evaluate $f(13)$ and $f'(13)$ and interpret your answers.

58. The National Center for Education Statistics shows the following trends in the percent of public school classrooms with Internet access.

Year	Percent of Classrooms with Internet Access
1994	3
1995	8
1996	14
1997	27
1998	51
1999	64
2000	77
2001	87
2002	92
2003	93

a. Find a power regression equation, $f(x) = ax^b$, that best models the data. Let x be the number of years after 1990.

b. Sketch the graph of $f(x)$.

c. How fast was the percent changing in 1998?

d. Evaluate $f(11)$ and $f'(11)$ and interpret your answers.

59. The total cost of producing x units of a product is given in the following table.

Quantity, x Units	Total Cost of Production
0	$12
1	$24
2	$27
3	$31
4	$39
5	$53
6	$73
7	$99

a. When 3 units are produced, use the table to determine the marginal cost.

b. When 6 units are produced, use the table to determine the marginal cost.

c. Find the cubic cost function using the regression features of your calculator.

d. Use the cost function from part c to determine $C'(3)$ and $C'(6)$.

e. Compare your answers in part d to your answers from parts a and b. How well does the cost function reflect the true marginal cost?

f. Predict the cost of producing 8 units.

60. The total cost of producing x units of a product is given in the following table.

Quantity, x Units	Total Cost of Production
0	$30
1	$55
2	$75
3	$85
4	$100
5	$120
6	$145
7	$185
8	$240
9	$310
10	$395

a. Use the table to find the marginal cost when 4 units are produced.

b. Use the table to find the marginal cost when 8 units are produced.

c. Find the quadratic cost function using the regression features of your calculator.

d. Use the cost function from part c to find $C'(4)$ and $C'(8)$.

e. Compare your answers in part d to your answers from parts a and b. How well does the cost function reflect the true marginal cost?

f. Predict the cost of producing 11 units.

TOPIC

10

Product, Quotient, and Chain Rules

IMPORTANT TOPICS IN TOPIC 10

- *Product Rule*
- *Quotient Rule*
- *Chain Rule: Power Form*

The Power Rule simplifies finding the derivative of functions of the form $f(x) = ax^n$. You also know that the derivative of the sum of two functions is the sum of the two derivatives and the derivative of the difference of two functions is the difference of the two derivatives. Derivatives of products of two functions and quotients of two functions, however, require special rules, which is the focus of this topic. Before starting your study of these special rules for derivatives, you should complete the Topic 10 Warm-up Exercises to brush up on the algebra and calculus skills needed.

TOPIC 10 WARM-UP EXERCISES

Be sure you can complete the following exercises before starting Topic 10.

Algebra Warm-up Exercises

Simplify each of the following expressions using the laws of exponents.

1. $x^2 \cdot x^5$ **2.** x^5/x^8 **3.** $(x^4)^3$

Expand each of the following expressions.

4. $(5x - 3)(2x + 7)$

5. $(x^2 - 5x)(2x^3 + 3x^2)$

Find the greatest common factor (GCF) of the following pairs.

6. x^2y^6 and $3x^4y^2$

7. $15x^3(y - 3)^2$ and $6x^2(y - 3)^3$

Factor each expression.

8. $x^2y^6 - 3x^4y^2$

9. $15x^3(y - 3)^2 + 6x^2(y - 3)^3$

Simplify each of the following rational expressions.

10. $\dfrac{3x^2y^4}{9x^6y}$

11. $\dfrac{8(x - 3)^2}{12(x - 3)^5}$

12. Expand and simplify $(2x^3 - 4)3x + 2x(3x^2)$.

13. Solve $\dfrac{3x - 6}{x^2 - 9} = 0$ for x.

14. Given $f(x) = x^3$ and $g(x) = 2x - 1$, find $f(g(x))$.

15. Find the equation of the line with slope $\frac{2}{3}$ passing through $(4, -1)$. Write the answer in slope–intercept form, $y = mx + b$.

Calculus Warm-up Exercises

1. Find the points of discontinuity for $f(x) = \dfrac{3x - 6}{x^2 - 9}$.

2. Use the limit definition to find the derivative of $f(x) = 3x^2 - 5x + 2$.

Find the derivative of each function using the Power Rule.

3. x^{12}

4. $5x^{-3}$

5. $\sqrt[4]{x}$

6. $4/x^7$

7. $2x^6 - 3x^2 + 7x - 4$

Answers to Warm-up Exercises

Algebra Warm-up Exercises

1. x^7 **2.** x^{-3} or $1/x^3$ **3.** x^{12}

4. $10x^2 + 29x - 21$ **5.** $2x^5 - 7x^4 - 15x^3$

6. x^2y^2 **7.** $3x^2(y - 3)^2$

8. $x^2y^2(y^4 - 3x^2)$

9. $3x^2(y - 3)^2[5x + 2(y - 3)]$ or $3x^2(y - 3)^2(5x + 2y - 6)$

10. $\dfrac{y^3}{3x^4}$ **11.** $\dfrac{2}{3(x - 3)^3}$

12. $6x^4 + 6x^3 - 12x$ **13.** $x = 2$

14. $f(g(x)) = (2x - 1)^3$

15. $y = \frac{2}{3}x - \frac{11}{3}$

Calculus Warm-up Exercises

1. $x = 3, -3$ **2.** $6x - 5$

3. $12x^{11}$ **4.** $-15x^{-4}$

5. $\dfrac{1}{4\sqrt[4]{x^3}}$ **6.** $-28/x^8$

7. $12x^5 - 6x + 7$

ou have learned the Power Rule, the Sum and Difference Properties, and the Constant Multiple Property to help find derivatives. We now consider three special cases of differentiation:

- Derivatives of products
- Derivatives of quotients
- Derivatives of composite functions

We then look at combinations of these cases.

Derivatives of Products

The Sum Property told us that if $y = f(x) + g(x)$, then $y' = f'(x) + g'(x)$. In other words, *the derivative of the sum of two functions is the sum of the derivatives of the individual functions*. The Difference Property told us that if $y = f(x) - g(x)$, then $y' = f'(x) - g'(x)$. In other words, *the derivative of the difference of two functions is the difference of the derivatives of the individual functions*. Unfortunately, derivatives of products of two functions and derivatives of quotients of two functions do not follow the same logical pattern.

Consider $y = (3x - 2)(x^2 + 5)$. The two functions being multiplied are $f(x) = 3x - 2$ and $g(x) = x^2 + 5$. For these functions, we know that $f'(x) = 3$ and $g'(x) = 2x$. If the derivative of $y = f(x) \cdot g(x)$ were simply the product of the derivatives of the individual factors, then y' would be simply $f'(x) \cdot g'(x)$ and, for our example, we would obtain a derivative of $3(2x) = 6x$. To show that this derivative is ***incorrect***, multiply the two factors together and then take the derivative with the Power Rule.

$$y = (3x - 2)(x^2 + 5) \rightarrow y = 3x^3 - 2x^2 + 15x - 10$$
$$\rightarrow y' = 9x^2 - 4x + 15$$

The **correct** derivative, $y' = 9x^2 - 4x + 15$, is very different from our hypothesized "guess" of $y' = 3(2x) = 6x$. Obviously, a special rule is needed for finding derivatives of products of two functions, the **Product Rule**.

Product Rule
If $y = f(x) \cdot g(x)$, then $y' = D_x[f(x)g(x)] = g(x) \cdot f'(x) + f(x) \cdot g'(x)$.

Tip: Rather than memorize the Product Rule symbolically, remember that the Product Rule says to *multiply each function by the derivative of the other function and add the products together.*

MATHEMATICS CORNER 10.1 

Proof of the Product Rule

Why does the Product Rule work?

Given $y = f(x) \cdot g(x)$, we will use the limit definition to determine y'.

$$y' = \lim_{h \to 0} \frac{f(x+h) \cdot g(x+h) - f(x) \cdot g(x)}{h}$$

The trick now is simplifying the difference quotient. Subtracting and adding $f(x)g(x+h)$ in the numerator yields:

$$y' = \lim_{h \to 0} \frac{f(x+h)g(x+h) - f(x)g(x+h) + f(x)g(x+h) - f(x)g(x)}{h}$$

Factor the numerator by grouping:

$$y' = \lim_{h \to 0} \frac{g(x+h)[f(x+h) - f(x)] + f(x)[g(x+h) - g(x)]}{h}$$

Separate into two fractions:

$$y' = \lim_{h \to 0} \left[\frac{g(x+h)[f(x+h) - f(x)]}{h} + \frac{f(x)[g(x+h) - g(x)]}{h} \right]$$

Separate into two limits:

$$y' = \lim_{h \to 0} \frac{g(x+h)[f(x+h) - f(x)]}{h} + \lim_{h \to 0} \frac{f(x)[g(x+h) - g(x)]}{h}$$

$$= \lim_{h \to 0} g(x+h) \cdot \lim_{h \to 0} \frac{[f(x+h) - f(x)]}{h} + \lim_{h \to 0} f(x) \cdot \lim_{h \to 0} \frac{[g(x+h) - g(x)]}{h}$$

Evaluate the limits. In the first fractions, $g(x+h) \to g(x)$ if $h \to 0$. In the second fraction, $f(x)$ is unchanged if $h \to 0$.

$$y' = g(x) \cdot \lim_{h \to 0} \frac{f(x+h) - f(x)}{h} + f(x) \cdot \lim_{h \to 0} \frac{g(x+h) - g(x)}{h}$$

The remaining limits are the definition of derivative:

$$y' = g(x) \cdot f'(x) + f(x) \cdot g'(x)$$

This result is called the **Product Rule.**

Example 1: Find $D_x[(3x - 2)(x^2 + 5)]$ using the Product Rule.

Solution: Let $f(x) = 3x - 2$ and $g(x) = x^2 + 5$, so $f'(x) = 3$ and $g'(x) = 2x$. Apply the Product Rule:

$$Dx\,[\underbrace{(3x - 2)}_{f(x)}\underbrace{(x^2 + 5)}_{g(x)}] = \underbrace{(x^2 + 5)}_{g(x)}\underbrace{(3)}_{f'(x)} + \underbrace{(3x - 2)}_{f(x)}\underbrace{(2x)}_{g'(x)}$$

$$= 3x^2 + 15 + 6x^2 - 4x \quad \text{Expand each term}$$

$$= 9x^2 - 4x + 15 \quad \text{Combine like terms}$$

The result is the same as the derivative found by multiplying the two functions together and using the Power Rule. ■

Example 2: Find $D_x\left[(x^3 - 2x)\left(4 + \frac{5}{x}\right)\right]$ using the Product Rule.

Solution: Use the Product Rule with $f(x) = x^3 - 2x$ and $g(x) = 4 + \frac{5}{x}$. Then

$$f'(x) = 3x^2 - 2 \quad \text{and} \quad g'(x) = -\frac{5}{x^2}.$$

Applying the Product Rule gives

$$D_x\left[\underbrace{(x^3 - 2x)}_{f(x)}\underbrace{\left(4 + \frac{5}{x}\right)}_{g(x)}\right] = \underbrace{\left(4 + \frac{5}{x}\right)}_{g(x)}\underbrace{(3x^2 - 2)}_{f'(x)} + \underbrace{(x^3 - 2x)}_{f(x)}\underbrace{\left(-\frac{5}{x^2}\right)}_{g'(x)}$$

$$= 12x^2 - 8 + 15x - \frac{10}{x} - 5x + \frac{10}{x} \quad \text{Expand each term}$$

$$= 12x^2 + 10x - 8 \quad \text{Combine like terms}$$

■

Example 3: Find $D_x[x^5(3x^3 - 2x + 5)]$ in two ways:

a. Using the Product Rule

b. Multiplying and using the Power Rule

Solution:

a. First, use the Product Rule:

$$D_x[x^5(3x^3 - 2x + 5)] = (3x^3 - 2x + 5)(5x^4) + x^5(9x^2 - 2) \quad f(x) = x^5,\ g(x) = 3x^3 - 2x + 5$$

$$= 15x^7 - 10x^5 + 25x^4 + 9x^7 - 2x^5 \quad \text{Expand each term}$$

$$= 24x^7 - 12x^5 + 25x^4 \quad \text{Combine like terms}$$

b. Second, multiply and use the Power Rule:

$$D_x x^5(3x^3 - 2x + 5) = D_x(3x^8 - 2x^6 + 5x^5) = 24x^7 - 12x^5 + 25x^4$$

As expected, both methods give the same result. ■

Derivatives of Quotients

A special rule was needed for derivatives of products of two functions. Because division is the inverse of multiplication, it seems logical that a special rule is also needed for derivatives of quotients of two functions. To develop this rule, we use two tools from the Tool Kit.

1. The Product Rule involved multiplying each function by the derivative of the other function and *adding* the products. A little logic should imply that derivatives of quotients would involve *subtraction*.
2. Recall that $D_x \frac{1}{x} = -\frac{1}{x^2}$. Notice the **negative sign** (we expected subtraction, remember) and the **square** in the denominator.

Combining these two results helps generate the **Quotient Rule**. A more rigorous proof of the Quotient Rule will be provided later after discussing composite functions.

Quotient Rule

If $y = \frac{f(x)}{g(x)}$, then $y' = D_x \frac{f(x)}{g(x)} = \frac{g(x)f'(x) - f(x)g'(x)}{[g(x)]^2}$.

Tip: **The derivative of a quotient is a quotient.** The numerator resembles the Product Rule, with subtraction instead of addition, and the denominator is the square of the original denominator. Because the numerator involves subtraction, the order in which the terms are subtracted is crucial. A simpler way to remember this rule is

$$\frac{(\text{denominator})(\text{derivative of numerator}) - (\text{numerator})(\text{derivative of denominator})}{(\text{denominator})^2}$$

or simply $\frac{D \cdot N' - N \cdot D'}{D^2}$.

Example 4: Find $D_x \frac{1}{x}$ using the Quotient Rule.

Solution: Let $f(x) = 1$ and $g(x) = x$, so $f'(x) = 0$ and $g'(x) = 1$. The Quotient Rule gives

$$D_x \frac{1}{x} = \frac{x(0) - 1(1)}{x^2} = -\frac{1}{x^2}$$

■

Example 5: Find $D_x \dfrac{x}{x^3 + 5}$ using the Quotient Rule.

Solution: Let $f(x) = x$ and $g(x) = x^3 + 5$, so $f'(x) = 1$ and $g'(x) = 3x^2$. The Quotient Rule gives

$$\begin{aligned} D_x \frac{x}{x^3 + 5} &= \frac{(x^3 + 5)(1) - x(3x^2)}{(x^3 + 5)^2} \\ &= \frac{x^3 + 5 - 3x^3}{(x^3 + 5)^2} \\ &= \frac{5 - 2x^3}{(x^3 + 5)^2} \end{aligned}$$

Tip: When using the Quotient Rule, it is *not* necessary to expand the denominator of the derivative (unless it is a monomial), but it is necessary to simplify the numerator.

Example 6: Find $D_x \dfrac{x^2 - 2}{5x + 1}$ using the Quotient Rule.

Solution: Let $f(x) = x^2 + 2$ and $g(x) = 5x + 1$, so $f'(x) = 2x$ and $g'(x) = 5$. Applying the Quotient Rule gives

$$\begin{aligned} D_x \frac{x^2 - 2}{5x + 1} &= \frac{(5x + 1)(2x) - (x^2 - 2)(5)}{(5x + 1)^2} \\ &= \frac{10x^2 + 2x - 5x^2 + 10}{(5x + 1)^2} \\ &= \frac{5x^2 + 2x + 10}{(5x + 1)^2} \end{aligned}$$

Check Your Understanding 10.1

For the functions in Exercises 1 and 2, find $f'(x)$ using an appropriate rule.

1. $(-2x + 7)(x^5 + 4x)$
2. $\dfrac{3 - 2x}{x^2 - 5}$
3. Find $D_x[x^3(2x^2 - 9)]$ in two ways.

Derivatives of Composite Functions

Composite functions are functions within functions, such as $(3x^4 - 5)^{10}$. Two functions are at work here. The "outer" function—the operation that would be performed last if evaluating—is the 10th power. Write this function as $f(u) = u^{10}$. The "inner" function—the base function contained within the grouping symbol—is the polynomial $3x^4 - 5$. Write this function as $u = 3x^4 - 5$. The **composition** of these two functions is the **composite function** $f(g(x)) = (3x^4 - 5)^{10}$. You should also see that functions such as $\sqrt{x^2 - 3x}$ and $\frac{3}{2 - 5x}$ are composite functions.

MATHEMATICS CORNER 10.2

Composite Functions

Be sure you see that $y = \sqrt{x^2 - 3x}$ and $y = \frac{3}{2 - 5x}$ are composite functions.

For $\sqrt{x^2 - 3x}$, let $f(u) = \sqrt{u}$ with $u = g(x) = x^2 - 3x$.

For $\frac{3}{2 - 5x}$, let $f(u) = \frac{3}{u}$ with $u = g(x) = 2 - 5x$.

How do we find the derivative of a composite function? We begin by considering the composite function $(3x^4 - 5)^2$. Expanding this function, we have $y = (3x^4 - 5)^2 = 9x^8 - 30x^4 + 25$. By the Power Rule, the derivative is $y' = 72x^7 - 120x^3$. Could we have found the derivative so easily had the function been $(3x^4 - 5)^{20}$? Obviously not, because expanding the 20th power is much more difficult than squaring. Apparently, there is a need for yet one more derivative rule.

To develop this rule, let's look again at $D_x(3x^4 - 5)^2$. The square on the outer function implies that the Power Rule could be applied: bring down the exponent "2" as the coefficient and reduce the exponent of the base function to degree 1, yielding $2(3x^4 - 5)$. Is that enough, though? Let's start with the derivative and work backward.

Factoring the derivative provides some insight into the rule:

$$D_x(3x^4 - 5)^2 = 72x^7 - 120x^3$$

Both terms contain the factor $24x^3$:

$$D_x(3x^4 - 5)^2 = 24x^3(3x^4 - 5)$$

You see that the derivative now contains the function $3x^4 - 5$ to the first power, which follows from the Power Rule. The Power Rule also indicates that there would be a factor of 2 (the original exponent) in the first factor. Continuing to factor the derivative gives

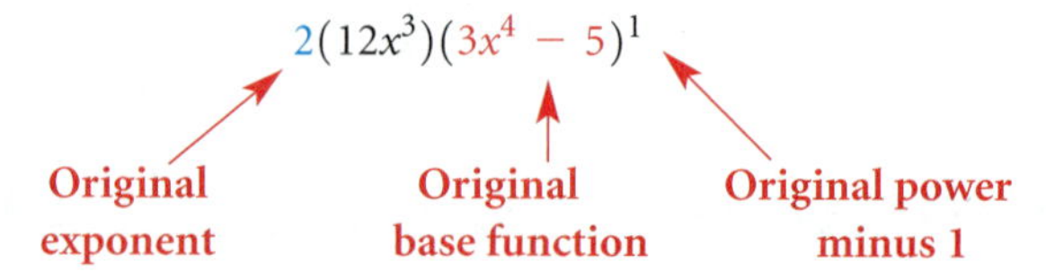

The 2 is the original exponent, and the $3x^4 - 5$ is the original base function with the exponent reduced by 1. Where, though, did the $12x^3$ come from? Here's a hint. What is the derivative of the inner function, $3x^4 - 5$? The $12x^3$ is the derivative of the base function $3x^4 - 5$!

Now we can easily find $D_x(3x^4 - 5)^{20}$. Using the same pattern, we apply the Power Rule and include the additional derivative of the base function to obtain a derivative of $20(12x^3)(3x^4 - 5)^{19}$.

We can now write the **General Power Rule,** sometimes called the **Chain Rule: Power Form**. The word *chain* is used because the derivatives are done in a chain-like fashion. In Topic 13, we go over two additional forms of the Chain Rule.

General Power Rule (Chain Rule: Power Form)

If $y = [f(x)]^n$, then

$$y' = D_x[f(x)]^n = n[f(x)]^{n-1} \cdot f'(x) \quad \text{or} \quad n \cdot f'(x) \cdot [f(x)]^{n-1}$$

The location of the factor $f'(x)$ is unimportant because of the multiplication.

Example 7: Find $D_x(x^2 - 3x + 7)^4$.

Solution:

$$f(x) = x^2 - 3x + 7, \quad n = 4, \quad \text{and} \quad f'(x) = 2x - 3$$

Using the Chain Rule,

$$D_x(x^2 - 3x + 7)^4 = 4(2x - 3)(x^2 - 3x + 7)^3$$

The derivative is in factored form, so no further simplification is necessary. ■

Warning! A common mistake in applying the General Power Rule is to replace the base of the expression with its derivative.

$$D_x(x^2 - 3x + 7)^4 \neq 4(2x - 3)^3$$

As shown in Example 7, the *correct* derivative is $4(2x - 3)(x^2 - 3x + 7)^3$.

Example 8: Find $D_x\sqrt{6x - 11}$.

Solution: First, recognize this expression as a composite function. The radical sign is treated as a grouping symbol, so the inner function is $6x - 11$ and the outer function is the square root. Second, rewrite the function using exponents. Then apply the Chain Rule.

$$
\begin{aligned}
D_x\sqrt{6x - 11} &= D_x(6x - 11)^{1/2} && \text{Rewrite using exponents} \\
&= \tfrac{1}{2}(6)(6x - 11)^{(1/2)-1} && \text{Apply the Chain Rule} \\
&= 3(6x - 11)^{-1/2} && \text{Simplify} \\
&= \frac{3}{(6x - 11)^{1/2}} && \text{Rewrite negative exponent as positive exponent} \\
&= \frac{3}{\sqrt{6x - 11}} && \text{Rewrite fraction exponent as radical}
\end{aligned}
$$

■

Tip: The simplified form of derivatives of radical function, such as that of Example 8, is much easier to work with when dealing with questions such as finding the points of discontinuity of the derivative. The points of discontinuity of $f'(x) = \dfrac{3}{\sqrt{6x - 11}}$ are easy to see because of the radical in the denominator; the points of discontinuity of $f'(x) = 3(6x - 11)^{-1/2}$ may not be as obvious.

Example 9: Find $D_x \dfrac{5}{(2x - 3)^7}$.

Solution: This expression is a composite function with $2x - 3$ as the inner function. The numerator is a constant, so the Quotient Rule can be avoided by rewriting the denominator using negative exponents.

$$D_x \frac{5}{(2x - 3)^7} = D_x[5(2x - 3)^{-7}] \quad \text{Rewrite using negative exponents}$$

$$= 5[-7(2)(2x - 3)^{-7-1}] \quad \text{Apply the Chain Rule}$$

$$= -70(2x - 3)^{-8} \quad \text{Simplify}$$

$$= -\frac{70}{(2x - 3)^8} \quad \text{Rewrite the negative exponent as positive exponent}$$

■

Warning! When rewriting a denominator that is a composite function, only the outer exponent is changed.

$$D_x \frac{5}{(2x - 3)^7} \neq D_x[5(2x^{-1} - 3)^{-7}]$$

As shown in Example 9, the *correct* derivative is $-\dfrac{70}{(2x - 3)^8}$.

Check Your Understanding 10.2

Find $f'(x)$ for each of the following functions.

1. $(3x^2 - 5)^{10}$ **2.** $\sqrt{3x^2 - 5}$ **3.** $\dfrac{6}{(3x^2 - 5)^4}$

Combining the Rules

The Chain Rule may also be used in conjunction with the Product Rule and the Quotient Rule. Simplifying derivatives obtained with combinations of these rules *always* involves factoring and algebraic simplification.

Example 10: Find $D_x[6x^4(x^3 - 2)^2]$.

Solution: This example involves the Product Rule because $6x^4$ and $(x^3 - 2)^2$ are being multiplied.

> **Tip:** Read the function aloud. "Times" means multiplication. An x came both before and after the word *times*, so the Product Rule is needed.

The Chain Rule will be needed to evaluate the derivative of $(x^3 - 2)^2$.

Let $f(x) = 6x^4$ and $g(x) = (x^3 - 2)^2$. Then $f'(x) = 24x^3$ using the Power Rule and $g'(x) = 2(3x^2)(x^3 - 2)$ using the Chain Rule.

Apply the Product Rule:

$$D_x \underbrace{6x^4}_{f(x)} \underbrace{(x^3 - 2)^2}_{g(x)} = \underbrace{(x^3 - 2)^2}_{g(x)} \underbrace{(24x^3)}_{f'(x)} + \underbrace{6x^4}_{f(x)} \cdot \underbrace{2(3x^2)(x^3 - 2)}_{g'(x)}$$

Rather than expand and multiply terms together, which can be a lot of work if the exponents are large, look for common factors and factor the expression.

The derivative has two terms: $(x^3 - 2)^2(24x^3)$ and $6x^4 \cdot 2(3x^2)(x^3 - 2)$, or $(x^3 - 2)^2 24x^3$ and $36x^6(x^3 - 2)$. The GCF of these two terms is $12x^3(x^3 - 2)$.

Factoring the GCF from the derivative gives:

$$D_x[6x^4(x^3 - 2)^2] = (x^3 - 2)^2(24x^3) + 6x^4 \cdot 2(3x^2)(x^3 - 2)$$

$$= 12x^3(x^3 - 2)[(x^3 - 2)2 + x(3x^2)] \quad \text{Factor GCF from both terms}$$

$$= 12x^3(x^3 - 2)(2x^3 - 4 + 3x^3) \quad \text{Simplify the expression in brackets}$$

$$= 12x^3(x^3 - 2)(5x^3 - 4) \quad \text{Combine terms in parentheses}$$ ■

As you know, it is always preferable to leave the derivative in factored form. In later topics, we set the derivative equal to zero and solve for x. Factored form makes that work much easier.

Example 11: If $y = \dfrac{2x^6}{(x^2 - 2)^3}$, for what value(s) of x does the derivative have a value of zero? Discuss the continuity of the derivative.

Solution: The function is a quotient, so the Quotient Rule applies. The denominator is a composite function, so the Chain Rule will be necessary to find its derivative.

Let $f(x) = 2x^6$ and $g(x) = (x^2 - 2)^3$. Then $f'(x) = 12x^5$, by the Power Rule, and $g'(x) = 3(2x)(x^2 - 2)^2$ by the Chain Rule.

So,

$$D_x \frac{2x^6}{(x^2-2)^3} = \frac{\overbrace{(x^2-2)^3}^{D} \cdot \overbrace{12x^5}^{N'} - \overbrace{2x^6}^{N} \cdot \overbrace{3(2x)(x^2-2)^2}^{D'}}{\underbrace{[(x^2-2)^3]^2}_{D^2}}$$

$$= \frac{(x^2-2)^3 \cdot 12x^5 - 12x^7(x^2-2)^2}{(x^2-2)^6}$$

To simplify, factor the GCF $12x^5(x^2-2)^2$ from the numerator.

$$D_x \frac{2x^6}{(x^2-2)^3} = \frac{12x^5(x^2-2)^2[x^2-2-x^2]}{(x^2-2)^6}$$ **Factor GCF from numerator**

$$= \frac{12x^5(x^2-2)^2[-2]}{(x^2-2)^6}$$ **Simplify within brackets**

Both numerator and denominator contain the factor $(x^2-2)^2$, so simplify the expression by dividing this factor from numerator and denominator:

$$D_x \frac{2x^6}{(x^2-2)^3} = \frac{-24x^5}{(x^2-2)^4}$$

Set the derivative equal to 0 and solve for x:

$$\frac{-24x^5}{(x^2-2)^5} = 0 \rightarrow -24x^5 = 0 \rightarrow x = 0$$

> **Warning!** Remember that denominators of fractions cannot equal zero, so the solution of a rational equation $\frac{f(x)}{g(x)} = 0$ must come from setting the numerator equal to zero.

The derivative has points of discontinuity at those value(s) of x for which the denominator equals zero. Because $(x^2-2)^5 = 0$ then $x^2 - 2 = 0$ and $x = \sqrt{2}, -\sqrt{2}$. The derivative has points of discontinuity at $x = \sqrt{2}, -\sqrt{2}$. ■

Check Your Understanding 10.3

Find $f'(x)$ for each of the following functions using the Product Rule, the Quotient Rule, the Chain Rule, or a combination of these rules.

1. $3x^3(2x+5)^4$

2. $(x^2-1)^3(2x+5)^4$

3. $\frac{3x^3}{(2x+5)^4}$

4. $\frac{(x^2-1)^3}{(2x+5)^4}$

Applications Using the Chain Rule: Power Form

Example 12: Find the equation of the tangent line to $f(x) = \sqrt{x^2 - 4x}$ at the point where $x = 5$.

Solution: The equation of a line is given by $y - y_1 = m(x - x_1)$. For this example,

The point (x_1, y_1) is the point $(5, f(5))$.
The slope is given by $f'(5)$ because it is a tangent line.

Setting $f(x) = (x^2 - 4x)^{1/2}$, the Chain Rule gives

$$f'(x) = \left(\tfrac{1}{2}\right)(x^2 - 4x)^{-1/2}(2x - 4) = \frac{x - 2}{\sqrt{x^2 - 4x}}$$

We now have

$$\begin{aligned} x_1 &= 5 \\ y_1 &= f(5) = \sqrt{5^2 - 4(5)} = \sqrt{5} \\ m &= f'(5) = \frac{5 - 2}{\sqrt{5^2 - 4(5)}} = \frac{3}{\sqrt{5}} = \frac{3\sqrt{5}}{5} \end{aligned}$$

The equation of the tangent line is found using $y - y_1 = m(x - x_1)$ and the above values for m and (x_1, y_1):

$$\begin{aligned} y - \sqrt{5} &= \frac{3\sqrt{5}}{5}(x - 5) \\ y &= \frac{3\sqrt{5}}{5}x - 2\sqrt{5} \end{aligned}$$

■

Example 13: The cost of producing x units of a product is $C(x) = 100\sqrt{0.2x + 3}$, where C is thousands of dollars. Evaluate $C(100)$ and $C'(100)$ and interpret your answers.

Solution: Because $C(x) = 100\sqrt{0.2x + 3}$ is a composite function, it can also be written as $C(x) = 100(0.2x + 3)^{1/2}$. The derivative can be found as follows:

$$\begin{aligned} C'(x) &= 100 \cdot \tfrac{1}{2}(0.2x + 3)^{-1/2} \cdot (0.2) \\ &= \frac{10}{\sqrt{0.2x + 3}} \end{aligned}$$

Evaluating gives

$$C(100) = 100\sqrt{0.2(100) + 3} = 479.583 \text{ thousand dollars and}$$

$$C'(100) = \frac{10}{\sqrt{0.2(100) + 3}} \approx 2.085 \text{ thousand dollars per unit}$$

If 100 units of the product are produced, the cost is approximately $479,583. The cost is increasing by approximately $2085 per unit produced. ■

MATHEMATICS CORNER 10.3

Proof of the Quotient Rule

The Quotient Rule is easily proven by rewriting the quotient as a product and then using the Product Rule and the Chain Rule:

$D_x \dfrac{f(x)}{g(x)} = D_x f(x) \cdot [g(x)]^{-1}$	Rewrite using negative exponents
$= [g(x)]^{-1} \cdot D_x f(x) + f(x) \cdot D_x [g(x)]^{-1}$	Apply the Product Rule
$= [g(x)]^{-1} \cdot f'(x) + f(x) \cdot (-1)[g(x)]^{-2} \cdot g'(x)$	Apply the Power Rule and the Chain Rule
$= \dfrac{f'(x)}{g(x)} - \dfrac{f(x)g'(x)}{[g(x)]^2}$	Rewrite negative exponents
$= \dfrac{f'(x)}{g(x)} \cdot \dfrac{g(x)}{g(x)} - \dfrac{f(x)g'(x)}{[g(x)]^2}$	Write with common denominators
$= \dfrac{g(x)f'(x)}{[g(x)]^2} - \dfrac{f(x)g'(x)}{[g(x)]^2}$	Simplify
$= \dfrac{g(x)f'(x) - f(x)g'(x)}{[g(x)]^2}$	Combine fractions

Check Your Understanding Answers

Check Your Understanding 10.1

1. $-12x^5 + 35x^4 - 16x + 28$
2. $\dfrac{2x^2 - 6x + 10}{(x^2 - 5)^2}$
3. $10x^4 - 27x^2$

   By the Product Rule,

   $$D_x[x^3(2x^2 - 9)] = x^3(4x) + 3x^2(2x^2 - 9) = 4x^4 + 6x^4 - 27x^2 = 10x^4 - 27x^2$$

   Expanding and using the Power Rule,

   $$D_x[x^3(2x^2 - 9)] = D_x(2x^5 - 9x^3) = 10x^4 - 27x^2$$

Check Your Understanding 10.2

1. $60x(3x^2 - 5)^9$
2. $\dfrac{3x}{\sqrt{3x^2 - 5}}$
3. $\dfrac{-144x}{(3x^2 - 5)^5}$

Check Your Understanding 10.3

1. $3x^2(2x + 5)^3(14x + 15)$
2. $2(x^2 - 1)^2(2x + 5)^3(10x^2 + 15x - 4)$
3. $\dfrac{3x^2(15 - 2x)}{(2x + 5)^5}$
4. $\dfrac{2(x^2 - 1)^2(2x^2 + 15x + 4)}{(2x + 5)^5}$

Topic 10 Review

10

This topic presented three additional rules—the Product Rule, the Quotient Rule, and the General Power Rule (or Chain Rule: Power Form)—for differentiation of algebraic functions and showed how to use them in combination.

CONCEPT	EXAMPLE
The **Product Rule** states that $$D_x[f(x)\cdot g(x)] = g(x)f'(x) + f(x)g'(x).$$ To find the derivative of the product of two functions, multiply each function by the derivative of the other function and add the products together.	$D_x[4x^2(5x^{-2} + 7x - 3)]$ $= (5x^{-2} + 7x - 3)\cdot D_x4x^2 + 4x^2\cdot D_x(5x^{-2} + 7x - 3)$ $= (5x^{-2} + 7x - 3)\cdot(8x) + 4x^2\cdot(-10x^{-3} + 7)$ $= 84x^2 - 24x$
The **Quotient Rule** states that $$D_x\frac{f(x)}{g(x)} = \frac{g(x)f'(x) - f(x)g'(x)}{[g(x)]^2}.$$ The derivative of the quotient of two functions is $\frac{D\cdot N' - N\cdot D'}{D^2}$, where N is the numerator $f(x)$ and D is the denominator $g(x)$. It is assumed that $g(x) \neq 0$.	$D_x\left[\frac{2x^3 - 5}{x^4 + 3x}\right] = \frac{(x^4 + 3x)(6x^2) - (2x^3 - 5)(4x^3 + 3)}{(x^4 + 3x)^2}$ $= \frac{(6x^6 + 18x^3) - (8x^6 - 14x^3 - 15)}{(x^4 + 3x)^2}$ $= \frac{-2x^6 + 32x^3 + 15}{(x^4 + 3x)^2}$
The **General Power Rule** (or **Chain Rule: Power Form**) states that $$D_x[f(x)]^n = n\cdot f'(x)\cdot[f(x)]^{n-1}$$ The derivative of a composite function is found by applying the Power Rule and multiplying by the derivative of the inner function.	$D_x(2x^3 - 5x + 1)^4 = 4(2x^3 - 5x + 1)^3(6x^2 - 5)$ $D_x\sqrt{x^2 - 4x} = D_x(x^2 - 4x)^{1/2} = \frac{x - 2}{\sqrt{x^2 - 4x}}$ $D_x\frac{6}{(2x - 7)^3} = D_x[6(2x - 7)^{-3}]$ $= 6(-3)(2x - 7)^{-4}(2)$ $= -36(2x - 7)^{-4}$
These three rules may be combined. Simplifying derivatives obtained using the Chain Rule combined with either the Product Rule or the Quotient Rule *always* requires factoring and algebraic simplification.	$D_x[(2x - 5)^3(3x + 2)^2]$ $= (3x + 2)^2\cdot D_x(2x - 5)^3 + (2x - 5)^3\cdot D_x(3x + 2)^2$ $= (3x + 2)^2\cdot 3(2)(2x - 5)^2 + (2x - 5)^3\cdot 2(3)(3x + 2)$ $= 6(3x + 2)(2x - 5)^2[(3x + 2) + (2x - 5)]$ $= 6(3x + 2)(2x - 5)^2(5x - 3)$

(continued)

$$D_x \frac{3x^5}{(x^2-4)^3}$$
$$= \frac{(x^2-4)^3 \cdot D_x 3x^5 - 3x^5 \cdot D_x(x^2-4)^3}{\left[[(x^2-4)^3]^2\right]}$$
$$= \frac{(x^2-4)^3(15x^4) - 3x^5 \cdot 3(2x)(x^2-4)^2}{[(x^2-4)^3]^2}$$
$$= \frac{3x^4(x^2-4)^2[5(x^2-4)-6x^2]}{(x^2-4)^6}$$
$$= \frac{3x^4(-x^2-20)}{(x^2-4)^4}$$

NEW TOOLS IN THE TOOL KIT:

- Using the Product Rule to find the derivative of the product of two functions
- Using the Quotient Rule to find the derivative of the quotient of two functions
- Using the Chain Rule (General Power Rule) to find the derivative of a composite function
- Combining the Chain Rule with the Product Rule, the Quotient Rule, or both

Topic 10 Exercises

10

Find $f'(x)$ for each of the following functions. Leave answers in factored form if possible.

1. $x^2(x^3 - 4x)$

2. $x^4(x^2 + 3x)$

3. $(2x - 5)(7x^2 + 4)$

4. $(7 - 2x)(x^2 + 5)$

5. $(x^3 - 2x)(4 - 5x^2)$

6. $(2x^3 - 5)(6 - x^4)$

7. $3x^4(2x^5 - 7x^2 + 5)$

8. $4x^3(3x^4 + 2x^3 - 7)$

9. $(x^2 - 3x + 2)(4x + 7)$

10. $(x^5 - 3x^3 + 4)(3x - 2)$

11. $(2x^3 - 5x + 7)(x^2 + 2x - 4)$

12. $(3x^4 - 5x^2 + 1)(x^3 - 2x + 5)$

13. $\sqrt{x}(3\sqrt{x} - 4)$

14. $\sqrt[3]{x}(\sqrt[3]{x^2} - 2)$

15. $(\sqrt{x} + 5)(\sqrt{x} - 5)$

16. $(2\sqrt{x} - 3)(2\sqrt{x} + 3)$

17. $(x^{2/3} - 2x^{1/3})(x^{4/3} + 3x)$

18. $(x^{3/4} - 3x^{1/4})(x^{5/4} + 2x)$

19. $(2x^{-3} - 5x^4)(x^2 + 3x^3)$

20. $(7x^{-4} + 4x^{-3})(x^{-5} - 2x^{-4})$

21. $\dfrac{x}{5x + 4}$

22. $\dfrac{3x}{2x - 7}$

23. $\dfrac{2x - 5}{x^2 + 7}$

24. $\dfrac{4 - 7x}{x^2 - 3}$

25. $\dfrac{-2x^2}{x^3 - 4}$

26. $\dfrac{3x^3}{x^2 + 5}$

27. $\dfrac{5x - 2}{x^3 - 4x}$

28. $\dfrac{2x - 3}{x^2 - 7x}$

29. $(7x - 2)^4$

30. $(5 - 3x)^7$

31. $(x^2 - 3x + 5)^3$

32. $(x^3 - x^2 + 6)^5$

33. $(3x^{-4} - 2x^{-1})^4$

34. $(-2x^{-5} + x^{-3})^5$

35. $(2x^3 - 5x + 2)^{-4}$

36. $(3x^5 + 7x^2 - 2)^{-8}$

37. $(x^{2/3} - 3x^{1/3})^5$

38. $(3x^{5/6} + 2x^{1/6})^3$

39. $x^6(x^2 - 4)^3$

40. $x^4(x^3 + 5)^2$

Find $f'(x)$ for each of the following functions. Write your answers in factored form.

41. $(3x - 5)^4(2x + 7)^3$

42. $(2x - 7)^3(5x + 4)^6$

43. $(x^2 - 4x)^5(x^3 - 3)^4$

44. $(x^4 - 3x^2)^3(2x^2 + 3)^4$

Find $f'(x)$ for each of the following functions.

45. $\dfrac{x^2}{(3x - 5)^3}$

46. $\dfrac{x^3}{(4 - 3x)^5}$

47. $\dfrac{(2x - 3)^4}{(x^2 + 7)^2}$

48. $\dfrac{(5 - 2x)^3}{(x^3 + 2)^4}$

In Exercises 49 and 50, find $f'(x)$ for each of the following functions in three ways:

a. Using the Chain Rule

b. Using the Product Rule

c. Expanding and using the Power Rule

Verify that your results are the same.

49. $(3x^2 - 5)^2$

50. $(7 - 3x^4)^2$

In Exercises 51 through 54, find $f'(x)$ for each of the following functions in two ways:

a. Using the Quotient Rule

b. Rewriting with negative exponents and using the Chain Rule

Verify that your results are the same.

51. $\dfrac{3}{x^2 - 5}$

52. $\dfrac{-4}{x^2 + 5x}$

53. $\dfrac{4x}{(x^2 - 5)^3}$

54. $\dfrac{3 - x}{(x^3 + 2x)^4}$

Find the equation of the tangent line to the given function at the specified point.

55. $f(x) = x^3(x^2 - 3)^4$ at $x = 2$

56. $f(x) = (3x^2 - 4)^2(2x + 5)^3$ at $x = 1$

57. $f(x) = \dfrac{x^3}{(x^2 - 4)^4}$ at $x = -1$

58. $f(x) = \dfrac{2x}{(x - 3)^5}$ at $x = -1$

For each of the following functions:

a. Find $f'(x)$.

b. Determine the value(s) of x for which $f'(x) = 0$.

c. Determine the value(s) of x for which $f'(x)$ is discontinuous.

59. $f(x) = \dfrac{x^3}{x^2 - 4}$

60. $f(x) = \dfrac{2x - 5}{x^3 + 4x}$

61. $f(x) = \dfrac{(x - 4)^2}{(x^2 + 5)^3}$

62. $f(x) = \dfrac{(x^3 - 2)^4}{(2x - 3)^7}$

63. The concentration of drug in a patient's bloodstream t hours after injection is given by $C(t) = \dfrac{0.4t}{t^2 + 1}$, where $C(t)$ is concentration in milligrams per cubic centimeter (mg/cm^3). At what rate is the concentration changing after one-half hour? After two hours? Interpret your answers.

64. The cost of manufacturing x units of a product is $C(x) = x\sqrt{2x + 5}$, where $C(x)$ is the cost in thousands of dollars and x is the number of units manufactured in hundreds. At what rate is cost changing if 200 units are manufactured? If 400 units are manufactured? Interpret your answers.

65. The number of CD players a store sells in a month is dependent on the price and can be modeled by $N(p) = \dfrac{2250}{p + 5}$, where N is the number sold and p is the price in dollars. Evaluate $N(25)$ and $N'(25)$ and interpret your answers.

66. The temperature of a patient during an illness is modeled by $T(t) = 98.6 + \dfrac{3t}{t^2 + 2}$, where T is degrees Fahrenheit and t is the number of hours after the onset of the illness. Evaluate $T(2)$ and $T'(2)$ and interpret your answers.

67. The cost of producing x units of a product is given by $C(x) = \sqrt{3x^2 + 400}$, where C is the cost in dollars. Find the marginal cost for producing 20 units and interpret your answer.

68. The cost of manufacturing x units of a product is $C(x) = x\sqrt{2x + 5}$, where $C(x)$ is the cost in thousands of dollars and x is the number of units manufactured in hundreds. Find the marginal cost for producing 100 units and interpret your answer.

69. The sum of \$5000 is deposited in an account paying r% interest compounded annually. After five years, the value of the account is given by $A(r) = 5000(1 + 0.01r)^5$. Evaluate $A(6)$ and $A'(6)$, and interpret your answers.

70. The sum of \$8000 is deposited into an annuity paying r% interest compounded quarterly. After four years, the value of the annuity is $V(r) = 8000(1 + 0.0025r)^{16}$. Evaluate $V(5)$ and $V'(5)$, and interpret your answers.

SHARPEN THE TOOLS

71. Develop a rule for the derivative of the product of three factors: $D_x[f(x)g(x)h(x)]$.

72. Use the rule derived in Exercise 71 to find the derivative of the following functions

a. $(x - 3)(x^2 + 5)(2x^3 - 3x + 1)$

b. $(2x + 5)^2(x^4 - 3)^3(x^2 - 5x + 2)^6$

73. Develop a rule for $D_x\dfrac{f(x)g(x)}{h(x)}$.

74. Use the rule derived in Exercise 73 to find the derivative of the following functions.

a. $\dfrac{(3 - 2x)(x^2 + 5)}{2x^3 - 7}$

b. $\dfrac{(3x^2 - 1)^4(5 - 8x)^3}{(x^3 - 2x + 1)^2}$

TOPIC

Higher-Order Derivatives

11

IMPORTANT TOOLS IN TOPIC 11

- *Second derivatives*
- *Other higher-order derivatives*
- *Symbols for higher-order derivatives*
- *Applications of second derivatives, including acceleration of objects in motion*

If a ball is thrown into the air, its height t seconds after being thrown can be expressed as a function $h(t) = -16t^2 + v_0t + h_0$, where v_0 is the initial velocity of the ball and h_0 is the initial height of the ball. You know that the derivative, $h'(t)$, gives the velocity of the ball at any time. You also know that the velocity changes as the ball moves through the air. The rate at which the velocity of the ball changes is called its acceleration and is the derivative of the velocity. In this topic, you learn how to find derivatives of derivatives, called higher-order derivatives. Before beginning this topic, you should complete the Topic 11 Warm-up Exercises to brush up on the algebra and calculus skills needed.

TOPIC 11 WARM-UP EXERCISES

Be sure you can successfully complete these exercises before starting Topic 11.

Algebra Warm-up Exercises

Write an equivalent expression using positive exponents.

1. x^{-2} **2.** $-4x^{-3}$

Write an equivalent expression using fraction exponents.

3. $\sqrt{x}$ **4.** $\sqrt[5]{x^3}$ **5.** $1/\sqrt[3]{x}$

Write the radical equivalent of each expression.

6. $x^{2/3}$ **7.** $x^{-3/4}$ **8.** $2x^{-1/2}$

Calculus Warm-up Exercises

Find $f'(x)$ for each of the following functions.

1. $f(x) = x^3 - 4x^2 + 6x - 5$

2. $f(x) = 2x - 7$

3. $f(x) = \sqrt{4x - 5}$

4. $f(x) = \dfrac{3}{x^2}$

5. $f(x) = \dfrac{6x}{3x - 1}$

6. $f(x) = (3x - 2)^4$

7. $f(x) = (x^2 - 5)(x^3 + 4x - 3)$

Answers to Warm-up Exercises

Algebra Warm-up Exercises

1. $1/x^2$ **2.** $-4/x^3$ **3.** $x^{1/2}$

4. $x^{3/5}$ **5.** $x^{-1/3}$ **6.** $\sqrt[3]{x^2}$

7. $1/\sqrt[4]{x^3}$ **8.** $2/\sqrt{x}$

Calculus Warm-up Exercises

1. $3x^2 - 8x + 6$ **2.** 2

3. $2/\sqrt{4x - 5}$ **4.** $-6/x^3$

5. $-\dfrac{6}{(3x - 1)^2}$ **6.** $12(3x - 2)^3$

7. $5x^4 - 3x^2 - 6x - 20$

ou now know that, given any algebraic function, you can find its derivative by applying an appropriate rule or combination of rules. You also know that the derivative is the instantaneous rate of change of the function for any value of x in the function's domain. For example,

$$f(x) = x^3 - 2x^2 + 5x - 1 \rightarrow f'(x) = 3x^2 - 4x + 5$$

You should notice that the derivative, $f'(x)$, is itself a function, so why not find its derivative?

Second Derivative

The derivative of $f'(x)$ is called the **second derivative** and is denoted in different ways:

$$f''(x) \quad \text{or} \quad y'' \quad \text{or} \quad D_x^2(f) \quad \text{or} \quad \frac{d^2y}{dx^2} \quad \text{or} \quad \frac{d^2}{dx^2}f(x)$$

The derivative $f'(x)$ is referred to as the **first derivative**. Thus, for the function above,

$$\begin{aligned} f(x) &= x^3 - 2x^2 + 5x - 1 \\ f'(x) &= 3x^2 - 4x + 5 \\ f''(x) &= 6x - 4 \end{aligned}$$

Example 1: Find $f''(x)$ for $f(x) = x^4 - 2x^2 + 6$.

Solution: This expression is a polynomial function, so apply the Power Rule:

$$\begin{aligned} f(x) &= x^4 - 2x^2 + 6 \\ f'(x) &= 4x^3 - 4x \\ f''(x) &= 12x^2 - 4 \end{aligned}$$

■

Example 2: Find $f''(x)$ for $f(x) = 3/x^2$.

Solution: Rewrite and use the Power Rule:

$f(x) = \dfrac{3}{x^2} \rightarrow f(x) = 3x^{-2}$ Rewrite using negative exponents

$\rightarrow f'(x) = -6x^{-3}$ Take first derivative

$\rightarrow f''(x) = 18x^{-4}$ Take second derivative

$= \dfrac{18}{x^4}$ Rewrite with positive exponent ■

Example 3: Find $f''(x)$ for $f(x) = 40\sqrt{x}$.

Solution: Rewrite and use the Power Rule:

$f(x) = 40\sqrt{x} \rightarrow f(x) = 40x^{1/2}$ Rewrite radical as an exponent

$\rightarrow f'(x) = 20x^{-1/2}$ Take first derivative

$$\rightarrow f''(x) = -10x^{-3/2} \quad \text{Take second derivative}$$
$$= -\frac{10}{\sqrt{x^3}} \quad \text{Rewrite as radical}$$
■

Example 4: Find $f''(x)$ for $f(x) = (5x - 2)^4$.

Solution: This expression is a composite function with $f(u) = u^4$ and $u = g(x) = 5x - 2$, so we apply the Chain Rule:

$$f(x) = (5x - 2)^4 \rightarrow f'(x) = 4(5x - 2)^3(5) = 20(5x - 2)^3$$
$$\rightarrow f''(x) = 20(3)(5x - 2)^2(5) = 300(5x - 2)^2$$
■

Example 5: Find $f''(x)$ for $f(x) = (x^2 - 4)^3$.

Solution: This expression is a composite function with $f(u) = u^3$ and $u = g(x) = x^2 - 4$, so we apply the Chain Rule:

$$f(x) = (x^2 - 4)^3 \rightarrow f'(x) = 3(x^2 - 4)^2(2x) = 6x(x^2 - 4)^2$$

The first derivative is a product of two functions of x, so we must apply the Product Rule to find the second derivative:

$$f''(x) = (x^2 - 4)^2 \cdot D_x(6x) + 6x \cdot D_x(x^2 - 4)^2$$
$$f''(x) = (x^2 - 4)^2(6) + 6x(2)(x^2 - 4)(2x)$$
$$= 6(x^2 - 4)^2 + 24x^2(x^2 - 4) \quad \text{Simplify}$$
$$= 6(x^2 - 4)[(x^2 - 4) + 4x^2] \quad \text{Factor out } 6(x^2 - 4)$$
$$= 6(x^2 - 4)(5x^2 - 4) \quad \text{Simplify within brackets}$$
■

Check Your Understanding 11.1

Find the second derivative of each of the following functions.

1. $f(x) = 3x^5 - 2x^3 + 7x^2$

2. $f(x) = -6/x^4$

3. $f(x) = 12\sqrt[3]{x}$

4. $f(x) = (2x - 3)^5$

5. $f(x) = (x^2 + 7)^3$

Applications of Higher-Order Derivatives

You already know that the first derivative of a function represents the rate of change of that function. The second derivative of the function represents the rate at which the first derivative is changing.

Example 6: Sales of a CD (in thousands of CDs) can be modeled by $S(t) = 400\sqrt{t}$, where t is the number of months after release of the CD. Evaluate the following sales and interpret the answers.

a. $S(4)$ **b.** $S'(4)$ **c.** $S''(4)$

Solution:

a. $S(4) = 400\sqrt{4} = 800$

Four months after the release of the CD, 800,000 CDs had been sold.

b. $S(t) = 400\sqrt{t} \rightarrow S'(t) = 200t^{-1/2} = \dfrac{200}{\sqrt{t}}$

$$S'(4) = \frac{200}{\sqrt{4}} = 100$$

Four months after the release of the CD, the number of CDs sold is growing by 100,000 CDs per month.

c. $S'(t) = 200t^{-1/2} \rightarrow S''(t) = -100t^{-3/2} = -\dfrac{100}{\sqrt{t^3}}$

$$S''(4) = -\frac{100}{\sqrt{4^3}} = -12.5$$

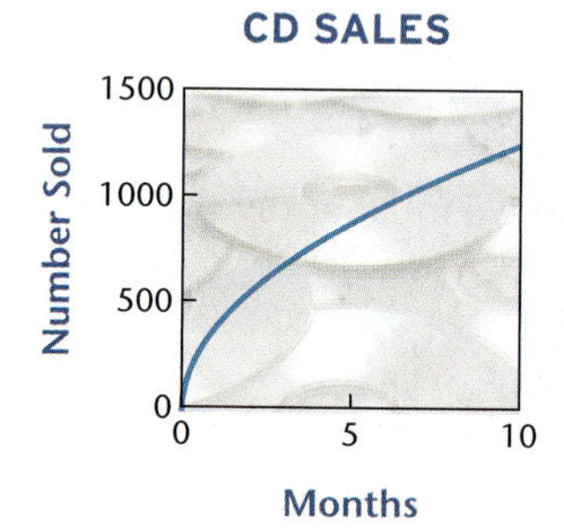

Figure 11.1

Four months after the release of the CD, the rate of growth in sales of CDs is decreasing by 12,500 CDs per month per month. Sales are still increasing, but at a slower rate. Figure 11.1 shows the graph of this function. You can see that sales will grow each month, which results in the positive value of the first derivative. All tangent lines will have a positive slope, but the value of the slopes of the tangent lines decreases in value as time increases. The rate of growth of sales slows down, as represented by the negative value for the second derivative. ■

Example 7: Union membership as a percentage of the labor force can be modeled by $M(x) = 0.003x^3 - 0.066x^2 + 4.64x - 81$, where M is membership as a percentage of the labor force and x is the number of years after 1900. At what rate was the rate of growth changing in 1950?

Solution: The key words in this question are *rate of growth* and *rate . . . changing*. The **rate of growth** is given by the first derivative. The rate at which that rate of growth changes is given by the second derivative. So,

$$M(x) = 0.003x^3 - 0.066x^2 + 4.64x - 81 \rightarrow M'(x) = 0.009x^2 - 0.132x + 4.64$$
$$\rightarrow M''(x) = 0.018x - 0.132$$

The year 1950 is 50 years after 1900, so $M''(50) = 0.018(50) - 0.132 = 0.768$. Thus, in 1950, the growth rate of union membership was increasing by 0.768% per year per year. ■

Projectile Motion

Recall that the height of a projectile above the ground at any time after it is launched can be modeled by $h(t) = at^2 + bt + c$. You also know that the **velocity** of the projectile at any time is given by $h'(t)$.

What about the second derivative? The second derivative, $h''(t)$, represents the rate at which the velocity changes, which is referred to as **acceleration**. (An automobile's gas pedal is called the accelerator because pressing down on the pedal or letting up on the pedal changes the rate at which the vehicle moves.)

Example 8: The height of a ball, in feet, t seconds after being thrown into the air is given by $h(t) = -16t^2 + 40t + 50$. Describe the position, velocity, and acceleration of the ball after one second.

Solution:

$h(1) = 74$ so after one second, the ball is 74 feet above the ground.

$h'(t) = -32t + 40$ so $h'(1) = 8$. After one second, the ball is rising at a rate of 8 feet per second.

$h''(t) = -32$. After one second, the ball's velocity is decreasing by 32 feet per second per second (which results from the pull of gravity on the ball). ■

Example 9: A quarterback fires a pass downfield to his receiver. The height of the ball, in yards, t seconds after its release is given by $h(t) = -0.64t^2 + 3.2t + 2$. Describe the position, velocity, and acceleration of the ball after three seconds.

Solution:

$h(3) = 5.84$. After three seconds, the ball is 5.84 yards above the ground.

$h'(3) = -0.64$. After three seconds, the ball is dropping at a rate of 0.64 yards per second. That is the ball's velocity.

$h''(3) = -1.28$. After three seconds, the velocity of the ball is decreasing by 1.28 yards per second per second. That is the ball's deceleration. ■

A Graphical Interpretation of Second Derivative

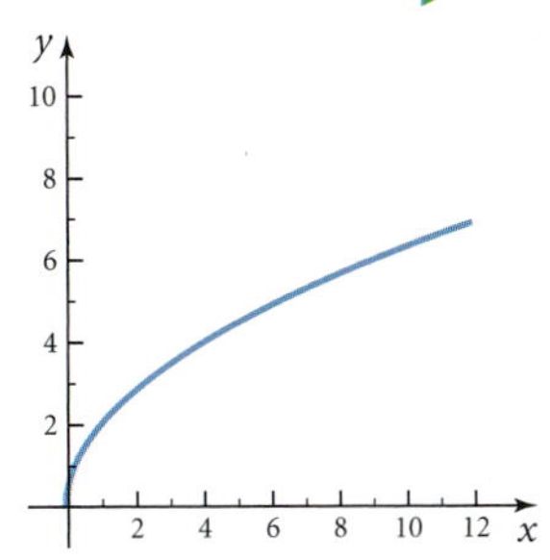

Figure 11.2

Consider the function $f(x) = 2\sqrt{x}$. The derivatives of this function are

$$f'(x) = 2\left(\frac{1}{2\sqrt{x}}\right) = \frac{1}{\sqrt{x}}$$

$$f''(x) = -\frac{1}{2}x^{-3/2} = -\frac{1}{2\sqrt{x^3}}$$

Look at the graph of $f(x) = 2\sqrt{x}$, shown in Figure 11.2. The graph of this function increases in value for all values of x. How do the tangent lines to the graph of this function behave? We will look at the tangent line at three specific points: $x = 1$, $x = 4$, and $x = 9$. The function and the tangent lines are shown in Figures 11.3 through 11.5.

Figure 11.3 shows the tangent line to $f(x) = 2\sqrt{x}$ at the point where $x = 1$. The slope of this tangent line is $f'(1) = 1$. Figure 11.4 shows the tangent line to $f(x) = 2\sqrt{x}$ at the point where $x = 4$. The slope of this tangent line is $f'(4) = \frac{1}{2}$. Figure 11.5 shows the tangent line to $f(x) = 2\sqrt{x}$ at the point where $x = 9$. The slope of this tangent line is $f'(9) = \frac{1}{3}$. As you can see from both the graph and

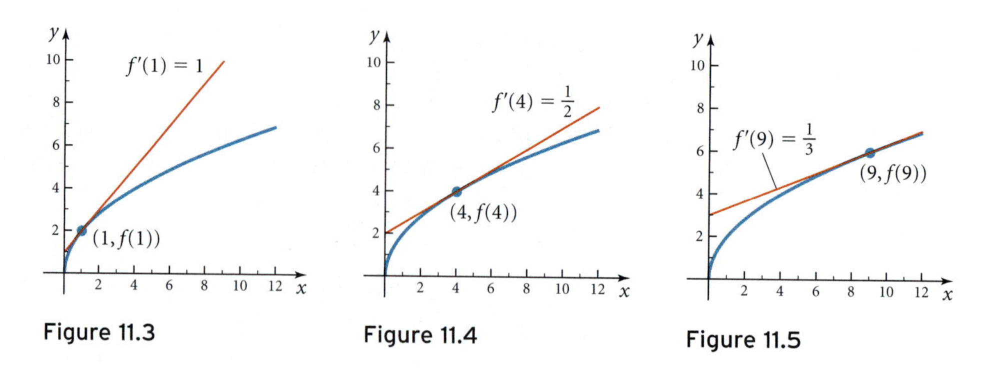

Figure 11.3 Figure 11.4 Figure 11.5

the actual slopes of the lines, the tangent line to $f(x) = 2\sqrt{x}$ at any point will have a positive slope, so $f'(x)$ is always positive. You should also see that the slopes of the tangent lines decrease in value for successively larger values of x. The negative value of the second derivative means that the tangent lines increase at a decreasing rate.

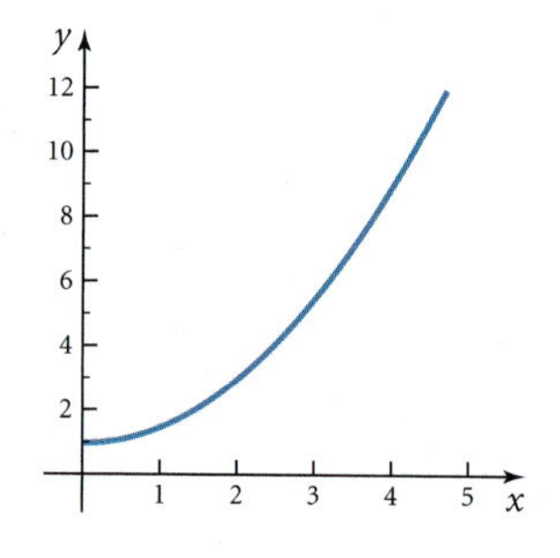

Figure 11.6

Now consider the function $f(x) = \frac{1}{2}x^2 + 1$ for $x > 0$. The derivatives of this function are

$$f'(x) = \tfrac{1}{2}(2x) = x$$
$$f''(x) = 1$$

The graph of $f(x) = \frac{1}{2}x^2 + 1$ is shown in Figure 11.6. The graph of this function increases in value for all values of $x > 0$. How do the tangent lines to the graph of this function behave? We will look at the tangent line at three specific points: $x = 1$, $x = 2$, and $x = 4$. The function and the tangent lines are shown in Figures 11.7 through 11.9.

Figure 11.7 shows the tangent line to $f(x) = \frac{1}{2}x^2 + 1$ at the point where $x = 1$. The slope of this tangent line is $f'(1) = 1$. Figure 11.8 shows the tangent line to $f(x) = \frac{1}{2}x^2 + 1$ at the point where $x = 2$. The slope of this tangent line is $f'(2) = 2$. Figure 11.9 shows the tangent line to $f(x) = \frac{1}{2}x^2 + 1$ at the point where $x = 4$. The slope of this tangent line is $f'(4) = 4$. As you can see from both the graph and the actual slopes of the lines, the tangent line to $f(x) = \frac{1}{2}x^2 + 1$ for any $x > 0$ will have a positive slope, so $f'(x)$ is always positive. You should also see that the slopes of the tangent lines increase in value for successively larger values of x. The positive value of the second derivative means that the tangent lines increase at a increasing rate. The graphical interpretation of the second derivative is explored in more detail in Topic 15.

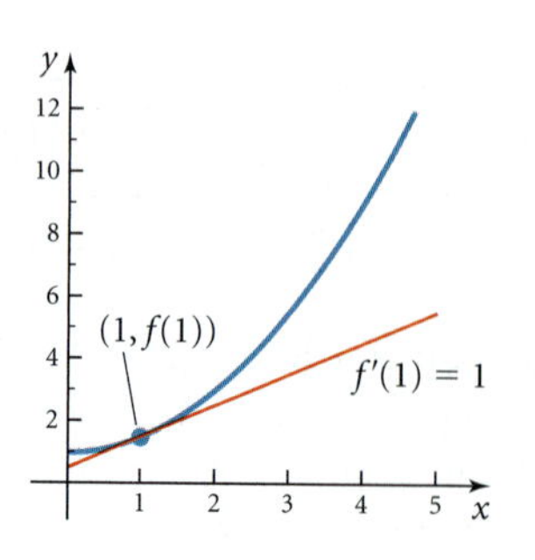

Figure 11.7

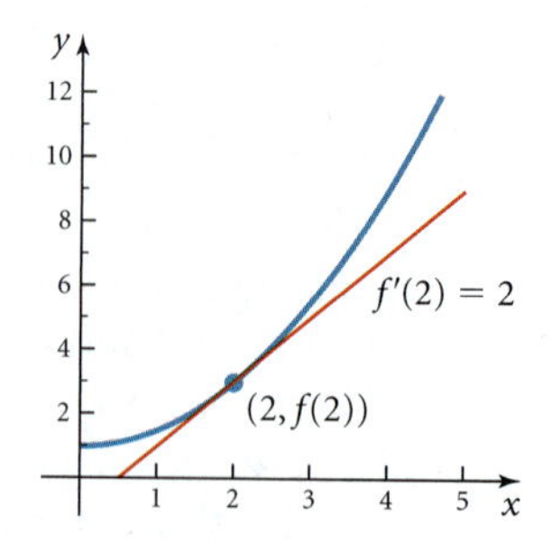

Figure 11.8

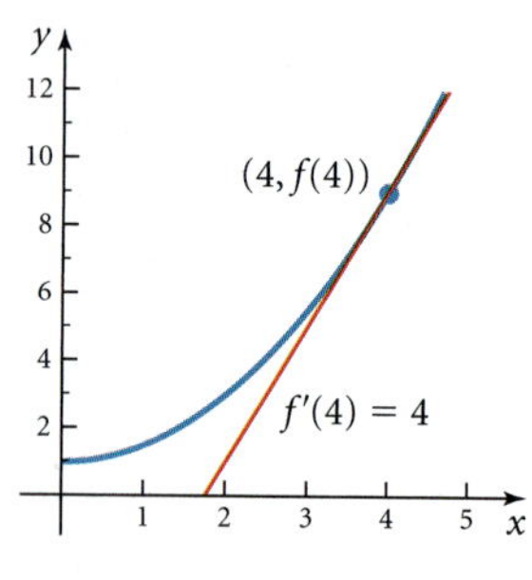

Figure 11.9

Other Higher-Order Derivatives

If it is possible to find second derivatives, obviously we can continue the process by finding third derivatives, fourth derivatives, or even higher-order derivatives. Third derivatives are denoted symbolically by

$$f'''(x) \quad \text{or} \quad y''' \quad \text{or} \quad \frac{d^3y}{dx^3} \quad \text{or} \quad D_x^3(f)$$

Derivatives beyond the third derivative are denoted as $f^{(n)}(x)$. So, for example, the fourth derivative is represented as $f^{(4)}(x)$.

Example 10: Given $f(x) = -4x^3 + 2x^2 - 7x + 8$, find $f'''(x)$.

Solution:

$$\begin{aligned} f(x) = -4x^3 + 2x^2 - 7x + 8 \rightarrow f'(x) &= -12x^2 + 4x - 7 \\ \rightarrow f''(x) &= -24x + 4 \\ \rightarrow f'''(x) &= -24 \end{aligned}$$

■

Example 11: Given $f(x) = 6 - 2x^3 + 5x^7$, find $f^{(4)}(x)$.

Solution:

$$\begin{aligned} f(x) = 6 - 2x^3 + 5x^7 \rightarrow f'(x) &= -6x^2 + 35x^6 \\ \rightarrow f''(x) &= -12x + 210x^5 \\ \rightarrow f'''(x) &= -12 + 1050x^4 \\ \rightarrow f^{(4)}(x) &= 4200x^3 \end{aligned}$$

■

An interesting application of the third derivative is a concept called **jerk.** Jerk is the rapid change in acceleration that causes an airplane to suddenly change altitudes, giving passengers the unpleasant sensation of dropping.

In the fall of 1972, President Richard Nixon appeared on television and reported to the American people the good news that "the rate of increase of inflation is going down." This marked the first time that a sitting president used the third derivative to improve his chances for re-election. Do you think that what he presented as good news was really good news?

Check Your Understanding Answers

Check Your Understanding 11.1

1. $60x^3 - 12x + 14$
2. $-\dfrac{120}{x^6}$
3. $-\frac{8}{3}x^{-5/3}$ or $-\dfrac{8}{\sqrt[3]{x^5}}$
4. $80(2x - 3)^3$
5. $6(x^2 + 7)(5x^2 + 7)$

Topic 11 Review

11

In this topic, you learned about second derivatives and their applications. Other higher-order derivatives were also introduced.

CONCEPT	EXAMPLE
The **second derivative** is the derivative of the **first derivative**, denoted symbolically as $f''(x)$ or y'' or $D_x^2 f$.	For $f(x) = 2x^5 - 7x^2 + 11x - 5$, the second derivative is $f''(x) = 40x^3 - 14$.
The second derivative gives the rate at which the first derivative is changing. A positive second derivative means that the first derivative's rate of increase or decrease is increasing. A negative second derivative means that the first derivative's rate of increase or decrease is decreasing.	For $f(x) = x^2 - 4$, $f''(x) = 2$, which means that the first derivative (the slope of the curve at any point) is always increasing. Total revenues from the sale of a new CD are given by $R(x) = 400\sqrt{x}$, where x is the number of weeks after the release of the CD. Four weeks after the release of the CD, the total revenue is \$800, or $R(4) = 800$. Revenue is increasing at a rate of \$100 per week, or $R'(4) = 100$. The rate of increase is decreasing by \$12.50 per week per week, or $R''(4) = -12.50$.
For projectile motion, the second derivative gives the **acceleration** of the object.	The path of a projectile is given by $f(x) = \sqrt{4x - x^2}$, where x is time in seconds and $f(x)$ is height above the ground in feet. Its acceleration after two seconds is $f''(2) = -\frac{1}{2}$, which means that the velocity of the projectile is decreasing by $\frac{1}{2}$ foot per second per second after two seconds.

NEW TOOLS IN THE TOOL KIT

- Finding the second derivative of a given function
- Interpreting acceleration as the second derivative of a projectile motion position function
- Finding other higher-order derivatives of functions

Topic 11 Exercises

11

Find $f''(x)$ for each of the following functions.

1. $f(x) = x^2 - 5x + 6$
2. $f(x) = 2x + 7$
3. $f(x) = 2 - 3x$
4. $f(x) = 2x - x^2 + 5$
5. $f(x) = 2x^3 - 3x^2 + 7x - 9$
6. $f(x) = -5x^3 + 4x^2 - 2x + 1$
7. $f(x) = -4x^5 + 7x^3 - 2x + 11$
8. $f(x) = 2x^6 - 5x^4 + 2x^3 - x$
9. $f(x) = -3x^{-2} + 7x - 2$
10. $f(x) = 7x^{-3} + 8x^{-1} - 5x$
11. $f(x) = \sqrt[3]{x}$
12. $f(x) = \sqrt[4]{x}$
13. $f(x) = -2/x^4$
14. $f(x) = 5/x^7$
15. $f(x) = \dfrac{3}{x - 2}$
16. $f(x) = -\dfrac{2}{x + 6}$
17. $f(x) = (2x - 5)^4$
18. $f(x) = (4 - 3x)^3$
19. $f(x) = (x^3 + 3)^5$
20. $f(x) = (2x^5 - 3)^4$
21. $f(x) = \sqrt{x^2 - 9}$
22. $f(x) = \sqrt{x^4 + 2x}$
23. $f(x) = x^3(2x - 5)^2$
24. $f(x) = -2x^4(7 - 2x)^3$
25. $f(x) = \dfrac{x^3}{(2x - 5)^2}$
26. $f(x) = \dfrac{-2x^4}{(7 - 2x)^3}$

Find the specified derivative for each of the following functions.

27. $f(x) = 2x^3 - 5x + 6$; find $f'''(x)$
28. $f(x) = 3x^5 - 2x^3 + 9x - 11$; find $f'''(x)$
29. $h(c) = \dfrac{2}{c^2} - \dfrac{3}{c}$; find $h'''(c)$
30. $m(d) = 2d^{2/3} - 3d^{1/3}$; find $m'''(d)$
31. $t(n) = 3n^{-1/3} + 2n^{4/3}$; find $t^{(4)}(n)$
32. $p(a) = 5a^{-2} + 6a^4 - 3a$; find $p^{(4)}(a)$
33. For a group of 10,000 people, the number who survive to age x is given by $N(x) = 1{,}000\sqrt{100 - x}$. Find the survival rate at age 36. At what rate is the survival rate changing at age 36?
34. Repeat Exercise 33 for age 96. Compare your answers.
35. After a category 4 hurricane, the number of customers without power is given by $P(x) = -1.5x^4 + 16.4x^3 - 51.3x^2 - 4.6x + 174.5$, where P is the number of customers without power, in thousands, and x is the number of days after the hurricane. Evaluate $P''(2)$ and interpret your answer.
36. Repeat Exercise 35 for $P''(4)$.
37. A woman on a weight-loss program weighs herself each week. Her weight each week is modeled by $W(t) = \dfrac{790 + 132t}{t + 4.4}$, where W is her weight in pounds and t is the number of weeks since starting the program. Evaluate $W(3)$, $W'(3)$, and $W''(3)$. Interpret your answers.
38. Repeat Exercise 37 for $W(5)$, $W'(5)$, and $W''(5)$.
39. The revenue generated from sales of a new software program is given by
$$R(x) = 10x^2 - 0.0033x^3,$$
where R is the revenue in dollars and x is the number of programs sold. How fast is the growth rate of sales changing when 25 programs are sold?
40. The profit generated from producing and selling x thousand textbooks is given by
$$P(x) = 2x + 0.025x^4,$$
where P is thousands of dollars. At what rate is the growth rate of the profit changing when 6000 books have been sold?

41. A company that manufactures cell phones finds that the cost of producing x phones is given by $C(x) = 250 + 40\sqrt{x}$. Evaluate $C(100)$, $C'(100)$, and $C''(100)$. Interpret your answers.

42. The cost of producing x calculators is given by $C(x) = \sqrt{3x^2 + 400}$. Evaluate $C(200)$, $C'(200)$, and $C''(200)$. Interpret your answers.

SHARPEN THE TOOLS

Find $f''(x)$ for each of the following functions.

43. $f(x) = 4x^6(x^2 - 3)^2$

44. $f(x) = \dfrac{4x^6}{(x^2 - 3)^2}$

45. Prove that for any polynomial function $P(x)$, $P^{(n)}(x) = 0$ for some value of n. What is that value of n?

COMMUNICATE

46. In July 2004, the National Endowment for the Arts released the study "Reading at Risk," which reported the results of a survey of more than 17,000 adults to determine how much reading that was not required various age groups had done over the past year. The report showed that "*the rate of reading is decreasing* among young people, but the downward trend is evident in every demographic group" (emphasis added). Specifically, the 2004 report stated that for young adults aged 18 to 24, less than 50% of them had done any nonrequired reading during the previous year. That proportion has dropped 28% since the 1982 survey.

a. Briefly explain how this statement relates to first and second derivatives.

b. Sketch a graph describing reading among young people, based on this statement.

TOPIC

12 Exponential and Logarithmic Functions

In Topics 1 through 3, we reviewed the basic algebraic functions: polynomial (linear, quadratic, and cubic), square root, and rational. Now consider the graph showing the increase in the number of Starbucks locations since 1971. You should see that a curve to model the data would not be algebraic. In this topic, we review exponential and logarithmic functions. This topic is a review, emphasizing what is needed for the study of calculus; it is *not* intended to provide in-depth instruction of exponential functions and logarithmic functions.

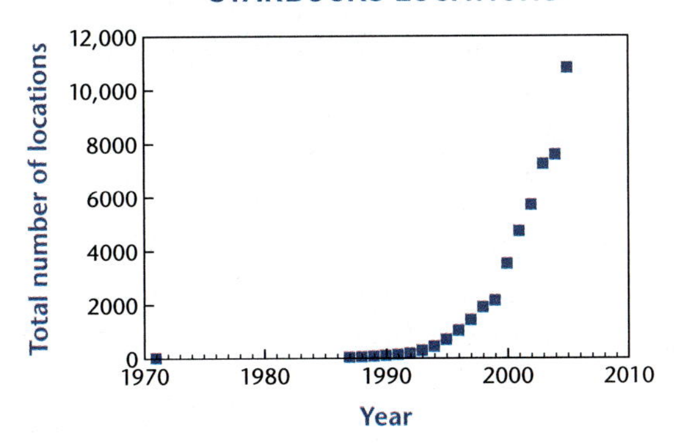

Before beginning the review of exponential and logarithmic functions, be sure to complete the Warm-up Exercises to sharpen your algebra and calculus skills.

IMPORTANT TOOLS IN TOPIC 12

- *Exponential functions*
- *Logarithmic functions*
- *Graphs of exponential and logarithmic functions*
- *Solving exponential and logarithmic equations*

TOPIC 12 WARM-UP EXERCISES

Algebra Warm-up Exercises

Be sure you can successfully complete these exercises before starting Topic 12.

1. Identify the base and the exponent in each of the following numerical expressions.

a. 4^3 **b.** -4^3 **c.** $(-4)^3$

2. Evaluate each of the following numerical expressions without using a calculator.

a. 4^3 **b.** 3^{-2} **c.** 7^0

d. -4^2 **e.** $(-4)^2$ **f.** 4^{-2}

g. $\left(\frac{1}{3}\right)^{-2}$ **h.** -4^{-1} **i.** 17^0

j. $9^{1/2}$ **k.** $9^{-1/2}$

3. Approximate the following numbers to three decimal places using a calculator.

a. $2^{1.7}$ **b.** $\left(\frac{1}{2}\right)^{-0.4}$ **c.** $3.75^{0.26}$

4. Simplify using the laws of exponents. Write your answers using only positive exponents.

a. $x^2 \cdot x^4$ **b.** $(x^2)^4$ **c.** x^2/x^5

d. $\dfrac{x^6 \cdot x^3}{x^5}$

e. $(2x^3y^4)(5x^{-2}y^{-6})$

f. $(3x^5y^{-2})^3(-2x^{-3}y^4)^2$

5. Solve $3x + 5 = 11$ for x.

Match each graph to the correct type of equation. Some equations may be used more than once.

6.

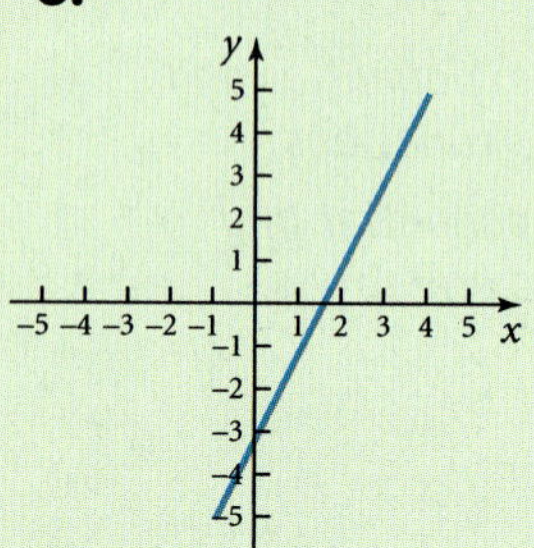

7.

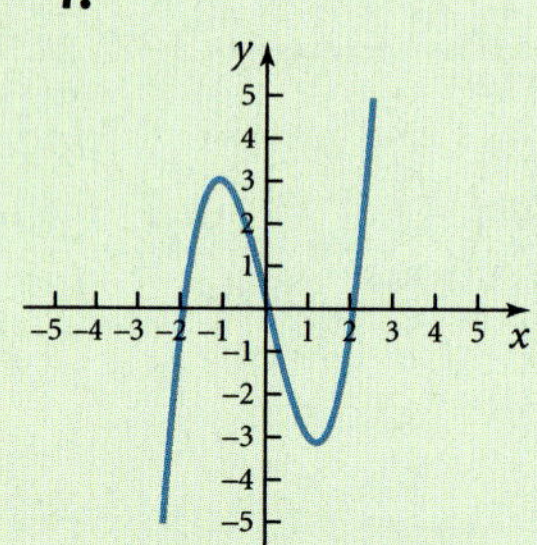

8.

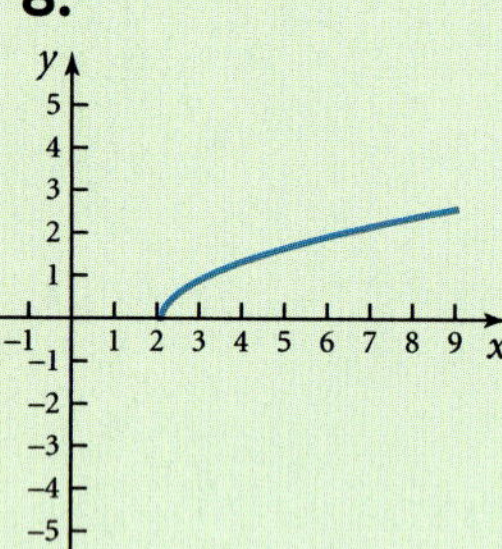

9.

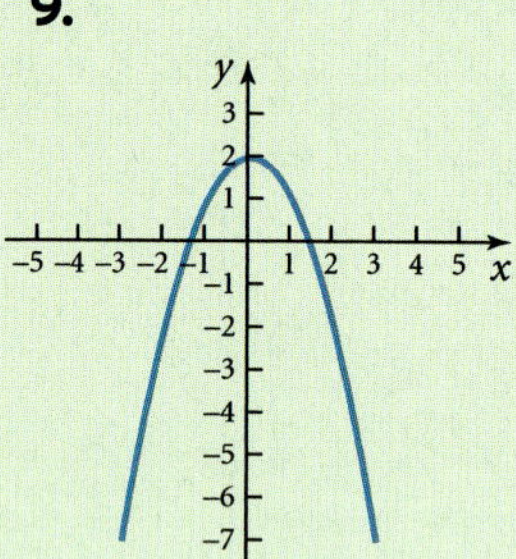

10.

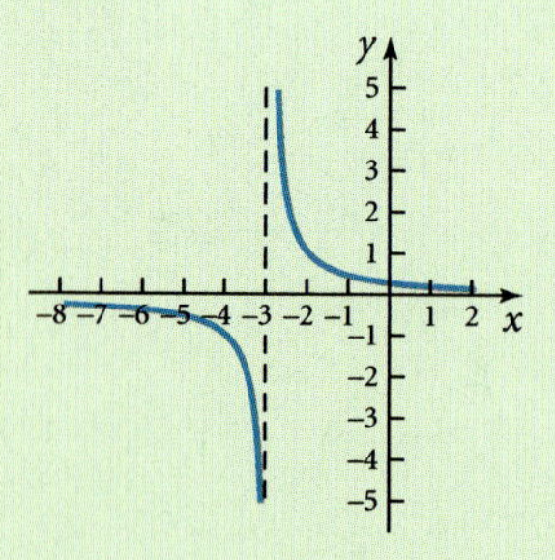

11.

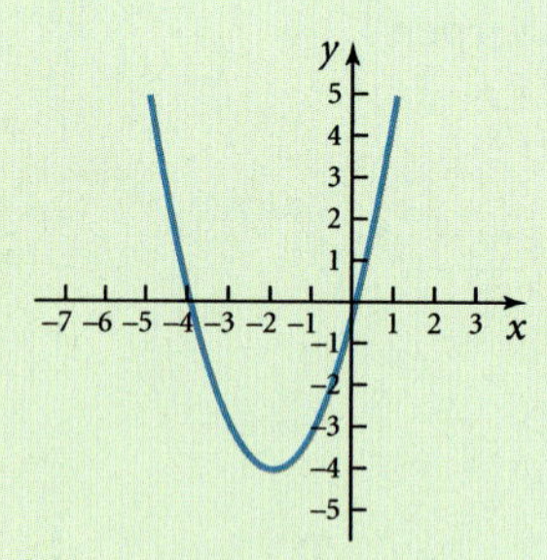

12.

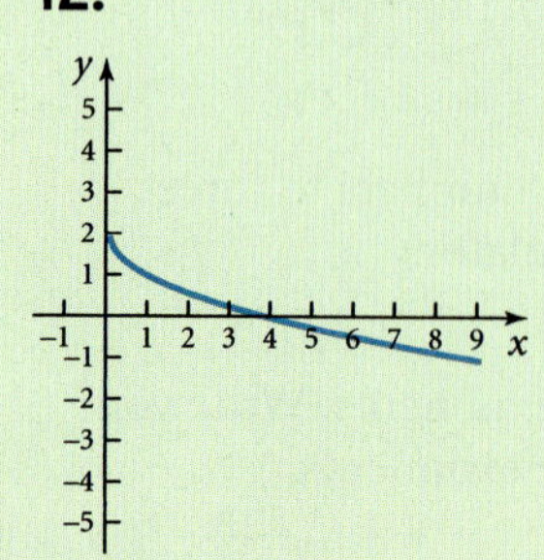

a. linear
b. quadratic
c. cubic
d. radical
e. rational

13. Are the following functions one to one?

a. $y = x^3$ **b.** $y = x^2$

c. $y = x^3 - 4x$ **d.** $y = |x|$

e. $y = 1/x$

14. Find the inverse of the given function. State the domain and range of $f(x)$ and of $f^{-1}(x)$.

a. $f(x) = 2x - 3$

b. $f(x) = \dfrac{3}{x + 5}$

c. $f(x) = \sqrt{x - 2}$

Calculus Warm-up Exercises

1. Evaluate the following limits.

a. $\lim\limits_{x \to -1} (3x^2 - 5x + 6)$

b. $\lim\limits_{x \to \infty} \dfrac{2}{x}$

c. $\lim\limits_{x \to \infty} \dfrac{x^2}{x + 1}$

2. Determine the horizontal and vertical asymptotes of each of the following functions.

a. $f(x) = 2/x$ **b.** $f(x) = \dfrac{-2x}{x + 7}$

c. $f(x) = \dfrac{3x}{x^2 - 4}$ **d.** $f(x) = \dfrac{x^2}{x - 5}$

3. Find the interval(s) for which the following functions are continuous.

a. $f(x) = \dfrac{3}{x + 4}$

b. $f(x) = x^3 - 4x$

c. $f(x) = \sqrt{x - 2}$

d.

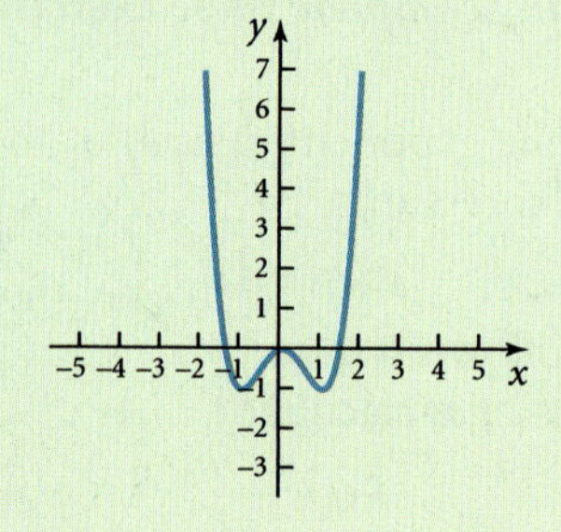

e.

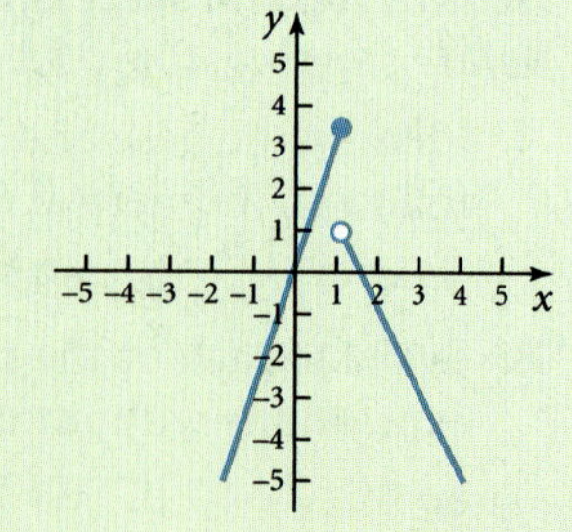

Answers to Warm-up Exercises

Algebra Warm-up Exercises

1. a. base 4, exponent 3
b. base 4, exponent 3
c. base −4, exponent 3

2. a. 64 **b.** $\frac{1}{9}$ **c.** 1
d. −16 **e.** 16 **f.** $\frac{1}{16}$
g. 9 **h.** $-\frac{1}{4}$ **i.** 1
j. 3 **k.** $\frac{1}{3}$

3. a. 3.249 **b.** 1.320 **c.** 1.410

4. a. x^6 **b.** x^8 **c.** $1/x^3$
d. x^4 **e.** $10x/y^2$ **f.** $108x^9y^2$

5. $x = 2$

6. A **7.** C **8.** D **9.** B
10. E **11.** B **12.** D

13. a. yes **b.** no **c.** no
d. no **e.** yes

14. a. $y = \dfrac{x+3}{2}$; domain and range are real numbers
b. $y = \dfrac{3}{x} - 5$; domain is $x \neq 0$, range is $y \neq -5$
c. $y = x^2 + 2$; domain is $x \geq 0$, range is $y \geq 2$

Calculus Warm-up Exercises

1. a. 14 **b.** 0
c. ∞, so does not exist

2. a. $x = 0, y = 0$ **b.** $x = -7, y = -2$
c. $x = 2, x = -2, y = 0$
d. $x = 5$, no horizontal

3. a. $(-\infty - 4) \cup (-4, \infty)$
b. all real numbers
c. $[2, \infty)$
d. all real numbers
e. $(-\infty, 1) \cup (1, \infty)$

In Topic 1, you learned that functions may be classified as algebraic or nonalgebraic.

- **Algebraic functions** include the polynomial functions, radical functions, and rational functions. Algebraic functions are constructed using the algebraic operations of addition, subtraction, multiplication, division, powers, and roots with polynomials.
- **Nonalgebraic (transcendental) functions** include the exponential and logarithmic functions and the trigonometric functions. "Transcendental" comes from the Latin word *transendere*, meaning to "cross over."

Exponential Functions

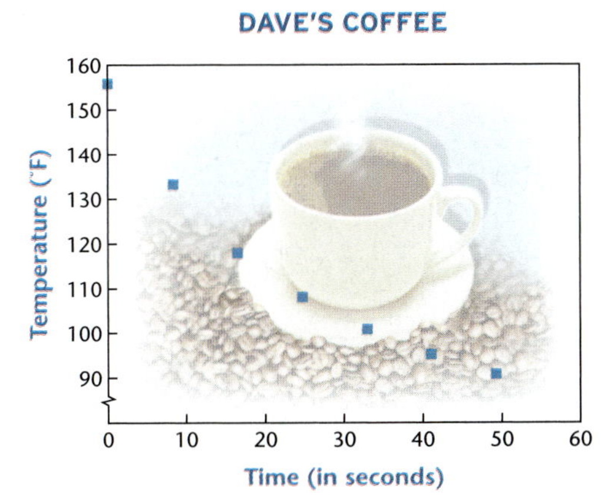

Figure 12.1

Many functions do not behave like polynomial, radical, or rational functions. Consider this example. Dave fixes a cup of coffee with cream and places it on the counter to cool. The temperature of the coffee at various times is shown in Figure 12.1.

Does the plot of points appear to be polynomial? The temperature drops quickly at first, but then more slowly over time. The change between each set of points is not the same, so it is not linear. How will the function behave as time increases? Obviously, the coffee will continue to cool, so the temperature will continue to decrease but at a slower rate. Because the temperature will not rise again at some point, the graph is not quadratic. Because the temperatures do not fluctuate between increasing and decreasing, the graph is not cubic. Rather, this graph is an example of a nonalgebraic function called an exponential function.

> **Definition:** An **exponential function** is any function of the form $f(x) = a^x$, where x is any real number and a is a positive real number such that $a \neq 1$.

The domain of an exponential function is all real numbers; the range is all positive real numbers, unless transformations are involved.

All the functions studied so far have a certain type of graph associated with them. Exponential functions are no exception. We consider the basic exponential function $f(x) = 2^x$. A table of values is shown below.

x		$f(x) = 2^x$
-2	$\frac{1}{4}$	$(2^{-2} = 1/2^2)$
-1	$\frac{1}{2}$	$(2^{-1} = 1/2^1)$
0	1	$(2^0 = 1)$
0.5	1.4	$(2^{0.5} = 2^{1/2} = \sqrt{2})$
1	2	
2	4	
3	8	

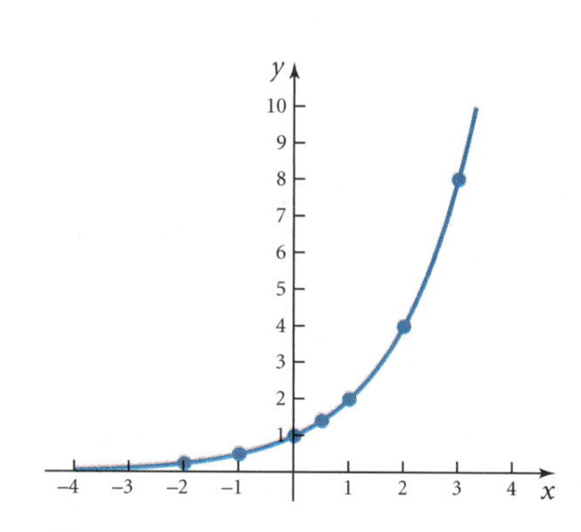

Figure 12.2

Using these values, the graph of $f(x) = 2^x$ is shown in Figure 12.2.

The graph of $f(x) = 2^x$ is continuous for all values of x, and it has three characteristics that help describe its shape:

- The graph passes through the point (0, 1).
- The graph is increasing for all values of x.
- The graph has an asymptote along the negative x-axis. As x approaches $-\infty$, the values of y get smaller and approach 0.

Now consider $f(x) = \left(\frac{1}{2}\right)^x$. A table of values and the graph of $f(x) = \left(\frac{1}{2}\right)^x$ are shown below and in Figure 12.3.

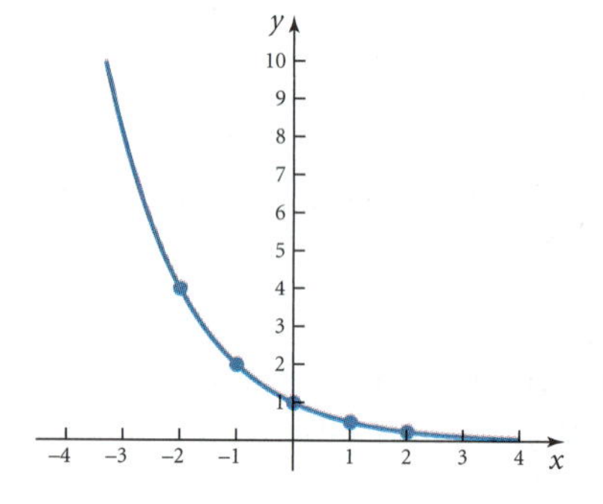

Figure 12.3

x	$f(x)$
-2	$\left(\frac{1}{2}\right)^{-2} = 2^2 = 4$
-1	$\left(\frac{1}{2}\right)^{-1} = 2^1 = 2$
0	$\left(\frac{1}{2}\right)^0 = 1$
1	$\frac{1}{2}$
2	$\frac{1}{4}$

The graph of $f(x) = \left(\frac{1}{2}\right)^x$ is continuous for all values of x, and it has three characteristics that help describe its shape:

- The graph passes through the point (0, 1).
- The graph is decreasing for all values of x.
- The graph has an asymptote along the positive x-axis. As x approaches ∞, the values of y get smaller and approach 0.

Graph of the Basic Exponential Function

The graph of $f(x) = a^x$ passes through the point (0, 1).

If $a > 1$, the graph is continuous and increasing and has an asymptote along the negative x-axis.

If $0 < a < 1$, the graph is continuous and decreasing and has an asymptote along the positive x-axis.

Recall the transformations that alter the location of a graph.

- **Horizontal translations:** $f(x + h)$ shifts the graph of $f(x)$ to the left h units; $f(x - h)$ shifts the graph of $f(x)$ to the right h units.
- **Vertical translations:** $f(x) + k$ shifts the graph of $f(x)$ up k units; $f(x) - k$ shifts the graph of $f(x)$ down k units.
- **Reflections:** $-f(x)$ reflects, or inverts, the graph of $f(x)$ about the x-axis; $f(-x)$ reflects, or inverts, the graph of $f(x)$ about the y-axis.

Transformations may be applied to graphs of exponential functions.

MATHEMATICS CORNER 12.1

Base of an Exponential Function

Exponential functions require a base number that is positive and not equal to 1. Why?

Suppose $a = 1$. Then $f(x) = 1^x = 1$, which is constant, not exponential.

Suppose $a < 0$. Two problems occur here. One is that a negative number raised to an odd integer power is negative, but a negative number raised to an even integer power is positive. Thus, we would have a function whose values fluctuate between positive and negative values and are not strictly increasing or decreasing.

The second problem with negative bases occurs with rational exponents. Consider, for instance, $(-2)^{1/2}$. Evaluating this expression leads to the square root of a negative number, which is imaginary.

Thus, we require that the base of an exponential function be positive and not equal to 1, ensuring a function that is defined for all values of x and that is strictly increasing or decreasing.

Example 1: Use the basic graph of $f(x) = 2^x$ and your knowledge of transformations to sketch the graph of each of the following functions without using a graphing calculator.

a. $f(x) = 2^{x-1}$

b. $f(x) = 2^x - 1$

c. $f(x) = -2^x$

d. $f(x) = 2^{-x}$

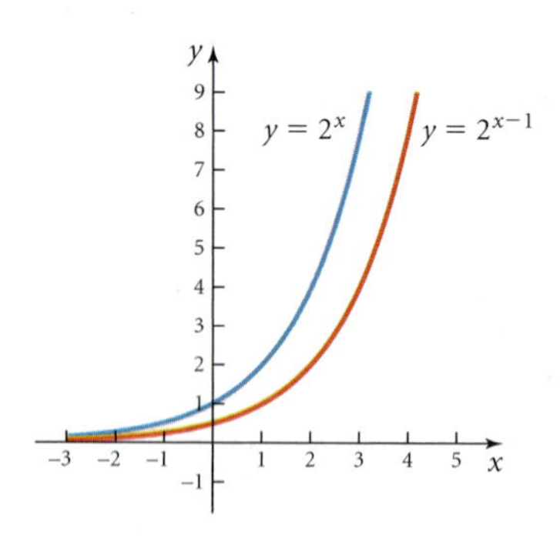

Figure 12.4

Solution:

a. $f(x) = 2^{x-1}$ is the graph of $f(x) = 2^x$ shifted to the right 1 unit. This graph passes through (1, 1) and increases with the asymptote along the negative x-axis. The domain is all real numbers, and the range is all positive real numbers ($y > 0$). See Figure 12.4.

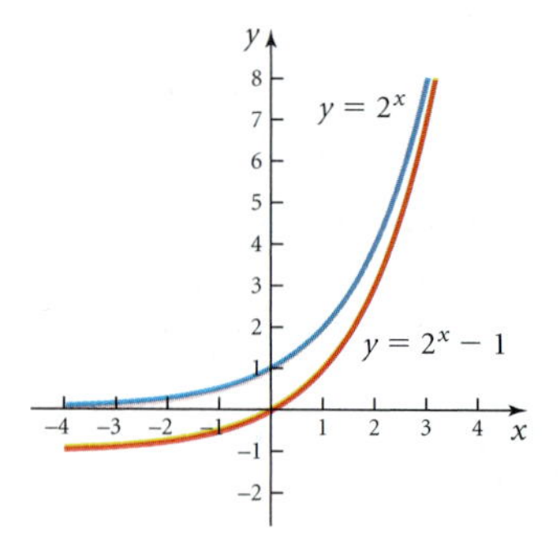

Figure 12.5

b. $f(x) = 2^x - 1$ is the graph of $f(x) = 2^x$ shifted downward 1 unit. This graph passes through (0, 0) and increases with the asymptote along the line $y = -1$. The domain is all real numbers, and the range is all $y > -1$. See Figure 12.5.

c. $f(x) = -2^x$ is the graph of $f(x) = 2^x$ inverted about the x-axis. The graph passes through $(0, -1)$ and decreases with the asymptote along the negative x-axis. The domain is all real numbers, and the range is all negative real numbers ($y < 0$). See Figure 12.6.

d. $f(x) = 2^{-x}$ is the graph of $f(x) = 2^x$ inverted about the y-axis. The graph passes through (0, 1) and decreases with the asymptote along the positive x-axis. The domain is all real numbers, and the range is all positive real numbers ($y > 0$). See Figure 12.7.

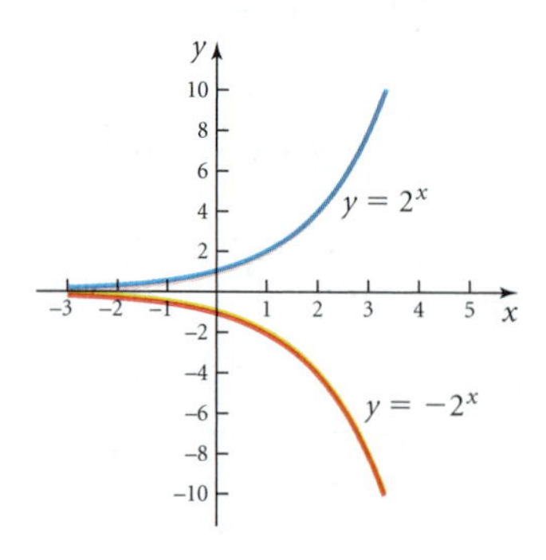

Figure 12.6

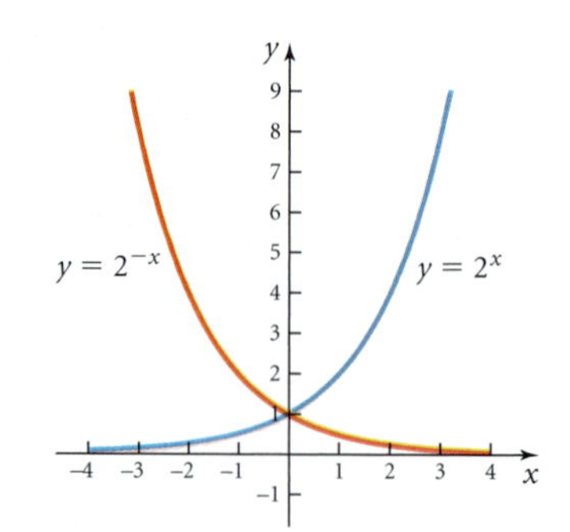

Figure 12.7

As you can see from the graphs in Example 1, graphs of exponential functions are **continuous** and are also **one to one**. (Recall that a one-to-one function has only one x value for each y value.)

MATHEMATICS CORNER 12.2

Graphs of $f(x) = 2^{-x}$ and $f(x) = \left(\frac{1}{2}\right)^x$

Did you see in Example 1d that the graph of $f(x) = 2^{-x}$ is the same as the graph of $f(x) = \left(\frac{1}{2}\right)^x$ shown in Figure 12.3? Why is it the same?

Using the laws of exponents, we can write $f(x) = \left(\frac{1}{2}\right)^x = (2^{-1})^x = 2^{-x}$, which explains why the graphs are the same.

Calculator Corner 12.1

How does changing the base of the exponential function affect the graph? Use your graphing calculator to draw the graphs of the following functions. How does each graph compare with the graph of $f(x) = 2^x$?

1. $f(x) = 4^x$ **2.** $f(x) = \left(\frac{1}{3}\right)^x$ **3.** $f(x) = 3^x$ **4** $f(x) = 10^x$

The Number e

One special base that frequently occurs in applications involving growth and decay is the number *e*. The number *e* is an irrational number (similar to π) that was first used by Leonhard Euler, a Swiss mathematician. It is called the **natural base**, and its value is approximately 2.718. This value can be approximated by examining $\lim_{n\to\infty}\left(1 + \frac{1}{n}\right)^n$. You can find *e* on your calculator by entering **2nd Ln 1 ENTER.**

Calculator Corner 12.2

Evaluate $\left(1 + \frac{1}{n}\right)^n$ by completing the table. Do not round the result your calculator gives. As the values of *n* increase, you should see the values of the expression approaching 2.718.

n	$\left(1 + \frac{1}{n}\right)^n$
10	
100	
1000	
100,000	
1,000,000	
10,000,000	
1,000,000,000	

Calculator Corner 12.3

Enter $y_1 = \left(1 + \frac{1}{x}\right)^x$ in your calculator. The TABLE feature of your calculator can also be used to show the limiting value of e. Under TBLSET, set TblStart $= 10$ and ΔTbl $= 200$. Then scroll down the table to see what the y values approach.

Using Your Calculator to Evaluate Exponential Functions

The exponential function e^x is best evaluated using your calculator rather than the approximation 2.718. To access the base e, enter **2nd Ln**. Then type in the desired exponent and press **ENTER**.

Example 2: Use your calculator to evaluate the following. Round to three decimal places.

a. $e^{1.2}$ **b.** $e^{-0.47}$ **c.** $3e^{-2}$

Solution:

a. Enter **2nd Ln** 1.2. The result is 3.320.

b. Enter **2nd Ln** −0.47. The result is 0.625.

c. Enter 3 **2nd Ln** −2. The result is 0.406. ■

Example 3: The growth in the total number of Starbucks locations each year since its beginning in 1971 can be modeled by the exponential function $f(x) = 0.2536e^{0.307x}$, where x represents the number of years after 1970. Use the model to predict the number of locations in 2007.

Solution: The year 2007 is 37 years after 1970, so the number of locations in 2007 is estimated by $f(37) = 0.2536e^{0.307(37)} \approx 21{,}742$. Check the Starbucks website at http://www.starbucks.com to see how accurate this estimate is. ■

Example 4: *Compound Interest* If an amount of money, P, is invested (or borrowed) at a specified annual rate, r, for a specified length of time, t, compounded n times per year, the amount of money available (or owed), A, at the end of that time period is found by

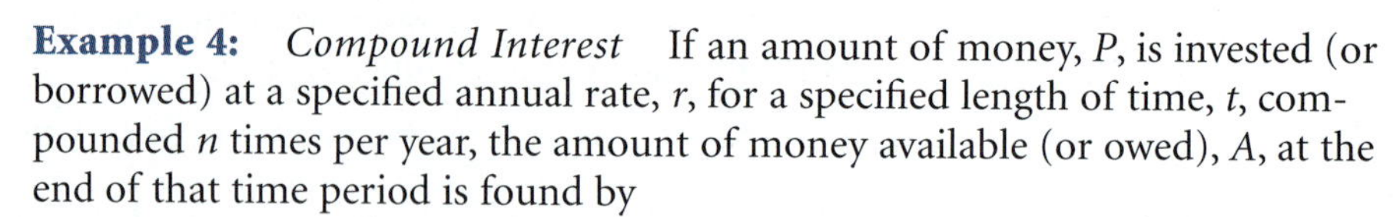

$$A = P\left(1 + \frac{r}{n}\right)^{nt}$$

where n is the number of compounding periods per year. Suppose \$2000 is invested at 6% compounded semiannually. How much money will there be after five years? How long will it take for the money to double?

Solution: Here, $P = \$2000$, $r = 0.06$, $n = 2$, and $t = 5$. (Be sure that the annual interest rate is always expressed as a decimal.) So,

$$A = 2000\left(1 + \frac{0.06}{2}\right)^{2(5)} \approx \$2687.83$$

Compounding	Number of Compounding Periods
Annual	1
Semiannual	2
Quarterly	4
Monthly	12
Daily	365

After five years, the account has about \$2687.83.

The graph of this function is shown in Figure 12.8. When the money doubles, the initial investment of \$2000 will be worth \$4000. Find the point on the graph where $y = \$4000$ and approximate the x value representing the time required to reach \$4000. It appears that the money will double after approximately 12 years (assuming no more money is deposited or withdrawn and the rate stays the same). We revisit this problem in Example 19 and show the algebraic way to find this same solution.

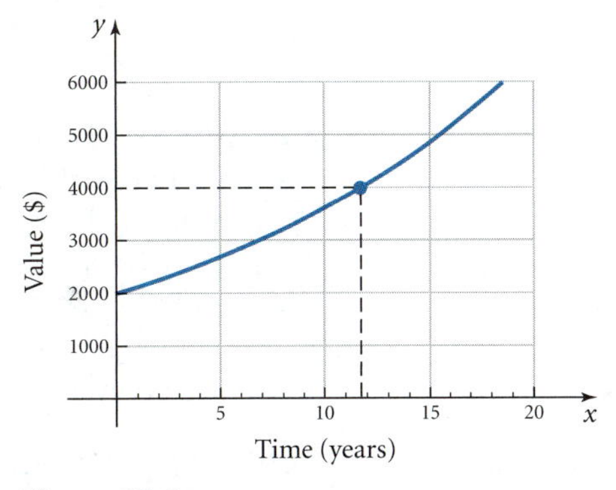

Figure 12.8

Now suppose interest is compounded continuously (which is very nearly what credit card companies do). There are then an infinite number of compounding periods. The amount available after t years is then found by

$$A = \lim_{n\to\infty} P_0\left(1 + \frac{r}{n}\right)^{nt}.$$

We know that $e = \lim_{n\to\infty}\left(1 + \frac{1}{n}\right)^{n}$. The exponent power, n, is the reciprocal of the fraction term $1/n$. With that format we could also write $e = \lim_{n\to\infty}\left(1+\frac{r}{n}\right)^{n/r}$.

We can now rewrite the compound interest formula as

$$\begin{aligned} A = \lim_{n\to\infty} P_0\left(1 + \frac{r}{n}\right)^{nt} &= P_0 \lim_{n\to\infty}\left(1 + \frac{r}{n}\right)^{(n/r)rt} \\ &= P_0 \lim_{n\to\infty}\left[\left(1 + \frac{r}{n}\right)^{(n/r)}\right]^{rt} \\ &= P_0 e^{rt} \end{aligned}$$

For continuous compounding, $A = P_0 e^{rt}$.

Example 5: *Continuous Compounding* If \$2000 is invested at 6% compounded continuously, how much money is available after five years?

Solution: Here, $A = 2000e^{0.06(5)} \approx 2699.72$. After five years, approximately \$2699.72 is in the account.

Compound Interest

$A = P\left(1 + \frac{r}{n}\right)^{nt}$ where A is the amount after t years, P is the initial amount, r is annual rate of interest, n is the number of compounding periods per year, and t is time in years.

$A = P_0e^{rt}$ for continuous compounding, where A is the amount after t years, P_0 is the initial amount, r is the annual rate of interest, and t is time in years.

Example 6: *Law of Growth* The model $P(t) = P_0e^{kt}$ is called the **exponential law of growth**. (In economics, it is referred to as the Malthusian law of growth, after Thomas Malthus, the economist who developed the law and the formula expressing the law.) In this model, P_0 represents the initial amount in the population and k is the rate of growth of the population, assuming $k > 0$.

A colony of mosquitoes grows according to $P(t) = 500e^{0.04t}$, where t is time in days. How many mosquitoes will there be after two weeks?

Solution: Two weeks is 14 days, so the number of mosquitoes will be

$$P(14) = 500e^{.04(14)} \approx 875.34 \text{ or } 876 \text{ mosquitoes}$$

■

Tip: In this case, it makes more sense to round up, because any fractional part of a mosquito is counted as another mosquito.

Example 7: *Logistic Growth* When modeling human populations, a better model is the **logistic model** of growth. This model of growth takes into account birth rates and death rates and allows for a tapering off of the population. In 1840, Belgian mathematician P. Verhulst proposed that the population of the United States could be modeled by

$$P(t) = \frac{500}{1 + 124e^{-0.024t}}$$

where P is the population in millions and t is the number of years after 1790. According to this model, what was the U.S. population in 1960?

Solution: The year 1960 is 170 years after 1790, so the population in 1960 is $P(170) \approx 161.47$, which says that the 1960 population should have been approximately 161.47 million people. (This estimate is way off the actual population of 179.7 million because it is based on an interval around 1790.)

In the graph of this logistic function in Figure 12.9, the population rises at first like an exponential function but then tapers off to a limiting value of 500 million. ■

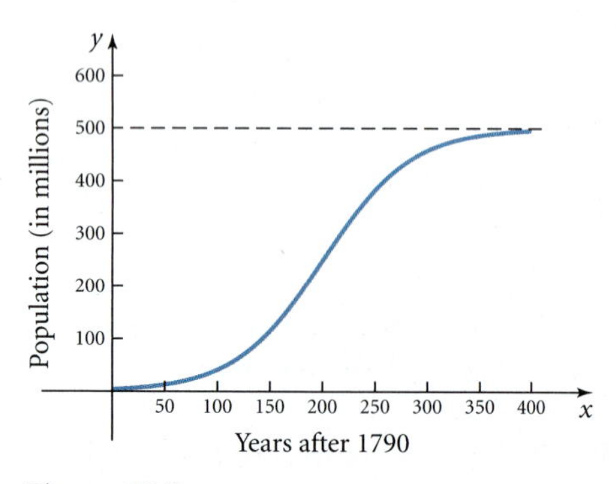

Figure 12.9

Calculator Corner 12.4

Graph $y = \dfrac{500}{1 + 124e^{-0.024t}}$. Use the Table feature of your calculator to show that the limiting value of the function is 500. Under Table Set, set TblStart at 50 with ΔTbl = 50. Then go to the table. As the values of time, denoted by x, increase, you should see the population, denoted by y, approach a value of 500.

Example 8: *Law of Decay* The model $P(t) = P_0e^{kt}$ can be used to model decay if $k < 0$. Suppose an insecticide is sprayed in an area where mosquitoes are breeding and the number of mosquitoes t hours after spraying is given by $P(t) = 6500e^{-0.3t}$. How many mosquitoes were there when the spraying started? How many are present after three hours?

Solution: Given $P(t) = 6500e^{-0.3t}$, we know that $P_0 = 6500$, so there were initially 6500 mosquitoes in the area. After three hours, there will be $P(3) = 6500e^{-0.3(3)} \approx 2642.7028$. Thus, after three hours, there are approximately 2643 mosquitoes still present. ■

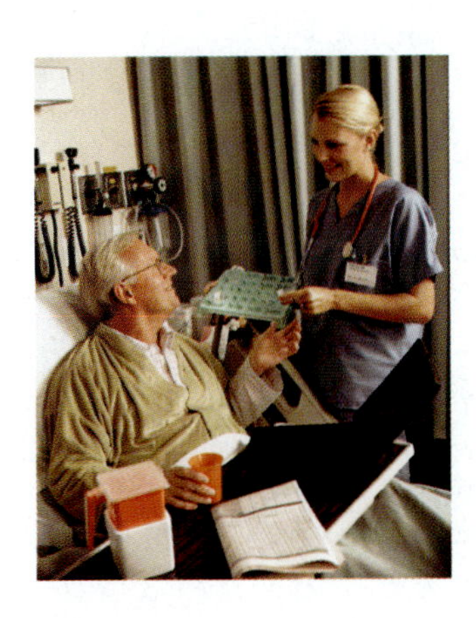

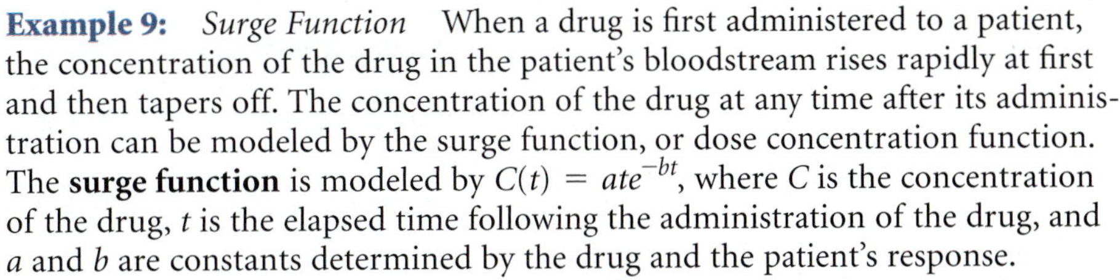

Example 9: *Surge Function* When a drug is first administered to a patient, the concentration of the drug in the patient's bloodstream rises rapidly at first and then tapers off. The concentration of the drug at any time after its administration can be modeled by the surge function, or dose concentration function. The **surge function** is modeled by $C(t) = ate^{-bt}$, where C is the concentration of the drug, t is the elapsed time following the administration of the drug, and a and b are constants determined by the drug and the patient's response.

A heart patient is given a dose of a medicine to control pulse rate. The concentration of the drug t hours after its administration is given by $C(t) = 0.8te^{-0.45t}$, where C is cubic centimeters.

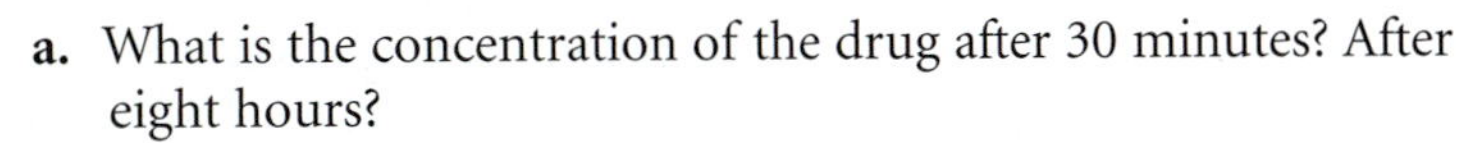

a. What is the concentration of the drug after 30 minutes? After eight hours?

b. Graph the function and estimate when the drug's concentration will reach its highest level?

c. A second injection is needed when the concentration drops below 0.25 unit per cubic centimeter. Use the graph to determine when that injection should be given.

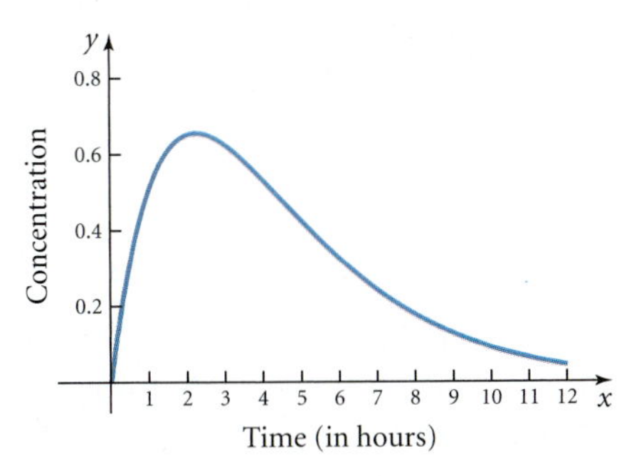

Figure 12.10

Solution:

a. After 30 minutes (or $\frac{1}{2}$ hour) the concentration is $C\left(\frac{1}{2}\right) \approx 0.32$ unit per cubic centimeter.

After eight hours, the concentration is $C(8) \approx 0.17$ unit per cubic centimeter. For now, the easiest way to answer the next two parts of the question is to examine the graph of the surge function.

b. The graph in Figure 12.10 shows that the highest value of the function occurs slightly after two hours. (In a later topic, we learn how to determine this value using calculus.)

c. The concentration will drop below 0.25 unit per cubic centimeter after approximately seven hours. ■

Calculator Corner 12.5

Graph $y = 0.8xe^{-0.45x}$ using a window of [0, 12] for x and [0, 1] for y. To find the value of the highest point, press **2nd TRACE** (CALC) and select option 4:maximum. Set the left bound at $x = 2$. Trace to a point slightly to the right of the maximum and press **ENTER** twice. The calculator shows a maximum of 0.654 unit per cubic centimeter after approximately 2.22 hours. To estimate when concentration is 0.25, use the Table feature, with TblStart = 6.5 and ΔTbl = .01. Scroll through the y column. It appears that the concentration is 0.25 units per cubic centimeter after approximately 6.87 hours.

The table summarizes the exponential models.

Type	Model
Exponential growth	$P(t) = P_0e^{kt}$, where $k > 0$
Exponential decay	$P(t) = P_0e^{kt}$, where $k < 0$
Logistic growth	$P(t) = \dfrac{c}{1 + ae^{-bt}}$
Surge function	$C(t) = ate^{-bt}$

Check Your Understanding 12.1

1. Use the graph of $f(x) = 3^x$ to sketch the following, without using a graphing calculator.

a. $f(x) = 3^{x+2}$ **b.** $f(x) = 3^x + 2$ **c.** $f(x) = 2 - 3^x$

2. Evaluate each of the following to three decimal places.

a. 1.7^4 **b.** $-2^{1.2}$ **c.** $e^{0.37}$ **d.** $5e^{-1.4}$

3. If \$2000 is invested at 4% compounded quarterly, what is the value of the investment after 3 years?

4. A colony of bacteria grows according to $P(t) = 350e^{0.025t}$, where t is in hours. Find the number of bacteria in the colony after two days.

The Logarithmic Function

The exponential function $f(x) = a^x$ is continuous and one to one. Recall from algebra that a continuous one-to-one function must have an inverse that is also a function. Let's find the inverse of the exponential function:

$$f(x) = a^x \rightarrow y = a^x$$

The inverse is found by interchanging x and y and then solving for y. Interchanging x and y gives

$$x = a^y$$

which now must be solved for y. The problem is that no algebraic technique works for solving this function for y, so we must define a new one. The y we are solving for is *the power to which a is raised to obtain the result x.* That is the definition of the logarithmic function.

Definition: The **logarithmic function** $f(x) = \log_a x$ is the inverse of the exponential function $f(x) = a^x$. In other words, $y = \log_a x$ if and only if $x = a^y$.

Remember that the domain of the inverse function is the range of the original function. Therefore, the domain of the logarithmic function is all positive real numbers and the range is all real numbers. As with the exponential function, the base number a is a positive real number not equal to 1. Just remember that $\log_a x$ means "the exponent placed on a to obtain x."

Example 10: Evaluate each of the following logarithms without a calculator.

a. $\log_3 9$ **b.** $\log_4 \frac{1}{16}$ **c.** $\log_{17} 1$ **d.** $\log_5(-5)$

Solution:

a. $\log_3 9$ means "the exponent placed on 3 to obtain 9." Because $3^2 = 9$, $\log_3 9 = 2$.

b. $\log_4 \frac{1}{16}$ means "the exponent placed on 4 to obtain $\frac{1}{16}$." Because $4^{-2} = \frac{1}{16}$, $\log_4 \frac{1}{16} = -2$.

c. $\log_{17} 1$ means "the exponent placed on 17 to obtain 1." Because $17^0 = 1$, $\log_{17} 1 = 0$.

d. $\log_5(-5)$ means "the exponent placed on 5 to obtain -5." Because any power of 5 must be a positive value, we know that $\log_5(-5)$ does not exist in the set of real numbers. ■

A logarithm is an exponent, so the laws of exponents can be rewritten for logarithms.

Properties of Logarithms

1. $\log_a 1 = 0$ because $a^0 = 1$
2. $\log_a a^x = x$ and $a^{\log_a x} = x$ because a^x and $\log_a x$ are inverses
3. $\log_a(m \cdot n) = \log_a m + \log_a n$ because $a^m \cdot a^n = a^{m+n}$
4. $\log_a \frac{m}{n} = \log_a m - \log_a n$ because $\frac{a^m}{a^n} = a^{m-n}$
5. $\log_a m^r = r \cdot \log_a m$ because $(a^m)^r = a^{m \cdot r}$

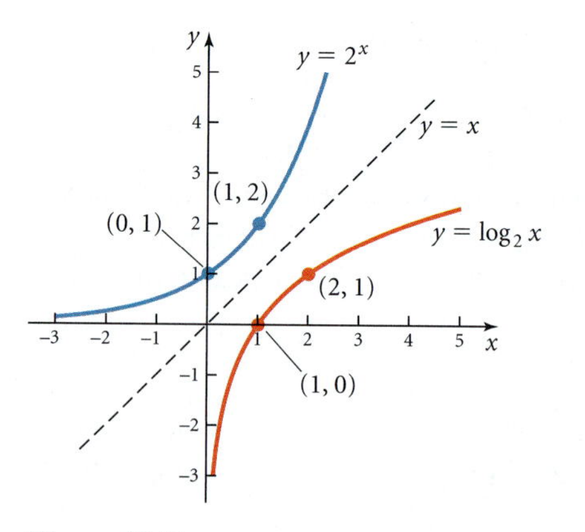

Figure 12.11

By now you know that every function has a graph associated with it. The graph of the logarithmic function shown in Figure 12.11 can be easily obtained by using the graph of the exponential function and the graphical properties of inverses.

From algebra, we know that the graphs of a function and its inverse will be symmetric about the line $y = x$, which means that a point (a, b) on the graph of the function will become the point (b, a) on the graph of the inverse. Because the exponential function passes through $(0, 1)$, its inverse must pass through $(1, 0)$.

Because the exponential function has an asymptote along the negative x-axis, its inverse must have an asymptote along the negative y-axis.

The domain of the logarithmic function is the positive real numbers (the range of the exponential function). The range of the logarithmic function is the set of real numbers (the domain of the exponential function).

The graph of a logarithmic function is continuous for all positive values of x. It has three characteristics that describe its shape:

- The graph passes through $(1, 0)$.
- The graph increases for all values of x.
- The graph has an asymptote along the negative y-axis.

As with all functions, transformations can be applied to graphs of logarithmic functions also to change their location.

Example 11: Use the graph of $\log_2 x$ and your knowledge of transformations to sketch each of the following functions without using a graphing calculator.

a. $f(x) = \log_2 x + 3$ **b.** $f(x) = \log_2(x - 3)$ **c.** $f(x) = -\log_2 x$

Solution:

a. $f(x) = \log_2 x + 3$ shifts the graph of $\log_2 x$ upward three units, so the curve increases and passes through $(1, 3)$. The asymptote remains along the y-axis. The domain is all positive real numbers, and the range is all real numbers. See Figure 12.12.

b. $f(x) = \log_2(x - 3)$ shifts the graph of $\log_2 x$ to the right three units, so the curve increases and passes through $(4, 0)$. The asymptotes shifts to $x = 3$. The domain is all $x > 3$, and the range is all real numbers. See Figure 12.13.

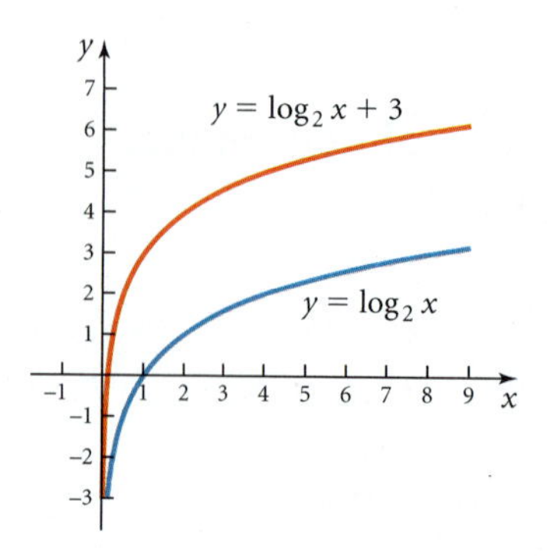

Figure 12.12

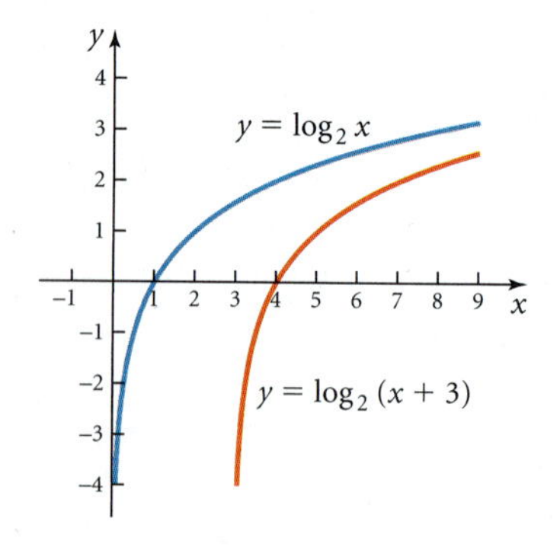

Figure 12.13

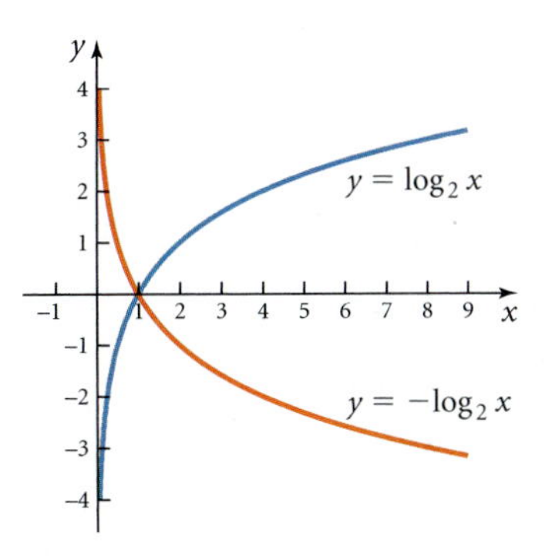

Figure 12.14

c. $f(x) = -\log_2 x$ reflects the graph of $\log_2 x$ about the x-axis. The curve passes through $(1, 0)$ and decreases. The asymptote remains along the y-axis, but along the positive side. The domain is all positive real numbers, and the range is all real numbers. See Figure 12.14. ■

Special Bases

Two special bases are frequently used when working with logarithms:

- **Common logarithms** use base 10 and are denoted **log *x***.
- **Natural logarithms** use base e and are denoted **ln *x***.

On your calculator, find the **log** key and the **ln** key. The **log** key is used to evaluate common logarithms, and the **ln** key is used to evaluate natural logarithms.

Example 12: Use your calculator to evaluate the following logarithms. Round to three decimal places.

a. $e^{-0.5}$ **b.** $\log 25$ **c.** $\log 0.8$ **d.** $\ln 1.7$

e. $\ln 800$ **f.** $\ln(-5)$

Solution:

a. Enter **2nd ln** -0.5 to obtain 0.607.

b. Enter **log** 25 to obtain 1.398. This result should seem logical because $\log 10 = 1$ and $\log 100 = 2$.

c. Enter **log** 0.8 to obtain -0.097. This result should seem logical because $\log 1 = 0$ and $\log 0.1 = -1$.

d. Enter **ln** 1.7 to obtain 0.531.

e. Enter **ln** 800 to obtain 6.685.

f. The domain of the ln function must be positive numbers, so $\ln(-5)$ does not exist. ■

To evaluate logarithms of bases other than 10 or e, we use the **Change of Base Formula**.

Change of Base Formula

If $y = \log_b x$, where $b \neq 10$ or e, then $\log_b x = \dfrac{\log x}{\log b}$ or $\dfrac{\ln x}{\ln b}$.

Example 13: Evaluate $\log_8 72$ to three decimal places.

Solution: Because 72 is not an exact power of 8, we use the Change of Base Formula:

$$\log_8 72 = \frac{\log 72}{\log 8} = \frac{\ln 72}{\ln 8} \approx 2.057$$

You should verify that using either base in the calculation yields the same result. ■

MATHEMATICS CORNER 12.3

Proof of Change of Base Formula

We use the laws of logarithms to prove the Change of Base Formula.

Let	$y = \log_b x$	Where $b \neq 10, e$
Then	$b^y = x$	Rewrite in exponential form
	$\log_a b^y = \log_a x$	Apply $\log_a$ to both sides, where $a = 10$ or e
	$y \cdot \log_a b = \log_a x$	Apply Logarithm Property 5
	$y = \dfrac{\log_a x}{\log_a b}$	Solve for y

Because y represented $\log_b x$, we can now write $\log_b x = \dfrac{\log x}{\log b}$ or $\dfrac{\ln x}{\ln b}$.

Calculator Corner 12.6

The Change of Base Formula can be used to graph logarithmic functions. For example, to draw the graph of $f(x) = \log_2 x$, enter $y1 = \dfrac{\log x}{\log 2}$ or $y2 = \dfrac{\ln x}{\ln 2}$. Both graphs should be the same.

Example 14: Walking speed in various cities is a function of the population of the city and is modeled by $W(P) = 0.35 \ln P + 2.74$, where W is the walking speed in feet per second and P is the population in thousands. How fast do residents of Philadelphia walk if the population of the city is 1,586,000?

Solution: P must be in thousands, so $W = 0.35 \ln 1586 + 2.74 \approx 5.32$. Residents of Philadelphia walk at a speed of about 5.32 feet per second. ■

Check Your Understanding 12.2

1. Given $f(x) = \log_3 x$, sketch each of the following graphs without using a graphing calculator.
   a. $f(x) = \log_3(x - 1)$ **b.** $f(x) = \log_3 x - 1$
2. Evaluate each logarithm without using a calculator.
   a. $\log_5 125$ **b.** $\log_7 \frac{1}{49}$ **c.** $\log_{1/8} 2$
3. Evaluate each logarithm using a calculator. Round each answer to three decimal places.
   a. $\log_4 27$ **b.** $\log_{18} 4$ **c.** $\log_{1/5} 47$

Solving Equations Involving Exponential and Logarithmic Functions

To be successful in calculus, you must be able to do three major things with exponential and logarithmic functions:

- Recognize and sketch the basic graphs.
- Evaluate using a graphing calculator.
- Solve equations involving exponential and logarithmic functions.

You have now reviewed the basic graphs of exponential functions and logarithmic functions and learned how to evaluate these functions on your calculator. We now turn to a discussion of solving exponential equations and logarithmic equations.

Some exponential equations can be solved using the following **Equality Property of Exponents.**

Equality Property of Exponents
If $a^m = a^n$, then $m = n$, where a is a positive real number not equal to 1.

Example 15: Solve the following equations by writing with like bases and using the Equality Property of Exponents.

a. $2^{3x} = 16$ **b.** $3^{x-1} + 2 = 11$ **c.** $4^{x+1} = 8$

Solution:

a. 16 is an exact power of 2, so we write

$2^{3x} = 2^4$ **Rewrite with like bases**

$3x = 4$ **Equality Property**

$x = \frac{4}{3}$ **Solve for *x***

b. First, we isolate the exponential expression.

$3^{x-1} + 2 = 11 \rightarrow 3^{x-1} = 9$ **Isolate the exponential expression**

$3^{x-1} = 3^2$ **Rewrite with like bases**

$x - 1 = 2$ **Equality Property**

$x = 3$ **Solve for *x***

c. Both bases are exact powers of 2, so we write the equation as

$4^{x+1} = 8 \rightarrow (2^2)^{x+1} = 2^3$ **Rewrite with like bases**

$2^{2x+2} = 2^3$ **Simplify exponent**

$2x + 2 = 3$ **Equality Property**

$x = \frac{1}{2}$ **Solve for *x*** ■

Remember that exponential functions and logarithmic functions are inverses of each other. To solve an exponential equation that cannot be solved with the Equality Property, we use logarithms.

Example 16: Solve the following equations, rounding each answer to three decimal places.

a. $e^x = 3.2$ **b.** $e^{2x-1} = 5.46$ **c.** $5e^{2-x} = 15$

d. $3e^{-2x} + 4 = 10$ **e.** $1.7^{4x} = 3$

Solution: Remember that $\log_a x$ and a^x are inverses. Logarithm Property 2 states that $\log_a a^x = x$, so $\ln e^x = \log_e e^x = x$.

a. $e^x = 3.2 \rightarrow \ln e^x = \ln 3.2$ Apply ln to each side

$x = \ln 3.2 \approx 1.163$ Apply Logarithm Property 2 and solve for x

Calculator Corner 12.7

Evaluate $e^{1.163}$ to verify that the result is approximately 3.2.

b. $e^{2x-1} = 5.46 \rightarrow \ln e^{2x-1} = \ln 5.46$ Apply ln to each side

$2x - 1 = \ln 5.46$ Apply Logarithm Property 2

$2x = 1 + \ln 5.46$ Solve for x

$$x = \frac{1 + \ln 5.46}{2}$$

$x \approx 1.349$

Warning! **Do not use the calculator or do any rounding until the final step.** If you round answers in each step of the solution process, the accuracy of the final solution be will affected.

c. $5e^{2-x} = 15 \rightarrow e^{2-x} = 3$ Isolate exponential expression

$\ln e^{2-x} = \ln 3$ Apply ln to each side

$2 - x = \ln 3$ Apply Logarithm Property 2

$-x = -2 + \ln 3$ Solve for x

$x = 2 - \ln 3$

$x \approx 0.901$

d. $3e^{-2x} + 4 = 10 \rightarrow 3e^{-2x} = 6$ Isolate exponential expression

$e^{-2x} = 2$

$\ln e^{-2x} = \ln 2$ Apply ln to each side

$-2x = \ln 2$ Apply Logarithm Property 2

$$x = -\frac{\ln 2}{2}$$ Solve for x

$x \approx -0.347$

e. This equation is not base e, but we will solve using inverses nonetheless. There are actually two methods that could be used to solve this equation.

Method 1: The base is 1.7, so we use the logarithm with the same base:

$$\log_{1.7} 1.7^{4x} = \log_{1.7} 3 \qquad \text{Apply } \log_{1.7} \text{ to each side}$$

$$4x = \log_{1.7} 3 \qquad \text{Apply Logarithm Property 2}$$

$$4x = \frac{\log 3}{\log 1.7} \quad \text{or} \quad 4x = \frac{\ln 3}{\ln 1.7} \qquad \text{Change of Base Formula}$$

$$x = \frac{\log 3}{4 \log 1.7} = \frac{\ln 3}{4 \ln 1.7} \qquad \text{Solve for } x$$

$$x \approx 0.518$$

Method 2: We can also apply ln to both sides of the equation:

$$\ln 1.7^{4x} = \ln 3$$

Warning! Be aware that $\ln 1.7^{4x} \neq 4x$ because ln has base e, not base 1.7.

$$4x \ln 1.7 = \ln 3 \qquad \text{Logarithm Property 5}$$

$$x = \frac{\ln 3}{4 \ln 1.7} \qquad \text{Solve for } x$$

$$x \approx 0.518$$

Both methods yield the same results, but method 2 is a bit simpler, so that is the method that we use in this textbook when solving equations of this type. ■

Calculator Corner 12.8

Graph $y_1 = 1.7^{4x}$. Trace to the point where $y = 3$. What is the x value?

Some logarithm equations can be solved using the following **Equality Property of Logarithms.**

Equality Property of Logarithms

If $\log_a m = \log_a n$, then $m = n$.

Example 17: Solve the following equations using the Equality Property of Logarithms. Round each answer to three decimal places.

a. $\ln(2x - 3) = \ln 7$

b. $\ln(x - 1) + \ln(x - 3) = \ln 8$

Solution: These are logarithmic equations with like bases in each term, so the Equality Property of Logarithms can be used to solve the equations:

a. $\ln(2x - 3) = \ln 7 \rightarrow 2x - 3 = 7$ **Equality Property of Logarithms**

$x = 5$ **Solve for x**

b. $\ln(x - 1) + \ln(x - 3) = \ln 8$

$\ln[(x - 1)(x - 3)] = \ln 8$ **Logarithm Property 3**

$(x - 1)(x - 3) = 8$ **Equality Property of Logarithms**

$x^2 - 4x + 3 = 8$ **Multiply left side**

$x^2 - 4x - 5 = 0$ **Solve for x**

$x = 5, -1$

Remember that the domain of the logarithm function must be positive numbers. If $x = -1$ is substituted into the original equation, it yields logarithms whose values do not exist. Thus, $x = -1$ is an extraneous solution. (An extraneous solution to an equation is a solution that is obtained algebraically from the equation but that does not satisfy the original equation.) The only solution to the equation is $x = 5$. ■

Warning! In Example 17, you should notice that $\ln(x - 1) \neq \ln x - \ln 1$ and $\ln(x - 3) \neq \ln x - \ln 3$. **The Distributive Property of Addition does not apply to the logarithm function.**

To solve logarithmic equations for which the Equality Property of Logarithms does not apply, convert to an exponential equation. Recall that $e^{\ln x} = x$. Also, the calculator will not be used until the final step.

Example 18: Use the properties of logarithms to solve the following equations.

a. $\ln x = 1.4$

b. $\ln(3x + 5) = 2$

c. $\log_2 x - \log_2(x - 2) = 3$

Solution:

a. $\ln x = 1.4 \rightarrow e^{\ln x} = e^{1.4}$ **Apply e to each side**

$x = e^{1.4}$ **Apply Logarithm Property 2 and solve for x**

$x \approx 4.055$

b. $\ln(3x + 5) = 2 \rightarrow e^{\ln(3x+5)} = e^2$ **Apply e to each side**

$3x + 5 = e^2$ **Apply Logarithm Property 2**

$3x = e^2 - 5$ **Solve for x**

$$x = \frac{e^2 - 5}{3}$$

$x \approx 0.796$

c. $\log_2 x - \log_2(x - 2) = 3$

$\log_2 \dfrac{x}{x - 2} = 3$ **Logarithm Property 4**

$$\frac{x}{x-2} = 2^3 \qquad \text{Rewrite in exponential form}$$

$$\frac{x}{x-2} = 8 \qquad \text{Simplify}$$

$$x = 8(x-2) \qquad \text{Solve for } x$$

$$x = \tfrac{16}{7}$$

Checking this answer in the original equation gives

$$\log_2 \tfrac{16}{7} - \log_2\left(\tfrac{16}{7} - 2\right) = \log_2 \tfrac{16}{7} - \log_2 \tfrac{2}{7} = 3$$

Because both $\log_2 \frac{16}{7}$ and $\log_2 \frac{2}{7}$ are defined, $x = \frac{16}{7}$ seems to be a logical solution. ■

Calculator Corner 12.9

Verify that $x = \frac{16}{7}$ is the solution of $\log_2 x - \log_2(x - 2) = 3$ by substituting $x = \frac{16}{7}$ for x in the left-hand expression.

Example 19: *Compound Interest Revisited* How long will it take for \$2000 invested at 6% compounded semiannually to double?

Solution: In Example 4, we estimated this time to be approximately 12 years. Now we will solve this problem algebraically. Use the compound interest formula with $A = 4000$, $P = 2000$, $r = 0.06$, and $n = 2$.

$$A = P\left(1 + \frac{r}{n}\right)^{nt}$$

$$4000 = 2000(1.03)^{2t}$$

$$2 = 1.03^{2t} \qquad \text{Divide both sides by 2000}$$

$$\ln 2 = \ln(1.03)^{2t} \qquad \text{Apply ln to both sides}$$

$$\ln 2 = 2t \ln 1.03 \qquad \text{Apply Logarithm Property 5}$$

$$\frac{\ln 2}{2 \ln 1.03} = t \qquad \text{Solve for } t \text{ by dividing by } 2 \ln 1.03$$

So, $t \approx 11.725$ or approximately 12 years, which agrees with our estimate in Example 4. ■

Example 20: The population of California was 29.76 million people in 1990 and was 33.87 million in 2000. Assuming the same rate of exponential growth continues, write a formula for the population of California at any time after 1990. Use your formula to estimate when the population of California will reach 50 million.

Solution: We must determine the function $P(t) = P_0e^{kt}$, knowing that the initial population was 29.76 million in 1990. With $P_0 = 29.76$, we have $P(t) = 29.76e^{kt}$.

Next, to determine the value of the growth constant k, we know that $P(10) = 33.87$. Thus,

$$P(10) = 29.76e^{10k} = 33.87$$

Solve this equation for k:

$$29.76e^{10k} = 33.87$$

$$e^{10k} = \frac{33.87}{29.76} \qquad \textbf{Divide both sides by 29.76}$$

$$10k = \ln\left(\frac{33.87}{29.76}\right) \qquad \textbf{Apply ln to both sides}$$

$$k = \frac{\ln\left(\frac{33.87}{29.76}\right)}{10} \approx 0.0129 \qquad \textbf{Solve for } \boldsymbol{k}$$

The population of California at any time after 1990 can be approximated by $P(t) = 29.76e^{0.0129t}$.

To estimate when the population of California will reach 50 million, we must determine the t value for which $P(t) = 50$:

$$50 = 29.76e^{0.0129t}$$

so

$$\frac{50}{29.76} = e^{0.0129t}$$

Solving for t gives

$$t = \frac{\ln\left(\frac{50}{29.76}\right)}{0.0129} \approx 40.22$$

The population of California will reach 50 million approximately 40 years after 1990, or in 2030. ■

Check Your Understanding 12.3

Solve the following equations by rewriting each side with like bases and then using the Equality Property.

1. $4^{3x-2} = 16$ **2.** $9^{4-x} = 27$

Solve without using a calculator.

3. $\log_4 64 = x$ **4.** $\log_x 9 = 2$ **5.** $\log 100 = x$

Solve using a calculator. Round each answer to three decimal places.

6. $5.3^x = 21$ **7.** $e^{x+2} = 1.7$ **8.** $3e^{x-1} = 12$

9. $\ln(3x) = 1.4$ **10.** $\log(2x + 3) = -2.8$

11. $\log_2(3x - 7) = 3$

12. How long will it take for \$2000 invested at 4% compounded quarterly to be worth \$5000?

Limits of Exponential Functions and Logarithmic Functions

Because exponential functions and logarithmic functions are continuous and either increasing or decreasing, we have the following limit rules for exponential functions.

Limits of Exponential Functions

$\lim_{x \to b} a^x = a^b$

$\lim_{x \to \infty} a^x = \infty$ if $a > 1$, and $\lim_{x \to \infty} a^x = 0$ if $0 < a < 1$

$\lim_{x \to \infty} a^{-x} = 0$ if $a > 1$, and $\lim_{x \to -\infty} a^{-x} = \infty$ if $0 < a < 1$

Example 21: Evaluate each limit, if it exists, rounding each answer to three decimal places.

a. $\lim_{x \to 2} 2e^{0.2x}$ **b.** $\lim_{x \to \infty} e^{-2x}$

Solution:

a. By the Substitution Property, $\lim_{x \to 2} 2e^{0.2x} = 2e^{0.4} \approx 2.984$.

b. Because e^{-2x} is a decreasing function, its value approaches the asymptote, which is $y = 0$. Thus, $\lim_{x \to \infty} e^{-2x} = 0$. ■

Limits of Logarithmic Functions

$\lim_{x \to b} (\log_a x) = \log_a b$

$\lim_{x \to \infty} (\log_a x) = \infty$ if $a > 1$

$\lim_{x \to \infty} (\log_a x) = -\infty$ if $0 < a < 1$

Example 22: Evaluate each limit, rounding to three decimal places if needed.

a. $\lim_{x \to 0.5} \ln(4 - 3x)$ **b.** $\lim_{x \to \infty} \ln(2x)$

Solution:

a. By the Substitution Property, $\lim_{x \to .5} \ln(4 - 3x) = \ln 2.5 \approx 0.916$.

b. Because $\ln(2x)$ increases for all values of x, $\lim_{x \to \infty} \ln(2x) = \infty$. ■

Example 23: For the logistic growth model $P(t) = \dfrac{c}{1 + ae^{-bt}}$, find $\lim_{t \to \infty} P(t)$.

Solution: If $t \to \infty$, then $e^{-bt} \to 0$, so $\lim_{t \to \infty} P(t) = \dfrac{c}{1 + a(0)} = c$. ■

Regression with Exponential and Logarithmic Functions

In Topic 4, you learned how to take a set of data and determine the function that best models that data using regression and your graphing calculator. Now we can add exponential functions and logarithmic models to the list of possible models that fit a set of data.

Example 24: Use the data from the beginning of this topic regarding Dave's cooling coffee and determine the exponential model that best describes the data.

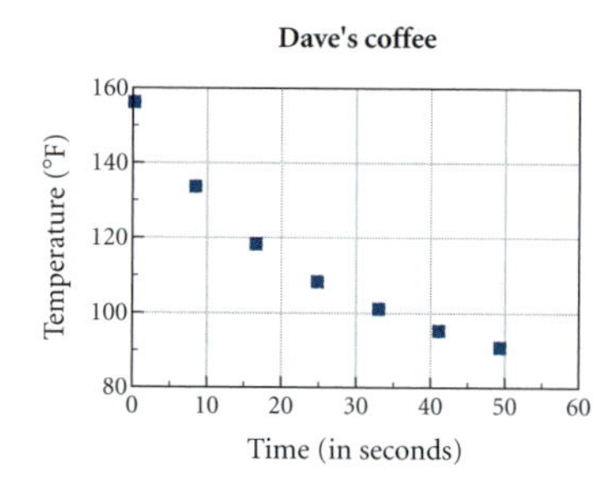

Figure 12.15

Time (seconds)	0.2	8.4	16.6	24.8	33.0	41.1	49.3
Temperature (°F)	155.8	133.2	117.9	107.9	100.7	94.9	90.5

Solution: The scatter plot is shown in Figure 12.15.

Using your graphing calculator, enter the time data in List 1 and the temperature data in List 2. Choose the ExpReg model. Your model should be $y = 146.9(0.989)^x$, where y is the temperature in degrees Fahrenheit and x is the time in seconds.

To convert $f(x) = a \cdot b^x$ to base e, use the property that $b = e^{\ln b}$ and rewrite $a \cdot b^x = a \cdot e^{(\ln b)x}$. Thus, $y = 146.9(0.989)^x$ could also be written as $y = 146.9e^{(\ln .989)x}$ or $y = 146.9e^{-0.0111x}$. ■

Convert from base *b* to base *e*:

$$y = a \cdot b^x = ae^{(\ln b)x}$$

Example 25: Enrollment at a college is given in the table shown. Draw the scatter plot of the data and determine the logarithmic function that models the data. Let x represent years after 1990.

Year	1995	1997	1999	2001	2003	2005
Enrollment	8128	10,255	12,126	13,320	14,028	14,988

Solution: The scatter plot is shown in Figure 12.16.

Enrollment grows quickly at first, but then the growth rate slows down. The enrollment does not fluctuate from increasing to decreasing. A logarithmic model seems to fit the data.

Using your graphing calculator, enter the time data in List 1 and the enrollment data in List 2. Choose the LnReg model. Your model should be $y = -1830.95 + 6245.1 \ln x$, where x is the number of years after 1990 and y is the enrollment.

Figure 12.16 shows the scatter plot and Figure 12.17 shows the logarithmic regression model for the data. As you can see from the graph, the logarithmic model provides a good estimate of the data for the domain $5 \le x \le 15$.

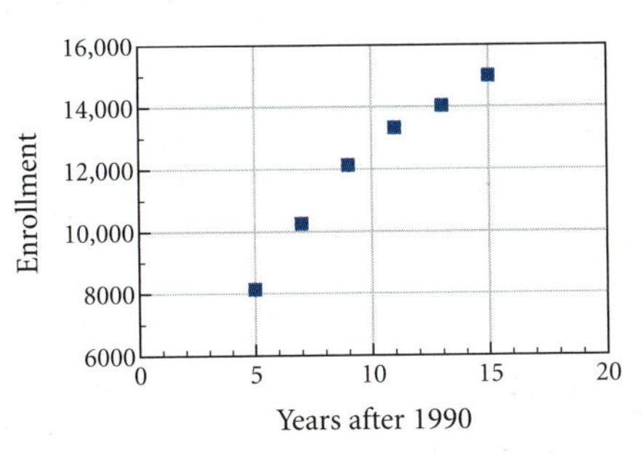

Figure 12.16

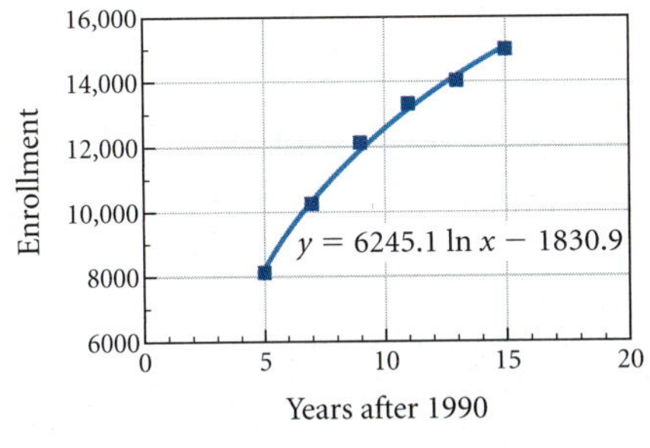

Figure 12.17

Check Your Understanding Answers

Check Your Understanding 12.1

1. a. **b.** **c.**

2. a. 8.352 **b.** −2.297
c. 1.448 **d.** 1.233

3. \$2253.65

4. 1163 bacteria

Check Your Understanding 12.2

1. a. **b.**

2. a. 3 **b.** −2 **c.** $-\frac{1}{3}$

3. a. 2.377 **b.** 0.480 **c.** −2.392

Check Your Understanding 12.3

1. $\frac{4}{3}$ **2.** $\frac{5}{2}$
3. 3 **4.** 3
5. 2 **6.** 1.826
7. −1.470 **8.** 2.386
9. 1.352 **10.** −1.499
11. 5 **12.** approximately 23 years

Topic 12 Review

12

This topic reviewed the basic concepts of exponential and logarithmic functions.

CONCEPT	EXAMPLE
An **exponential function** is any function of the form $f(x) = a^x$, where x is any real number and a is a positive real number not equal to 1.	$f(x) = 3^x$ and $f(x) = \left(\frac{2}{3}\right)^x$ are exponential functions.
If $a > 1$, the graph of the exponential function $f(x) = a^x$ is an increasing continuous curve that passes through (0, 1) and has an asymptote along the negative x-axis.	The graph of $f(x) = 3^x$ is 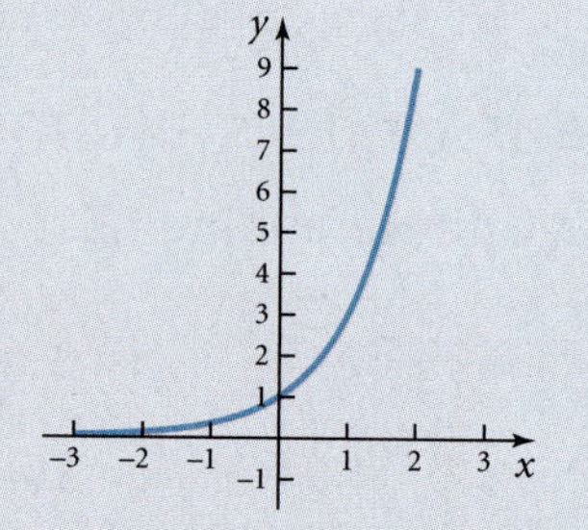
If $0 < a < 1$, the graph of the exponential function $f(x) = a^x$ is a decreasing continuous curve that passes through (0, 1) and has an asymptote along the positive x-axis.	The graph of $f(x) = \left(\frac{2}{3}\right)^x$ is 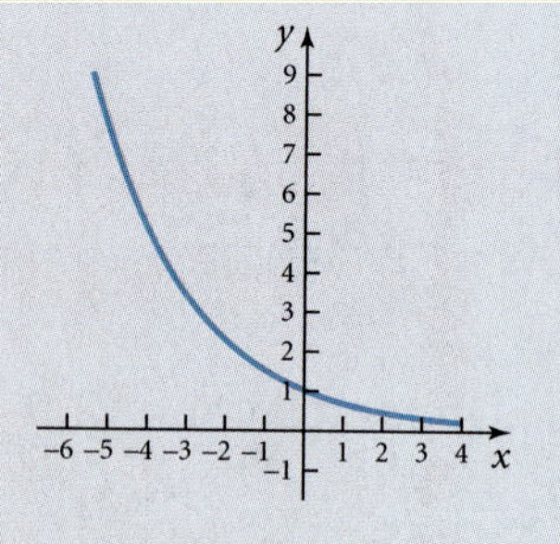
The inverse of an exponential function is a logarithmic function. A **logarithmic function** is any function of the form $f(x) = \log_a x$, where x is a positive real number and a is a positive real number not equal to 1.	The inverse of $f(x) = 3^x$ is $f(x) = \log_3 x$.

The graph of the basic logarithmic function is symmetric about $y = x$ to the graph of the basic exponential function. If $a > 1$, the logarithm graph is an increasing continuous curve that passes through $(1, 0)$ and has an asymptote along the negative y-axis.	The graph of $f(x) = \log_3 x$ is 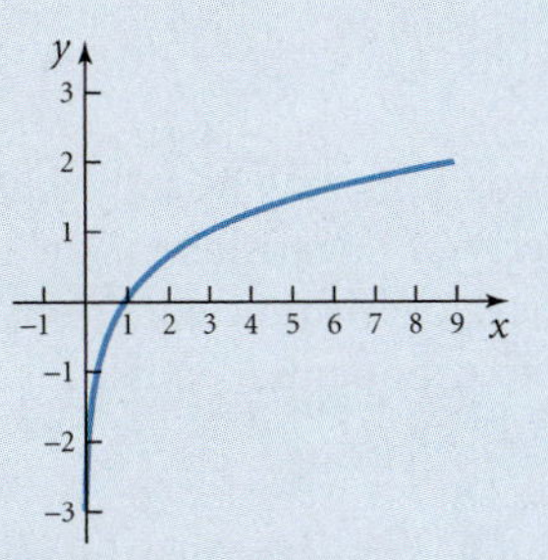
If $0 < a < 1$, the logarithm graph is a decreasing continuous curve passing through $(1, 0)$ with an asymptote along the positive y-axis.	The graph of $f(x) = \log_{0.3} x$ is 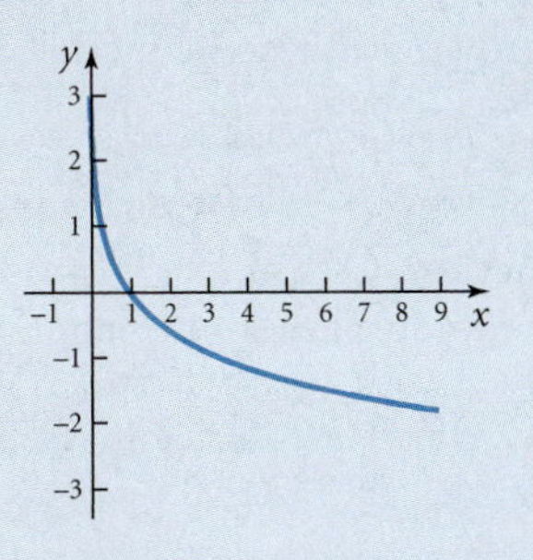
Exponential functions and logarithmic functions are inverses. In other words, $y = \log_a x$ if and only if $x = a^y$. Any logarithmic equation $y = \log_a x$ can be written exponentially as $x = a^y$ and vice versa.	If $4^2 = 16$, then $\log_4 16 = 2$. If $\log_5 \frac{1}{125} = -3$, then $5^{-3} = \frac{1}{125}$.
The two common bases for logarithms are base 10, written $\log x$, and base e, written $\ln x$.	$\log 100$ is a **common logarithm**, referring to the exponent placed on 10 to obtain 100. $\ln 100$ is a **natural logarithm**, referring to the exponent placed on e to obtain 100.
Logarithms using bases other than 10 or e can be evaluated by the **Change of Base Formula**: $\log_b x = \dfrac{\log x}{\log b} = \dfrac{\ln x}{\ln b}$.	$\log_{13} 502 = \dfrac{\log 502}{\log 13} = \dfrac{\ln 502}{\ln 13} \approx 2.424$
The **Equality Property of Exponents** states that if $a^m = a^n$, where a is a positive real number not equal to 1, then $m = n$.	If $3^{x-1} = 3^2$, then $x - 1 = 2$.
Exponential equations may be solved using the Equality Property of Exponents if the expressions can be written with like bases. Otherwise, exponential equations are solved using logarithms.	If $3^{x-1} = 3^2$, then $x - 1 = 2$ and $x = 3$. If $4^{2x-1} = 7$, then $\ln(4^{2x-1}) = \ln 7$. Solving gives $2x - 1 = \dfrac{\ln 7}{\ln 4}$, so $x = \dfrac{\ln 7/\ln 4 + 1}{2} \approx 1.202$

(continued)

The **Equality Property of Logarithms** states that if $\log_a m = \log_a n$, then $m = n$.	If $\ln(2x - 1) = \ln 7$, then $2x - 1 = 7$.
Logarithmic equations may be solved using the Equality Property of Logarithms if the bases are the same. Otherwise, logarithmic equations are solved by converting to exponential equations.	If $\ln(2x - 1) = \ln 7$, then $2x - 1 = 7$ and $x = 4$. If $\ln(2x - 1) = 1.7$, then $2x - 1 = e^{1.7}$ and $x = \frac{1 + e^{1.7}}{2} \approx 3.237$.
$P(t) = P_0e^{kt}$ models the **exponential law of growth** if $k > 0$ and exponential decay if $k < 0$. Exponential growth can also be modeled by the **logistic model,** $P(t) = \frac{c}{1 + ae^{-bt}}$. The **surge function,** $C(t) = ate^{-bt}$, models quantities that grow rapidly at first but then taper off.	*Exponential growth:* The population of a colony of mosquitoes grows according to $P(t) = 500e^{0.25t}$, where t is in days. *Exponential decay:* After spraying with insecticide, the population of the colony decreases according to $P(t) = 500e^{-0.68t}$, where t is in days. *Logistic:* The proportion of public school classrooms with Internet access is modeled by $P(t) = \frac{98}{1 + 678e^{-0.8t}}$, where t is the number of years since 1990. *Surge:* The concentration (in milligrams per cubic centimeters) of a drug in a patient's bloodstream t hours after injection is given by $C(t) = 4te^{-0.5t}$.
Limits of exponential functions and logarithmic functions for specific points within their domains can be determined by evaluating the function at that point: $\lim_{x \to a} b^x = b^a$ and $\lim_{x \to a} (\log_b x) = \log_b a$ for $a > 0$.	$\lim_{x \to 2} 1.65^x = 1.65^2$ $\lim_{x \to 7} (\log_4 x) = \log_4 7$

NEW TOOLS IN THE TOOL KIT

- Exponential functions
- Logarithmic functions
- Graphs of exponential and logarithmic functions
- Evaluating exponential and logarithmic expressions
- Evaluating logarithms using the Change of Base Formula
- Using the exponential law of growth, exponential decay, logistic growth, and the surge function
- Solving exponential and logarithmic equations
- Using regression to find exponential and logarithmic models of data
- Evaluating limits of exponential and logarithmic functions

Topic 12 Exercises

12

Match the equation to the correct graph.

1. $y = 3^x$
2. $y = 3^{x-1}$
3. $y = 3^x - 1$
4. $y = -3^x$
5. $y = 3^{x+1} - 2$
6. $y = 3^{-x}$
7. $y = \log_3 x$
8. $y = \log_3(x - 1)$
9. $y = \log_3 x - 1$
10. $y = -\log_3 x$

a.

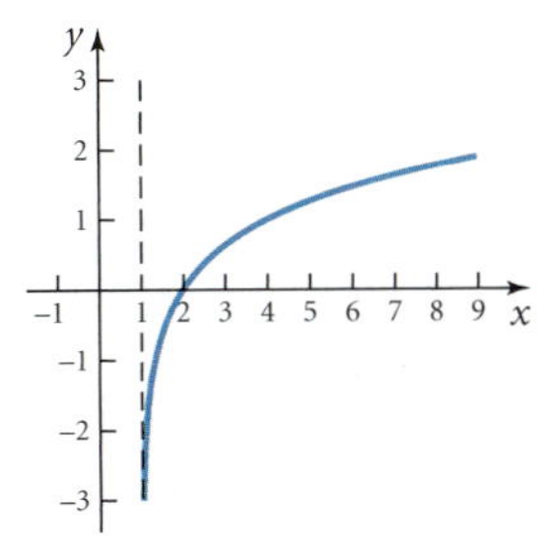

b.

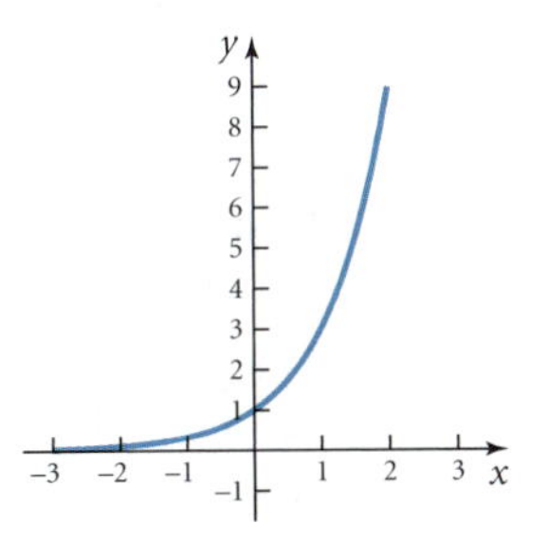

c.

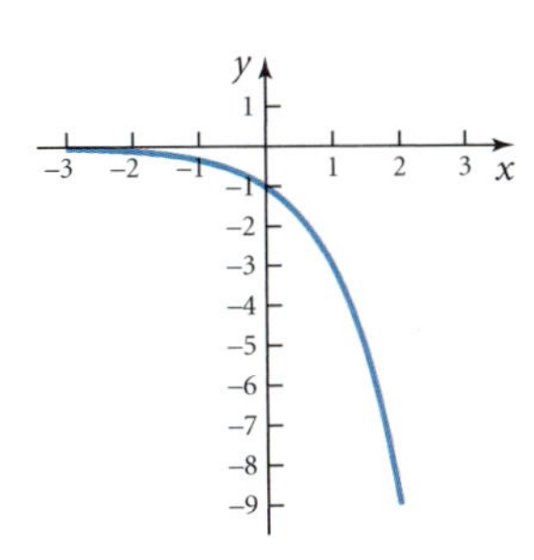

d.

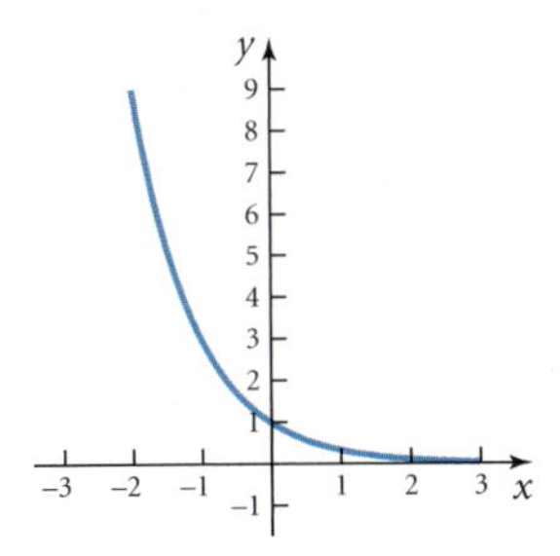

e.

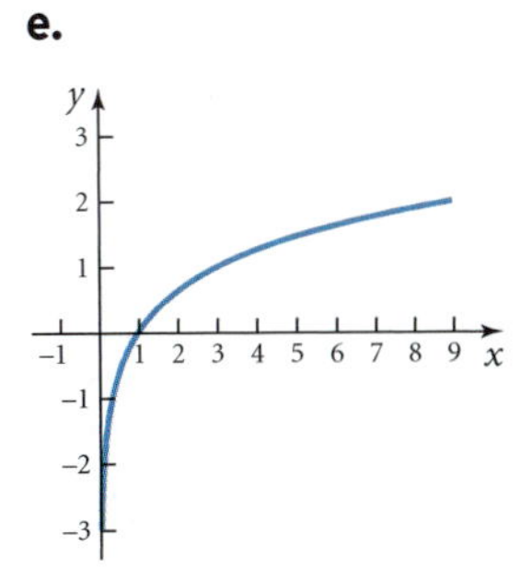

f.

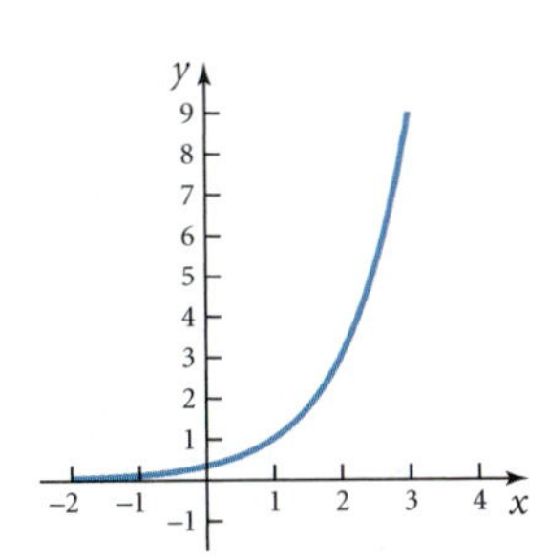

g.

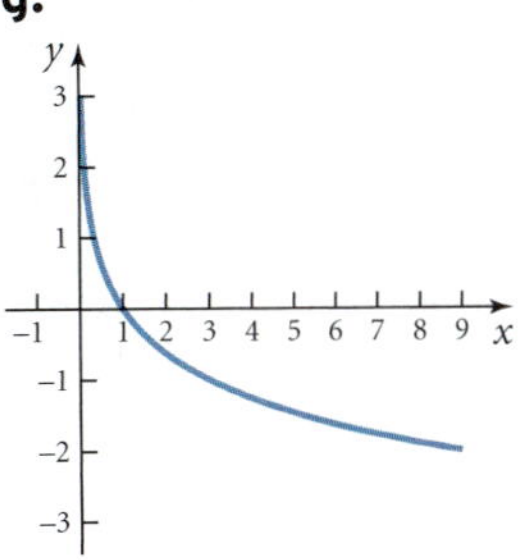

h.

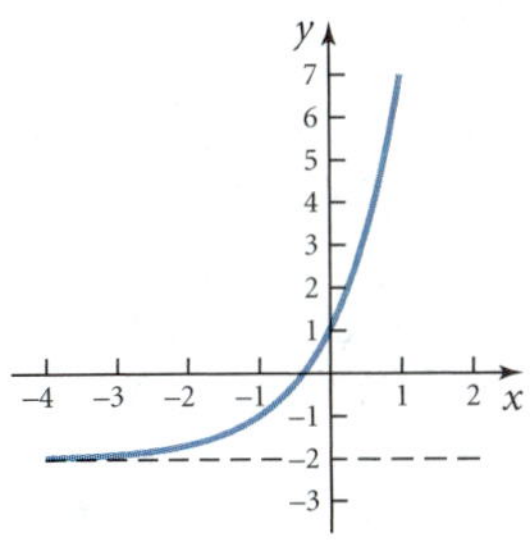

i.

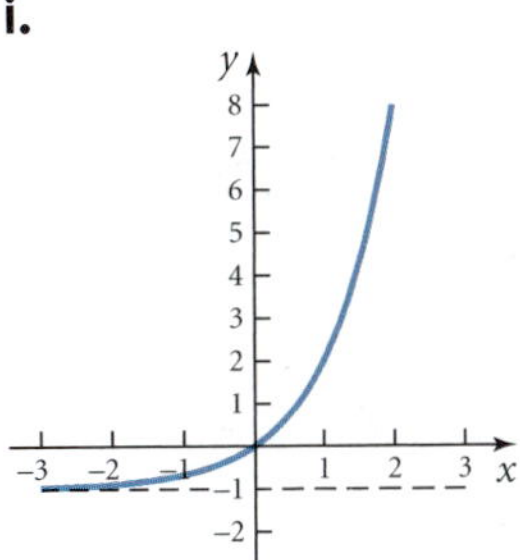

j.

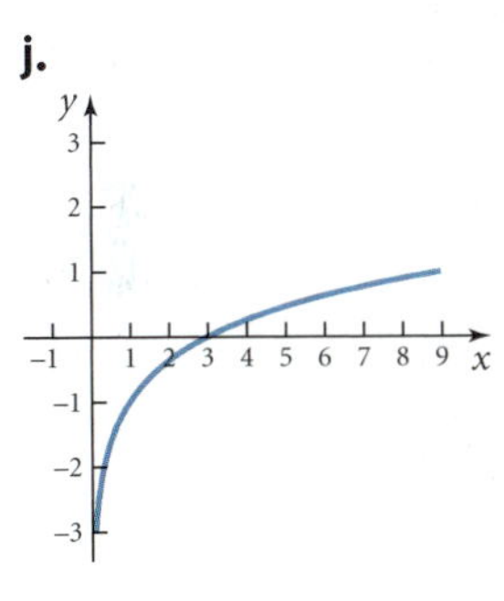

Evaluate each numerical expression exactly, without using a calculator.

11. 5^0
12. -7^0
13. 2^{-3}
14. 3^{-4}
15. 4^2
16. 9^2
17. -4^{-3}
18. -9^{-2}
19. $\log 1$
20. $\log 10$
21. $\ln e$
22. $\ln e^4$
23. $\log_4 16$
24. $\log_5 625$
25. $\log_{1/16} 4$
26. $\log_5 \frac{1}{5}$
27. $\log_3 \frac{1}{9}$
28. $\log_{1/2} \frac{1}{8}$
29. $\log_7 7^2$
30. $\log_8 8^{-3}$

Evaluate each numerical expression using a calculator. Give each answer to three decimal places.

31. $e^{1.2}$
32. $e^{-2.5}$
33. $e^{-0.56}$
34. $e^{0.73}$
35. $\ln 4$
36. $\ln 11$
37. $\ln 0.3$
38. $\ln 0.24$
39. $\log 19$
40. $\log 72$
41. $\log 0.5$
42. $\log 0.3$
43. $\log_6 14$
44. $\log_{11} 46$
45. $\log_{23} 15$
46. $\log_8 2.3$
47. $\log_{1/2} \frac{2}{3}$
48. $\log_{1/5} \frac{3}{4}$

49. $\log_3 120$
50. $\log_7 5$
51. $\log_6(-6)$
52. $\ln(-e)$

Solve each of the following equations without using a calculator.

53. $7^{x-1} = 49$
54. $6^{2-x} = 36$
55. $8^{3x-1} = 32$
56. $9^{2x+1} = 27$
57. $\log_4 x = \frac{1}{2}$
58. $\log_x 64 = 3$
59. $\log_4 x = -2$
60. $\log_4 \frac{1}{2} = x$

Solve each of the following equations using a calculator. Round each answer to three decimal places.

61. $e^x = 8$
62. $e^{-x} = 4$
63. $e^{2x-5} = 6$
64. $e^{3x+2} = 2$
65. $3e^{2x} + 5 = 11$
66. $2e^{x-1} - 7 = 9$
67. $2^{3x} = 13$
68. $5^{2x} = 21$
69. $7^{2-x} = 10$
70. $11^{4-3x} = 50$
71. $5^{x+1} - 3 = 6$
72. $6^{2-x} + 5 = 12$
73. $\log_2(3x) = 4$
74. $\log_5(2x) = 2$
75. $2\log_5(2x - 3) = 2$
76. $3\log_2(5x + 2) = 9$
77. $\ln(x - 2) = 6.1$
78. $\ln(4x - 3) = 1.3$
79. $\ln(3 - 2x) = -2$
80. $\ln(2x - 7) = -0.4$

Use properties of logarithms to solve the following equations. Round each answer to three decimal places as necessary.

81. $\log x + \log(x - 3) = 1$
82. $\ln(2 + x) + \ln x = \ln 3$
83. $\log_4 x - \log_4(x + 3) = -2$
84. $\log_3 x + \log_3(x + 8) = 2$
85. $\ln(2x) + \ln(x - 5) = \ln 3$
86. $\log_5(2x) - \log_5(x - 3) = 1$
87. If \$3000 is invested at 5% compounded semiannually, find the value of the investment after two years.
88. If \$5000 is invested at 6% compounded quarterly, find the value of the investment after five years.
89. Refer to Exercise 87. How long will it take for the initial investment of \$3000 to reach a value of \$5000?
90. Refer to Exercise 88. How long will it take for the initial investment of \$5000 to double?
91. Refer to Exercise 87. Suppose the compounding is done continuously. How much money is in the account after two years?
92. Refer to Exercise 88. Suppose the compounding is done continuously. How much money is in the account after five years?
93. The number of cell phone subscribers, in millions, is given by $N(t) = 6.052(1.378)^t$, where N is the number of subscribers in millions and t is the number of years after 1990.

a. How many subscribers were there in 2005?

b. When will there be one billion subscribers?

94. The cost of a 30-second commercial during the Super Bowl can be modeled by $C(x) = 55.66(1.114)^x$, where C is the cost in thousands and x is the number of years after 1967.

a. What did a commercial cost in the 2004 Super Bowl?

b. When will the cost of commercial be \$5,000,000?

95. The value of a new car depreciates approximately 20% per year for the first five years. The value of the car after t years is modeled by $V(t) = P(0.8)^t$, where P is the initial price of the car. Mark buys a new car for \$25,960.

a. Find the value of Mark's car after three years.

b. When will Mark's car be worth \$8000?

96. The percentage of information retained x weeks after learning it is approximately $P(x) = 80e^{-0.5x} + 20$.

a. Evaluate $P(3)$ and interpret your answer.

b. How much information is retained after three months?

97. The number of bacteria in a culture is approximated by $P(t) = 750e^{0.012t}$, where t is in days.

a. How many bacteria are in the culture after one week?

b. When will there be 5000 bacteria?

98. Atmospheric pressure, in pounds per square inch, a miles above sea level is given by $P(a) = 14.5e^{-0.21a}$. Find the atmospheric pressure for the following locations.

a. Mount McKinley, elevation 3.85 miles above sea level

b. Denver, the "mile-high" city

c. Death Valley, 632 feet below sea level

99. The population of a city, in millions, is given by $P(x) = 1.225e^{0.031x}$, where x is the number of years after 1990. Assume the growth rate continues.

a. What will the population of the city be in 2008?

b. When will the population be 2.5 million?

100. The population of a city was 38,880 in 1980 and 45,010 in 1990.

a. Find a model for the population of the city t years after 1980.

b. Use your model to predict the population in 2008.

101. Carbon-14, a radioactive element, has a half-life of 5760 years. The amount of carbon-14 remaining after x years is given by $P(x) = P_0e^{-0.00012x}$, where P_0 is the original amount. Suppose a fossilized bone is uncovered and testing shows that 80% of the original amount of carbon-14 remains. How old is the bone?

102. Repeat Exercise 101 for a fossilized bone with 70% of the original amount of carbon-14 remaining.

103. The probability of an accident at a given blood alcohol (BAC) level is given by $P(b) = e^{21.5b}$, where P is the percent probability ($P\%$) of having an accident and b is the BAC level.

a. In most states, the maximum legal BAC limit when driving is 0.08. What is the probability of an accident at that level?

b. At what level is the probability of an accident 50%?

104. In 1957, a nuclear reactor accident occurred in the Soviet Union. The amount of plutonium remaining from an initial amount of 10 pounds after t years is given by $N(t) = 10e^{-0.00002845t}$. (Plutonium is a radioactive element with a half-life of 24,360 years.) How much plutonium remains after 50 years? Does your answer explain why the Soviet government set aside a large region near the accident as a permanent reserve?

105. The population of Charlottesville, Virginia, was 131,107 in 1990 and 159,576 in 2000. Assuming the exponential growth rate continues, find a formula for the population of Charlottesville at any time after 1990. Use your formula to predict when the population will reach 200,000.

106. The school-age population of Douglas County, Colorado, which encompasses the city of Denver, has grown from 13,000 in 1990 to 35,000 in 2000. Assuming the exponential growth rate continues, find a formula for the school-age population of Douglas County at any time after 1990. Use your formula to predict when the school-age population will reach 70,000.

Newton's Law of Cooling states that the rate at which an object cools is proportional to the original temperature and the temperature of the medium in which it is placed. The model is $T(t) = M + (P - M)e^{-kt}$, where T is the temperature of the object at any time, t is in minutes, M is the temperature of the medium in which the object is placed, and P is the initial temperature. Use Newton's law of cooling to solve Exercises 107 and 108.

107. A casserole is taken out of a 375°F oven and placed on the counter to cool. The room temperature is 70°F. After 10 minutes, the temperature of the casserole is 300°F.

a. What is the temperature of the casserole after 30 minutes?

b. Sketch the graph of the function using an appropriate window.

c. When will the temperature of the casserole be 120°F?

108. A freshly poured cup of coffee has a temperature of 200°F. Don poured the coffee at 7 a.m. The room temperature is 72°F. After five minutes, the temperature of the coffee is 180°F. If Don likes to drink his coffee when its temperature is between 120°F and 140°F, during what time interval should he drink his coffee?

109. Walking speed in a city is given by $W = 0.35\ln P + 2.74$, where W is the walking speed in feet per second and P is the population of the city, in thousands. How much faster do residents of Orlando (population 1,752,192) walk than residents of Buffalo (population 328,000)?

110. Use the function in Exercise 109 to estimate the population of a city if the walking speed is 5.8 feet per second.

111. The equation $f(t) = 82 - 14\ln(t + 1)$ models the retention of information using test scores on that information, where $f(t)$ is the average test score t months after learning the information.

a. Evaluate $f(1)$, $f(2)$, and $f(4)$ and interpret your answers.

b. Graph the function, using a window of $[0, 9]$ for x and $[0, 90]$ for y. What does the graph indicate about the amount of information being retained?

112. The equation $f(x) = 1 + 5.5\ln(x + 1)$ models the average number of consecutive free throws a basketball player can make, where x is the number of consecutive days of practice for two hours.

a. After seven days of two-hour practices, estimate the number of consecutive free throws the player can make.

b. How much practice is needed to be able to make 10 consecutive free throws?

113. The percentage of the population living in metropolitan areas is increasing and can be modeled by $P(t) = 25.5\ln t - 44.7$, where t is the number of years after 1890.

a. Estimate the percentage of the population living in metropolitan areas in 2005.

b. In what year did 50% of the population live in metropolitan areas?

114. The proportion of students in a psychology experiment who could remember an eight-digit number correctly for t minutes is given by $N(t) = 0.9 - 0.2\ln t$.

a. Estimate the proportion of students who could remember the number after 20 minutes.

b. After how long will only 10% of the students remember the number?

115. The percentage of the U.S. labor force that is 16 years of age or older can be modeled by $y = \dfrac{76.4}{1 + 3.75e^{-0.018x}}$, where x is the number of years after 1900.

a. Estimate the percentage in 2000.

b. When will the percentage be 50%?

c. What is the limiting value of the percentage?

116. The growth in the number of two-year colleges in the United States can be modeled by $y = \dfrac{1259}{1 + 121.7e^{-0.77x}}$, where $x = 1$ represents the decade from 1901 to 1910.

a. Estimate the number of two-year colleges in the United States during the 1980s.

b. In what decade will there be 1250 two-year colleges?

117. The concentration of a drug in a patient's bloodstream t hours after the drug was injected is given by $C(t) = 1.6te^{-0.5t}$ mg/cm^3.

a. What is the concentration after one hour?

b. What is the concentration after 12 hours?

118. Refer to Exercise 117.

a. Graph the function using a window of $[0, 12]$ for x and $[0, 2]$ for y.

b. Use the graph to estimate when the concentration will be 1 mg/cm^3.

c. Use the graph to estimate the time at which concentration is the highest.

SHARPEN THE TOOLS

119. A calculus professor dies suddenly and the body is placed in a cooler with a temperature of 40°F. A zealous student finds the body and calls in the coroner to estimate the time of death. The body's temperature is 92.4°F at 3 p.m. and 90.8°F at 4 p.m. Assuming normal body temperature of 98.6°F at the time of death, when did the professor die?

120. Earthquake intensity is measured using the *Richter scale*. The Richter number is given by $R = \log \dfrac{I}{I_0}$, where I is the intensity of the earthquake and I_0 is a constant representing standard intensity. (Standard intensity means that a certain amount of shifting of the plates beneath the surface of the earth is acceptable. I_0 is a measure of that standard amount of acceptable movement.)

a. If an earthquake measures 5.8 on the Richter scale, how intense is the earthquake?

b. The 1906 San Francisco earthquake measured 8.3 on the Richter scale; another earthquake in San Francisco in 1989 measured 7.8 on the Richter scale. A newscaster at the time reported that the 1906 earthquake was three times more intense than the 1989 one. Justify his statement.

121. Solve $e^{0.2x} = 3 - \ln x$ using a graphing calculator.

a. Graph $y_1 = e^{0.2x}$ and $y_2 = 3 - \ln x$. Then use ISECT to determine the x value at the point of intersection.

b. Now graph $e^{0.2x} - (3 - \ln x) = 0$ and solve for x by determining the x-intercept. How does this result compare with your answer from part a?

c. Why can't this equation be solved using algebraic methods?

122. Solve $4 - x^2 = x \ln x$ using the two approaches given in Exercise 119. Why can't this equation be solved using algebraic methods?

123. Solve $x^2 = 2^x$. State the window you used for the graph.

124. Solve $x^3 = 3^x$. State the window you used for the graph. What conjecture might you make about the solutions of $x^n = n^x$ for n a positive integer?

CALCULATOR CONNECTION

125. The growth in the total number of Starbucks locations each year since its beginnings in 1971 is shown in the table at the top of the right column.

a. Draw a scatter plot of the data.

b. Use the regression feature of your calculator to find an exponential growth model for the data. Let x represent the number of years after 1970.

c. Use the regression feature of your calculator to find a logistic growth model for the data. Let x be the number of years after 1970.

d. Use the two models you obtained in parts b and c to estimate the total number of Starbucks locations in 1994 and in 2004.

e. Based on your answers in part d, which of the two models do you think gives the truest prediction and why?

f. Use the two models you obtained to predict the number of Starbucks locations in 2010.

Year	Number of Locations
1971	1
1987	17
1988	33
1989	55
1990	84
1991	116
1992	165
1993	272
1994	425
1995	676
1996	1,015
1997	1,412
1998	1,886
1999	2,135
2000	3,501
2001	4,709
2002	5,688
2003	7,225
2004	7,569
2005	10,801

Source: Starbucks.

126. The U.S. population, in millions, for various years is shown in the table.

Year	U.S. Population	Year	U.S. Population
1840	17.1	1940	132.6
1850	23.2	1950	151.7
1860	31.4	1960	179.7
1870	38.6	1970	203.7
1880	50.2	1980	226.7
1890	63.0	1990	248.7
1900	76.2	2000	281.4
1910	92.2	2004	295.0
1920	106.1	2006	300.0
1930	123.4		

Data from: U.S. Census Bureau

a. Determine a logistic model for the U.S. population growth.

b. Use your model to estimate the population in 1900 and in 2000. How close is the estimate to the true population?

c. Use your model to predict the U.S. population in 2010.

127. *Hurricane Aftermath* After a major category 4 hurricane, a power company provided the following information regarding the number of customers without power in the service area.

Day	Number Without Power
1	2,560,000
2	1,240,000
3	697,982
4	392,294
5	225,316

a. Find a logarithmic model for the data. Let N be the number of customers, in millions, without power and $x = 1$ be the day of the hurricane.

b. Use your model to estimate the number of customers without power on day 2 and day 4. How close did the estimate come to the true number?

c. Use your model to estimate when there were fewer than 1000 customers without power.

128. The table shown at the top of the right column gives the total number of two-year colleges in the United States for each decade, beginning with 1901.

a. Draw a scatter plot of the data.

b. Until 1960, the data grew exponentially; after 1960, however, it appears more logarithmic. Use the regression feature of your calculator to write a piecewise-defined function describing the total number of two-year colleges at any time t.

c. Use your model to predict the total number of two-year colleges in the 2001–2010 decade.

Decade	x	Total Number of Colleges
1901–1910	1	25
1911–1920	2	74
1921–1930	3	180
1931–1940	4	238
1941–1950	5	330
1951–1960	6	412
1961–1970	7	909
1971–1980	8	1,058
1981–1990	9	1,106
1991–2000	10	1,155

Data from: American Association of Community Colleges

129. The percent of the U.S. population living in metropolitan areas since 1900 is shown in the table.

Year	Percent Living in Metropolitan Areas
1900	25.5
1910	28.3
1920	34.0
1930	44.6
1940	47.8
1950	56.1
1960	63.3
1970	69.0
1980	74.8
1990	77.5
1996	79.8
2000	80.3

Data from: U.S. Census Bureau

a. Find a logistic growth model for the data. Let x be the number of years after 1900.

b. Use your model to estimate the percentage of the population living in metropolitan areas in 2005.

c. What is the limiting value of your model? In other words, find $\lim_{x \to \infty}$. Does your answer seem logical?

130. The 2000 U.S. census provided the following information regarding average household size.

Year	Average Household Size
1900	4.60
1930	4.01
1940	3.68
1950	3.38
1960	3.29
1970	3.11
1980	2.75
1990	2.63
2000	2.59

a. Find an exponential model for the data. Let x be the number of years after 1900.

b. Convert the model from part a to base e.

c. Use your model to estimate the average household size in 2008.

d. What is the limiting value of the model? Does this answer seem logical? Realistically, what is the limiting value of the data?

131. In 1965, just four years after the first planar integrated circuit was discovered, Gordon Moore made a famous observation. Moore's law predicted that the number of transistors on a computer chip would double approximately every two years. In April 2005, Intel estimated that this pattern would continue until the end of 2010.

a. Use the law of growth to create a formula $P(t) = P_0 e^{kt}$ for a quantity that doubles every two years.

b. The following table gives the actual number of transistors. Find an exponential model for the data, converting to $y = a \cdot e^{bt}$. Compare your model with the model in part a. How good was Moore's 1965 prediction?

Year	Number of Transistors
1971	2,250
1972	2,500
1974	5,000
1978	29,000
1982	120,000
1985	275,000
1989	1,180,000
1993	3,100,000
1997	7,500,000
1999	24,000,000
2000	42,000,000
2002	220,000,000
2003	410,000,000

Source: www.intel.com/technology/mooreslaw.htm.

COMMUNICATION

132. The definitions of exponential and logarithmic functions require that the base, a, be a positive real number not equal to 1. Briefly explain why this restriction is necessary.

133. Briefly explain how exponential functions and polynomial functions differ. How do their graphs compare?

134. The exponential law of growth, the logistic model, and the surge function are three models of exponential growth. Discuss the appropriate use of each model. Give some real examples that might be modeled by each of these functions.

GROUP CALCULATOR PROJECTS

These projects provide two opportunities to generate an exponential function. Divide students into groups of three. Project 1 requires each group to have 40 dice. Project 2 requires small bags of M&M candies.

Group Project 12.1

Roll 40 dice. Remove all the dice with four dots facing up and record the number of dice left. Roll the remaining dice and again remove the 4s and record the number left. Continue for about six rolls or until there are fewer than six dice left. Now complete the table.

Roll Number	Dice Remaining
0	40
1	
2	
3	
4	
5	
6	
7	

Graph the data, using the roll number along the horizontal axis and the number of dice on the vertical axis. Create an exponential model for the data.

Group Project 12.2

Put 40 M&Ms in a cup. Pour them out on a flat surface and remove the ones with an *M* facing up. Record the number remaining. Put the remaining candies back in the cup and pour them out again, this time removing the ones with an *M* facing up. Continue for five rolls.

Complete the table.

Trial Number	Number of M&Ms Remaining
0	40
1	
2	
3	
4	
5	

Graph the data, using the trial number as x and the number of candies as y.

Create an exponential model for the data.

TOPIC 13

Derivatives of Exponential and Logarithmic Functions

The value of Steve's new car decreases over time and can be modeled by the exponential function $V(t) = \$28{,}250(0.8)^t$, where t is the age of the new car in years. At what rate is the value of Steve's car changing after one year? You know that rates of change are found using derivatives, but so far you only know how to find derivatives of algebraic functions. In this topic, you learn how to find derivatives of exponential and logarithmic functions. Before starting your study of these derivatives, be sure to complete the Warm-up Exercises to sharpen the algebra and calculus tools needed.

IMPORTANT TOOLS IN TOPIC 13

- *Derivative of e^x*
- *Derivative of $\ln x$*
- *Chain Rule: Exponential Form*
- *Chain Rule: Logarithmic Form*

TOPIC 13 WARM-UP EXERCISES

Be sure you can successfully complete the following exercises before starting Topic 13.

Algebra Warm-up Exercises

1. Given $f(x) = -2x^2 + 3x - 5$, evaluate the difference quotient $\dfrac{f(a+h) - f(a)}{h}$.

2. Given $f(x) = x^5$ and $g(x) = 4 - 3x$, write the composite function $f(g(x))$.

Solve the following equations. Round each answer to three decimal places as needed.

3. $e^{x-3} = 14$

4. $8 - 2^{3x} = 5$

5. $\ln(3x + 5) = 2.4$

Calculus Warm-up Exercises

1. Evaluate each limit, if it exists.

a. $\lim\limits_{x \to -1} (x^3 - 4x + 2)$

b. $\lim\limits_{x \to 6} \dfrac{4}{x - 6}$

c. $\lim\limits_{x \to 6} \dfrac{4x}{x - 3}$

d. $\lim\limits_{x \to 6} \dfrac{4x - 24}{x^2 - 36}$

2. Given $f(x) = -2x^2 + 3x - 5$ if, use the limit definition to find $f'(x)$.

Find $f'(x)$ for each of the following functions using an appropriate derivative rule.

3. $f(x) = -2x^2 + 3x - 5$

4. $f(x) = (5 - 2x)(3x^2 - 4)$

5. $f(x) = \dfrac{5x - 7}{x^3 + 2}$

6. $f(x) = (-2x^2 + 3x - 5)^4$

7. Find $f'(x)$ for the following function. Write your answer in factored form.

$$f(x) = (x^3 - 5)^2(2x^4 + 3)^5$$

Find $f''(x)$ for each of the following functions.

8. $f(x) = -2x^2 + 3x - 5$

9. $f(x) = (2x^2 - 5)^4$

10. Find the equation of the tangent line to $f(x) = -2x^2 + 3x - 5$ at the point where $x = 1$.

Sketch the graph of the following without using a graphing calculator.

11. $f(x) = e^{x-2}$

12. $f(x) = 2 - e^x$

13. $f(x) = \ln(x + 3) - 2$

Answers to Warm-up Exercises

Algebra Warm-up Exercises

1. $-4a - 2h + 3$
2. $f(g(x)) = (4 - 3x)^5$
3. 5.639
4. 0.528
5. 2.008

Calculus Warm-up Exercises

1. **a.** 5 **b.** does not exist
   c. 8 **d.** $\frac{1}{3}$
2. $f'(x) = \lim_{h \to 0}(-4x - 2h + 3) = -4x + 3$
3. $-4x + 3$
4. $-18x^2 + 30x + 8$
5. $\dfrac{-10x^3 + 21x^2 + 10}{(x^3 + 2)^2}$
6. $4(-2x^2 + 3x - 5)^3(-4x + 3)$
7. $2x^2(x^3 - 5)(2x^4 + 3)^4(26x^4 - 100x + 9)$
8. -4
9. $16(2x^2 - 5)^2(14x^2 - 5)$
10. $y = -x - 3$

11.

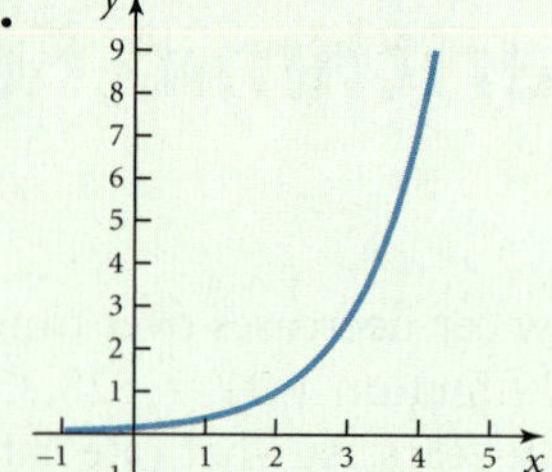

12.

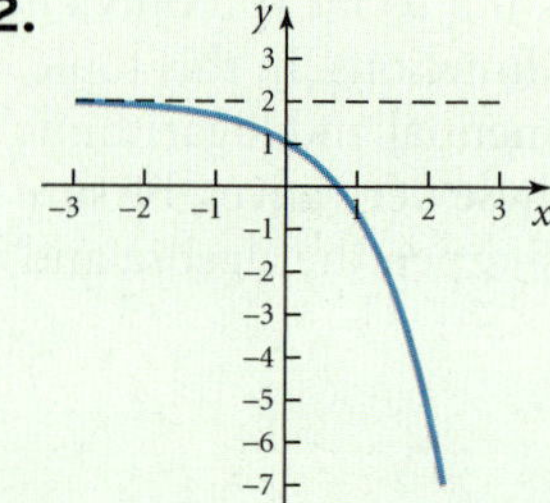

13.

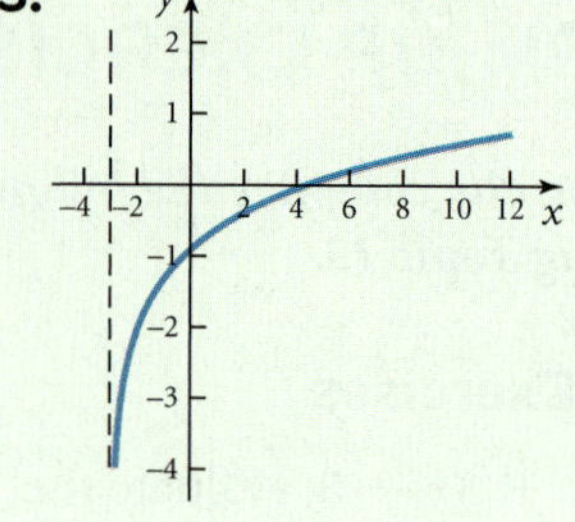

s we return to the discussion of derivatives, you need to remember that the derivative of a function represents the *rate of change* of the function for some value in its domain or the *slope* of the tangent line to the graph of the function at some point. You have learned the following rules for determining derivatives of algebraic functions:

- Power Rule
- Product Rule
- Quotient Rule
- Chain Rule: Power Form

Exponential and logarithmic functions are not algebraic, so how do we find their derivatives? The remainder of this topic answers that question.

Warning! $D_x e^x \neq xe^{x-1}$. The Power Rule only applies if the base is the variable and the exponent is a constant.

Finding the Derivative of $f(x) = e^x$

First, be sure you understand why the Power Rule does not apply to finding the derivative of $f(x) = e^x$. The Power Rule involves constant powers of a variable x, but in the exponential function, the base is the constant and the power is the variable. Because the Power Rule does not apply, we resort to the limit definition of derivatives:

$$f'(x) = \lim_{h \to 0} \frac{f(x+h) - f(x)}{h}$$

$$= \lim_{h \to 0} \frac{e^{x+h} - e^x}{h} \quad \text{Substitute } f(x+h) = e^{x+h} \text{ and } f(x) = e^x$$

$$= \lim_{h \to 0} \frac{e^x e^h - e^x}{h} \quad \text{Laws of exponents}$$

$$= \lim_{h \to 0} \frac{e^x(e^h - 1)}{h} \quad \text{Factor } e^x \text{ from numerator}$$

$$= e^x \cdot \lim_{h \to 0} \frac{e^h - 1}{h} \quad e^x \text{ is a constant because it is not a function of } h$$

At this point, evaluating $\lim_{h \to 0} \frac{e^h - 1}{h}$ yields the indeterminate form 0/0, which implies that some type of algebraic simplification needs to be done. Because there is no more factoring to do, we now approach the limit numerically. Consider the following table, choosing values for h that approach zero.

h	$\frac{e^h - 1}{h}$
0.1	1.05170918076
0.01	1.00501670842
0.001	1.00050016670
0.0001	1.0000500017
↓	↓
0	1

From the values in the table, it appears that $\lim_{h \to 0} \frac{e^h - 1}{h} = 1$.

Calculator Corner 13.1

You can also use the Table feature of your calculator to see $\lim_{h \to 0} \frac{e^h - 1}{h}$. Enter $y_1 = \frac{e^x - 1}{x}$. Then set the table with Table Start $= 0.1$ and ΔTbl $= -0.001$. Scroll through the table and you will see the y values approach 1.

Returning to our limit definition,

$$D_x e^x = \lim_{h \to 0} \frac{e^{x+h} - e^x}{h}$$
$$= e^x \lim_{h \to 0} \frac{e^h - 1}{h}$$
$$= e^x \cdot 1$$
$$= e^x$$

So, we get $\boxed{D_x e^x = e^x}$. In other words, *the exponential function e^x is its own derivative!*

Calculator Corner 13.2

Graph $y_1 = e^x$ using Zoom Decimal. Evaluate dy/dx, where $x = 1$, using your calculator. You should notice a value of $dy/dx = 2.71828$.

TI-83/84: Graph the function. Then **2nd TRACE** (CALC) 6: *dy/dx* and trace to the point where $x = 1$. Notice the y value at this point. Then press **ENTER** and notice the value of the derivative!

Repeat for different values of x and notice that the value of the derivative is always the same as the value of the function at any given point.

Example 1: Find $f'(x)$ for each of the following functions.

a. $3e^x$ **b.** $e^x - 3x^6$ **c.** e^x/x **d.** x^3e^x **e.** $(e^x - 4)^3$

Solution:

a. $D_x(3e^x) = 3 \cdot D_x(e^x) = 3e^x$

b. $D_x(e^x - 3x^6) = D_x(e^x) - D_x(3x^6)$
$= e^x - 18x^5$

c. By the Quotient Rule,

$$D_x\left(\frac{e^x}{x}\right) = \frac{x \cdot D_x e^x - e^x \cdot D_x x}{x^2}$$
$$= \frac{xe^x - e^x}{x^2}$$
$$= \frac{e^x(x - 1)}{x^2}$$

d. By the Product Rule,

$$D_x(x^3e^x) = e^x \cdot D_x x^3 + x^3 \cdot D_x e^x$$
$$= e^x(3x^2) + x^3e^x$$
$$= x^2e^x(3 + x)$$

Factored form is preferred

e. By the Chain Rule,

$$D_x(e^x - 4)^3 = 3(e^x - 4)^2 \cdot D_x(e^x - 4)$$
$$= 3(e^x - 4)^2 \cdot e^x \quad \text{or} \quad 3e^x(e^x - 4)^2$$

We now know that $D_x e^x = e^x$. What if the x were replaced by a function of x, creating the composite function $e^{f(x)}$? How would the derivative look? Before answering, let's review the Power Rule and the resulting Chain Rule when x was replaced by $f(x)$.

Power Rule	Chain Rule: Power Form
$D_x x^n = nx^{n-1}$	$D_x[f(x)]^n = n[f(x)]^{n-1} \cdot f'(x)$

When the x is replaced by $f(x)$ to create a composite function, the Power Rule still applied, but an additional factor of $f'(x)$ appeared in the derivative (see Topic 10). We know that an exponential function is its own derivative and that, for composite functions, an additional factor of $f'(x)$ must appear in the derivative. So, we can write a second form of the Chain Rule for derivatives of exponential functions.

Chain Rule: Exponential Form

$$D_x e^{f(x)} = e^{f(x)} \cdot f'(x)$$

Example 2: Find dy/dx for each of the following functions.

a. e^{3x-2} **b.** e^{x^4}

Solution: Each of these functions is a composite function because the exponents are functions of x.

a. By the Chain Rule: Exponential Form,

$$\begin{aligned} D_x e^{3x-2} &= e^{3x-2} \cdot D_x(3x - 2) \\ &= e^{3x-2}(3) \quad \text{or} \quad 3e^{3x-2} \end{aligned}$$

b. $D_x e^{x^4} = e^{x^4} \cdot D_x x^4 = e^{x^4} \cdot 4x^3 \quad \text{or} \quad 4x^3 e^{x^4}$ ■

Example 3: Find $f''(x)$ for $f(x) = e^{x^4}$.

Solution: From part b of Example 2, you know that

$$f(x) = e^{x^4} \rightarrow f'(x) = 4x^3 e^{x^4}$$

The Product Rule is now required for the second derivative. (Why? Read the function aloud: $f'(x)$ is the product of two functions of x.)

To apply the Product Rule, let $g(x) = 4x^3$ and $h(x) = e^{x^4}$. Then

$$g'(x) = 12x^2 \quad \text{and} \quad h'(x) = 4x^3 e^{x^4}$$

By the Product Rule,

$$\begin{aligned} f''(x) &= D_x[g(x)h(x)] \\ &= h(x)g'(x) + g(x)h'(x) \end{aligned}$$

Substituting, we now have

$$\begin{aligned} f''(x) &= e^{x^4} \cdot D_x 4x^3 + 4x^3 \cdot D_x e^{x^4} \\ &= e^{x^4}(12x^2) + 4x^3 \cdot e^{x^4} \cdot 4x^3 \\ &= 4x^2 e^{x^4}(3 + 4x^4) \end{aligned}$$ ■

Example 4: The population of a city t years after 1995 can be modeled by $P(t) = 62.4e^{0.02t}$, where P is the population in thousands. Evaluate $P'(10)$ and interpret your answer.

Solution:

$$P(t) = 62.4e^{0.02t} \rightarrow P'(t) = 62.4e^{0.02t}(0.02) = 1.248e^{0.02t}$$

Thus,

$$P'(10) = 1.248e^{0.02(10)} \approx 1.524$$

In 2005, the population of the city was increasing by approximately 1524 people per year. ■

Check Your Understanding 13.1

Find $f'(x)$ for each of the following functions.

1. $f(x) = e^{7x}$

2. $f(x) = e^{x^2} - x^2$

3. $f(x) = 3e^{-2x}$

4. $f(x) = (e^{3x} - 4)^2$

5. $f(x) = x^4e^{5x}$

6. $f(x) = e^{5x}/x^4$

7. $f(x) = e^7$

8. $f(x) = x^{7e}$

Finding the Derivative of $f(x) = \ln x$

Like exponential functions, the logarithmic function is also not an algebraic function. This time, however, the limit definition of the derivative can be avoided by making use of the inverse relationship between exponential functions and logarithmic functions.

Let

$$e^{\ln x} = x \qquad \textbf{Logarithm Property 2}$$

Then

$$D_x e^{\ln x} = D_x x \qquad \textbf{Take the derivative of each side}$$

$$e^{\ln x} \cdot D_x \ln x = 1 \qquad \textbf{Chain Rule: Exponential Form}$$

$$x \cdot D_x \ln x = 1 \qquad \textbf{Logarithm Property 2}$$

$$\boxed{D_x \ln x = \frac{1}{x}} \qquad \textbf{Solve for } D_x \ln x$$

To agree with the domain of the logarithm function, the assumption here is that $x > 0$.

Example 5: Find $f'(x)$ for each of the following functions.

a. $f(x) = 4\ln x$

b. $f(x) = x^2 + \ln x$

c. $f(x) = x^3 \ln x$

d. $f(x) = (8x - 3\ln x)^5$

Solution:

a. $D_x 4\ln x = 4D_x \ln x = 4(1/x) = 4/x$

b. $D_x(x^2 + \ln x) = D_x x^2 + D_x \ln x = 2x + 1/x$

c. By the Product Rule,

$$\begin{aligned} D_x(x^3 \ln x) &= (\ln x) \cdot D_x x^3 + x^3 \cdot D_x \ln x \\ &= (\ln x) \cdot 3x^2 + x^3\left(\frac{1}{x}\right) \\ &= 3x^2 \ln x + x^2 \\ &= x^2(3\ln x + 1) \end{aligned}$$

d. By the Chain Rule,

$$\begin{aligned} D_x(8x - 3\ln x)^5 &= 5(8x - 3\ln x)^4 \cdot D_x(8x - 3\ln x) \\ &= 5(8x - 3\ln x)^4 \cdot \left(8 - \frac{3}{x}\right) \end{aligned}$$

■

Before discussing the derivative of logarithms involving composite functions, let's review what we already know.

Composite Function	Derivative
$[f(x)]^n$	Chain Rule: Power Form $n[f(x)]^{n-1} \cdot f'(x)$
$e^{f(x)}$	Chain Rule: Exponential Form $e^{f(x)} \cdot f'(x)$

Now we consider composite functions of the form $\ln f(x)$. We know that $D_x \ln x = 1/x$ and that, for composite functions, an additional factor of $f'(x)$ must appear in the derivative. Replacing x by $f(x)$ in the derivative definition and including the additional factor, we obtain the following definition.

Chain Rule: Logarithmic Form

$$D_x \ln f(x) = \frac{1}{f(x)} \cdot f'(x) = \frac{f'(x)}{f(x)}$$

where $f(x) > 0$ and $f'(x)$ exists.

MATHEMATICS CORNER 13.1

General Form of the Chain Rule

Let $y = f(u)$, where $u = g(x)$. Then $y = f(g(x))$. Now, $y' = f'(g(x)) \cdot g'(x)$, assuming $f'(u)$ and $g'(x)$ exist.

Example 6: Find $f'(x)$ for each of the following functions.

a. $f(x) = \ln(5x)$ **b.** $f(x) = \ln(x + 5)$ **c.** $f(x) = \ln x^4$

d. $f(x) = (\ln x)^4$ **e.** $f(x) = 3x^2 \ln(4x)$

Solution:

a. $D_x \ln(5x) = 1/5x \cdot 5 = 1/x$

b. $D_x \ln(x + 5) = \dfrac{1}{x + 5} \cdot 1 = \dfrac{1}{x + 5}$

> **Warning!** $D_x \ln(x + 5) \neq D_x(\ln x) + D_x(5)$. The Distributive Property of addition does not apply to the logarithm function.

c. $D_x \ln x^4 = \dfrac{1}{x^4} \cdot 4x^3 = \dfrac{4x^3}{x^4} = \dfrac{4}{x}$

This derivative can also be evaluated using logarithm properties:

$$D_x \ln x^4 = D_x(4 \ln x) = 4 \cdot D_x \ln x = 4 \cdot \frac{1}{x} = \frac{4}{x}$$

d. By the Chain Rule: Power Form,

$$D_x(\ln x)^4 = 4(\ln x)^3 \cdot \left(\frac{1}{x}\right) = \frac{4(\ln x)^3}{x}$$

e. By the Product Rule,

$$\begin{aligned} D_x(3x^2 \ln(4x)) &= \ln(4x) \cdot D_x 3x^2 + 3x^2 \cdot D_x \ln(4x) \\ &= \ln(4x) \cdot 6x + 3x^2 \cdot \frac{1}{4x} \cdot 4 \\ &= 6x \ln(4x) + 3x \\ &= 3x[2 \ln(4x) + 1] \end{aligned}$$

■

Example 7: For what value(s) of x does $f(x) = x^2 \ln x$ have a horizontal tangent line?

Solution: A horizontal line has a slope of zero, and tangent lines are determined by the first derivative. Thus, we must determine the value(s) of x for which $f'(x) = 0$. So,

$$f(x) = x^2 \ln x \rightarrow f'(x) = (\ln x)(2x) + x^2\left(\frac{1}{x}\right) = 2x \ln x + x$$

$$\begin{aligned} f'(x) = 0 \rightarrow 2x \ln x + x &= 0 \\ x(2 \ln x + 1) &= 0 \\ x = 0 \quad \text{or} \quad 2 \ln x + 1 &= 0 \end{aligned}$$

Because $x = 0$ is not in the domain of the function, the solution must come from $2 \ln x + 1 = 0$:

$$2 \ln x + 1 = 0 \rightarrow \ln x = -\frac{1}{2} \rightarrow x = e^{-1/2} \approx 0.607$$

■

Calculator Corner 13.3

Graph $y_1 = x^2 \ln x$ using a window of $[-2, 6]$ for x and $[-1, 3]$ for y. Verify that the tangent line is horizontal when $x \approx 0.607$.

Example 8: Costs to produce a new machine are $C(x) = 200 \ln(x + 2)$, where C is the cost in dollars and x is the number of units produced, in thousands.

a. Find the marginal cost if 2000 units are produced.

b. Find the marginal cost if 8000 units are produced.

Solution: Marginal cost is given by $C'(x)$, so

$$C'(x) = 200 \cdot \frac{1}{x + 2} \cdot 1 = \frac{200}{x + 2}$$

a. $C'(2) = \dfrac{200}{4} = 50$

If 2000 units are produced, the additional cost of producing 1000 more units is \$50.

b. $C'(8) = \dfrac{200}{10} = 20$

If 8000 units are produced, the additional cost of producing 1000 more units is \$20. ■

Example 9: Find the equation of the line tangent to $f(x) = 4x + \ln x$ at the point where $x = 2$.

Solution: The equation of the tangent line is given by $y - y_1 = m(x - x_1)$, where (x_1, y_1) is the point of tangency and m is the slope of the tangent line, which is determined by the first derivative. For this function,

$$f(x) = 4x + \ln x \rightarrow f'(x) = 4 + \frac{1}{x}$$

Thus,

$$x_1 = 2$$
$$y_1 = f(2) = 4(2) + \ln 2 = 8 + \ln 2 \approx 8.69$$
$$m = f'(2) = 4 + \frac{1}{2} = 4.5$$

The equation is

$$y - 8.69 = 4.5(x - 2)$$
$$y - 8.69 = 4.5x - 9.0$$
$$y = 4.5x - 0.31$$

■

Calculator Corner 13.4

Graph $y_1 = 4x + \ln x$ and $y_2 = 4.5x - 0.31$ using a window of $[-1, 6]$ for x and $[-10, 30]$ for y. Verify that y_2 is tangent to y_1 at $x = 2$.

Check Your Understanding 13.2

Find $f'(x)$ for each function.

1. $f(x) = 8\ln x$

2. $f(x) = 8 + \ln x$

3. $f(x) = 8x \ln x$

4. $f(x) = (8x^2 - \ln x)^5$

5. $f(x) = 4\ln(2x)$

6. $f(x) = \ln(3x^5 - 7x)$

7. $f(x) = 2x^3\ln(4x)$

8. $f(x) = e^{8x}/\ln x$

Derivatives of General Exponential and Logarithmic Functions

We have previously shown that $D_x e^{f(x)} = e^{f(x)} \cdot f'(x)$. Because $e^{\ln x} = x$, we can create a formula for $D_x a^x$.

First, write

$$a^x = (e^{\ln a})^x = e^{(\ln a)x} \qquad \text{Logarithm Property 2}$$

Then

$$\begin{aligned} D_x a^x &= D_x e^{(\ln a)x} && \text{Take derivative of each side} \\ &= e^{(\ln a)x} \cdot D_x(\ln a)x && \text{Chain Rule: Exponential Form} \\ &= e^{(\ln a)x} \cdot (\ln a) && \text{Remember that } \ln a \text{ is a constant} \\ &= a^x \cdot \ln a && \text{Logarithm Property 2} \end{aligned}$$

Thus,

$$\boxed{D_x a^x = a^x \cdot (\ln a)}$$

Example 10: Find $f'(x)$ for $f(x) = 6^x$.

Solution:

$$f(x) = 6^x \rightarrow f'(x) = 6^x \cdot \ln 6$$ ■

If $D_x a^x = a^x \cdot \ln a$, we can find the derivative of the composite function $a^{f(x)}$ by following the same methods used to obtain the Chain Rule.

Derivative of the General Exponential Function

$$D_x a^{f(x)} = a^{f(x)} \cdot \ln a \cdot f'(x)$$

Example 11: Find $f'(x)$ for $f(x) = 6^{3x-5}$.

Solution:

$$f(x) = 6^{3x-5} \rightarrow f'(x) = 6^{3x-5} \cdot \ln 6 \cdot 3 \quad \text{or} \quad 3(\ln 6) \cdot 6^{3x-5}$$

Finally, we can write these last two rules for differentiation of logarithmic functions.

Derivative of the General Logarithmic Function

$$D_x \log_a x = \frac{1}{(\ln a)x}$$

where $x > 0$.

$$D_x \log_a f(x) = \frac{1}{(\ln a)f(x)} \cdot f'(x)$$
$$= \frac{f'(x)}{(\ln a)f(x)}$$

where $f(x) > 0$.

Be sure you see the similarities to the preceding derivative rules for $\ln x$ and $\ln f(x)$.

MATHEMATICS CORNER 13.2

Proof of the Derivative of the General Logarithmic Function

The Change of Base Formula can be used to prove that $D_x \log_a f(x) = \frac{f'(x)}{(\ln a)f(x)}$:

$$\log_a f(x) = \frac{\ln f(x)}{\ln a}$$

so

$$D_x \log_a f(x) = D_x \frac{\ln f(x)}{\ln a}$$
$$= \frac{1}{\ln a} \cdot D_x \ln f(x)$$
$$= \frac{1}{\ln a} \cdot \frac{f'(x)}{f(x)}$$

Example 12: Find $f'(x)$ for the following functions.

a. $\log_3 x$

b. $\log_8(3x - 2)$

Solution:

a. $D_x \log_3 x = \dfrac{1}{(\ln 3)x}$

b. $D_x \log_8(3x - 2) = \dfrac{1}{(\ln 8)(3x - 2)} \cdot 3 = \dfrac{3}{(\ln 8)(3x - 2)}$ ■

The rules for derivatives of exponential and logarithmic functions are summarized as follows.

Derivatives of Exponential Functions and Logarithmic Functions

$D_x e^x = e^x$

$D_x e^{f(x)} = e^{f(x)} \cdot f'(x)$

$D_x \ln x = \dfrac{1}{x}$, where $x > 0$

$D_x \ln f(x) = \dfrac{f'(x)}{f(x)}$, where $f(x) > 0$

$D_x a^x = a^x \cdot (\ln a)$

$D_x a^{f(x)} = a^{f(x)} \cdot (\ln a) \cdot f'(x)$

$D_x \log_a x = \dfrac{1}{(\ln a)x}$, where $x > 0$

$D_x \log_a f(x) = \dfrac{f'(x)}{(\ln a)f(x)}$, where $f(x) > 0$

Check Your Understanding 13.3

Find $f'(x)$ for each of the following functions.

1. $f(x) = 3^x$
2. $f(x) = 3^{2x}$
3. $f(x) = (3^x - 2)^4$
4. $f(x) = \log_5 x$
5. $f(x) = \log_3(2x^2 + 5)$
6. $f(x) = x^2 \cdot 2^x$

Check Your Understanding Answers

Check Your Understanding 13.1

1. $7e^{7x}$

2. $2xe^{x^2} - 2x$

3. $-6e^{-2x}$

4. $2(e^{3x} - 4) \cdot 3e^{3x} = 6e^{3x}(e^{3x} - 4)$

5. $x^3 e^{5x}(5x + 4)$

6. $\dfrac{e^{5x}(5x - 4)}{x^5}$

7. 0

8. $7ex^{7e-1}$

Check Your Understanding 13.2

1. $8/x$
2. $1/x$
3. $8 + 8\ln x$
4. $5(8x^2 - \ln x)^4\left(16x - \frac{1}{x}\right)$
5. $4/x$
6. $\frac{15x^4 - 7}{3x^5 - 7x}$
7. $2x^2(1 + 3\ln 4x)$
8. $\frac{e^{8x}\left(8\ln x - \frac{1}{x}\right)}{(\ln x)^2}$

Check Your Understanding 13.3

1. $3^x \cdot (\ln 3)$
2. $2(\ln 3) \cdot 3^{2x}$
3. $4(3^x - 2)^3(\ln 3)3^x$
4. $\frac{1}{(\ln 5)x}$
5. $\frac{4x}{(\ln 3)(2x^2 + 5)}$
6. $x \cdot 2^x[(\ln 2)x + 2]$

Topic 13 Review

13

In this topic, you learned how to find derivatives of exponential and logarithmic functions.

CONCEPT	EXAMPLE
Derivative of $f(x) = e^x$	If $f(x) = e^x$, then $f'(x) = e^x$.
Derivative of $f(x) = \ln x$	If $f(x) = \ln x$, then $f'(x) = 1/x$.
The **Chain Rule** can also be applied to exponential and logarithmic functions. **Chain Rule: Exponential Form** $D_x e^{f(x)} = e^{f(x)} \cdot f'(x)$ **Chain Rule: Logarithmic Form** $D_x \ln f(x) = \frac{f'(x)}{f(x)}$	$D_x e^{2x-5} = 2e^{2x-5}$ $D_x \ln(2x - 5) = \frac{2}{2x - 5}$
Generic form of the Chain Rule for composite functions	If $y = f(u)$, where $u = g(x)$, then $y = f(g(x))$. Then $y' = f'(g(x)) \cdot g'(x)$, assuming $f'(u)$ and $g'(x)$ exist.

(continued)

Derivatives of the general exponential function $y = a^x$ and the general logarithmic function $y = \log_a f(x)$ were also defined.

$D_x a^x = a^x \cdot \ln a$	$D_x 8^x = 8^x \cdot \ln 8$
$D_x a^{f(x)} = a^{f(x)} \cdot (\ln a) \cdot f'(x)$	$D_x 8^{2x+3} = 8^{2x+3} \cdot (\ln 8) \cdot 2$
$D_x \log_a x = \dfrac{1}{(\ln a)x}$ for $x > 0$	$D_x \log_8 x = \dfrac{1}{(\ln 8)x}$
$D_x \log_a f(x) = \dfrac{f'(x)}{(\ln a)f(x)}$ for $f(x) > 0$	$D_x \log_8(2x + 3) = \dfrac{2}{(\ln 8)(2x + 3)}$

NEW TOOLS IN THE TOOL KIT

- Derivatives of e^x and $\ln x$
- Derivatives of $e^{f(x)}$ and $\ln f(x)$
- Derivatives of a^x and $\log_a x$
- Derivatives of $a^{f(x)}$ and $\log_a f(x)$

Topic 13 Exercises

13

Are the following statements true or false?

1. $\dfrac{d}{dx}(e^x) = xe^{x-1}$

2. $\dfrac{d}{dx}(x^e) = ex^{e-1}$

3. $\dfrac{d}{dx}(\ln x) = \dfrac{1}{x}$

4. $\dfrac{d}{dx}\ln(2x) = \dfrac{1}{2x}$

Find $f'(x)$ for each of the following functions.

5. e^{3x}

6. e^{-2x}

7. $4e^{x^2-5}$

8. e^{2x^3+5}

9. $x^2 - e^{5x}$

10. $e^{7x} - x^3$

11. x^2e^{3x}

12. x^3e^{-7x}

13. e^{-5x}/x^4

14. x^3/e^{2x}

15. x^{3e}

16. $7x^{-4e}$

17. e^4

18. $9e^{-3}$

19. $\ln(x - 3)$

20. $\ln(x^2 + 4)$

21. $\ln x^2$

22. $\ln x^{-6}$

23. $(\ln x)^2$

24. $(\ln x)^{-6}$

25. $x^2\ln(4x)$

26. $x^3\ln(2x)$

27. $\dfrac{\ln x}{x^4}$

28. $\dfrac{5\ln x}{x^3}$

29. $-\ln 6$

30. $\ln 8$

31. $[x^3 - 3e^x - \ln(2x)]^5$

32. $\left(2e^{5x} - \dfrac{2}{x^2} + 3\ln x\right)^4$

33. $(x^5 - 3e^x)(2x + 5)$

34. $(x^3 + 5e^x)(3 - 4x)$

35. $\dfrac{3x^5 - e^{2x}}{2x + 5}$

36. $\dfrac{4x^3 + e^{7x}}{3 - 4x}$

37. $(x^5 - 3e^x)^2(2x + 5)^3$

38. $(x^3 + 5e^x)^5(3 - 4x)^2$

39. For $y = x^2 - 3\ln x$,

a. Find the slope of the tangent line to $f(x)$ at $x = 2$.

b. Find the equation of the tangent line to $f(x)$ at $x = 2$.

40. For $y = 3e^{-4x} + x - 2$,

a. Find the slope of the tangent line to $f(x)$ at $x = 1$.

b. Find the equation of the tangent line to $f(x)$ at $x = 1$.

41. The population of a city is approximated by $P(t) = 5{,}000e^{0.012t}$, where t is the number of years after 1980. How fast was the population growing in 2002?

42. After spraying an insecticide, the number of mosquitoes near a lake is approximated by $N(t) = 12{,}000e^{-0.03t}$, where t is the number of hours after spraying. Find $N'(12)$ and interpret your answer.

43. A casserole is taken out of an oven and placed on a counter to cool. Its temperature after t minutes is modeled by $T(t) = 70 + 305e^{-0.07t}$, where T is degrees Fahrenheit and t is in minutes. Evaluate $T'(30)$ and interpret your answer.

44. The probability of having an accident at a given blood alcohol concentration (BAC) level is modeled by $P(b) = e^{21.5b}$, where P is the whole number percent (P%) and b is the blood alcohol level ($0 \le b \le 0.40$). In most states, the maximum legal BAC limit when driving is $b = 0.08$. Evaluate $P(0.08)$ and $P'(0.08)$ and interpret your answers.

45. The percent of information retained t months after being tested on that information is given by $f(t) = 82 - 14\ln(t + 1)$. Evaluate $f(2)$ and $f'(2)$ and interpret your answers.

46. A list of 40 words is to be memorized. The number of words learned after x days of study is modeled by $N(x) = 40\ln(0.2x + 1)$, where N is the number of words memorized. At what rate are the words being memorized after three days? After 10 days?

47. Refer to Exercise 45. At what rate is the retention rate changing after two months?

48. Refer to Exercise 46. At what rate is the memorization rate changing after 10 days?

49. The proportion of the U.S. population living in metropolitan areas can be modeled by $y = \dfrac{94.54}{1 + 2.98e^{-0.03x}}$, where x is the number of years after 1900. At what rate was the proportion changing in 1930? In 2000?

50. The proportion of the U.S. labor force 16 years of age and older can be modeled by $y = \dfrac{76.4}{1 + 3.75e^{-0.018t}}$, where t is the number of years after 1900. At what rate was the proportion changing in 1950? In 2000?

51. The number of bacteria in a colony at any time can be modeled by $P(t) = 450(1.099)^t$, where t is in days. How fast is the colony growing after one week?

52. The population of a city, in thousands, over the past decade has decreased and can be modeled by $P(t) = 152(0.9844)^t$, where t is the number of years since 1995. Evaluate $P(6)$ and $P'(6)$ and interpret your answers.

53. The concentration of a drug in a patient's bloodstream t hours after the drug was injected is approximated by $C(t) = 1.6te^{-0.5t}$ mg/cm^3. Evaluate $C(3)$, $C'(3)$, and $C''(3)$ and interpret your answers.

54. The number of transistors on a computer chip has grown according to $N(t) = 1.335(1.4257)^t$, where N is the number of transistors in thousands and t is the number of years after 1971. Evaluate $N(35)$, $N'(35)$, and $N''(35)$ and interpret your answers.

Find $f'(x)$ for each of the following functions.

55. 12^{3x}

56. -8^{5x}

57. $4 \cdot 2^{x-1}$

58. $7 + 13^{2+x}$

59. $4x \cdot 2^{x^3}$

60. $6x^2 \cdot 4^{x+1}$

61. $\log_6 x$

62. $\log_{12} x$

63. $5\log_7(x - 2)$

64. $8\log_{12}(3x + 5)$

65. $x^2 \log_{11} x$

66. $x^4 \log_2 x$

Find $f''(x)$ for each of the following functions.

67. e^{6x}

68. e^{-3x}

69. e^{6x^2}

70. e^{2x^3}

71. $x^{-2}e^{3x}$

72. x^4e^{2x}

73. e^{3x}/x^2

74. e^{5x}/x^4

75. 6^x

76. -3^x

77. 4^{x^2}

78. 9^{x^4}

79. $\ln(5 - 2x)$

80. $2\ln(5x + 3)$

81. $\log_5(x - 2)$

82. $\log_6(2x + 3)$

COMMUNICATE

83. Explain in your own words why $\frac{d}{dx}e^x \neq xe^{x-1}$.

84. Explain why $\frac{d}{dx}\ln(3x - 5) \neq \frac{d}{dx}\ln(3x) - \frac{d}{dx}(5)$.

SHARPEN YOUR SKILLS

85. Prove that $D_x\ln(kx) = 1/x$ for k a constant. Assume $x > 0$.

86. Prove that $D_x\ln(x^k) = k/x$ for k a constant. Assume $x > 0$.

CALCULATOR CONNECTION

87. Given $f(x) = \dfrac{x\ln x}{x^2 + 2}$,

a. Use your calculator's derivative function to evaluate $f'(3)$.

b. Verify your answer to part a using appropriate derivative rules.

c. Graph $f(x)$ using a window of $[0, 14]$ for x and $[-0.5, 0.5]$ for y. Verify by inspection that the derivative obtained in parts a and b seems logical for the graph.

d. Evaluate $\lim_{x\to 0} f(x)$ by inspecting the graph.

e. Evaluate $\lim_{x\to\infty} f(x)$ by inspecting the graph.

88. Given $f(x) = x^3e^{-2x}\ln(x^4)$,

a. Use your calculator's derivative function to evaluate the derivative at $x = -\frac{1}{2}$, at $x = \frac{1}{2}$, and at $x = 3$. Could these answers be verified using appropriate derivative rules?

b. Evaluate $\lim_{x\to 0} f(x)$, $\lim_{x\to\infty} f(x)$, and $\lim_{x\to-\infty} f(x)$.

c. Graph $f(x)$ using a window of $[-2, 6]$ for x and $[-1, 3]$ for y.

d. Examine the graph closely about $x = 0$. What is $f(0)$?

e. Estimate the point(s) where the graph has a horizontal tangent line.

89. *Marginal Analysis* The cost and revenue functions for producing x units of a certain product are $C(x) = 6000 + 15x$ and $R(x) = 1811\ln(1.2x + 1)$.

a. Find the marginal cost if 200 units are produced.

b. Find the marginal revenue if 200 units are produced.

c. Find the marginal profit if 200 units are produced.

d. Graph the two functions and use your graphing calculator to determine the break-even point(s).

e. To make a profit, how many units should be produced?

90. *Marginal Analysis* The cost and revenue functions for producing x units of a certain product are $C(x) = 4500e^{0.008x}$ and $R(x) = 1500\sqrt{0.73x}$.

a. Find the marginal cost if 125 units are produced.

b. Find the marginal revenue if 125 units are produced.

c. Find the marginal profit if 125 units are produced.

d. Graph the two functions and use your graphing calculator to determine the break-even point(s).

e. To make a profit, how many units should be produced?

UNIT 1

Review

Unit 1 presented the introductory concepts of calculus: the limit and the derivative.

TOPIC 5	■ Limits of algebraic functions ■ Limits of rational functions ■ Limits of piecewise defined functions ■ Evaluating limits from graphs of functions ■ Evaluating limits of difference quotients
TOPIC 6	■ Continuity of functions at a point ■ Determining points of discontinuity from a function ■ Determining points of discontinuity from a graph
TOPIC 7	■ Average rate of change ■ Instantaneous rate of change ■ Slope of a curve
TOPIC 8	■ The derivative as the limit of a difference quotient ■ Nonexistence of derivatives
TOPIC 9	■ Rules for differentiation of algebraic functions ■ Power Rule ■ Sum and Difference Properties ■ Constant Multiple Property ■ Marginal analysis ■ Equations of tangent lines ■ Velocity of objects in motion
TOPIC 10	■ Product Rule ■ Quotient Rule ■ Chain Rule: Power Form
TOPIC 11	■ Higher-order derivatives ■ Acceleration of objects in motion
TOPIC 12	■ Review of exponential functions ■ Review of logarithmic functions ■ Exponential growth and decay ■ Logistic growth models ■ Surge functions
TOPIC 13	■ Derivatives of exponential functions ■ Derivatives of logarithmic functions

Having completed Unit 1, you should now be able to:

1. Evaluate limits numerically.
2. Evaluate limits graphically.
3. Evaluate limits algebraically.
4. Evaluate left-hand and right-hand limits.
5. Determine the existence of limits.
6. Evaluate limits of difference quotients.
7. Evaluate limits at infinity.
8. Evaluate limits of piecewise defined functions.
9. Determine intervals for which a function is continuous.
10. Determine intervals for which the graph of a function is continuous.
11. Identify points of discontinuity.
12. Understand and apply the three conditions for continuity of a function.
13. Determine the average rate of change of a function between two points.
14. Determine the instantaneous rate of change of a function at a specific point or at any point.
15. Find the slope of a curve at a specific point or at any point.
16. Determine the velocity of an object in motion.
17. Find the derivative of an algebraic function using the limit definition.
18. Find the derivative of an algebraic function using an appropriate rule (Power Rule, Product Rule, Quotient Rule, Chain Rule).
19. Interpret the meaning of the derivative.
20. Evaluate marginal cost, marginal revenue, and marginal profit.
21. Find the equation of the tangent line to a function at a point.
22. Find the second derivative of a function.
23. Interpret the meaning of the second derivative.
24. Determine the acceleration of an object in motion.
25. Evaluate exponential and logarithmic expressions with and without a calculator.
26. Sketch graphs of exponential and logarithmic functions using the basic graphs and transformations.
27. Solve exponential and logarithmic equations.
28. Solve problems involving exponential growth, exponential decay, and logistic growth models.
29. Use regression to fit exponential or logarithmic models to a set of data.
30. Find the derivative of exponential and logarithmic functions using appropriate rules.

UNIT 1 TEST

Evaluate each of the following limits, if they exist.

1. $\lim_{x\to 1}(x^2 - 4)$ **2.** $\lim_{x\to 4} 7$

3. $\lim_{x\to -2}\sqrt{5 - 2x}$ **4.** $\lim_{x\to -1}\frac{3x}{x^2 - 4}$

5. $\lim_{x\to 2}\frac{3x}{x^2 - 4}$ **6.** $\lim_{x\to 2}\frac{3x - 6}{x^2 - 4}$

7. $\lim_{x\to\infty}\frac{2x}{x - 3}$ **8.** $\lim_{x\to\infty}\frac{3}{x^2 + 1}$

9. $\lim_{x\to\infty}\frac{x^2}{x + 4}$

10. Given $f(x) = \begin{cases} 2x & \text{if } x < -1 \\ x + 3 & \text{if } -1 \le x < 2 \\ x^2 + 1 & \text{if } x \ge 2 \end{cases}$

Evaluate the following limits.

a. $\lim_{x\to -1} f(x)$ **b.** $\lim_{x\to 0} f(x)$

c. $\lim_{x\to 2^-} f(x)$ **d.** $\lim_{x\to 2^+} f(x)$

e. $\lim_{x\to 2} f(x)$

11. Given $f(x)$ as shown in the graph, evaluate the following limits, if they exist.

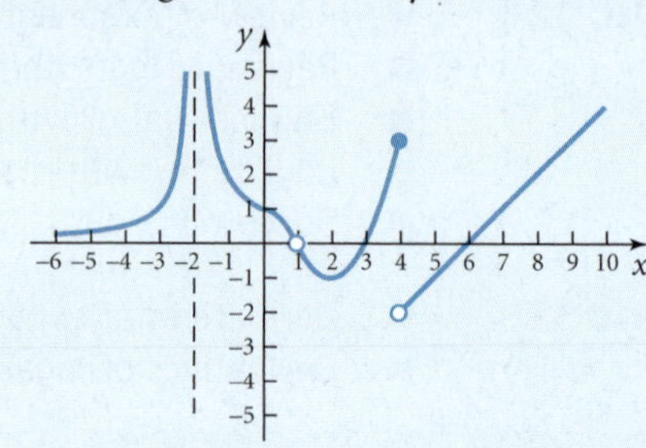

a. $\lim_{x \to -2} f(x)$ **b.** $\lim_{x \to 0} f(x)$

c. $\lim_{x \to 1} f(x)$ **d.** $\lim_{x \to 4^-} f(x)$

e. $\lim_{x \to 4^+} f(x)$ **f.** $\lim_{x \to 4} f(x)$

12. Given $f(x)$ as shown in the graph, determine if the following statements true or false.

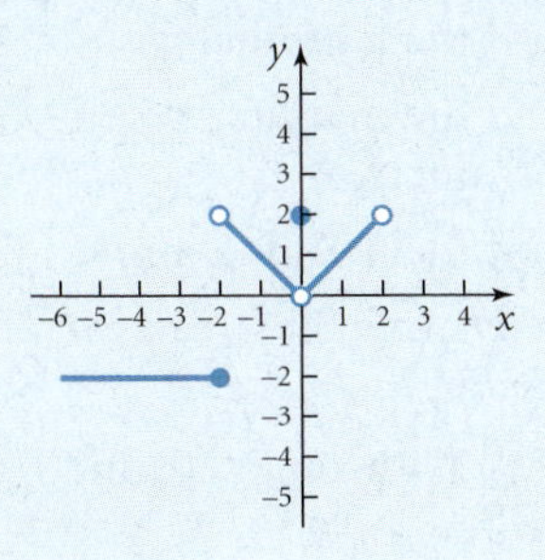

a. $\lim_{x \to -2^+} f(x) = 1$

b. $\lim_{x \to -2^-} f(x) = -2$

c. $\lim_{x \to -2^-} f(x) = \lim_{x \to -2^+} f(x)$

d. $\lim_{x \to -2} f(x)$ exists

e. $\lim_{x \to -2} f(x) = 2$

f. $\lim_{x \to 0} f(x) = 0$

13. Given $f(x) = x^2 - 3x + 4$, evaluate

$$\lim_{h \to 0} \frac{f(x + h) - f(x)}{h}.$$

In exercises 14 through 16, state the interval(s) of x for which each function is continuous.

14. $f(x) = \sqrt{4 - x}$ **15.** $f(x) = \dfrac{5}{x^2 - 1}$

16. $f(x) = 2x^3 - 5x + 7$

17. Given $f(x)$ as shown in the graph, determine if $f(x)$ is continuous at each of the indicated points. If not, give the mathematical condition that is violated.

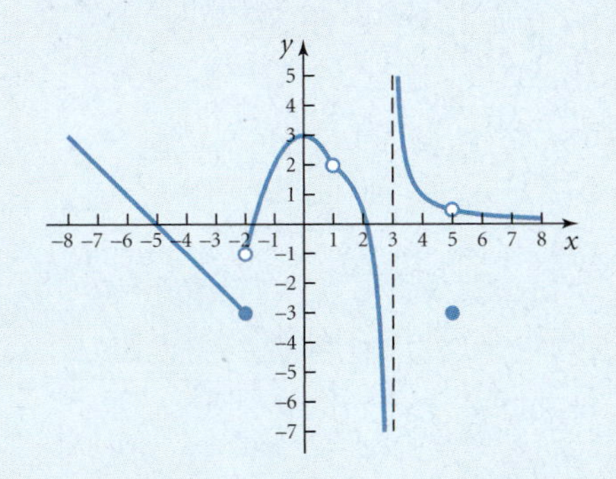

a. $x = -2$

b. $x = 0$

c. $x = 1$

d. $x = 3$

e. $x = 5$

18. Given $f(x)$ as shown in the graph, determine if the following statements are true or false.

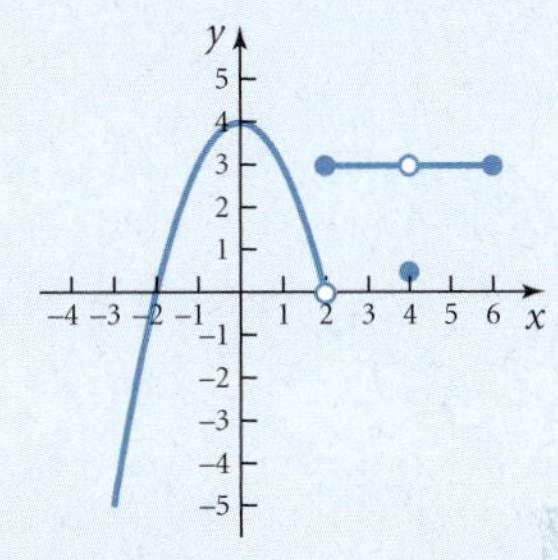

a. $f(x)$ is continuous at $x = 4$

b. $f(x)$ is continuous at $x = 0$

c. $\lim_{x \to 3} f(x) = \lim_{x \to 5} f(x)$

d. $f(x)$ is continuous at $x = 2$

19. Health care spending, in millions of dollars, in the United States since 1970 is shown in the table.

Year	Per Capita Spending on Health Care
1970	\$362
1980	\$1005
1988	\$2012
1993	\$3250
1997	\$3981
2000	\$4670
2001	\$4995
2002	\$5440

a. Determine the average rate of change in spending between 1988 and 2000 and interpret your answer.

b. If per capita spending on health care is modeled by $S(x) = 4.14x^2 + 23x + 356.8$, where x is the number of years after 1970, determine the instantaneous rate of change in spending in 2000 and interpret your answer.

20. Use the limit definition of derivative to find $f'(x)$ for $f(x) = x^2 - 3x + 7$.

For Exercises 21 through 30, find $f'(x)$ for each of the following functions using an appropriate rule.

21. $f(x) = 2x^3 - 5x + 11$

22. $f(x) = 3x^{-6}$

23. $f(x) = \sqrt[4]{x}$

24. $f(x) = -\dfrac{4}{x^3} + 5x$

25. $f(x) = (x^4 - 3)(x^5 + 7x - 2)$

26. $f(x) = \dfrac{3x - 5}{x^2 + 2}$

27. $f(x) = (x^4 - 3)^7$

28. $f(x) = \sqrt{3 - 2x}$

29. $f(x) = \dfrac{5}{(3x - 2)^3}$

30. $f(x) = (2x^3 - 5)^4(3x^4 + 5)^3$

31. Find the slope of the curve $f(x) = -2x^3 + 6x^2 - 7x + 5$ at the point where $x = -1$.

32. Find the equation of the tangent line to $f(x) = \sqrt{7 - 3x}$ at the point where $x = 1$.

33. A rock is thrown upward from a platform. Its height above the ground, in feet, t seconds after being thrown is given by $h(t) = -16t^2 + 40t + 5$.

a. What is the velocity of the rock after one second? Interpret your answer.

b. What is the velocity of the rock after three seconds? Interpret your answer.

c. When is the velocity of the rock zero?

d. When does the rock hit the ground?

34. The percentage of the U.S. population living in poverty can be modeled by $P(t) = 0.04t^2 - 0.91t + 14$, where t is the number of years after 1990. At what rate was the percentage changing in 2005?

35. A woman on a weight-loss program weighs herself each week. Her weight at any time is given by $W(t) = \dfrac{790 + 132t}{t + 4.4}$, where W is her weight in pounds and t is the number of weeks after starting the program. Find $W'(1)$ and $W'(6)$ and interpret your answers.

36. Bill's Barber Shop has weekly costs of $C(x) = 480 + 10x$ and weekly revenues of $R(x) = 0.2x^2 + 14x$, where x is the number of customers per week.

a. Find the marginal cost for 65 customers.

b. Find the marginal revenue for 65 customers.

c. Find the marginal profit for 65 customers.

d. Determine the break-even point.

For Exercises 37 through 40, find $f''(x)$ for the given function.

37. $f(x) = 3x^5 - 2x^4 + 15x - 11$

38. $f(x) = \dfrac{3}{2x + 5}$

39. $f(x) = \sqrt{3 - 2x}$

40. $f(x) = (3x^2 - 2)^4$

41. The number of traffic accidents per month at a busy intersection has decreased since a traffic light was installed. The number of accidents per month can be modeled by $N(t) = -0.17t^3 + 2.22t^2 - 9.46t + 82.83$, where t is the number of months after installation of the light. Evaluate $N(8)$, $N'(8)$, and $N''(8)$ and interpret your answers.

42. The population of a city has grown according to $P(x) = 2000\sqrt{3.8x + 5000}$, where P is the population in hundreds and x is the number of years after the city's incorporation. At what rate is the growth rate changing after five years?

Match each function to its graph.

43. $f(x) = e^x$

44. $f(x) = e^{-x}$

45. $f(x) = e^x + 2$

46. $f(x) = e^x - 2$

47. $f(x) = -e^x$

48. $f(x) = e^{x+2}$

49. $f(x) = e^{x-2}$

50. $f(x) = \ln x + 1$

51. $f(x) = \ln(x + 1)$

52. $f(x) = -\ln x$

a.

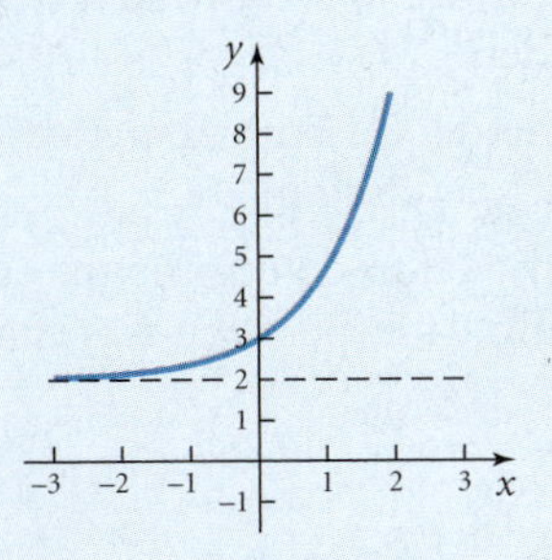

b.

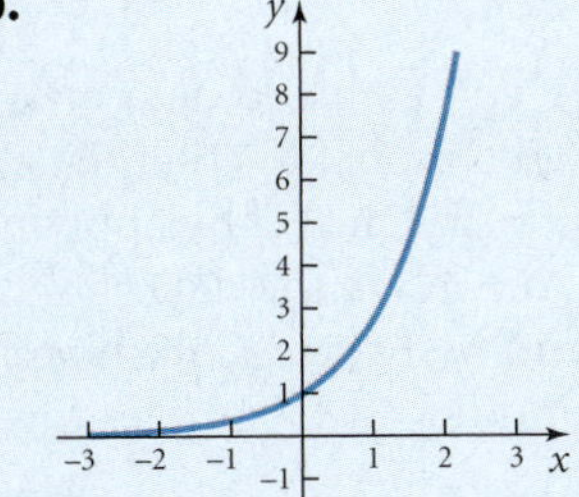

c.

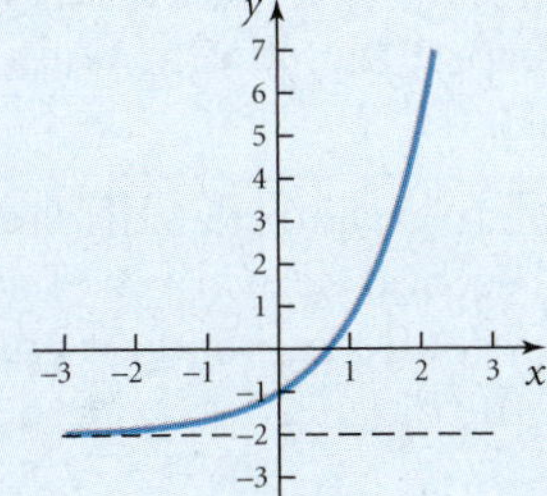

d.

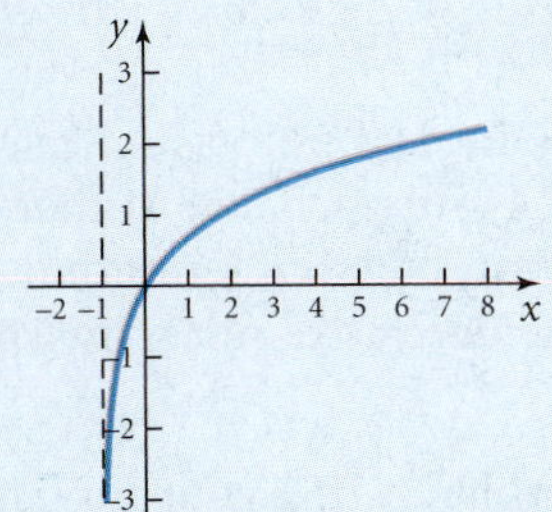

e.

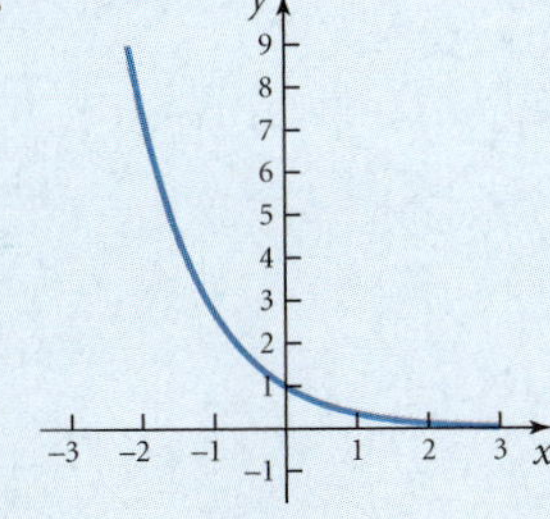

f.

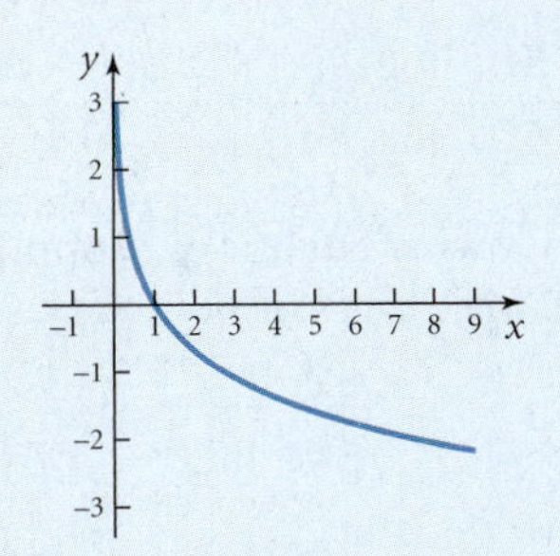

g.

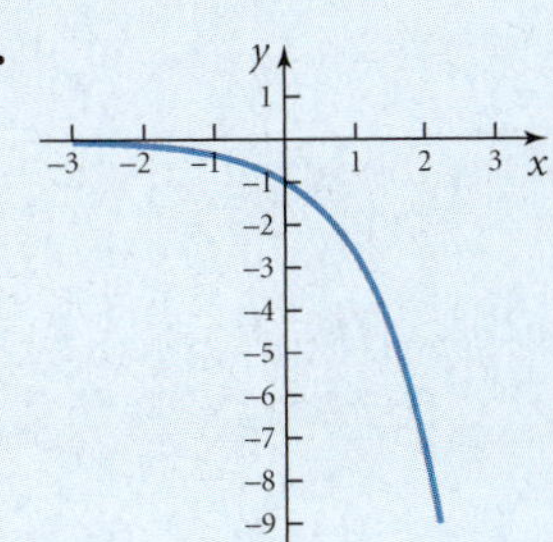

h.

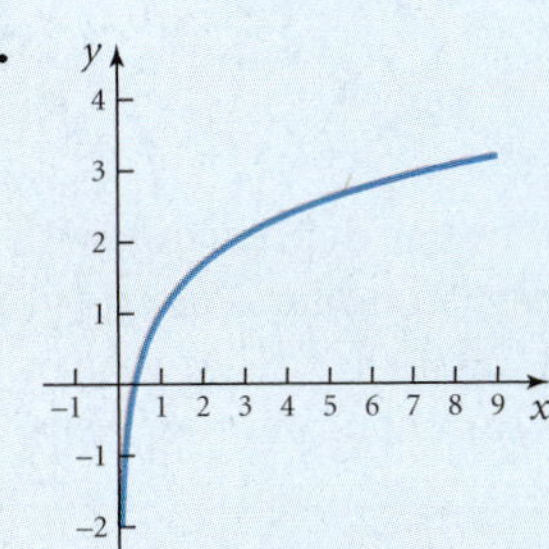

i.

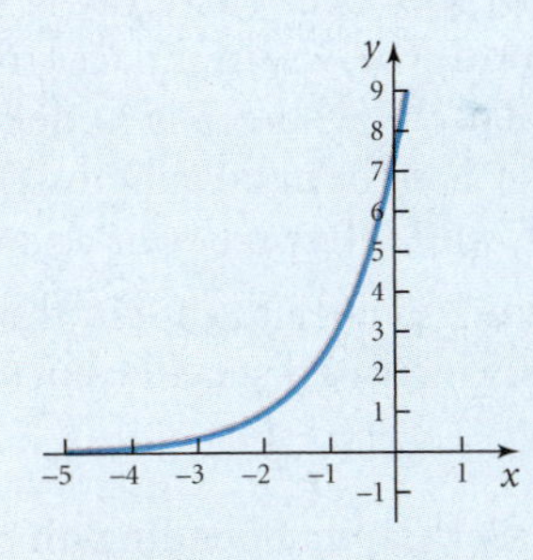

j.

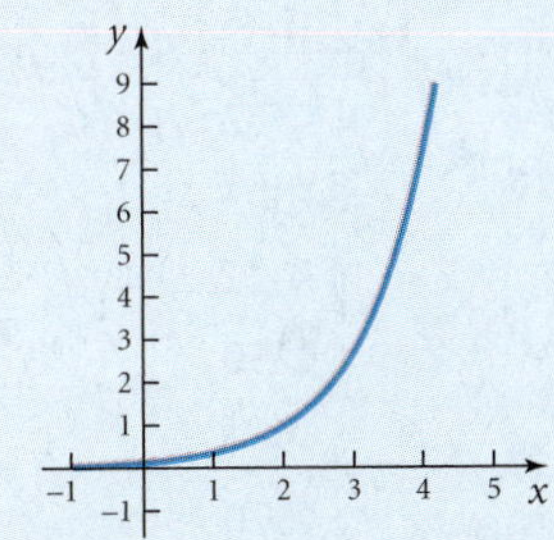

Evaluate the following expressions. Round each answer to three decimal places, if necessary.

53. e^2

54. $e^{-0.27}$

55. $\ln 2$

56. $\log 7$

57. $e^{\ln 17}$ **58.** $\log_3 9$

59. $\ln e^{-4}$ **60.** $\log_4 29$

Solve each of the following equations. Round each answer to three decimal places, if necessary.

61. $3e^x = 12$ **62.** $e^{2x-3} - 6 = 4$

63. $3^{-x} = 11$ **64.** $\ln(x - 3) = 1.4$

65. $\log_5(2x + 1) = 2$

66. $\log_3 x + \log_3(x - 2) = 1$

Find $f'(x)$ for each of the following functions.

67. $3e^x - 8x + 5$ **68.** $\ln(5x + 3)$

69. e^{4-3x^2} **70.** $(\ln x^2)^4$

71. $x^{4e} + e^{4x}$ **72.** $x^5 e^{-3x}$

73. $\dfrac{3e^{2x}}{8x}$

74. $\left(2x^3 - 6e^{3x} - \dfrac{5}{x}\right)^4$

75. The value of a new car depreciates with age. The value of a car (in dollars) after t years is given by $V(t) = 25{,}000e^{-0.2t}$. At what rate is the value of the car changing two years after purchase?

76. A chilled mug of root beer currently at 40°F left in a 70°F room will warm up according to $T(t) = 70 - 30e^{-1.5t}$, where T is in degrees Fahrenheit and t is in hours. Evaluate $T(2)$, $T'(2)$ and $T''(2)$ and interpret your answers.

77. Per capita spending on health care in the United States has increased steadily since 1970.

Year	Per Capita Spending on Health Care
1970	\$362
1980	\$1005
1990	\$2012
1993	\$3250
1997	\$3981
2000	\$4670
2001	\$4995
2002	\$5440

Data from: OECD Health Data, 2006 Newsletter

a. Draw a scatter plot of the data.

b. Find an exponential regression model for the data. Be sure to state what x represents in the model.

c. Use your model to estimate per capita health care spending in 2005.

d. Use your model to determine the rate at which health care spending was changing in 2000.

UNIT 1 PROJECT

In the Unit 0 Project, you gathered data on a topic of interest and created a model for that data. This project continues with that model by applying calculus techniques to analyze the data.

Using the first and best regression model you determined in the Unit 0 Project, answer the following questions.

1. Find the first and second derivatives of your function and interpret them in the context of your data.
2. Choose an x value from your data list and evaluate the first and second derivatives of your function at that point. Interpret your answers in the context of the data.
3. Choose any two of your data points and calculate the average rate of change between those points. Interpret your answer.
4. Choose one of your data points and calculate the instantaneous rate of change at that point, using both the limit of a difference quotient definitionand the derivative rules. Interpret your answer.
5. Write a word problem using your function that would require the use of a derivative to solve. Include the solution and interpret the answers.

Applications of the Derivative

UNIT 2

A magazine article chronicled the impressive growth and increased revenues of a local medical supply company.

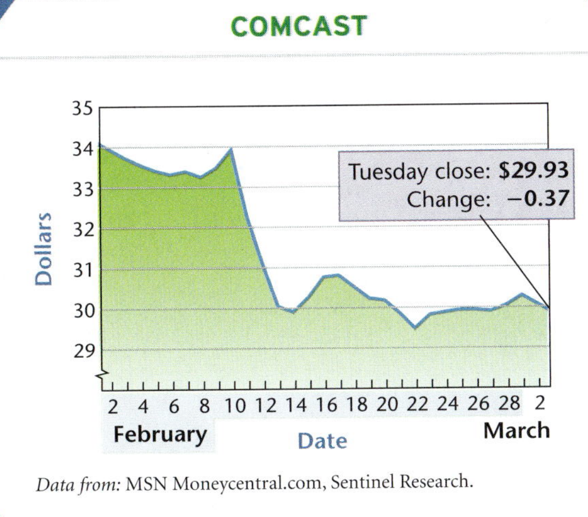

Data from: MSN Moneycentral.com, Sentinel Research.

You may have read or heard similar things about the rapid growth of Starbucks, the decline of videotape rentals at Blockbuster, the changes in passenger traffic at your local airport, or the fluctuations of the stock market. The most likely asked questions were of this type:

- At what rate were the company's revenues growing in 2005?
- At what rate was the airport's passenger traffic declining in 2004?
- When was the rate of decrease in Comcast stock the greatest?

In this unit, you learn to use the various derivatives now at your disposal as mathematical tools to analyze changes in a function by exploring increasing and decreasing behaviors and extreme points of the function. We also generate

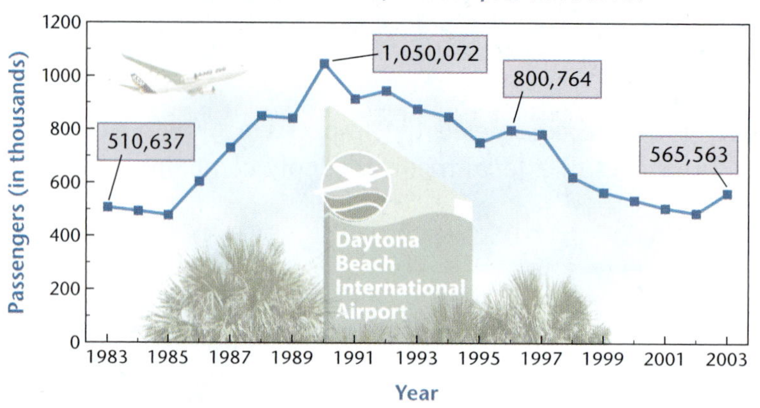

Data from: Daytona Beach International Airport.

calculus techniques to optimize the behavior of a function. In Unit 1, you built the foundation of the house of calculus. In Unit 2, we build on that foundation.

TOPIC

First Derivative Test and Graphs of Functions

14

Calculus was defined as the mathematical tool used to analyze rates of change in variable quantities. Now that you have learned the rules for finding derivatives of algebraic, exponential, and logarithmic functions, let's put them to use to study the behavior of those functions. How can derivatives be used to determine whether a function increases or decreases? How does the derivative help locate maximum and minimum values of functions? In this topic, we answer both of those questions. Before starting your study of function behavior, complete the Topic 14 Warm-up Exercises to sharpen your algebra and calculus skills.

IMPORTANT TOOLS IN TOPIC 14

- *Increasing and decreasing functions*
- *Extreme points of functions*
- *Classifying extreme points as maximum or minimum points*
- *Classifying extreme points as absolute or relative*
- *Using the First Derivative Test to determine intervals of increasing and decreasing behavior and to locate and categorize extreme points*

TOPIC 14 WARM-UP EXERCISES

Be sure you can successfully complete the following exercises before starting Topic 14.

Algebra Warm-up Exercises

1. State the domain of each of the following functions.

a. $f(x) = x^2 - 4x + 11$

b. $f(x) = \dfrac{x}{x^2 - 9}$

c. $f(x) = \sqrt{2x + 8}$

d. $f(x) = \dfrac{x}{\sqrt{2x + 8}}$

2. Find the x-intercept(s) and y-intercept of each of the following functions.

a. $f(x) = 2x^2 - x - 6$

b. $f(x) = \dfrac{2x - 1}{x^2 + 4}$

c. $f(x) = x^3 - 5x^2 + 4x$

3. Find the vertical and horizontal asymptotes of each function.

a. $f(x) = \dfrac{x - 1}{x^2 - 9}$

b. $f(x) = \dfrac{-2x^2 + 1}{x^2 - 9}$

c. $f(x) = e^{2x} - 4$

4. Solve $1.3x^2 - 0.4x - 3.5 = 0$ using the Quadratic Formula. Round your answer to three decimal places.

Calculus Warm-up Exercises

Find $f'(x)$ for each of the following functions.

1. $f(x) = 3x^2 - 5x + 6$

2. $f(x) = 4x^3 - x^4$

3. $f(x) = \dfrac{x}{x^2 - 9}$

4. $f(x) = e^{2x} - 4$

5. $f(x) = \ln(3x - 4)$
6. $f(x) = \sqrt{8 - 2x}$
7. Given $f'(x) = x^2 - 4x$, determine the value(s) of x for which $f'(x) = 0$.
8. Given $f'(x) = x^2 - 4x$, determine if $f'(x)$ has a positive or negative value over the indicated intervals.
   a. $x < 0$ **b.** $x > 4$ **c.** $0 < x < 4$
9. Find the slope of $f(x) = x^2 - 3x$ at the point where $x = -1$.

State the point(s) of discontinuity for the following functions.

10. $f(x) = \dfrac{x}{x^2 - 9}$
11.

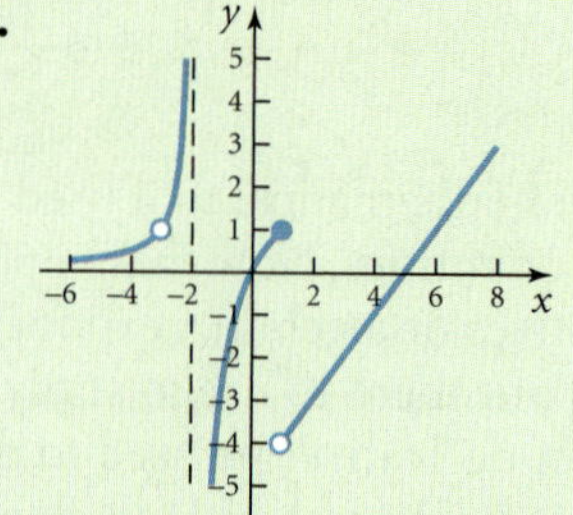

Answers to Warm-up Exercises

Algebra Warm-up Exercises

1. **a.** $(-\infty, \infty)$
   b. $x \neq 3, -3$ or $(-\infty, -3) \cup (-3, 3) \cup (3, \infty)$
   c. $x \geq -4$ or $[-4, \infty)$
   d. $x > -4$ or $(-4, \infty)$
2. **a.** $(0, -6), (2, 0), \left(-\frac{3}{2}, 0\right)$ **b.** $\left(0, -\frac{1}{4}\right), \left(\frac{1}{2}, 0\right)$
   c. $(0, 0), (4, 0), (1, 0)$
3. **a.** $x = 3, x = -3, y = 0$
   b. $x = 3, x = -3, y = -2$
   c. $y = -4$
4. $x \approx -1.494, 1.802$

Calculus Warm-up Exercises

1. $f'(x) = 6x - 5$
2. $f'(x) = 12x^2 - 4x^3$
3. $f(x) = \dfrac{-x^2 - 9}{(x^2 - 9)^2}$
4. $f'(x) = 2e^{2x}$
5. $f'(x) = \dfrac{3}{3x - 4}$
6. $f'(x) = -1/\sqrt{8 - 2x}$
7. $x = 0, 4$
8. **a.** positive **b.** positive **c.** negative
9. -5
10. discontinuous at $x = 3$ and $x = -3$
11. discontinuous at $x = -3$, $x = -2$, and $x = 1$

n this topic, we use the tools of differentiation for a more in-depth analysis of function behavior.

Increasing and Decreasing Functions and Extreme Points

One way to analyze the behavior of a function is to study the graph of that function. There are many ways to describe the graph of a function. We can describe the *location* of the graph on the coordinate plane, the *direction* in which the graph moves, and the *shape*, or *curvature*, of the graph. Your Tool Kit already contains three ways to discuss the **location** of a graph: the domain, the intercepts, and the asymptotes or other points of discontinuity. (Location, direction, and shape are the author's terms.)

Example 1: Identify the domain, the intercepts, and the asymptotes of the following functions.

a. $f(x) = x^2 - 4x + 3$ **b.** $f(x) = \dfrac{x + 1}{x^2 - 9}$

Solution:

a. The function $f(x) = x^2 - 4x + 3$ is a polynomial function. The domain is the set of all real numbers, so there are no asymptotes or points of discontinuity. $f(0) = 3$, so the y-intercept is $(0, 3)$.
If $x^2 - 4x + 3 = 0$ then $x = 1, 3$. The x-intercepts are $(1, 0)$ and $(3, 0)$.

b. The function $f(x) = \dfrac{x + 1}{x^2 - 9}$ is a rational function. The domain is all $x \neq 3, -3$. The asymptotes are $x = 3, x = -3$, and $y = 0$. Because $f(0) = -\frac{1}{9}$, the y-intercept is $\left(0, -\frac{1}{9}\right)$. Because $\dfrac{x + 1}{x^2 - 9} = 0$ if $x = -1$, the x-intercept is $(-1, 0)$. ■

Although these concepts provide some information about the graph, they do not give a complete picture. The intercepts provide several points that belong to the graph. The domain and the asymptotes are the boundaries that define where the graph goes, but they don't describe how the dots get connected within those boundaries.

A second way to describe a graph is to analyze its **direction**. As x increases in value, do the values of $f(x)$ increase, decrease, or remain constant? You probably have an intuitive concept of what it means for a quantity (or function) to *increase* or *decrease*. If the graph of $f(x)$ rises from left to right over some interval $[a, b]$, we say that $f(x)$ increases over that interval. If the graph of $f(x)$ falls from left to right over some interval $[a, b]$, we say that $f(x)$ decreases over that interval.

Let's review the formal definition of increasing and decreasing from algebra.

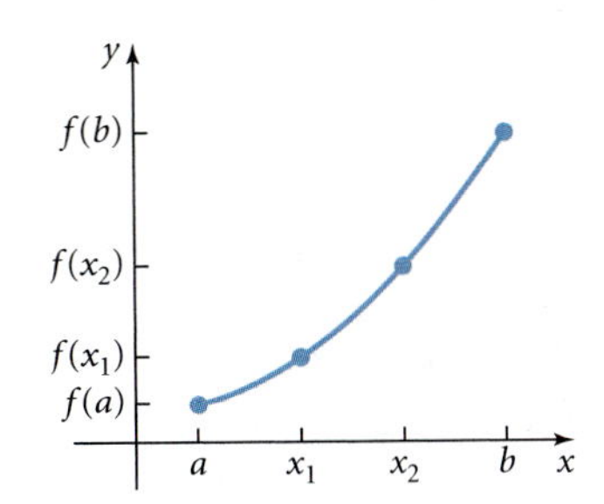

Figure 14.1

Definition: On the interval $[a, b]$,

$f(x)$ is **increasing** if $f(a) < f(x_1) < f(x_2) < f(b)$ for all $a < x_1 < x_2 < b$.

$f(x)$ is **decreasing** if $f(a) > f(x_1) > f(x_2) > f(b)$ for all $a < x_1 < x_2 < b$.

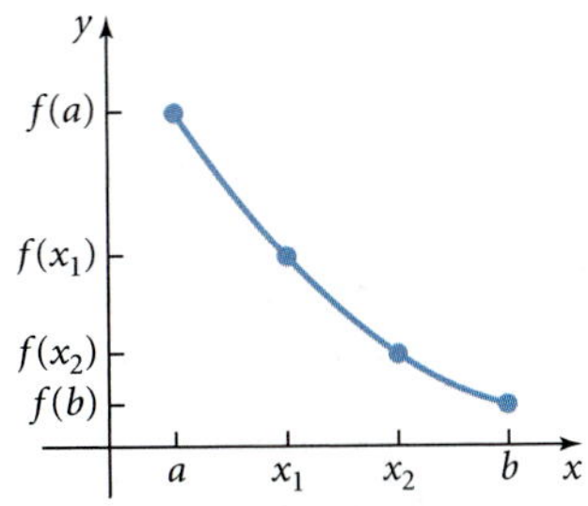

Figure 14.2

The graph in Figure 14.1 increases because $f(a) < f(x_1) < f(x_2) < f(b)$ for $a < x_1 < x_2 < b$.

The graph in Figure 14.2 decreases because $f(a) > f(x_1) > f(x_2) > f(b)$ for $a < x_1 < x_2 < b$.

Example 2: Are the following functions increasing or decreasing?

a.

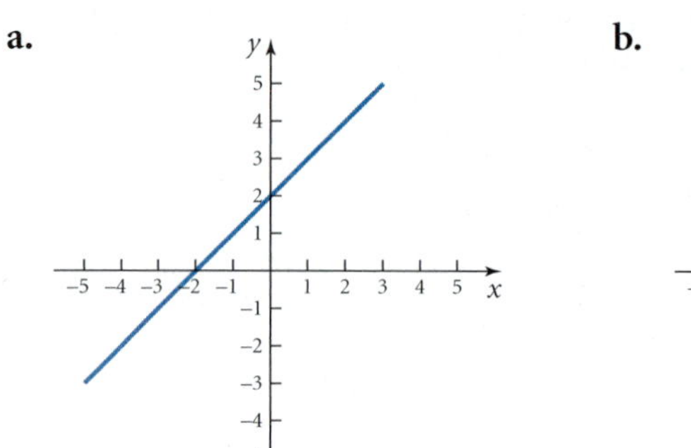

b.

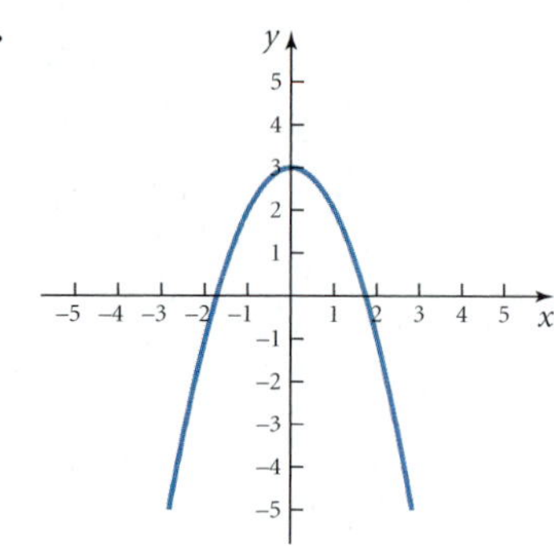

Solution:

a. The graph shows a linear function whose values constantly increase, so $f(x)$ is increasing for all values of x.

b. The graph shows a quadratic function. The values increase until reaching the vertex, and then the values decrease. We describe the direction in which the function moves using x values. Thus, $f(x)$ increases for $x < 0$ and decreases for $x > 0$. ■

As shown in Example 2, a graph may both increase and decrease within its domain. The point on a graph where its direction changes is an *extreme point*, or turning point. Extreme points may be either maximum points or minimum points.

- A *minimum point* occurs if the graph changes from decreasing to increasing.
- A *maximum point* occurs if the graph changes from increasing to decreasing.

Definition: An **extreme point** of a graph is the point where the graph changes direction. If the direction changes from increasing to decreasing, the extreme point is a **maximum point.** If the direction changes from decreasing to increasing, the extreme point is a **minimum point**.

Example 3: Identify the extreme points of the following graphs.

a. **b.** **c.**

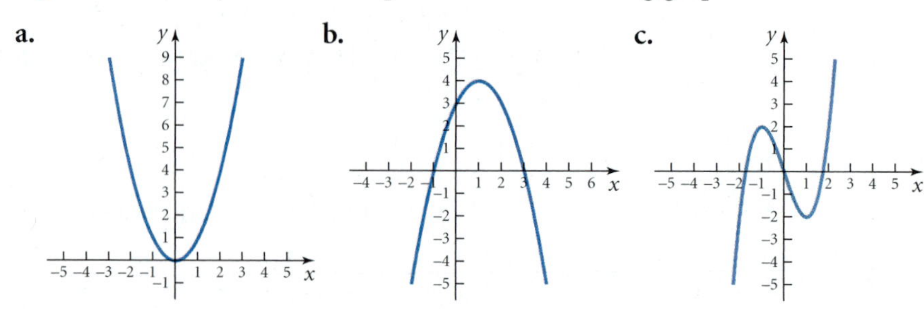

Solution:

a. This function decreases for $x < 0$ and increases for $x > 0$, so $f(x)$ has a minimum at $x = 0$.

b. This function increases for $x < 1$ and decreases for $x > 1$, so $f(x)$ has a maximum at $x = 1$.

c. This function increases for $x < -1$ decreases for $-1 < x < 1$, and increases for $x = 1$. The function has a maximum at the point where $x = -1$ and a minimum at the point where $x = -1$. ■

Figure 14.3

Look carefully at the minimum points identified in Examples 3a and 3c. Both minimum points represented the location on a graph where a decreasing graph changed to an increasing graph, but they are different types of extreme points.

In Figure 14.3, showing the graph from Example 3a, the extreme point at $(0, 0)$ is the *only* extreme point of the graph. Because it is the lowest point on the graph, $f(0) = 0$ is the smallest value the function ever has. We call this type of extreme point an **absolute extreme point**.

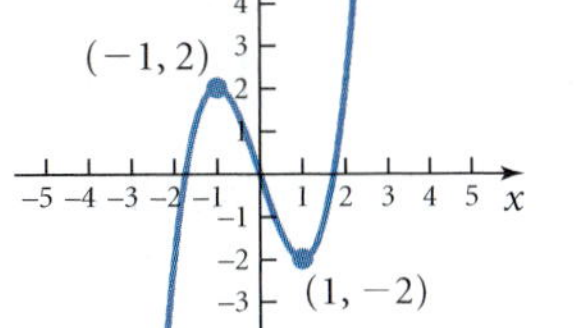

Figure 14.4

In Figure 14.4, showing the graph from Example 3c, the extreme point at $(1, f(1))$ is a turning point because the direction of the graph changes at that point. In this graph, $f(1)$ is not the smallest value for $f(x)$. It is, however, the smallest value of $f(x)$ in an open interval containing $(1, f(1))$. We call this type of extreme point a **relative**, or **local**, **extreme point**. Similarly, $(-1, f(-1))$ is also a relative extreme point.

Thus, turning points of graphs may be either maximum or minimum points and may be further classified as absolute or relative extreme points.

Definition: Given the graph of $f(x)$ with $(c, f(c))$ as a turning point of the graph,

$f(c)$ is an **absolute maximum** if $f(c) \geq f(x)$ for all x in the domain of $f(x)$.

$f(c)$ is an **absolute minimum** if $f(c) \leq f(x)$ for all x in the domain of $f(x)$.

$f(c)$ is a **relative maximum** if $f(c) \geq f(x)$ for some open interval containing $(c, f(c))$.

$f(c)$ is a **relative minimum** if $f(c) \leq f(x)$ for some open interval containing $(c, f(c))$.

Example 4: For the following graphs, locate the extreme points and identify them as maximum or minimum and absolute or relative.

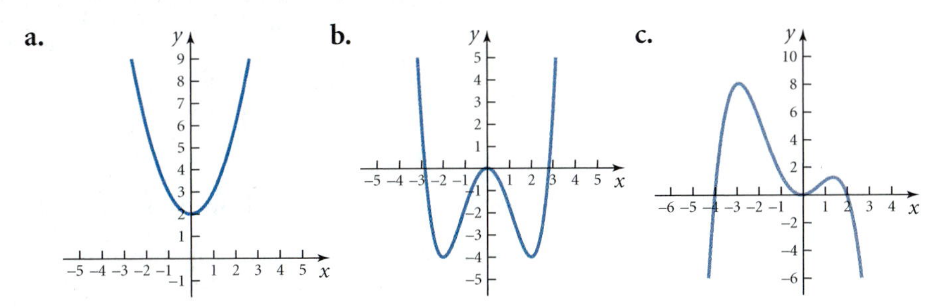

Solution:

a. The graph has a turning point at $x = 0$, which is also the lowest point of the graph. Therefore, the function has an absolute minimum at $(0, f(0))$.

b. The graph has three turning points: a minimum at $x = -2$, a maximum at $x = 0$, and a minimum at $x = 2$. The two minimum points are the lowest points on the graph (they have the same y value), so they are both absolute minimum points. The maximum point, however, is not the highest point on the graph, so $(0, f(0))$ is a relative maximum point.

c. The graph has three turning points: a maximum at $x = -3$, a minimum at $x = 0$, and a maximum at $x \approx 1.5$. The maximum at $x = -3$ is the highest point on the graph, so $(-3, f(-3))$ is the absolute maximum. There is a relative minimum at $(0, f(0))$ and a relative maximum at $(1.5, f(1.5))$. ■

Warning! Extreme points are points and should be stated as such. It is incorrect to say that the maximum is $x = -3$. Rather, you should say that the maximum occurs at the point $(-3, f(-3))$ or that $f(-3)$ is the maximum value of the function.

First Derivative Test

Now let's see what role derivatives play in the discussion of a graph's direction (increasing or decreasing) and extreme points (maximum or minimum). We already know that the derivative of $f(x)$ tells us the slope of the function at any point or the rate of change of the function. How is the derivative, $f'(x)$, related to the graph of $f(x)$? What can the derivative tell us about the graph of a function?

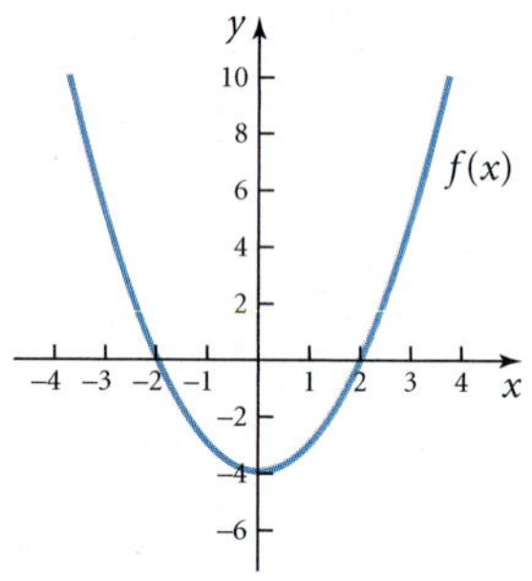

Figure 14.5

Figure 14.6

Example 5: What can you tell about the function $f(x) = x^2 - 4$?

Solution: From the graph of $f(x)$ in Figure 14.5, we see that

$f(x)$ is decreasing for $x < 0$.

$f(x)$ is increasing for $x > 0$.

$f(x)$ has an extreme point at $x = 0$.

What can you tell about derivative of the function $f'(x) = 2x$?

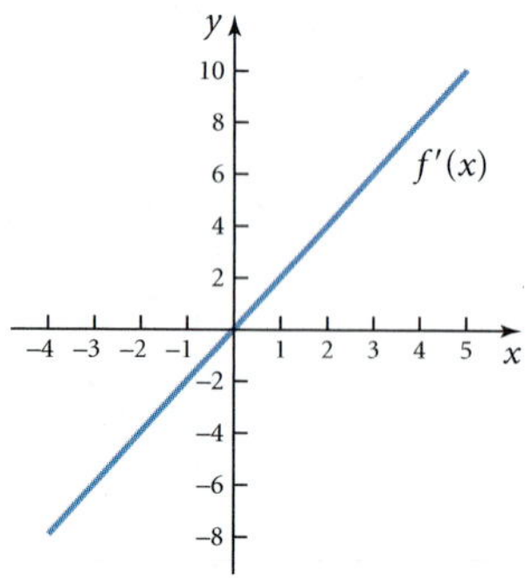

Solution: From the graph of $f'(x)$ in Figure 14.6, we see that

$f'(x)$ is above the x-axis for $x > 0$.

$f'(x)$ is below the x-axis for $x < 0$.

$f'(x)$ has an x-intercept at $x = 0$.

Comparing the results in Figure 14.7, we can reach these conclusions:

- When $f(x)$ is increasing, the graph of $f'(x)$ is above the x-axis.
- When $f(x)$ is decreasing, the graph of $f'(x)$ is below the x-axis.
- $f(x)$ has an extreme point where the graph of $f'(x)$ has an x-intercept.

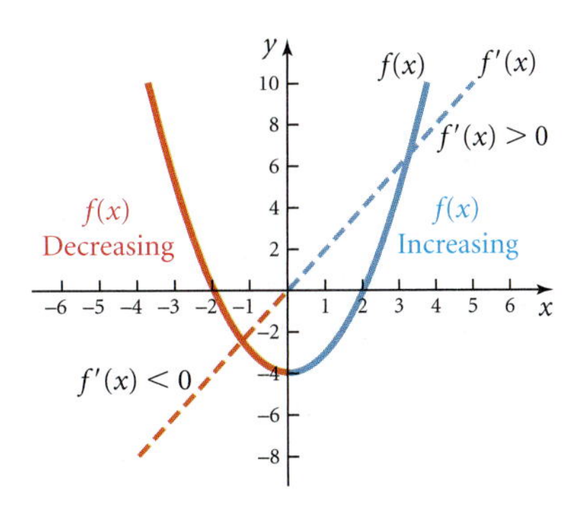

Figure 14.7

To determine if the conclusions from Example 5 hold for all functions, let's examine another function.

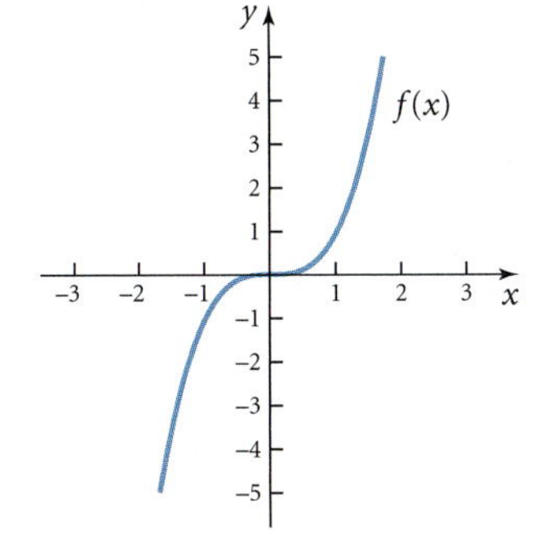

Figure 14.8

Example 6: What can you tell about the function $f(x) = x^3$?

Solution: From the graph of $f(x)$ in Figure 14.8, we see that

$f(x)$ is increasing for all values of x.
$f(x)$ is never decreasing.
$f(x)$ has no extreme point (remember that $(0, 0)$ is a point of inflection).

What can you tell about the derivative of the function, $f'(x) = 3x^2$?

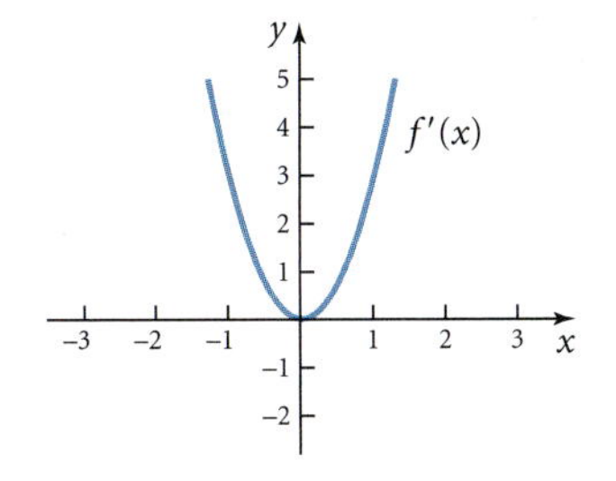

Figure 14.9

Solution: From the graph of $f'(x)$ in Figure 14.9, we see that

$f'(x)$ is above the x-axis for $x > 0$ and $x < 0$.
$f'(x)$ is never below the x-axis.
$f'(x)$ has an x-intercept at $x = 0$.

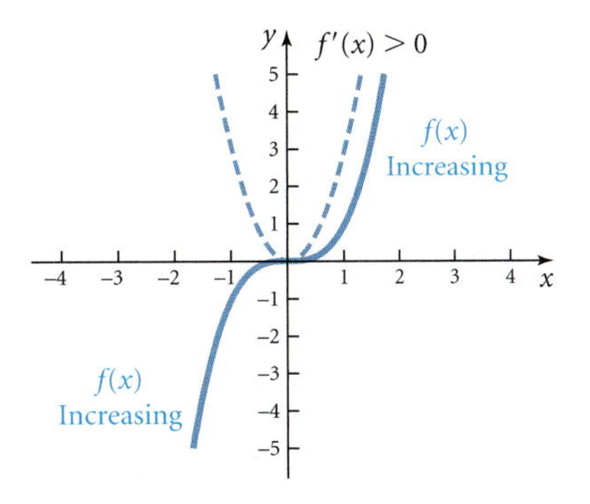

Figure 14.10

Again, when $f(x)$ is increasing, the graph of $f'(x)$ is above the x-axis. Yet even though $f'(x)$ had an x-intercept, $f(x)$ had no extreme point. See Figure 14.10. ■

Before attempting to establish any final conclusions, we consider one more function.

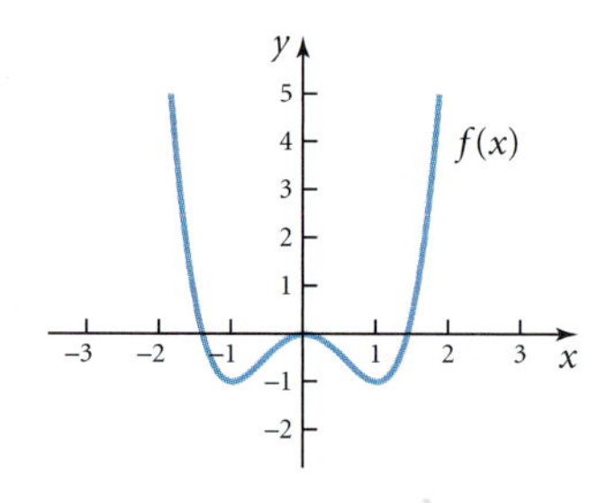

Figure 14.11

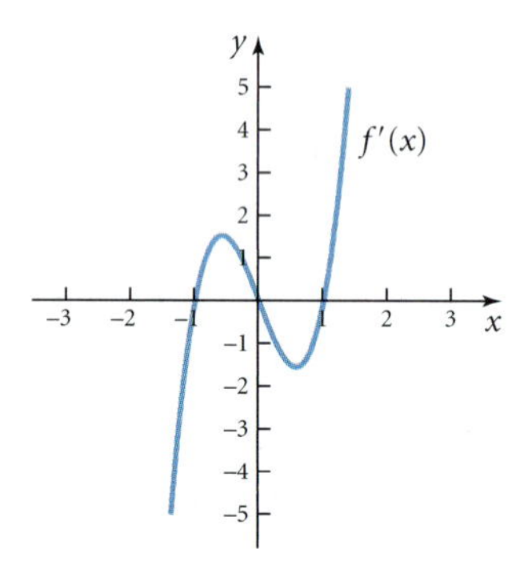

Figure 14.12

Example 7: What can you tell about the function $f(x) = x^4 - 2x^2$?

Solution: From the graph in Figure 14.11, we see that

$f(x)$ is increasing for $-1 < x < 0$ and $x > 1$.
$f(x)$ is decreasing for $x < -1$ and $0 < x < 1$.
$f(x)$ has extreme points at $x = -1, x = 0$, and $x = 1$.

What can you tell about the derivative, $f'(x) = 4x^3 - 4x$?

Solution: From the graph of $f'(x)$ in Figure 14.12, we see that

$f'(x)$ is above the x-axis for $-1 < x < 0$ and $x > 1$.
$f'(x)$ is below the x-axis for $x < -1$ and $0 < x < 1$.
$f'(x)$ has x-intercepts at $x = -1, x = 0$, and $x = 1$.

Comparing results in Figure 14.13, we reach these conclusions:

- When $f(x)$ is increasing, the graph of $f'(x)$ is above the x-axis.
- When $f(x)$ is decreasing, the graph of $f'(x)$ is below the x-axis.
- $f(x)$ has an extreme point where the graph of $f'(x)$ has an x-intercept.

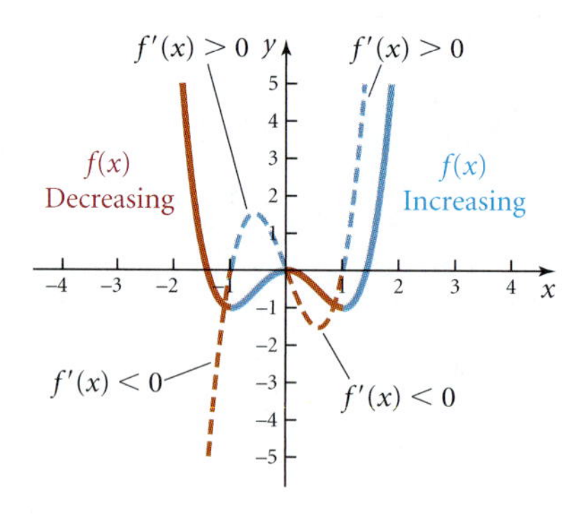

Figure 14.13

Based on the observations from Examples 5 through 7, we can make the following conclusions:

- The graph of a function $f(x)$ is increasing when the graph of its derivative $f'(x)$ is above the x-axis.
- The graph of a function $f(x)$ is decreasing when the graph of its derivative $f'(x)$ is below the x-axis.
- If $f(x)$ has an extreme point, it will occur at the point where the graph of $f'(x)$ has an x-intercept.

Now we mathematically state these conclusions and formalize the result.

1. *The graph of a function $f(x)$ is increasing when the graph of its derivative $f'(x)$ is above the x-axis.*

   What does it mean to say that a graph is above the x-axis? Any point that is above the x-axis has a y value greater than zero. If the graph of $f'(x)$ is above the x-axis, then $f'(x) > 0$.

2. *The graph of a function $f(x)$ is decreasing when the graph of its derivative $f'(x)$ is below the x-axis.*

   What does it mean to say that a graph is below the x-axis? Any point that is below the x-axis has a y value less than zero. If the graph of $f'(x)$ is below the x-axis, then $f'(x) < 0$.

3. *If $f(x)$ has an extreme point, it will occur at the point where the graph of $f'(x)$ has an x-intercept.*

   What does it mean to say that a graph has an x-intercept? The x-intercept is a point on the x-axis, so its y value must be zero. If the graph of $f'(x)$ has an x-intercept, that x value was obtained by solving $f'(x) = 0$.

Restating these conclusions mathematically yields the following important result, the **First Derivative Test**:

First Derivative Test

Given a function $f(x)$:

$f(x)$ is increasing when $f'(x) > 0$.
$f(x)$ is decreasing when $f'(x) < 0$.

If $f(x)$ has an extreme point, it will occur when $f'(x) = 0$.

If $(c, f(c))$ represents an extreme point of $f(x)$, then

$f(c)$ is a minimum if $f(x)$ changes from decreasing to increasing at $(c, f(c))$.
$f(c)$ is a maximum if $f(x)$ changes from increasing to decreasing at $(c, f(c))$.

The First Derivative Test can be used to find extreme points of functions.

Example 8: Use the First Derivative Test to locate and describe the extreme point(s) of $f(x) = x^2 - 4$.

Solution:

Step 1: Find $f'(x)$. $f'(x) = 2x$
Step 2: Set $f'(x) = 0$. $f'(x) = 2x = 0 \rightarrow x = 0$

The value(s) of x for which either $f'(x) = 0$ or $f'(x)$ is undefined is (are) the **critical value(s)**. Critical values for which $f(x)$ is defined are the possible extreme points of the function.

Definition: The value(s) of x for which either $f'(x) = 0$ or $f'(x)$ is undefined is (are) the **critical value(s)** of the function.

Step 3: Draw a number line, plot the critical value(s), and determine the values of $f'(x)$ in each interval on the line.

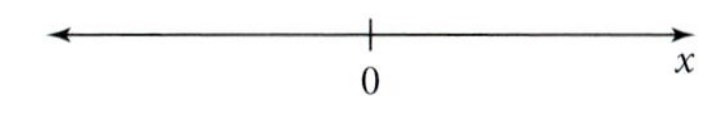

The critical value here is 0, so partition the number line into two intervals, $x < 0$ and $x > 0$. Next, select a number in the interval $x < 0$ and substitute that value into the derivative to determine if the derivative is positive or negative on that interval. Then select a number in the interval $x > 0$ to determine if the derivative is positive or negative on that interval.

Test $x = -1$: $f'(-1) = 2(-1) = -2 < 0$

$f'(x) < 0$ for $x < 0$, so we know that $f(x)$ is decreasing on the interval $x < 0$.

Test $x = 1$: $f'(1) = 2(1) = 2 > 0$

$f'(x) > 0$ for $x > 0$, so we know that $f(x)$ is increasing on the interval $x > 0$.

On the number line, we use a plus sign for positive values of the derivative and a minus sign for negative values of the derivative.

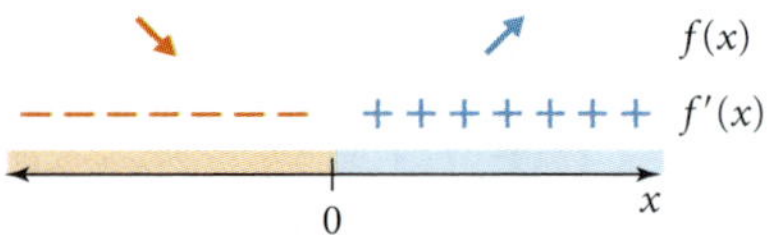

> **Tip:** You should see that selecting *any* $x < 0$ will yield a negative result for $f'(x)$ and selecting *any* $x > 0$ will yield a positive result for $f'(x)$.

Step 4: Determine the extreme point(s).

Because $f(x)$ changes from decreasing to increasing at $x = 0$, we know that $(0, f(0))$ is a minimum point for $f(x)$. Furthermore, this point must be *absolute* because it is the only turning point of the function. ■

Example 9: Use the First Derivative Test to find and describe the extreme point(s) of $f(x) = x^3 - 2$. See Figure 14.14.

Solution:

Step 1: $f'(x) = 3x^2$ — **Find the derivative**

Step 2: $f'(x) = 0 \rightarrow 3x^2 = 0 \rightarrow x = 0$ — **Find the critical value(s)**

Step 3: 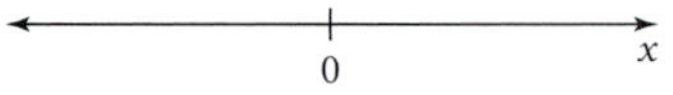 — **Draw the number line**

Test $x = -1$: $f'(-1) = 3(-1)^2 = 3 > 0$

$f(x)$ increases for $x < 0$

Test $x = 1$: $f(1) = 3(1)^2 = 3 > 0$

$f(x)$ increases for $x > 0$

Evaluate $f'(x)$ at test values

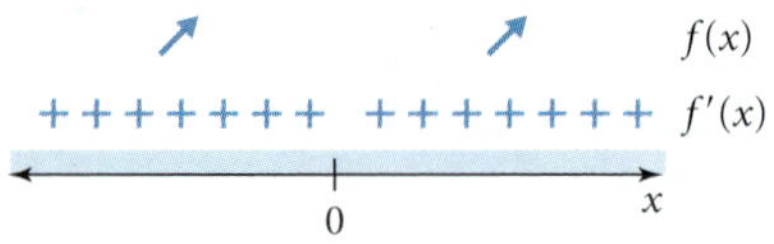

Step 4: $f(x)$ increases for all x, so there is no extreme point. Figure 14.14 verifies this result. — **Determine extreme points.** ■

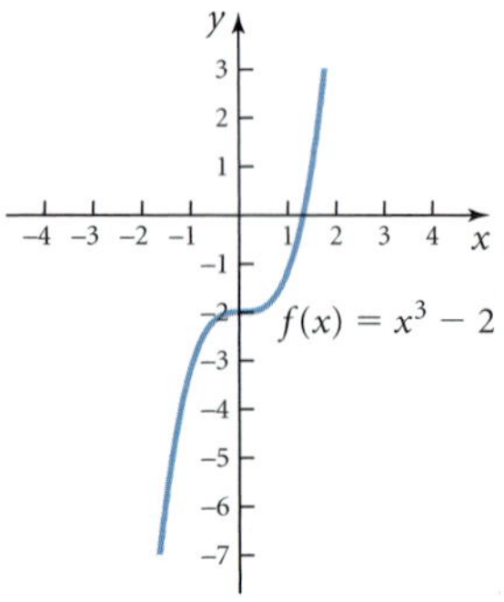

Figure 14.14

Example 10: Use the First Derivative Test to find and describe the extreme point(s) of $f(x) = x^4 - 2x^2$.

Solution:

Step 1: $f'(x) = 4x^3 - 4x$

Step 2: $f'(x) = 0 \rightarrow 4x^3 - 4x = 0$

$\rightarrow 4x(x^2 - 1) = 0$

$\rightarrow 4x(x - 1)(x + 1) = 0$

$\rightarrow x = 0, 1, -1$ are the critical values

Step 3:

Test $x = -2$: $f'(-2) = 4(-2)(-2 - 1)(-2 + 1)$
$= (-8)(-3)(-1) < 0$

> **Tip:** Be aware that the important result here is the *sign* of the answer, not the actual number!

Test $x = -0.5$: Evaluating $f'(-0.5)$ shows that $4x < 0, x - 1 < 0$, and $x + 1 > 0$. Thus, the product will be greater than zero. The actual numbers are $(-2), (-1.5)$, and (0.5), but we are only seeking the sign of the result.

Test $x = 0.5$: Evaluating $f'(0.5)$ shows that $4x > 0, x - 1 < 0$, and $x + 1 > 0$, so the product will be less than zero.

Test $x = 2$: $f'(2) = 4(2)(2 - 1)(2 + 1) = 8(1)(3) > 0$

Thus,

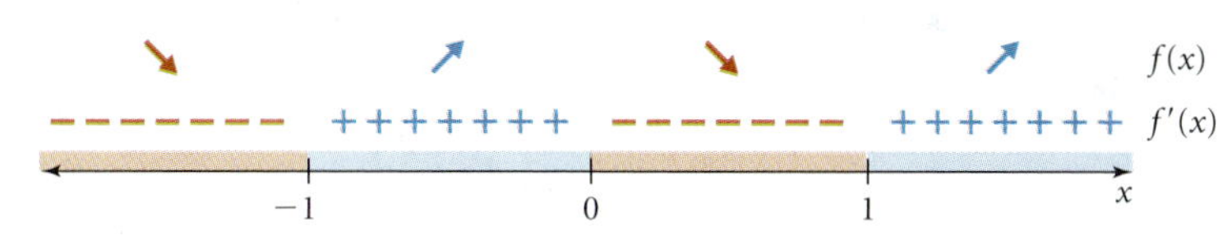

Step 4: The number line shows that $f(x)$ decreases for $x < -1$, increases for $-1 < x < 0$, decreases for $0 < x < 1$, and increases for $x > 1$. Thus, the extreme points occur at $x = -1, x = 1$ and $x = 0$. The points $(-1, f(-1))$ and $(1, f(1))$ are minimum points, and $(0, f(0))$is the maximum point.

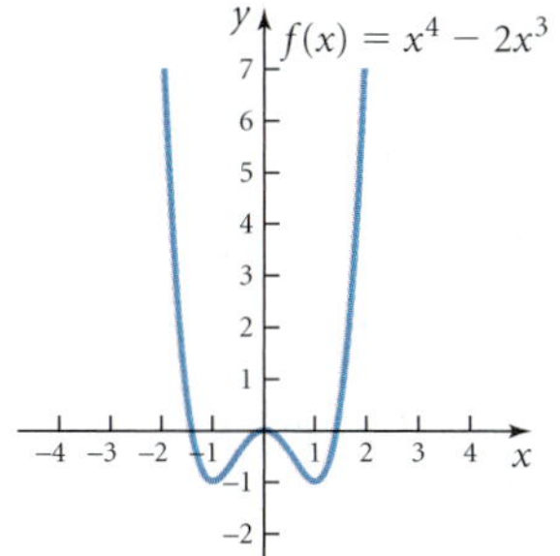

Figure 14.15

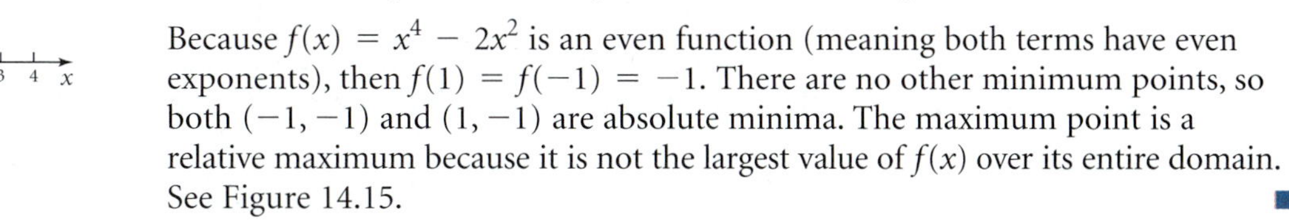

Because $f(x) = x^4 - 2x^2$ is an even function (meaning both terms have even exponents), then $f(1) = f(-1) = -1$. There are no other minimum points, so both $(-1, -1)$ and $(1, -1)$ are absolute minima. The maximum point is a relative maximum because it is not the largest value of $f(x)$ over its entire domain. See Figure 14.15. ■

Example 11: For what interval(s) does $f(x) = x^4 - 4x^3$ decrease?

Solution:

$f'(x) = 4x^3 - 12x^2$

$f'(x) = 0 \rightarrow 4x^3 - 12x^2 = 0$

$4x^2(x - 3) = 0 \rightarrow x = 0, 3$

The critical values are $x = 0$ and $x = 3$

Test $x = -1$: $f'(-1) = 4(-1)^2(-1 - 3) = -16 < 0$
Test $x = 1$: $f'(1) = 4(1)^2(1 - 3) = -8 < 0$
Test $x = 4$: $f'(4) = 4(4)^2(4 - 3) = 64 > 0$

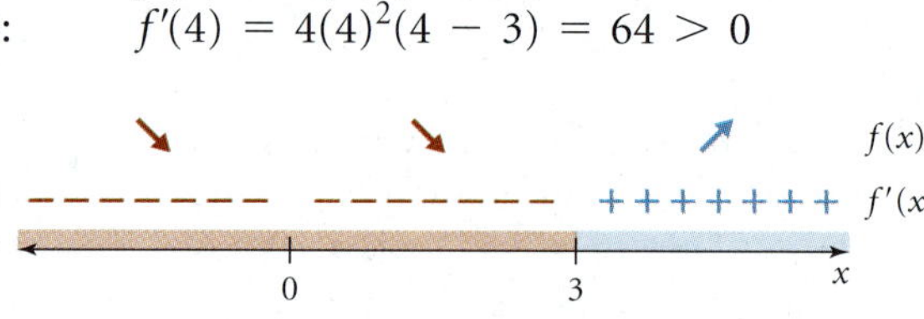

Figure 14.16

$f(x) = x^4 - 4x^3$

In this example, $f(x)$ decreases if $f'(x) < 0$, so $f(x)$ is decreasing on $(-\infty, 3)$. The graph in Figure 14.16 verifies the results.

Example 12: For what intervals does $f(x) = \dfrac{-4}{x + 5}$ increase?

Solution:

$$f'(x) = \frac{4}{(x + 5)^2}$$

There are no values of x for which $f'(x) = 0$. Because $f'(x)$ is undefined at $x = -5$, the critical value is $x = -5$. This critical value is indicated on the number line as a dashed line to show that it is an asymptote and not a member of the domain of the function.

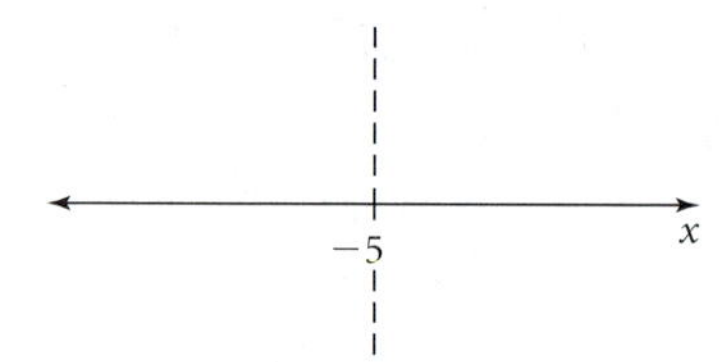

Test $x = -6$: $f'(-6) = \dfrac{4}{(-6 + 5)^2} = 4 > 0$

Test $x = 6$: $f'(6) = \dfrac{4}{(6 + 5)^2} = \dfrac{4}{121} > 0$

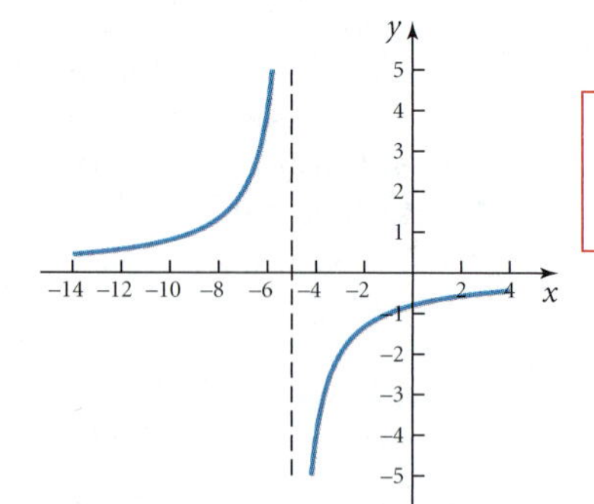

Figure 14.17

$f(x) = \dfrac{-4}{x + 5}$

Tip: Because the numerator is a positive constant and the denominator is the square of an expression, the value of the derivative will always be positive.

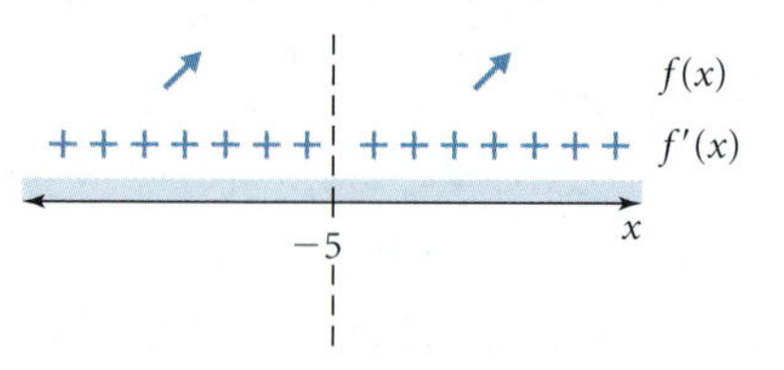

Thus, $f(x)$ increases on $(-\infty, -5)$ and $(-5, \infty)$. The graph in Figure 14.17 verifies the results.

Check Your Understanding 14.1

1. Find the critical value(s) of each of the following functions.
   a. $f(x) = 3x - x^2$ **b.** $f(x) = x^3 - 3x^2$
   c. $f(x) = \dfrac{5}{x - 2}$
2. Determine the interval(s) for which each function is increasing and decreasing.
   a. $f(x) = 3x - x^2$ **b.** $f(x) = x^3 - 3x^2$
   c. $f(x) = \dfrac{5}{x - 2}$
3. Determine the extreme points of each function. Identify each extreme point as maximum or minimum and classify it as absolute or relative.
   a. $f(x) = 3x - x^2$ **b.** $f(x) = x^3 - 3x^2$
   c. $f(x) = \dfrac{5}{x - 2}$

Applications of the First Derivative Test

The First Derivative Test can be used in applications involving real data and real situations.

Example 13: Union membership as a percentage of the total workforce can be modeled by $P(x) = 0.0003x^3 - 0.066x^2 + 4.64x - 81$, where x is the number of years after 1900 and P is the percentage of the total workforce. When did union membership as a percentage of the total workforce reach its maximum value?

Solution:

$$P(x) = 0.0003x^3 - 0.066x^2 + 4.64x - 81$$
$$\text{so } P'(x) = 0.0009x^2 - 0.132x + 4.64$$

The maximum value will occur when $P'(x) = 0$. Because $P'(x) = 0.0009x^2 - 0.132x + 4.64$, we must solve $0.0009x^2 - 0.132x + 4.64 = 0$. The Quadratic Formula gives solutions of $x \approx 88.2$ and $x \approx 58.4$.

Calculator Corner 14.1

To solve $0.0009x^2 - 0.132x + 4.64 = 0$ on your TI-84 Plus Silver, press **APPS PolySmlt ENTER ENTER**. Select 1: PolyRootFinder. Then enter Degree = 2 **ENTER**. On the next screen, enter the coefficients and press **SOLVE**.

Placing the critical values on the number line yields

The variable x represents time, so the domain of the function is $[0, \infty)$ and the number line begins at 0.

Test $P = 50$: $P'(50) = 0.29 > 0$
Test $P = 60$: $P'(60) = -0.04 < 0$
Test $P = 100$: $P'(100) = 0.44 > 0$

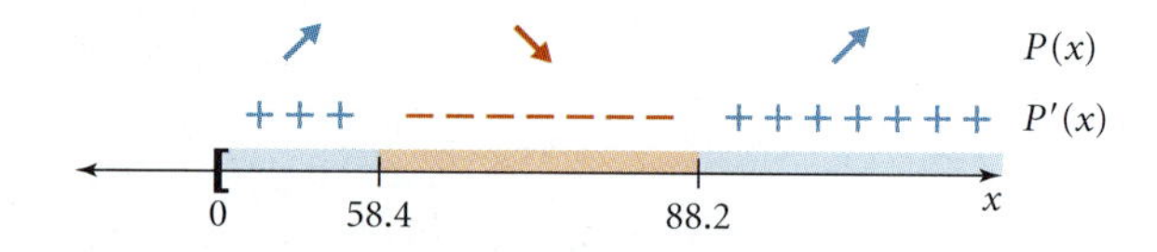

Thus, the function increases on $(0, 58.4)$ and decreases on $(58.4, 88.2)$.

The maximum occurs at 58.4. Because x is "years after 1900," the maximum membership occurred during 1958. ■

Calculator Corner 14.2

Graph $y_1 = 0.0003x^3 - 0.066x^2 + 4.64x - 81$. Set the x window as $[0, 120]$ with xScl $= 10$ and the y window as $[0, 60]$ with yScl $= 10$. Graph the function. Then press **2nd TRACE** (Calc) and select 3: minimum. Use the left or right arrows to trace to a point slightly to the left of the minimum point. Press **ENTER**. Then use the left or right arrows to trace to a point slightly to the right of the minimum point and press **ENTER ENTER**. Your calculator shows the minimum point as $x = 88.2$ and $y = 20.66$, which means that the minimum percentage of 20.66% occurred after approximately 88 years.

To find the maximum value, press **2nd TRACE** (Calc), select 4: maximum and follow the same steps as above.

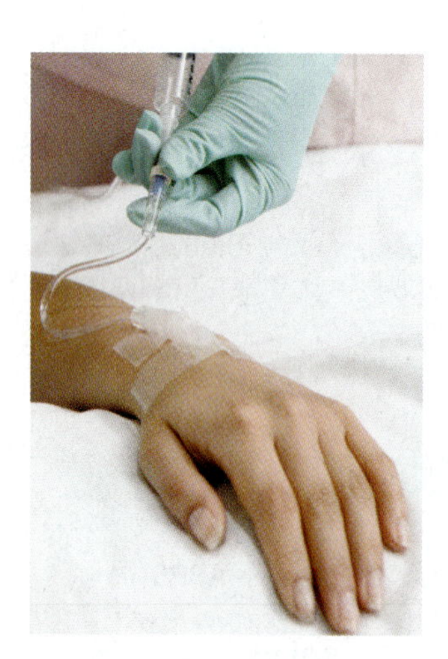

Example 14: The concentration of drug in a patient's bloodstream t hours after injection is given by $C(t) = \frac{0.4t}{t^2 + 1}$, where $C(t)$ is concentration in milligrams per cubic centimeter. When is the concentration at its maximum level? What is the maximum concentration?

Solution: The domain of $C(t)$ is $[0, \infty)$ because t represents time. By the Quotient Rule,

$$C'(t) = \frac{0.4(1 - t^2)}{(t^2 + 1)^2}$$

The maximum level will occur when $C'(t) = 0$:

$$\frac{0.4(1 - t^2)}{(t^2 + 1)^2} = 0 \rightarrow t = 1, -1$$

Because the domain is $[0, \infty)$, $t = 1$ is the only possible critical value.

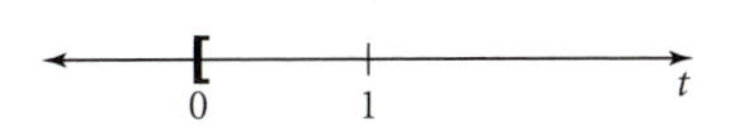

Testing $t = 0.5$ yields $C'(.5) > 0$. Testing $t = 2$ yields $C'(2) < 0$.

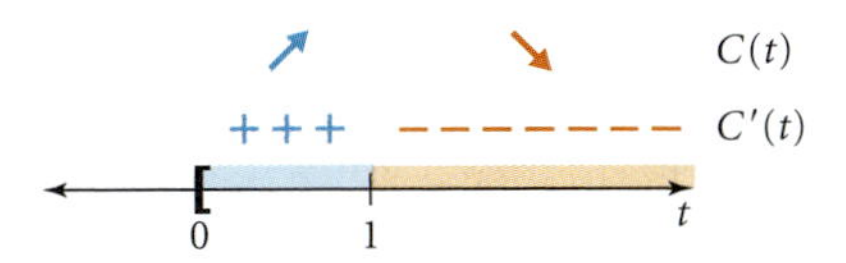

The concentration increases initially, reaching its maximum level after one hour. The maximum concentration is $C(1) = 0.2 \text{ mg/cm}^3$. ■

Calculator Corner 14.3

Graph $y_1 = \dfrac{0.4t}{t^2 + 1}$ using an x window of $[0, 6]$ and a y window of $[0, 0.04]$. Verify that the maximum occurs at the point $(1, 0.2)$.

Example 15: The total cost of producing x thousand units of a new video game is $C(x) = 1.28x^3 - 12.6x^2 + 32x + 17$, where C is in thousands of dollars. Determine the number of units that should be produced for minimum cost. What is the minimum cost?

Solution: The domain of the function is $[0, \infty)$. The extreme points are determined by the first derivative

$$C'(x) = 3.84x^2 - 25.2x + 32$$

The minimum occurs if $C'(x) = 0 \rightarrow 3.84x^2 - 25.2x + 32 = 0$. Using the Quadratic Formula (or the TI-84), we obtain critical values of $x \approx 1.72, 4.84$.

Test $x = 1$: $C'(1) = 10.64 > 0$, so costs are increasing between 0 and 1.72.

Test $x = 2$: $C'(2) = -3.04 < 0$, so costs are decreasing between 1.72 and 4.84.

Test $x = 5$: $C'(5) = 2 > 0$, so costs are increasing after 4.84.

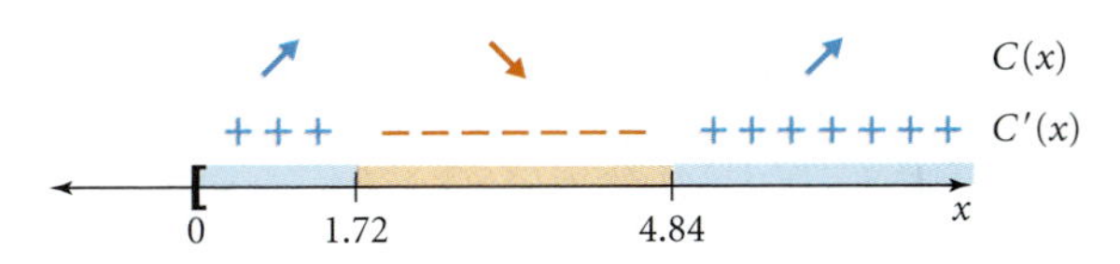

The minimum cost will occur at $x \approx 4.84$, or if 4840 units are produced.

To find the actual cost, evaluate $C(4.84) \approx 21.844$. The cost will be approximately \$21,844 to produce 4840 units. ■

Check Your Understanding Answers

Check Your Understanding 14.1

1. **a.** $x = 1.5$ or $\frac{3}{2}$ **b.** $x = 0, x = 2$
   c. $x = 2$
2. **a.** increases for $(-\infty, 1.5)$, decreases for $(1.5, \infty)$
   b. increases for $(-\infty, 0)$ and $(2, \infty)$; decreases for $(0, 2)$
   c. decreases for $(-\infty, 2)$ and $(2, \infty)$
3. **a.** There is an absolute maximum at $x = 1.5$.
   b. There is a relative maximum at $x = 0$ and a relative minimum at $x = 2$.
   c. no extreme points.

Topic 14 Review

14

This topic discussed two ways of describing the graph of a function: its *location* on the coordinate plane and the *direction* in which it moves.

CONCEPT	EXAMPLE
The **location** of a graph is described by its domain, its x- and y-intercepts, and its asymptotes, if any.	For $f(x) = \dfrac{2x - 5}{x + 3}$, we know the following: 1. The domain is all $x \neq -3$. 2. The asymptotes are $x = -3$ and $y = 2$. 3. The intercepts are $\left(\frac{5}{2}, 0\right)$ and $\left(0, -\frac{5}{3}\right)$.
The **direction** of a graph may be described as **increasing** or **decreasing**. A graph increases if it rises from left to right. A graph decreases if it falls from left to right. On the interval $[a, b]$, $f(x)$ is increasing if $f(a) < f(x_1) < f(x_2) < f(b)$ for all $a < x_1 < x_2 < b$.	$f(x) = e^x$ increases for all x. 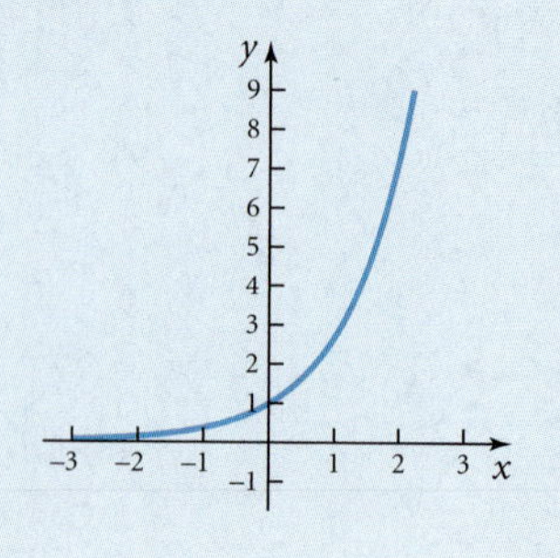

$f(x)$ is decreasing if

$$f(a) > f(x_1) > f(x_2) > f(b)$$

for all $a < x_1 < x_2 < b$.

$f(x) = 3 - (x + 1)^3$ decreases for all x.

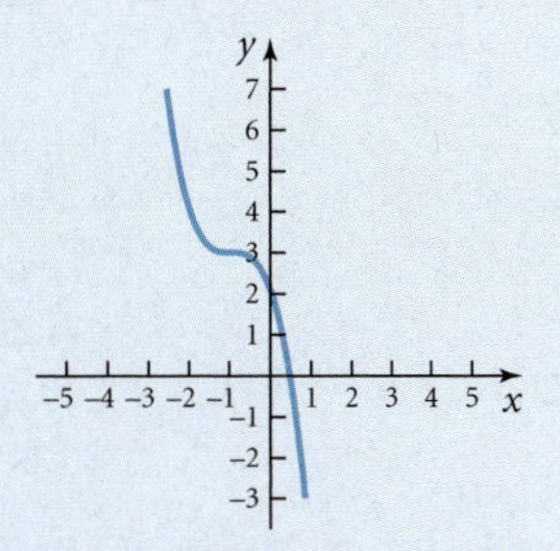

$f(x) = 4 - x^2$ increases for $x < 0$ and decreases for $x > 0$.

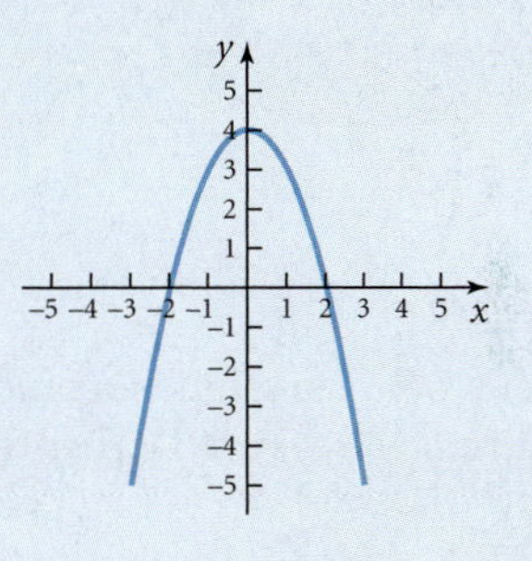

The point where the direction of a graph changes is an **extreme point** of the graph. A **maximum point** occurs if increasing changes to decreasing as x increases in value. A **minimum point** occurs if decreasing changes to increasing as x increases in value.

The function $f(x) = 3x - x^3$ has a maximum at $(1, f(1))$ and a minimum at $(-1, f(-1))$.

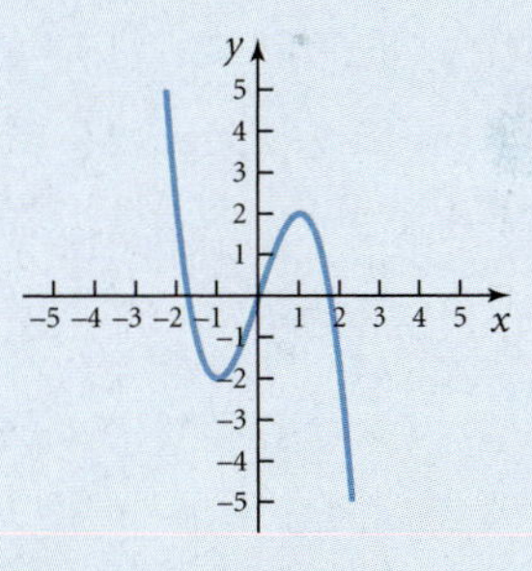

An **absolute extreme point** is the highest or lowest point in the entire domain of the function.

$f(x) = x^2 - 3$ has an absolute minimum at $(0, -3)$.

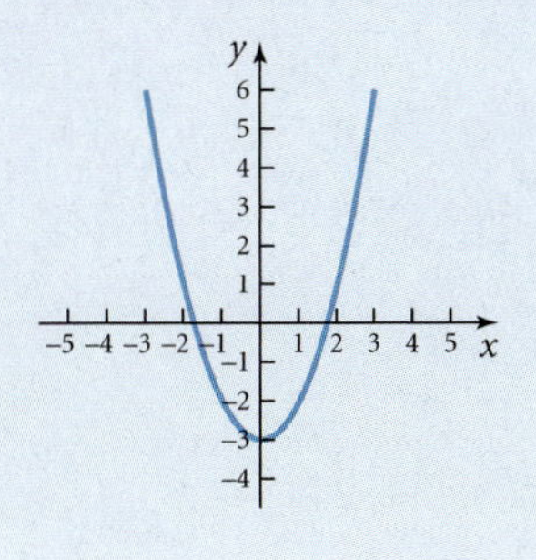

(continued)

A **relative extreme point** is a turning point of the graph.	The function $f(x) = 3x - x^3$ has a relative maximum at $(1, f(1))$ and a relative minimum at $(-1, f(-1))$. 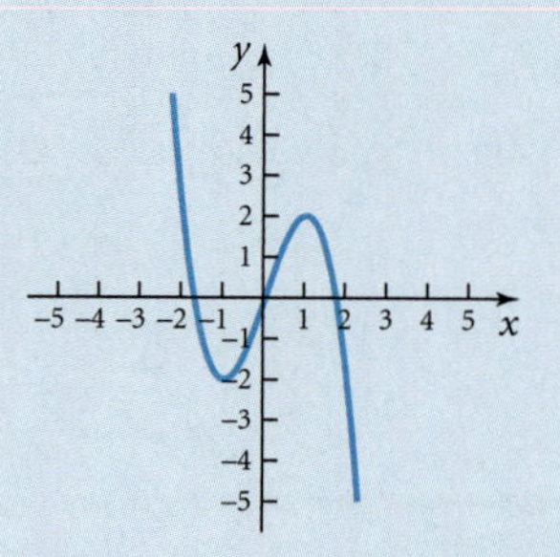
The **critical value(s)** of a function are the value(s) of x for which either $f'(x) = 0$ or $f'(x)$ is undefined.	The critical values of $f(x) = 3x - x^3$ are $x = -1$ and $x = 1$ because $f'(-1) = f'(1) = 0$. The critical value of $f(x) = \frac{2}{x - 3}$ is $x = 3$ because $f'(3)$ is undefined.
The relationship between a function and its first derivative is given by the **First Derivative Test.** Given a function $f(x)$, $f(x)$ is increasing when $f'(x) > 0$. $f(x)$ is decreasing when $f'(x) < 0$. If $f(x)$ has an extreme point, it occurs when $f'(x) = 0$. If $(c, f(c))$ represents an extreme point of $f(x)$, then $f(c)$ is a minimum if $f(x)$ changes from decreasing to increasing at $(c, f(c))$ and $f(c)$ is a maximum if $f(x)$ changes from increasing to decreasing at $(c, f(c))$.	To analyze $f(x) = 3x - x^3$: $f'(x) = 3 - 3x^2$ $3 - 3x^2 = 0 \rightarrow x = -1, 1$ The critical values are $x = -1$ and $x = 1$. Draw a number line and place the critical values on the line. Test values in each interval to determine if $f'(x)$ is positive or negative. 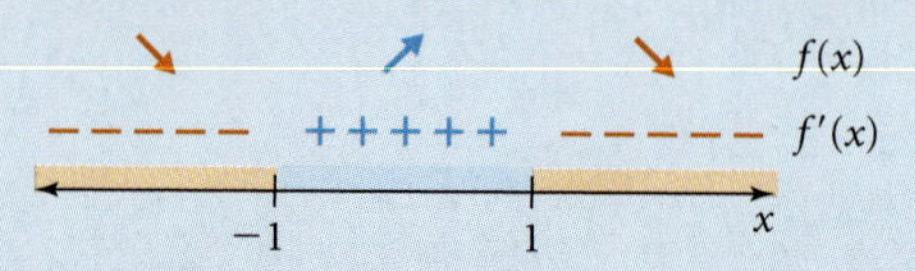 The function decreases for $x < -1$, increases for $-1 < x < 1$, and decreases for $x > 1$. There is a relative minimum point at $(-1, f(-1))$ and a relative maximum point at $(1, f(1))$.

NEW TOOLS IN THE TOOL KIT

- Describing the graph of a function by identifying the domain, intercepts, and asymptotes
- Describing the intervals of x for which the graph of a function is increasing or decreasing
- Identifying extreme points
- Classifying extreme points as maximum or minimum points
- Classifying extreme points as absolute or relative
- Using the First Derivative Test to determine intervals of increasing and decreasing and extreme points

Topic 14 Exercises 14

For each graph, state the interval(s) of x for which

a. $f(x)$ is increasing. **b.** $f(x)$ is decreasing.

1.

2.

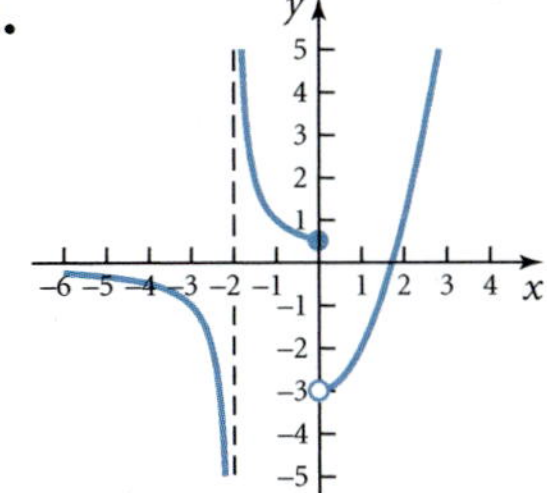

For each graph, state the interval(s) of x for which

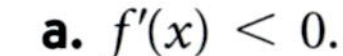
a. $f'(x) < 0$.

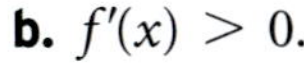
b. $f'(x) > 0$.

3.

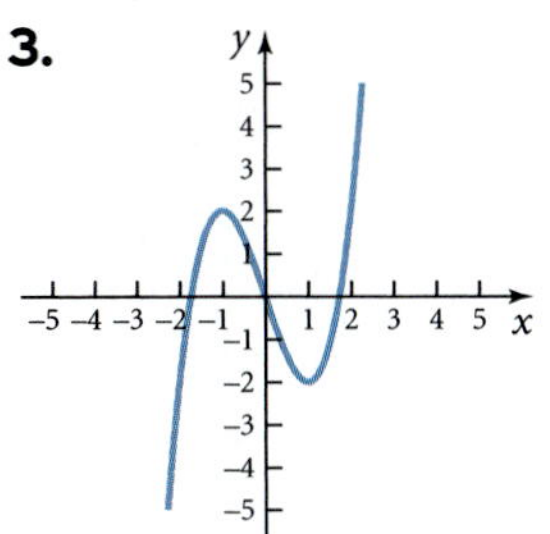

4.

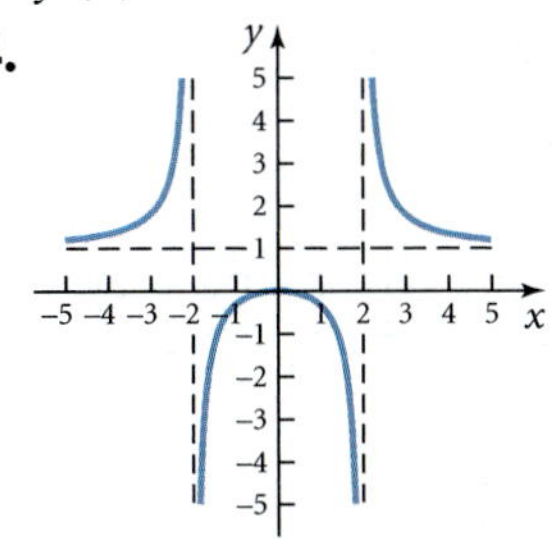

Given $f(x)$ in Exercises 5 and 6, identify each of the following points.

a. relative maximum **b.** relative minimum

c. absolute maximum **d.** absolute minimum

e. x-intercept **f.** y-intercept

5.

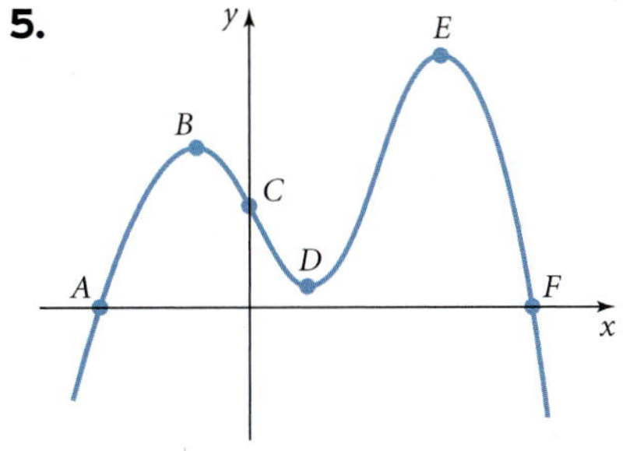

6.

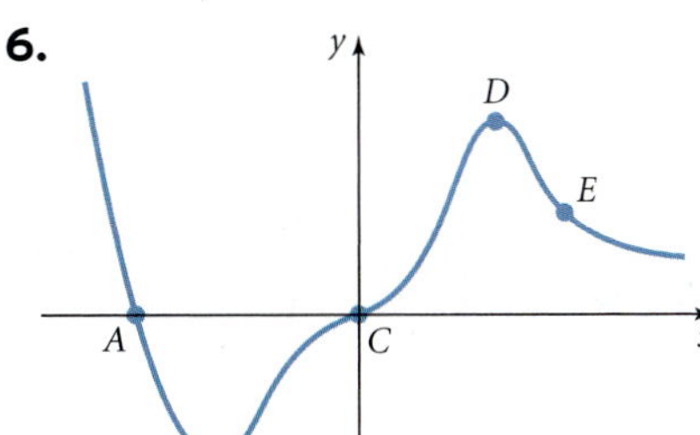

For Exercises 7 through 16, use the graph of $f(x)$ to determine if each of the following statements is true or false.

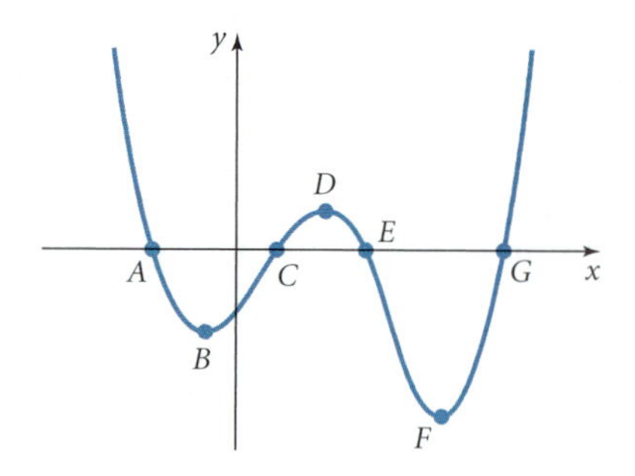

7. B is a relative minimum
8. D is an absolute extreme point
9. D is an absolute maximum
10. F is a relative minimum
11. A, C, E, G are x-intercepts
12. B is the y-intercept
13. $f(x)$ decreases between D and F
14. $f(x)$ increases between B and D
15. $f'(x) > 0$ between A and B
16. $f'(x) < 0$ between F and G

In Exercises 17 through 20, the curve in black represent functions and the curve in red represent the derivatives of those functions. Match each curve to its derivative.

17. 18.

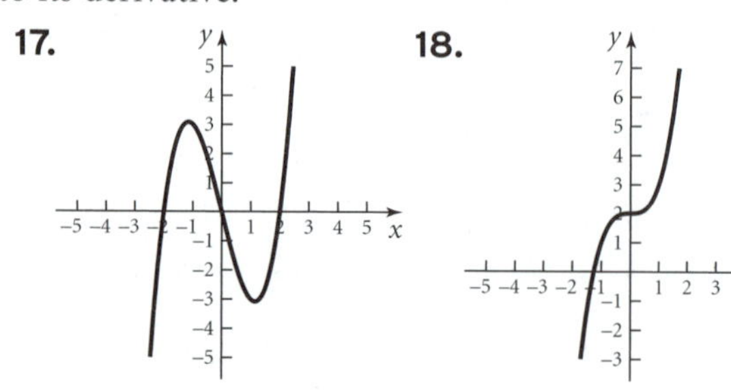

19. 20.

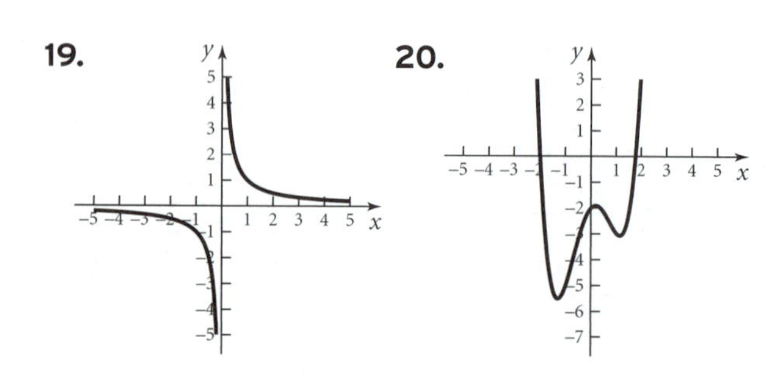

a. b.

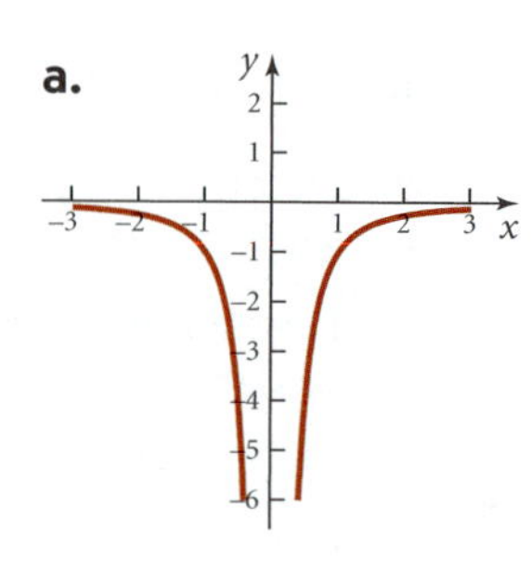

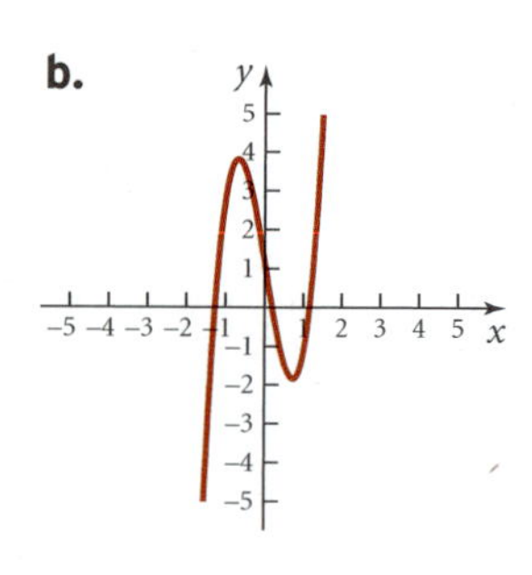

c. d.

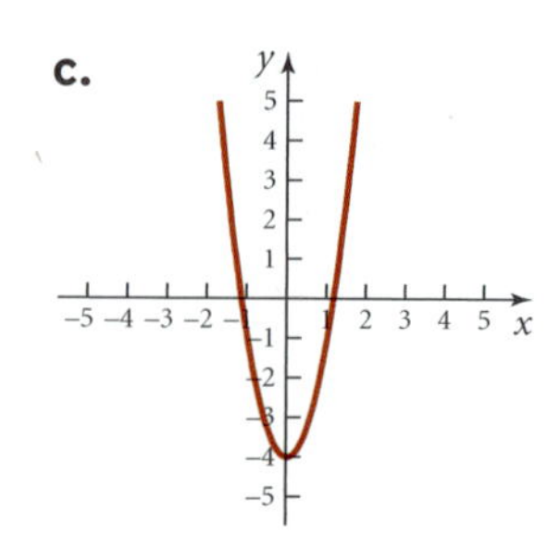

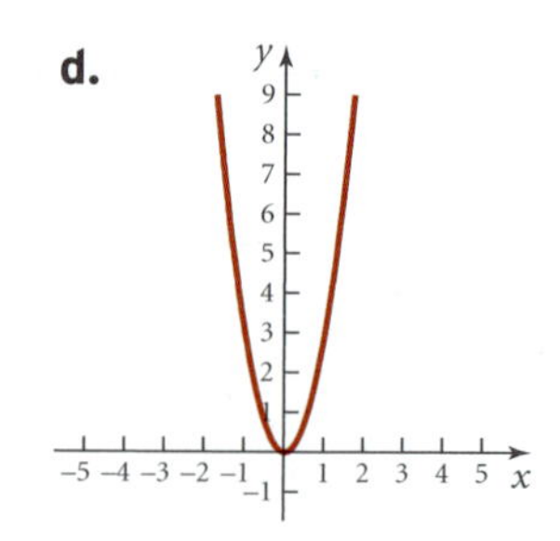

Find the critical values for each function.

21. $f(x) = x^3 - 3x^2 - 9x + 4$
22. $f(x) = -2x^3 + 3x^2 + 12x - 2$
23. $f(x) = x^4 - 4x^3$
24. $f(x) = 5x^3 - 3x^5$
25. $f(x) = \dfrac{x}{x - 3}$
26. $f(x) = \dfrac{-3x}{x + 5}$
27. $f(x) = \dfrac{-2x}{x^2 - 4}$
28. $f(x) = \dfrac{x}{x^2 - 9}$
29. $f(x) = (3x - 2)^5$
30. $f(x) = (x^2 - 4x + 3)^3$

For each of the following functions, determine

a. the interval(s) of x for which $f(x)$ increases
b. the interval(s) of x for which $f(x)$ decreases
c. extreme points

31. $f(x) = x^3 - 12x$
32. $f(x) = x^3 - 3x$
33. $f(x) = x^4 - 4x^3$
34. $f(x) = 5x^3 - 3x^5$
35. $f(x) = x^4 - 4x^3 - 8x^2 + 5$
36. $f(x) = 2x^4 - 8x^3 - 16x^2$
37. $f(x) = x^4 + 4x^3 - 20x^2 - 12$
38. $f(x) = -x^4 - 4x^3 - 4x^2 + 1$

39. $f(x) = (3x - 2)^5$

40. $f(x) = (x^2 - 4x + 3)^3$

41. $f(x) = \dfrac{2x}{x - 5}$

42. $f(x) = \dfrac{-4x}{x + 3}$

43. $f(x) = \dfrac{x}{x^2 - 1}$

44. $f(x) = \dfrac{2x}{x^2 - 4}$

45. $f(x) = \dfrac{x}{x^2 + 9}$

46. $f(x) = \dfrac{-5x}{x^2 + 1}$

47. $f(x) = \dfrac{-2x^2}{x^2 + 16}$

48. $f(x) = \dfrac{3x^2}{x^2 + 25}$

49. $f(x) = \dfrac{x^2}{x^2 - 4}$

50. $f(x) = \dfrac{2x^2}{x^2 - 9}$

51. $f(x) = x(2x - 3)^2$

52. $f(x) = -2x(3x - 1)^3$

53. $f(x) = x^3(x - 5)^2$

54. $f(x) = x^2(2 - x)^3$

55. $f(x) = \sqrt{2x - 5}$

56. $f(x) = \sqrt{6 - 3x}$

57. $f(x) = \sqrt{x^2 - 16}$

58. $f(x) = \sqrt{x^2 + 4x}$

59. $f(x) = e^x$

60. $f(x) = e^{-0.3x}$

61. $f(x) = xe^x$

62. $f(x) = xe^{-x}$

63. $f(x) = \ln x$

64. $f(x) = 4 - \ln x$

65. $f(x) = x \ln x$

66. $f(x) = 2x - x \ln x$

67. According to the Florida Agency for Workforce Innovation in March 2004, the monthly average number of unemployment claims in Volusia County is given by $N(t) = 21.26t^2 - 330.62t + 2155$, where t is the number of years after 1990.

a. During what years did the number of claims decrease?

b. Find the extreme point and interpret it in the context of this problem.

68. According to the Florida Department of Highway Safety and Motor Vehicles, the percent of motorists refusing to take an alcohol breath test before being issued citations for driving under the influence of alcohol has climbed according to $R(t) = 3.635t^2 - 23.1t + 53.54$, where t is the number of years after 1998.

a. During what years was there a decrease in the percent refusing the breath test?

b. Find the extreme point and interpret it in the context of this problem.

69. Luis hits a baseball into the air. Its height in feet above the ground after t seconds is given by $h(t) = -10.11t^2 + 67.6t + 6.79$.

a. How long was the ball rising?

b. When will the ball hit the ground?

c. Find the maximum height the ball reaches and when that maximum height occurs.

70. A rocket is launched from a platform, and its height in feet above the ground after t seconds is given by $h(t) = -16t^2 + 120t + 80$.

a. When will the rocket hit the ground?

b. For what time interval does the rocket rise?

c. For what time interval does the rocket fall?

d. What is the maximum height the rocket reaches? At what time after launch does the maximum height occur?

71. The cost of producing n units of a new DVD is $C(n) = 3 + 5\ln(n + 1) - n$, where C is cost in thousands of dollars and n is the number of units produced in thousands.

a. What are the fixed costs?

b. Find the extreme point and interpret it in the context of the problem.

72. The number of new cases being reported during a flu epidemic at a college is given by $N(t) = 50 + 120te^{-0.2t}$, where N is the number of cases and t is the number of days after the first cases were reported.

a. How many cases were initially reported?

b. For how many days did the number of new cases increase?

c. Find the extreme point and interpret it in the context of the problem.

73. Given the following information about $f(x)$, which of the following graphs best describes the function?

$f(x)$ is continuous for all real numbers.

$f(x)$ has an absolute maximum at $(3, 3)$, a relative maximum at $(-1, 1)$, and a relative minimum at $(0, 0)$.

a.

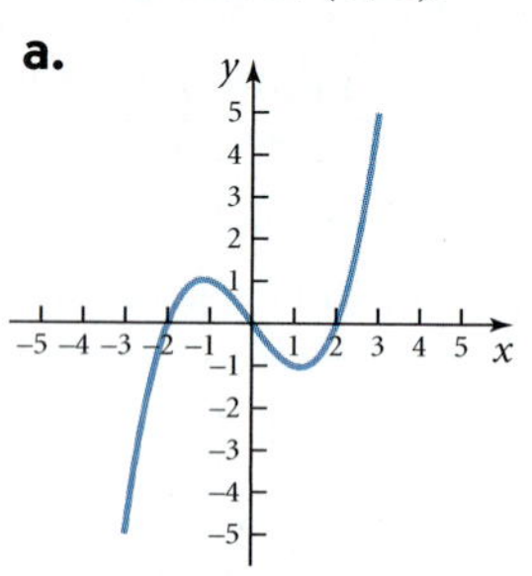

b.

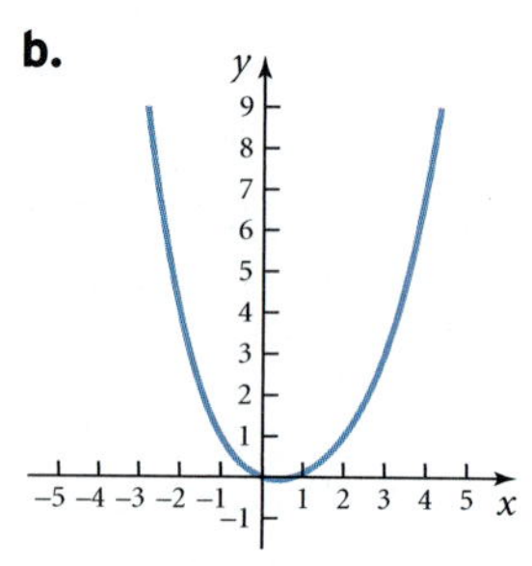

c. **d.**

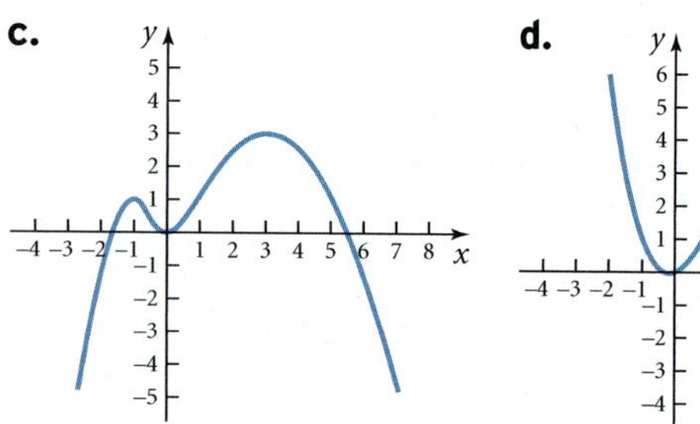

74. Given the following information about $f(x)$, which of the following graphs best describes the function?

$f(x)$ is discontinuous at $x = 2$ and $x = -2$.

$f(x)$ has a horizontal asymptote at $y = 1$.

$f(x)$ has a relative maximum at $(0, 0)$.

$f(x)$ increases for $(-\infty, -2)$ and $(-2, 0)$ and decreases for $(0, 2)$ and $(2, \infty)$.

a.

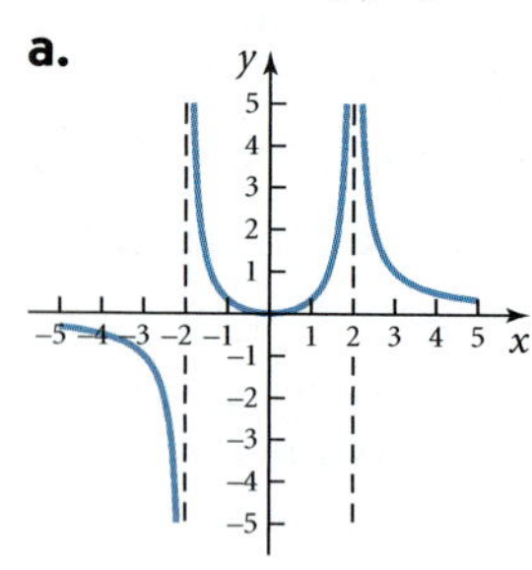

b.

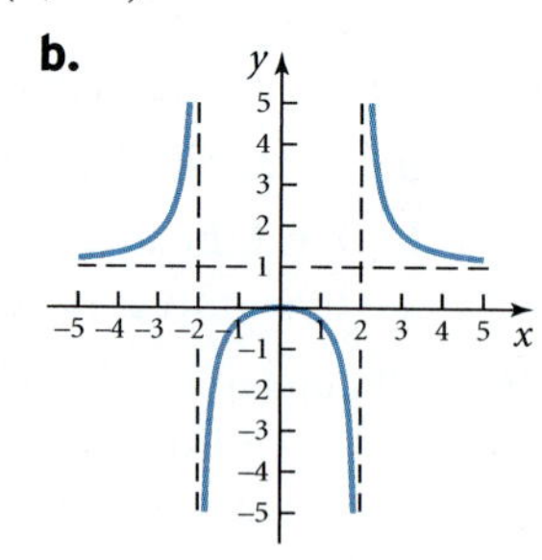

c.

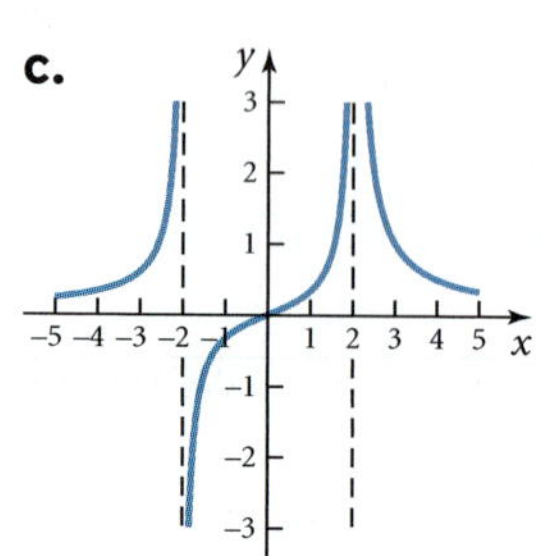

d.

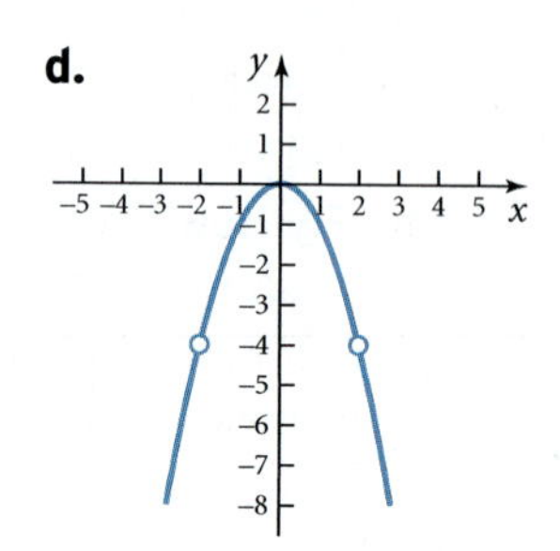

SHARPEN THE TOOLS

For each of the following functions, determine

a. the interval(s) of x for which $f(x)$ increases

b. the interval(s) of x for which $f(x)$ decreases

c. the extreme points

75. $f(x) = \dfrac{2x^2}{x^3 - 4x}$ **76.** $f(x) = \dfrac{-2x^3}{x^3 - 12x}$

77. $f(x) = x^2e^{-3x}$ **78.** $f(x) = x^2 \ln x$

The following graph shows the graph of the *derivative*, $f'(x)$. Use the graph to solve Exercises 79 and 80 for *function* $f(x)$.

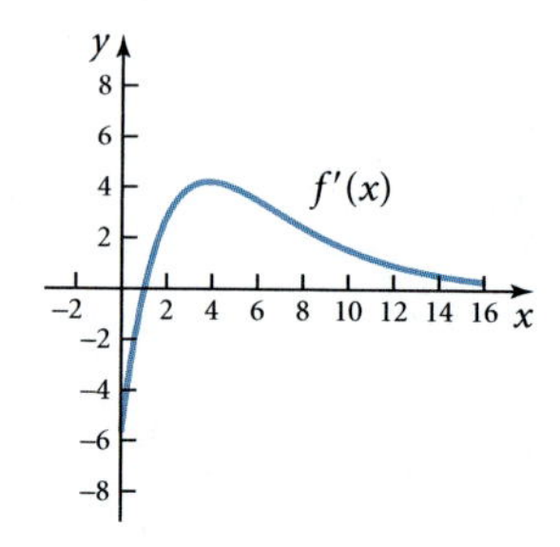

79. For what interval(s) of x does $f(x)$ increase and decrease?

80. Identify the extreme points of $f(x)$.

CALCULATOR CONNECTION

For the functions in Exercises 81 through 84, answer the following questions.

a. Use your graphing calculator to draw a good graph of the function. What is the window used for the graph?

b. What is the domain of the function?

c. What are the extreme points of the graph?

d. For what interval(s) does the function increase? For what interval(s) does the function decrease?

81. $f(x) = x^2e^{-x}$

82. $f(x) = x^2 \ln x$

83. $f(x) = e^{-x} - x \ln x$

84. $f(x) = x^3e^{2x} \ln(x^4)$

85. Passenger traffic at Daytona Beach International Airport for selected years is as given in the table.

Year	Number of Passengers
1983	540,637
1985	486,735
1988	823,155
1990	1,050,072
1992	973,224
1994	805,736
1996	800,764
1998	602,845
2000	562,815
2002	495,878
2003	565,563
2004	631,038

Data from: Daytona Beach International Airport.

a. Determine a cubic regression equation that models the data. Let x be the number of years after 1980.

b. Determine the intervals for which the function increases and decreases and interpret these intervals in the context of the problem.

c. Find the extreme points and interpret them in the context of the problem.

86. Annual revenue in million dollars at a theme park for recent years is as given in the table.

Year	Revenue, in Million Dollars
1990	35.63
1992	39.46
1994	48.71
1996	51.56
1998	50.86
2000	45.13
2002	42.76
2004	48.91
2006	53.74

a. Find a cubic regression function that models the data. Let x represent the number of years after 1990.

b. Determine the intervals for which the function increases and decreases and interpret these intervals in the context of the problem.

c. Find the extreme points and interpret them in the context of the problem.

TOPIC

15

Second Derivative Test and Graphs of Functions

IMPORTANT TOOLS IN TOPIC 15

- *Concavity*
- *Points of inflection*
- *Second Derivative Test*
- *Using the First and Second Derivative Tests to sketch graphs of functions*

You now know that the first derivative can be used to describe the behavior of a graph. The First Derivative Test shows that a function increases when its first derivative is positive and decreases when its first derivative is negative. These changes in direction tell where the function has maximum or minimum points. The second derivative can also help describe the behavior of the graph of a function. In this topic, you learn the Second Derivative Test. Before starting this topic, complete the Warm-up Exercises to sharpen your algebra and calculus skills.

TOPIC 15 WARM-UP EXERCISES

Be sure you can successfully complete the following exercises before starting Topic 15.

Algebra Warm-up Exercises

Solve each of the following equations.

1. $x^2 - 3x - 4 = 0$ **2.** $x^3 - 4x = 0$

3. $x^4 - 3x^3 = 0$ **4.** $\dfrac{4}{x - 3} = 0$

5. $\dfrac{4x}{x - 3} = 0$

Calculus Warm-up Exercises

Find the second derivative of each of the following functions.

1. $f(x) = x^2 - 3x - 4$ **2.** $f(x) = x^3 - 12x$

3. $f(x) = x^4 - 2x^3$ **4.** $f(x) = \dfrac{4}{x - 3}$

5. $f(x) = \dfrac{4x}{x - 3}$

6–10. For exercises 1 through 5, state the interval(s) of x for which $f(x)$ increases and decreases and identify the extreme point(s).

Answers to Warm-up Exercises

Algebra Warm-up Exercises

1. $x = 4, -1$ **2.** $x = 0, 2, -2$

3. $x = 0, 3$ **4.** no solution

5. $x = 0$

Calculus Warm-up Exercises

1. 2 **2.** $6x$ **3.** $12x^2 - 12x$

4. $\dfrac{8}{(x - 3)^3}$ **5.** $\dfrac{24}{(x - 3)^3}$

6. decreases for $(-\infty, 1.5)$, increases for $(1.5, \infty)$; absolute minimum at $x = 1.5$

7. increases for $(-\infty, -2)$ and $(2, \infty)$, decreases for $(-2, 2)$; relative maximum at $x = -2$, relative minimum at $x = 2$

8. decreases for $(-\infty, 1.5)$, increases for $(1.5, \infty)$; absolute minimum at $x = 1.5$

9. decreases for $(-\infty, 3)$ and $(3, \infty)$; no extreme points

10. decreases for $(-\infty, 3)$and $(3, \infty)$; no extreme points

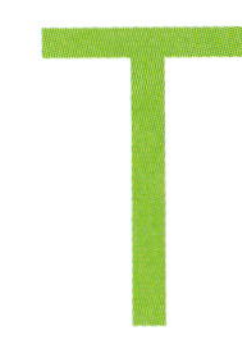

Topic 14 began the discussion of describing graphs of functions.

- The **location** of a graph on the coordinate plane is indicated by its domain, its x- and y-intercepts, and its asymptotes, where applicable.
- The **direction** of a graph, whether it increases or decreases, is determined by the value of the first derivative of the function. Points where graphs change direction are called **extreme points** and may be either **maximum points** or **minimum points**.

Concavity

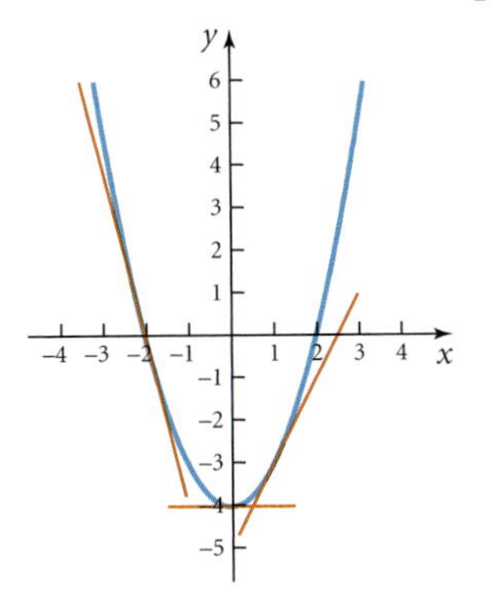

Figure 15.1

Besides the location and direction, the third and final way to describe the graph of a function is by analyzing its **shape**, or how it bends. A graph's **concavity** can be up or down.

> **Definition:** A graph is said to be **concave up** on an interval if all the tangent lines to the curve in that interval are below the curve.

Consider the function $f(x) = x^2 - 4$. It should be apparent from Figure 15.1 that the tangent line to $f(x)$ at any point will always be below the curve, no matter what the x value at the point of tangency. This graph is concave up for all values of x.

> **Tip:** We sometimes say that a concave up graph will "hold water."

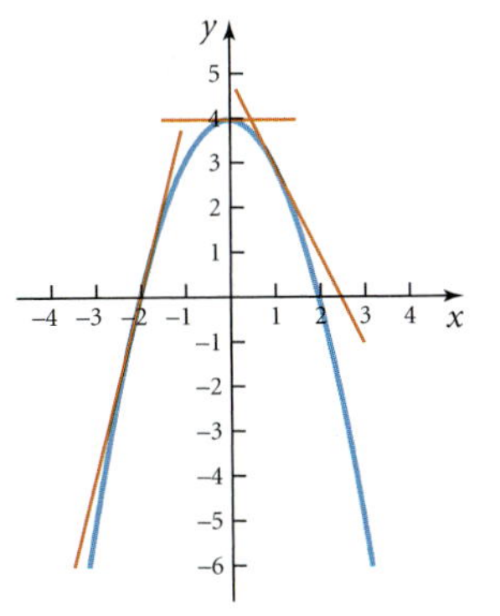

Figure 15.2

> **Definition:** A graph is said to be **concave down** on an interval if all the tangent lines to the curve in that interval are above the curve.

Consider the function $f(x) = 4 - x^2$. As shown in Figure 15.2, the tangent line to $f(x)$ at any point will always be above the curve, no matter what the x value at the point of tangency. This graph is concave down for all values of x.

> **Tip:** A concave down graph does not "hold water."

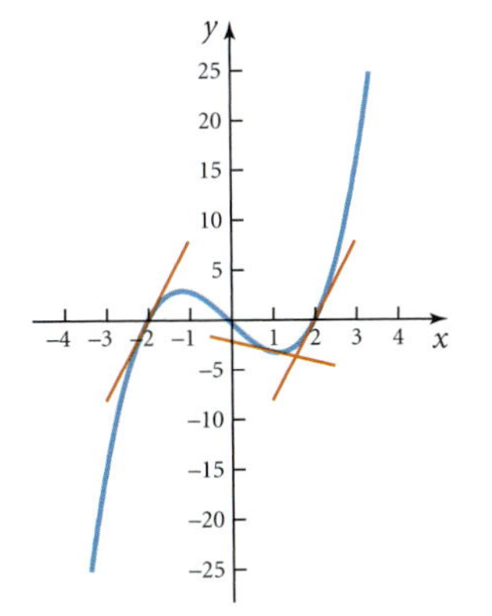

Figure 15.3

Now consider the function $f(x) = x^3 - 4x$. By examining the tangent lines and the general shaping of the curve, you can see in Figure 15.3 that the graph is concave down for $x < 0$ and concave up for $x > 0$. The point where the concavity of a graph changes is called the **point of inflection**.

> **Definition:** Any point where the concavity of a graph changes is a **point of inflection**.

Example 1: Determine the interval(s) for which each graph is concave up and concave down.

a.

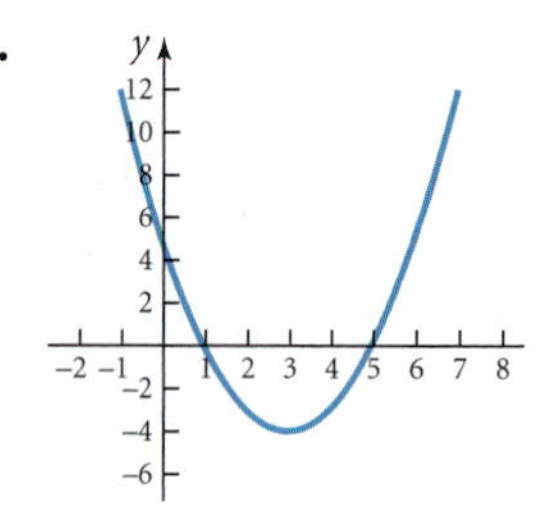

b.

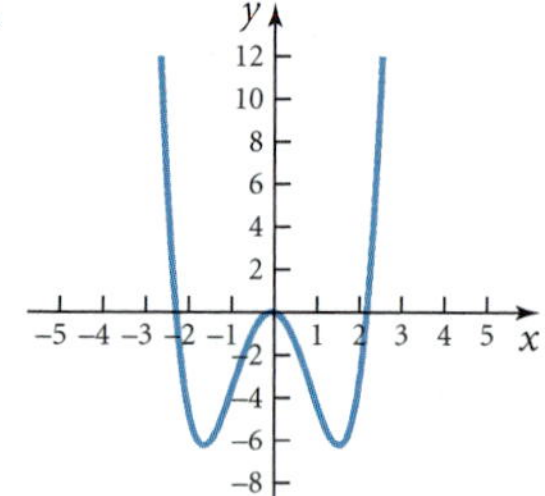

c.

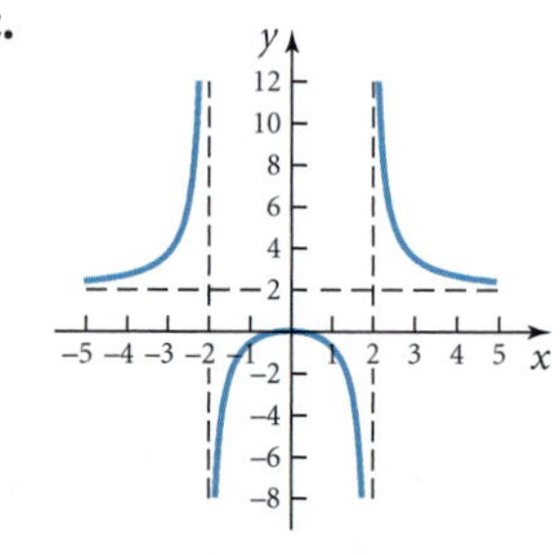

Solution: Remember that x values are used to describe the intervals.

a. This graph is concave up for all values of x.

b. This graph appears to be concave up for $(-\infty, -1.5)$, concave down for $(-1.5, 1.5)$, and concave up for $(1.5, \infty)$. There are points of inflection at $(-1.5, f(-1.5))$ and $(1.5, f(1.5))$ because the concavity changes at approximately those points.

c. This graph is concave up for $(-\infty, -2)$ and $(2, \infty)$ and concave down for $(-2, 2)$. Even though the concavity changes at $x = -2$ and $x = 2$, there are no points of inflection there because the graph is undefined at those values. ■

Second Derivative Test

We saw in Topic 14 that the first derivative determined the intervals where a graph increased or decreased and the graph's extreme points. What relationship exists between derivatives and the concavity and points of inflection of a graph?

Consider a curve that is concave up, such as $f(x) = x^2 - 4$. Examine the slopes of the tangent lines to the curve as x increases in value. You should see from Figure 15.4 that

- Where $x < 0$, the slopes of the tangent lines are negative.
- Where $x = 0$, the slope of the tangent line is zero.
- Where $x > 0$, the slopes of the tangent lines are positive.

Figure 15.4

In addition, as x increases in value, the slopes of the tangent lines also increase in value. Putting all this together, we can say that *a function is concave up if the slopes of the tangent lines to f(x) are increasing.*

Remember two things:

1. Slopes of tangent lines to the graph of a function are determined by the first derivative of the function.
2. A function is increasing where its derivative is positive.

So, "a function is concave up if the slopes of the tangent lines to $f(x)$ are increasing" means that the first derivative $f'(x)$ is increasing.

For $f'(x)$ to increase, we know that *its* derivative must be positive. The derivative of $f'(x)$ is $f''(x)$. So,

$$\text{"a function is concave up if } \underbrace{\text{the slopes of the tangent lines to } f(x)}_{f'(x)} \text{ are increasing"}$$

In other words,

$$f'(x) \quad \text{is increasing}$$

which means that

$$\frac{d}{dx}f'(x) > 0$$

or

$$f''(x) > 0$$

Thus, a function is concave up if $f''(x) > 0$.

Now consider a graph that is concave down, such as $f(x) = 4 - x^2$. Examine the slopes of the tangent lines to the curve as x increases in value. You should see from Figure 15.5 that

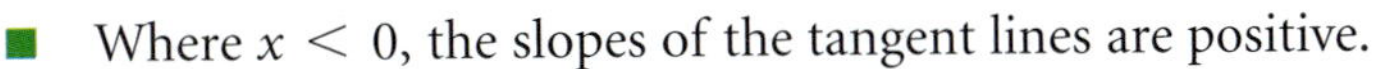

- Where $x < 0$, the slopes of the tangent lines are positive.
- Where $x = 0$, the slope of the tangent line is zero.
- Where $x > 0$, the slopes of the tangent lines are negative.

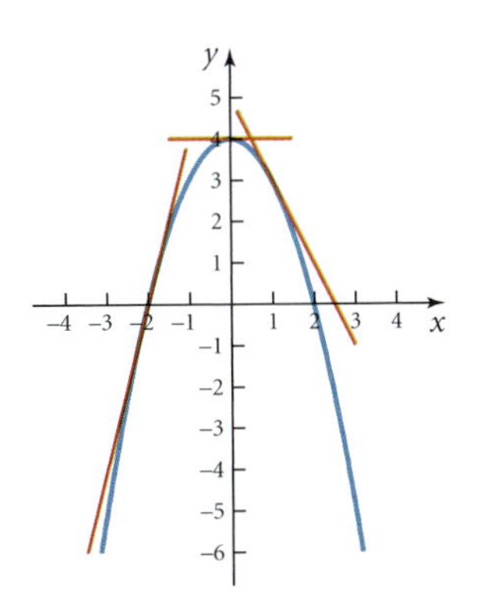

Figure 15.5

Thus, as x increases in value, the *slopes of the tangent lines to f(x) decrease in value.* We also know that a function is decreasing if its derivative is negative.

Following the same logic as before,

$$\underbrace{\text{slopes of the tangent lines to } f(x)}_{f'(x)} \text{ are decreasing}$$

In other words,

$$f'(x) \quad \text{is decreasing}$$

which means that

$$\frac{d}{dx}f'(x) < 0$$

or

$$f''(x) < 0$$

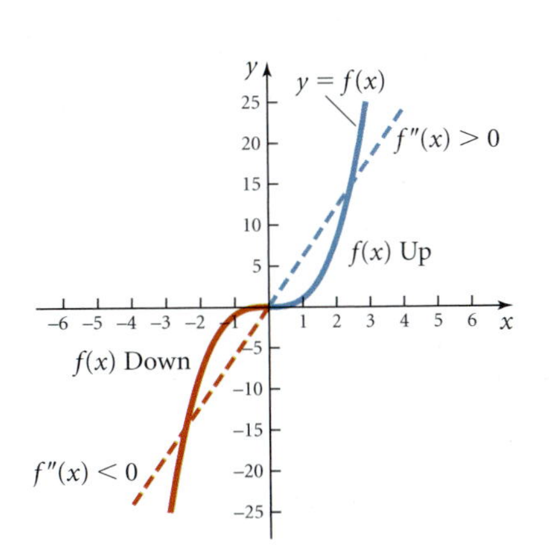

Figure 15.6

Thus, a function is concave down if $f''(x) < 0$.

Let's look at a graphic interpretation of concavity before summarizing. Consider $f(x) = x^3$and its second derivative $f''(x) = 6x$, shown in Figure 15.6.

Here, $f(x) = x^3$ is concave down for $x < 0$ and concave up for $x > 0$. Where $f(x) = x^3$ is concave down, the graph of $f''(x)$ is below the x-axis, or $f''(x) < 0$, as determined in the preceding discussion. Where $f(x) = x^3$ is concave up, the graph of $f''(x)$ is above the x-axis, or $f''(x) > 0$, as was also seen in the preceding discussion. The point of inflection for $f(x) = x^3$ is at the origin, which is where the graph of $f''(x)$ crosses the axis, or where $f''(x) = 0$.

These results can be summarized in the **Second Derivative Test**.

Second Derivative Test

Given a function $f(x)$,

- $f(x)$ is concave up if $f''(x) > 0$.
- $f(x)$ is concave down if $f''(x) < 0$.
- If $f(x)$ has a point of inflection, then $f''(x) = 0$.

If $(c, f(c))$ is an extreme point of $f(x)$, then

- $(c, f(c))$ is a minimum point if $f''(c) > 0$.
- $(c, f(c))$ is a maximum point if $f''(c) < 0$.

Example 2: Discuss the concavity and find the points of inflection for $f(x) = x^2 - 4$.

Solution:

Step 1: Find $f''(x)$.

$$f(x) = x^2 - 4$$
$$f'(x) = 2x$$
$$f''(x) = 2$$

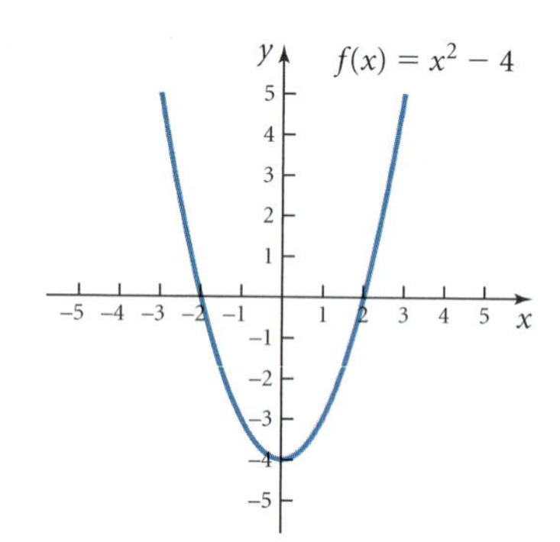

Figure 15.7

Step 2: Set $f''(x) = 0$ to find the possible points of inflection.

Possible points of inflection occur only at those values of x for which either $f''(x) = 0$ or $f''(x)$ is undefined. Because $f''(x) = 2 > 0$ for all x, $f(x)$ is concave up for all values of x. The concavity never changes, so there are no points of inflection. The graph in Figure 15.7 verifies the result. ■

Example 3: Discuss the concavity and find the points of inflection for $f(x) = x^3 - 4x$.

Solution:

Step 1: Find $f''(x)$.

$$f(x) = x^3 - 4x$$
$$f'(x) = 3x^2 - 4$$
$$f''(x) = 6x$$

Step 2: Set $f''(x) = 0$ to find the possible points of inflection.

$$f''(x) = 0 \rightarrow 6x = 0$$
$$x = 0$$

Step 3: Draw a number line, plot the possible point(s) of inflection, and determine the sign of $f''(x)$ in each interval on the line.

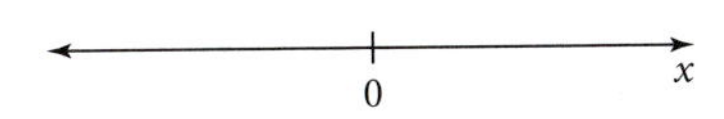

There is one possible point of inflection, so the number line is partitioned into two intervals, $x < 0$ and $x > 0$. Select an x value in each interval and evaluate $f''(x)$ for that number to determine if $f''(x)$ is positive or negative in that interval. Place either a plus sign or a minus sign above the interval to indicate the value of $f''(x)$ in that interval.

Test $x = -1$. $\quad f''(-1) = 6(-1) < 0$

$f''(x) < 0$ for $x < 0$, so we know that $f(x)$ is concave down on the interval $x < 0$.

Test $x = 1$. $\quad f''(1) = 6(1) > 0$

$f''(x) > 0$ for $x > 0$, so we know that $f(x)$ is concave up on the interval $x > 0$.

We denote the conclusions on the number line as follows.

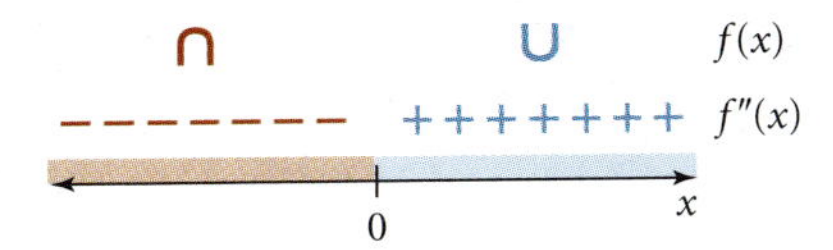

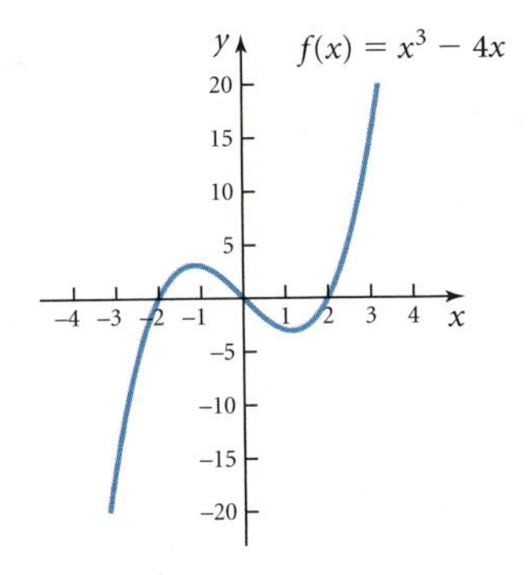

Figure 15.8

Tip: As with the First Derivative Test, we are only interested in the **sign** of the derivative in each interval, not the actual value of the derivative.

Step 4: Determine the point(s) of inflection. Points of inflection occur where concavity changes. Thus, the point of inflection for this curve is $(0, f(0))$ or $(0, 0)$. The graph in Figure 15.8 verifies the results. ■

Example 4: Discuss the concavity and find the points of inflection for $f(x) = x^4 - 4x^3$.

Solution:

Step 1:

$$f(x) = x^4 - 4x^3$$
$$f'(x) = 4x^3 - 12x^2$$
$$f''(x) = 12x^2 - 24x$$

Step 2:

$$f''(x) = 0 \rightarrow 12x^2 - 24x = 0$$
$$\rightarrow 12x(x - 2) = 0$$
$$\rightarrow \quad x = 0 \quad \text{and} \quad x = 2$$

Step 3:

Test $x = -1$. $f''(-1) = 12(-1)(-1 - 2) > 0$

$f''(x) > 0$ for $x < 0$, so we know that $f(x)$ is concave up for the interval $x < 0$.

Test $x = 0.5$. $f''(0.5) = 12(0.5)(0.5 - 1) < 0$

$f''(x) < 0$ for $0 < x < 2$, so we know that $f(x)$ is concave down for the interval $0 < x < 2$.

Test $x = 3$. $f''(3) = 12(3)(3 - 2) > 0$

$f''(x) > 0$ for $x > 2$, so we know that $f(x)$ is concave up for the interval $x > 2$.

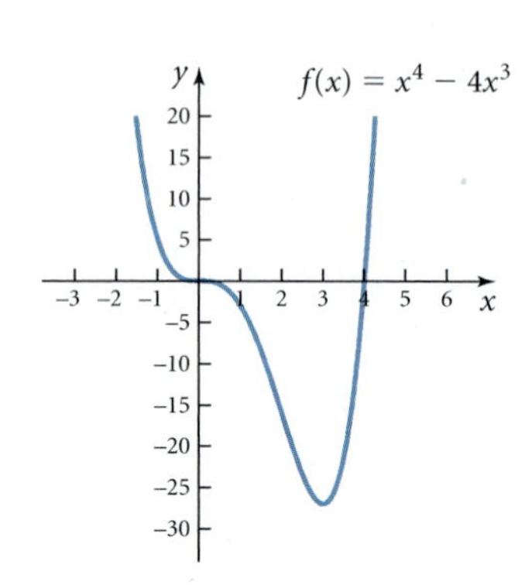

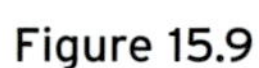
Figure 15.9

Placing these results on the number line yields

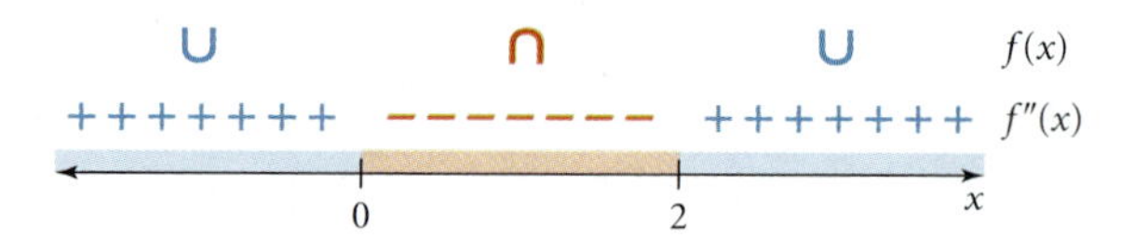

Step 4: The concavity changes at $x = 0$ and $x = 2$, so the points of inflection are $(0, f(0))$and $(2, f(2))$, or $(0, 0)$ and $(2, -16)$. The graph in Figure 15.9 verifies the results. ■

Example 5: Discuss the concavity and find the point of inflection for $f(x) = \dfrac{4}{x - 3}$.

Solution: The domain of $f(x)$ is all real numbers such that $x \neq 3$.

Step 1: $f''(x) = \dfrac{8}{(x - 3)^3}$

Step 2: Because $\dfrac{8}{(x - 3)^3} \neq 0$, the only possible inflection point is where $x = 3$, the value for which $f''(x)$ is undefined.

Step 3: Test $x = 0$. $f''(0) = \dfrac{8}{(0 - 3)^3} < 0$

$f(x)$ is concave down for $x < 3$.

Test $x = 4$. $f''(4) = \dfrac{8}{(4 - 3)^3} > 0$

$f(x)$ is concave up for $x > 3$.

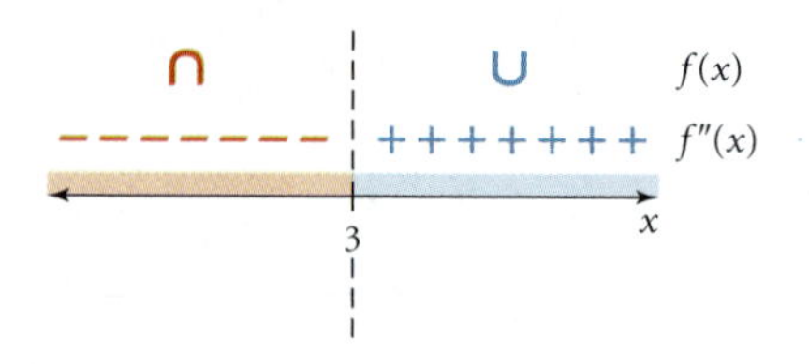

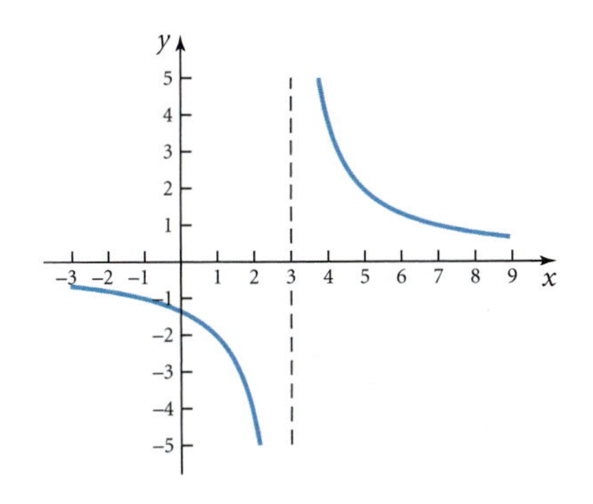

Figure 15.10

Step 4: The concavity changes at $x = 3$, but that is an asymptote of the graph. Because the function is undefined at the asymptote, there is no point of inflection. The graph in Figure 15.10 verifies the results. ■

Example 6: Discuss the concavity and find the point of inflection for $f(x) = \ln(x + 2)$.

Solution: The domain of the function is all x such that $x > -2$.

Step 1:

$$f(x) = \ln(x + 2)$$

$$f'(x) = \frac{1}{x + 2}$$

$$f''(x) = -\frac{1}{(x + 2)^2}$$

Step 2: Because $-\frac{1}{(x + 2)^2} \neq 0$, the only possible inflection point is at $x = -2$, which is not in the domain of the function.

Step 3: Test $x = 0$.

$$f''(x) = -\frac{1}{(0 + 2)^2} < 0.$$

$f(x)$ is concave down for all $x > -2$.

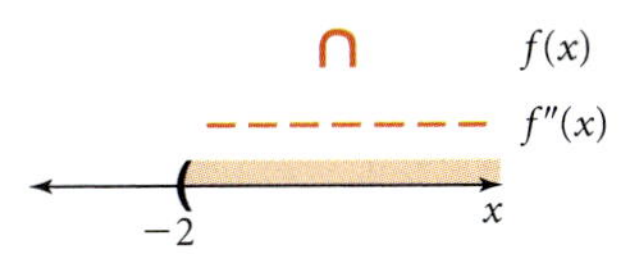

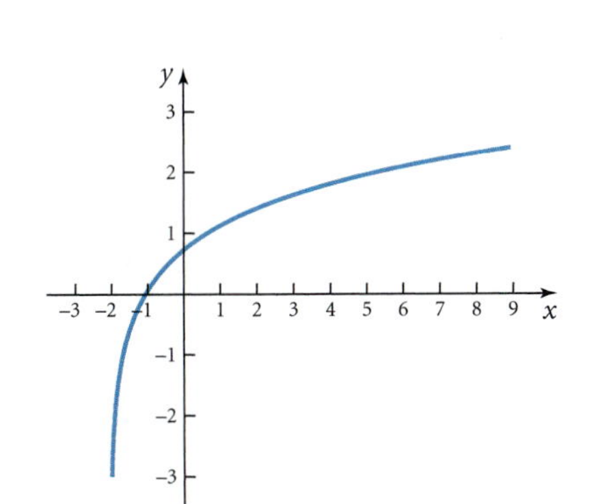

Figure 15.11

Step 4: The concavity never changes, so there is no point of inflection. The graph in Figure 15.11 verifies the results. ■

Check Your Understanding 15.1

Discuss the concavity and determine the point(s) of inflection for the following functions using the Second Derivative Test.

1. $f(x) = x^3 - 2$

2. $f(x) = x^4 - 6x^2$

3. $f(x) = e^{-x}$

4. $f(x) = \frac{2}{x - 3}$

Graphing Functions Using Derivative Tests

The graph of a function $f(x)$ can be analyzed as follows.

Function	Information Provided
$f(x)$	Domain, intercepts, asymptotes
$f'(x)$	Intervals of increasing and decreasing, extreme point(s)
$f''(x)$	Intervals of concavity, point(s) of inflection

Once all the information has been gathered, plot the intercepts, the extreme point(s), and the point(s) of inflection along with the asymptotes, if any. Then draw the curve between these points using the information about direction and shaping. The following table summarizes what we know about the value of the derivatives and the behavior and shaping of the graph.

Derivative Value	Shape of Curve	Behavior of Function
$f'(x) > 0$ ($f(x)$ increases) $f''(x) > 0$ (concave up)	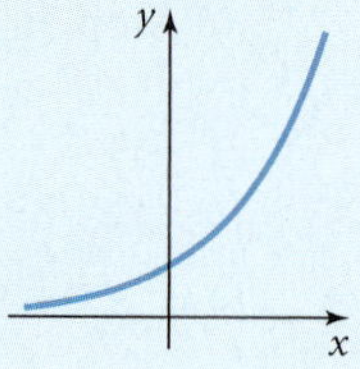 y, x	$f(x)$ increases at a faster rate
$f'(x) > 0$ ($f(x)$ increases) $f''(x) < 0$ (concave down)	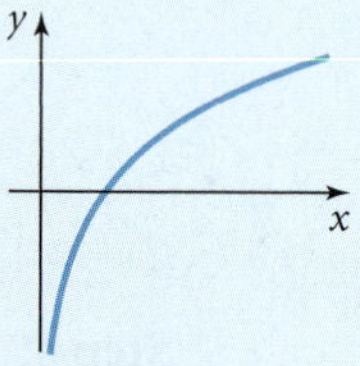 y, x	$f(x)$ increases at a slower rate
$f'(x) < 0$ ($f(x)$ decreases) $f''(x) > 0$ (concave up)	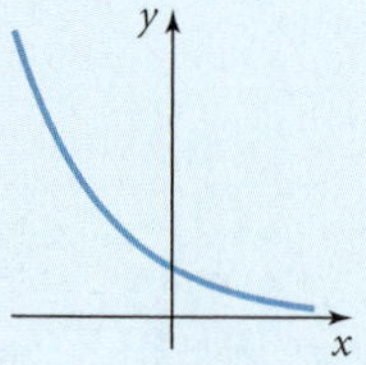 y, x	$f(x)$ decreases at a slower rate
$f'(x) < 0$ ($f(x)$ decreases) $f''(x) < 0$ (concave down)	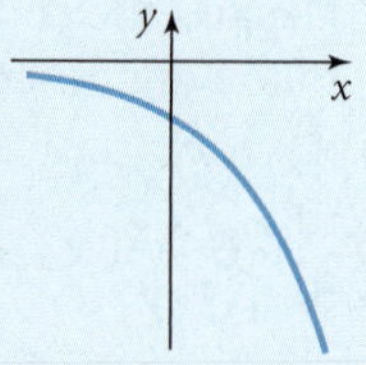 y, x	$f(x)$ decreases at a faster rate

Example 7: Use the Derivative Tests to analyze and sketch the graph of $f(x) = 4x^3 - x^4$.

Solution:

Step 1: Because $f(x)$ is a polynomial, the domain is the set of all real numbers and there are no asymptotes. To determine the y-intercept, set $x = 0$. Because $f(0) = 0$, the y-intercept is $(0, 0)$. To determine the x-intercept, set $f(x) = 0$:

$$0 = 4x^3 - x^4 \rightarrow 0 = x^3(4 - x) \rightarrow x = 0, 4$$

The x-intercepts are $(0, 0)$ and $(4, 0)$.

> **Warning!** Intercepts are points on the graph, so they should be stated as ordered pairs.

Step 2: Because $f(x) = 4x^3 - x^4$, $f'(x) = 12x^2 - 4x^3$. To determine the critical values, set $f'(x) = 0$:

$$12x^2 - 4x^3 = 0 \rightarrow 4x^2(3 - x) = 0$$

So, $x = 0, 3$ are the critical values. The number line analysis yields

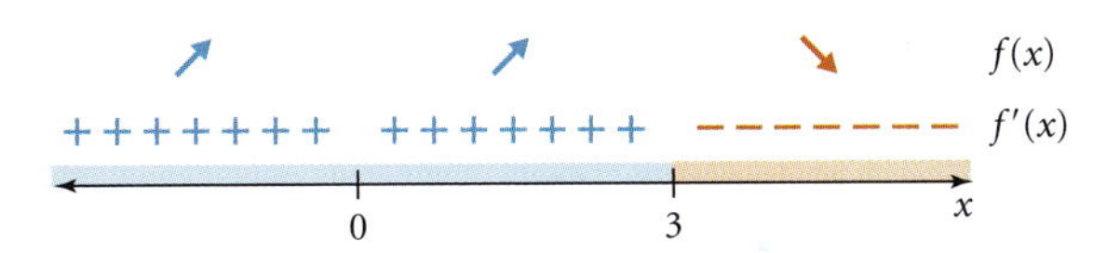

Thus, $f(x)$ increases for $(-\infty, 3)$ and decreases for $(3, \infty)$. There is an absolute maximum at $x = 3$, and $f(3) = 27$ is the maximum value of the function.

Step 3: $f(x) = 4x^3 - x^4 \rightarrow f''(x) = 24x - 12x^2$

Setting $f''(x) = 0$ gives

$$24x - 12x^2 = 0 \rightarrow 12x(2 - x) = 0$$

So $x = 0, 2$ are the possible inflection points. The number line analysis yields

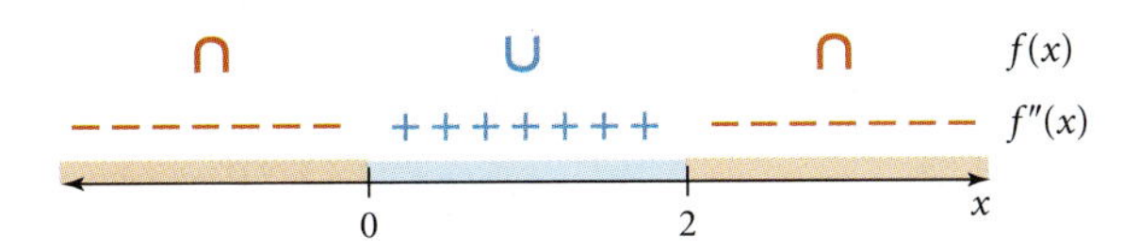

Thus, $f(x)$ is concave down for $(-\infty, 0)$ and $(2, \infty)$ and concave up for $(0, 2)$. The points of inflection are at $x = 0$ and $x = 2$, or $(0, 0)$ and $(2, 16)$.

Tip:

To test for extreme points and intervals of increasing and decreasing, substitute the test numbers into the first derivative.

To test for concavity and points of inflection, substitute the test numbers into the second derivative.

To determine the y-value of the extreme points or points of inflection, substitute into the original function.

Step 4: Plot the intercepts, extreme point, and points of inflection. See Figure 15.12.

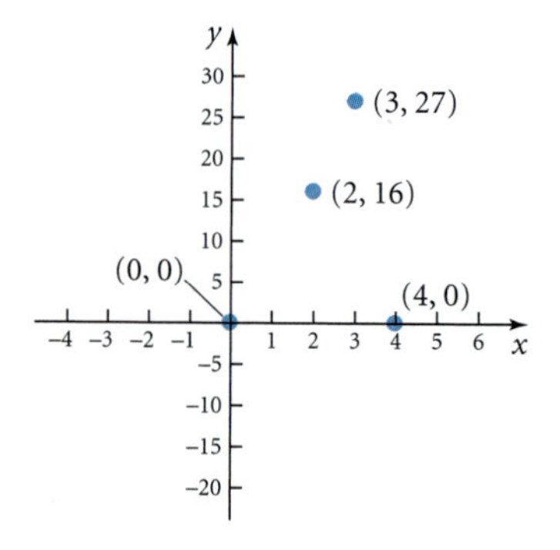

Figure 15.12

Now connect the dots, using the information about direction and concavity. See Figure 15.13.

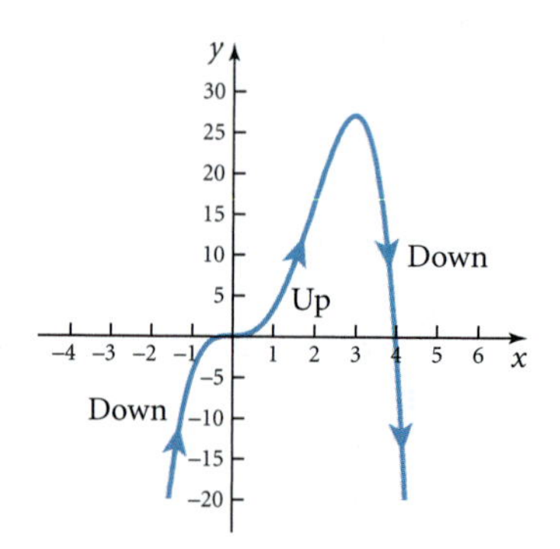

Figure 15.13 ■

Example 8: Use the Derivative Tests to analyze and sketch the graph of $f(x) = x^3 - 12x + 2$.

Solution:

Step 1: Because $f(x)$ is a polynomial, the domain is all real numbers and there are no asymptotes. To determine the y-intercept, set $x = 0$. Because $f(0) = 2$, the y-intercept is $(0, 2)$. To determine the x-intercept, set $f(x) = 0$. The equation $0 = x^3 - 12x + 2$ does not factor easily, but by using your calculator you can obtain approximate solutions of $x \approx 0.167, 3.38, -3.54$. Thus, the x-intercepts are $(0.167, 0)$, $(3.38, 0)$, and $(-3.54, 0)$.

Calculator Corner 15.1

To solve $x^3 - 12x + 2 = 0$ on the TI-84+ Silver, press **APPS** and select :PolySmlt from the alphabetized menu. Press **ENTER ENTER ENTER**. Enter 3 as Degree of Polynomial. Then enter the coefficients: $a_3 = 1, a_2 = 0, a_1 = -12, a_0 = 2$. Press **SOLVE** and the solutions will appear on the home screen.

Step 2: Because $f(x) = x^3 - 12x + 2, f'(x) = 3x^2 - 12$. To determine the critical values, set $f'(x) = 0$:

$$3x^2 - 12 = 0 \rightarrow 3(x + 2)(x - 2) = 0$$

So, $x = -2, 2$ are the critical values. The number line analysis yields

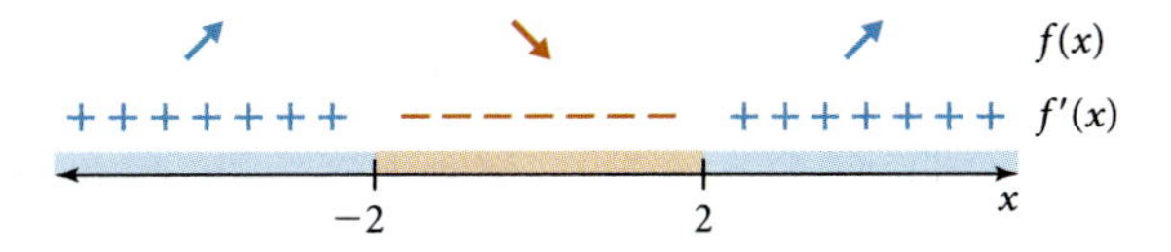

Thus, $f(x)$ increases for $(-\infty, -2)$ and $(2, \infty)$ and decreases for $(-2, 2)$. There are extreme points at $x = -2$ and $x = 2$. The point $(-2, 18)$ is a relative maximum, and the point $(2, -14)$ is a relative minimum.

Step 3: $f(x) = x^3 - 12x + 2 \rightarrow f''(x) = 6x$

Setting $f''(x) = 0$ gives $6x = 0 \rightarrow x = 0$. The number line analysis yields

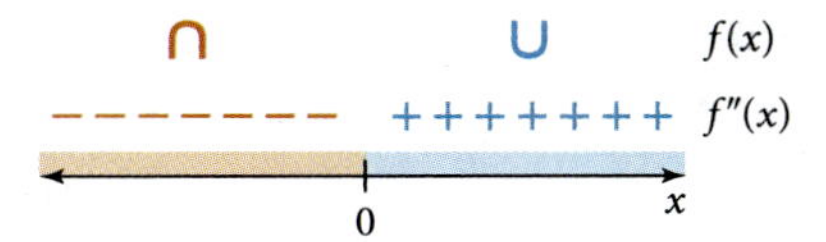

Thus, $f(x)$ is concave down for $(-\infty, 0)$ and concave up for $(0, \infty)$. The point of inflection is $(0, 2)$.

Step 4: Plot the intercepts, the extreme point, and the points of inflection. See Figure 15.14.

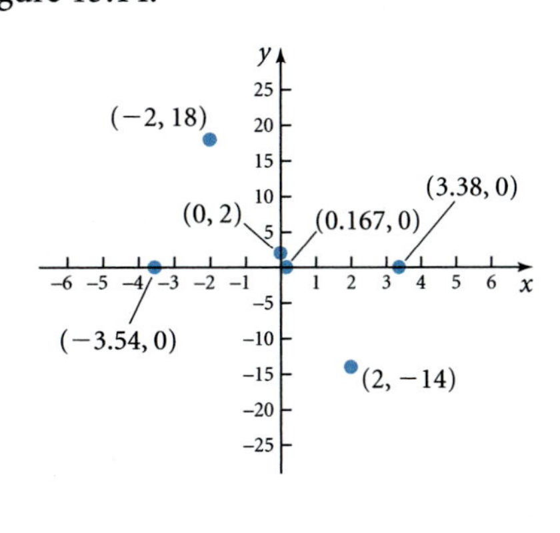

Figure 15.14

Now connect the dots, as shown in Figure 15.15, using the information on direction and shaping obtained above.

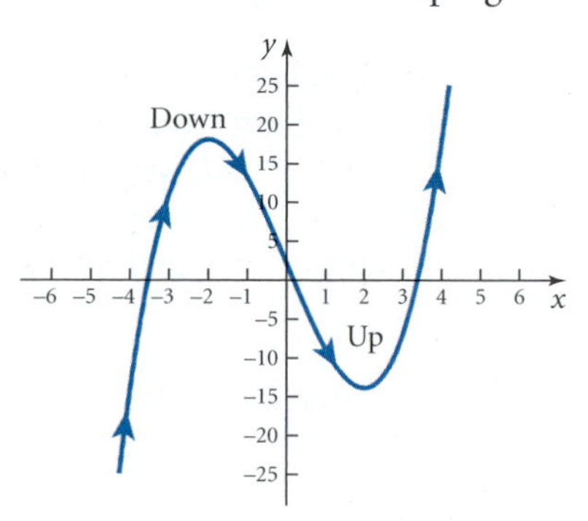

Figure 15.15 ■

Example 9: Use the Derivative Tests to analyze and sketch the graph of $f(x) = \frac{2}{x+3}$.

Solution:

Step 1: The domain is all $x \neq -3$. Because $f(0) = \frac{2}{3}$, the y-intercept is $\left(0, \frac{2}{3}\right)$. Because $f(x) = \frac{2}{x+3} \neq 0$, there is no x-intercept. The asymptotes are $x = -3$ and $y = 0$.

Step 2: $f(x) = \frac{2}{x+3} \rightarrow f'(x) = \frac{-2}{(x+3)^2}$

Because $f'(x) = \frac{-2}{(x+3)^2} \neq 0$, the only critical value is $x = -3$. The number line analysis yields

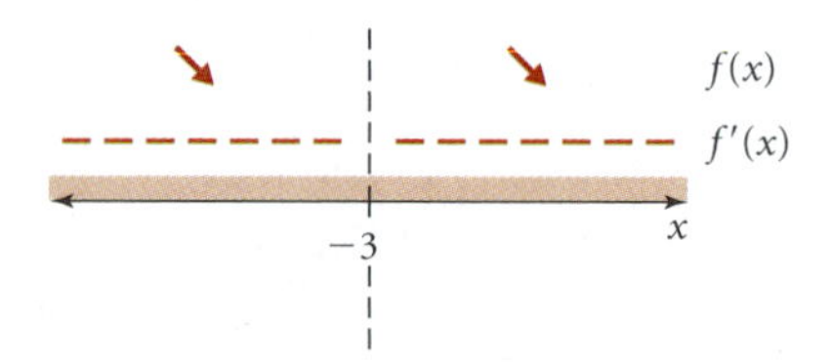

Thus, $f(x)$ decreases for $(-\infty, -3)$ and $(3, \infty)$. There is no extreme point because the direction does not change.

Step 3: $f(x) = \frac{2}{x+3} \rightarrow f''(x) = \frac{4}{(x+3)^3}$

Because $f''(x) = \frac{4}{(x+3)^3} \neq 0$, the only possible inflection point is at $x = -3$. The number line analysis yields

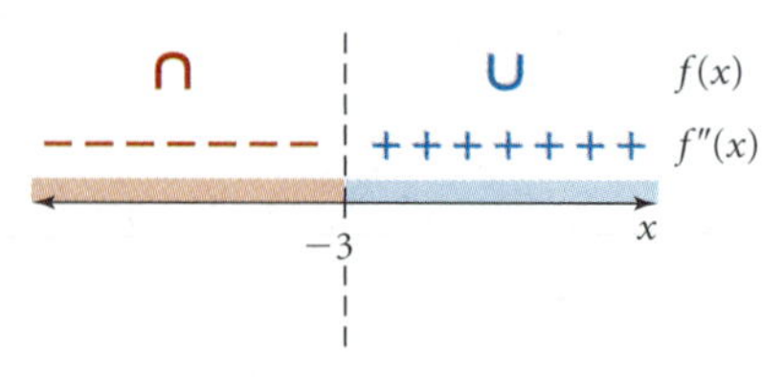

Thus, $f(x)$ is concave down for $(-\infty, -3)$ and concave up for $(3, \infty)$. There is no point of inflection because $f(x)$ is discontinuous at $x = -3$.

Step 4: The only items to plot for this graph are the asymptotes and the intercept, as shown in Figure 15.16.

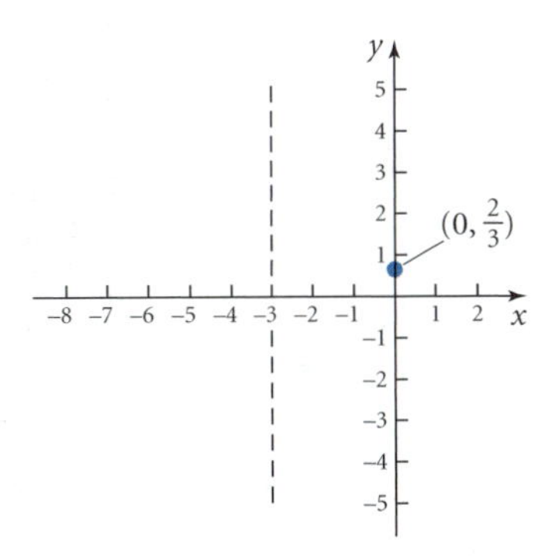

Figure 15.16

When $x > -3$, the curve must decrease, be concave up, and fit between the asymptotes. When $x < -3$, the curve must decrease, be concave down, and fit between the asymptotes. There is no point available for the left side of the asymptote, so evaluating $f(-4)$ and plotting a point at $(-4, -2)$ may help in sketching. See Figure 15.17.

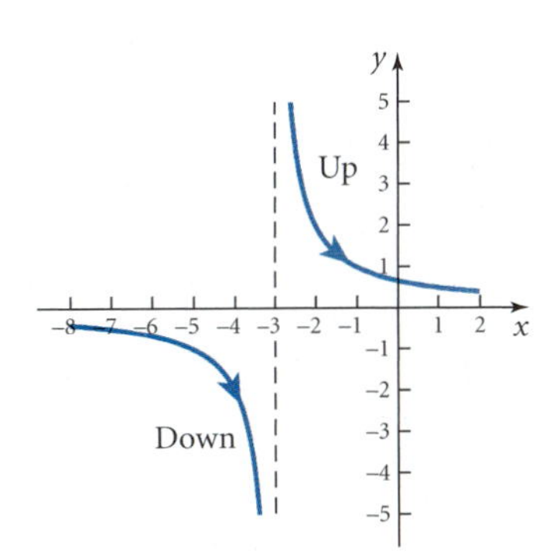

Figure 15.17 ■

Example 10: Use the Derivative Tests to analyze and sketch $f(x) = \frac{2}{x^2 - 4}$.

Solution

Step 1: The domain is all $x \neq -2, 2$. Because $f(0) = -\frac{1}{2}$, $\left(0, -\frac{1}{2}\right)$ is the y-intercept. Because $f(x) = \frac{2}{x^2 - 4} \neq 0$, there is no x-intercept. The asymptotes are $x = -2$, $x = 2$, and $y = 0$.

Step 2: $f(x) = \frac{2}{x^2 - 4} \rightarrow f'(x) = -\frac{4x}{(x^2 - 4)^2}$

$$f'(x) = -\frac{4x}{(x^2 - 4)^2} = 0 \rightarrow x = 0$$

Thus, the critical values are $x = 0, -2, 2$. The number line analysis yields

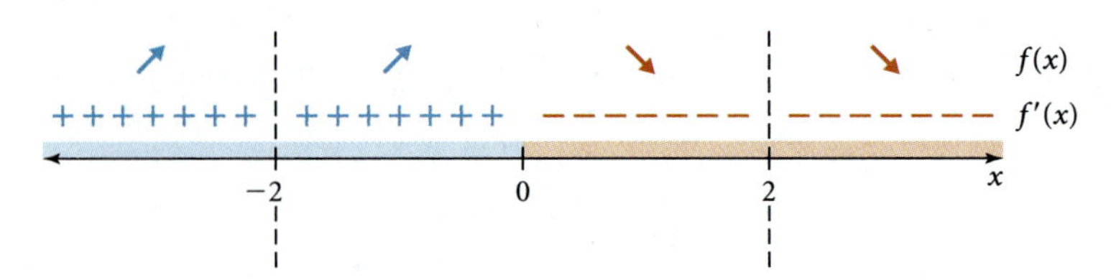

Thus, $f(x)$ increases for $(-\infty, -2)$ and $(-2, 0)$ and decreases for $(0, 2)$ and $(2, \infty)$. There is an extreme point at $x = 0$. The point $\left(0, -\frac{1}{2}\right)$ is a relative maximum.

Step 3: Using the Quotient Rule,

$$f(x) = \frac{2}{x^2 - 4} \rightarrow f''(x) = \frac{4(4 + 3x^2)}{(x^2 - 4)^3}$$

Because $f''(x) = \frac{4(4 + 3x^2)}{(x^2 - 4)^3} \neq 0$, the only possible inflection points are at the asymptotes, $x = -2$ and $x = 2$. The number line analysis yields

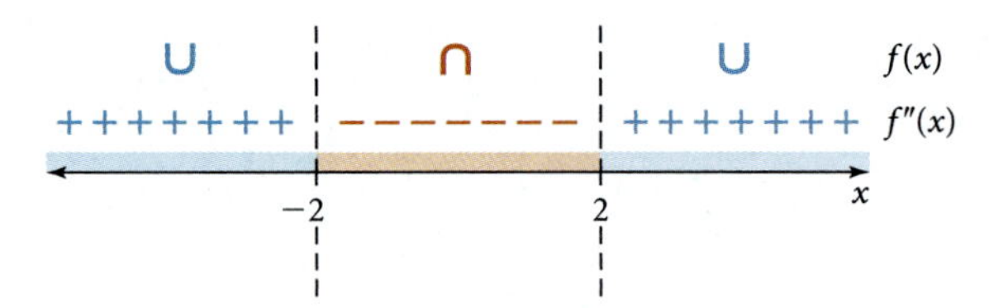

Thus, $f(x)$ is concave up for $(-\infty, -2)$ and $(2, \infty)$ and concave down for $(-2, 2)$. There are no points of inflection.

Step 4: The only items to plot are the asymptotes and the y-intercept. Evaluating $f(3)$ and $f(-3)$ may help in graphing. We obtain $f(3) = 0.4$ and $f(-3) = 0.4$. See Figure 15.18.

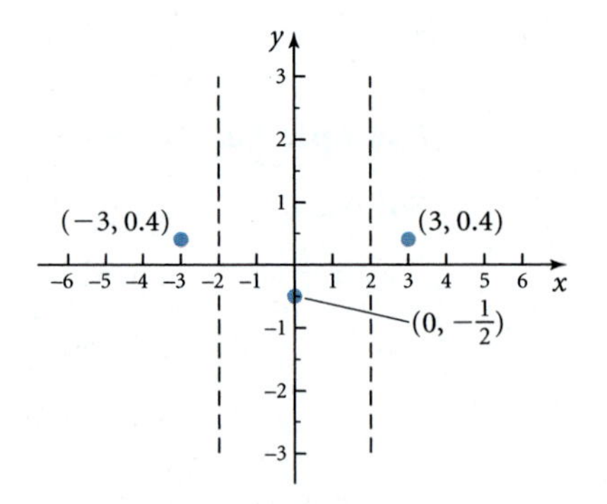

Figure 15.18

To connect the dots, we know that where $x < -2$, the graph must increase, be concave up, and fit between the asymptotes. Where $x > 2$, the graph must decrease, be concave up, and fit between the asymptotes. Where $-2 < x < 2$, the graph will increase for $(-2, 0)$ and decrease for $(0, 2)$ with a relative maximum point at $\left(0, -\frac{1}{2}\right)$, as shown in Figure 15.19.

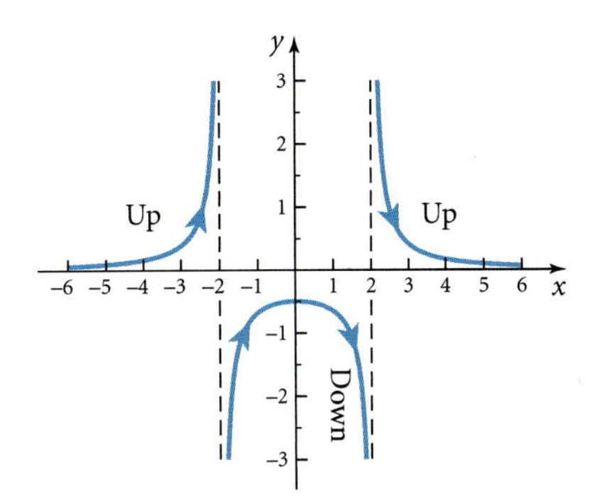

Figure 15.19 ■

Example 11: Use the Derivative Tests to analyze and sketch $f(x) = xe^{-x}$.

Solution:

Step 1: The domain is all real numbers. Because $f(0) = 0$, the y-intercept is $(0, 0)$. There is an asymptote along the positive x-axis because of e^{-x}.

Step 2: $f(x) = xe^{-x} \rightarrow f'(x) = e^{-x}(1 - x)$.

If $f'(x) = e^{-x}(1 - x) = 0$

then $x = 1$ is the critical value. The number line analysis yields

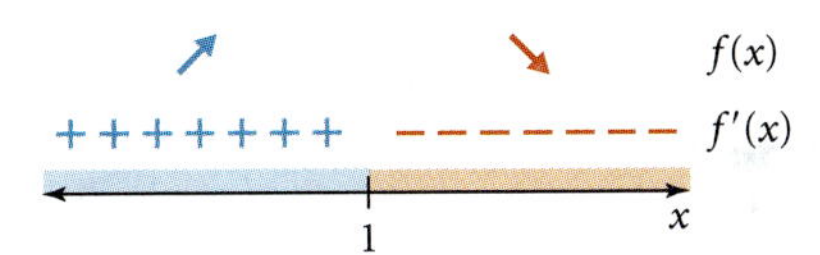

Thus, $f(x)$ increases for $(-\infty, 1)$ and decreases for $(1, \infty)$. The extreme point is at $x = 1$, so $(1, e^{-1}) \approx (1, 0.4)$ is the absolute maximum.

Step 3: $f(x) = xe^{-x} \rightarrow f''(x) = e^{-x}(x - 2)$

If $f''(x) = e^{-x}(x - 2) = 0$, then $x = 2$.

The number line analysis yields

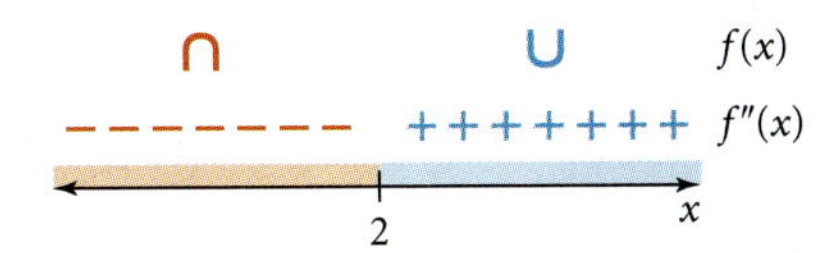

Thus, $f(x)$ is concave down for $(-\infty, 2)$ and concave up for $(2, \infty)$. The concavity changes at $x = 2$, so the point of inflection is $(2, 2e^{-2}) \approx (2, 0.27)$.

Step 4: Plot the intercept, the extreme point, and the point of inflection, as shown in Figure 15.20.

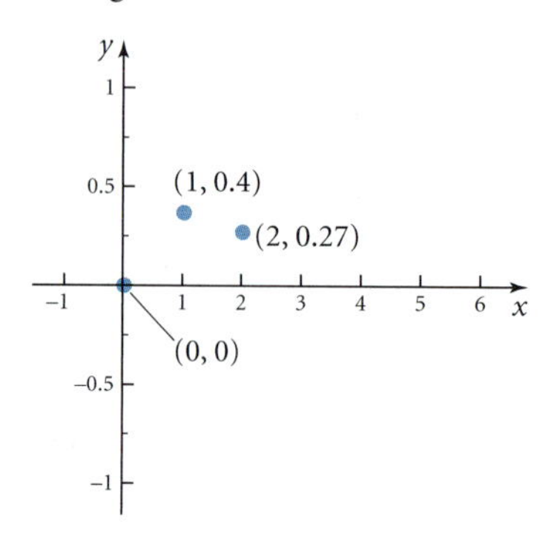

Figure 15.20

Connecting the dots using the direction and shaping information gives the graph in Figure 15.21.

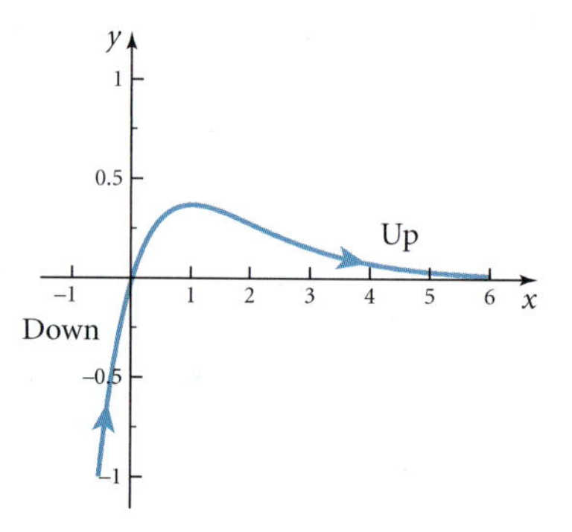

Figure 15.21

Example 12: *Point of Diminishing Returns* Sales of a new electronic game t months after its release are modeled by $S(t) = \dfrac{80{,}000}{1 + 4{,}000e^{-0.4t}}$, where S is the total number of games sold. The curve is logistic, which means that sales rise rapidly at first and then taper off to a limiting value of 80,000. Therefore,

$$S'(t) = \frac{128{,}000{,}000e^{-0.4t}}{(1 + 4{,}000e^{-0.4t})^2}$$

and $S'(t) > 0$ for all values of t (why?), so we know that sales continually increase and there is no maximum. Then

$$S''(t) = \frac{-51{,}200{,}000e^{-0.4t}(1 - 4000e^{-0.4t})}{(1 + 4000e^{-0.4t})^3}$$

To find the point of inflection, we must set $S''(t) = 0$:

$$S''(t) = \frac{-51{,}200{,}000e^{-0.4t}(1 - 4000e^{-0.4t})}{(1 + 4000e^{-0.4t})^3} = 0$$

which means that $-51{,}200{,}000e^{-0.4t}(1 - 4000e^{-0.4t}) = 0$ because the denominator cannot have a value of zero.

We also know, however, that $-51{,}200{,}000e^{-0.4t} \neq 0$. Thus, we must solve

$$1 - 4000e^{-0.4t} = 0$$

$$e^{-0.4t} = \frac{1}{4000} \quad \text{Isolate the exponential expression}$$

$$-0.4t = \ln\left(\frac{1}{4000}\right) \quad \text{Apply ln to each side}$$

$$t = \frac{\ln\left(\frac{1}{4000}\right)}{-0.4} \approx 20.74 \quad \text{Solve for } t$$

Sales are concave up for approximately the first 21 months and then become concave down. In other words, the point of inflection represents the time at which the growth in sales begins to taper off. Economists refer to this point, where sales continue to grow but at a slower rate, as the **point of diminishing returns.**

Figure 15.22 shows the logistic curve describing sales of the game. You can see the rapid increase in sales at the beginning and then the tapering off of the sales. Sales continue to increase and approach the limiting value represented by the asymptote. Figure 15.23 shows the second derivative of the logistic function describing game sales. During the first 21 months, the positive second derivative indicates that the increase in sales increases. After 21 months, the negative second derivative indicates the slowing down of the increase in sales.

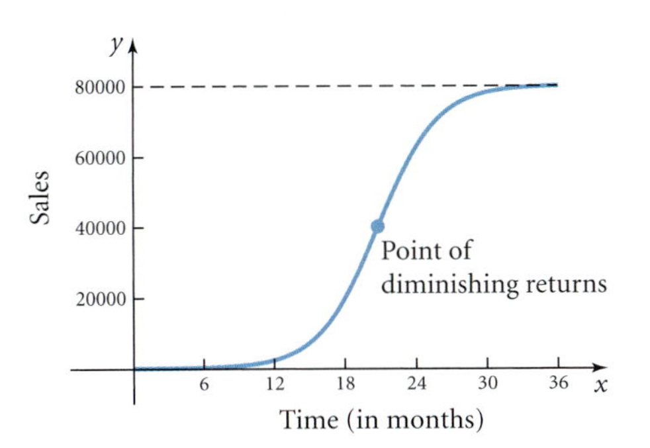

Figure 15.22
Graph of $S(t)$.

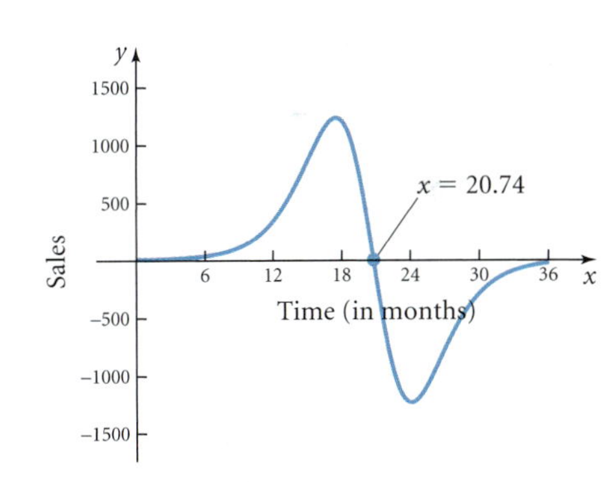

Figure 15.23
Graph of $S''(t)$.

Check Your Understanding 15.2

Use the Derivative Tests to analyze and sketch the graph of each function.

1. $f(x) = x^4 - 4x^3$ **2.** $f(x) = \dfrac{x}{x^2 - 9}$

Check Your Understanding Answers

Check Your Understanding 15.1

1. concave down for $(-\infty, 0)$ and concave up for $(0, \infty)$; point of inflection is at $x = 0$, or the point $(0, -2)$
2. concave up for $(-\infty, -1)$ and $(1, \infty)$; concave down for $(-1, 1)$; points of inflection are at $x = -1$ and $x = 1$, or the points $(-1, -5)$ and $(1, -5)$.
3. concave up for all x; no point of inflection
4. concave down for $(-\infty, 3)$ and concave up for $(3, \infty)$; no point of inflection because $x = 3$ is the asymptote

Check Your Understanding 15.2

1. Domain is all real numbers. Intercepts are $(0, 0)$ and $(4, 0)$. Decreases for $(-\infty, 3)$ and increases for $(3, \infty)$. Direction changes at $x = 3$, so the absolute minimum is $(3, -27)$. Concave up for $(-\infty, 0)$ and $(2, \infty)$; concave down for $(0, 2)$. Concavity changes at $x = 0$ and $x = 2$, so the points of inflection are $(0, 0)$ and $(2, -16)$.

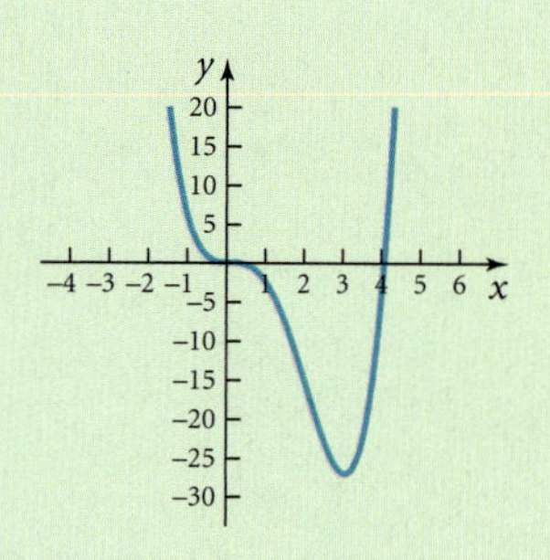

2. Domain is all $x \neq -3, 3$. Intercept is $(0, 0)$. Asymptotes are $x = -3, x = 3, y = 0$. Decreases for $(-\infty, -3)$, $(-3, 3)$, and $(3, \infty)$. No extreme points. Concave down for $(-\infty, -3)$ and $(0, 3)$. Concave up for $(-3, 0)$ and $(3, \infty)$. Concavity changes at $x = 0$, so $(0, 0)$ is the point of inflection.

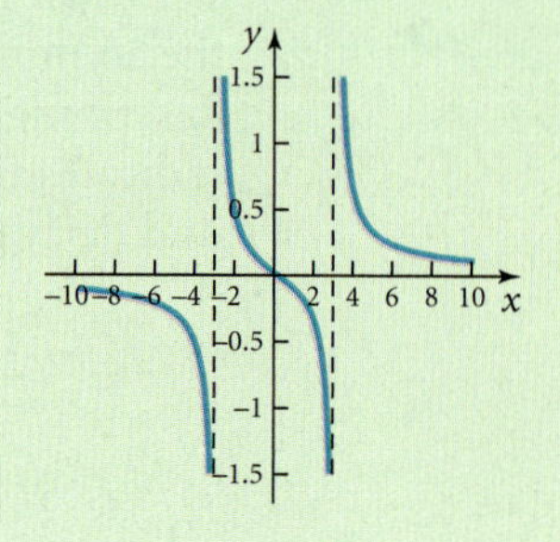

Topic 15 Review

15

This topic discussed the third way of describing the graph of a function: through its shaping, or bend. You were introduced to concavity, points of inflection, and the Second Derivative Test.

CONCEPT	EXAMPLE
The shaping, or bend, of a graph is called the **concavity** of the graph. A graph is **concave up** on an interval if all the tangent lines to the curve on that interval are below the graph. A graph is **concave down** on an interval if all the tangent lines to the curve on that interval are above the graph.	The graph of $f(x) = x^2 - 2$ is concave up for all values of x. 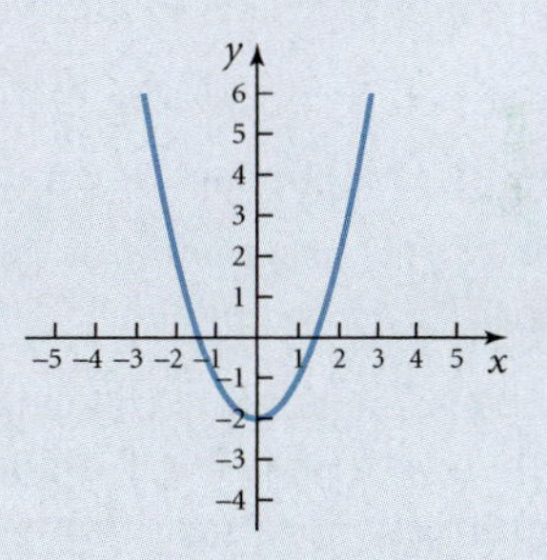 The graph of $f(x) = \ln x$ is concave down for all $x > 0$. 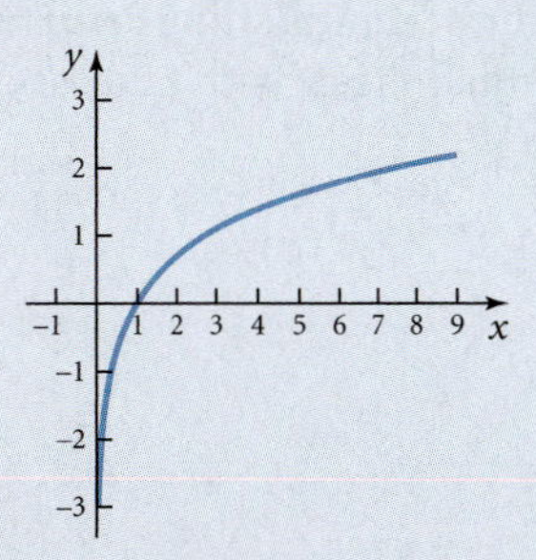 The graph of $f(x) = x^3 - 3x$ is concave down for $x < 0$ and concave up for $x > 0$. 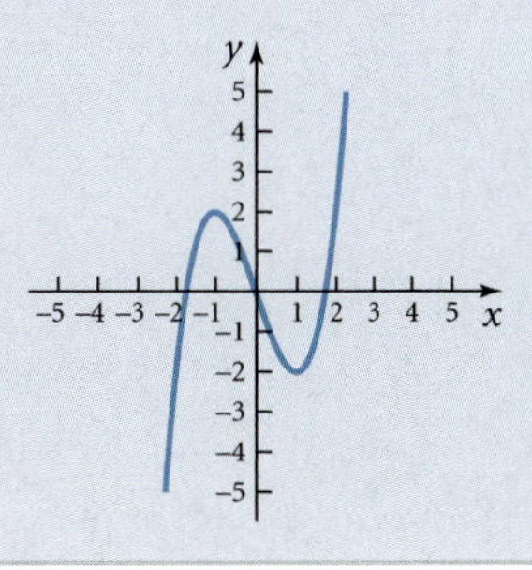

(continued)

Any point where the concavity of a graph changes is called a **point of inflection.**

The graph of $f(x) = x^3 - 3x$ is concave down for $x < 0$ and concave up for $x > 0$.

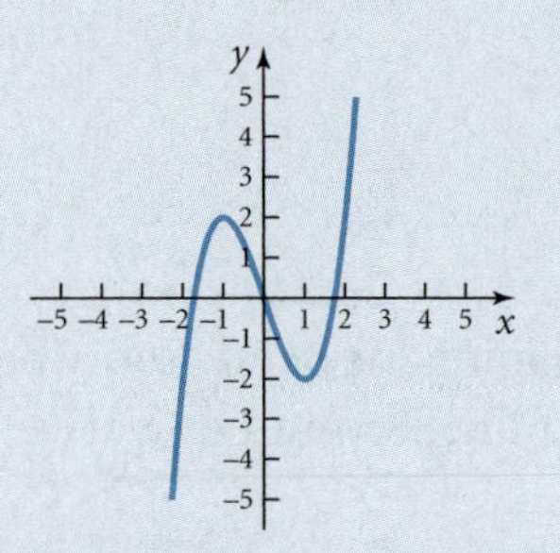

The point of inflection is (0, 0).

The relationship between a function and its second derivative is given by the **Second Derivative Test.**

- The graph of $f(x)$ is concave up if $f''(x) > 0$.
- The graph of $f(x)$ is concave down if $f''(x) < 0$.
- If the graph of $f(x)$ has a point of inflection, then $f''(x) = 0$ or $f''(x)$ is undefined.

Suppose $(c, f(c))$ is a possible extreme point of $f(x)$. If $f''(c) > 0$, then $(c, f(c))$ is a **minimum point.** If $f''(c) < 0$, then $(c, f(c))$ is a **maximum point.**

Given $f(x) = x^3 - 3x, f'(x) = 3x^2 - 3 = 0$. So $x = -1, 1$ are the critical values.

$f''(x) = 6x$

$f''(-1) = -6 < 0$, so $(-1, f(-1))$ or $(-1, 2)$ is the relative maximum.

$f''(1) = 6 > 0$, so $(1, f(1))$ or $(1, -2)$ is the relative minimum.

In economics, one application of the second derivative is the **point of diminishing returns,** which is the point at which the growth of a function begins to slow down.

Total sales of a new product are modeled by $S(x) = \dfrac{500}{1 + 10e^{-0.1x}}$, where x is the number of weeks after the release of the product.

$$S''(x) = \frac{50e^{-0.1x}(10e^{-0.1x} - 1)}{(1 + 10e^{-0.1x})^3}$$

$$S''(x) = 0 \rightarrow x \approx 23$$

Sales are concave up for the first 23 weeks and then become concave down. After 23 weeks, sales continue to grow, but at a slower rate.

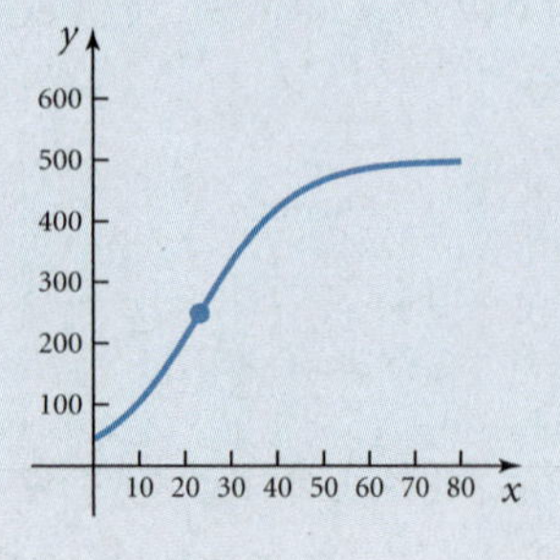

NEW TOOLS IN THE TOOL KIT

- Describing a graph using its concavity
- Identifying intervals for which the graph of a function is concave up or concave down
- Determining points of inflection
- Interpreting a point of inflection
- Using the Second Derivative Test to find intervals of concavity and points of inflection
- Using the Second Derivative Test to determine whether a possible extreme point is a maximum point or a minimum point
- Sketching the graph of a function using information about its location, direction, extreme points, concavity, and points of inflection

Topic 15 Exercises 15

For each graph in Exercises 1 and 2, state the intervals for which

a. $f(x)$ is concave up. **b.** $f(x)$ is concave down.

1.

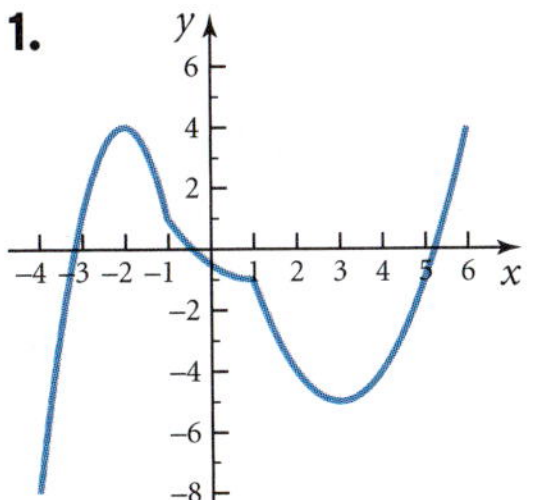

2.

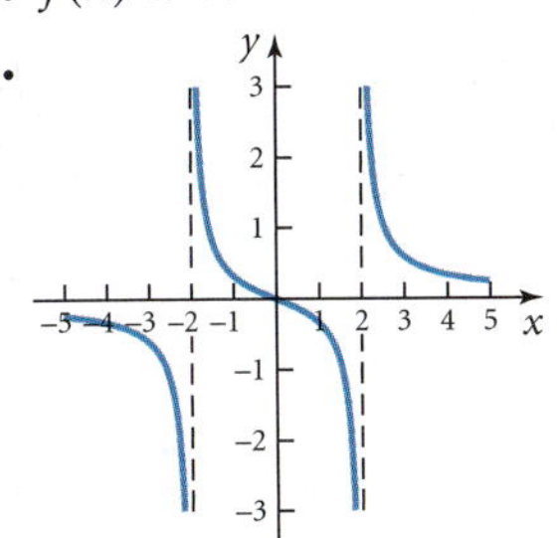

For the graphs in Exercises 3 and 4, identify the intervals for which

a. $f''(x) > 0$. **b.** $f''(x) < 0$.

3.

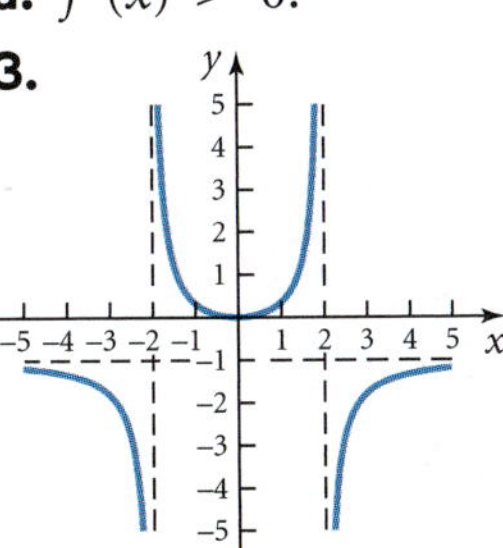

4.

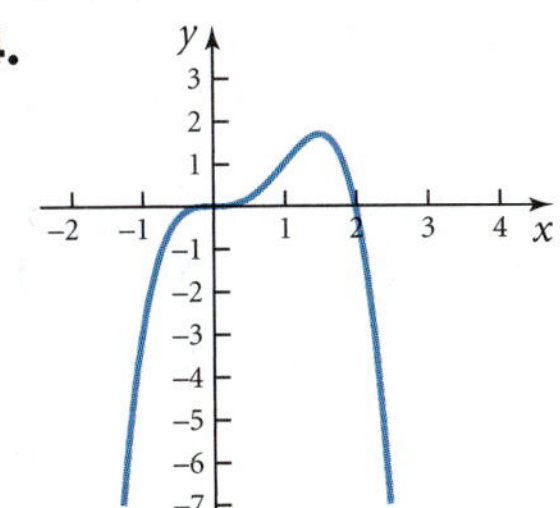

In Exercises 5 and 6, identify the points of inflection for each graph.

5.

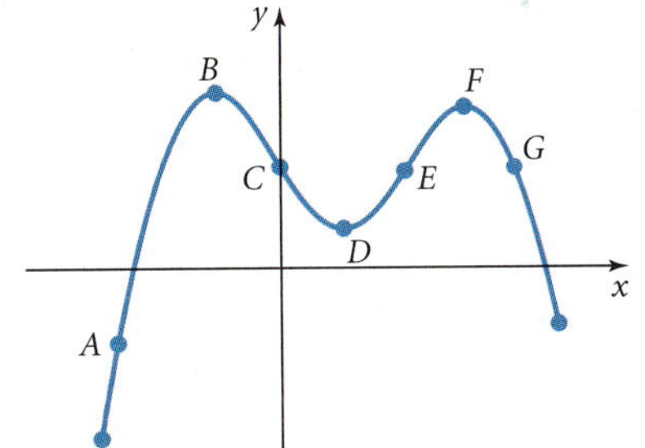

6.

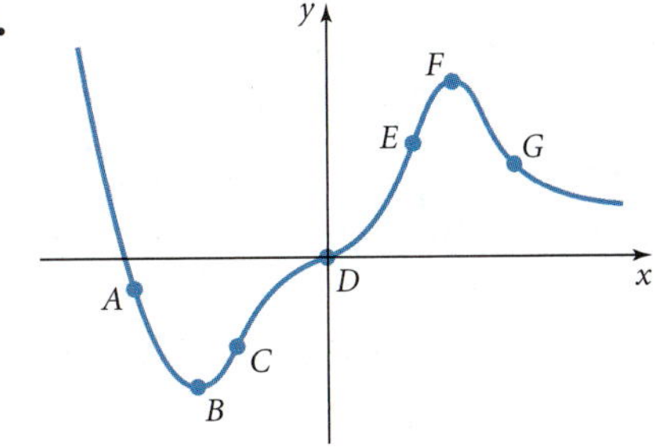

Use the graph to determine if the following statements are true or false.

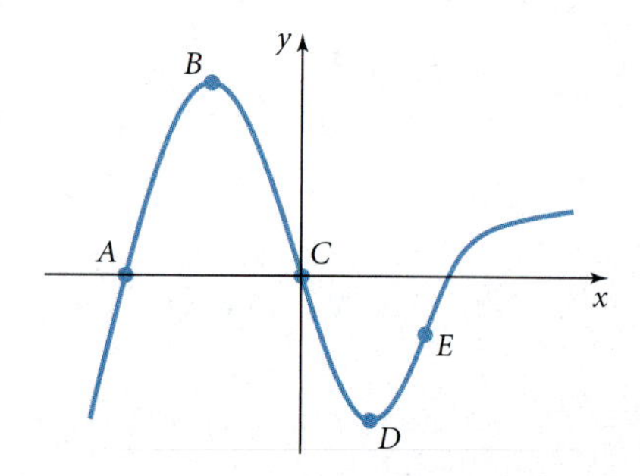

7. C is a point of inflection

8. $f''(x) < 0$ between A and C

9. $f''(D) < 0$

10. $f''(E) = 0$

11. $f''(A) = 0$

12. $f''(B) < 0$

13. $f''(x) > 0$ between C and E

14. A is a point of inflection

For the functions in Exercises 15 through 40, use the Second Derivative Test to find

a. The interval(s) for which $f(x)$ is concave up.

b. The interval(s) for which $f(x)$ is concave down.

c. The point(s) of inflection.

15. $x^3 - 12x$

16. $x^3 - 3x$

17. $x^4 - 16x^3$

18. $x^4 - x^5$

19. $x^3 - 3x^2 - 9x + 7$

20. $2x^3 + 4x^2 - 5x + 1$

21. $x^4 + 8x^3 + 18x^2 + 8$

22. $x^4 + 2x^3 - 36x^2 + 6x - 5$

23. $10x^3 - 3x^5$

24. $3x^5 - 5x^4$

25. $(2x - 6)^4$

26. $(3x + 2)^5$

27. $(x^2 - 4)^3$

28. $(x^2 - 1)^4$

29. $\dfrac{2}{x - 5}$

30. $-\dfrac{3}{x + 1}$

31. $\dfrac{2x}{x + 5}$

32. $\dfrac{4x}{2x + 3}$

33. $\dfrac{x}{x^2 + 9}$

34. $\dfrac{-3x}{x^2 + 1}$

35. $\sqrt{x - 5}$

36. $\sqrt[3]{x + 4}$

37. x^2e^x

38. $x^2 \ln x$

39. $2x \ln x$

40. xe^{-x}

For Exercises 41 through 44,

a. Determine the possible extreme points.

b. Use the Second Derivative test to determine what type of extreme point exists.

41. $x^4 - 16x^3$

42. $x^4 - x^5$

43. x^2e^x

44. $x^2 \ln x$

In Exercises 45 through 60, use the graphing strategy presented in this topic to analyze and sketch the graph of the function.

45. $x^4 - 16x^3$

46. $x^4 - x^5$

47. $x^3 - 3x^2 - 9x + 7$

48. $x^3 - 6x^2 + 9x - 4$

49. $3x^5 - 5x^4$

50. $10x^3 - 3x^5$

51. $(2x - 6)^4$

52. $(3x + 2)^5$

53. $\dfrac{2x^2}{x^2 - 9}$

54. $\dfrac{x}{x^2 - 4}$

55. $\dfrac{3x}{x^2 + 4}$

56. $\dfrac{3}{x^2 + 1}$

57. $\ln(x^2 + 2)$

58. $\ln(x^2 - 1)$

59. $x^2e^{-0.5x}$

60. xe^{-x}

In Exercises 61 through 64, use the information provided to sketch the graph of the function.

61. $f(0) = 0, f(-4) = 0, f(4) = 0$
$f(-2) = 4, f(2) = -4$
increases for $(-\infty, -2)$ and $(2, \infty)$
decreases for $(-2, 2)$
concave up for $(0, \infty)$
concave down for $(-\infty, 0)$
point of inflection at $(0, 0)$

62. $f(0) = 0$
asymptotes at $x = -3, x = 3, y = 0$
increases for $(-\infty, -3), (-3, 3), (3, \infty)$
no extreme points
concave up for $(-\infty, -3)$ and $(0, 3)$
concave down for $(-3, 0)$ and $(3, \infty)$
point of inflection at $(0, 0)$

63. $f(1) = 0$
asymptotes at $x = 0, y = 0$
increases for $(-\infty, 0)$ and $(2, \infty)$
decreases for $(0, 2)$
absolute minimum at $(2, -3)$
concave up for $(-\infty, 0)$ and $(0, 3)$
concave down for $(3, \infty)$
point of inflection at $(3, -1)$

64. $f(-3) = 0$
asymptote at $y = 0$
absolute maximum at $(0, 3)$
point of inflection at $(1, 2)$
increases for $(-\infty, 0)$
decreases for $(0, \infty)$
concave up for $(1, \infty)$
concave down for $(-\infty, 1)$

65. After an outbreak of measles, the number of cases reported each week grows according to the following graph. Use the graph to estimate when the Centers for Disease Control could say that the rate of growth in new cases was beginning to taper off.

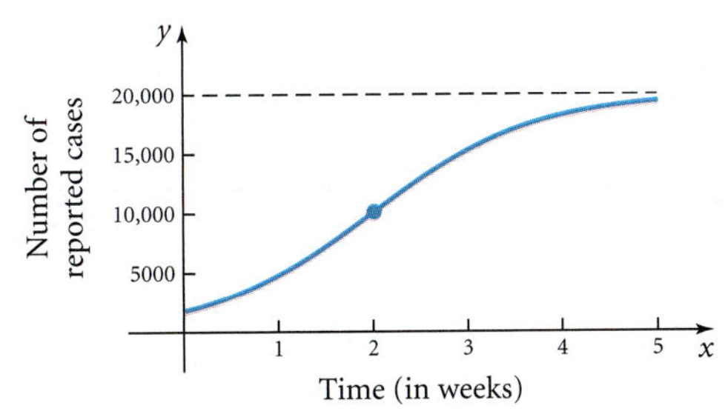

66. The annual sales, in billions, of a large grocery chain is shown in the graph. Use the graph to estimate when the rate of change of sales was at its maximum. When was the rate of change of sales at its minimum? Assume $t = 0$ represents the year 2000.

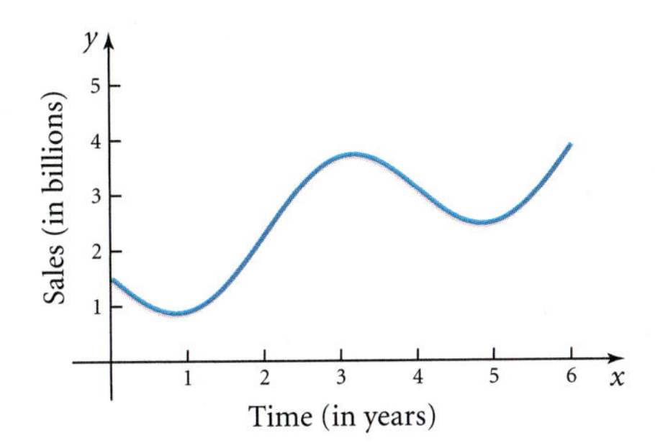

67. The concentration in milligrams per cubic centimeters of a drug in a patient's bloodstream t hours after injection is $C(t) = 0.8te^{-0.45t}$.

a. For what time interval does the concentration increase?

b. For what time interval does the concentration decrease?

c. For what time interval is the concentration concave up? Interpret in the context of this problem.

68. The concentration in milligrams per cubic centimeters of a drug in a patient's bloodstream t hours after injection is $C(t) = 2.6te^{-0.3t}$.

a. For what time interval does the concentration increase?

b. For what time interval does the concentration decrease?

c. For what time interval is the concentration concave down? Interpret in the context of this problem.

69. The proportion of the U.S. workforce that is 16 years of age or older has increased continually since 1900 and can be modeled by $P(t) = \dfrac{76.4}{1 + 3.5e^{-0.018t}}$, where t is the number of years since 1900. When did the growth rate start to taper off?

70. Total sales, in thousands, of a new John Grisham best seller t weeks after its release are given by $S(t) = \dfrac{94.54}{1 + 2.98e^{-0.1t}}$. When did the rate of sales begin to taper off?

SHARPEN THE TOOLS

The graph below represents the first derivative, $f'(x)$, of a function. Use the graph to answer Exercises 71 through 76.

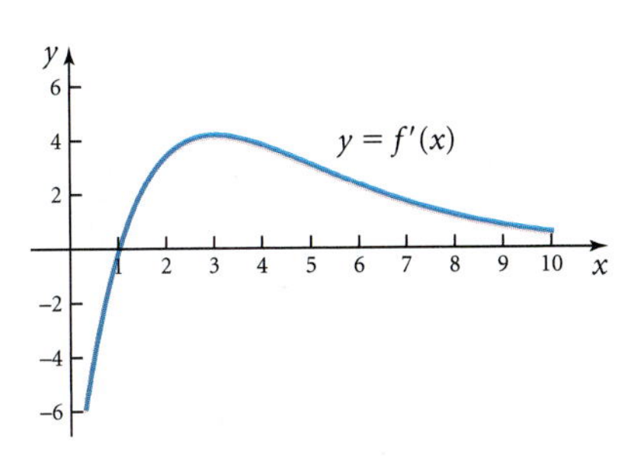

71. For what value(s) of x is $f(x)$ concave down?

72. For what value(s) of x does $f(x)$ increase?

73. For what value(s) of x does $f(x)$ decrease?

74. For what value(s) of x is $f(x)$ concave up?

75. For what value(s) of x does $f(x)$ have a point of inflection?

76. For what value(s) of x does $f(x)$ have an extreme point?

For Exercises 77 through 84, use the graphing strategy to analyze and sketch the graph of each function.

77. $f(x) = x^5 - 5x^4 + 5x^3 - 23$

78. $f(x) = x^2e^{-0.5x}$

79. $f(x) = \sqrt[3]{x^2} - 1$

80. $f(x) = 8x - 10x^{4/5}$

81. $f(x) = \dfrac{2x^2}{x^2 + 9}$

82. $f(x) = x^2 \ln x$

83. $f(x) = \dfrac{1}{x} - \dfrac{1}{x^2}$

84. $f(x) = \dfrac{x^3 + x + 4}{x^2 + 1}$

CALCULATOR CONNECTION

For Exercises 85 and 86, use the graphing strategy and your calculator to analyze and sketch a graph of each function.

85. $f(x) = \dfrac{e^x - 2}{x - 1}$

86. $f(x) = \dfrac{3}{\ln(x^2)}$

The small town of DeBary, Florida, has seen a boom of growth in recent years. A January 25, 2004, article in the *Daytona Beach News-Journal* stated that "DeBary is growing rapidly. . . . However, since 2002, people are still drawn to DeBary, but at a slower rate."

87. Explain the relationship of first and second derivatives to this statement.

88. The actual population data for DeBary is given in the table.

Year	Population
1990	7,176
2000	15,559
2002	15,925
2003	16,127
2004	17,027
2005	17,052

Data from: Daytona Beach News-Journal (January 25, 2004, p. 3I)

Find a logistic equation that models the data. When did the growth rate begin to taper off? Does your result support the article's statement?

TOPIC

Absolute Extrema

16

Consider the cubic function $R(p) = -5p^3 + 55p^2 + 60p$, which has the set of all real numbers for its domain. Using the First and Second Derivative Tests, you can show that the relative minimum is at approximately $(-0.51, R(-0.51))$ and the relative maximum is at approximately $(7.84, R(7.84))$ (see Figure 16.1). Suppose, though, $R(p) = -5p^3 + 55p^2 + 60p$ represented the revenue of a product sold for $\$p$. It would no longer make sense to use the set of real numbers as the domain. Rather, the domain would be $p \geq 0$. How would that change in the domain affect the extreme points? Figure 16.2 shows that $(-0.51, R(-0.51))$ could no longer be the minimum point because it is outside the domain of the function. In this topic, you learn a technique for determining absolute extreme points of functions that are defined on specific intervals. Before starting your study of absolute extrema, complete the Warm-up Exercises to sharpen your algebra and calculus skills.

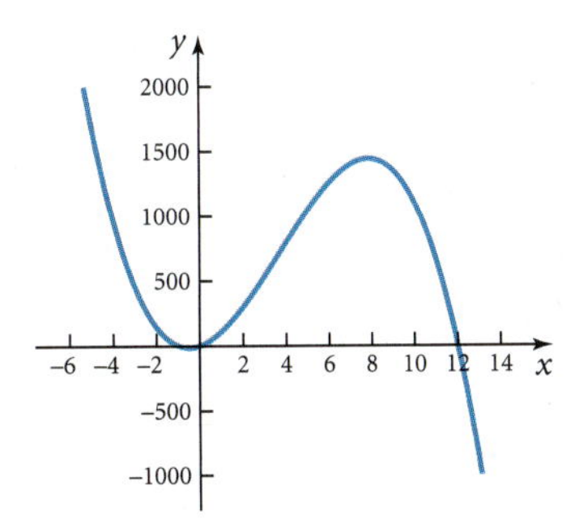

Figure 16.1
Graph of $R(p)$ with domain all real numbers.

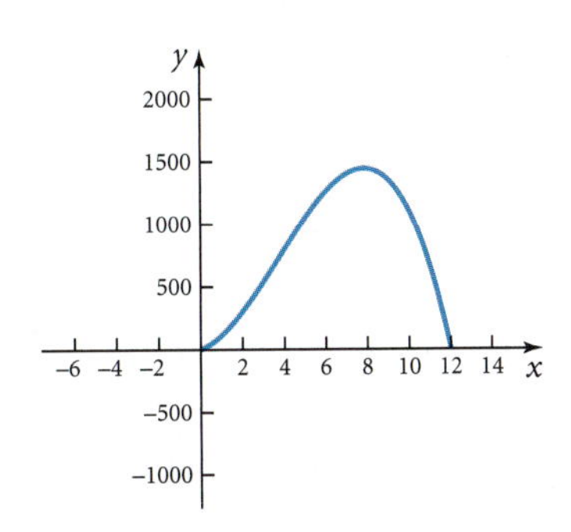

Figure 16.2
Graph of $R(p)$ with $p \geq 0$, $R(p) \geq 0$.

IMPORTANT TOOLS IN TOPIC 16

- *Absolute maximum value of a function on an interval*
- *Absolute minimum value of a function on an interval*
- *Elasticity of demand*
- *Maximum revenue*

TOPIC 16 WARM-UP EXERCISES

Be sure you can successfully complete these exercises before starting Topic 16.

Algebra Warm-up Exercises

1. Given $f(x) = 2x^3 - 5x^2 + 7x - 4$, find each of the following values.

a. $f(1)$ **b.** $f(0)$ **c.** $f(-3)$

Sketch the graph of each function on the indicated domain.

2. $f(x) = x^3 - 4x$

a. on real numbers **b.** on $[-1, 2]$

3. $f(x) = \sqrt{x + 2}$

a. on $[-2, \infty)$ **b.** on $[0, 7]$

Solve each of the following equations for x.

4. $2x^2 - x - 6 = 0$

5. $3x^3 - 12x = 0$

6. $\dfrac{x^2 - 1}{x^2 + 6} = 0$

7. $2 - \dfrac{8}{x^2} = 0$

8. $3x^2 + 11x + 2 = 0$

9. For what x interval does $f(x) = -4x^2 + 28x + 120$ have a positive value?

Calculus Warm-up Exercises

For each of the following functions, identify the interval(s) where $f(x)$ increases and decreases and identify the extreme points.

1. $f(x) = x^3 - 12x$

2. $f(x) = \dfrac{x}{x^2 - 4}$

Evaluate the following limits.

3. $\lim\limits_{x \to 0^+} \left(3x^2 - \dfrac{4}{x}\right)$

4. $\lim\limits_{x \to \infty} \left(3x^2 - \dfrac{4}{x}\right)$

5. Identify the extreme points in the following graph. Classify each extreme point as maximum or minimum, and absolute or relative.

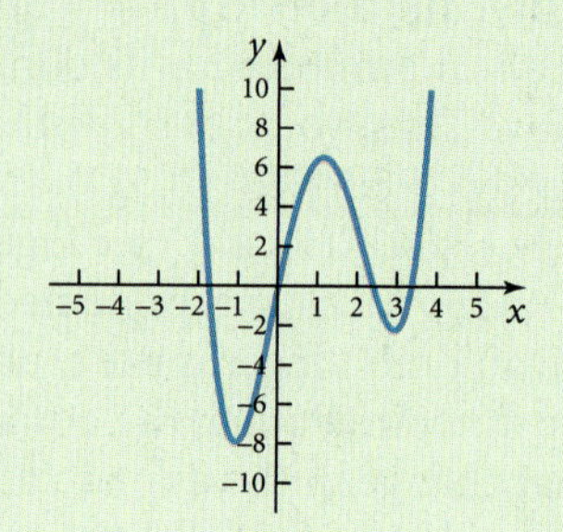

Answers to Warm-up Exercises

Algebra Warm-up Exercises

1. a. 0 **b.** -4 **c.** -124

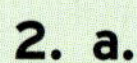

2. a.

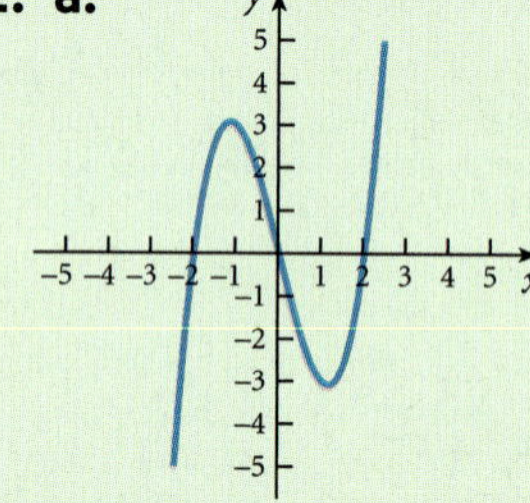

b.

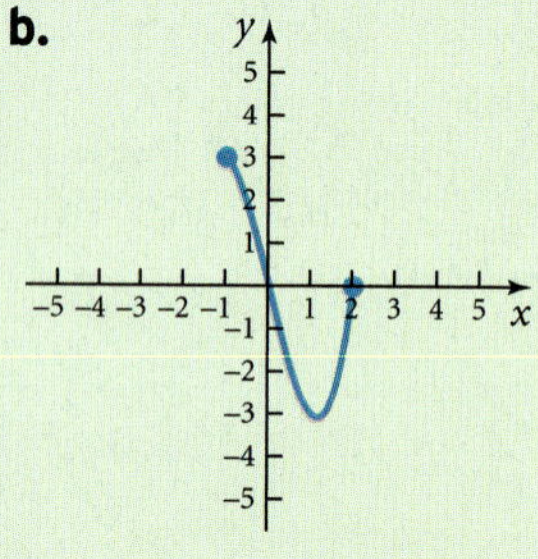

3. a.

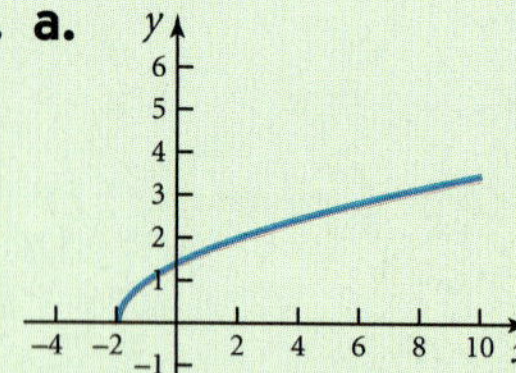

b.

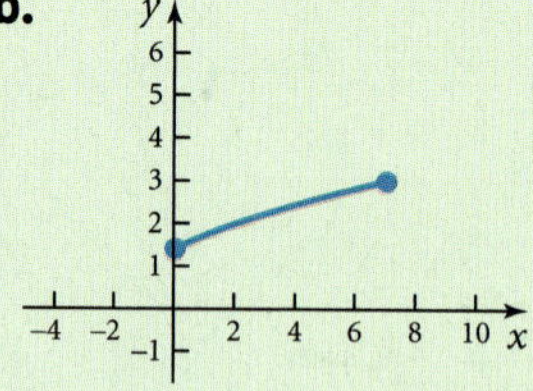

4. $x = -\frac{3}{2}, 2$

5. $x = 0, 2, -2$

6. $x = 1, -1$

7. $x = 2, -2$

8. $x = \dfrac{-11 + \sqrt{97}}{6} \approx -0.192,$

$x = \dfrac{-11 - \sqrt{97}}{6} \approx -3.475$

9. $-3 < x < 10$

Calculus Warm-up Answers

1. increases for $(-\infty, -2)$ and $(2, \infty)$; decreases for $(-2, 2)$; relative minimum at $x = 2$; relative maximum at $x = -2$

2. decreases for $(-\infty, -2)$, $(-2, 2)$, and $(2, \infty)$; no extreme points

3. $-\infty$

4. ∞

5. absolute minimum at $x \approx -1$; relative minimum at $x \approx 3$; relative maximum at $x = 1$

Absolute Extrema

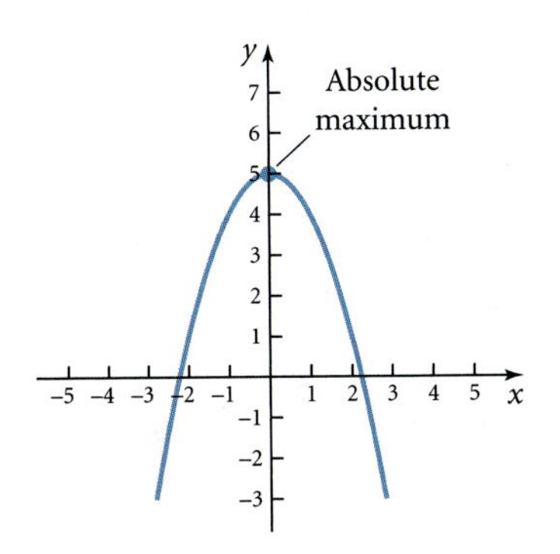

Figure 16.3

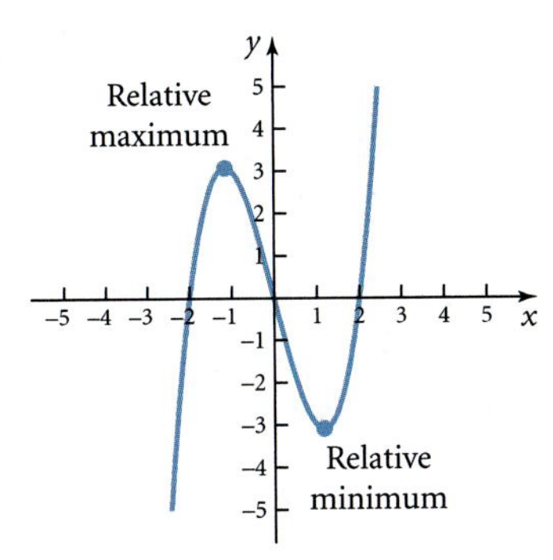

Figure 16.4

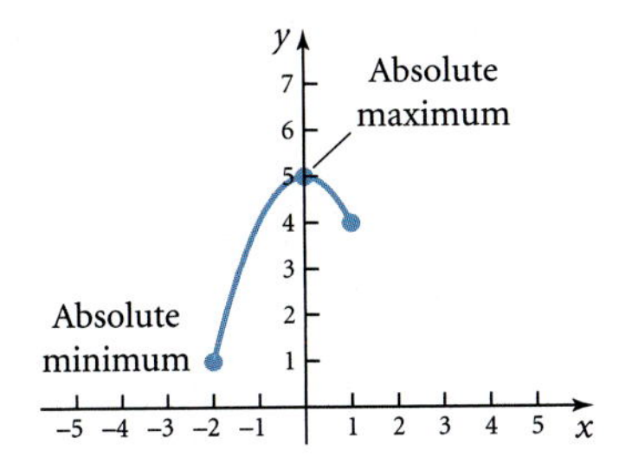

Figure 16.5

In Topic 14, you saw that a graph's **extreme points** occur when the direction of the graph changes. The direction changes from increasing to decreasing at a **maximum point**, and the direction changes from decreasing to increasing at a **minimum point**. Furthermore, you learned that extreme points could be either **absolute** or **relative**.

> **Definition:** The **absolute maximum value** of a function, if it exists, is the largest value of the function over its domain. The **absolute minimum value** of a function, if it exists, is the smallest value of the function over its domain.

Example 1: The graph in Figure 16.3 has an absolute maximum point. The graph is increasing for $x < 0$ and decreasing for $x > 0$. The point $(0, 5)$ is the highest point on the graph. There is no minimum point for this graph. ■

Example 2: The graph in Figure 16.4 has two turning points. The graph increases for $x < -1$, decreases for $-1 < x < 1$, and increases for $x > 1$. Thus, there is a maximum point at $(-1, 3)$ and a minimum point at $(1, -3)$. Both of these points are relative extrema because they are not the highest or lowest points on the entire graph. ■

Example 3: Suppose the parabola from Figure 16.3 had its domain restricted to $[-2, 1]$. In that case, what is the absolute maximum? What is the absolute minimum?

Solution: According to the graph in Figure 16.5, the absolute maximum is still the turning point at $(0, 5)$. Because the domain was restricted to a specific interval, however, the absolute minimum is now at the end point, $(-2, 1)$. ■

Notice that the **absolute extreme points** of functions over a specified interval in the domain may occur at either a turning point or at an interval endpoint.

> **Warning!** Absolute extrema of functions on a specific interval in the domain are not necessarily the extreme points obtained from the first derivative.

Strategy for Determining Absolute Extrema of $f(x)$ Over $[a, b]$

Find $f'(x)$ and determine all critical values, c, contained within $[a, b]$.

Evaluate $f(c)$ for all possible critical values.

Evaluate $f(a)$ and $f(b)$.

The absolute minimum is the smallest of the above values; the absolute maximum is the largest of the above values. It is assumed that $f(x)$ is continuous on the interval $[a, b]$.

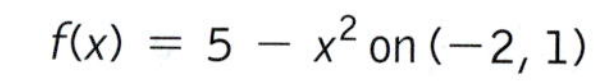

Absolute Extreme Points on an Interval

A function may not have an absolute minimum or an absolute maximum on an interval, as shown in these graphs.

$f(x) = 5 - x^2$ on $[-2, 1]$

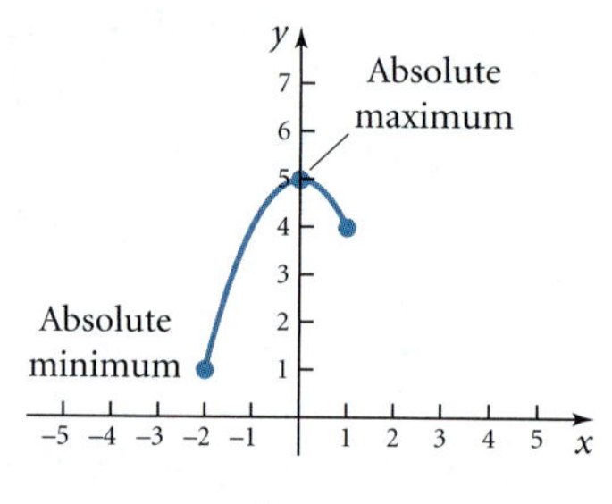

$f(x) = 5 - x^2$ on $(-2, 1)$

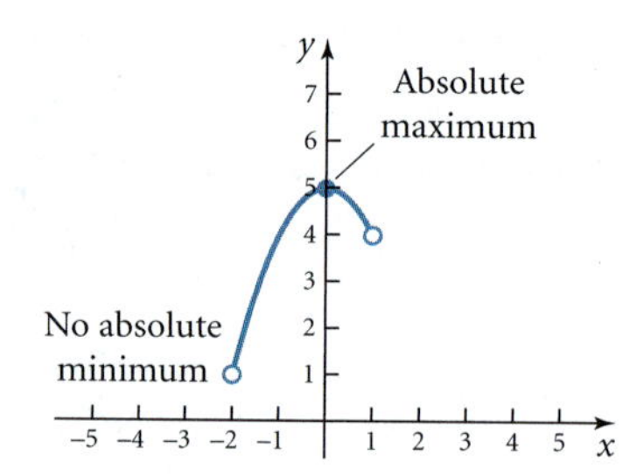

$f(x) = 5 - x^2, x \neq 0$, on $(-2, 1)$

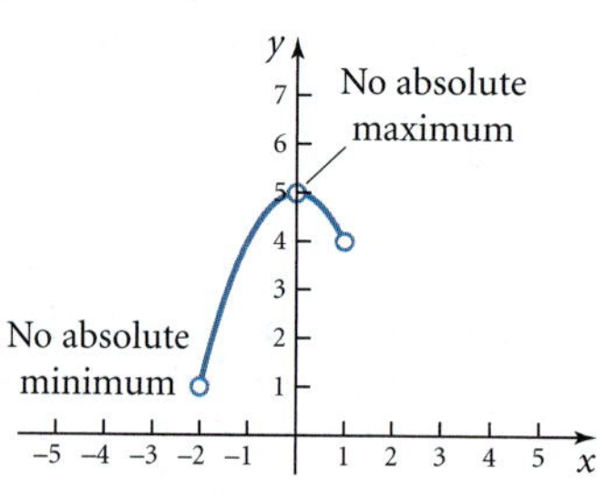

$f(x) = 5 - x^2, x \neq 0$, on $[-2, 1]$

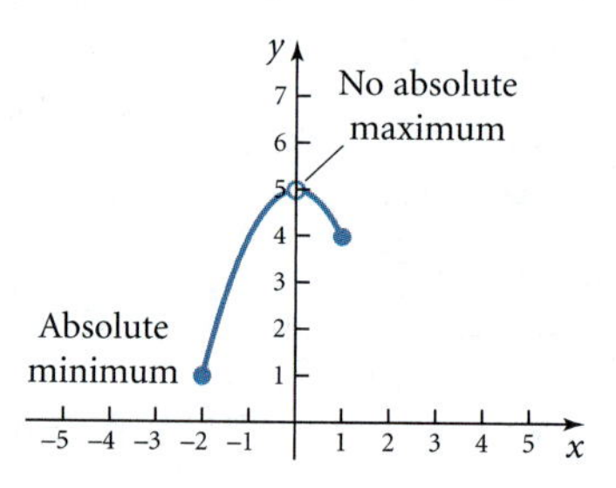

Example 4: Find the absolute extrema of $f(x) = x^3 - 6x^2 + 22$ on

a. $[-1, 7]$ **b.** $[-1, 2]$

Solution:

a. $f(x) = x^3 - 6x^2 + 22 \rightarrow f'(x) = 3x^2 - 12x$

To find the critical values, set

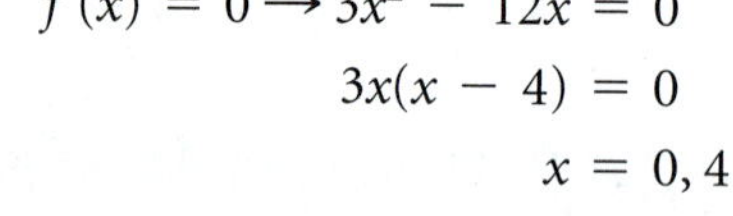

$$f'(x) = 0 \rightarrow 3x^2 - 12x = 0$$
$$3x(x - 4) = 0$$
$$x = 0, 4$$

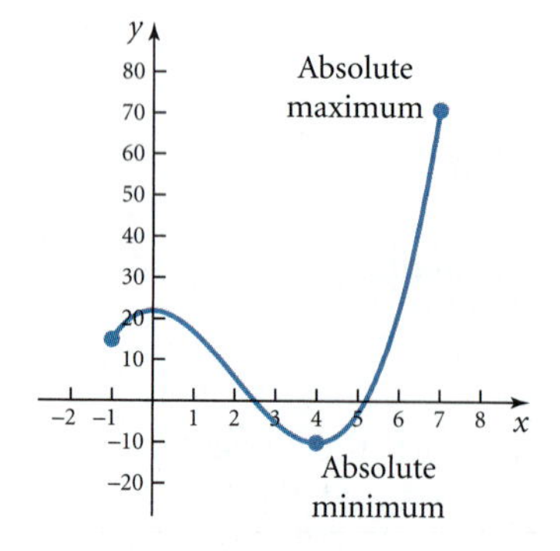

Figure 16.6

So, the critical values are 0 and 4, and both critical values are contained within $[-1, 7]$.

Now evaluate:

$f(0) = 22 \qquad f(-1) = 15$

$f(4) = -10 \qquad f(7) = 71$

The absolute maximum value is 71 at $x = 7$. The absolute minimum value is -10 at $x = 4$. The graph in Figure 16.6 verifies the results.

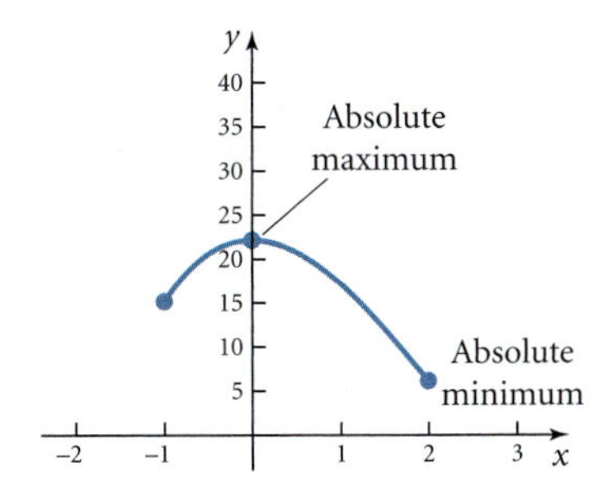

Figure 16.7

b. The critical values remain the same, $x = 0, 4$. Because $x = 4$ is not contained in the interval $[-1, 2]$, it will not be included as a possible absolute extreme point on this interval.

We evaluate $f(x)$ at the critical value $x = 0$ and at the two end points:

$$f(0) = 22$$
$$f(-1) = 15$$
$$f(2) = 6$$

The absolute maximum value is 22 if $x = 0$. The absolute minimum value is 6 if $x = 2$. The graph in Figure 16.7 verifies the results. ■

Example 5: Determine the absolute extrema of $f(x) = 12 - x - \frac{9}{x}$ on $(0, \infty)$.

Solution: First, find the critical values:

$$f(x) = 12 - x - \frac{9}{x} \rightarrow f'(x) = -1 + \frac{9}{x^2}$$

$$-1 + \frac{9}{x^2} = 0 \rightarrow x = 3, -3$$

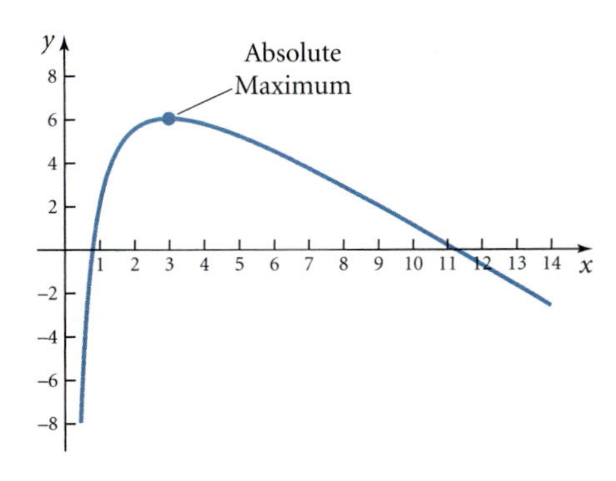

Figure 16.8

The critical values are $x = 3$ and $x = -3$. Only $x = 3$, however, is included in the interval $(0, \infty)$. Evaluating $f(x)$ at the critical value yields

$$f(3) = 6$$

Because $(0, \infty)$ is an open interval, it is not possible to directly evaluate $f(x)$ at each end point. We will use limits instead:

$$\lim_{x \to 0^+} f(x) = 12 - 0 - \infty = -\infty$$
$$\lim_{x \to \infty} f(x) = 12 - \infty - 0 = -\infty$$

Thus, the absolute maximum value is 6 when $x = 3$. There is no absolute minimum value. The graph in Figure 16.8 verifies the results. ■

Check Your Understanding 16.1

Find the absolute extrema of each function on the specified interval.

1. $f(x) = x^4 - 2x^2$ on $[-3, 2]$.

2. $f(x) = x^4 - 2x^2$ on $[0, 2]$.

3. $f(x) = \frac{x^2}{x - 2}$ on $(2, \infty)$.

Applications Involving Absolute Extrema

In some applications, you must find extreme points of functions over specific intervals in the domain rather than over the entire domain. For instance, you may need to analyze costs or production output over a certain time interval. The procedure is similar to the strategy presented in previous examples.

Example 6: Fuel economy (in miles per gallon) of the average American compact car is modeled by $E(x) = -0.015x^2 + 1.14x + 8.3$, where $20 \leq x \leq 60$ mph. At what speed is fuel economy the greatest? How many miles per gallon does the car achieve at that speed?

Solution: The word *greatest* implies the maximum value over [20, 60]. First, find the critical value:

$$E(x) = -0.015x^2 + 1.14x + 8.3 \rightarrow E'(x) = -0.030x + 1.14$$

$$-0.030x + 1.14 = 0 \rightarrow x = 38$$

So, the critical value is 38.

Now evaluate:

$$E(38) = 29.96$$
$$E(20) = 25.1$$
$$E(60) = 22.7$$

Thus, maximum fuel economy is 29.96 mpg when the car travels at 38 mph. (You should also see that the minimum fuel economy is 22.7 mpg at 60 mph.) ■

Example 7: The value of a timber forest, in dollars per acre, after t years is estimated by $V(t) = 480\sqrt{t} - 40t$, where $0 \leq t \leq 50$. What is the maximum value of the forest?

Solution:

$$V(t) = 480\sqrt{t} - 40t \rightarrow V'(t) = \frac{240}{\sqrt{t}} - 40$$

Find the critical value:

$$\frac{240}{\sqrt{t}} - 40 = 0 \rightarrow t = 36$$

So, the critical value is 36.

Now evaluate:

$$V(36) = \$1440$$
$$V(0) = \$0$$
$$V(50) = \$1394$$

Thus, the maximum value of the forest is \$1440 per acre after 36 years. ■

Calculator Corner 16.1

Graph $y_1 = 480\sqrt{x} - 40x$ using a window of [0, 50] for x and [0, 1500] for y. Verify that the maximum value occurs at $x = 36$.

Elasticity of Demand

In Topic 9, we discussed the supply of a product and the demand for that product as functions of the price of the product. Usually, as the price of the product increases, supply will increase and demand will decrease. **Equilibrium** occurs when the price is such that the quantity supplied is the same as the quantity demanded. The point on the graph for which supply and demand are equal is called the **equilibrium point.**

Suppose we want to examine how a change in the price of the product affects the demand for that product. This sensitivity of demand to changes in price will vary depending on the product. A $1 increase in the price of a gallon of gasoline will have a large impact on the demand for the gasoline; a $1 increase in the annual cost of automobile insurance will not have a significant effect on demand for insurance. Economists measure this sensitivity to price changes using **elasticity**. Because the actual change in price and demand vary from product to product, elasticity uses the *percent change* rather than the *actual change* to measure the sensitivity to price changes.

Let p be the price of a product and q be the demand for that product. We use Δp to denote the change in price and Δq to denote the change in demand. The fractional change in price is $\Delta p/p$, and the fractional change in demand is $\Delta q/q$. Because Δp and Δq usually have opposite signs (demand goes down when the price goes up), we will use absolute value to obtain positive values for elasticity.

Definition: The **elasticity of demand** for a product is the ratio of the percent change in demand to the percent change in price that caused the change in demand:

$$E = \frac{\%\text{ change in demand}}{\%\text{ change in price}} = \left|\frac{\Delta q/q}{\Delta p/p}\right| = \left|\frac{\Delta q}{q}\cdot\frac{p}{\Delta p}\right| = \left|\frac{p}{q}\cdot\frac{\Delta q}{\Delta p}\right|$$

Example 8: Suppose the price of a product increases from $9 to $10 and demand drops from 150 units to 110 units. Find the elasticity of demand.

Solution:

$$p = \$9 \qquad q = 150$$
$$\Delta p = \$1 \qquad \Delta q = -40$$

Thus,

$$E = \left|\frac{9}{150}\cdot\frac{-40}{1}\right| = |-2.4| = 2.4$$

An elasticity of 2.4 means that increasing the price of the product by 1% creates a 2.4% drop in demand. ■

Economists are not interested in only the elasticity number. Elasticity of demand is used to see how sensitive the demand for a good is to a price change. The higher the elasticity, the more sensitive consumers are to price changes. A very low elasticity implies that changes in price have little influence on demand.

Ratio	Type	Meaning
$\lvert E \rvert > 1$	Elastic	Demand is sensitive to changes in price. The larger the elasticity, the more sensitive demand is to price changes.
$0 \leq \lvert E \rvert < 1$	Inelastic	Demand is not sensitive to price changes.
$\lvert E \rvert = 1$	Unit elastic	The percent changes in price and demand are about equal.

In Example 8, $E = 2.4$ implies that the demand is highly sensitive to the price of the product.

Example 9: The price of a ticket to an athletic event rose from \$45 to \$50. Average attendance dropped from 18,500 to 17,600. Find the elasticity of demand and interpret your answer.

Solution:

$$p = \$45 \qquad q = 18{,}500$$
$$\Delta p = \$5 \qquad \Delta q = -900$$

Thus,

$$E = \left| \frac{45}{18{,}500} \cdot \frac{-900}{5} \right| \approx 0.438$$

Demand for the athletic event is inelastic, which implies that the demand for tickets is not highly sensitive to changes in the price of the ticket (at least not if $p = \$45$!). ■

In our definition of elasticity, Δp measured the *exact* change in the price of a product and Δq measured the *exact* change in the demand of that product. If the exact values are not available but the demand function $q = D(p)$ is known, then $\Delta q / \Delta p \approx dq/dp = D'(p)$.

Definition: The **elasticity of demand** for a demand function $q = D(p)$ is

$$E = \left| \frac{p}{q} \cdot \frac{dq}{dp} \right| = \left| \frac{p \cdot D'(p)}{D(p)} \right|$$

If $|E| > 1$, the demand is **elastic.** If $0 \leq |E| < 1$, the demand is **inelastic.**

Example 10: The demand for a product is $q = -5p^2 + 55p + 60$, where p is the price. Find the demand elasticity at $p = \$7$ and $p = \$11$ and interpret your answers.

Solution:

$$q = -5p^2 + 55p + 60 \qquad p = 7 \rightarrow q(7) = 200 \qquad p = 11 \rightarrow q(11) = 60$$
$$\frac{dq}{dp} = D'(p) = -10p + 55 \quad p = 7 \rightarrow D'(7) = -15 \quad p = 11 \rightarrow D'(11) = -55$$

For $p = 7$,

$$E = \left|\frac{p}{q} \cdot \frac{dq}{dp}\right| = \left|\frac{7}{200} \cdot (-15)\right| = 0.525$$

The demand if $p = \$7$ is inelastic. A 1% increase in price creates a 0.525% decrease in demand.

For $p = 11$,

$$E = \left|\frac{p}{q} \cdot \frac{dq}{dp}\right| = \left|\frac{11}{60} \cdot (-55)\right| \approx 10.08$$

The demand if $p = \$11$ is elastic. A 1% increase in price creates a 10.08% decrease in demand. ■

Relationship of Elasticity of Demand and Revenue

Total revenue can be determined as the product of the price per unit and the number of units demanded.

Definition: For the demand function $D(p)$, **total revenue** is given by $R(p) = p \cdot D(p)$.

In Example 10, with $D(p) = -5p^2 + 55p + 60$, we saw that the demand was inelastic for $p = \$7$ yet was elastic for $p = \$11$. What is the effect of the demand elasticity on the revenue?

Elasticity of Demand and Revenue

If $|E| > 1$, the demand is elastic and revenue is decreasing.

If $0 \leq |E| < 1$, the demand is inelastic and revenue is increasing.

If $|E| = 1$, the demand is **unit elastic** and the revenue is neither increasing nor decreasing; in other words, **maximum revenue** occurs where $|E| = 1$.

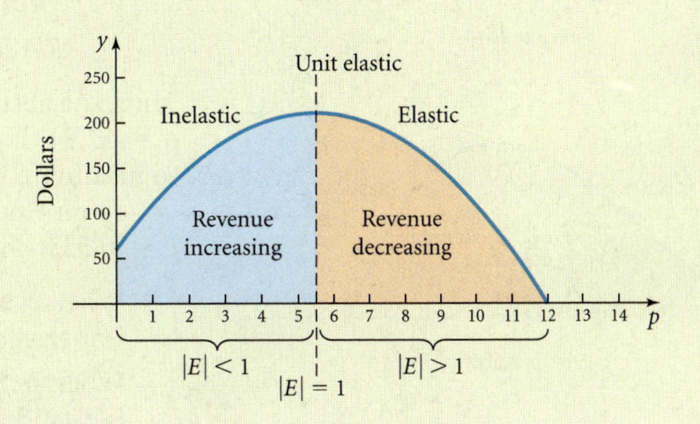

To increase revenue:

Raise prices if $|E| < 1$.
Lower prices if $|E| > 1$.

Example 11: Demand for a product is $D(p) = -5p^2 + 55p + 60$. For what range of prices is the demand elastic? Inelastic? Unit elastic? What effect does the demand elasticity have on the revenue?

Solution: In this example, p represents price and price cannot be negative, so we know that $p \geq 0$. Because $D(p)$ is the number of units demanded, we also know that $D(p) \geq 0$. So,

$$\begin{aligned} D(p) = -5p^2 + 55p + 60 &\geq 0 \\ -5(p^2 - 11p - 12) &\geq 0 \\ -5(p - 12)(p + 1) &\geq 0 \end{aligned}$$

which occurs if $-1 \leq p \leq 12$. Here $p \geq 0$, however, so the domain of $D(p) = -5p^2 + 55p + 60$ is $0 \leq p \leq 12$.

Now evaluate:

$$E = \left|\frac{p \cdot D'(p)}{D(p)}\right| = \frac{p}{-5p^2 + 55p + 60} \cdot |-10p + 55|$$

Total revenue is

$$\begin{aligned} R(p) = p \cdot D(p) \rightarrow R(p) &= p(-5p^2 + 55p + 60) \\ R(p) &= -5p^3 + 55p^2 + 60p \end{aligned}$$

where $0 \leq p \leq 12$.

To find the maximum revenue, we must determine the critical values:

$$\begin{aligned} R(p) = -5p^3 + 55p^2 + 60p \rightarrow R'(p) &= -15p^2 + 110p + 60 \\ R'(p) = 0 \rightarrow -15p^2 + 110p + 60 &= 0 \end{aligned}$$

Using either the Quadratic Formula or a calculator, we obtain $p \approx -0.5$ or $p \approx 7.84$. The domain is $[0, 12]$, so only $p \approx 7.84$ is a meaningful answer.

Testing the critical value and the end points yields

$$\begin{aligned} R(7.84) &\approx 1441.6 \\ R(0) &= 0 \\ R(12) &= 0 \end{aligned}$$

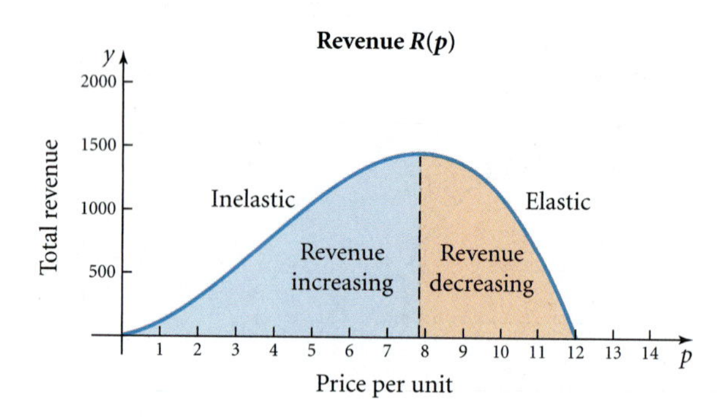

Figure 16.9

Thus, the absolute maximum revenue is \$1441.60 when $p = \$7.84$. If $p = \$7.84$, then $E(7.84) = |0.99774| \approx 1$, so maximum revenue occurs when $|E| = 1$. See Figure 16.9.

From Example 10, we already know that for $p = \$7$, $E = 0.525$, and for $p = \$11$, $E = 10$. So,

When $p < \$7.84$, you see that $|E| < 1$. The demand is inelastic, and revenue is increasing.

When $p > \$7.84$, you see that $|E| > 1$. The demand is elastic, and revenue is decreasing. ■

Example 12: Demand for a product is $D(p) = -4p^2 + 28p + 120$. The current price is \$5, but the manufacturer is considering increasing the price to \$6. What effect would this increase have on the revenue? Should the price be raised?

Solution: Revenue is given by

$$R = p(-4p^2 + 28p + 120)$$
$$= -4p^3 + 28p^2 + 120p.$$

Because $R(p) \geq 0$, we know that

$$-4p^3 + 28p^2 + 120p \geq 0$$
$$-4p(p^2 - 7p - 30) \geq 0$$
$$-4p(p - 10)(p + 3) \geq 0$$

so $p \leq -3$ or $0 \leq p \leq 10$. Here $p \geq 0$, however, so the domain is $0 \leq p \leq 10$.

Maximum revenue occurs if $R'(p) = 0 \rightarrow -12p^2 + 56p + 120 = 0$. By either the Quadratic Formula or a calculator, we obtain $p \approx 6.26$ or $p \approx -1.60$. Because $p \geq 0$, only $p \approx 6.26$ is a meaningful answer. Furthermore, $R''(p) = -24p + 56$, so $R''(6.26) = -24(6.26) + 56 < 0$ and $p \approx 6.26$ is the maximum. Thus, maximum revenue is reached at a price of \$6.26. Raising the price of the product from \$5 to \$6 would increase revenue and therefore is probably a good move for the manufacturer. ■

Calculator Corner 16.2

Graph $y_1 = -4x^3 + 28x^2 + 120x$ using a window of $[0, 10]$ for x and $[0, 1200]$ for y. Verify that the maximum occurs if $x = 6.26$.

Check Your Understanding 16.2

1. Initial claims for unemployment compensation in an area can be modeled by $N(t) = 31t^2 - 560t + 3007$, where t is the number of years after 1990. Find the absolute extrema between 1998 and 2003 $(8 \leq t \leq 13)$.
2. Determine the demand elasticity of the following situations and interpret your answers.
   a. Price increases from \$6 per unit to \$7 per unit; demand drops from 1000 units to 900 units.
   b. Price drops from \$70 per unit to \$60 per unit; demand rises from 10,500 units to 12,000 units.
3. The demand for a product is given by $D(p) = -p^2 + 100p$. Find the elasticity of demand for $p = \$35$ and $p = \$80$ and interpret your answers.
4. Find the maximum revenue of the demand function $D(p) = -p^2 + 5p + 50$.

Check Your Understanding Answers

Check Your Understanding 16.1

1. The absolute maximum value is 63 at $x = -3$. The absolute minimum value is -1 at $x = 1$ or $x = -1$.
2. The absolute maximum value is 8 at $x = 2$. The absolute minimum value is -1 at $x = 1$.
3. The absolute minimum value is 8 at $x = 4$. There is no absolute maximum value.

Check Your Understanding 16.2

1. The absolute minimum is approximately 478 claims when t is approximately 9 (1999). The absolute maximum is approximately 966 claims when t is 13 (2003).
2. **a.** $E = 0.6$, so demand is inelastic and is not sensitive to changes in price.
   b. $E = 1$, so demand is unit elastic.
3. If $p = \$35, E = 0.46$. Demand at the \$35 price is inelastic and would not be affected much by changes in price. If $p = \$80, E = 3$. Demand at the \$80 price is elastic and would be sensitive to changes in price.
4. Maximum revenue occurs if $p = \$6.08$.

Topic 16 Review

16

This topic explored the concept of absolute extreme points in detail and introduced elasticity of demand.

CONCEPT	EXAMPLE
The **absolute maximum value** of a function over an interval is the largest value of the function over that interval.	The absolute maximum value of $f(x) = 4x - x^2$ is 4 when $x = 2$. 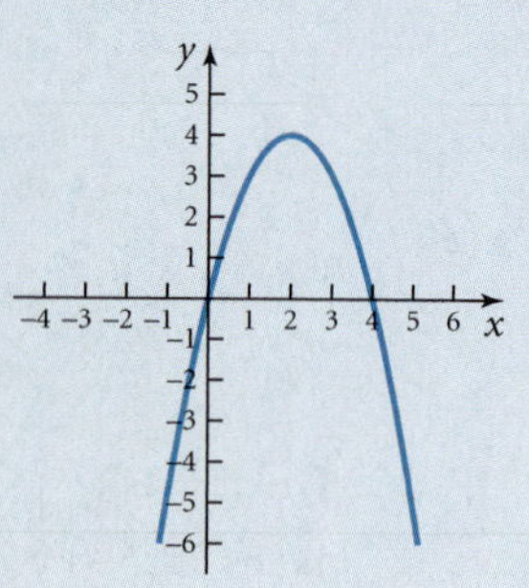

The **absolute minimum value** of a function over an interval is the smallest value of the function over that interval.	The absolute minimum value of $f(x) = 4x - x^2$ on $[-1, 3]$ is -5 when $x = -1$. 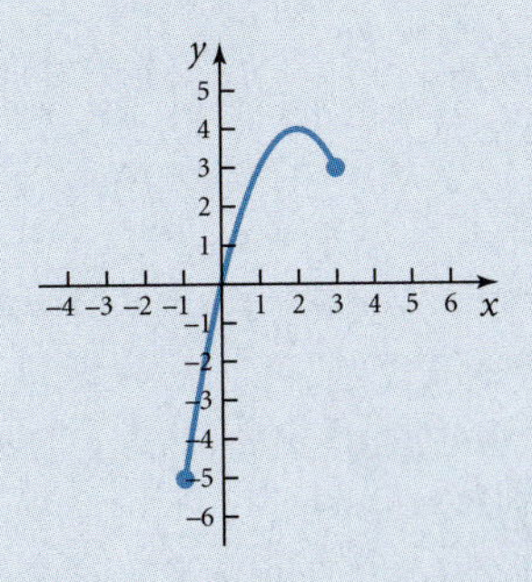
Absolute extreme points of a function over a specific interval may occur at either a turning point or at an end point of the interval.	The absolute minimum value of $f(x) = 4x - x^2$ on $[-1, 3]$ is -5 when $x = -1$, the left end point of the interval. The absolute maximum value is 4 at the turning point $x = 2$. 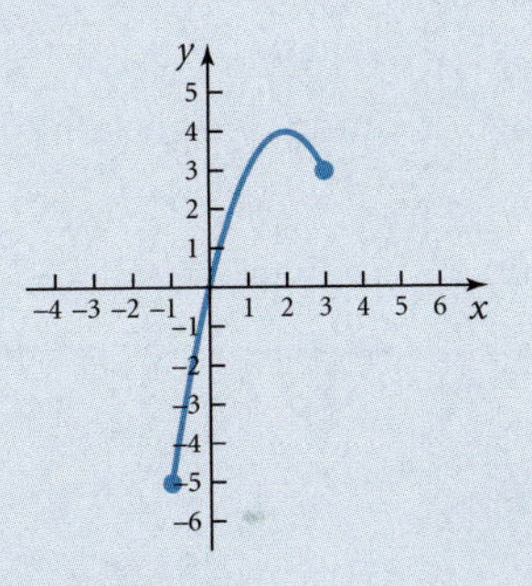
To determine the absolute extrema of a function over a specific interval: • Find the critical values of the function contained within the interval. • Evaluate the function at the critical value(s) and at the end points. • The absolute maximum is the largest of those values, and the absolute minimum is the smallest of those values.	For $f(x) = 4x - x^2$ on $[-1, 3]$, $f'(x) = 4 - 2x$, so $f'(x) = 0$ if $x = 2$. Evaluating gives $f(2) = 4$ $f(-1) = -5$ $f(3) = 3$ The absolute maximum value is 4 if $x = 2$, and the absolute minimum value is -5 if $x = -1$.
For the demand function $D(p)$, sensitivity to price changes is measured by **elasticity of demand**, denoted as E. Elasticity compares the percent change in demand with the percent change in price: $E = \dfrac{\text{\% change in demand}}{\text{\% change in price}} = \left\| \dfrac{p}{q} \cdot \dfrac{\Delta q}{\Delta p} \right\|$	If the price of a product increases from \$20 to \$22, the demand drops from 100 units to 95 units. The elasticity of demand is $E = \left\| \dfrac{20}{100} \cdot \dfrac{-5}{2} \right\| = \|-0.5\| = 0.5$

(continued)

Elasticity of demand for a demand function $q = D(p)$ is $$E = \left\lvert \frac{p}{q} \cdot \frac{dq}{dp} \right\rvert = \left\lvert \frac{p \cdot D'(p)}{D(p)} \right\rvert$$	Demand for a product is $q = -5p^2 + 55p + 60$. If $p = \$10$, then $q(10) = 110$ and $dq/dp = -10p + 55$. Then $$E = \left\lvert \frac{10}{110} \cdot (-45) \right\rvert \approx 4.09$$
Elasticity measures the sensitivity of demand to price changes. • If $\lvert E \rvert > 1$, the demand is **elastic**. Demand for the product is sensitive to changes in price. • If $0 \le \lvert E \rvert < 1$, the demand is **inelastic**. Demand for the product is not greatly affected by changes in price. • If $\lvert E \rvert = 1$, the demand is **unit elastic**. The percent changes in demand and in price are approximately the same.	If the price of a product increases from \$20 to \$22, the demand drops from 100 units to 95 units. The elasticity of demand is 0.5. Demand is inelastic, so demand for this product is not greatly affected by price changes. Demand for a product is $q = -5p^2 + 55p + 60$. If $p = \$10$, then $E = 4.09$. Demand is elastic, so demand for this product is sensitive to price changes.
Total revenue of a product is given by $R(p) = p \cdot D(p)$. The **maximum revenue** occurs where $\lvert E \rvert = 1$ or where $R'(p) = 0$.	If demand for a product is $D(p) = -2p^2 + 12p + 72$, total revenue is $R(p) = -2p^3 + 12p^2 + 72p$. Because $R'(p) = -6p^2 + 24p + 72$, $R'(p) = 0$ if $p = 6$ or $p = -2$. Maximum revenue occurs if $p = \$6$. If $p = \$6$, then $q = 72$ and $D'(6) = -12$, and $$E = \left\lvert \frac{6}{72} \cdot (-12) \right\rvert = 1$$

NEW TOOLS IN THE TOOL KIT

- Finding the absolute maximum value of $f(x)$ over an interval
- Finding the absolute minimum value of $f(x)$ over an interval
- Determining elasticity of demand
- Determining maximum revenue

Topic 16 Exercises

16

Use the following graph to decide if the statements in Exercises 1 through 12 are true or false.

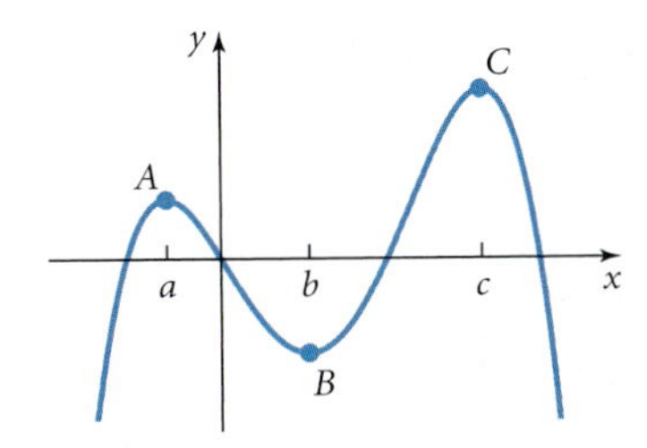

1. C is the absolute maximum of $f(x)$.
2. A is the absolute maximum of $f(x)$.
3. B is the absolute minimum of $f(x)$.
4. A is a relative maximum of $f(x)$.
5. B is the absolute minimum of $f(x)$ on $[0, c]$.
6. C is the absolute maximum of $f(x)$ on $[a, b]$.
7. A is a relative maximum of $f(x)$ on $[a, c]$.
8. B is the absolute minimum of $f(x)$ on $[a, c]$.
9. $f'(c) > 0$
10. $f'(b) = 0$
11. $f'(a) = 0$
12. $f(a) \geq f(x)$ for all real number values of x

Find the absolute maximum and absolute minimum of each of the following functions on the indicated interval.

13. $f(x) = x^3 - 6x^2$ on $[-1, 5]$
14. $f(x) = x^3 - 3x^4$ on $[-2, 6]$
15. $f(x) = x^3 - 6x^2$ on $[-2, 2]$
16. $f(x) = x^3 - 3x^4$ on $[1, 5]$
17. $f(x) = x^4 - 4x^3 - 8x^2 + 7$ on $[-2, 4]$
18. $f(x) = 3x^4 - 8x^3 - 18x^2 - 11$ on $[-2, 4]$
19. $f(x) = x^4 - 4x^3 - 8x^2 + 7$ on $[1, 5]$
20. $f(x) = 3x^4 - 8x^3 - 18x^2 - 11$ on $[-2, 1]$
21. $f(x) = xe^{-x}$ on $[-1, 2]$
22. $f(x) = x\ln(x^2)$ on $[0.1, 2]$
23. $f(x) = 2x(x - 4)^3$ on $[0, 3]$
24. $f(x) = 5x(5 - x)^4$ on $[-1, 2]$
25. $f(x) = \dfrac{x^2}{x - 3}$ on $(3, \infty)$
26. $f(x) = \dfrac{x^2}{x - 3}$ on $(-\infty, 3)$
27. $f(x) = x - 3 - \dfrac{4}{x^2}$ on $(-\infty, 0)$
28. $f(x) = x + 3 + \dfrac{4}{x^2}$ on $(0, \infty)$
29. The fatality rate on U.S. highways is modeled by $f(t) = 0.0038t^2 - 0.35t + 8.3$, where F is the fatality rate per 100 million vehicle miles traveled and t is the driver's age. Find the absolute extrema over $[30, 60]$ and interpret your answers.
30. The percentage of the U.S. population that lives in poverty is estimated by $P(x) = 0.04x^2 - 0.91x + 17$, where x is the number of years after 1990. Find the absolute extrema over $[8, 15]$ and interpret your answers.
31. The number of bachelor's degrees in engineering awarded anually in the United States is estimated by $B(x) = -0.07x^2 + 2.98x + 46.8$, where B is in thousands and x is the number of years after 1970. Find the absolute extrema over $[1, 32]$ and interpret your answers.
32. The number of bachelor's degrees in foreign languages awarded annually in the United States is modeled by $L(x) = 0.025x^2 - 0.96x + 20.8$, where L is in thousands and x is the number of years after 1970. Find the absolute extrema over $[1, 32]$ and interpret your answers.
33. Daily high temperatures in New York City in November can be modeled by $T(x) = -0.012x^3 + 0.62x^2 - 8.85x + 85.6$, where T is degrees Fahrenheit and $1 \leq x \leq 30$. Find the absolute extreme points and interpret your answers.

34. The average monthly temperature for Daytona Beach can be modeled by $T(x) = 0.014x^4 - 0.433x^3 + 3.78x^2 - 7.24x + 62.11$, where T is degrees Fahrenheit and $1 \leq x \leq 12$. Find the absolute extreme points and interpret your answers.

35. The concentration of a drug in the bloodstream t hours after being administered is modeled by $C(t) = 120te^{-0.4t}$, where C is in milligrams per cubic centimeter and $0 \leq t \leq 24$. When is the concentration at its maximum level?

36. The number of pounds lost per week by a woman on a weight-loss program is modeled by $y = 12te^{-0.5t}$, where y is the number of pounds lost in the xth week. What is her maximum weekly weight loss, and when does it occur?

Are the statements in Exercises 37 through 42 true or false?

37. If $|E| < 1$, demand is elastic.

38. Demand is inelastic if $|E| = 1$.

39. If $|E| > 1$, revenue can be increased by lowering prices.

40. If $|E| < 1$, revenue is increasing.

41. Revenues are at their maximum level if $|E| = 1$.

42. Elasticity compares the exact change in demand with the exact change in price.

43. A major cereal producer decides to lower the price of one of its popular cereals from \$3.60 to \$3 per 14-ounce box. If the quantity demanded increases by 18%, what is the elasticity of demand?

44. Suppose that Karin always buys exactly five granola bars each week, regardless of whether they are regularly priced at \$1 or on sale for \$0.50. Based on this information, what is the elasticity of demand for the granola bars?

45. When tolls on the Dulles Airport Greenway were reduced from \$1.75 to \$1, traffic increased from 10,000 to 26,000 trips a day. Assuming all changes in quantity were due to the change in price, what is the elasticity of demand for the Dulles Airport Greenway?

46. Suppose Calvin buys 42 quarts of bottled water each month when the price per quart is \$1, but when water goes on sale for \$0.60 per quart, he buys 60 quarts per month. What is the elasticity of demand for water?

For Exercises 47 through 52, calculate the elasticity of demand and state whether it is elastic, inelastic, or unit elastic. Will revenue rise, decline, or stay the same?

47. The price of a ticket to Boston Red Sox baseball game rises from \$10 to \$12 per game. The quantity of tickets sold falls from 160,000 to 144,000.

48. The price of a calculus textbook falls from \$75 to \$70. The number of books sold rises from 1000 to 1075.

49. The price of a water bed rises from \$500 to \$600. The quantity demanded falls from 100,000 to 80,000.

50. A firm can sell 1200 units at \$14 per unit and 2000 units at \$10 per unit.

51. A store can sell 100 units at \$10 per unit and 99 units at \$11 per unit.

52. Admission prices fall from \$12 to \$5, and daily attendance rises from 60,000 to 100,000.

53. Demand for a new video game, in hundreds, is $D(p) = -1.6p^2 + 100p + 50$. Find the elasticity of demand for $p = \$15$ and $p = \$50$ and interpret your answers.

54. Demand for a new microwave oven, in thousands, is $D(p) = -0.2p^2 + 70p + 43$. Find the elasticity of demand for $p = \$125$ and $p = \$275$ and interpret your answers.

55. Demand for tickets at a sporting event, based on average daily attendance, is

$$D(p) = -7.7p^2 + 495.8p + 10{,}000.$$

The current ticket price is \$36, but the manager is considering a \$5 increase in prices. Based on elasticity of demand, should the manager raise the price? Justify your answer.

56. Demand for tickets to a theme park, based on average daily attendance, is

$$D(p) = -10p^2 + 678p + 7500.$$

The current admission price is \$52, but the park is considering a \$5 decrease in ticket prices. Based on elasticity of demand, should the park lower the price? Justify your answer.

57. Refer to Exercise 53. What is the maximum revenue?

58. Refer to Exercise 54. What is the maximum revenue?

59. Refer to Exercise 55. For what ticket price will revenue be a maximum?

60. Refer to Exercise 56. For what ticket price will revenue be a maximum?

SHARPEN THE TOOLS

61. The number of bachelor's degrees in mathematics awarded annually in the United States is modeled by $B(x) = -0.004x^3 + 0.24x^2 - 3.74x + 28.6$, where B is in thousands, and x is the number of years after 1970. Find the absolute extrema over $[1, 32]$ and interpret your answers.

62. The number of bachelor's degrees in business management awarded annually in the United States is modeled by $B(x) = 0.028x^4 - 0.18x^3 + 3.45x^2 - 13.3x + 125$, where B is in thousands and x is the number of years after 1970. Find the absolute extrema over $[1, 32]$ and interpret your answers.

63. The concentration of a drug t hours after its administration is approximated by $C(t) = \dfrac{230t}{t^2 + 9}$, where C is in milligrams per cubic centimeter and $0 \leq t \leq 24$. What are the absolute extrema of the concentration over this interval?

64. As manager of a hotel, you want to increase the number of occupancies by 12%. It has been determined that the price elasticity of demand for rooms in your hotel is 0.2. What should you do to the cost of rooms to increase occupancy?

For Exercises 65 and 66, determine the absolute extrema of each of the following functions over the indicated interval.

65. $f(x) = 2x - 3x^{2/3}$ on $[0, 3]$

66. $f(t) = 0.8te^{-0.45t}$ on $[1, 5]$

TOPIC

17

Optimization

IMPORTANT TOOLS IN TOPIC 17

- *Solving optimization problems*
- *Writing the objective function*
- *Determining the constraint equation*
- *Using the optimization strategy to determine the maximum or minimum value*

So far in our discussion of extreme points of functions, you have been given a function describing some quantity and have then used derivatives to generate the desired maximum or minimum value for the function. Quite often, an additional step is needed. Suppose, for instance, you are fencing in a storage area and need to know what dimensions will result in the minimum cost for the fencing. When no function is given, it becomes necessary to create your own function subject to the given limitations. In this topic, you learn a strategy for solving such optimization problems. Before starting this topic, complete the Warm-up Exercises to sharpen your algebra and calculus skills.

TOPIC 17 WARM-UP EXERCISES

Algebra Warm-up Exercises

Be sure you can successfully complete these exercises before starting Topic 17.

1. Find the perimeter and the area of the following geometric figures.

a. A rectangle measuring 17 inches by 11 inches

b. A square measuring 3.5 meters by 3.5 meters

c. A right triangle with sides of 9 feet, 12 feet, and 15 feet

2. True or false: A square is a rectangle.

3. Find the circumference and area of a circle whose radius is 3 units.

4. Find the volume and surface area of a right circular cylinder with radius 2 units and height 6 units.

5. Find the distance between the following points.

a. $(3, 2)$ and $(-1, 5)$

b. (x, y) and $(3, 0)$

6. Solve each equation for the indicated variable.

a. $10 = 2l + 2w$ for l

b. $50 = lw$ for w

c. $100 = \pi r^2 h$ for h

7. Solve each of the following equations.

a. $l^2 = 2500$ **b.** $w^2 - 4w - 5 = 0$

c. $3 - \dfrac{75}{l^2} = 0$ **d.** $2l - \dfrac{432}{l^2} = 0$

e. $4x^2 - 3 = 0$ **f.** $3x(x^2 - 4) = 0$

Calculus Warm-up Exercises

1. Find $f'(x)$ and $f''(x)$ for each of the following functions.

a. $f(x) = 3x^2 - 7x + 4$ **b.** $f(x) = 4x + \dfrac{36}{x}$

c. $f(x) = (3x^2 - 5)^4$

2. Determine the extreme points of each of the following functions.

a. $f(x) = 3x^2 - 7x + 4$ **b.** $f(l) = 4l + \dfrac{36}{l}$

c. $f(x) = x^3 - 6x^2$

3. Verify that $l = 5$ is the minimum point of $A(l) = 3l + \dfrac{75}{l}$, with $l > 0$, using the following tests.

a. The First Derivative Test

b. The Second Derivative Test

Answers to Warm-up Exercises

Algebra Warm-up Exercises

1. a. perimeter $= 56$ inches, area $= 187$ square inches

b. perimeter $= 14$ meters, area $= 12.25$ square meters

c. perimeter $= 36$ feet, area $= 54$ square feet

2. true

3. $C = 6\pi \approx 18.85$ units, $A = 9\pi \approx 28.27$ units^2

4. $V = 24\pi \approx 75.4$ units^3, $SA = 32\pi \approx 100.53$ units^2

5. a. $d = 5$

b. $d = \sqrt{(x-3)^2 + y^2}$

6. a. $l = 5 - w$

b. $w = 50/l$

c. $h = 100/\pi r^2$

7. a. $l = \pm 50$

b. $l = 5, -1$

c. $l = \pm 5$

d. $l = 6$

e. $x = \pm\dfrac{\sqrt{3}}{2}$

f. $x = 0, 2, -2$

Calculus Warm-up Exercises

1. a. $f'(x) = 6x - 7$, $f''(x) = 6$

b. $f'(x) = 4 - \dfrac{36}{x^2}$, $f''(x) = \dfrac{72}{x^3}$

c. $f'(x) = 24x(3x^2 - 5)^3$, $f''(x) = 24(3x^2 - 5)^2(21x^2 - 5)$

2. a. $x = \frac{7}{6}$ is the absolute minimum.

b. $l = 3$ is the relative minimum; $l = -3$ is the relative maximum

c. $x = 4$ is the relative minimum; $x = 0$ is the relative maximum

3. a. $A'(l) = 3 - \dfrac{75}{l^2}$

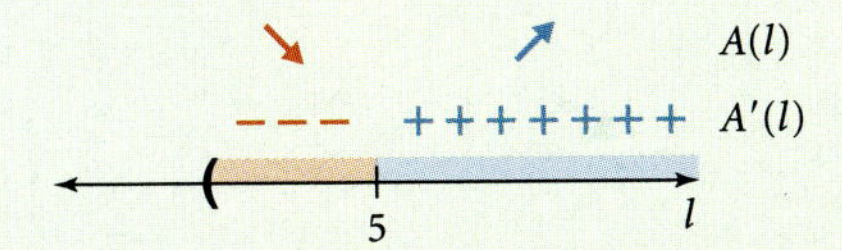

Decreasing changes to increasing, so $c = 5$ is the minimum.

b. $A''(l) = \dfrac{150}{l^3}$, so $A''(5) = 150/5^3 > 0$. Concavity is positive, so $l = 5$ is the minimum point.

s noted in the introduction to this topic, in many real-life situations a maximum or minimum value is desired yet a function may not be readily available. In these situations, you will need to derive that function based on certain known or given information. Such problems are called **optimization problems.**

See the Geometry Formulas table for a review of geometry formulas.

Geometry Formulas

Triangle	**Rectangle**
$P = a + b + c$ $A = \dfrac{1}{2}bh$ (figure labels: a, h, c, b)	$P = 2l + 2w$ $A = lw$ (figure labels: w, l)

(continued)

Square $P = 4s$ $A = s^2$	**Trapezoid** $A = \frac{1}{2}h(b_1 + b_2)$
Circle $C = 2\pi r$ $A = \pi r^2$	**Sphere** $V = \frac{4}{3}\pi r^3$
Cylinder $V = \pi r^2 h$ $A = 2\pi r^2 + 2\pi rh$	**Cone** $V = \frac{1}{3}\pi r^2 h$

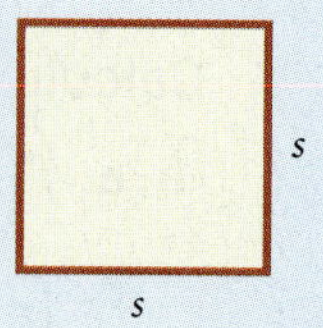

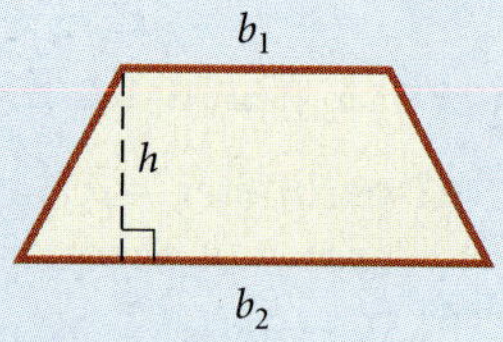

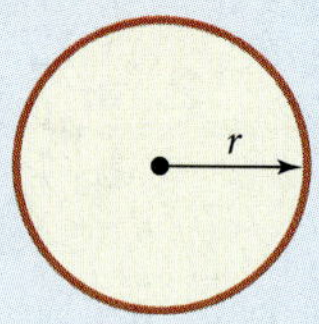

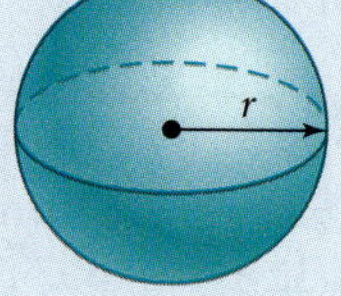

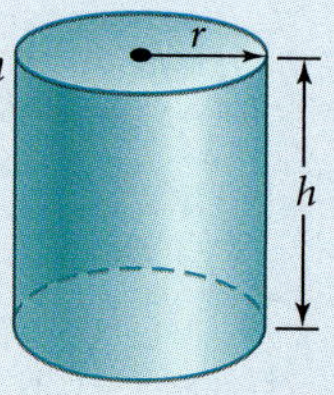

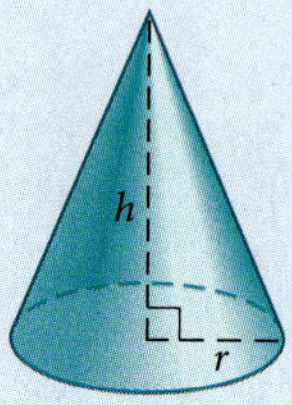

Solving Optimization Problems

Optimization problems involve

- **Variables**, which represent the unknowns
- An **objective function**, which is the function to be maximized or minimized
- **Constraints**, which impose restrictions on the values of the variables

Solving optimization problems requires thought and organization. Consider the following **strategy for solving optimization problems.**

Strategy for Solving Optimization Problems

Read the problem carefully. Be sure you understand the situation being described and the terms being used. **Draw a picture** or diagram if possible.

Identify the unknowns and represent them as variables. Draw a diagram and label the variables.

Write the constraint equation. Express the relationship between the unknowns as an equation. Solve this equation for one of the unknowns.

Determine the objective function. What is to be optimized? Express it as a function in terms of one variable only, using the

(continued)

constraint equation as necessary. Be sure to state the domain of the function.

Determine the critical value(s). Use the first derivative to find the critical value(s) of the objective function. Use either the First Derivative Test or the Second Derivative Test to verify that this critical value is the desired extreme value.

Substitute the critical values into the constraint equation to determine the value of each variable. Determine the actual maximum or minimum value, if requested.

Example 1: Suppose the perimeter of a rectangle is fixed at 70 inches.

a. What are some possible dimensions for the rectangle?

b. How do the different dimensions affect the area of the rectangle?

c. What dimensions provide maximum area?

Solution: The perimeter of a rectangle is given by $P = 2l + 2w$.

a. Some possibilities for dimensions are shown in Figure 17.1.

Figure 17.1

Because length is a measurement, you know that $l > 0$, and because there are only 70 inches total for the four sides, $l < 35$.

b. If the dimensions are 34×1, the area is 34 square inches.

If the dimensions are 30×5, the area is 150 square inches.

If the dimensions are 20×15, the area is 300 square inches.

It appears that the area increases as the length and width become more equal in size.

c. Let l = length and w = width.

Constraint: Perimeter is 70 inches, so $2l + 2w = 70$. Solving for w gives $w = 35 - l$.

Objective: Maximum area, where $A = l \cdot w$.

To find the derivative, we must rewrite the area formula using only one variable, so substitute the expression from the constraint equation into the area formula. Here, $A = l \cdot w$ but $w = 35 - l$, so

$$A = l(35 - l)$$
$$A(l) = 35l - l^2$$
$$A'(l) = 35 - 2l$$

where $0 < l < 35$. The extreme value occurs if $35 - 2l = 0 \rightarrow l = 17.5$.

Because the domain is an open interval, there are two ways to verify that $l = 17.5$ provides the maximum area. A number line analysis could be done, using the First Derivative Test.

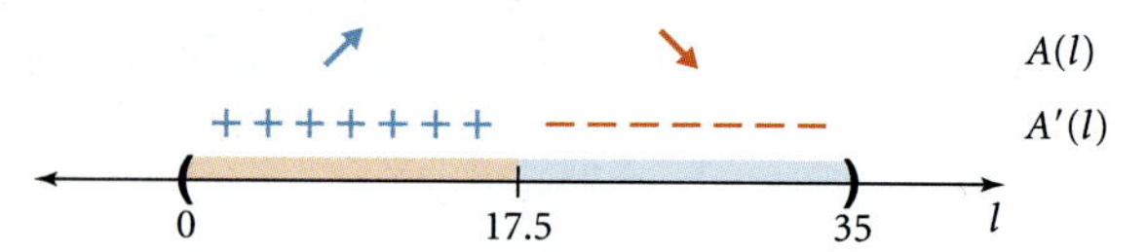

Testing $l = 1$ gives $A'(1) = 33 > 0$, so area is increasing in the interval $(0, 17.5)$.

Testing $l = 20$ gives $A'(20) = -5 < 0$, so area is decreasing in the interval $(17.5, 35)$.

Thus, the maximum point of the function over the specified interval occurs if $l = 17.5$.

A second way to verify that the maximum occurs at $l = 17.5$ is to use the Second Derivative Test:

$$A(l) = 35l - l^2$$
$$A'(l) = 35 - 2l$$
$$A''(l) = -2 < 0$$

The second derivative is negative, so the concavity is downward and the extreme point must be a maximum.

Finally, we substitute $l = 17.5$ into the constraint equation to solve for w:

$$w = 35 - l$$

so $w = 35 - 17.5 = 17.5$. The area of the rectangle is maximized if the dimensions are 17.5 inches by 17.5 inches. (Yes, a square is a rectangle!) The graph of the area function in Figure 17.2 confirms the result. ■

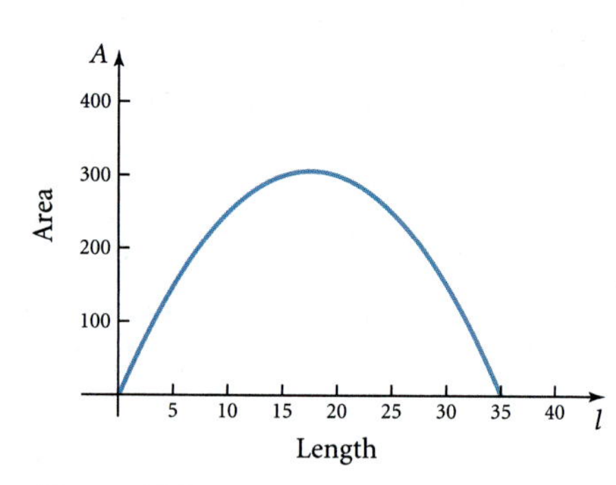

Figure 17.2

Calculator Corner 17.1

Graph $y_1 = 35x - x^2$ using a window of $[0, 35]$ for x and $[0, 350]$ for y. Verify that $x = 17.5$ is the maximum value by pressing **2nd TRACE** (CALC). Select option 4:Maximum. Trace to a point slightly to the left of the maximum and press **ENTER**; then trace to a point slightly to the right of the maximum and press **ENTER ENTER**.

Figure 17.3

Example 2: A farmer has 1800 feet of fence to use to enclose a rectangular grazing area, as shown in Figure 17.3. What dimensions will provide maximum grazing area?

Solution: Let l = length and w = width.

Constraint: Perimeter is 1800 feet, so $2l + 2w = 1800$. Solving for l gives $l = 900 - w$.

Objective: Maximum area, where $A = l \cdot w$.

To find the derivative, we must rewrite the area formula using only one variable, so substitute the expression from the constraint equation into the area formula. Here, $A = l \cdot w$ but $l = 900 - w$, so

$$A = (900 - w)w$$
$$A(w) = 900w - w^2$$
$$A'(w) = 900 - 2w$$

where $0 < w < 900$. The extreme point occurs if $900 - 2w = 0 \rightarrow w = 450$.

Now verify that the desired extreme value occurs if $w = 450$. First, $A'(w) = 900 - 2w$, so $A''(w) = -2 < 0$. The second derivative is always negative, so the concavity is downward and $(450, A(450))$ is the maximum point.

Finally, substitute $w = 450$ into the constraint equation to solve for l:

$$l = 900 - w$$

so $l = 900 - 450 = 450$. The grazing area is maximized if the dimensions are 450 feet by 450 feet. ■

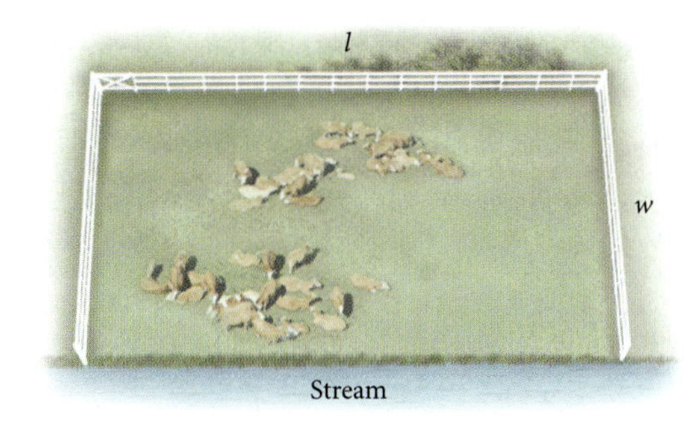

Figure 17.4

Example 3: Suppose the grazing area described in Example 2 is beside a stream, and that side does not need to be fenced. (See Figure 17.4.) Also assume the stream is straight. What dimensions provide maximum grazing area?

Solution: Let l = length and w = width.

Constraint: Perimeter is 1800 feet, but only one length is fenced, so $l + 2w = 1800$. Solving for l gives $l = 1800 - 2w$.

Objective: Maximum area, where $A = l \cdot w$.

To find the derivative, we must rewrite the area formula using only one variable, so substitute the expression from the constraint equation into the area formula. Here, $A = l \cdot w$ but $l = 1800 - 2w$, so

$$A = (1800 - 2w)w$$
$$A(w) = 1800w - 2w^2$$
$$A'(w) = 1800 - 4w$$

where $0 < w < 900$. The extreme point occurs if $1800 - 4w = 0 \rightarrow w = 450$.

Now, verify that the desired extreme value occurs if $w = 450$. First, $A'(w) = 1800 - 4w$, so $A''(w) = -4 < 0$. Because the second derivative is always negative, the concavity is downward and $A(450)$ is the maximum value.

Finally, substitute $w = 450$ into the constraint equation to solve for l:

$$l = 1800 - 2w$$

so $l = 1800 - 2(450) = 900$. The grazing area is maximized if the dimensions are 900 feet by 450 feet, where the 900-foot side runs parallel to the stream. ■

Example 4: Suppose the area of a rectangle is fixed at 100 square inches.

a. What are some possible dimensions for the rectangle?

b. How do the different dimensions affect the perimeter of the rectangle?

c. What dimensions would allow for minimum perimeter?

Solution: We know that area = (length) × (width), or $A = lw$.

a. Some possible dimensions are shown in Figure 17.5.

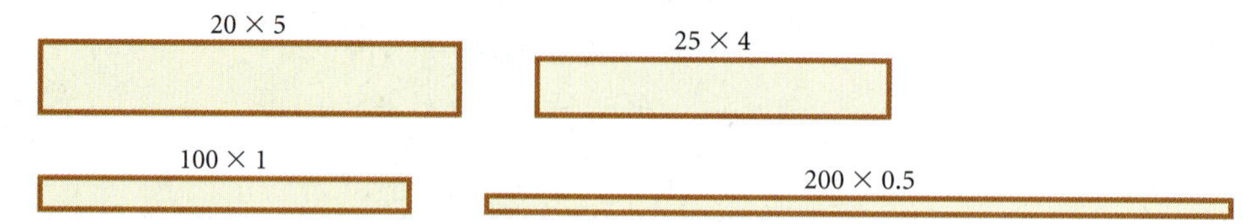

Figure 17.5

As the length increases, the width decreases, so the only restriction is that $l > 0$.

b. If the dimensions are 20 × 5, the perimeter is $2(20) + 2(5) = 50$ inches.
If the dimensions are 25 × 4, the perimeter is $2(25) + 2(4) = 58$ inches.
If the dimensions are 100 × 1, the perimeter is $2(100) + 2(1) = 202$ inches.
If the dimensions are 200 × 0.5, the perimeter is $2(200) + 2(0.5) =$ 401 inches.

It appears that the perimeter decreases as the length and width become more equal in value.

c. Let l = length and w = width.

Constraint: Area is 100 square inches, so $l \cdot w = 100$. Solving for w gives $w = 100/l$.
Objective: Minimum perimeter, where $P = 2l + 2w$.

To find the derivative, we must rewrite the perimeter formula using only one variable, so substitute the expression from the constraint equation into the perimeter formula. Here, $P = 2l + 2w$, but $w = \dfrac{100}{l}$, so

$$P = 2l + 2\left(\frac{100}{l}\right)$$

$$P(l) = 2l + \frac{200}{l}$$

$$P'(l) = 2 - \frac{200}{l^2}$$

where $l > 0$. The extreme point occurs if

$$2 - \frac{200}{l^2} = 0 \rightarrow 2l^2 - 200 = 0$$
$$\rightarrow l^2 = 100$$
$$\rightarrow l = 10, -10$$

We know that $l > 0$, so $l = 10$ is the only meaningful answer.

To verify that the minimum occurs if $l = 10$, use the Second Derivative Test:

$$P'(l) = 2 - \frac{200}{l^2}$$

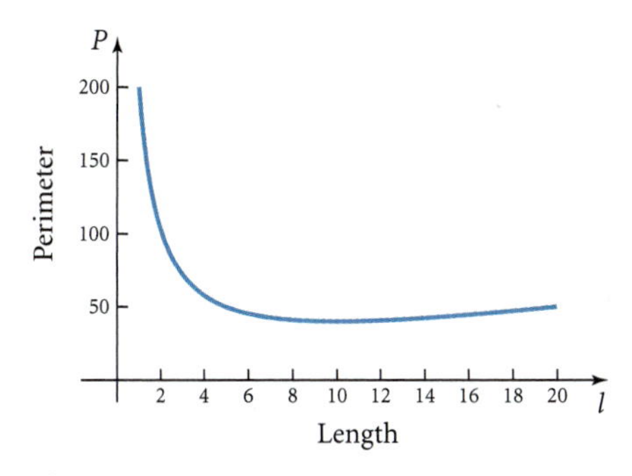

Figure 17.6

so $P''(l) = 400/l^3$ and $P''(10) = 0.4 > 0$. Because the second derivative is positive if $l = 10$, we know that concavity is upward and the point is a minimum.

Finally, substitute $l = 10$ into the constraint equation to find the width:

$$w = \frac{100}{l}$$

so $w = 10$. The perimeter of the rectangle is minimized if the dimensions are 10 inches by 10 inches. The graph of the perimeter function in Figure 17.6 verifies the result. ■

Calculator Corner 17.2

Graph $y_1 = 2x + \dfrac{200}{x}$ using a window of [0, 30] for x and [0, 100] for y. Verify that $x = 10$ is the minimum value by pressing **2nd TRACE** (CALC). Select option 3:Minimum. Trace to a point slightly to the left of the minimum and press **ENTER**; then trace to a point slightly to the right of the minimum and press **ENTER ENTER**.

Check Your Understanding 17.1

Use the optimization strategy to solve the following problems.

1. Find two positive numbers whose sum is 20 and whose product is a maximum.
2. The perimeter of a rectangle is 300 yards. Find the dimensions that provide maximum area.
3. The area of a rectangle is 400 square meters. Find the dimensions that provide minimum perimeter.

Example 5: A building supply store wishes to fence in a rectangular storage area next to its building. (See Figure 17.7.) The area to be enclosed is 1280 square feet. The side along the building needs no fence. Along the front, the amount of fencing needed costs \$8 per foot, and the other two sides use fencing costing \$5 per foot. What dimensions will minimize the total cost? What is the minimum cost?

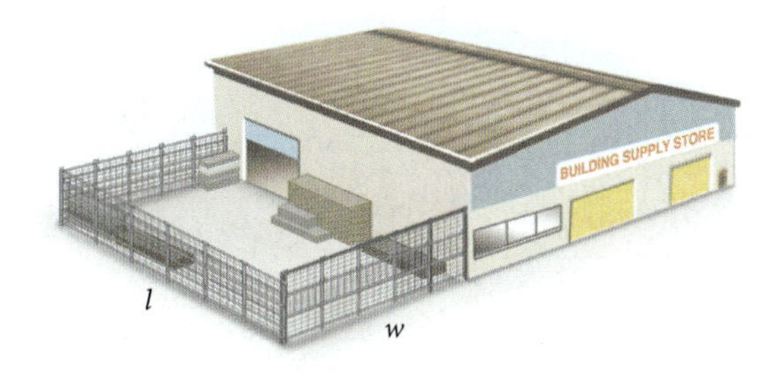

Figure 17.7

Solution: Let l = length and w = width.

Constraint: Area is 1280 square feet, so $l \cdot w = 1280$. Solving for w gives $w = 1280/l$. We know that $l > 0$ because length is a dimension.

Objective: Minimum cost, where the cost per side is determined by cost per side = (cost per foot) · (number of feet). So, for the front fence, cost = $\$8l$. For the side fences, cost = $2(\$5)(w) = \$10w$. The total cost is $C = 8l + 10w$.

To find the derivative, we must rewrite the cost function using only one variable, so substitute the expression from the constraint equation into the cost function. Here, $C = 8l + 10w$ but $w = 1280/l$, so

$$C = 8l + 10\left(\frac{1280}{l}\right)$$

$$C(l) = 8l + \frac{12{,}800}{l}$$

$$C'(l) = 8 - \frac{12{,}800}{l^2}$$

where $l > 0$. The extreme point occurs if

$$8 - \frac{12{,}800}{l^2} = 0$$

$$8l^2 - 12{,}800 = 0$$

$$l^2 = 1600$$

$$l = 40, -40$$

Because $l > 0$, $l = 40$ is the only meaningful answer.

To verify that the minimum occurs if $l = 40$, use the Second Derivative Test:

$$C'(l) = 8 - \frac{12{,}800}{l^2}$$

so $C''(l) = 25{,}600/l^3$ and $C''(40) = 25{,}600/64{,}000$. Because $C''(40) > 0$, the concavity is upward and the point is a minimum.

Finally, substitute $l = 40$ into the constraint equation to find w:

$$w = \frac{1280}{l}$$

so $w = 1280/40 = 32$. The cost is minimized if the front fence ($8 per foot) is 40 feet and the other two sides are 32 feet.

The total cost is

$$C = 8l + 10w$$

$$C = 8(40) + 10(32)$$

$$C = 320 + 320$$

$$C = \$640$$

Do you see that the cost of the more expensive fence equals the cost of the two less expensive fences? That is why the total cost is minimized with these dimensions. ■

Figure 17.8

Example 6: A $10\frac{3}{4}$-ounce can of Scrumptious Tomato Soup is to contain 12 cubic inches of soup. (See Figure 17.8.) The cost of materials for the top and bottom of the can is 5 cents per square inch, and the cost of materials for the side of the can is 3 cents per square inch. To minimize cost, what should the dimensions of the can be?

Solution: Let r = radius of the can and h = height of the can.

Constraint: The volume is 12 cubic inches, so $\pi r^2 h = 12$. Solving for h yields $h = 12/\pi r^2$.

To determine the cost, we will need the surface area of the top, bottom, and side of the can. (See Figure 17.9.)

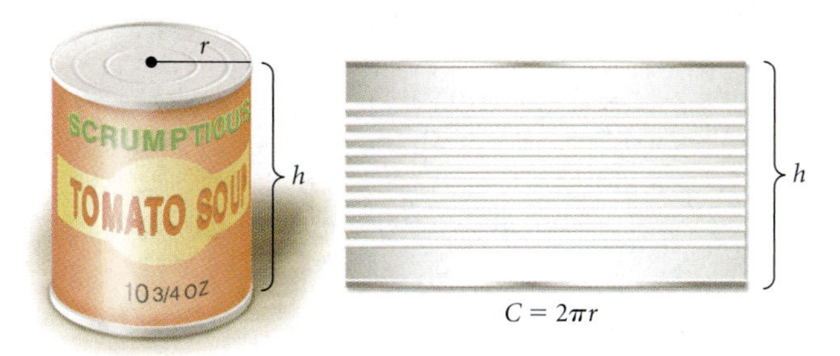

Figure 17.9

1. The top of the can is a circle with radius r, so the area is πr^2 square inches. The bottom of the can also has an area of πr^2 square inches.
2. To figure the surface area of the side of the can, imagine cutting the can from top to bottom and laying the side out as a rectangle. The width of the rectangle will be the height of the can. The length of the rectangle will be the circumference of the top (or bottom) of the can. Thus, the surface area of the side is $2\pi rh$ square inches.

Objective: Minimum cost, where cost is (dollar amount per square inch) $\cdot$ (number of square inches).

So,

$$\text{total cost} = (\text{cost of top}) + (\text{cost of bottom}) + (\text{cost of side})$$
$$C = 0.05(\pi r^2) + 0.05(\pi r^2) + 0.03(2\pi rh)$$
$$C = 0.1\pi r^2 + 0.06\pi rh$$

where $r > 0$ and $h > 0$. To find the derivative, we must rewrite the cost function using only one variable. Because $h = 12/\pi r^2$, we have

$$C = 0.1\pi r^2 + 0.06\pi r\left(\frac{12}{\pi r^2}\right)$$

So

$$C(r) = 0.1\pi r^2 + \frac{0.72}{r}$$

$$C'(r) = 0.2\pi r - \frac{0.72}{r^2}$$

where $r > 0$. The extreme value occurs if

$$0.2\pi r - \frac{0.72}{r^2} = 0$$
$$0.2\pi r^3 - 0.72 = 0$$
$$r^3 = \frac{0.72}{0.2\pi} \approx 1.1459$$
$$r = \sqrt[3]{\frac{0.72}{0.2\pi}} \approx 1.046$$

To verify that the minimum value occurs if $r = 1.046$, use the Second Derivative Test:

$$C'(r) = 0.2\pi r - \frac{0.72}{r^2}$$

so $C''(r) = 0.2\pi + 1.44/r^3$ and $C''(1.046) = 0.2\pi + 1.44/1.046^3$. Because $C''(1.046) > 0$, the concavity is upward and $(1.046, C(1.046))$ is the minimum point.

Finally, substitute $r = 1.046$ into the constraint equation to find h:

$$h = \frac{12}{\pi r^2}$$

but $r = 1.046$, so $h = 12/\pi(1.046)^2 \approx 3.49$. To minimize cost, the can should have a radius of approximately 1.046 inches and a height of 3.49 inches. (Check a $10\frac{3}{4}$-ounce soup can. What are its dimensions?) ■

Calculator Corner 17.3

To evaluate a cube root on the TI-83/84, go to **MATH** and select option 4:$\sqrt[3]{\ }$(. Press **ENTER** and then type in the number for which you desire the cube root. Press **ENTER** to see the root.

Example 7: Find the point on $f(x) = x^2$ that is closest to the point $(0, 1)$.

Solution: Let (x, y) be the point on $f(x)$. (See Figure 17.10.) The distance from (x, y) to the point $(0, 1)$ is then calculated using the distance formula:

$$d = \sqrt{(x-0)^2 + (y-1)^2} = \sqrt{x^2 + (y-1)^2}$$

Constraint: (x, y) is a point on the graph of $f(x)$, so $y = x^2$. Thus,

$$d = \sqrt{x^2 + (y-1)^2} = \sqrt{x^2 + (x^2-1)^2}$$

Objective: Minimize distance, where $d = \sqrt{x^2 + (x^2-1)^2}$.

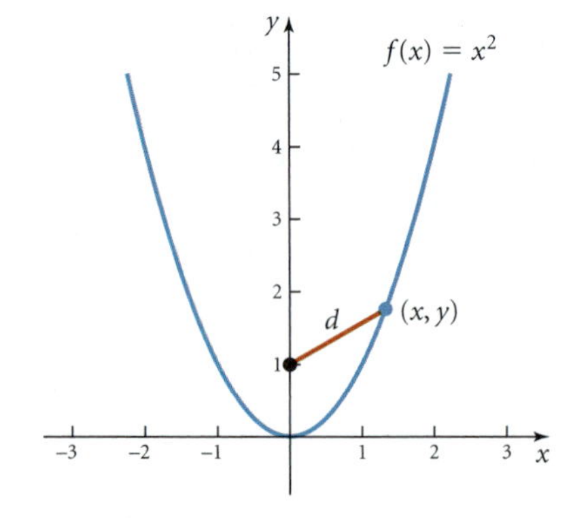

Figure 17.10

Minimizing the expression underneath the radical will minimize the value of the radical itself. Thus, we must find the minimum value of $d(x) = x^2 + (x^2 - 1)^2$:

$$d'(x) = 2x + 2(x^2 - 1)(2x) \quad \text{Power Rule and Chain Rule}$$
$$= 2x[1 + 2(x^2 - 1)] \quad \text{Factor } 2x \text{ from both terms}$$
$$= 2x(2x^2 - 1) \quad \text{Simplify within brackets}$$

To find the extreme points, solve $2x(2x^2 - 1) = 0$, yielding $x = 0, 1/\sqrt{2}$, and $-1/\sqrt{2}$ as the critical values. By the First Derivative Test, we can do the following number line.

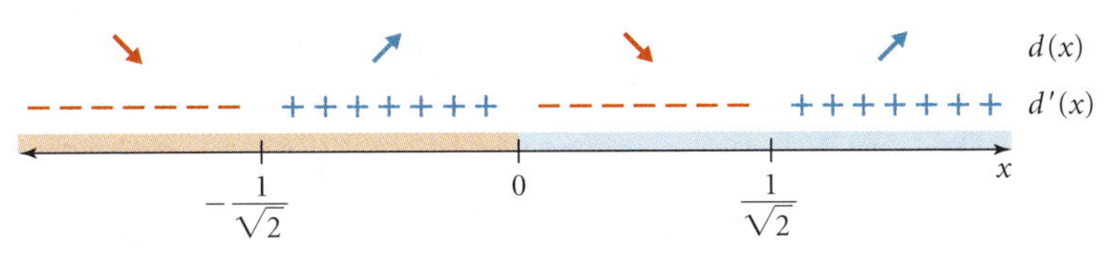

Thus, minimums occur at both $x = 1/\sqrt{2}$ and $x = -1/\sqrt{2}$.

To determine the y value, substitute $x = 1/\sqrt{2}$ and $x = -1/\sqrt{2}$ into the constraint equation, $f(x) = x^2$. Because $f(1/\sqrt{2}) = f(-1/\sqrt{2}) = \frac{1}{2}$, there are two minimum points. The points on $y = x^2$ that are closest to $(0, 1)$ are $\left(\frac{1}{\sqrt{2}}, \frac{1}{2}\right)$ and $\left(-\frac{1}{\sqrt{2}}, \frac{1}{2}\right)$. ■

MATHEMATICS CORNER 17.1

Perpendicularity of Tangent Line

An interesting geometric relationship with the tangent line occurs at the two points on $f(x) = x^2$ that are closest to the point $(0, 1)$, as found in Example 7.

Consider the point $\left(\frac{1}{\sqrt{2}}, \frac{1}{2}\right)$. At $x = 1/\sqrt{2}$, the slope of the tangent line to $f(x) = x^2$ is

$$m_1 = f'\left(\frac{1}{\sqrt{2}}\right) = 2\left(\frac{1}{\sqrt{2}}\right) = \frac{2}{\sqrt{2}}$$

The slope of the line between $\left(\frac{1}{\sqrt{2}}, \frac{1}{2}\right)$ and $(0, 1)$ is

$$m_2 = \frac{1 - \frac{1}{2}}{0 - 1/\sqrt{2}} = \frac{\frac{1}{2}}{-1/\sqrt{2}} = -\frac{\sqrt{2}}{2}$$

The slopes of these two lines are negative reciprocals ($m_1 m_2 = -1$), so the two lines are perpendicular.

The same perpendicularity occurs at the point $\left(-\frac{1}{\sqrt{2}}, \frac{1}{2}\right)$.

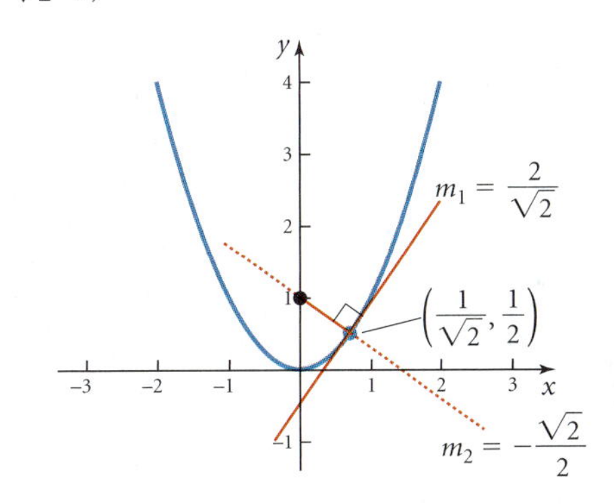

You should be able to verify that the tangent line to $f(x) = x^2$ at $\left(-\frac{1}{\sqrt{2}}, \frac{1}{2}\right)$ is perpendicular to the line between $(0, 1)$ and $\left(-\frac{1}{\sqrt{2}}, \frac{1}{2}\right)$

Example 8: Brian is designing a storage container to store wrapping paper. To accommodate rolls of paper, the container must be 3 feet long. (See Figure 17.11.) The container must also have 14 square feet of surface area. What dimensions would provide the maximum storage space (volume)?

Solution:

Constraint: Surface area must be 14 square feet. For this box,

$$\text{total surface area} = \text{area of top} + \text{area of bottom} + \text{area of four sides}$$
$$14 = 3x + 3x + xh + xh + 3h + 3h$$
$$14 = 6x + 2xh + 6h$$
$$7 = 3x + xh + 3h$$

Figure 17.11

Solving for x gives

$$7 = 3x + xh + 3h$$
$$7 - 3h = x(3 + h)$$
$$x = \frac{7 - 3h}{h + 3}$$

where $h > 0$.

Objective: Maximize volume, where $V = 3xh$.

Substituting $x = \dfrac{7 - 3h}{h + 3}$, we obtain

$$V(h) = 3\left(\frac{7 - 3h}{h + 3}\right)h \quad \text{or} \quad V(h) = \frac{21h - 9h^2}{h + 3}$$

We know that $h > 0$ and that volume must also be positive. Thus, $21h - 9h^2 > 0$. Factoring, we have $3h(7 - h) > 0$, which says that $0 < h < 7$.

To find the extreme point, we use the derivative.

$$V(h) = \frac{21h - 9h^2}{h + 3} \rightarrow V'(h) = \frac{(h + 3)(21 - 18h) - (21h - 9h^2)(1)}{(h + 3)^2}$$
$$= \frac{-9h^2 - 54h + 63}{(h + 3)^2}$$

The extreme point(s) occur where

$$\frac{-9h^2 - 54h + 63}{(h + 3)^2} = 0$$

The denominator cannot be zero, so

$$-9h^2 - 54h + 63 = 0$$
$$-9(h^2 + 6h - 7) = 0$$
$$-9(h + 7)(h - 1) = 0$$
$$h = -7, 1$$

Because $0 < h < 7$, the only possible critical value is $h = 1$. By the First Derivative Test, we can do the following number line analysis.

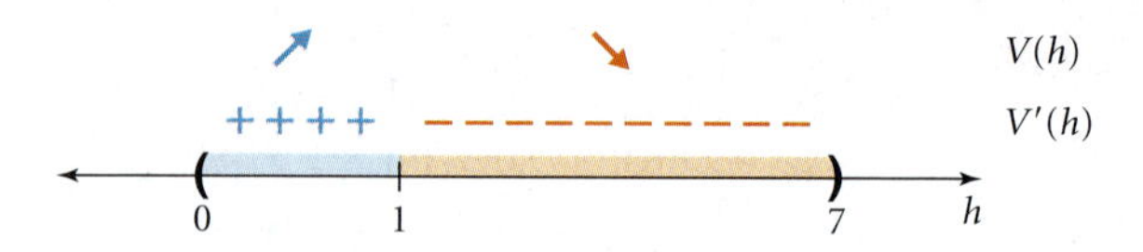

So, the volume is maximized if $h = 1$. Substituting $h = 1$ into the constraint equation $x = \dfrac{7 - 3h}{h + 3}$, we see that the width of the container should be $x = \dfrac{7 - 3(1)}{1 + 3} = 1$. So, maximum storage capacity is reached if the container is 3 feet by 1 foot by 1 foot. ■

Check Your Understanding 17.2

Solve using the optimization strategy.

1. A rectangular storage area of 500 square feet is to be fenced along all four sides. One side uses fencing that costs \$6 per foot, and the other sides use fencing that costs \$4 per foot. Find the dimensions that will minimize the total cost of the fence.
2. Find the point(s) on $f(x) = -x^2$ closest to the point $(0, -2)$.
3. Rework Example 5 by solving the constraint equation for l. Verify that the same extreme point results.

Check Your Understanding Answers

Check Your Understanding 17.1

1. 10 and 10
2. 75 yards by 75 yards
3. 20 m by 20 m

Check Your Understanding 17.2

1. The side with the \$6 fence is 20 feet long, and the other side is 25 feet.
2. $\left(\frac{\sqrt{6}}{2}, -\frac{3}{2}\right)$ and $\left(-\frac{\sqrt{6}}{2}, -\frac{3}{2}\right)$
3. Here, $lw = 1280 \rightarrow l = 1280/w$. Then

$$C(w) = \frac{10{,}240}{w} + 10w$$

$$C'(w) = -\frac{10{,}240}{w^2} + 10$$

$$C'(w) = 0 \rightarrow w = 32$$

If $w = 32$, then $l = 1280/w = 1280/32 = 40$. The cost is minimized if the front fence is 40 feet and the other two sides are 32 feet.

Topic 17 Review

17

This topic discussed strategies for solving optimization problems.

CONCEPT	EXAMPLE
An **optimization problem** involves determining the minimum or maximum value in a situation for which no function is given.	A storage area next to a building is to be constructed using 2000 feet of fencing. No fence is needed along the building. Determine the dimensions that will allow for the largest amount of area.
An optimization problem involves variables, constraints, and an objective function. • The **variables** are the unknowns in the problem. • **Constraints** are limitations on the value of the variables. • The **objective function** is the quantity to be optimized.	In the above example, • the variables are the length and width of the storage area. • the constraint is the 2000 feet of fencing available. • the objective function is the area that is to be maximized.
The **strategy for solving optimization problems** is as follows. • Read the problem. • Identify the unknowns and represent them as variables. • Write the constraint equation. • Express the objective function in terms of one variable. • Find the critical value(s). • Substitute the critical value into the constraint equation to determine the value of all variables.	In the above example, the variables are length, l, and width, w. The constraint is the amount of fencing available: $l + 2w = 2000$. Objective function: Maximize $A = lw$. Solving $l + 2w = 2000 \rightarrow l = 2{,}000 - 2w$. Substitute $l = 2000 - 2w$ into $A = lw$: $A(w) = (2000 - 2w)w = 2000w - 2w^2.$ $A'(w) = 2000 - 4w$ $A'(w) = 0 \rightarrow w = 500$ is the critical value. Because, $A''(w) = -4$, the extreme point is a maximum. If $w = 500$, then $l = 1000$. The area is maximized if the dimensions are 500 feet by 1000 feet, with the 1000-foot side running parallel to the building.

NEW TOOLS IN THE TOOL KIT:

- Solving optimization problems using the optimization strategy
- Writing a constraint equation
- Expressing an objective function in terms of one variable
- Minimizing perimeter, costs, distance, and surface area
- Maximizing area and volume

Topic 17 Exercises

17

Solve the following optimization problems. Assume $x > 0$ and $y > 0$.

1. Maximize $A = xy$ if $x + y = 12$.
2. Maximize $A = xy$ if $2x + y = 16$.
3. Minimize $P = 2x + 2y$ if $xy = 36$.
4. Minimize $P = 8x + 10y$ if $xy = 100$.
5. Minimize $F = x^2 + y^2$ if $x + 2y = 20$.
6. Minimize $F = x^2 + y$ if $x^2y = 16$.
7. Minimize $C = 4x + 10y$ if $xy = 10$.
8. Minimize $C = 5x + 2y$ if $x^2y = 10$.
9. Find two numbers whose sum is 40 and whose product is a maximum.
10. Find two numbers whose sum is 500 and whose product is a maximum.
11. Find two numbers whose difference is 40 and whose product is a minimum.
12. Find two numbers whose difference is 12 and whose product is a minimum.
13. The product of two positive numbers is 64. Find the numbers that have the least sum.
14. The product of two positive numbers is 45. Find the numbers that have the least sum.

Solve the following problems using the optimization strategy.

15. A board that is 25 feet long is to be cut into two pieces. Where should the cut be made so that the product of the two lengths will be a maximum?
16. The perimeter of a rectangle is 60 yards. What dimensions will provide for the maximum area?
17. A farmer has 1500 feet of fencing to use to enclose a rectangular grazing area. What dimensions will provide the maximum grazing area?
18. A farmer is fencing a rectangular grazing area beside a stream with 1200 feet of fencing. If the side along the stream is not to be fenced, what dimensions will result in the maximum grazing area?
19. A boarding kennel is building pens as indicated in the diagram. The area to be enclosed is 640 square feet. The side fences cost \$8 per foot, and the front fence costs \$12.50 per foot. The side along the building will not be fenced. What dimensions will provide minimum total cost for the fence?

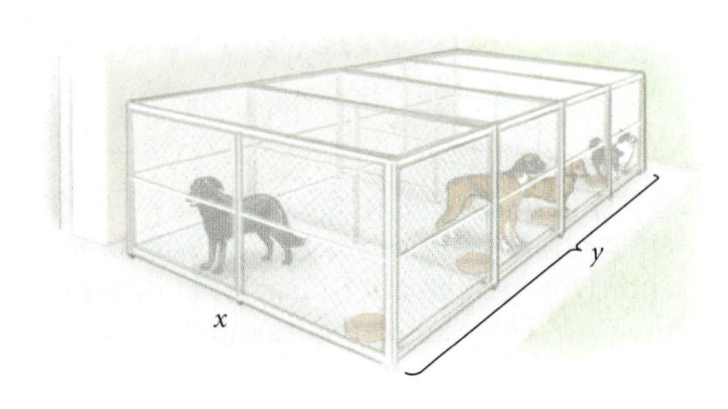

20. Haley is planning to fence in her vegetable garden with 640 feet of fence. Her garden has three sections, with the middle section twice as wide as the other two sections. What dimensions will provide the maximum area for the garden?

21. A rectangle has 225 square centimeters of area. What dimensions result in the least perimeter?

22. Don has \$3000 to purchase fencing to enclose a storage area. The side along the driveway needs a fence costing \$15 per foot, and the other three sides will use a fence costing \$5 per foot. What dimensions will maximize the total enclosed area?

23. Mario is enclosing a rectangular storage area next to his building. The area to be enclosed is 2880 square feet. Along the front, the fencing used costs \$12 per foot; the other sides use fencing costing \$8 per foot. The side next to the building is not fenced. What dimensions will minimize the total cost of the storage? What is the cost?

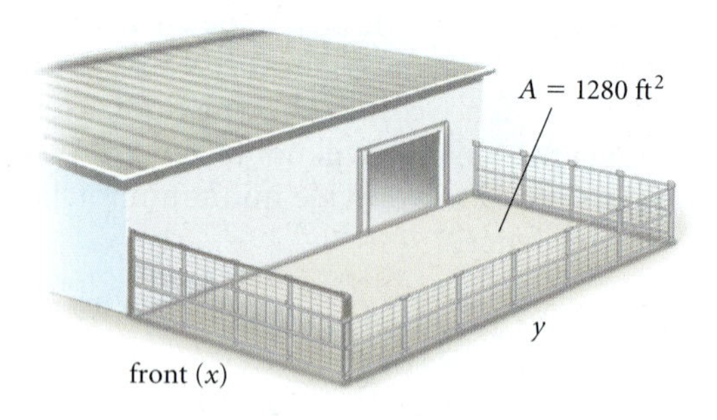

24. A page contains 54 square inches of print. The top and bottom margins are $1\frac{1}{2}$ inches, and the left and right margins are 1 inch. What dimensions of the page will result in the least amount of paper being used? See figure at top of next column.

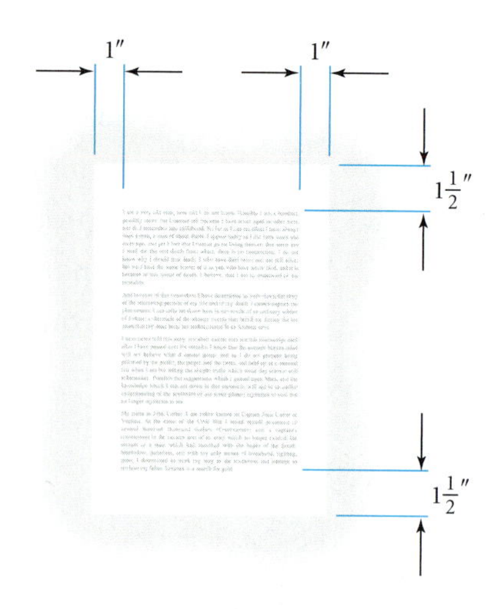

25. A page of 96 square inches has 1-inch margins at the top and bottom and $1\frac{1}{2}$-inch margins on the left and right. What dimensions of the page will provide maximum printed area?

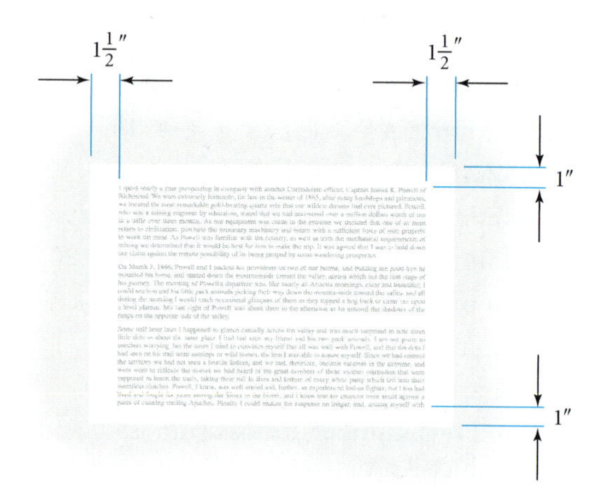

26. The infield of a racetrack is in the shape of a rectangle with semicircles at each end. The perimeter (one length around the track) is one-half mile, or 2640 feet. What dimensions for the rectangle will maximize the infield area?

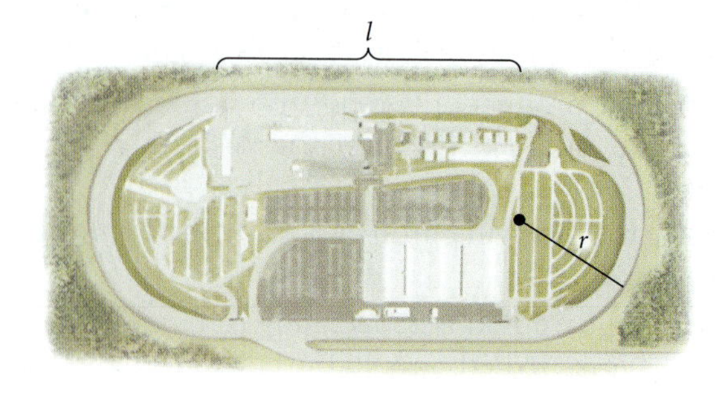

27. A Norman window has the shape of a rectangle topped by a semicircle. If the area of the rectangular window is 50 square feet, what dimensions will minimize the perimeter of the entire window? What is the total area of the window?

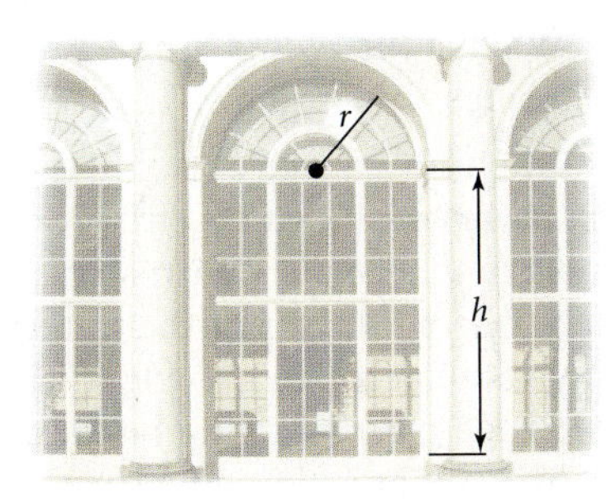

28. A box with an open top and a square base is to have a surface area of 108 square inches. Determine the dimensions that will maximize the volume of the box.

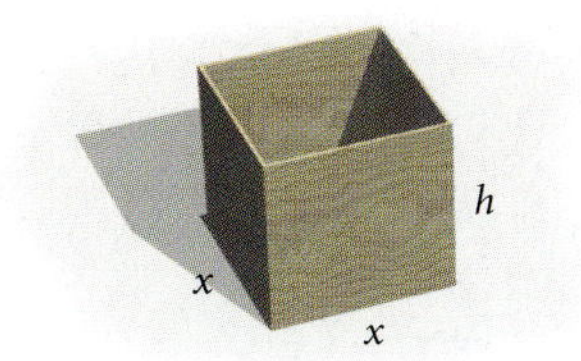

29. A box with an open top and a square base is to have a volume of 108 cubic inches. What dimensions allow for the least amount of material to construct the box?

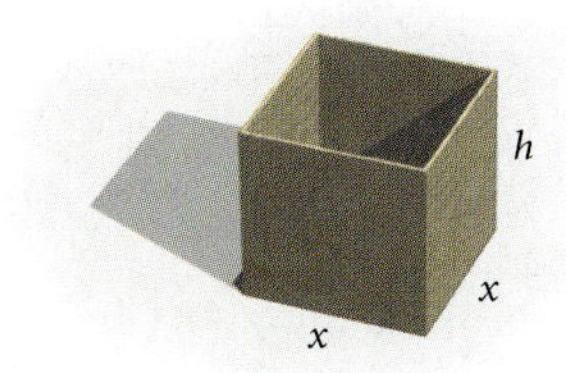

30. A bus stop shelter has two square sides, a back, and a roof. The volume is 256 cubic feet. What dimension will allow for the least amount of material to be used?

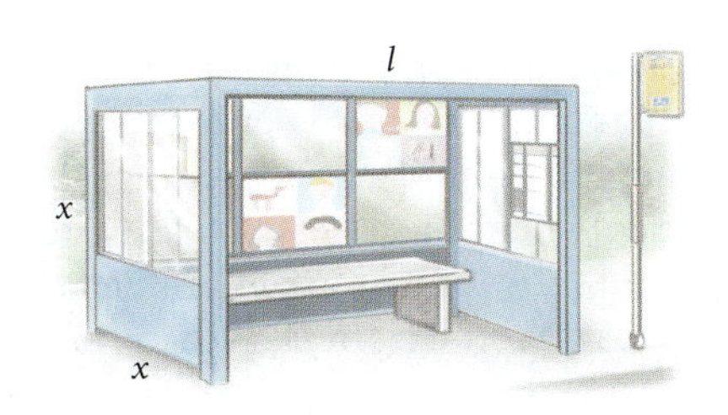

31. Scrumptious Soup Company makes a soup can with a volume of 250 cm^3. What dimensions will allow for the minimum amount of metal to produce the can?

32. A jumbo-size can of baked beans has a volume of 600 cm^3. What dimensions will allow for the minimum amount of metal to produce the can?

33. The surface area of a can of chunked chicken requires 60 square inches of material. What dimensions allow for maximum volume?

34. A large can of tuna requires a surface area of 100 square inches. What dimensions provide the maximum volume?

35. Find the point in the first quadrant on the graph of $y = 4 - x^2$ that is closest to the point $(0, 2)$.

36. Find the point in the first quadrant on the graph of $y = x^2 - 2$ that is closest to the point $(0, 1)$.

37. Two posts, one 12 feet high and the other 28 feet high, are 30 feet apart. They are anchored by two wires running from the top of each pole to a single stake in the ground at a point between the two posts. Where should the stake be placed so that the minimum amount of wire is used?

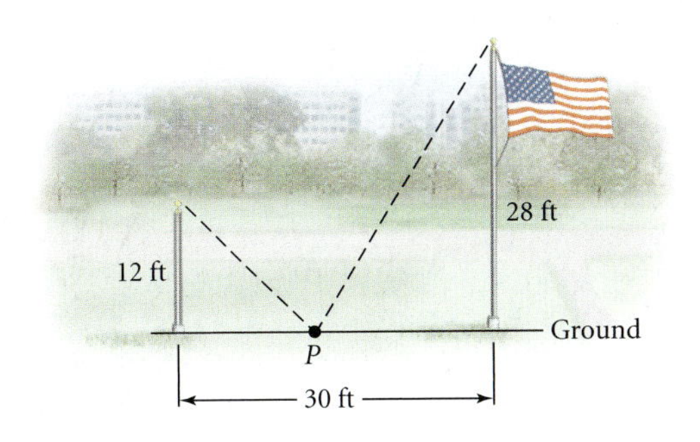

38. A factory must run a pipe across a lake to a warehouse. It costs \$2 per foot to run the pipe along the ground and \$3 per foot to run the pipe under water. The lake is 300 feet wide. Where should P be located to determine the least expensive route from factory to warehouse?

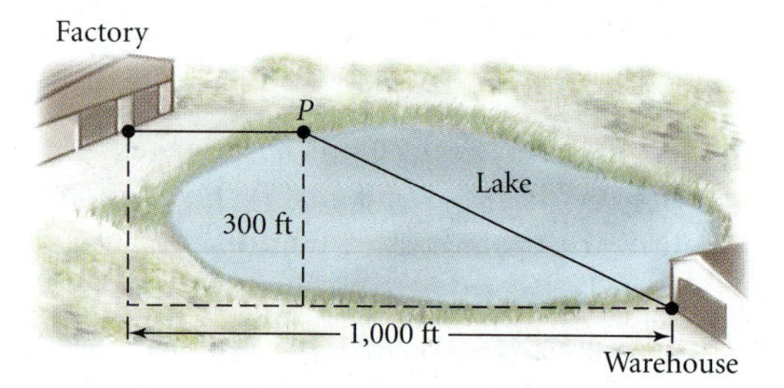

SHARPEN THE TOOLS

39. Find the points on $y = x^2 - 4$ that are closest to $(1, 2)$. Round coordinates to the nearest tenth.

40. Find the points on $y = x^3 - 4x$ that are closest to $(-1, 0)$. Round coordinates to the nearest hundredth.

41. A 100-inch-long piece of string is to be cut into two pieces. One piece will form a circle, and the other piece will form a square. How long should each piece of string be to minimize the total area? What is the radius of the circle? How long is each side of the square?

42. A 100-inch-long piece of string is cut into three pieces. One piece forms a circle, one piece forms a square, and one piece forms an equilateral triangle. If the perimeter of the triangle is equal in length to the perimeter of the square, how long should each piece of string be to minimize the total area? What is the radius of the circle? How long is each side of the square? How long is each side of the triangle?

TOPIC

Business Applications

18

IMPORTANT TOOLS IN TOPIC 18

- *Maximizing revenue*
- *Maximizing yield*
- *Minimizing costs of inventory control*
- *Cobb–Douglas productivity models*

Maurice owns a small plant that manufactures ride-on lawn mowers. How does he decide how many mowers to produce and at what price to sell them to maximize his profit? What constraints should he consider when answering these questions? In this topic, we look at ways calculus can be used in constrained optimization applications from the business area, such as maximizing revenues and inventory control. Before you start this topic, be sure to complete the Warm-up Exercises to sharpen your algebra and calculus skills.

TOPIC 18 WARM-UP EXERCISES

Be sure you can successfully complete these exercises before starting Topic 18.

Algebra Warm-up Exercises

Simplify each of the following expressions

1. $\left(\frac{2}{x^{0.3}}\right)^4$ **2.** $\sqrt[4]{\frac{12(7)^4}{5}}$

Solve each of the following equations.

3. $-5p^2 + 55p + 60 = 0$

4. $3m^2 - 4m - 9 = 0$

5. Solve for p: $p^2 - 4p \geq 0$.

6. For what values of x does $-5x^2 + 55x + 60$ have a positive value?

7. Solve each of the following equations.

a. $4 = y^{1/2}$ **b.** $4 = y^{2/3}$

c. $4 = y^{0.2}$

Calculus Warm-up Exercises

A rectangular storage area of 500 square feet is to be fenced on all four sides. One side uses fencing that costs \$6 per foot, and the other three sides use fencing that costs \$4 per foot.

1. Find the dimensions that will minimize the total cost of the fence.

2. Verify that these dimensions provide the minimum total cost.

3. What is the minimum total cost?

Answers to Warm-up Exercises

Algebra Warm-up Exercises

1. $16/x^{1.2}$

2. $7\sqrt[4]{\frac{12}{5}} \approx 8.713$

3. $p = 12, -1$

4. $m = \frac{2 \pm \sqrt{31}}{3} \approx 2.523, -1.189$

5. $p \leq 0$ or $p \geq 4$ **6.** $-1 \leq x \leq 12$

7. a. $4^2 = 16$ **b.** $4^{3/2} = 8$

c. $4^5 = 1024$

Calculus Warm-up Exercises

1. The side with the \$6 fence is 20 feet long, and the other dimension is 25 feet.
2. First Derivative Test:

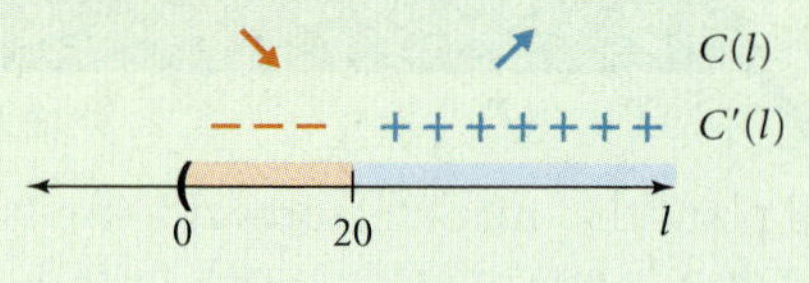

Decreasing changes to increasing, so this result is the minimum.

Second Derivative Test: $C''(20) = 8000/20^3 > 0$. Concavity is upward, so this result is the minimum.

3. Cost is $C = 10l + 8w = 10(20) + 8(25) = 200 + 200 = \$400$.

Maximizing Revenue and Profit

In Topic 17 you learned a strategy for solving optimization problems.

Strategy for Solving Optimization Problems

Read the problem carefully. Draw a picture or diagram if possible.

Identify the unknowns and represent them as variables.

Write the constraint equation. Solve this equation for one of the unknowns.

Determine the objective function. Express this as a function in terms of one variable only, using the constraint equation as necessary.

Determine the critical value(s). Verify that this critical value provides the desired extreme.

Substitute the critical value into the constraint equation to determine the value of all the variables.

We now apply this strategy to several types of business applications involving revenue, maximizing profit, and maximizing yield. **Revenue** is the number of items sold times the price per item, or $R = p \cdot q$. **Maximizing revenue** involves determining the price at which a quantity should be sold in order to obtain maximum revenue. **Profit** is revenue minus cost. **Maximizing profit** involves determining the price at which a quantity should be sold in order to obtain maximum profit.

Example 1: *Maximum Revenue* Rosie's Discount Mart sells paperback books. At price \$$p$, Rosie can sell $q(p) = -5p^2 + 55p + 60$ books. What price would give Rosie the greatest revenue? How many books would she have to sell to generate the maximum revenue?

Solution: Let p be the price and $q(p)$ be the demand at price p.

Constraint: We know that $q(p) = -5p^2 + 55p + 60$ and that revenue is $R = p \cdot q$. We also know that $p \geq 0$ and $q \geq 0$.

Objective: Maximize revenue, $R = p \cdot q$.

Substituting the constraint, $R = p(-5p^2 + 55p + 60)$.

The demand must be nonnegative, so $-5p^2 + 55p + 60 \geq 0$

$$-5(p^2 - 11p - 12) \geq 0$$
$$-5(p - 12)(p + 1) \geq 0$$

which yields

$$-1 \leq p \leq 12$$

We also know, however, that $p \geq 0$, so the domain is $0 \leq p \leq 12$.

To determine the maximum, we use the derivative of the objective function:

$$R = p(-5p^2 + 55p + 60) \quad \text{or} \quad R = -5p^3 + 55p^2 + 60p$$

Then $R'(p) = -15p^2 + 110p + 60$.

Now we find the extreme points:

$$-15p^2 + 110p + 60 = 0$$
$$-5(3p^2 - 22p - 12) = 0$$
$$p \approx 7.84, -0.51$$

Because the quadratic expression $3p^2 - 22p - 12$ does not factor, the Quadratic Formula was used to solve the above equation. You should be able to verify the result. Because $0 \leq p \leq 12$, $p = 7.84$ is the only meaningful critical value. Furthermore,

$$R''(p) = -30p + 110$$

so $R''(7.84) = -30(7.84) + 110 < 0$. Therefore, the maximum occurs if $p = 7.84$ and the maximum revenue will occur if each paperback book is priced at \$7.84. At that price, the demand is $q(7.84) = -5(7.84)^2 + 55(7.84) + 60 = 183.872$ or approximately 184 books. The revenue generated will be $R = \$7.84(184) = \$1442.56$. So, if Rosie prices each book at \$7.84, 184 books will be demanded with a revenue of \$1442.56. ■

Calculator Corner 18.1

The TI-84 Plus Silver Edition will solve polynomial equations. Press **APPS** and scroll down to: PolySmlt. Then press **ENTER** three times. Enter the degree of the polynomial (2) and press **ENTER**. Then enter the coefficients and press **SOLVE**.

Example 2: *Maximum Profit* Mark's Restaurant can produce one chicken sandwich for \$2. The sandwiches sell for \$5 each, and, at this price, his customers buy 1200 sandwiches each month. Because of rising costs from suppliers, the restaurant is planning to raise the price of the sandwich. Based on the results of previous price increases, Mark estimates that he will sell 120 fewer sandwiches each month for each \$1 increase in the price. At what price should the sandwiches be sold to maximize Mark's profit? What is the maximum profit?

Solution: Let x be the new price and $P(x)$ be the new profit.

Constraint: Total profit is a function of both price and the number of sandwiches sold:

profit = (number sold) · (profit per sandwich)

Because 1200 sandwiches sell if the price is \$5 and 120 fewer are sold each month for each \$1 price increase, we know that

$$\begin{aligned} \text{number sold} &= 1200 - 120(\text{number of \$1 increases}) \\ &= 1200 - 120(x - \$5) \\ &= 1200 - 120x + 600 \\ &= 1800 - 120x \\ &= 120(15 - x) \end{aligned}$$

where $5 \le x \le 15$. Because the cost per sandwich is \$2, the profit per sandwich is $x - 2$ dollars. Substituting $120(15 - x)$ for the number of sandwiches and $x - 2$ for the profit per sandwich into the profit equation gives

$$P(x) = 120(15 - x)(x - 2)$$

where $5 \le x \le 15$.

Tip: The domain is $5 \le x \le 15$. You should see that if $x > 15$, the number sold would be negative.

Objective: Maximum profit, where $P(x) = 120(15 - x)(x - 2)$.

Using the Product Rule,

$$P'(x) = 120[(15 - x)(1) + (x - 2)(-1)]$$
$$P'(x) = 120(17 - 2x)$$

To determine the extreme value, set $120(17 - 2x) = 0$ so $x = 17/2 = 8.5$. To verify that the maximum occurs if $x = 8.5$, use the Second Derivative Test:

$$P'(x) = 120(17 - 2x)$$

so $P''(x) = 120(-2) < 0$, and the maximum occurs if $x = 8.5$. If the new price of the sandwich is \$8.50, the maximum possible profit is

$$P(8.5) = 120(15 - 8.5)(8.5 - 2) = \$5070.$$

■

The method outlined in Example 2 is valid for any situation in which an increase in one variable causes a decrease in another related variable. **Maximizing yield** involves optimization situations in which an increase in one variable causes a decrease in another related variable.

Example 3: *Maximum Yield* Taylor's Orchards has always planted 40 trees per acre, with a yield of 300 apples per tree. For each additional tree planted per acre, the yield drops by 5 apples per tree. How many trees should be planted per acre for maximum yield?

Solution: Let n represent the number of trees per acre and T represent the total yield. The total yield per acre is given by (number of trees) • (yield per tree).

Constraint: For each additional tree beyond 40, the yield per tree drops by 5. The yield per tree is 300 minus 5 for each additional tree, which is expressed as

$$\text{yield per tree} = 300 - 5(n - 40)$$

Objective: Maximize total yield, where the total yield per acre is given by (number of trees) • (yield per tree). So,

$$T = n[300 - 5(n - 40)]$$
$$= n(500 - 5n)$$

where $n \geq 40$. So, $40 \leq n \leq 100$ because if $n > 100$, the total yield would be negative.

Now let's find the extreme point:

$$T(n) = n(500 - 5n) \quad \text{or} \quad T(n) = 500n - 5n^2 \text{ and } T'(n) = 500 - 10n$$

Because $T'(n) = 0$ if $500 - 10n = 0$, $n = 50$ is the critical value. Because $T''(n) = -10 < 0$, the concavity is always downward and the extreme point is a maximum.

Maximum yield occurs if 50 trees are planted. The yield per tree would be $300 - 5(50 - 40) = 250$ apples per tree for a maximum yield of 50(250) or 12,500 apples per acre. ■

Check Your Understanding 18.1

1. Determine the maximum revenue if $q(p) = -p^2 + 66p$.
2. If an electronic game sells for $40, the store can sell 60 games per month. For each increase of $1 in the price of the game, the store sells three fewer games per month. If each unit costs the store $25, what price will maximize the store's profit?

Inventory Control

Another type of problem frequently encountered in business applications is **inventory control**. Suppose a manufacturer intends to produce and sell a large number of a particular item over the next year. The manufacturer could produce all the items at once at the beginning of the year. This method would hold down production costs, but it would require large storage costs to store the unsold items during the year. Another way is to spread out the production over the year. Although this method would reduce storage costs, it would also significantly increase the cost of production because it costs more to manufacture small quantities many times than to produce large quantities fewer times. Most manufacturers seek a middle ground, determining the number of production runs that will minimize the cost of production, which includes the cost of storage.

Example 4: MAC Boats anticipates a demand for 12,000 fishing boats over the next year. The start-up costs of each production run are $5000, and it costs the company $40 to store each boat during the year. How many boats should be made during each production run to minimize total costs? How many production runs should there be?

Solution: We assume demand is uniform over the year and costs do not change during the year.

Let b be the number of boats produced in each production run and n be the number of production runs.

Constraints: There are n production runs of b boats each, and the total number of boats produced must be 12,000, so $n \cdot b = 12{,}000$. Solving for n yields $n = 12{,}000/b$.

The total costs for the year are

$$\text{total costs} = \text{set-up costs} + \text{storage costs}$$

Set-up costs are \$5000n. What about storage costs? If all the boats are produced at the beginning, storage costs would be \$40(12,000) or \$480,000. See Figure 18.1.

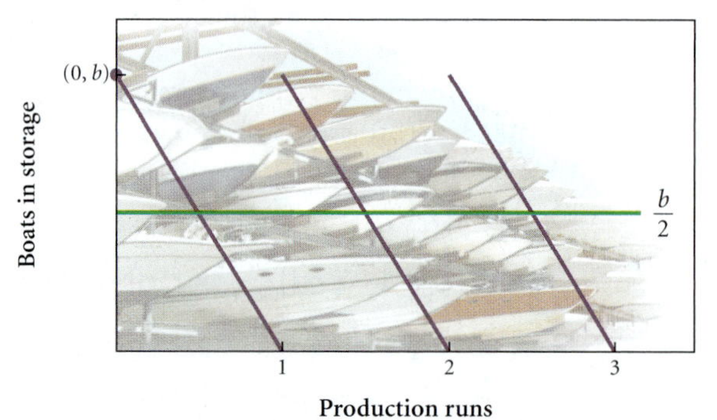

Figure 18.1

Assuming demand is uniform, the number of boats in storage between production runs decreases from b to 0. The average number stored each day is $b/2$. It costs \$40 per boat for storage, so the total storage costs will be \$40($b/2$) or $20b$.

Objective: Minimize costs.

Going back to the original equation:

$$\text{total costs} = \text{set-up costs} + \text{storage costs}$$
$$C = 5000n + 20b$$

Substituting $n = 12{,}000/b$ gives

$$C(b) = 5000\left(\frac{12{,}000}{b}\right) + 20b$$

$$C(b) = \frac{60{,}000{,}000}{b} + 20b$$

$$C'(b) = -\frac{60{,}000{,}000}{b^2} + 20$$

where $0 < b \leq 12{,}000$. To find the extreme value, solve

$$-\frac{60{,}000{,}000}{b^2} + 20 = 0$$
$$-60{,}000{,}000 + 20b^2 = 0$$
$$b^2 = 3{,}000{,}000$$
$$b \approx \pm 1732.05$$

Because $0 < b \leq 12{,}000$, the only meaningful solution is $b \approx 1732$.

To prove that $b = 1732$ is the minimum, use the Second Derivative Test:

$$C'(b) = -\frac{60{,}000{,}000}{b^2} + 20$$

$$C''(b) = \frac{120{,}000{,}000}{b^3}$$

Thus, $C''(1732) = 120{,}000{,}000/1732^3 > 0$, so at $b = 1732$, the concavity is upward and the extreme point is a minimum. Therefore, approximately 1732 boats should be produced in each production run.

How many production runs should there be? From the constraint equation, we know that $n = 12{,}000/b$, so $n = 12{,}000/1732 \approx 6.93$. Costs will be minimized if there are seven production runs of approximately 1732 boats each. ■

Cobb–Douglas Productivity Model

Productivity of a plant or factory is measured by the total number of units produced. It usually depends on both the amount of labor and the amount of capital. Labor refers to the number of employees or workers hired to do the job. Capital is the operating budget, which includes, for example, the cost of the building and the equipment. The **Cobb–Douglas productivity model** describes productivity as a function of both labor and capital.

Cobb–Douglas Productivity Model

The productivity of a plant or factory is given by

$$P = Kx^a y^{1-a}$$

where P is the number of units produced, x is the number of employees, and y is the operating budget or capital. K and a are constants that are determined by each individual factory or plant, with $0 < a < 1$.

Example 5: Brian's Beach Shop manufactures surfboards. Daily operating costs are \$80 per employee and \$25 per machine. The number of surfboards produced each day is given by $q = 4.5x^{0.8}y^{0.2}$, where x is the number of employees and y is the number of machines. If Brian wants to produce 90 surfboards each day at minimum cost, how many employees and how many machines should he use?

Solution: Let x be the number of employees and y be the number of machines.

Constraint: $q = 4.5x^{0.8}y^{0.2}$, where $x > 0$ and $y > 0$ and $q = 90$. Thus, $90 = 4.5x^{0.8}y^{0.2}$. Solving for y yields

$$90 = 4.5x^{0.8}y^{0.2}$$

$$20 = x^{0.8}y^{0.2}$$

$$\frac{20}{x^{0.8}} = y^{0.2}$$

To solve for y, we raise each side of the equation to the $1/0.2$ power. Because $1/0.2 = 5$, it is equivalent to raising each side to the fifth power:

$$\left(\frac{20}{x^{0.8}}\right)^5 = (y^{0.2})^5$$

$$\left(\frac{20^5}{x^4}\right) = y$$

Objective: Minimum costs, where cost is \$80 per employee and \$25 per machine. Costs are given by $C = 80x + 25y$. Substituting $(20^5/x^4) = y$ gives

$$C = 80x + 25\left(\frac{20^5}{x^4}\right) \quad \text{and} \quad C'(x) = 80 - \frac{100(20^5)}{x^5}$$

Next, find the extreme points:

$$80 - \frac{100(20^5)}{x^5} = 0 \rightarrow 80x^5 - 100(20^5) = 0$$

$$x^5 = \frac{100(20^5)}{80} = \frac{5(20^5)}{4}$$

Taking the fifth root of each side gives

$$x = \sqrt[5]{\frac{5(20^5)}{4}} = 20\sqrt[5]{\frac{5}{4}} \approx 20.9$$

Thus, costs are minimized if Brian uses approximately 21 employees. Because $(20^5/x^4) = y$, we have $y = 20^5/21^4 \approx 16.45$. So, Brian will minimize his costs if he uses 21 employees and 16 machines. ■

Check Your Understanding 18.2

1. Repeat Example 4 with 20,000 boats demanded, production costs of \$6000 per run, and storage costs of \$50 per boat per year.
2. Repeat Example 5 if labor costs are \$100 per employee and \$40 per machine and the desired output is 75 surfboards. The Cobb–Douglas productivity model is $q = 5x^{0.75}y^{0.25}$.

Check Your Understanding Answers

Check Your Understanding 18.1

1. $R = \$42{,}592$
2. The price of each game should be \$42.50, which gives a maximum profit of \$918.75.

Check Your Understanding 18.2

1. There should be nine production runs of approximately 2222 boats each.
2. There should be 16 employees and 12 machines.

Topic 18 Review

18

This topic discussed strategies for solving business-related optimization problems. Recall that an optimization problem involves

- Variables
- An objective function, which is the quantity to be optimized
- Constraints, or the limitations on the value of the variables

Also recall that the strategy for solving optimization problems is as follows:

- Read the problem.
- Identify the unknowns and represent them as variables.
- Write the constraint equation and solve for one of the variables.
- Determine the objective function and write it as a function of only one variable.
- Find the critical value(s).
- Substitute the critical value(s) into the constraint equation to determine the value of all variables.

CONCEPT	EXAMPLE
Revenue is the number of items sold times the price per item, or $R = p \cdot q$. **Maximizing revenue** involves determining the price at which a quantity should be sold to obtain maximum revenue.	Lindsay owns a tattoo parlor. If she charges \$$p$ for a basic tattoo, the weekly demand is $D(p) = -0.02p^2 + 0.03p + 1825$ What price would give the greatest revenue?
Maximizing profit involves determining the price at which a quantity should be sold to obtain maximum profit.	If it costs Lindsay \$25 per tattoo, what price will generate the maximum profit?
Maximizing yield involves optimization situations in which an increase in one variable causes a decrease in another related variable.	Jerry rents snowmobiles at a ski lodge. If the rental fee is \$25 per half hour, Jerry knows he can expect 50 rentals per day. For each \$1 increase in the rental fee, he loses 3 rentals per day. What should Jerry charge per half hour to maximize his revenue?

(continued)

Inventory control problems involve determining how many production runs should be done to minimize production costs.	Jerry's friend Pat sells snowmobiles and expects to sell 50 each month. Each snowmobile ordered costs Pat \$7000, and there is a fixed charge of \$250 per order. It costs Pat \$50 to store a snowmobile for a year. How many snowmobiles should Pat order and how often should the orders be placed to minimize inventory costs?
The **Cobb–Douglas productivity model** relates productivity to labor and capital. The model is $P = Kx^a y^{1-a}$, where P is the number of units produced, x is the number of employees, and y is the operating budget or capital. K and a are constants, where $0 < a < 1$, and are determined by each individual factory or plant.	A snowmobile plant has a Cobb–Douglas productivity model of $q = 0.14x^{0.3}y^{0.7}$, where x is the number of employees and y is the daily operating budget. Daily operating costs amount to \$2500 per employee. If the plant wants to produce 120 snow mobiles per day at minimum cost, how many employees should be hired?

NEW TOOLS IN THE TOOL KIT

- Maximizing revenue
- Maximizing profit
- Maximizing yield
- Minimizing costs of inventory control problems
- Using the Cobb–Douglas productivity model to maximize production

Topic 18 Exercises

18

In Exercises 1 through 4, find the price p that maximizes revenue if the demand is given by the indicated function.

1. $q(p) = -p^2 + 24p$
2. $q(p) = -p^2 + 120p$
3. $q(p) = -p^2 + 5p + 50$
4. $q(p) = -4p^2 + 28p + 120$
5. Demand for a video game, in hundreds, is given by $q(p) = -1.6p^2 + 100p + 50$. What price will maximize revenue?
6. Demand for a microwave oven, in thousands, is given by $q(p) = -0.2p^2 + 70p + 43$. What price will maximize revenue?
7. Demand for tickets to a professional football game is given by $q(p) = -7.7p^2 + 495.8p + 10{,}000$. What price will maximize revenue?
8. Demand for tickets to a theme park is given by $q(p) = -10p^2 + 678p + 7500$. What price will maximize revenue?

9. Marge is planning a casino bus trip. If 100 people sign up, the cost is $300 per person. For each additional person above 100, the cost per person is reduced by $2 per person. To maximize Marge's revenue, how many people should go on the trip? What is the cost per person?

10. A charter dinner cruise boat holds 50 people. The company will charter the boat for 35 or more people. If 40 people are on board, the cost per person is $150. For each additional person, the cost per person is reduced by $3. How large a group should be on the cruise to maximize the revenue? What is the cost per person?

11. A peach orchard has an average yield of 90 bushels per tree if there are 20 trees per acre. For each additional tree per acre, the yield decreases by 3 bushels per tree. How many trees should the orchard plant per acre to maximize the yield? What is the total yield?

12. An orange grove plants 25 trees per acre and gets a yield of 116 bushels of oranges per tree. For each additional tree planted per acre, the yield decreases by 4 bushels per tree. How many trees should be planted per acre to maximize the yield? What is the total yield?

13. Matt's Top 40 rents movies. If the rental fee is $4 each, Matt knows he can rent 100 movies per week. For each additional $1 increase in the rental fee, Matt loses 10 rentals per week. What rental fee should Matt charge for a movie to maximize his revenue?

14. Barb's Babysitting charges $8 per hour and, at that rate, averages 20 jobs each week. For each additional $1 charge per hour, the number of jobs per week declines by two. What should Barb charge per hour to maximize revenue?

15. Refer to Exercise 13. If each movie costs Matt $2, what should his rental price be to maximize profit?

16. Refer to Exercise 14. If Barb spends $2 per job on supplies, what should she charge per hour to maximize her profits?

17. When Jerry's Jalopies charges $20 to do an oil change, there are 80 customers per month. For each additional $1 charge, the number of customers per month drops by four. If it costs Jerry $5 per customer for the supplies, what should he charge for an oil change to maximize profits? How many customers will there be each month?

18. Missy's Tutoring charges $35 per hour for a tutoring session and has 60 clients each week. For each additional $1 charge, there are two fewer clients each week. It costs $12 per client for supplies. What should Missy charge per hour to maximize profits? How many clients will there be each week?

19. A bicycle plant assembles 2000 bicycles per month. Each production run costs $1200, and it costs $20 to store a bicycle for a month. How many production runs should the plant use to minimize inventory costs? How many bicycles are assembled in each production run?

20. A soda bottling company bottles 20,000 cases of lime soda each year. Each production run costs $1400, plus a storage cost of $18 per case. How many production runs should the company use to minimize inventory costs? How many cases are bottled in each production run?

21. A textbook publisher estimates that the demand for a new calculus book will be 6000 copies. Each book costs $12 to print, and set-up costs are $1800 for each printing run. Storage costs $3 per book per year. How many books should be printed per printing run and how many printings should there be to minimize inventory costs?

22. A car dealer expects to sell 500 new convertibles this year. Each convertible costs the dealer $16,000 plus a fixed $5000 delivery fee per order. It costs $500 to store each car for a year. How many orders should be made and how many cars should there be in each order to minimize inventory costs?

23. A golf club manufacturer finds that production of golf clubs follows the model $q = 25x^{0.25}y^{0.75}$, where x is the number of employees and y is the number of machines. If the manufacturer must produce 2000 clubs and it costs $70 per employee and $20 per machine, how many employees and how many machines will minimize costs?

24. Refer to Exercise 23. If production is increased to 3000 clubs and costs increase to $30 per machine, how many employees and how many machines will minimize costs?

25. Tammy manages a test laboratory for a pharmaceutical company. The number of tests she can

run each month has a Cobb–Douglas model of $C = 2.5x^{0.3}y^{0.7}$, where x is the number of employees and y is the number of lab stations. The company needs to run 50 tests per month. If employees earn an average of $2500 per month and it costs $850 per lab station per month, how many employees should Tammy hire to minimize the monthly operating cost? How many lab stations will she need?

26. Refer to Exercise 25. If the company decides that Tammy must run 80 tests per month and lab costs increase to $900 per month, how many employees will Tammy need to hire to minimize the monthly operating cost?

SHARPEN THE TOOLS

27. A 150-room resort hotel is filled at a room rate of $125 per day. For each $5 increase in the room rate, three fewer rooms are rented. What room rate will result in maximum daily revenue? How many rooms will be rented at that rate?

28. Refer to Exercise 27. If it costs $10 per day to clean each occupied room, what room rate will maximize daily profit? How many rooms will be rented at that rate?

29. Two years ago, Burrows Orchards had 50 trees planted per acre with a yield of 60 bags of fruit per tree. Last year, they removed 10 trees per acre and saw the yield increase to 80 bags of fruit per tree. How many trees should be planted per acre to maximize the yield?

30. Because weather forecasters are predicting a bad winter, a company that manufactures generators for home use expects a demand for 20,000 generators this winter. Each production run costs $20,000, and it costs the company $350 to store each generator during the year.

a. How many generators should be produced in each production run to minimize the costs? How many production runs should there be?

b. If the winter is not as bad as predicted, the demand is reduced to 7500 generators. For this quantity, how many generators should be produced in each production run and how many runs should there be?

c. As the plant production manager, how would you decide how many generators to produce?

CALCULATOR CONNECTION

31. The following table shows the yield of a vineyard.

Number of Vines per Acre	Bottles per Acre
100	320
120	320
140	320
150	320
155	316
160	312
165	308
170	304

To maximize the yield, how many vines should be planted per acre?

32. Data for a soft-drink bottling company show the relationship between labor hours (in thousands), capital (net assets, in millions), and number of bottles produced (in thousands) per month, as given in the table.

Labor Hours (x)	Capital (y)	Production
100	11	340
100	13	360
110	14	395
125	16	445
133	17	475
140	20	525
151	23	570
152	23	575
160	24	600
166	26	635

The Cobb–Douglas productivity model is $q = 6.92x^{0.678}y^{0.322}$. Corporate headquarters requests that 700,000 bottles be produced each month, but labor may not exceed 180,000 hours and capital outlay may not exceed $30 million. Labor costs average about $60 per hour. As production manager, how would you meet the request while still maintaining the constraints of the plant?

TOPIC

Implicit Differentiation and Related Rates

19

All the functions studied so far have been expressed as equations in the form $y = f(x)$. The derivative gives a function's instantaneous rate of change for any value of x or the slope of the graph of the function at any point. Not all equations are expressed in the form $y = f(x)$, however. Consider, for example, the equation of a circle, $x^2 + y^2 = r^2$. In this topic, you learn a method to find the rate of change of functions not in the form $y = f(x)$, called implicit differentiation. Before beginning this topic, complete the Warm-up Exercises to sharpen your required algebra and calculus skills.

IMPORTANT TOOLS IN TOPIC 19

- *Explicit and implicit equations*
- *Implicit differentiation*
- *Related rates problems*

TOPIC 19 WARM-UP EXERCISES

Be sure you can successfully complete the following exercises before starting Topic 19.

Algebra Warm-up Exercises

1. Solve for x:

$$\frac{400}{x-3} = 20$$

Determine the value of x in each right triangle.

2.

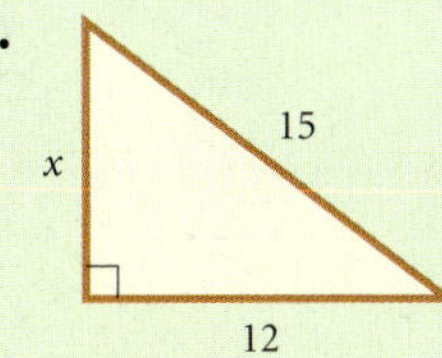

3.

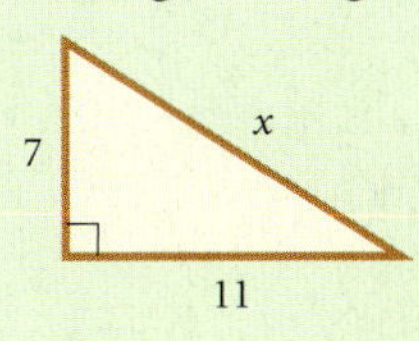

4.

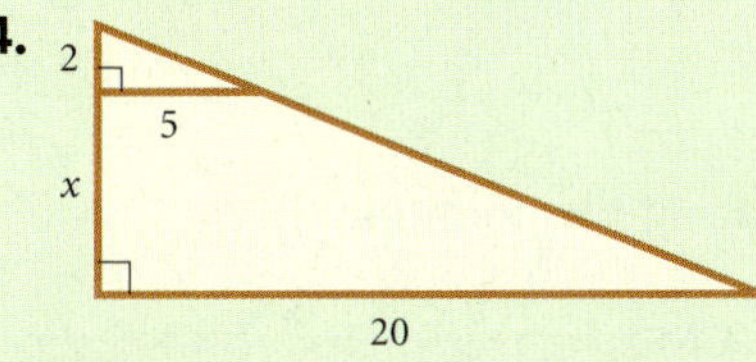

Calculus Warm-up Exercises

Find $f'(x)$ for each function in Exercises 1 through 4.

1. $f(x) = 4x^3 - 5x^2 + 7x - 2/x^4$

2. $f(x) = 3x^2(4x - 5)^3$

3. $f(x) = \sqrt{x^2 - 4}$

4. $f(t) = 2t^2e^{-0.3t}$

5. Maximize $A = x^2y$ if $3x + y = 10$.

Answers to Warm-up Exercises

Algebra Warm-up Exercises

1. $x = 23$

2. $x = 9$

3. $x = \sqrt{170} \approx 13.04$

4. $x = 6$

Calculus Warm-up Exercises

1. $f'(x) = 12x^2 - 10x + 7 + 8/x^5$

2.
$$\begin{aligned} f'(x) &= (4x-5)^3(6x) + 3x^2 \cdot 3(4x-5)^2(4) \\ &= 6x(4x-5)^2(6x + 4x - 5) \\ &= 6x(4x-5)^2(10x - 5) \\ &= 30x(4x-5)^2(2x-1) \end{aligned}$$

3. $f'(x) = \dfrac{1}{2}(x^2 - 4)^{-1/2}(2x) = \dfrac{x}{\sqrt{x^2 - 4}}$

4.
$$\begin{aligned} f'(t) &= 4te^{-0.3t} + 2t^2(-0.3e^{-0.3t}) \\ &= 2te^{-0.3t}(2 - 0.3t) \end{aligned}$$

5. Maximum is $x = 20/9$.

ll the functions studied so far have been expressed as equations in the form $y = f(x)$. Functions of this form are called **explicit functions**. Not all equations are expressed explicitly. The equation of a circle, $x^2 + y^2 = r^2$, for example, is not an explicit equation because the left side of the equation contains both x and y. It is not solved explicitly for y.

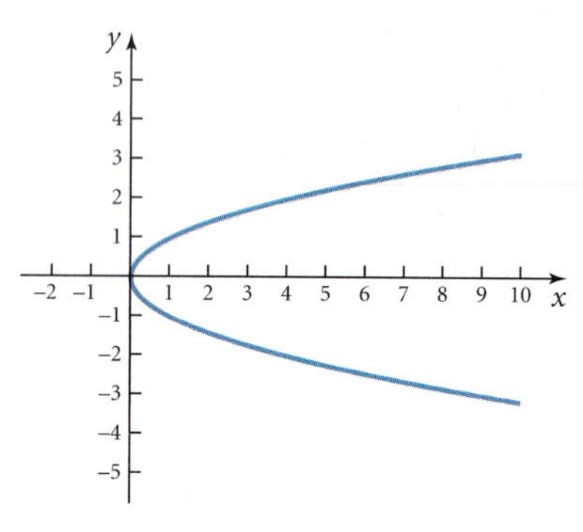

Figure 19.1

Consider $x - y^2 = 0$. Equations of the form $F(x, y) = 0$ are referred to as **implicit equations**. The implicit assumption is that the equation could be solved for y and turned into an explicit function. This particular equation can be expressed as an explicit function by solving for y and obtaining either $y = \sqrt{x}$ or $y = -\sqrt{x}$, which are both explicit functions that can be differentiated. By examining the graph of $x - y^2 = 0$ in Figure 19.1, you should see that a slope exists at all points on the curve except the vertex $(0, 0)$, where the tangent line is vertical. Thus, it must be possible to find the derivative at those points and determine the slope.

For instance, what is the slope at the point $(4, 2)$? The point $(4, 2)$ is on the upper half of the parabola, so we use the function $y = \sqrt{x}$ to determine the derivative. Here, $y = \sqrt{x}$, so $dy/dx = \dfrac{1}{2\sqrt{x}}$ and the derivative at $(4, 2)$ is $y'(4) = \dfrac{1}{2\sqrt{4}} = \frac{1}{4}$. The slope of at the point $(4, 2)$ is $\frac{1}{4}$.

Implicit Differentiation

How would you find dy/dx for $y = x - y^2$? This equation is not easily solved for y, so we find the derivative through a differentiation technique called **implicit differentiation**. This process is used to find dy/dx for equations of the form $F(x, y) = 0$ without having to solve the equation for y first. The basic assumption is that the y variable must be treated as $f(x)$ when differentiating.

For example, we use the General Power Rule (the Chain Rule: Power Form) when evaluating $\dfrac{d}{dx}(y^n)$.

Implicit Differentiation of y^n

If $y = f(x)$, then

$$\frac{d}{dx}(y^n) = n \cdot y^{n-1} \cdot \frac{dy}{dx}$$

Example 1: Find $\dfrac{dy}{dx}$ for $x - y^2 = 0$. Evaluate at the point $(4, 2)$.

Solution: Differentiate each side with respect to x:

$$\frac{d}{dx}(x - y^2) = \frac{d}{dx}(0)$$

$$\frac{d}{dx}(x) - \frac{d}{dx}(y^2) = \frac{d}{dx}(0)$$

Evaluate each derivative:

$$1 - 2y \cdot \frac{dy}{dx} = 0$$

Solve for dy/dx:

$$1 = 2y \cdot \frac{dy}{dx}$$

$$\frac{1}{2y} = \frac{dy}{dx}$$

At the point $(4, 2)$,

$$\frac{dy}{dx} = \frac{1}{2(2)} = \frac{1}{4}$$

which agrees with the previous result obtained by first solving for y. ■

Example 2: Find dy/dx for $y = x - y^2$. Evaluate at the point $(2, 1)$.

Solution: Differentiate each side with respect to x:

$$\frac{d}{dx}(y) = \frac{d}{dx}(x - y^2)$$

Evaluate the derivatives:

$$\frac{dy}{dx} = 1 - 2y \cdot \frac{dy}{dx}$$

Solve for $\frac{dy}{dx}$:

$$\frac{dy}{dx} + 2y \cdot \frac{dy}{dx} = 1$$

$$\frac{dy}{dx}(1 + 2y) = 1$$

$$\frac{dy}{dx} = \frac{1}{1 + 2y}$$

At the point $(2, 1)$,

$$\frac{dy}{dx} = \frac{1}{1 + 2(1)} = \frac{1}{3}$$

■

Example 3: Find dy/dx for $y^2 = y^3 - x^2$ and evaluate at the point $(-2, 2)$.

Solution: Differentiate each side with respect to x:

$$\frac{d}{dx}(y^2) = \frac{d}{dx}(y^3 - x^2)$$

Evaluate the derivatives:

$$2y \cdot \frac{dy}{dx} = 3y^2 \cdot \frac{dy}{dx} - 2x$$

Solve for dy/dx:

$$2y \cdot \frac{dy}{dx} - 3y^2 \cdot \frac{dy}{dx} = -2x$$

$$(2y - 3y^2)\frac{dy}{dx} = -2x$$

$$\frac{dy}{dx} = -\frac{2x}{2y - 3y^2}$$

At the point $(-2, 2)$,

$$\frac{dy}{dx} = -\frac{2(-2)}{2(2) - 3(2)^2} = -\frac{-4}{4 - 12} = -\frac{1}{2}$$

■

Example 4: Find dy/dx for $xy^3 = 54$ and evaluate at the point (2, 3).

Solution: Differentiate each side with respect to x:

$$\frac{d}{dx}(xy^3) = \frac{d}{dx}(54)$$

To evaluate the left side, we must use the Product Rule because both factors (x and y^3) are functions of x:

$$\frac{d}{dx}(xy^3) = \frac{d}{dx}(54)$$

$$y^3 \cdot \frac{d}{dx}(x) + x \cdot \frac{d}{dx}(y^3) = 0$$

$$y^3(1) + x \cdot 3y^2 \cdot \frac{dy}{dx} = 0$$

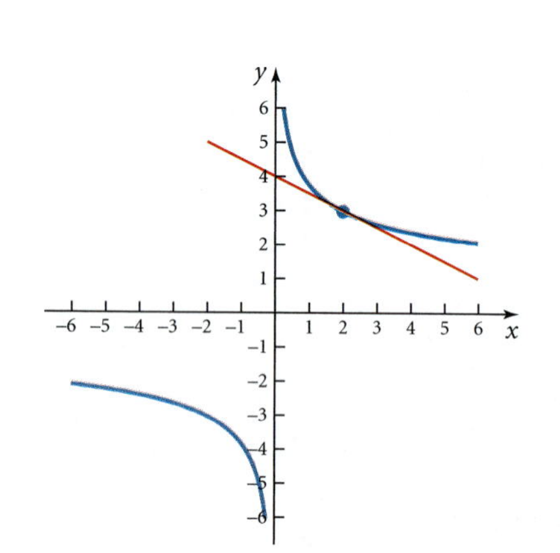

Figure 19.2

Solve for $\frac{dy}{dx}$:

$$\frac{dy}{dx} = -\frac{y^3}{3xy^2} = -\frac{y}{3x}$$

At the point (2, 3),

$$\frac{dy}{dx} = -\frac{3}{3(2)} = -\frac{3}{6} = -\frac{1}{2}$$

The graph of the function (Figure 19.2) verifies that a slope of $-\frac{1}{2}$ at (2, 3) is reasonable.

■

Example 5: A factory uses both full-time and part-time workers in its labor force. Factory output is given by $Q = 4x^3 + 4xy^2 + y^3$, where x is the number of hours full-time workers are used per week and y is the number of hours part-time workers are used per week. Current factory output is 164,000 units with full-time workers used for 30 hours per week and part-time workers used for 20 hours per week.

a. Find dy/dx.

b. Evaluate dy/dx at the current output level and interpret.

Solution:

a. The current output is $Q = 164{,}000$, so $164{,}000 = 4x^3 + 4xy^2 + y^3$. Take the derivative of each side with respect to x:

$$\frac{d}{dx}164{,}000 = \frac{d}{dx}(4x^3 + 4xy^2 + y^3)$$

$$0 = \frac{d}{dx}(4x^3) + \frac{d}{dx}(4xy^2) + \frac{d}{dx}(y^3)$$

(Remember to use the Product Rule for the second term!) Then,

$$0 = 12x^2 + 4x \cdot 2y\frac{dy}{dx} + 4 \cdot y^2 + 3y^2\frac{dy}{dx}$$

Solve for $\dfrac{dy}{dx}$:

$$-12x^2 - 4y^2 = (8xy + 3y^2)\frac{dy}{dx}$$

$$\frac{-12x^2 - 4y^2}{8xy + 3y^2} = \frac{dy}{dx}$$

b. Current output is 164,000 if $x = 30$ and $y = 20$, so

$$\frac{dy}{dx} = \frac{-12(30)^2 - 4(20)^2}{8(30)(20) + 3(20)^2} = -\frac{12{,}400}{6000} \approx -2.07$$

The current output level will be maintained if the number of hours worked by full-time workers increases by 1 hour per week and the number of hours worked by part-time workers decreases by approximately 2.07 hours per week. ■

Check Your Understanding 19.1

Use implicit differentiation to find dy/dx for each of the following implicit equations. Evaluate each derivative at the indicated point.

1. $x^2 + 2y^2 = 22$ at the point $(-2, 3)$
2. $x^3y^2 = 8$ at the point $(2, -1)$

Related Rates

In many instances, the variables of an equation will themselves be functions of another variable, usually time. Problems of this type are called **related rates problems**. Implicit differentiation is used in solving such problems.

Example 6: An object falling into a pool sends out circular ripples. If the radius of the outermost ripple grows at a rate of 3 feet per second, how fast is the area changing when the radius of the outermost ripple is 6 feet? (See Figure 19.3.)

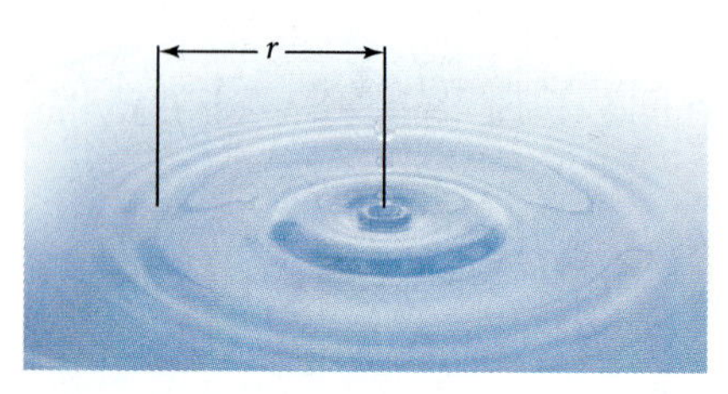

Figure 19.3

Solution: The area of the ripple is given by $A = \pi r^2$, but because the radius is changing over time, both the radius, r, and the area, A, are functions of time, t.

Constraint: The radius grows at a rate of 3 feet per second, so $dr/dt = 3$.

Objective: Find $\frac{dA}{dt}$ when $r = 6$.

Differentiate the area function $A = \pi r^2$ with respect to time:

$$\frac{d}{dt}(A) = \frac{d}{dt}(\pi r^2)$$

$$\frac{dA}{dt} = 2\pi r \cdot \frac{dr}{dt}$$ Product Rule is not needed because π is a constant

We know that $dr/dt = 3$, however, so

$$\frac{dA}{dt} = 2\pi r \cdot 3 = 6\pi r$$

which gives us the rate of change of the area of the outermost ripple for any radius. If the radius is 6 feet, then $dA/dt = 6\pi(6) = 36\pi \approx 113.097$. Thus, when the radius of the outermost circular ripple is 6 feet, the area of that ripple is increasing by approximately 113 square feet per second. ■

Warning! The derivative must be found *before* the known values are substituted!

Example 7: *The Sliding Ladder Problem* A 6-foot ladder is leaning against a wall. José is standing at the top of the ladder, and Mario is standing on the ground at a point 5 feet from the base of the wall, as shown in Figure 19.4. The ladder begins to slide down the wall at a rate of 3 feet per minute. How fast is the bottom of the ladder moving when it hits Mario?

Figure 19.4

Solution: Let x be the distance from the base of the ladder to the wall and y be the height of the top of the ladder above the ground. From the Pythagorean Theorem, we know that $x^2 + y^2 = 6^2$. We also know that $dy/dt = -3$ (a negative rate because the ladder is sliding *down* the wall). We want to find dx/dt when $x = 5$.

Differentiating with respect to t gives:

$$\frac{d}{dt}(x^2 + y^2) = \frac{d}{dt}(6^2)$$

$$2x\frac{dx}{dt} + 2y\frac{dy}{dt} = 0$$

Solve for $\frac{dx}{dt}$:

$$\frac{dx}{dt} = -\frac{y}{x}\frac{dy}{dt}$$

If $x = 5$, then $5^2 + y^2 = 6^2 \rightarrow y^2 = 36 - 25 = 11$, so $y = \sqrt{11}$.

Now substitute $dy/dt = -3$, $x = 5$, and $y = \sqrt{11}$:

$$\frac{dx}{dt} = -\frac{\sqrt{11}}{5}(-3) \approx 1.99$$

The bottom of the ladder is moving at a rate of approximately 1.99 feet per minute away from the base of the wall at the instant it hits Mario. ■

Example 8: The number of trout in a lake is related to the amount of bacteria in the lake. The trout population is modeled by $y = \frac{3200}{x + 5}$, where x is the bacteria count per cubic centimeter of water and y is the number of trout in the lake. After some time, the bacteria count is rising at a rate of 4 bacteria per cubic centimeter per day. Find the rate of change in the number of trout if there are 160 trout in the lake.

Solution: Both bacteria count and number of trout are related to time.

We know

$$y = \frac{3200}{x + 5} \quad \text{and} \quad \frac{dx}{dt} = 4$$

We want to find $\frac{dy}{dt}$ if $y = 160$. First, differentiate each side:

$$\frac{d}{dt}(y) = \frac{d}{dt}\left(\frac{3200}{x + 5}\right)$$

$$\frac{dy}{dt} = -\frac{3200}{(x + 5)^2}\frac{dx}{dt}$$

Substitute $dx/dt = 4$ and $y = 160$. The equation, however, also requires a value for x. Because $y = \frac{3200}{x + 5}$, we have $160 = \frac{3200}{x + 5}$, so $x = 15$. Next, substitute $dx/dt = 4$ and $x = 15$:

$$\frac{dy}{dt} = -\frac{3200}{(15 + 5)^2}(4) = -32$$

At the current bacteria growth rate, when there are 160 trout in the lake, the number of trout is decreasing by 32 trout per day. ■

Example 9: Past company records show that revenue for the number of video games produced and sold daily is given by $R = 420x - x^2$, where R is the generated revenue in dollars and x is the number of units produced and sold daily. It is also known that games sell at a rate of six per day. How fast is revenue changing when 60 games are being produced and sold daily?

Solution: In this example, both revenue and number of games sold are functions of time.

We know

$$R = 420x - x^2 \quad \text{and} \quad \frac{dx}{dt} = 6 \text{ games per day}$$

We want to find dR/dt if $x = 60$.

First, find the derivative:

$$\frac{d}{dt}(R) = \frac{d}{dt}(420x - x^2)$$

$$\frac{dR}{dt} = 420\frac{dx}{dt} - 2x\frac{dx}{dt}$$

Now substitute the known values, $dx/dt = 6$ and $x = 60$:

$$\frac{dR}{dt} = 420(6) - 2(60)(6) = 1800$$

When 60 games are being produced and sold daily, the generated revenue is increasing by $1800 per day. ■

Check Your Understanding 19.2

1. The radius of a snowball in the form of a sphere melts at a rate of 0.25 inch per minute. How fast is the snowball's volume changing when its radius is 2 inches?
2. The revenue from sales of a certain type of television is given by $R = 250x - x^2$, where x is the number of televisions sold. If the televisions are selling at a rate of 15 sets per week, how fast is the revenue changing in a week when 50 sets are sold?

Check Your Understanding Answers

Check Your Understanding 19.1

1. $dy/dx = -x/2y$, so at the point $(-2,3)$, $dy/dx = \frac{1}{3}$.
2. $dy/dx = -3y/2x$, so at the point $(2,-1)$, $dy/dx = \frac{3}{4}$.

Check Your Understanding 19.2

1. $dV/dt = -4\pi \approx -12.57$. The volume is decreasing by approximately 12.57 cubic inches per minute.
2. $dR/dt = 2250$. Revenue is increasing by $2250 per week.

Topic 19 Review

19

This topic introduced implicit differentiation and a type of application problem called related rates. Related rates problems are solved using implicit differentiation.

CONCEPT	EXAMPLE
A function that is expressed in the form $y = f(x)$ is an **explicit function.**	$y = x^4 - 2x^2 + 5x - 7$ and $y = \dfrac{2x}{3x + 5}$ are examples of explicit functions.
A function that is expressed in the form $F(x, y) = 0$ is an **implicit equation.**	$x = y^2 - 3$ and $2x^3y = 5x + 3y^2$ are examples of implicit equations.
Implicit equations can sometimes be differentiated by first solving for y and then finding dy/dx.	To differentiate $x = y^2 - 3$, first solve for y: $$y = \sqrt{x + 3} \quad \text{or} \quad y = -\sqrt{x + 3}$$ Then $$\frac{dy}{dx} = \frac{1}{2\sqrt{x + 3}} \quad \text{or} \quad \frac{dy}{dx} = -\frac{1}{2\sqrt{x + 3}}$$
Implicit differentiation is a process used to find dy/dx for implicit equations without first solving the equation for y. Find d/dx for both sides of the equation, remembering that $d/dx(y^n) = n \cdot y^{n-1} \cdot dy/dx$. Then solve for dy/dx.	To differentiate $2x^3y = 5x + 3y^2$: $$\frac{d}{dx}(2x^3y) = \frac{d}{dx}(5x + 3y^2)$$ $$2x^3\frac{dy}{dx} + 6x^2y = 5 + 6y\frac{dy}{dx}$$ $$(2x^3 - 6y)\frac{dy}{dx} = 5 - 6x^2y$$ $$\frac{dy}{dx} = \frac{5 - 6x^2y}{2x^3 - 6y}$$
Related rates problems involve two or more variables that are functions of some other variable, usually time.	The sides of a square are shrinking at a rate of 0.5 inch per minute. When the length of a side is 4 inches, at what rate is the area of the square changing? **Solution:** Use the formula $A = s^2$. Find dA/dt if $s = 4$ and $ds/dt = -0.5$: $$\frac{d}{dt}(A) = \frac{d}{dt}(s^2)$$ $$\frac{dA}{dt} = 2s\frac{ds}{dt}$$

(continued)

	If $s = 4$ and $ds/dt = -0.5$, then $$\frac{dA}{dt} = 2(4)(-0.5) = -4$$ When the length of a side is 4 inches, the area is decreasing by 4 square inches.

NEW TOOLS IN THE TOOL KIT

- Distinguishing between explicit and implicit equations
- Solving implicit equations for y and differentiating to find dy/dx
- Differentiating with implicit differentiation to find dy/dx
- Solving related rate problems using implicit differentiation

Topic 19 Exercises

19

Are the following statements true or false?

1. $y = x^2 - 3$ is an implicit equation.
2. $x^2y = 3$ is an implicit equation.
3. $x^2 + y^2 = 25$ is an implicit equation.
4. $y = \sqrt{x - 3}$ is an implicit equation.
5. If $x^2 + y^2 = 4$, then $dy/dx = -x/y$.
6. If $x^3 + 3y^2 = 10$, then $dy/dx = -2y/x^3$.
7. If $x^2y^2 = 4$, then $dy/dx = -4xy$.
8. If $2xy^3 = 61$, then $dy/dx = 61/6xy^2$.
9. If $A = \pi r^2$, then $dr/dt = 1/2\pi r \cdot dA/dt$.
10. If $V = \frac{4}{3}\pi r^3$, then $dV/dt = (4\pi r^2)dr/dt$.

In Exercises 11 and 12, find dy/dx using implicit differentiation. Then check by solving for y and differentiating.

11. $x^2 - 3y = 0$
12. $x^3y = 6$

In Exercises 13 through 22, find dy/dx using implicit differentiation.

13. $4x^2 - 3y^3 = 0$
14. $2x^3 + 5y^2 = 0$
15. $3x - 2y^2 = y$
16. $y^2 = 4x^2 + 3y$
17. $x^3y^2 = 10$
18. $5x^4y^3 = 11$
19. $x^2 - y^2 = x^3$
20. $x^3 + y^3 = x^2$
21. $2x^4 - 3x^2y + y^3 = 0$
22. $4x^3 + 2x^2y^2 - y^3 = 0$
23. Find the slope of the line tangent to $2x^2 - 3y^2 = 6$ at the point $(3, -2)$.
24. Find the slope of the line tangent to $x^2 - 4y^2 = 4$ at the point $(4, \sqrt{3})$.
25. Find the equation of the line tangent to $3x^2 + y^2 = 12$ at the point $(-1, -3)$.
26. Find the equation of the line tangent to $2x^2 + 3y^2 = 14$ at the point $(-1, 2)$.

27. Find the equation of the line tangent to $x^3y^2 = 4x$ at the point $(2, -1)$.

28. Find the equation of the line tangent to $y^3 = y^2 + 3x^4$ at the point $(-2, 4)$.

29. Factory output of a certain type of sneaker is based on the number of full-time and part-time workers used. Output is given by $Q = 2x^4 + 3x^2y^2 + y^3$, where x is the number of hours worked by full-time employees each week and y is the number of hours worked by part-time employees each week. Current factory output is 1,427,000 units with full-time workers used for 20 hours each week and part-time workers used for 30 hours each week. Evaluate dy/dx at the current output level and interpret your answer.

30. Repeat Exercise 29 if $Q = 2x^3 + xy^2 + 3y^2$ and current output is 94,300 units with full-time workers used for 35 hours each week and part-time workers used for 15 hours each week. Evaluate dy/dx at the current output level and interpret your answer.

31. The area of a circular wound heals at a rate of 0.5 cm^2 per day. How fast is the radius of the wound changing when the area is 2 cm^2?

32. In calm water, an oil spill spreads out in a circular fashion. Suppose the radius of the spill is increasing by 3 feet per minute. How fast is the area of the spill growing when the radius is 10 feet?

33. A point moves along the curve $y = e^{0.5x}$. Its x-coordinate changes at a rate of 0.2 unit per day. How fast is the y-coordinate changing when $x = 1.2$?

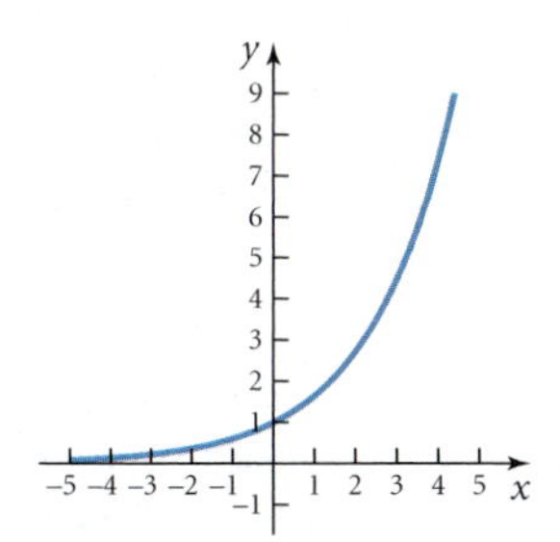

34. A point moves along the curve $y = 3/x$. Its y-coordinate changes at a rate of -0.4 unit per second. How fast is the x-coordinate changing when $y = 0.8$? See graph at top of next column.

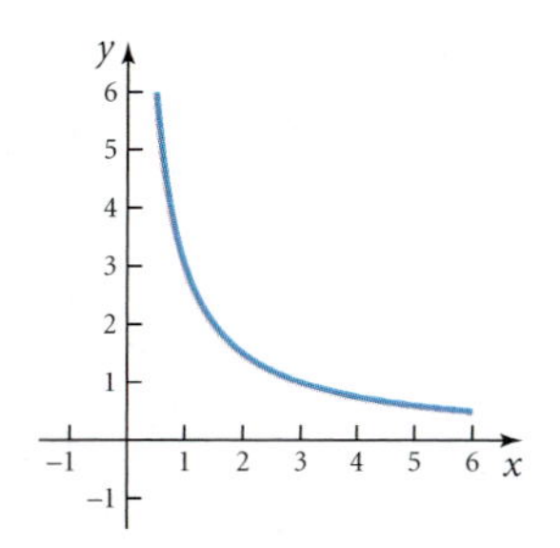

35. Lynn and Charlie work in offices that are 40 miles apart. Lynn leaves home and drives to her office going due east at a rate of 50 mph. At the same time, Charlie leaves home and drives to his office going due north at a rate of 60 mph. How fast is the distance between Lynn and Charlie changing when Lynn has driven 10 miles?

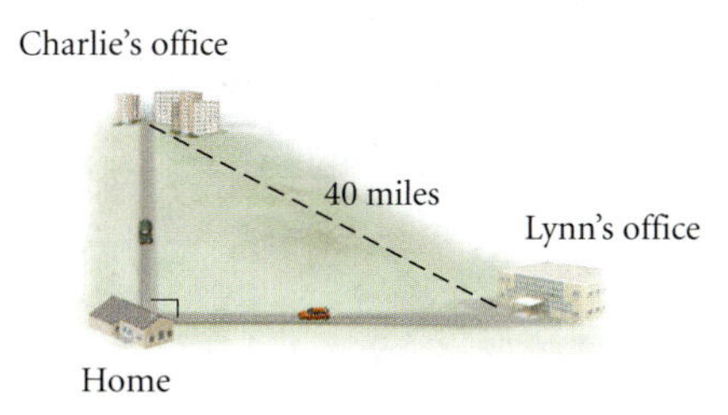

36. Two delivery drivers leave a store. Larry drives due north at 45 mph, and Lucy goes due east at 35 mph. How fast is the distance between the two drivers changing after 30 minutes?

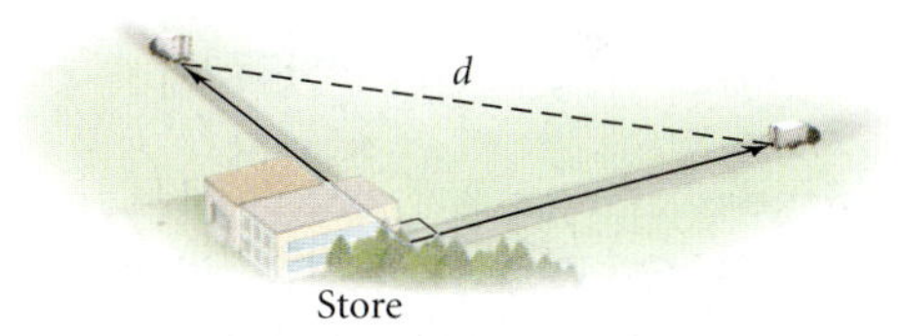

37. A spherical balloon is being inflated at a rate of 500 cm^3 per minute. How fast is the radius of the balloon changing when the radius is 20 cm?

38. The sides of a cube are expanding at a rate of 1.5 inches per second. When the length of a side is 5 inches, find the rate at which

a. The volume is changing.

b. The surface area is changing.

39. Andy placed a 25-foot ladder against his house so he could clean the gutters. If the base of the ladder begins to slide away from the house at a rate of 2 feet per second, how fast is the ladder sliding down the house when the base of the ladder is 12 feet from the house?

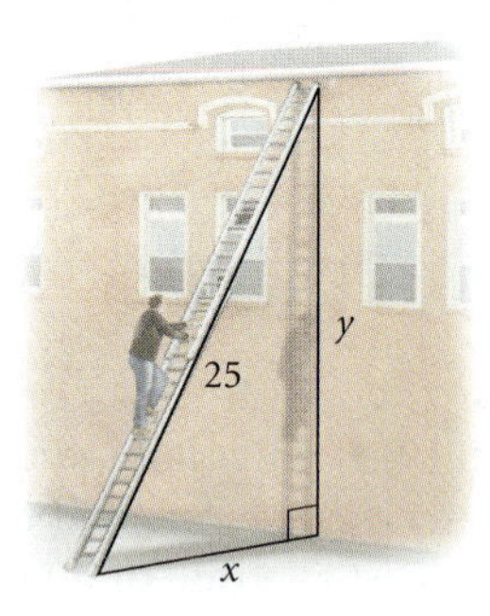

40. A firefighter places a 40-foot extension ladder against a building, with the base of the ladder 6 feet from the building. The firefighter decreases the length of the ladder at a rate of 2 feet per second. At what rate is the ladder sliding down the building when the ladder is 30 feet long?

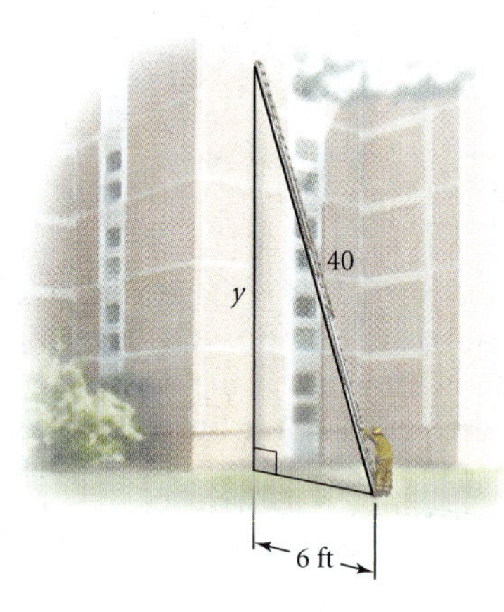

41. Sand is falling into a conical pile at a rate of 8 cubic feet per minute. The diameter of the pile is equal to the altitude. At what rate is the height of the pile changing when the pile is 10 feet high?

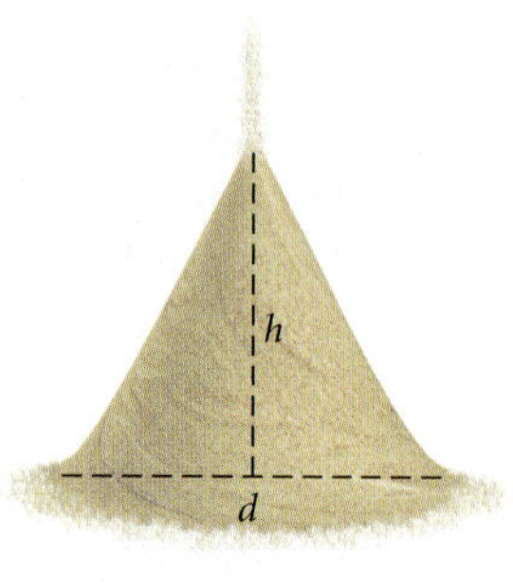

42. A plane at an altitude of 30,000 feet is flying at a rate of 300 mph toward an air traffic control tower on the ground. How fast is the distance from the plane to the tower changing when the horizontal distance is 5000 feet?

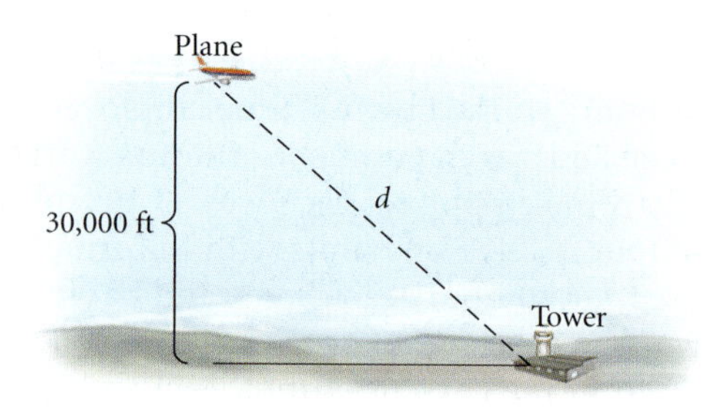

43. The number of home sales for a national real estate company is related to mortgage rates. The relationship is given by $0.3N^3 + 4.4r = 50$, where N is the number of sales, in millions, and r is the annual interest rate (r%). At what rate is the number of home sales changing if the current mortgage rate is 7.5% and that mortgage rate is increasing at a rate of 0.5% per year?

44. Repeat Exercise 43 if the relationship is given by $0.35N^2 + 5.2r = 60$, the current mortgage rate is 8.25%, and the mortgage rate is decreasing at a rate of 0.5% per year.

45. A stainless-steel supplier will produce x thousand units of a certain type of pipe each week when the wholesale price is $\$p$ per pipe. Company research shows that the quantity of pipes produced, x, in thousands, and the price per pipe, p, are related by $x^2 - 2xp + 4p^2 = 525$. Currently, 5000 pipes are produced at a price of \$12.50. If the price is rising \$0.25 per week, how fast is the supply changing?

46. If q thousand units of a product are produced, the total cost, in thousand dollars, is given by $C^2 - 2q^3 = 3500$. When 1500 units are produced and the number produced is increasing at a rate of 200 units per week, at what rate are costs changing?

SHARPEN THE TOOLS

47. The equation for a curve called the *folium of Descartes* is $x^3 + y^3 = 6xy$.

a. Find the location of all horizontal tangent lines.

b. Besides the origin, where does the curve intersect the line $y = x$?

c. Try to draw the graph.

48. The equation for a *cardioid* (see graph below) is

$$x^2 + y^2 = \sqrt{x^2 + y^2} + 4x.$$

Find the location of the horizontal tangent lines.

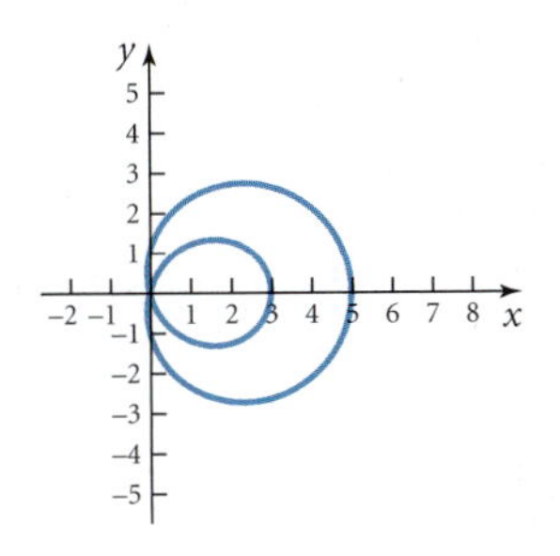

49. Determine $\dfrac{dy}{dx}$ for $\dfrac{x^3}{y^3 - 1} = \dfrac{x + 1}{y - 1}$.

50. Find the equation of the line perpendicular to the graph of $(xy^3 + y)^2 = x^2 + 25$ at the point $(0, 5)$.

51. Find the equation of the line tangent to the graph of $y^3 + y^2 = xy + 1$ at the point where $y = 1$.

52. Find the maximum and minimum points on the rotated ellipse $2x^2 - xy + y^2 = 12$. Draw the graph of the ellipse.

UNIT

2 Review

Unit 2 presented strategies for analyzing the behavior of a function, for graphing functions, and for solving optimization problems.

TOPIC 14	■ Determining the domain, intercepts, and asymptotes of a graph ■ First Derivative Test ■ Extreme points of functions (maximum or minimum, absolute or relative) ■ Describing where a function or graph is increasing and decreasing
TOPIC 15	■ Describing the concavity of a graph ■ Points of inflection ■ Second Derivative Test ■ Drawing graphs of functions using information on location, direction, and shaping
TOPIC 16	■ Extreme points of functions over an interval ■ Elasticity of demand
TOPIC 17	■ Optimization problems and a strategy for solving them
TOPIC 18	■ Business applications, including inventory control and Cobb–Douglas productivity functions ■ Maximizing revenue and minimizing costs
TOPIC 19	■ Implicit differentiation ■ Related rates problems and a strategy for solving related rates problems using implicit differentiation

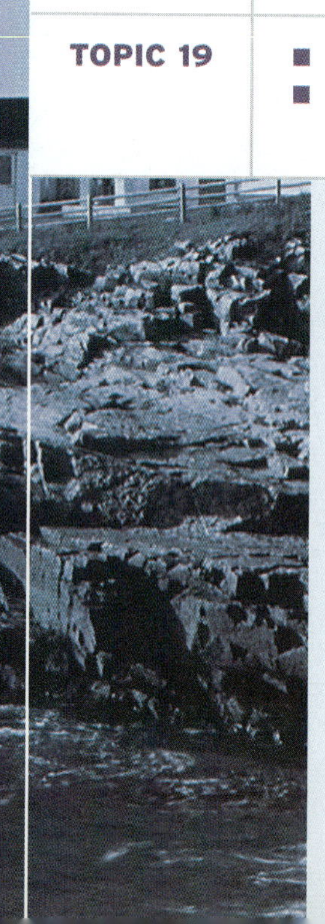

Having completed Unit 2, you should now be able to:

1. Describe the location of a graph by stating the domain, the intercepts, and the asymptotes.
2. Use the First Derivative Test to determine the intervals for which a function is increasing or decreasing.
3. Determine the extreme points of a function.
4. Classify the extreme points as maximum or minimum.
5. Classify the extreme points as absolute or relative.
6. Use the Second Derivative Test to determine the intervals for which a function is concave up or concave down.
7. Determine the points of inflection of a function.
8. Draw the graph of a function using information on location, direction, and shaping.
9. Determine the absolute extreme points of a function over a specified interval in its domain.
10. Determine elasticity of demand for a demand function and classify as elastic, inelastic, or unit elastic.
11. Solve application problems involving optimizing a function with the function given.
12. Solve application problems involving optimizing a function by writing an objective function subject to a constraint.
13. Distinguish between implicit and explicit equations.
14. Determine dy/dx for implicit equations using implicit differentiation.
15. Solve related rates problems using implicit differentiation.

UNIT 2 TEST

The graphs in Exercises 1 through 5 represent the graph of the derivative, $f'(x)$, of a function. For each graph, answer the following questions.

a. For what interval(s) of x does $f(x)$ increase?
b. For what interval(s) of x does $f(x)$ decrease?
c. For what value(s) of x does $f(x)$ have a minimum point?
d. For what value(s) of x does $f(x)$ have a maximum point?
e. For what interval(s) of x is $f(x)$ concave up?
f. For what interval(s) of x is $f(x)$ concave down?
g. For what value(s) of x does $f(x)$ have a point of inflection?

1.

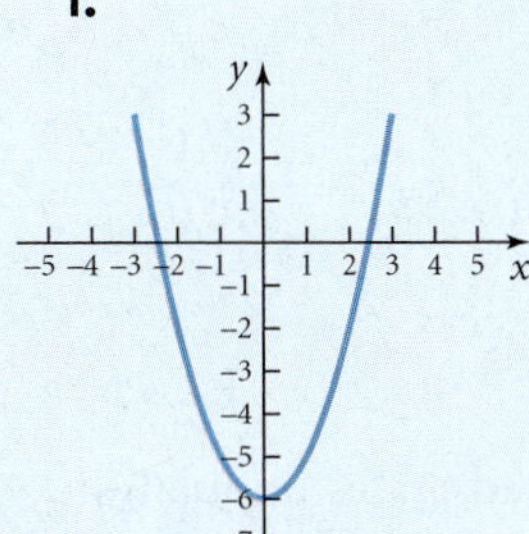

2.

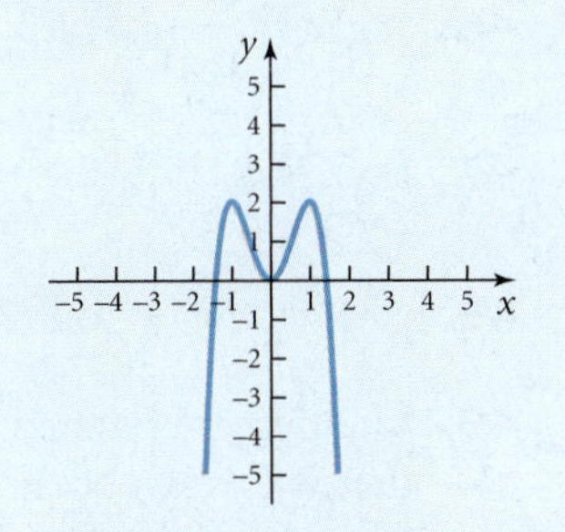

3.

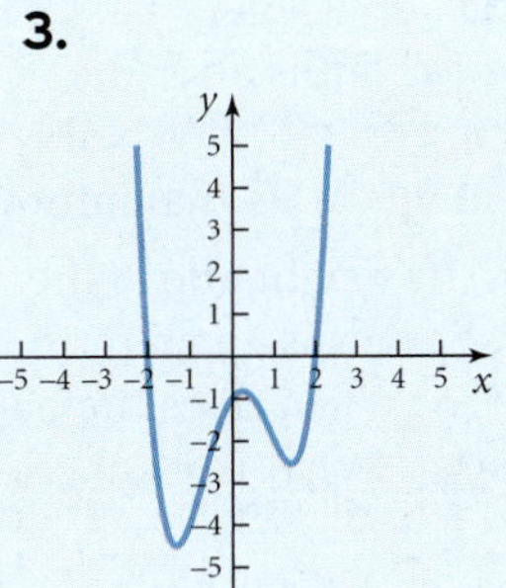

4.

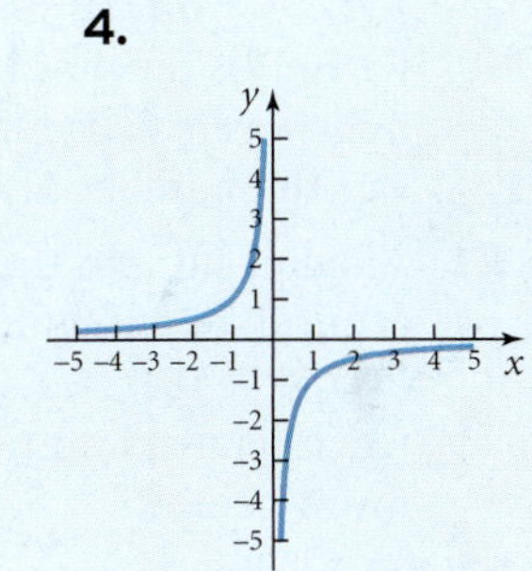

5.

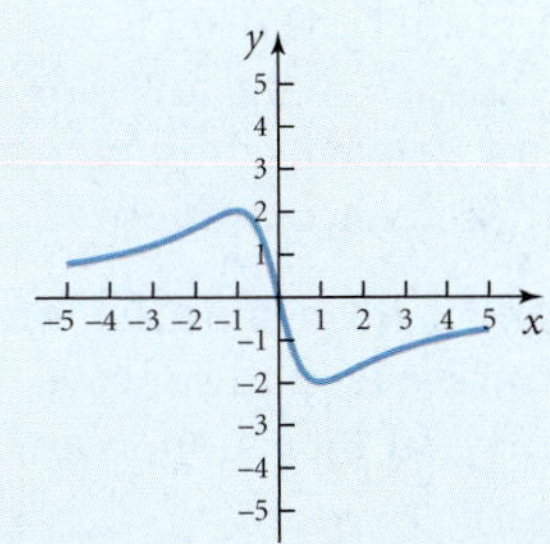

For Exercises 6 through 10, state the domain, intercepts, and asymptotes of the functions.

6. $f(x) = x^2 - 6x$

7. $f(x) = \sqrt{x - 3}$

8. $f(x) = \dfrac{5x}{x + 2}$

9. $f(x) = \dfrac{x^2}{x^3 - 4x}$

10. $f(x) = e^{-3x}$

For Exercises 11 through 18, determine the intervals for which $f(x)$ is increasing and decreasing. Identify the extreme points and classify them as minimum or maximum, relative or absolute.

11. $f(x) = x^2 - 8x + 3$

12. $f(x) = 12x^2 - x^3$

13. $f(x) = x^4 - 4x^3 - 8x^2 + 64$

14. $f(x) = (x^2 - 2x - 8)^2$

15. $f(x) = 6x - 10x^{3/5}$

16. $f(x) = xe^{-x}$

17. $f(x) = \dfrac{3x}{x^2 - 4}$

18. $f(x) = \dfrac{x^2}{\ln(x)}$

19. The percent of the United States population living in poverty can be modeled by $P(t) = 0.04t^2 - 0.91t + 17$, where t is the number of years after 1990. When did the percent living in poverty decrease? What was the minimum percent?

20. Traffic accidents per month at a busy intersection have decreased since a traffic light was installed. The number of accidents per month at the intersection can be modeled by $A(t) = -0.17t^3 + 2.22t^2 - 9.46t + 82.83$, where t is the number of months after the light was installed and $0 \leq t \leq 12$. When was the number of accidents at its maximum?

21. A ball is hit into the air. Its height above the ground, in feet, after t seconds is given by $h(t) = -16t^2 + 64t + 6$. When does the ball reach its maximum height? What is its maximum height?

22. The concentration of a drug in a patient's bloodstream t hours after the drug is injected is given by $C(t) = 2.4te^{-0.6t}\ \text{mg/cm}^3$. For what time period does the concentration increase? What is the maximum concentration?

For Exercises 23 through 32, determine the intervals for which $f(x)$ is concave up and concave down. Identify the point(s) of inflection, if any.

23. $f(x) = x^2 - 8x + 3$

24. $f(x) = 12x^2 - x^3$

25. $f(x) = x^4 - 6x^2 - 5$

26. $f(x) = 6x - 10x^{3/5}$

27. $f(x) = xe^{-x}$

28. $f(x) = (x^2 - 2x - 8)^2$

29. $f(x) = (x - 1)^5$

30. $f(x) = \dfrac{4}{x - 3}$

31. $f(x) = \dfrac{4x}{x + 1}$

32. $f(x) = \dfrac{x^2 + 1}{x^2 - 4}$

For Exercises 33 through 40, use the graphing strategy to draw the graph of the function.

33. $f(x) = 12x^2 - x^3$

34. $f(x) = x^4 - 6x^2 - 5$

35. $f(x) = (x - 1)^5$

36. $f(x) = 6x - 10x^{3/5}$

37. $f(x) = xe^{-x}$

38. $f(x) = \dfrac{4}{x - 3}$

39. $f(x) = \dfrac{4x}{x^2 + 1}$

40. $f(x) = \dfrac{x^2 + 1}{x^2 - 4}$

41. Traffic accidents per month at a busy intersection have decreased since a traffic light was installed. The number of accidents per month at the intersection can be modeled by $A(t) = -0.17t^3 + 2.22t^2 - 9.46t + 82.83$, where t is the number of months after the light was installed and $0 \leq t \leq 12$. When was the rate of decrease in the number of accidents at its maximum?

42. Monthly sales, in thousands, of a new laptop computer grow according to the surge model $N(t) = 4.7te^{-0.3t}$, where t is the number of months after the introduction of the computer. Determine the point of diminishing returns and interpret your answer.

For Exercises 43 through 48, determine the absolute extreme points of the function over the indicated interval.

43. $f(x) = x^3 - 6x^2$ on $[1, 5]$

44. $f(x) = 3x^4 - 8x^3 - 18x^2 + 5$ on $[-3, 2]$

45. $f(x) = xe^{-3x}$ on $[-2, 3]$

46. $f(x) = \dfrac{x^2}{\ln x}$ on $[1.5, 4]$

47. $f(x) = x + 3 + \dfrac{4}{x}$ on $(0, \infty)$

48. $f(x) = \dfrac{x^2}{x - 3}$ on $(3, \infty)$

49. The percent of the United States population that is foreign born is given by

$$P(t) = 0.00006t^3 - 0.00675t^2 + 0.0523t + 14.15,$$

where t is the number of years after 1900. Find the absolute extrema on the interval [50, 90] and interpret your answers.

50. The average annual number of unemployment claims filed in a county is given by $U(t) = 21.3t^2 - 331t + 2155$, where t is the number of years after 1990. Find the absolute extrema on the interval [5, 12] and interpret your answers.

For Exercises 51 through 53, determine the elasticity of demand and classify as elastic or unelastic.

51. If the price of a calculator drops from \$125 to \$100, the demand increases from 25,500 to 27,800.

52. Steve can sell 150 home theater systems at a price of \$6500 and 120 systems at a price of \$8000.

53. The demand for a new television is given by $D(p) = -7p^2 + 354p - 1748$, where p is the price in hundreds of dollars. Find the elasticity for $p = \$1400$ and $p = \$2700$ and interpret your answers.

54. For Exercise 53, determine the maximum revenue.

55. Maximize $A = 4xy$ if $2x + 2y = 60$.

56. Minimize $C = 6x + 15y$ if $xy = 40$.

57. The sum of two numbers is 30. Find the two numbers that provide for a maximum product.

58. The product of two numbers is 30. Find the two numbers that provide for the minimum sum.

59. Alexis is fencing in a rectangular flower garden with 140 feet of fence. Find the dimensions of the garden providing maximum area if

a. All four sides are fenced.

b. Only three of the four sides are fenced.

60. Harry must enclose a rectangular area beside his auto repair shop that will provide 1280 square feet of work space. If the front fence costs \$10 per foot and the other two sides costs \$8 per foot, leaving the side next to the shop unfenced, what dimensions will allow for the most economical fence?

61. A box with a square base and an open top has a volume of 256 cubic inches. What dimensions will provide for minimum surface area?

62. A can of hair spray in the shape of a right circular cylinder has a total surface area of 75 square inches. What dimensions give maximum volume?

63. Find the point on the curve $y = x^2 - 1$ that is closest to the point (0, 3).

64. Yukari is organizing a bus tour. If 50 people sign up, the cost per person is \$198. For each additional person, the cost per person decreases by \$3. How large a group should Yukari have to maximize her revenue? What is the cost per person?

65. A cherry orchard yields an average of 88 quarts of cherries per tree if 40 trees are planted. For each additional tree, the yield per tree decreases by 2 quarts. How many trees should be planted to provide the maximum yield?

66. At a cost of \$10 per haircut, Bill's Barber Shop can expect 150 customers per week. If Bill raises the price of a haircut by \$1, there will only be 140 customers per week.

a. What should Bill charge to maximize his revenue?

b. If it costs Bill \$4 per haircut for supplies and labor, what should he charge to maximize his profit?

67. Jean's Candle Company expects to produce 12,000 boxes of candles this year. Each production run costs \$670, and it costs \$1 to store each box for a year. How many production runs and how many boxes per run will minimize inventory costs?

68. A bicycle assembly plant has a Cobb–Douglas productivity model of $q = 3.2x^{0.25}y^{0.75}$, where x is the number of employees and y is the number of machines. The plant must assemble 6000 bicycles with costs of \$65 per employee and \$30 per machine. How many employees and how many machines should be used to minimize costs?

For Exercises 69 through 72, find dy/dx using implicit differentiation.

69. $x^3 - 2y = 0$

70. $2x^4 - 3y^3 = y$

71. $3x^3y^2 = 4x^2$

72. $x^3 - y^3 = 3xy$

73. Find the slope of the line tangent to $x^2 + 2y^2 = 8$ at the point $(-2, \sqrt{2})$.

74. Find the equation of the line tangent to $x^2 + y^2 = 16$ at the point $(\sqrt{7}, 3)$.

75. A factory's output is modeled by $Q = 2x^2 + 3xy + y^2$, where x is the number of hours per week worked by full-time employees and y is the number of hours per week worked by part-time employees. Current output is 56,200 units with 30 hours per week worked by full-time employees and 20 hours per week worked by part-time employees. Find dy/dx and interpret the answer.

76. An observer is tracking a small plane flying at an altitude of 5000 feet. The plane flies directly over the observer at a rate of 1000 feet per minute. Find the rate of change of the distance from the plane to the observer when the plane is 12,000 feet horizontally from the observer.

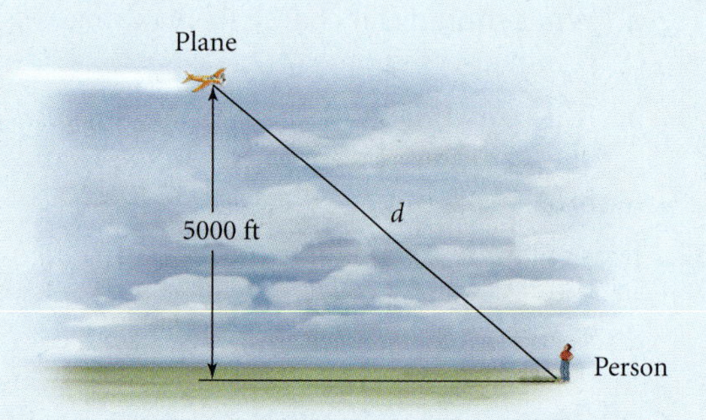

77. The area of a circular sinkhole increases at a rate of 420 square yards per day. How fast is the radius of the sinkhole growing when its radius is 50 yards?

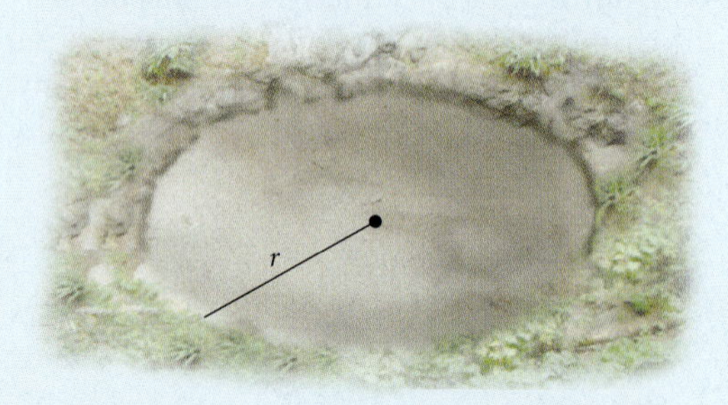

78. Two boats leave a dock, one heading due east at 40 mph and one heading due north at 50 mph. How fast is the distance between the boats changing after 1.5 hours?

UNIT 2 PROJECT

In the Unit 0 Project, you gathered data on a topic of interest and created a model for that data using regression. In the Unit 1 Project, you analyzed the rates of change of that model. This project continues the analysis of that model by exploring its extreme points.

Using the data and best regression model you determined in the Unit 0 Project, answer the following questions.

1. Determine the intervals for which your function increases and decreases. Interpret in the context of the data.
2. What are the extreme points of your model? Interpret in the context of the data.
3. What are the absolute extreme points of your model over its domain? Interpret in the context of the data.
4. Determine the intervals for which your function is concave up and concave down. Interpret in the context of the data.
5. What is the point of inflection for your model? Interpret in the context of the data.
6. Using your data and the function describing it, write a word problem that requires the derivative for solving. Include the solution and explain the answer in the context of the data.

The Integral and Its Applications

UNIT 3

Chronic Care Solutions, now CCS Medical, is a mail-order medical supply company in Clearwater, Florida, whose annual revenues have grown rapidly in recent years.

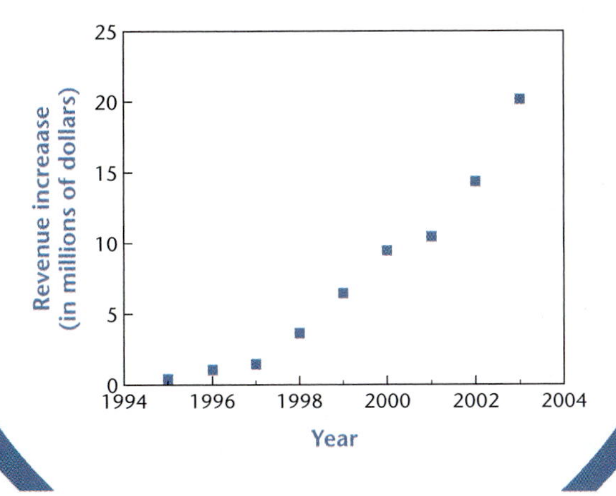

(*Source: Tampa Bay Magazine*, September/October 2003, page 170)

The graph above shows the annual increase in the company's revenues from 1994 to 2003.

Knowing the rate at which the company's revenue was growing, could you determine the revenue in 2006? Could you predict the revenue in 2010? What are the total revenues generated by the company from 2000 to 2007?

These are the types of questions you will learn how to answer in this unit. The previous units focused on finding the derivative of a given function and using that derivative to analyze the function. In this unit we will be working backwards, using the derivative of a function to determine the function.

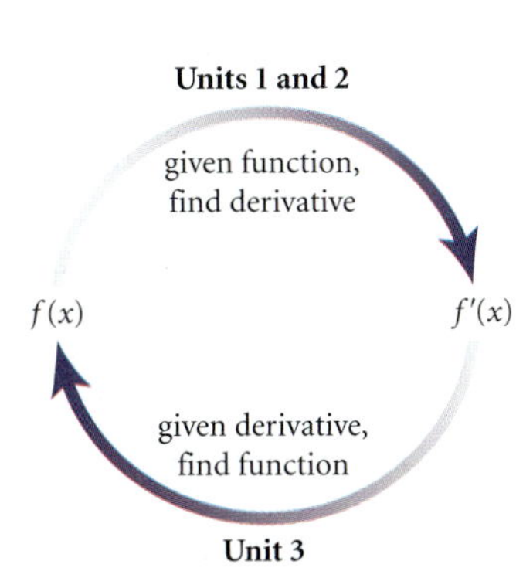

TOPIC

20 Antiderivatives and Integrals

IMPORTANT TOOLS IN TOPIC 20

- *Antiderivatives and integrals*
- *Power Rule for Integration*
- *Properties of integrals*
- *Determining the value of the constant of integration*

Suppose that you are going to a job interview and the interviewer has given you directions from your home to the office. Once the interview is complete, how would you get back to your home? Naturally, you would simply reverse the directions you had been given. You have invested a considerable amount of time learning how to find derivatives of functions. In this topic you will see that the process of differentiation can be reversed—if we know the derivative of a function, we can work backwards and determine that function. Before beginning this topic, be sure to complete the Warm-up Exercises and sharpen the algebra and calculus skills you will need.

TOPIC 20 WARM-UP EXERCISES

Be sure you can successfully complete the following exercises before starting Topic 20.

Algebra Warm-up Exercises

Simplify each of the following, leaving answers as rational numbers.

1. $\frac{1}{3} + 1$ **2.** $\frac{2}{5} - 1$ **3.** $-3 + 1$

4. $-5 - 1$ **5.** $\dfrac{1}{5/3}$ **6.** $8^{2/3}$

Rewrite each of the following in the form x^n.

7. $\dfrac{1}{x^7}$ **8.** $\sqrt[4]{x^3}$ **9.** $\dfrac{4}{3x^2}$

Calculus Warm-up Exercises

Find $f'(x)$ for each of the following functions.

1. $f(x) = x^{12}$ **2.** $f(x) = 5x^3$

3. $f(x) = -2x^5 + 3x^4 - 5x + 10$

4. $f(x) = 3 - 2x^{-4}$ **5.** $f(x) = \dfrac{7}{x^3}$

6. $f(x) = \sqrt[3]{x}$ **7.** $f(x) = x^{-2/5}$

8. Find the slope of $f(x) = \frac{1}{2}x^4 - 3x^2 + 3$ at the point $(2, -1)$.

9. Find the equation of the line tangent to $f(x) = 3x^2 - 5x + 7$ at the point where $x = 2$.

10. The cost, in thousands of dollars, of producing x hundred units of a product is $C(x) = -0.5x^2 + 6x + 10$. Find the marginal cost if 500 units are produced.

Answers to Warm-up Exercises

Algebra Warm-up Exercises

1. $\frac{4}{3}$ 2. $-\frac{3}{5}$ 3. -2
4. -6 5. $\frac{3}{5}$ 6. 4
7. x^{-7} 8. $x^{3/4}$ 9. $\frac{4}{3}x^{-2}$

Calculus Warm-up Exercises

1. $12x^{11}$ 2. $15x^2$ 3. $-10x^4 + 12x^3 - 5$
4. $8x^{-5}$ 5. $-\frac{21}{x^4}$ 6. $\dfrac{1}{3\sqrt[3]{x^2}}$
7. $-\dfrac{2}{5}x^{-7/5}$ 8. slope is 4.
9. $y = 7x - 5$
10. Marginal cost is 1, which means that when 500 units are produced, the additional cost of producing one hundred more units is $1,000.

very mathematical operation has an inverse operation that "undoes" the operation. Consider the following examples.

- $n + 3 - 3$ reverts back to n because subtraction undoes addition. Let $n = 4$.

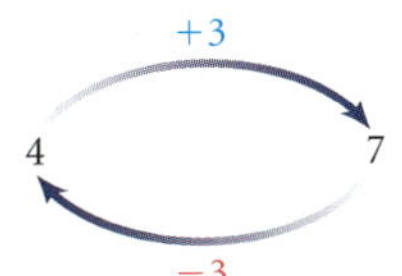

- $(\sqrt{a})^2$ reverts back to a because the square undoes the square root. Let $a = 4$.

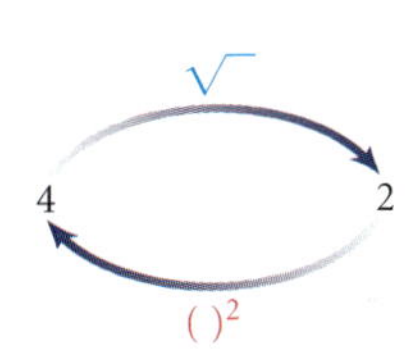

- $\ln(e^n)$ reverts back to n because ln and e are inverses of each other. Let $n = 4$.

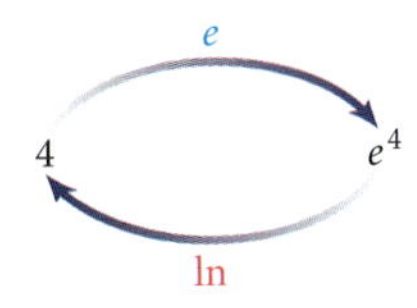

In algebra, after you learned to multiply polynomials together, then you learned factoring, which broke the polynomial back down into its original factors.

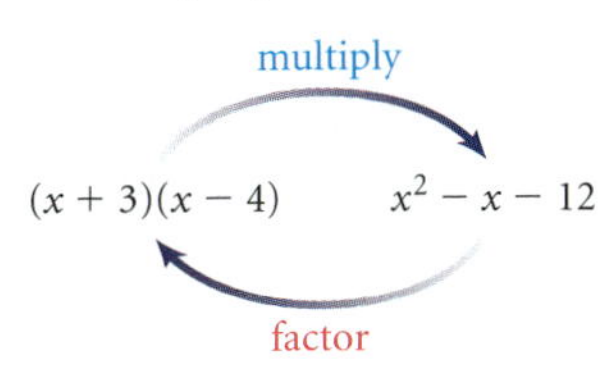

Knowing the derivative of a function, it should be possible to work backwards and determine the original function. If we know the rate at which sales of a product are growing, for instance, can we determine the sales at any time? Or if we know the slope of a curve at any point and a point through which the curve passes, can we determine the equation of the curve? The answer, of course, is yes!

Antiderivatives

The rate at which a function changes is given by the derivative of that function, and the process of finding a derivative is called differentiation. The process of converting a derivative back to the original function is called finding an ***antiderivative*** or ***antidifferentiation***.

Definition: An **antiderivative** of $f(x)$ is a function $F(x)$ such that $F'(x) = f(x)$.

For instance, $F(x) = 2x - 5$ is an antiderivative of $f(x) = 2$ because $F'(x) = 2 = f(x)$. An easy way to organize the discussion of finding antiderivatives is to look at some of the derivative rules you already know and rewrite them as antiderivative rules.

- We know that the derivative of a constant function $f(x) = k$ is 0, so the antiderivative of 0 is a constant k.
- We know that the derivative of a linear function $f(x) = kx$ is the constant k, so an antiderivative of a constant k is a linear function kx.

What would be an antiderivative of the linear function $f(x) = x$? We know that the antiderivative will include x^2, but the derivative of x^2 is $2x$ and our function is just $1x$. Thus, we will divide our antiderivative by 2 (or multiply by $\frac{1}{2}$) and obtain an antiderivative of $\frac{1}{2}x^2$ or $\frac{x^2}{2}$.

Let's review what we have so far.

Function	Antiderivative
0	Constant, k
Constant, k	Linear, kx
Linear, x	Quadratic, $\frac{1}{2}x^2$ or $\frac{x^2}{2}$

What about an antiderivative of x^2? Following the same logic, we know that the antiderivative will include x^3, but the derivative of x^3 is $3x^2$ and our function is just $1x^2$. As before, we will divide the antiderivative by 3 (or multiply by $\frac{1}{3}$) and obtain an antiderivative of $\frac{1}{3}x^3$ or $\frac{x^3}{3}$. Let's add this result to the table.

Function	Antiderivative
0	Constant, k
Constant, k	Linear, kx
Linear, x	Quadratic, $\frac{1}{2}x^2$ or $\frac{x^2}{2}$
Quadratic, x^2	Cubic, $\frac{1}{3}x^3$ or $\frac{x^3}{3}$

You can probably guess what an antiderivative of the cubic function x^3 is. If you said $\frac{1}{4}x^4$ or $\frac{x^4}{4}$ you would be correct! By now you should see a pattern developing. The exponent of an antiderivative is just one more than the exponent of the function, and the coefficient is one divided by that new exponent. Thus, an antiderivative of x^n is $\frac{1}{n+1}x^{n+1}$ or $\frac{x^{n+1}}{n+1}$.

But now consider the following derivatives:

$$D_x(x^2) = 2x$$
$$D_x(x^2 - 5) = 2x$$
$$D_x(x^2 + 13) = 2x$$

Which quadratic function is an antiderivative of $2x$? Actually, there are infinitely many antiderivatives, because $D_x(x^2 + C) = 2x$ for any value of the constant, C. Thus, we must say that the antiderivative of $2x$ is $x^2 + C$, where C is an arbitrary constant called the **constant of integration**. Then the antiderivative of x^n is

$$\frac{1}{n+1}x^{n+1} + C \quad \text{or} \quad \frac{x^{n+1}}{n+1} + C$$

Before continuing any further, we need a symbol to indicate finding an antiderivative. That symbol is called an *integral* sign.

Definition: The symbol used to indicate finding an antiderivative is called the **integral** sign and is denoted as $\int f(x)dx$.

The dx is part of the symbol and indicates that the antidifferentiation is to be done with respect to the variable x. Its meaning will be discussed in more detail in a later topic.

The process of finding an antiderivative is called integration.

The function $f(x)$ is called the integrand.

We can now write our first rule for integration.

Power Rule for Integration

$$\int x^n dx = \frac{1}{n+1}x^{n+1} + C = \frac{x^{n+1}}{n+1} + C$$

where $n \neq -1$.

An integral of this type is called an **indefinite integral**, because it specifies a family of functions. The constant term, C, is called the **constant of integration** and is unknown.

Note that the only restriction on the exponent is that $n \neq -1$.

 MATHEMATICS CORNER  20.1 

Integration as the Inverse of Differentiation

Compare the Power Rule for Integration to the Power Rule for Derivatives and you will see the inverse process of mathematics at work.

To apply the Power Rule for Derivatives, we multiply by the exponent and then subtract 1 from the exponent.

To apply the Power Rule for Integration, start at the end and reverse the steps. We would undo the subtraction by adding 1 to the exponent and then reverse the multiplication by dividing by the exponent.

Example 1: Find each antiderivative.

a. $\int x^{13} dx$ **b.** $\int x^{-5} dx$ **c.** $\int \sqrt{x} dx$ **d.** $\int \frac{1}{x^3} dx$

Solution: Using the Power Rule for Integration, we obtain

a. $\int (x^{13}) dx = \frac{1}{13+1}x^{13+1} + C = \frac{1}{14}x^{14} + C \text{ or } \frac{x^{14}}{14} + C$

b. $\int x^{-5} dx = \frac{1}{-5+1}x^{-5+1} + C = -\frac{1}{4}x^{-4} + C$

The final answer will be left in the same form as the original function.

c. First we must write $\sqrt{x}$ as a power of x in order to apply the Power Rule.

$$\int \sqrt{x} dx = \int x^{1/2} dx = \frac{1}{1/2+1}x^{1/2+1} + C$$ Rewrite as x^n and apply Power Rule

$$= \frac{1}{3/2}x^{3/2} + C$$ Simplify fractions

$$= \frac{2}{3}x^{3/2} + C$$ Simplify coefficient

$$= \frac{2}{3}\sqrt{x^3} + C$$ Rewrite in original form

d. Again we must rewrite $\frac{1}{x^3}$ as a power of x in order to apply the Power Rule.

$$\int \frac{1}{x^3}dx = \int x^{-3}dx = \frac{1}{-3+1}x^{-3+1} + C = -\frac{1}{2}x^{-2} + C = -\frac{1}{2x^2} + C$$

The answer is expressed in the same form as the original function. ■

Properties of Integrals

The following properties of integrals, similar to the properties of derivatives, will simplify our work.

Properties of Integrals

Constant Multiple Property:

$$\int c \cdot f(x)dx = c\int f(x)dx$$

Sum Property:

$$\int [f(x) + g(x)]dx = \int f(x)dx + \int g(x)dx$$

Difference Property:

$$\int [f(x) - g(x)]dx = \int f(x)dx - \int g(x)dx$$

Example 2: Evaluate $\int 2x^4dx$.

Solution: By the Constant Multiple Property,

$$\int 2x^4dx = 2\int x^4dx = 2 \cdot \frac{1}{5}x^5 + C = \frac{2}{5}x^5 + C$$ ■

Example 3: Evaluate $\int \left(p^6 - 4p + 7\right)dp$.

Solution: Using the Sum and Difference Properties, we have

$$\int \left(p^6 - 4p + 7\right)dp = \int p^6dp - \int 4pdp + \int 7dp$$

$$= \frac{1}{7}p^7 - 4 \cdot \frac{1}{2}p^2 + 7p + C$$

$$= \frac{1}{7}p^7 - 2p^2 + 7p + C$$ ■

Example 4: Evaluate $\int\left(\frac{3}{w^2} - 5w^2\right)dw$.

Solution: Use the Difference Property.

$$\int\left(\frac{3}{w^2} - 5w^2\right)dw = \int\left(3w^{-2} - 5w^2\right)dw \qquad \text{Rewrite as a power of } w$$

$$= \int 3w^{-2}dw - \int 5w^2dw \qquad \text{Use the Difference Property}$$

$$= 3 \cdot \frac{1}{-2+1}w^{-2+1} - 5 \cdot \frac{1}{2+1}w^{2+1} + C \qquad \text{Apply the Power Rule}$$

$$= -3w^{-1} - \frac{5}{3}w^3 + C \qquad \text{Simplify}$$

$$= -\frac{3}{w} - \frac{5}{3}w^3 + C \qquad \text{Rewrite in original form}$$

Solutions to integration problems can always be checked. Because the integrand is the derivative of the solution, just take the derivative of the solution and be sure to end up with the original integrand. For instance, in Example 4,

$$D_w\left(-\frac{3}{w} - \frac{5}{3}w^3 + C\right) = D_w\left(-3w^{-1} - \frac{5}{3}w^3 + C\right)$$

$$= -3\left(-1w^{-1-1}\right) - \frac{5}{3}\left(3w^2\right) + 0$$

$$= 3w^{-2} - 5w^2 = \frac{3}{w^2} - 5w^2$$

which was the original integrand. ■

Check Your Understanding 20.1

Find the antiderivative of each of the following.

1. $\int 5dx$ **2.** $\int x^{10}dx$ **3.** $\int 3a^5da$ **4.** $\int (c^{-4} - c^3)dc$

5. $\int\left(\sqrt[4]{x^3}\right)dx$ **6.** $\int\left(2n^7 - 3n^5 + 5n - \frac{4}{n^5}\right)dn$

Finding the Value of the Constant of Integration, C

There are instances in which the value of the unknown constant, C, can be determined.

Example 5: Find the equation of the curve passing through the point $(2, -3)$ if the slope of the curve at any point is given by $\frac{dy}{dx} = 2x^3 + 6x - 5$.

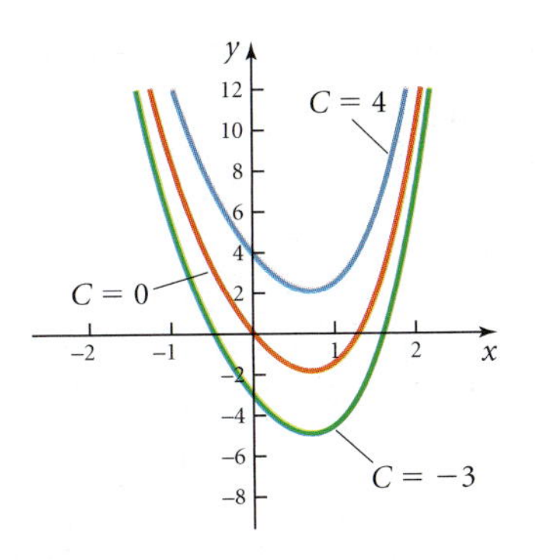

Figure 20.1

Solution: In this problem, we are given two pieces of information about the curve: its slope at any point and a specific point on the curve, called an **initial condition**. Because the slope at any point is given by the *derivative* and we desire the *original function*, the first step is to find the antiderivative.

$$y = \int \left(2x^3 + 6x - 5\right)dx = \frac{1}{2}x^4 + 3x^2 - 5x + C$$

The antiderivative is actually a family of functions with different y-intercepts. Three members of the family are shown in the graph in Figure 20.1. You should see that each curve has the same shape but a different y-intercept.

We must find the one member of that family that contains the point $(2, -3)$. To do that, substitute $x = 2$ and $y = -3$ into the antiderivative and solve for C.

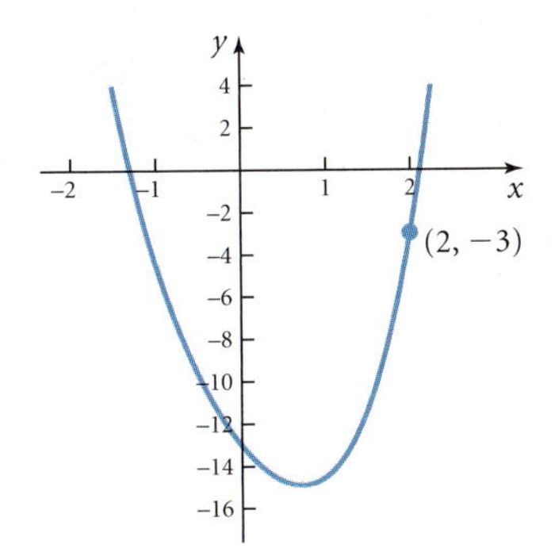

Figure 20.2

$$y = \frac{1}{2}x^4 + 3x^2 - 5x + C$$

so

$$-3 = \frac{1}{2}(2)^4 + 3(2)^2 - 5(2) + C$$
$$-3 = 8 + 12 - 10 + C$$
$$-3 = 10 + C$$
$$C = -13$$

Thus, the curve we needed to find has the equation $y = \frac{1}{2}x^4 + 3x^2 - 5x - 13$. See Figure 20.2. ■

Example 6: An art collection purchased for \$20,000 increases in value at a rate of $300\sqrt{t}$ dollars per year. What is the value of the collection after 25 years?

Solution: What do we know here? We know the original value of the collection—that's the initial condition—and the *rate* at which it increases in value—that's the derivative of the function representing the value. We begin with the antiderivative.

$$V(t) = \int 300\sqrt{t}\,dt = \int 300t^{1/2}dt = 300 \cdot \frac{2}{3}t^{3/2} + C = 200\sqrt{t^3} + C$$

Tip: Be sure you see that the variable in Example 6 is time, being represented by t, so the integral was written in the form $\int f(t)dt$.

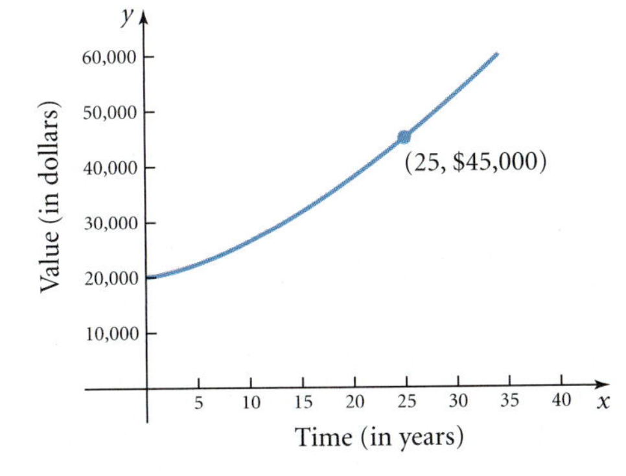

Figure 20.3

We know that the original value is \$20,000, or $V(0) = 20{,}000$. Substitute $t = 0$ and $V(0) = 20{,}000$ into the antiderivative.

$$V(t) = 200\sqrt{t^3} + C$$

so $20{,}000 = 200\sqrt{0} + C$, and $C = 20{,}000$.

Thus, the value of the art collection after t years is given by

$$V = 200\sqrt{t^3} + 20{,}000.$$ See Figure 20.3.

After 25 years the collection is worth

$$V(25) = 200\sqrt{25^3} + 20{,}000 = \$45{,}000$$ ■

Example 7: Alyssa's Bridal Shop creates custom bridal gowns. The monthly marginal cost of producing x gowns is given by $M'(x) = 21x^{4/3} - 6x^{1/2} + 50$. The fixed monthly costs are \$5,000. Find the cost of producing 12 gowns per month.

Solution: In Topic 9 you learned that marginal cost is the derivative of the cost function. Alyssa's costs are given by

$$\begin{aligned} M(x) &= \int \left(21x^{4/3} - 6x^{1/2} + 50\right)dx \\ &= 21 \cdot \frac{3}{7}x^{7/3} - 6 \cdot \frac{2}{3}x^{3/2} + 50x + C \\ &= 9x^{7/3} - 4x^{3/2} + 50x + C \end{aligned}$$

We also know that costs are determined by the sum of the variable costs and the fixed costs. The fixed costs are \$5,000, so $M(0) = 5{,}000$. Thus, we have a cost function of

$$M(x) = 9x^{7/3} - 4x^{3/2} + 50x + 5{,}000$$

If Alyssa makes 12 gowns per month, her costs will be $M(12) \approx \$8{,}400.82$. ■

Projectile Motion Revisited

You already know that the function $h(t) = at^2 + v_0t + h_0$ gives the height of an object in motion t seconds after its release.

- The quadratic coefficient, a, is the acceleration of the object at any time. The quadratic term is negative because gravity pulls the object down.
- The linear coefficient, v_0, is the initial velocity of the object, or $v(0)$.
- The constant term, h_0, is the initial height of the object, or $h(0)$.

You also know that $h'(t) = v(t)$ represents the velocity of the object at any time, and that $h''(t) = v'(t) = a(t)$ is the acceleration of the object. Knowing the acceleration of an object and its initial height and initial velocity, it is now possible to work backwards and determine the function describing the height of the object at any time.

Given the acceleration, $-a$, and knowing that $a(t) = v'(t)$ and $v(0) = v_0$, we can find the velocity by integrating.

$$v(t) = -\int a\,dt = -at + v_0$$

Knowing that $h'(t) = v(t)$ and $h(0) = h_0$, we can now find the height function by integrating again.

$$h(t) = \int v(t)dt = \int (-at + v_0)dt = -\frac{1}{2}at^2 + v_0t + h_0$$

Example 8: A ball is thrown upward from a 3-foot platform at 60 feet per second. Assume that acceleration is -32 feet/sec^2. How high will the ball be after 2 seconds?

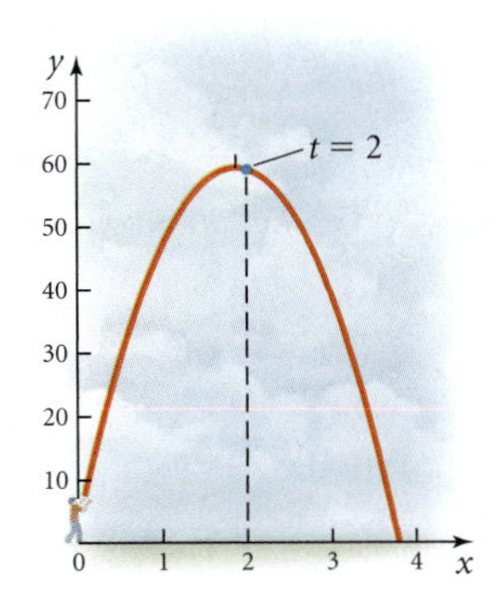

Figure 20.4

Solution: We know $a(t) = -32$, so $v(t) = \int -32dt = -32t + C$.

In this equation the constant term represents the initial velocity, which we know is 60 feet per second. This means that the velocity is

$$v(t) = -32t + 60$$

Then $h(t) = \int v(t)dt = \int(-32t + 60)dt = -16t^2 + 60t + C$.

Here the constant term represents the initial height, which we know is 3 feet. Thus, $h(t) = -16t^2 + 60t + 3$.

The height of the ball after 2 seconds is $h(2) = 59$, so after 2 seconds the ball is 59 feet above the ground. See Figure 20.4.

The same height function could have been obtained by substituting $a = 32$, $v_0 = 60$, and $h_0 = 3$ into the function $h(t) = -\frac{1}{2}at^2 + v_0t + h_0$. ■

Check Your Understanding 20.2

1. Find the equation of the curve passing through $(-1, 3)$ if the slope of the curve is given by $\frac{dy}{dx} = 3x^5 - 4x + 7$.
2. The rate at which the area of an oil spill grows is given by $A'(t) = 2t^{0.6} + 5$ square feet per hour. If the area is 10 square feet 1 hour after the spill began, how large is the spill after 24 hours?
3. A small rocket is launched from the ground with an initial velocity of 90 feet per second. How high is the rocket after 5 seconds? Assume that $a = -32$ feet per second per second.

Check Your Understanding Answers

Check Your Understanding 20.1

1. $5x + C$
2. $\frac{1}{11}x^{11} + C$
3. $\frac{1}{2}a^6 + C$
4. $-\frac{1}{3}c^{-3} - \frac{1}{4}c^4 + C$
5. $\frac{4}{7}\sqrt[4]{x^7} + C$
6. $\frac{1}{4}n^8 - \frac{1}{2}n^6 + \frac{5}{2}n^2 + \frac{1}{n^4} + C$

Check Your Understanding 20.2

1. $y = \frac{1}{2}x^6 - 2x^2 + 7x + 11.5$
2. Area after 24 hours is ≈ 325.7 square feet.
3. After 5 seconds the rocket is 50 feet above the ground.

Topic 20 Review

20

In this topic we introduced the concept of finding an antiderivative, which reverses differentiation and converts a derivative back to the original function.

CONCEPT	EXAMPLE
An **antiderivative** of $f(x)$ is a function $F(x)$ such that $F'(x) = f(x)$.	$F(x) = 2x^2 + 3x - 8$ is an antiderivative of $f(x) = 4x + 3$, because $F'(x) = f(x)$.
The symbol for antiderivative is the **integral**, $\int f(x)dx$.	$\int (4x + 3)dx$ means to find the antiderivative of $4x + 3$.
The **Power Rule for Integration** is $\int x^n dx = \frac{1}{n+1}x^{n+1} + C$, where $n \neq -1$. This type of integral is called an **indefinite integral** because the antiderivative is a **family of functions** all having the same shape but different y-intercepts if graphed.	$\int x^3 dx = \frac{1}{4}x^4 + C$ Here are several members of the family of functions $F(x) = \frac{1}{4}x^4 + C$, with $C = 0$, $C = -2$, and $C = 4$. 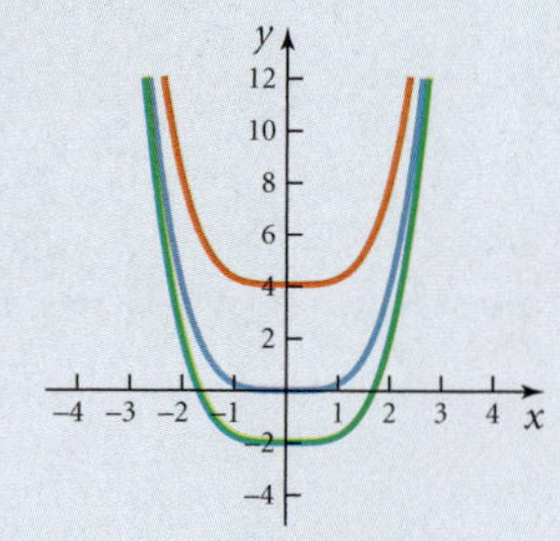
Three properties of integration: **Constant Multiple Property:** $\int c \cdot f(x)dx = c\int f(x)dx$ **Sum Property:** $\int [f(x) + g(x)]dx = \int f(x)dx + \int g(x)dx$ **Difference Property:** $\int [f(x) - g(x)]dx = \int f(x)dx - \int g(x)dx$	By the Constant Multiple Property: $\int 3x^7 dx = 3\int x^7 dx = \frac{3}{8}x^8 + C$ By the Sum and Difference Properties: $\int (x^5 - 3x^2 + 5)dx = \int x^5 dx - \int 3x^2 dx + \int 5dx = \frac{1}{6}x^6 - x^3 + 5x + C$

Given an antiderivative and an **initial condition**, it is possible to determine the value of C in that antiderivative. • Find the antiderivative. • Substitute the initial condition to determine the value of C.	Sales of a product grow at a rate of $S'(t) = 4t + 5$ units per week. If 25 units were sold the second week, how many units were sold the 10th week? $S(t) = \int (4t + 5)dt = 2t^2 + 5t + C$ But $S(2) = 25$, so $25 = 2(2)^2 + 5(2) + C$ and $C = 7$. Sales at any time are given by $S(t) = 2t^2 + 5t + 7$. Thus, in the 10th week, sales were $S(10) = 257$ units.
Given the acceleration a, initial velocity v_0, and initial height h_0 of an object in motion, the height $h(t)$ of the object at any time can be written as $h(t) = -\frac{1}{2}at^2 + v_0t + h_0$, using integration and knowing that $a(t) = v'(t)$ and $v(t) = h'(t)$.	A ball is thrown upward from a 6-foot wall with initial velocity of 40 feet per second. Assume that acceleration is -32 feet/sec^2. How high is the ball after 1 second? $a(t) = -32 \rightarrow v(t) = -32t + v_0$ $v(t) = -32t + 40$ $h(t) = \int (-32t + 40)dt = -16t^2 + 40t + h_0$ so $h(t) = -16t^2 + 40t + 6$ Then $h(1) = 30$ feet above the ground.

NEW TOOLS IN THE TOOL KIT

- Antiderivative as the inverse of derivative
- Integral sign as the symbol of antidifferentiation
- Power Rule for Integration
- Determining the value of C given an initial condition
- Determining the height of an object in motion given the acceleration, initial velocity, and initial height

Topic 20 Exercises

20

Are the equations in Exercises 1 through 4 true or false?

1. $\int (x^2 - 3x^4)dx = \int x^2dx - \int 3x^4dx$

2. $\int ((x-1)(x+3))dx = \int (x-1)dx \cdot \int (x+3)dx$

3. $\int \frac{x^2 - 3}{x + 1}dx = \frac{\int (x^2 - 3)dx}{\int (x + 1)dx}$

4. $\int \left(2x^3 + \frac{5}{x^2}\right)dx = \int 2x^3dx + \int \frac{5}{x^2}dx$

For Exercises 5 through 30, find the antiderivative.

5. $\int 4dx$ **6.** $\int 11dt$

7. $\int b^8db$ **8.** $\int x^{14}dx$

9. $\int 3x^{11}dx$ **10.** $\int 5x^8dx$

11. $\int (x^2 - 3x + 5)dx$ **12.** $\int (x^4 - 2x^3 + 6)dx$

13. $\int (2x^5 - 3x^{-2})dx$ **14.** $\int (4x^3 - 2x^{-4} + 6)dx$

15. $\int \frac{2}{x^3}dx$ **16.** $\int \frac{3}{x^4}dx$

17. $\int \frac{3}{2x^2}dx$ **18.** $\int \frac{6}{5x^4}dx$

19. $\int 6\sqrt{x}dx$ **20.** $\int 8\sqrt[3]{x}dx$

21. $\int \left(8x^3 + 4x + \frac{5}{x^2}\right)dx$

22. $\int \left(6x^4 + 3x - \frac{2}{x^5}\right)dx$

23. $\int (x^{3/4} - 3x^2)dx$

24. $\int (x^{2/5} - 4x^{-2})dx$

25. $\int \left(\frac{24}{\sqrt[5]{x^2}} + 21\sqrt[5]{x^2}\right)dx$

26. $\int \left(\frac{6}{\sqrt[3]{x^2}} + 8\sqrt[3]{x^2}\right)dx$

27. $\int (x + 3)^2dx$ **28.** $\int \left(x^2 - 5\right)^2 dx$

29. $\int \frac{x^3 - 2x^2 + 3}{x^2}dx$ **30.** $\int \frac{2x^5 + 3x^4 - 5x}{x^4}dx$

In Exercises 31 through 34, find $f(x)$ such that

31. $f'(x) = 3x - 5$ and $f(4) = -3$.

32. $f'(x) = 9 - 2x$ and $f(3) = -6$.

33. $f'(x) = \frac{1}{2}x^3 - 2x + 3$ and $f(1) = 3$.

34. $f'(x) = -2x^2 + 3x - 1$ and $f(-2) = 5$

35. Find the equation of the curve passing through the point $(2, -5)$ with slope $\frac{dy}{dx} = 7 - 4x^3$.

36. Find the equation of the curve passing through the point $(1, 3)$ with slope $\frac{dy}{dx} = 5x^2 - 2x + 1$.

37. Find the equation of the curve passing through the point $(1, -4)$ with slope $\frac{dy}{dx} = 3\sqrt{x} - 2x$.

38. Find the equation of the curve passing through the point $(1, 1)$ with slope $\frac{dy}{dx} = 2x^{2/3} + \frac{1}{x^2}$.

39. Sales of a new plasma TV, in thousands, grow at a rate of $N'(x) = 0.06x^2 - 0.8x + 3$ thousand sets per month, where x is the number of months after the initial introduction. If 1000 TVs were sold initially, how many were sold during the fourth month?

40. Hotel industry profit is growing at a rate of $P'(x) = 2.2x - 5.97$ billion dollars per year, where x is the number of years after 2000. If the industry saw a \$22 billion profit in 2000, what is the expected profit in 2007?

41. The unemployment rate in the United States has changed at a rate of $R'(x) = 0.003x^2 - 0.11x + 0.723$ percent per year, where x is the number of years after 1970. If the unemployment rate in 1970 was 5%, what was the unemployment rate in 2006?

42. The number of consumer bankruptcy filings, in thousands, has changed at a rate of $B'(t) = 20.7t^2 - 380.84t + 1744$ thousand filings per year, where t is the number of years after 1995. If there were 891,500 bankruptcy filings in 2000, how many filings were there in 2006?

43. The area of a small surface wound heals at a rate of $A'(t) = -\frac{5}{t^2}$ cm^2 per day. If the area of the wound on the first day was 8 cm^2, what was the area on the third day?

44. Passenger traffic at a small airport has changed at a rate of $P'(x) = 1.2x^2 - 38.46x + 266.44$ thousand passengers per year, where x is the number of years after 1980. If there were 562,815 passengers in 2000, how many will there be in 2007?

45. Marginal revenue, in thousand dollars, is $R'(x) = -0.6x + 7$, where x is the number of units produced, in hundreds. Find the revenue if 350 units are produced. Assume $R(0) = 0$.

46. Marginal cost, in hundred dollars, is $C'(x) = 0.6\sqrt{x} + 3$, where x is the number of units. If the fixed costs are \$350, find the cost of producing 25 units.

47. The marginal cost, in thousands, for producing x hundred cell phones is given by $C'(x) = \frac{20}{\sqrt{x}}$. If the fixed costs are \$250, find the cost of producing 400 phones.

48. The marginal profit, in thousands, for producing x hundred cell phones is given by $P'(x) = 0.2 + 0.5x$. Find the profit if 400 cell phones are produced. Assume that $P(0) = 0$.

49. A rocket is launched from the ground at 300 feet per second. How high is the rocket after 1 second?

50. A rock is dropped from a 30-foot platform. What is the height of the rock after 1 second?

51. Blaise hit a baseball into the air at 90 feet per second. Assume that the bat was 4 feet above the ground when the ball was hit.

a. How high is the ball after 2 seconds?

b. What is the velocity of the ball after 2 seconds?

c. What is the maximum height the ball reaches?

d. When will the ball hit the ground?

52. A projectile is launched from a 10-meter platform with an initial velocity of 20 meters per second and an acceleration of 18 meters per second per second.

a. How high is the projectile after 1 second?

b. What is the velocity of the projectile after 1 second?

c. What is the maximum height the projectile reaches?

d. When will the projectile hit the ground?

SHARPEN THE TOOLS

In Exercises 53 through 58, find the antiderivative.

53. $\int (x^2 - 2)^3 dx$

54. $\int (2x - 3)^4 dx$

55. $\int x^3(x^2 + 1)^2 dx$

56. $\int (x^3 - 2x + 1)^2 dx$

57. $\int \frac{x^3 - 64}{x - 4} dx$

58. $\int \frac{x^4 - 16}{x + 2} dx$

59. Betty's Craft Corner sells handmade baskets. If Betty produces x hundred baskets each month, her marginal cost in hundred dollars is $C'(x) = -1.2x + 5$, and her marginal revenue in hundred dollars is $R'(x) = 8.4 + \frac{20}{\sqrt{x}}$. Betty's fixed costs per month are \$800. What is her profit if she makes 200 baskets per month?

60. The longest home run in major league baseball history was hit in 1960 by Mickey Mantle of the New York Yankees in a game in Detroit against the Detroit Tigers. The rate at which the height of the ball in feet, y, changed with respect to its horizontal distance in feet from home plate, x, is given by $\frac{dy}{dx} = 0.9 - 0.0028x$. Assume an initial height of 4 feet.

a. What is the height of the ball at any time?

b. What is the height of the ball when it is 100 feet from home plate?

c. What is the maximum height of the ball?

d. How far from home plate did the ball travel before hitting the ground?

COMMUNICATE

61. How are derivatives and integrals related to each other?

62. Briefly explain why $\int f(x)dx$ is an "indefinite" integral. Include a graphical interpretation in your explanation.

CALCULATOR CONNECTION

63. The table below gives the change in total box office revenues, in million dollars, for a new movie during the first 12 weeks after its release.

Week	Change in Total Revenue, in Million Dollars
1	1.3
2	1.8
3	2.6
4	2.0
5	1.2
6	1.8
7	1.0
8	0.6
9	0.8
10	0.4
11	0.2
12	0.1

a. Find a cubic regression model, $R'(t)$, for the change in total revenues in any given week. Round coefficients to four decimal places.

b. Use your model to estimate the change in total revenue in week 4 and in week 8. How close was your estimate to the true change as given in the table?

c. Use integration to find a model for the total revenue in any given week, $R(t)$. Use $R(1) = 1.3$.

d. Use your model to estimate the total revenue in week 4 and in week 8.

64. Sales of a new DVD, in thousands, change each week as given in the table below.

Week	Change in Sales, in Thousands
1	1.58
2	0.66
3	0.50
4	0.42
5	0.38
6	0.33
7	0.31
8	0.29
9	0.27
10	0.26

a. Find a power regression model for the change in total sales in any given week, $S'(t)$. Round coefficients to four decimal places.

b. Use your model to estimate the change in sales in week 3 and in week 8. How close was your estimate to the actual change as given in the table?

c. Use integration to find a model for the total sales in any week, $S(t)$. Use $S(1) = 1.58$.

d. Use your model to estimate the total sales in week 3 and in week 8.

TOPIC

More Rules for Integration

21

Officials at a local college have determined that the college's enrollment since 1990 has grown at a rate of $E'(t) = \dfrac{6245}{t}$ students per year. They wish to predict the college's enrollment for 2010, knowing that the enrollment for 2007 was 8128. You know that you need to find an antiderivative to make the prediction, but the Power Rule cannot be applied to this function because of the exponent of one in the denominator. In this topic you will learn how to find antiderivatives of nonalgebraic functions. Before beginning this topic, you should complete the Warm-up Exercises to sharpen the algebra and calculus skills you will need.

IMPORTANT TOOLS IN TOPIC 21

- *Antiderivative of* e^x
- *Antiderivative of* e^{kx}
- *Antiderivative of* $\dfrac{1}{x}$
- *Antiderivative of* a^x

TOPIC 21 WARM-UP EXERCISES

Be sure you can successfully complete the following exercises before starting Topic 21.

Algebra Warm-up Exercises

Evaluate each of the following expressions.

1. 3^0 **2.** $\frac{0}{7}$ **3.** $\frac{5}{0}$ **4.** 2^{-1}

5. $\ln(e^6)$ **6.** $e^{\ln 3}$ **7.** $(e^x)^2$

Calculus Warm-up Exercises

Find $f'(x)$ for each of the following functions.

1. e^x **2.** e^{-4x}

3. $\ln x$ **4.** $\ln(x^3 + 5)$

5. 3^x **6.** 7^{3x+1}

7. $\log_8 x$ **8.** $\log_4(5x - 2)$

Find each antiderivative.

9. $\int x^{10} dx$ **10.** $\int (3x^5 - 8x + 4)dx$

11. $\int \frac{4}{a^7} da$ **12.** $\int 4q^{2/3} dq$

13. The population of a bacteria culture is given by $P(t) = 520e^{0.25t}$, where t is in days. How fast is the population changing after 1 week?

Answers to Warm-up Exercises

Algebra Warm-up Exercises

1. 1 **2.** 0 **3.** undefined **4.** $\frac{1}{2}$

5. 6 **6.** 3 **7.** e^{2x}

Calculus Warm-up Exercises

1. e^x **2.** $-4e^{-4x}$ **3.** $\frac{1}{x}$ **4.** $\dfrac{3x^2}{x^3 + 5}$

5. $3^x \cdot \ln 3$ **6.** $3(\ln 7) \cdot 7^{3x+1}$

7. $\dfrac{1}{x(\ln 8)}$ **8.** $\dfrac{5}{(5x - 2)\ln 4}$

9. $\frac{1}{11}x^{11} + C$ **10.** $\frac{1}{2}x^6 - 4x^2 + 4x + C$

11. $-\dfrac{2}{3a^6} + C$ **12.** $\frac{12}{5}q^{5/3} + C$

13. The number of bacteria is increasing by about 748 bacteria per day after 1 week.

In Topic 20 you learned the Power Rule for Integration.

$$\int x^n dx = \frac{1}{n+1}x^{n+1} + C \qquad \text{for } n \neq -1$$

Why the restriction on the value of n? Why must $n \neq -1$? To find out why, let's try to apply the Power Rule when $n = -1$.

If $n = -1$, then $x^n = x^{-1} = \frac{1}{x}$.

$$\int \frac{1}{x}dx = \int x^{-1}dx$$

$$= \frac{1}{-1+1}x^{-1+1} + C$$

$$= \frac{1}{0}x^0 + C$$

But you know that $\frac{1}{0}$ is undefined, which is why the Power Rule doesn't work for $\int \frac{1}{x}dx$ and explains the restriction for the Power Rule that $n \neq -1$. The antiderivative of x^{-1} or $\frac{1}{x}$ is not an algebraic function. So how do we find $\int \frac{1}{x}dx$?

Derivatives of e^x and ln x

Recall the derivative rules you learned for nonalgebraic functions.

$$D_x \ln x = \frac{1}{x} \qquad \text{for } x > 0$$

$$D_x e^x = e^x$$

Any derivative rule can be reversed and written as an antiderivative rule, so we now add two more integration rules to the Tool Kit.

Antiderivative Rules for Nonalgebraic Functions

Logarithmic Rule: $\int \frac{1}{x}dx = \ln x + C \qquad \text{for } x > 0$

Exponential Rule: $\int e^x dx = e^x + C$

You will also recall that the domain of $f(x) = \ln x$ is all $x > 0$. In this text we will assume that $x > 0$ when finding $\int \frac{1}{x}dx$.

Example 1: Find each antiderivative.

a. $\int \frac{6}{x}dx$ **b.** $\int \left(\frac{3}{g} - \frac{3}{g^2}\right)dg$ **c.** $\int \frac{5}{2v}dv$

Solution:

a. By rewriting and using the Constant Multiple Property, we have

$$\int \frac{6}{x}dx = \int 6 \cdot \frac{1}{x}dx = 6\int \frac{1}{x}dx = 6\ln x + C$$

We assume here that $x > 0$.

b. Rewriting and using the Difference Property, we have

$$\int \left(\frac{3}{g} - \frac{3}{g^2}\right)dg = \int \frac{3}{g}dg - \int \left(\frac{3}{g^2}dg\right) = \int \frac{3}{g}dg - \int 3g^{-2}dg$$

$$= 3\ln g + \frac{3}{g} + C$$

Warning! $\int \frac{3}{g^2}dg \neq 3\ln g^2 + C$

The natural logarithm applies only when the denominator of an integrand is a single variable with an exponent of one.

Warning! $\frac{3}{g} \neq 3\ln g$

The natural logarithm applies only when *integrating* $\frac{3}{g}$: $\int \frac{3}{g}dg = 3\ln g + C$.

c. $$\int \frac{5}{2v}dv = \int \frac{5}{2} \cdot \frac{1}{v}dv$$

$$= \frac{5}{2}\int \frac{1}{v}dv$$

$$= \frac{5}{2}\ln v + C$$

Warning! $\int \frac{5}{2v}dv \neq 5\ln(2v) + C$

■

Example 2: Find each antiderivative.

a. $\int 3e^x dx$ **b.** $\int (3 + e^x)dx$ **c.** $\int (2e^x + x^{2e} + 2x)dx$

Solution:

a. Using the Constant Multiple Property, we have

$$\int 3e^x dx = 3\int e^x dx = 3e^x + C$$

b. Using the Sum Property, we have

$$\int (3 + e^x)dx = \int 3dx + \int e^x dx = 3x + e^x + C$$

c. We will use the Sum Property here, but be careful! The first term ($2e^x$)is an exponential function, the second term (x^{2e}) is algebraic (just a power of x, because e is a constant), and the third term ($2x$) is linear.

$$\int (2e^x + x^{2e} + 2x)dx = \int 2e^x dx + \int x^{2e} dx + \int 2xdx$$

$$= 2e^x + \frac{1}{2e + 1}x^{2e+1} + x^2 + C$$ ■

Composite Exponential Functions

You also know, by the Chain Rule, that $D_x e^{kx} = ke^{kx}$. In other words, the derivative of e^{kx} is e^{kx} multiplied by k, the derivative of kx. Reversing this process and applying it to antiderivatives, we obtain another integration rule.

Integration of e^{kx}

$$\int e^{kx} dx = \frac{1}{k}e^{kx} + C$$

You should see the inverse process here. The derivative of e^{kx} involves *multiplying* by k, so the antiderivative of e^{kx} must *divide* by k.

Example 3: Find each antiderivative.

a. $\int e^{3x} dx$ **b.** $\int 2e^{0.25t} dt$ **c.** $\int \left(4e^{-3a} - \frac{3a}{4} + \frac{3}{a}\right)da$

Solution:

a. With $k = 3$, we have $\int e^{3x} dx = \frac{1}{3}e^{3x} + C$.

b. With $k = 0.25$, we have

$$\int 2e^{0.25t}dt = \frac{2}{0.25}e^{0.25t} + C$$
$$= 8e^{0.25t} + C$$

c. The first term ($4e^{-3a}$) uses the exponential property. The second term $-\frac{3a}{4}$ is just a linear expression with a coefficient of $-\frac{3}{4}$. The third term $\left(\frac{3}{a}\right)$ will involve ln.

Using the Sum and Difference Properties, we have

$$\int\left(4e^{-3a} - \frac{3a}{4} + \frac{3}{a}\right)da = \int 4e^{-3a}da - \int \frac{3}{4}ada + \int \frac{3}{a}da$$
$$= \frac{4}{-3}e^{-3a} - \frac{3}{4}\cdot\frac{1}{2}a^2 + 3\ln a + C$$
$$= -\frac{4}{3}e^{-3a} - \frac{3a^2}{8} + 3\ln a + C$$

■

Let's summarize the rules of integration you have learned so far.

Rules of Integration

$$\int x^n dx = \frac{1}{n+1}x^{n+1} + C \qquad \text{for } n \neq -1$$

$$\int \frac{1}{x}dx = \ln x + C \qquad \text{for } x > 0$$

$$\int e^x dx = e^x + C$$

$$\int e^{kx}dx = \frac{1}{k}e^{kx} + C$$

Check Your Understanding 21.1

Find each antiderivative.

1. $\int 7dx$ **2.** $\int 3x^4 dx$ **3.** $\int 2x^{-5}dx$

4. $\int \sqrt[3]{x}dx$ **5.** $\int (x^2 - 6x + 5)dx$ **6.** $\int \frac{8}{x^3}dx$

7. $\int \frac{8}{x}dx$ **8.** $\int 5e^x dx$ **9.** $\int e^{5x}dx$

Example 4: A bacteria culture grows at a rate of $P'(t) = 45e^{0.3t}$ bacteria per hour. If there were initially 120 bacteria in the culture, how many will there be after 1 day?

Solution: Here we are given the *rate* at which the population changes but asked to evaluate the actual population at a given time. We must use the integral to convert the rate of change of the population back to actual population function.

$$P(t) = \int 45e^{0.3t}dt = \frac{45}{0.3}e^{.3t} + C = 150e^{0.3t} + C$$

To determine the population at a given time, we must determine the value of C. Using the fact that the initial population was 120 (the initial condition), we know that $P(0) = 120$. Substituting $P(0) = 120$ into $P(t) = 150e^{0.3t} + C$ gives us the value of C.

$$P(t) = 150e^{0.3t} + C \rightarrow 120 = 150e^{0} + C$$
$$-30 = C$$

The population of the culture at any time is given by $P(t) = 150e^{0.3t} - 30$. After 1 day, or 24 hours, there will be $P(24) \approx 200{,}885$ bacteria in the culture. ■

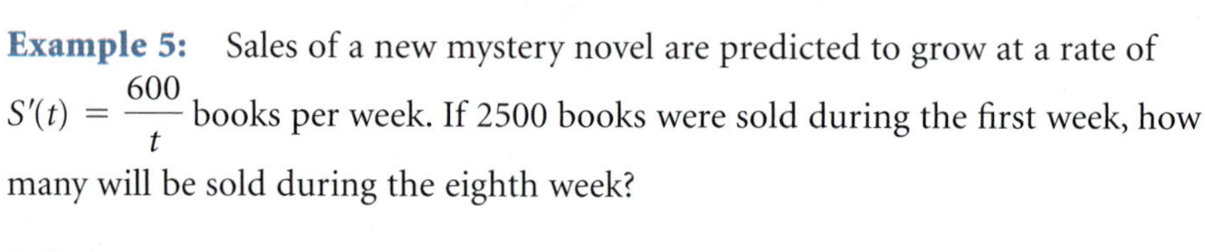

Example 5: Sales of a new mystery novel are predicted to grow at a rate of $S'(t) = \frac{600}{t}$ books per week. If 2500 books were sold during the first week, how many will be sold during the eighth week?

Solution: Again we are given the rate at which sales are changing, but we must determine actual sales, so we will use the integral.

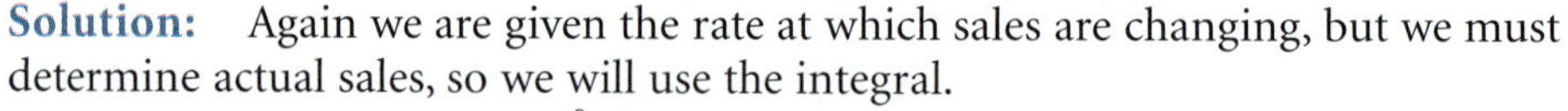

The expression $S(t) = \int \frac{600}{t}dt = 600 \ln t + C$ gives the sales at any time.

We know that $S(1) = 2500$, so $2500 = 600 \ln 1 + C$, which gives $C = 2500$ because $\ln 1 = 0$. Thus, the sales at any time are given by $S(t) = 600 \ln t + 2500$. Sales during the eighth week will be $S(8) = 600 \ln 8 + 2500 \approx 3748$ books. ■

Integrals of General Exponential Functions

Because we know that $e^{\ln a} = a$, we can easily write all exponential functions in the form $e^{f(x)}$.

Example 6: Evaluate $\int 4^x dx$.

Solution:

$$e^{\ln a} = a, \text{ so } a^x = (e^{\ln a})^x = e^{(\ln a)x}$$

In this integral, $a = 4$, so $\int 4^x dx = \int e^{(\ln 4)x}dx = \frac{1}{\ln 4}e^{(\ln 4)x} + C$

$$\approx 0.72135e^{1.3863x} + C, \text{ or } 0.72135(4^x) + C.$$ ■

Example 7: Annual per capita spending on health care in the United States since 1970 has grown at a rate of $S'(x) = 34.3(1.087)^x$ dollars per year, where x is the number of years after 1970. If annual per capita spending in 1980 was $1005, how much was spent in 2005?

Solution: We know both the rate of spending since 1970 and the amount spent in 1980. The amount spent in any given year is found using the antiderivative:

$$\begin{aligned} S(x) &= \int 34.3(1.087)^x dx \\ &= \int 34.3e^{(\ln 1.087)x} dx \\ &= \frac{34.3}{\ln 1.087} e^{(\ln 1.087)x} + C \\ &\approx 411.2e^{0.0834x} + C \quad \text{or} \quad 411.2(1.087^x) + C \end{aligned}$$

To find the value of C, we know that for 1980, $S(10) = 1005$. Thus, we have

$$\begin{aligned} 1005 &= 411.2e^{0.0834*10} + C \\ 1005 &= 946.79 + C \\ C &\approx 58 \end{aligned}$$

The amount spent in any given year is $S(x) = 411.2e^{(\ln 1.087)x} + 58$ or $S(x) = 411.2(1.087^x) + 58$.

The amount spent in 2005 is $S(35) \approx \$7680$. ■

Check Your Understanding 21.2

Find each antiderivative, knowing that $a^x = e^{(\ln a)x}$.

1. $\int 6^x dx$ **2.** $\int 4 \cdot 3^x dx$ **3.** $\int \left(8^x - 4x^8 + \frac{8}{x} - \frac{x}{8}\right) dx$

4. Evaluate $\int 2.7(1.5)^x dx$ by converting to $e^{f(x)}$ and knowing that $e^{\ln a} = a$.

Check Your Understanding Answers

Check Your Understanding 21.1

1. $7x + C$
2. $\frac{3}{5}x^5 + C$
3. $-\frac{1}{2}x^{-4} + C$
4. $\frac{3}{4}\sqrt[3]{x^4} + C$
5. $\frac{1}{3}x^3 - 3x^2 + 5x + C$
6. $-\frac{4}{x^2} + C$
7. $8\ln x + C$
8. $5e^x + C$
9. $\frac{1}{5}e^{5x} + C$

Check Your Understanding 21.2

1. $\frac{6^x}{\ln 6} + C$
2. $\frac{4}{\ln 3} \cdot 3^x + C$
3. $\frac{8^x}{\ln 8} - \frac{4}{9}x^9 + 8\ln x - \frac{x^2}{16} + C$
4. $\int 2.7e^{(\ln 1.5)x} dx = \frac{2.7}{\ln 1.5} e^{(\ln 1.5)x} + C$
$= \frac{2.7}{\ln 1.5}(1.5)^x + C$ or $6.659(1.5)^x + C$

Topic 21 Review

21

This topic presented rules of integration of exponential and logarithmic functions.

CONCEPT	EXAMPLE
The **integral rules for exponential functions** are $\int e^x dx = e^x + C$ and $\int e^{kx} dx = \frac{1}{k}e^{kx} + C$.	$\int e^{2x} dx = \frac{1}{2}e^{2x} + C$ $\int e^{-0.2t} dt = \frac{1}{-0.2}e^{-0.2t} + C = -5e^{-0.2t} + C$
The **integral rule for the natural logarithmic function** is $\int \frac{1}{x} dx = \ln x + C$ for $x > 0$.	$\int \frac{6}{x} dx = 6 \ln x + C$ $\int -3x^{-1} dx = -3 \ln x + C$
The **integral rule for the general exponential function** is $\int a^x dx = \int e^{(\ln a)x} dx = \frac{1}{\ln a}e^{(\ln a)x} + C$ or $= \frac{1}{\ln a}a^x + C$	$\int 11^x dx = \frac{1}{\ln 11}(11)^x + C$

NEW TOOLS IN THE TOOL KIT

- Finding $\int e^x dx$ and $\int e^{kx} dx$
- Finding $\int \frac{1}{x} dx$
- Finding $\int a^x dx$

Topic 21 Exercises

21

Are Exercises 1 though 8 true or false?

1. $\int 3e^x dx = 3e^x + C$

2. $\int \frac{1}{x} dx = \ln x + C$

3. $\int e^x dx = \frac{1}{x+1} e^{x+1} + C$

4. $\int e^{x^2} dx = \frac{1}{2x} e^{x^2} + C$

5. $\int \frac{1}{x^2} dx = \ln x^2 + C$

6. $\int e^{5x} dx = \frac{1}{5} e^{5x} + C$

7. $\int \frac{3}{5x} dx = 3 \ln |5x| + C$

8. $\int -\frac{2}{3x} dx = -2 \ln |3x| + C$

For Exercises 9 though 26, find each antiderivative.

9. $\int e^{8x} dx$

10. $\int e^{-2x} dx$

11. $\int 3e^{-4x} dx$

12. $\int 7e^{5x} dx$

13. $\int 12e^{0.3t} dt$

14. $\int 200e^{-0.5t} dt$

15. $\int \frac{4}{x} dx$

16. $\int -\frac{8}{x} dx$

17. $\int \left(\frac{2}{a} - \frac{5}{a^3}\right) da$

18. $\int \left(\frac{2}{a^4} - \frac{11}{a}\right) da$

19. $\int (e^x - 3)^2 dx$

20. $\int \left(\frac{2}{x} - 5\right)^2 dx$

21. $\int \left(e^{-3x} - x^3 + \frac{3}{x}\right) dx$

22. $\int \left(4x^3 - \frac{4}{x^3} + e^{4x}\right) dx$

23. $\int \frac{x^3 - 3x^2 + 5x}{x^2} dx$

24. $\int \frac{2x^2 - 3x^3 + 5x^5}{x^3} dx$

25. $\int \frac{x^2(x-4)^3}{x^4 - 4x^3} dx$

26. $\int \frac{x(x+3)^4}{x^3 + 3x^2} dx$

(Hint for Exercises 25 and 26: Factor the denominator and simplify before integrating.)

In Exercises 27 through 32, find each antiderivative, knowing that $a^x = e^{(\ln a)x}$.

27. $\int 7^x dx$

28. $\int -8^x dx$

29. $\int (3 \cdot 2^x - x^2) dx$

30. $\int (x^4 + 4 \cdot 5^x) dx$

31. $\int \left(6^x - e^{6x} + \frac{6}{x} + \frac{x}{6}\right) dx$

32. $\int \left(4^x - e^{4x} + \frac{4}{x} - \frac{x^2}{4}\right) dx$

In Exercises 33 through 38, find $f(x)$ using the given information.

33. $f'(x) = e^{3x} + 4$ with $f(0) = 2$

34. $f'(x) = 2 - e^{-x}$ with $f(0) = -1$

35. $f'(x) = -\frac{2}{x} + x$ with $f(1) = 3$

36. $f'(x) = x^2 + \frac{3}{x}$ with $f(1) = 2$

37. $f'(x) = 3x^2 - 5^x + 1$ with $f(1) = -2$

38. $f'(x) = x^3 - 2^x - 3$ with $f(1) = 3$

39. Find the equation of the curve passing through $(1, 2)$ if the slope is given by $\frac{dy}{dx} = \frac{3}{x} - 4$.

40. Find the equation of the curve passing through $(e, -4)$ if the slope is given by $\frac{dy}{dx} = \frac{2}{x}$.

41. Find the equation of the curve passing through (0, 1) if the slope is given by $\frac{dy}{dx} = 6e^{-0.2x} - x + 3$.

42. Find the equation of the curve passing through (0, −3) if the slope is given by $\frac{dy}{dx} = 0.5e^{x} + 2$.

43. The number of cell phone subscribers, in millions, since 1990 has grown at a rate of $N'(t) = 1.94e^{0.32t}$ million subscribers per year, where t is the number of years after 1990. If there were 149.4 million subscribers in 2000, how many subscribers were there in 2006?

44. The population of a city, in millions, since 1990 has grown at a rate of $P'(t) = 0.38e^{0.031t}$ million people per year, where t is the number of years after 1990. If there were 1.67 million in 2000, estimate the population in 2007.

45. The number of mosquitoes in a lake area after an insecticide spraying decreases at a rate of $M'(t) = -360e^{-0.03t}$ mosquitoes per hour. If there were 12,000 mosquitoes initially, how many will there be after 3 hours?

46. Maria takes a cake out of a 375°F oven and places it on a counter. The rate at which the cake cools is given by $T'(t) = -21.35e^{-0.07t}$ degrees per minute. If the cake needs to be not more than 150°F before it can be iced, how long will Maria have to wait before icing the cake?

47. Arthur pours a cup of coffee and places it on the table, where it cools at a rate of $T'(t) = -1.6(0.989)^t$ degrees Fahrenheit per minute. If the coffee was initially 180°F, what will the temperature be after 5 minutes? (Remember that $a^x = e^{(\ln a)x}$.)

48. Average household size in the United States since 1900 has decreased at a rate of $H'(t) = -0.029(0.9939)^t$ persons per year, where t is the number of years after 1900. If average household size was 2.75 in 1990, predict average household size in 2010. (Remember that $a^x = e^{(\ln a)x}$.)

49. Enrollment of a college since 1990 has grown at a rate of $E'(t) = \frac{6245}{t}$ students per year, where t is the number of years after 1990. If there were 8128 students in 1995, how many students will there be in 2008?

50. The percentage of the U.S. population living in metropolitan areas since 1890 has grown at a rate of $P'(t) = \frac{25.5}{t}$ percent per year, where t is the number of years after 1890. In 2000, 75% of the U.S. population lived in a metropolitan area. Estimate the percentage for 2008.

51. Sales of a new computer are growing at a rate of $S'(x) = 50e^{0.2x} + 70x - 40$ computers per month, where x is the number of months after the computer was put on the market. If $S(0) = 0$, what are the sales during the eighth month?

52. Sales of a new sailboat grow at a rate of $S'(x) = 900x + \frac{250}{x}$ boats per month. If 450 boats were sold during the first month, how many were sold during the sixth month?

COMMUNICATE

53. Explain why $\int e^{x^2} dx$ cannot be integrated using the rules learned in this topic.

54. Explain why $\int x \ln x dx$ cannot be integrated using the rules learned in this topic.

CALCULATOR CONNECTION

55. In the introduction to this unit, you were given a graph of the growth in revenues of CCS Medical, a medical supply company in Clearwater, Florida. The table gives the actual annual increase in revenues, in million dollars.

Year	Increase in Revenue, in Million Dollars
1994	0.05
1995	0.35
1996	1.05
1997	1.4
1998	3.6
1999	6.4
2000	9.4
2001	10.4
2002	14.3
2003	20.1

a. Use the regression feature of your calculator to determine an exponential function that models the growth in revenue. Let x represent the number of years after 1990.

b. Use your model to estimate the company's revenues in 2006. Let $R(4) = 0.05$.

c. Use your model to predict the company's revenues in 2010.

56. The total number of Starbucks locations has grown rapidly since its beginnings in 1971. The table gives the annual increase in the total number of locations.

(*Source:* www.starbucks.com)

Year	Number of New Locations
1971	1
1987	16
1988	16
1989	22
1990	29
1991	32
1992	49
1993	107
1994	153
1995	251
1996	339
1997	397
1998	474
1999	249
2000	1366
2001	1208
2002	979
2003	1537
2004	344
2005	1912

a. Use the regression feature of your calculator to determine an exponential function that models the growth in the number of new Starbucks locations. Let x represent the number of years after 1970.

b. Use your model to predict the total number of Starbucks locations in 2004 if there were 2135 locations in 1999.

c. In 2004 there were actually 7,569 Starbucks locations. How well did your model predict the actual number?

TOPIC

22

Substitution Techniques for Integration

IMPORTANT TOOLS IN TOPIC 22

- *Computing differentials*
- *Substitution to find antiderivatives of some composite functions*

Your integration Tool Kit now contains the Power Rule for Integration, the Sum and Difference properties, and the exponential and logarithmic rules. Using these rules, you know, for instance, that $\int x^{26}\,dx = \frac{x^{27}}{27} + C$ and $\int \frac{6}{n}\,dn = 6\ln n + C$. In this topic you will use these rules and properties to find antiderivatives of composite functions such as $(x - 5)^{10}$ and $\frac{6}{2n - 5}$ using a process called substitution. Before beginning this topic, be sure to complete the Warm-up Exercises to sharpen the algebra and calculus skills you will need.

TOPIC 22 WARM-UP EXERCISES

Be sure you can successfully complete the following exercises before starting Topic 22.

Algebra Warm-up Exercises

Expand each of the following expressions.

1. $(x - 5)^2$ **2.** $(x - 5)^3$

3. Factor the polynomial so that the cubic coefficient is 1: $\frac{1}{2}x^3 - 3x^2 + 6x - 4$.

Calculus Warm-up Exercises

Find $f'(x)$ for each of the following functions.

1. $f(x) = e^{2x}$ **2.** $f(x) = (x^2 - 4)^3$

3. $f(x) = \frac{2}{x^4}$ **4.** $f(x) = -\frac{3}{x}$

Simplify as necessary and perform each integration.

5. $\int (3x - 2)(x + 4)\,dx$

6. $\int \frac{x^3 - 5x^2 + 3x - 4}{x^2}\,dx$

7. $\int (5x - 4)^2\,dx$ **8.** $\int x(x^2 - 3)^2\,dx$

9. $\int \sqrt{x}\,dx$ **10.** $\int e^{2x}\,dx$

11. $\int \frac{4}{x}\,dx$

Answers to Warm-up Exercises

Algebra Warm-up Exercises

1. $x^2 - 10x + 25$

2. $x^3 - 15x^2 + 75x - 125$

3. $\frac{1}{2}(x^3 - 6x^2 + 12x - 8)$

Calculus Warm-up Exercises

1. $2e^{2x}$ **2.** $3(x^2 - 4)^2(2x) = 6x(x^2 - 4)^2$

3. $-\frac{8}{x^5}$ **4.** $\frac{3}{x^2}$

5. $x^3 + 5x^2 - 8x + C$

6. $\frac{1}{2}x^2 - 5x + 3\ln x + \frac{4}{x} + C$

7. $\frac{25}{3}x^3 - 20x^2 + 16x + C$

8. $\frac{1}{6}x^6 - \frac{3}{2}x^4 + \frac{9}{2}x^2 + C$

9. $\frac{2}{3}\sqrt{x^3} + C$

10. $\frac{1}{2}e^{2x} + C$

11. $4\ln x + C$

In your Tool Kit you now have several rules and properties for differentiation:

- Power Rule
- Sum and Difference Properties
- Exponential and logarithmic derivative rules
- Product and Quotient rules
- Chain Rule for composite functions

For integration, you have learned these rules and properties:

- Power Rule
- Rules for integrating exponential and logarithmic functions
- Sum and Difference Properties

However, there is no Product Rule, no Quotient Rule, and no Chain Rule for integration. Other techniques must be developed for integrating products, quotients, and composite functions.

In this topic we will consider integration of certain types of composite functions using a technique called *substitution*. We will demonstrate the technique first in an example and then express the technique in general terms.

Integration by Substitution

Suppose we needed to integrate $\int (x - 5)^2\,dx$. Because the integrand is raised to the second power, you know that the antiderivative will involve a third power. One way to integrate $\int (x - 5)^2\,dx$ is to expand the integrand and use the Power Rule.

$$\int (x - 5)^2\,dx = \int (x^2 - 10x + 25)\,dx$$

$$= \frac{1}{3}x^3 - 5x^2 + 25x + C$$

Now factor $\frac{1}{3}$ from the first three terms

$$= \frac{1}{3}(x^3 - 15x^2 + 75x) + C_1, \text{ where } C_1 = \frac{1}{3}C$$

We know that $(x - 5)^3 = x^3 - 15x^2 + 75x - 125$, which you should see almost matches the cubic expression in parentheses above. We will now subtract 125 from that expression so that it can be written in factored form as $(x - 5)^3$.

To maintain an equivalent expression, we must then add $\frac{1}{3}(125)$. With these computations, we now have

$$= \frac{1}{3}(x^3 - 15x^2 + 75x - 125) + C_1 + \frac{125}{3}$$

We know that $(x - 5)^3 = x^3 - 15x^2 + 75x - 125$. We also know that C_1 represents a constant term, so $C_1 + \frac{125}{3}$ is still a constant, which we will write as C_2. Thus, we have a final result of

$$= \frac{1}{3}(x - 5)^3 + C_2$$

where

$$C_2 = C_1 + \frac{125}{3}$$

Even though there is nothing wrong with this integration technique, it wouldn't work too well with an integral like $\int (x - 5)^{10}\,dx$. Expanding $(x - 5)^{10}$ is a lot more difficult to perform. A better method for integrating some composite functions is the **substitution** technique, which uses an expression called a **differential**.

> **Definition:** For $y = f(x)$, we define the **differential** as $dy = f'(x)\,dx$, which gives the change in y for an infinitely small change in x.

Example 1: Find the differential for each of the following.

a. $y = 5 - 7x$ **b.** $u = 5x^2 + 3$

Solution:

a. Here we have $f(x) = 5 - 7x$, so $f'(x) = -7$. The differential is $dy = -7dx$.

b. Here we have $f(x) = 5x^2 + 3$, so $f'(x) = 10x$. The differential is $du = 10xdx$. ■

Now let's rework $\int (x - 5)^2\,dx$ using the substitution technique. You should see that $(x - 5)^2$ is the composite function $f(g(x)) = (x - 5)^2$, so $g(x) = x - 5$ and $f(x) = x^2$.

STEPS	CALCULATION
Step 1: Let $u = g(x)$.	$u = x - 5$
Step 2: Compute the differential $du = g'(x) \cdot dx$.	$du = 1dx$
Step 3: Rewrite $\int (f(g(x)))\,dx = \int f(u)\,du$.	$\int (x - 5)^2\,dx = \int u^2\,du$
Step 4: Integrate $\int f(u)\,du$.	$\int u^2\,du = \frac{1}{3}u^3 + C$
Step 5: Rewrite step 4 as a function of x by substituting $u = g(x)$.	$= \frac{1}{3}(x - 5)^3 + C$

Let's formalize the process of integration by substitution.

Substitution Technique for Integration of Some Composite Functions

To integrate a composite function $\int f(g(x))\,g'(x)\,dx$:

1. Let $u = g(x)$.
2. Compute the differential $du = g'(x)\,dx$.
3. Substitute and rewrite $\int f(g(x))\,g'(x)\,dx$ as $\int f(u)\,du$.
4. Integrate and obtain the antiderivative $F(u) + C$.
5. Substitute $u = g(x)$ and write the antiderivative as $F(g(x)) + C$.

Example 2: Integrate $\int (x-5)^{10}\,dx$.

Solution: $(x-5)^{10}$ is a composite function with $g(x) = x - 5$.

Let $u = x - 5$. Then $du = dx$.

Substituting $u = x - 5$ and $du = dx$, we have

$$\int (x-5)^{10}\,dx = \int u^{10}\,du.$$

Integrating gives $= \frac{1}{11}u^{11} + C.$

Now substitute $u = x - 5$ $= \frac{1}{11}(x-5)^{11} + C.$ ■

Example 3: Integrate $\int \sqrt{2x-5}\,dx$.

Solution: Let $u = 2x - 5$. Then $du = 2dx$.

Because the integrand contains only "dx," we must solve $du = 2dx$ for dx, so $\frac{1}{2}du = dx$.

Substituting $u = 2x - 5$ and $\frac{1}{2}du = dx$, we have

$$\begin{aligned}
\int \sqrt{2x-5}\,dx &= \int \sqrt{u}\left(\frac{1}{2}du\right) \\
&= \int \frac{1}{2}\sqrt{u}\,du \\
&= \int \frac{1}{2}u^{1/2}\,du \\
&= \frac{1}{2}\cdot\frac{2}{3}u^{3/2} + C \\
&= \frac{1}{3}u^{3/2} + C
\end{aligned}$$

Substituting $u = x - 5$ gives

$$= \frac{1}{3}(2x-5)^{3/2} + C$$

Writing in original form, we have

$$= \frac{1}{3}\sqrt{(2x - 5)^3} + C$$ ■

Remember that solutions to integration problems can always be checked by differentiating the antiderivative and showing that the original function is obtained. Let's check the answer to Example 3.

$$D_x\frac{1}{3}\sqrt{(2x - 5)^3} = D_x\frac{1}{3}(2x - 5)^{3/2}$$

$$= \frac{1}{3} \cdot \frac{3}{2}(2x - 5)^{1/2}(2) \text{ by the Chain Rule}$$

$$= (2x - 5)^{1/2} \quad \text{or} \quad \sqrt{2x - 5}$$

Example 4: Integrate $\int \frac{8}{(5 - 4x)^3}dx$.

Solution: Let $u = g(x) = 5 - 4x$. Then $du = g'(x)\,dx = -4dx$. As in Example 3, the integrand contains only "dx."
Solving $du = -4dx$, we have $-\frac{1}{4}du = dx$.

Substituting $u = 5 - 4x$ and $-\frac{1}{4}du = dx$, we have

$$\int \frac{8}{(5 - 4x)^3}dx = \int \frac{8}{u^3}\left(-\frac{1}{4}du\right)$$

$$= \int -2u^{-3}\,du$$

$$= -2 \cdot -\frac{1}{2}u^{-2} + C$$

$$= \frac{1}{u^2} + C$$

Substituting $u = 5 - 4x$, we have $\frac{1}{(5 - 4x)^2} + C$. ■

Example 5: Evaluate $\int x^2(x^3 - 5)^6\,dx$.

Solution: Let $u = g(x) = x^3 - 5$.

> **Tip:** How do you know which function to use for the u substitution, x^2 or $x^3 - 5$? Choose the function with the higher power, so that the differential $du = g'(x)dx$ will have the correct power.

Then $du = 3x^2\,dx$. In this integrand the "x^2" must be included with the "dx" when substituting, so that all functions of x are converted to functions of u. Thus, if $du = 3x^2\,dx$, then $\frac{1}{3}du = x^2\,dx$.

Substituting $u = x^3 - 5$ and $\frac{1}{3}du = x^2\,dx$, we have

$$\int x^2(x^3 - 5)^6\,dx = \int (x^3 - 5)^6 x^2\,dx = \int u^6\left(\frac{1}{3}du\right)$$

Integrating gives

$$= \frac{1}{3} \cdot \frac{1}{7}u^7 + C$$

$$= \frac{1}{21}u^7 + C$$

Substituting $u = x^3 - 5$ gives

$$= \frac{1}{21}(x^3 - 5)^7 + C$$

■

Warning! ***Not every integral of a composite function can be evaluated using the substitution method.***

For instance, had the integral in Example 5 been $\int x(x^3 - 5)^6\,dx$, the substitution technique would not have worked. Letting $u = x^3 - 5$ gives a differential involving $x^2\,dx$, and the substitution would have yielded $du = 3x^2\,dx$ or $x\,dx = \frac{1}{3x}du$. Because the differential is not strictly a function of u, the integration cannot be completed. There are other integration techniques that could be used, but they are beyond the scope of this text.

Check Your Understanding 22.1

Integrate each of the following integrals using the substitution technique.

1. $\int (7x + 3)^5\,dx$ **2.** $\int \frac{7}{(3 - 2x)^4}dx$ **3.** $\int \sqrt{7x + 3}\,dx$

4. $\int \frac{4}{5x - 2}dx$ **5.** $\int \frac{4x}{(x^2 - 5)^2}dx$ **6.** $\int xe^{-x^2}\,dx$

Determining the Value of the Constant of Integration C

As in Topic 21, given an **initial condition**, the value of the constant C in the antiderivative can be determined.

- Determine the antiderivative.
- Substitute the initial condition.
- Solve for C.

Example 6: Find $f(x)$ if $f'(x) = \dfrac{6x^2}{(x^3 - 4)^2}$ and $f(2) = -1$.

Solution:

$$f(x) = \int \frac{6x^2}{(x^3 - 4)^2}\,dx$$

Using substitution, let $u = x^3 - 4$. Then $du = 3x^2\,dx$. Because the integrand contains $6x^2$, we will multiply by 2 and substitute $2du = 6x^2\,dx$.

Substituting $u = x^3 - 4$ and $2du = 6x^2\,dx$, we have

$$f(x) = \int \frac{6x^2}{(x^3 - 4)^2}\,dx = \int \frac{1}{u^2}(2du)$$

Integrating gives

$$= \frac{2}{-1}u^{-1} + C$$

$$= -\frac{2}{u} + C$$

Substituting $u = x^3 - 4$ gives

$$f(x) = -\frac{2}{x^3 - 4} + C$$

Thus, $f(x) = -\dfrac{2}{x^3 - 4} + C$.

Substituting the initial condition $f(2) = -1$, we have

$$f(2) = -\frac{2}{2^3 - 4} + C = -1$$

so $-\frac{1}{2} + C = -1$ and $C = -\frac{1}{2}$.

Thus, $f(x) = -\dfrac{2}{x^3 - 4} - \dfrac{1}{2}$.

> **Warning!** Be sure to convert the antiderivative back to a function of x before substituting the initial condition.

■

Example 7: Find the equation of the curve passing through $(-1, 2)$ whose slope is given by $\dfrac{dy}{dx} = 6x(x^2 - 2)^5$.

Solution: Because we are given the slope of the curve, the equation can be found by

$$y = \int 6x(x^2 - 2)^5\,dx$$

Let $u = x^2 - 2$. Then $du = 2xdx$.

The integrand contains $6xdx$, so we will multiply by 3 and substitute $3du = 6xdx$.

Substituting $u = x^2 - 2$ and $3du = 6xdx$, we have

$$y = \int 6x(x^2 - 2)^5\,dx = \int u^5(3du)$$

Integrating gives

$$= 3 \cdot \frac{1}{6}u^6 + C = \frac{1}{2}u^6 + C$$

Substituting $u = x^2 - 2$, we have

$$y = \frac{1}{2}(x^2 - 2)^6 + C$$

We know that the curve must pass through $(-1, 2)$, so to find the value of C, let $x = -1$ and $y = 2$.

$$y = \frac{1}{2}(x^2 - 2)^6 + C$$

so

$$2 = \frac{1}{2}(1) + C \rightarrow C = \frac{3}{2}$$

The equation of the curve is $y = \frac{1}{2}(x^2 - 2)^6 + \frac{3}{2}$. ■

Example 8: A woman on a weight loss program loses weight at the rate of $W'(t) = -\dfrac{8}{0.4t + 2}$ pounds per week, where t is the number of weeks after starting the program. If her initial weight was 170 pounds, what is her weight after 4 weeks?

Solution: We know the rate at which her weight changes, so her weight at any time is

$$W(t) = \int -\frac{8}{0.4t + 2}\,dt$$

Let $u = 0.4t + 2$. Then $du = 0.4dt$. Solving for dt gives $\dfrac{1}{0.4}du = dt$.

Substituting $u = 0.4t + 2$ and $\dfrac{1}{0.4}du = dt$, we have

$$W(t) = \int -\frac{8}{0.4t + 2}\,dt = \int -\frac{8}{u} \cdot \frac{1}{0.4}\,du$$

$$= \int -\frac{20}{u}\,du$$

Integrating gives

$$W(t) = -20\ln u + C$$

Substituting $u = 0.4t + 2$, we have $W(t) = -20\ln(0.4t + 2) + C$.

We know that $W(0) = 170$, so $170 = -20\ln 2 + C$.

Then $C = 170 + 20\ln 2 \approx 183.86$.

Her weight at any time is given by $W(t) = -20\ln(0.4t + 2) + 183.86$.

Her weight after 4 weeks is $W(4) = -20\ln(0.4(4) + 2) \approx 158$.

After 4 weeks her weight is about 158 pounds. ■

Example 9: Marginal revenue, in thousand dollars, generated by producing x hundred units of a product is given by $R'(x) = xe^{0.2x^2} + 1$. Find the revenue generated by producing 500 units of the product.

Solution: Revenue is

$$R(x) = \int \left(xe^{0.2x^2} + 1\right) dx$$

The integrand is a sum, so we can write $R(x) = \int \left(xe^{0.2x^2}\right) dx + \int 1dx$.

The first integral requires substitution, so let's look at it first.

Let $u = 0.2x^2$. Then $du = 0.4xdx$ and $\frac{1}{0.4}du = xdx$.

Substituting $u = 0.2x^2$ and $\frac{1}{0.4}du = xdx$, we have

$$\int xe^{0.2x^2} dx = \int e^{u}\left(\frac{1}{0.4}du\right)$$

Integrating gives

$$= 2.5e^{u} + C$$

Substituting $u = 0.2x^2$, we have

$$= 2.5e^{0.2x^2} + C$$

Now, let's go back to the revenue function.

$$R(x) = \int \left(xe^{0.2x^2}\right) dx + \int 1dx$$

$$= 2.5e^{0.2x^2} + x + C$$

Initial revenue would be 0, so $R(0) = 0$. Thus,

$$R(0) = 2.5e^{0} + 0 + C = 0 \rightarrow C = -2.5$$

The revenue is given by $R(x) = 2.5e^{0.2x^2} + x - 2.5$.

Revenue for 500 units would be $R(5) \approx 373.533$. If 500 units are produced, revenue is about \$373,533. ■

Check Your Understanding 22.2

1. Determine the value of C given that $f'(x) = 6x(x^2 + 1)^3$ and $f(1) = 5$.
2. Find the equation of the curve passing through $\left(1, \frac{1}{2}\right)$ if the slope of the curve is given by $\dfrac{dy}{dx} = 5x - \dfrac{4}{(2x - 1)^2}$.

Check Your Understanding Answers

Check Your Understanding 22.1

1. $\frac{1}{42}(7x + 3)^6 + C$
2. $\dfrac{7}{6(3 - 2x)^3} + C$
3. $\frac{2}{21}\sqrt{(7x + 3)^3}\,dx$
4. $\frac{4}{5}\ln(5x - 2) + C$
5. $-\dfrac{2}{x^2 - 5} + C$
6. $-\frac{1}{2}e^{-x^2} + C$

Check Your Understanding 22.2

1. $C = -7$
2. $y = \dfrac{5}{2}x^2 + \dfrac{2}{2x - 1} - 4$

Topic 22 Review

22

This topic presented integration by substitution, a technique for integrating some composite functions.

CONCEPT	EXAMPLE
If $y = f(x)$, the **differential** is defined as $dy = f'(x)\,dx$.	The differential of $y = 3x^2 - 5$ is $dy = 6xdx$.
The **substitution** technique can be used to evaluate integrals of some types of composite functions. Given a composite function $f(g(x))$, let $u = g(x)$. Compute the differential $du = g'(x)dx$, solving for dx. Substitute so that $\int f(g(x))g'(x)\,dx = \int f(u)\,du$. Integrate $\int f(u)\,du$ and obtain the antiderivative $F(u) + C$. Substitute $u = g(x)$ to obtain the antiderivative $F(g(x)) + C$.	Integrate $\dfrac{4x}{(x^2 - 5)^3}dx$. Let $u = x^2 - 5$ so $du = 2xdx$ and $2du = 4xdx$. Substituting, we have $\dfrac{4x}{(x^2 - 5)^3}dx = \displaystyle\int \frac{2}{u^3\,du}$ $= \displaystyle\int 2u^{-3}\,du$

(continued)

	$= -u^{-2} + C$ $= -\dfrac{1}{u^2} + C$ $= -\dfrac{1}{(x^2-5)^2} + C$
If an **initial condition** is given, the value of the constant term C can be determined. After finding the antiderivative $F(g(x)) + C$, substitute the initial condition and solve for C.	Find the equation of the curve passing through $(2, -1)$ if the slope is given by $\dfrac{dy}{dx} = 4x - 3$. $\dfrac{dy}{dx} = 4x - 3 \rightarrow y = 2x^2 - 3x + C$ $y(2) = -1 = 2(2)^2 - 3(2) + C$ so $C = -3$. Thus, the equation of the curve is $y = 2x^2 - 3x - 3$.
The answer to the integration can be checked by differentiating the answer and verifying that the original integrand is obtained.	$\dfrac{4x}{(x^2-5)^3}dx = -\dfrac{1}{(x^2-5)^2} + C$ To check the answer, $D_x\left(-\dfrac{1}{(x^2-5)^2} + C\right)$ $= D_x\left(-1(x^2-5)^{-2}\right) + D_x(C)$ $= 2(x^2-5)^{-3}(2x) + 0$ $= \dfrac{4x}{(x^2-5)^3}$ which was the original integrand.

NEW TOOLS IN THE TOOL KIT

- Computing differentials
- Using the substitution technique to integrate some composite functions
- Checking the answer of the integration by differentiating and verifying that the original integrand is obtained
- Determining the value of the constant term C when substitution has been used for integration

Topic 22 Exercises

22

Are the following statements true or false?

1. $\int \sqrt{x^2+4}\,dx$ can be integrated using the substitution technique.

2. $\int e^{3x}\,dx$ can be integrated using the substitution technique.

3. $\int \frac{x}{x^2-4}\,dx$ can be integrated using the substitution technique.

4. $\int e^{x^2}\,dx$ can be integrated using the substitution technique.

5. $\int \frac{3}{4x-1}\,dx = 3\ln|4x-1| + C.$

6. $\int e^{x^2}\,dx = \frac{1}{2x}e^{x^2} + C.$

In Exercises 7 through 10 compute the differential of the given function.

7. $y = x^4 - 3x^2 + 6$
8. $y = 2x^3 - 5\ln(x^4)$
9. $u = 2e^{-3x}$
10. $u = \frac{4}{x}$

In Exercises 11 and 12 perform each integration in two ways:

a. using the exponential rule
b. using substitution

11. $\int e^{0.4x}\,dx$
12. $\int e^{-3x+1}\,dx$

In Exercises 13 through 18 perform each integration in two ways and verify that the two results are equivalent:

a. expanding and using the Power Rule
b. using substitution

13. $\int (x-3)^2\,dx$
14. $\int (x-2)^3\,dx$
15. $\int (5x-4)^2\,dx$
16. $\int (2x+7)^2\,dx$
17. $\int x(x^2-3)^2\,dx$
18. $\int x^2(x^3+2)^2\,dx$

In Exercises 19 through 50, perform the integration using substitution.

19. $\int (3x-5)^7\,dx$
20. $\int (7-2x)^9\,dx$
21. $\int \frac{dx}{x+5}$
22. $\int \frac{dx}{4-x}$

(Hint for Exercise 21 and 22: $\int \frac{dx}{x} = \int \frac{1}{x}dx$.)

23. $\int \frac{b^2}{b^3-4}\,db$
24. $\int \frac{2a^3}{a^4+1}\,da$
25. $\int \frac{5x^2}{(x^3-4)^3}\,dx$
26. $\int \frac{x^3}{(x^4+1)^5}\,dx$
27. $\int a\sqrt{a^2-3}\,da$
28. $\int 3c\sqrt{4-c^2}\,dc$
29. $\int 3xe^{x^2+1}\,dx$
30. $\int 2x^3e^{x^4}\,dx$
31. $\int \frac{4p}{\sqrt{p^2+1}}\,dp$
32. $\int \frac{3p^3}{\sqrt{p^4-2}}\,dp$
33. $\int e^{x/3}\,dx$
34. $\int e^{-2x/5}\,dx$
35. $\int \frac{e^x}{e^x-3}\,dx$
36. $\int \frac{e^{2/x}}{x^2}\,dx$
37. $\int t^2e^{-t^3}\,dt$
38. $\int 3t^5e^{-2t^6}\,dt$
39. $\int \frac{\ln(5v)}{v}\,dv$
40. $\int \frac{(\ln(3q))^4}{q}\,dq$
41. $\int x^3(x^4+2)^6\,dx$
42. $\int x^2(x^3+4)^5\,dx$
43. $\int \frac{dx}{x\ln(x^2)}$
44. $\int \frac{dx}{x\ln(x^3)}$

45. $\int \frac{3x^2 - 4}{(x^3 - 4x + 1)^2} dx$

46. $\int \frac{6x^2 + 4}{(x^3 + 2x - 4)^3} dx$

47. $\int \left(3x^5 + \frac{4}{(3x - 2)^5}\right) dx$

48. $\int (2xe^{-x^2} + 6x^2)\, dx$

49. $\int \frac{e^x - 3}{(e^x - 3x)^2} dx$ **50.** $\int \frac{4x - 2e^{-x}}{(e^{-x} + x^2)^3} dx$

In Exercises 51 through 54 find $f(x)$ given the indicated initial condition.

51. $f'(x) = 5x(x^2 - 3)^4$ and $f(2) = \frac{3}{2}$.

52. $f'(x) = \frac{x}{\sqrt{x^2 + 1}}$ and $f(0) = -2$.

53. $f'(x) = xe^{-x^2}$ and $f(0) = 1$.

54. $f'(x) = \frac{x^2}{x^3 + 1}$ and $f(0) = -1$.

In Exercises 55 through 58 find the equation of the curve.

55. Passing through $(2, -1)$ with $\frac{dy}{dx} = \frac{4x}{x^2 - 3}$

56. Passing through $(0, 2)$ with $\frac{dy}{dx} = x^3e^{-x^4}$

57. Passing through $(0, 1)$ with $\frac{dy}{dx} = x\sqrt{x^2 + 1} - x$

58. Passing through $(-1, 0)$ with $\frac{dy}{dx} = x^3(x^4 - 2)^2 + 2x$

59. In still water, the radius of an oil spill spreads at a rate of $R'(t) = \frac{6}{\sqrt{0.3t + 2}}$ feet per minute. How large is the radius of the spill after 30 minutes? Assume $R(0) = 0$.

60. Total sales of a new DVD increase at a rate of $S'(t) = 20\sqrt{4t + 1}$ thousand DVDs per month. What were the total sales during the third month? Assume $S(0) = 0$.

61. Marginal demand, in thousands, for a new gas barbecue grill is $D'(p) = -\frac{4p}{\sqrt{10 - p^2}}$, where the price of the grill is p hundred dollars. If there is a demand for 15,000 grills when the price is \$125, find the demand if the price is \$275.

62. Repeat Exercise 61 if the price is \$225. Sketch a graph of the demand function.

63. Total box office revenues for the hit movie *Star Wars Episode III: Revenge of the Sith*, x weeks after its release in May 2005, changed at a rate of $R'(x) = 75.445xe^{-0.125x^2}$ million dollars per week. If the total revenues during the third week were \$307.9 million, what were the total revenues during week 4?

64. Refer to Exercise 63. A news release (June 5, 2005) stated that Fox expected the movie to gross over \$400 million and perhaps rival the \$431 million from *Star Wars Episode I.* Assuming that the rate of change is as indicated in Exercise 63, how long will it take for the total revenues to be \$400 million? How long will it take for the total revenues to reach \$431 million?

65. A factory built along a river discharges pollution into the river at a rate of $P'(t) = \frac{4t}{t^2 + 2}$ thousand tons per year, where t is the number of years since the factory was built. Assuming there was zero initial pollution, how much pollution was the factory dumping into the river after two years?

66. Newsom Car Sales holds a week-long Labor Day Tent Sale every year. Records show that cars sell at a rate of $S'(x) = \frac{15}{x + 3}$ cars per day, where x is the number of days since the sale started. If 30 cars were sold on the first day of the sale, how many cars were sold on the fifth day of the sale?

SHARPEN THE TOOLS

The integral $\int x^m(x + b)^n\, dx$ can also be evaluated using substitution.

Let $u = x + b$ and $du = dx$.

Solving $u = x + b$ gives $x = u - b$.

For instance, to integrate $\int x(x+5)^2\,dx$, let $u = x + 5$, $x = u - 5$, and $du = dx$. Then we have

$$\int x(x+5)^2\,dx = \int (u-5)u^2\,du \quad \text{Substitute } u = x+5 \text{ and } x = u-5$$

$$= \int (u^3 - 5u^2)\,du \quad \text{Multiply the polynomials}$$

$$= \frac{1}{4}u^4 - \frac{5}{3}u^3 + C \quad \text{Integrate using the Power Rule}$$

$$= \frac{1}{4}(x+5)^4 - \frac{5}{3}(x+5)^3 + C \quad \text{Substitute } u = x+5$$

Perform the integrations in Exercises 67 through 70 using this technique.

67. $\int x^2(x+1)^3\,dx$

68. $\int x^3(x-2)^4\,dx$

69. $\int \frac{x}{x+1}\,dx$

70. $\int \frac{x^3}{x^2-3}\,dx$

COMMUNICATE

71. Prove that $\int e^{mx+b}\,dx = \frac{1}{m}e^{mx+b} + C$.

72. Prove that $\int \frac{k}{mx+b}\,dx = \frac{k}{m}\ln|mx+b| + C$, for k a constant.

73. Explain why $\int \frac{2}{(x^2-3)^4}\,dx$ cannot be evaluated with the substitution technique.

74. Explain why $\int \sqrt{x^2-4}\,dx$ cannot be evaluated with the substitution technique.

CALCULATOR CONNECTION

75. Refer to Exercise 65.

a. Graph the rate function $P'(t) = \frac{4t}{t^2+2}$ using a window of $[0, 8]$ for x and $[0, 2]$ for y. Use the **TRACE** feature of the calculator to determine the value of $P'(2)$. Interpret the result.

b. Graph the pollution function $P(t) = 2\ln(t^2+2) - 1.386$ using a window of $[0, 8]$ for x and $[0, 10]$ for y. Use the derivative feature of your calculator to evaluate $P'(2)$. Interpret the result.

c. Are the results of parts a and b the same? Explain your answer.

Calculator Corner 22.1

After graphing the desired function, press **2nd TRACE (CALC)**. Select option **6: *dy/dx***. Type in the desired x value and press **ENTER**. The calculator will display the value of the derivative at the selected point.

76. Refer to Exercise 63.

a. Graph the rate function $R'(x) = 75.445xe^{-0.125x^2}$ using a window of $[0, 8]$ for x and $[0, 120]$ for y. Use the **TRACE** feature of the calculator to determine the value of $R'(4)$. Interpret the result.

b. Graph the total revenue function $R(x) = -301.78e^{-0.125x^2} + 405.87$ using a window of $[0, 8]$ for x and $[0, 500]$ for y. Use the derivative feature of your calculator to evaluate $R'(4)$. Interpret the result.

c. Are the results of parts a and b the same? Explain your answer.

TOPIC

23

Definite Integrals

Suppose a local electronics store is selling a new computer and knows the rate at which the computers have been selling since being introduced. Using an indefinite integral and the knowledge that initial sales were zero, you could determine a function that describes sales of the computer at any time. The store could then predict sales at some specific time. Suppose, though, that the store wanted to determine the total sales during the first quarter or the average sales for the first year. An indefinite integral will not provide the answer to these questions. In this topic you will learn about definite integrals, which give information about functions over a time interval rather than at one specific point. Before starting this topic, be sure to complete the Warm-up Exercises to sharpen the algebra and calculus skills you will need.

IMPORTANT TOOLS IN TOPIC 23

- *Definite integral*
- *Fundamental Theorem of Calculus*
- *Average value*
- *Evaluating definite integrals using substitution*

TOPIC 23 WARM-UP EXERCISES

Be sure you can successfully complete the following exercises before starting Topic 23.

Algebra Warm-up Exercises

1. Given $f(x) = 2e^{-3x}$, evaluate

a. $f(2)$ **b.** $f(0)$ **c.** $f(2) - f(0)$

2. Given $f(x) = x^3 - \frac{1}{2}x^2 + 3x - 1$, evaluate

a. $f(3)$ **b.** $f(-1)$ **c.** $f(3) - f(-1)$

Use the properties of logarithms to simplify each of the following. Do not evaluate.

3. $\ln 8 - \ln 2$ **4.** $\ln 4 + \ln 6$ **5.** $3 \ln 2$

Calculus Warm-up Exercises

Perform each integration.

1. $\int (x^2 - 4x + 6)dx$

2. $\int e^{3x}dx$

3. $\int 10\sqrt[4]{x}\,dx$

4. $\int \left(5x^2 - \frac{3}{x}\right)dx$

5. Find $f(x)$ given that $f'(x) = 2x^3 - 4x + \frac{1}{x^2}$ and $f(1) = -3$.

6. The number of new cases in a flu epidemic grows at the rate of $150\sqrt{t}$ new cases per day. If there were 25 new cases on the first day, how many new cases were there on the fourth day?

Perform each integration using substitution.

7. $\int 3x(x^2 - 5)^4dx$

8. $\int 4xe^{-x^2}dx$

9. $\int \frac{6}{4 - 3x}dx$

10. $\int \frac{6x^3}{(x^4 - 2)^2}dx$

Answers to Warm-up Exercises

Algebra Warm-up Exercises

1. a. $2e^{-6} \approx 0.005$ b. 2
   c. $2(e^{-6} - 1) \approx -1.995$
2. a. 30.5 b. −5.5 c. 36
3. ln 4 4. ln 24 5. ln 8

Calculus Warm-up Exercises

1. $\frac{1}{3}x^3 - 2x^2 + 6x + C$
2. $\frac{1}{3}e^{3x} + C$
3. $8\sqrt[4]{x^5} + C$
4. $\frac{5}{3}x^3 - 3\ln x + C$
5. $f(x) = \frac{1}{2}x^4 - 2x^2 - \frac{1}{x} - \frac{1}{2}$
6. 725 new cases
7. $\frac{3}{10}(x^2 - 5)^5 + C$
8. $-2e^{-x^2} + C$
9. $-2\ln|4 - 3x| + C$
10. $-\dfrac{3}{2(x^4 - 2)} + C$

Definite Integrals

You now have two useful tools for antidifferentiation in your Tool Kit.

- Converting the derivative of a function back to the original function using an indefinite integral.
- Determining the value of the unknown constant in the antiderivative by substituting a given initial condition.

Suppose, though, that we are interested only in the value of an antiderivative over some specific interval in its domain. For instance, given the rate at which sales of a product are growing, a business manager may want to predict the total sales in the next quarter. An integral is still needed, because we need to convert a derivative back to the original function. But an indefinite integral, which will provide the sales at one specific point in time, will not work in this case where we need to predict sales over a period of time.

The **definite integral** of $f(x)$ on the interval $[a, b]$ is denoted by

$$\int_a^b f(x)dx$$

and represents the net change in the value of the antiderivative on the interval $[a, b]$. The numbers a and b are called the **limits of integration**.

How do we evaluate a definite integral?

- The expression $\int_a^b f(x)dx$ is an integral, so we first integrate using an appropriate method to find the antiderivative.
- Then we evaluate the antiderivative for b and for a and determine the difference.

This procedure results from the **Fundamental Theorem of Calculus.**

Fundamental Theorem of Calculus
Given $f(x)$ continuous on the interval $[a, b]$ and $F(x)$ an antiderivative of $f(x)$, then $\int_a^b f(x)dx = F(x)|_a^b = F(b) - F(a)$.

Example 1: Evaluate $\int_3^5 (2x - 7)dx$.

Solution: First, find the antiderivative.

$$\int_3^5 (2x - 7)dx = (x^2 - 7x + C)|_3^5$$

Here we have $F(x) = x^2 - 7x + C$ that must be evaluated between $a = 3$ and $b = 5$. Next evaluate $F(b) - F(a)$.

$$\begin{aligned}\int_3^5 (2x - 7)dx &= (x^2 - 7x + C)|_3^5 \\ &= F(5) - F(3) \\ &= (5^2 - 7(5) + C) - (3^2 - 7(3) + C) \\ &= (-10 + C) - (-12 + C) \\ &= 2\end{aligned}$$

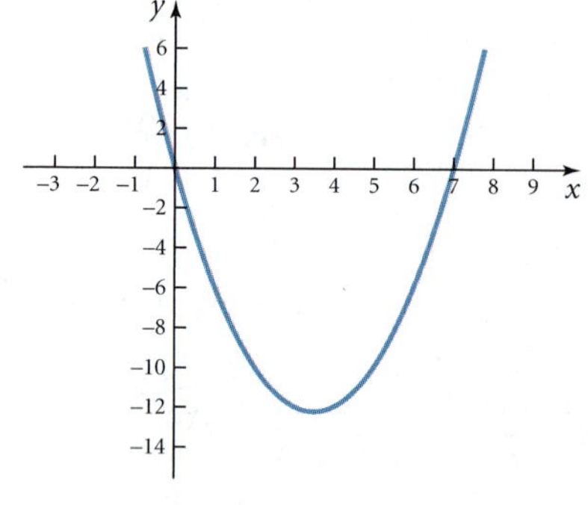

Figure 23.1

Did you notice that the C disappeared? The C will always cancel out in the subtraction, so it is not necessary to include it when writing the antiderivative. You should also note that the result of a definite integral is a numerical value, not a family of functions as was the case with indefinite integrals.

To explore graphically what is happening with this definite integral, look at the graph of the antiderivative $F(x) = x^2 - 7x$, with $C = 0$, in Figure 23.1.

Focus on the interval $[3, 5]$, as shown in Figure 23.2, and you will see that the net change in the value of the function between $F(3)$ and $F(5)$ is two units.

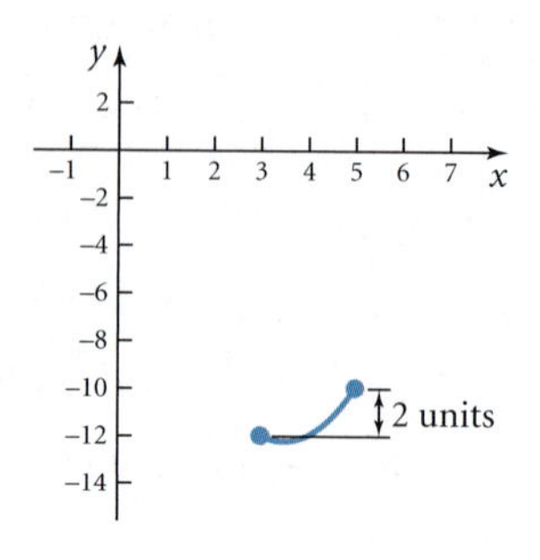

Figure 23.2

Tip: Graph $y = x^2 - 7x + C$ for several values of C. You should be able to see that the net change on [3, 5] is two units, regardless of the C value chosen. Graphed here in Figure 23.3 are $y = x^2 - 7x + 6$, $y = x^2 - 7x$, and $y = x^2 - 7x - 8$.

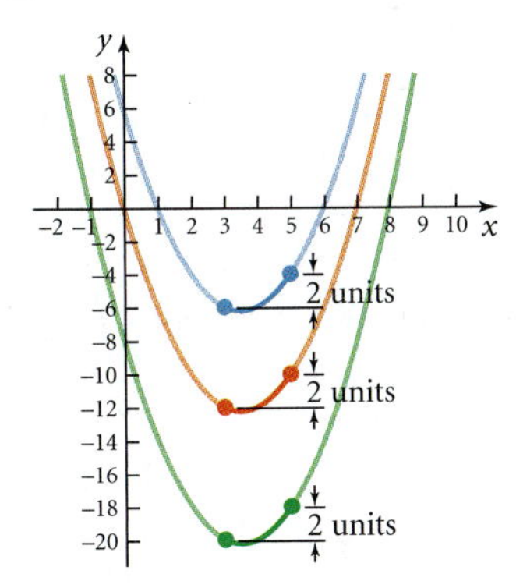

Figure 23.3

Example 2: Evaluate $\int_{-1}^{2} (2x^3 - 6x^2)dx$.

Solution:

$$\int_{-1}^{2} (2x^3 - 6x^2)dx = \left(\frac{1}{2}x^4 - 2x^3\right)\Bigg|_{-1}^{2}$$

$$= \left(\frac{1}{2}(2)^4 - 2(2)^3\right) - \left(\frac{1}{2}(-1)^4 - 2(-1)^3\right)$$

$$= -8 - 2\frac{1}{2}$$

$$= -10\frac{1}{2}$$

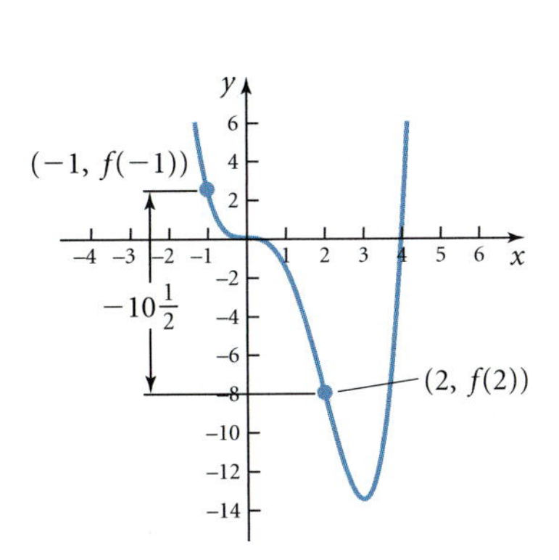

Figure 23.4

The graph in Figure 23.4 shows clearly that the net change can be negative, because the value decreases from $f(-1)$ to $f(2)$.

Example 3: Evaluate $\int_{-2}^{0} e^{-2x}dx$.

Solution:

$$\int_{-2}^{0} e^{-2x}dx = -\frac{1}{2}e^{-2x}\Bigg|_{-2}^{0}$$

$$= -\frac{1}{2}(e^0 - e^4)$$

$$= -\frac{1}{2}(1 - e^4) \approx 26.8$$

Tip: When the antiderivative involves a constant multiple, as was the case with $-\frac{1}{2}e^{-2x}$ in Example 3, it is sometimes easier to factor out the constant and multiply it last, as was done in the example, rather than include it in each term.

The expression $-\frac{1}{2}e^{-2x}\big|_{-2}^{0} = \left(-\frac{1}{2}e^{0}\right) - \left(-\frac{1}{2}e^{4}\right)$, which is the same as $-\frac{1}{2}\left(e^{0} - e^{4}\right)$.

Example 4: Evaluate $\int_2^7 \frac{6}{c}\,dc$.

Solution:

$$\int_2^7 \frac{6}{c}\,dc = 6\ln c\big|_2^7$$

$$= 6(\ln 7 - \ln 2) = 6\ln\frac{7}{2} \approx 7.52$$

Calculator Corner 23.1

Definite integrals can be evaluated on the TI-83/84. Press **MATH** and select option **9: FNINT(**. Enter the integral as follows: fnInt(function, x, lower limit, upper limit) then press **ENTER**. Example 4 would be entered as fnInt($6/x$, x, 2, 7).

Example 5: Production costs of a new machine grow at a rate of $C'(t) = 200\sqrt{t}$ dollars per week. Find the total production costs during the first four weeks of production. What are the total production costs during the next four weeks?

Solution: We are given the rate at which costs grow, so an integral will be used to determine the actual cost. A definite integral will be used because costs must be determined over an interval of time, rather than at one specific time. The first four weeks indicate the interval [0, 4], from the beginning through the end of the fourth week.

$$\int_0^4 200\sqrt{t}\,dt = 200 \cdot \frac{2}{3}t^{3/2}\Big|_0^4$$

$$= \frac{400}{3}\left(4^{3/2} - 0\right)$$

$$= \frac{3200}{3} \approx 1066.67$$

Total production costs for the first four weeks are about \$1066.67.

For the next four weeks, we must begin at the end of the fourth week and go until the end of the eighth week, indicating the interval [4, 8].

$$\int_4^8 200\sqrt{t}dt = 200 \cdot \frac{2}{3}t^{3/2}\Big|_4^8$$
$$= \frac{400}{3}\left(8^{3/2} - 4^{3/2}\right)$$
$$\approx 1950.32$$

Total production costs for the second month are about $1950.32. ■

Example 6: The number of new cases in a flu epidemic during the first week grows at a rate of $N'(x) = 4e^{0.8x}$ new cases per day. How many cases were reported between the end of the second day and the end of the fifth day?

Solution:

$$\int_2^5 4e^{0.8x}dx = \frac{4}{0.8}e^{0.8x}\Big|_2^5$$
$$= 5(e^{0.8(5)} - e^{0.8(2)})$$
$$\approx 248.23$$

From the end of the second day to the end of the fifth day there were about 248 new cases of flu reported. ■

Check Your Understanding 23.1

Evaluate each integral.

1. $\int_0^2 (x^3 - 6x^2 + 4)dx$ **2.** $\int_1^8 \sqrt[3]{x}dx$

3. $\int_1^4 \frac{3}{x}dx$ **4.** $\int_0^1 e^{x/2}dx$

5. In Example 6 determine the total number of new cases from the end of the fifth day to the end of the seventh day.

Average Value of a Function

Suppose a local electronics store is selling a new computer. After x months the total number of computers sold is given by $S(x) = 40\sqrt{x}$. How might we determine the average number of computers sold over the first four months?

One approach might be to consider sales at the beginning and sales at the end, or in other words, from $S(0) = 0$ to $S(4) = 80$. Then you might be tempted to say that the average number of computers sold during the four-month period is

$\frac{0 + 80}{2}$ or 40 computers. However, if you look at sales each month during the first four months, you will see what is wrong with this average.

$$S(0) = 0$$
$$S(1) = 40$$
$$S(2) \approx 57$$
$$S(3) \approx 69$$
$$S(4) = 80$$

You now see that sales quickly reached 40 computers after only one month, but it took three more months to sell 40 more computers, so 40 does not really represent the average number sold. If we average the five data values we now have, the average number of computers sold during the four-month period is $\frac{0 + 40 + 57 + 69 + 80}{5}$ or about 49 computers.

Obviously, we could continue by generating even more data values by shortening the intervals and increasing the number of intervals and then calculating the average. (Imagine being able to evaluate $f(x)$ for every value of x in $[0, 4]$ and then calculate the average.) Instead, we will make a definition.

Definition: Given a function $f(x)$ continuous on $[a, b]$, the **average value** of $f(x)$ on $[a, b]$ is given by

$$\frac{1}{b - a}\int_a^b f(x)dx$$

Example 7: Suppose a local electronics store is selling a new computer. After x months the total number of computers sold is given by $S(x) = 40\sqrt{x}$. Determine the average number of computers sold over the first four months.

Solution: "Over the first four months" means a time interval of $[0, 4]$. Thus, the average value is given by

$$\frac{1}{4 - 0}\int_0^4 40\sqrt{x}\,dx$$

Integrating, we have

$$\frac{1}{4 - 0}\int_0^4 40\sqrt{x}\,dx = \frac{1}{4} \cdot 40 \cdot \frac{2}{3}x^{3/2}\Big|_0^4$$
$$= \frac{20}{3}(4^{3/2} - 0) = 53\frac{1}{3}$$

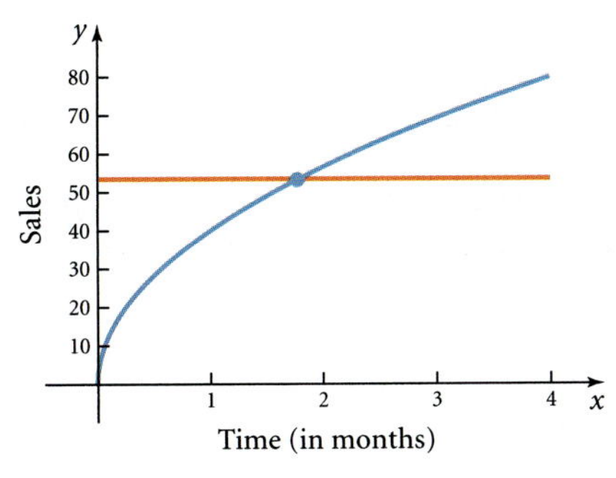

Figure 23.5

The average number of computers sold over the first four months was about 53 computers per month.

The graph of $S(x) = 40\sqrt{x}$ in Figure 23.5 along with the line $S = 53\frac{1}{3}$ representing the average provides some additional insight. The average line, $S = 53\frac{1}{3}$, divides the vertical dimension of the graph of $S(x) = 40\sqrt{x}$ roughly in the middle. About half of the S values are below $53\frac{1}{3}$, and about half of the S values are above $53\frac{1}{3}$. ■

Example 8: In any given year, the percentage of the U.S. population who were foreign born is given by $P(x) = 0.00006x^3 - 0.00675x^2 + 0.0523x + 14.15$, where x is the number of years after 1900. Find the average annual percentage of the population who were foreign born between 1920 and 1940; between 1960 and 1980.

Solution: 1920 is 20 years after 1900 and 1940 is 40 years after 1900 so the interval is [20, 40]. The average annual percentage who were foreign born between 1920 and 1940 is given by

$$\frac{1}{40 - 20}\int_{20}^{40}(0.00006x^3 - 0.00675x^2 + 0.0523x + 14.15)dx$$

$$= \frac{1}{20}(0.000015x^4 - 0.00225x^3 + 0.02615x^2 + 14.15x)\Big|_{20}^{40}$$

$$= \frac{1}{20}(502.24 - 277.86)$$

$$\approx 11.2\%$$

The average annual percentage who were foreign born between 1960 and 1980 is given by

$$\frac{1}{80 - 60}\int_{60}^{80}(0.00006x^3 - 0.00675x^2 + 0.0523x + 14.15)dx$$

$$= \frac{1}{20}(0.000015x^4 - 0.00225x^3 + 0.02615x^2 + 14.15x)\Big|_{60}^{80}$$

$$= \frac{1}{20}(761.76 - 651.54)$$

$$\approx 5.51\%$$ ■

Check Your Understanding 23.2

1. Find the average value of $f(x) = x^3$ on [0, 2].
2. In Example 8 determine the average annual percentage of the U.S. population who were foreign born between 1980 and 2000.

Using Substitution to Evaluate Definite Integrals

The substitution technique discussed in Topic 22 can also be used to evaluate definite integrals. Find the antiderivative $F(x)$ and then evaluate $F(b) - F(a)$.

Example 9: Evaluate $\int_2^4 \frac{x}{x^2 - 1} dx$.

Solution: The expression $\frac{x}{x^2 - 1}$ is a composite function so we will use the substitution technique.

Let $u = x^2 - 1$. Then $du = 2xdx$ and $\frac{1}{2}du = xdx$.

Substitute $u = x^2 - 1$ and $\frac{1}{2}du = xdx$.
Then

$$\int_2^4 \frac{x}{x^2 - 1} dx = \int_{x=2}^{x=4} \frac{1}{2} \cdot \frac{1}{u} du$$

Tip: The limits of integration on $\int_2^4 \frac{x}{x^2 - 1} dx$ are values of x. When making the substitution converting the integrand to a function of u, it is a good idea to indicate that the limits of integration are still in terms of x.

Integrate

$$= \frac{1}{2} \ln|u| \Big|_{x=2}^{x=4}$$

Substitute $u = x^2 - 1$

$$= \frac{1}{2} \ln|x^2 - 1| \Big|_2^4$$

Evaluate $F(4) - F(2)$

$$= \frac{1}{2}(\ln 15 - \ln 3) = \frac{1}{2} \ln 5 \approx 0.805$$

Warning! Be sure to convert the antiderivative to a function of x before substituting the limits of integration.

■

MATHEMATICS CORNER 23.2 

Substitution to Evaluate Definite Integrals

Another way to evaluate definite integrals that involve substitution is to convert the limits of integration to values of u. Let's rework Example 9, $\int_2^4 \frac{x}{x^2 - 1}dx$. Let $u = x^2 - 1$ and $\frac{1}{2}dx = xdx$ as in Example 9. The interval of integration is [2, 4], so convert these x values to u values using the substitution $u = x^2 - 1$.

If $x = 2$, then $u = 3$. If $x = 4$, then $u = 15$.

Now substitute:

$$\int_2^4 \frac{x}{x^2 - 1}dx = \int_3^{15} \frac{1}{2} \cdot \frac{1}{u}du$$

$$= \frac{1}{2}\ln u \Big|_3^{15} = \frac{1}{2}(\ln 15 - \ln 3)$$

which is the same result obtained in Example 9.

Example 10: Evaluate $\int_0^1 \frac{18x^2}{(x^3 + 1)^4}dx$.

Solution: Let $u = x^3 + 1$. Then $du = 3x^2\,dx$ and $6du = 18x^2\,dx$. Then

$$\int_0^1 \frac{18x^2}{(x^3 + 1)^4}dx = \int_{x=0}^{x=1} \frac{6}{u^4}du$$

$$= \int_0^1 6u^{-4}du = 6 \cdot \frac{1}{-3}u^{-3}\Big|_{x=0}^{x=1}$$

$$= -\frac{2}{u^3}\Big|_0^1 = -\frac{2}{(x^3 + 1)^3}\Big|_0^1$$

$$= -\frac{2}{8} - \frac{-2}{1}$$

$$= 1\frac{3}{4}$$

■

Example 11: A woman on a weight loss program loses weight at the rate of $W'(t) = -\frac{8}{0.04t + 2}$ pounds per week. How much weight did she lose during the first four weeks of the program?

Solution: The first four weeks means the interval of integration is [0, 4]. The integral to be evaluated is $\int_0^4 -\frac{8}{0.04t + 2}dt$.

Using substitution with $u = 0.04t + 2$ and $du = 0.04dt$, we have

$$\int_0^4 -\frac{8}{0.04t + 2}dt = \int_{t=0}^{t=4} \frac{-8}{0.04} \cdot \frac{1}{u}du$$
$$= -200 \ln|u|\Big|_{t=0}^{t=4} = -200 \ln|0.04t + 2|\Big|_0^4$$
$$= -200(\ln 3.6 - \ln 2) \approx -15.4$$

Why is the result negative? Definite integrals measure net change, and the net change in her weight is a loss, or a decrease. During the first four weeks of the program, the woman lost about 15 pounds. ■

Check Your Understanding 23.3

Evaluate each integral using substitution.

1. $\int_1^3 (3x - 5)^2 dx$
2. $\int_{-2}^0 \frac{7}{3 - 2x}dx$
3. $\int_2^6 \sqrt{2x - 3}\, dx$
4. $\int_0^1 4xe^{-x^2}dx$

Check Your Understanding Answers

Check Your Understanding 23.1

1. -4
2. $\frac{45}{4}$
3. $-3\ln 4$
4. ≈ 1.3
5. about 1079 total cases

Check Your Understanding 23.2

1. 2 units
2. 8.24%

Check Your Understanding 23.3

1. 8
2. ≈ 2.966
3. $\frac{26}{3}$
4. ≈ 1.264

Topic 23 Review

23

This topic presented definite integrals, that provide information about a function over an interval in the domain.

CONCEPT	EXAMPLE
The **definite integral** of $f(x)$ on $[a, b]$ is denoted by $\int_a^b f(x)dx$. The definite integral gives the net change in the value of an antiderivative over the interval $[a, b]$ in its domain. The numbers a and b are called the **limits of integration**.	The expression $\int_1^3 x^3 dx$ is a definite integral. When evaluated, the result is the net change in the value of the antiderivative $\frac{x^4}{4} + C$ on the interval $[1, 3]$.
Definite integrals are evaluated using the **Fundamental Theorem of Calculus** $\int_a^b f(x)dx = F(x)\vert_a^b = F(b) - F(a).$	$\int_1^3 x^3 dx = \left.\frac{x^4}{4}\right\vert_1^3 = \frac{3^4}{4} - \frac{1^4}{4} = 20$
The **average value** of a function over an interval in its domain is found by evaluating $\frac{1}{b-a}\int_a^b f(x)dx.$	The average value of $f(x) = x^3$ on $[0, 3]$ is $\frac{1}{3}\int_0^3 x^3 dx = \frac{1}{3}\left(\frac{x^4}{4}\right)_0^3 = \frac{1}{3}\left(\frac{3^4}{4} - 0\right) = \frac{1}{3}\left(\frac{81}{4}\right) = \frac{27}{4}.$
Substitution can be used when evaluating definite integrals involving some composite functions.	$\int_2^5 \frac{x}{x^2 - 1}dx = \frac{1}{2}\int_{x=2}^{x=5} \frac{1}{u}du = \frac{1}{2}\ln u \Big\vert_{x=2}^{x=5}$ $= \frac{1}{2}\ln(x^2 - 1)_2^5$ $= \frac{1}{2}(\ln 24 - \ln 3)$ $= \frac{1}{2}\ln 8 \approx 1.04$

NEW TOOLS IN THE TOOL KIT

- Evaluating definite integrals
- Determining the net change in the value of an antiderivative over a specific interval in its domain using a definite integral
- Determining the average value of a function over a specific interval in its domain
- Evaluating definite integrals using substitution

Topic 23 Exercises

23

Evaluate each definite integral.

1. $\int_{-2}^{3} 6\,dx$

2. $\int_{-1}^{5} -3\,dx$

3. $\int_{1}^{3} 4x\,dx$

4. $\int_{-1}^{2} x^2\,dx$

5. $\int_{2}^{4} (9x^2 - 3)\,dx$

6. $\int_{1}^{4} (4 - 5x)\,dx$

7. $\int_{0}^{1} (2x^3 - 4x + 5)\,dx$

8. $\int_{-2}^{0} (3x^5 + 2x - 7)\,dx$

9. $\int_{1}^{3} \left(2x^{-2} - \frac{5}{x}\right)dx$

10. $\int_{1}^{2} \left(3x^{-5} + \frac{2}{x}\right)dx$

11. $\int_{-4}^{-2} -\frac{3}{x^4}\,dx$

12. $\int_{-3}^{-1} \frac{2}{x^5}\,dx$

13. $\int_{1}^{8} 8\sqrt[3]{x}\,dx$

14. $\int_{0}^{1} 14\sqrt[4]{x^3}\,dx$

15. $\int_{0}^{2} 4e^{-2x}\,dx$

16. $\int_{0}^{3} 6e^{4x}\,dx$

17. $\int_{-1}^{2} (x - 3)^2\,dx$

18. $\int_{-2}^{1} (2x + 5)^2\,dx$

19. $\int_{1}^{3} \left(21x^{3/4} - 8x^{-1/3}\right)dx$

20. $\int_{2}^{4} \left(10x^{2/3} + 6x^{-1/4}\right)dx$

21. $\int_{1}^{3} \frac{2x^2 - 5x + 3}{2x}\,dx$

22. $\int_{1}^{4} \frac{4x^3 - 2x^2 + 7x}{4x^3}\,dx$

Evaluate each integral using substitution.

23. $\int_{0}^{1} x(x^2 - 1)^3\,dx$

24. $\int_{-1}^{1} x^3(x^4 - 3)^2\,dx$

25. $\int_{-3}^{1} \sqrt{3 - 2x}\,dx$

26. $\int_{0}^{8} \sqrt{5x + 9}\,dx$

27. $\int_{1}^{3} \frac{4x^2}{x^3 + 2}\,dx$

28. $\int_{3}^{6} \frac{8x}{x^2 - 3}\,dx$

29. $\int_{-1}^{2} \frac{8x}{(x^2 + 3)^2}\,dx$

30. $\int_{2}^{4} -\frac{6x^2}{(x^3 - 1)^3}\,dx$

31. $\int_{0}^{2} x^2\sqrt{x^3 + 4}\,dx$

32. $\int_{-2}^{0} x\sqrt{4 - x^2}\,dx$

Find the average value of the given function over the indicated interval.

33. $f(x) = x^2$ on $[-1, 3]$

34. $f(x) = 0.5x^3$ on $[-2, 0]$

35. $f(x) = e^{2x}$ on $[0, 2]$

36. $f(x) = e^{-x/2}$ on $[1, 3]$

37. $f(x) = \frac{4}{x}$ on $[2, 5]$

38. $f(x) = -\frac{2}{x}$ on $[1, 4]$

39. $f(x) = \frac{4}{x^2}$ on $[2, 5]$

40. $f(x) = -\frac{2}{x^3}$ on $[1, 4]$

41. $f(x) = x^3 - 4x + 5$ on $[1, 4]$

42. $f(x) = 3x^2 - 5x + 2$ on $[-2, 1]$

43. $f(x) = 4x\sqrt{6 - x^2}$ on $[0, 2]$

44. $f(x) = 3x(x^2 - 5)^4$ on $[0, 2]$

45. $f(x) = 1.5xe^{-0.5x^2}$ on $[0, 3]$

46. $f(x) = \frac{3x^2}{2x^3 + 5}$ on $[1, 3]$

47. Hotel industry profit has grown at a rate of $P'(x) = 2.2x - 5.97$ billion dollars per year since 1990. Find the total profits between 2000 and 2004.

48. Sales of a new plasma TV are growing at a rate of $S'(t) = 0.06t^2 - 0.8t + 3$ thousand sets per month, where t is the number of months after the TV was first offered for sale. Find the total number of plasma TVs sold between the end of the third and sixth months.

49. The number of consumer bankruptcy filings, in thousands, is changing at a rate of $N'(x) = 20.7x^2 - 380.84x + 1744$ thousand filings per year, where x is the number of years after 1995. Find the total number of bankruptcy filings between 2000 and 2005.

50. Passenger traffic at a regional airport since 1980 has changed at a rate of $1.2x^2 - 38.46x + 266.44$ thousand passengers per year, where x is the number of years after 1980. How many total passengers used the airport between 2002 and 2005?

51. Sales of a new novel are growing at a rate of $S'(t) = \frac{600}{t}$ books per week. How many copies of the novel were sold from the end of the third week to the end of the sixth week?

52. Costs of producing a machine change at a rate of $C'(x) = 8.4 + 20\sqrt{x}$ hundred dollars per week. Find the total costs during the first four weeks of production.

53. Sales of a new computer have grown at a rate of $S'(x) = 50e^{0.2x} + 70x - 40$ computers per month since the introduction of the computer. Find the total sales between the end of the fifth and eighth months.

54. Sales of a new sailboat have grown at a rate of $S'(x) = 900x + \frac{250}{x}$ boats per month. How many boats were sold from the end of the second month to the end of the sixth month?

55. Weekly box office revenues for *Star Wars Episode III: Revenge of the Sith* grew at a rate of $R'(x) = 75.445xe^{-0.125x^2}$ million dollars per week after the movie was released. What were the total box office revenues between the end of the second and the end of the sixth weeks?

56. Since the release of a new DVD, its sales have changed at a rate of $S'(t) = 20\sqrt{4t + 1}$ thousand DVDs per month. How many DVDs were sold during the first six months?

57. Newsom Car Sales holds a week-long Labor Day Tent Sale every year. Records show that cars sell at a rate of $S'(x) = \frac{150}{x + 3}$ cars per day, where x

is the number of days since the sale started. Find the total number of cars sold during the first three days of the sale.

58. A factory built along a river dumps pollution into the river at a rate of $P'(t) = \dfrac{4t}{t^2 + 2}$ thousand tons per year, where t is the number of years since the factory was built. How much pollution did the factory dump into the river during the first five years?

59. The cost, in thousand dollars, for producing x units of a product is given by $C(x) = 0.6\sqrt{x} + 3$. Find the average cost for the first 100 units.

60. The cost, in hundred dollars, for producing x units of a product is given by $C(x) = 4\sqrt{x} + 5$. Find the average cost for the first 250 units.

61. The revenue, in thousand dollars, generated from producing x hundred units of a new product is given by $R(x) = -0.1x^2 + 7x$. Find the total revenue for the first 1000 units sold.

62. The profit, in thousand dollars, generated from producing x hundred units of a new product is given by $P(x) = 0.2x + 0.25x^2$. Find the total profit for the first 500 units sold.

63. Melissa invests \$2500 into an account paying 5% compounded quarterly. Find the average value of the account over the first two years.

64. Horacio invests \$4000 into an account paying 4.2% compounded monthly. Find the average value of the account over the first three years.

65. Repeat Exercise 61 if the compounding is continuous.

66. Repeat Exercise 62 if the compounding is continuous.

67. The amount that workers contribute monthly for health insurance premiums is given by $A(t) = 0.07t^3 - 3.4t^2 + 53.7t - 240$, where t is the number of years after 1980. Find the average monthly premium between 1995 and 2000.

68. The height of a rocket, in feet, t seconds after launch is given by $h(t) = -3.4t^2 + 150t + 50$. Find the average height of the rocket between 20 seconds and 40 seconds.

69. The concentration of a drug (in mg/cm^3) t hours after its injection is given by $C(t) = 8te^{-0.15t^2}$. What is the average concentration during the first two hours?

70. The concentration of a drug (in mg/cm^3) t hours after its injection is given by $C(t) = \dfrac{0.8t}{t^2 + 1}$. What is the average concentration during the first three hours?

COMMUNICATE

71. Compare and contrast definite and indefinite integrals. How are they similar? How are they different? Explain how to determine which integral to use in a given application problem.

72. Explain why $\displaystyle\int_{-2}^{2} \frac{7}{(3x - 2)^2}\,dx$ cannot be integrated using the Fundamental Theorem of Calculus.

Find the mistake in each of the following integration problems. Correct the mistake and give the correct answer.

73.
$$\int_1^3 (3x^2 - 2x)dx = (x^3 - x^2)\Big|_1^3 = 3^3 - 1^2 = 26$$

74.
$$\int_1^2 \frac{4}{(3x - 2)^3}dx = \int_1^2 \frac{4}{u^3}\cdot\frac{1}{3}du = \frac{4}{3}\int_1^2 u^{-3}du = \left(\frac{4}{3}\cdot\frac{1}{-2}u^{-2}\right)\Big|_1^2 = -\frac{2}{3}(2^{-2} - 1^{-2}) = \frac{1}{2}$$

SHARPEN THE TOOLS

Evaluate each definite integral using substitution.

75. $\displaystyle\int_1^2 x^3(x^2 - 1)^3dx$

76. $\displaystyle\int_4^7 \frac{x^2}{(x - 3)^5}dx$

Use the Fundamental Theorem of Calculus to prove the following Properties of Definite Integrals. Demonstrate each property using a specific function and interval of integration.

77. $\int_a^a f(x)dx = 0$

78. $\int_a^b f(x)dx = -\int_b^a f(x)dx$

79. $\int_a^b f(x)dx = \int_a^c f(x)dx + \int_c^b f(x)dx$

for $a < c < b$

80. $\int_a^b [f(x) \pm g(x)]dx = \int_a^b f(x)dx \pm \int_a^b g(x)dx$

CALCULATOR CONNECTION

81. The table gives the change in total box office revenues, in million dollars, for a new movie during the first 12 weeks after its release.

Week	Change in Total Revenue, in Million Dollars
1	1.3
2	1.8
3	2.6
4	2.0
5	1.2
6	1.8
7	1.0
8	0.6
9	0.8
10	0.4
11	0.2
12	0.1

a. Find a cubic regression model, $R'(t)$, for the change in total revenues in any given week. Round coefficients to four decimal places.

b. Use your model to estimate the total revenues for the first four weeks.

82. Sales of a new DVD, in thousands, change each week as shown in the table.

Week	Change in Sales, in Thousands
1	1.58
2	0.66
3	0.50
4	0.42
5	0.38
6	0.33
7	0.31
8	0.29
9	0.27
10	0.26

a. Find a power regression model for the change in total sales in any given week, $S'(t)$. Round coefficients to four decimal places.

b. Use your model to estimate the total sales from the end of the first week to the end of the fourth week.

83. Temperatures on a day in May in St. Louis are given in the table.

Time of Day	Temperature (°F)
7 a.m.	49
8 a.m.	58
9 a.m.	66
10 a.m.	72
11 a.m.	76
12 noon	79
1 p.m.	80
2 p.m.	78
3 p.m.	78
4 p.m.	74
5 p.m.	69
6 p.m.	63

a. Use the regression feature of your calculator to determine a quadratic function that models the temperature at any time. Let t represent the number of hours after midnight.

b. Use your model to estimate the average daytime temperature between 8 a.m. and noon.

c. Use your model to estimate the average daytime temperature between 2 p.m. and 5 p.m.

84. The total number of Starbucks locations has grown rapidly since its beginnings in 1971. The table below gives the annual increase in the number of new Starbucks locations.

Year	Number of New Starbucks Locations
1971	1
1987	16
1988	16
1989	22
1990	29
1991	32
1992	49
1993	107
1994	153
1995	251
1996	339
1997	397
1998	474
1999	249
2000	1366
2001	1208
2002	979
2003	1537
2004	344
2005	1912

Source: www.starbucks.com

a. Use the regression feature of your calculator to determine an exponential function that models the growth in the number of new Starbucks locations. Let x represent the number of years after 1970.

b. Use your model to estimate the average number of new Starbucks locations between 1990 and 1995.

c. Use your model to estimate the average number of new Starbucks locations between 2001 and 2005.

Use your calculator's function integrator to evaluate each of the following integrals. Explain why these integrals cannot be evaluated using the methods in your Tool Kit.

85. $\int_0^1 e^{-x^2/2}dx$

86. $\int_2^5 \frac{3}{x^2+1}dx$

87. $\int_2^6 \ln x\, dx$

88. $\int_3^5 x^2 \ln x\, dx$

89. $\int_{-3}^2 \sqrt{x^2+1}\, dx$

90. $\int_1^4 \frac{\sqrt{x^3+2}}{x}dx$

TOPIC 24

Areas and Definite Integrals

Given a function $f(x)$, you learned in previous units that the derivative, $f'(x)$, has two major applications. First, the derivative gives the rate at which the function is changing at any point in its domain. The derivative also gives the slope of the graph of the function at any point. You also know that, given the rate at which a function changes, the integral returns the function itself. A definite integral gives the net change in the value of an antiderivative over some specific interval. You will learn in this topic that the definite integral also has a geometric interpretation. Before starting this topic, be sure to complete the Warm-up Exercises to sharpen the algebra and calculus skills you will need.

IMPORTANT TOOLS IN TOPIC 24

- *Using definite integrals to find the area between a curve and the x-axis*
- *Using definite integrals to find the area between two curves*
- *Finding areas under curves using Riemann Sums*

TOPIC 24 WARM-UP EXERCISES

Be sure you can successfully complete the following exercises before starting Topic 24.

Algebra Warm-up Exercises

Find the area of the indicated geometric figures using an appropriate formula.

1.

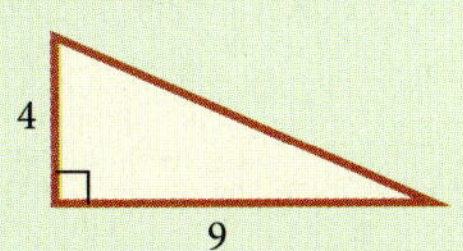

2.

3.

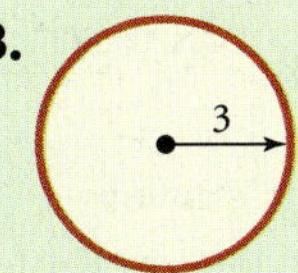

4.

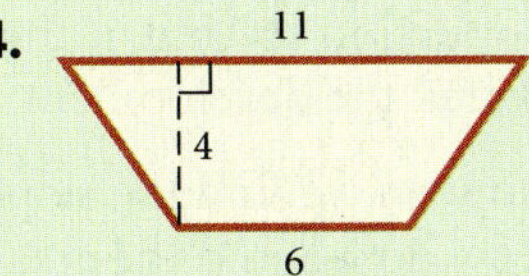

5.

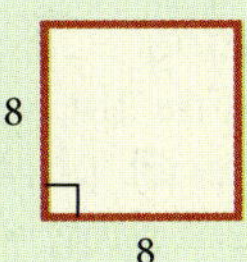

6. Approximate the area of the region below if each grid is one square unit.

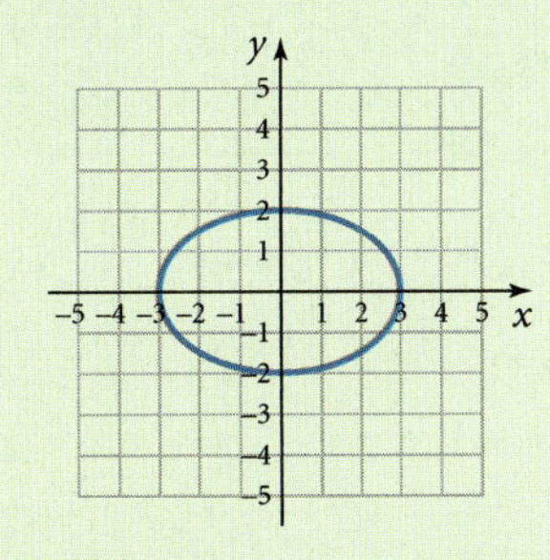

7. How are the graphs of $y = x^2$ and $y = -x^2$ related to each other?

8. Solve each equation for x.

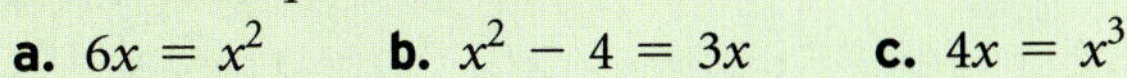

a. $6x = x^2$ **b.** $x^2 - 4 = 3x$ **c.** $4x = x^3$

Calculus Warm-up Exercises

Evaluate the following definite integrals.

1. $\int_{-1}^{2} (x^2 - 3x + 4)dx$ **2.** $\int_{-3}^{-1} \frac{4}{x^2}dx$

3. $\int_0^3 e^{-2x}dx$

4. $\int_1^4 (\sqrt{x} + 3)dx$

Evaluate each integral using substitution.

5. $\int_2^5 \frac{4}{3x - 1}dx$

6. $\int_0^1 x(x^2 - 5)^3 dx$

7. $\int_{-1}^0 xe^{-x^2}dx$

Evaluate using your graphing calculator's function integrator tool. Explain why the integral cannot be integrated with any other method currently in your Tool Kit.

8. $\int_0^2 \frac{1}{(x^3 + 1)^2}dx$

9. $\int_0^2 \sqrt{4 - x^2}dx$

Answers to Warm-up Exercise

Algebra Warm-up Exercises

1. 18 square units
2. 48 square units
3. 9π square units
4. 34 square units
5. 64 square units
6. about 18 square units (answers may vary)
7. $y = -x^2$ is the graph of $y = x^2$ reflected, or inverted, about the x-axis.
8. a. $x = 0, 6$
b. $x = 4, -1$
c. $x = 0, 2, -2$

Calculus Warm-up Exercises

1. $10\frac{1}{2}$ or 10.5
2. $2\frac{2}{3}$ or 2.67
3. $-\frac{1}{2}(e^{-6} - 1) \approx 0.499$
4. $13\frac{2}{3}$ or 13.67
5. $\frac{4}{3}(\ln 14 - \ln 5) \approx 1.373$
6. -46.125
7. $-\frac{1}{2}(1 - e^{-1}) \approx -0.316$
8. 0.8007 If $u = x^3 + 1$ and $du = 3x^2dx$, then the substitution for dx is still a function of x, so the integration cannot be completed.
9. 3.1416 If $u = 4 - x^2$ and $du = 2xdx$, then the substitution for dx is still a function of x, so the integration cannot be completed.

Figure 24.1

In the Introduction to Unit 1, there were two major questions that motivated the development of the calculus:

- How do we find the slope of a nonlinear curve?
- How do we find the area of a nongeometric region?

You have now learned that the slope question is answered by the derivative. Now we focus on the second question.

Suppose that a developer needs to determine the area of a lake on a piece of property. Using an aerial photograph overlaid on a grid, he obtains the picture shown in Figure 24.1.

Figure 24.2

If each grid represents 1000 square feet, how might the developer estimate the area? First, count all the grids completely enclosed within the lake's boundaries—no grid may extend outside the lake's perimeter. These grids have been shaded in the diagram. There are 22 such grids, each representing 1000 square feet, as shown in Figure 24.2.

Next, look at the remaining unshaded partial grids and estimate about how many more complete grids are enclosed. You might estimate about eight more total grids. Thus, there are about 22 + 8, or 30, grids within the lake's boundaries, so the developer could estimate the lake's area to be about $30 \cdot 1000$ or 30,000 square feet.

The method of grid counting was used here because the area enclosed was not a geometric region. A method similar to this was used by Archimedes and will be discussed later in this topic. But first let's see how integrals play a role in determining an area.

Areas Above the *x*-Axis

You already know that the definite integral $\int_a^b f(x)dx$ provides two important pieces of information.

- The expression $\int_a^b f(x)dx$ gives the net change in the value of the antiderivative $F(x)$ on the interval [a, b] in its domain.
- The expression $\frac{1}{b-a}\int_a^b f(x)dx$ gives the average value of the function $f(x)$ on the interval [a, b].

Now let's explore the relationship between definite integrals and area.

Example 1: You know that $\int_0^2 (4 - 2x)dx = (4x - x^2)\big|_0^2 = 4$.

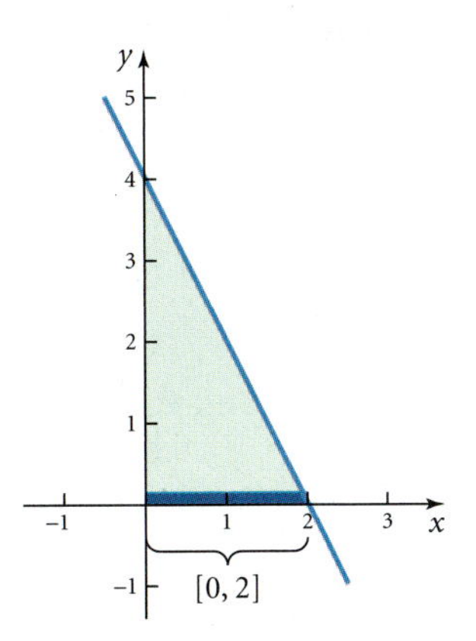

Figure 24.3

Look at the graph of $y = 4 - 2x$ on the interval [0, 2], shown in Figure 24.3. Shade the interval [0, 2] on the x-axis. The interval end point $x = 2$ is also a point on the graph. The interval end point $x = 0$ is not a point on the graph, so draw a vertical line from (0, 0) to the point $(0, f(0))$.

You have now enclosed a triangular geometric region: one boundary is the interval [0, 2] on the x-axis; a second boundary is the line from (0, 0) to $(0, f(0))$; the third boundary is the piece of the graph between $f(0)$ and $f(2)$. The region is triangular, so we use the formula $A = \frac{1}{2}bh$ to find its area.

The base of this triangle is the interval along the x-axis, and its length is $2 - 0 = 2$ units. The height of this triangle is the line from (0, 0) to $(0, f(0))$; its length is $f(0) = 4$ units. The area is $A = \frac{1}{2}(2)(4) = 4$ square units.

The area enclosed by the region is the same as the value obtained from the definite integral. Is this always the case? To find out, let's consider a second example. ■

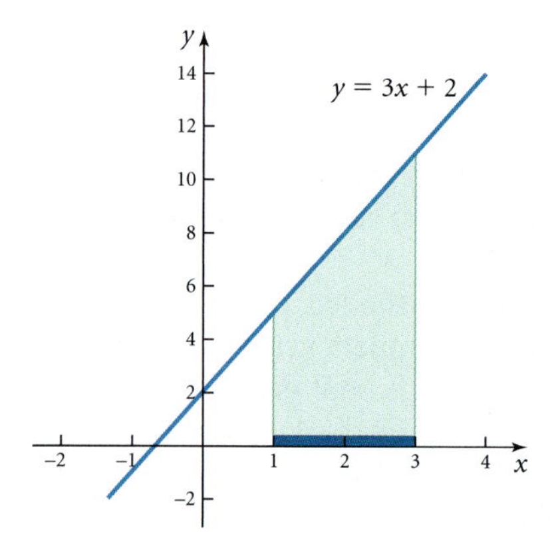

Figure 24.4

Example 2: You know that $\int_1^3 (3x + 2)dx = \left(\frac{3}{2}x^2 + 2x\right)\Big|_1^3 = \left(\frac{27}{2} + 6\right) - \left(\frac{3}{2} + 2\right) = 16$. Look at the graph of $y = 3x + 2$ on the interval $[1, 3]$, shown in Figure 24.4. Shade the interval $[1, 3]$ on the x-axis. Neither interval end point is a point on the graph, so extend vertical lines from $(1, 0)$ to $(1, f(1))$ and $(3, 0)$ to $(3, f(3))$.

As in Example 1, you have enclosed a geometric region. The boundaries are the interval $[1, 3]$ on the x-axis, the line from $(1, 0)$ to $(1, f(1))$, the line from $(3, 0)$ to $(3, f(3))$, and the piece of the graph between $f(1)$ and $f(3)$.

The region forms a trapezoid whose area is given by $A = \frac{1}{2}h(b_1 + b_2)$, where b_1 and b_2 are the bases (the two parallel sides) and h is the altitude (the perpendicular distance between the two parallel sides).

In this example we have

$$b_1 = f(1) = 5$$
$$b_2 = f(3) = 11$$
$$h = 3 - 1 = 2$$

The area is $A = \frac{1}{2}(2)(5 + 11) = 16$ square units. As in Example 1, the value of the definite integral is the same as the area of the region enclosed by the graph of the function and the x-axis. ■

Even though these two examples are by no means a proof, they intuitively suggest the following generalization.

Area Rule 1

Given $f(x) \geq 0$ on [a, b], the area of the region enclosed by $f(x)$ and the x-axis on [a, b] is given by

$$A = \int_a^b f(x)dx$$

It is assumed that $f(x)$ is continuous over the interval [a, b].

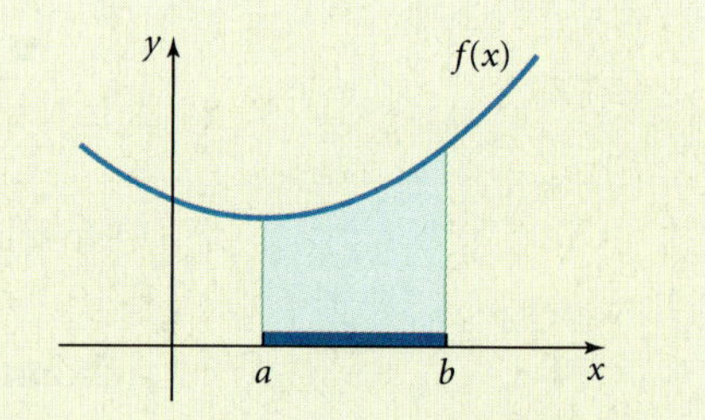

Example 3: Find the area of the region between $f(x) = x^3 + 1$ and the x-axis on $[0, 2]$.

Solution: Draw the graph and determine the region. (See Figure 24.5.) This region is not "geometric" because one of the boundaries is a curve and no geometric shape is formed for which an area formula exists. We will use Area Rule 1.

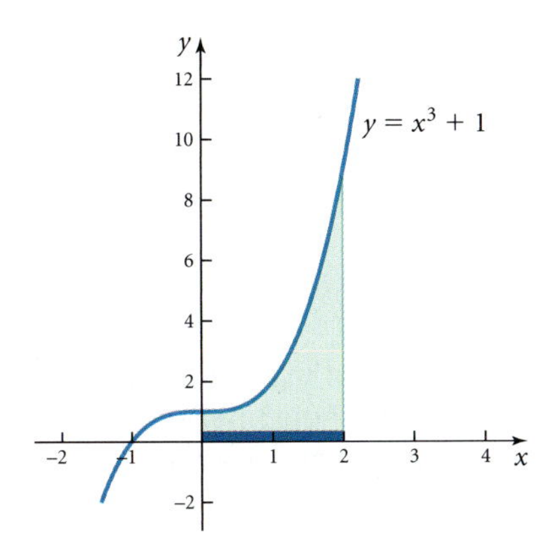

Figure 24.5

$$A = \int_0^2 (x^3 + 1)dx = \left(\frac{1}{4}x^4 + x\right)\Bigg|_0^2$$

$$= \left(\frac{1}{4}(16) + 2\right) - 0 = 6 \text{ square units}$$

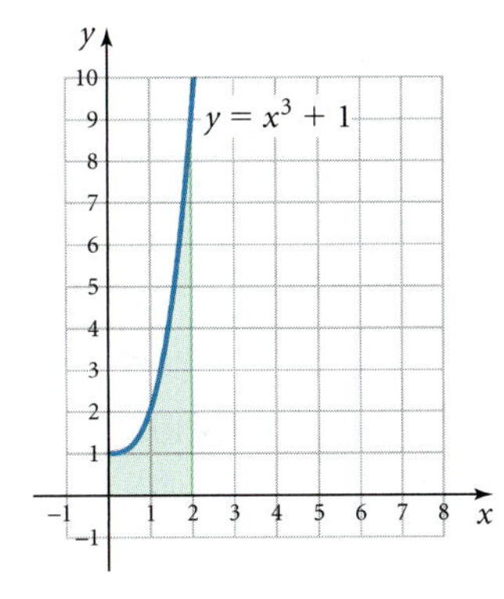

Figure 24.6

By counting unit squares and estimating, six square units seems a reasonable area for this region. (See Figure 24.6.) ■

Example 4: Find the area of the region between $f(x) = \sqrt{x} + 1$ and the x-axis on $[1, 4]$.

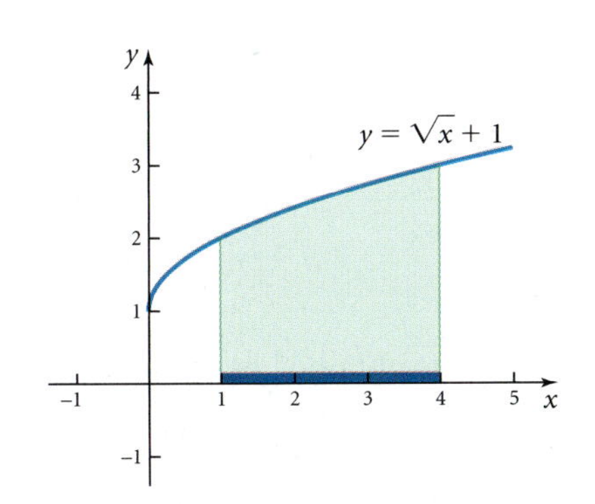

Figure 24.7

Solution: Graph the function and determine the region. (See Figure 24.7.) Using Area Rule 1, the area of the region is

$$A = \int_1^4 (\sqrt{x} + 1)dx = \left(\frac{2}{3}x^{3/2} + x\right)\Bigg|_1^4$$

$$= \left(\frac{2}{3}(4)^{3/2} + 4\right) - \left(\frac{2}{3}(1)^{3/2} + 1\right)$$

$$= \frac{23}{3} \text{ or } 7\frac{2}{3} \text{ square units}$$ ■

Calculator Corner 24.1

Graph $y_1 = \sqrt{x} + 1$. The area integral $\int_1^4 (\sqrt{x} + 1)dx$ can be evaluated on your calculator as follows.

Press **2nd TRACE (CALC)** and select option **7:** $\int$ ***f*****(*****x*****)*****dx***.

Type 1 as the lower limit and press **ENTER**.

Type 4 as the upper limit and press **ENTER**.

The calculator will then shade the region and give the value of the integral.

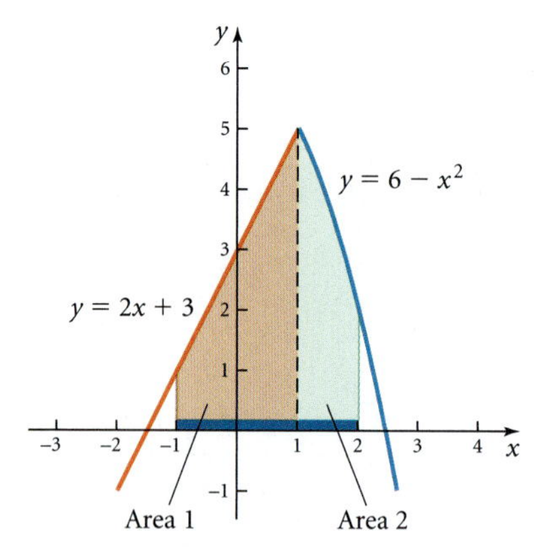

Figure 24.8

Example 5: Find the area of the region between $f(x)$ and the x-axis on $[-1, 2]$ if $f(x)$ is the piecewise defined function

$$f(x) = \begin{cases} 2x + 3 & x < 1 \\ 6 - x^2 & x \geq 1 \end{cases}$$

Solution: The graph of the piecewise defined function in Figure 24.8 shows the desired area. The function is continuous over the interval $[-1, 2]$. However, the upper boundary of the region for the interval $[-1, 1]$ is different from the upper boundary of the region for the interval $[1, 2]$. We will consider two separate areas:

The area below $y = 2x + 3$ on the interval $[-1, 1]$

The area below $y = 6 - x^2$ on the interval $[1, 2]$

Area 1: $$A_1 = \int_{-1}^{1} (2x + 3)dx = (x^2 + 3x)\Big|_{-1}^{1} = 4 - (-2) = 6$$

Area 2: $$A_2 = \int_{1}^{2} (6 - x^2)dx = \left(6x - \tfrac{1}{3}x^3\right)\Big|_{1}^{2}$$
$$= \left(12 - \tfrac{8}{3}\right) - \left(6 - \tfrac{1}{3}\right) = 3\tfrac{2}{3}$$

The total area of the region is $A_1 + A_2$ or $6 + 3\frac{2}{3} = 9\frac{2}{3}$ square units. ■

Areas Below the x-Axis

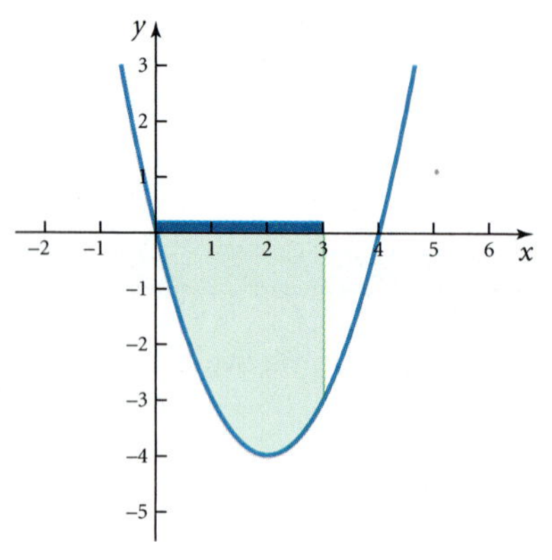

Figure 24.9

Beware of the temptation to conclude that the value obtained from any definite integral is always an area. Here's a simple example why it is not.

$$\int_0^3 (x^2 - 4x)dx = \left(\frac{1}{3}x^3 - 2x^2\right)\Big|_0^3 = (9 - 18) - 0 = -9$$

Because area can never be negative, it should be obvious that -9 does not represent the area of a region.

To see what happened here, look at the graph of $y = x^2 - 4x$ on $[0, 3]$. (See Figure 24.9.) From the graph you can see that the area enclosed by $y = x^2 - 4x$ and the x-axis on $[0, 3]$ is *below* the x-axis. In order to apply Area Rule 1, the area must be *above* the x-axis. We know from algebra that $y = -f(x)$ inverts the graph of $y = f(x)$ about the x-axis. Thus, $y = -f(x) = -(x^2 - 4x)$ will invert the graph of $y = x^2 - 4x$ about the x-axis, allowing the area to be above the x-axis.

So, the area enclosed by $y = x^2 - 4x$ and the x-axis on $[0, 3]$ is given by

$$A = -\int_0^3 (x^2 - 4x)dx = -\left(\frac{1}{3}x^3 - 2x^2\right)\Big|_0^3 = -(-9) = 9$$

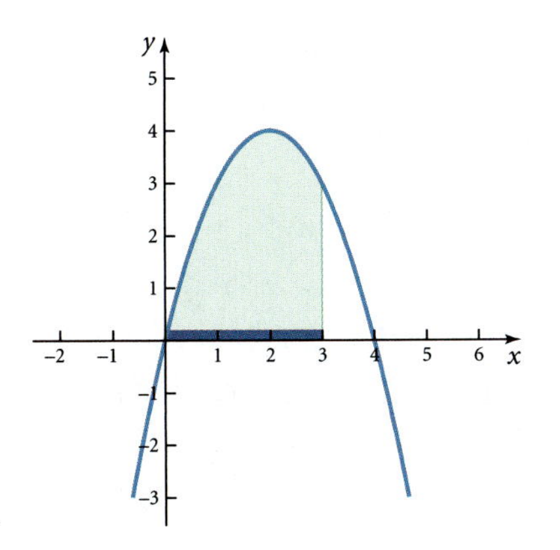

Figure 24.10

Remember that the negative placed in front of the integral inverts the graph so that the area will be positive, as shown in Figure 24.10.

We now state this result formally.

Area Rule 2

Given $f(x) \leq 0$ on [a, b], the area of the region enclosed by $f(x)$ and the x-axis on [a, b] is given by

$$A = -\int_a^b f(x)dx$$

It is assumed that $f(x)$ is continuous over the interval [a, b].

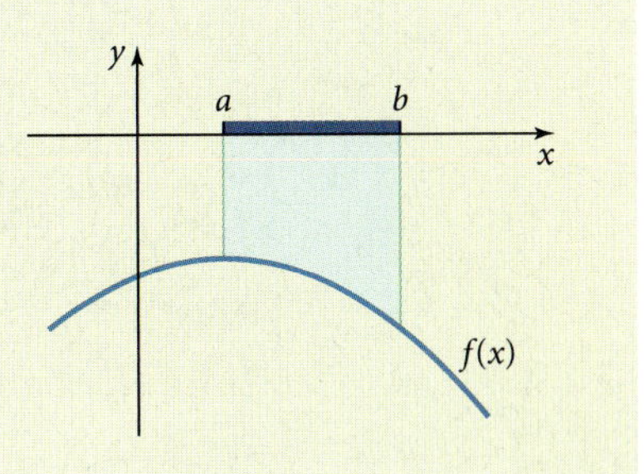

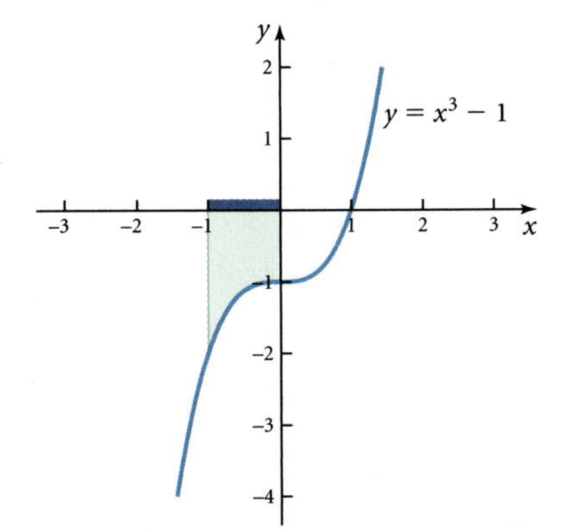

Figure 24.11

Example 6: Find the area of the region between $f(x) = x^3 - 1$ and the x-axis on $[-1, 0]$.

Solution: Graph the function and indicate the region. (See Figure 24.11.) The area to be evaluated is below the x-axis, so Area Rule 2 will be used.

$$A = -\int_{-1}^{0}(x^3 - 1)dx = -\left(\frac{1}{4}x^4 - x\right)\Bigg|_{-1}^{0}$$

$$= -\left[0 - \left(\frac{1}{4} - (-1)\right)\right] = -\left(-\frac{5}{4}\right) = \frac{5}{4} \text{ square units}$$

By now you may recognize the importance of graphing the function before proceeding with an Area Rule. The need to see the graph of the function is especially apparent in the next example.

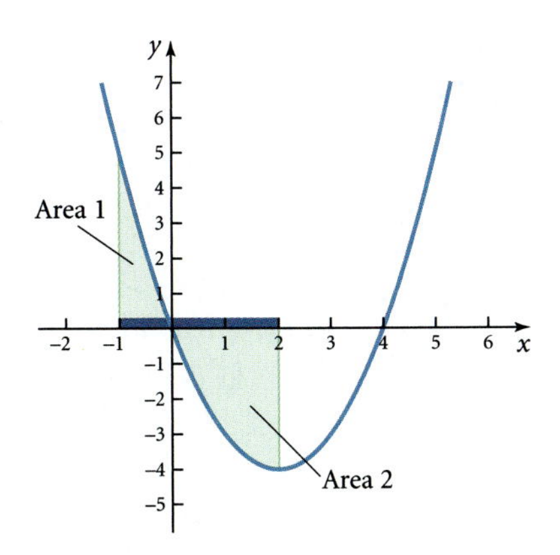

Figure 24.12

Example 7: Find the area of the region between $f(x) = x^2 - 4x$ and the x-axis on $[-1, 2]$.

Solution: Graph the function and determine the region. (See Figure 24.12.) From the graph you can see that the area enclosed here is really two areas:

One above the x-axis on $[-1, 0]$

The other below the x-axis on $[0, 2]$

We will need to find each area separately and then add them to get the total area.

Area I is above the x-axis on $[-1, 0]$, so by Area Rule 1 we have

$$A_1 = \int_{-1}^{0}\left(x^2 - 4x\right)dx = 2\frac{1}{3}$$

Area 2 is below the x-axis on $[0, 2]$, so by Area Rule 2 we have

$$A_2 = -\int_0^2 (x^2 - 4x)dx = -\left(-5\frac{1}{3}\right) = 5\frac{1}{3}$$

The total area is $A = A_1 + A_2 = 2\frac{1}{3} + 5\frac{1}{3} = 7\frac{2}{3}$ square units. ■

In Example 7, had you simply evaluated $\int_{-1}^{2} (x^2 - 4x)dx$ without looking at the graph, you would have obtained a value of -3. Area cannot be negative, but in this case the total area was not just $-(-3) = 3$.

MATHEMATICS CORNER 24.1

Area Versus Net Change

Evaluating $\int_{-1}^{2}(x^2 - 4x)dx$ gives a value of -3, which cannot be an area because it is negative. A common mistake among calculus students is to assume that the correct area is 3, but you know from Example 7 that the true area is $7\frac{2}{3}$ square units.

Recall that the definite integral gives the net change in the value of an antiderivative on some interval. The -3 obtained by evaluating $\int_{-1}^{2}(x^2 - 4x)dx$ is the net difference in the two areas you saw graphically.

$A_1 = 2\frac{1}{3}$ and $A_2 = 5\frac{1}{3}$, so $A_1 - A_2 = 2\frac{1}{3} - 5\frac{1}{3} = -3$, which is what the definite integral is actually measuring.

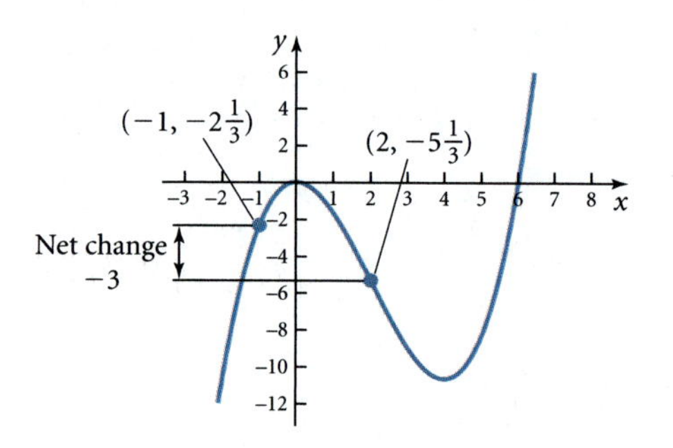

Graph the antiderivative $y = \frac{1}{3}x^3 - 2x^2$. The net change on $[-1, 2]$ is the change from $f(-1)$ to $f(2)$ which is -3 units.

Check Your Understanding 24.1

Find the area of each indicated region. Sketch a graph of the function and shade the region.

1. Between $f(x) = \dfrac{4}{x}$ and the x-axis on $[1, 2]$
2. Between $f(x) = x^3$ and the x-axis on $[0, 3]$
3. Between $f(x) = 1 - x^2$ and the x-axis on $[1, 3]$
4. Between $f(x) = e^{2x} - 4$ and the x-axis on $[-2, 0]$
5. Between $f(x) = 1 - x^2$ and the x-axis on $[-3, 0]$

Areas Between Two Curves

You have now learned two rules for evaluating the area of a region that has one boundary along the x-axis. Suppose that the area to be found is enclosed between two curves, as shown in Figure 24.13. How would you find the area?

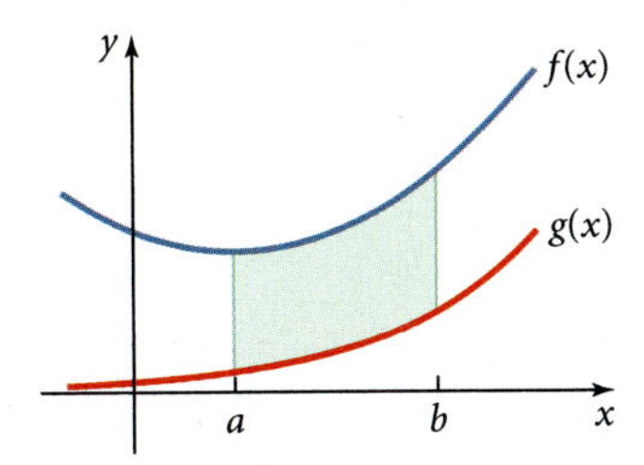

Figure 24.13

Using the rules you already have, you know that

$\int_a^b f(x)dx$ gives the area between $f(x)$ and the x-axis on [a, b].

$\int_a^b g(x)dx$ gives the area between $g(x)$ and the x-axis on [a, b].

Subtracting the area under $g(x)$ from the area under $f(x)$ and using the properties of integrals gives the area between the two curves.

$$A = \int_a^b f(x)dx - \int_a^b g(x)dx = \int_a^b [f(x) - g(x)]dx$$

Area Rule 3
If $f(x) \geq g(x)$ on [a, b], then the area of the region enclosed between the two curves is given by

$$A = \int_a^b [f(x) - g(x)]dx$$

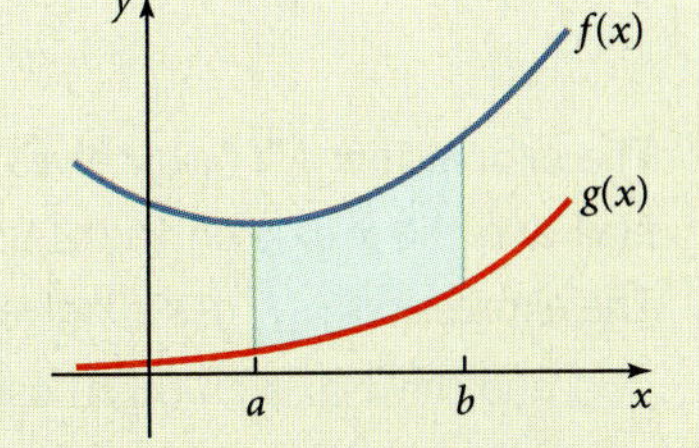

It is assumed that $f(x)$ and $g(x)$ are continuous on [a, b]. The location of the two curves with respect to the x-axis does not matter when using Area Rule 3. See Mathematics Corner 24.2 for an explanation.

MATHEMATICS CORNER 24.2

Area Between Two Curves

To show that the location of the x-axis is unimportant in Area Rule 3, consider the following three cases.

Case 1: $f(x) \geq g(x)$ with $f(x) \geq 0$ and $g(x) \geq 0$.

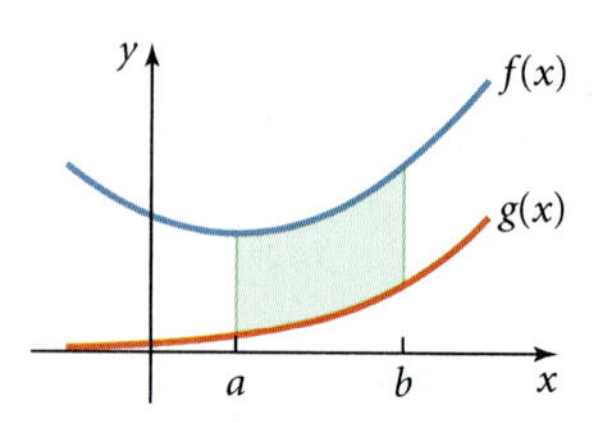

The expression $\int_a^b f(x)dx$ gives the area between $f(x)$ and the x-axis on [a, b], and $\int_a^b g(x)dx$ gives the area between $g(x)$ and the x-axis on [a, b]. By subtracting the area under $g(x)$ from the area under $f(x)$ and using the properties of integrals, the area between the curves is given by

$$A = \int_a^b f(x)dx - \int_a^b g(x)dx$$

$$= \int_a^b [f(x) - g(x)]dx$$

Case 2: $f(x) \geq g(x)$ with $f(x) \geq 0$ and $g(x) \leq 0$.

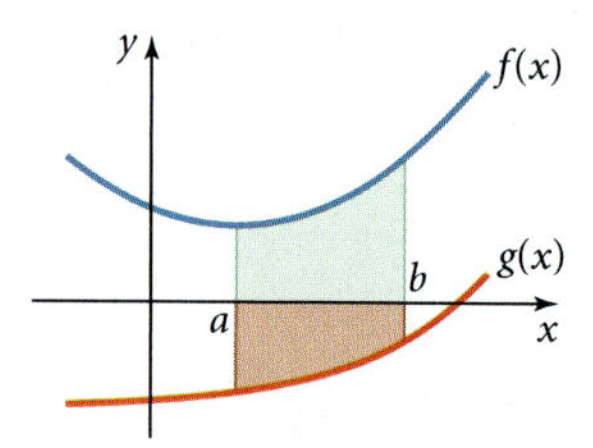

The expression $\int_a^b f(x)dx$ gives the area between $f(x)$ and the x-axis on [a, b] using Area Rule 1. The expression $-\int_a^b g(x)dx$ gives the area between $g(x)$ and the x-axis on [a, b] using Area Rule 2.

The total area is the sum of these two areas.

$$A = \int_a^b f(x)dx + \left[-\int_a^b g(x)dx\right]$$

$$= \int_a^b [f(x) - g(x)]dx$$

Case 3: $f(x) \geq g(x)$ with $f(x) \leq 0$ and $g(x) \leq 0$.

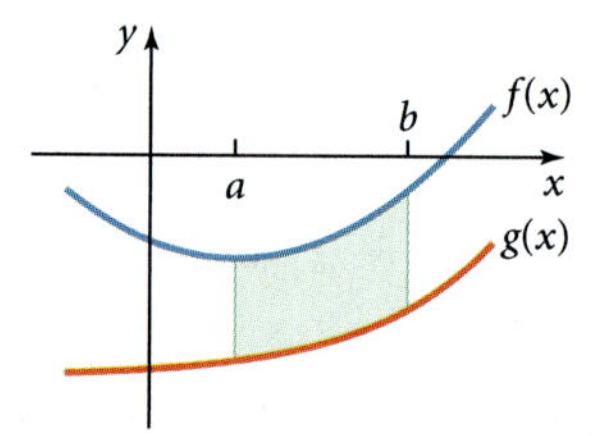

The expression $-\int_a^b f(x)dx$ gives the area between $f(x)$ and the x-axis on [a, b] using Area Rule 2.

The expression $-\int_a^b g(x)dx$ gives the area between $g(x)$ and the x-axis on [a, b] using Area Rule 2.

Subtracting the smaller area from the larger area gives

$$A = -\int_a^b g(x)dx - \left[-\int_a^b f(x)dx\right]$$

$$= -\int_a^b g(x)dx + \int_a^b f(x)dx$$

$$= \int_a^b f(x)dx - \int_a^b g(x)dx = \int_a^b [f(x) - g(x)]dx$$

All three cases yield the same result, so the x-axis plays no role in determining area between two curves.

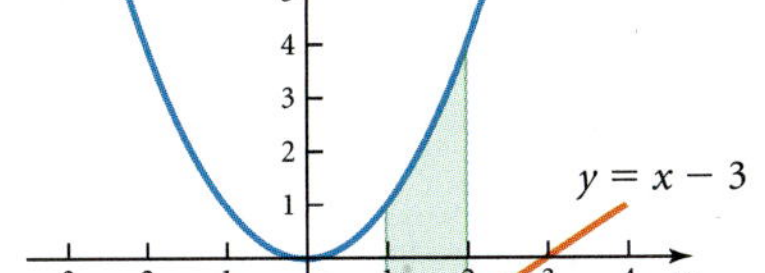
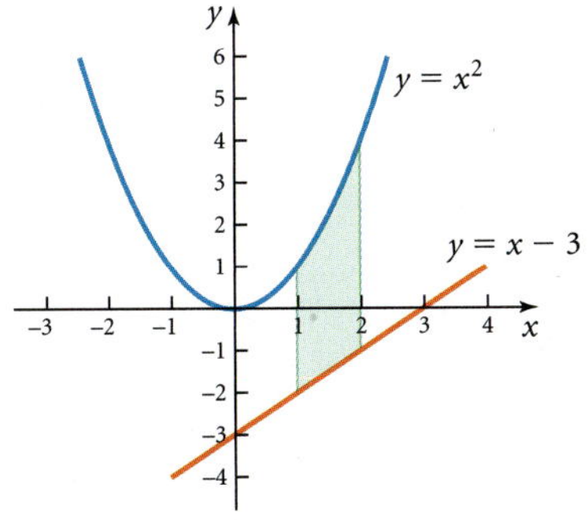

Figure 24.14

Example 8: Find the area of the region between $y = x^2$ and $y = x - 3$ on $[1, 2]$.

Solution: First, draw the graph of the two functions and indicate the region. (See Figure 24.14.) Which function is $f(x)$ and which function is $g(x)$? Area Rule 3 says that $f(x) \geq g(x)$, which means that $f(x)$ is the top function and $g(x)$ is the bottom function. For this example, $f(x) = x^2$ and $g(x) = x - 3$. The area is then found by

$$A = \int_1^2 [x^2 - (x - 3)]dx \quad \text{Apply Area Rule 3}$$

$$= \int_1^2 (x^2 - x + 3)dx \quad \text{Simplify the integrand}$$

$$= \left(\frac{1}{3}x^3 - \frac{1}{2}x^2 + 3x\right)\Bigg|_1^2 \quad \text{Integrate}$$

$$= \left(\frac{8}{3} - 2 + 6\right) - \left(\frac{1}{3} - \frac{1}{2} + 3\right) \quad \text{Apply the Fundamental Theorem of Calculus}$$

$$= 3\frac{5}{6} \text{ or about 3.83 square units} \quad \text{Simplify}$$

■

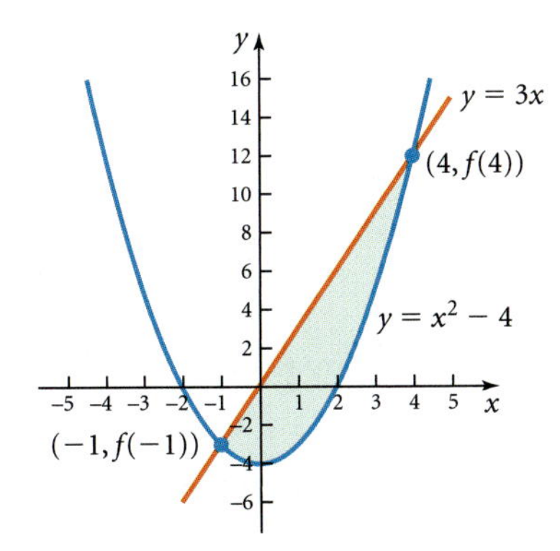

Figure 24.15

Example 9: Find the area of the region enclosed by $y = x^2 - 4$ and $y = 3x$.

Solution: In this example, no interval was given, so we must graph the functions to find the region that is enclosed by these two curves. (See Figure 24.15.) In the graph, the two functions intersect each other twice and enclose a region. To calculate the area of this shaded region using Area Rule 3, we have $f(x) = 3x$ as the top function and $g(x) = x^2 - 4$ as the bottom function. The interval [a, b] is determined by the two points of intersection of the two functions. Those values can be found by setting $f(x) = g(x)$.

$$x^2 - 4 = 3x \rightarrow x^2 - 3x - 4 = 0$$
$$(x - 4)(x + 1) = 0$$
$$x = -1, 4$$

The intersection points are the boundaries of the interval for which the area is to be found, so the interval of integration is $[-1, 4]$.

The area is found by Area Rule 3.

$$A = \int_{-1}^4 [3x - (x^2 - 4)]dx \quad \text{Apply Area Rule 3}$$

$$= \int_{-1}^4 (3x - x^2 + 4)dx \quad \text{Simplify the integrand}$$

$$= \left(\frac{3}{2}x^2 - \frac{1}{3}x^3 + 4x\right)\Bigg|_{-1}^4 \quad \text{Integrate}$$

$$= \left(24 - \frac{64}{3} + 16\right) - \left(\frac{3}{2} + \frac{1}{3} - 4\right)$$ Apply the Fundamental Theorem of Calculus

$$= 20\frac{5}{6} \text{ or } 20.833 \text{ square units}$$ Simplify

Check Your Understanding 24.2

1. Find the area of the region between $y = x^2 - 1$ and $y = x + 5$ on $[-1, 2]$. Draw the graphs and indicate the region.
2. Find the area of the region enclosed by $y = x^2 - 1$ and $y = x + 5$. Draw the graphs and indicate the region.

Areas Using Riemann Sums

Archimedes (287–212 BC)

Estimating areas of nongeometric regions by using geometric approximations was one of the major contributions of Greek mathematicians. Archimedes (287–212 BC) used areas of rectangles to approximate areas of nongeometric regions.

Consider the region enclosed by $y = 4 - x^2$ and the x-axis on $[0, 2]$. Divide the region into two rectangles by partitioning the interval $[0, 2]$ into two equal subintervals, creating two rectangles whose heights are determined by the value of the function at the left-hand end point of each subinterval. (See Figure 24.16.)

The width of each rectangle is $\frac{2 - 0}{2} = 1$ unit. The heights of the rectangles are $f(0) = 4$ and $f(1) = 3$.

The area of the first rectangle is $1 \times 4 = 4$ square units.

The area of the second rectangle is $1 \times 3 = 3$ square units.

The total area is the sum of the areas of the two rectangles: $4 + 3 = 7$ square units.

Figure 24.16

You can see that this area estimate is larger than the true area, because some of the area enclosed by the two rectangles is outside the true area enclosed by the curve.

Consider using four rectangles, instead of only two, and repeat the process. Divide the region into four rectangles by partitioning the interval $[0, 2]$ into four equal subintervals, creating four rectangles whose heights are determined by the value of the function at the left-hand end point of each subinterval. (See Figure 24.17.)

The width of each rectangle is $\frac{2 - 0}{4} = 0.5$ unit. The heights of the rectangles are $f(0) = 4, f(0.5) = 3.75, f(1) = 3$, and $f(1.5) = 1.75$.

The total area is the sum of the areas of the four rectangles: $0.5(4) + 0.5(3.75) + 0.5(3) + 0.5(1.75) = 6.25$ square units.

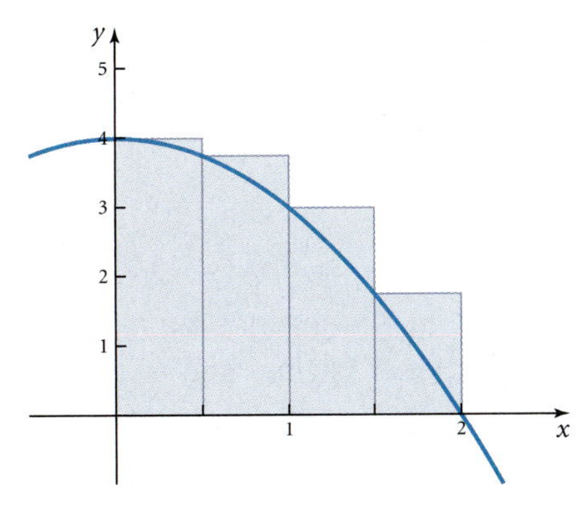

Figure 24.17

The extraneous area outside the curve is now smaller, so you know that the area estimate of 6.25 square units is getting closer to the true area.

Now consider using eight rectangles. (See Figure 24.18.)

The width of each rectangle is $\frac{2-0}{8} = 0.25$. The height of each rectangle is the value of the function at the left-hand end point of each subinterval.

The total area is given by

$$\begin{aligned} A &= 0.25(f(0) + f(.25) + f(.50) + f(.75) + f(1) \\ &\quad + f(1.25) + f(1.5) + 5(1.75)) \\ &= 0.25(4 + 3.9375 + 3.75 + 3.4375 + 3 + 2.4375 \\ &\quad + 1.75 + 0.9375) \\ &= 5.8125 \text{ square units} \end{aligned}$$

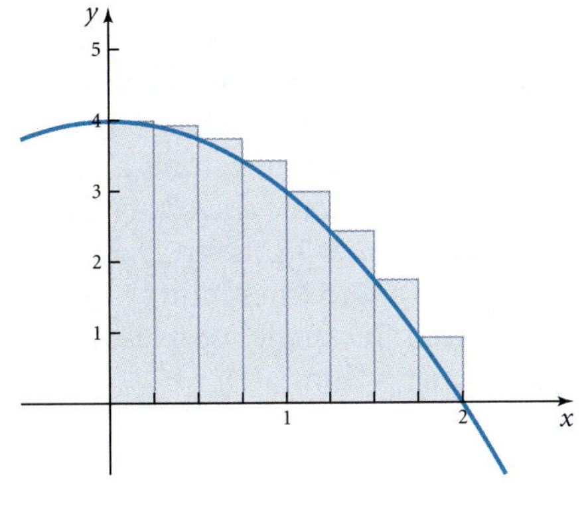

Figure 24.18

As before, the width of each rectangle gets smaller and the extraneous area outside the curve continues to shrink.

Do you see that calculus is at work in this method? As the number of rectangles increases, the width of each rectangle decreases and approaches zero. As the number of rectangles increases, the extraneous area decreases, so that the estimated area approaches the true area. Limits are at work here.

For Greek mathematicians, this was fairly sophisticated thinking. It wasn't until the development of calculus in the 1600s that mathematicians had the mathematical tools to formalize Archimedes' method.

Find the area enclosed by $y = f(x)$ and the x-axis on [a, b]. Let $a = x_0$ and $b = x_n$. We will denote the individual end points between a and b as $x_1, x_2, x_3, \ldots x_i, \ldots x_{n-1}$. The width of each rectangle is $\frac{b-a}{n} = \Delta x_i$. The height of each rectangle is the value of $f(x)$ at the left-hand end point of each subinterval. (See Figure 24.19.)

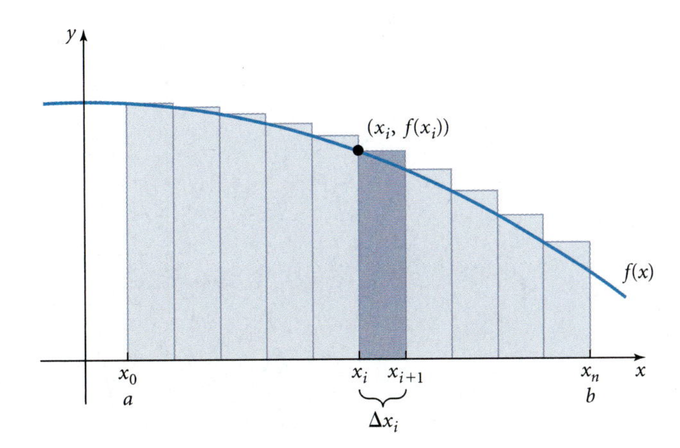

Figure 24.19

Consider one rectangle under this curve.

The width of the shaded rectangle is $\dfrac{b-a}{n} = \Delta x_i$.

The height of the shaded rectangle is $f(x_i)$.

The area of the shaded rectangle is $A_i = f(x_i)\Delta x_i$.

The total area enclosed on [a, b] is the sum of all n such rectangles, which is denoted mathematically using Σ, which is the Greek letter "sigma" and is called a **summation symbol.**

$$A = \sum_{i=0}^{n-1} f(x_i)\Delta x_i$$

Georg Friedrich Bernhard Riemann (1826–1866)

This summation symbol indicates to calculate the product $f(x_i)\Delta x_i$ for all n rectangles and then sum the products. The sum is referred to as a **Riemann Sum,** for the mathematician Georg Friedrich Bernhard Riemann (1826–1866) who formalized Archimedes' method of exhaustion by applying calculus tools.

How can we use the Riemann Sum to evaluate the area of the enclosed region? We showed previously that, in order to get closer and closer to the true area, we needed more and more rectangles, so n gets larger and approaches infinity. As the number of rectangles increases, the width of each rectangle decreases and approaches zero. That means the limit of the Riemann Sum as $n \to \infty$ (or as $\Delta x_i \to 0$) will give the true area of the region.

$$A = \lim_{\Delta x_i \to 0} \sum_{i=0}^{n} f(x_i)\Delta x_i$$

It can then be proven, using calculus that is beyond the scope of this book, that this limit is the same as the definite integral we have been using to find areas of enclosed regions.

Areas Using Riemann Sums

Given the region enclosed by $y = f(x)$ and the x-axis on [a, b], the area of the region is given by

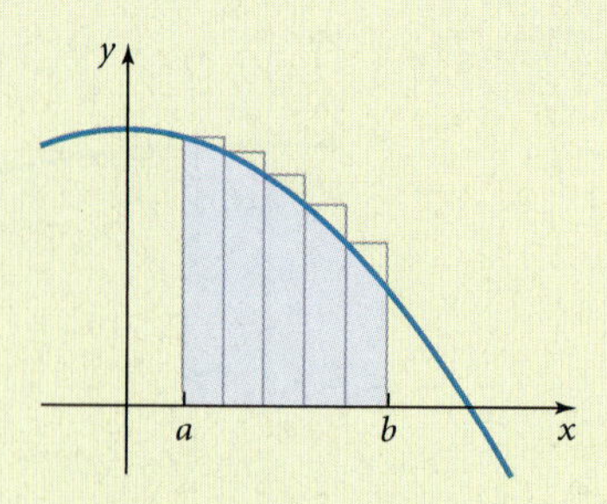

$$A = \lim_{\Delta x_i \to 0} \sum_{i=0}^{n} f(x_i)\Delta x_i = \int_a^b f(x)dx$$

The Riemann Sum definition of area also gives a geometric explanation of the role of dx in the integrand of a definite integral. The definite integral denotes area. The $f(x)$ denotes the height of each rectangle in Archimedes' method, and the dx represents Δx, which is the width of each rectangle used in Archimedes' method. Writing a definite integral as $\int_a^b f(x)$, without the dx, is asking for an area but only providing one dimension—the height of the rectangle.

Return now to the area enclosed by $y = 4 - x^2$ and the x-axis on $[0, 2]$. Our estimated areas were

A = 7 square units, using two rectangles

A = 6.25 square units, using four rectangles

A = 5.8125 square units, using eight rectangles

The true area is given by $A = \int_0^2 (4 - x^2)dx = 5\frac{1}{3}$ square units. Had we continued with more and more rectangles, the estimated area would have gotten extremely close to the true area of $5\frac{1}{3}$.

MATHEMATICS CORNER 24.3

Areas by Riemann Sums

Approximating the area of a region using Riemann Sums involves constructing rectangles under the curve. **Circumscribed rectangles** are drawn in such a way that they include extraneous area outside the region. The estimated area will be larger than the true area, but as the number of rectangles increases, the estimated area approaches the true area.

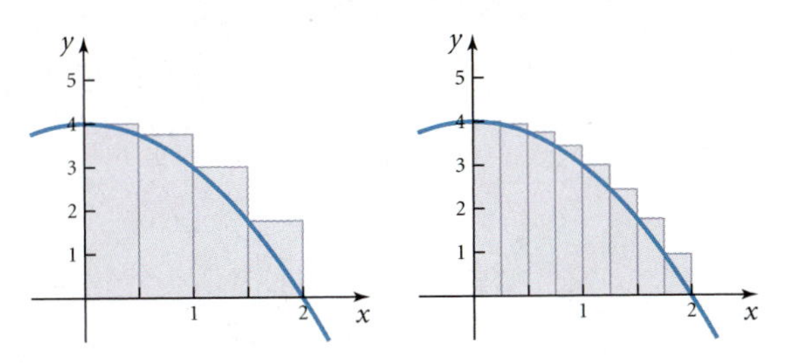

Inscribed rectangles are drawn within the enclosed region. Here the estimated area will be smaller than the true area, but as the number of rectangles increases, the estimated area will approach the true area.

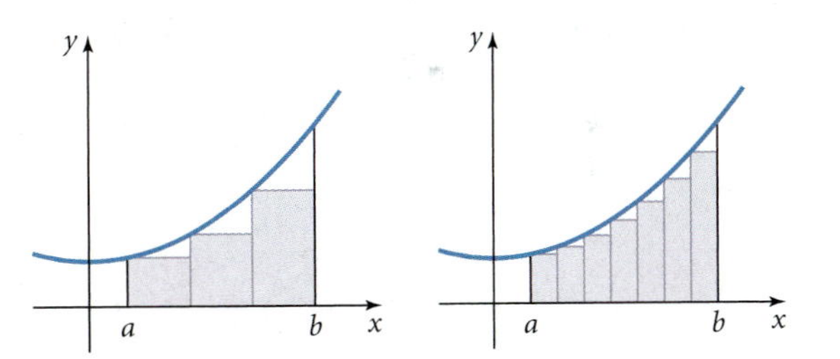

Riemann Sums can also be constructed using **midpoints** of each of the subintervals.

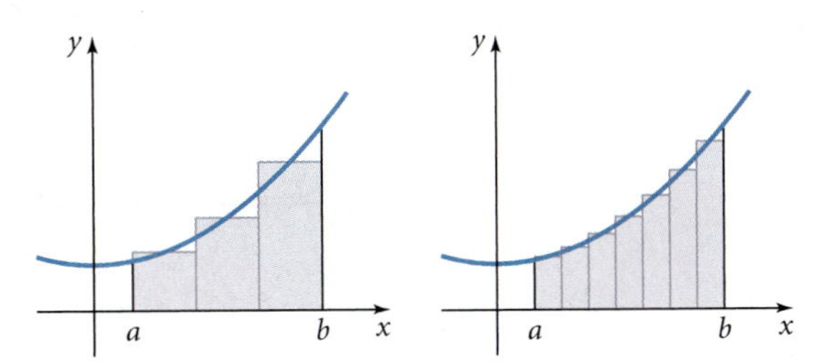

As with the other two methods, as the number of rectangles increases, the estimated area approaches the true area of the region.

Check Your Understanding Answers

Check Your Understanding 24.1

1. Area about 2.773 square units

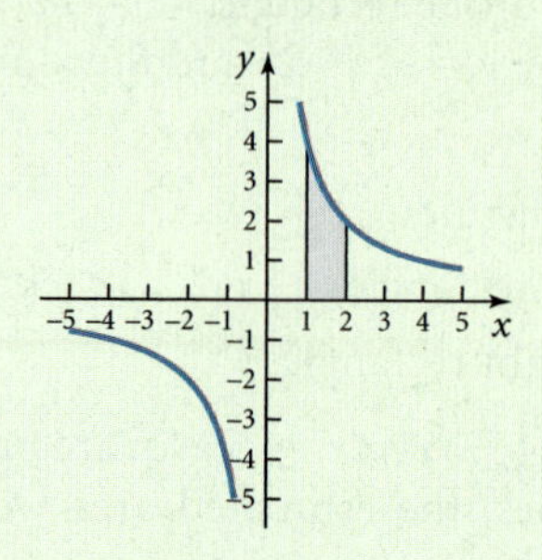

2. Area is 20.25 square units

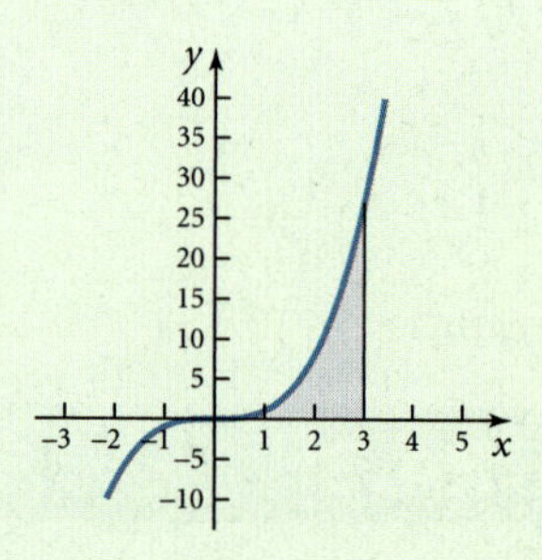

3. Area is $6\frac{2}{3}$ square units

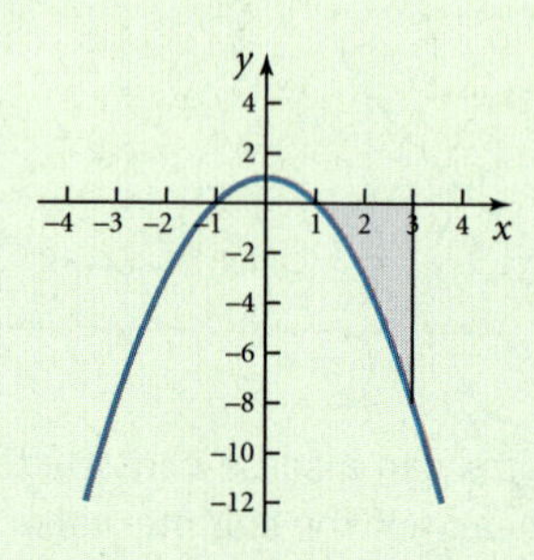

4. Area is about 7.509 square units

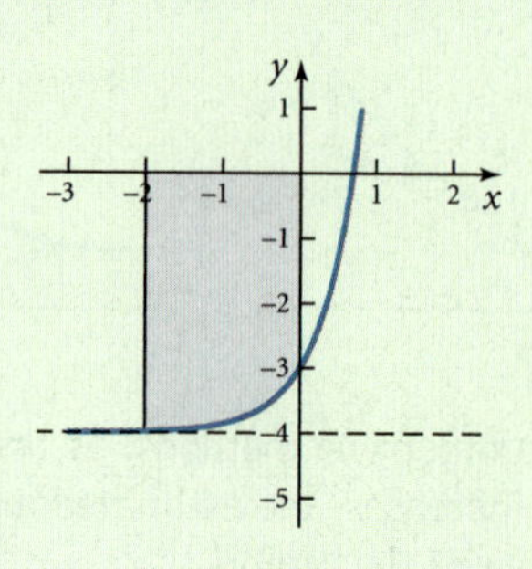

5. Area is $7\frac{1}{3}$ square units

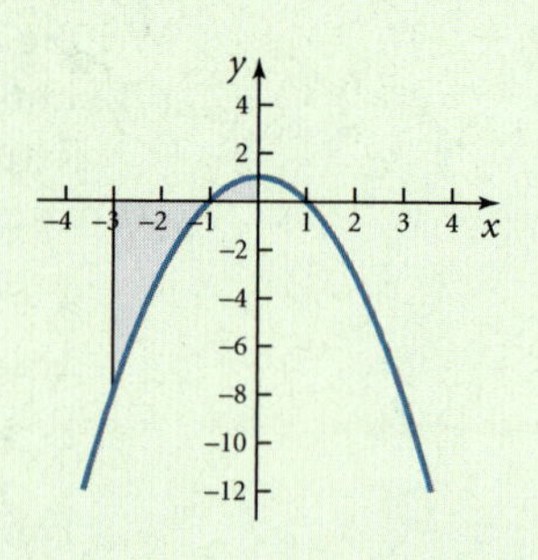

Check Your Understanding 24.2

1. Area is 16.5 square units

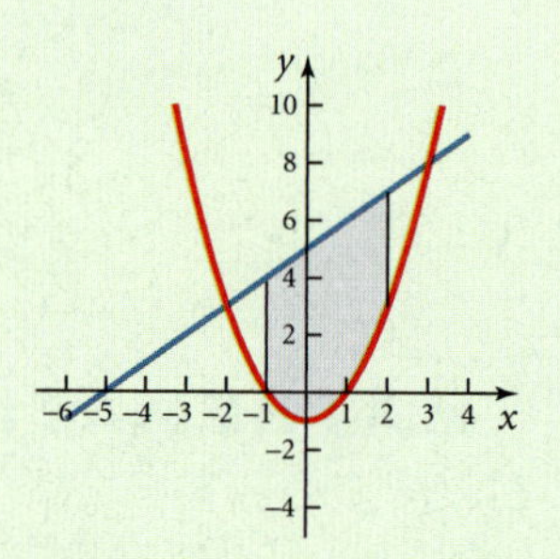

2. Area is $20\frac{5}{6}$ square units

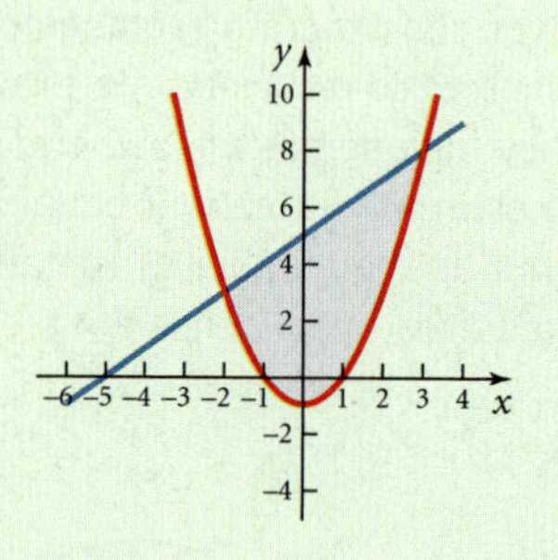

Topic 24 Review

24

In this topic you learned how to use definite integrals to determine the area of nongeometric regions.

CONCEPT	EXAMPLE
Area Rule 1 stated that for $f(x) \geq 0$ on [a, b], the area of the region between $f(x)$ and the x-axis on the interval [a, b] is given by $\int_a^b f(x)dx$.	The area between $f(x) = x^2 + 1$ on $[-2, 1]$ and the x-axis is given by $\int_{-2}^{1}(x^2 + 1)dx$. 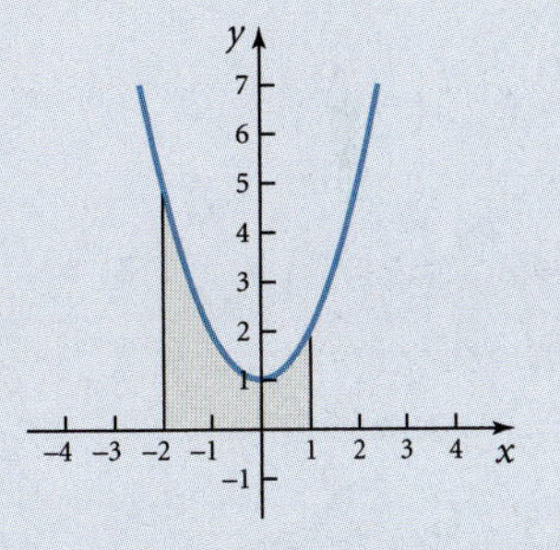
Area Rule 2 stated that for $f(x) \leq 0$ on [a, b], the area of the region between $f(x)$ and the x-axis on the interval [a, b] is given by $-\int_a^b f(x)dx$.	The area between $f(x) = x^2 - 4$ on $[-2, 1]$ and the x-axis is given by $-\int_{-2}^{1}(x^2 - 4)dx$. 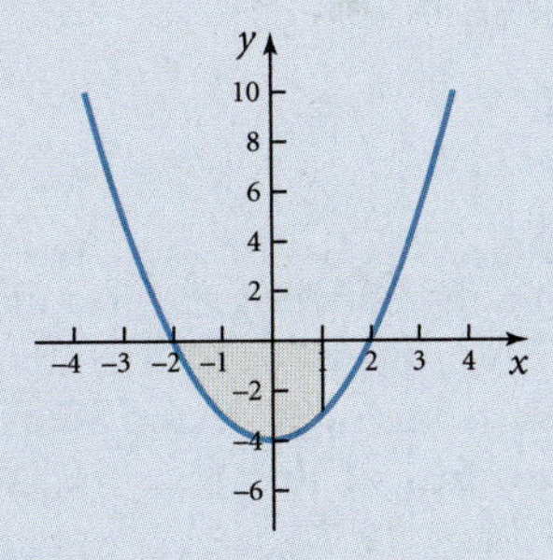
Area Rule 3 stated that for $f(x) \geq g(x)$ on [a, b], the area of the region between the two curves on [a, b] is given by $\int_a^b [f(x) - g(x)]dx$.	The area between $f(x) = x^2 - 4$ and $f(x) = 6 - x$ is given by $\int_{-2}^{1}\big((6 - x) - (x^2 - 4)\big)dx$.

(continued)

In this rule, the location of the x-axis is not relevant to the definition.

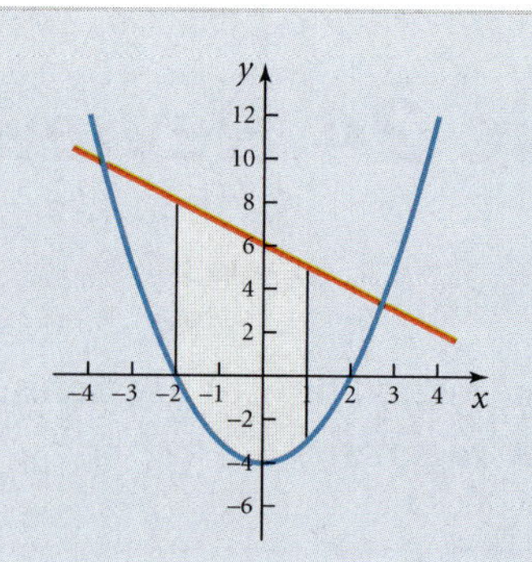

Area Rule 1 and Area Rule 2 may be combined if the enclosed region is both above and below the x-axis.

The area between $f(x) = x^2 - 4$ and the x-axis on

$$\int_{-3}^{-2}(x^2 - 4)dx + \left[-\int_{-2}^{1}(x^2 - 4)dx\right]$$

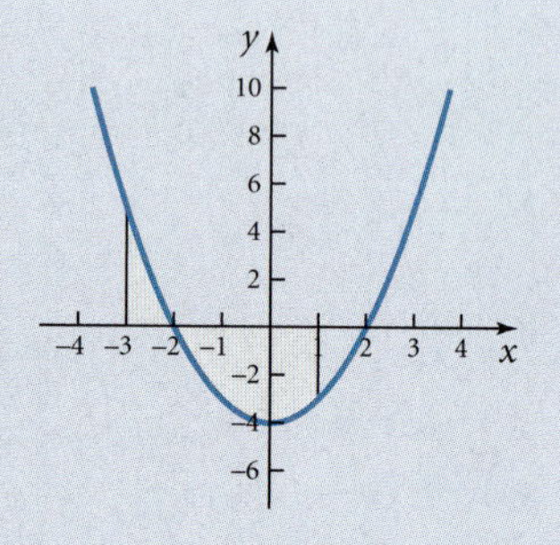

Areas under continuous piecewise defined functions can be determined by dividing the enclosed area into two or more regions, calculating each area, and then adding the areas together.

The area between the x-axis and

$f(x) = \begin{cases} x + 3 & x \leq -1 \\ x^2 + 1 & x > -1 \end{cases}$ on $[-3, 1]$ is given by

$$A = \int_{-3}^{-1}(x + 3)dx + \int_{-1}^{1}(x^2 + 1)dx$$

The concept of approximating the area under a curve using rectangles was first used by Archimedes and was later formalized using **Riemann Sums**. It can be shown that $\int_a^b f(x)dx = \lim_{\Delta x_i \to 0} \sum_{i=0}^{n} f(x_i)\Delta x_i$ gives the area under the curve $y = f(x)$ on the interval [a, b], assuming $f(x)$ is continuous on [a, b].

Given $f(x) = x^2 + 1$, the area between $f(x)$ and the x-axis on $[0, 2]$ can be approximated using inscribed rectangles.

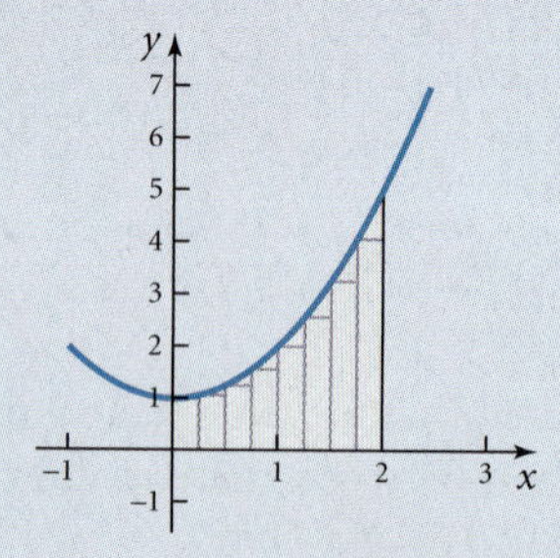

The integral $\int_0^2(x^2 + 1)dx$ gives the actual area.

NEW TOOLS IN THE TOOL KIT

- Using definite integrals and area rules to find the area between $y = f(x)$ and the x-axis on an interval [a, b]
- Using definite integrals and area rules to find the area enclosed by continuous piecewise defined functions on [a, b]
- Using definite integrals to find the area of regions enclosed between two curves
- Using Riemann Sums to approximate areas of enclosed regions

Topic 24 Exercises 24

In Exercises 1 through 32, determine the area of the region between the given function and the x-axis on the specified interval.

1. $f(x) = 2x + 5$ on $[-1, 3]$
2. $f(x) = 3 - 4x$ on $[-2, 0]$
3. $f(x) = x^2$ on $[1, 3]$
4. $f(x) = 2x^2$ on $[1, 4]$
5. $f(x) = 2 - x^2$ on $[-1, 1]$
6. $f(x) = 3 + 2x - x^2$ on $[0, 3]$
7. $f(x) = x^3$ on $[0, 2]$
8. $f(x) = -x^3$ on $[-3, -1]$
9. $f(x) = 4 - x^2$ on $[2, 4]$
10. $f(x) = 2x - x^2$ on $[3, 5]$
11. $f(x) = x^3 - 4x$ on $[1, 3]$
12. $f(x) = 2x - x^3$ on $[-1, 1]$
13. $f(x) = 4 - x^2$ on $[0, 3]$
14. $f(x) = 1 - x^3$ on $[-1, 2]$
15. $f(x) = 6x - x^2$ on $[4, 7]$
16. $f(x) = x^3 - 4x$ on $[1, 4]$
17. $f(x) = e^x$ on $[0, 3]$
18. $f(x) = e^{-x}$ on $[-2, 1]$
19. $f(x) = \dfrac{3}{x}$ on $[-3, -1]$
20. $f(x) = -\dfrac{5}{x}$ on $[1, 3]$
21. $f(x) = \sqrt{x} + 1$ on $[0, 4]$
22. $f(x) = 4 - \sqrt{x}$ on $[0, 4]$
23. $f(x) = e^{-0.5x}$ on $[-1, 2]$
24. $f(x) = e^{0.3x}$ on $[-3, 3]$
25. $f(x) = 2xe^{-0.5x^2}$ on $[0, 4]$
26. $f(x) = 0.4xe^{-0.2x^2}$ on $[0, 2]$
27. $f(x) = \dfrac{2x}{x^2 + 1}$ on $[2, 5]$
28. $f(x) = \dfrac{x^2}{x^3 + 1}$ on $[0, 3]$
29. $f(x) = x(x^2 - 2)^3$ on $[0, 1]$
30. $f(x) = x^2(3 - x^3)^2$ on $[0, 2]$
31. $f(x) = 4x\sqrt{x^2 - 1}$ on $[1, 3]$
32. $f(x) = x\sqrt{x^2 + 1}$ on $[1, 4]$

In Exercises 33 through 52 find the area of the region enclosed between the two curves.

33. $y = 2x - 3$ and $y = x^2$ on $[-1, 2]$
34. $y = x^2 + 2$ and $y = 3x - 2$ on $[-2, 1]$
35. $y = \sqrt{x}$ and $y = -2x$ on $[0, 3]$

36. $y = \sqrt{x} + 1$ and $y = -2x$ on $[1, 4]$
37. $y = x^3$ and $y = 3 - x$ on $[-2, 0]$
38. $y = x^3 + 1$ and $y = 2x - 3$ on $[-1, 1]$
39. $y = e^x$ and $y = \frac{1}{x}$ on $[1, 2]$
40. $y = e^{0.2x}$ and $y = \frac{1}{x^2}$ on $[1, 3]$
41. $y = x^2$ and $y = e^{0.5x}$ on $[-1, 2]$
42. $y = e^{-0.5x}$ and $y = -x^2$ on $[-2, 1]$
43. $y = x^3$ and $y = x^2$
44. $y = x^5$ and $y = x^2$
45. $y = x^2$ and $y = 2x$
46. $y = x^2$ and $y = -3x$
47. $y = x^3$ and $y = 4x$
48. $y = x^3$ and $y = x$
49. $y = 4 - x^2$ and $y = x + 2$
50. $y = 4x - x^2$ and $y = 4 - x$
51. $y = 2 - x^2$ and $y = x^2 - 2$
52. $y = 4x - x^2$ and $y = x^2 - 4x$

Find the area of the region enclosed by $f(x)$ and the x-axis for the given function over the specified interval.

53. $f(x) = \begin{cases} x^2 + 1 & x \le 1 \\ 2x & x > 1 \end{cases}$ on $[-1, 3]$

54. $f(x) = \begin{cases} \sqrt{x} & 0 \le x \le 4 \\ 6 - x & x > 4 \end{cases}$ on $[1, 5]$

55. $f(x) = \begin{cases} e^{-x} & x < 0 \\ x + 1 & 0 \le x < 2 \\ \frac{4}{x} + 1 & x \ge 2 \end{cases}$ on $[-1, 4]$

56. $f(x) = \begin{cases} x^3 + 1 & x \le 1 \\ \frac{2}{x} & 1 < x \le 2 \\ \frac{x}{2} & x > 2 \end{cases}$ on $[0, 3]$

In Exercises 57 through 60 evaluate $\int_0^5 f(x)dx$ for the indicated regions.

57.

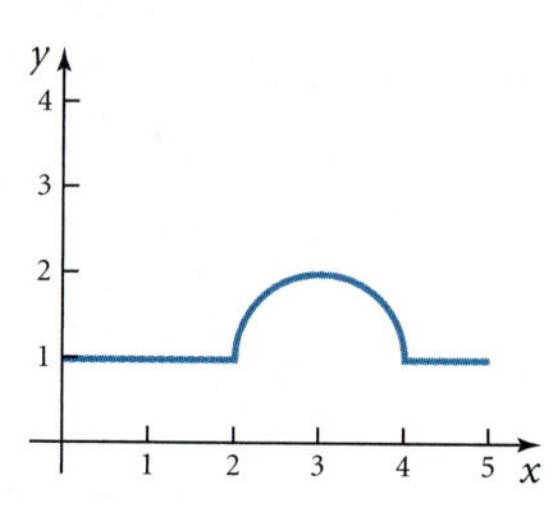

58.

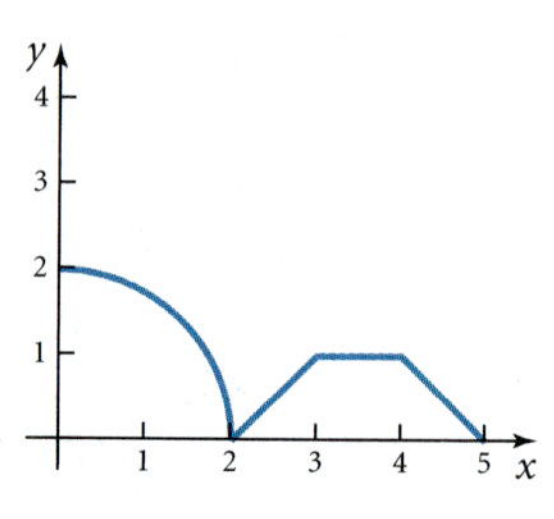

59.

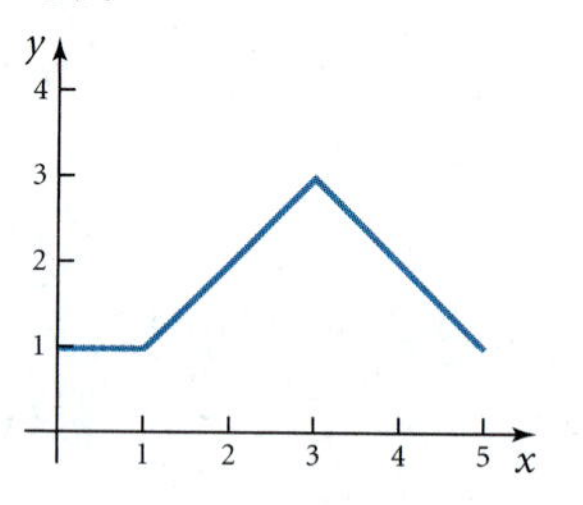

60.

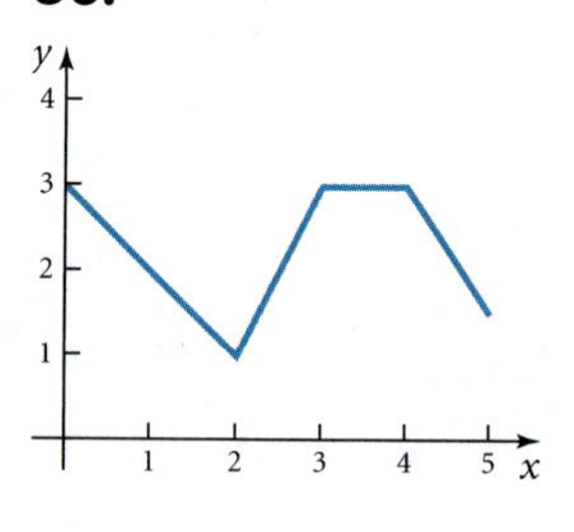

In Exercises 61 through 64, estimate the area of each region using Riemann Sums with the indicated number of rectangles.

61. between $y = \ln x$ and the x-axis on $[1, 4]$ using four rectangles; using eight rectangles.
62. between $y = x\ln x$ and the x-axis on $[1, 3]$ using four rectangles; using eight rectangles.
63. between $y = e^{-x^2/2}$ and the x-axis on $[-1, 2]$ using four rectangles; using eight rectangles.
64. between $y = \sqrt{x^3 + 4x}$ and the x-axis on $[0, 3]$ using four rectangles; using eight rectangles.
65. Trade for a small country has declined in recent years. Since 2002, annual imports for the country are given by $I(t) = 20.93t^{0.668}$ billion dollars and annual exports are given by $E(t) = 22.79t^{0.334}$ billion dollars, where t is the number of years after 2002.

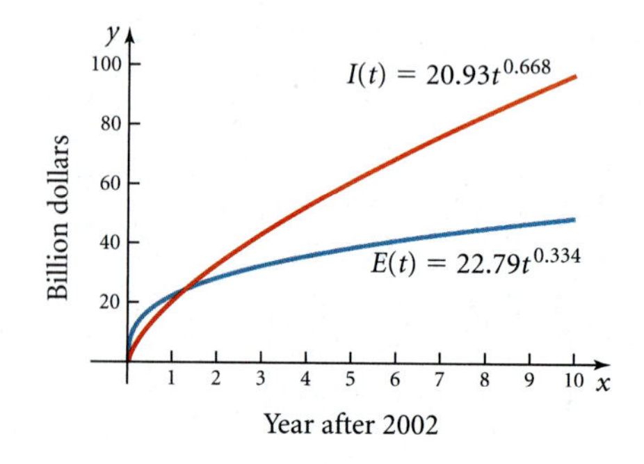

When imports exceed exports, the country is said to have a trade deficit. Find the total trade deficit from 2004 to 2008.

66. Trade in a neighboring country has improved since 2002. Annual imports for the country are given by $I(t) = 18.62t^{0.381}$ billion dollars, and annual exports are given by $E(t) = 16.3t^{0.594}$ billion dollars, where t is the number of years after 2002.

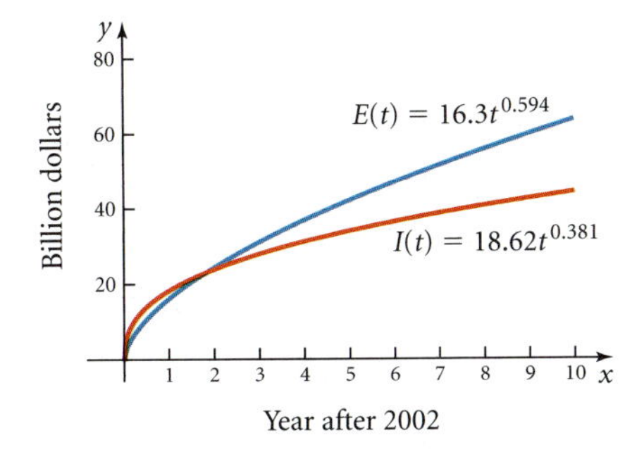

When exports exceed imports, the country is said to have a trade surplus. Find the total trade surplus from 2004 to 2008.

67. Marginal revenue, in thousand dollars, for x hundred new sports cars is given by $MR(x) = \dfrac{400}{x}$. Marginal cost, in thousand dollars, is given by $MC(x) = \dfrac{450}{x^2}$.

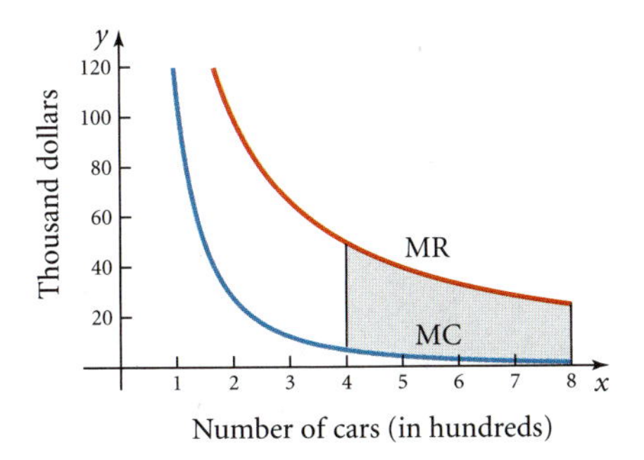

Find the total profit for $x = 400$ to $x = 800$ cars.

68. A union is negotiating its next contract with the employer. The employer offers to pay $20{,}000e^{0.02t}$ billion dollars per year, but the union is demanding $20{,}000e^{0.05t}$ billion dollars per year.

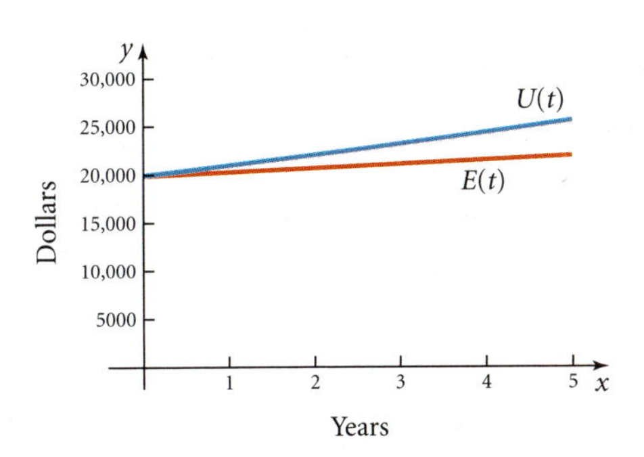

What is the accumulated salary difference over the five-year life of the contract?

SHARPEN THE TOOLS

For each of the following, draw the graph and shade the region specified. Then find the area of the region.

69. Between $f(x) = x^3 - 4x$ and the x-axis on $[-3, 1]$

70. Between $f(x) = 2x - x^3$ and the x-axis on $[-2, 2]$

71. Between $y = x + 3$, $y = 5 - 2x$, and $y = 1 - 0.5x$

72. Between $y = x^2$, $y = 3 - 2x$, and $y = 2 - x$

73. Between $y = x^3 - 4x$ and $y = x - 2$

74. Between $y = x^4 - 4x^2$ and $y = x - 1$

75. Between $y = x^3 - 4x^2$ and $y = x^2 - 4x - 3$

76. Between $y = x^3 - 4x^2 + 4$ and $y = 2 - x$

77. Between $y = e^{-0.5x}$, $y = e^{0.2x}$, and $y = x + 4$

78. Between $y = \sqrt{x}$, $y = x - 3$, and $y = 4 - x$

The natural logarithm $\ln a$, for $a > 0$, is sometimes defined as the area under $y = \dfrac{1}{x}$ on the interval $[1, a]$.

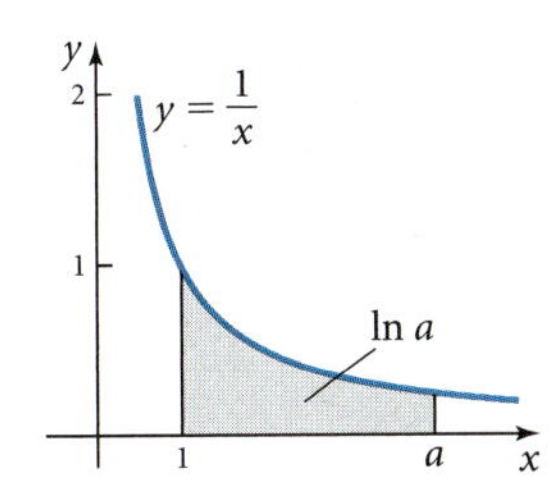

79. Prove that $\displaystyle\int_1^a \frac{1}{x}\,dx = \ln a$ for all $a > 0$.

80. Verify that the definition is correct for $a = 6$ by first evaluating $\displaystyle\int_1^6 \frac{1}{x}\,dx$ and then by evaluating $\ln 6$ on your calculator.

CALCULATOR CONNECTION

81. Marginal revenue for x hundred camera cell phones is given by $MR(x) = 15\ln(x + 1)$ thousand dollars. Marginal costs are given by $MC(x) = 20e^{-0.2(x-5)^2}$ thousand dollars.

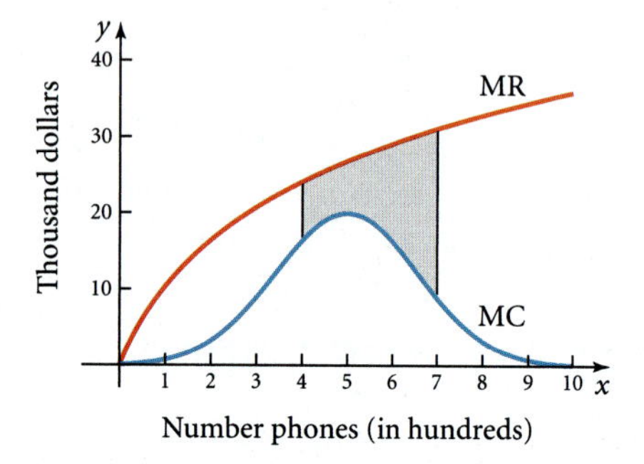

Find the total accumulated profit for $x = 400$ to $x = 700$ phones.

82. Annual imports for a country since 2000 are given by $I(t) = 9te^{0.05t}$ billion dollars. Annual exports are given by $E(t) = 25\ln(t + 1)$ billion dollars.

a. For what years did the country have a trade surplus?

b. Find the accumulated trade deficit from 2005 to 2008.

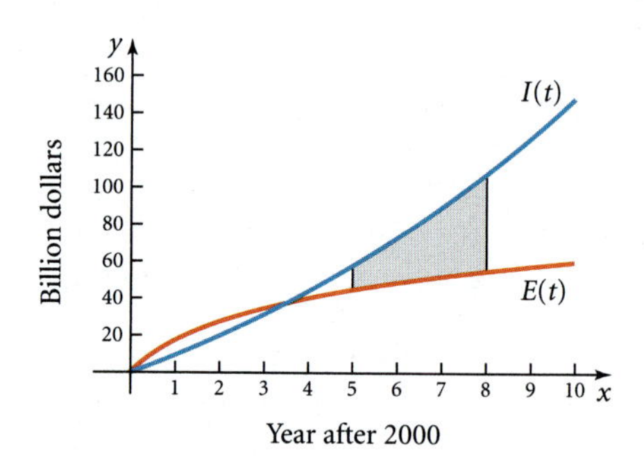

TOPIC

Applications of Definite Integrals

25

You now know how to evaluate definite integrals and how to determine the area under a curve and the area between curves. We now look at several applications from business and economics that make use of definite integrals and area under curves—Lorenz curves and the Gini index, producer and consumer surplus, and probability density functions. Before starting your study of these applications, be sure to complete the Warm-up Exercises to sharpen the algebra and calculus skills you will need.

IMPORTANT TOOLS IN TOPIC 25

- *Lorenz curves*
- *Gini Index to measure income inequality*
- *Producer surplus*
- *Consumer surplus*
- *Market surplus*
- *Probability density functions*

TOPIC 25 WARM-UP EXERCISES

Be sure you can successfully complete the following exercises before starting Topic 25.

Algebra Warm-up Exercises

Solve each equation for x.

1. $3.2x^2 - 5.12x = 1.152$

2. $2x^3 - 3x^2 + x = 0$

3. Find the equilibrium point if $S(p) = 2p + 3$ and $D(p) = -p^2 + 5p + 7$.

4. Find the equation of the line between (10, 22) and (30, 8). Write the equation in slope-intercept form, $y = mx + b$.

Calculus Warm-up Exercises

Evaluate each definite integral. Round answers to three decimal places.

1. $\int_0^1 \left(x - 0.6x^{1.72}\right)dx$ **2.** $\int_3^6 3e^{-0.2x}dx$

Find the area of the region.

3. Between $y = x^2 + 1$ and the x-axis on $[1, 3]$

4. Between $y = x^2 + 1$ and $y = x + 3$

Answers to Warm-up Exercises

Algebra Warm-up Exercises

1. $x = -0.2, 1.8$ **2.** $x = 0, \frac{1}{2}, 1$

3. $p = 4, S(4) = D(4) = 11$, so equilibrium point is (4, 11)

4. $y = -0.7x + 29$

Calculus Warm-up Exercises

1. 0.279

2. 3.714

3. $10\frac{2}{3}$ square units

4. 4.5 square units

Lorenz Curves and the Gini Index

Today the average fast-food restaurant cashier earns $6.35 per hour. The average doctor earns $210,000 per year, which is $4038 per week (assuming the doctor works 52 weeks a year) and $67.31 per hour (assuming 60 hours per week). The average NFL player makes about $1.5 million per season, which is $75,000 per game (assuming 20 games per season) and $15,000 per hour (assuming five hours per game). In 2005 there were 691 billionaires in the United States, yet nearly 36 million people were living in poverty. The issue of income inequality is of vital importance to any society.

One way to measure the degree of income inequality in a society is to rank everyone in that society by his or her income, then divide that ranking into quintiles (each representing 20% of the population), and compute what share of the total income that quintile earns. In an "ideal" society, each quintile would have the same share (20%) of the total income because everyone would have the same income. Realistically, of course, that is not the case. In the United States, for instance, in 2005 the top 5% of the population earned nearly 40% of total income.

Consider the distribution of income for the United States in 2004.

Income Quintile	Percent (%) of Total Income	Cumulative Percent (%) of Total Income
Lowest 20% (lower class)	3.4% or 0.034	3.4% or 0.034
Second 20% (lower-middle class)	8.7% or 0.087	3.4 + 8.7 = 12.1% or 0.121
Middle 20% (middle class)	14.7% or 0.147	12.1 + 14.7 = 26.8% or 0.268
Fourth 20% (upper-middle class)	23.2% or 0.232	26.8 + 23.2 = 50.0% or 0.500
Highest 20% (upper class)	50.0% or 0.500	100% or 1.000

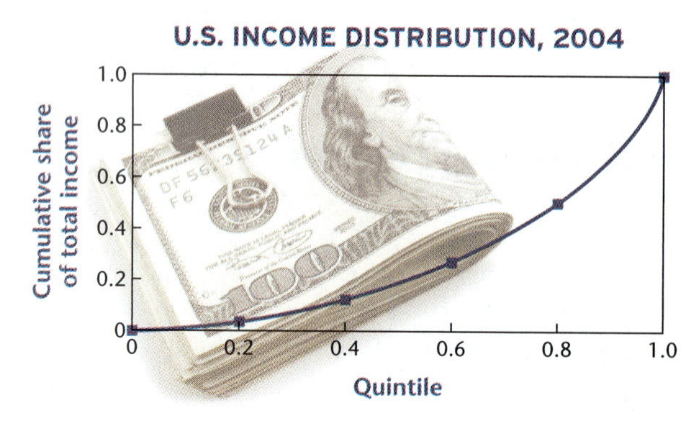

Figure 25.1

We plot this data using a **Lorenz curve**, which is a geometric representation of the income distribution among the five quintiles. The horizontal axis represents the five *cumulative* quintiles. The vertical axis represents the *cumulative* share of the total income for each quintile. Percents are expressed in decimal form, so $0 \leq x \leq 1$ and $0 \leq y \leq 1$, where a value of one means 100%.

Figure 25.1 shows the Lorenz curve for the 2004 income data. To determine the equality or inequality of a distribution, we compare it to the ideal distribution in which each quintile has the same share (20%) of the total income. Graphically, the ideal distribution is represented by the line $y = x$, and the actual distribution is represented by its Lorenz curve.

MATHEMATICS CORNER 25.1

Lorenz Functions

Lorenz curves were introduced in 1905 by the American mathematician Max Otto Lorenz as a graphical representation of income distribution.

Lorenz curves represent functions that have the following properties:

$L(0) = 0 \quad L(1) = 1$

$0 < L(x) < x < 1$

$L'(x) > 0$ so Lorenz curves are always increasing

$L''(x) > 0$ so Lorenz curves are always concave up

A Lorenz curve always begins at the point (0, 0) and ends at the point (1, 1). The curve is increasing and concave up and will always be below the line $y = x$.

The area between the ideal distribution and the actual distribution is called the **area of inequality**. The **Gini Index** or **Gini Coefficient** is used to measure the inequality. The Gini Index is a ratio comparing the area of inequality to the total area. An example is shown in Figure 25.2.

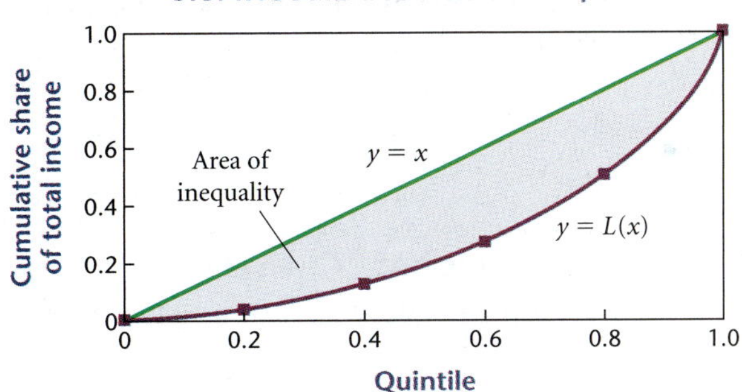

Figure 25.2

Definition: Given the Lorenz Curve $L(x)$ as indicated and area of inequality A.

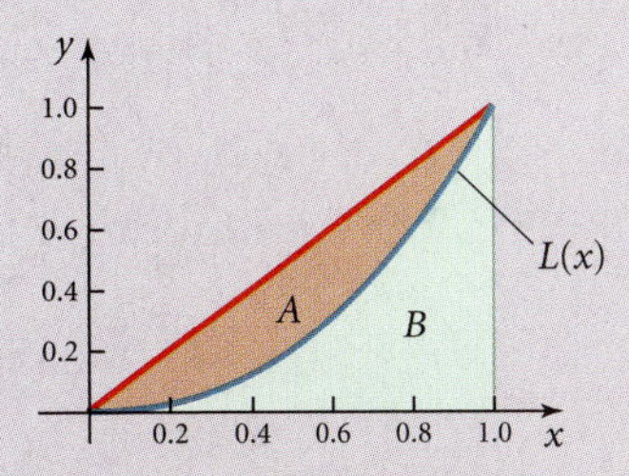

The Gini Index (or Gini Coefficient) is given by $G = \dfrac{A}{A + B}$.

MATHEMATICS CORNER 25.2

Gini Index

The Gini Index was developed by the Italian statistician Corrado Gini (1884–1965). Gini was also a sociologist and leading fascist theorist. The Gini Index is a number between 0 and 1, where 0 is perfect equality (everyone has the same income) and 1 is perfect inequality (one person has all the income).

The Gini Index is mostly used to measure income inequality within a society, but it can also be used to compare populations (for instance, to compare income distribution in the United States to that of another country) or to compare segments within a population (for instance, to compare income distributions of men and women).

A Gini Index of 0 is perfect equality, because if income is equally distributed, then the Lorenz curve is the same as the line $y = x$, so the area of inequality is $A = 0$. If one person has all the income, then the Gini Index is 1, because $A = A + B$. The closer the Gini Index is to 0, the more equal the distribution of income; the closer the Gini Index is to 1, the more unequal the distribution of income.

To calculate the Gini Index we must determine the values of areas A and $A + B$.

The area $A + B$ is the area of a triangle with base 1 and height 1, so its area is $\frac{1}{2}$. Thus, we know that $A + B = \frac{1}{2}$.

The area of inequality A is the area of the region between the line $y = x$ and the Lorenz curve $y = L(x)$. Thus, $A = \int_0^1 \left(x - L(x)\right)dx$.

The Gini Index is the ratio of these two areas.

Definition: Given the Lorenz Curve $L(x)$ as indicated and area of inequality A.

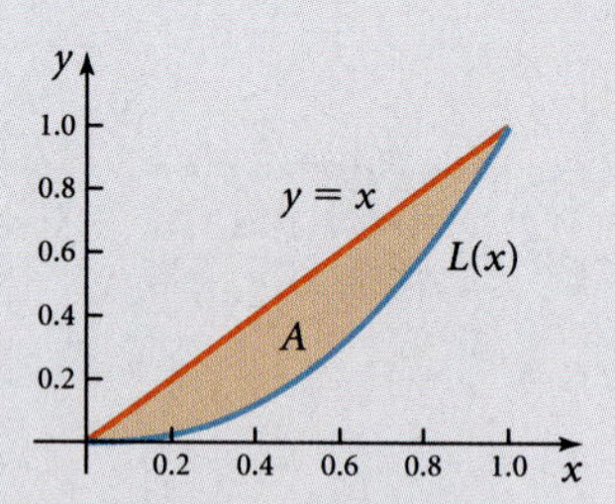

The **Gini Index** (or **Gini Coefficient**) is given by

$$G = \frac{\int_0^1 \left(x - L(x)\right)dx}{\frac{1}{2}} = 2\int_0^1 \left(x - L(x)\right)dx$$

Example 1: Calculate the Gini Index for income distribution in the United States in 2004 if $L(x) = 0.845x^{2.05}$.

Solution:

$$G = 2\int_0^1 \left(x - 0.845x^{2.05}\right)dx$$

$$= 2\left(\frac{1}{2}x^2 - \frac{0.845}{3.05}x^{3.05}\right)\Bigg|_0^1$$

$$= 2\left(\frac{1}{2} - \frac{0.845}{3.05}\right) = 0.446$$

The income inequality of the United States in 2004, represented by the Gini Index of 0.446, is better than the Gini Index of some countries, such as Brazil (0.591) or South Africa (0.593), but it is worse than many other countries, such as Denmark (0.247) or Japan (0.249). ■

Example 2: Economists often state that the inequality of income distribution in the United States has worsened since 1970. Consider the U.S. income distribution for 1968. Calculate the Gini Index for 1968 and compare it to that of 2004.

Income Quintile	Percent (%) of Total Income	Cumulative Percent (%) of Total Income
Lowest 20% (lower class)	4.2%	4.2% or 0.042
Second 20% (lower-middle class)	11.1%	15.3% or 0.153
Middle 20% (middle class)	17.5%	32.8% or 0.328
Fourth 20% (upper-middle class)	24.4%	57.2% or 0.572
Highest 20% (upper class)	42.8%	100% or 1

The Lorenz curve for this data (1968) is $L(x) = 0.924x^{1.94}$.

Solution:

$$G = 2\int_0^1 \left(x - 0.924x^{1.94}\right)dx$$

$$= 2\left(\frac{1}{2}x^2 - \frac{0.924}{2.94}x^{2.94}\right)\Bigg|_0^1$$

$$= 2\left(\frac{1}{2} - \frac{0.924}{2.94}\right) = 0.371$$

The Gini Index for 1968 was 0.371. From Example 1, we know that the Gini Index for 2004 was 0.446. Because $0.371 < 0.442$, the distribution of income in 1968 was more equal than in 2004. ■

The Gini Index is mostly used to measure income inequality within a society, but it can also be used to compare populations.

Example 3: Government data show that income distribution for doctors is given by the Lorenz curve $L_d(x) = 0.85x^{1.76}$, and income distribution for engineers is given by the Lorenz curve $L_e(x) = 0.78x^2 + 0.22x$. For which profession is the distribution of income more equal?

Solution: For doctors:

$$G_d = 2\int_0^1 \left(x - 0.85x^{1.76}\right)dx = 0.384$$

For engineers:

$$G_e = 2\int_0^1 \left(x - (0.78x^2 + 0.22x)\right)dx = 2\int_0^1 (0.78x - 0.78x^2)dx = 0.26$$

Income distribution for engineers is more equal than that of doctors. There are some doctors who are highly specialized and earn extremely high incomes, which causes the distribution of incomes among doctors to be more unequal. ■

Check Your Understanding 25.1

Determine the Gini Index for each of the following Lorenz curves.

1. $L(x) = 0.527x^{1.809}$ **2.** $L(x) = 1.27x^2 - 0.27x$

Producer and Consumer Surplus

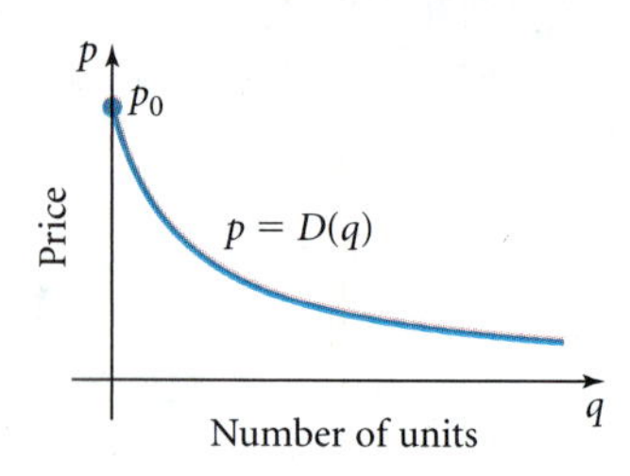

Figure 25.3

Consider the demand curve $p = D(q)$, giving the price at which q units are demanded. The highest price customers are willing to pay is denoted as p_0, as shown in Figure 25.3.

When a new item first comes on the market, few consumers are willing to buy it at the high price. As the price decreases, more will be sold at the lower prices. The **demand price** is what consumers are willing to pay. **Market price** is what they do pay. At market price p_m consumers will demand q_m units. (See Figure 25.4.)

Consumer surplus measures the savings the consumer gets from buying at a price below what the consumer would have been willing to pay. Consumer surplus is represented by the area between the demand curve and the market price.

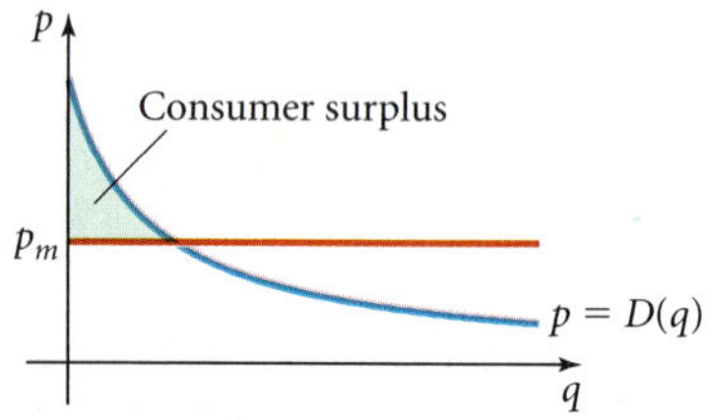

Figure 25.4

Definition: **Consumer surplus** is the area between the demand curve $p = D(q)$ and the market price p_m and is calculated by

$$CS = \int_0^{q_m} \left(D(q) - p_m\right)dq.$$

Example 4: The demand function for q hundred new computers is $p = \frac{30}{q+1}$, where p is the price in hundred dollars. When the price of the computer is \$1250, what is the consumer surplus?

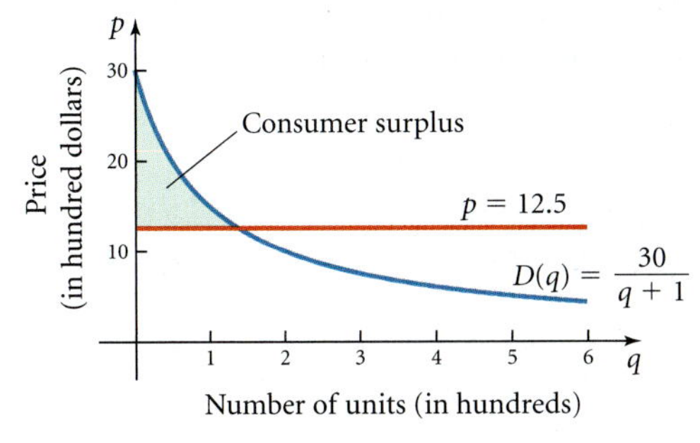

Figure 25.5

Solution: If the price is \$1250, then $p = 12.5$, so $12.5 = \frac{30}{q+1}$ and $q = 1.4$. Consumer surplus is then given by

$$\begin{aligned} CS &= \int_0^{1.4} \left(\frac{30}{q+1} - 12.5 \right) dq = \left(30\ln(q+1) - 12.5q\right)\Big|_0^{1.4} \\ &= 8.764 \text{ hundred dollars} \end{aligned}$$

Consumers gain about \$876.40 buying at \$1250 rather than at the price they would have been willing to pay. See Figure 25.5. ■

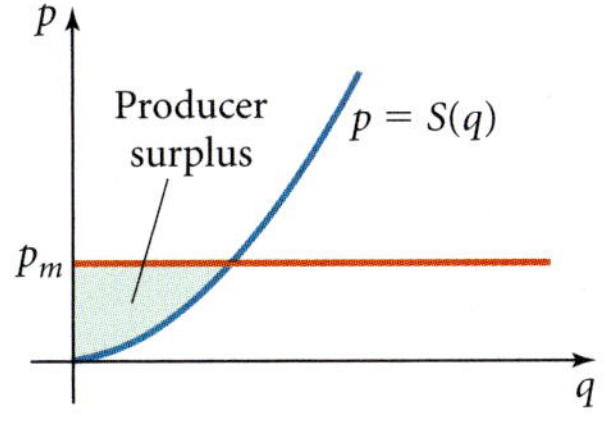

Figure 25.6

Now consider the supply function $p = S(q)$, the price at which producers are willing to supply q units of a product. Producers are more willing to supply a product as the price of the product increases. See Figure 25.6.

At market price p_m, producers are willing to supply q_m units. If the market price is above the price given by the supply function, the producers experience a benefit called *producer surplus.* Producer surplus is the suppliers' gain from selling at the current price rather than one they would have been willing to accept. Producer surplus is the area between the market price p_m and the supply function $p = S(q)$.

> **Definition:** **Producer surplus** is represented by the area between the market price p_m and the supply function $p = S(q)$ and is calculated by
>
> $$PS = \int_0^{q_m} \left(p_m - S(q)\right) dq.$$

Example 5: The supply function for q hundred new computer is given as $p = 2q^2 + q$, where p is the price in hundred dollars. If the market price is \$1250, what is the producer surplus?

Solution: If the price is \$1250, then $p = 12.5$, so $12.5 = 2q^2 + q$ and $q = 2.26$. The producer surplus is given by

$$\begin{aligned} PS &= \int_0^{2.26} \left(12.5 - (2q^2 + q)\right) dq \\ &= \left(12.5q - \frac{2}{3}q^3 - \frac{1}{2}q^2\right)\Big|_0^{2.26} \\ &= 18 \text{ hundred dollars} \end{aligned}$$

Producers gain about \$1800 supplying the computer at a price of \$1250 rather than the price at which they would have been willing to supply the computers. ■

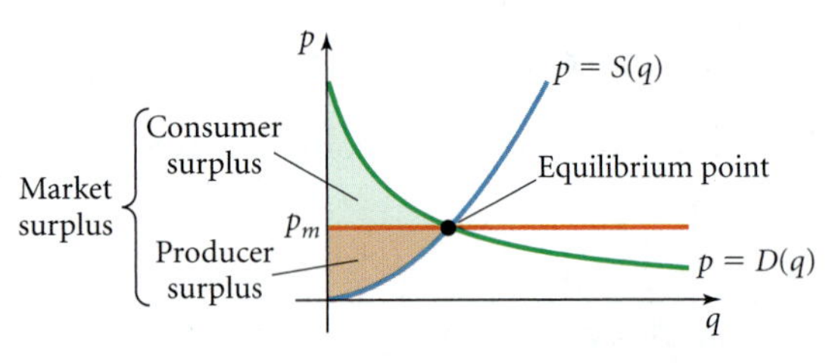

Figure 25.7

The price for which supply equals demand is called the **equilibrium price**. Graphically, the point where supply equals demand is called the **equilibrium point**. Market surplus is the total benefit to both consumer and producer. See Figure 25.7.

Market surplus is the sum of consumer surplus and producer surplus up to the equilibrium price and is also the area between the demand and supply curves.

Definition: **Market surplus** is represented by the area between the demand and supply curves up to the equilibrium point and is calculated by

$$MS = CS + PS = \int_0^{q_m} \big(D(q) - S(q)\big)dq$$

Example 6: Demand for q hundred new computers is $p = \frac{30}{q+1}$, where p is the price in hundred dollars. The supply function for q hundred new computers is given $p = 2q^2 + q$, where p is the price in hundred dollars. Determine the equilibrium point and find the market surplus up to that point.

Solution: The equilibrium point occurs for the quantity at which

$$\frac{30}{q+1} = 2q^2 + q$$
$$\rightarrow 2q^3 + 3q^2 + q - 30 = 0$$
$$\rightarrow q = 2$$

Calculator Corner 25.1

On the TI-84 Plus Silver Edition, the equation $2q^3 + 3q^2 + q - 30 = 0$ can be solved by pressing **APPS** and selecting **:PolySmlt**. Press **ENTER** three times. Enter 3 as the degree of the polynomial. On the next screen enter the coefficients and press **SOLVE**. The calculator returns the three roots, only one of which is real ($q = 2$).

If $q = 2$, then $p = \frac{30}{2+1} = 2(2)^2 + 2 = 10$. The equilibrium point is (2, 10), which means that at a price of \$1000, supply and demand both equal 200 computers. (See Figure 25.8.)

Market surplus is given by

$$\text{MS} = \int_0^2 \left(\frac{30}{q+1} - (2q^2 + q)\right)dq$$
$$= \left(30\ln(q+1) - \frac{2}{3}q^3 - \frac{1}{2}q^2\right)_0^2$$
$$\approx 25.625 \text{ hundred dollars or } \$2562.50$$

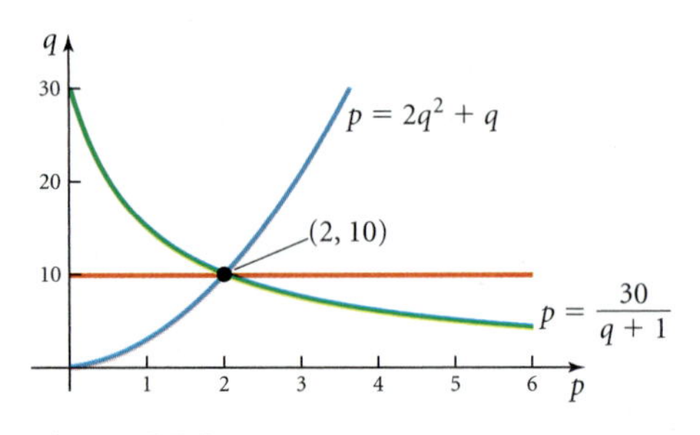

Figure 25.8

Market surplus can also be found as the sum of consumer surplus and producer surplus.

At $p = 10$, we have

$$CS = \int_0^2 \left(\frac{30}{q + 1} - 10\right) dq \approx 12.958$$

and

$$PS = \int_0^2 \left(10 - (2q^2 + q)\right) dq \approx 12.667.$$

Market surplus is then

$$MS = CS + PS$$

$$MS \approx 12.958 + 12.667$$

$$MS \approx 25.625 \text{ hundred dollars or } \$2562.50$$

■

Check Your Understanding 25.2

The demand function for a product is $p = -0.3q + 330$, and the supply function is $p = 0.2q + 75$, where p is the price in dollars.

1. Find the consumer surplus if $p = \$\,255$.
2. Find the producer surplus if $p = \$155$.
3. Determine the equilibrium point.
4. Find the market surplus up to the equilibrium point.

Probability Density Functions

The study of probability began in the 1600s as rich aristocrats with too much time on their hands got tired of losing money in the gaming parlors and turned to the mathematicians for advice on winning strategies. Since that time probability has evolved into a major branch of mathematics with widespread applications. We now take a brief look at how calculus applies to probability.

First we need a little background terminology. Suppose that x represents the number of minutes a bank customer waits in line before getting to the teller and that wait time is known to be in the interval $[0, 15]$, which means that customers may wait anywhere from 0 to 15 minutes. There are infinitely many possible waiting times contained in this interval. The variable x is called a continuous random variable because it can represent any value in the interval. What is the probability that the next customer will wait between five and eight minutes? Questions of this sort are answered using a probability density function.

A **probability density function** of a continuous random variable x over an interval I is a function with these properties:

1. $f(x) \geq 0$ for all x in the interval I.
2. The total area under the graph of $y = f(x)$ is 1, where $1 = 100\%$. (See Figure 25.9.)

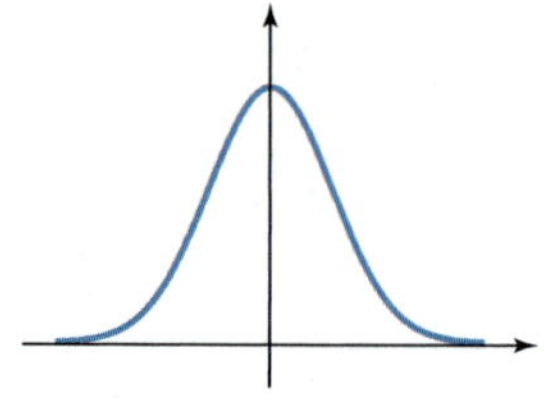

Figure 25.9

3. The probability that a selected value of the random variable lies in the interval [a, b] is given by $P(a \leq x \leq b) = \int_a^b f(x)dx$. (See Figure 25.10.)

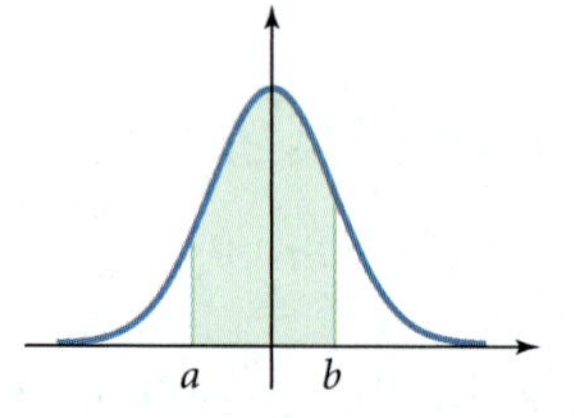

Figure 25.10

4. Because the total area under the curve is 1, $0 \leq P(a \leq x \leq b) \leq 1$. Probability must be a number between 0 and 1, or between 0% and 100%.

There are several types of probability density functions. We will look at two types. The first type is an **exponential probability density function,** $p(x) = ae^{-ax}$, which is shown in the next example.

Example 7: After a new copy machine is installed, the number of months before it first needs repairs has the exponential probability density function $f(x) = 0.02e^{-0.02x}$, where x is time in months. What is the probability that the machine will first need repairs

a. during the first year?
b. during the second year?

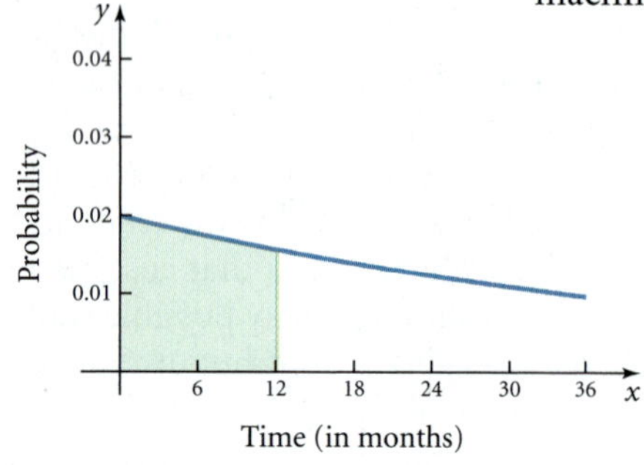

Figure 25.11

Solution: Figure 25.11 shows the graph of the function.

a. During the first year means during the first 12 months, or the interval [0, 12]. The probability that the machine needs repairs at some time during the first 12 months is given by

$$P(0 \leq x \leq 12) = \int_0^{12} \left(0.02e^{-0.02x}\right)dx = 0.213$$

b. During the second year means from the end of the first year to the end of the second year, or the interval [12, 24]. The probability that the machine needs repairs at some time during the second year is given by

$$P(12 \leq x \leq 24) = \int_{12}^{24} \left(0.02e^{-0.02x}\right)dx = 0.168$$

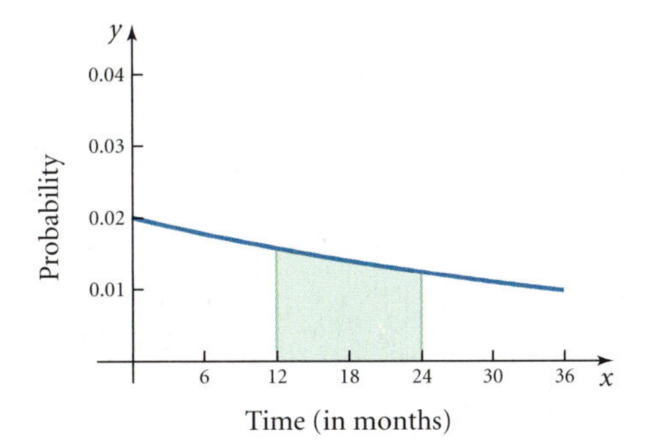

Figure 25.12

The most widely used probability density function is the **normal distribution**, whose graph is the bell-shaped curve. Normal distributions are symmetric, which means that the values in the distribution cluster around the average and then taper off gradually toward the tails of the graph. Some examples of distributions that are normal are:

- heights of men and women. Most men are between 67 inches and 72 inches tall. As heights increase (or decrease), there are fewer men at those heights. How many men do you know that are 80 inches (7' 6") tall?
- SAT or ACT scores. SAT scores average about 1030 and are spread fairly symmetrically toward the end points of 200 and 1600. Very few students score toward the lower or higher end of the scale; most students score between 900 and 1200. ACT scores follow the same symmetric pattern, with an average of about 22.5.
- the quantity of liquid in a 1-liter bottle. Most 1-liter bottles of water probably contain 1 liter of water, give or take a few milliliters.

The **normal distribution curve** is given by $f(x) = \dfrac{1}{\sigma\sqrt{2\pi}}e^{-(x-\mu)^2/2\sigma^2}$ and is used to determine the proportion of the distribution lying between two specific values. In this definition the symbol μ represents the mean of the distribution, and the symbol σ represents the standard deviation of the distribution. The mean is the average of all the values in the distribution. The standard deviation is a statistical measure of the standard amount by which each value in the distribution varies from that average. The larger the standard deviation is, the more spread out the distribution.

Example 8: Heights of American men are normally distributed with an average of 70 inches and a standard deviation of 2.9 inches. In order to apply for a police academy, men must be at least 66 inches tall and not more than 74 inches tall. What proportion of American men qualify, based strictly on height?

Solution: The average is $\mu = 70$ and the standard deviation is $\sigma = 2.9$, so the distribution looks like that shown in Figure 25.13.

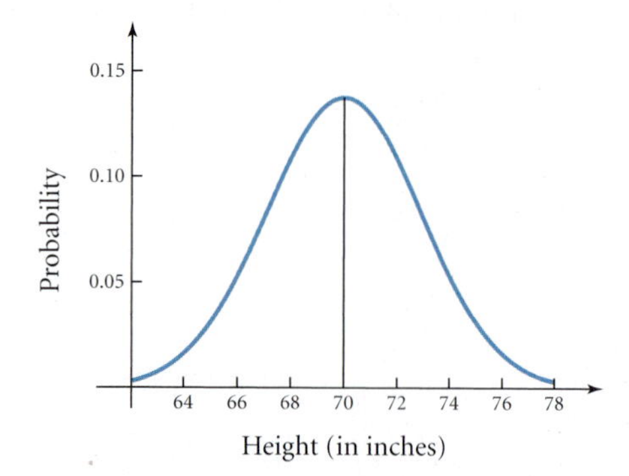

Figure 25.13

Substituting $\mu = 70$ and $\sigma = 2.9$, the normal distribution function for heights of American men is given by

$$f(x) = \frac{1}{2.9\sqrt{2\pi}} e^{-(x-70)^2/2(2.9)^2}$$

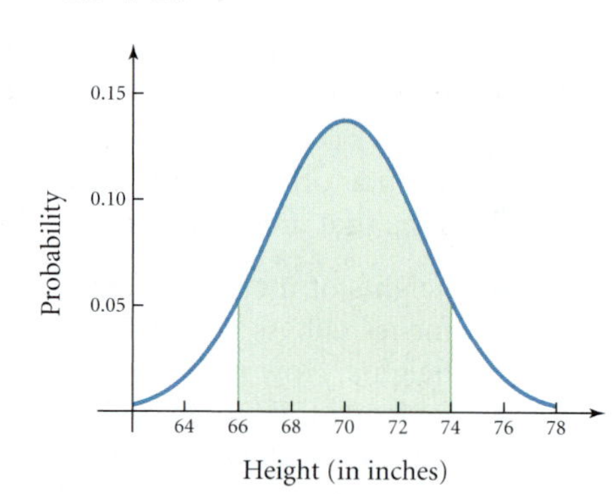

Figure 25.14

To determine the probability that a particular man's height falls in the interval [66, 74], we must integrate $\int_{66}^{74} \frac{1}{2.9\sqrt{2\pi}} e^{-(x-70)^2/2(2.9)^2} dx$, as shown in Figure 25.14.

However, this function is not integrable with substitution or any other method discussed in this unit. It can be integrated using either Riemann Sums or the TI-84 calculator.

Solution Using Riemann Sums: Partition the interval [66, 74] into subintervals of width one unit. Use the left-hand end point of each subinterval to determine the height of each rectangle. The area under the curve can be estimated by

$$\begin{aligned} \text{Area} &= 1\big(f(66) + f(67) + f(68) + f(69) + f(70) \\ &\quad + f(71) + f(72) + f(73)\big) \\ &\approx 1(0.05313 + 0.08056 + 0.10845 + 0.1296 + 0.1376 \\ &\quad + 0.1296 + 0.10845 + 0.08056) \\ &\approx 0.82796 \end{aligned}$$

About 0.83, or 83%, of American men would meet the height restriction.

Solution Using the TI-84:

a. Graph $y_1 = \dfrac{1}{2.9\sqrt{2\pi}}e^{-(x-70)^2/2(2.9)^2}$ using a window of [60, 80] for x and [0, 0.5] for y. Then press **2nd TRACE (CALC)** and select option **7:** $\int$ ***f*(*x*)*dx***. Enter 66 as the Lower Limit and press **ENTER**. Enter 74 as the Upper Limit and press **ENTER**. The calculator shades the desired region and gives an area of about 0.8322.

b. Press **MATH** and select option **9: FNINT(**. Enter the function on the screen as fnInt $(f(x), x, 66, 74)$. Press **ENTER**. The calculator returns a value of about 0.8322. ■

MATHEMATICS CORNER 25.3

Standard Normal Distribution

The standard normal distribution has a mean (average) of 0 and a standard deviation of 1. The graph is a bell-shaped curve centered around 0.

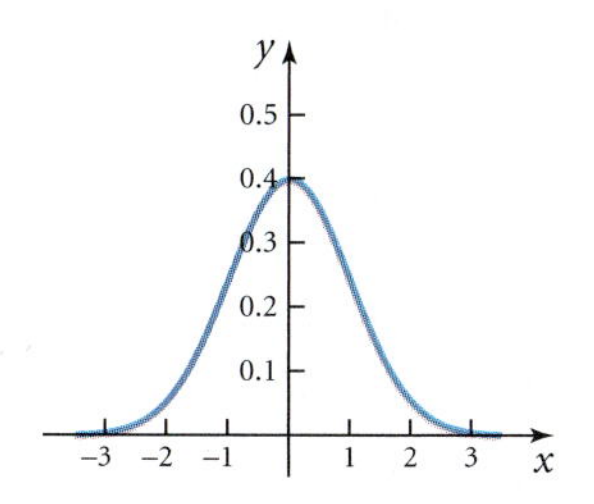

Any value of a random variable from a normal distribution can be converted to a standard score using the formula $z = \dfrac{x - \mu}{\sigma}$.

Example 8 could have been done as follows.

$$P(66 \leq x \leq 74) = P(-1.3793 \leq z \leq 1.3793)$$
$$= \int_{-1.3793}^{1.3793} \frac{1}{\sqrt{2\pi}} e^{-x^2/2} dx$$
$$= 0.83219772 \text{ or about } 0.8322$$

The obvious advantage of converting to a standard score is that the function to be integrated is slightly simpler to evaluate.

Check Your Understanding 25.3

1. Given the exponential density function $f(x) = 0.06e^{-0.06x}$, find the probability that the random variable x is in between 1 and 6. In other words, evaluate $P(1 \leq x \leq 6)$.
2. Use the information in Example 8 about heights of American men. What proportion of American men are at least 65 inches tall but not more than 72 inches tall?

Check Your Understanding Answers

Check Your Understanding 25.1

1. 0.625 **2.** 0.423 **3.** $q = 510$ and $p = \$177$ **4.** \$65,025

Check Your Understanding 25.2

1. \$9375 **2.** \$16,000

Check Your Understanding 25.3

1. 0.244 **2.** 0.712

Topic 25 Review

25

This topic presented three applications of definite integrals and area from the fields of business and economics.

CONCEPT	EXAMPLE
A **Lorenz curve** is used to describe income distribution in a population. The area between the line $y = x$, which represents ideal equality, and the Lorenz curve is the **area of inequality**.	For the Lorenz curve $L(x) = x^{2.3}$, the area of inequality is shown below. 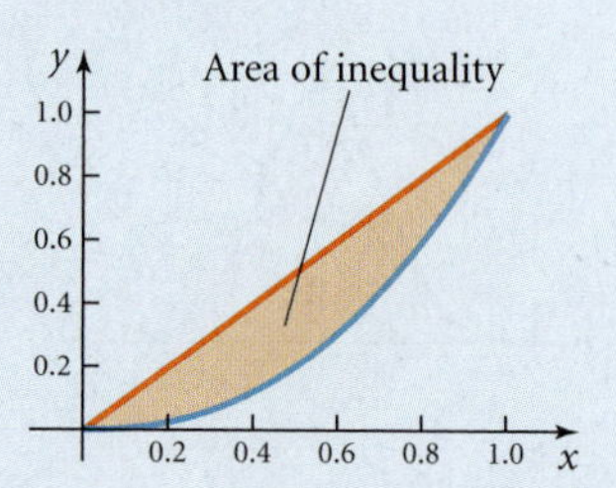
The **Gini Index**, or **Gini Coefficient**, is the ratio of the area of inequality to the total area. The Gini Index is found using $G = 2\int_0^1 \left(x - L(x)\right)dx$. The Gini Index is a number between 0 and 1. The closer the Gini Index is to 0, the more equitable the distribution. The closer the Gini Index is to 1, the more unequal the income distribution.	The Gini Index for the area of inequality described by the Lorenz curve $L(x) = x^{2.3}$ is given by $G = 2\int_0^1 (x - x^{2.3})dx = 0.39$. The Gini Index for the area of inequality described by the Lorenz curve $L(x) = x^{1.8}$ is given by $G = 2\int_0^1 (x - x^{1.8})dx = 0.286$.

(continued)

Consumer surplus is represented by the area between the demand curve $p = D(q)$ and the market price p_m and is calculated by $CS = \int_0^{q_m} (D(q) - p_m)dq$.

Demand for q hundred units of a product is $p = \frac{15}{q + 3}$, where p is the price in hundred dollars. If $p = \$250$, the consumer surplus is given by

$$CS = \int_0^3 \left(\frac{15}{q + 3} - 2.5\right)dq.$$

Producer surplus is represented by the area between the market price p_m and the supply function $p = S(q)$ and is calculated by $PS = \int_0^{q_m} (p_m - S(q))dq$.

The supply function for q hundred units of a product is $p = q^2 + 2q$, where p is the price in hundred dollars. If the market price is \$1500, the producer surplus is $PS = \int_0^3 (15 - (q^2 + 2q))dq$.

Market surplus is represented by the area between the demand and supply curves up to the equilibrium point and is calculated by $MS = CS + PS = \int_0^{q_m} (D(q) - S(q))dq$.

Demand for q hundred units of a product is $p = \frac{15}{q + 3}$, where p is the price in hundred dollars. The supply function for q hundred units of a product is $p = q^2 + 2q$, where p is the price in hundred dollars. The equilibrium point is $p \approx 3.6$ if $q \approx 1.15$. Market surplus is given by

$$MS = \int_0^{1.15} \left(\frac{15}{q + 3} - \left(q^2 + 2q\right)\right)dq.$$

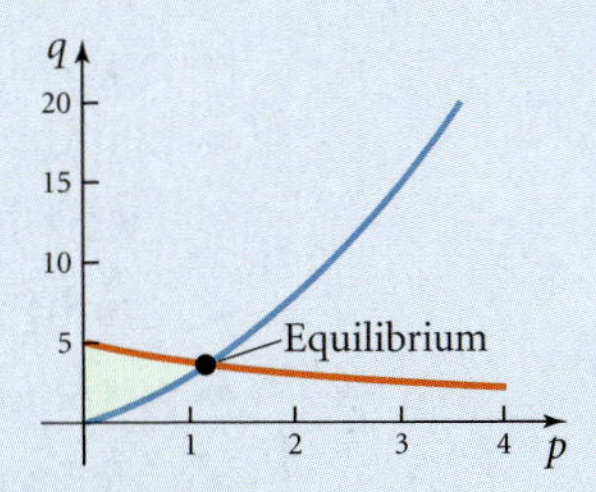

A **probability density function** of a continuous random variable x over an interval I is a function with these properties:

- $f(x) \geq 0$ for all x in the interval I.
- The total area under the graph of $y = f(x)$ is 1.
- The probability that a selected value of the random variable lies in the interval [a, b] is given by $P(a \leq x \leq b) = \int_a^b f(x)dx$.

The number of years until a new car needs repairs is given by the exponential probability distribution $f(x) = 0.25e^{-0.25x}$ on [0, 2.5].

The probability that the car needs repairs during the first six months is given by $\int_0^{0.5} 0.25e^{-0.25x}dx = 0.118$.

The probability that the car needs repairs during the first two years is given by $\int_0^2 0.25e^{-0.25x}dx = 0.393$.

(continued)

• Because the total area under the curve is 1, $0 \leq P(a \leq x \leq b) \leq 1$. Two common types of probability density functions are the **exponential probability density function** and the **normal distribution.**	The probability that the car needs repairs during the second year is given by $\int_1^2 0.25e^{-0.25x}dx = 0.172$.
The **normal distribution curve** is given by $f(x) = \frac{1}{\sigma\sqrt{2\pi}}e^{-(x-\mu)^2/2\sigma^2}$ and is widely used to determine the proportion of the distribution lying between two specific values. In this definition the symbol μ represents the mean or average of the distribution, and the symbol σ represents the standard deviation of the distribution. The graph of this curve is the bell-shaped curve.	Scores on a math placement test are normally distributed with mean of 34.2 and standard deviation of 3.7. The normal probability distribution function for scores is given by $f(x) = \frac{1}{3.7\sqrt{2\pi}}e^{-(x-34.2)^2/2(3.7)^2}$. The probability that a student scores between 30 and 40 is given by $\int_{30}^{40}\left(\frac{1}{3.7\sqrt{2\pi}}e^{-(x-34.2)^2/2(3.7)^2}\right)dx = 0.81$.

NEW TOOLS IN THE TOOL KIT

- Lorenz Curves to graphically represent income distribution
- Gini Index or Gini Coefficient to measure the inequality of income distribution
- Consumer surplus
- Producer surplus
- Market surplus
- Probability density functions and their properties
- Exponential density functions
- Normal distributions and probabilities

Topic 25 Exercises

1. Use the Lorenz curve below to answer the following questions.

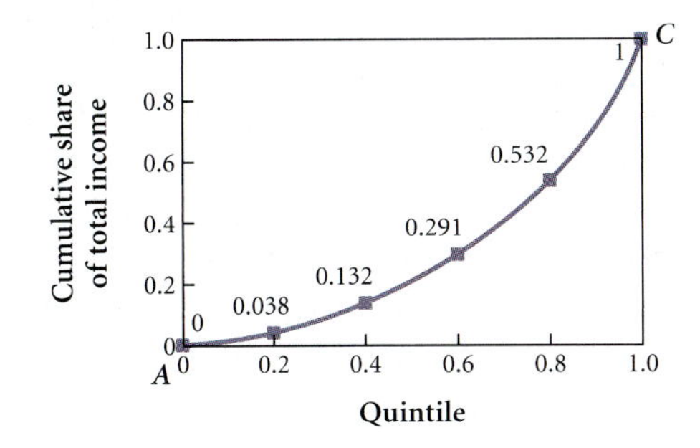

a. What percent of the total income does the top 20% of the population earn?

b. What percent of the total income does the top 40% of the population earn?

c. What percent of the total income does the bottom 40% of the population earn?

d. Describe points A and C. Why is the Lorenz curve always anchored at these two points?

2. Use the graph to answer the following questions.

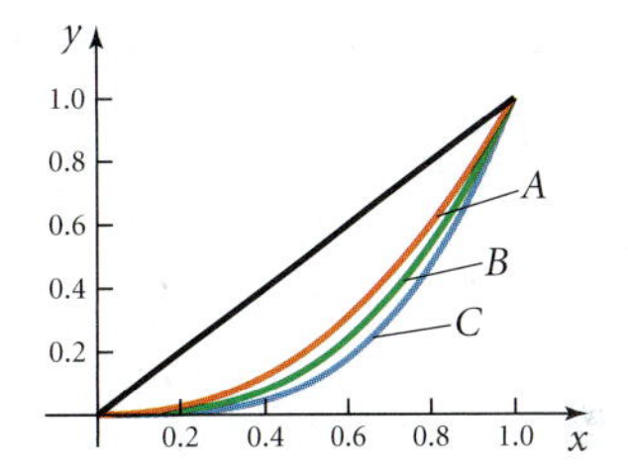

a. Which curve has the most equal income distribution?

b. Which curve has the least equal income distribution?

3. The following table gives the percentage of total income earned by three professions within a certain society. Use the table to answer the following questions.

Income Quintile	Accountants	Lawyers	Doctors
Lowest 20%	5%	7%	2%
Second 20%	8%	10%	6%
Middle 20%	10%	25%	9%
Fourth 20%	20%	25%	19%
Highest 20%	57%	33%	64%
Lorenz Curve	$y = 0.72x^{1.757}$	$y = 0.96x^{1.684}$	$y = 0.71x^{2.306}$

a. Draw the Lorenz curve for each profession.

b. Calculate the Gini Index for each profession. Round to three decimal places.

c. Which profession has the most equitable income distribution and why?

d. Which profession has the least equitable income distribution and why?

4. Refer to the graph at the right. The Gini Index is represented by which of the following?

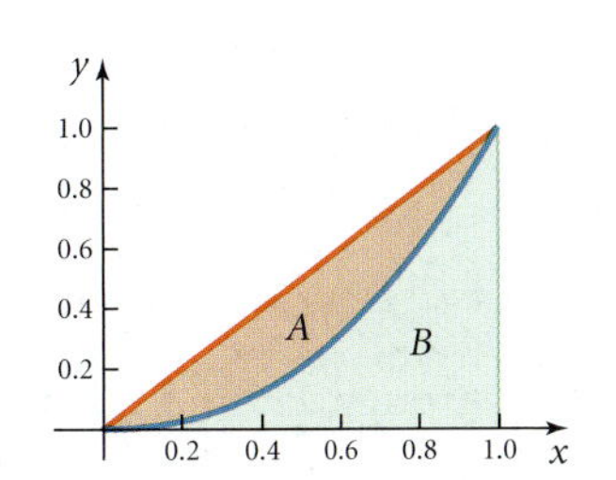

a. $\frac{A}{B}$ b. $\frac{A}{A+B}$ c. A d. $\frac{B}{A+B}$

Find the Gini Index for each of the following Lorenz curves. Round to three decimal places.

5. $L(x) = x^{3.1}$ **6.** $L(x) = x^{2.7}$

7. $L(x) = 0.37x^{1.9}$ **8.** $L(x) = 0.58x^{2.03}$

9. $L(x) = 0.2x^2 + 0.8x$ **10.** $L(x) = 0.6x^2 + 0.4x$

11. Income distribution in Bangladesh is given by $y = 0.892x^{1.44}$. Income distribution in India is given by $y = 1.2x^2 - 0.2x$. Find the Gini Index for each country. Which country has the most equitable income distribution?

12. Income distribution in Mexico is given by $y = 1.8x^2 - 0.8x$. Income distribution in India is given by $y = 1.2x^2 - 0.2x$. Find the Gini Index for each country. Which country has the most equitable income distribution?

13. A country's current income distribution is given by $y = 0.766x^{1.9}$. The country is proposing sweeping tax reforms that are intended to improve the income distribution. Income distribution after the tax reforms is supposed to be $y = 1.5x^2 - 0.5x$.

a. Calculate the Gini Index for income distribution before and after the tax reforms.

b. Based on the Gini Index, did the reforms improve income distribution? Explain your answer.

14. Income distribution for professional athletes is given by $y = 0.876x^{2.15}$, and income distribution for public school classroom teachers is given by $y = 0.7x^2 + 0.3x$.

a. Calculate the Gini Index for each profession.

b. Based on the Gini Index, for which profession is income most equally distributed? Explain your answer.

In Exercises 15 through 18 find the consumer surplus for the given demand curve at the indicated value.

15. $p = D(q) = 8 - 0.007q$ if $p = \$6.60$

16. $p = D(q) = 17.5 - 0.001q$ if $p = \$16.70$

17. $p = D(q) = 18e^{-0.003q}$ if $q = 650$

18. $p = D(q) = 400 - 10\sqrt{q}$ if $q = 400$

In Exercises 19 through 22, find the producer surplus for the given supply curve at the indicated value.

19. $p = S(q) = 0.005q + 2.5$ if $p = \$5$

20. $p = S(q) = 0.23q + 1.75$ if $p = \$42$

21. $p = S(q) = (0.1q - 2)^2$ if $q = 150$

22. $p = S(q) = 50\ln(0.2q)$ if $q = 720$

23. Supply and demand for a product are given by $p = S(q) = 0.005q + 2.5$ and $p = D(q) = 4 - 0.0075q$. Find the market surplus for 120 units of the product.

24. Supply and demand for a product are given by $p = S(q) = 5\sqrt{q} + 12$ and $p = D(q) = \frac{6200}{q + 1}$. Find the market surplus for 100 units of the product.

25. Market supply and demand, in hundreds, for cases of bottled water are given by $p = S(q) = 4q - 1.5$ and $p = D(q) = 10.5 - 2q$. Equilibrium price is \$6.50 per case.

a. Find the market surplus up to equilibrium using the integral definition.

b. Verify the market surplus by calculating $MS = CS + PS$.

26. Market supply and demand for a new luxury car are given by $p = S(q) = 2q + 8$ and $p = D(q) = 80 - 1.6q$, where p is the price in thousand dollars, and equilibrium price is \$48,000.

a. Find the market surplus up to equilibrium using the integral definition.

b. Verify the market surplus by calculating $MS = CS + PS$.

27. Market supply and demand for a new computer are given by $p = S(q) = 0.02q + 1.015$ and $p = D(q) = 20e^{-0.002q}$, where $0 \leq q \leq 1000$ and p is in hundred dollars.

a. Sketch the supply and demand curves on the same set of axes.

b. Determine the equilibrium point.

c. Find the consumer surplus up to equilibrium.

d. Find the producer surplus up to equilibrium.

e. Find the market surplus up to equilibrium.

28. Market supply and demand for a new video game are given by $p = S(q) = 4\sqrt{q} + 10$ and $p = D(q) = 100e^{-0.008q}$, where $0 \leq q \leq 300$.

a. Sketch the supply and demand curves on the same set of axes.

b. Determine the equilibrium point.

c. Find the consumer surplus up to equilibrium.

d. Find the producer surplus up to equilibrium.

e. Find the market surplus up to equilibrium.

29. Waiting time in the drive-through lane at Benny's Burgers has the exponential probability density function $p(x) = 0.12e^{-0.12x}$, where x is the number of minutes the customer waits in line before placing an order. What is the probability that a customer will wait in line between two and five minutes?

30. The number of months before a new car needs repairs has the exponential probability density function $p(x) = 0.05e^{-0.05x}$, where x is the age of the car in months. What is the probability that a new car will need repairs during the first year?

31. Benny's Burgers (see Exercise 29) claims that customers never wait in the drive-through line more than five minutes to place an order. What is the probability that a customer waits in line more than five minutes? Is Benny's claim justified?

32. Refer to Exercise 30. What is the probability that a new car needs repairs after the second year?

In Exercises 33 through 40, use the standard normal distribution $y = \frac{1}{\sqrt{2\pi}}e^{-\frac{x^2}{2}}$ to determine the probabilities. Round answers to four decimal places.

33. $P(-1 \leq x \leq 1)$ **34.** $P(0 \leq x \leq 2)$

35. $P(-3 \leq x \leq 0)$ **36.** $P(-2 \leq x \leq 3)$

37. $P(-1.4 \leq x \leq 2.6)$

38. $P(-3.1 \leq x \leq -0.4)$

39. $P(0.62 \leq x \leq 1.84)$

40. $P(-1.47 \leq x \leq 2.03)$

41. SAT scores are normally distributed with $\mu = 1026$ and $\sigma = 180$. What is the probability that a test taker scored between 1000 and 1300?

42. ACT scores are normally distributed with $\mu = 24.6$ and $\sigma = 3.4$. What is the probability that a test taker scored between 18 and 26?

43. A person's intellectual ability is measured by the Intelligence Quotient, or IQ. IQs are calculated by $IQ = \frac{MA}{CA} \cdot 100$, where MA is the person's mental age and CA is the person's chronological age. IQs are normally distributed with $\mu = 100$ and $\sigma = 15$. What proportion of people have IQs between 80 and 120?

44. The quantity of liquid in a 1-liter bottle of water is normally distributed with $\mu = 1$ liter and $\sigma = 0.03$ liter. What proportion of 1-liter bottles have between 0.95 and 1.05 liters of liquid?

45. Heights of American women are normally distributed with $\mu = 64.5$ inches and $\sigma = 2.75$ inches. To be a Rockette at the Radio City Music Hall in New York City, a woman must be between 68 and 72 inches tall. What proportion of American women qualify, based strictly on height?

46. Most players in the National Basketball Association are between 72 and 84 inches tall. Heights of American men are normally distributed with $\mu = 70$ inches and $\sigma = 2.9$ inches. What proportion of American men are between 72 and 84 inches tall?

SHARPEN THE TOOLS

EMPIRICAL RULE

Given the standard normal distribution $y = \frac{1}{\sqrt{2\pi}}e^{-\frac{x^2}{2}}$, the Empirical Rule is often used to estimate the proportion of a distribution that lies within certain standard deviations of the mean. The Empirical Rule states that

- About 68% of a normal distribution lies within one standard deviation of the mean.
- About 95% of a normal distribution lies within two standard deviations of the mean.
- About 99.7% of a normal distribution lies within three standard deviations of the mean.

Use the Empirical Rule to answer Exercises 47 through 52.

47. Verify that $P(-1 \leq x \leq 1) \approx 68\%$.

48. Verify that $P(-2 \leq x \leq 2) \approx 95\%$.

49. Verify that $P(-3 \leq x \leq 3) \approx 99.7\%$.

50. Verify that $P(-2 \leq x \leq 1) = P(-1 \leq x \leq 2) \approx 81.5\%$.

51. Any value that is more than two deviations from the mean is considered to be unusual. (A woman who is six feet tall is unusual, because her height is 2.7 deviations above the average female height of 64.5 inches.) What proportion of a normal distribution falls in this category?

52. Any value that is more than three deviations from the mean is considered an outlier. (Shaquille O'Neal, at 7′ 6″, is an outlier for men's height, because his height is more than six deviations above the average male height of 70 inches.) What proportion of a normal distribution falls in this category?

PROPERTIES OF NORMAL DISTRIBUTIONS

Property 1: The normal distribution curve is symmetric about its mean, meaning that the areas given by $P(-a \leq x \leq \mu)$ and $P(\mu \leq x \leq a)$ are the same.

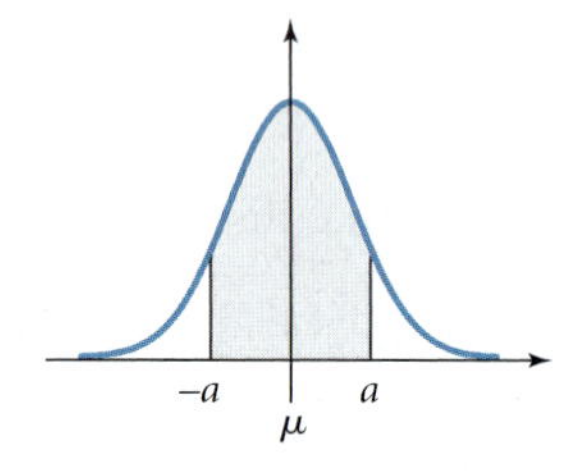

Property 2: Knowing that the total area under the curve is 1, the area outside $P(a \leq x \leq b)$ is given by $1 - P(a \leq x \leq b)$.

Property 3: The area to the left of $x = a$ is denoted as $P(x < a)$, and the area to the right of $x = b$ is denoted as $P(x > b)$. Thus, we have $P(x < a) + P(x > b) = 1 - P(a \leq x \leq b)$.

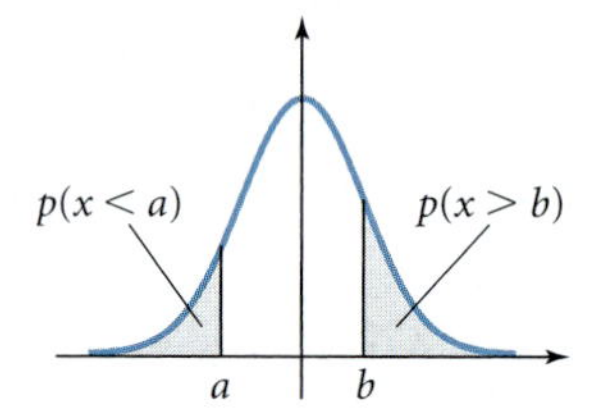

Use these three properties to answer the following questions about IQ scores, which are normally distributed with $\mu = 100$ and $\sigma = 15$.

53. Find $P(80 \leq x \leq 120)$. Use the result and the above properties to find

a. $P(x < 80)$

b. $P(x > 120)$

54. MENSA requires an IQ above 130 to be considered for membership. What proportion of adults would qualify?

CALCULATOR CONNECTION

The table below gives the income distribution by quintile for the United States in selected years.

Quintile	1947	1959	1968	1979	1988	1990	1992	1995	2001
0.2	5	4.9	4.2	4.2	3.8	3.9	3.8	3.7	3.5
0.4	11.8	12.3	11.1	10.3	9.6	9.6	9.4	9.1	8.8
0.6	17	17.9	17.5	16.9	16	15.9	15.8	15.2	14.5
0.8	23.1	23.8	24.4	24.6	24.3	24	24.2	23.3	23.1
1	43.1	41.1	42.8	44	46.3	47.6	46.8	48.7	50.1

Use the Power Regression feature of your calculator to find the Gini Index for the years in Exercises 55 through 62.

55. 1947 **56.** 1959 **57.** 1968

58. 1979 **59.** 1988 **60.** 1990

61. 1992 **62.** 1995

63. For which year was the income distribution in the United States most equitable?

64. For which year was the income distribution in the United States least equitable?

65. It is often said that "the rich get richer and the poor get poorer." Do the Gini Indexes support this statement? Explain your answer.

66. John Galbreath, a noted economist, stated in 2002 that the United States is getting richer faster than any other nation and that the wealth gap has steadily worsened since the 1970s. Do the Gini Indexes support this statement? Explain your answer.

The table below gives the income distribution by quintile for three selected countries. Use the table for Exercises 67 through 70.

Quintile	India	Czechoslovakia	Mexico
0.2	8.5	10.5	4.1
0.4	12.1	13.9	7.8
0.6	15.8	16.9	12.5
0.8	21.1	21.3	20.2
1	42.5	37.4	55.4

67. Use the table to answer these questions.

a. Use the regression feature of your calculator to find a power regression function to model income distribution for each country.

b. Calculate the Gini Index for each country.

c. Which country has the most equitable distribution?

d. Which country has the least equitable distribution?

68. Use the table to answer these questions.

a. Use the regression feature of your calculator to find a quadratic regression function to model income distribution for each country.

b. Calculate the Gini Index for each country.

c. Which country has the most equitable distribution?

d. Which country has the least equitable distribution?

69. Use the table to answer these questions.

a. Use the regression feature of your calculator to find a quartic regression function to model income distribution for each country.

b. Calculate the Gini Index for each country.

c. Which country has the most equitable distribution?

d. Which country has the least equitable distribution?

70. Use the table to answer these questions.

a. Use the regression feature of your calculator to find a cubic regression function to model income distribution for each country.

b. Calculate the Gini Index for each country.

c. Which country has the most equitable distribution?

d. Which country has the least equitable distribution?

Use the normal distribution curve $y = \frac{1}{\sigma\sqrt{2\pi}}e^{\frac{-(x-\mu)^2}{2\sigma^2}}$ for Exercises 71 and 72.

71. SAT scores are normally distributed with $\mu = 1026$ and $\sigma = 180$. Estimate the interval of scores earned by the middle 50% of the distribution. In other words, find the scores a and b such that $P(a \leq x \leq b) = 0.50$, where $p(a \leq x \leq \mu) = p(\mu \leq x \leq b) = 0.25$.

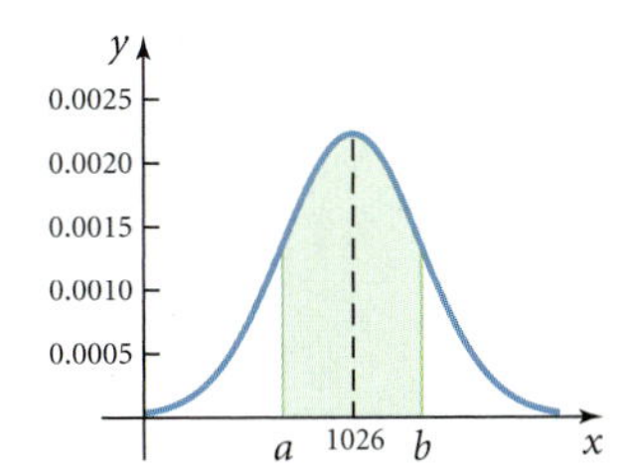

72. Heights of American men are normally distributed with $\mu = 70$ inches and $\sigma = 2.9$ inches. Estimate the height interval for the middle 60% of American men.

World distribution of income for two selected years is given in this table. Use this table for Exercises 73 through 78.

Cumulative Percentage of World Population	Cumulative Percentage of World Income, 1988	Cumulative Percentage of World Income, 1993
10	0.9%	0.8%
20	2.3%	2%
50	9.6%	8.5%
75	25.9%	22.3%
85	41%	37.1%
90	53.1%	49.2%
95	69.8%	66.3%
99	91.7%	91.5%

73. Use your calculator to find a power regression function for world distribution of income in 1988.

74. Use your calculator to find a power regression function for world distribution of income in 1993.

75. Calculate the Gini Index for world income distribution in 1988.

76. Calculate the Gini Index for world income distribution in 1993.

77. The Gini Index for world income distribution in 1979 was 0.662, and the Gini Index for world income distribution in 2000 was 0.637. Use the Gini Index you calculated in Exercise 75 to briefly describe how the world income distribution has changed since 1979.

78. The Gini Index for world income distribution in 1979 was 0.662, and the Gini Index for world income distribution in 2000 was 0.637. Use the Gini Index you calculated in Exercise 76 to briefly describe how the world income distribution has changed since 1979.

TOPIC

Other Methods of Integration

26

You now have a rather extensive set of integration techniques in your Tool Kit. However, consider how you might find these integrals.

- $\int x \ln x dx$
- $\int x^2 e^x dx$
- $\int \frac{2}{\sqrt{9 + x^2}} dx$

You should be able to determine that none of the integration techniques you have learned can be applied to these integrals. In this topic you will learn some additional methods for integration. Before starting this topic, be sure to complete the Warm-up Exercises to sharpen the algebra and calculus skills you will need.

IMPORTANT TOOLS IN TOPIC 26

- *Integration by parts*
- *Continuous income flow*
- *Tables of integration*

TOPIC 26 WARM-UP EXERCISES

Calculus Warm-up Exercises

Use an appropriate integration technique to find each of the following integrals.

1. $\int e^{3x} dx$ **2.** $\int \frac{4}{x} dx$

3. $\int \frac{4}{x^2} dx$ **4.** $\int (x - 3)^2 dx$

5. $\int_1^4 6\sqrt{x} dx$ **6.** $\int_0^2 xe^{x^2} dx$

Find the differential for each of the following.

7. $u = e^{-3x}$ **8.** $v = \ln(2x)$

9. $u = x^2 + 4x - 3$ **10.** $v = \frac{3}{x^4}$

Given the differential, integrate both sides to find v.

11. $dv = e^{5x} dx$ **12.** $dv = x^2 dx$

Evaluate each integral using your calculator's function integrator. Write answers correct to three decimal places.

13. $\int_5^8 \frac{x}{(x - 4)^2} dx$ **14.** $\int_1^2 x^2 e^x dx$

Answers to Warm-up Exercises

1. $\frac{1}{3}e^{3x} + C$ **2.** $4 \ln x + C$ **3.** $-\frac{4}{x} + C$

4. $\frac{1}{3}(x - 3)^3 + C$ using substitution; $\frac{1}{3}x^3 - 3x^2 + 9x + C$ by expanding and using the Power Rule

5. 28

6. $\frac{1}{2}(e^4 - 1) \approx 26.799$

7. $du = -3e^{-3x}dx$

8. $dv = \frac{1}{x}dx$

9. $du = (2x + 4)dx$

10. $dv = -\frac{12}{x^5}dx$

11. $v = \frac{1}{5}e^{5x}$

12. $v = \frac{1}{3}x^3$

13. 4.386

14. 12.060

The integration techniques you have learned so far are not universal—they cannot be applied to any integral. In fact, there are numerous integration techniques, many of which are beyond the scope of this book. In this topic you will learn one more useful method of integration, called **integration by parts**. You will also learn how to use a table of integrals to evaluate integrals not easily evaluated by the other methods presented in this unit.

Integration by Parts

One factor that complicates integration is the lack of a product or quotient rule for integration. Integration by parts is, in some sense, a product rule for integration. It is based on the idea of breaking apart the integrand and replacing it with another integrand that is equivalent yet hopefully simpler to integrate.

Integration by Parts

If $u(x)$ and $v(x)$ are differentiable functions, then $\int u\,dv = uv - \int v\,du$.

MATHEMATICS CORNER 26.1

Proof of Integration by Parts

Given $u(x)$ and $v(x)$, the differentials are $du = u'dx$ and $dv = v'dx$.

According to the Product Rule for Differentiation,

$$\frac{d}{dx}(uv) = u'v + uv'$$

Next integrate each side with respect to x.

$$\int\left(\frac{d}{dx}(uv)\right)dx = \int(u'v + uv')dx$$

Because differentiation and integration are inverse operations, the left side of the equation is uv. We apply the Sum Property to the right side of the equation.

$$uv = \int u'v\,dx + \int uv'\,dx$$

Now simplify, knowing that $du = u'dx$ and $dv = v'dx$.

$$uv = \int v\,du + \int u\,dv$$

Finally, solving for $\int u\,dv$, we get $\int u\,dv = uv - \int v\,du$, which is the formula for Integration by Parts.

Example 1: Use integration by parts to find $\int x \ln x dx$.

Solution: Let $\boldsymbol{u = \ln x}$ and $\boldsymbol{dv = xdx}$. Then $\boldsymbol{du = \frac{1}{x}dx}$ and $\boldsymbol{v = \int xdx = \frac{1}{2}x^2}$. Substitute into the integration by the parts formula.

$$\int udv = \quad u\ v \quad - \int \quad v \qquad du$$
$$\downarrow\downarrow \qquad\quad \downarrow \qquad \downarrow$$
$$= (\ln x)\frac{x^2}{2} - \int\left(\frac{x^2}{2} \cdot \frac{1}{x}dx\right)$$
$$= \frac{x^2}{2}\ln x - \int \frac{x}{2}dx$$
$$= \frac{x^2}{2}\ln x - \frac{1}{4}x^2 + C$$

As you can see here, the integral obtained using integration by parts is much easier to integrate than the original integral.

> **Tip:** How do you know which factor to choose for u and which factor to choose for dv? Two general guidelines for choosing are:
>
> Choose u so that u' is simpler than u.
>
> Let dv be the more complicated factor, but one that you know how to integrate.

■

Example 2: Use integration by parts to find $\int xe^{4x}dx$.

Solution: Both x and e^{4x} are easy to integrate, but x becomes simpler when differentiated, so we will let $u = x$ and $dv = e^{4x}dx$. Then, $du = dx$ and $v = \frac{1}{4}e^{4x}$. Substituting into the integration by parts formula gives

$$\int xe^{4x}dx = x\left(\frac{1}{4}e^{4x}\right) - \int \frac{1}{4}e^{4x}dx$$
$$= \frac{1}{4}xe^{4x} - \frac{1}{16}e^{4x} + C$$

■

Some integrals may require repeated applications of integration by parts, as you will see in Example 3.

Example 3: Use integration by parts to find $\int x^2 e^x dx$.

Solution: Let $u = x^2$ and $dv = e^x dx$.
Then $du = 2xdx$ and $v = e^x$.
Substituting into the integration by parts formula gives

$$\int x^2 e^x dx = x^2 e^x - \int 2xe^x dx \qquad \textbf{Equation 1}$$

The integral that we must now integrate, $\int 2xe^x dx$, is somewhat simpler than the original but still requires integration by parts to simplify. Using integration by parts again, this time we have $u = 2x$ and $dv = e^x dx$. Then, $du = 2dx$ and $v = e^x$. Thus,

$$\int 2xe^x dx = 2xe^x - \int 2e^x dx \qquad \textbf{Equation 2}$$

Now substitute the result from Equation 2 into Equation 1:

$$\begin{aligned}
\int x^2 e^x dx &= x^2 e^x - \int 2xe^x dx && \textbf{Equation 1} \\
&= x^2 e^x - \left[2xe^x - \int 2e^x dx\right] && \textbf{Substitute Equation 2} \\
&= x^2 e^x - 2xe^x + 2e^x + C && \textbf{Integrate} \\
&= e^x(x^2 - 2x + 2) + C && \textbf{Factor } e^x
\end{aligned}$$

■

Remember that the solution to an integration problem can always be checked by differentiating the answer to see if you get the original integrand. (If you don't, then you know your integration is incorrect.) Checking the solution to Example 3, we have

$$D_x(e^x(x^2 - 2x + 2) + C) = D_x(e^x(x^2 - 2x + 2)) + 0$$

Using the Product Rule,

$$\begin{aligned}
&= e^x(2x - 2) + e^x(x^2 - 2x + 2) \\
&= e^x(2x - 2 + x^2 - 2x + 2) \\
&= e^x x^2
\end{aligned}$$

which was the original integrand.

Check Your Understanding 26.1

Use integration by parts to simplify each of the following integrals. Verify your solution using differentiation.

1. $\int 4xe^{2x} dx$ **2.** $\int 4x^2 e^{2x} dx$

Integration by parts can also be used to evaluate definite integrals.

Example 4: Evaluate $\int_4^7 x\sqrt{x-3}\,dx$.

Solution: This integral can also be simplified with substitution, but here we will use integration by parts.

Let $u = x$ and $dv = \sqrt{x-3}\,dx$.

Then $du = dx$ and $v = \frac{2}{3}\sqrt{(x-3)^3}$ or $\frac{2}{3}(x-3)^{3/2}$.

Using integration by parts, we have

$$\int_4^7 x\sqrt{x-3}\,dx = x\left(\frac{2}{3}(x-3)^{3/2}\right) - \int \frac{2}{3}(x-3)^{3/2}dx \qquad \text{Integration by parts}$$

$$= \frac{2}{3}x(x-3)^{3/2} - \frac{2}{3}\cdot\frac{2}{5}(x-3)^{5/2}\Bigg|_4^7 \qquad \text{Integrate}$$

$$= \left[\frac{2}{3}(7)(4)^{3/2} - \frac{4}{15}(4)^{5/2}\right] - \left[\frac{2}{3}(4)(1)^{3/2} - \frac{4}{15}(1)^{5/2}\right] \qquad \text{Fundamental Theorem of Calculus}$$

$$= \left(\frac{14}{3}(8) - \frac{4}{15}(32)\right) - \left(\frac{8}{3} - \frac{4}{15}\right) \qquad \text{Simplify}$$

$$= \frac{132}{5} = 26\frac{2}{5} = 26.4$$

■

Continuous Income Flow

Suppose a steady flow of money is being put into an account that earns interest over a specific time period. The **future value** of the income flow is the total amount of money accumulated over the time period (the money transferred into the account plus the interest earned). Definite integrals are used to determine the future value.

Future Value

Let $f(t)$ be the rate at which money flows into an account. Let r be the annual interest rate, which is compounded continuously. Let T be the time period, with $0 \le t \le T$. The **future value** of the income flow after T years is given by

$$FV = e^{rT}\int_0^T f(t)e^{-rt}dt$$

If the money flows in at a constant rate, then the integral to be evaluated is of the form $\int_0^T ke^{-rt}dt$,x which is easily integrated with methods currently in the Tool Kit.

Example 5: Money is transferred into an account at a rate of \$2500 per year. The account earns 6% annual interest, compounded continuously. What is the value of the account after three years?

Solution: In this example we have $f(t) = 2500$, $r = 0.06$, and $T = 3$. The future value is

$$
\begin{aligned}
FV &= e^{0.06(3)} \int_0^3 2500e^{-0.06t}dt \\
&= e^{0.18}\left(\frac{2500}{-0.06}e^{-0.06t}\right)\Bigg|_0^3 \\
&= e^{0.18}\left(\frac{2500}{-0.06}\right)(e^{-0.18} - e^0) \\
&\approx \$8217.39
\end{aligned}
$$

The future value of the account after three years is about \$8,217.39. ■

If the money flows into the account at a variable rate, then integration by parts will be needed to evaluate the integral and determine the future value of the account.

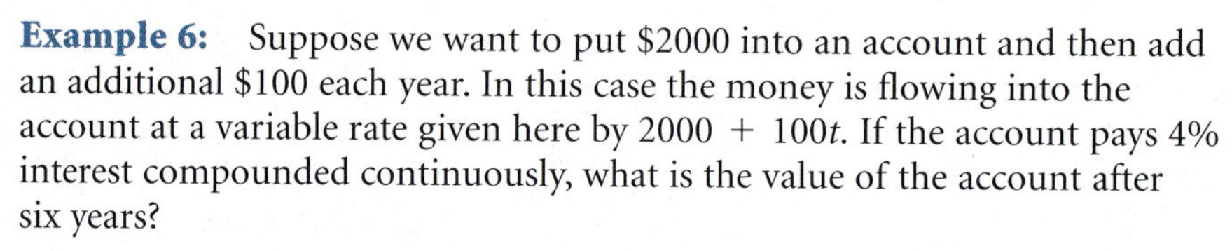

Example 6: Suppose we want to put \$2000 into an account and then add an additional \$100 each year. In this case the money is flowing into the account at a variable rate given here by $2000 + 100t$. If the account pays 4% interest compounded continuously, what is the value of the account after six years?

Solution: In this example we have $f(t) = 2000 + 100t$, $r = 0.04$, and $T = 6$. The future value is given by

$$FV = e^{0.04(6)} \int_0^6 (2000 + 100t)e^{-0.04t}dt$$

Now we must use integration by parts to evaluate the integral.
Let $u = 2000 + 100t$ and $dv = e^{-0.04t}dt$.

Then $du = 100dt$ and $v = -\frac{1}{0.04}e^{-0.04t} = -25e^{-0.04t}$.

Using integration by parts we have

$$
\begin{aligned}
FV &= e^{0.04(6)} \int_0^6 (2000 + 100t)e^{-0.04t}dt \\
&= e^{0.24}\left[(2000 + 100t)(-25e^{-0.04t}) - \int -25e^{-0.04t}(100)dt\right]_0^6 \\
&= e^{0.24}\left[(2000 + 100t)(-25e^{-0.04t}) - \frac{2500}{0.04}e^{-0.04t}\right]_0^6 \\
&= e^{0.24}[(2600(-25e^{-0.24}) - 62{,}500e^{-0.24}) - (2000(-25) - 62{,}500)] \\
&= e^{0.24}(12{,}204.95) \\
&\approx \$15{,}515.53
\end{aligned}
$$

The future value of the account after six years is about \$15,515.53. ■

Check Your Understanding 26.2

Evaluate each integral using integration by parts.

1. $\int_0^1 (xe^{2x})dx$ **2.** $\int_1^4 x(x - 2)^6 dx$

3. Repeat Example 6 if $f(t) = 500 + 50t$ and the time period is 10 years.

Integration Using Integral Tables

Most of the integrals you will encounter can be handled with the techniques now in your Tool Kit. However, you may run into an integral that cannot be evaluated using one of these methods. When that happens, you may wish to consult a **table of integrals**.

The integral tables provide a list of the most common integral forms, which have been evaluated for you using integration by parts or some other integration technique. The table provides you with a formula with which you can then easily evaluate the integral by making a few substitutions.

A brief table of integrals follows on the next pages of this book and in Appendix B. The table has been separated into sections according to the form of the integrand to make it easier for you to decide which formula to choose. You will see sections for linear forms $ax + b$, the radical forms $\sqrt{ax + b}$ and $\sqrt{x^2 \pm a^2}$, and forms involving e^x and $\ln x$. The a and b in the formulas are constants, and you will also see n used to denote powers.

To use the table, you must first determine the form of the integrand to be evaluated. Scan through the formulas in that section to find the one that matches that integrand. Then substitute the appropriate values of a, b, and n into the formula to determine the value of the integral.

A Brief Table of Integrals

Basic Formulas

1. $\int x^n dx = \frac{1}{n + 1}x^{n+1} + C$ where $n \neq -1$

2. $\int \frac{1}{x} dx = \ln|x| + C$

3. $\int e^x dx = e^x + C$

4. $\int e^{kx} dx = \frac{1}{k}e^{kx} + C$

(continued)

A Brief Table of Integrals *(continued)*

Integration by Parts Formula

5. $\int u\,dv = uv - \int v\,du$

Forms Involving $ax + b$

6. $\int \frac{x}{ax+b}dx = \frac{x}{a} - \frac{b}{a^2}\ln|ax+b| + C$

7. $\int \frac{1}{(ax+b)(cx+d)}dx = \frac{1}{ad-bc}\ln\left|\frac{ax+b}{cx+d}\right| + C$

8. $\int \frac{x}{(ax+b)(cx+d)}dx = \frac{1}{ad-bc}\left(\frac{d}{c}\ln|cx+d| - \frac{b}{a}\ln|ax+b|\right) + C$

9. $\int \frac{1}{x^2(ax+b)}dx = -\frac{1}{b}\left(\frac{1}{x} + \frac{a}{b}\ln\left|\frac{x}{ax+b}\right|\right) + C$

Forms Involving $\sqrt{ax + b}$

10. $\int \frac{x}{\sqrt{ax+b}}dx = \frac{2ax-4b}{3a^2}\sqrt{ax+b} + C$

11. $\int \frac{1}{x\sqrt{ax+b}}dx = \frac{1}{\sqrt{b}}\ln\left|\frac{\sqrt{ax+b}-\sqrt{b}}{\sqrt{ax+b}+\sqrt{b}}\right| + C$

Forms Involving $\sqrt{a^2 + x^2}$

12. $\int \sqrt{a^2+x^2}\,dx = \frac{x}{2}\sqrt{a^2+x^2} + \frac{a^2}{2}\ln\left|x + \sqrt{a^2+x^2}\right| + C$

13. $\int \frac{1}{\sqrt{a^2+x^2}}dx = \ln\left|x + \sqrt{a^2+x^2}\right| + C$

14. $\int \frac{1}{x\sqrt{a^2+x^2}}dx = -\frac{1}{a}\ln\left|\frac{\sqrt{a^2+x^2}+a}{x}\right| + C$

15. $\int x^2\sqrt{a^2+x^2}\,dx = \frac{x}{8}(a^2+2x^2)\sqrt{a^2+x^2} - \frac{a^4}{8}\ln\left|x + \sqrt{a^2+x^2}\right| + C$

Forms Involving $\sqrt{a^2 - x^2}$

16. $\int \frac{1}{x\sqrt{a^2-x^2}}dx = -\frac{1}{a}\ln\left|\frac{a+\sqrt{a^2-x^2}}{x}\right| + C$

17. $\int \frac{1}{x^2\sqrt{a^2-x^2}}dx = \frac{-\sqrt{a^2-x^2}}{a^2x} + C$

18. $\int \frac{1}{a^2-x^2}dx = \frac{1}{2a}\ln\left|\frac{a+x}{a-x}\right| + C$

19. $\int \frac{\sqrt{a^2-x^2}}{x}dx = \sqrt{a^2-x^2} - a\ln\left|\frac{a+\sqrt{a^2-x^2}}{x}\right| + C$

A Brief Table of Integrals *(continued)*

Forms Involving $\sqrt{x^2 - a^2}$

20. $\int \sqrt{x^2 - a^2}\,dx = \frac{x}{2}\sqrt{x^2 - a^2} - \frac{a^2}{2}\ln\left|x + \sqrt{x^2 - a^2}\right| + C$

21. $\int \frac{\sqrt{x^2 - a^2}}{x^2}\,dx = -\frac{\sqrt{x^2 - a^2}}{x} + \ln\left|x + \sqrt{x^2 - a^2}\right| + C$

22. $\int \frac{1}{\sqrt{x^2 - a^2}}\,dx = \ln\left|x + \sqrt{x^2 - a^2}\right| + C$

23. $\int \frac{1}{x^2\sqrt{x^2 - a^2}}\,dx = \frac{\sqrt{x^2 - a^2}}{a^2 x} + C$

24. $\int \frac{1}{x^2 - a^2}\,dx = \frac{1}{2a}\ln\left|\frac{x - a}{x + a}\right| + C$

Forms Involving e^{ax} and $\ln x$

25. $\int xe^{ax}\,dx = \frac{1}{a^2}(ax - 1)e^{ax} + C$

26. $\int \ln x\,dx = x\ln|x| - x + C$

27. $\int \frac{1}{x\ln x}\,dx = \ln|\ln x| + C$

28. $\int x^n \ln x\,dx = \frac{x^{n+1}}{n+1}\left(\ln|x| - \frac{1}{n+1}\right) + C$ where $n \neq -1$

Reduction Formulas

29. $\int x^n e^{ax}\,dx = \frac{1}{a}x^n e^{ax} - \frac{n}{a}\int x^{n-1}e^{ax}\,dx + C$

30. $\int (\ln x)^n\,dx = x(\ln x)^n - n\int (\ln x)^{n-1}\,dx + C$ where $n \neq -1$

31. $\int x^n\sqrt{a + bx}\,dx = \frac{2}{b(2n+3)}\left[x^n(a + bx)^{3/2} - na\int x^{n-1}\sqrt{a + bx}\,dx\right] + C$

where $n \neq -\frac{3}{2}$

Example 7: Find $\int \frac{x}{3x + 2}\,dx$.

Solution: The denominator of this integrand involves the linear form $ax + b$; the numerator is x. Look at the section of the Table of Integrals headed "Forms Involving $ax + b$" to find the formula whose integrand matches the integrand in the example. Do you see that Formula 6 has the same form as this integrand?

Formula 6:

$$\int \frac{x}{ax + b}\,dx = \frac{x}{a} - \frac{b}{a^2}\ln|ax + b| + C$$

To evaluate the integral, use Formula 6 with $a = 3$ and $b = 2$.

$$\int \frac{x}{3x + 2}\,dx = \frac{x}{3} - \frac{2}{9}\ln|3x + 2| + C$$ ■

Example 8: Find $\int \frac{2}{\sqrt{9 + x^2}}\,dx$.

Solution: The denominator of the integrand involves the form $\sqrt{a^2 + x^2}$; the numerator is a constant. Look at the section of the Table of Integrals headed "Forms Involving $\sqrt{a^2 + x^2}$" to find the formula whose integrand matches the integrand in the example. You should see that Formula 13 has the same form as this integrand.

Formula 13:

$$\int \frac{1}{\sqrt{a^2 + x^2}}\,dx = \ln\left|x + \sqrt{a^2 + x^2}\right| + C$$

To evaluate the integral, use Formula 13 with $a = 3$. The constant in the numerator will be factored out and treated as a constant multiple.

$$\int \frac{2}{\sqrt{9 + x^2}}\,dx = 2\left(\ln\left|x + \sqrt{9 + x^2}\right|\right) + C$$

$$= 2\ln\left|x + \sqrt{9 + x^2}\right| + C$$ ■

Example 9: Find $\int x^4 \ln x\,dx$.

Solution: The integrand involves $\ln x$. Use Formula 28 with $n = 4$.

Formula 28:

$$\int x^n \ln x\,dx = \frac{x^{n+1}}{n + 1}\left(\ln x - \frac{1}{n + 1}\right) + C \quad \text{where } n \neq -1$$

$$\int x^4 \ln x\,dx = \frac{x^5}{5}\left(\ln x - \frac{1}{5}\right) + C$$ ■

MATHEMATICS CORNER 26.2  

Forms Involving $\sqrt{a^2 + bx^2}$

Suppose that the integral in Example 8 had been $\int \frac{2}{\sqrt{9 + 4x^2}}dx$.

Formula 13 can still be used, but the radicand must be simplified so that the quadratic coefficient is 1.

$$\int \frac{2}{\sqrt{9 + 4x^2}}dx = \int \frac{2}{\sqrt{4\left(\frac{9}{4} + x^2\right)}}dx$$

$$= \int \frac{2}{2\sqrt{\frac{9}{4} + x^2}}dx$$

$$= \int \frac{1}{\sqrt{\frac{9}{4} + x^2}}dx$$

Now use Formula 13 with $a^2 = \frac{9}{4}$ and $a = \frac{3}{2}$.

$$= \ln\left|x + \sqrt{\frac{9}{4} + x^2}\right| + C$$

This same technique works with any of the radicands involving radical expressions.

Example 10: Find $\int (\ln x)^2 dx$.

Solution: This integrand involves $\ln x$ raised to a power, so we must use Reduction Formula 30 with $n = 2$. **Reduction formulas** reduce the integrand to a simpler integrand, by reducing the exponent. The resulting integrand may or may not require the use of another formula, depending on the integrand.

Formula 30:

$$\int (\ln x)^n dx = x(\ln x)^n - n\int (\ln x)^{n-1}dx + C \quad \text{where } n \neq -1$$

Using Formula 30 with $n = 2$:

$$\int (\ln x)^2 dx = x(\ln x)^2 - 2\int \ln x\, dx + C$$

To integrate $\int \ln x dx$, use Formula 26.

Formula 26:

$$\int \ln x dx = x \ln x - x + C$$

$$= x(\ln x)^2 - 2(x \ln x - x) + C$$

Check Your Understanding 26.3

Use the integral tables to find each integral.

1. $\int \frac{x}{\sqrt{2x + 5}} dx$ **2.** $\int \frac{\sqrt{x^2 - 16}}{x^2} dx$

3. $\int x^2 e^{3x} dx$

Check Your Understanding Answers

Check Your Understanding 26.1

1. $2xe^{2x} - e^{2x} + C$

2. $e^{2x}(2x^2 - 2x + 1) + C$

Check Your Understanding 26.2

1. $\frac{1}{4}e^2 + \frac{1}{4} \approx 2.097$ **2.** $\frac{3849}{56} \approx 68.732$

3. $9017.33

Check Your Understanding 26.3

1. $\frac{x - 5}{3}\sqrt{2x + 5} + C$

2. $-\frac{\sqrt{x^2 - 16}}{x} + \ln\left|x + \sqrt{x^2 - 16}\right| + C$

3. $e^{3x}\left(\frac{1}{3}x^2 - \frac{2}{9}x + \frac{2}{27}\right) + C$

Topic 26 Review

26

This topic presented two other techniques of integration—integration by parts and tables of integrals. An application to continuous income flow was also discussed.

CONCEPT	EXAMPLE
Integration by parts rewrites the integrand into an integral that is equivalent and simpler to integrate. The formula for integration by parts is $$\int udv = uv - \int vdu$$ To decide which factor is u and which factor is dv, consider the following: **1.** Choose u so that u' is simpler than u. **2.** Let dv be the more complicated factor, but one that you know how to integrate.	To integrate $\int 2xe^{3x}dx$ using integration by parts, let $u = 2x$ and $dv = e^{3x}dx$. Then $du = 2dx$ and $v = \frac{1}{3}e^{3x}$. Then by integration by parts we have $$\int 2xe^{3x}dx = 2x\left(\frac{1}{3}e^{3x}\right) - \int \frac{1}{3}e^{3x}(2dx) = \frac{2}{3}xe^{3x} - \frac{2}{3}\left(\frac{1}{3}e^{3x}\right) + C = \frac{2}{3}xe^{3x} - \frac{2}{9}e^{3x} + C$$
Integration by parts can be used to evaluate both indefinite and definite integrals.	Evaluate $\int_0^1 3x(x + 2)^2dx$. Let $u = 3x$ and $dv = (x + 2)^2dx$. Then $du = 3dx$ and $v = \frac{1}{3}\left(x + 2\right)^3$. Then $$\int_0^1 3x(x + 2)^2dx = 3x\left(\frac{1}{3}(x + 2)^3\right)\Big\vert_0^1 - \int_0^1 \frac{1}{3}(x + 2)^3 3dx = \left[x(x + 2)^3 - \frac{1}{4}(x + 2)^4\right]_0^1 = 10.75$$

(continued)

Let $f(t)$ be the rate at which money flows into an account. Let r be the annual interest rate, which is compounded continuously. Let T be the time period, with $0 \leq t \leq T$. The **future value** of the income flow after T years is given by $$FV = e^{rT}\int_0^T f(t)e^{-rt}dt$$	Money is transferred into an account at a rate of \$3000 per year, with an additional \$200 added each year. The account earns 4% annual interest compounded continuously. After three years, the value of the account is given by $$FV = e^{(0.04)(3)}\int_0^3 (3000 + 200t)e^{-0.04t}dt \approx \$10{,}499.37$$
For integrals that cannot be evaluated using any of the techniques in the Tool Kit, use a **table of integrals**. Tables of integrals are arranged by the form of the integrand.	The integrals $\int\sqrt{9 + x^2}dx$ and $\int_1^2 x\ln x\,dx$ must be integrated using the Table of Integrals.

NEW TOOLS IN THE TOOL KIT

- Using integration by parts as an integration technique for both indefinite and definite integrals
- Calculating continuous income flow
- Using the table of integrals to evaluate an integral

Topic 26 Exercises

26

Use integration by parts to evaluate each of the following integrals.

1. $\int xe^{-2x}dx$
2. $\int 3xe^{3x}dx$
3. $\int \ln x\,dx$
4. $\int (\ln x)^2dx$
5. $\int a\ln(3a)da$
6. $\int a^2\ln(4a)da$
7. $\int b\sqrt{b - 2}\,db$
8. $\int \frac{4p}{\sqrt{p + 2}}dp$
9. $\int x^3\ln x\,dx$
10. $\int x^4\ln x\,dx$
11. $\int x^2e^x dx$
12. $\int x\ln x^3dx$
13. $\int 3x(x - 2)^4dx$
14. $\int x(x + 1)^6dx$
15. $\int \frac{x}{e^x}dx$
16. $\int \frac{\ln x}{x^3}dx$
17. $\int_1^4 x^2\ln\sqrt{x}\,dx$
18. $\int_0^3 \frac{x^2}{e^x}dx$

19. $\int_0^2 \sqrt{2x^2 + 1}\,dx$ **20.** $\int_1^3 \sqrt{3x^2 - 2}dx$

21. $\int_2^5 x^2 \ln x\,dx$ **22.** $\int_1^4 x \ln \sqrt{x}\,dx$

Find the area enclosed by the x-axis and the given function on the indicated interval.

23. $f(x) = \dfrac{2 \ln x}{x^3}$ on the interval $[1, e]$

24. $f(x) = x^3 e^{-x}$ on the interval $[0, 1]$

Simplify each of the following integrals using (a) integration by parts and (b) substitution. Which method is easier to use and why?

25. $\int x\sqrt{x + 8}\,dx$ **26.** $\int \dfrac{3x}{(x - 2)^4}dx$

Use the Table of Integrals to simplify each of the following integrals.

27. $\int \dfrac{4x}{\sqrt{2x - 3}}dx$ **28.** $\int \dfrac{3}{\sqrt{x^2 + 16}}dx$

29. $\int \dfrac{x}{(3x - 2)(5x + 7)}dx$

30. $\int \dfrac{7}{16 - x^2}dx$ **31.** $\int \sqrt{x^2 + 16}\,dx$

32. $\int \dfrac{6}{x \ln x}dx$ **33.** $\int \dfrac{2}{x\sqrt{9 - x^2}}dx$

34. $\int \dfrac{4}{x\sqrt{5x + 3}}dx$ **35.** $\int_3^4 \dfrac{6}{x \ln x}dx$

36. $\int_5^{10} \sqrt{x^2 - 25}\,dx$ **37.** $\int_1^2 \dfrac{1}{x\sqrt{9 - x^2}}dx$

38. $\int_1^4 x^2 \ln x dx$ **39.** $\int_0^2 x^3 e^{2x} dx$

40. $\int_2^5 (\ln x)^3 dx$ **41.** $\int x^2 e^{3x} dx$

42. $\int (\ln x)^2 dx$

43. Find the equation of the curve passing through $(1, 2)$ if the slope of the tangent line is $\dfrac{dy}{dx} = (x - 1)e^{-2x}$.

44. Find the equation of the curve passing through $(4, -2)$ if the slope of the tangent line is $\dfrac{dy}{dx} = 2x \ln \sqrt{x}$.

45. If x thousand units are produced and sold, the marginal profit in thousand dollars is given by $P'(x) = \dfrac{400 \ln(x + 1)}{(x + 1)^2}$. Find the total profit over $0 \le x \le 5$. Assume that $P(0) = 0$.

46. If q units of a product are produced, the marginal cost is $C'(q) = (0.2q + 3)e^{0.02q}$ dollars per unit. The total cost of 10 units is \$400. Find the total cost of the first 25 units.

47. After a drug is administered, the concentration of the drug in the patient's bloodstream is approximated by $C(t) = 1.2te^{-0.2t}$ mg/cm^3 per hour. Find the average concentration of the drug during the first four hours after it was injected.

48. The population of a city is approximated by $P(t) = 0.5t^2 e^{-0.01t^2}$ thousand people, where t is the number of years after 2000. What was the average population of the city from 2005 to 2008?

49. Money is transferred into an account at a rate of $200 + 25t$ dollars per year. The account earns 6% annual interest, compounded continuously. What is the value of the account after four years?

50. Money is transferred into an account at a rate of $500 + 100t$ dollars per year. The account earns 5% annual interest, compounded continuously. What is the value of the account after three years?

51. Money is transferred into an account at a rate of $1200te^{-0.25t}$ dollars per year. The account earns 4% annual interest, compounded continuously. What is the value of the account after five years?

52. Money is transferred into an account at a rate of $250te^{-0.3t}$ dollars per year. The account earns 5% annual interest, compounded continuously. What is the value of the account after 10 years?

CALCULATOR CONNECTION

Use your calculator to evaluate each integral. Verify your solution using either integration by parts or an appropriate formula from the Table of Integrals. Round answers to three decimal places.

53. $\int_1^3 x^5 \ln x^2 dx$ **54.** $\int_2^5 \frac{x^2}{\ln(x^2)} dx$

55. $\int_1^2 \frac{3}{(2x - 1)(5x + 4)} dx$

56. $\int_1^3 3x^2\sqrt{2x^2 + 7}\, dx$

SHARPEN THE TOOLS

Use integration by parts to simplify each integral.

57. $\int x(\ln x)^3 dx$ **58.** $\int x^2(\ln x)^2 dx$

59. $\int_0^2 x^2\sqrt{4x + 1}\, dx$ **60.** $\int_0^1 x^2\sqrt{2x^2 + 5}dx$

61. $\int x^3\sqrt{x + 2}\, dx$ **62.** $\int x^4 e^{-x} dx$

Use integration by parts to prove each of the following formulas from the Table of Integrals.

63. Formula 18 **64.** Formula 21

65. Formula 29 **66.** Formula 30

67. Jamal is putting money aside for his newborn son's college fund. He has two options.

Option 1: Tenth National Bank offers an account paying 6% compounded continuously. Money will flow into the account with an initial investment of $1200 and annual increases of $150 per year.

Option 2: Midtown Savings and Loan offers an account paying 7% compounded continuously with money flowing into the account at a constant rate of $2000 per year.

Which account provides Jamal a larger amount of money after 18 years for his son's college education?

68. Mario is 30 years old and is setting up a retirement account. He has two plans to consider.

Option 1: An account is offered paying 4.5% interest compounded continuously with money flowing in at a rate of $200 per month ($2400 per year).

Option 2: An account is offered paying 4% interest compounded continuously with money flowing in at a continuous rate of $100 per month with $10 monthly increases each year ($1200 + $120*t*).

Which plan will provide Mario with the largest amount of money when he retires at age 65?

TOPIC

Differential Equations

27

Newton's Law of Cooling states that *the rate at which an object cools is proportional to the difference between the temperature of the object and the temperature of the medium in which the object is placed.* The exponential Law of Growth states that *the rate at which a population grows is proportional to the population at any time.* You have already learned the model that describes the temperature, T, of an object at any time t after being placed in a medium of temperature $M(T(t) = M + (P + M)e^{kt})$ and the model that describes the size of a population at any time ($P(t) = P_0e^{kt}$). In this topic you will learn about differential equations, which explain how those models were derived. Before starting your study of differential equations, be sure to complete the Warm-up Exercises to sharpen the algebra and calculus skills you will need.

IMPORTANT TOOLS IN TOPIC 27

- *Differential equations*
- *General and particular solutions*
- *Separation of variables*
- *Implicit and explicit solutions*

TOPIC 27 WARM-UP EXERCISES

Be sure you can successfully complete the following exercises before starting Topic 27.

Algebra Warm-up Exercises

Use the Laws of Exponents to simplify each of the following expressions.

1. $x^4 \cdot x^3$ **2.** $(x^4)^3$ **3.** $e^x \cdot e^2$

4. Write as a product: e^{x+y}.

5. Find the exponential function $y = Ce^{kx}$ that passes through the points (0, 2) and (4, 4).

Simplify each expression using the Properties of Logarithms.

6. $\ln x + \ln y$

7. $5 \ln x$

8. $\ln x^2 - \ln y^3$

9. $2 \ln x + \ln y - 3 \ln z$

Are the following given in explicit form or implicit form?

10. $y = \sqrt{x + 1}$ **11.** $x^2 + y^2 = 16$

12. $4x^3y^2 = 16$ **13.** $y = e^{-2x}$

Calculus Warm-up Exercises

Find each of the following integrals.

1. $\int (x^2 - 3x)dx$

2. $\int e^{5x}\,dx$ **3.** $\int \frac{4}{x}dx$

4. $\int (x^2 - 3x)dx$ if $y(6) = 20$

5. Demand for a new product is given by $D(p) = -0.2p^3 + 90p + 43$, where p is the price. Find the elasticity of demand for $p = \$5$.

Answers to Warm-up Exercises

Algebra Warm-up Exercises

1. x^7 **2.** x^{12}

3. e^{x+2} **4.** $e^x \cdot e^y$

5. $y = 2e^{(\ln 2/4)x}$ or $y = 2e^{0.1733x}$

6. $\ln(xy)$ **7.** $\ln x^5$

8. $\ln \dfrac{x^2}{y^3}$ **9.** $\ln \dfrac{x^2 y}{z^3}$

10. explicit **11.** implicit

12. implicit **13.** explicit

Calculus Warm-up Exercises

1. $\frac{1}{3}x^3 - \frac{3}{2}x^2 + C$ **2.** $\frac{1}{5}e^{5x} + C$

3. $4\ln x + C$ **4.** $\frac{1}{3}x^3 - \frac{3}{2}x^2 + 2$

5. $E = \left|\dfrac{-p \cdot D'(p)}{D(p)}\right| = 0.8$

Differential Equations

In Topic 12 you learned the exponential Law of Growth model $P(t) = P_0e^{kt}$. The actual theory behind this model, developed by Thomas Malthus, says that "the rate at which a population grows is proportional to the population at any time." If $P(t)$ denotes the population at any time t, then the Law of Growth can be written as $\dfrac{dP}{dt} = kP$.

Equations of this type are known as **differential equations**, because they involve the derivative of some function that is yet to be determined. Differential equations are important in modeling and occur in a number of applications in business and economics as well as in the social and life sciences.

A **differential equation** is an equation involving an unknown function and one or more of its derivatives. The following are examples of differential equations.

$$\frac{dy}{dx} = e^{2x} \qquad y' = x^2 - 3x + 2 \qquad (y')^2 - 3y' = 0 \qquad y'' = 2x^2y^3$$

The **solution** of a differential is the unknown function that is determined to satisfy the equation.

The **general solution** is the family of all possible solutions to the differential equation.

A **particular solution** is a member of the general solution that satisfies specific given conditions called initial conditions.

The simplest differential equations are those of the form $y' = g(x)$, which are easily solved by integrating each side of the equation with respect to the independent variable.

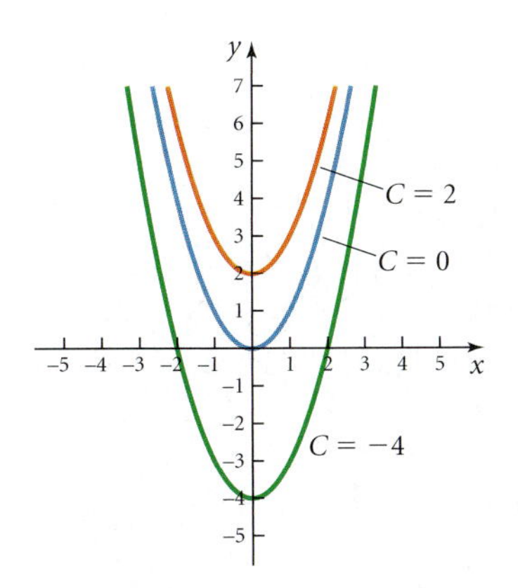

Figure 27.1

Example 1: Solve $y' = 2x$.

Solution: Integrate each side with respect to x:

$$\int y'dx = \int 2xdx$$
$$y = x^2 + C$$

The general solution is $y = x^2 + C$, which is a family of parabolas. Three members of this family are shown in Figure 27.1. ■

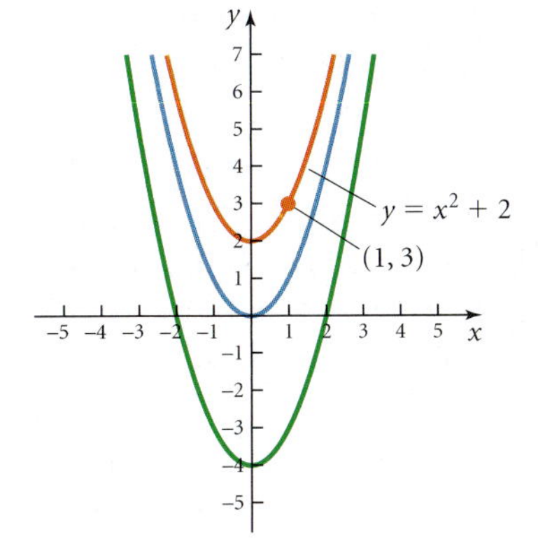

Figure 27.2

Example 2: Solve $y' = 2x$ given that $y(1) = 3$.

Solution: This example is an initial value problem, because we must find the solution to a differential equation subject to the initial condition, $y(1) = 3$. In other words, we are seeking the equation that satisfies the differential equation and whose graph passes through the point $(1, 3)$.

First, find the *general solution*:

$$\int y'dx = \int 2x\,dx$$
$$y = x^2 + C$$

To find the particular solution, we know that $y(1) = 3$.

$$y = x^2 + C \quad \text{but} \quad y(1) = 3, \text{so } 3 = 1^2 + C, \text{which gives } C = 2$$

The *particular solution* (the only member of the general solution that passes through the point $(1, 3)$) is $y = x^2 + 2$. See Figure 27.2. ■

Example 3: Solve $\dfrac{dy}{dx} = 3x^2 + 6x - 5$. Find the particular solution that passes through the point $(-1, 2)$.

Solution: $y' = 3x^2 + 6x - 5$, so

$$\int y'dx = \int (3x^2 + 6x - 5)\,dx$$
$$y = x^3 + 3x^2 - 5x + C$$

If $y(-1) = 2$, then $2 = (-1)^3 + 3(-1)^2 - 5(-1) + C$ and $C = -5$. The solution is $y = x^3 + 3x^2 - 5x - 5$. ■

Calculator Corner 27.1

Draw the graph of $y = x^3 + 3x^2 - 5x - 5$. Verify that the graph passes through the point $(-1, 2)$ by pressing **2nd TRACE (CALC)** and selecting option **1:VALUE**. Type -1 for the x value and press **ENTER**.

Check Your Understanding 27.1

Find the general solution of each differential equation.

1. $y' = 2x^3$
2. $y' = 3e^{-2x}$
3. Find the particular solution of $y' = \dfrac{2}{x^3}$ given that $y(1) = 4$.

Separation of Variables

Some differential equations can be solved by a method called **separation of variables.** This method rewrites the equation so that each side is a function of only one of the variables. Then each side is integrated with respect to that variable.

Separation of Variables

To solve $\dfrac{dy}{dx} = \dfrac{f(x)}{g(y)}$:

1. Rewrite the equation—separate the variables—as $g(y)dy = f(x)dx$.
2. Integrate each side of the equation: $\int g(y)dy = \int f(x)dx$.

Example 4: Solve $\dfrac{dy}{dx} = 2xy$ and verify the solution.

Solution: Separate the variables by multiplying by dx and dividing by y.

$$\frac{1}{y}dy = 2xdx$$

Integrate each side.

$$\int \frac{1}{y}dy = \int 2xdx$$

$$\ln y = x^2 + C, \text{ where } y > 0$$

Solve for y.

$$y = e^{x^2+C}$$

Apply Laws of Exponents.

$$y = e^{x^2} \cdot e^C$$

It may not be immediately apparent, but the factor e^C represents a constant, so we can write the solution as

$$y = C_1 e^{x^2}, \text{ where } C_1 = e^C$$

To verify that $y = C_1e^{x^2}$ is a solution of $\frac{dy}{dx} = 2xy$, find the derivative of the solution and then substitute that solution and its derivative into the differential equation.

$$y = C_1e^{x^2}, \text{ so}$$

$$y' = C_1e^{x^2}(2x)$$

Substituting $y = C_1e^{x^2}$ into $\frac{dy}{dx} = 2xy$, we have $\frac{dy}{dx} = 2x\left(C_1e^{x^2}\right)$.

It should be apparent that both results are equal: $C_1e^{x^2}(2x) = 2x\left(C_1e^{x^2}\right)$. ■

In the next example, we will derive the exponential Law of Growth model using separation of variables.

Example 5: The rate at which a population grows is proportional to the population at any time t. If $P(0) = P_0$ is the initial population, find a formula for the population at any time.

Solution:

$$\underbrace{\textit{The rate at which } p\ (t) \textit{ grows}}_{\frac{dP}{dt}} \quad \underbrace{\textit{is proportional to}}_{=k} \quad \underbrace{\textit{the population at any time}}_{P}$$

k is called the constant of proportionality, which refers to the growth rate of the population. This constant will appear in the solution we are about to find. We now solve $\frac{dP}{dt} = kP$ by separation of variables.

Separate variables.

$$\frac{1}{P}dP = k\,dt$$

Integrate both sides.

$$\int \frac{1}{P}dP = \int k\,dt$$

Solve for P.

$$\ln P = kt + C$$
$$P = e^{kt+C}$$

Simplify.

$$P = e^{kt} \cdot e^{C} = C_1e^{kt}$$

But we know that $P(0) = P_0$, so $C_1 = P_0$ and therefore

$$P(t) = P_0e^{kt}$$

which is the exponential Law of Growth. ■

You learned in Topic 17 that equations can be expressed explicitly or implicitly. To refresh your memory, $y = x^3 - 5x + e^x$ is in explicit form $y = f(x)$, but $x^3y + \frac{x}{2y} = 0$ is in implicit form $F(x, y) = C$. Implicit form means that the equation has not been solved for y but the assumption is that the equation could be solved for y.

Solutions to differential equations may also be expressed in **explicit form** $(y = f(x))$ or in **implicit form** $(F(x, y) = C)$.

Example 6: Solve $\frac{dy}{dx} = \frac{2x}{y}$.

Solution: Separate variables.

$$y\,dy = 2x\,dx$$

Integrate both sides.

$$\int y\,dy = \int 2x\,dx$$

$$\frac{1}{2}y^2 = x^2 + C$$

$$\text{or} \quad y^2 = 2x^2 + C_1, \quad \text{where } C_1 = 2C$$

The **implicit** solution is $y^2 - 2x^2 = C_1$.

The **explicit** solution is $y = \pm\sqrt{2x^2 + C_1}$.

The implicit solution is a family of hyperbolas, but the equation is not readily graphed. The explicit solution is also a family of hyperbolas but is in a form that is more easily graphed. Both solutions are equivalent and yield the same graph, so the choice is merely one of simplicity or ease of computation. (See Figure 27.3.) ■

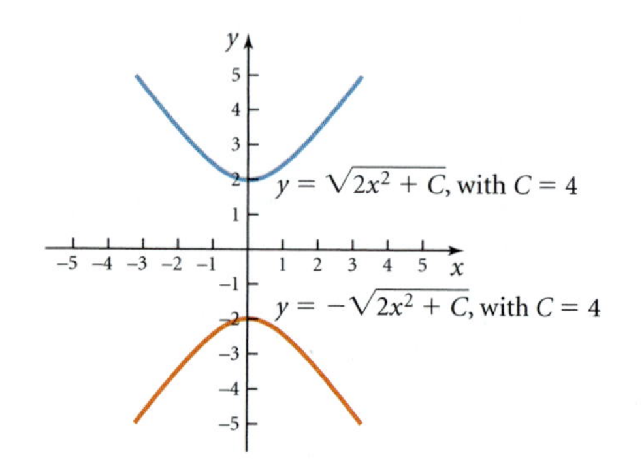

Figure 27.3

Example 7: Solve $y^2y' + 4x = 0$ if $y(0) = 3$.

Solution: Write y' as $\frac{dy}{dx}$.

$$y^2\frac{dy}{dx} + 4x = 0$$

Separate variables.

$$y^2dy = -4x\,dx$$

Integrate both sides.

$$\int y^2dy = \int -4x\,dx$$

$$\frac{1}{3}y^3 = -2x^2 + C$$

Simplify **implicitly**:

$$y^3 = -6x^2 + C_1 \quad \text{where } C_1 = 3C$$
$$6x^2 + y^3 = C_1$$

Substitute the initial condition $y(0) = 3$:

$$6(0)^2 + 3^3 = C_1$$
$$C_1 = 27$$

Implicit solution: $6x^2 + y^3 = 27$

Simplify **explicitly**:

$$y^3 = -6x^2 + C_1 \text{ where } C_1 = 3C$$
$$y = \sqrt[3]{-6x^2 + C_1}$$

$$3 = \sqrt[3]{-6(0)^2 + C_1}$$
$$3 = \sqrt[3]{C_1} \text{ so } C_1 = 27$$

Explicit solution: $y = \sqrt[3]{-6x^2 + 27}$

Check Your Understanding 27.2

1. Solve by separation of variables. Write the answer both implicitly and explicitly.

$$\frac{dy}{dx} = \frac{4x}{y}$$

2. Solve by separation of variables.

$$\frac{dy}{dx} = 6x^2y \text{ if } y(0) = 3$$

Applications

Example 8: The rate at which a rumor spreads through a college dormitory is proportional to the number of people who haven't heard the rumor. Express the number of people who have heard the rumor as a function of time.

Solution: Let $N(t)$ represent the number in the dormitory who have heard the rumor and let B denote the total number of people in the dormitory.

The rate at which people hear the rumor is $\frac{dN}{dt}$.

The number who have not heard the rumor is $B - N$.

The differential equation to represent the rate at which the rumor spread is

$$\underbrace{\frac{dN}{dt}}_{\textit{rate at which rumor spreads}} \underbrace{=-k}_{\textit{is proportional to}} \underbrace{B - N}_{\textit{number who haven't heard}}$$

Now solve $\frac{dN}{dt} = k(B - N)$ using separation of variables to determine $N(t)$, the number who have heard the rumor at any time t.

Separate variables.

$$\frac{1}{B - N}dN = kdt$$

Integrate both sides.

$$\int \frac{1}{B-N}dN = \int kdt$$

$$-\ln(B-N) = kt + C, \text{ where } B > N$$

$$\ln(B-N) = -kt + C_1 \text{ where } C_1 = -C$$

Simplify.

$$B - N = e^{-kt+C_1} = C_2e^{-kt} \text{ where } C_2 = e^{C_1}$$

Solve for N.

$$-N = -B + C_2e^{-kt}$$

$$N = B + C_3e^{-kt}, \text{ where } C_3 = -C_2$$

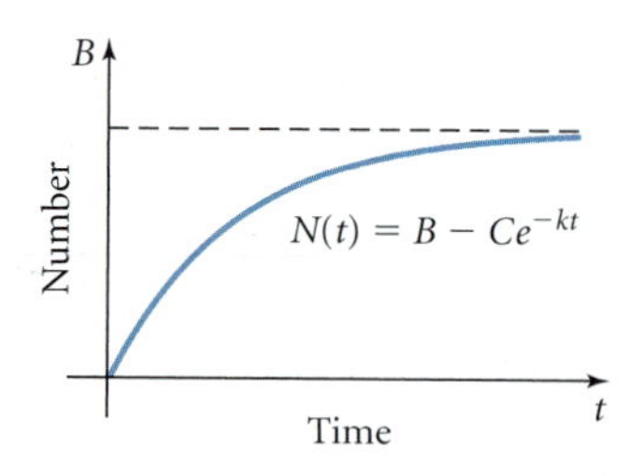

Figure 27.4

This equation is sometimes referred to as a learning curve. Graphically the total number of people, B, is the horizontal asymptote. (See Figure 27.4.) ■

Example 9: Elasticity of demand for a certain product is 0.8. Find the demand function $x = D(p)$.

Solution: Elasticity of demand is given by $E = -\frac{pD'(p)}{D(p)}$ or $E = -\frac{p}{D(p)} \cdot D'(p)$.

With $E = 0.8$ and $D(p) = x$, we have $0.8 = -\frac{p}{x} \cdot \frac{dx}{dp}$.

Separate variables.

$$\frac{0.8}{p}dp = -\frac{1}{x}dx$$

Integrate both sides.

$$\int \frac{0.8}{p}dp = -\int \frac{1}{x}dx$$

$$0.8\ln p = -\ln x + C$$

Simplify.

$$\ln x + 0.8\ln p = C$$

$$\ln x + \ln p^{0.8} = C$$

$$\ln(xp^{0.8}) = C$$

Solve for x.

$$xp^{0.8} = e^C = C_1$$

$$x = \frac{C_1}{p^{0.8}}$$

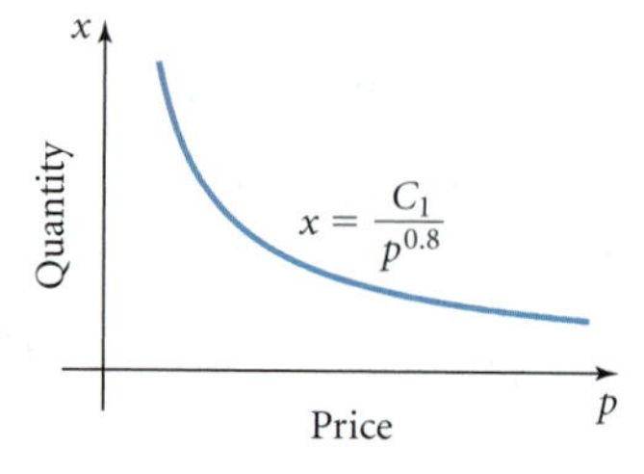

Figure 27.5

The graph in Figure 27.5 shows that as the price of the product increases, the demand for the product decreases. ■

Calculator Corner 27.2

For the demand function $x = \frac{C_1}{p^{0.8}}$, how do various values of the constant affect the graph?

Graph $y_1 = \frac{5}{x^{0.8}}$, $y_2 = \frac{10}{x^{0.8}}$, and $y_3 = \frac{20}{x^{0.8}}$. Use a window of [0, 10] for both x and y.

Example 10: The population of the United States was 282.2 million in 2000 and 296.4 million in 2005. Assume that the Law of Growth applies.

a. Find a formula for the population of the United States at any time after 2000.

b. Predict the population of the United States in 2010.

c. Estimate when the population of the United States hit 300 million.

Solution:

a. We assume that the Law of Growth applies, so we know that the solution is of the form $P(t) = P_0e^{kt}$. Letting P_0 represent the population in 2000, we can then write $P(t) = 282.2e^{kt}$.

Next we must find the value of k, the constant of proportionality. Using the fact that the population in 2005 was 296.4 million, we know that $P(5) = 296.4$. Thus, $P(5) = 282.2e^{k(5)} = 296.4$.

Now we solve for k.

$$282.2e^{k(5)} = 296.4$$

$$e^{k(5)} = \frac{296.4}{282.2} \qquad \text{Divide both sides by 282.2}$$

$$k(5) = \ln\left(\frac{296.4}{282.2}\right) \qquad \text{Write as logarithmic equation}$$

$$k = \frac{\ln\left(\frac{296.4}{282.2}\right)}{5} \approx 0.00982 \qquad \text{Solve for } k \text{ by dividing by 5}$$

Thus, the population of the United States at any time after 2000 can be approximated by $P(t) = 282.2e^{0.00982t}$.

b. To predict the population in 2010, we must evaluate $P(10)$.

$$P(10) = 282.2e^{0.00982(10)} \approx 311.32 \text{ million people}$$

c. To estimate when the population was 300 million, we must solve

$$300 = 282.2e^{0.00982t}$$

$$e^{0.00982t} = \frac{300}{282.2}$$

$$0.00982t = \ln\left(\frac{300}{282.2}\right)$$

$$t = \frac{\ln\left(\frac{300}{282.2}\right)}{0.00982} \approx 6.2$$

The U.S. population reached 300 million in 2006 (six years after 2000). In fact, this milestone was reached in October 2006 amid much fanfare. ■

Check Your Understanding Answers

Check Your Understanding 27.1

1. $y = \frac{1}{2}x^4 + C$

2. $y = -\frac{3}{2}e^{-2x} + C$

3. $y = -\frac{1}{x^2} + 5$

Check Your Understanding 27.2

1. $4x^2 - y^2 = C$ or $y = \pm\sqrt{4x^2 + C}$

2. $y = 3e^{2x^3}$

Topic 27 Review 27

This topic introduced differential equations, a solution method called separation of variables, and several related applications.

CONCEPT	EXAMPLE
A **differential equation** is an equation involving an unknown function and one or more of its derivatives.	The equations $\frac{dy}{dx} = 3e^{-2x}$ and $y' = x^2 + 2x - 5$ are examples of differential equations.
The **solution** of a differential equations is the unknown function that is found to satisfy the equation. The **general solution** is the family of all possible solutions. A **particular solution** is a member of the family that satisfies specific initial conditions.	The general solution of $\frac{dy}{dx} = 3e^{-2x}$ is $y = -\frac{3}{2}e^{-2x} + C$. Two particular solutions are $y = -\frac{3}{2}e^{-2x}$ and $y = -\frac{3}{2}e^{-2x} + 6$.

(continued)

Differential equations of the form $\frac{dy}{dx} = f(x)$ can be solved by integrating both sides with respect to the independent variable.	To solve $\frac{dy}{dx} = 3e^{-2x}$, first write the equation as $dy = 3e^{-2x}dx$. Then integrate each side. $\int dy = \int 3e^{-2x}dx$ so $y = -\frac{3}{2}e^{-2x} + C$
Differential equations of the form $\frac{dy}{dx} = \frac{f(x)}{g(y)}$ can be solved by **separation of variables**: Rewrite the equation (separate the variables) as $g(y)dy = f(x)dx$. Integrate each side: $\int g(y)dy = \int f(x)dx$	To solve $\frac{dy}{dx} = 4x^3y$: First write the equation as $\frac{dy}{y} = 4x^3\,dx$. Then integrate both sides. $\int \frac{dy}{y} = \int 4x^3\,dx$; $\ln y = x^4 + C$ or $y = e^{x^4+C} = C_1e^{x^4}$
Solutions to differential equations may be written in **explicit form** or **implicit form.**	In the solution above to $\frac{dy}{dx} = 4x^3y$, $\ln y = x^4 + C$ is an implicit solution and $y = e^{x^4+C} = C_1e^{x^4}$ is an explicit solution.

NEW TOOLS IN THE TOOL KIT

- Recognizing differential equations
- Solving differential equations by integrating each side
- Solving differential equations by separation of variables
- Writing solutions to differential equations in explicit form and in implicit form
- Solving application problems involving differential equations

Topic 27 Exercises

27

Solve each differential equation.

1. $y' = x + 4$
2. $y' = 2 - x$
3. $y' = 3x^2 + x$
4. $y' = 4x - x^2$
5. $y' = \dfrac{4x}{x^2 + 1}$
6. $y' = xe^{-x^2}$
7. $y' = 4e^{-x/2}$
8. $y' = \dfrac{e^x}{e^x + 4}$
9. $y' = \dfrac{x + 3}{x}$
10. $y' = \dfrac{6}{x}$
11. $y' = x\sqrt{x - 3}$
12. $y' = 4x\sqrt{5 - x}$

In Exercises 13 through 16, write each of the following statements as a differential equation. Then solve the equation, assuming that $C = 0$. See Mathematics Corner 27.1 for an explanation of variation.

MATHEMATICS CORNER 27.1

Variation

There are three types of variation:

Direct variation, which says that $y = kx$.

Inverse variation, which says that $y = \dfrac{k}{x}$.

Joint variation, which says that $y = k(xw)$.

In each variation, k is called the **constant of proportionality.**

13. The rate of change of V with respect to x varies directly to the square root of x.
14. The rate of change of P with respect to t varies directly to the square root of $4 - t$.
15. The rate of change of N with respect to t is inversely proportional to t^2.
16. The rate of change of A with respect to y varies jointly with y and $3 - y$.

Solve each differential equation using separation of variables.

17. $\dfrac{dy}{dx} = y + 4$
18. $\dfrac{dy}{dx} = 2 - y$
19. $\dfrac{dy}{dx} = \dfrac{3x}{y}$
20. $\dfrac{dy}{dx} = \dfrac{x^2}{y}$
21. $\dfrac{dy}{dx} = x\sqrt{y}$
22. $\dfrac{dy}{dx} = x(1 + y)$
23. $4y\dfrac{dy}{dx} = 3e^x$
24. $x\dfrac{dy}{dx} = y\ln x$
25. $(x^2 + 1)\dfrac{dy}{dx} - 2xy = 0$
26. $xy + \dfrac{dy}{dx} = 10x$

Solve each differential equation and find the member of the general solution that passes through (1, 2).

27. $\dfrac{dy}{dx} = \dfrac{x}{4}$
28. $\dfrac{dy}{dx} = \dfrac{4}{x}$
29. $\dfrac{dy}{dx} = \dfrac{x}{y}$
30. $\dfrac{dy}{dx} = \dfrac{y}{x}$

Find the particular solution of each of the following differential equations.

31. $y(x + 1) + \dfrac{dy}{dx} = 0$ if $y(-2) = 1$
32. $2xy' - \ln x^2 = 0$ if $y(1) = 3$
33. $\dfrac{dy}{dx} = \dfrac{4x}{9y}$ if $y(-1) = 2$
34. $\dfrac{dy}{dx} = -\dfrac{3y}{2x}$ if $y(1) = 2$
35. The rate at which the population of a city grows is proportional to the population at any time. The population in 1990 was 8713, and the population in 2000 was 12,546.
    a. Find a formula for the population of the city at any time after 1990.
    b. Predict the population of the city in 2010.

36. The half-life of a radioactive element is the amount of time needed for half of the element to disintegrate. Carbon-14 is an element used in a process called carbon dating and has a half-life of 5760 years.

a. Find a formula $P(t) = P_0e^{-kt}$ for the amount of C-14 present at any time t.

b. Suppose a fossil is unearthed and through testing it is found that the fossil contains 40% of the original amount of C-14. How old is the fossil?

Newton's Law of Cooling states that *the rate at which an object cools is proportional to the difference between the temperature of the object and the temperature of the medium in which the object is placed.* If T is the temperature of the object at any time t and T_m is the temperature of the medium, then $\frac{dT}{dt} = k(T - T_m)$. Use this differential equation to solve Exercises 37 through 40.

37. Suppose you remove a cake from a 350°F oven and place it on a counter to cool. Room temperature is 75°F. After 10 minutes the temperature of the cake is 225°F.

a. Find a formula for the temperature of the cake at any time t.

b. If the cake must be 150°F in order to be iced, how long must you wait before icing the cake?

38. A casserole is taken out of a 400°F oven and placed on the table. Room temperature is 72°F. After 10 minutes the temperature of the casserole is 300°F.

a. Find a formula for the temperature of the casserole at any time t.

b. The casserole tastes best when the temperature is below 180°F. How long will it take before the casserole can be eaten?

39. Newton's Law of Cooling can also be used to measure the rate at which an object heats up. Terry takes soda from a 35°F refrigerator and places it on a picnic table in 80°F heat. After 15 minutes the temperature of the drink is 50°F.

a. Find a formula for the temperature of the drink at any time t.

b. What is the temperature of the drink after 30 minutes?

40. Cheryl takes a gallon of ice cream from a grocery store's 28°F freezer and places it in her grocery cart. The temperature of the store is 70°F. After 10 minutes the temperature of the ice cream is 40°F. For health reasons, the ice cream should be put in Cheryl's home freezer before it reaches 60°F. How long does Cheryl have to get home and put the ice cream away?

COMMUNICATE

41. Suppose $y = f(x)$ is a solution of a differential equation, $y'(x) = e^{kx}$. Explain why $y = f(x) + C$ is also a solution. Verify your answer by providing an example.

42. Suppose $y = f(x)$ is a solution of a differential equation, $y'(x) = e^{kx}$. Explain why $y = Cf(x)$ is also a solution. Verify your answer by providing an example.

SHARPEN THE TOOLS

43. Newton's Law of Cooling states that the rate at which an object cools is proportional to the difference between the temperature of the object and the temperature of the medium in which it is placed. If $T(t)$ is the temperature of the object at any time, T_m is the temperature of the surrounding medium, and the initial temperature is $T(0) = T_0$, find a formula for the temperature of the object at any time. (Solve $\frac{dT}{dt} = k(T - T_m)$.)

44. The Logistic Growth model assumes that the rate at which a population changes is proportional to the product of the population at any time and the difference between the rate at which things enter or leave the population. If $P(t)$ is the population at any time, a is the constant rate at which things enter the population, bP is the rate at which things leave the population, and $P(0) = P_0$, find a formula for the population at any time. (Solve $\frac{dP}{dt} = kP(a - bP)$.)

Two families of curves are said to be orthogonal if each member of the first family is perpendicular to each member of the second family. Each curve in the first family is said to be the **orthogonal trajectory** of each curve in the second family. The graph shows two curves that are orthogonal—the tangent lines at their point of intersection are perpendicular, meaning that the slopes of the tangent lines are negative reciprocals.

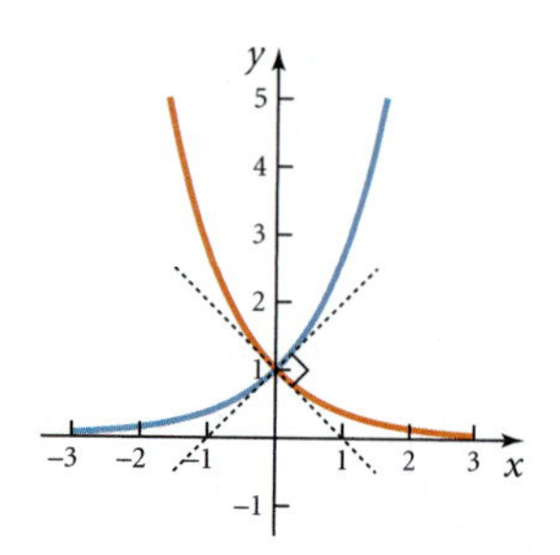

Suppose that $y_1(x)$ is the first family. To find the orthogonal trajectory,

1. Find $\frac{dy_1}{dx}$, which represents the slope of each curve in the family.
2. Find the negative reciprocal of this slope.
3. Solve $\frac{dy_2}{dx}$ = the result from step 2 to find the second family.

Suppose $y = \frac{k}{x}$ is the first family. Then $xy = k$, so by the Product Rule and implicit differentiation, we have $x\frac{dy_1}{dx} + y = 0$. Solving for $\frac{dy_1}{dx}$ gives $\frac{dy_1}{dx} = -\frac{y}{x}$, which is the slope of each curve in the family.

The orthogonal trajectory must be a curve whose slope is the negative reciprocal of the first family. The first curve's slope is given by $\frac{dy_1}{dx} = -\frac{y}{x}$. Thus, we know that $\frac{dy_2}{dx} = \frac{x}{y}$, the negative reciprocal.

Solve to find the orthogonal family.

$$\frac{dy_2}{dx} = \frac{x}{y} \text{ so } y\,dy = x\,dx$$

$$\int y\,dy = \int x\,dx$$

$$\frac{1}{2}y^2 = \frac{1}{2}x^2 + C$$

$$y^2 - x^2 = C$$

The orthogonal trajectories of $y = \frac{k}{x}$ are given by $y^2 - x^2 = C$.

Sketch one member of each family to see the orthogonality (perpendicularity). The graphs of $y = \frac{4}{x}$ and $y^2 - x^2 = 4$ are presented.

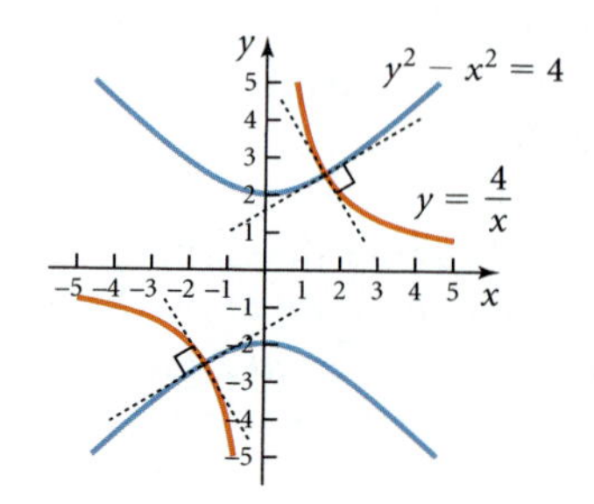

Find the orthogonal trajectory of the following families. Sketch a representative curve from each family to show the orthogonality.

45. $x^2 + y^2 = C$

46. $y^2 = kx$

47. $x^2 = ky$

48. $y = Ce^x$

CALCULATOR CONNECTION

A puppy that weighs 2 pounds at birth gains weight at a rate of $\frac{dW}{dt} = k(A - W)$ pounds per year.

49. Solve the differential equation if $k = 0.5$ and $A = 50$. Graph the solution and estimate the maximum weight of the dog.

50. Solve the differential equation if $k = 0.8$ and $A = 50$. Graph the solution and estimate the maximum weight of the dog.

51. Solve the differential equation if $k = 0.9$ and $A = 80$. Graph the solution and estimate the maximum weight of the dog.

52. Solve the differential equation if $k = 1$ and $A = 80$. Graph the solution and estimate the maximum weight of the dog.

UNIT 3

Review

Unit 3 presented the integral and several applications of the integral.

TOPIC 20	■ Antiderivatives as the reverse of derivatives ■ The integral as the symbol of antidifferentiation ■ The Power Rule for Integration ■ Properties of Integration: Constant Multiple Property, Sum Property, Difference Property ■ Indefinite integrals and families of solutions ■ Determination of the value of the unknown constant using an initial condition ■ Solving application problems to determine the value of a function at a specific point, given an initial condition
TOPIC 21	■ Integral rules with exponential functions as integrands ■ Integral rules resulting in the natural logarithm function
TOPIC 22	■ Substitution technique for integrating certain composite functions ■ Differentials
TOPIC 23	■ Definite integrals as the net change in the value of an antiderivative over an interval ■ Using the Fundamental Theorem of Calculus to evaluate definite integrals ■ Average value of a function over an interval
TOPIC 24	■ Using definite integrals to find the area enclosed by a nongeometric region ■ Estimating areas using Riemann Sums ■ Solving application problems to determine the value of a function over an interval, given its rate of change
TOPIC 25	■ Lorenz curves and the Gini Index ■ Producer, consumer, and market surplus ■ Probability density functions
TOPIC 26	■ Integration by parts ■ Continuous income flow ■ Tables of integrals
TOPIC 27	■ Differential equations ■ General and particular solutions to differential equations ■ Separation of variables to solve differential equations

Having completed Unit 3, you should now be able to

1. Simplify integrals using the Power Rule for Integration.
2. Use the Constant Multiple Property, the Sum Property, and the Difference Property to simplify integrals.
3. Determine the family of solutions to an indefinite integral.
4. Determine the value of the unknown constant given an initial condition.
5. Find the equation of a curve with a given slope passing through a specified point.
6. Use indefinite integrals to solve application problems requiring the value of a function at a specific point given the rate of change of the function.
7. Determine the height of an object in motion given its acceleration, initial velocity, and initial height.
8. Simplify integrals whose integrand is of the form e^x or e^{kx}.
9. Simplify integrals whose integrand is of the form $\frac{k}{x}$.
10. Use the substitution technique to simplify integrals of certain composite functions.
11. Calculate the future value of a continuous income flow.
12. Evaluate definite integrals using the Fundamental Theorem of Calculus.
13. Interpret definite integrals as net change.
14. Use definite integrals to determine the average value of a function on an interval.
15. Use definite integrals to determine the area enclosed by nongeometric regions—between $f(x)$ and the x-axis on an interval or between two curves.
16. Estimate the area of a nongeometric region using a Riemann Sum.
17. Use definite integrals to solve application problems requiring the value of function over an interval in its domain, given the rate of change of the function.
18. Determine whether to use an indefinite integral or a definite integral when solving an application problem.
19. Draw Lorenz curves and determine the area of inequality.
20. Calculate the Gini Index and interpret the result.
21. Find producer surplus, consumer surplus, and market surplus, and interpret the results.
22. Evaluate exponential density functions.
23. Evaluate normal probability density functions.
24. Simplify integrals using integration by parts.
25. Simplify integrals using the Table of Integrals.
26. Solve differential equations of the form $y' = f(x)$ by integrating both sides.
27. Solve differential equations of the form $\frac{dy}{dx} = \frac{f(x)}{g(y)}$ using separation of variables.
28. Find particular solutions of differential equations, given an initial condition.
29. Write solutions to differential equations in either implicit or explicit form.

UNIT 3 TEST

Use an appropriate method to find each of the following integrals.

1. $\int (x^3 - 4x + 6)\,dx$
2. $\int \frac{8}{x^3}\,dx$
3. $\int e^{-3x}\,dx$
4. $\int (x^{3/4} - x^{-4})\,dx$
5. $\int_1^9 6\sqrt{x}\,dx$
6. $\int_2^4 \left(\frac{2}{x} + x^2\right)dx$
7. $\int_0^{10} 100e^{-0.025t}\,dt$
8. $\int_1^4 \frac{x^2 - 3x + 5}{x^2}\,dx$

Find $f(x)$ such that

9. $f'(x) = 7x - \frac{2}{x^3}$ and $f(2) = 3$

10. $f'(x) = e^{0.2x} + 3$ and $f(0) = -2$

Find the equation of the curve passing through the given point with the indicated slope.

11. $\frac{dy}{dx} = x^{2/3} + 2x$; $(1, -1)$

12. $\frac{dy}{dx} = \frac{3}{x} - 4x$; $(1, 1)$

Use substitution to find each of the following integrals.

13. $\int xe^{x^2}\,dx$

14. $\int_0^1 \frac{x}{(x^2-4)^3}\,dx$

15. $\int_{-2}^{0} x^2\sqrt{1-x^3}\,dx$

16. $\int 4x^2(2x^3-5)^6\,dx$

17. $\int \frac{(\ln x)^2}{x}\,dx$

18. $\int_2^3 \frac{x^4}{x^5-1}\,dx$

Use integration by parts to evaluate each of the following integrals.

19. $\int_0^2 te^{2t}\,dt$

20. $\int x(x+3)^4\,dx$

21. $\int x^3 \ln x\,dx$

22. $\int_4^7 \frac{2m}{\sqrt{m-3}}\,dm$

Use the Table of Integrals to find each of the following integrals.

23. $\int \frac{x}{\sqrt{2x-5}}\,dx$

24. $\int \frac{x}{2x-5}\,dx$

25. $\int \frac{1}{9-x^2}\,dx$

26. $\int \frac{3}{\sqrt{x^2+4}}\,dx$

27. $\int \sqrt{x^2-9}\,dx$

28. $\int x^6 \ln x\,dx$

29. $\int \frac{6}{x\sqrt{16-x^2}}\,dx$

30. $\int x\sqrt{x+2}\,dx$

For Exercises 31 through 36, find the area of the indicated region. Sketch the graph and shade the region.

31. Between $y = 2 - x^3$ and the x-axis on $[-1, 1]$

32. Between $y = e^x - 3$ and the x-axis on $[0, 1]$

33. Between $y = x^2 - 5x - 6$ and the x-axis on $[-2, 2]$

34. Between $y = \frac{2}{x} + x$ and the x-axis on $[1, 4]$

35. Between $y = 1 - x^2$ and $y = \frac{2}{x}$ on $[1, 3]$

36. Between $y = x^2 - 5x - 6$ and $y = x - 11$

37. A college's enrollment, in thousands, since 1998 is approximated by $E(t) = 0.076t^3 - 0.528t^2 + 0.28t + 12$. Find the average annual enrollment of the college between 2000 and 2004.

38. The concentration of a drug in a patient's bloodstream, in mg/cm^3, t hours after injection is approximated by $C(t) = 2te^{-0.3t^2}$. Find the average concentration of the drug during the first six hours after its injection.

39. The rate at which a measles epidemic spreads through an elementary school is approximated by $M'(t) = 90t - 3t^2$ new cases per day. If there were 30 cases reported on the first day, how many new cases were reported on the seventh day?

40. Sales of a new best seller are growing at a rate of $N'(t) = \frac{25}{t}$ thousand books per week, where t is the number of weeks after the release of the book. If 12,700 books were sold during the first week, how many books were sold during the fourth week?

41. The rate at which a measles epidemic spreads through an elementary school is given by $M'(t) = 90t - 3t^2$ new cases per day. How many total new cases were reported between the seventh and 14th days?

42. Sales of a new best seller are growing at a rate of $N'(t) = \frac{25}{t}$ thousand books per week, where t is the number of weeks after the release of the book. How many books were sold during the second month (from the end of the fourth week to the end of the eighth week)?

43. Total box office revenue for a new movie for any week t changed at a rate of $R'(t) = -24.75te^{-0.15t^2}$ million dollars per week. Find the total box office revenue during the third week. Assume $R(1) = 71$.

44. Money is being transferred into an account at a rate of $\$5000 + 250t$ per year. If the account pays 5.5% compounded continuously, what is the future value of the account after four years?

Find the Gini Index for each of the following Lorenz curves.

45. $L(x) = 0.86x^{2.35}$

46. $L(x) = 0.75x^2 + 0.25x$

47. Income distribution for accountants is given by $A(x) = 1.375x^2 - 0.375x$. Income distribution for lawyers is given by $L(x) = 0.01x^2 + 0.283x + 0.125$. Income distribution for doctors is given by $D(x) = 0.71x^{2.31}$.

a. Find the Gini Index for all three professions.

b. Which profession has the most equitable income distribution?

c. Which profession has the least equitable income distribution?

48. Find the consumer surplus for $p = D(q) = -0.035q + 24.5$ if $p = \$18$.

49. Find the producer surplus for $p = S(q) = 12e^{0.0025q}$ if $q = 425$ units.

50. Using the supply and demand curves in Exercises 48 and 49, find the market surplus up to equilibrium.

51. Waiting time in a doctor's office has the exponential probability density function $p(x) = 0.05e^{-0.05x}$, where x is the number of minutes the patient waits for the doctor. What is the probability that a patient will wait between 15 and 30 minutes?

52. A new light bulb is advertised to have an average life of 1200 hours with a standard deviation of 15 hours. If the life of the bulb is normally distributed, what is the probability that one of these light bulbs will last between 1180 and 1210 hours?

Solve each differential equation.

53. $y' = 5 - 3x$

54. $y' = \dfrac{2x}{x^2 + 3}$

55. $y' = 4y$

56. $y' = 2x(y + 3)$

57. $xy' = y$

58. $2y^2y' = x^2 + 6$

Find the particular solution of each differential equation.

59. $y'(x + 1) = 2y$ if $y(1) = 3$

60. $xy' = \ln x^2$ if $y(1) = 2$

Chronic Care Solutions (now CCS Medical) was discussed in the Unit 3 Introduction. The table below gives the increase in the number of patients each year, in thousands, since the company's founding in 1994.

Year	Increase in Number of Patients, in Thousands
1994	0.1
1995	0.525
1996	1.375
1997	1.1
1998	5.45
1999	5.5
2000	6.1
2001	9.9
2002	12.5
2003	22.5

61. Use your calculator to find an exponential model for the growth in the number of patients, $P'(t)$. Let t represent the number of years since 1990.

62. Use the model from Exercise 61 to estimate the average annual increase in the number of new patients between 2000 and 2003.

63. Use the model from Exercise 61 to find an exponential model for the number of patients each year, $P(t)$.

64. Use the model from Exercise 63 to estimate the average annual number of patients between 2000 and 2003.

UNIT 3 PROJECT

In the Unit 0 Project, you gathered data on a topic of interest and created a best model for that data using regression. In the Unit 1 Project, you used that model to analyze the rates of change of the model. In the Unit 2 Project, you analyzed the behavior of the model and determined the extrema for the model. This project continues the analysis of the data.

Using the data and best regression model you determined in the Unit 0 Project, answer the following questions.

1. Choose two data points and determine the average value of your model over the interval between those two points. Interpret your answer.
2. State the first derivative of your model.
3. Using the first derivative and one data point as an initial condition, estimate the value of your model at another point in the domain.
4. Using the first derivative and two data points, evaluate the definite integral over that interval and interpret your answer.
5. For your data, write a word problem that would require the use of an indefinite integral to solve. Include the solution to the problem.
6. For your data, write a word problem that would require the use of a definite integral to solve. Include the solution to the problem.

Multivariable Calculus

UNIT 4

The calculus you have learned up to this point is referred to as *single-variable calculus*, because all the functions were of the form $y = f(x)$.

The dependent variable was a function of a single independent variable. But there are many situations in which the dependent variable is a function of two or more independent variables. Consider these examples.

- A factory production manager pays full-time employees \$35 per hour and part-time employees \$22 per hour. If full-time employees are used x hours per week and part-time employees are used y hours per week, then the total cost for salaries per week is given by $C = 35x + 22y$.
- On an achievement test, correct answers are worth two points but 0.25 points are deducted for each incorrect answer. A test taker's score is $S = 2r - 0.25w$, where r is the number of correct answers and w is the number of incorrect answers.

- The volume and surface area of a soda can in the shape of a right circular cylinder are given by $V = \pi r^2 h$ and $SA = 2\pi r^2 + 2\pi rh$, where r is the radius of the base of the can and h is the height of the can.
- The area of a trapezoid is $A = \frac{1}{2}h(b_1 + b_2)$, where b_1 and b_2 are the bases of the trapezoid (the two parallel sides) and h is the altitude (the perpendicular distance between the two parallel sides).

These examples are just a few of the many applications in which the quantity being represented depends on more than one other related quantity.

In this final unit we extend our study of calculus to functions involving more than one independent variable. The calculus of functions of more than one variable is called *multivariable calculus*.

TOPIC 28

Introduction to Functions of More than One Variable

A factory production manager pays his full-time employees \$35 per hour and his part-time employees \$22 per hour. If full-time employees are used for x hours a week and part-time employees are used for y hours a week, the total cost for salaries for a week is given by $C = 35x + 22y$. This function is called a function of two variables, because the total cost of salaries depends on two variables, the number of hours worked by full-time employees and the number of hours worked by part-time employees. In this topic you will learn more about functions of more than one variable. Before starting the topic, be sure you can complete the warm-up exercises to sharpen the algebra and calculus skills you will need.

IMPORTANT TOOLS IN TOPIC 28

- *Evaluating functions of more than one variable*
- *Identifying the domain of functions of more than one variable*
- *Sketching the graph of some functions of two variables*
- *Determining level curves for graphs of functions of two variables*

TOPIC 28 WARM-UP EXERCISES

Be sure you can successfully complete these exercises before starting Topic 28.

Algebra Warm-Up Exercises

State the domain of each of the following functions.

1. $f(x) = 3x + 2$

2. $f(x) = \dfrac{3}{x + 2}$

3. $f(x) = \sqrt{2x + 5}$

4. $f(x) = \ln(x - 3)$

5. $f(x) = e^{-2x}$

6. $f(x) = \dfrac{4x}{x^2 - 9}$

7. $f(x) = \dfrac{2}{\sqrt{x - 3}}$

8. $f(x) = \dfrac{4x}{\sqrt{x^2 - 9}}$

9. Given $f(x) = 2x^3 - 4x^3 + \dfrac{5}{x} - 3$, evaluate $f(2)$.

10. Evaluate $x^2y + \dfrac{3x}{y} - \sqrt{x}$ if $x = 4$ and $y = 2$.

Sketch the graph of each function.

11. $f(x) = 3x + 2$

12. $f(x) = \dfrac{3}{x - 2}$

13. $f(x) = \sqrt{2x + 5}$

14. $f(x) = e^{-2x}$

15. $f(x) = \ln(x - 3)$

Calculus Warm-up Exercises

1. The output of a factory is given by $q = 6x^{0.5}y^{0.5}$, where x is the number of employees and y is the number of machines used. Daily operating costs are \$75 per employee and \$27 per machine. Output goal is 50 units per day. How many employees and how many machines should be used to minimize daily operating cost?

2. Express $2x^2 - y^2 = 8$ in explicit form.

3. Express $y = \dfrac{4}{\sqrt{2 - x^2}}$ in implicit form.

Answers to Warm-up Exercises

Algebra Warm-up Exercises

1. all real numbers, or $(-\infty, \infty)$
2. $x \neq -2$, or $(-\infty, -2) \cup (-2, \infty)$
3. $x \geq -\frac{5}{2}$, or $[-\frac{5}{2}, \infty]$
4. $x > 3$, or $(3, \infty)$
5. all real numbers, or $(-\infty, \infty)$
6. $x \neq 3, -3$, or $(-\infty, -3) \cup (-3, 3) \cup (3, \infty)$
7. $x > 3$, or $(3, \infty)$
8. $x < -3, x > 3$, or $(-\infty, -3) \cup (3, \infty)$
9. -16.5
10. 36

11. **12.**

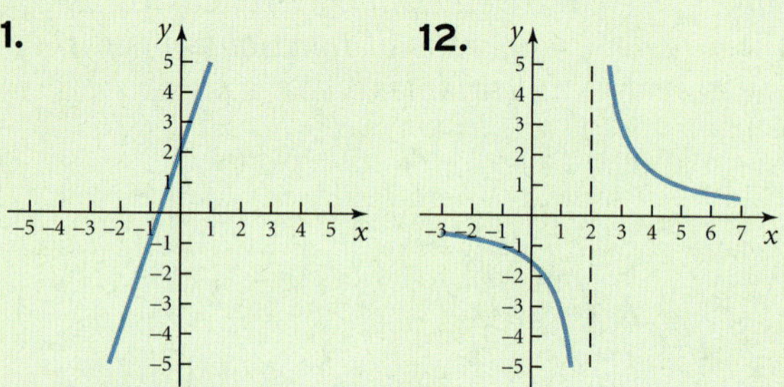

13. **14.**

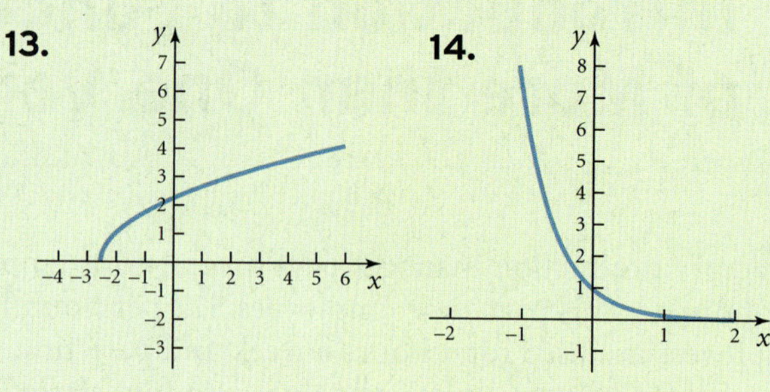

15.

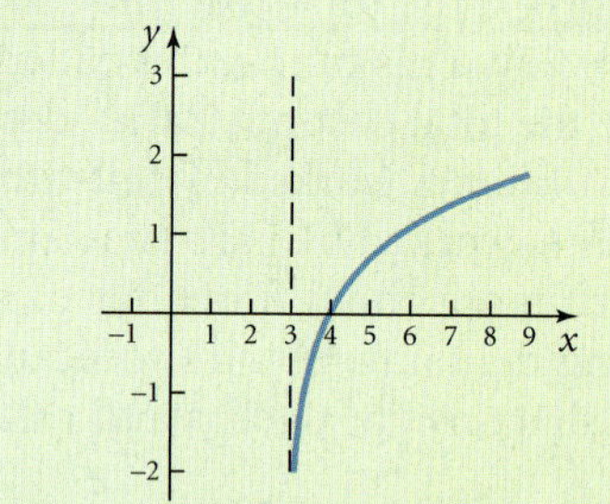

Calculus Warm-up Exercises

1. 5 employees and 14 machines
2. $y = \pm\sqrt{2x^2 - 8}$
3. $y^2(2 - x^2) = 16$

The calculus of functions of more than one variable is called **multivariable calculus**. In this topic we will focus primarily on functions with two independent variables, and we will also present several examples of functions of more than two variables.

> **Definition:** A **function of two variables, $f(x, y)$,** is a rule that assigns to each ordered pair (x, y) in the domain one unique real number $f(x, y)$ in the range. x and y are the **independent variables**; $z = f(x, y)$ is the **dependent variable.**

With single-variable functions, you already know how to

- Evaluate the function.
- Graph the function.
- Find derivatives of the function.
- Integrate the function.

The same operations can be defined for functions of more than one variable.

Evaluating Functions of Two Variables

The domain of a function of two variables is a set of ordered pairs (x, y). The domain may be the entire xy-plane or it may be restricted to some part of the xy-plane, as we will see in the following examples. Given a specific point (x_1, y_1) in the domain, substituting these two values for x and y in the function will produce the corresponding range value, as is shown in the next three examples.

Example 1: Given $f(x, y) = \dfrac{3x^2 - y^2}{x + y}$:

a. State the domain of f.

b. Evaluate $f(2, -1)$.

Solution:

a. $f(x, y) = \dfrac{3x^2 - y^2}{x + y}$ is a rational function so you know that the denominator cannot equal zero. The domain then is all ordered pairs (x, y) such that $x + y \neq 0$, or $y \neq -x$. Graphically, the domain is all points in the plane except those on the line $y = -x$. The shaded region in the graph in Figure 28.1 represents the domain of the function.

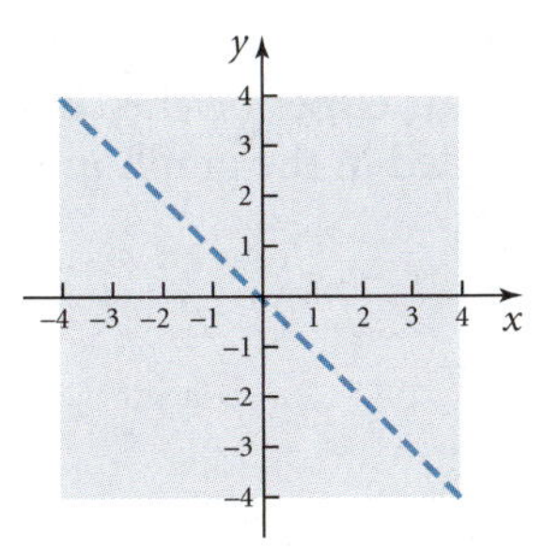

Figure 28.1

b. To evaluate $f(2, -1)$, substitute $x = 2$ and $y = -1$ into the function.

$$f(2, -1) = \frac{3(2)^2 - (-1)^2}{2 + (-1)} = \frac{12 - 1}{2 - 1} = \frac{11}{1} = 11$$

Example 2: Given $f(x, y) = x^2 - x \ln y$:

a. State the domain of f.

b. Evaluate $f(4, 2)$.

Solution:

a. $\ln y$ is defined only if $y > 0$, so the domain is all ordered pairs (x, y) such that $y > 0$. Graphically, the domain is all points in the half-plane above the x-axis. The domain is shaded in the graph in Figure 28.2.

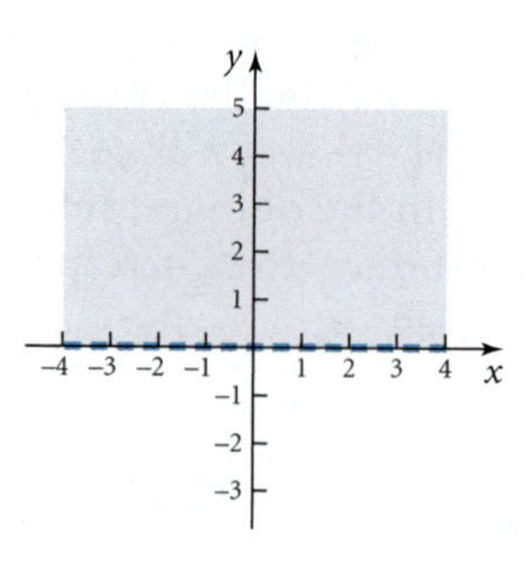

Figure 28.2

b. To evaluate $f(4, 2)$, substitute $x = 4$ and $y = 2$.

$$f(4, 2) = 4^2 - 4\ln 2 = 16 - 4\ln 2 \approx 13.227$$

Example 3: Given $f(x, y) = \sqrt{9 - x^2 - y^2}$:

a. State the domain of f.

b. Evaluate $f(2, 1)$.

Solution:

a. $\sqrt{9 - x^2 - y^2}$ is defined only if $9 - x^2 - y^2 \geq 0$. The domain is the set of all ordered pairs (x, y) such that $x^2 + y^2 \leq 9$. Graphically, the domain is all points either on the circle $x^2 + y^2 = 9$ or in its interior. The domain is shaded in the graph in Figure 28.3.

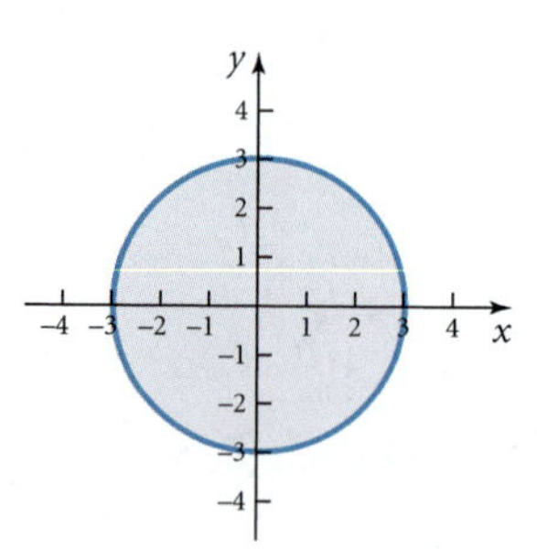

Figure 28.3

b. To evaluate $f(2, 1)$, substitute $x = 2$ and $y = 1$.

$$f(2, 1) = \sqrt{9 - 2^2 - 1^2} = \sqrt{4} = 2$$

Let's look at one example with more than two independent variables.

Example 4: A fence along a garden border consists of three pieces with lengths denoted by a, b, and c. All dimensions are in feet. The cost of the length a is

\$10 per foot, the cost of the length b is \$12 per foot, and the cost of the length c is \$8 per foot.

a. Express the total cost as a function of a, b, and c.

b. Evaluate $C(10, 6, 2)$ and interpret the answer.

Solution:

a. The total cost is given by $C = 10a + 12b + 8c$.

b. $C(10, 6, 2) = 10(10) + 12(6) + 8(2) = 100 + 72 + 16 = 188.$
To build a fence with first piece of length 10 feet, second piece of length 6 feet, and third piece of length 2 feet, the total cost is \$188. ■

There are many practical applications of functions of two variables.

Example 5: A board foot is a unit of measure equal to 1 square foot and 1 inch of thickness. It is used to measure logs and lumber. The *Doyle Log Rule* determines the lumber yield of a log (in board feet) based on the diameter d (in inches) and the length L (in feet) of the log. The number of board feet is given by $B(d, L) = L\left(\frac{d-4}{4}\right)^2$.

a. State the domain of B.

b. If a log is 20 feet long and 30 inches in diameter, find the number of board feet.

c. Evaluate $B(25, 12)$ and interpret your answer.

Solution:

a. Because diameter and length are dimensions, they must be positive values. The domain is all ordered pairs (d, L) such that $d > 0$ and $L > 0$. These values are contained within the first quadrant of the graph in Figure 28.4.

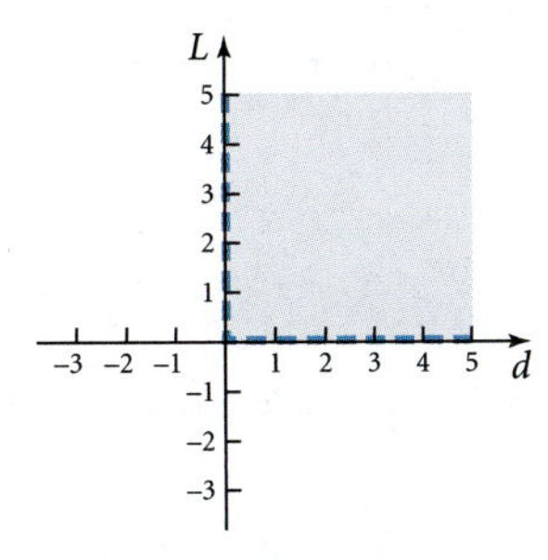

Figure 28.4

b. With $L = 20$ and $d = 30$, we have

$$B(30, 20) = 20\left(\frac{30-4}{4}\right)^2 = 845 \text{ board feet}$$

c. $B(25, 12) = 12\left(\frac{25-4}{4}\right)^2 = 330.75$

A log that is 12 feet long and 25 inches in diameter contains 330.75 board feet. ■

Example 6: *Cobb–Douglas Productivity Model.* Brian's Beach Shop manufactures surfboards. The number produced each day is given by $q = 4.5x^{0.8}y^{0.2}$, where x is the number of employees and y is the number of machines used.

a. State the domain of q.

b. Evaluate $q(20, 10)$ and interpret.

c. Evaluate $q(40, 20)$ and interpret.

Solution:

a. The number of employees and the number of machines must be positive values, so the domain is all ordered pairs (x, y) such that $x > 0$ and $y > 0$. Graphically, as illustrated in Figure 28.5, the domain is all points in the first quadrant of the xy-plane.

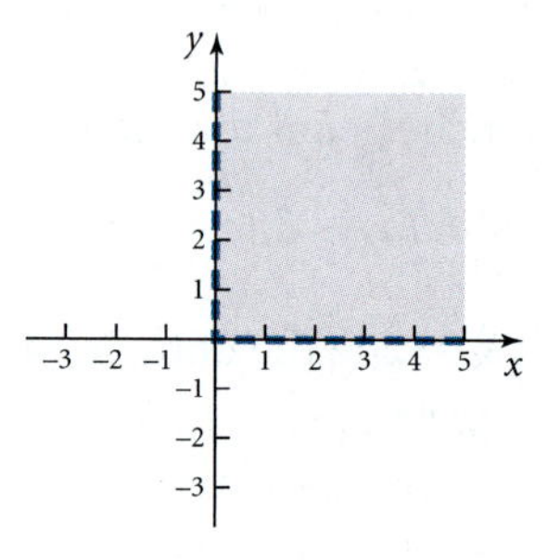

Figure 28.5

b. $q(20, 10) = 4.5(20)^{0.8}(10)^{0.2} = 78.3$. If Brian uses 20 employees and 10 machines, he can produce about 78 surfboards.

c. $q(40, 20) = 4.5(40)^{0.8}(20)^{0.2} = 156.7$. With 40 employees and 20 machines, Brian can produce about 157 boards. In other words, if the number of employees and the number of machines are both doubled, the output is also doubled. (See Exercise 50.) ■

Check Your Understanding 28.1

For each of the following functions:

a. State the domain of the function.

b. Evaluate the function at the indicated values of the independent variables.

1. $f(x, y) = \sqrt{x^2 - y^2}$; find $f(3, -2)$

2. $P(t, d) = \dfrac{2d}{t^2}$; find $P(4, 6)$

3. $A(b_1, b_2, h) = \frac{1}{2}h(b_1 + b_2)$; find $A(6, 14, 2)$

Graphs of Functions of Two Variables

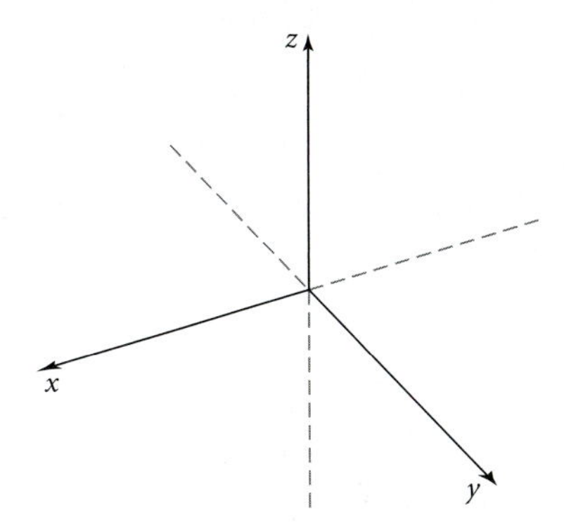

Figure 28.6

The graph of a function of two variables, $z = f(x, y)$, is a set of **ordered triples** (x, y, z) in three dimensions. To create a coordinate system in three dimensions, picture the two-dimensional xy-plane and imagine drawing a third axis perpendicular to the plane through the origin. The usual representation of the three-dimensional coordinate system is shown in Figure 28.6.

Example 7: Plot the point (2, 3, 1).

Solution: In Figure 28.7, to plot (2, 3, 1) begin at the origin and count 2 units along the x-axis. Then count 3 units parallel to the positive y-axis and 1 unit up.

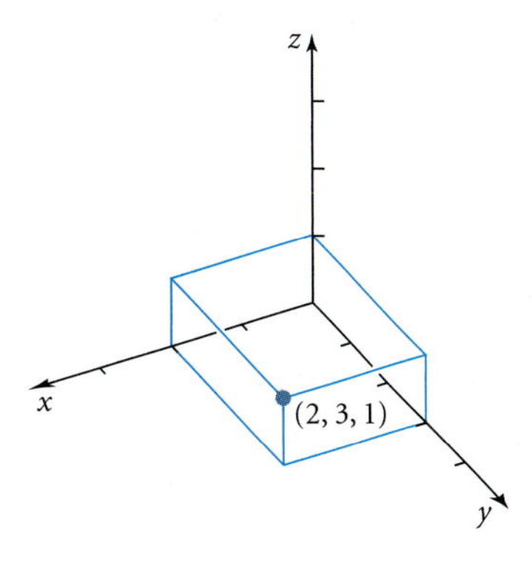

Figure 28.7

The point (2, 3, 1) is the corner point of a box 2 units by 3 units by 1 unit high. ■

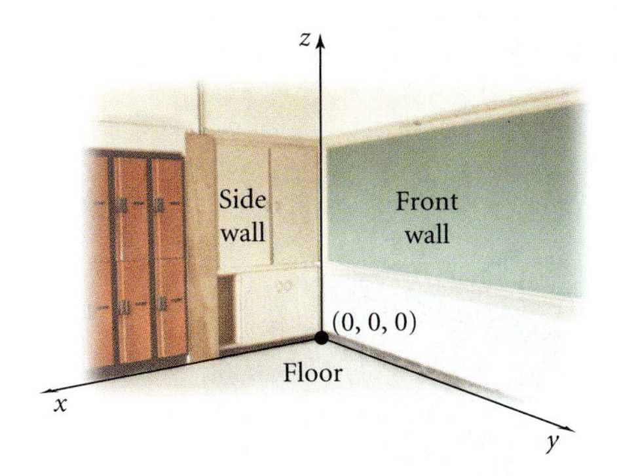

Figure 28.8

To better visualize the three-dimensional coordinate system, focus on the lefthand corner of your classroom, as you see it sitting at your desk. The point on the floor where the floor, the front wall, and the side wall meet represents the origin as shown in Figure 28.8. The baseboard along the side of the room is the x-axis, the baseboard along the front is the y-axis, and the vertical seam where the walls meet is the z-axis. The room as we have just described it represents the first **octant** of the three-dimensional coordinate system, where $x > 0, y > 0$, and $z > 0$. (In two dimensions, there are four quadrants. In three dimensions, there are eight octants, four above the xy-plane and four below the xy-plane.) If $x < 0$, you are in the room or hallway behind the front wall. If $y < 0$, you are in the room or hallway beside the side wall. If $z < 0$, you are underneath the floor.

The floor represents the xy-plane, the side wall represents the xz-plane, and the front wall represents the yz-plane.

To "see" the ordered triple (2, 3, 1), you would start in the corner of the room (the origin) and walk 2 units along the side wall, then turn and walk 3 units parallel to the front wall. Then step up 1 unit and you are there—suspended in the air 1 unit above the floor.

Definition: The graph of $z = f(x, y)$ is a set of ordered triples (x, y, z) with domain a region in the xy-plane. The graph is called a **surface.**

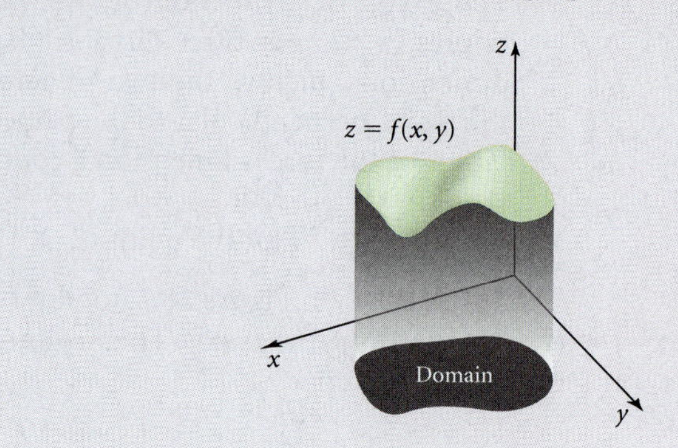

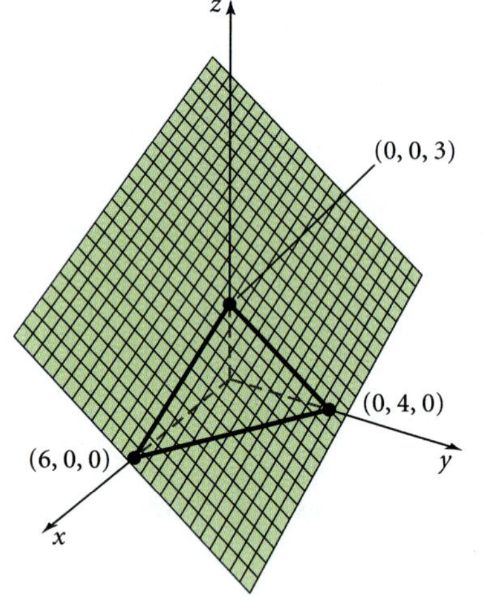

Figure 28.9

Although it is not the purpose of this topic to teach you to become proficient at drawing graphs of various functions of two variables, some of the more common surfaces are presented in the following examples. You should be able to recognize the surface associated with each function. A summary is provided after Example 9.

Example 8: The graph of a linear equation $z = ax + by + c$ is a **plane**, or **planar surface.** If $2x + 3y + 4z = 12$, then the x-intercept is $(6, 0, 0)$, the y-intercept is $(0, 4, 0)$, and the z-intercept is $(0, 0, 3)$. Figure 28.9 shows the graph of the planar surface $z = f(x, y) = 3 - \frac{1}{2}x - \frac{3}{4}y$. ■

Example 9: The graph of $z = x^2 + y^2$, as shown in Figure 28.10, is a **circular paraboloid.** In the xz-plane, the cross section, as shown in Figure 28.11, is the parabola $z = x^2$. In the yz-plane, the cross section, as

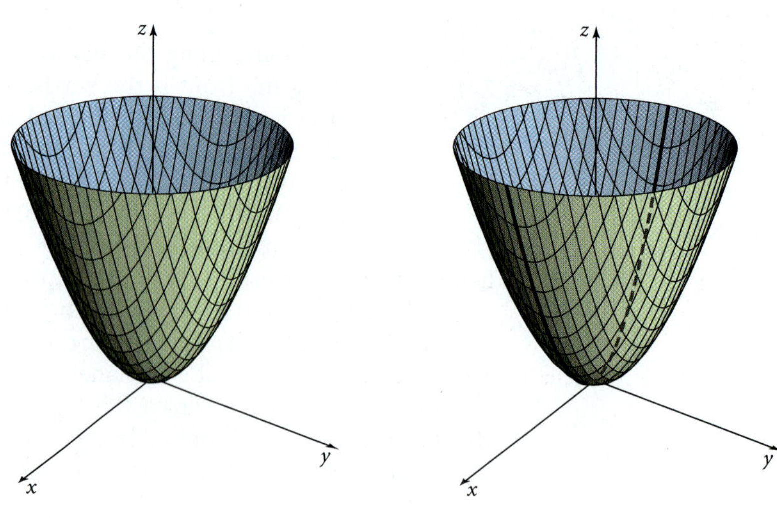

Figure 28.10

Figure 28.11

shown in Figure 28.12, is the parabola $z = y^2$. Any cross section of the paraboloid parallel to the xy-plane, as shown in Figure 28.13, is a circle, $x^2 + y^2 = k$.

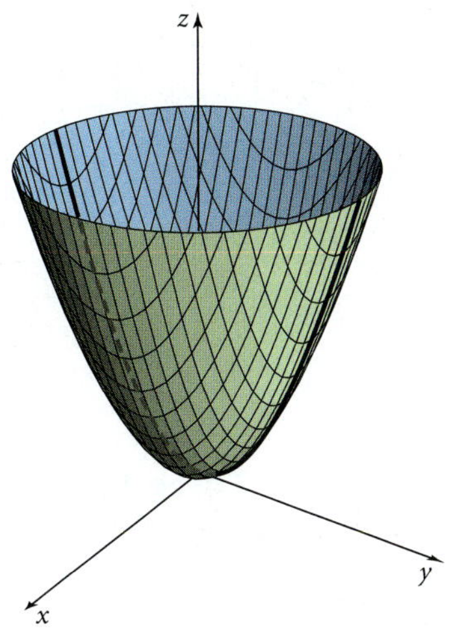

Figure 28.12

Figure 28.13

Common Graphs in Two Dimensions

In two dimensions, the **conic sections** are:

the *circle*, given by $x^2 + y^2 = r^2$;

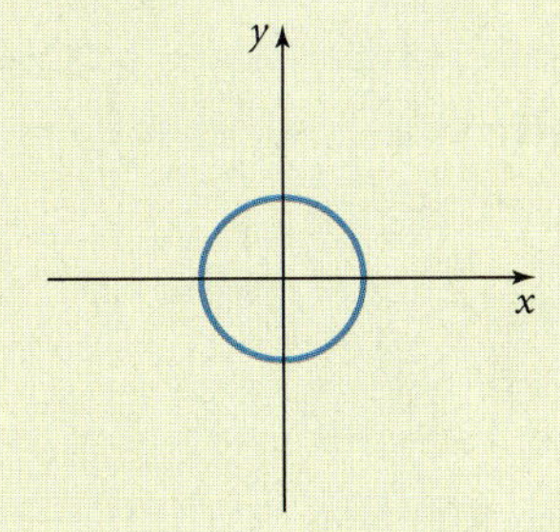

the *ellipse*, given by $\dfrac{x^2}{a^2} + \dfrac{y^2}{b^2} = 1$;

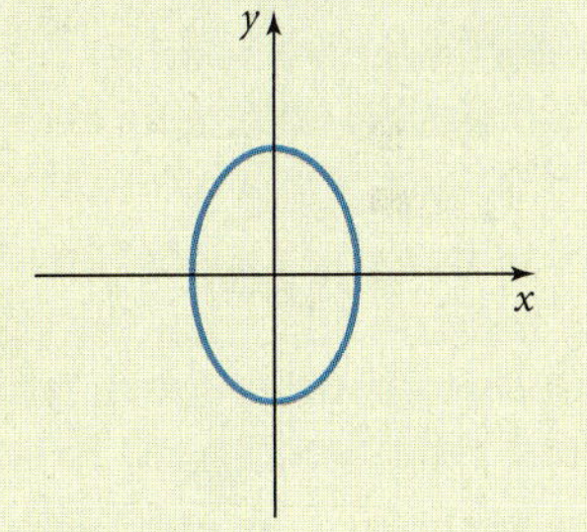

the *parabola*, given by $ax^2 + bx + k = y$;

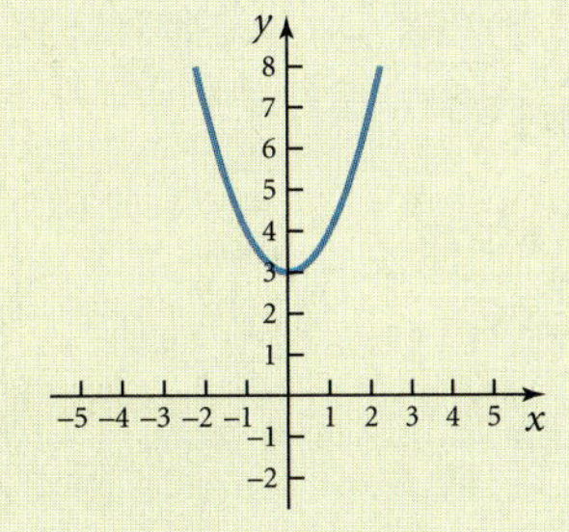

with $a = 1, b = 0,$ and $k = 3,$ or

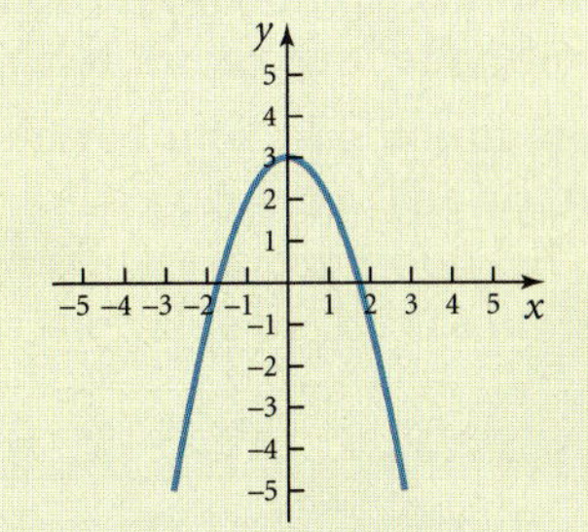

with $a = -1, b = 0,$ and $k = 3;$

(continued)

the *hyperbola*, given by $\frac{x^2}{a^2} - \frac{y^2}{b^2} = 1$ or $\frac{y^2}{a^2} - \frac{x^2}{b^2} = 1.$

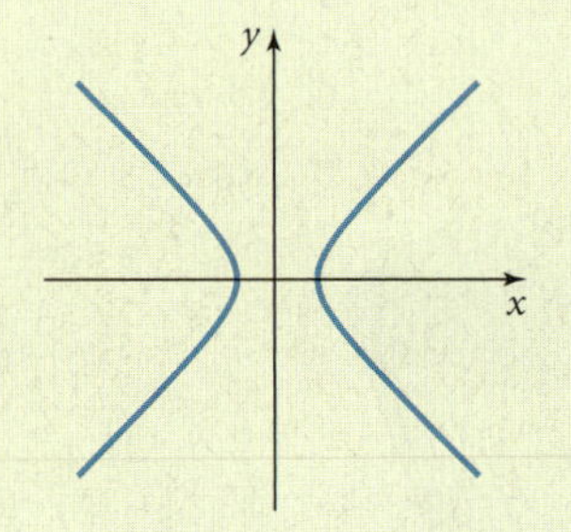

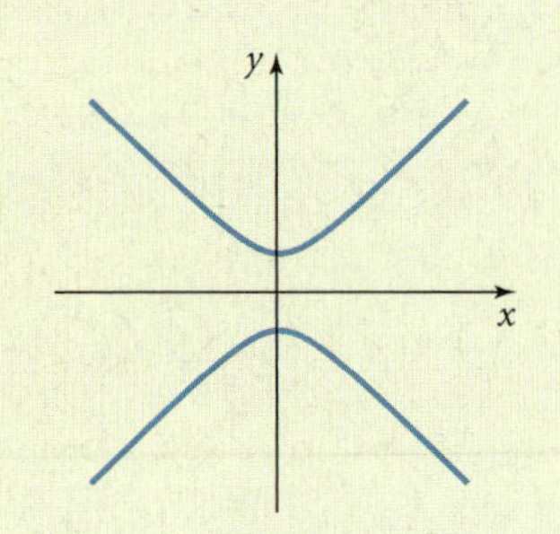

Common Surfaces in Three Dimensions

In three dimensions, the common **surfaces** are:

the *sphere*, given by $x^2 + y^2 + z^2 = r^2$;

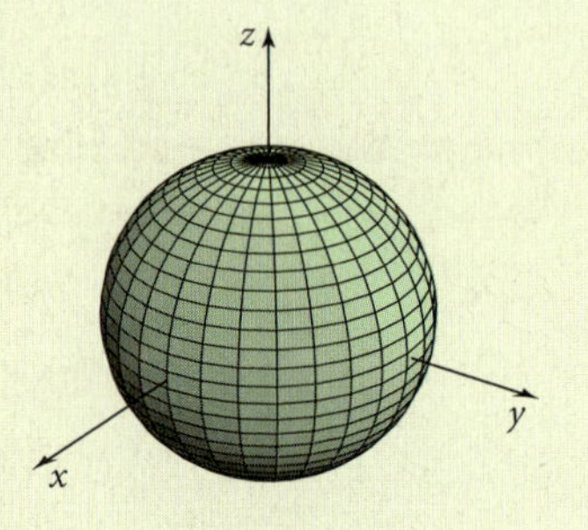

the *ellipsoid*, given by $\frac{x^2}{a^2} + \frac{y^2}{b^2} + \frac{z^2}{c^2} = 1$;

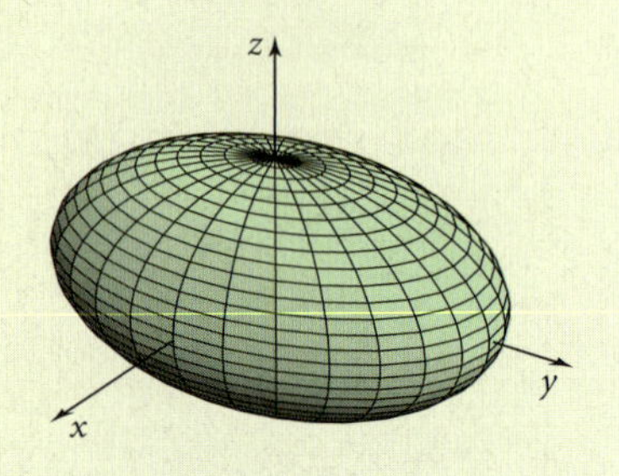

the *paraboloid*, given by $z = ax^2 + by^2$;

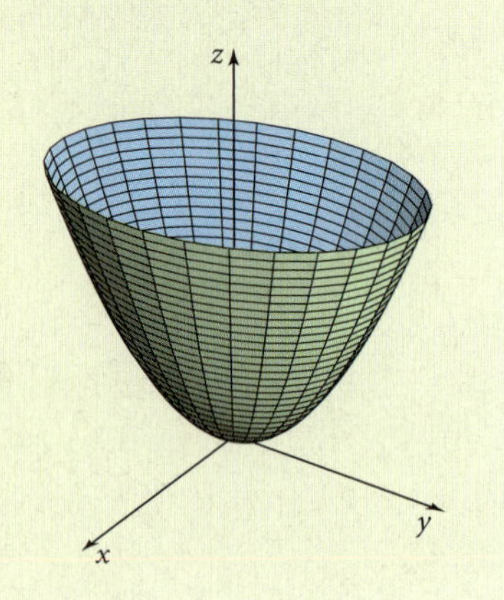

the *hyperboloid of one sheet*, given by $\frac{x^2}{a^2} + \frac{y^2}{b^2} - \frac{z^2}{c^2} = 1$;

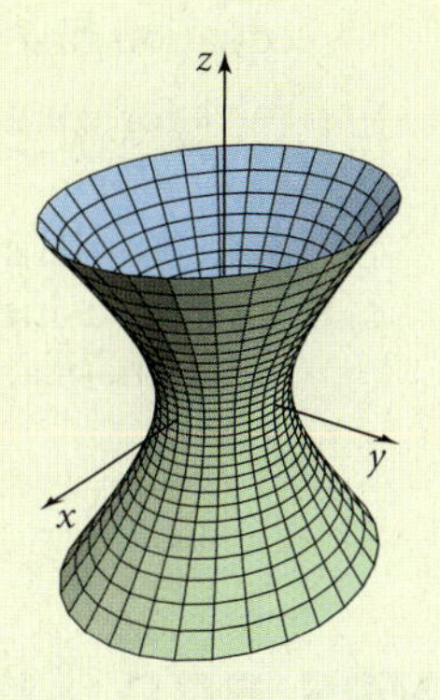

the *hyperboloid of two sheets*, given by $\frac{x^2}{a^2} - \frac{y^2}{b^2} - \frac{z^2}{c^2} = 1$;

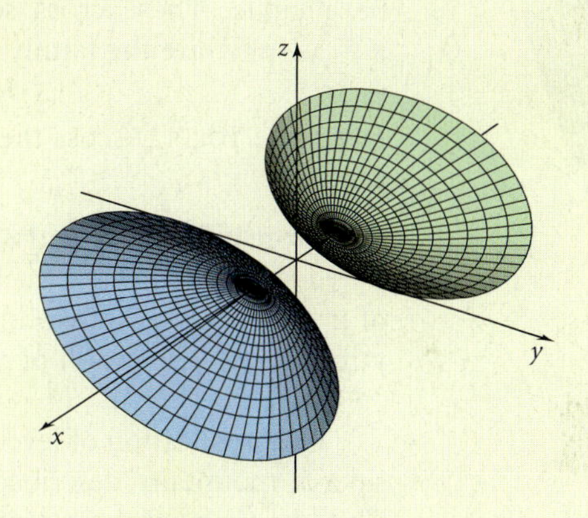

the *hyperbolic paraboloid* (the "saddle"), given by $z = x^2 - y^2$.

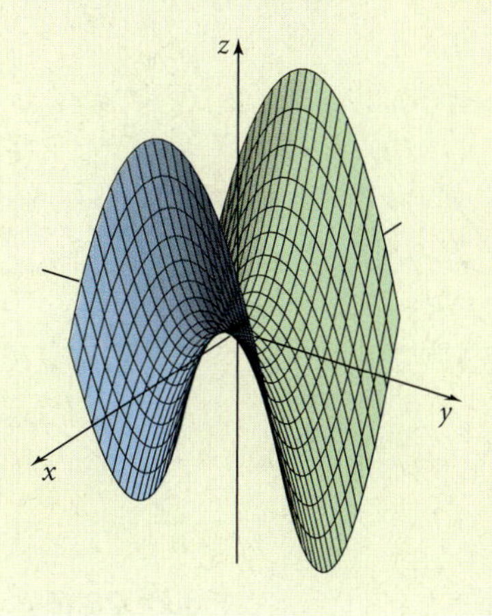

Calculator Corner 28.1

To graph the conic sections on the TI-83/84, it is necessary to solve the equation first for y. For instance, to graph the ellipse $2x^2 + y^2 = 1$, solve for y and get $y = \pm\sqrt{1 - 2x^2}$. Then enter in the calculator $y_1 = \sqrt{1 - 2x^2}$ and $y_2 = -\sqrt{1 - 2x^2}$ to see the graph. The three-dimensional surfaces cannot be graphed on the TI-83/84. They can be graphed using a TI-89 or using a computer algebra system such as Mathematica, Derive, Maple, or Converge.

Level Curves and Their Applications

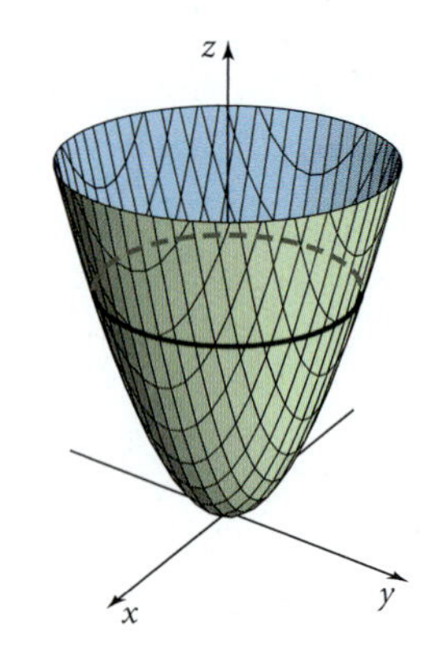

Figure 28.14

Consider the paraboloid $z = x^2 + y^2$ from Example 9. Looking at the cross section of the graph in Figure 28.14 where $z = 4$, you should see a circle whose equation is $x^2 + y^2 = 4$. In fact, for any $z > 0$, the cross section parallel to the xy-plane will be a circle. These cross sections are called **level curves**. The level curves of $z = x^2 + y^2$ are the family of circles $x^2 + y^2 = k$.

Example 10: Discuss the level curves of the hyperbolic paraboloid $z = x^2 - y^2$.

Solution: The level curves of $z = x^2 - y^2$ are the family of hyperbolas $x^2 - y^2 = k$. If $k > 0$, then the hyperbolas have x-intercepts. The graph of $x^2 - y^2 = 4$ is shown in Figure 28.15. If $k < 0$, then the hyperbolas have y-intercepts. The graph of $x^2 - y^2 = -4$ is shown in Figure 28.16. ■

One application of level curves is **topographical maps**, or **contour maps**. Consider a "mountain" described by the elliptic paraboloid $z = 2000 - 3x^2 - 2y^2$. The "elevation" of the mountain at any point is given by $f(x, y)$. The level curve at that point is $f(x, y) = k$. The family of all level curves is a topographical, or contour, map. For example, in Figure 28.17 the closer you are to the top of the mountain, the smaller the level curve is.

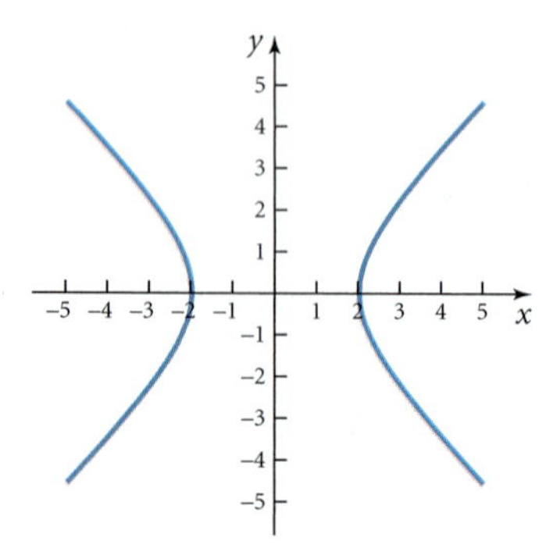

Figure 28.15

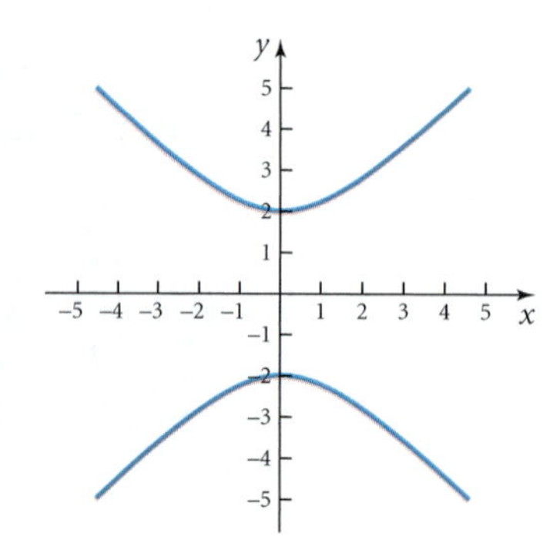

Figure 28.16

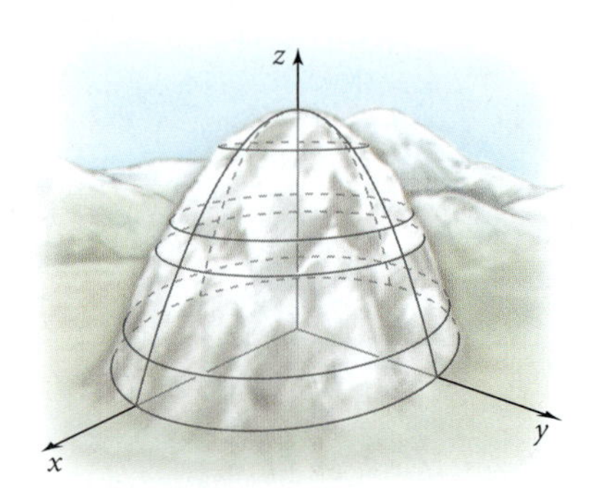

Figure 28.17

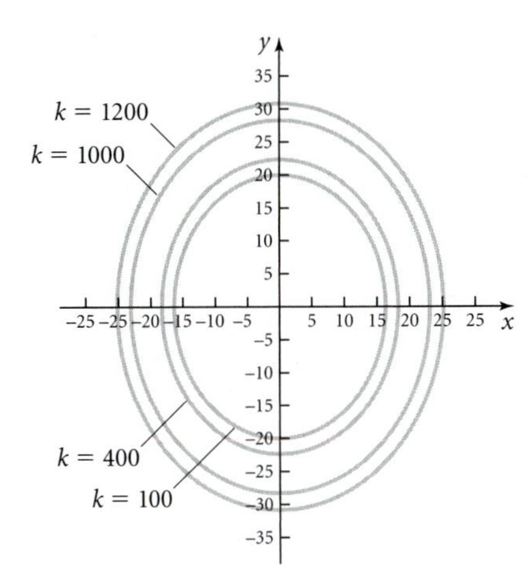

Figure 28.18

The level curves for $k = 1200, k = 1000, k = 400$, and $k = 100$ are shown in Figure 28.18.

To find the level curves, set $f(x, y) = k$. For instance, if $k = 1200$, the level curve is $1200 = 2000 - 3x^2 - 2y^2$, which is the ellipse $\frac{3x^2}{800} + \frac{y^2}{400} = 1$.

Level curves are also used in business and economics. An **isoquant curve** represents the combination of production factors that result in equal amounts of output ("iso" means "equal"). For example, if 25 employees and 10 machines create an output of 3000 units, then to maintain the same output, if 30 employees were used, we would need fewer than 10 machines. In other words, as the value of one factor—the number of employees in this case—increases, then the value of the other factor—the number of machines in this case—must decrease to keep the output constant. An **isoquant map** shows a set of isoquant curves for different levels of output.

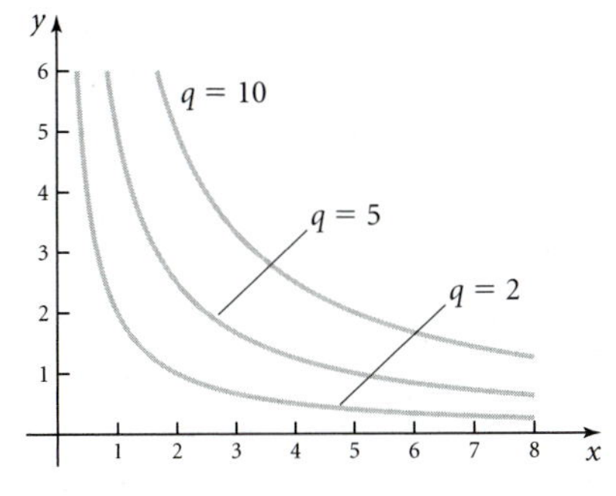

Figure 28.19

Example 11: Suppose that output for a factory is given by $q = xy$. In Figure 28.19, isoquant curves are presented for several different levels of output. Each isoquant curve is given by $y = \frac{q}{x}$. The farther away from the x-axis and y-axis the curve is, the greater the output. ■

Check Your Understanding 28.2

Describe the level curves for each surface.

1. $f(x,y) = x^2 - y$ for $k = 1, 3, -4$.
2. $f(x,y) = \frac{x}{y}$ for $k = 1, 3, -0.5$.

Check Your Understanding Answers

Check Your Understanding 28.1

1. a. The domain is all ordered pairs (x, y) such that $x^2 - y^2 \geq 0$, or $x^2 \geq y^2$.
   b. $f(3, -2) = \sqrt{3^2 - (-2)^2} = \sqrt{5}$
2. a. The domain is all ordered pairs (t, d) such that $t \neq 0$.
   b. $P(4, 6) = \frac{2(6)}{4^2} = \frac{12}{16} = \frac{3}{4}$
3. a. The domain is all ordered triples (b_1, b_2, h) such that $b_1 > 0, b_2 > 0, h > 0$.
   b. $A(6, 14, 2) = \frac{1}{2}(2)(6 + 14) = 20$

Check Your Understanding 28.2

1. $k = 1 \rightarrow y = x^2 - 1$
   $k = 3 \rightarrow y = x^2 - 3$
   $k = -4 \rightarrow y = x^2 + 4$
   The level curves are the family of parabolas with vertex at $(0, -k)$.
2. $k = 1 \rightarrow y = x$
   $k = 3 \rightarrow y = \frac{1}{3}x$
   $k = -0.5 \rightarrow y = -2x$
   The level curves are the family of lines passing through the origin with slope of $m = \frac{1}{k}$.

Topic 28 Review

28

This topic introduced the concept of functions of more than one independent variable, including determining the domain, sketching the graph, and finding level curves.

CONCEPT	EXAMPLE
A **function of two variables, $f(x, y)$**, is a rule that assigns to each ordered pair (x, y) in the domain one unique real number $f(x, y)$ in the range. x and y are the **independent variables**; $z = f(x, y)$ is the **dependent variable.**	$f(x, y) = x\sqrt{y}$ is a function of two variables. The independent variables are x and y. The dependent variable is $z = f(x, y)$. The domain of f is the set of all (x, y) such that $y \geq 0$.
The graph of a function of two variables is a **surface** in three dimensions whose domain is an ordered pair in the xy-plane and whose range is a real number. The points on that surface, (x, y, z), are called **ordered triples.**	The graph of $f(x, y) = x^2$ is 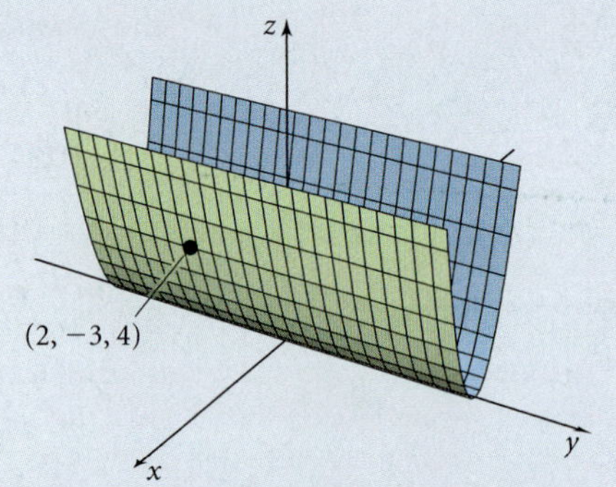 The domain is the set of all (x, y). The range is $z \geq 0$. An ordered triple on the surface is $(2, -3, 4)$.
The two-dimensional cross section of a surface at a specified range value is called a **level curve**. The equation of the level curves is $f(x, y) = k$.	Letting $k = 4$, one level curve for $f(x, y) = x^2 - y$ is the parabola $4 = x^2 - y$. 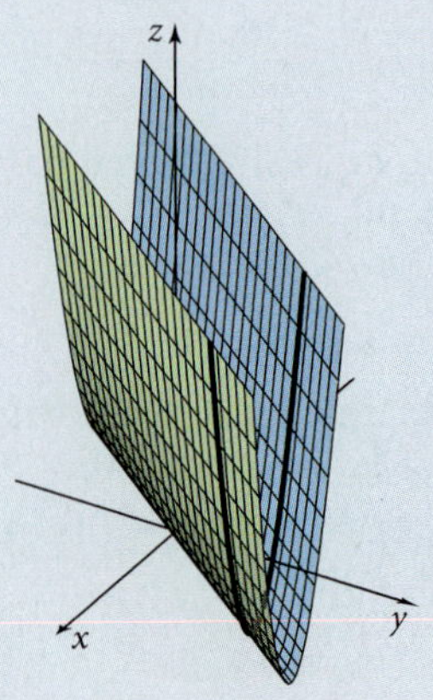

Two applications of level curves are **contour maps** and **isoquant graphs**.	A mountain has a shape described by $z = 4000 - 2x^2 - y^2$. At an elevation of $k = 1000$, the level curve is the ellipse $2x^2 + y^2 = 3000$. At an elevation of $k = 3000$, the level curve is $2x^2 + y^2 = 1000$. As the elevation increases, the size of the level curve decreases.

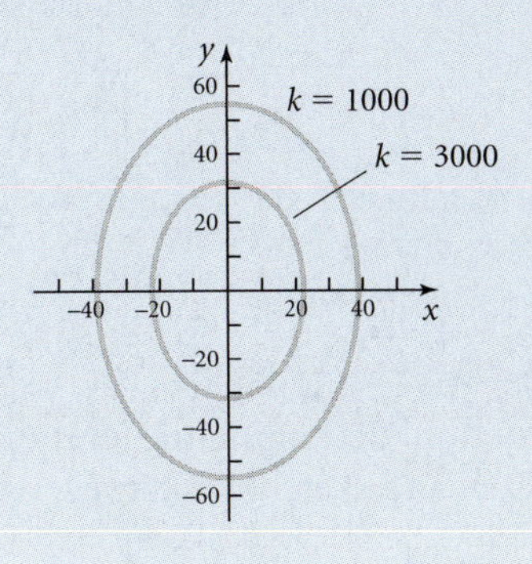

NEW TOOLS IN THE TOOL KIT

- Evaluating functions of two or more variables
- Identifying the domain of a function of two variables
- Sketching the graph of some functions of two variables
- Determining level curves for graphs of functions of two variables

Topic 28 Exercises

28

Evaluate each function for the indicated values of the independent variables.

1. For $f(x, y) = 2x^2 + x^3y - 4y$, find $f(2, -1)$.
2. For $f(x, y) = (x^2y^3 + 4x - 5y)^3$, find $f(-1, 1)$.
3. For $P(t, n) = 2t\sqrt{n} + \frac{4}{t^2} - 7n$, find $P(6, 4)$.
4. For $N(c, d) = \frac{c}{d} + \sqrt{cd}$, find $N(4, 9)$.
5. For $V(r, h) = \frac{4}{3}\pi r^3 h$, find $V(10, 8)$.
6. For $A(r, h) = 2\pi r(r + h)$, find $A(6, 12)$.
7. For $f(x, y) = e^{xy} - x^2y + \ln(3x)$, find $f(1, 2)$.
8. For $f(x, y) = xe^y - ye^x + xy$, find $f(2, -1)$.
9. For $A(P, r, t) = P\left(1 + \frac{r}{4}\right)^{4t}$, find $A(2000, 0.04, 6)$.
10. For $V(x, y, z) = 2x^2 + 3y^2 - z^2$, find $V(3, 4, 5)$.
11. For $Q(a, b, c) = a^3b^2 - a^2bc + 3c$, find $Q(3, -1, 4)$.
12. For $N(m, n, p) = \frac{m^2n}{p} - \frac{mn}{p^3}$, find $N(3, -2, 4)$.

State the domain of each function. Sketch the graph of the domain of the function in the xy-plane.

13. $f(x, y) = 3x - 2y + 6$

14. $f(x, y) = x^2 + y^2$

15. $f(x, y) = \dfrac{2x^2}{x - y}$

16. $f(x, y) = \sqrt{4 - x^2 - y^2}$

17. $f(x, y) = 2\sqrt{x^2 - y}$

18. $f(x, y) = \dfrac{e^{-2x}}{y}$

19. $f(x, y) = x \ln(y + 2)$

20. $f(x, y) = \dfrac{x}{\sqrt{x^2 - y^2}}$

For each function, determine the level curve at the indicated value and sketch the curve.

21. $f(x, y) = -x^2 + y$ for $k = 2$

22. $f(x, y) = x^2 - 4x - y$ for $k = -4$

23. $f(x, y) = \sqrt{16 - x^2 - y^2}$ for $k = 3$

24. $f(x, y) = e^x - y$ for $k = -2$

Describe the family of level curves of each function. Sketch several representative curves.

25. $f(x, y) = xy$

26. $f(x, y) = x^2 + y$

27. $f(x, y) = 2x + 3y$

28. $f(x, y) = xe^y$

Match each function to its surface.

29. $f(x, y) = 2x + 3y - 6$

30. $f(x, y) = \sqrt{4 - x^2 - y^2}$

31. $f(x, y) = 4 - x^2 - y^2$

32. $f(x, y) = x^2 + y^2 - 4$

33. $f(x, y) = x^2 - y^2$

34. $f(x, y) = -\sqrt{x^2 + y^2}$

a.

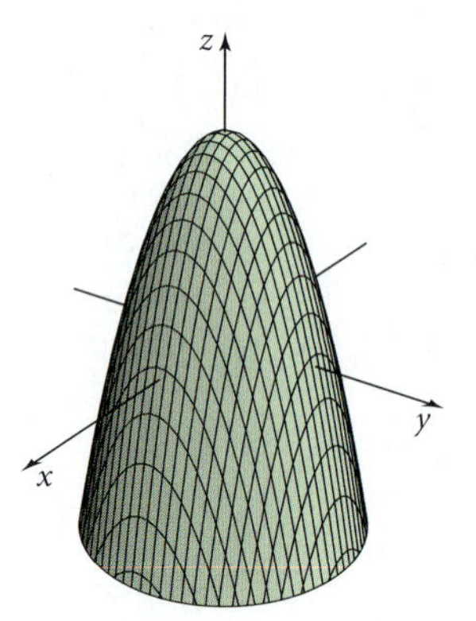

b.

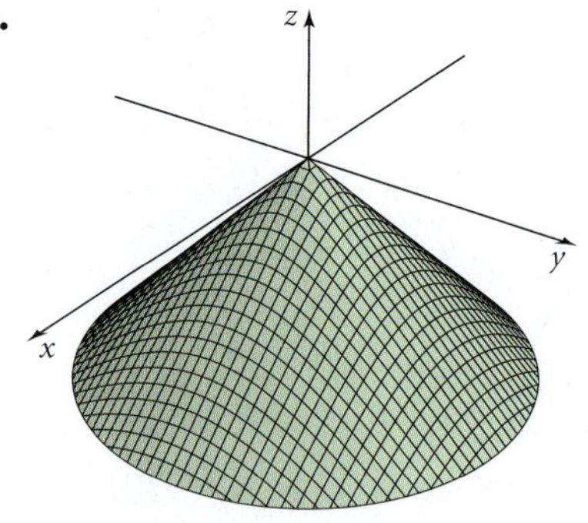

c.

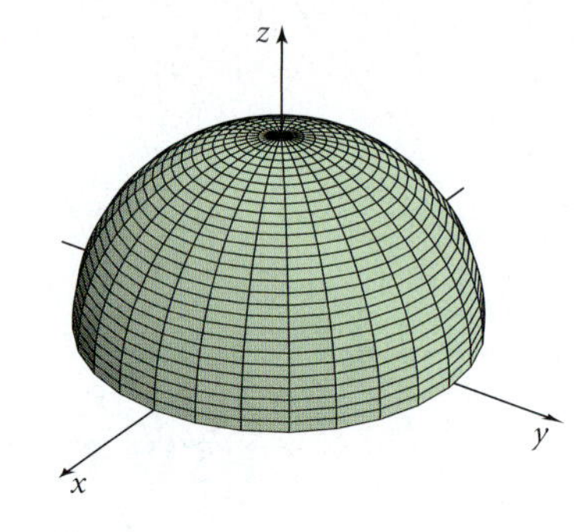

d.

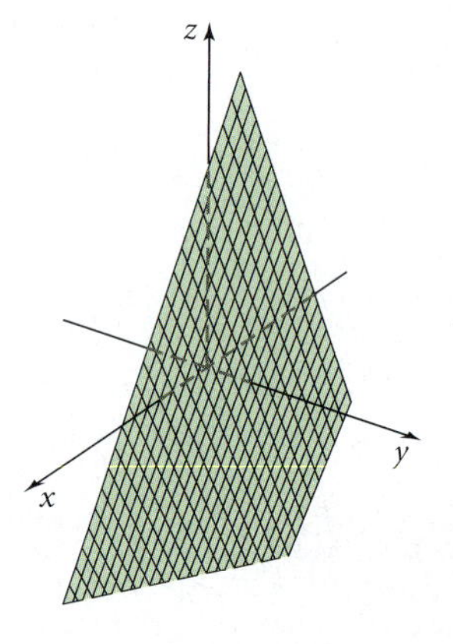

e.

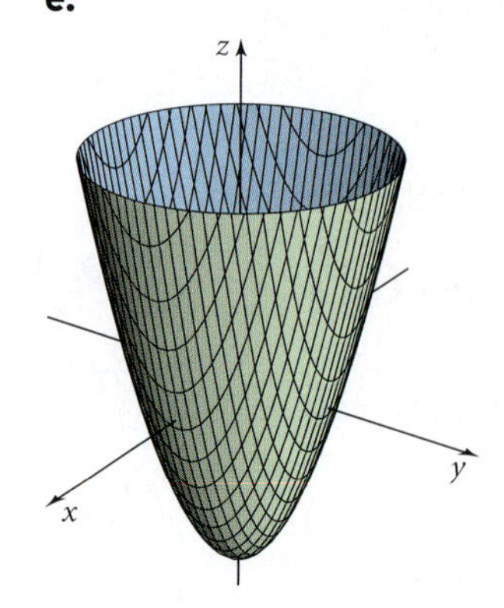

f.

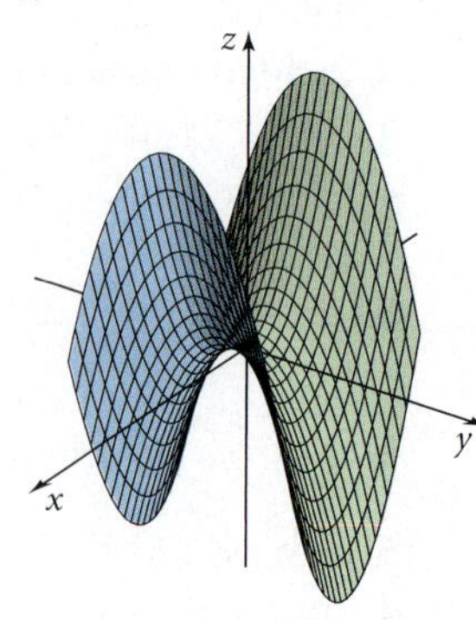

35. It costs Fifi's Dog Groomers \$12 to groom a small dog and \$19 to groom a large dog. Fixed monthly costs are \$1950.

a. Write a cost function $C(x, y)$ for grooming x small dogs and y large dogs in a month.

b. Evaluate $C(45, 72)$ and interpret the answer.

36. Refer to Exercise 35. The revenue is \$35 for a small dog and \$50 for a large dog.

a. Write a profit function $P(x, y)$ for grooming x small dogs and y large dogs in a month.

b. Evaluate $P(45, 72)$ and interpret the answer.

37. A college bookstore sells two brands of calculators, the MT-56 and the TR-98. The MT-56 sells for \$$a$ and the TR-98 sells for \$$b$. At those prices, the demand for the MT-56 is given by $D_a = 300 - 5a + 10b$ calculators and the demand for the TR-98 is given by $D_b = 125 + 15a - 15b$ calculators.

a. Determine the bookstore's total revenue as a function of a and b.

b. Find the total revenue if $a = \$150$ and $b = \$120$.

38. A large theme park sells a one-day pass and a discounted four-day pass. The one-day pass costs \$$x$ and the four-day pass costs \$$y$. On any given weekday, demand for the one-day pass is given by $D_1 = 850 - 30x + 50y$ passes and demand for the four-day pass is given by $D_2 = 600 + 45x - 10y$ passes.

a. Determine the park's total daily revenue as a function of x and y.

b. Find the total daily revenue if $x = \$65$ and $y = \$210$.

39. The duration of a scuba dive, in minutes, is given by $T(v, D) = \dfrac{33v}{D + 33}$, where v is the volume of air in the tank (at sea level pressure) and D is the depth of the dive in feet. Evaluate $T(90, 30)$ and $T(80, 40)$.

40. A person's intelligence quotient is based on the ratio of mental age, m, to chronological age, c, and is given by $Q(m, c) = \dfrac{m}{c} \cdot 100$. Evaluate $Q(15, 18)$ and $Q(12, 10)$.

The *Wilson lot-size formula*, $Q = \sqrt{\dfrac{2CN}{h}}$, provides the optimal quantity, Q, of a product that a business should order, based on the following factors: the cost of placing an order, C; the number of items that the store sells per week, N; and the weekly holding cost per item, h. Use the Wilson lot-size formula for the next two questions.

41. How many tennis rackets should Paulo's Pro Shop order each week if it costs \$15 to place the order, \$0.50 to hold each racket, and Paulo expects to sell 50 rackets per week?

42. Lee's Car Sales sells a luxury model convertible. It costs \$225 to place the order, \$15 to hold each car, and Lee expects to sell 25 convertibles. How many cars should she order?

The Cobb–Douglas Productivity Model is given by $Q(x, y) = Kx^a y^{1-a}$, where x is the number of employees, and y is the annual operating cost, in thousand dollars. K and a are positive constants. Use this productivity model to answer the next two questions.

43. If $K = 100$ and $a = 0.4$, determine the number of items produced with 100 employees and an annual operating budget of \$500,000.

44. If $K = 1000$ and $a = 0.5$, determine the number of items produced with 500 employees and an annual operating budget of \$1,000,000.

45. Since 1998 the National Football League has used a rating system to compare its quarterbacks on their passing ability. The formula used is

$$R(A, C, Y, T, I) = \left(\frac{\frac{C}{A} - 0.3}{0.2} + \frac{\frac{Y}{A} - 3}{4} + \frac{\frac{T}{A}}{0.05} + \frac{0.095 - \frac{I}{A}}{0.04} \right) \cdot \frac{100}{6}$$

where A is the number of passes attempted, C is the number of passes completed, Y is the total number of yards gained passing, T is the number of touchdown passes, and I is the number of interceptions.

The performance statistics for four quarterbacks during the 2004–2005 NFL season are provided in the table. Find the rating of each quarterback, rounded to the nearest tenth.

	Name/Team	Passes Attempted	Passes Completed	Yards Gained Passing	Touchdown Passes	Interceptions
a.	Tom Brady, Patriots	474	288	3692	28	14
b.	Donovan McNabb, Eagles	469	300	3875	31	8
c.	Daunte Culpepper, Vikings	548	379	4717	39	11
d.	Peyton Manning, Colts	497	336	4557	49	10
e.	Brett Favre, Packers	540	346	4088	30	17

46. A person's body mass can be calculated using the body mass index function $B(w, h) = \frac{w}{h^2} \cdot 703$, where w is weight in pounds and h is height in inches. If $B < 18.5$, the person is considered underweight; if $18.5 \leq B \leq 24.9$, the person is considered normal; if $25 \leq B \leq 29.9$, the person is overweight; if $B \geq 30$, the person is considered obese. Calculate the body mass index of each of the following and classify as underweight, normal, overweight, or obese.

a. an adult woman weighing 165 pounds who is 66 inches tall

b. an adult man weighing 180 pounds who is 72 inches tall

c. a teenage boy weighing 215 pounds who is 70 inches tall

d. a teenage girl weighing 115 pounds who is 67 inches tall

47. In November 2001 the National Weather Service published an updated Wind Chill Temperature Index, which uses air temperature in degrees Fahrenheit (°F) and wind speed in miles per hour to measure how cold it actually feels. Assuming temperatures below 40°F and wind speeds between 5 and 60 mph, the wind chill is approximated by $W(T, v) = 35.74 + 0.6215T - 35.75v^{0.16} + 0.4275Tv^{0.16}$, where T is air temperature (°F) and v is wind speed (mph). Calculate the following wind chills and interpret the answer.

a. $W(25, 10)$ **b.** $W(-5, 15)$

c. On a winter day in Chicago, the wind speed was 30 mph with a wind chill of -10°F. What was the temperature?

d. On a cold winter day in Minneapolis, the temperature was 0°F with a wind chill of -25°F. What was the wind speed?

48. The National Weather Service also has a Heat Index table that uses air temperature, T, in degrees Fahrenheit (°F) and relative humidity, r, to measure how hot it actually feels. Assuming temperatures of at least 80°F and relative humidity of at least 40%, the heat index is approximated by

$$\begin{aligned} H(T, r) = &-42.4 + 2.049T + 10.143r \\ &- 0.2248Tr - 0.00684T^2 - 0.0548r^2 \\ &+ 0.00123T^2r + 0.00085Tr^2 \\ &- 0.00000199T^2r^2 \end{aligned}$$

Calculate the following heat indexes and interpret the answer.

a. $H(96, 50)$

b. $H(104, 70)$

c. On a summer day in Orlando, the temperature was 94°F with a heat index of 107°F. What was the relative humidity?

d. On another summer day in Richmond, the relative humidity was 49% with a heat index of 108°F. What was the temperature?

SHARPEN THE TOOLS

49. Last year, using 2000 employees and an operating budget of $1,000,000, the D'Angelo Machine Works produced 250 milling machines. This year the budget was cut to $250,000 and 75 milling machines were produced.

a. Find the Cobb–Douglas Productivity Model for D'Angelo. Round K and a to two decimal places.

b. Use your model from part a to predict the output for next year if D'Angelo goes back to an operating budget of $1 million but cuts the workforce to 1,000 employees.

50. For the Cobb–Douglas Productivity Model $Q = Kx^a y^{1-a}$, prove that if the labor force and the operating budget are both doubled, the output is also doubled. In other words, prove that $Q(2x, 2y) = 2Q(x, y)$.

51. The monthly payments for a mortgage are given by the function

$$M(P, i, n) = P\left(\frac{i}{1 - (1 + i)^{-n}}\right)$$

where P is the amount borrowed, i is the monthly interest rate $\left(i = \frac{r}{12}\right)$, and n is the total number of months over which the loan must be repaid.

The Shimmels are buying a home and need to finance $220,000. They have two options: a 30-year mortgage with an annual rate of 7.5% or a 15-year mortgage with an annual rate of 6.75%.

a. Calculate the monthly payment for each option.

b. What is the total amount paid back under each option? (Multiply the monthly payment by the total number of months.)

c. How much interest will they pay with each option? (Interest = total amount repaid − amount borrowed.)

d. In the long run, which option is better and why?

52. The monthly payments for a car loan are given by the function

$$M(P, i, n) = P\left(\frac{i}{1 - (1 + i)^{-n}}\right)$$

where P is the amount borrowed, i is the monthly interest rate $\left(i = \frac{r}{12}\right)$, and n is the total number of months to repay the loan.

Jerry is buying a car and needs to finance $25,000. He can take a 6-year loan with an annual rate of 8.25% or a 5-year loan with an annual rate of 7%.

a. Calculate the monthly payment for each option.

b. What is the total amount paid back under each option? (Multiply the monthly payment by the total number of months.)

c. How much interest will be paid with each option? (Interest = total amount repaid − amount borrowed.)

d. In the long run, which option is better and why?

CALCULATOR CONNECTION

53. Given the Cobb–Douglas Productivity Model $Q(x, y) = 6x^{0.4}y^{0.6}$, graph the level curves corresponding to $k = 3$, $k = 6$, and $k = 12$ using a window of $[0, 5]$ for both x and y.

54. A large corporation has 10 plants across the country. In 2005, the data for each plant gave the number of labor hours (in thousands), capital (in millions of dollars), and total number of units produced, as shown in the table.

Labor	250	270	300	320	320	350	400	440	450	450
Capital	25	35	42	45	50	56	65	71	75	78
Output	124	141	160	170	175	192	220	245	250	253

The plants all use the same technology, and the Cobb–Douglas Productivity Model is $Q(x, y) = 0.82x^{0.78}y^{0.22}$, where x is the number of employees (in thousands), y is the capital (in millions of dollars), and Q is the output (in hundreds of units).

a. Graph isoquants for $Q = 100$, $Q = 150$, $Q = 200$, $Q = 250$, and $Q = 300$. Use a window of $[0, 400]$ for both x and y. The isoquant curves are given by $y = \dfrac{Q}{x}$, so you will first need to solve $Q(x, y) = 0.82x^{0.78}y^{0.22}$ for y and then substitute the output values.

b. The CEO wants to keep labor fixed at 320,000 employees. If labor is fixed at 320,000, use the graph from part a to approximate how much capital is required for an output of 200 units.

c. The CFO wants to keep capital fixed at \$60 million. If capital is fixed at \$60 million, use the graph from part a to approximate how much labor is required for an output of 200 units.

TOPIC

Partial Derivatives

29

Brian's Beach Shop manufactures surfboards. The output each week is given by $Q(x, y) = 4.5x^{0.8}y^{0.2}$, where x is the number of employees and y is the number of machines used. Suppose Brian wanted to know the rate of change in output if the number of employees changes but the number of machines stays fixed. You know that a derivative is needed to find rate of change, but how do you find derivatives of functions of more than one variable? The derivative of a function of more than one variable is called a partial derivative. In this topic you will learn how to find partial derivatives and how to interpret their meaning. Before starting the topic, be sure to complete the Warm-up Exercises to sharpen the algebra and calculus skills you will need.

IMPORTANT TOOLS IN TOPIC 29

- *Finding first partial derivatives of functions of more than one variable*
- *Partial derivatives as a rate of change in a specific direction*
- *Partial derivatives as a slope along a surface in the x or y direction*
- *Second partial derivatives*

TOPIC 29 WARM-UP EXERCISES

Be sure you can successfully complete the following exercises before starting Topic 29.

Algebra Warm-up Exercises

Solve each system of equations.

1. $x - 2y = 4$
$3x + y = 6$

2. $y = x^2$
$x = y^2$

Calculus Warm-up Exercises

Find $f'(x)$ for each of the following functions.

1. $f(x) = 5$

2. $f(x) = 5x^3$

3. $f(x) = e^{5x}$

4. $f(x) = \frac{5}{x^2}$

5. $f(x) = 5x^2 \ln(2x)$

6. $f(x) = \frac{2x^3}{(3x - 5)^2}$

7. Use the limit definition of derivative to find $f'(x)$ for $f(x) = 3x^2 - 5x + 4$.

8. The output of a shoe factory is given by the Cobb–Douglas Model $Q(x, y) = 2.5x^{0.7}y^{0.3}$, where x is the number of employees (in hundreds), y is the capital (in thousand dollars), and Q is the output (in thousands of units).

a. Evaluate $Q(5, 225)$ and interpret your answer.

b. Suppose the capital is fixed at \$500,000. How many employees are needed to produce an output of 7500 units?

Use implicit differentiation to find $\frac{dy}{dx}$.

9. $x^2 + 5y = 6$

10. $2x^3y^2 = 5 - \frac{y}{x}$

Answers to Warm-up Exercises

Algebra Warm-up Exercises

1. $\left(\frac{16}{7}, -\frac{6}{7}\right)$

2. $(0, 0)$ and $(1, 1)$

Calculus Warm-up Exercises

1. $f'(x) = 0$

2. $f'(x) = 15x^2$

3. $f'(x) = 5e^{5x}$

4. $f'(x) = -\frac{10}{x^3}$

5. $f'(x) = 5x(1 + 2\ln(2x))$

6. $f'(x) = \dfrac{6x^2(x-5)}{(3x-5)^3}$

7. $f'(x) = \lim\limits_{h\to 0} \dfrac{(3(x+h)^2 - 5(x+h) + 4) - (3x^2 - 5x + 4)}{h}$

$= \lim\limits_{h\to 0} \dfrac{6xh + 3h^2 - 5h}{h}$

$= \lim\limits_{h\to 0} (6x + 3h - 5) = 6x - 5$

8. a. $Q(5, 225) = 39.1628$

With 500 employees and $225,000 of capital, the factory produces 39,163 shoes.

b. about 33 employees

9. $\dfrac{dy}{dx} = -\dfrac{2x}{5}$

10. $\dfrac{dy}{dx} = \dfrac{y - 6x^4y^2}{4x^5y + x}$

Partial Derivatives

You already know that the first derivative of a function $f(x)$ gives the rate at which the function changes with respect to changes in the independent variable x. You also know that the derivative was defined as a limit:

$$f'(x) = \lim_{h\to 0} \frac{f(x+h) - f(x)}{h}$$

assuming the limit exists.

What about rates of change for functions of two variables? For instance, in a Cobb–Douglas Productivity Model, you know that output is a function of both labor and capital. How will changes in the amount of labor affect the amount of output? We can answer this question by holding capital constant and then allowing labor to vary and examining the effect of that variation on the output. By holding labor fixed and allowing capital to vary, we can see how changes in the amount of capital affect the amount of output.

We use a similar process to find the rate of change of $f(x, y)$ with respect to one of its independent variables. Derivatives of this sort are called **partial derivatives**, because the derivative is taken with respect to only one of the independent variables. The process of finding the derivatives of a function of two or more variables is called **partial differentiation**.

Definition: If $z = f(x, y)$, then the **first partial derivatives** of $f(x, y)$ with respect to x and y are

$$f_x = \lim_{h\to 0} \frac{f(x+h, y) - f(x, y)}{h} \text{ and } f_y = \lim_{h\to 0} \frac{f(x, y+h) - f(x, y)}{h}$$

assuming that the limits exist.

First partial derivatives can also be symbolized as $\dfrac{\partial f}{\partial x}$ and $\dfrac{\partial f}{\partial y}$. The "$\partial$" is not a Greek letter but simply a mathematical symbol meaning partial derivative.

To apply the definition, we will treat one of the independent variables as a constant and then use derivative rules on the other independent variable.

Example 1: For $f(x, y) = 3x^2 + x^3y^2 - 5y$, find f_x and f_y.

Solution: To find f_x, treat y as a constant.

$$f_x = 6x + 3x^2y^2 - 0 = 6x + 3x^2y^2$$

Tip: Differentiating the term x^3y^2 did not require the Product Rule, because the y variable was considered constant.

To find f_y, treat x as a constant.

$$f_y = 0 + 2x^3y - 5 = 2x^3y - 5$$

Tip: Differentiating the term x^3y^2 did not require the Product Rule, because the x variable was considered constant.

■

Example 2: For $f(x, y) = \frac{x^2}{2y} + \frac{4y^2}{x}$, find $f_x(2, -1)$ and $f_y(2, -1)$.

Solution: Holding y constant,

$$f(x, y) = \left(\frac{1}{2y}\right)x^2 + 4y^2\left(\frac{1}{x}\right), \text{ so } f_x = \left(\frac{1}{2y}\right)2x + 4y^2\left(-\frac{1}{x^2}\right)$$

$$f_x = \frac{x}{y} - \frac{4y^2}{x^2}$$

Thus,

$$f_x(2, -1) = \frac{2}{-1} - \frac{4(-1)^2}{2^2} = -2 - 1 = -3$$

Holding x constant,

$$f(x, y) = x^2\left(\frac{1}{2y}\right) + \left(\frac{1}{x}\right)4y^2, \text{ so } f_y = x^2\left(-\frac{1}{2y^2}\right) + \left(\frac{1}{x}\right)(8y)$$

$$f_y = -\frac{x^2}{2y^2} + \frac{8y}{x}$$

Thus,

$$f_y(2, -1) = -\frac{2^2}{2(-1)^2} + \frac{8(-1)}{2} = -2 - 4 = -6$$

■

Graphical Interpretation of First Partial Derivatives

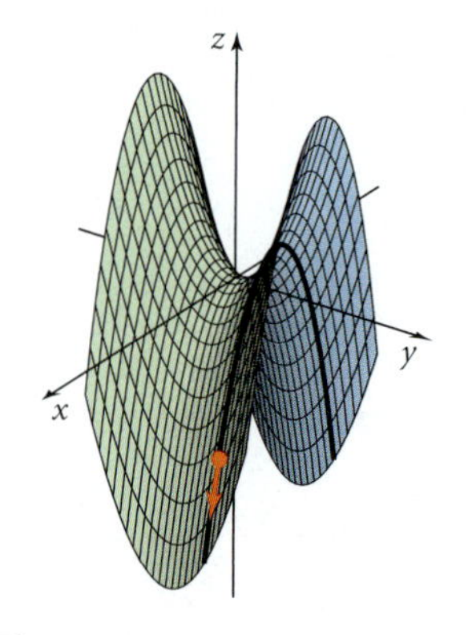

Figure 29.1

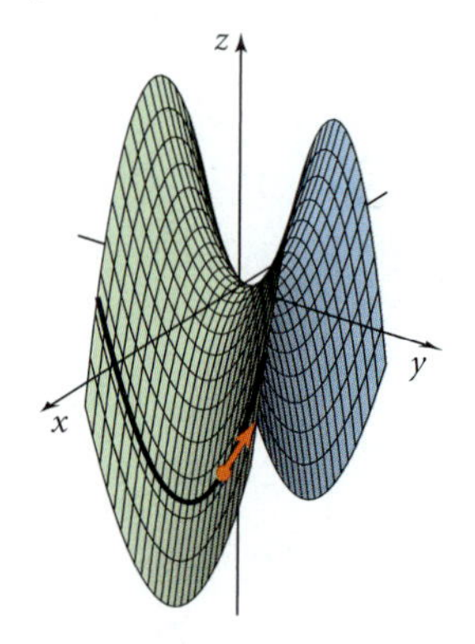

Figure 29.2

As with ordinary derivatives, there is both a graphical interpretation of the first partial derivative and a functional interpretation of the first partial derivative. We first consider the graphical interpretation.

Example 3: For $f(x, y) = y^2 - x^2$, find $f_x(2, 1)$ and $f_y(2, 1)$.

Solution: $f_x = -2x$ and $f_y = 2y$. Evaluating with $x = 2$ and $y = 1$, we have $f_x(2, 1) = -2(2) = -4$ and $f_y(2, 1) = 2(1) = 2$.

Let's explore the graphical interpretation of these first partial derivatives. $f(2, 1) = (1)^2 - 2^2 = 1 - 4 = -3$. This means that $(2, 1, -3)$ is a point on the graph of $f(x, y) = y^2 - x^2$. Hold y constant and move from the point $(2, 1, -3)$ along the curve in the x-direction.

From Figure 29.1 you see that the values of the function decrease as the point moves along the curve. The expression $f_x(2, 1) = -4$ means that the value of $f(x, y)$ changes at a rate of $-4 \dfrac{\textit{units of } f}{\textit{units of } x}$, or, in other words, the value of $f(x, y)$ decreases four units for a one-unit increase in x. We say that the slope of $f(x, y)$ in the x-direction is -4.

Now hold x constant and move from the point $(2, 1, -3)$ along the curve in the y-direction. From Figure 29.2 you see that the values of the function increase as the point moves along the curve. The expression $f_y(2, 1) = 2$ means that the value of $f(x, y)$ changes at a rate of $2 \dfrac{\textit{units of } f}{\textit{units of } y}$, or, the value of $f(x, y)$ increases two units for each unit increase in y. We say that the slope of $f(x, y)$ in the y-direction is 2. ■

Functional Interpretation of First Partial Derivatives

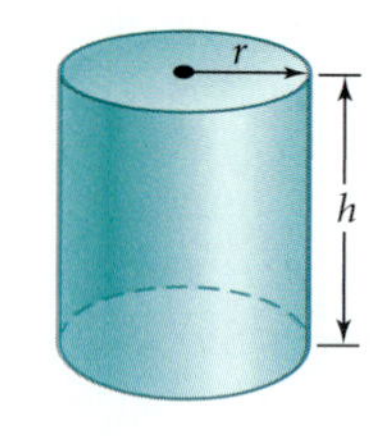

Figure 29.3

Now we turn to some functional interpretations of first partial derivatives.

Example 4: The volume of a right circular cylinder (a soda can) is given by $V(r, h) = \pi r^2 h$, where r is the radius and h is the height as shown in Figure 29.3.

a. Find V_r and V_h.

b. Evaluate V_r and V_h if $r = 4$ inches and $h = 10$ inches and interpret your answers.

Solution:

a. $V_r = 2\pi rh$ and $V_h = \pi r^2$.

b. $V_r(4, 10) = 2\pi(4)(10) = 80\pi$. If the height of the cylinder is fixed at 10 inches, the volume increases at a rate of 80π cubic inches per 1 inch increase in the radius. $V_h(4, 10) = \pi(4)^2 = 16\pi$. If the radius of the cylinder is fixed at 4 inches, the volume increases at a rate of 16π cubic inches per 1 inch increase in the height. ■

Example 5: *Marginal Revenue.* A college bookstore sells two kinds of calculators. The price of the MT-56 is given by $a = \frac{1}{5}x + \frac{2}{15}y - 76\frac{2}{3}$ dollars. The price of the TR-98 is given by $b = \frac{1}{5}x + \frac{1}{15}y - 68\frac{1}{3}$ dollars, where x is the number of MT-56 calculators sold and y is the number of TR-98 calculators sold this month.

a. Find the revenue function $R(x, y)$.

b. Find R_x and R_y.

c. Evaluate $R_x(350, 480)$ and interpret your answer.

Solution:

a. Revenue is price times quantity: $R(x, y) = a \cdot x + b \cdot y$.

$$R(x, y) = \left(\frac{1}{5}x + \frac{2}{15}y - 76\frac{2}{3}\right) \cdot x + \left(\frac{1}{5}x + \frac{1}{15}y - 68\frac{1}{3}\right) \cdot y$$

$$= \frac{1}{5}x^2 + \frac{1}{3}xy - 76\frac{2}{3}x + \frac{1}{15}y^2 - 68\frac{1}{3}y$$

b. $R_x = \frac{2}{5}x + \frac{1}{3}y - 76\frac{2}{3}$ and $R_y = \frac{1}{3}x + \frac{2}{15}y - 68\frac{1}{3}$.

c. $R_x(350, 480) = \frac{2}{5}(350) + \frac{1}{3}(480) - 76\frac{2}{3} = 223\frac{1}{3}$.

When 350 MT-56 calculators and 480 TR-98 calculators have been sold, the bookstore's revenue is increasing at a rate of $223.33 for each additional MT-56 calculator sold. ■

Definition: For the Cobb–Douglas Productivity Model

$$Q(x, y) = Kx^a y^{1-a},$$

where x represents labor units and y represents capital, we define the following:

- $Q_x(x, y)$ gives the change in output per unit change in labor, holding capital fixed. This change is called **marginal productivity of labor.**
- $Q_y(x, y)$ gives the change in output per unit change in capital, holding labor fixed. This change is called **marginal productivity of capital.**

Example 6: Brian's Beach Shop manufactures surfboards. The output each week is given by $Q(x, y) = 4.5x^{0.8}y^{0.2}$, where x is the number of employees and y is the number of machines used. Compute the marginal productivity of labor and the marginal productivity of capital if $x = 25$ and $y = 10$. Interpret the answers.

Solution: $Q_x = 3.6x^{-0.2}y^{0.2}$ and $Q_y = 0.9x^{0.8}y^{-0.8}$.

For $x = 25$ and $y = 10$, we evaluate $Q_x(25, 10)$, which gives

$$Q_x(25, 10) = 3.6(25)^{-0.2}(10)^{0.2} \approx 2.997$$

This answer tells us that when the shop uses 25 employees and 10 machines and keeps the number of machines fixed at 10, weekly output increases by about three boards if one additional employee is working.

For $x = 25$ and $y = 10$, we evaluate $Q_y(25, 10)$, which gives

$$Q_y(25, 10) = 0.9(25)^{0.8}(10)^{-0.8} \approx 1.873$$

This answer tells us that when the shop uses 25 employees and 10 machines and keeps the number of employees fixed at 25, output increases by about two boards if one additional machine is used. ■

A logical question often asked in business is whether output increases more rapidly if labor is increased (with capital fixed) or if capital is increased (with labor fixed). To approximate the change in productivity, we define the following.

Change in Productivity
Given the Cobb–Douglas Productivity Model $Q(x, y) = Kx^a y^{1-a}$, where x represents labor units and y represents capital:

- The **change in output with respect to labor** is Q_x times the change in labor.
- The **change in output with respect to capital** is Q_y times the change in capital.

Example 7: Brian's Beach Shop currently uses 25 employees and 10 machines (see Example 6). Would output increase more if the number of employees were increased to 30 or if three more machines were added?

Solution: From Example 6 you know that output each week is given by $Q(x, y) = 4.5x^{0.8}y^{0.2}$, where x is the number of employees and y is the number of machines used. You also know that $Q_x = 3.6x^{-0.2}y^{0.2}$ and $Q_y = 0.9x^{0.8}y^{-0.8}$.

The shop is considering changing labor by adding five employees, keeping the number of machines fixed. The other consideration is to add three additional machines, keeping the number of employees fixed.

We must compare $Q_x(25, 10) \cdot 5$ and $Q_y(25, 10) \cdot 3$.

Additional labor: $Q_x(25,10) \cdot 5 = 2.997(5) = 14.985$

Additional machine: $Q_y(25,10) \cdot 3 = 1.873(3) = 5.619$

If five additional employees were hired and the shop still used 10 machines, then the weekly output would increase by about 15 surfboards. If three new machines are added, still using 25 employees, then the weekly output increases by about six boards. It is clear that adding five more employees and keeping the same number of machines would result in a greater increase in output. ■

Let's summarize what you have learned about the meaning of first partial derivatives.

$y = f(x)$	$z = f(x, y)$
$f'(x)$ is the rate at which $f(x)$ changes with respect to changes in x.	f_x is the rate at which $f(x, y)$ changes with respect to x, holding y constant. f_y is the rate at which $f(x, y)$ changes with respect to y, holding x constant.
$f'(x)$ is the slope of the curve $y = f(x)$ at any point.	f_x is the slope of $f(x, y)$ in the x-direction, with y held constant. f_y is the slope of $f(x, y)$ in the y-direction, with x held constant.

Check Your Understanding 29.1

Find f_x and f_y for each of the following functions.

1. $f(x, y) = x^3y^2 - e^{xy} + \frac{4}{x}$
2. $f(x, y) = \sqrt{4 - 2x^2 - y^2}$
3. $f(x, y) = \frac{4x}{3x - 2y}$
4. In Exercise 1, evaluate $f_x(2, 0)$.
5. In Exercise 3, evaluate $f_y(-3, 2)$.
6. In Example 5, evaluate $R_y(200, 420)$ and interpret.

Higher-Order Partial Derivatives

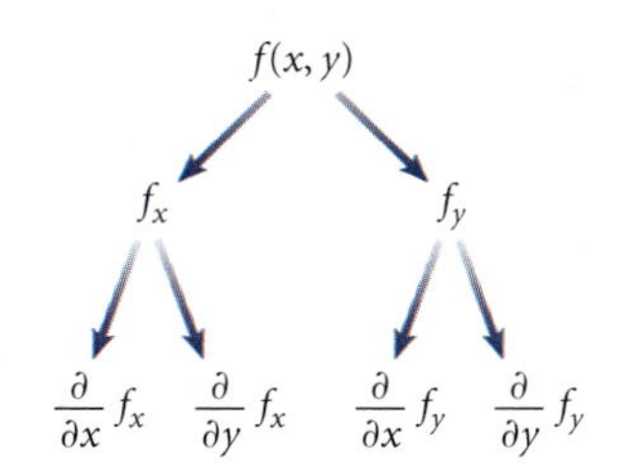

Figure 29.4

You already know that for the first derivative of $f(x)$, you can also find its second derivative, if it exists. Finding second derivatives for functions of two variables is a little more complicated, because there are two first partial derivatives for $f(x, y)$, one for each independent variable. That means there are four second partial derivatives, two second partial derivatives for each first partial derivative as shown in Figure 29.4.

Definition: For $f(x, y)$ the **second partial derivatives** are

$$f_{xx} = \frac{\partial}{\partial x} f_x = \frac{\partial}{\partial x}\frac{\partial f}{\partial x} = \frac{\partial^2 f}{\partial x^2} \qquad f_{xy} = \frac{\partial}{\partial y} f_x = \frac{\partial}{\partial y}\frac{\partial f}{\partial x} = \frac{\partial^2 f}{\partial y \partial x}$$

$$f_{yy} = \frac{\partial}{\partial y} f_y = \frac{\partial}{\partial y}\frac{\partial f}{\partial y} = \frac{\partial^2 f}{\partial y^2} \qquad f_{yx} = \frac{\partial}{\partial x} f_y = \frac{\partial}{\partial x}\frac{\partial f}{\partial y} = \frac{\partial^2 f}{\partial x \partial y}$$

The notation f_{xy} means find f_x and then differentiate it with respect to y. The first subscript is the first derivative, and the second subscript is the second derivative.

Example 8: Find the second partial derivatives of $f(x, y) = x^3 + 2x^4y^2 - \frac{3}{y}$.

Solution: First we find the first partial derivatives.

$$f_x = 3x^2 + 8x^3y^2 \quad \text{and} \quad f_y = 4x^4y + \frac{3}{y^2}$$

From these first partial derivatives, the second partial derivatives are

$$f_x = 3x^2 + 8x^3y^2 \qquad f_y = 4x^4y + \frac{3}{y^2}$$

$$\text{so } f_{xx} = 6x + 24x^2y^2 \qquad \text{so } f_{yy} = 4x^4 - \frac{6}{y^3}$$

$$\text{and } f_{xy} = 16x^3y \qquad \text{and } f_{yx} = 16x^3y$$

■

The second partial derivatives f_{xy} and f_{yx} are called **mixed partial derivatives,** because the first derivative and the second derivative are done with respect to different variables. You should see that $f_{xy} = f_{yx}$ in Example 8. While it is not the case that $f_{xy} = f_{yx}$ for all functions of two variables, the equality is true if f_{xy} and f_{yx} are both continuous, which is usually the case in most practical applications.

Example 9: Evaluate $f_{xy}(2, -3)$ for $f(x, y) = ye^{xy}$.

Solution: The first partial derivative with respect to x is

$$f_x = ye^{xy}(y) = y^2e^{xy}$$

The second partial derivative with respect to y requires the Product Rule, because both factors, y^2 and e^{xy}, are functions of the variable y.

$$f_{xy} = y^2 \cdot \frac{\partial}{\partial y}e^{xy} + e^{xy} \cdot \frac{\partial}{\partial y}y^2 \quad \text{Apply the Product Rule}$$

$$f_{xy} = y^2 \cdot xe^{xy} + 2ye^{xy} \quad \text{Find derivatives}$$

$$f_{xy} = ye^{xy}(xy + 2) \quad \text{Simplify}$$

Then $f_{xy}(2, -3) = (-3)e^{2(-3)}(2(-3) + 2) = -3e^{-6}(-4) = 12e^{-6}$. ■

Check Your Understanding 29.2

Find the second partial derivatives of each function.

1. $f(x, y) = 2x^3y + 4x - 3y$
2. $f(x, y) = \dfrac{x^2}{2y - 3}$
3. $f(x, y) = x^2 - y^2\ln(4x - 3)$
4. For $f(x, y) = 4x^3y^2 - \dfrac{3}{x^2} + e^y$ evaluate $f_{xy}(1, -2)$ and $f_{yx}(1, -2)$.

Check Your Understanding Answers

Check Your Understanding 29.1

1. $f_x = 3x^2y^2 - ye^{xy} - \dfrac{4}{x^2}$ $\quad f_y = 2x^3y - xe^{xy}$
2. $f_x = \dfrac{-2x}{\sqrt{4 - 2x^2 - y^2}}$ $\quad f_y = \dfrac{-y}{\sqrt{4 - 2x^2 - y^2}}$
3. $f_x = \dfrac{-8y}{(3x - 2y)^2}$ $\quad f_y = \dfrac{8x}{(3x - 2y)^2}$
4. $f_x(2, 0) = -1$
5. $f_y(-3, 2) = -\dfrac{24}{169}$
6. $R_y(200, 420) = 54\frac{1}{3}$
   When 200 MT-56 calculators and 420 TR-98 calculators have been sold, the bookstore's revenue increases an additional \$54.33 for each additional TR-98 sold.

Check Your Understanding 29.2

1. $f_{xx} = 12xy \quad f_{yy} = 0$
 $f_{xy} = 6x^2 \quad f_{yx} = 6x^2$

2. $f_{xx} = \dfrac{2}{2y - 3} \quad f_{yy} = \dfrac{8x^2}{(2y - 3)^3}$
 $f_{xy} = \dfrac{-4x}{(2x - 3)^2} \quad f_{yx} = \dfrac{-4x}{(2y - 3)^2}$

3. $f_{xx} = 2 + \dfrac{16y^2}{(4x - 3)^2} \quad f_{yy} = -2\ln(4x - 3)$
 $f_{xy} = \dfrac{-8y}{4x - 3} \quad f_{yx} = \dfrac{-8y}{4x - 3}$

4. $f_{xy}(1, -2) = f_{yx}(1, -2) = -48$

Topic 29 Review

29

This topic introduced partial derivatives to differentiate functions of two or more variables.

CONCEPT	EXAMPLE
If $z = f(x, y)$, then the **first partial derivatives** of $f(x, y)$ with respect to x and y are $\dfrac{\partial f}{\partial x} = f_x = \lim_{h \to 0} \dfrac{f(x + h, y) - f(x, y)}{h}$ $\dfrac{\partial f}{\partial y} = f_y = \lim_{h \to 0} \dfrac{f(x, y + h) - f(x, y)}{h}$ assuming that the limits exist. To determine f_x, treat the y variable as a constant and use regular differentiation rules. To determine f_y, treat the x variable as a constant and use regular differentiation rules.	Given $f(x, y) = 3x^2 - 2xy^3 + 7y$, the first partial derivatives are $f_x = 6x - 2y^3$ and $f_y = -6xy^2 + 7$. Note that the Product Rule was not necessary when differentiating $-2xy^3$. To find f_x, y was treated as a constant. To find f_y, x was treated as a constant.
The Cobb–Douglas Productivity Model is $Q(x, y) = Kx^a y^{1-a}$, where x represents labor units and y represents capital. $Q_x(x, y)$ gives the change in output per unit change in labor, holding capital fixed. This change is called **marginal productivity of labor**. $Q_y(x, y)$ gives the change in output per unit change in capital, holding labor fixed. This change is called **marginal productivity of capital**.	A factory's output per week, in hundreds, is given by $Q(x, y) = 5x^{0.7}y^{0.3}$, where x is the number of employees and y is the number of machines used. **a.** The marginal productivity for 30 employees and 10 machines is $Q_x(30, 10) = 2.52$ and $Q_y(30, 10) = 3.24$. Output changes at a rate of 252 units if there are 10 machines and the number of employees increases by one. Output changes at a rate of 324 units if the number of

(continued)

The **change in output with respect to labor** is Q_x times the change in labor.

The **change in output with respect to capital** is Q_y times the change in capital.

employees stays fixed and the number of machines is increased by one.

b. The company must decide whether to hire five more employees or add two new machines. The change in output with respect to labor is Q_x times the change in labor: $(2.52)(5) = 12.6$. The change in output with respect to capital is Q_y times the change in capital: $(3.24)(2) = 6.48$. The greater change in output occurs if five additional employees are hired.

Graphically, given point (x_1, y_1, z_1) on the surface $z = f(x, y)$, then $f_x(x_1, y_1)$ is the slope in the x-direction and $f_y(x_1, y_1)$ is the slope in the y-direction.

For $f(x, y) = x^2 + y^2$, we have $f_x(2, 1) = 4$ and $f_y(2, -1) = -2$. Graphically, at the point $(2, -1, 5)$:

If y is held constant the value of f increases at a rate of four units per one unit change in x.

If x is held constant the value of f decreases at a rate of two units per one unit change in y.

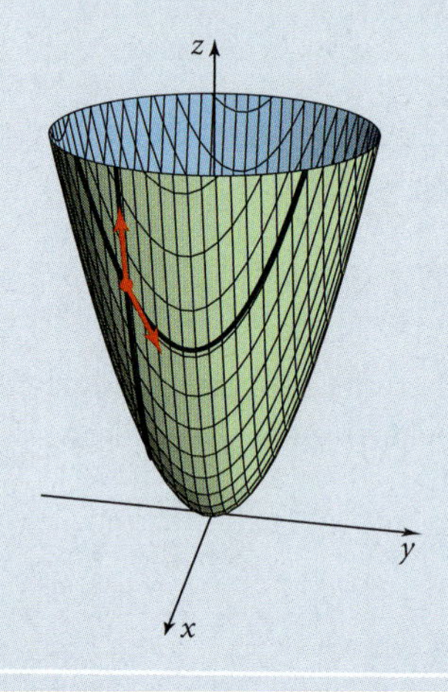

For $f(x, y)$, the **second partial derivatives** are

$$f_{xx} = \frac{\partial}{\partial x} f_x$$

$$f_{xy} = \frac{\partial}{\partial y} f_x$$

$$f_{yy} = \frac{\partial}{\partial y} f_y$$

$$f_{yx} = \frac{\partial}{\partial x} f_y$$

Given $f(x, y) = 3x^2 - 2xy^3 + 7y$, the second partial derivatives are

$$f_{xx} = 6$$

$$f_{xy} = -6y^2$$

$$f_{yy} = -12xy$$

$$f_{yx} = -6y^2$$

f_{xy} and f_{yx} are called **mixed partial derivatives.** The mixed-order partial derivatives are equal if both f_{xy} and f_{yx} are continuous, which is usually the case in most practical applications.

Given $f(x, y) = 3x^2 - 2xy^3 + 7y$, the mixed partial derivatives are $f_{xy} = -6y^2$ and $f_{yx} = -6y^2$. In this case $f_{xy} = f_{yx}$ because both f_{xy} and f_{yx} are continuous.

NEW TOOLS IN THE TOOL KIT

- Finding first partial derivatives of functions of two variables
- Evaluating first partial derivatives at specific points
- Interpreting first partial derivatives as a rate of change or as a slope along the surface in either the x-direction or the y-direction
- Finding second partial derivatives of functions of two variables
- Evaluating second partial derivatives at specific points

Topic 29 Exercises

29

Find f_x and f_y for each of the following functions.

1. $f(x, y) = 5x - 2y + 7$

2. $f(x, y) = x^2 + 5y + 9x$

3. $f(x, y) = x^3 - 2xy + 4y^2$

4. $f(x, y) = x^2y - xy^3$

5. $f(x, y) = y\sqrt{x}$

6. $f(x, y) = \sqrt{xy}$

7. $f(x, y) = e^{xy}$

8. $f(x, y) = e^{x/y}$

9. $f(x, y) = ye^{xy}$

10. $f(x, y) = x^2ye^{3y}$

11. $f(x, y) = \sqrt{x^2 - y^2}$

12. $f(x, y) = \sqrt{4 - x^2 - y^2}$

13. $f(x, y) = \ln(2x - 3y)$

14. $f(x, y) = \ln(x^2 - y^2)$

15. $f(x, y) = \dfrac{x^4}{3y} - \dfrac{2y^2}{5x}$

16. $f(x, y) = \dfrac{x^2}{2x - 3y}$

17. $f(x, y) = \dfrac{2xy}{x - y}$

18. $f(x, y) = \dfrac{x^2 - y^2}{2x^2 + 3y}$

19. $f(x, y) = y^3\ln(4x - 3)$

20. $f(x, y) = x^2y^2 + \ln(4x - 3)$

For each of the following functions, find the slope in the x-direction and the slope in the y-direction at the indicated point.

21. $f(x, y) = 9 - x^2 - 3y^2$ at $(1, 1, 5)$

22. $f(x, y) = x^2 - 2y^2$ at $(2, -1, 2)$

23. $f(x, y) = x^2e^{3y}$ at $(3, 0, 9)$

24. $f(x, y) = \dfrac{xy}{x + y}$ at $\left(-2, 5, \dfrac{10}{3}\right)$

For what values of x and y are both $f_x = 0$ and $f_y = 0$ simultaneously?

25. $f(x, y) = x^2 - 6xy + y^2 + 4x - 16y + 2$

26. $f(x, y) = x^3 - 15xy - y^3$

27. $f(x, y) = xy - \dfrac{1}{x} - \dfrac{1}{y}$

28. $f(x, y) = \ln(2 - x^2 - y^2)$

Find the four second partial derivatives of each of the following functions. Verify that the mixed partial derivatives are equal.

29. $f(x, y) = 2x^2 - 5xy + y^2$

30. $f(x, y) = x^4 + 4x^2y^2 - y^4$

31. $f(x, y) = \sqrt{x^2 - y^2}$

32. $f(x, y) = \sqrt{4 - x^2 - y^2}$

33. $f(x, y) = \ln(x^2 - y^2)$

34. $f(x, y) = y\ln(2x + 5)$

35. $f(x, y) = x^2e^y$ **36.** $f(x, y) = xye^y$

A function $f(x, y)$ satisfies Laplace's equation, named for Pierre Laplace (1749–1827), if $f_{xx} + f_{yy} = 0$ for all (x, y) in the domain of f. Do the following functions satisfy Laplace's equation? Justify your answer.

37. $f(x, y) = 8xy$ **38.** $f(x, y) = \dfrac{1}{\sqrt{x + y}}$

39. $f(x, y) = xe^y - ye^x$ **40.** $f(x, y) = x^2 - y^2$

41. The surface area of a right circular cylinder is given by $S(r, h) = 2\pi r^2 + 2\pi rh$, where r is the radius and h is the height.

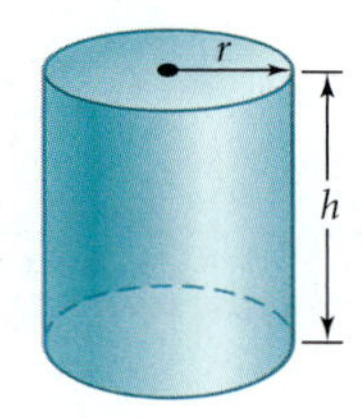

a. Find S_r and S_h.

b. Evaluate S_r and S_h if the radius is 4 inches and the height is 10 inches. Interpret your answers.

42. The volume of a cone is given by $V(r, h) = \frac{1}{3}\pi r^2h$, where r is the radius and h is the height.

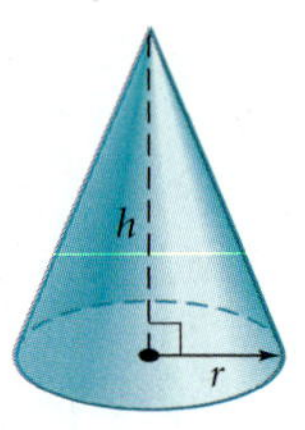

a. Find V_r and V_h.

b. Evaluate V_r and V_h if the radius is 4 inches and the height is 10 inches. Interpret your answers.

43. A college spends x thousand dollars a year on newspaper and television advertising and y thousand dollars on mailings and brochures. The total cost, in thousand dollars, of these advertising venues is given by $C(x, y) = 4x^2 + 5y$.

a. Find C_x and C_y.

b. Evaluate C_x and C_y at (20, 35) and interpret your answers.

44. The local Kiwanis Club sponsors a pancake breakfast each spring. Supplies and labor are donated. Total sales from the breakfast depend on the amount spent on advertising, a, and the number of tickets sold, t. Sales from last year's breakfast are given by $S(t, a) = 850 - 0.05a^2 + 3t$.

a. Find S_t and S_a.

b. Evaluate S_t and S_a if advertising cost \$300 and 1650 tickets were sold. Interpret your answers.

45. The duration of a scuba dive, in minutes, is given by $T(v, D) = \dfrac{33v}{D + 33}$, where v is the volume of air in the tank (at sea level pressure) and D is the depth of the dive in feet. Evaluate $T(90, 30)$, $T_v(90, 30)$, and $T_D(90, 30)$ and interpret your answers.

46. A person's body mass index (BMI) depends on weight, in pounds, and height, in inches. BMI is given by $B(w, h) = \dfrac{w}{h^2} \cdot 703$. Evaluate $B(175, 70)$, $B_w(175, 70)$, and $B_h(175, 70)$ and interpret your answers.

47. In November 2001 the National Weather Service published an updated Wind Chill Temperature Index that uses air temperature in degrees Fahrenheit (°F) and wind speed in miles per hour to measure how cold it actually feels. Assuming temperatures below 40°F and wind speeds between 5 and 60 mph, the wind chill is approximated by $W(T, v) = 35.74 + 0.6215T - 35.75v^{0.16} + 0.4275Tv^{0.16}$, where T is air temperature (°F) and v is wind speed (mph). Calculate the following and interpret the answer.

a. $W(15, 10)$ **b.** $W_T(15, 10)$

c. $W_v(15, 10)$

48. The National Weather Service also has a Heat Index table that uses air temperature, T, in degrees Fahrenheit (°F) and relative humidity, r, to measure how hot it actually feels. Assuming temperatures of at least 80°F and relative humidity of at least 40%, the heat index is approximated by

$$\begin{aligned} H(T, r) = {} & -42.4 + 2.049T + 10.143r \\ & - 0.2248Tr - 0.00684T^2 \\ & - 0.0548r^2 + 0.00123T^2r \\ & + 0.00085Tr^2 - 0.00000199T^2r^2 \end{aligned}$$

Calculate the following and interpret the answer.

a. $H(96, 60)$

b. $H_T(96, 60)$

c. $H_r(96, 60)$

49. A bowling ball manufacturer has the Cobb–Douglas Productivity Model $Q(x, y) = 12x^{0.25}y^{0.75}$, where x is labor, measured in thousands of employees; y is capital, measured in thousand dollars; and Q is the number of bowling balls produced, in thousands.

a. Find Q_x and Q_y.

b. If the factory currently uses 2200 employees and $750,000 of capital, find the marginal productivity of labor and the marginal productivity of capital.

c. Would production increase more rapidly with 500 more employees or $250,000 additional capital?

50. Jo Jo's Jelly Beans has 10 plants across the country all using the same technology. Output is given by $Q(x, y) = 0.725x^{0.65}y^{0.35}$, where x is the number of employees in thousands, y is the capital in million dollars, and Q is the number of bags of jelly beans in millions.

a. Find Q_x and Q_y.

b. If the factory currently uses 13,500 employees and $1,200,000 of capital, find the marginal productivity of labor and the marginal productivity of capital.

c. Would production increase more rapidly with 2500 more employees or $200,000 additional capital?

51. Gary's Overhead Door Company has output modeled by $Q(x, y) = 1.4x^{0.72}y^{0.28}$, where x is the number of employees in hundreds, y is the capital in thousand dollars, and Q is the number of garage doors in hundreds.

a. Find Q_x and Q_y.

b. If the factory currently has 6500 employees and $820,000 of capital, find the marginal productivity of labor and the marginal productivity of capital.

c. Would production increase more rapidly with 100 more employees or $100,000 additional capital?

52. Betty's Crafts makes unique wicker gift baskets. The Cobb–Douglas model is $Q(x, y) = 0.36x^{0.81}y^{0.19}$, where x is the number of employees, y is the capital in thousand dollars, and Q is the number of baskets in hundreds.

a. Find Q_x and Q_y.

b. If Betty currently has 45 employees and uses $36,000 of capital, find the marginal productivity of labor and the marginal productivity of capital.

c. Would production increase more rapidly if Betty hired one more employee or used an additional $5000 in capital?

53. A large appliance store sells two models of refrigerators. The freezer-on-top model sells for $\$a$ and the side-by-side model sells for $\$b$. At those prices, the demand for the freezer-on-top model is $D_1 = 500 - 0.4a + 0.75b$ and the demand for the side-by-side model is $D_2 = 425 + 0.85a - 0.2b$.

a. Find the revenue function $R(a, b)$.

b. Evaluate $R(a, b)$, $R_a(a, b)$, and $R_b(a, b)$ if $a = \$925$ and $b = \$1475$ and interpret your answers.

54. A large theme park sells one-day passes for $\$x$ and four-day passes for $\$y$. At those prices the demand for the one-day pass is $D_1 = 850 - 30x + 50y$ and the demand for the four-day pass is $D_2 = 600 + 45x - 10y$.

a. Find the daily revenue function $R(x, y)$.

b. Evaluate $R(x, y)$, $R_x(x, y)$, and $R_y(x, y)$ if $x = \$65$ and $y = \$210$ and interpret your answers.

55. Refer to the appliance store in Exercise 53. The cost of the refrigerators to the store is given by

$$C(a, b) = \frac{a + b}{2} + 8\sqrt{a + b}.$$

a. Find the profit function $P(a, b)$.

b. Evaluate $P(a, b)$, $P_a(a, b)$, and $P_b(a, b)$ if $a = \$925$ and $b = \$1475$ and interpret your answers.

56. Refer to the theme park in Exercise 54. The park's daily costs are given by

$$C(x, y) = 3x^2 + e^{0.05y} + \frac{xy}{10}.$$

a. Find the daily profit function $P(x, y)$.

b. Evaluate $P(x, y)$, $P_x(x, y)$, and $P_y(x, y)$ if $x = \$65$ and $y = \$210$ and interpret your answers.

SHARPEN THE TOOLS

Find the second partial derivatives of each of the following functions. Verify that the mixed partial derivatives are equal.

57. $f(x, y) = x^2e^{xy}$

58. $f(x, y) = y^3e^{x^2y}$

59. $f(x, y) = \dfrac{xy}{2x - 3y}$

60. $f(x, y) = \dfrac{x^2}{x^2 + y^2}$

Partial derivatives can also be defined for functions of more than two variables. To find the first partial derivative with respect to one of the independent variables, treat all the other independent variables as constants. For example, if $P(a, b, c) = 2a^2b - 3b + 5abc$, the first partial derivatives are $P_a = 4ab + 5bc$, $P_b = 2a^2 - 3 + 5ac$, and $P_c = 5ab$.

61. Given $V(a, b, c) = 2a^3 + 3a^2b + 2b^2c + c^3$, find V_a, V_b, and V_c.

62. Given $f(x, y, z) = \dfrac{xy}{z} + \dfrac{xz}{y} + \dfrac{yz}{x}$, find f_x, f_y, and f_z.

Third partial derivatives can also be defined for functions of two variables. In subscript notation, the order of differentiation moves from left to right. P_{aba} means find the first derivative with respect to a, then the second derivative of P_a with respect to b, and finally the third derivative of P_{ab} with respect to a. For example, if $P(a, b, c) = 2a^2b - 3b + 5abc$, then $P_a = 4ab + 5bc$, $P_{ab} = 4a + 5c$, and $P_{aba} = 4$.

63. Given $f(x, y) = x^3y + 2x^2y^2 - 4y^3$, find f_{xxx} and f_{yxy}.

64. Given $P(r, t) = e^{rt} - \dfrac{t}{r} + 4rt$, find P_{rrt} and P_{trt}.

65. A company manufactures two types of refrigerators, a freezer-on-top model and a side-by-side model. The cost of producing x freezer-on-top models and y side-by-side models is given by $C(x, y) = 8\sqrt{xy} + 75x + 150y + 1200$ dollars.

a. Determine the marginal cost C_x and C_y for $x = 60$ and $y = 80$. Interpret your answers.

b. If additional productivity is required, which model causes the cost to rise at a faster rate? Justify your answer.

66. The number of applicants, in hundreds, to a small college is given by $N(t, r) = \dfrac{400}{t\sqrt{r}}$, where t is the tuition in thousand dollars and r is room and board in thousand dollars.

a. The cost of room and board is currently \$5000 and tuition is \$12,000. Evaluate N, N_t, and N_r for those values and interpret your answers.

b. Which will change the number of applicants more rapidly, changing the tuition or changing the cost of room and board?

67. The monthly payments for a 30-year mortgage are given by the function $M(P, i) = P\left(\dfrac{i}{1 - (1 + i)^{-360}}\right)$, where P is the amount borrowed and i is the monthly interest rate $\left(i = \dfrac{r}{12}\right)$.

a. Find M_P.

b. Evaluate M_P for a mortgage of \$220,000 with an annual interest rate of 7%. Interpret your answer.

c. For a mortgage of \$220,000 with an annual interest rate of 7%, use the result of part b to determine how much the payment would increase if the amount borrowed increased by \$5000.

68. The monthly payments for a 5-year car loan are given by the function $M(P, i) = P\left(\dfrac{i}{1 - (1 + i)^{-60}}\right)$, where P is the amount borrowed and i is the monthly interest rate $\left(i = \dfrac{r}{12}\right)$.

a. Find M_P.

b. Evaluate M_P for a loan of \$25,000 with an annual interest rate of 8%. Interpret your answer.

c. For a loan of \$25,000 with an annual interest rate of 8%, use the result of part b to determine how much the payment would increase if the amount borrowed increased by \$1000.

TOPIC 30

Optimization and Lagrange Multipliers

Suppose that a company sells two different models of a product. The company's sales revenue is a function of how many of each model is sold. How can the company determine how many of each model should be sold in order to maximize the company's profit? Optimization problems similar to this occur regularly in business. This topic will present two methods for optimizing functions of two variables. Before starting this topic, be sure to complete the Warm-up Exercises to sharpen the algebra and calculus skills you will need.

IMPORTANT TOOLS IN TOPIC 30

- *Extreme points of functions of two variables*
- *Second Partial Derivative Test to classify critical points as a relative maximum, a relative minimum, or a saddle point*
- *Lagrange multipliers to solve constrained optimization problems*

TOPIC 30 WARM-UP EXERCISES

Be sure you can successfully complete these exercises before starting Topic 30.

Algebra Warm-up Exercises

Solve each system of equations.

1. $3x + 2y = 10$, $5y - 2x = -4$

2. $2x + 5y^2 = 0$, $3x - 3y = 0$

3. $2x + y^2 = 0$, $x^2 - 2y = 0$

4. $x - 2y = 0$, $3y + 2z = 0$, $x + y + z = 6$

Solve each equation for x.

5. $2x + 3y = 12$

6. $3x - y^2 = 0$

7. $0.2x - 0.3y + 15 = 0$

Calculus Warm-up Exercises

For each of the following functions, determine the critical values. Then use the second derivative to determine the relative extreme points of each function.

1. $f(x) = 2x^2 - 8x + 3$

2. $f(x) = 2x^3 - 6x^2$

3. $f(x) = \dfrac{4}{x - 1}$

4. $f(x) = \dfrac{4x^2}{x^2 - 1}$

5. A rectangular storage area is to be fenced in along all four sides to enclose an area of 500 square feet. One side uses fencing that costs \$6 per foot; the other three sides use fencing that costs \$4 per foot. Find the dimensions that will minimize the cost.

For each of the following functions, find f_x and f_y. Evaluate each derivative at the given point.

6. $f(x, y) = 3x^2 - 7xy + 2y^2 - 6y + 8$ at $(2, -1)$

7. $f(x, y) = 3xe^{xy^2}$ at $(3, 0)$

8. $f(x, y) = \dfrac{x - 3y}{x^2 + y^2}$ at $(1, -2)$

9. Find the second partial derivatives of $f(x, y) = x^3 - \dfrac{7x}{y} + \dfrac{2y}{x} + y^3$.

Answers to Warm-up Exercises

Algebra Warm-up Exercises

1. $\left(\frac{58}{19}, \frac{8}{19}\right)$
2. $(0, 0)$ and $\left(-\frac{2}{5}, -\frac{2}{5}\right)$
3. $(0, 0)$ and $(-2, 2)$
4. $(8, 4, -6)$
5. $x = 6 - \frac{3}{2}y$
6. $x = \dfrac{y^2}{3}$
7. $x = \frac{3}{2}y - 75$

Calculus Warm-up Exercises

1. Critical value is $c = 2$.
   $f''(2) = 4 > 0$, so $f(2)$ is a relative minimum.
2. Critical values are $c = 0, 2$.
   $f''(0) = -12 < 0$, so $f(0)$ is a relative maximum.
   $f''(2) = 12 > 0$, so $f(2)$ is a relative minimum.
3. Critical value is $c = 1$. Because there is an asymptote at $x = 1$, there is no extreme point for this function.
4. Critical values are $c = 0, -1$, and 1.
   $f''(0) = -8 < 0$, so $f(0)$ is a relative maximum.
   There are asymptotes at $x = 1$ and $x = -1$.
5. The side with the \$6 fence is 20 feet long; the other dimension is 25 feet.
6. $f_x = 6x - 7y$ so $f_x(2, -1) = 19$; $f_y = -7x + 4y - 6$ so $f_y(2, -1) = -24$
7. $f_x = 3e^{xy^2}(xy^2 + 1)$ so $f_x(3, 0) = 3$; $f_y = 6x^2ye^{xy^2}$ so $f_y(3, 0) = 0$
8. $f_x = \dfrac{y^2 - x^2 + 6xy}{(x^2 + y^2)^2}$ so $f_x(1, -2) = -\dfrac{9}{25}$;
   $f_y = \dfrac{-3x^2 + 3y^2 - 2xy}{(x^2 + y^2)^2}$ so $f_y(1, -2) = \dfrac{13}{25}$
9. $f_{xx} = 6x + \dfrac{4y}{x^3}$, $f_{yy} = 6y - \dfrac{14x}{y^3}$,
   $f_{xy} = \dfrac{7}{y^2} - \dfrac{2}{x^2}$, and $f_{yx} = \dfrac{7}{y^2} - \dfrac{2}{x^2}$

Extreme Points of Functions of Two Variables

By examining the graphs of a few of the more common surfaces, you see that functions of two variables do have extreme points.

Figure 30.1 shows the graph of $f(x, y) = x^2 + y^2$, which is a paraboloid. Any cross-sectional curve perpendicular to the xy plane and passing through $(0, 0, 0)$ is a parabola with a minimum point at $(0, 0)$. Thus, the graph of $f(x, y) = x^2 + y^2$ has a *minimum point* at the point $(0, 0, 0)$.

Figure 30.2 shows the graph of $f(x, y) = 4 - x^2 - y^2$, which is an inverted paraboloid. Any cross-sectional curve perpendicular to the xy plane and passing

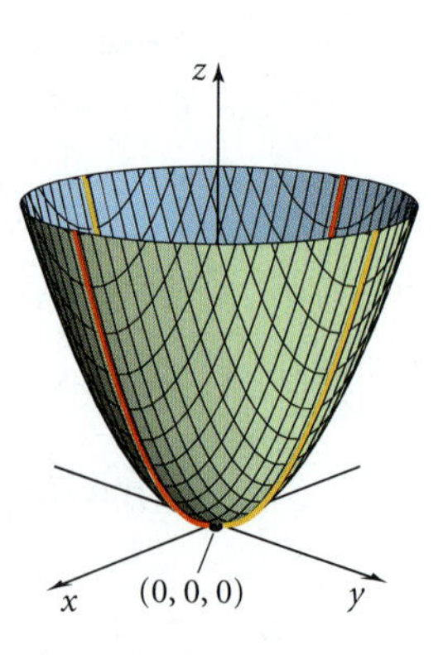

Figure 30.1

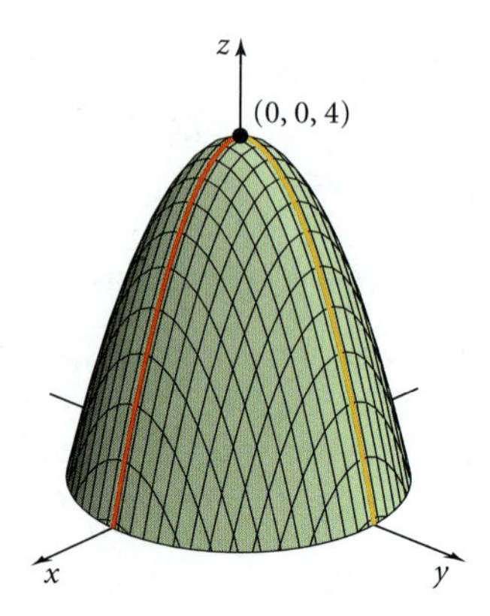

Figure 30.2

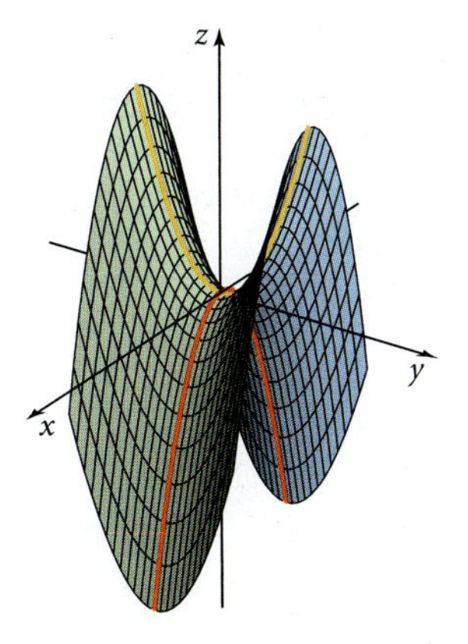

Figure 30.3

through (0, 0, 4) is a parabola with a maximum point at (0, 4). Thus, the graph of $f(x, y) = 4 - x^2 - y^2$ has a *maximum point* at (0, 0, 4).

Figure 30.3 shows the graph of $f(x, y) = y^2 - x^2$, which is a hyperbolic paraboloid. This graph does not have a minimum point or a maximum point. The cross-sectional curve for $x = 0$ is the parabola $z = y^2$, which has a minimum at (0, 0). The cross-sectional curve for $y = 0$ is the parabola $z = -x^2$, which has a maximum at (0, 0). Points for which one cross-sectional curve has a minimum and another cross-sectional curve has a maximum are called *saddle points*. The graph of $f(x, y) = x^2 + y^2$ has a saddle point at (0, 0, 0).

To begin the discussion of how to find extreme points of functions of two variables, we review the tools currently in the Tool Kit for extreme points of functions of one variable.

Extreme Points of Functions of One Variable $y = f(x)$

For $y = f(x)$ with c in the interval (a, b) in the domain of $f(x)$,

$x = c$ is a **critical value** if $f'(c) = 0$ or if $f'(c)$ is undefined.
$(c, f(c))$ is a **relative minimum** if $f(c) \leq f(x)$ for all x in the interval.
$(c, f(c))$ is a **relative maximum** if $f(c) \geq f(x)$ for all x in the interval.

Using the second derivative, you also know that

$(c, f(c))$ is a relative minimum if $f''(c) > 0$.
$(c, f(c))$ is a relative maximum if $f''(c) < 0$.

The critical value of $f(x)$ is a member of the domain for which the first derivative either equals zero or is undefined. But the domain of functions of two variables is a set of ordered pairs in the xy-plane, so $f(x, y)$ will have critical points, rather than critical values. Extreme points for $f(x, y)$ are defined in a manner similar to that which you already know for functions of one variable.

Definition: For $z = f(x, y)$ and the point (a, b) contained in some small region R of the domain,

(a, b) is a **critical point** if $f_x(a, b) = 0$ *and* $f_y(a, b) = 0$.
$f(a, b)$ is a **relative maximum value** if $f(a, b) \geq f(x, y)$ for all (x, y) contained in the region R.
$f(a, b)$ is a **relative minimum value** if $f(a, b) \leq f(x, y)$ for all (x, y) contained in the region R.

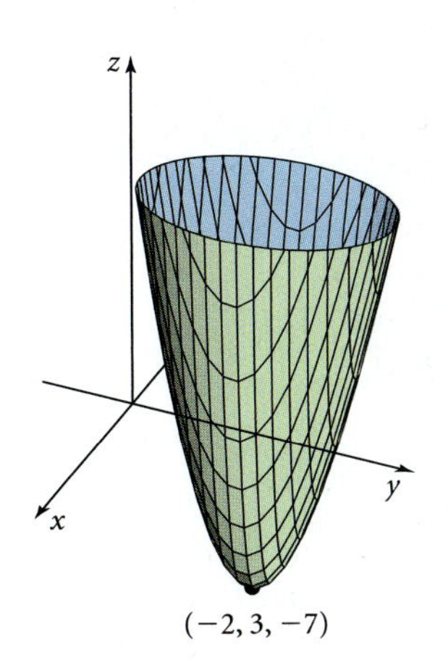

Figure 30.4

Before introducing the second derivative test to determine if the critical point is a relative extreme point, you must first learn how to find the critical points.

Example 1: Find the critical points of $f(x, y) = 2x^2 + y^2 + 8x - 6y + 10$.

Solution: The critical points occur where both $f_x = 0$ *and* $f_y = 0$. For the given function, the first derivatives are $f_x = 4x + 8$ and $f_y = 2y - 6$. If $4x + 8 = 0$, then $x = -2$. If $2y - 6 = 0$, then $y = 3$. That means the critical point is $(-2, 3)$.

The graph as shown in Figure 30.4 of $f(x, y) = 2x^2 + y^2 + 8x - 6y + 10$ clearly shows a minimum point at $(-2, 3, -7)$. ■

MATHEMATICS CORNER 30.1

By completing the square, you can see that $(-2, 3, f(-2, 3))$ is the vertex of the surface $f(x, y) = 2x^2 + y^2 + 8x - 6y + 10$ from Example 1.

$$\begin{aligned} f(x, y) &= 2x^2 + y^2 + 8x - 6y + 10 \\ &= 2x^2 + 8x + y^2 - 6y + 10 && \text{Place like variables together} \\ &= 2(x^2 + 4x) + (y^2 - 6y) + 10 && \text{Factor 2 from terms containing } x \\ &= 2(x^2 + 4x + 4) + (y^2 - 6y + 9) + 10 - 8 - 9 && \text{Complete the square} \\ &= 2(x + 2)^2 + (y - 3)^2 - 7 && \text{Simplify} \end{aligned}$$

So the vertex of the paraboloid is at $(-2, 3, -7)$, which agrees with the answer found in Example 1.

Example 2: Find the critical points of $f(x, y) = x^3 - 4xy + 2y^2 - 1$.

Solution: For the given function, the first derivatives are $f_x = 3x^2 - 4y$ and $f_y = -4x + 4y$. The critical points occur where $3x^2 - 4y = 0$ and $-4x + 4y = 0$. If $-4x + 4y = 0$, then $y = x$. Substituting $y = x$ into $3x^2 - 4y = 0$ gives $3x^2 - 4x = 0$. Solving $3x^2 - 4x = 0$, we have $x(3x - 4) = 0$, so $x = 0$ or $x = \frac{4}{3}$. Because $y = x$, the critical points are $(0, 0)$ and $\left(\frac{4}{3}, \frac{4}{3}\right)$.

By looking at the graph from two different angles, you can see that $(0, 0, -1)$ is a **saddle point** (Figure 30.5) and $\left(\frac{4}{3}, \frac{4}{3}, \frac{-59}{27}\right)$ is a relative minimum point (Figure 30.6).

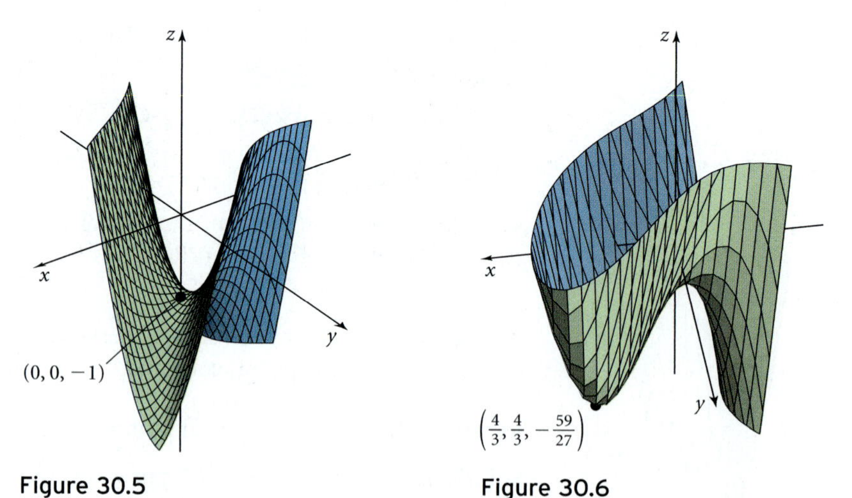

Figure 30.5

Figure 30.6

How can you determine the relative extreme points of a function of two variables without looking at its graph? The answer lies, as it did before, with the function's second derivative.

Second Partial Derivative Test

Given $z = f(x, y)$ such that all four second partial derivatives exist, let (a, b) be a critical point. Then we define $D = f_{xx}(a, b) \cdot f_{yy}(a, b) - (f_{xy}(a, b))^2$ and

if $D > 0$ and $f_{xx}(a, b) > 0$, then f has a relative minimum point at (a, b).

if $D > 0$ and $f_{xx}(a, b) < 0$, then f has a relative maximum point at (a, b).

if $D < 0$, then f is a saddle point at (a, b).

if $D = 0$, no conclusion can be made.

Example 3: Use the Second Partial Derivative Test to prove that $f(-2, 3)$ is a relative minimum of $f(x, y) = 2x^2 + y^2 + 8x - 6y + 10$. (See Example 1.)

Solution: The first partial derivatives are $f_x = 4x + 8$ and $f_y = 2y - 6$. From Example 1 you know that the critical point is $(-2, 3)$.

The second partial derivatives are $f_{xx} = 4, f_{xy} = 0, f_{yy} = 2$, and $f_{yx} = 0$. For the critical point $(-2, 3)$ we have $f_{xx}(-2, 3) = 4, f_{yy}(-2, 3) = 2$, and $f_{xy}(-2, 3) = 0$. Thus, $D = 4(2) - 0^2 = 8$. Because $D > 0$, and $f_{xx}(-2, 3) > 0$, we know that the critical point is a relative minimum for $f(x, y)$. ■

Example 4: Identify the relative extreme points of $f(x, y) = x^3 - 4xy + 2y^2 - 1$. (See Example 2.)

Solution: The first partial derivatives are $f_x = 3x^2 - 4y$ and $f_y = -4x + 4y$. From Example 2 you know that the critical points are $(0, 0)$ and $\left(\frac{4}{3}, \frac{4}{3}\right)$.

The second partial derivatives are $f_{xx} = 6x, f_{xy} = -4, f_{yy} = 4$, and $f_{yx} = -4$.

For the critical point $(0, 0)$, $f_{xx}(0, 0) = 6(0) = 0, f_{yy}(0, 0) = 4$, and $f_{xy}(0, 0) = -4$. Thus, $D = 0(4) - (-4)^2 = -16$. Because $D < 0$, we know that the critical point $(0, 0)$ is a saddle point for $f(x, y)$.

For the critical point $\left(\frac{4}{3}, \frac{4}{3}\right)$, $f_{xx}\left(\frac{4}{3}, \frac{4}{3}\right) = 6 \cdot \frac{4}{3} = 8, f_{yy}\left(\frac{4}{3}, \frac{4}{3}\right) = 4$, and $f_{xy}\left(\frac{4}{3}, \frac{4}{3}\right) = -4$. Thus, $D = 8(4) - (-4)^2 = 16$. Because $D > 0$ and $f_{xx}\left(\frac{4}{3}, \frac{4}{3}\right) > 0$, we know that the critical point $\left(\frac{4}{3}, \frac{4}{3}\right)$ is a relative minimum for $f(x, y)$. ■

Example 5: Determine the critical points and identify the relative extrema of $f(x, y) = 3x^2 + y^2 + 2xy + 6y - 4$.

Solution: First we find the critical points. The first partial derivatives are $f_x = 6x + 2y$ and $f_y = 2y + 2x + 6$.

Setting the first derivatives equal to zero gives $6x + 2y = 0 \rightarrow y = -3x$ and $2y + 2x + 6 = 0 \rightarrow y + x + 3 = 0$.

Substitute $y = -3x$ into $y + x + 3 = 0$ and solve for x:

$$-3x + x + 3 = 0$$

$$-2x + 3 = 0$$

$$x = \frac{3}{2}$$

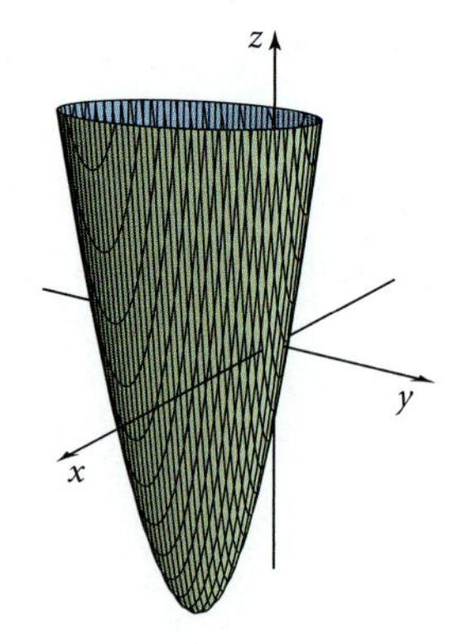

Figure 30.7

Now solve for y: $y = -3x$ so $y = -3\left(\frac{3}{2}\right) = -\frac{9}{2}$.

The critical point is $\left(\frac{3}{2}, -\frac{9}{2}\right)$.

Next use the Second Partial Derivative Test to determine the type of extreme point: $f_x = 6x + 2y$ and $f_y = 2y + 2x + 6$, so $f_{xx} = 6$, $f_{yy} = 2$, and $f_{xy} = f_{yx} = 2$.

At the critical point we have $f_{xx}\left(\frac{3}{2}, -\frac{9}{2}\right) = 6$, $f_{yy}\left(\frac{3}{2}, -\frac{9}{2}\right) = 2$, and $f_{xy}\left(\frac{3}{2}, -\frac{9}{2}\right) = 2$. Thus, $D = 6(2) - 2^2 = 8$. Because $D > 0$ and $f_{xx}\left(\frac{3}{2}, -\frac{9}{2}\right) > 0$, we know that the critical point is a relative minimum for $f(x, y)$.

The graph in Figure 30.7 of $f(x, y) = 3x^2 + y^2 + 2xy + 6y - 4$ clearly shows a minimum point at $\left(\frac{3}{2}, -\frac{9}{2}, -\frac{35}{2}\right)$. ■

Example 6: Determine the critical points and identify the relative extrema of $f(x, y) = \frac{1}{x} - \frac{4}{y} - 2xy$.

Solution: First we find the critical points. The first partial derivatives are $f_x = -\frac{1}{x^2} - 2y$ and $f_y = \frac{4}{y^2} - 2x$.

Setting $f_x = 0$:

$$-\frac{1}{x^2} - 2y = 0 \rightarrow y = -\frac{1}{2x^2}$$

Substitute $y = -\frac{1}{2x^2}$ into $f_y = \frac{4}{y^2} - 2x = 0$ and solve for x:

$$f_y = \frac{4}{y^2} - 2x = 0 \rightarrow \frac{4}{\left(-\frac{1}{2x^2}\right)^2} - 2x = 0$$

$$\frac{4}{\frac{1}{4x^4}} - 2x = 0$$

$$16x^4 - 2x = 0$$

$$2x(8x^3 - 1) = 0 \text{ so } x = 0 \text{ or } x = \frac{1}{2}$$

But $x \neq 0$ so the only possible critical value is $x = \frac{1}{2}$.

Now solve for y: $y = -\frac{1}{2x^2}$ so $y = -\frac{1}{2\left(\frac{1}{2}\right)^2} = -\frac{1}{\frac{1}{2}} = -2$.

The critical point is $\left(\frac{1}{2}, -2\right)$.

Next use the Second Partial Derivative Test to determine the type of extreme point: $f_x = -\frac{1}{x^2} - 2y$ and $f_y = \frac{4}{y^2} - 2x$, so $f_{xx} = \frac{2}{x^3}$, $f_{yy} = -\frac{8}{y^3}$, and $f_{xy} = f_{yx} = -2$.

At the critical point we have $f_{xx}\left(\frac{1}{2}, -2\right) = 16, f_{yy}\left(\frac{1}{2}, -2\right) = 1$ and, $f_{xy}\left(\frac{1}{2}, -2\right) = -2$. Thus, $D = 16(1) - (-2)^2 = 12$. Because $D > 0$ and $f_{xx}\left(\frac{1}{2}, -2\right) > 0$, we know that the critical point is a relative minimum for $f(x, y)$. ■

Example 7: The weekly profit from the sale of x backpacks and y calculators at a local college bookstore is given by $P(x, y) = -0.2x^2 - 0.3y^2 - 0.2xy + 25x + 30y + 10$ dollars. How many backpacks and calculators must the store sell in order to maximize the store's profit?

Solution: First we must determine the critical point. The first partial derivatives are $P_x = -0.4x - 0.2y + 25$ and $P_y = -0.6y - 0.2x + 30$.

Now set the first derivatives equal to zero: $-0.4x - 0.2y + 25 = 0$ and $-0.6y - 0.2x + 30 = 0$.

Multiplying both equations by 5, we have $y = 125 - 2x$ and $-3y - x + 150 = 0$. Substituting $y = 125 - 2x$ gives $-3(125 - 2x) - x + 150 = 0$.

Solve for x:

$$-375 + 5x + 150 = 0$$
$$5x = 225$$
$$x = 45$$

Now find y: $y = 125 - 2x$ and $x = 45$, so $y = 125 - 2(45) = 125 - 90 = 35$.

The critical point is $(45, 35)$.

Next we use the second partial derivatives to determine if the critical point is the maximum for $P(x, y)$. $P_x = -0.4x - 0.2y + 25$ and $P_y = -0.6y - 0.2x + 30$, so $P_{xx} = -0.4, P_{yy} = -0.6$, and $P_{xy} = P_{yx} = -0.2$.

At the critical point, $P_{xx}(45, 35) = -0.4, P_{yy}(45, 35) = -0.6$, and $P_{xy}(45, 35) = -0.2$. Then $D = (-0.4)(-0.6) - (-0.2)^2 = 0.20$. Because $D > 0$ and $P_{xx}(45,35) < 0$, the critical point is a relative maximum for $P(x, y)$.

Weekly maximum profit occurs if 45 backpacks and 35 calculators are sold. The maximum profit is given by $P(45, 35) = -0.2(45)^2 - 0.3(35)^2 - 0.2(45)(35) + 25(45) + 30(35) + 10 = \$1097.50$. ■

Check Your Understanding 30.1

For each of the following,

- **a.** Determine the critical point(s).
- **b.** Identify the critical point as a relative minimum point, a relative maximum point, or a saddle point.

1. $f(x, y) = x^2 - y^2 - 2x + 6y + 4$
2. $f(x, y) = x^2 + y^2 + xy - 3y$
3. $f(x, y) = -x^3 + 6xy - y^3$

Constrained Optimization and Lagrange Multipliers

Many of the optimization problems that we encounter require optimizing some quantity subject to constraints. For example, the maximum profit a company receives from its two best-selling products may be subject to shipping constraints or manufacturing constraints. Or the maximum area of an enclosed region may be subject to the amount of fencing available. Problems of this type are called **constrained optimization problems**. The components of a constrained optimization problem are

- The variables
- The objective function—that is, the quantity that is to be optimized
- The constraints
- The solution

Example 8: A rectangular area of 500 square feet is to be enclosed. Fencing for one side costs \$6 per foot; fencing for the other three sides costs \$4 per foot. Find the dimensions that provide for minimum cost to enclose the area.

Solution: Let x = length and y = width.

The objective function is cost: $C(x, y) = 6x + 4x + 2(4y) = 10x + 8y$. The constraint is the area: $xy = 500$.

The method of solution in Topic 17 was to solve the constraint equation for one of the variables and then substitute into the objective function, thus converting the objective function to a function of one variable. Then we used the first derivative to find the extreme point. Doing the same in this example, we have the following.

Solving constraint equation for y: $y = \dfrac{500}{x}$.

Substituting into objective function: $C(x) = 10x + 8\left(\dfrac{500}{x}\right) = 10x + \dfrac{4000}{x}$, where $x > 0$.

Derivative: $C'(x) = 10 - \dfrac{4000}{x^2}$.

Solve $C'(x) = 0$:

$$10 - \frac{4000}{x^2} = 0$$
$$x^2 = 400$$
$$x = 20$$

Verify: $C''(x) = \dfrac{8000}{x^3}$, so

$C''(20) = \dfrac{8000}{20^3} > 0$, which means the critical value is a minimum.

Find y: $y = \dfrac{500}{x}$ and $x = 20$, so

$$y = \frac{500}{20} = 25.$$

The minimum cost will be achieved if the \$6 fence is 20 feet long and the other dimension is 25 feet. ■

Lagrange Multipliers

An alternate method for solving constrained optimization problems is the method of **Lagrange multipliers**. Why is another method needed? There may be times that the constraint equation cannot easily be solved. Also, we need a method suitable for functions of more than two independent variables or with more than one constraint equation. The method of Lagrange multipliers fits these scenarios quite nicely. Let's outline the method.

Solving Constrained Optimization Problems with Lagrange Multipliers

To optimize $f(x, y)$ subject to $g(x, y) = c$,

1. Write the **Lagrangian** $F(x, y) = f(x, y) + \lambda(c - g(x, y))$. The λ factor is the **Lagrange multiplier**.
2. Find F_x, F_y, and F_λ.
3. Set $F_x = 0, F_y = 0$, and $F_\lambda = 0$ and solve for the critical point (a, b).
4. Evaluate $f(a, b)$ to determine the optimal solution.

The method of Lagrange multipliers is due to Joseph Lagrange (1736–1813), a prominent 18th-century French mathematician. The method assumes that the extreme point exists and then provides the critical point.

Joseph Lagrange (1736–1813)

Example 9: Rework Example 7 using Lagrange multipliers.

Solution: The problem is to minimize $C(x, y) = 10x + 8y$ subject to $xy = 500$. The variables x and y are lengths of fence, so we also know that $x > 0$ and $y > 0$. The Lagrangian is $F(x, y) = 10x + 8y + \lambda(500 - xy)$.

Differentiating, $F_x = 10 - \lambda y, F_y = 8 - \lambda x$, and $F_\lambda = 500 - xy$.

Set the derivatives equal to zero:

$$10 - \lambda y = 0$$
$$8 - \lambda x = 0$$
$$500 - xy = 0$$

Solve and find the critical point:

$$10 - \lambda y = 0, \text{ so } \lambda = \frac{10}{y}$$

$$8 - \lambda x = 0, \text{ so } \lambda = \frac{8}{x}$$

Equating the two gives $\frac{10}{y} = \frac{8}{x}$ or $10x = 8y$, so $y = \frac{5}{4}x$.

Substitute $y = \frac{5}{4}x$ into $500 - xy = 0$:

$$500 - x\left(\frac{5}{4}x\right) = 0$$

$$500 - \frac{5}{4}x^2 = 0$$

$$500 = \frac{5}{4}x^2$$

$$400 = x^2$$
$$20 = x$$

$x = -20$ is not considered because the domain is $x > 0$. $y = \frac{5}{4}x$ and $20 = x$, so $y = \frac{5}{4}(20) = 25$.

The critical point is $(20, 25)$.
The minimum cost is $f(20, 25) = 10(20) + 8(25) = \$400$. ■

The Lagrange multiplier, λ, gives the extra "whatever is being optimized" that could be obtained if one more unit of "what constrains the choice" were available. In Example 9, we knew that $\lambda = \frac{10}{y}$ and that $y = 25$, so $\lambda = \frac{10}{25} = \frac{2}{5}$ or 0.40. With dimensions of 20 feet by 25 feet, the additional cost of one more square foot of area is \$0.40.

Example 10: Maximize $f(x, y) = 2xy - 4x$ if $x + y = 8$.

Solution: The Lagrangian is $F(x, y) = 2xy - 4x + \lambda(8 - x - y)$.

The first partial derivatives are $F_x = 2y - 4 - \lambda$, $F_y = 2x - \lambda$, and $F_\lambda = 8 - x - y$.

Setting the derivatives equal to zero and solving gives

$$2y - 4 - \lambda = 0 \text{ or } \lambda = 2y - 4$$
$$2x - \lambda = 0 \text{ or } \lambda = 2x$$

Thus, $2y - 4 = 2x$, so $y - 2 = x$.

Substituting $y - 2 = x$ into $F_\lambda = 0$:

$$8 - (y - 2) - y = 0$$
$$10 - 2y = 0$$
$$y = 5$$

We know that $y - 2 = x$ and $y = 5$, so $x = 5 - 2 = 3$. The critical point is $(3, 5)$. The maximum is $f(3, 5) = 2(3)(5) - 4(3) = 18$. ■

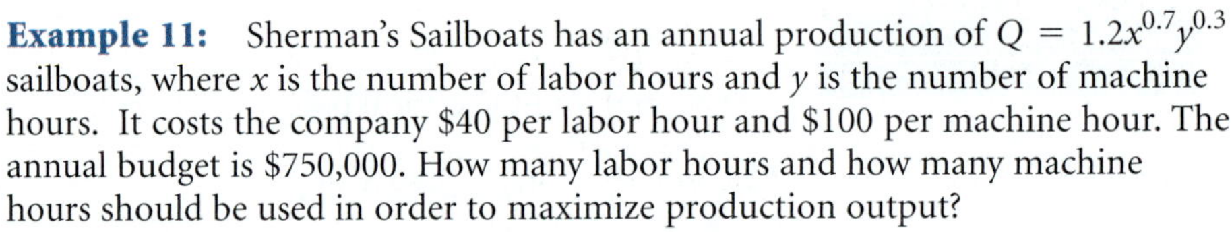

Example 11: Sherman's Sailboats has an annual production of $Q = 1.2x^{0.7}y^{0.3}$ sailboats, where x is the number of labor hours and y is the number of machine hours. It costs the company \$40 per labor hour and \$100 per machine hour. The annual budget is \$750,000. How many labor hours and how many machine hours should be used in order to maximize production output?

Solution: The problem is to maximize $Q = 1.2x^{0.7}y^{0.3}$ subject to $40x + 100y = 750{,}000$.

The Lagrangian is $F(x,y) = 1.2x^{0.7}y^{0.3} + \lambda(750{,}000 - 40x - 100y)$.

The first partial derivatives are $F_x = 0.84x^{-0.3}y^{0.3} - 40\lambda$, $F_y = 0.36x^{0.7}y^{-0.7} - 100\lambda$, and $F_\lambda = 750{,}000 - 40x - 100y$.

Setting the derivatives equal to zero and solving gives $0.84x^{-0.3}y^{0.3} - 40\lambda = 0$ and $0.36x^{0.7}y^{-0.7} - 100\lambda = 0$.

Solve each equation for λ and equate the two results:

$$0.84x^{-0.3}y^{0.3} - 40\lambda = 0, \text{ so } \lambda = \frac{0.84x^{-0.3}y^{0.3}}{40} = \frac{0.84y^{0.3}}{40x^{0.3}}$$

$$0.36x^{0.7}y^{-0.7} - 100\lambda = 0, \text{ so } \lambda = \frac{0.36x^{0.7}y^{-0.7}}{100} = \frac{0.36x^{0.7}}{100y^{0.7}}$$

Thus,

$$\frac{0.84y^{0.3}}{40x^{0.3}} = \frac{0.36x^{0.7}}{100y^{0.7}}.$$

$$84y = 14.4x \text{ or } y = \frac{6}{35}x.$$

Substitute $y = \frac{6}{35}x$ into $750{,}000 - 40x - 100y = 0$:

$$750{,}000 - 40x - 100\left(\frac{6}{35}x\right) = 0$$

$$26{,}250{,}000 - 1400x - 600x = 0$$

$$26{,}250{,}000 = 2000x$$

$$13{,}125 = x$$

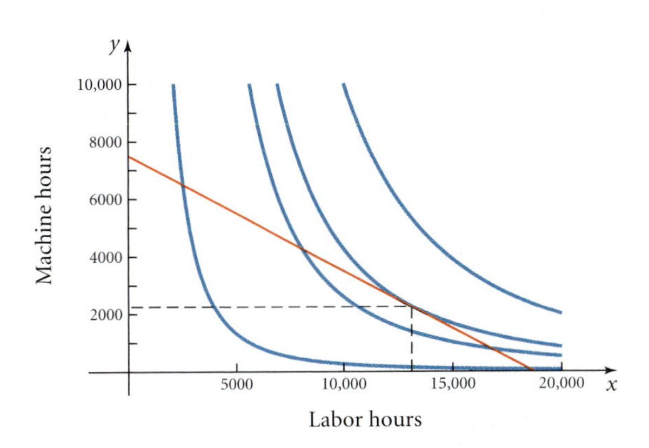

Figure 30.8

Then $y = \frac{6}{35}(13{,}125) = 2250$. The critical point is $(13{,}125, 2250)$.

Maximum production is $Q(13{,}125, 2250) = 1.2(13{,}125)^{0.7}(2250)^{0.3} = 9279$. If 13,125 labor hours and 2250 machine hours are used, the maximum production is 9279 sailboats.

The graph in Figure 30.8 shows the level curves for four different production levels, plus the constraint equation. The optimal solution of (13,125, 2250) occurs when the constraint equation is tangent to the level curve corresponding to a production level of $Q = 9279$. ■

Check Your Understanding 30.2

Use Lagrange multipliers to optimize each of the following.

1. Maximize $A = xy$ subject to $6x + 8y = 360$.
2. Minimize $C = 50x + 35y$ subject to $xy = 490$.

Check Your Understanding Answers

Check Your Understanding 30.1

1. a. $(1, 3)$
   b. $D < 0$, so $f(1, 3)$ is a saddle point.
2. a. $(-1, 2)$
   b. $D > 0$ and $f_{xx}(-1, 2) > 0$, so $f(-1, 2)$ is a relative minimum.
3. a. $(0, 0)$ and $(2, 2)$
   b. $f(0, 0)$ is a saddle point because $D < 0$. $f(2, 2)$ is a relative maximum because $D > 0$ and $f_{xx}(2, 2) < 0$.

Check Your Understanding 30.2

1. Critical point is $(30, 22.5)$, so maximum is $A(30, 22.5) = 675$.
2. Critical point is $(7\sqrt{7}, 10\sqrt{7}) \approx (18.52, 26.458)$, so minimum is $C(7\sqrt{7}, 10\sqrt{7}) = 50(7\sqrt{7}) + 35(10\sqrt{7}) = 700\sqrt{7} \approx 1852$.

Topic 30 Review

30

In this topic you learned how to find extreme points of functions of two variables, using either the Second Partial Derivative Test or Lagrange multipliers.

CONCEPT

Functions of two variables may have **relative minimum** points, **relative maximum** points, or **saddle points.**

$f(x, y)$ has a **relative minimum** at (a, b) if the cross-sectional curves through (a, b) in any direction have minimum points at (a, b).

$f(x, y)$ has a **relative maximum** at (a, b) if the cross-sectional curves through (a, b) in any direction have maximum points at (a, b).

$f(x, y)$ has a **saddle point** at (a, b) if the cross-sectional curve in one direction has a minimum point at (a, b) and the cross-sectional curve in another direction has a maximum point at (a, b).

EXAMPLE

$f(x, y) = x^2 + y^2 - 2$ has a minimum point at $(0, 0, -2)$.

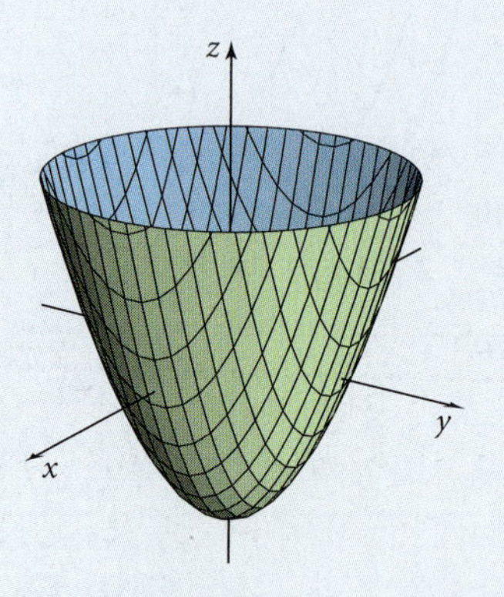

$f(x, y) = 2 - x^2 - y^2$ has a maximum point at $(0, 0, 2)$.

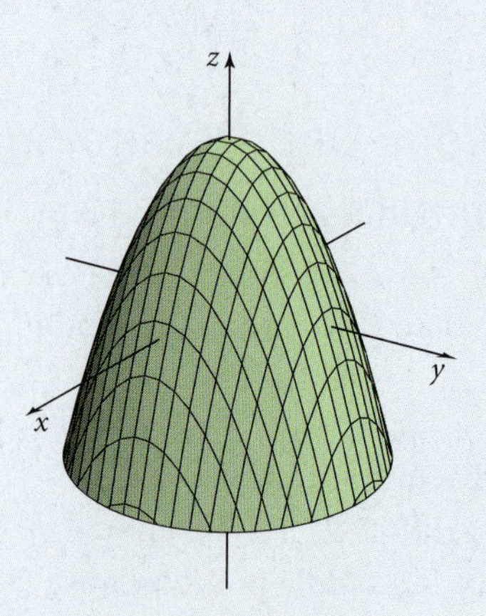

$f(x, y) = x^2 - y^2$ has a saddle point at $(0, 0, 0)$.

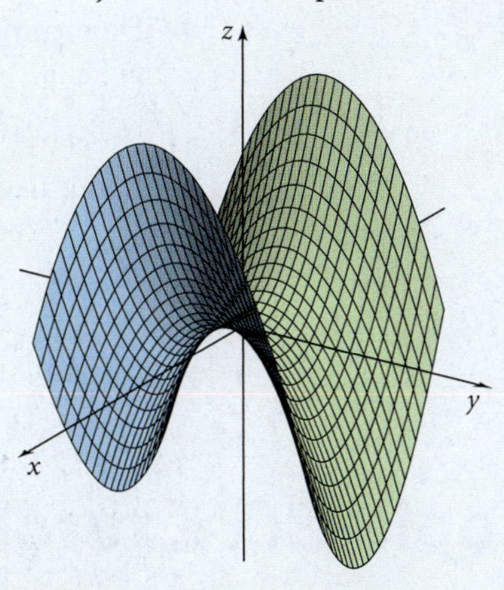

The **critical point** of $f(x, y)$ is the point (a, b) in the domain of f such that $f_x = 0$ and $f_y = 0$.

For $f(x, y) = x^2 - y^2, f_x = 2x$ and $f_y = -2y$. $f_x = 0$ and $f_y = 0$ if $x = 0$ and $y = 0$. The critical point is $(0, 0)$.

To determine if a critical point is a relative minimum, a relative maximum, or a saddle point, we define

$$D = f_{xx}(a, b) \cdot f_{yy}(a, b) - (f_{xy}(a, b))^2.$$

The **Second Partial Derivative Test** says that

- If $D > 0$ and $f_{xx}(a, b) > 0$, then f has a relative minimum at (a, b).
- If $D > 0$ and $f_{xx}(a, b) < 0$, then f has a relative maximum point at (a, b).
- If $D < 0$, then f has a saddle point at (a, b).
- If $D = 0$, no conclusion can be drawn.

For $f(x, y) = x^2 - y^2$, the critical point is $(0, 0)$. We have $f_x = 2x$ and $f_y = -2y$, so $f_{xx} = 2, f_{yy} = -2$, and $f_{xy} = f_{yx} = 0$. Then $D = (2)(-2) - 0^2 = -4$. Because $D < 0$, the critical point is a saddle point for $f(x, y)$.

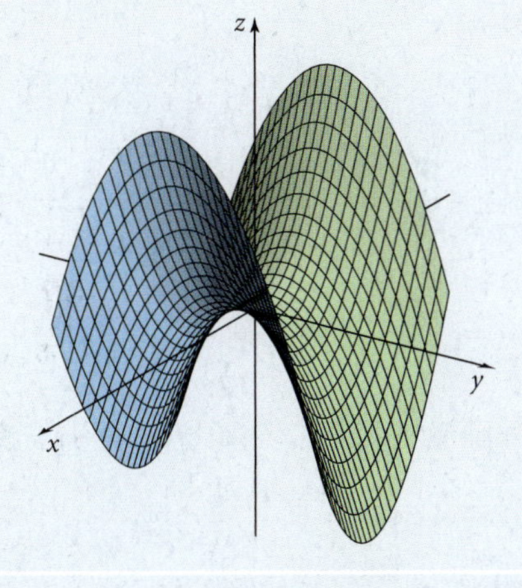

To solve **constrained optimization** problems, we use the method of **Lagrange multipliers.** To use Lagrange multipliers to optimize $f(x, y)$ subject to $g(x, y) = c$, define the **Lagrangian** $F(x, y) = f(x, y) + \lambda(c - g(x, y))$. Then set $F_x = 0, F_y = 0$, and $F_\lambda = 0$. Solve the resulting system of equations to determine the critical point.

Maximize $A = 4xy$ subject to $x + 2y = 10$. The Lagrangian is $F(x, y) = 4xy + \lambda(10 - x - 2y)$, so $F_x = 4y - \lambda, F_y = 4x - 2\lambda$, and $F_\lambda = 10 - x - 2y$. Thus, $\lambda = 4y = 2x$, so $y = \frac{x}{2}$. Substitute $y = \frac{x}{2}$ into $10 - x - 2y = 0$: $10 - x - 2\left(\frac{x}{2}\right) = 0$, which gives $x = 5$. Then $y = \frac{5}{2}$. The maximum is $A\left(5, \frac{5}{2}\right) = 4(5)\left(\frac{5}{2}\right) = 50$.

NEW TOOLS IN THE TOOL KIT

- Determining critical points of functions of two variables
- Using the Second Partial derivative Test to determine if a critical point (a, b) is a relative maximum, a relative minimum, or a saddle point
- Using Lagrange multipliers to solve constrained optimization problems

Topic 30 Exercises

30

For each of the following functions, determine the critical point(s) and use the Second Partial Derivative Test to determine if the critical point is a relative maximum, a relative minimum, or a saddle point.

1. $f(x, y) = 4 - x^2 - y^2$

2. $f(x, y) = x^2 + y^2 - 2$

3. $f(x, y) = 2x^2 - y^2$

4. $f(x, y) = x^2 - 4y^2$

5. $f(x, y) = x^2 + y^2 + 4x - 2y + 5$

6. $f(x, y) = x^2 + y^2 - 2x + 6y - 2$

7. $f(x, y) = x^2 + y^2 + 4xy - 3y$

8. $f(x, y) = 6xy - x^2 - y^2 + 8x$

9. $f(x, y) = x^3 + y^3 - 2xy$

10. $f(x, y) = x^3 + y^3 + 2xy$

11. $f(x, y) = e^{x^2+y^2}$

12. $f(x, y) = e^{x^2-y^2}$

13. $f(x, y) = e^{-(x^2+y^2)}$

14. $f(x, y) = e^{x^2y}$

15. $f(x, y) = 4xy - \frac{2}{x} + \frac{8}{y}$

16. $f(x, y) = 2x + 4y - \frac{8}{xy}$

17. $f(x, y) = x^3 + y^3 + xy$

18. $f(x, y) = x^2 + y^2 - xy^2$

Solve each of the following constrained optimization problems using the method of Lagrange multipliers.

19. Maximize $A = 4xy$ if $2x + 2y = 80$.

20. Maximize $A = 10x^2y$ if $3x + 5y = 45$.

21. Minimize $C = 20x + 15y$ if $xy = 540$.

22. Minimize $C = 6xy - 0.2x^2 - 0.8y^2$ if $xy = 200$.

23. Trudy's Camping Supply Store has received an order for 3000 camping tents. She will fill the order with x hundred tents from the Atlanta plant and y hundred tents from the Kansas City plant. The total cost of the order is given by $C(x, y) = 0.5x^2 - 6x + 0.2y^2 - 10y + 240$ thousand dollars. How many tents should be shipped from each plant to minimize the cost?

24. If Trudy's Camping Supply Store (see Exercise 23) sells x hundred tents from the Atlanta plant and y hundred tents from the Kansas City plant, the profit is given by $P(x, y) = -2x^2 - 3y^2 + 2xy + 45x + 20y$ thousand dollars. How many tents should be shipped from each plant to maximize her profit?

25. A rectangular area is to be enclosed using 1000 feet of fencing. Find the dimensions that allow for the maximum area to be enclosed.

26. A rectangular area is to be enclosed using 1000 feet of fencing. Find the dimensions that allow for the maximum area, if one of the sides is unfenced.

27. A carton with a square base is to be constructed with a volume of 640 cubic feet. The cost of the bottom is \$6 per square foot. The cost of the sides and top is \$4 per square foot. What dimensions will minimize the cost?

28. Brian's Beach Shop sells two models of surfboard. The price of a 6-foot board is $\$a$ and the price of a 9-foot board is $\$b$. The revenue is given by $R(a, b) = 150a + 350b + 2.5ab - 2a^2 - 1.5b^2$. What prices will maximize the shop's revenue?

29. Maximize $Q = x^{2/3}y^{1/3}$ subject to $2x + 3y = 12$.

30. Maximize $Q = 80x^{0.2}y^{0.8}$ subject to $50x + 40y = 100{,}000$.

31. Trudy's Camping Supply Store must ship x hundred tents from the Atlanta plant and y hundred tents from the Kansas City plant. The cost of shipping from the Atlanta plant is $C_1 = 2.5x^2 + 10$, and the cost of shipping from the Kansas City plant is $C_2 = 5y^2 + 12$. She needs to ship 3000 tents. How many tents should be shipped from each plant to minimize the cost?

32. The number of students a tutoring center can enroll depends on the number of teachers, t, and the number of staff members, s, and is given by $N(t, s) = 10t^{0.6}s^{0.4}$. Teachers are paid \$40,000 annually and staff members are paid \$25,000 annually. The annual budget is \$600,000. How many teachers and staff members should be hired to maximize the number of students that can be enrolled?

33. Sherman's Sailboats operates two manufacturing plants. The cost of producing x sailboats at the Tampa plant is given by $C_1 = 0.04x^2 + 5x + 325$ thousand dollars. The cost of producing y sailboats at the Virginia Beach plant is given by $C_2 = 0.06y^2 + 9y + 180$ thousand dollars. Each sailboat sells for \$15,000.

a. Find the profit function $P(x, y)$.

b. How many boats should be manufactured at each plant to maximize the profit?

34. Brian's Beach Shop sells two models of surfboards. The revenue from a 6-foot boards and b 9-foot boards is given by $R(a, b) = -a^2 - 2b^2 + 2ab + 150a + 200b$. How many of each board should Brian manufacture to maximize revenue, if his production cannot exceed 500 boards?

35. A production editor has been given \$80,000 to spend on the production of a new calculus textbook. The editor estimates that if a thousand dollars are spent on development and b thousand dollars are spent on marketing, about $N(a, b) = 35a^{1.2}b$ copies of the book will be sold. How much money should be allotted to development and how much to marketing to maximize sales?

36. Refer to Exercise 35. Suppose the editor is given an additional \$10,000. How will the additional money in the budget affect the amount allotted to development and marketing?

SHARPEN THE TOOLS

37. A factory must lay a pipe from Building A along a river to point P and then under the river to Building B. The river is 500 feet wide and the two buildings are 3000 feet apart. It costs \$10 per foot to lay the pipe along the ground and \$25 per foot to lay the pipe under the water.

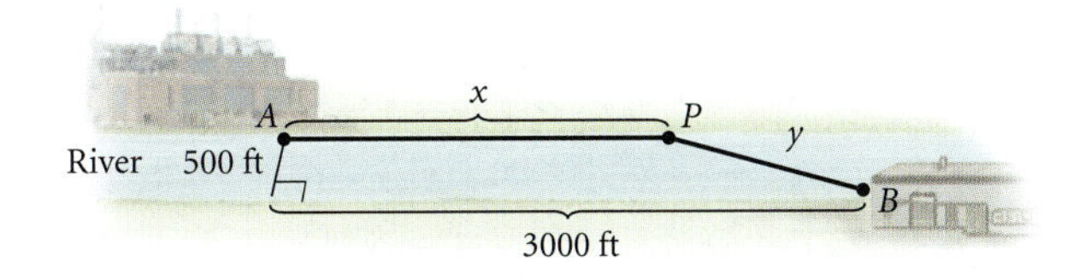

Where should point P be located to minimize the cost of laying the pipe?

38. Common blood types are determined genetically by the alleles A, B, and O. (An allele is any of a group of possible mutational forms of a gene.) A person whose blood type is AA, BB, or OO is said to be homozygous. A person whose blood type is AB, AO, or BO is said to be heterozygous. The *Hardy–Weinberg Law* states that the proportion of the population that are heterozygous (of blood type AB, AO, or BO) is given by $P(a, b, r) = 2ab + 2ar + 2br$, where a is the percent who are Type A, b is the percent who are Type B, and r is the percent that are Type O. If $a + b + r = 1$, what is the maximum proportion that can be Type AB, AO, or BO?

39. *Heron's Formula* for the area of a triangle states that $A = \sqrt{s(s - a)(s - b)(s - c)}$, where s is the *semiperimeter* given by $s = \frac{a + b + c}{2}$. The sides of the triangle are given by a, b, and c.

a. Given a fixed total perimeter, find the dimensions a, b, and c to maximize the area.

b. If the total perimeter of the triangle is 60 cm, what are the dimensions of the triangle and what is the area?

40. *Heron's Formula* for the area of a triangle states that $A = \sqrt{s(s - a)(s - b)(s - c)}$, where s is the semiperimeter given by $s = \frac{a + b + c}{2}$. The sides of the triangle are given by a, b, and c.

a. Given a fixed area, find the dimensions a, b, and c to minimize the total perimeter.

b. If the area is 60 square inches, what are the dimensions of the triangle and what is the total perimeter?

TOPIC

Double Integrals

31

So far in your brief study of multivariable calculus, you have learned to

- Evaluate functions of two or more variables
- Determine level curves of graphs of functions of two variables
- Find first and second partial derivatives of functions of two variables
- Determine extreme points of functions of two variables using the Second Partial Derivative Test
- Optimize functions of two variables using Lagrange multipliers

Now we turn our attention to the final calculus tool of this book, integration of functions of two variables. Before starting this topic, be sure to complete the Warm-up Exercises to sharpen the algebra and calculus skills you will need.

IMPORTANT TOOLS IN TOPIC 31

- *Integration of functions of two variables*
- *Evaluating double integrals over rectangular and nonrectangular regions in the plane*
- *Sketching the region of integration*
- *Reversing the order of integration*
- *Average value of a function of two variables*
- *Volumes of solid regions*

TOPIC 31 WARM-UP EXERCISES

Be sure you can successfully complete these exercises before starting Topic 31.

Algebra Warm-up Exercises

1. Write the equation of the line passing through $(2, -3)$ and $(-1, 4)$ in slope–intercept form.

2. Expand and combine terms: $x^2(2x - 3)^2 - x(x + 4)^2$

3. Simplify: $x^2\left(\frac{y^2}{2}\right)^2$ if $y = \frac{x}{2}$.

Calculus Warm-up Exercises

Evaluate each of the following integrals.

1. $\int\left(3x^5 - 2x + \frac{7}{x}\right)dx$ **2.** $\int\left(2e^{-y} + \frac{3}{y^2}\right)dy$

3. $\int_1^4\left(2\sqrt{x} - 3x^2\right)dx$ **4.** $\int_1^8\left(x^{2/3} - 6x^{-4}\right)dx$

5. Find the area of the region between $f(x) = x^2 + 3$ and the x-axis on the interval $[-2, 3]$.

6. Find the area of the region between $f(x) = x^2 + 3$ and $f(x) = 4x$.

7. Find the average value of $f(x) = x^3 - 4x$ on the interval $[-2, 1]$.

8. Given $Q(x, y) = 40x^{0.2}y^{0.8}$, find

a. Q_x **b.** Q_y **c.** Q_{xx} **d.** Q_{yx}

Answers to Warm-up Exercises

Algebra Warm-up Exercises

1. $y = -\frac{7}{3}x + \frac{5}{3}$ **2.** $4x^4 - 13x^3 + x^2 - 16x$ **3.** $\frac{x^6}{64}$

Calculus Warm-up Exercises

1. $\frac{1}{2}x^6 - x^2 + 7\ln x + C$ **2.** $-2e^{-y} - \frac{3}{y} + C$

3. $-53\frac{2}{3}$ **4.** 16.604 **5.** $26\frac{2}{3}$ square units

6. $1\frac{1}{3}$ square units **7.** $\frac{3}{4}$

8. a. $Q_x = 8x^{-0.8}y^{0.8}$ **b.** $Q_y = 32x^{0.2}y^{-0.2}$

c. $Q_{xx} = -6.4x^{-1.8}y^{0.8}$ **d.** $Q_{yy} = 6.4x^{-0.8}y^{-0.2}$

Integrating Functions of Two Variables

For a function of one variable, $f(x)$, you evaluated both indefinite and definite integrals. The integral had several meanings and uses.

- Net rate of change in the value of an antiderivative over an interval
- Average value of a function over an interval
- Area of a nongeometric region

Integration can also be defined for functions of two (or more) variables. Consider the function $f(x, y) = 6xy^2$. Just as there are two first partial derivatives for $f(x, y)$, there are also two ways to integrate $f(x, y)$.

- Integrating with respect to the variable x, we treat y as a constant:

$$\int 6xy^2 dx = \int (6y^2)x dx$$

$$= (6y^2)\left(\frac{x^2}{2}\right) = 3y^2x^2 + C(y)$$

  Because y was treated as a constant, the constant of integration is a function of y.

- Integrating with respect to the variable y, we treat x as a constant:

$$\int 6xy^2 dx = \int (6x)y^2 dy$$

$$= (6x)\left(\frac{y^3}{3}\right) = 2xy^3 + C(x)$$

  Because x was treated as a constant, the constant of integration is a function of x.

Remember that you can check an integration problem by differentiating the solution and verifying that you obtain the original integrand. For example,

$\frac{\partial}{\partial x}(3y^2x^2 + C(y)) = (3y^2)2x + 0 = 6xy^2$, because $\frac{\partial}{\partial x}C(y) = 0$ if y is a constant.

$\frac{\partial}{\partial y}(2xy^3 + C(x)) = 2x(3y^2) + 0 = 6xy^2$, because $\frac{\partial}{\partial y}C(x) = 0$ if x is a constant.

Definite integrals of functions of two variables can also be evaluated by combining the technique outlined above with the Fundamental Theorem of Calculus you learned in Topic 23.

Example 1: Evaluate $\int_2^y \left(2x^3y - \frac{3}{y}\right)dx$.

Solution: The upper limit of integration is $x = y$; the lower limit of integration is $x = 2$.

First, integrate with respect to x, treating y as a constant:

$$\int_2^y \left(2x^3y - \frac{3}{y}\right)dx = \int_2^y \left((2y)x^3 - \frac{3}{y}\right)dx \qquad \textbf{2y and } \frac{3}{y} \textbf{ are constants}$$

$$= \left(2y\left(\frac{x^4}{4}\right) - \left(\frac{3}{y}\right)x\right)_2^y \qquad \textbf{Integrate with respect to } x$$

$$= \left[\frac{1}{2}x^4y - \left(\frac{3}{y}\right)x\right]_2^y \qquad \textbf{Simplify expressions}$$

Next, apply the Fundamental Theorem of Calculus; let $x = y$ and then let $x = 2$:

$$= \left[\frac{1}{2}y^4y - \frac{3}{y}(y)\right] - \left[\frac{1}{2}(2)^4y - \frac{3}{y}(2)\right]$$

Simplify:

$$= \frac{1}{2}y^5 - 3 - 8y + \frac{6}{y}$$

You see that the result is still a function of y, because y was treated as a constant throughout the integration process. ■

Double Integrals

We know that the graph of $f(x,y)$ is a surface in three dimensions whose domain is some region R in the xy-plane. Just as the definite integral $\int_a^b f(x)dx$ was used to determine the area of regions in the plane and the average value of functions of one variable on an interval, double integrals can be used to determine volumes under surfaces in three dimensions and average value of functions of two variables over a region in the plane.

The integral $\iint_R f(x,y)dA$ is called a **double integral** or an **iterated integral.**

The **limits of integration** on the two integrals define the boundaries of the **region of integration** in the plane.

If R is a **rectangular region** bounded by $a \le x \le b$ and $c \le y \le d$, the double integral is written as either $\int_a^b \int_c^d f(x,y)dydx$ or $\int_c^d \int_a^b f(x,y)dxdy$.

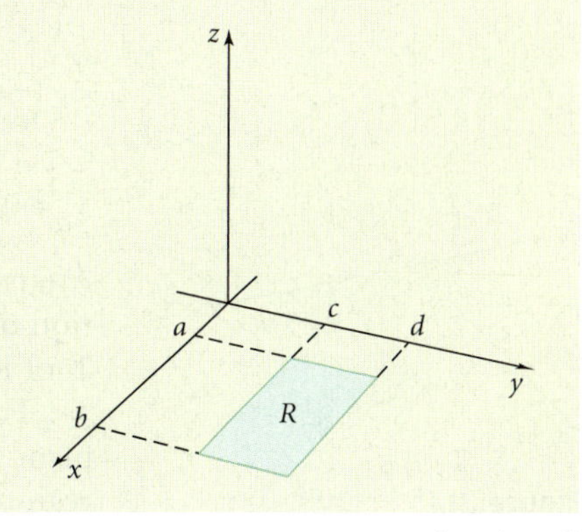

(continued)

If R is a **nonrectangular region** bounded by $a \le x \le b$ and $g_1(x) \le y \le g_2(x)$, the double integral is written as $\int_a^b \int_{g_1(x)}^{g_2(x)} f(x, y)dydx$.

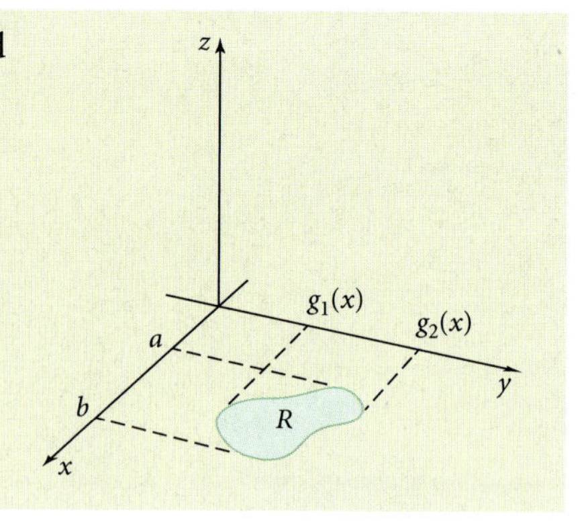

To evaluate a double integral, begin with the innermost integral and then work to the outermost integral, as shown in Example 2.

Example 2: Evaluate $\int_0^4 \int_{-1}^3 (3x + 2y - 6)dydx$.

Solution: Begin with the innermost integral and integrate with respect to y, treating x as a constant:

$$\int_0^4 \int_{-1}^3 (3x + 2y - 6)dydx = \int_0^4 \left[\int_{-1}^3 (3x + 2y - 6)dy \right] dx$$

$$= \int_0^4 [3xy + y^2 - 6y]_{-1}^3 dx \quad \text{3x and 6 are constants}$$

$$= \int_0^4 [(9x + 9 - 18) - (-3x + 1 + 6)]dx \quad \text{Let } y = 3 \text{ and } y = -1$$

$$= \int_0^4 (12x - 16)dx \quad \text{Simplify integrand}$$

$$= (6x^2 - 16x)_0^4 \quad \text{Evaluate integral}$$

$$= 32$$

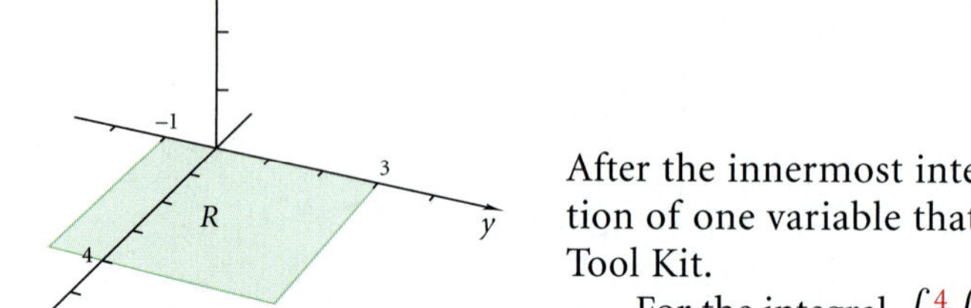

Figure 31.1

After the innermost integration is complete, the remaining integral is a function of one variable that can be evaluated using the methods already in the Tool Kit.

For the integral $\int_0^4 \int_{-1}^3 (3x + 2y - 6)dydx$, the region of integration in the plane is the rectangular region defined by $-1 \le y \le 3$ and $0 \le x \le 4$. This region is shown in the graph in Figure 31.1. ■

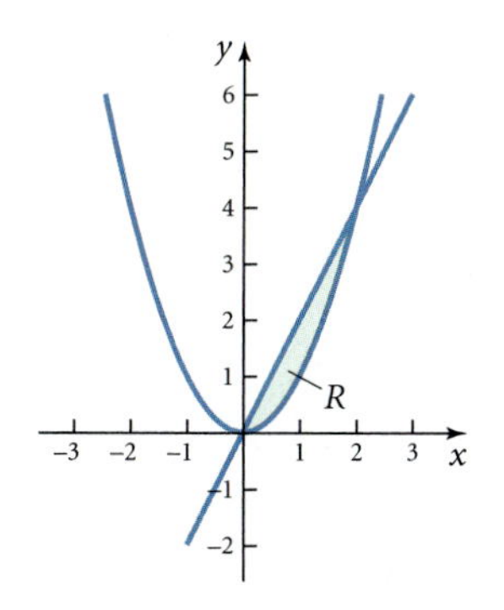

Figure 31.2

Example 3: Evaluate $\int_0^2 \int_{x^2}^{2x} (2x^3y - 6y + 3x)dydx$ and sketch the region of integration in the plane.

Solution: The region of integration is defined by $x^2 \leq y \leq 2x$ and $0 \leq x \leq 2$. The graph of this region is shown in Figure 31.2.

$$\int_0^2 \int_{x^2}^{2x} (2x^3y - 6y + 3x)dydx = \int_0^2 \left[\int_{x^2}^{2x} (2x^3y - 6y + 3x)dy \right] dx$$

$$= \int_0^2 \left[x^3y^2 - 3y^2 + 3xy\right]_{x^2}^{2x} dx$$

Treat x as a constant

Letting $y = 2x$ and $y = x^2$, we have

$$= \int_0^2 [(x^3(2x)^2 - 3(2x)^2 + 3x(2x)) - (x^3(x^2)^2 - 3(x^2)^2 + 3x(x^2))]dx$$

Simplifying the integrand gives

$$= \int_0^2 [(4x^5 - 12x^2 + 6x^2) - (x^7 - 3x^4 + 3x^3)]dx$$

$$= \int_0^2 (-x^7 + 4x^5 + 3x^4 - 3x^3 - 6x^2)dx$$

Integrating:

$$= \left(-\frac{x^8}{8} + \frac{2}{3}x^6 + \frac{3}{5}x^5 - \frac{3}{4}x^4 - 2x^3\right)_0^2$$

$$= \frac{28}{15} \approx 1.87$$

■

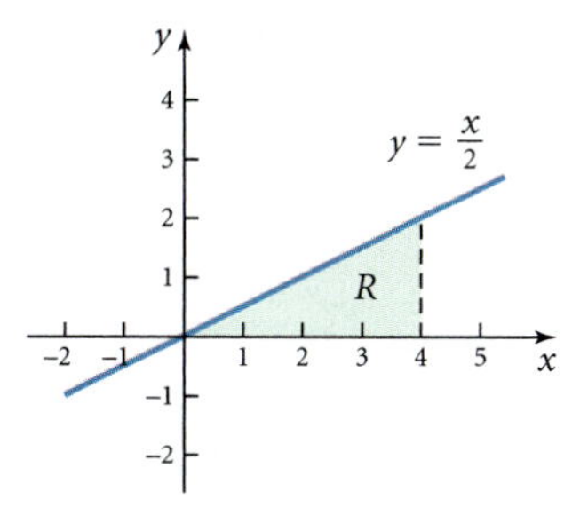

Figure 31.3

Example 4: Compute $\iint_R x^2y\, dA$ where R is the region shown in Figure 31.3.

Solution: First it is necessary to define the boundaries of R. The upper boundary is the line $y = \frac{x}{2}$ and the lower boundary is $y = 0$ (the x-axis), where $0 \leq x \leq 4$.

Next use these boundaries to construct the integral to be evaluated:

$$\iint_R x^2y\, dA = \int_0^4 \int_0^{x/2} x^2y\, dydx$$

> **Tip:** The limits of integration for the innermost integral are the boundaries for y from the region of integration. The lower boundary is the lower limit; the upper boundary is the upper limit. The limits of integration for the outermost integral are the boundaries for x from the region of integration. The outermost limits of integration will always be constants.

Integrate as in the previous examples.

$$\int_0^4\int_0^{x/2} x^2y\,dydx = \int_0^4\left[x^2\left(\frac{y^2}{2}\right)\right]_0^{x/2} dx \qquad x^2 \text{ is treated as a constant}$$

$$= \int_0^4\left(\frac{x^2}{2}\cdot(y)^2\right)_0^{x/2} dx$$

$$= \int_0^4\left(\frac{x^2}{2}\cdot\left(\frac{x}{2}\right)^2 - 0\right)dx \qquad \text{Let } y = \frac{x}{2} \text{ and } y = 0$$

$$= \int_0^4\left[\frac{x^2}{2}\cdot\frac{x^2}{4} - 0\right]dx$$

$$= \int_0^4\frac{x^4}{8}dx$$

$$= \frac{128}{5} = 25.6$$

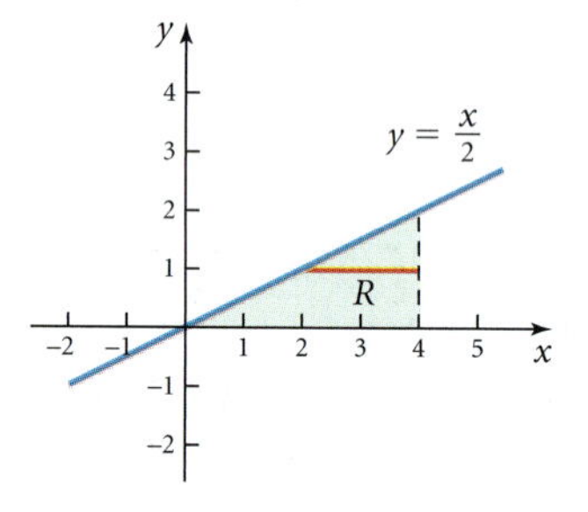

Figure 31.4

The order of integration can also be written as $\iint_R x^2y\,dxdy$. Then the innermost limits of integration must be functions of y, and the outermost limits of integration must be constant values of y. Let's demonstrate by reversing the order of integration for the region R in Example 4 as shown in Figure 31.4.

The line $y = \dfrac{x}{2}$ can also be written as $x = 2y$. The boundaries for x are now $2y \le x \le 4$, and the boundaries for y are $0 \le y \le 2$.

The integral to be evaluated is now $\int_0^2\int_{2y}^4 x^2y\,dxdy$. Evaluating this integral we have

$$\int_0^2\int_{2y}^4 x^2y\,dxdy = \int_0^2\left[\frac{x^3}{3}\cdot y\right]_{2y}^4 dy \qquad \text{Treat } y \text{ as a constant}$$

$$= \int_0^2\left(\frac{64y}{3} - \frac{8y^4}{3}\right)dy \qquad \text{Let } x = 4 \text{ and } x = 2y$$

$$= \left(\frac{32y^2}{3} - \frac{8y^5}{15}\right)_0^2$$

You should see that the same answer is obtained regardless of the order in which the integration is performed. ■

Check Your Understanding 31.1

Find each of the following integrals.

1. $\int(2xe^y - x^3y)dy$

2. $\int(2xe^y - x^3y)dx$

Evaluate each of the following double integrals. Sketch the region of integration in the plane.

3. $\int_1^3 \int_{-2}^1 (2x + 5y^2 - 3)dxdy$ **4.** $\int_1^4 \int_3^{x+2} \frac{y}{x} dydx$

5. Sketch the region in the plane and reverse the order of integration of $\int_0^2 \int_0^{\sqrt{x}} f(x, y)dydx$.

Applications of Double Integrals

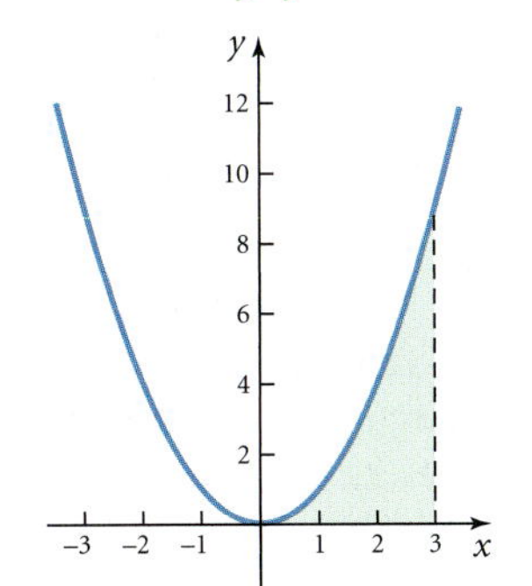

Figure 31.5

In Topic 24 you learned that $\int_0^3 x^2dx$ gives the area between $y = x^2$ and the x-axis on the interval $0 \leq x \leq 3$. See Figure 31.5.

What does the definite double integral mean graphically?

Definition: $f(x, y) \geq 0$ over a region R in the plane. The **volume** of the solid region that lies above R and below the surface $f(x, y)$ is given by

$$V = \iint_R f(x, y)dA.$$

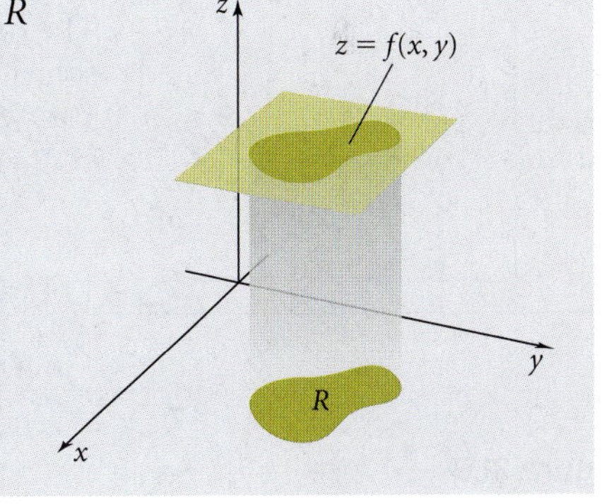

In Example 4 we found that $\iint_R x^2y\, dA$, with R the region shown in Figure 31.6, had a value of 25.6. This means that the volume in Figure 31.7 of the solid region above this triangular region in the plane and below the surface $z = x^2y$ is 25.6 cubic units.

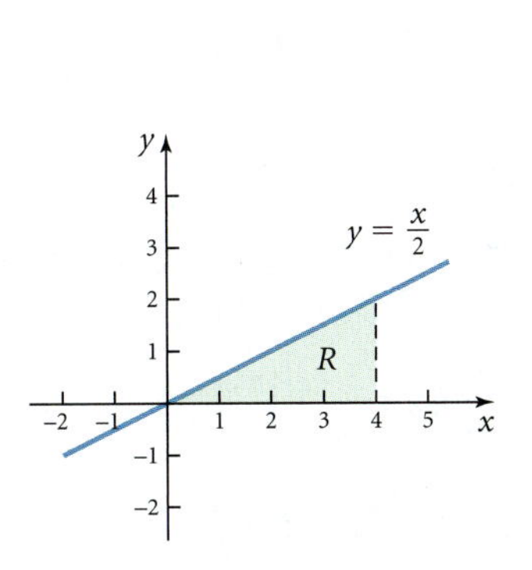

Figure 31.6

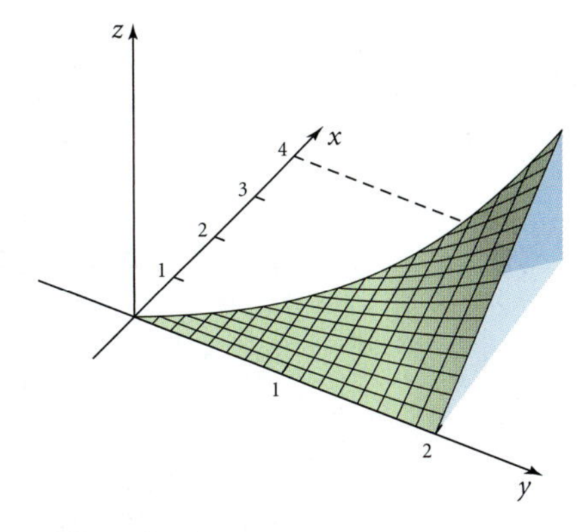

Figure 31.7

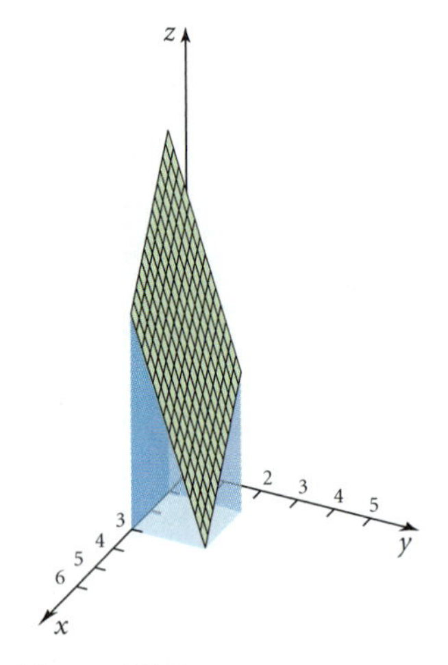

Figure 31.8

Example 5: Find the volume of the solid region above R and below $2x + 3y + z = 12$, if R is the rectangular region bounded by $1 \le x \le 3$ and $0 \le y \le 2$.

Solution: The upper surface of the solid region in Figure 31.8 is the function $z = f(x, y)$, so we must first solve $2x + 3y + z = 12$ for z, obtaining $z = f(x, y) = 12 - 2x - 3y$. Then the volume of the solid region is given by

$$
\begin{aligned}
V &= \int_1^3 \int_0^2 (12 - 2x - 3y)\,dydx \\
&= \int_1^3 \left(12y - 2xy - \frac{3}{2}y^2\right)\Big|_0^2 dx \\
&= \int_1^3 (18 - 4x)\,dx \\
&= (18x - 2x^2)_1^3 \\
&= (54 - 18) - (18 - 2) \\
&= 36 - 16 = 20 \text{ cubic units}
\end{aligned}
$$

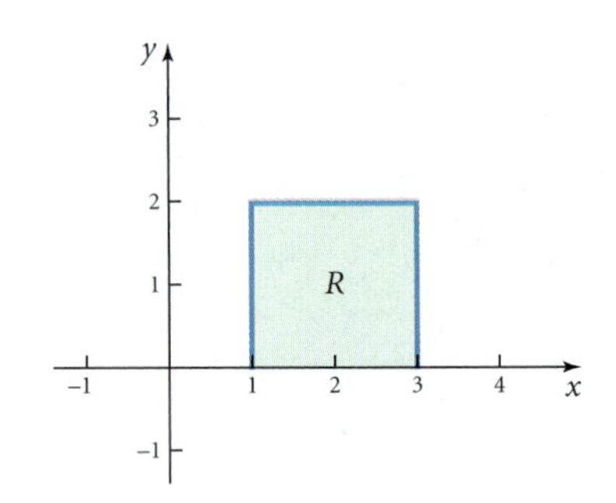

Figure 31.9

The limits of integration in Example 5 can be reversed easily and the same solution will result because the region of integration is rectangular, bounded by $1 \le x \le 3$ and $0 \le y \le 2$ as shown in Figure 31.9.

Reversing the order of integration gives

$$
\begin{aligned}
V &= \int_0^2 \int_1^3 (12 - 2x - 3y)\,dxdy \\
&= \int_0^2 (12x - x^2 - 3yx)_1^3\,dy \\
&= \int_0^2 ((36 - 9 - 9y) - (12 - 1 - 3y))\,dy \\
&= \int_0^2 (16 - 6y)\,dy \\
&= (16y - 3y^2)_0^2 = (32 - 12) - 0 = 20 \text{ cubic units}
\end{aligned}
$$

■

The average value of a function of one variable, $f(x)$, over the interval [a, b] is given by $\frac{1}{b - a}\int_a^b f(x)\,dx$. The average value of a function of two variables, $f(x, y)$, can also be defined.

Definition: The **average value** of $f(x, y)$ on region R in the plane is given by

$$\bar{f} = \frac{1}{A}\iint_R f(x, y)\,dA$$

where A is the area of the region R in the plane.

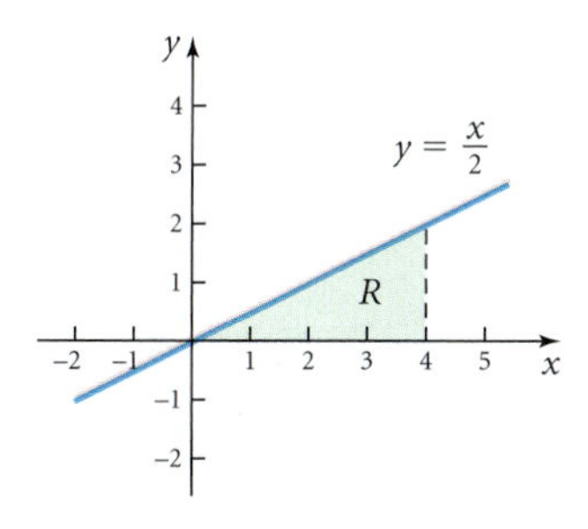

Figure 31.10

Example 6: Find the average value of $f(x, y) = x^2y$ if R is the region shown in Figure 31.10. (See Example 4.)

Solution: R is the triangular region bounded by $y = \frac{x}{2}$ and the x-axis for $0 \le x \le 4$. The area of the triangular region can be found by integrating $\int_0^4 \frac{x}{2}\,dx$ or by using the geometric formula for the area of a triangle with a base of 4 units and height 2 units. The area of R is 4 square units.

We also know from Example 4 that $\int_0^4 \int_0^{x/2} x^2y\,dydx = 25.6$. The average value is

$$\bar{f} = \frac{1}{A}\iint_R x^2y\,dA = \frac{1}{4}\int_0^4\int_0^{x/2} x^2y\,dydx$$
$$= \frac{1}{4}(25.6)$$
$$= 6.4$$

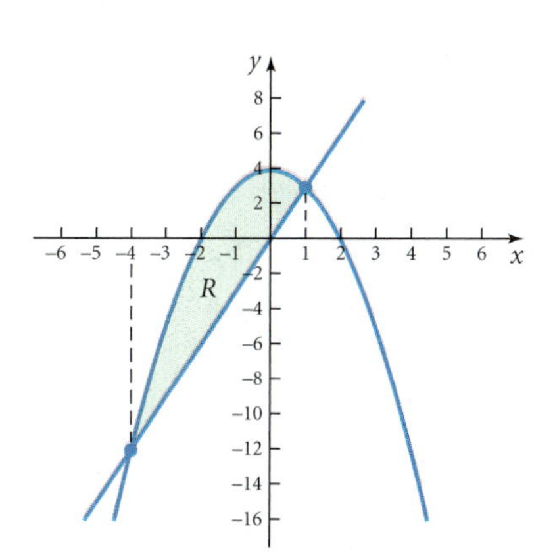

Figure 31.11

Example 7: Find the average value of $z = xy$ on the region R shown in Figure 31.11.

Solution: The region R is bounded by the curves $y = 4 - x^2$ and $y = 3x$, so $3x \le y \le 4 - x^2$. To determine the boundaries of x we must determine the points where the two curves intersect.

Setting $4 - x^2 = 3x$, we obtain $x^2 + 3x - 4 = 0$.
Factoring we have $(x + 4)(x - 1) = 0$.
Solving gives $x = -4, 1$.

The area of the region R is given by

$$A = \int_{-4}^{1}(4 - x^2 - 3x)\,dx$$
$$= \left(4x - \frac{x^3}{3} - \frac{3x^2}{2}\right)_{-4}^{1}$$
$$= 20\frac{5}{6} \text{ or } \approx 20.83$$

If $A = \frac{125}{6}$, then $\frac{1}{A} = \frac{6}{125}$, and the average value of $f(x, y)$ on this region R is then found by evaluating

$$\bar{f} = \frac{6}{125}\int_{-4}^{1}\int_{3x}^{4-x^2}(xy)\,dydx$$
$$= \frac{6}{125}\int_{-4}^{1}\left(\frac{xy^2}{2}\right)_{3x}^{4-x^2}dx$$
$$= \frac{6}{125}\cdot\frac{1}{2}\int_{-4}^{1}(x(4 - x^2)^2 - x(3x)^2)\,dx$$

$$= \frac{3}{125}\int_{-4}^{1}(x(16 - 8x^2 + x^4) - x(9x^2))\,dx$$
$$= \frac{3}{125}\int_{-4}^{1}(16x - 17x^3 + x^5)\,dx$$
$$= \frac{3}{125}\left(8x^2 - \frac{17x^4}{4} + \frac{x^6}{6}\right)_{-4}^{1}$$
$$= \frac{3}{125}\left(\left(8 - \frac{17}{4} + \frac{1}{6}\right) - \left(128 - 1088 + 682\frac{2}{3}\right)\right)$$
$$= \frac{3}{125}(281.25)$$
$$= 6.75$$

Example 8: Brian's Beach Shop is planning to price its new deluxe surfboards between \$250 and \$400. Marketing reports that at price $\$p$ per board, the demand for the board will be between $q = 500 - p$ and $q = 800 - p$ boards sold during the first year. What is the average possible revenue the shop can expect during the first year from the sale of the new deluxe board?

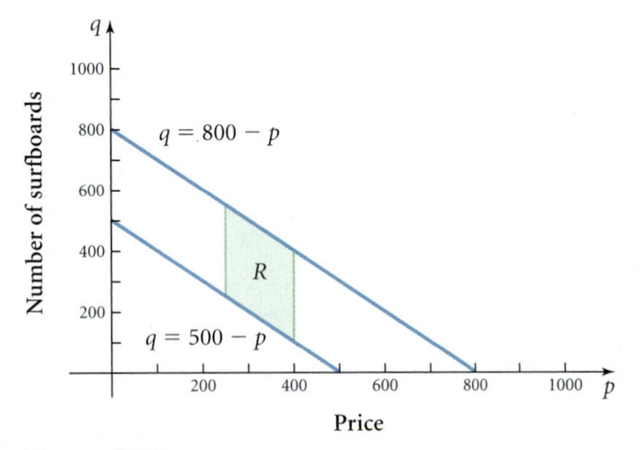

Figure 31.12

Solution: Revenue is price times quantity: $R = pq$.

The constraints are $250 \le p \le 400$ and $500 - p \le q \le 800 - p$. The region of integration is shown in Figure 31.12. Average revenue is found by

$$\overline{R} = \frac{1}{A}\int_{250}^{400}\int_{500-p}^{800-p} pq\,dq\,dp$$

The area of the region in the plane is given by

$$A = \int_{250}^{400}((800 - p) - (500 - p))\,dp$$
$$= \int_{250}^{400} 300dp$$
$$= (300p)_{250}^{400}$$
$$= 300(400 - 250)$$
$$= 45{,}000$$

The average revenue is

$$\overline{R} = \frac{1}{A}\int_{250}^{400}\int_{500-p}^{800-p} pq\,dq\,dp = \frac{1}{45{,}000}\int_{250}^{400}\int_{500-p}^{800-p} pq\,dq\,dp$$
$$= \frac{1}{45{,}000}\int_{250}^{400}\left.\frac{pq^2}{2}\right|_{500-p}^{800-p}dp$$
$$= \frac{1}{45{,}000}\cdot\frac{1}{2}\int_{250}^{400}(p(800 - p)^2 - p(500 - p)^2)\,dp$$

$$= \frac{1}{90{,}000}\int_{250}^{400}(p(640{,}000 - 1600p + p^2) - p(250{,}000 - 1000p + p^2))\,dp$$

$$= \frac{1}{90{,}000}\int_{250}^{400}(390{,}000p - 600p^2\,dp)$$

$$= \frac{1}{90{,}000}(195{,}000p^2 - 200p^3)_{250}^{400}$$

$$= \frac{1}{90{,}000}(9{,}337{,}500{,}000)$$

$$= \$103{,}750$$

The average possible expected revenue during the first year from the sale of the deluxe surfboard is $103,750. ■

Check Your Understanding 31.2

In the first two questions, find the volume of the solid region below the surface $x + y + z = 8$ and above the indicated region.

1. the rectangular region bounded by $1 \leq x \leq 4$ and $0 \leq y \leq 1$
2. the region bounded by $y = x^2$ and $y = 2x$
3. Find the average value of $x + y + z = 8$ on the region indicated in question 2.

Check Your Understanding Answers

Check Your Understanding 31.1

1. $2xe^y - \frac{x^3y^2}{2} + C(x)$

2. $x^2e^y - \frac{x^4y}{4} + C(y)$

3. 106

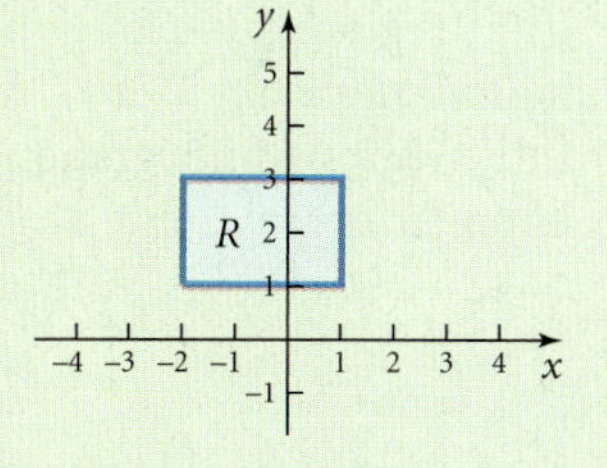

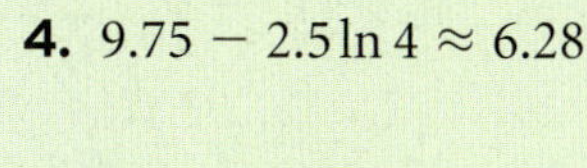

4. $9.75 - 2.5\ln 4 \approx 6.28$

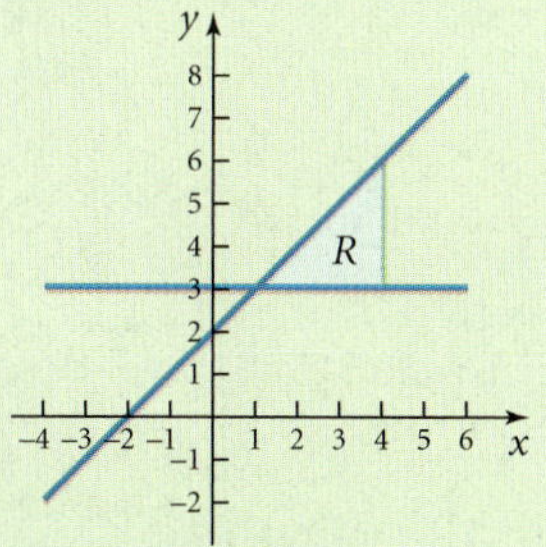

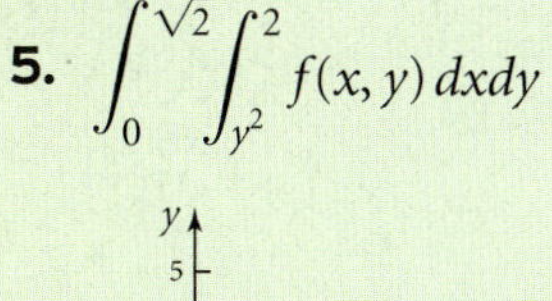

5. $\int_0^{\sqrt{2}}\int_{y^2}^{2} f(x, y)\,dxdy$

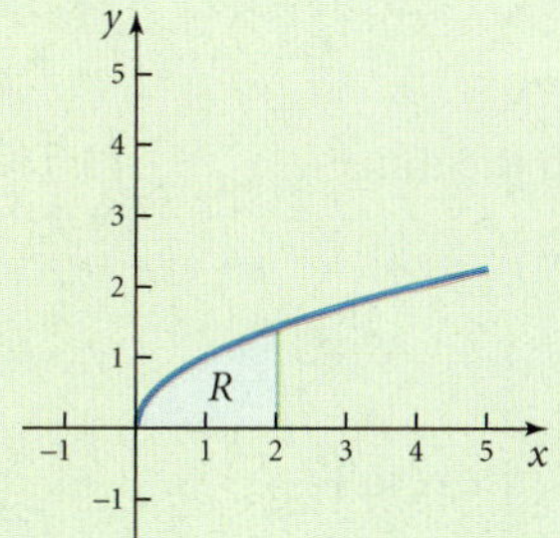

Check Your Understanding 31.2

1. 15 cubic units

2. $\frac{36}{5} = 7.2$ cubic units

3. $\bar{f} = \frac{1}{4/3}\int_0^2\int_{x^2}^{2x}(8 - x - y)\,dydx$

$$= \frac{3}{4}\left(\frac{36}{5}\right) = 5.4$$

Topic 31 Review

31

This topic presented integration for functions of two variables.

CONCEPT	EXAMPLE
Functions of two variables can be integrated with respect to either of the variables. To integrate $\int f(x, y)\,dy$, treat x as a constant. To integrate $\int f(x, y)\,dx$, treat y as a constant.	To integrate $\int 3x^2y^5\,dx$, treat y as a constant. Thus, we have $$\int 3x^2y^5\,dx = x^3y^5 + C(y)$$ To integrate $\int 3x^2y^5\,dy$, treat x as a constant. Thus, we have $$\int 3x^2y^5\,dy = \frac{1}{2}x^2y^6 + C(x)$$
The integral $\int_a^b \int_{g_1(x)}^{g_2(x)} f(x, y)\,dydx$ is a **double integral** or an **iterated integral**. The **limits of integration** on the integrals define the **region of integration** in the plane. If the region of integration is **rectangular**, the boundaries are $a \le x \le b$ and $c \le y \le d$. If the region of integration is **nonrectangular**, the boundaries are $a \le x \le b$ and $g_1(x) \le y \le g_2(x)$.	$\int_1^4 \int_{-2}^3 f(x, y)\,dydx$ is a double integral. The region of integration is rectangular, with boundaries $1 \le x \le 4$ and $-2 \le y \le 3$. $\int_1^4 \int_{-2x}^{3-x^2} f(x, y)\,dydx$ is a double integral. The region of integration is nonrectangular, with boundaries $-2x \le y \le 3 - x^2$ and $1 \le x \le 4$.
To evaluate a double integral, begin with the innermost integral: $$\int_a^b \int_{g_1(x)}^{g_2(x)} f(x, y)\,dydx = \int_a^b \left[\int_{g_1(x)}^{g(x)} f(x, y)\,dy \right] dx$$ After simplifying the innermost integral, the resulting integral is a function of one variable.	$$\int_0^1 \int_{x^2}^{x} 3xy^2 dydx = \int_0^1 \left[\int_{x^2}^{x} 3xy^2 dy \right] dx = \int_0^1 xy^3 \Big\|_{x^2}^{x} dx = \int_0^1 [x(x)^3 - x(x^2)^3]\,dx = \int_0^1 (x^4 - x^7)\,dx = \frac{x^5}{5} - \frac{x^8}{8} \Big\|_0^1 = \frac{3}{40}$$

The order of integration can be reversed: $$\int_a^b \int_{g_1(x)}^{g_2(x)} f(x,y)\,dydx = \int_c^d \int_{h_1(y)}^{h_2(y)} f(x,y)\,dxdy$$ The region of integration is then given by $h_1(y) \leq x \leq h_2(y)$ and $c \leq y \leq d$.	$$\int_0^1 \int_{x^2}^{x} 3xy^2 dydx = \int_0^1 \int_y^{\sqrt{y}} 3xy^2 dxdy$$ $$= \frac{3}{2}\int_0^1 y^2x^2 \Big\vert_y^{\sqrt{y}} dy$$ $$= \frac{3}{2}\int_0^1 (y^2(\sqrt{y})^2 - y^2y^2)\,dy$$ $$= \frac{3}{2}\int_0^1 (y^3 - y^4)\,dy$$ $$= \frac{3}{2}\left(\frac{y^4}{4} - \frac{y^5}{5}\right)\Big\vert_0^1$$ $$= \frac{3}{2}\left(\frac{1}{20}\right) = \frac{3}{40}$$
The **average value** of $f(x,y)$ on a region R in the plane is given by $\bar{f} = \frac{1}{A}\iint_R f(x,y)\,dA$, where A is the area of the region in the plane.	For $\int_0^1 \int_{x^2}^{x} 3xy^2\,dydx$, the area of the region in the plane is given by $A = \int_0^1 (x - x^2)\,dx = \frac{1}{6}$. The average value of $\int_0^1 \int_{x^2}^{x} 3xy^2\,dydx$ is then given by $\frac{1}{\frac{1}{6}}\int_0^1 \int_{x^2}^{x} 3xy^2\,dydx = 6\left(\frac{3}{40}\right) = \frac{9}{20}$.
Given $f(x,y) \geq 0$ over a region R in the plane, the **volume** of the solid lying above R and below the surface $z = f(x,y)$ is given by $V = \iint_R f(x,y)\,dA$, where dA may be done as $dydx$ or $dxdy$.	Let R be the region in the plane bounded by $0 \leq y \leq 1 - x$ and $0 \leq x \leq 1$. The volume of the solid lying above R and below $z = 4 - x^2$ is given by either $$V = \int_0^1 \int_0^{1-x} (4 - x^2)\,dydx \text{ or}$$ $$V = \int_0^1 \int_0^{1-y} (4 - x^2)\,dxdy.$$

NEW TOOLS IN THE TOOL KIT

- Integrating a function of two variables with respect to one of the variables
- Evaluating double integrals of functions of two variables over rectangular regions in the plane
- Evaluating double integrals of functions of two variables over nonrectangular regions in the plane
- Reversing the order of integration
- Determining the average value of a function of two variables over a region in the plane
- Determining the volume of a solid below a surface and above a region in the plane

Topic 31 Exercises

31

Find each integral with respect to the indicated variable.

1. $\int (x + y)\,dy$

2. $\int (x - 3y)\,dx$

3. $\int (x^2 - y^2)\,dx$

4. $\int x^3y^2\,dy$

5. $\int x^3e^{2y}\,dy$

6. $\int x^3e^{2y}\,dx$

7. $\int \frac{3}{xy}\,dx$

8. $\int \sqrt{xy}\,dy$

Evaluate each double integral.

9. $\int_0^3 \int_0^2 (x + 2y)\,dydx$

10. $\int_1^3 \int_{-1}^2 (3y - 4x)\,dydx$

11. $\int_0^2 \int_0^1 (x^2 + y^2)\,dydx$

12. $\int_0^1 \int_0^2 (2x^2 - y^2)\,dydx$

13. $\int_0^1 \int_0^3 (x^2e^{2y})\,dxdy$

14. $\int_{-1}^2 \int_0^3 (4ye^{-x})\,dxdy$

15. $\int_{-1}^2 \int_0^1 \left(\frac{4xy}{4 + x^2}\right)dxdy$

16. $\int_0^2 \int_2^3 xy(x^2 - 4)^3\,dxdy$

17. $\int_1^4 \int_x^4 \left(\frac{2x}{y^2} + \frac{3y}{x}\right)dydx$

18. $\int_1^2 \int_{x^2}^{x+2} \frac{xy + 1}{x^4}\,dydx$

19. $\int_0^1 \int_{y-1}^0 e^{x+y}\,dxdy$

20. $\int_0^4 \int_{y/2}^{\sqrt{y}} xy^2\,dxdy$

21. $\int_0^1 \int_{-x}^{x^2} y\,dydx$

22. $\int_0^3 \int_{x^2}^{3x} \frac{4}{x}\,dydx$

23. $\int_0^1 \int_0^{2x} e^{x^2}\,dydx$

24. $\int_1^3 \int_{1/x}^{x} xy\,dydx$

For each of the following double integrals, sketch the region of integration and reverse the order of integration. Do not integrate.

25. $\int_{-2}^{2}\int_{0}^{4-2x} f(x, y)\, dydx$

26. $\int_{0}^{2}\int_{x}^{2} f(x, y)\, dydx$

27. $\int_{0}^{2}\int_{x^2}^{2x} f(x, y)\, dydx$

28. $\int_{1}^{4}\int_{1}^{\sqrt{x}} f(x, y)\, dydx$

29. $\int_{0}^{2}\int_{y^2}^{4} f(x, y)\, dxdy$

30. $\int_{0}^{1}\int_{0}^{\sqrt{1-y}} f(x, y)\, dxdy$

In Exercises 31 through 36, find the volume of the given solid over the indicated region of integration.

31. $f(x, y) = xy$, where R is the region given by $0 \leq x \leq 1$ and $0 \leq y \leq 4$

32. $f(x, y) = 3 - x - y$, where R is the region given by $0 \leq x \leq 2$ and $0 \leq y \leq 1$

33. $f(x, y) = 4 - x^2$, where R is the region given by $0 \leq x \leq 2$ and $0 \leq y \leq 4 - x^2$

34. $f(x, y) = x + y$, where R is the region given by $0 \leq x \leq 2$ and $2x \leq y \leq 4$

35. $f(x, y) = 6 - y^2$, where R is the region given by $0 \leq x \leq 2$ and $0 \leq y \leq x$

36. $f(x, y) = 4 - x^2 - y^2$, where R is the region given by $0 \leq x \leq 1$ and $x^2 \leq y \leq x$

In Exercises 37 through 40, find the average value of each function over the indicated region in the plane.

37. $f(x, y) = x^2$, where R is the region given by $0 \leq x \leq 4$ and $0 \leq y \leq 2$

38. $f(x, y) = xy$, where R is the region given by $0 \leq x \leq 3$ and $-1 \leq y \leq 2$

39. $f(x, y) = x^2 + y^2$, where R is the region given by $-1 \leq x \leq 2$ and $1 \leq y \leq x + 2$

40. $f(x, y) = e^{x+y}$, where R is the region given by $0 \leq x \leq 2$ and $-x \leq y \leq 3$

41. Trudy's Camping Supply Store is preparing to launch the sales of a deluxe six-person tent, which will be priced between \$800 and \$1200. Marketing data suggest that at a price of \$$p$ per tent the annual demand for the tent will be between $q = 1250 - p$ and $q = 1600 - p$ tents. What is the average possible revenue that the company can expect during the first year from sales of the new tent?

42. Trudy's Camping Supply Store is planning to introduce a four-person tent. The demand is expected to be between 3000 and 4000 tents per year if the price is between $p = 0.2q + 50$ and $p = 0.3q + 20$ tents. What is the average possible revenue that the store can expect during the first year from sales of the new tent?

43. The total number of sailboats manufactured by Sherman's Boats each month is given by $Q = 50x^{0.6}y^{0.4}$, where x is the monthly capital, in \$1000 units, and y is the labor in worker hours. Monthly capital investment must be between \$10,000 and \$12,000. Monthly labor hours must be between 2800 and 3200 worker hours. What is the average monthly number of sailboats produced?

44. Repeat Exercise 43 if monthly output is given by $Q = 80x^{0.7}y^{0.3}$, the monthly capital investment is between \$8000 and \$10,000, and the monthly labor hours are between 2400 and 3000 worker hours.

COMMUNICATE

45. Given $\iint_R e^{x^2}dA$, explain whether $\iint_R f(x, y)dydx$ or $\iint_R f(x, y)dxdy$ is the easier method of integration. Then evaluate the integral over the region bounded by $0 \leq x \leq 1$ and $0 \leq y \leq x$.

46. Given $\iint_R xe^{xy}dA$, explain whether $\iint_R f(x, y)dydx$ or $\iint_R f(x, y)dxdy$ is the easier method of integration. Then evaluate the integral over the region bounded by $0 \leq x \leq 2$ and $0 \leq y \leq 1$.

SHARPEN THE TOOLS

The iterated integral $\int_a^b \int_{g_1(x)}^{g_2(x)} \int_{h_1(x,y)}^{h_2(x,y)} f(x, y, z)\, dz\, dydx$ is called a *triple integral.* The limits of integration define the solid region of integration in three dimensions. To evaluate a triple integral, begin with the innermost integral and work toward the outermost integral. For example,

$$\int_1^3 \int_0^1 \int_1^2 (x + y - z)dzdydx = \int_1^3 \int_0^1 \left[\int_1^2 (x + y - z)dz \right] dydx \quad \text{Integrate first with respect to } z$$

$$= \int_1^3 \int_0^1 \left(xz + yz - \frac{z^2}{2} \right)_1^2 dydx \quad \text{Treat } x \text{ and } y \text{ as constants}$$

$$= \int_1^3 \int_0^1 \left(x + y - \frac{3}{2} \right) dydx \quad \text{Let } z = 2 \text{ and } z = 1$$

The remaining integral is a double integral that can be integrated using the methods presented in this topic.

$$= \int_1^3 \left(xy + \frac{y^2}{2} - \frac{3}{2}y \right)_0^1 dx \quad \text{Integrate with respect to } y$$

$$= \int_1^3 (x - 1)dx \quad \text{Let } y = 1 \text{ and } y = 0$$

$$= \left(\frac{x^2}{2} - x \right)_1^3 = 2$$

Evaluate each triple integral.

47. $\displaystyle\int_0^4 \int_1^2 \int_{-1}^3 (2x + 3y - z)dzdydx$

48. $\displaystyle\int_0^2 \int_0^3 \int_0^1 (xyz)dzdydx$

49. $\displaystyle\int_0^1 \int_0^x \int_0^{xy} xdzdydx$

50. $\displaystyle\int_1^4 \int_0^1 \int_0^x (2e^{-x^2})dzdydx$

51. $\displaystyle\int_0^2 \int_{2x}^4 \int_0^{y^2-4x^2} 4dzdydx$

52. $\displaystyle\int_0^3 \int_0^x \int_0^{6-x-y} 3dzdydx$

UNIT 4

Review

Unit 4 presented the basic concepts of multivariable calculus, including partial derivatives, double integrals, optimization, and Lagrange multipliers.

TOPIC 28	■ Functions of two variables ■ Domains of functions of two variables ■ Graphs of functions of two variables ■ Level curves
TOPIC 29	■ First partial derivatives ■ Interpreting first partial derivatives as slope in the x-direction or slope in the y-direction ■ Second partial derivatives
TOPIC 30	■ Determining critical points of functions of two variables ■ Using the Second Partial Derivative Test to determine if the critical point is a relative maximum, a relative minimum, or a saddle point ■ Solving constrained optimization problems using Lagrange multipliers
TOPIC 31	■ Evaluating double integrals over rectangular regions ■ Evaluating double integrals over nonrectangular regions ■ Volume of a solid over a region R in the plane and under a surface ■ Average value of a function of two variables over a region R

Having completed Unit 4, you should now be able to

1. Evaluate a function of two or more variables.
2. State the domain of a function of two variables and sketch the domain as a region in the xy-plane.
3. Determine the family of level curves for a function of two variables and sketch the curve in the xy-plane.
4. Recognize the graphs of the basic surfaces.
5. Determine the first partial derivative of a function of two or more variables.
6. Evaluate the first partial derivative at a specific point and interpret the answer.
7. Determine the second partial derivative of a function of two or more variables.
8. Find the critical point(s) of a function of two variables.
9. Use the Second Partial Derivative Test to determine if the critical point is a relative maximum, a relative minimum, or a saddle point.
10. Solve constrained optimization problems using Lagrange multipliers.
11. Evaluate double integrals over rectangular regions in the plane.
12. Evaluate double integrals over nonrectangular regions in the plane.
13. Determine the region of integration for a double integral and reverse the order of integration.
14. Find the volume of a solid region under a surface and over a region in the plane.
15. Find the average value of a function of two variables using double integrals.
16. Solve application problems involving functions of two variables.

UNIT 4 TEST

Evaluate each function at the indicated point.

1. Evaluate $f(2, -3)$ for $f(x, y) = 2x^3 - 3xy^2 + 5y$.
2. Evaluate $Q(250, 15)$ for $Q(x, y) = 85x^{0.63}y^{0.37}$.
3. Evaluate $V(2, 1, 6)$ for $V(a, b, c) = \sqrt{a^2b + ab^2 + bc^2}$.
4. Evaluate $g(-3, 4, 2)$ for $$g(x, y, z) = \frac{xy}{x + z} + \frac{xz}{x + y} + \frac{yz}{x + y}.$$

State the domain of each function. Sketch the domain as a region in the plane.

5. $f(x, y) = \ln(x^2 + y)$
6. $f(x, y) = 4x + \sqrt{x - 3y}$

For each function, determine the level curve at the indicated point and sketch the curve in the plane.

7. $f(x, y) = 6 - x^2 - y^2$ for $k = 2$
8. $f(x, y) = e^{-x} - y$ for $k = 1$

Describe the family of level curves of each of the following functions.

9. $f(x, y) = \sqrt{6 - x^2 - y^2}$
10. $f(x, y) = 6xy$
11. Tickets to a church supper cost \$5 for adults and \$2 for children. It costs the church \$3.25 to prepare each adult meal and \$1.50 to prepare each child's meal. Fixed costs are \$275.
    a. Write a revenue function $R(a, c)$ if a adult tickets and c child's tickets are sold.
    b. Evaluate $R(80, 64)$ and interpret the answer.
    c. Write a cost function $C(a, c)$ for the cost of serving a adult meals and c child's meals.
    d. Evaluate $C(80, 64)$ and interpret the answer.
    e. Write a profit function $P(a, c)$ for selling a adult tickets and c child's tickets.
    f. Evaluate $P(80, 64)$ and interpret your answer.
    g. If the church social hall can seat up to 300 for a dinner, how many adults and how many children should be served for the church to break even on the meal?
12. For a church holiday dinner, the price of an adult ticket is \$$x$ and the price of a child's ticket is \$$y$. At these prices, the demand for adult tickets is $D_1 = 200 - 3x + 8y$, and the demand for children's tickets is $D_2 = 250 + 6x - 40y$.

a. Determine the revenue function $R(x, y)$.

b. Find the total revenue if $x = \$10$ and $y = \$4$.

13. The monthly output of a factory is $Q = 90x^{0.6}y^{0.4}$, where x is the number of employees and y is the number of machines. How many units are produced with 50 employees and 16 machines?

For each of the following functions, find f_x and f_y.

14. $f(x, y) = e^{-x} - y$

15. $f(x, y) = e^{x^2 y}$

16. $f(x, y) = \ln(3y - 2x)$

17. $f(x, y) = \sqrt{6 - x^2 - y^2}$

18. $f(x, y) = \dfrac{6xy}{x + 3y}$

19. $f(x, y) = x^3 e^{2x/y}$

Find the slope in the x-direction and the slope in the y-direction at the indicated point.

20. $f(x, y) = 6 - x^2 - y^2$ at $(1, 1, 4)$

21. $f(x, y) = \dfrac{6xy}{x + 3y}$ at $(2, -1, 12)$

Find the second partial derivatives of each of the following functions.

22. $f(x, y) = e^{-x} - y$

23. $f(x, y) = \ln(3y - 2x)$

24. $f(x, y) = e^{xy^2}$

25. A race track is in the shape of a rectangle with semicircles on each end. The area of the track is $A(x, r) = 2xr + \pi r^2$.

a. Find A_x and A_r.

b. Evaluate A_x and A_r if $x = 500$ feet and $r = 60$ feet. Interpret your answers.

26. The monthly output of a factory is $Q = 90x^{0.6}y^{0.4}$, where x is the number of employees and y is the number of machines.

a. Find Q_x and Q_y.

b. If the factory currently uses 50 employees and 16 machines, find the marginal productivity of labor (employees) and the marginal productivity of capital (machines).

c. At the current level of 50 employees and 16 machines, would production increase more rapidly with five more employees or two more machines?

27. A local school is hosting a barbeque. The price of an adult ticket is $\$x$ and the price of a child's ticket is $\$y$. At these prices, the demand for adult tickets is $D_1 = 200 - 3x + 8y$ and the demand for children's tickets is $D_2 = 250 + 6x - 40y$.

a. Write the revenue function $R(x, y)$

b. Find R_x and R_y.

c. Evaluate R_x and R_y if $x = \$10$ and $y = \$4$. Interpret your answers.

28. The cost to the school for preparing the barbecue (see Exercise 27) is

$$C(x, y) = 3x^2 + \frac{xy}{8} + y^2.$$

a. Write the profit function $P(x, y)$.

b. Find P_x and P_y.

c. Evaluate P_x and P_y if $x = \$10$ and $y = 4$. Interpret your answers.

For each function, find the critical point(s) and use the Second Partial Derivative Test to determine if the critical point is a relative maximum, a relative minimum, or a saddle point.

29. $f(x, y) = 6 - x^2 - y^2$

30. $f(x, y) = x^2 + 6x + y^2 - 4y + 8$

31. $f(x, y) = x^3 + y^3 - 6xy$

32. $f(x, y) = 8x + 2y - \dfrac{4}{xy}$

33. $f(x, y) = e^{x^2 - y^2}$

Solve each of the following constrained optimization problems using Lagrange multipliers.

34. Maximize $A = 8xy$ if $2x + y = 80$.

35. Minimize $C = 40x + 25y$ if $xy = 320$. Assume $x > 0$ and $y > 0$.

36. Maximize $Q = 90x^{0.6}y^{0.4}$ if $3x + 5y = 10$.

37. Jon's Marina has received an order for 120 sailboats. Jon can fill the order with x boats from the California boatyard and y boats from the Virginia boatyard. The total cost is $C(x, y) = 0.5x^2 - 8x + 0.2y^2 - 6y + 650$. How many boats should be transported from each boatyard to minimize the cost?

38. Jon's Marina sells two models of sailboats. The Windsurfer sells for x thousand dollars and the Pelican sells for y thousand dollars. The revenue from the sale of the boats is $R(x, y) = 50x + 75y + 5xy - 3x^2 - 4y^2$. To the nearest dollar, what prices for the two boats will maximize the revenue?

Find each of the following integrals with respect to the indicated variable.

39. $\int x^2 e^{3y}\,dx$ **40.** $\int \left(\frac{6}{xy^2}\right) dy$

Find each double integral.

41. $\int_2^3 \int_{-1}^4 (3x - 2y + 6)\,dydx$

42. $\int_2^5 \int_1^2 \left(\frac{2x}{y^3} + \frac{4y}{x^2}\right) dxdy$

43. $\int_1^4 \int_1^x \frac{x + y^3}{xy}\,dydx$

44. $\int_{-2}^1 \int_x^{2-x^2} 4xy\,dydx$

45. $\int_0^1 \int_y^{\sqrt{y}} 4x\sqrt{y}\,dxdy$

46. $\int_1^3 \int_0^1 \left(\frac{6xy}{x^2 + 4}\right) dxdy$

For each of the following double integrals, sketch the region of integration. Then reverse the order of integration. Do not perform the integration.

47. $\int_1^2 \int_0^{\ln x} f(x, y)\,dydx$

48. $\int_0^4 \int_{y/2}^{\sqrt{y}} f(x, y)\,dxdy$

49. Find the volume of the solid under the surface $f(x, y) = 6 - x^2 - y^2$ and over the region R in the plane described by $0 \le x \le 2$ and $1 \le y \le 2$.

50. Find the volume of the solid under the surface $f(x, y) = 2x - 3y + 8$ and over the region R in the plane described by $0 \le x \le 4$ and $0 \le y \le \sqrt{x}$.

51. Find the average value of $f(x, y) = 4x + y$ over the region R in the plane described by $0 \le x \le 2$ and $0 \le y \le 1$.

52. Find the average value of $f(x, y) = x^2 + y^2$ over the region R described by $1 \le x \le 2$ and $0 \le y \le x^2$.

53. The Marina's annual output of sailboats is $Q = 30x^{0.7}y^{0.3}$, where x is the annual labor cost, in thousand dollars, and y is the annual capital, in thousand dollars. Find The Marina's average annual output if annual labor costs must be between \$650,000 and \$850,000 and annual capital must be between \$500,000 and \$600,000.

Geometry Formulas

Triangle

$P = a + b + c$

$A = \frac{1}{2}bh$

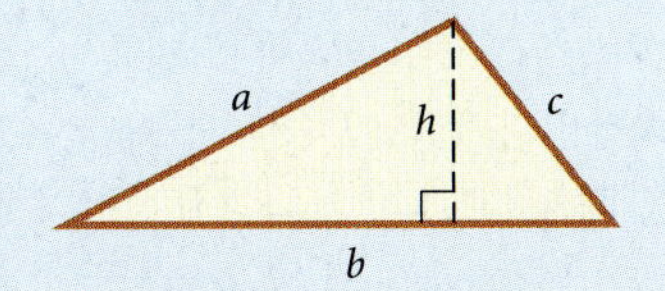

Rectangle

$P = 2l + 2w$

$A = lw$

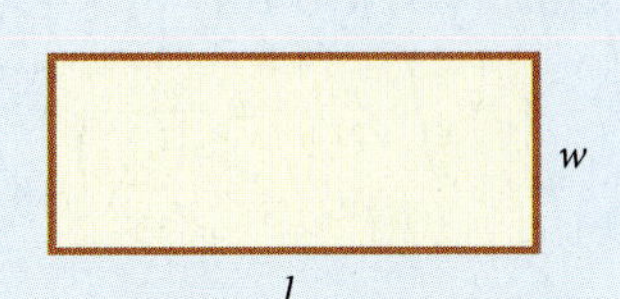

Square

$P = 4s$

$A = s^2$

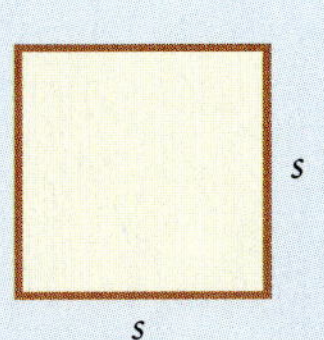

Trapezoid

$A = \frac{1}{2}h(b_1 + b_2)$

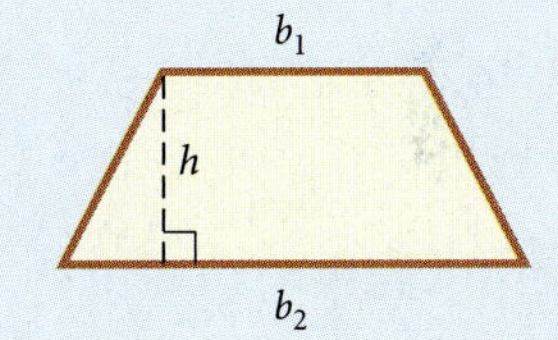

Circle

$C = 2\pi r$

$A = \pi r^2$

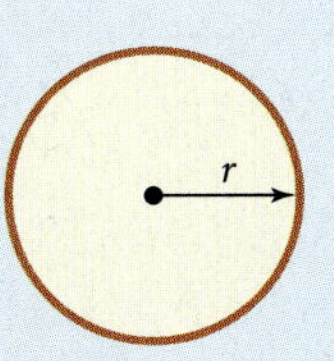

Sphere

$V = \frac{4}{3}\pi r^3$

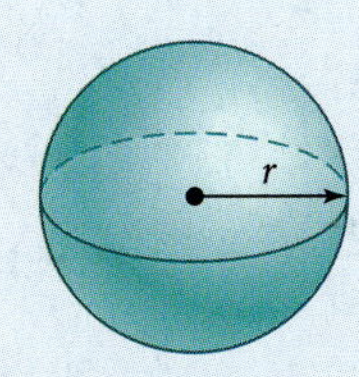

Cylinder

$V = \pi r^2 h$

$A = 2\pi r^2 + 2\pi rh$

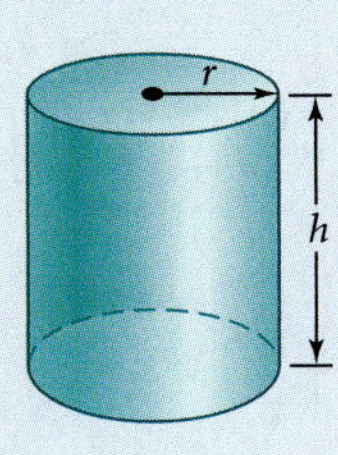

Cone

$V = \frac{1}{3}\pi r^2 h$

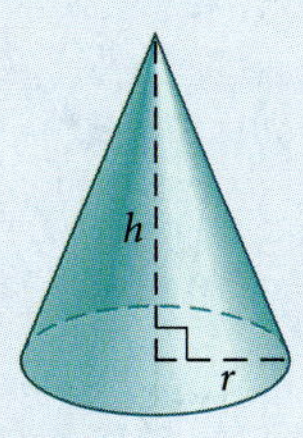

APPENDIX

B A Brief Table of Integrals

Basic Formulas

1. $$\int x^n dx = \frac{1}{n+1}x^{n+1} + C \quad \text{where } n \neq -1$$

2. $$\int \frac{1}{x} dx = \ln|x| + C$$

3. $$\int e^x dx = e^x + C$$

4. $$\int e^{kx} dx = \frac{1}{k}e^{kx} + C$$

Integration by Parts Formula

5. $$\int u\,dv = uv - \int v\,du$$

Forms Involving $ax + b$

6. $$\int \frac{x}{ax+b} dx = \frac{x}{a} - \frac{b}{a^2}\ln|ax+b| + C$$

7. $$\int \frac{1}{(ax+b)(cx+d)} dx = \frac{1}{ad-bc}\ln\left|\frac{ax+b}{cx+d}\right| + C$$

8. $$\int \frac{x}{(ax+b)(cx+d)} dx = \frac{1}{ad-bc}\left(\frac{d}{c}\ln|cx+d| - \frac{b}{a}\ln|ax+b|\right) + C$$

9. $$\int \frac{1}{x^2(ax+b)} dx = -\frac{1}{b}\left(\frac{1}{x} + \frac{a}{b}\ln\left|\frac{x}{ax+b}\right|\right) + C$$

Forms Involving $\sqrt{ax + b}$

10. $$\int \frac{x}{\sqrt{ax+b}} dx = \frac{2ax-4b}{3a^2}\sqrt{ax+b} + C$$

11. $$\int \frac{1}{x\sqrt{ax+b}} dx = \frac{1}{\sqrt{b}}\ln\left|\frac{\sqrt{ax+b}-\sqrt{b}}{\sqrt{ax+b}+\sqrt{b}}\right| + C$$

Forms Involving $\sqrt{a^2 + x^2}$

12. $$\int \sqrt{a^2 + x^2}\,dx = \frac{x}{2}\sqrt{a^2 + x^2} + \frac{a^2}{2}\ln|x + \sqrt{a^2 + x^2}| + C$$

13. $$\int \frac{1}{\sqrt{a^2 + x^2}}\,dx = \ln|x + \sqrt{a^2 + x^2}| + C$$

14. $$\int \frac{1}{x\sqrt{a^2 + x^2}}\,dx = -\frac{1}{a}\ln\left|\frac{\sqrt{a^2 + x^2} + a}{x}\right| + C$$

15. $$\int x^2\sqrt{a^2 + x^2}\,dx = \frac{x}{8}(a^2 + 2x^2)\sqrt{a^2 + x^2} - \frac{a^4}{8}\ln|x + \sqrt{a^2 + x^2}| + C$$

Forms Involving $\sqrt{a^2 - x^2}$

16. $$\int \frac{1}{x\sqrt{a^2 - x^2}}\,dx = -\frac{1}{a}\ln\left|\frac{a + \sqrt{a^2 - x^2}}{x}\right| + C$$

17. $$\int \frac{1}{x^2\sqrt{a^2 - x^2}}\,dx = -\frac{\sqrt{a^2 - x^2}}{a^2x} + C$$

18. $$\int \frac{1}{a^2 - x^2}\,dx = \frac{1}{2a}\ln\left|\frac{a + x}{a - x}\right| + C$$

19. $$\int \frac{\sqrt{a^2 - x^2}}{x}\,dx = \sqrt{a^2 - x^2} - a\ln\left|\frac{a + \sqrt{a^2 - x^2}}{x}\right| + C$$

Forms Involving $\sqrt{x^2 - a^2}$

20. $$\int \sqrt{x^2 - a^2}\,dx = \frac{x}{2}\sqrt{x^2 - a^2} - \frac{a^2}{2}\ln|x + \sqrt{x^2 - a^2}| + C$$

21. $$\int \frac{\sqrt{x^2 - a^2}}{x^2}\,dx = -\frac{\sqrt{x^2 - a^2}}{x} + \ln|x + \sqrt{x^2 - a^2}| + C$$

22. $$\int \frac{1}{\sqrt{x^2 - a^2}}\,dx = \ln|x + \sqrt{x^2 - a^2}| + C$$

23. $$\int \frac{1}{x^2\sqrt{x^2 - a^2}}\,dx = \frac{\sqrt{x^2 - a^2}}{a^2x} + C$$

24. $$\int \frac{1}{x^2 - a^2}\,dx = \frac{1}{2a}\ln\left|\frac{x - a}{x + a}\right| + C$$

Forms Involving e^{ax} and $\ln x$

25. $$\int xe^{ax}\,dx = \frac{1}{a^2}(ax - 1)e^{ax} + C$$

26. $$\int \ln x\,dx = x\ln x - x + C$$

27. $\displaystyle\int \frac{1}{x \ln x} dx = \ln|\ln x| + C$

28. $\displaystyle\int x^n \ln x \, dx = \frac{x^{n+1}}{n+1}\left(\ln x - \frac{1}{n+1}\right) + C$ where $n \neq -1$

Reduction Formulas

29. $\displaystyle\int x^n e^{ax} dx = \frac{1}{a} x^n e^{ax} - \frac{n}{a}\int x^{n-1} e^{ax} dx + C$

30. $\displaystyle\int (\ln x)^n dx = x(\ln x)^n - n\int (\ln x)^{n-1} dx + C$ where $n \neq -1$

31. $\displaystyle\int x^n \sqrt{a + bx}\, dx = \frac{2}{b(2n+3)}\left[x^n(a+bx)^{3/2} - na\int x^{n-1}\sqrt{a+bx}\, dx\right] + C$

where $n \neq -\dfrac{3}{2}$

Glossary of Important Terms

Absolute Extreme Points. The **absolute maximum value** of a function, if it exists, is the largest value of the function over its domain. The **absolute minimum value** of a function, if it exists, is the smallest value of the function over its domain. Given the graph of $f(x)$ with $(c, f(c))$ a turning point of the graph, $f(c)$ is an absolute maximum if $f(c) \geq f(x)$ for all x in the domain of $f(x)$. $f(c)$ is an absolute minimum if $f(c) \leq f(x)$ for all x in the domain of $f(x)$.

Absolute Value Function. An **absolute value function** has the form $f(x) = |p(x)|$. The domain is all real numbers and the range is all non-negative numbers.

Algebraic Functions. Algebraic functions include the polynomial functions (linear, quadratic, cubic, and higher), radical functions and rational functions. Algebraic functions are constructed using the algebraic operations of addition, subtraction, multiplication, division, powers, and roots with polynomials.

Antiderivative. An **antiderivative** of $f(x)$ is a function $F(x)$ such that $F'(x) = f(x)$. Antidifferentiation is the reverse of differentiation.

Area of Inequality. The **area of inequality** graphically represents the area between the ideal income distribution, $y = x$, and the actual income distribution given by the Lorenz curve $y = L(x)$.

Area Rule 1. Given $f(x) \geq 0$ on $[a, b]$, the area of the region enclosed by $f(x)$ and the x-axis on $[a, b]$ is given by $A = \int_a^b f(x)dx$. It is assumed that $f(x)$ is continuous over the interval $[a, b]$.

Area Rule 2. Given $f(x) \leq 0$ on $[a, b]$, the area of the region enclosed by $f(x)$ and the x-axis on $[a, b]$ is given by $A = -\int_a^b f(x)dx$. It is assumed that $f(x)$ is continuous over the interval $[a, b]$.

Area Rule 3. If $f(x) \geq g(x)$ on an interval $[a, b]$, then the area of the region enclosed between the two curves is given by $A = \int_a^b [f(x) - g(x)]dx$. It is assumed that $f(x)$ and $g(x)$ are continuous on $[a, b]$. The location of the two curves with respect to the x-axis does not matter when using Area Rule 3.

Areas Using Riemann Sums. Consider the region enclosed by $y = f(x)$ and the x-axis on $[a, b]$. Divide $[a, b]$ into n rectangles. The width of each rectangle is given by $\Delta x_i = \frac{b - a}{n}$ and the height of each rectangle is given by $f(x_i)$. The area of one rectangle is $A_i = f(x_i)\Delta x_i$. The total area of all n rectangles is given by the **Riemann Sum** $\sum_{i=0}^{n} f(x_i)\Delta x_i$. As the number of rectangles increases, the total area of the rectangles approaches the actual area of the region. The area of the region is given by

$$A = \lim_{\Delta x_i \to 0} \sum_{i=0}^{n} f(x_i)\Delta x_i = \int_a^b f(x)dx.$$

Asymptotes of Rational Functions. The **asymptotes** of a rational function are lines that the curve approaches but never touches. Given $f(x) = \frac{p(x)}{q(x)} = \frac{ax^n + \cdots + c}{bx^m + \cdots + d}$, where $f(x)$ is in simplified form and $q(x) \neq 0$. The **vertical asymptote**(s) is (are) $x = k$, where $q(k) = 0$. The **horizontal asymptote** represents the behavior of the function as $|x|$ gets quite large and is determined by examining the degrees of $p(x)$ and $q(x)$:

If $n < m$, the horizontal asymptote is $y = 0$.

If $n = m$, the horizontal asymptote is $y = \frac{a}{b}$.

If $n > m$, there is no horizontal asymptote.

Average Rate of Change. The **average rate of change** of a quantity over an interval is given by the ratio

$\dfrac{\text{amount of change in the quantity over the interval}}{\text{amount of change in the interval}}$.
If $f(x)$ represents the quantity, the average rate of change on $[a, b]$ is given by $\dfrac{f(b) - f(a)}{b - a}$.

Average Value of a Function. Given a function $f(x)$ continuous on $[a, b]$ the **average value** of $f(x)$ on $[a, b]$ is given by $\dfrac{1}{b-a}\displaystyle\int_a^b f(x)dx$.

Average Value of a Function of Two Variables. The **average value** of $f(x, y)$ on region R in the plane is given by $\bar{f} = \dfrac{1}{A}\displaystyle\iint_R f(x, y)dA$, where A is the area of the region R in the plane.

Break-even Point. The **break-even point** gives the number of items for which costs and revenues are equal. To determine the break-even point, set $C(x) = R(x)$ and solve for x.

Calculus. **Calculus** deals with the behavior of functions and is the mathematical tool used to analyze changes in variable quantities. The three major concepts of calculus are the limit, the derivative, and the integral.

Chain Rule: Exponential Form. The **Chain Rule: Exponential Form** is used to differentiate composite exponential functions: $D_x e^{f(x)} = e^{f(x)} \cdot f'(x)$.

Chain Rule: Logarithmic Form. The **Chain Rule: Logarithmic Form** is used to differentiate composite logarithmic functions:

$$D_x \ln f(x) = \frac{1}{f(x)} \cdot f'(x) = \frac{f'(x)}{f(x)},$$

where $f(x) > 0$ and $f'(x)$ exists.

Chain Rule: Power Form (General Power Rule). The **Chain Rule: Power Form** is used to find the derivative of composite algebraic functions. If $y = [f(x)]^n$, then $y' = D_x[f(x)]^n = n[f(x)]^{n-1} \cdot f'(x)$ or $n \cdot f'(x) \cdot [f(x)]^{n-1}$.

Change of Base Formula. The **Change of Base Formula** is used to evaluate logarithms that are not common logs or natural logs. If $y = log_b x$, where $b \neq 10$ or e, then $log_b x = \dfrac{\log x}{\log b}$ or $\dfrac{\ln x}{\ln b}$.

Cobb-Douglas Productivity Model. The productivity of a plant or factory is given by $P = Kx^a y^{1-a}$ where P is the number of units produced, x is the number of employees, and y is the operating budget or capital. K and a are constants that are determined by each individual factory or plant, with $0 < a < 1$.

Concave Down. A graph is said to be **concave down** on an interval if all the tangent lines to the curve in that interval are above the curve.

Concave Up. A graph is said to be **concave up** on an interval if all the tangent lines to the curve in that interval are below the curve. Sometimes we say that a concave up graph will "hold water."

Constant Multiple Property for Derivatives. The **Constant Multiple Property for Derivatives** states: $D_x(c \cdot f(x)) = c \cdot f'(x)$.

Constant Multiple Property of Integrals. The **Constant Multiple Property of Integrals** states: $\int c \cdot f(x)dx = c\int f(x)dx$

Consumer Surplus. **Consumer surplus** is the area between the demand curve $p = D(q)$ and the market price p_m and is calculated by $CS = \int_o^{qm}(D(q) - p_m)dq$. Consumer surplus measures the savings the consumer gets from buying at a price below what they would have been willing to pay.

Continuity. Intuitively, a function $f(x)$ is **continuous** if its graph can be drawn without lifting the pencil. Mathematically, a function is **continuous at a point** in its domain if the graph has no holes, asymptotes, or jumps at that point. In other words, a function $f(x)$ is continuous at a point $(a, f(a))$ if

1. $f(a)$ is defined,
2. $\lim_{x \to a} f(x)$ exists, and
3. $f(a) = \lim_{x \to a} f(x)$.

Continuous Income Flow and Future Value. Suppose a steady flow of money is being put into an account which earns interest over a specific time period. The **future value** of the income flow is the total amount of money accumulated over the time period. Let $f(t)$ be the rate at which money flows into an account. Let r be the annual interest rate, which is compounded continuously.

Let T be the time period, with $0 \le t \le T$. The future value of the income flow after T years is given by $FV = e^{rT}\int_0^T f(t)e^{-rt}dt$.

Critical Point for a Function of Two Variables. For $z = f(x, y)$and the point (a, b) contained in some small region R of the domain:

(a, b) is a **critical point** if $f_x(a, b) = 0$ and $f_y(a, b) = 0$.

$f(a, b)$ is a **relative maximum value** if $f(a, b) \ge f(x, y)$ for all (x, y) contained in the region R.

$f(a, b)$ is a **relative minimum value** if $f(a, b) \le f(x, y)$ for all (x, y) contained in the region R.

Critical Value. The value(s) of x for which either $f'(x) = 0$ or $f'(x)$ is undefined is (are) the **critical value(s)** of the function.

Cubic Function. Cubic functions are of the form $f(x) = ax^3 + bx^2 + cx + d$ or $f(x) = a(x - h)^3 + k$. A cubic function has degree three. The domain and range of cubic functions are the set of real numbers.

Decreasing Functions. A function decreases if its graph falls from left to right. Algebraically we say that on the interval $[a, b]$, $f(x)$ is **decreasing** if $f(a) > f(x_1) > f(x_2) > f(b)$ for all $a < x_1 < x_2 < b$.

Definite Integral. The **definite integral** of $f(x)$ on the interval $[a, b]$ is denoted by $\int_a^b f(x)dx$ and denotes the net change in the value of the antiderivative on the interval $[a, b]$. The numbers a and b are called the **limits of integration.**

Demand Function. The **demand function $D(p)$** refers to the quantity of units demanded by consumers at price p.

Derivative. The **derivative** of $f(x)$ is found by evaluating $\lim_{h \to 0} \frac{f(x + h) - f(x)}{h}$, if the limit exists. If the derivative exists, then $f(x)$ is said to be **differentiable**. The process of finding a derivative is called **differentiation**. Symbolically, the derivative of $f(x)$ can be written in several ways: $f'(x), y', \frac{dy}{dx}, D_x(f)$, and $\frac{d}{dx}f$. The derivative of a function gives the **instantaneous rate of change** of the function or the **slope** of the graph of the function.

Difference Quotients. The **difference quotient** is an algebraic expression which has the form $\frac{f(a + h) - f(a)}{h}$.

Differential. For $y = f(x)$, the **differential** is $dy = f'(x)dx$, which gives the change in y for an infinitely small change in x.

Differential Equations. A **differential equation** is an equation involving an unknown function and one or more of its derivatives. The **solution** of a differential is the unknown function that is determined to satisfy the equation. The **general solution** is the family of all possible solutions to the differential equation. A **particular solution** is a member of the general solution that satisfies specific given conditions called **initial conditions.**

Discontinuous. A function is **discontinuous** at a point $(a, f(a))$ if there is an asymptote, a hole, or a break in the graph at that point. Any point where a function is discontinuous is called a **point of discontinuity**.

Domain. The **domain** of a function is the set of all possible replacement values for the independent variable. The domain is sometimes referred to as the input variable.

Double Integral. The integral $\iint_R f(x, y)dA$ is called a **double integral** or an **iterated integral**. The limits of integration on the two integrals define the boundaries of the **region of integration** in the plane. If R is a **rectangular region** bounded by $a \le x \le b$ and $c \le y \le d$, then the double integral is written as either $\int_a^b \int_c^d f(x, y)dydx$ or $\int_a^b \int_c^d f(x, y)dxdy$. If R is a **non-rectangular region** bounded by $a \le x \le b$ and $g_1(x) \le y \le g_2(x)$, then the double integral is written as $\int_a^b \int_{g_1(x)}^{g_2(x)} f(x, y)dydx$.

The Number e. One special base that occurs frequently in applications involving growth and decay is the base e. The number $\boldsymbol{e}$ is an irrational number that was first used by the Swiss mathematician Leonhard Euler. e is called the **natural base** and its value is about 2.718. This value can be approximated by examining $\lim_{n\to\infty}\left(1+\frac{1}{n}\right)^n$. You can find e on your calculator by entering **2nd ln 1 ENTER.**

Elasticity of Demand. Economists measure sensitivity to price changes using **elasticity**. The **elasticity of demand** for a product is the ratio of the percent change in demand to the percent change in price that caused the change in demand.

$$E = \frac{\%\text{ change in demand}}{\%\text{ change in price}} = \left|\frac{\frac{\Delta q}{q}}{\frac{\Delta p}{p}}\right| = \left|\frac{\Delta q}{q}\cdot\frac{p}{\Delta p}\right| = \left|\frac{p}{q}\cdot\frac{\Delta q}{\Delta p}\right|$$

The **elasticity of demand** for a demand function $q = D(p)$ is $E = \left|\frac{p}{q}\cdot\frac{dq}{dp}\right| = \left|\frac{p\cdot D'(p)}{D(p)}\right|$.

If $|E| > 1$, the demand is **elastic**, meaning that demand is sensitive to changes in price. If $0 \le |E| < 1$, the demand is **inelastic**, meaning that demand is not sensitive to price changes.

Elasticity of Demand and Revenue. If $|E| > 1$, the demand is elastic and revenue is decreasing. If $0 \le |E| < 1$, the demand is inelastic and revenue is increasing. If $|E| = 1$ the demand is unit elastic and the revenue is neither increasing nor decreasing; the revenue is at a maximum.

Equilibrium Point. The **equilibrium point** is the price for which the quantity producers supply equals the quantity demanded by buyers. To determine the equilibrium point, set $S(p) = D(p)$ and solve for p.

Explicit Function. Functions of the form $y = f(x)$ are called **explicit functions**.

Exponential Function. An **exponential function** is any function of the form $f(x) = a^x$, where x is any real number and a is a positive real number such that $a \neq 1$. The domain of an exponential function is all real numbers; the range is all positive real numbers, unless transformations are involved. The graph of $f(x) = a^x$ passes through the point (0, 1). If $a > 1$, the graph is continuous and increasing and has an asymptote along the negative x-axis. If $0 < a < 1$, the graph is continuous and decreasing and has an asymptote along the positive x-axis.

Extraneous Solution. An **extraneous solution** to an equation is a solution that is obtained algebraically from the equation but does not satisfy the original equation.

Extreme Point. An **extreme point** of a graph is the point where the graph changes direction. If the direction changes from increasing to decreasing, the extreme point is a **maximum point**. If the direction changes from decreasing to increasing, the extreme point is a **minimum point**.

First Derivative Test. Given a function $f(x)$, then $f(x)$ is increasing when $f'(x) > 0$ and $f(x)$ is decreasing when $f'(x) < 0$. If $f(x)$ has extreme points, they occur when $f'(x) = 0$. If $(c, f(c))$ represents an extreme point of $f(x)$, then $f(c)$ is a minimum if $f(x)$ changes from decreasing to increasing at $(c, f(c))$. $f(c)$ is a maximum if $f(x)$ changes from increasing to decreasing at $(c, f(c))$.

First Partial Derivative. If $z = f(x, y)$, then the **first partial derivatives** of $f(x, y)$ with respect to x and y are

$$f_x = \lim_{h\to 0}\frac{f(x+h, y) - f(x, y)}{h} \quad \text{and}$$

$$f_y = \lim_{h\to 0}\frac{f(x, y+h) - f(x, y)}{h}$$

assuming that the limits exist. First partial derivatives can also be symbolized as $\frac{\partial f}{\partial x}$ and $\frac{\partial f}{\partial y}$. To find f_x, treat y as a constant. To find f_y, treat x as a constant.

Function. A **function** is a correspondence between two sets that assigns to each member of the first set (the **domain** or the **independent variable**) one and only one element from the second set (the **range** or the **dependent variable**).

Function Notation. To denote that an equation represents a function, replace y by $f(x)$. The notation $f(x)$ means the same thing as y and tells us that the equation to follow describes a function. For instance, $f(x) = 2x - 3$ indicates that the equation $y = 2x - 3$ is a function.

Function of Two Variables. A **function of two variables, $f(x, y)$**, is a rule that assigns to each ordered pair (x, y) in the domain one unique real number $f(x, y)$ in the range. x and y are the **independent variables**; $z = f(x, y)$ is the **dependent variable**.

Fundamental Theorem of Calculus. Given $f(x)$ continuous on the interval $[a, b]$ and $F(x)$ an antiderivative of $f(x)$, then

$$\int_a^b f(x)dx = F(x)\big|_a^b = F(b) - F(a).$$

Gini Index (Gini Coefficient). The **Gini Index** or **Gini Coefficient** is used to measure the area of inequality of an income distribution. The Gini Index is a ratio comparing the area of inequality to the total area. Given a Lorenz Curve $L(x)$, the Gini Index (or Gini Coefficient) is given by

$$G = \frac{\int_0^1 (x - L(x))dx}{\frac{1}{2}} = 2\int_0^1 (x - L(x))dx.$$

Horizontal Translation. Given the graph of $f(x)$, $f(x - h)$ shifts the graph of $f(x)$ to the right h units and $f(x + h)$ shifts the graph of $f(x)$ to the left h units.

Implicit Differentiation. **Implicit differentiation** is a process used to find $\frac{dy}{dx}$ for equations of the form $F(x, y) = 0$ without having to solve the equation for y first. The basic assumption is that the y variable must be treated as $f(x)$ when differentiating. If $y = f(x)$, then

$$\frac{d}{dx}(y^n) = n \cdot y^{n-1} \cdot \frac{dy}{dx}.$$

Implicit Equation. Equations of the form $F(x, y) = 0$ are referred to as **implicit equations.** The implicit assumption is that the equation could be solved for y and turned into an explicit function.

Increasing Functions. A function increases if its graph rises from left to right. Algebraically we say that on the interval $[a, b]$, $f(x)$ is **increasing** if $f(a) < f(x_1) < f(x_2) < f(b)$ for all $a < x_1 < x_2 < b$.

Indefinite Integral. An integral of the form $\int f(x)\,dx$ is called an **indefinite integral,** because it specifies a family of functions. $\int f(x)\,dx = F(x) + C$, where the constant term, C, is called the **constant of integration.**

Indeterminate Form. The fraction $\frac{0}{0}$ is **indeterminate.** There is no unique answer, because any number multiplied by 0 is 0.

Initial Condition. To determine the value of the constant of integration, some specific point on the graph of the function must be given. This point is called an **initial condition.**

Initial Value Problem. An initial value problem is a differential equation which must be solved subject to an initial condition. In other words we are seeking the equation that satisfies the differential equation and whose graph passes through the point specified point.

Instantaneous Rate of Change. The **instantaneous rate of change** of $f(x)$ when $x = a$ is given by $\lim_{h \to 0} \frac{f(a + h) - f(a)}{h}$.

Integral. The symbol used to indicate finding an antiderivative is called the **integral sign** and is denoted as $\int f(x)\,dx$. The process of finding an antiderivative is called **integration.** The function $f(x)$ is called the **integrand.**

Integration by Parts. **Integration by parts** is, in some sense, a product rule for integration. It's based on the idea of breaking apart the integrand and replacing it with another integrand that is equivalent yet hopefully simpler to integrate. If $u(x)$ and $v(x)$ are differentiable functions, then $\int udv = uv - \int vdu$.

Inventory Control. A problem frequently encountered in business applications is **inventory control.** In such problems, we must determine the number of production runs that will minimize the cost of production, including storage costs.

Lagrange Multipliers. **Lagrange multipliers** can be used to solve constrained optimization problems. To optimize $f(x, y)$ subject to $g(x, y) = c$:

1. Write the **Lagrangian** $F(x, y) = f(x, y) + \lambda(c - g(x, y))$. The λ factor is the Lagrange multiplier.

2. Find F_x, F_y, and F_λ.
3. Set $F_x = 0$, $F_y = 0$, and $F_\lambda = 0$ and solve for the critical point (a, b).
4. Evaluate $f(a, b)$ to determine the optimal solution.

Law of Decay. The model $P(t) = P_0e^{kt}$ can be used to model decay if $k < 0$. P_0 represents the initial amount and k is the rate of decay.

Law of Growth. The model $P(t) = P_0e^{kt}$ is called the **exponential law of growth.** P_0 represents the initial amount in the population and k is the rate of growth of the population, assuming $k > 0$.

Left-hand Limit. The **left-hand limit** of $f(x)$ is the value that $f(x)$ approaches as x approaches a from the left using only values of x that are smaller than a. Symbolically, the left-hand limit is written as $\lim_{x \to a^-} f(x)$.

Level Curve. For a function $z = f(x, y)$, the **level curves** are given by $k = f(x, y)$.

Limit of a Function. The **limit of a function** refers to how the y values of the function behave as x approaches a certain value. Symbolically, the limit of a function is denoted as $\lim_{x \to a} f(x)$.

Linear Function. **Linear functions** can be written in the form $f(x) = mx + b$. Linear functions have degree one, meaning that the independent variable is raised to the first power. The domain and range of a linear function are always the set of real numbers. The graph of a linear function is a straight line.

Logarithm Function. The **logarithmic function** $f(x) = \log_a x$ is the inverse of the exponential function $f(x) = a^x$. In other words, $y = \log_a x$ if and only if $x = a^y$. **Common logs** use base 10 and are denoted by ***log x***. **Natural logs** use base e and are denoted by ***ln x***.

Logistic Growth Model. When modeling human populations, a better model is the **logistic model** of growth. This model of growth takes into account birth rates and death rates and allows for a tapering off of the population. The logistic model is $P(t) = \dfrac{c}{1 + ae^{-bt}}$.

Lorenz Curve. A **Lorenz curve** is a geometric representation of the income distribution of a population. The graph of a Lorenz curve always begins at the point $(0, 0)$ and ends at the point $(1, 1)$. The curve is increasing and concave up and will always be below the line $y = x$.

Marginal Analysis. In business and economics, **productivity** is defined as the ratio of output units to input units, such as labor or materials. **Marginal productivity** is the additional output resulting from adding one more unit of input. **Marginal cost** is the additional cost of adding one more input unit; **marginal revenue** is the additional revenue generated from the addition of one more input unit. If $C(x)$ is cost, $R(x)$ is revenue, and $P(x)$ is profit, then Marginal cost $= C'(x)$, Marginal revenue $= R'(x)$, and **Marginal profit** is $P'(x) = R'(x) - C'(x)$.

Marginal Productivity. For the Cobb Douglas Productivity Model $Q(x, y) = Kx^ay^{1-a}$, where x represents labor units and y represents capital, $Q_x(x, y)$ gives the change in output per unit change in labor, holding capital fixed, called the **marginal productivity of labor**, and $Q_y(x, y)$ gives the change in output per unit change in capital, holding labor fixed, called the **marginal productivity of capital**. The **change in output with respect to labor** is Q_x times the change in labor. The **change in output with respect to capital** is Q_y times the change in capital.

Market Surplus. **Market surplus** is the sum of consumer surplus and producer surplus up to the equilibrium price and is also the area between the demand and supply curves. Market surplus is calculated by

$$MS = CS + PS = \int_0^{q_m} (D(q) - S(q))\,dq.$$

Multi-variable Calculus. **Multi-variable calculus** deals with functions of more than one variable. The domain of $z = f(x, y)$ is a set of ordered pairs (x, y).

Non-algebraic Functions. **Non-algebraic functions,** sometimes referred to as **transcendental functions**, include the exponential and logarithmic functions and the trigonometric functions. The terms of a non-algebraic function cannot be expressed as a real number power of x.

Optimization Problems. **Optimization problems** involve ***variables***, which represent the unknowns, an ***objective function***, which is the function to be maximized or minimized, and ***constraints***, which impose restrictions on the values of the variables.

Partial Derivatives. Derivatives of a function of two or more variables are called **partial derivatives**, because the derivative can be taken with respect to either variable. The process of finding the derivatives of a function of two or more variables is called **partial differentiation**.

Piecewise defined Function. A **piecewise defined function** is one which has a different formula for different intervals in the domain.

Point of Diminishing Returns. In business applications, the point of inflection represents the time at which the growth in sales begins to taper off. This point is referred to as the **point of diminishing returns.** Sales continue to grow, but at a slower rate.

Point of Inflection. Any point for which the concavity, or shaping, of a function changes is called a **point of inflection.**

Power Rule for Derivatives. If $f(x) = x^n$, then $f'(x) = D_x(x^n) = nx^{n-1}$, where n is any real number.

Power Rule for Integration.

$$\int x^n\,dx = \frac{1}{n+1}x^{n+1} + C = \frac{x^{n+1}}{n+1} + C$$

where $n \neq -1$.

Probability Density Function. A **probability density function** of a continuous random variable x over an interval I is a function with these properties:

1. $f(x) \geq 0$ for all x in the interval I.
2. The total area under the graph of $y = f(x)$ is 1, where $1 = 100\%$.
3. The probability that a selected value of the random variable lies in the interval $[a, b]$ is given by $P(a \leq x \leq b) = \int_a^b f(x)\,dx$.
4. Because the total area under the curve is 1, $0 \leq P(a \leq x \leq b) \leq 1$. Probability must be a number between 0 and 1, or between 0% and 100%.

An ***exponential probability density function*** is of the form $f(x) = Ae^{-Ax}$ for $x \geq 0$ and A a positive constant.

The most widely used probability density function is the **normal distribution**, whose graph is the bell-shaped curve. The normal distribution curve is given by

$$f(x) = \frac{1}{\sigma\sqrt{2\pi}}e^{-(x-\mu)^2/2\sigma^2}.$$

Producer Surplus. Producer Surplus is represented by the area between the market price p_m and the supply function $p = S(q)$ and is calculated by $PS = \int_0^{q_m}(p_m - S(q))dq$. Producer surplus is the suppliers' gain from selling at the current price rather than one they would have been willing to accept.

Product Rule for Derivatives. If $y = f(x) \cdot g(x)$ then $y' = D_x[f(x)g(x)] = g(x) \cdot f'(x) + f(x) \cdot g'(x)$. In other words, the derivative of the product of two functions is found by multiplying each function by the derivative of the other and adding those products.

Projectile Motion. Projectile motion refers to functions describing an object in motion. The path of such objects in motion is parabolic, and the object's height above the ground after t seconds is given by $h(t) = at^2 + vt + c$. The quadratic coefficient, a, is negative and accounts for the pull of gravity causing the object to return to the ground. The linear coefficient, v, is the rate at which the object was put into motion and is referred to as the **initial velocity**. The constant term, c, represents the **initial height** of the object.

Power Function. A **power function** is any function of the form $f(x) = x^p$, where p is a positive real number.

Quadratic Function. Quadratic functions can be written as $f(x) = ax^2 + bx + c$, called polynomial form, or as $f(x) = a(x - h)^2 + k$, called vertex form. A quadratic function has degree two. The domain of a quadratic function is the set of all real numbers. The range is restricted and is determined by the y-coordinate of the vertex. The graph of a quadratic function is a **parabola**. The turning point of the parabola is called the **vertex** and is the point where the direction of the graph changes.

Quotient Rule for Derivatives. If $y = \dfrac{f(x)}{g(x)}$, then

$$y' = D_x\frac{f(x)}{g(x)} = \frac{g(x)f'(x) - f(x)g'(x)}{[g(x)]^2}.$$

A simpler way to remember this rule is $\frac{D \cdot N' - N \cdot D'}{D^2}$ or

$$\frac{(\text{denominator})\left(\begin{matrix}\text{derivative of}\\ \text{numerator}\end{matrix}\right) - (\text{numerator})\left(\begin{matrix}\text{derivative of}\\ \text{numerator}\end{matrix}\right)}{(\text{denominator})^2}$$

Range. The **range** of a function is the set of all possible replacement values of the dependent variable. The range is sometimes referred to as the output variable.

Rational Function. A **rational function** is a ratio of two polynomial functions, written in the form $f(x) = \frac{p(x)}{q(x)}$. The domain of a rational function $f(x) = \frac{p(x)}{q(x)}$ is all values of x for which $q(x) \neq 0$.

The graph of the basic rational function $f(x) = \frac{1}{x}$ is a **hyperbola**.

Reflection. Given the graph of $f(x)$, $-f(x)$ reflects the graph of $f(x)$ about the x-axis and $f(-x)$ reflects the graph of $f(x)$ about the y-axis.

Regression. **Regression** is a mathematical tool that transforms real data into a function which models the data. Regression can be done on a graphing calculator or using Excel.

Related Rates. **Related rate problems** involve equations for which the variables are functions of another variable, usually time. Implicit differentiation is used to solve related rate problems.

Relative Extreme Points. Given the graph of $f(x)$ with $(c, f(c))$ a turning point of the graph, $f(c)$ is a **relative maximum** if $f(c) > f(x)$ for some open interval containing $(c, f(c))$. $f(c)$ is a **relative minimum** if $f(c) < f(x)$ for some open interval containing $(c, f(c))$.

Right-hand Limit. The **right-hand limit** of $f(x)$ is the value that $f(x)$ approaches as x approaches a from the right using only values of x that are larger than a. Symbolically, the right-hand limit is written as $\lim_{x \to a^+} f(x)$.

Second Derivative. The derivative of $f'(x)$ is called the **second derivative** and is denoted by $f''(x)$ or y'' or $D_x^2(f)$ or $\frac{d^2y}{dx^2}$ or $\frac{d^2}{dx^2}f(x)$. $f'(x)$ is referred to as the **first derivative**. The second derivative gives the rate at which the rate of increase or decrease is changing. For projectile motion, the second derivative, $h''(t)$, represents the rate at which the velocity changes, which is referred to as **acceleration**.

Second Derivative Test. Given a function $f(x)$, $f(x)$ is concave up if $f''(x) > 0$ and $f(x)$ is concave down if $f''(x) < 0$. If $f(x)$ has a point of inflection, then $f''(x) = 0$. If $(c, f(c))$ is an extreme point of $f(x)$, then $(c, f(c))$ is a minimum point if $f''(c) > 0$ and $(c, f(c))$ is a maximum point if $f''(c) < 0$.

Second Partial Derivative Test. Given $z = f(x, y)$ such that all four second partial derivatives exist. Let (a, b) be a critical point. Then we define $D = f_{xx}(a, b) \cdot f_{yy}(a, b) - (f_{xy}(a, b))^2$.

If $D > 0$ and $f_{xx}(a, b) > 0$, then f has a relative minimum point at (a, b).

If $D > 0$ and $f_{xx}(a, b) < 0$, then f has a relative maximum point at (a, b).

If $D < 0$, then f has a saddle point at (a, b).

If $D = 0$, then no conclusion can be made.

Separation of Variables. Some differential equations can be solved by a method called **separation of variables.** This method rewrites the equation so that each side is a function of only one of the variables. Then each side is integrated with respect to that variable. To solve $\frac{dy}{dx} = \frac{f(x)}{g(y)}$, first rewrite the equation—**separate the variables**—as $g(y)dy = f(x)dx$ and then integrate each side of the equation: $\int g(y)dy = \int f(x)dx$.

Single Variable Calculus. **Single variable calculus** deals only with functions of the form $y = f(x)$.

Slope of a Curve at a Point. The **slope of a curve** at a point on the curve is the slope of the tangent line to the curve at that point and is given by $m = \lim_{h \to 0} \frac{f(a + h) - f(a)}{h}$.

Slope of a Curve at any Point. The **slope of a curve** $f(x)$ **at any point**, x, is the slope of the tangent line to the curve at that point and is given by $m = \lim_{h \to 0} \frac{f(x + h) - f(x)}{h}$.

Slope of a Line. The **slope of the line** $y = mx + b$ is m. The slope represents the steepness of the line, as measured by the ratio of the change in y to the change in x.

Slope in the x-direction and Slope in the y-direction. $f_x(a, b) = k$ gives the **slope in the x-direction**, which means that the value of $f(x, y)$ changes at a rate of $k \frac{\text{units of } f}{\text{units of } x}$, or, in other words, the value of $f(x, y)$ changes k units for a one unit increase in x. Similarly, $f_y(a, b) = k$ gives the **slope in the y-direction**, which means that the value of $f(x, y)$ changes at a rate of $k \frac{\text{units of } f}{\text{units of } y}$, or, the value of $f(x, y)$ changes k units for each unit increase in y.

Square Root Function. A square root function has the form $f(x) = \sqrt{p(x)}$, where the expression $p(x)$ is called the **radicand**. The domain of a **square root function** is restricted, because the radicand must be non-negative. The range of a square root function is also restricted because evaluating a square root always yields a non-negative number, the principal square root.

Step Functions. A **step function** is similar to a piecewise defined function, but the formulas which define each piece of the function are constants.

Straight Line Depreciation. **Straight line depreciation** estimates the value of an asset as it loses value, or depreciates, with use over time. Given its original price P, its expected lifetime L, and its scrap value C (its value at the end of its expected lifetime), the value V of the asset at any time t during its lifetime is modeled by

$$\text{Value} = \text{Price} - \left(\frac{\text{Price} - \text{Scrap Value}}{\text{Lifetime}}\right) \cdot t$$

or

$$V = P - \left(\frac{P - C}{L}\right) \cdot t.$$

The slope of this equation represents the rate of depreciation and the intercept is the original value of the asset.

Substitution Technique for Integration of Some Composite Functions. To integrate a composite function $\int f(g(x))g'(x)dx$ using **substitution**:

1. Let $u = g(x)$.
2. Compute the differential $du = g'(x)dx$.
3. Substitute and rewrite $\int f(g(x))g'(x)dx$ as $\int f(u)du$.
4. Integrate and obtain the antiderivative $F(u) + C$.
5. Substitute $u = g(x)$ and write the antiderivative as $F(g(x)) + C$.

Sum and Difference Rule for Derivatives. The **Sum Rule for Derivatives** states that the derivative of the sum of two functions is the sum of the individual derivatives:

$$D_x(f(x) + g(x)) = f'(x) + g'(x).$$

The **Difference Rule for Derivatives** states that the derivative of the difference of two functions is the difference of the individual derivatives:

$$D_x(f(x) - g(x)) = f'(x) - g'(x).$$

Sum and Difference Properties for Integrals. The **integral of the sum of two functions** is the sum of the integrals of the individual functions:

$$\int [f(x) + g(x)]\,dx = \int f(x)dx + \int g(x)dx.$$

The **integral of the difference two functions** is the difference of the integrals of the individual functions:

$$\int [f(x) - g(x)]\,dx = \int f(x)dx - \int g(x)dx.$$

Supply Function. The **supply function $S(p)$** refers to the quantity of units sellers are willing to supply (or sell) at price p.

Surface. The graph of $z = f(x, y)$ is a set of ordered triples (x, y, z) with domain a region in the xy-plane. The graph is called a **surface.**

Surge Function. The **surge function** is modeled by $C(t) = ate^{-bt}$, where t is the elapsed time following the administration of a drug, C is the concentration of the drug, and a and b are constants determined by the drug and the patient's response.

Transformations. The basic graph of any function can be moved to a different location in the coordinate plane by using **transformations**. The three basic transformations are horizontal translations, vertical translations, and reflections.

Undefined Expressions. The fraction $\frac{k}{0}$, with k some non-zero constant, is **undefined**, because any number multiplied by zero is zero, not the constant.

Velocity. For an object in motion, the instantaneous rate of change of the object's position is called the **velocity** of the object. Positive velocity indicates the rate at which an object rises; negative velocity indicates the rate at which an object falls. The **speed** of an object represents only how fast it is moving, not its direction, and is $|velocity|$.

Vertex of a Quadratic Function. The turning point of a parabola is called the **vertex** and is the point where the direction of the graph changes. If $f(x) = a(x - h)^2 + k$ the vertex is at the point (h, k). If $f(x) = ax^2 + bx + c$, then the vertex (h, k) is given by $h = -\frac{b}{2a}$ and $k = f(h)$.

Vertical Line Test. If every vertical line drawn through any point on the graph of a relation intersects the graph only once, then the equation represented by that graph denotes a function.

Vertical Translation. Given the graph of $f(x)$, $f(x) + k$ shifts the graph of $f(x)$ up k units and $f(x) - k$ shifts the graph of $f(x)$ down k units.

Volume of a Solid. Given $f(x, y) \geq 0$ over a region R in the plane. The **volume** of the solid region that lies above R and below the surface $f(x, y)$ is given by $V = \iint_R f(x, y)\, dA$.

x-Intercept of a Graph. The **x-intercept** of a graph is the point where the graph passes through the horizontal axis. To find the x-intercept, let $y = 0$ and solve the equation for x.

y-Intercept of a Graph. The **y-intercept** of a graph the point where the graph passes through the vertical axis. To find the y-intercept, let $x = 0$. The **y-intercept** of the line $y = mx + b$ is the point $(0, b)$.

Answers

Unit 0

TOPIC 1

1. yes **3.** no **5.** yes **7.** yes **9.** no **11.** no
13. yes **15.** 1 **17.** 2 **19.** $-2a^2 + a - 5$
21. $-2a^2 - 4ah - 2h^2 + a + h - 5$
23. $2ah + h^2 - 3h$ **25.** 3 **27.** $2a + h$
29. $4a + 2h - 3$ **31.** $3a^2 + 3ah + h^2$
33. C **35.** D **37.** E **39.** B

41. slope 3; $(0, -4)$

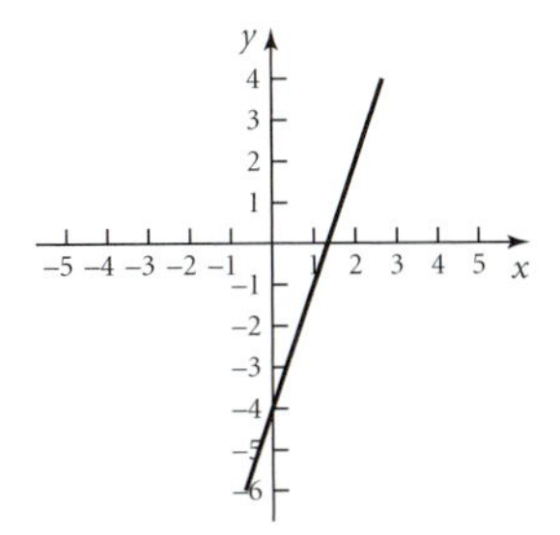

43. slope $\frac{2}{3}$; $(0, -4)$

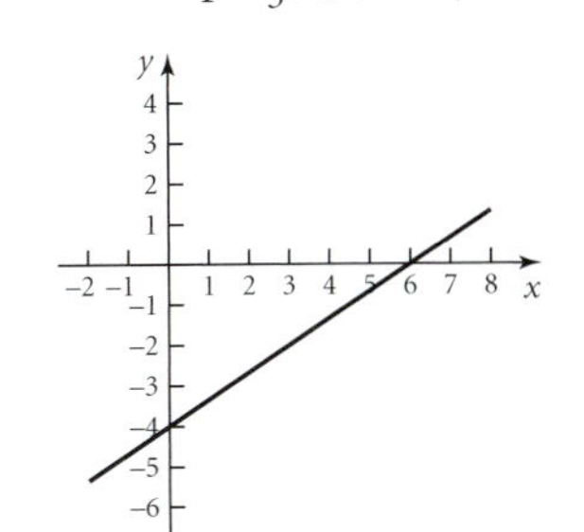

45. no slope; no y-intercept

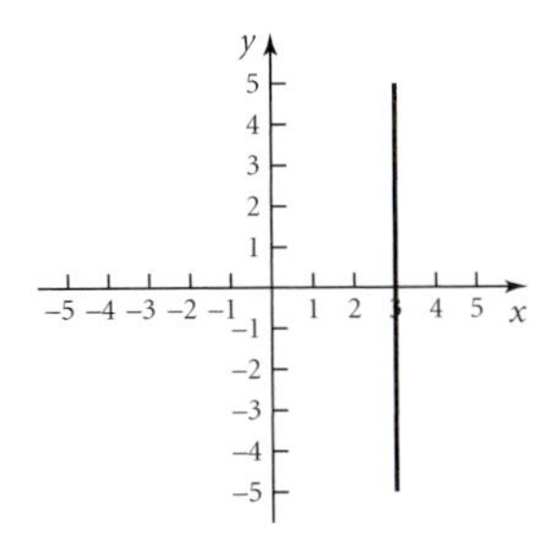

47. slope 0; $(0, 6)$

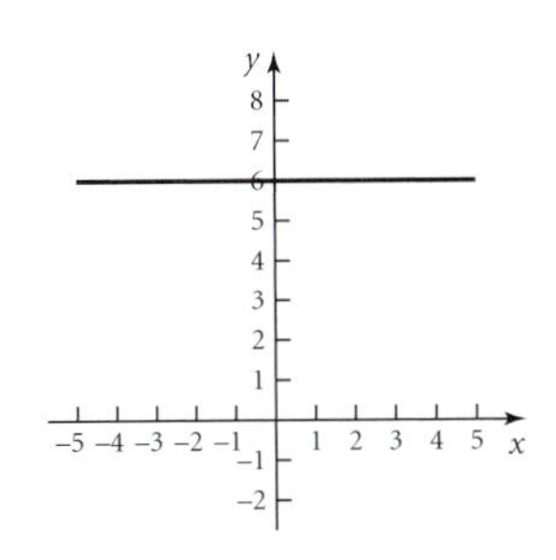

49. C **51.** A **53.** F

55. domain: reals
range: $y \geq 0$

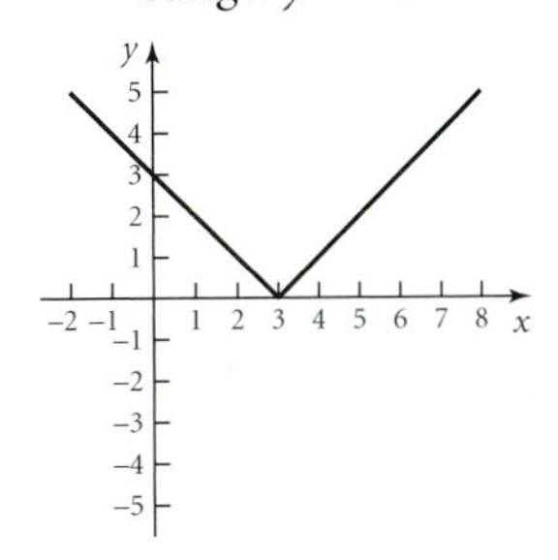

57. domain: reals
range: $y \geq -3$

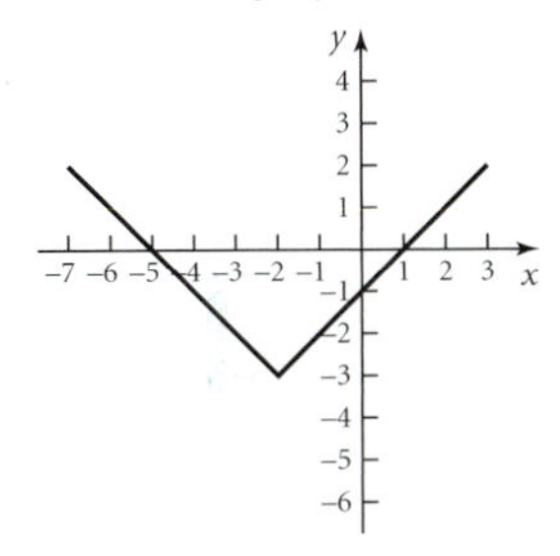

59. domain: reals
range: $y \leq 3$

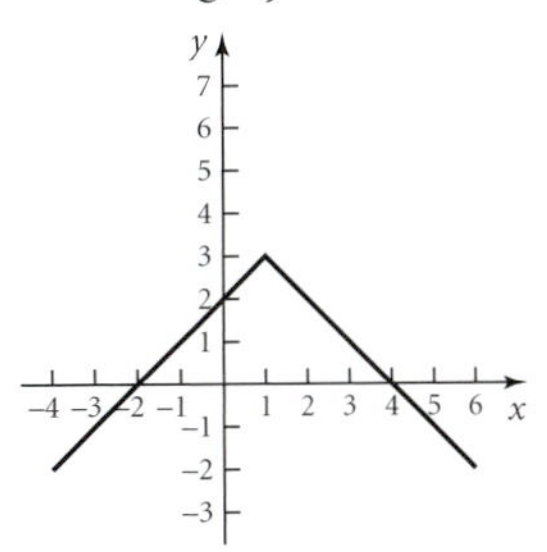

61. a. $y = \$125 + \$0.02x$ **b.** \$237
63. a. 25.84 years
b.

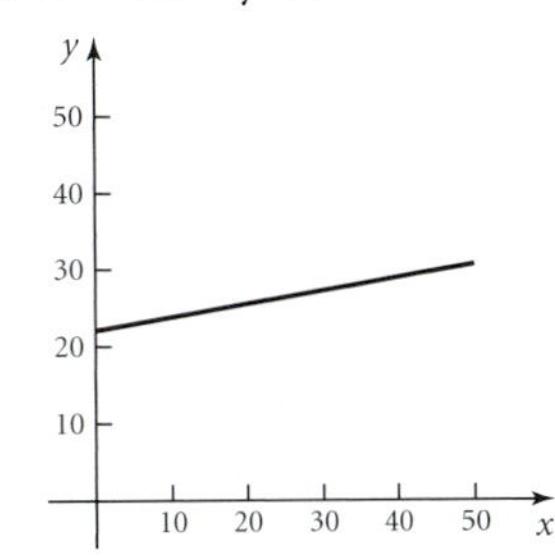

c. 34.7 years after 1980, so around 2015

65. a. 72 years **b.** The slope is 0.203. Life expectancy for men increases by 0.203 year per year for each birth year after 1960. **67.** $F = 50 + 10h$; \$110
69. $V = 950{,}000 - 35{,}000t$, with domain $0 \leq t \leq 25$; \$600,000; approximately 24 years from when the press was purchased **71.** Cost and revenue will be equal if 20 systems are installed. **73.** Supply equals demand if the price is \$45.
75. a. The intercepts are $(-4, 0)$ and $(4, 0)$.

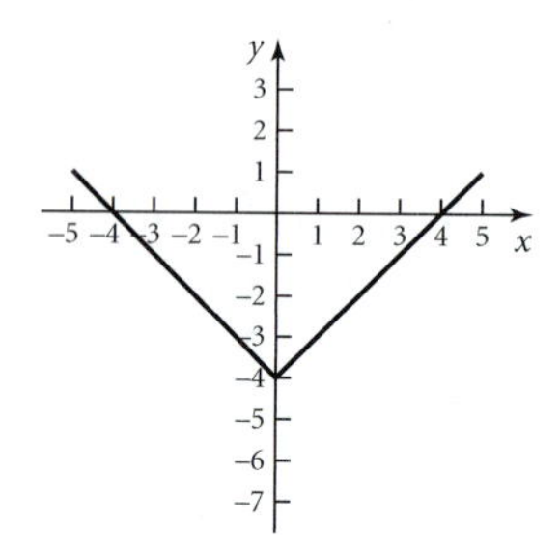

b. $x = 4, -4$ **c.** The x-intercepts of the graph of $f(x)$ are the same as the solutions to $f(x) = 0$.

77. a.

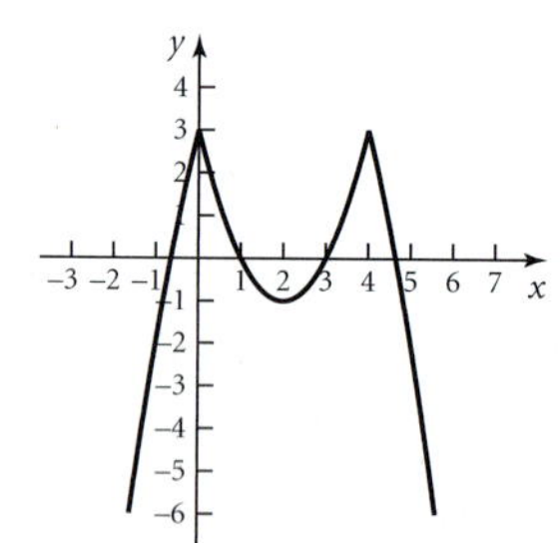

b. $x \approx -0.646, 1, 3, 4.646$

TOPIC 2

1. D **3.** A **5.** E **7.** I **9.** C **11.** L

13. vertex $(0, -2)$
domain: reals
range: $y \geq -2$

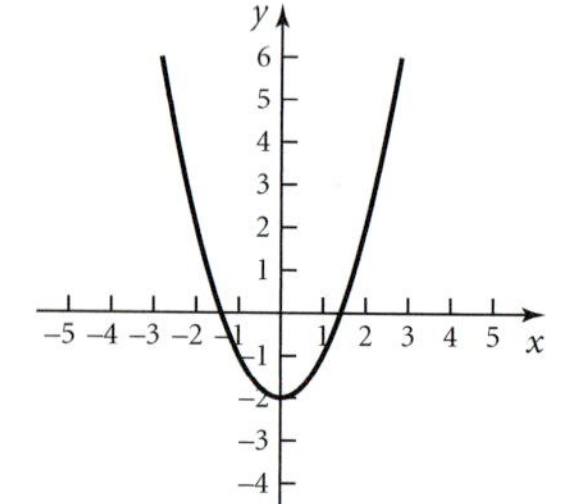

15. vertex $(0, 3)$
domain: reals
range: $y \leq 3$

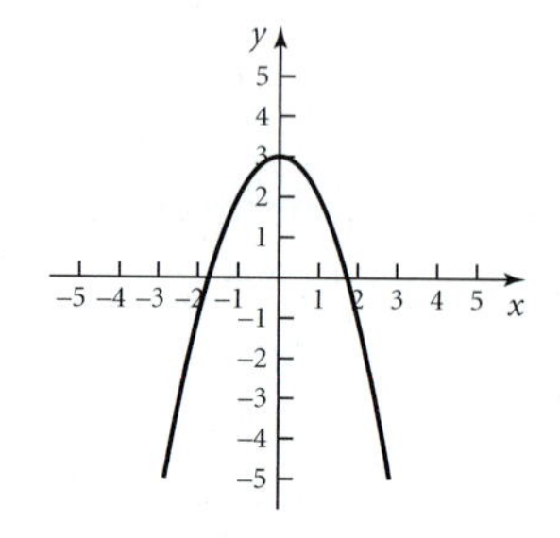

17. vertex $(-4, 3)$
domain: reals
range: $y \geq 3$

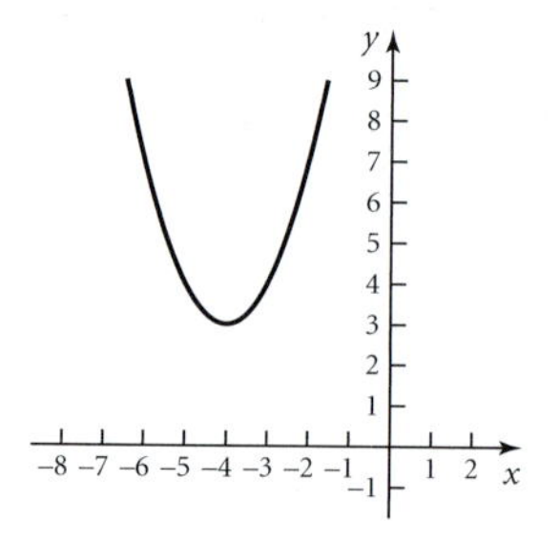

19. vertex $(1, 5)$
domain: reals
range: $y \leq 5$

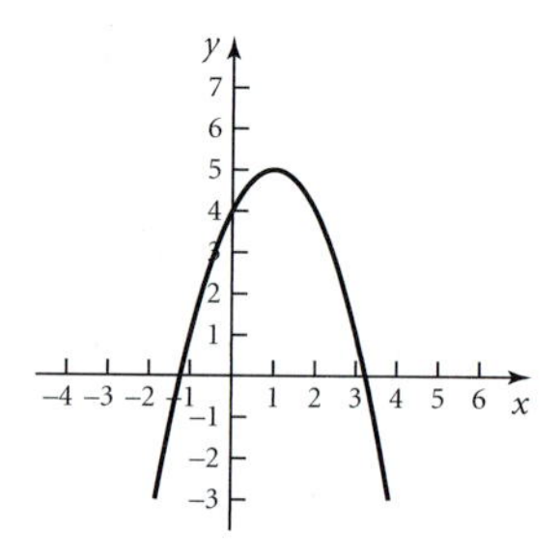

21. $(0, -2)$

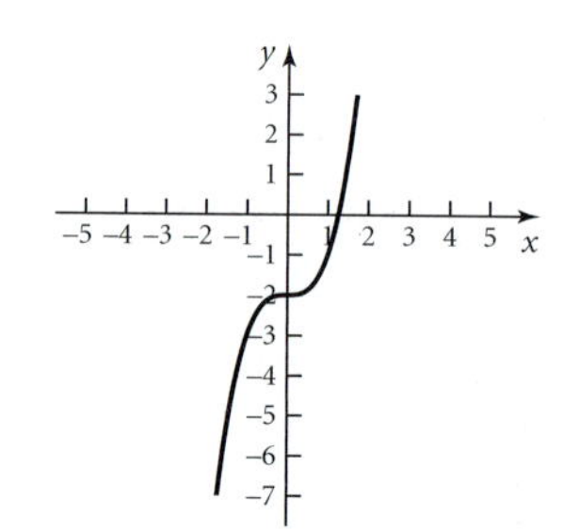

23. $(2, 0)$

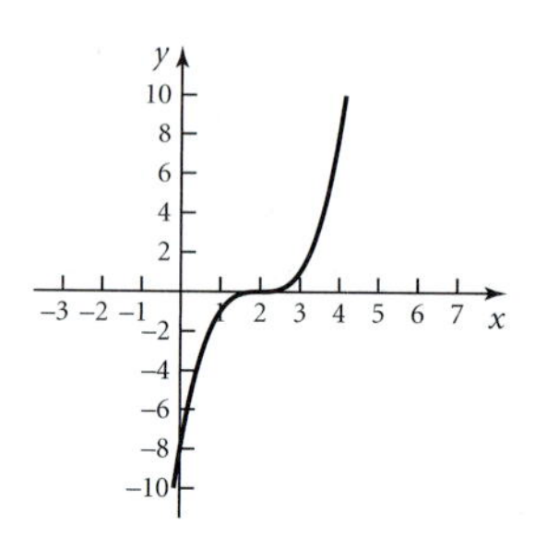

25. $(1, -2)$

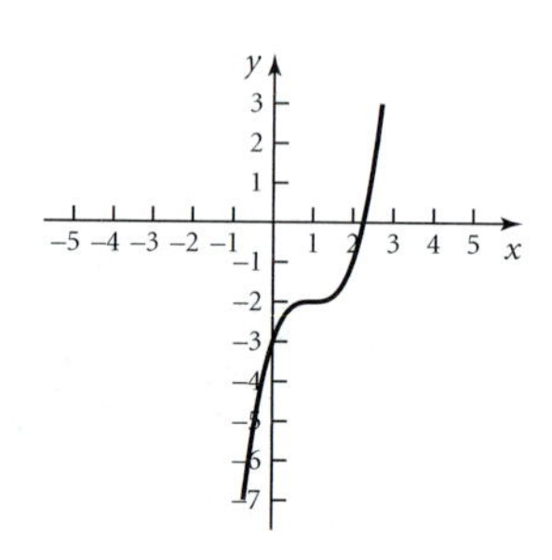

27. vertex $(2, -4)$
domain: reals
range: $y \geq -4$

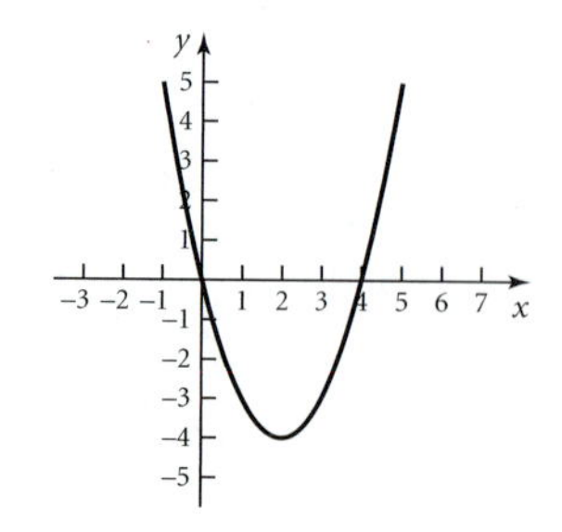

29. vertex
domain: reals
range: $y \geq -5$

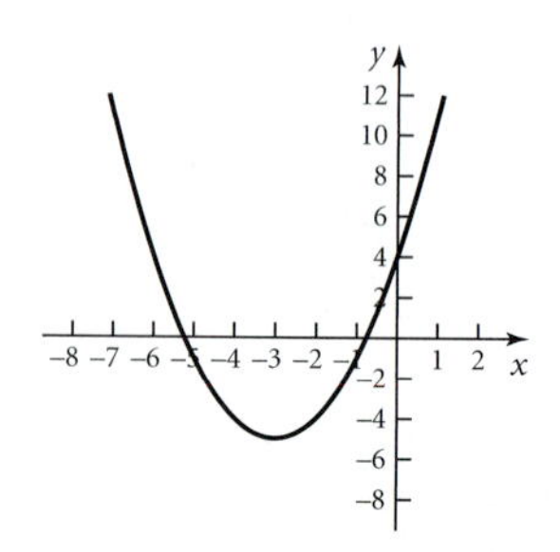

31. (0, 1)
domain: reals
range: reals

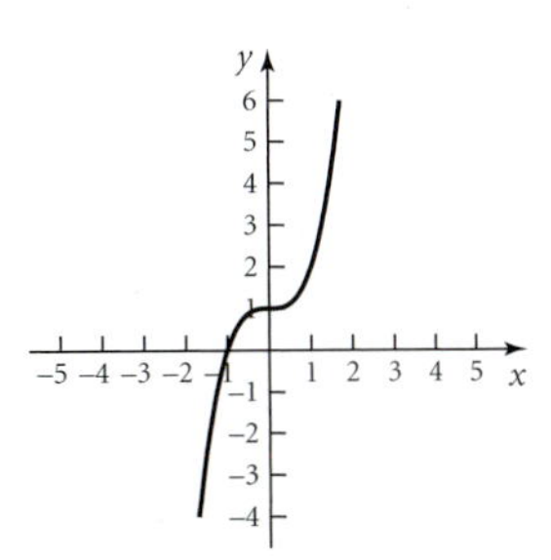

33. (1, 2)
domain: reals
range: reals

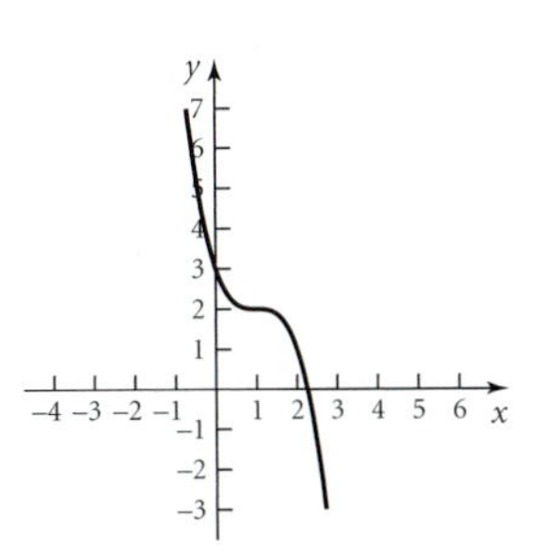

35. domain: $x \geq 0$
range: $y \geq -3$

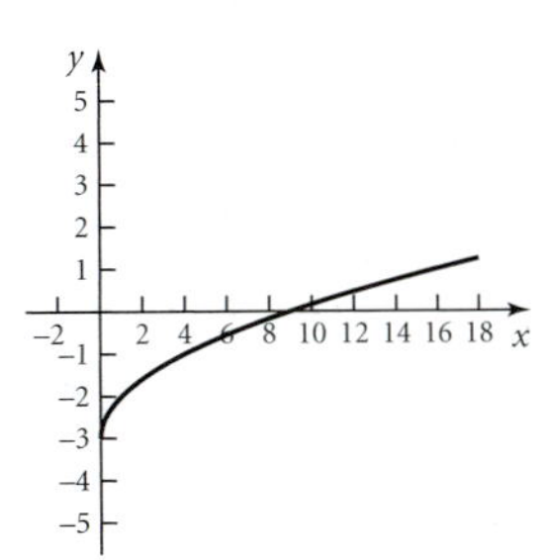

37. domain: $x \geq 0$
range: $y \geq 2$

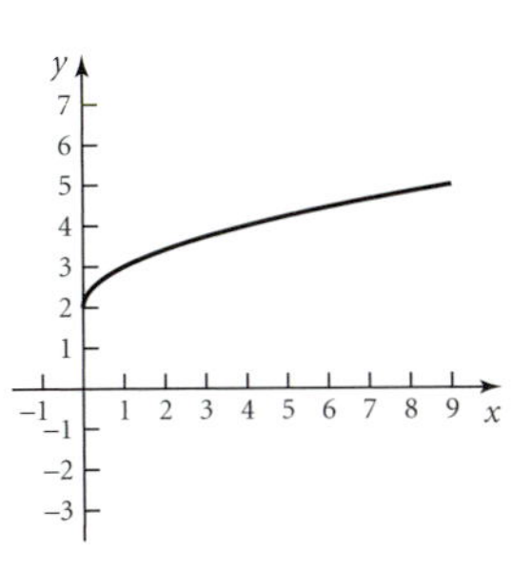

39. domain: $x \geq 0$
range: $y \leq 4$

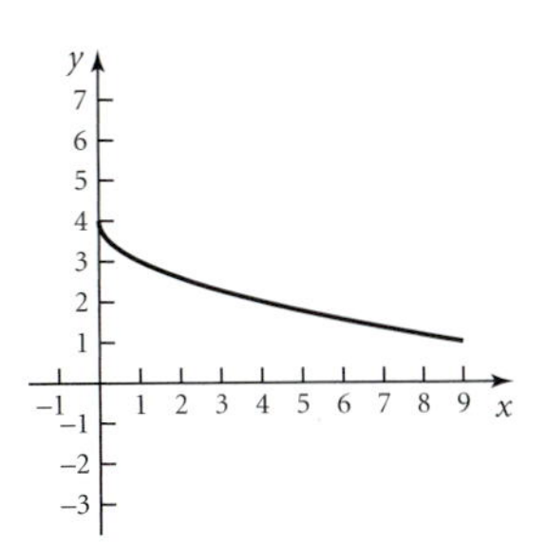

41. domain: $x \geq 5$
range: $y \geq 0$

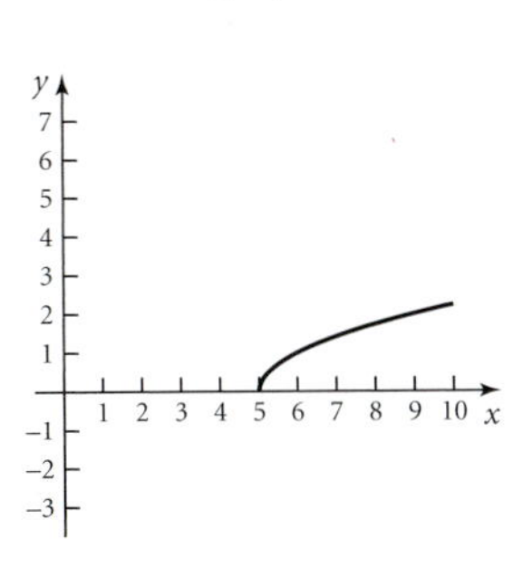

43. domain: $x \geq 3$
range: $y \geq 1$

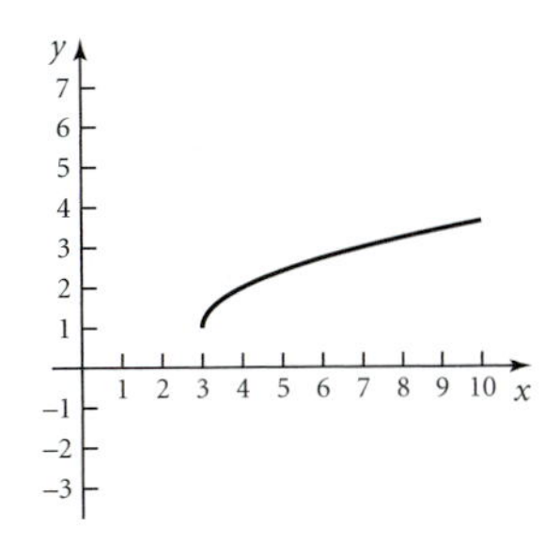

45. domain: $x \geq -3$
range: $y \leq 2$

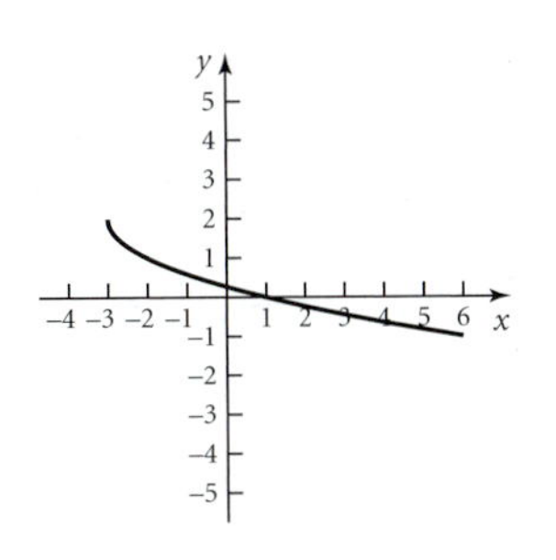

47. domain: $x \leq 2$
range: $y \geq 3$

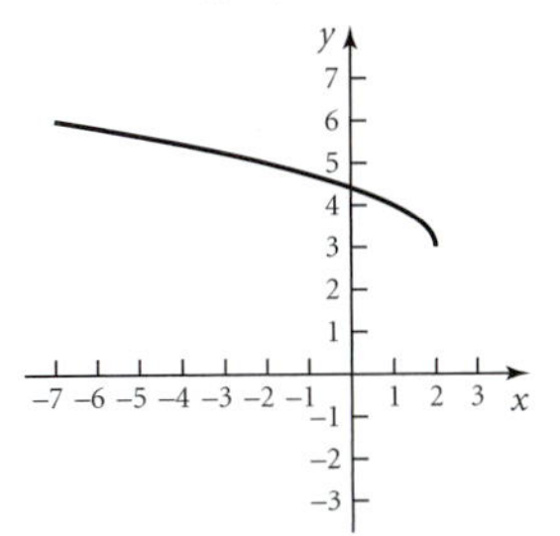

49. a.

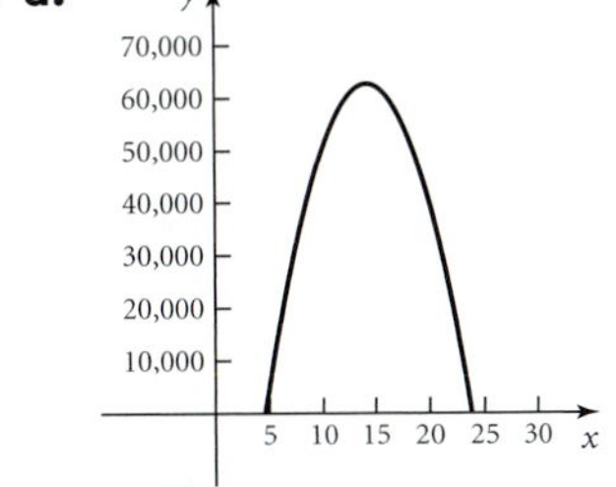

b. The vertex is (14.2, 62734.2). The maximum number of new cases was 62,734, which occurred in 1994.
51. a. \$34.21 billion **b.** The minimum profit was about \$14 billion in 2003.
53. 1950: $P(50) = 19.02$

1980: $P(80) = 74.61$

2005: $P(105) = 215.05$

An item costing \$19.02 in 1950 cost \$74.61 in 1980 and \$215.05 in 2009.
55. a. 1,368,200 filings **b.** 2 million filings in approximately 2009

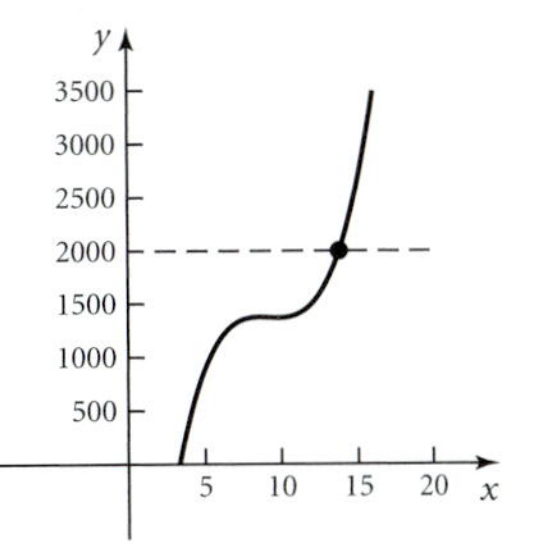

57. 10,000 **59.** Cost equals revenue if either 100 or 200 motors are manufactured.

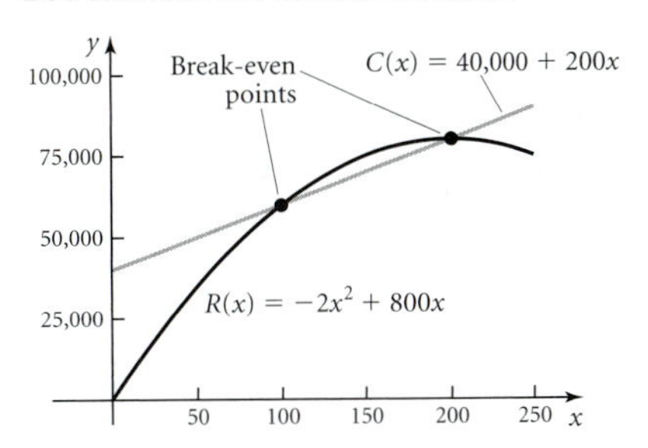

61. maximum profit of $5000 if 150 motors are manufactured **63. a.** $x(x + 2)(x - 2)$ **b.** $x = 0, -2, 2$ **c.** x-intercepts are $x = -2, 0, 2$

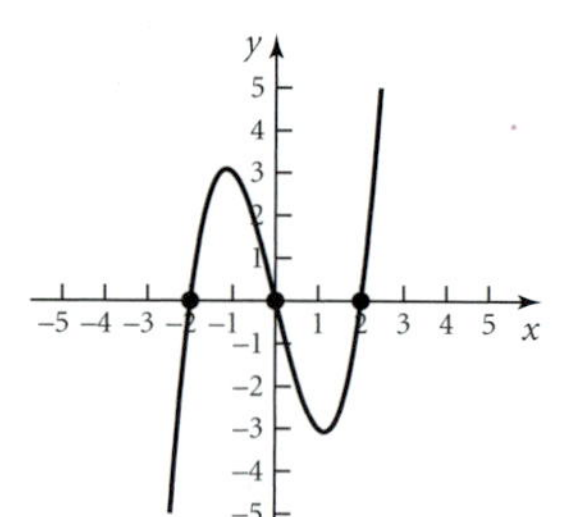

d. Given $f(x)$, the solutions of $f(x) = 0$ are the x-intercepts of the graph of $y = f(x)$.

65. a.

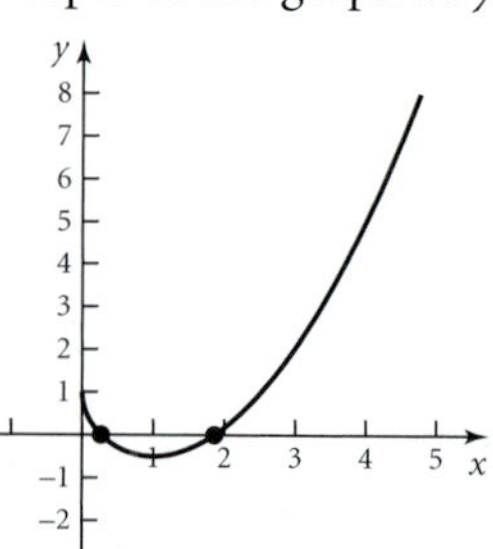

b. $x \approx 0.268, 1.858$

67. domain: $(-\infty, 0] \cap [4, \infty)$
range: $[0, \infty)$

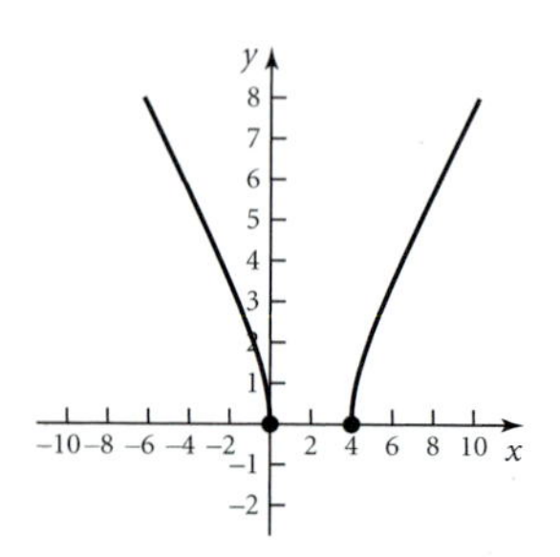

69. Supply equals demand at either \$2.46 or \$54.21.

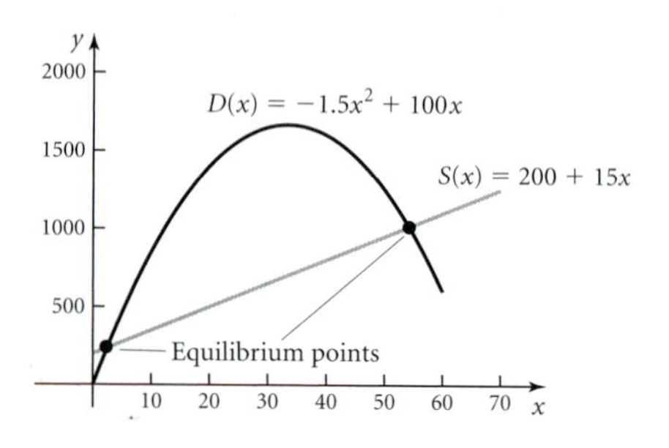

TOPIC 3

1. $x = 2, y = 0$ **3.** $x = -\frac{7}{3}, y = -\frac{2}{3}$
5. $x = 4, -4; y = 5$ **7.** $x = 2, -2; y = 0$
9. no vertical; $y = 1$ **11.** no vertical; $y = 0$

13. domain: $x \neq 3$
range: $y \neq 0$
asymptotes:
$x = 3, y = 0$

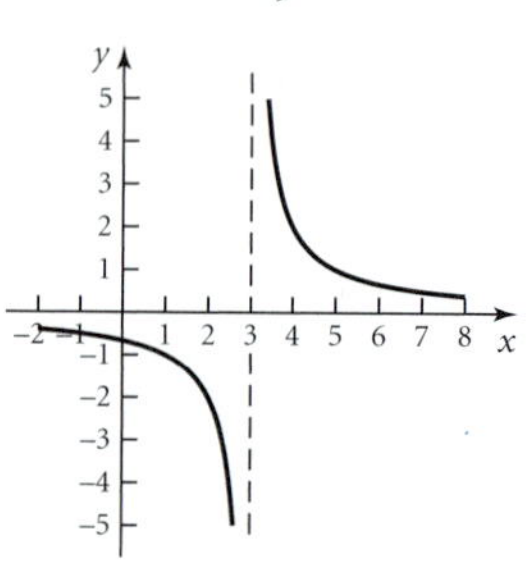

15. domain: $x \neq -1$
range: $y \neq -3$
asymptotes:
$x = -1, y = -3$

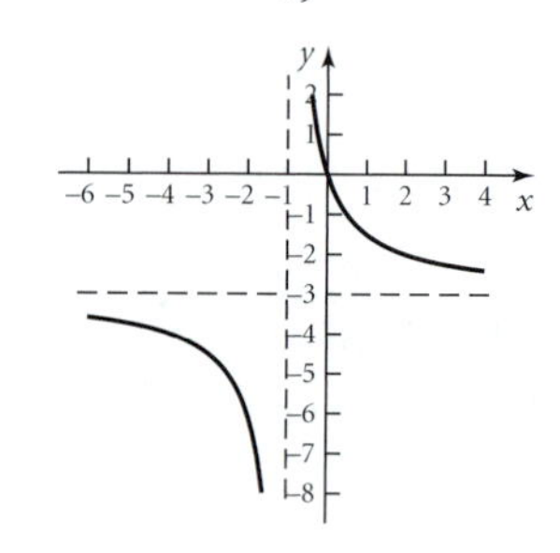

17. domain: $x \neq 3, -3$
range: reals
asymptotes:
$x = 3, x = -3, y = 0$

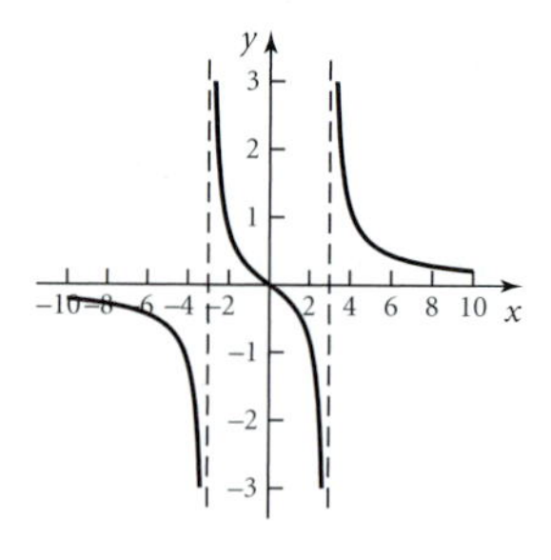

19. domain: $x \neq 3, -3$
range
$(-\infty, 0] \cup (1, \infty)$
asymptotes:
$(-\infty, -1) \cup [0, \infty)$
$x = 3, x = -3, y = 1$

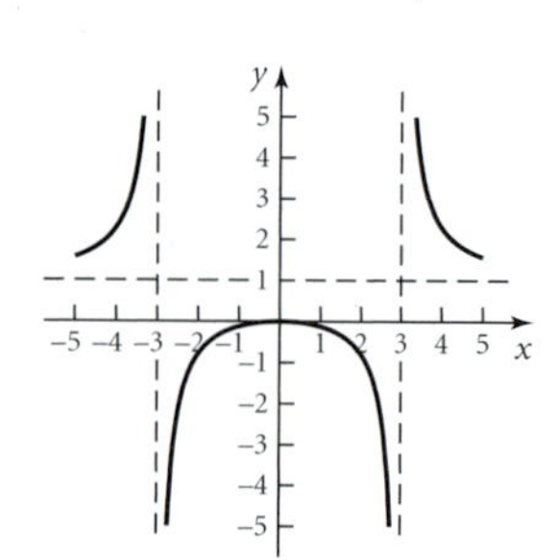

21. domain: reals
range: $\left(0, \frac{1}{3}\right]$
asymptotes: no vertical, $y = 0$

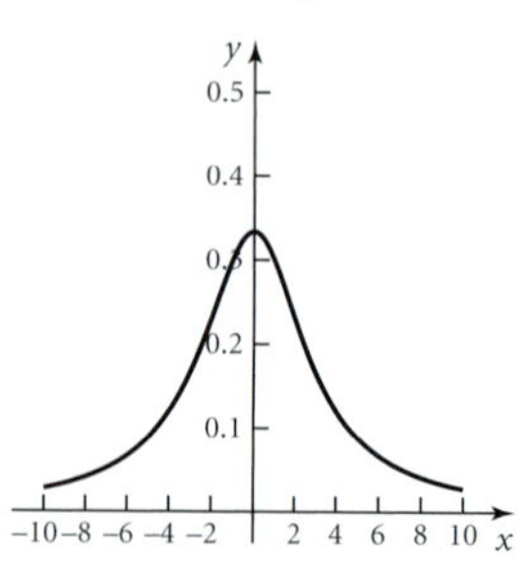

23. a. -6 **b.** 1 **c.** 16 **25. a.** 3 **b.** 7 **c.** 9 **d.** -1 **27. a.** 5 **b.** 2 **c.** 5 **29. a.** 3 **b.** 5 **c.** 2 **31. a.** 4 **b.** 4 **c.** 1

33.

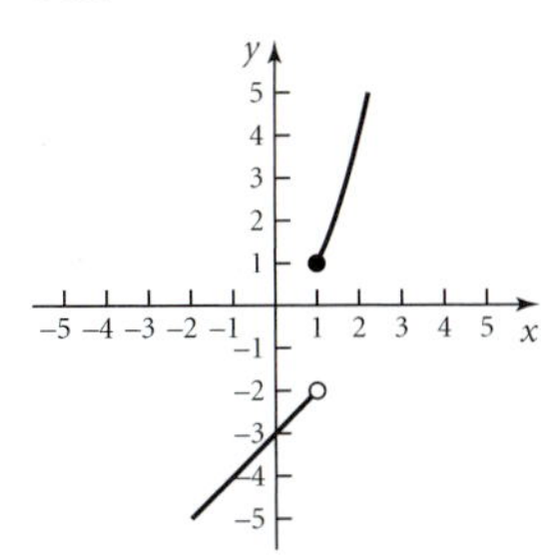

35.

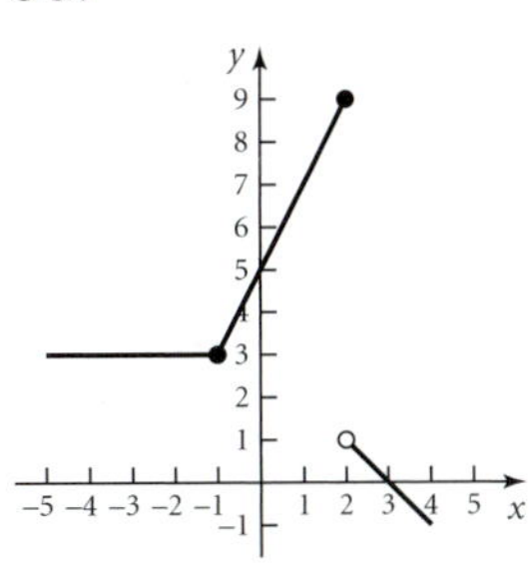

37.

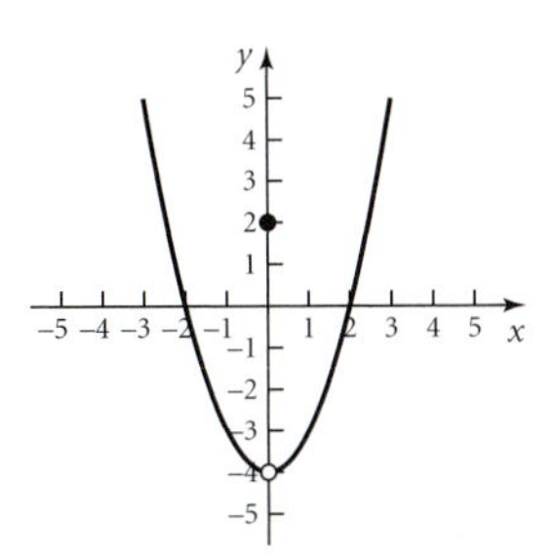

39.

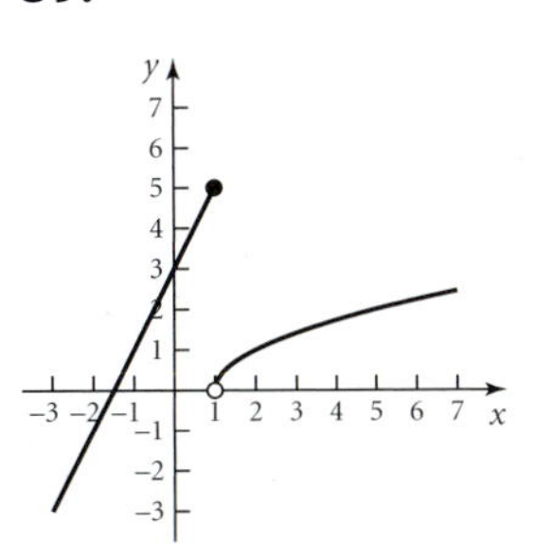

41.

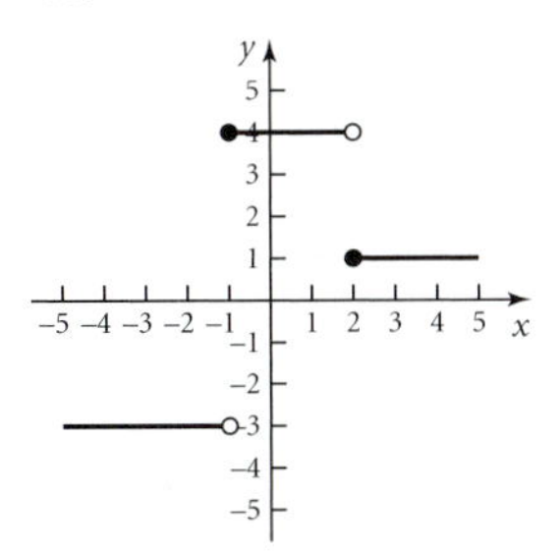

43. a. \$6.70 **b.** \$11.20

c. $y = \begin{cases} \$2.20 & x \leq 1 \\ \$4.70 & 1 < x \leq 2 \\ \$5.20 & 2 < x \leq 3 \\ \$6.70 & 3 < x \leq 4 \\ \$8.20 & 4 < x \leq 5 \\ \$9.70 & 5 < x \leq 6 \end{cases}$

d.

45. a. \$1640 **b.** \$10,430 **c.** \$47,449.50

47. a. $AveP(x) = -0.6x + 5.97 + \frac{22.1}{x}$
b. $AveP(10) =$ \$2.18 billion

c.

49. a. $AveC(x) = \frac{5600}{x} + 70 + 30x$
b. $AveC(10) =$ \$930 cost per year after 10 years
$AveC(30) \approx$ \$1157 cost per year after 30 years

c.

d. after approximately 14 years

51. a. $(0, \infty)$

b.

c. $x \approx 63.25$ feet

53. a. \$0.45 **b.** \$1.25 **c.** \$1.70

d. $y = \begin{cases} 0.45 & x \leq 2 \\ 0.85 & 2 < x \leq 3 \\ 1.25 & 3 < x \leq 4 \\ 1.70 & 4 < x \leq 30 \end{cases}$

55. a. $y = \begin{cases} 0.10d & d \le 10{,}450 \\ 1045 + 0.15(d - 10{,}450) & 10{,}450 < d \le 39{,}800 \\ 5447.50 + 0.25(d - 39{,}800) & 39{,}800 < d \le 102{,}800 \\ 21{,}197.50 + 0.28(d - 102{,}800) & 102{,}800 < d \le 166{,}450 \\ 39{,}019.50 + 0.33(d - 166{,}450) & 166{,}450 < d \le 326{,}450 \\ 91{,}819.50 + 0.35(d - 326{,}450) & d > 326{,}450 \end{cases}$ **b.** \$2477.50 **c.** \$8047.50

TOPIC 4

1. A **3.** B **5.** D **7.** C

Answers to regression equations in exercises 9 through 22 may vary slightly due to rounding.

9. a.

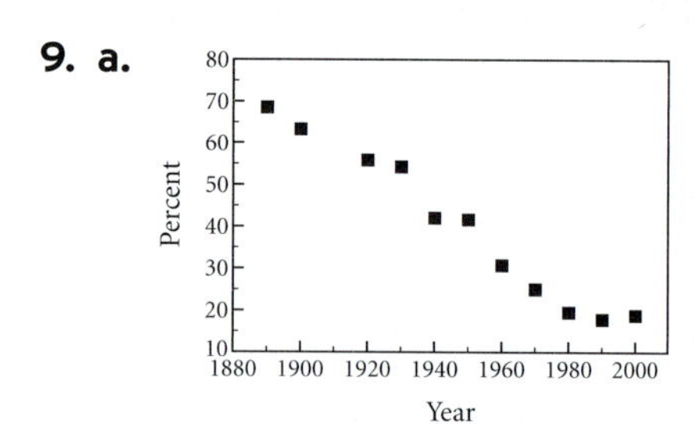

b. $y = -0.51x + 69.23$ **c.** 33.53%, which is fairly close to 30.5% **d.** approximately 9.05%

11. a.

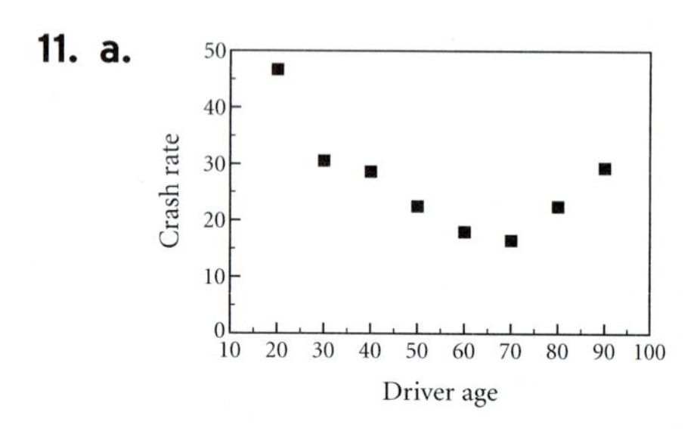

b. $y = 0.0147x^2 - 1.86x + 76.8$ **c.** 25.9, which is fairly close to 28.5

13. a.

b. $y = -0.012x^3 + 0.619x^2 - 8.85x + 85.6$
c. 53°F, which is close to 51°F **d.** 47.8°F **e.** 36.4°F

15. a.

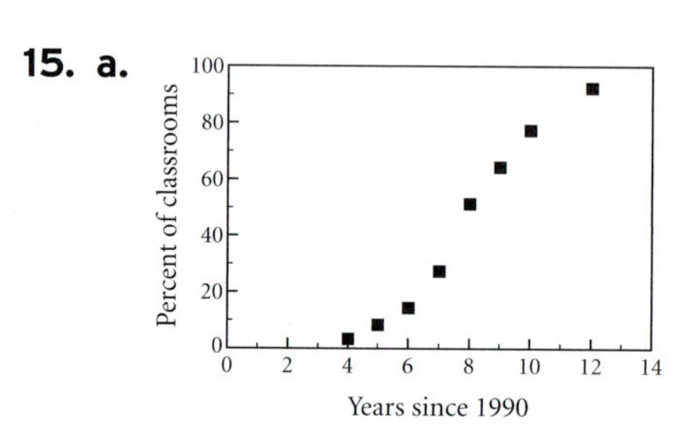

b. $y = 0.0405x^{3.2742}$ **c.** 23.7%, which is very close to 27%

17. a.

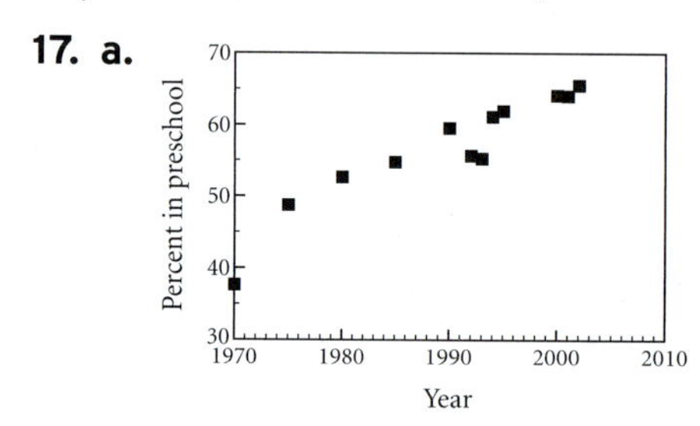

b. Letting x represent the number of years after 1970, we obtain the following:

$y = 0.73x + 42.24 \quad r^2 = 0.89$

$y = -0.011x^2 + 1.084x + 40.43 \quad r^2 = 0.91$

$y = 0.0019x^3 - 0.1092x^2 + 2.3665x + 38.037 \quad r^2 = 0.95$

The cubic model has the highest r^2 value. **c.** For 1975 ($x = 5$), linear predicts 46%, quadratic predicts 46%, and cubic predicts 47.4%. The true value was 48.6%, so the cubic model was the best. For 1993 ($x = 23$), linear predicts 59%, quadratic predicts 60%, and cubic predicts 58%. The true value was 55.1%, so cubic was still the closest. **d.** Using the cubic model, we predict 76.9% for 2009.

19. a.

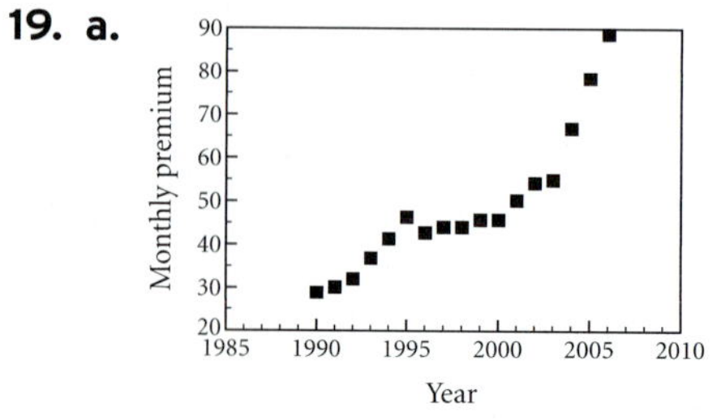

b. Letting x represent the number of years after 1980, we obtain the following:

$y = 2.9108x - 3.7353 \quad r^2 = 0.82$

$y = 0.0495x^3 - 2.4669x^2 + 41.492x - 192.45 \quad r^2 = 0.98$

$y = 2.91x^{0.9717} \quad r^2 = 0.87$

The domain is $1990 \leq x \leq 2006$. Based on r^2 values, cubic is most likely the best predictor. **c.** For 2009 $(x = 29)$, cubic predicts \$143.41.

21. a. $D(x) = 3437.54x + 12{,}914.6$, where x represents years after 2000. **b.** $S(x) = -72.6x^2 + 2018.43x + 112{,}558.6$. **c.** The shortage in 2008 is $D(8) - S(8) \approx 16{,}355$. The shortage in 2015 is $D(15) - S(15) \approx 37{,}978$.

Unit O Test

1. no **2.** yes, Domain is set of all reals. Range is set of all reals. **3.** yes, Domain is all $x \geq 2$. Range is all $y \geq 0$. **4.** no **5.** no **6.** yes, Domain is $\{-1, 3, 6\}$. Range is $\{4, 5\}$ **7. a.** -13 **b.** -9 **c.** $-2a^2 + 3a - 4$ **d.** $-2a^2 - 4ah - 2h^2 + 3a + 3h - 4$ **8. a.** $m = -3; (0, 5)$

b. $m = \frac{5}{2}; (0, -5)$

9. domain: reals
range: $y \geq -3$
vertex $(0, -3)$

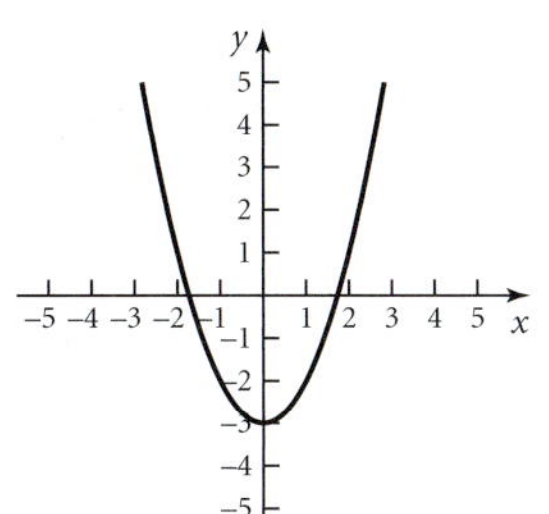

10. domain: reals
range: $y \leq 4$
vertex $(-1, 4)$

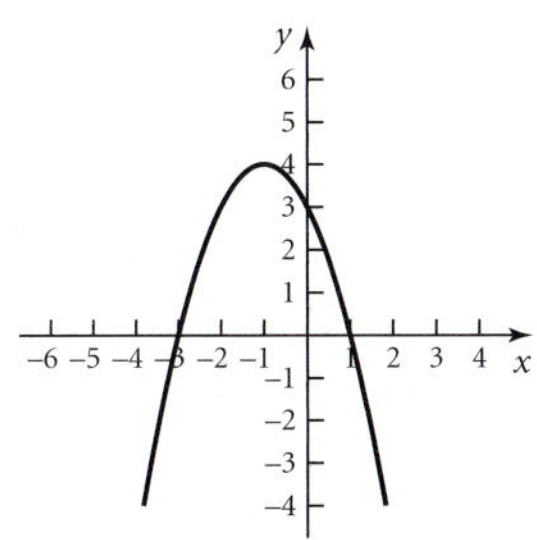

11. domain: reals
range: reals
point of inflection $(0, 2)$

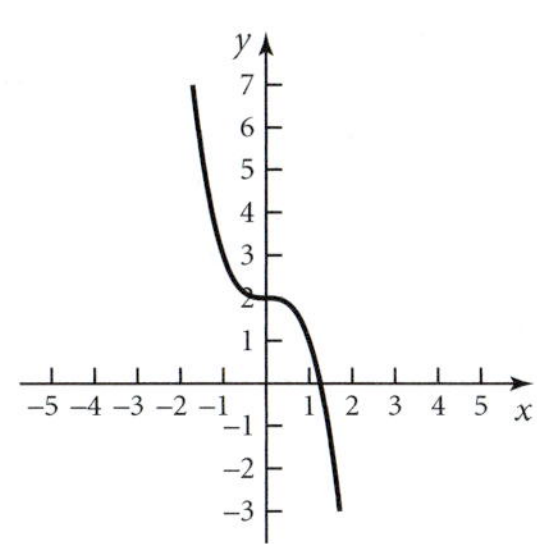

12. domain: reals
range: reals
point of inflection $(-1, -4)$

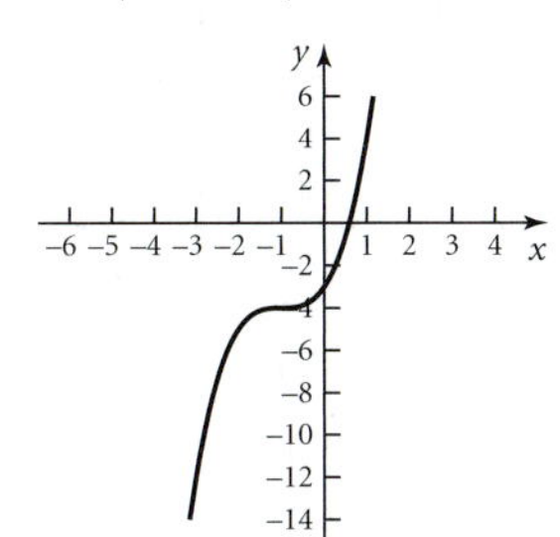

13. domain: $x \geq 2$
range: $y \geq 0$
End point at $(2, 0)$

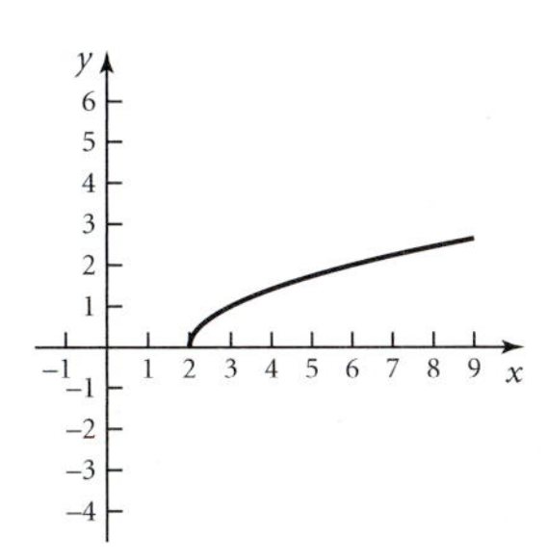

14. domain: $x \neq 3$; Vertical asymptote $x = 3$ range: $y \neq 0$; Horizontal asymptote $y = 0$

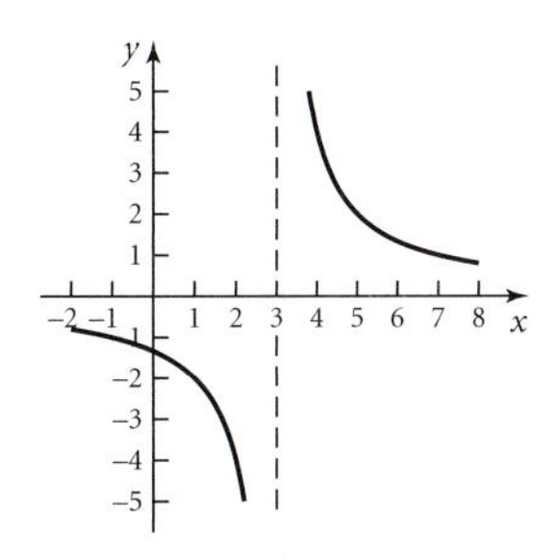

15. domain: reals
range: $y \geq -3$
corner point at $(-4, -3)$

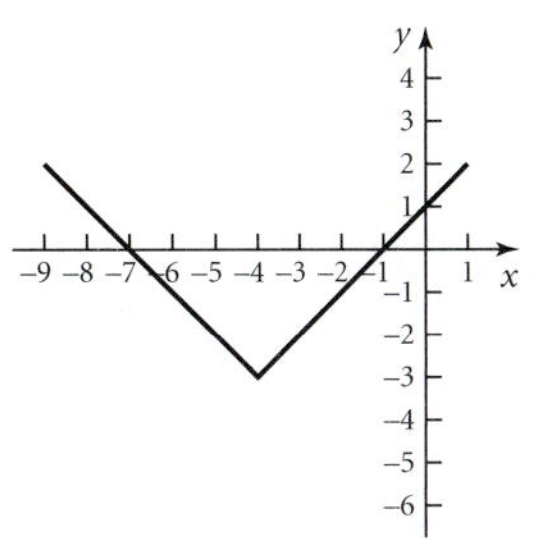

16. domain: reals
range: $y \geq -\frac{1}{4}$

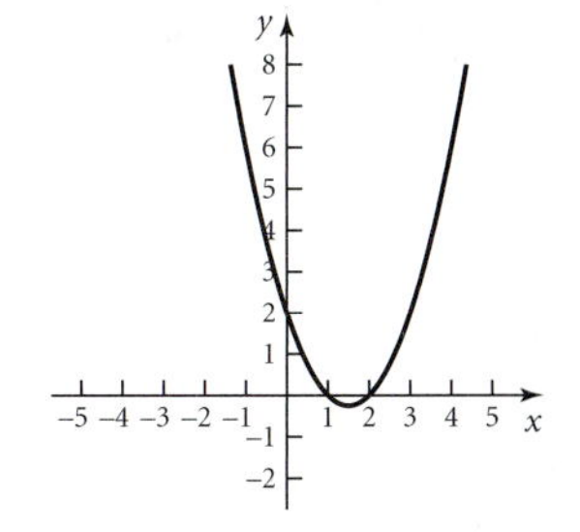

17. domain: reals
range: reals

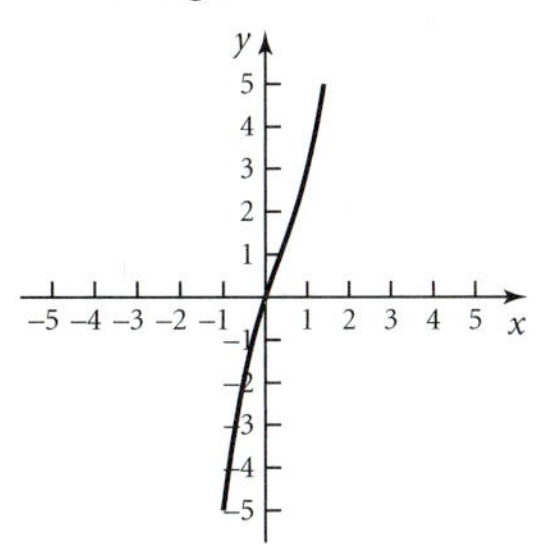

18. a. 4 **b.** 1 **c.** 4 **d.** -7

e.

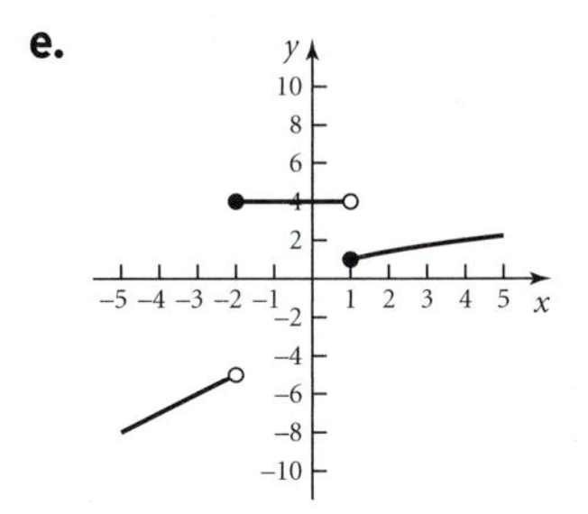

19. $x = 3, y = 0$ **20.** $x = 4, y = 6$ **21.** no vertical, $y = 3$ **22.** $x = 1, x = -1$, no horizontal **23.** $35,900 **24. a.** $C = 65 + 30h$ **b.** $125 **25. a.** $V = 325{,}000 - 13{,}750t$, where $0 \le t \le 20$ **b.** $228,750 **c.** approximately 16 to 17 years **26.** Cost equals revenue with either 8 or 25 customers. **27.** Maximum profit occurs with either 16 or 17 customers and is $72. **28.** Supply equals demand if the price is $466.67. **29. a.** $y = 0.3943x + 5.018$, where x is the number of years after 1970. **b.** 14.48% **c.** 20.8% **30. a.** $y = 22.03x^2 - 1028.9x + 24{,}254.46$, where x is the number of years after 1970. **b.** approximately 13,778 degrees in 1985 **c.** approximately 17,635 degrees in 2009

Unit 1

TOPIC 5

1. 1 **3.** -4 **5.** does not exist **7.** does not exist **9.** 2 **11.** 2 **13.** 2 **15.** does not exist **17.** ∞ **19. a.** 3 **b.** -1 **c.** does not exist **21. a.** 2 **b.** -2 **c.** does not exist **23. a.** 2 **b.** does not exist **c.** -1 **d.** does not exist **e.** does not exist **f.** does not exist **25.** B **27. a.** true **b.** true **c.** false **d.** true **e.** false **f.** true **29.** 7 **31.** 42 **33.** -6 **35.** 3 **37.** 1 **39.** does not exist **41.** $\frac{3}{2}$ **43.** $\frac{1}{2}$ **45.** does not exist **47.** does not exist **49.** $4a^2 - 3a$

51. a. -3
b. 9
c. does not exist

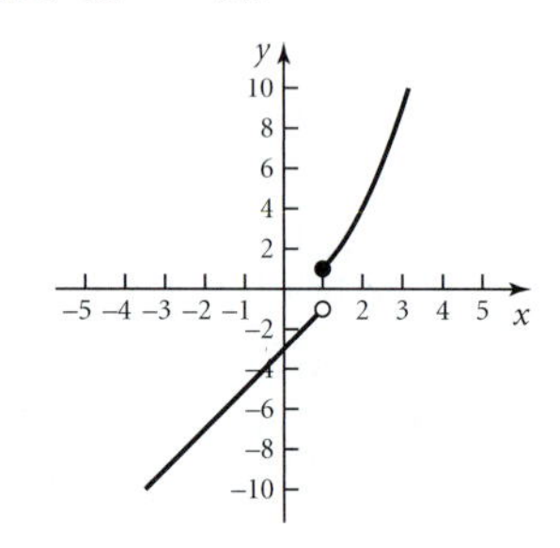

53. a. -3 **b.** does not exist **c.** does not exist **55. a.** 5 **b.** 5 **57. a.** $2a + h$ **b.** $2a$ **59. a.** 0 **b.** 0 **61. a.** $-4a - 2h + 7$ **b.** $-4a + 7$

63. a. $-\dfrac{2}{(a+1)(a+h+1)}$ **b.** $-\dfrac{2}{(a+1)^2}$

65. a. $3a^2 + 3ah + h^2$ **b.** $3a^2$

67. a. $\dfrac{1}{\sqrt{a+h} + \sqrt{a}}$ **b.** $\dfrac{1}{2\sqrt{a}}$

69. ∞ **71.** $-\infty$ **73.** ∞ **75.** 2 **77.** 0 **79.** 0 **81.** 5 **83.** ∞ **85.** 0 **87.** $-\frac{2}{3}$

89. a. $44,000 **b.** 2 years **c.** limit does not exist; there is a salary increase after 3 years **91. a.** $W(5) = 154.3$ pounds; $W(9) = 147.6$ pounds; both are approximately 1 pound over her actual weights at those times **b.** 132 pounds is the limit. Her weight will gradually approach 132 pounds. **93. a.** $6.70 **b.** $6.70 **c.** $8.20 **d.** does not exist

95.

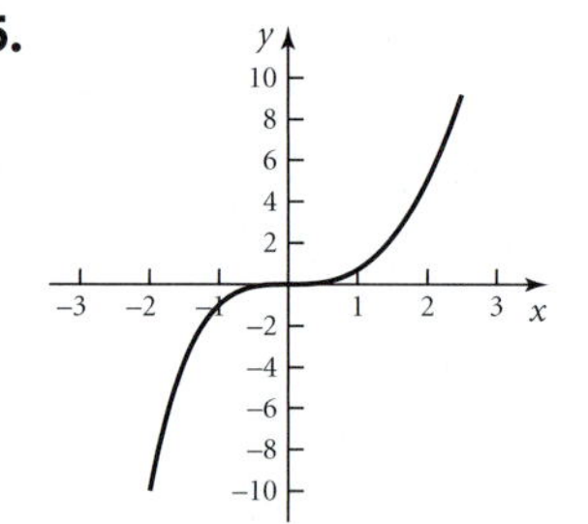

$$\lim_{x \to \infty} \frac{5x^3}{x + 6} = \infty$$

97.

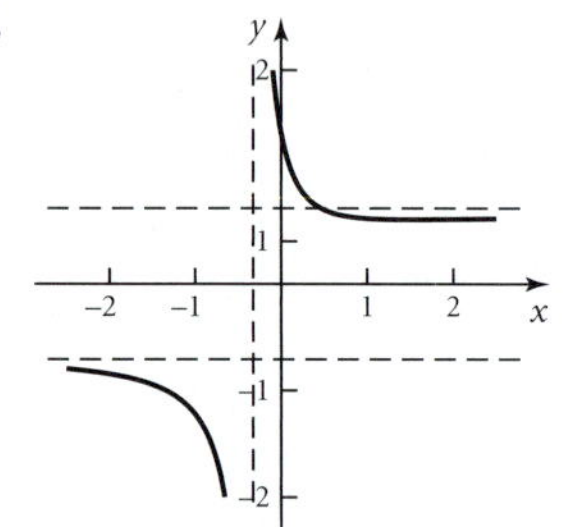

$$\lim_{x \to \infty} \frac{\sqrt{4x^2 + 2}}{3x + 1} = \frac{2}{3}$$

99. a.
$$T(d) = \begin{cases} 0.10d & d \le \$14{,}600 \\ \$1460 + 0.15(d - \$14{,}600) & \$14{,}600 < d \le \$59{,}400 \\ \$8180 + 0.25(d - \$59{,}400) & \$59{,}400 < d \le \$119{,}950 \\ \$23{,}317.50 + 0.28(d - \$119{,}950) & \$119{,}950 < d \le \$182{,}800 \\ \$40{,}915.50 + 0.33(d - \$182{,}800) & \$182{,}800 < d \le \$326{,}450 \\ \$88{,}320 + 0.35(d - \$326{,}450) & d > \$326{,}450 \end{cases}$$

b. $T(\$92{,}340) = \$16{,}415$ **c.** $9580 **d.** $8180

TOPIC 6

1. a. 1 **b.** does not exist **c.** no **3. a.** 4 **b.** 4 **c.** yes **5. a.** -1 **b.** 1 **c.** -1 **d.** does not exist **e.** yes **f.** no **7. a.** 6 **b.** does not exist **c.** no **9. a.** 4 **b.** 9 **c.** no **11.** $x = 0$; $f(0)$ is undefined **13.** $x = 0$; $f(0) \neq \lim_{x\to 0} f(x)$ **15.** $x = 2$; $\lim_{x\to 2} f(x)$ does not exist, $f(2)$ is undefined **17.** $x = 1$; $\lim_{x\to 1} f(x)$ does not exist **19.** $x = 2$; $f(2)$ is undefined, $\lim_{x\to 2} f(x)$ does not exist; $x = 3$; $f(3)$is undefined **21. a.** false **b.** true **c.** true **d.** false **e.** true **f.** true **g.** true **h.** false **i.** true **j.** false **k.** true **23. a.** 1 **b.** 4 **c.** discontinuous because $f(-4) \neq \lim_{x\to -4} f(x)$ **25. a.** does not exist **b.** undefined **c.** discontinuous because $f(-1)$ is undefined and $\lim_{x\to -1} f(x)$ does not exist **27. a.** does not exist **b.** 6 **c.** discontinuous because $\lim_{x\to 0} f(x)$ does not exist **29. a.** 0 **b.** 0 **c.** continuous **31.** all real numbers, or $(-\infty, \infty)$ **33.** all real numbers, or $(-\infty, \infty)$ **35.** $x \leq \frac{1}{2}$, or $(-\infty, \frac{1}{2}]$ **37.** $a \neq -3$, or $(-\infty, -3) \cup (-3, \infty)$ **39.** $x \neq 2, -2, 1$ or $(-\infty, -2) \cup (-2, 2) \cup (2, \infty)$

41. $x > 2$, or $(2, \infty)$ **43.** all real numbers, or $(-\infty, \infty)$

45. $z \leq -4$, $z \geq 4$, or $(-\infty, -4] \cup [4, \infty)$

47. all real numbers, or $(-\infty, \infty)$

49. a **51.** c **53.** b

55. a.

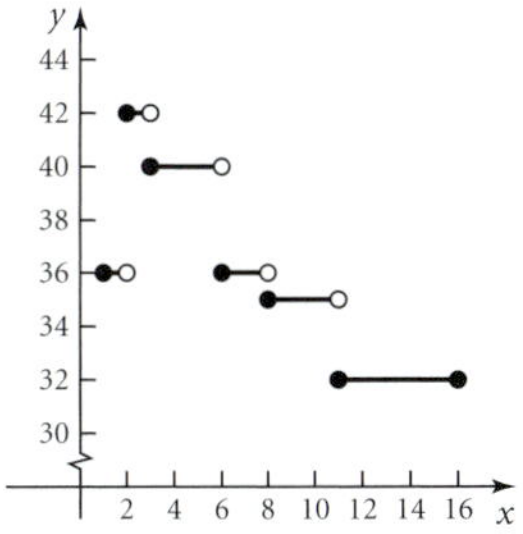

b. 1, 2, 5, 7, and 10; students adding or dropping the class

59. $x \geq 0$, or $[0, \infty)$

61. $x < 0$, $x > 9$, or $(-\infty, 0) \cup (9, \infty)$

TOPIC 7

1. a. 24. The average price of a new car increased by \$24 per year between 1930 and 1940. **b.** 189.67. The average price of a new car increased by \$189.67 per year between 1950 and 1980. **c.** 601.65. The average price of a new car increased by \$601.65 per year between 1970 and 1990. **3. a.** -24.8. The average annual income decreased by \$24.80 per year between 1930 and 1940. **b.** 307.05. The average annual income increased by \$307.05 per year between 1950 and 1970. **c.** 2226. The average annual income increased by \$2226 per year between 1990 and 2004. **5. a.** -322.10. The average cost of a new home decreased by \$322.10 per year between 1930 and 1940. **b.** 747.50. The average cost of a new home increased by \$747.50 per year between 1950 and 1970. **c.** 5428.6. The average cost of a new home increased by \$5428.60 per year between 1980 and 1990. **7. a.** 0.805. Union membership as a percent of the labor force increased by 0.805% per year between 1930 and 1950. **b.** -0.295. Union membership as a percent of the labor force decreased by 0.295% per year between 1960 and 1980. **c.** -0.265. Union membership as a percent of the labor force decreased by 0.265% per year between 1980 and 1997. **9. a.** 32. The height of the projectile increased by 32 feet per second between 1 second and 2 seconds after its launch. The projectile is rising at a rate of 32 feet per second. **b.** 8. The height of the projectile increased by 8 feet per second between 2 seconds and 2.5 seconds after its launch. The projectile is rising at a rate of 8 feet per second. **c.** -32. The height of the projectile decreased by 32 feet per second between 3 seconds and 4 seconds after its launch. The projectile is falling at a rate of 32 feet per second. **d.** 6.04 seconds after launch **11. a.** 48. The projectile is rising at a rate of 48 feet per second after 1 second. **b.** -48. The projectile is falling at a rate of 48 feet per second after 4 seconds. **c.** 0. The projectile is at its turning point after 2.5 seconds. **13. a.** 23.5. In 1950, union membership was 23.5% of the labor force. **b.** 0.29. In 1950, union membership was increasing by 0.29% per year. **c.** -0.16. In 1980, union membership was decreasing by 0.16% per year. **15. a.** -91. In 1992, the number of motorcyclists killed decreased by approximately 91 deaths per year. **b.** Approximately 3668 were killed in 2002. **c.** 309. In 2002, the number of motorcyclists killed increased by approximately 309 deaths per year. **d.** approximately 3977 deaths in 2003 **17. a.** -5 **b.** -5 **19. a.** 1 **b.** -9

21. $y = 2x + 1$

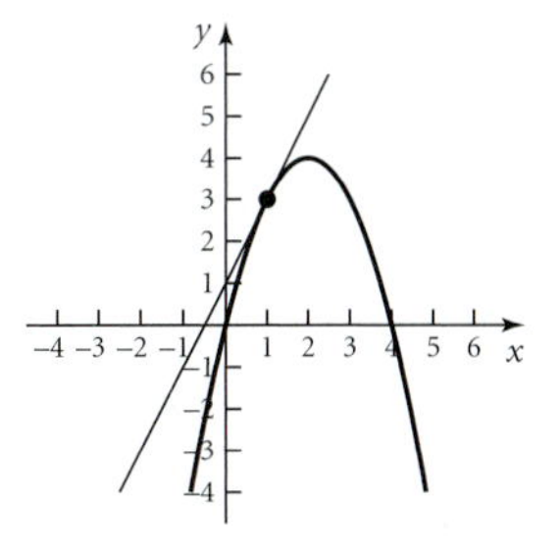

23. $m = 3$ **25.** $m = 2x$ **27.** $m = 4x + 5$
29. $m = 3 - 2x$ **31.** $m = 3x^2$

33. $m = \dfrac{1}{2\sqrt{x+1}}$ **35.** $m = -\dfrac{2}{(x+1)^2}$

37. a. $y = 15.9x^2 - 1344x + 26{,}625.5$, where x represents years after 1900 and $0 \le x \le 110$ **b.** 1960: \$3225.50, which is close to the actual amount of \$3199; 1980: \$20,865.50, which is a little higher than the actual amount of \$17,173; 2000: \$51,225.50, which is less than the actual amount of \$56,644 **c.** Between 1990 and 2004, the model gives an increase of \$1740.60, which is close to the increase of \$2226 from Exercise 3. **d.** The model is fairly good estimating the rate of change but is not very good at giving the actual income each year.
39. 1960: average annual income increased at a rate of \$564 per year; 1980: average annual income increased at a rate of \$1200 per year; 2000: average annual income increased at a rate of \$1836 per year.
41. a. $y = 0.0008x^3 - 0.046x^2 + 0.633x + 5.3$, where x is the number of years after 1970
b. $IRC = 0.0024x^2 - 0.092x + 0.633$ **c.** In 1980, unemployment is decreasing by 0.047% per year. In 2000, unemployment is increasing by 0.033% per year.
d. For 1981, predict an unemployment rate of approximately 7.053%. For 2001, predict an unemployment rate of approximately 4.033%. (For these calculations, the actual data values from the table were used for 1980 and 2000 rates.)

TOPIC 8

1. 3 **3.** $2x$ **5.** 0 **7.** $-4x$ **9.** $4x - 5$
11. $4 - 3x^2$ **13.** 11 **15.** $2x + 6$
17. $-\dfrac{3}{x^2}$ **19.** 8 **21.** 6 **23.** $\frac{1}{4}$ **25.** 2
27. $x = 0, 3, 4$ **29. a.** positive **b.** negative **c.** zero
31. a. $P'(t) = 0.00086t + 0.0117$ **b.** $P'(15) = 0.0246$. In 1995, the price of a gallon of gasoline was increasing at a rate of \$0.0246 per year. **33. a.** 20.7875. In 1985, union membership was 20.7875% of the labor force. **b.** -0.0775. In 1985, union membership was decreasing by 0.0775% per year. **c.** 20.71%
35. a. $h'(t) = -32t + 96$ gives the rate at which the object rises or falls at any time. **b.** After 1 second, the ball is rising at a rate of 64 feet per second. After 2.5 seconds, the ball is rising at a rate of 16 feet per second. **c.** 3 seconds. The ball reaches its highest point after 3 seconds.
37. a. $(-\infty, -1) \cup (1, \infty)$ **b.** $(-1, 1)$
c. $x = -1, 1$ **d.** $f'\left(\frac{1}{2}\right)$

39. a.

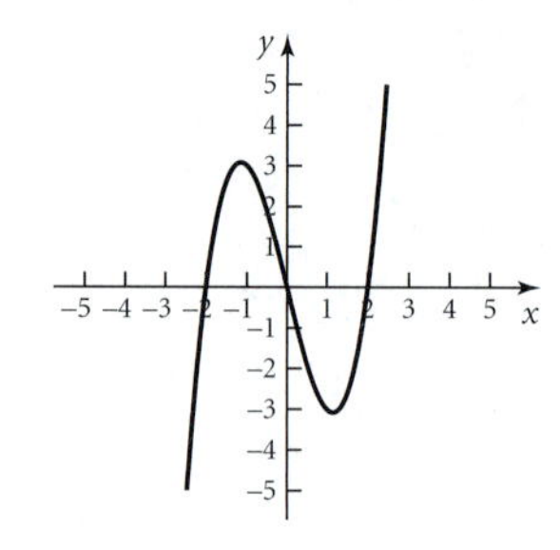

b. $f'(x) = 3x^2 - 4$

c. $f'(1) = -1; f'(-2) = 8$ **d.**

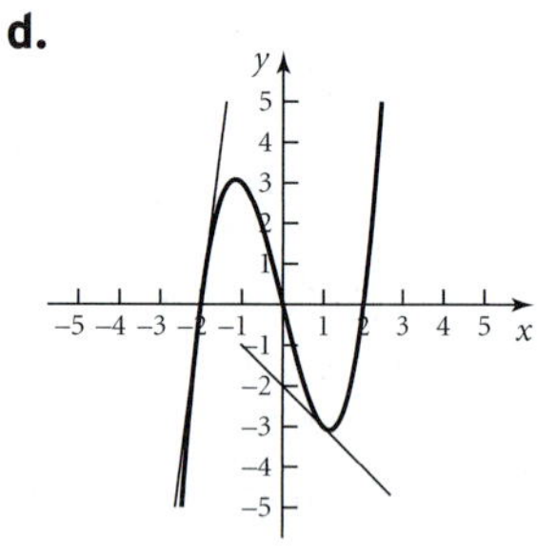

41. a.

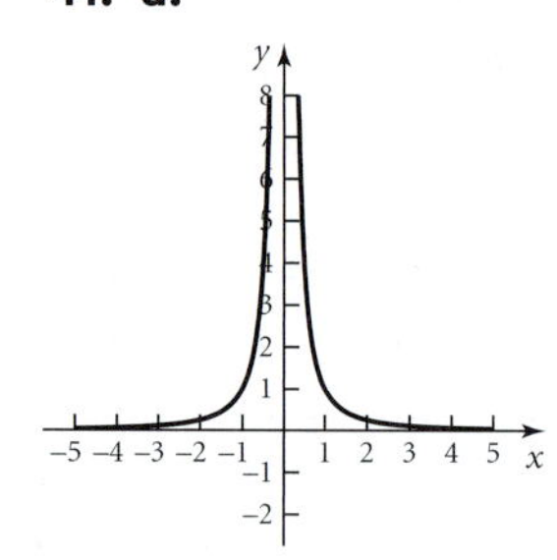

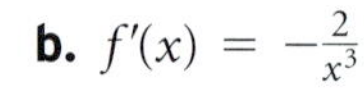
b. $f'(x) = -\frac{2}{x^3}$

c. $f'(1) = -2; f'(-2) = \frac{1}{4}$ **d.**

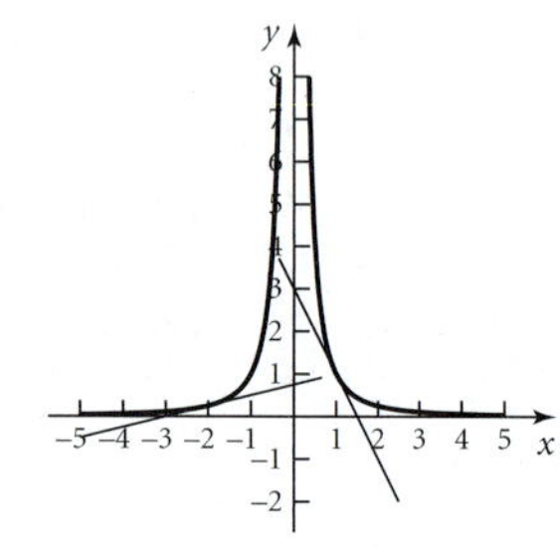

43. a.

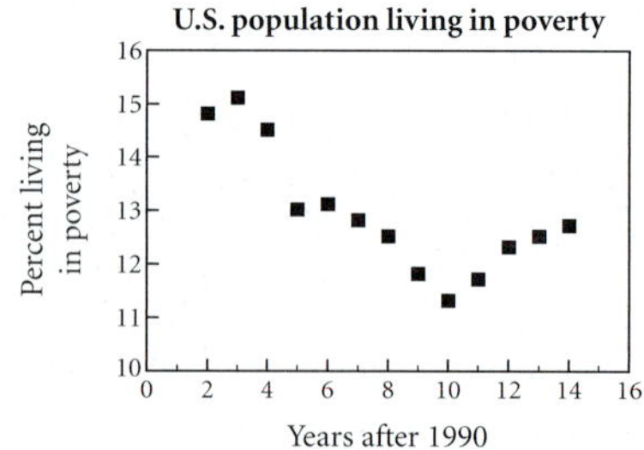

b. $y = 0.05x^2 - 1.034x + 17.3$, where x is the number of years since 1990; domain is $[0, 14]$ **c.** 1998: 12.23%, which is just slightly lower than the actual 12.5%; 2004: 12.624%, which is very close to the actual 12.7% **d.** $f'(x) = 0.10x - 1.034$ **e.** 1998: -0.234. In 1998, the percentage living in poverty was decreasing at a rate of 0.234% per year; 2004: 0.366. In 2004, the percentage living in poverty was increasing at a rate of 0.366% per year. **f.** decreasing by 0.134% in 1999 **g.** about 12.99%

TOPIC 9

1. $12x^{11}$ **3.** $-4x^{-5}$ **5.** $\frac{1}{5\sqrt[5]{x^4}}$ **7.** $\frac{2}{3}x^{-1/3}$

9. $1.7x^{0.7}$ **11.** $-\frac{6}{x^4}$ **13.** $\frac{3x^2}{2}$ or $\frac{3}{2}x^2$ **15.** $-\frac{8}{3x^5}$

17. -3 **19.** $12x^3 - 6x^2 + 7$ **21.** $3x^2 - 4 - \frac{6}{x^2}$

23. $\frac{3}{2}x^{-1/4} - 5x^{-2}$ **25.** $\frac{2}{\sqrt{x}} + 3.9x^{-2.3}$

27. $x - \frac{1}{4}$ **29.** $30x^5 - 140x^3$ or $10x^3(3x^2 - 14)$

31. $\frac{17}{3\sqrt[3]{x^2}} - \frac{20}{x^5} - 6x$ **33.** $50x - 30$ or $10(5x - 3)$

35. 6 **37.** $\frac{3}{4}$ **39.** $\frac{1}{4}$ **41.** 0 **43. a.** slope 4; $y = 4x - 3$ **b.** slope 25; $y = 25x - 27$

45. a. 18 feet per second **b.** -46 feet per second **c.** 1.5625 seconds **47. a.** increasing by \$100,000 per month **b.** increasing about \$63,246 per month

49. a. 6013 bacteria after 4 days **b.** 2810.9; after 4 days, the number of bacteria is increasing at a rate of approximately 2811 per day **51.** increasing at a rate of approximately 5102 customers per day after 1 week

53. a. $P(x) = 0.01x^2 + 0.3x - 30$ **b.** $C(50) = 140$; $R(50) = 150$; $P(50) = 10$. The cost of producing 50 units is \$140; the revenue generated from the sale of 50 units is \$150, and the profit earned from selling 50 units is \$10. **c.** $C'(x) = 2$; $R'(x) = 0.02x + 2.3$; $P'(x) = 0.02x + 0.3$ **d.** $C'(50) = 2$; $R'(50) = 3.3$; $P'(50) = 1.3$. When 50 units are produced and sold, cost increases by \$2 per unit; revenue increases by \$3.30 per unit, and profit increases by \$1.30 per unit.

55. a. $C'(x) = \frac{20}{\sqrt{x}}$ **b.** $C'(100) = 2$. When 100 phones are produced, the cost of producing one more is \$2.

57. a. $f(x) = 193.8x^3 - 4288.3x^2 + 34{,}003x - 16{,}719$, where x is the number of years after 1990.

b.

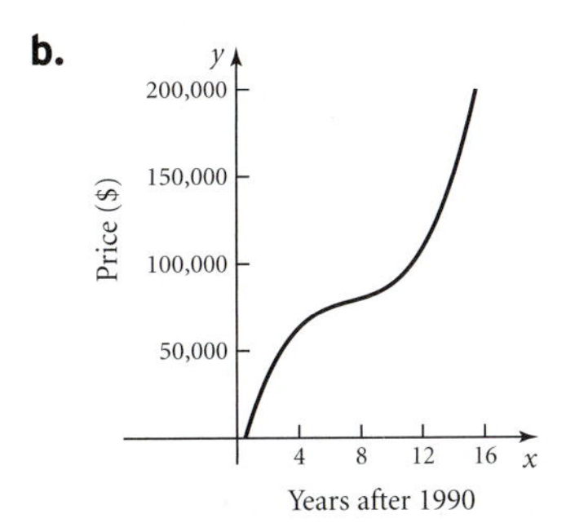

c. increasing at a rate of approximately \$14,805.40 per year **d.** $f(13) = \$126{,}375.90$; $f'(13) = \$20{,}763.80$. In 2003, the median sales price of existing homes was \$126,375.90 and the price was increasing at a rate of \$20,763.80 per year. **59. a.** \$8 **b.** \$26 **c.** $C(x) = 0.4x^3 - 2.5x^2 + 10.33x + 13.23$ **d.** $C'(3) = 6.13$; $C'(6) = 23.53$ **e.** Both are fairly close to the true value. **f.** $\$99 + C'(7) = \$99 + \$34.13 = \$133.13$

TOPIC 10

1. $5x^4 - 12x^2 = x^2(5x^2 - 12)$
3. $42x^2 - 70x + 8 = 2(21x^2 - 35x + 4)$
5. $-25x^4 + 42x^2 - 8$ **7.** $6x^3(9x^5 - 21x^2 + 10)$
9. $12x^2 - 10x - 13$
11. $10x^4 + 16x^3 - 39x^2 - 6x + 34$

13. $3 - \frac{2}{\sqrt{x}}$ **15.** 1 **17.** $2x + \frac{5}{3}x^{2/3} - 8x^{1/3}$

19. $-2x^{-2} - 30x^5 - 105x^6$

21. $\frac{4}{(5x + 4)^2}$ **23.** $\frac{-2x^2 + 10x + 14}{(x^2 + 7)^2}$

25. $\frac{2x(x^3 + 8)}{(x^3 - 4)^2}$ **27.** $\frac{-10x^3 + 6x^2 - 8}{(x^3 - 4x)^2}$

29. $28(7x - 2)^3$ **31.** $3(2x - 3)(x^2 - 3x + 5)^2$
33. $4(-12x^{-5} + 2x^{-2})(3x^{-4} - 2x^{-1})^3$
35. $-4(6x^2 - 5)(2x^3 - 5x + 2)^{-5}$
37. $5\left(\frac{2}{3}x^{-1/3} - x^{-2/3}\right)\left(x^{2/3} - 3x^{1/3}\right)^4$
39. $12x^5(x^2 - 4)^2(x^2 - 2)$
41. $6(3x - 5)^3(2x + 7)^2(7x + 9)$
43. $2(x^2 - 4x)^4(x^3 - 3)^3(11x^4 - 34x^3 - 15x + 30)$

45. $\frac{-x(3x + 10)}{(3x - 5)^4}$ **47.** $\frac{4(2x - 3)^3(3x + 14)}{(x^2 + 7)^3}$

49. $12x(3x^2 - 5)$ **51.** $-\frac{6x}{(x^2 - 5)^2}$

53. $\frac{-4(5 + 5x^2)}{(x^2 - 5)^4}$ **55.** $y = 140x - 272$

57. $y = \frac{17}{243}x + \frac{14}{243}$ **59. a.** $f'(x) = \dfrac{x^4 - 12x^2}{(x^2 - 4)^2}$
b. $x = 0, 2\sqrt{3}, -2\sqrt{3}$ **c.** $x = 2, -2$
61. a. $f'(x) = \dfrac{2(x - 4)(-2x^2 + 12x + 5)}{(x^2 + 5)^4}$
b. $x = 4, -0.3912, 6.3912$ **c.** There are no points of discontinuity. **63.** After one-half hour, the concentration is increasing at a rate of 0.192 mg/cm^3 per hour. After two hours, the concentration is decreasing at a rate of 0.048 mg/cm^3 per hour. **65.** $N(25) = 75$. If the price is $25, the store sells 75 CD players. $N'(25) = -2.5$. If the price is $25, the number of CD players sold decreases at a rate of 2.5 CD players per dollar.
67. $C'(20) = 1.5$. If 20 units are produced, the additional cost of producing one more unit is $1.50.
69. $A(6) = \$6691.13$, which means that the amount in the account will be $6691.13 if the annual compound rate is 6%. $A' = \$315.62$, which means that the amount in the account is increasing at a rate of $315.62 per year when the rate is 6%. **71.** $D_x[f(x)g(x)h(x)] = f(x)g(x)h'(x) + f(x)g'(x)h(x) + f'(x)g(x)h(x)$ **73.** $D_x\left[\dfrac{f(x)g(x)}{h(x)}\right] = \dfrac{f(x)g'(x)h(x) + f'(x)g(x)h(x) - f(x)g(x)h'(x)}{[h(x)]^2}$

TOPIC 11

1. 2 **3.** 0 **5.** $12x - 6$ **7.** $-80x^3 + 42x$
9. $-18x^{-4}$ **11.** $-\frac{2}{9}x^{-5/3}$ or $-\dfrac{2}{9\sqrt[3]{x^5}}$
13. $-\dfrac{40}{x^6}$ **15.** $\dfrac{6}{(x - 2)^3}$ **17.** $48(2x - 5)^2$
19. $30x(x^3 + 3)^3(7x^3 + 3)$ **21.** $-\dfrac{9}{\sqrt{(x^2 - 9)^3}}$.
23. $80x^3 - 240x^2 + 150x = 10x(8x^2 - 24x + 15)$
25. $\dfrac{150x}{(2x - 5)^4}$ **27.** 12 **29.** $-48/c^5 + 18/c^4$
31. $\frac{280}{27}n^{-13/3} + \frac{80}{81}n^{-8/3}$
33. $N'(36) = -62.5; N''(36) = -0.49$
35. $P''(2) = 22.2$, which means that the rate at which the number of customers without power is changing is increasing by 22,200 people per day per day on the second day. **37.** $W(3) = 160.3; W'(3) = -3.82; W''(3) = 1.03$. After three weeks on the program, the woman weighs 160.3 pounds. She is losing weight at a rate of 3.82 pounds per week, and the rate of decrease is increasing by 1.03 pounds per week per week.
39. $R''(25) = 19.505$ **41.** $C(100) = 650$; $C'(100) = 2; C''(100) = -0.01$. The cost of producing 100 phones is $650. The cost is increasing at a rate of $2 per phone. The increase in the cost is decreasing by $0.01 per phone per phone. **43.** $24x^4(15x^4 - 56x^2 + 45)$
45. Answers vary.

TOPIC 12

1. B **3.** I **5.** H **7.** E **9.** J **11.** 1 **13.** $\frac{1}{8}$
15. 16 **17.** $-\frac{1}{64}$ **19.** 0 **21.** 1 **23.** 2
25. $-\frac{1}{2}$ **27.** -2 **29.** 2 **31.** 3.320 **33.** 0.571
35. 1.386 **37.** -1.204 **39.** 1.279 **41.** -0.301
43. 1.473 **45.** 0.864 **47.** 0.585 **49.** 4.358
51. undefined **53.** 3 **55.** $\frac{8}{9}$ **57.** 2 **59.** $\frac{1}{16}$
61. 2.079 **63.** 3.396 **65.** 0.347 **67.** 1.233
69. 0.817 **71.** 0.365 **73.** $\frac{16}{3} \approx 5.333$ **75.** 4
77. 447.858 **79.** 1.432 **81.** 5 **83.** $\frac{1}{5}$ **85.** 5.284
87. $3311.44 **89.** approximately 10.34 years
91. $3315.51 **93. a.** approximately 742.4 million subscribers **b.** 16 years, or 2006 **95. a.** $13,291.52
b. approximately five years **97. a.** approximately 816 bacteria **b.** approximately 158 days
99. a. 2.14 million **b.** 23 years, or approximately 2013
101. 1860 years **103. a.** 5.58% **b.** 0.18
105. $P(t) = 131{,}107e^{0.0197t}$; 21.5 years after 1990, or about 2011 or 2012 **107. a.** approximately 201.7°F

b.

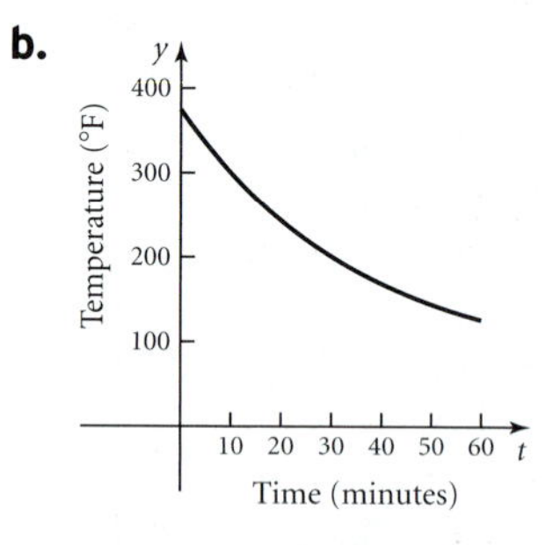

c. approximately 65 minutes
109. Orlando residents walk approximately 1.123 times faster than Buffalo residents. **111. a.** $f(1) = 72.3$; $f(2) = 66.6; f(4) = 59.5$. After one month, the average test score on the information is 72.3; after two months, the average test score is 66.6; and after four months, the average test score is 59.5. Over time, the amount of information retained is decreasing.

b.

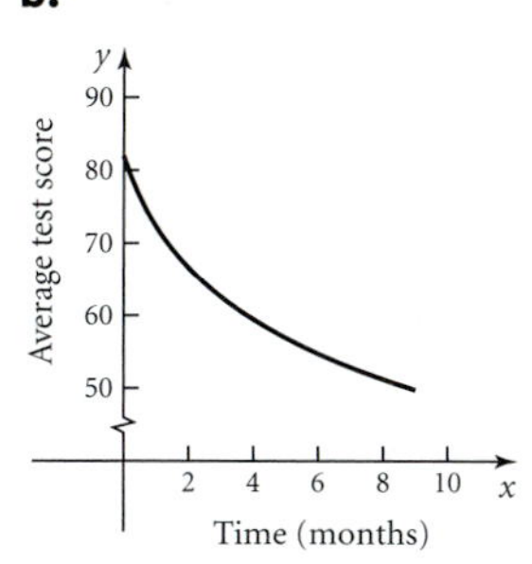

113. a. 76.3% **b.** 41 years, or approximately 1931
115. a. 47.2% **b.** approximately 2009 **c.** 76.4%
117. a. $0.97\,\text{mg/cm}^3$ **b.** $0.048\,\text{mg/cm}^3$
119. 11:24 a.m.
121. a. intersects at $x \approx 3.114$

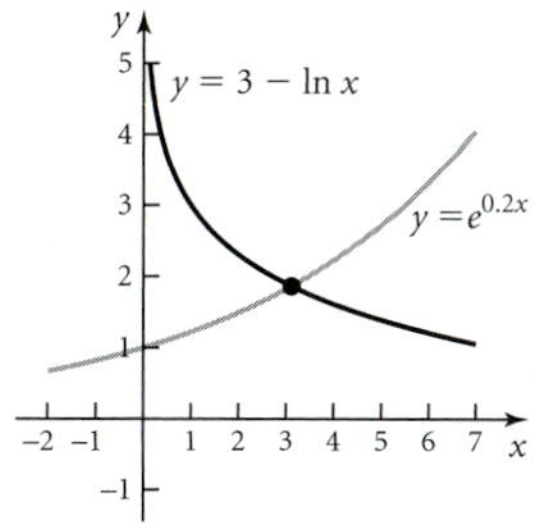

b. $x = 3.114$ **c.** The equation mixes algebraic and nonalgebraic functions.

123. $x = -0.7667, 2, 4$

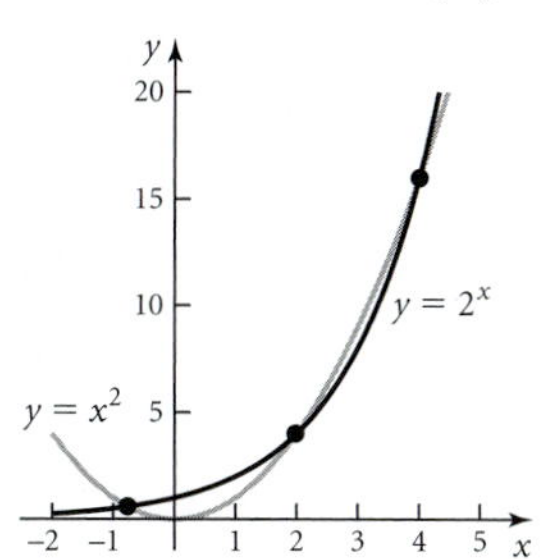

125. a.

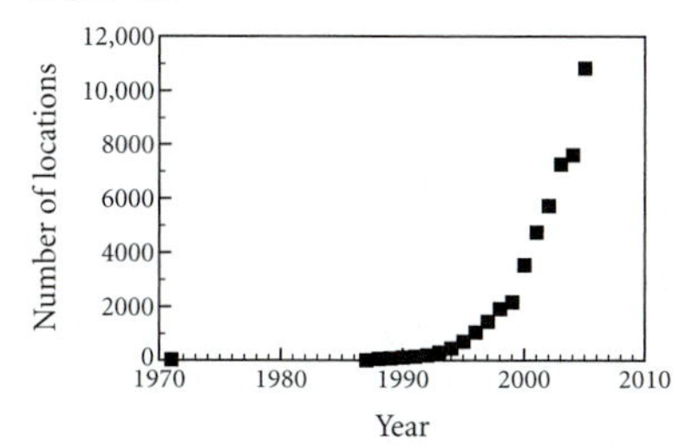

b. $y = 0.2536(1.3594)^x$, where x is the number of years after 1970. **c.** $y = \dfrac{21{,}014.3}{1 + 99{,}221.7e^{-0.3276x}}$, where x is the number of years after 1970. **d.** 1994: exponential gives 402 locations and logistic gives 536 locations; 2004: exponential gives 8669 and logistic gives 8602. **e.** For 1994, the exponential seems better, but for later years, logistic is better. **f.** exponential: 54,707 locations; logistic: 17,481 locations

127. a. $y = 2.425 - 1.464 \ln x$ **b.** day 2: 1.41 million; day 4: 0.395 million. Both are quite close to the actual numbers **c.** 5.24, or by the sixth day

129. a. $y = \dfrac{92.5}{1 + 2.918e^{-0.03x}}$, where x is the number of years after 1900 **b.** 82.2% **c.** 92.5%, which seems logical **131. a.** $P(t) = P_0e^{0.34657t}$ **b.** $y = 0.001335(1.42567)^x = 0.001335e^{0.35464x}$, where x in years after 1970, which is quite close to the doubling model in part a. **133.** Answers vary.

TOPIC 13

1. false **3.** true **5.** $3e^{3x}$ **7.** $8xe^{x^2-5}$
9. $2x - 5e^{5x}$ **11.** $xe^{3x}(3x + 2)$
13. $\dfrac{e^{-5x}(-5x - 4)}{x^5}$ **15.** $3ex^{3e-1}$
17. 0 **19.** $\dfrac{1}{x - 3}$
21. $2/x$ **23.** $\dfrac{2 \ln x}{x}$ **25.** $x[1 + 2\ln(4x)]$
27. $\dfrac{1 - 4\ln x}{x^5}$ **29.** 0
31. $5\left(3x^2 - 3e^x - \dfrac{1}{x}\right)[x^3 - 3e^x - \ln(2x)]^4$
33. $12x^5 + 25x^4 - 6xe^x - 21e^x$
35. $\dfrac{24x^5 + 75x^4 - 4xe^{2x} - 8e^{2x}}{(2x + 5)^2}$
37. $2(x^5 - 3e^x)(2x + 5)^2(13x^5 + 25x^4 - 6xe^x - 24e^x)$
39. a. $m = 2.5$ **b.** $y = 2.5x - 3.08$
41. 78.13 people per year **43.** $T'(30) = -2.6$. After 30 minutes, the temperature is decreasing at a rate of 2.6°F per minute. **45.** $f(2) = 66.6\%$. After two months, 66.6% of the information is retained; $f'(2) = -4.7$. After two months, the amount of information retained is decreasing by 4.7% per month. **47.** After two months, the retention rate is increasing by 1.6 words per day per day. **49.** In 1930, the proportion was increasing by 0.7% per year. In 2000, the proportion was increasing by 0.32% per year. **51.** $P'(t) = 82.26$. After one week, the colony is growing at a rate of approximately 82 bacteria per day. **53.** $C(3) \approx 1.07$. After three hours there is $1.07\,\text{mg/cm}^3$ of drug in the bloodstream; $C'(3) \approx -0.18$. The

concentration is decreasing by approximately 0.18 mg/cm^3 per hour after three hours; $C''(3) \approx -0.09$. The rate of decrease is decreasing by approximately 0.09 mg/cm^3 per hour per hour after three hours.
55. $(3 \ln 12) \cdot 12^{3x}$ **57.** $4 \cdot 2^{x-1} \cdot \ln 2$
59. $4 \cdot 2^{x^3}(3x^3 \ln 2 + 1)$ **61.** $\frac{1}{x \ln 6}$
63. $\frac{5}{(x-2)\ln 7}$ **65.** $x\left(\frac{1}{\ln 11} + 2 \log_{11} x\right)$
67. $36e^{6x}$ **69.** $12e^{6x^2}(1 + 12x^2)$
71. $3e^{3x}(3x^{-2} - 4x^{-3} + 2x^{-4})$
73. $\frac{3e^{3x}(3x^2 - 4x + 2)}{x^4}$ **75.** $6^x(\ln 6)^2$
77. $(2 \ln 4)4^{x^2}(2x^2 \ln 4 + 1)$ **79.** $-\frac{4}{(5-2x)^2}$
81. $-\frac{1}{\ln 5(x-2)^2}$ **83.** Answers vary
85. Answers vary **87. a.** 0.02735 **b.** By the Product Rule

$$\begin{aligned} f'(3) &= \frac{(x^2+2)(x \cdot \frac{1}{x} + \ln x) - (x \ln x)(2x)}{(x^2+2)^2} \\ &= \frac{(x^2+2)(1+\ln x) - x^2 \ln x}{(x^2+2)^2} \\ &= \frac{x^2 + x^2 \ln x + 2 + 2\ln x - 2x^2 \ln x}{(x^2+2)^2} \\ &= \frac{x^2 - x^2 \ln x + 2 + 2 \ln x}{(x^2+2)^2} \\ \text{So } f'(3) &= \frac{9 - 9 \ln 3 + 2 + 2 \ln 3}{(9+2)^2} \\ &= \frac{11 - 7 \ln 3}{121} \approx 0.02735 \end{aligned}$$

c.

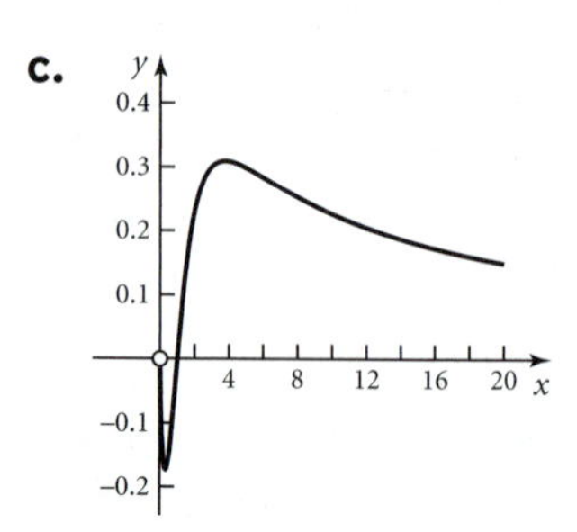

d. 0 **e.** 0 **89. a.** \$15 per unit **b.** \$9.02 per unit **c.** −\$5.98 per unit **d.** Break-even points are $x \approx 28$ units or $x \approx 318$ units.

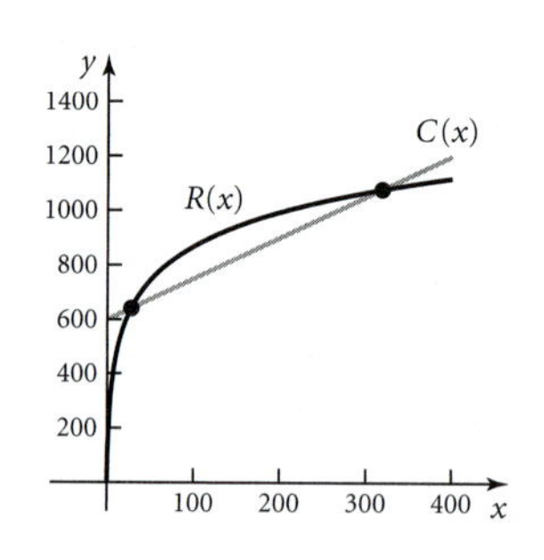

e. To make a profit, between 28 units and 318 units should be produced.

UNIT 1 TEST

1. −3 **2.** 7 **3.** 3 **4.** 1 **5.** does not exist **6.** $\frac{3}{4}$ **7.** 2 **8.** 0 **9.** ∞ **10. a.** does not exist **b.** 3 **c.** 5 **d.** 5 **e.** 5 **11. a.** does not exist (∞) **b.** 1 **c.** 0 **d.** 3 **e.** −2 **f.** does not exist **12. a.** false **b.** true **c.** false **d.** false **e.** false **f.** true **13.** $2x - 3$ **14.** $x \leq 4$ or $(-\infty, 4]$ **15.** $x \neq 1, -1$, or $(-\infty, -1) \cup (-1, 1) \cup (1, \infty)$ **16.** all real numbers **17. a.** discontinuous because $\lim_{x \to -2} f(x)$ does not exist **b.** continuous **c.** discontinuous because $f(1)$ is undefined **d.** discontinuous because $\lim_{x \to 3} f(x)$ does not exist and $f(3)$ is undefined **e.** discontinuous because $\lim_{x \to 5} f(x) \neq f(5)$ **18. a.** false **b.** true **c.** true **d.** false **19. a.** Per capita health care spending increased an average of \$221.50 per year between 1988 and 2000. **b.** The instantaneous rate of change in 2000 is $y'(30) = \$271.40$. In 2000, per capita health care spending is increasing at a rate of \$271.40 per year. **20.** $2x - 3$ **21.** $6x^2 - 5$ **22.** $-18x^{-7}$ **23.** $\frac{1}{4\sqrt[4]{x^3}}$ **24.** $\frac{12}{x^4} + 5$ **25.** $9x^8 + 20x^4 - 8x^3 - 21$ **26.** $\frac{-3x^2 + 10x + 6}{(x^2+2)^2}$ **27.** $28x^3(x^4 - 3)^6$ **28.** $\frac{-1}{\sqrt{3 - 2x}}$ **29.** $-\frac{45}{(3x-2)^4}$ **30.** $12x^2(2x^3 - 5)^3(3x^4 + 5)^2(12x^4 - 15x + 10)$ **31.** −25 **32.** $y = -\frac{3}{4}x + 2\frac{3}{4}$ **33. a.** The rock is rising at a rate of 8 feet per second. **b.** The rock is falling at a rate of 56 feet per second. **c.** 1.25 seconds **d.** approximately 3.4 seconds **34.** increasing by 0.29% per year **35.** $W'(1) = -7.17$. After one week, the woman is losing weight at a rate of approximately 7 pounds per week; $W'(6) = -1.93$. After six weeks the woman is losing weight at a rate of approximately 2 pounds per week. **36. a.** \$10 **b.** \$40 **c.** \$30 **d.** 40 customers, \$880 **37.** $f''(x) = 60x^3 - 24x^2$ **38.** $f''(x) = \dfrac{24}{(2x+5)^3}$ **39.** $f''(x) = -1/\sqrt{(3-2x)^3}$ **40.** $f''(x) = 24(3x^2-2)^2(21x^2-2)$ **41.** $N(8) = 62.19$. There are approximately 62 accidents the eighth month; $N'(8) = -6.58$. In the eighth month, the number of accidents is decreasing by approximately 6.58 accidents per month; $N''(8) = -3.72$. In the eighth month, the rate at which accidents are decreasing is decreasing by approximately 3.72 accidents per month per month. **42.** $P''(5) = -0.02$. After five years, the growth rate of the population is decreasing at a rate of approximately 2 people per year per year. **43.** B **44.** E **45.** A **46.** C **47.** G **48.** I **49.** J **50.** H **51.** D **52.** F **53.** 7.389 **54.** 0.763 **55.** 0.693 **56.** 0.845 **57.** 17 **58.** 2 **59.** −4 **60.** 2.429 **61.** 1.386 **62.** 2.651 **63.** −2.183 **64.** 7.055 **65.** 12 **66.** 3 **67.** $3e^x - 8$ **68.** $\dfrac{5}{5x+3}$ **69.** $-6xe^{4-3x^2}$ **70.** $\dfrac{8(\ln x^2)^3}{x}$ **71.** $4ex^{4e-1} + 4e^{4x}$ **72.** $x^4e^{3x}(-3x+5)$ **73.** $\dfrac{3e^{2x}(2x-1)}{8x^2}$ **74.** $4\left(2x^3 - 6e^{3x} - \dfrac{5}{x}\right)^3\left(6x^2 - 18e^{3x} + \dfrac{5}{x^2}\right)$ **75.** $V'(2) = -3351.6$. After two years, the car is decreasing in value at a rate of approximately \$3352 per year. **76.** $T(2) = 68.5$°F. After two hours, the temperature of the root beer is 68.5°F; $T'(2) = 2.24$. After two hours, the temperature of the root beer is increasing at a rate of 2.24°F degrees per hour; $T''(2) = -3.36$. After two hours, the rate at which the temperature is increasing is decreasing by approximately 3.36°F per hour per hour.

77. a.

Health care spending

Per capita spending

Years after 1970

b. $S(x) = 398.07(1.0872)^x = 398.07e^{0.0836x}$, where x is the number of years after 1970 **c.** $S(35) = \$7425$ **d.** $S'(30) = 408.67$. In 2000, per capita spending was increasing at a rate of approximately \$409 per year.

Unit 2

TOPIC 14

1. a. $(-\infty,-1) \cup [0,2)$ **b.** $(2,\infty)$ **3. a.** $(-1,1)$ **b.** $(-\infty,-1)\cup(1,\infty)$ **5. a.** B **b.** D **c.** E **d.** none **e.** A, F **f.** C **7.** true **9.** false **11.** true **13.** true **15.** false **17.** c **19.** a **21.** 3, −1 **23.** 0, 3 **25.** 3 **27.** 2, −2 **29.** $\frac{2}{3}$ **31. a.** $(-\infty,-2)$ and $(2,\infty)$ **b.** $(-2,2)$ **c.** relative minimum at $(2,-16)$, relative maximum at $(-2,16)$ **33. a.** $(3,\infty)$ **b.** $(-\infty,3)$ **c.** absolute minimum at $(3,-27)$ **35. a.** $(-1,0)$ and $(4,\infty)$ **b.** $(-\infty,-1)$ and $(0,4)$ **c.** relative minimum at $(-1,2)$, absolute minimum at $(4,-123)$, relative maximum at $(0,5)$ **37. a.** $(-5,0)$ and $(2,\infty)$ **b.** $(-\infty,-5)$ and $(0,2)$ **c.** absolute minimum at $(-5,-387)$, relative minimum at $(2,-44)$, relative maximum at $(0,-12)$ **39. a.** $(-\infty,\infty)$ **b.** none **c.** none **41. a.** none **b.** $(-\infty,5)\cup(5,\infty)$ **c.** none **43. a.** none **b.** $(-\infty,-1)\cup(-1,1)\cup(1,\infty)$ **c.** none **45. a.** $(-3,3)$ **b.** $(-\infty,-3)$ and $(3,\infty)$ **c.** absolute minimum at $(-3,-\frac{1}{6})$, absolute maximum at $(3,\frac{1}{6})$ **47. a.** $(-\infty,0)$ **b.** $(0,\infty)$ **c.** absolute maximum at $(0,0)$ **49. a.** $(-\infty,-2)\cup(-2,0)$ **b.** $(0,2)\cup(2,\infty)$ **c.** relative maximum at $(0,0)$ **51. a.** $(-\infty,\frac{1}{2})$ and $(\frac{3}{2},\infty)$ **b.** $(\frac{1}{2},\frac{3}{2})$ **c.** relative minimum at $(\frac{3}{2},0)$ relative maximum at $(\frac{1}{2},2)$ **53. a.** $(-\infty,3)$ and $(5,\infty)$ **b.** $(3,5)$ **c.** relative minimum at $(5,0)$; relative maximum at $(3,108)$ **55. a.** $[\frac{5}{2},\infty)$ **b.** none **c.** absolute minimum at $(\frac{5}{2},0)$ **57. a.** $[4,\infty)$ **b.** $(-\infty,-4]$ **c.** absolute minimum at $(-4,0)$, absolute minimum at $(4,0)$ **59. a.** $(-\infty,\infty)$ **b.** none **c.** none **61. a.** $(-1,\infty)$ **b.** $(-\infty,-1)$ **c.** absolute minimum at $(-1,-e^{-1})$ **63. a.** $(0,\infty)$ **b.** none **c.** none **65. a.** $(e^{-1},\infty)\approx(0.368,\infty)$ **b.** $(0,e^{-1})\approx(0,0.368)$ **c.** absolute minimum at $(e^{-1},-e^{-1})$ **67. a.** decreases for $(0,7.8)$, or from 1990 to 1998 **b.** The absolute minimum is at $x = 7.8$, so claims hit their lowest number in 1998, when there were approximately 870 claims ($x = 8$). **69. a.** from 0 to approximately 3.34 seconds **b.** hits the ground after approximately 6.8 seconds **c.** Maximum height is approx. 119.8 feet after 3.34 seconds. **71. a.** \$3000 **b.** Maximum costs of \$7047 occur if 4000 DVDs are produced. **73.** c **75. a.** none **b.** $(-\infty,-2)\cup(-2,0)\cup(0,2)\cup(2,\infty)$ **c.** none **77. a.** $(0,\frac{2}{3})$ **b.** $(-\infty,0)$ and $(\frac{2}{3},\infty)$ **c.** absolute minimum at $(0,0)$, relative maximum at $(\frac{2}{3},\frac{4}{9}e^{-2})$ **79. a.** increases for $(1,\infty)$ **b.** decreases for $(0,1)$

81. a.

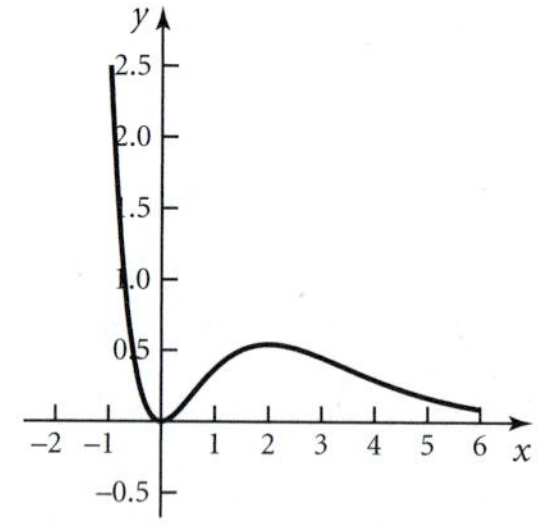

using $[-5, 10]$ for x and $[-1, 1]$ for y **b.** $(-\infty, \infty)$ **c.** absolute minimum at $x = 0$; relative maximum at $x = 2$ **d.** increases for $(0, 2)$, decreases for $(-\infty, 0)$ and $(2, \infty)$

83. a.

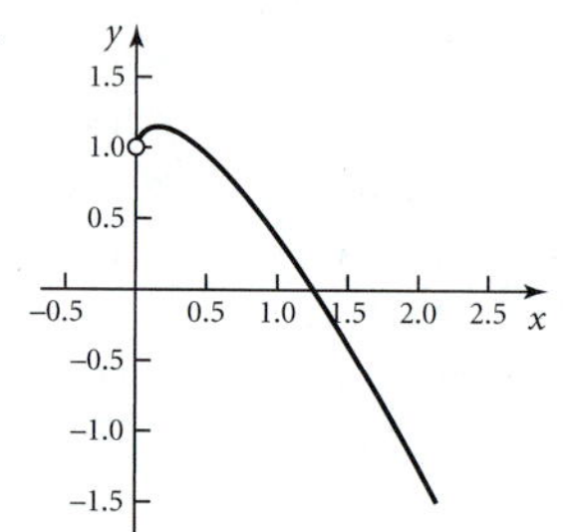

using $[0, 2]$ for x and $[-2, 2]$ for y **b.** $(0, \infty)$ **c.** absolute maximum at $x \approx 0.1564$ **d.** increases for $(0, 0.1564)$, decreases for $(0.1564, \infty)$ **85. a.** $y = 406.178x^3 - 19{,}717.25x^2 + 273{,}355x - 241{,}112$. **b.** The function increases for $(0, 10.1)$ and $(22.3, \infty)$ and decreases for $(10.1, 22.3)$. Passenger traffic increased from 1980 to 1990, decreased from 1990 to 2002, and then increased after 2002. **c.** The relative minimum is at $x = 22.3$, and the relative maximum is at $x = 10.1$. Passenger traffic was lowest in 2002 and highest in 1990.

TOPIC 15

1. a. $(-1, 1) \cup (1, \infty)$ **b.** $(-\infty, -1)$ **3. a.** $(-2, 2)$ **b.** $(-\infty, -2)$ and $(2, \infty)$ **5.** C, E **7.** true **9.** false **11.** false **13.** true **15. a.** $(0, \infty)$ **b.** $(-\infty, 0)$ **c.** $x = 0$ **17. a.** $(-\infty, 0)$ and $(8, \infty)$ **b.** $(0, 8)$ **c.** $x = 0$ and $x = 8$ **19. a.** $(1, \infty)$ **b.** $(-\infty, 1)$ **c.** $x = 1$ **21. a.** $(-\infty, -3)$ and $(-1, \infty)$ **b.** $(-3, -1)$ **c.** $x = -3, x = -1$ **23. a.** $(-\infty, -1)$ and $(0, 1)$ **b.** $(-1, 0)$ and $(1, \infty)$ **c.** $x = -1, x = 0, x = 1$ **25. a.** $(-\infty, \infty)$ **b.** none **c.** none **27. a.** $(-\infty, -2), \left(-\sqrt{\frac{4}{5}}, \sqrt{\frac{4}{5}}\right)$, $(2, \infty)$ **b.** $\left(-2, -\sqrt{\frac{4}{5}}\right)$ and $\left(\sqrt{\frac{4}{5}}, 2\right)$ **c.** $x = 2, x = \sqrt{\frac{4}{5}}, x = -\sqrt{\frac{4}{5}}, x = -2$ **29. a.** $(5, \infty)$ **b.** $(-\infty, 5)$ **c.** none **31. a.** $(-\infty, -5)$ **b.** $(-5, \infty)$ **c.** none

33. a. $(-3\sqrt{3}, 0)$ and $(3\sqrt{3}, \infty)$ **b.** $(-\infty, -3\sqrt{3})$ and $(0, 3\sqrt{3})$ **c.** $x = 0, x = -3\sqrt{3}, x = 3\sqrt{3}$ **35. a.** none **b.** $[5, \infty)$ **c.** none **37. a.** $(-\infty, -3.4)$ and $(-0.6, \infty)$ **b.** $(-3.4, -0.6)$ **c.** $x \approx -3.4, x \approx -0.6$ **39. a.** $(0, \infty)$ **b.** none **c.** none **41. a.** $x = 0, x = 12$ **b.** $(12, -6912)$ is a minimum **43. a.** $x = 0, x = -2$ **b.** $(0, 0)$ is a minimum, $(-2, 4e^{-2})$ is a maximum **45.** intercepts at $(0, 0)$ and $(16, 0)$; increases for $(12, \infty)$; decreases for $(-\infty, 12)$; absolute minimum at $(12, -6912)$; concave up for $(-\infty, 0)$ and $(8, \infty)$; concave down for $(0, 8)$; points of inflection at $(0, 0)$ and $(8, -4096)$

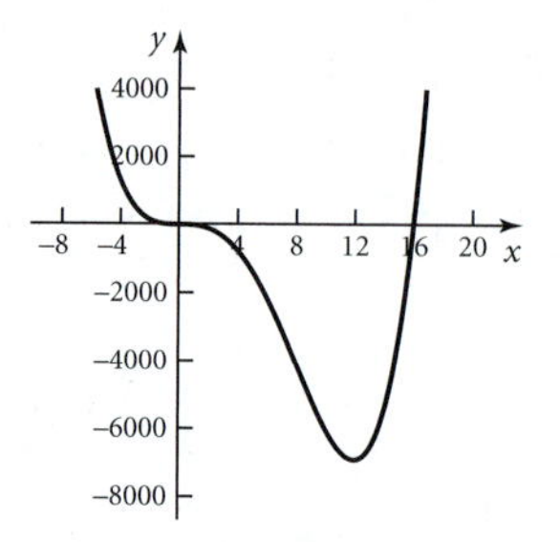

47. intercepts at $(0, 7)$, $(0.66, 0)$, $(-2.28, 0)$, and $(4.62, 0)$; increases for $(-\infty, -1)$ and $(3, \infty)$; decreases for $(-1, 3)$; relative minimum at $(3, -20)$; relative maximum at $(-1, 12)$; concave up for $(1, \infty)$; concave down for $(-\infty, 1)$; point of inflection at $(1, -4)$

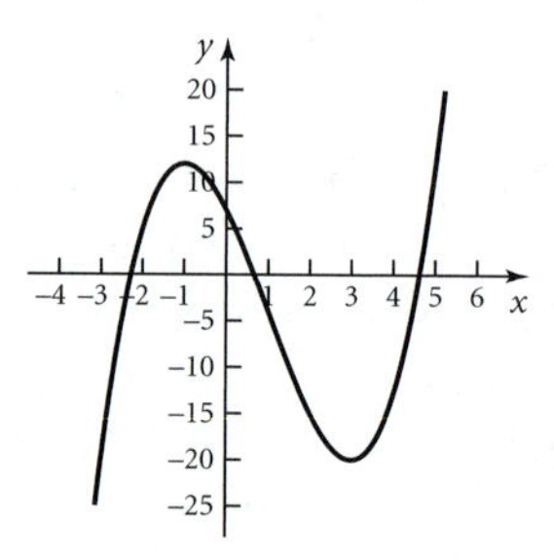

49. intercepts at $(0, 0)$ and $\left(\frac{5}{3}, 0\right)$; increases for $(-\infty, 0)$ and $\left(\frac{4}{3}, \infty\right)$; decreases for $\left(0, \frac{4}{3}\right)$; relative minimum at $\left(\frac{4}{3}, -3.16\right)$; relative maximum at $(0, 0)$; concave up for $(1, \infty)$; concave down for $(-\infty, 1)$; point of inflection at $(1, -2)$

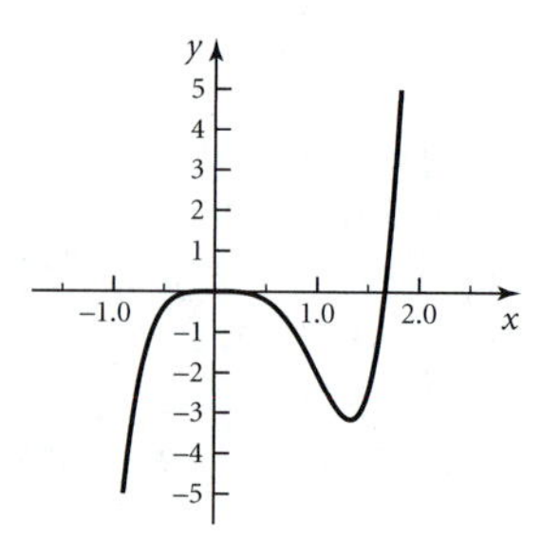

51. intercepts at (0, 1296) and (3, 0); increases for $(3, \infty)$; decreases for $(-\infty, 3)$; concave up for $(-\infty, \infty)$; absolute minimum at (3, 0)

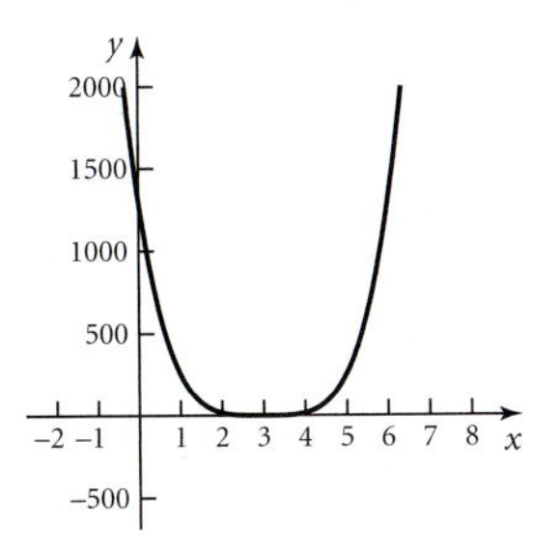

53. intercept at (0, 0); increases for $(-\infty, -3) \cup (-3, 0)$; decreases for (0, 3) and $(3, \infty)$; relative maximum at (0, 0); asymptotes at $x = 3, x = -3$, and $y = 2$; concave up for $(-\infty, -3)$ and $(3, \infty)$; concave down for $(-3, 3)$

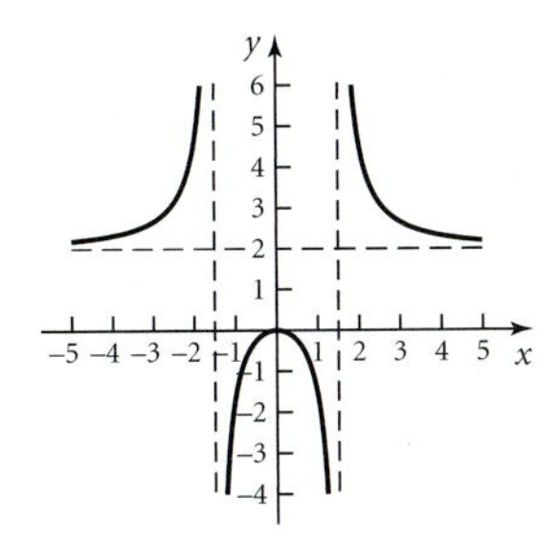

55. intercept at (0, 0); increases for $(-2, 2)$; decreases for $(-\infty -2)$ and $(2, \infty)$; absolute minimum at $\left(-2, -\frac{3}{4}\right)$; absolute maximum at $\left(2, \frac{3}{4}\right)$; asymptote at $y = 0$; concave up for $(-\sqrt{12}, 0)$ and $(\sqrt{12}, \infty)$; concave down for $(-\infty, -\sqrt{12})$ and $(0, \sqrt{12})$; points of inflection at $(-\sqrt{12}, -0.65)$, $(\sqrt{12}, 0.65)$, and (0, 0)

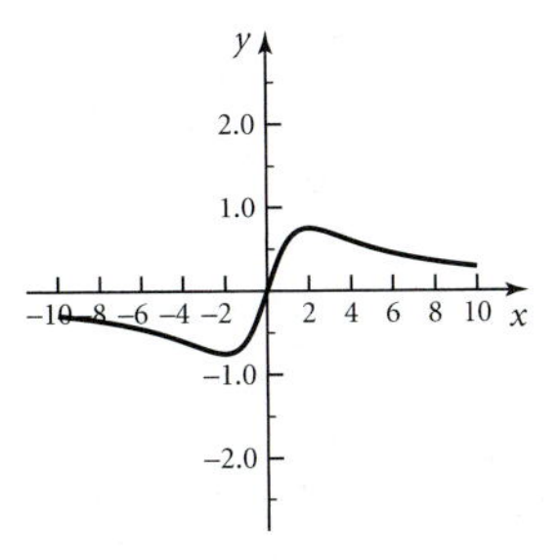

57. intercept at $(0, \ln 2)$; increases for $(0, \infty)$; decreases for $(-\infty, 0)$; absolute minimum at $(0, \ln 2)$; asymptote at $y = 0$; concave up for $(-\sqrt{2}, \sqrt{2})$; concave down for $(-\infty, -\sqrt{2})$ and $(\sqrt{2}, \infty)$; points of inflection at $(-\sqrt{2}, \ln 4)$ and $(\sqrt{2}, \ln 4)$

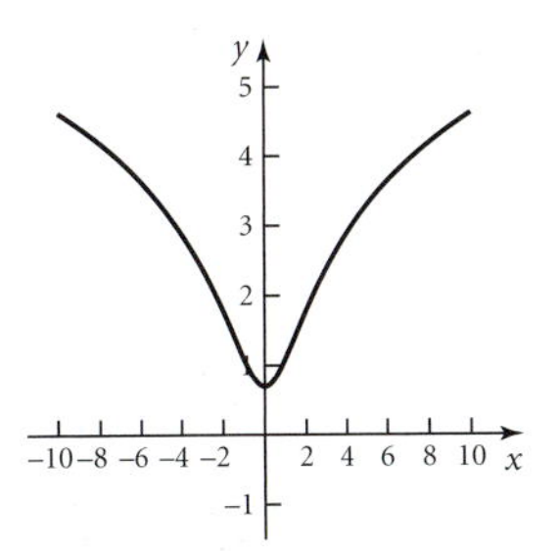

59. intercept at (0, 0); increases for (0, 4); decreases for $(-\infty, 0)$ and $(4, \infty)$; absolute minimum at (0, 0); relative maximum at (4, 2.165); asymptote at $y = 0$; concave up for $(-\infty, 1.17)$ and $(6.83, \infty)$; concave down for (1.17, 6.83); points of inflection at (1.17, 0.76) and (6.83, 1.53)

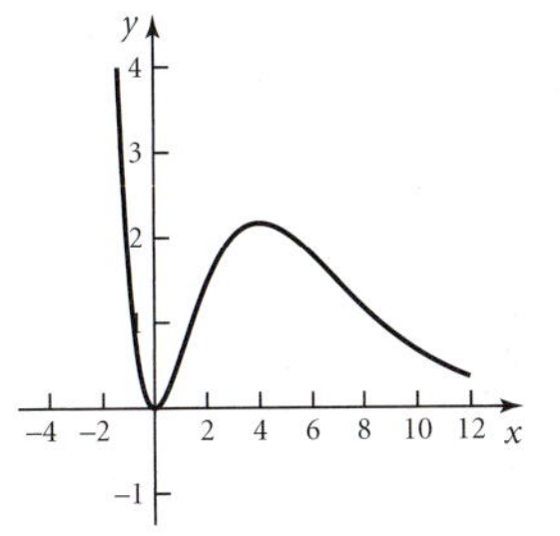

61. **63.**

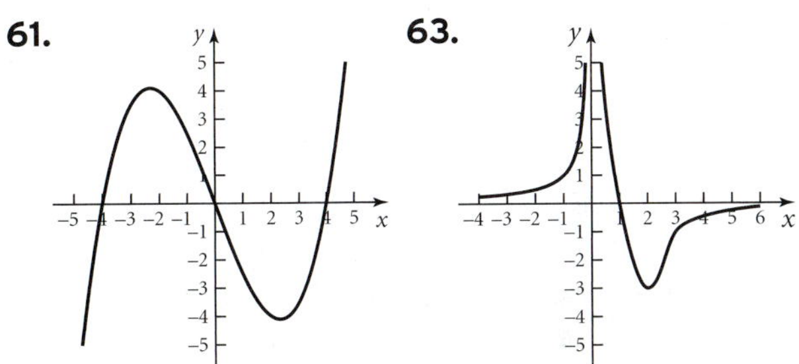

65. approximately the 2nd week **67. a.** (0, 2.2) **b.** $(2.2, \infty)$ **c.** concave up for $(4.4, \infty)$. The rate of decrease of concentration increases after approximately 4.4 hours. **69.** around 1970 **71.** $(3, \infty)$ **73.** (0, 1) **75.** $x = 3$ **77.** intercepts at (0, −23) and (3.79, 0); increases for $(-\infty, 1)$ and $(3, \infty)$; decreases for (1, 3); relative minimum at (3, −50); relative maximum at (1, −22); concave up for (0, 0.63) and $(2.4, \infty)$; concave down for $(-\infty, 0)$ and (0.63, 2.4); points of inflection at (0, −23), (0.63, −22.44), and (2.4, −40.14)

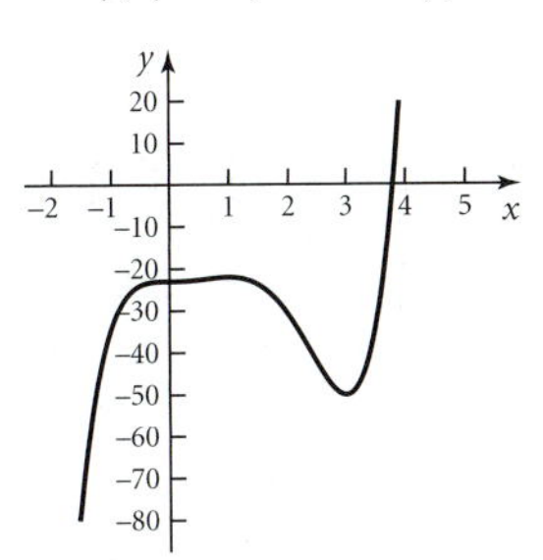

79. intercepts at $(0, -1)$, $(1, 0)$, and $(-1, 0)$; increases for $(0, \infty)$; decreases for $(-\infty, 0)$; concave down for $(-\infty, \infty)$; absolute minimum at $(0, -1)$

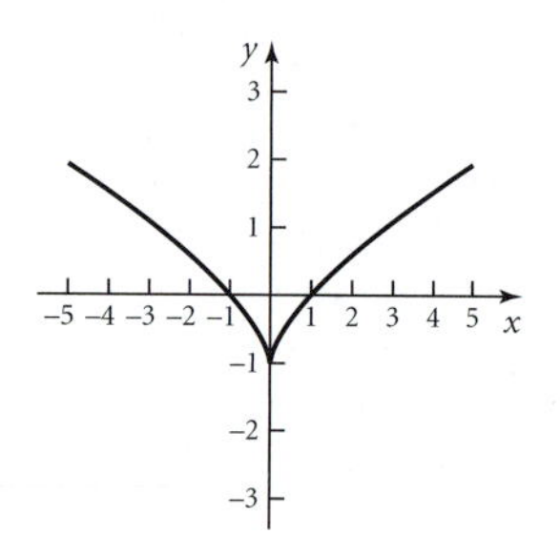

81. intercept at $(0, 0)$; increases for $(0, \infty)$; decreases for $(-\infty, 0)$; absolute minimum at $(0, 0)$; asymptote at $y = 2$; concave up for $(-\sqrt{3}, \sqrt{3})$; concave down for $(-\infty, -\sqrt{3})$ and $(\sqrt{3}, \infty)$; points of inflection at $\left(-\sqrt{3}, \frac{1}{2}\right)$ and $\left(\sqrt{3}, \frac{1}{2}\right)$

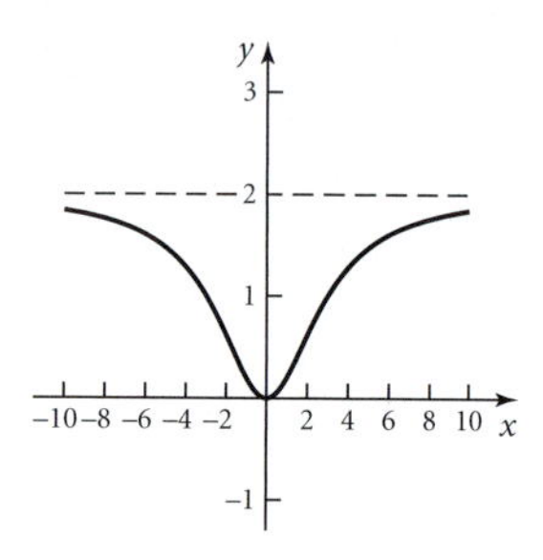

83. intercept at $(1, 0)$; increases for $(0, 2)$; decreases for $(-\infty, 0)$ and $(2, \infty)$; absolute maximum at $\left(2, \frac{1}{4}\right)$; domain $x \neq 0$; concave up for $(3, \infty)$; concave down for $(-\infty, 0)$ and $(0, 3)$; point of inflection at $\left(3, \frac{2}{9}\right)$

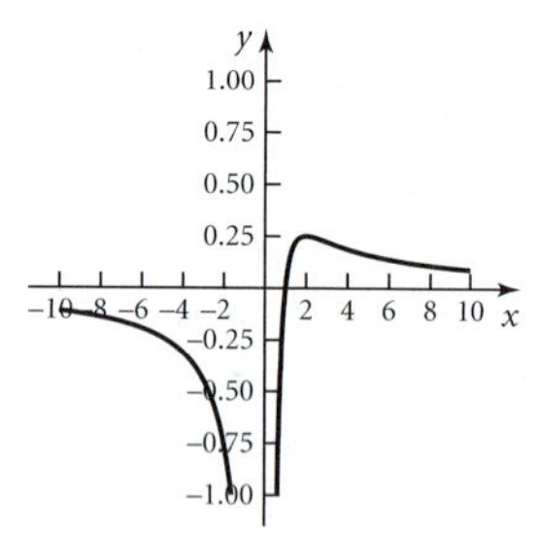

85. intercepts at $(0, 1)$ and $(\ln 2, 0)$; increases for $(-\infty, 0)$ and $(1.59, \infty)$; decreases for $(0, 1)$ and $(1, 1.59)$; relative minimum at $(1.59, 4.92)$; relative maximum at $(0, 1)$; asymptote at $x = 1$; concave up for $(-\infty, -0.78)$ and $(1, \infty)$; concave down for $(-0.78, 1)$; point of inflection at $(-0.78, 0.87)$

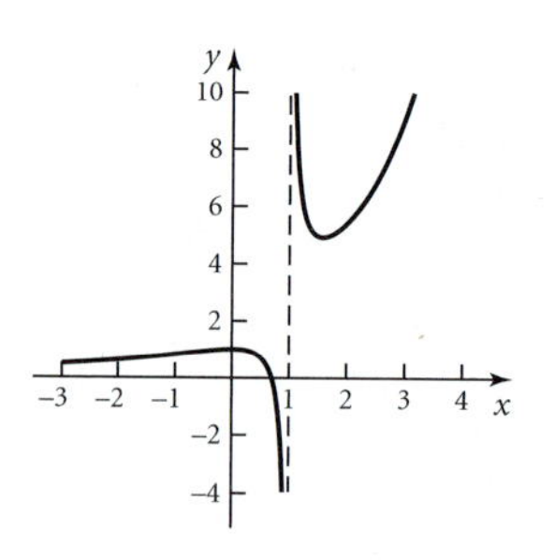

87. The population at any time is $P(t)$. The rate of growth in population is $P'(t)$ and the rate at which the growth rate changes is $P''(t)$. Since DeBary is still growing, we know that $P'(t) > 0$, but since the growth rate started to slow down around 2002, we know that $P''(t) > 0$ before 2002 and $P''(t) < 0$ after 2002.

TOPIC 16

1. true **3.** false **5.** true **7.** false **9.** false **11.** true **13.** Absolute minimum is -32 when $x = 4$. Absolute maximum is 0 when $x = 0$. **15.** Absolute minimum is -32 when $x = -2$. Absolute maximum is 0 when $x = 0$. **17.** Absolute minimum is -121 when $x = 4$. Absolute maximum is 23 when $x = -2$. **19.** Absolute minimum is -121 when $x = 4$. Absolute maximum is -4 when $x = 1$. **21.** Absolute minimum is $-e \approx -2.718$ when $x = -1$. Absolute maximum is $e^{-1} \approx 0.368$ when $x = 1$. **23.** Absolute minimum is -54 when $x = 1$. Absolute maximum is 0 when $x = 0$. **25.** Absolute minimum is 12 when $x = 6$. **27.** Absolute maximum is -6 when $x = -2$. **29.** Absolute minimum fatality rate is 0.24 accident per million vehicle miles traveled when driver age is 46. Absolute maximum fatality rate is 1.22 accidents per million vehicle miles traveled when driver age is 30. **31.** Absolute minimum number of engineering degrees is 49,710 in 1971. Absolute maximum number of engineering degrees is 78,510 in 1991. **33.** Absolute minimum daily high temperature is 47.1°F on November 10. Absolute maximum daily high temperature is 77.4°F on November 1. **35.** 2.5 hours **37.** false **39.** true **41.** true **43.** 1.08 **45.** 3.73 **47.** 0.5; demand is inelastic; revenue increases **49.** 1; demand is unit elastic; revenue stays same **51.** 0.1; demand is inelastic; revenue increases **53.** At \$15, $|E| = 0.66$; demand is inelastic. At \$50, $|E| = 2.86$; demand is elastic. Demand is sensitive to changes in price. **55.** Demand is inelastic; raise the price. **57.** approximately \$59,959.93 if $p = \$41.92$

59. \$51.36 **61.** Absolute minimum number of mathematics degrees was approximately 11,176 in 1981. Absolute maximum number of mathematics degrees was approximately 25,096 in 1971. **63.** Absolute minimum concentration is 0 mg/cm^3 at the beginning ($t = 0$). Absolute maximum concentration is 38.3 mg/cm^3 after three hours. **65.** Absolute minimum value is -1 at $x = 1$. Absolute maximum value is $x = 0$ at $x = 0$.

TOPIC 17

1. Maximum is 36 if $x = y = 6$. **3.** Minimum is 24 if $x = y = 6$. **5.** Maximum if 80 if $x = 4$ and $y = 8$. **7.** Minimum is 40 if $x = 5$ and $y = 2$. **9.** 20 and 20 **11.** 20 and -20 **13.** 8 and 8 **15.** cut in the middle so that each piece is 12.5 feet long **17.** 375 feet by 375 feet **19.** 45.25 feet by 14.14 feet **21.** 15 centimeters by 15 centimeters **23.** 33.94 feet by 84.85 feet **25.** 8 inches by 12 inches **27.** 6 feet wide by 8 feet high for an area of approximately 65.3 square feet **29.** 6 inches by 6 inches for the base, 3 inches high **31.** radius 3.4 cm, height 6.88 cm **33.** radius approximately 1.78 inches, height approximately 3.58 inches **35.** $\left(\sqrt{3/2}, \frac{5}{2}\right)$ **37.** The stake should be placed 9 feet from the 12-foot-high post. **39.** $(-2.3, 1.29)$ and $(2.4, 1.76)$ **41.** The piece that forms the circle should be approximately 44 inches, and the piece that forms the square should be approximately 56 inches. The radius of the circle is approximately 7 inches and each side of the square is approximately 14 inches.

TOPIC 18

1. $p = \$16$ **3.** $p = \$6.08$ **5.** $p = \$41.92$ **7.** $p = \$51.36$ **9.** 125 people at \$250 per person **11.** 25 trees per acre and 75 bushels per tree for a total yield of 1875 bushels **13.** \$7 **15.** \$8 **17.** \$22.50, 70 customers **19.** approximately 4 runs per month of approximately 500 bikes each **21.** approximately 2 runs of 3000 books each **23.** approximately 14 employees and 143 machines **25.** approximately 5 employees and 36 lab stations **27.** $p = \$187.50$, 112 or 113 rooms **29.** 40 trees **31.** approximately 275 vines

TOPIC 19

1. false **3.** true **5.** true **7.** false **9.** true **11.** $dy/dx = 2x/3$ **13.** $dy/dx = 8x/9y^2$ **15.** $\dfrac{dy}{dx} = \dfrac{3}{4y + 1}$ **17.** $dy/dx = -3y/2x$ **19.** $\dfrac{dy}{dx} = \dfrac{2x - 3x^2}{2y}$ **20.** $\dfrac{dy}{dx} = \dfrac{2x - 3x^2}{3y^2}$ **21.** $\dfrac{dy}{dx} = \dfrac{8x^3 - 6xy}{3x^2 - 3y^2}$ **23.** -1 **25.** $y = -x - 4$ **27.** $y = \frac{1}{2}x - 2$ **29.** $dy/dx = -2.3$. Current output level is maintained if the number of full-time work hours increases by 1 hour per week and the number of part-time work hours decreases by approximately 2.3 hours per week. **31.** $dr/dt = -0.1$. The radius of the wound decreases at a rate of 0.1 centimeter per day. **33.** $dy/dt = 0.1e^{0.6} \approx 0.1822$. The y-coordinate increases approximately 0.1822 unit per day. **35.** approximately 78.1 mph **37.** approximately 0.0995 cm per minute **39.** approximately -1.094 feet per second **41.** $8/25\pi \approx 0.1$ feet per minute **43.** $dN/dt = -0.166$. Sales are dropping approximately 166,000 homes per year. **45.** $dx/dt = 1.5$. Supply increases 1500 per week. **47. a.** Horizontal tangent lines occur if $x = 0$ or $x = \sqrt[3]{16} \approx 2.520$. The points of tangency are $(0, 0)$ and $(2.5198, 3.1748)$. **b.** $(3, 3)$

c.

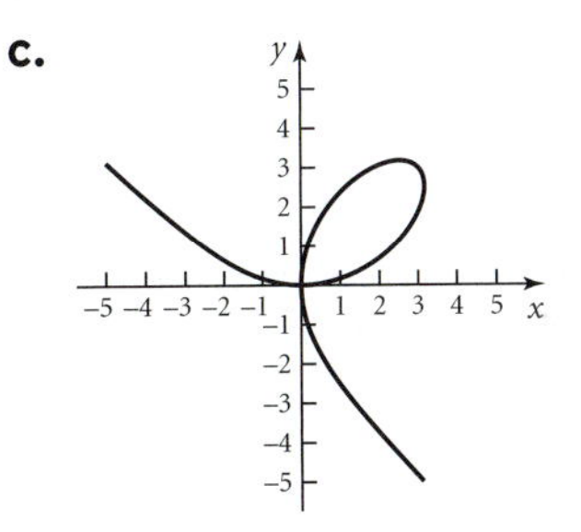

49. $\dfrac{dy}{dx} = \dfrac{-3x^2y + 3x^2 + y^3 - 1}{x^3 - 3xy^2 - 3y^2}$ **51.** $y = x/4 + 3/4$

UNIT 2 TEST

1. a. $(0, \infty)$ **b.** $(-\infty, 0)$ **c.** absolute minimum at $x = 0$ **d.** none **e.** $(-\infty, \infty)$ **f.** none **g.** none **2. a.** $(-\infty, -1) \cup (0, 1)$ **b.** $(-1, 0) \cup (1, \infty)$ **c.** relative minimum at $x = 0$ **d.** absolute maximum at $x = -1, x = 1$ **e.** $(-0.5, 0.5)$ **f.** $(-\infty, -0.5) \cup (0.5, \infty)$ **g.** $x = -0.5$ and $x = 0.5$ **3. a.** $(-1.5, 0) \cup (1.5, \infty)$ **b.** $(-\infty, -1.5) \cup (0, 1.5)$ **c.** relative minimum at $x = 1.5$; absolute minimum at $x = -1.5$ **d.** relative maximum at $x = 0$ **e.** $(-\infty, -1) \cup (1, \infty)$ **f.** $(-1, 1)$ **g.** $x = -1$ and $x = 1$ **4. a.** $(-\infty, 0) \cup (0, \infty)$ **b.** none **c.** none **d.** none **e.** $(-\infty, 0)$ **f.** $(0, \infty)$

g. none **5. a.** $(-\infty, -1) \cup (1, \infty)$ **b.** $(-1, 1)$ **c.** absolute minimum at $x = 1$ **d.** absolute maximum at $x = -1$ **e.** $(-\infty, -2) \cup (0, 2)$ **f.** $(-2, 0) \cup (2, \infty)$ **g.** $x = -2, x = 0$, and $x = 1$ **6.** domain $(-\infty, \infty)$; intercepts $(0, 0)$ and $(6, 0)$ **7.** domain $[3, \infty)$; intercepts $(3, 0)$ **8.** domain $x \neq -2$; intercepts $(0, 0)$; asymptotes $x = -2$ and $y = 5$ **9.** domain $x \neq 0, -2, 2$; asymptotes $x = 2, x = -2$, and $y = 0$; hole at $(0, 0)$ **10.** domain $(-\infty, \infty)$; intercepts $(0, 1)$; asymptote $y = 0$ **11.** increases for $(4, \infty)$; decreases for $(-\infty, 4)$; absolute minimum $x = 4$ **12.** increases for $(0, 8)$; decreases for $(-\infty, 0) \cup (8, \infty)$; relative minimum $x = 0$; relative maximum $x = 8$ **13.** increases for $(-1, 0) \cup (4, \infty)$; decreases for $(-\infty, -1) \cup (0, 4)$; relative minimum $x = -1$; absolute minimum $x = 4$; relative maximum $x = 0$ **14.** increases for $(-2, 1) \cup (4, \infty)$; decreases for $(-\infty, -2) \cup (1, 4)$; absolute minimum $x = -2$ and $x = 4$; relative maximum $x = 1$ **15.** increases for $(-\infty, -1) \cup (1, \infty)$; decreases for $(-1, 1)$; relative minimum $x = 1$; relative maximum $x = -1$ **16.** increases for $(-\infty, 1)$; decreases for $(1, \infty)$; absolute maximum $x = 1$ **17.** decreases for $(-\infty, -2) \cup (-2, 2) \cup (2, \infty)$; no extreme points **18.** increases for $(e^{0.5}, \infty)$; decreases for $(0, 1)(1, e^{0.5})$; relative maximum: $x = e^{0.5} \approx 1.65$ **19.** The percent decreased from 1990 to 2001. The minimum percent was approximately 11.8% in 2001. **20.** At $t = 0$, with 83 accidents. **21.** after 2 seconds; maximum height 70 feet **22.** Increases from 0 to 1.7 hours; maximum concentration is 1.47 mg/cm^3 after 1.7 hours. **23.** concave up for $(-\infty, \infty)$; no points of inflection **24.** concave up for $(-\infty, 4)$; concave down for $(4, \infty)$; points of inflection at $x = 4$ **25.** concave up for $(-\infty, -1)$ and $(1, \infty)$; concave down for $(-1, 1)$; points of inflection at $x = -1$ and $x = 1$ **26.** concave up for $(0, \infty)$; concave down for $(-\infty, 0)$; points of inflection at $x = 0$ **27.** concave up for $(2, \infty)$; concave down for $(-\infty, 2)$; points of inflection at $x = 2$ **28.** concave up for $(-\infty, -0.73)$ and $(2.73, \infty)$; concave down for $(-0.73, 2.73)$ points of inflection at $x = -0.73$ and $x = 2.73$ **29.** concave up for $(1, \infty)$; concave down for $(-\infty, 1)$; points of inflection at $x = 1$ **30.** concave up for $(3, \infty)$ concave down for $(-\infty, 3)$; no points of inflection **31.** concave up for $(-\infty, -1)$; concave down for $(-1, \infty)$; no points of inflection **32.** concave up for $(-\infty, -2)$ and $(2, \infty)$; concave down for $(-2, 2)$; no points of inflection **33.** intercept at $(0, 0)$ and at $(12, 0)$; increases for $(0, 8)$; decreases for $(-\infty, 0)$ and $(8, \infty)$; relative minimum at $(0, 0)$; relative maximum at $(8, 256)$; concave up for $(-\infty, 4)$; concave down for $(4, \infty)$; points of inflection at $(4, 128)$

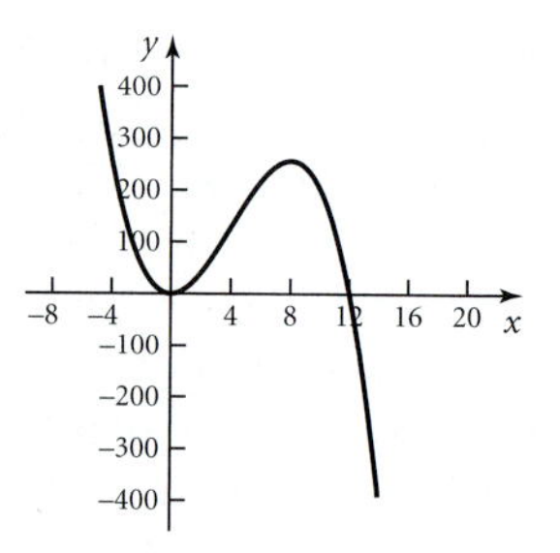

34. intercepts at $(0, -5)$, $(-2.6, 0)$, $(2.6, 0)$; increases for $(-\sqrt{3}, 0)$ and $(\sqrt{3}, \infty)$; decreases for $(-\infty, \sqrt{3})$ and $(0, \sqrt{3})$; absolute minimum at $(\sqrt{3}, -14)$ and $(-\sqrt{3}, -14)$; relative maximum at $(0, -5)$; concave up for $(-\infty, -1)$ and $(1, \infty)$; concave down for $(-1, 1)$; points of inflection at $(1, -10)$ and $(-1, -10)$

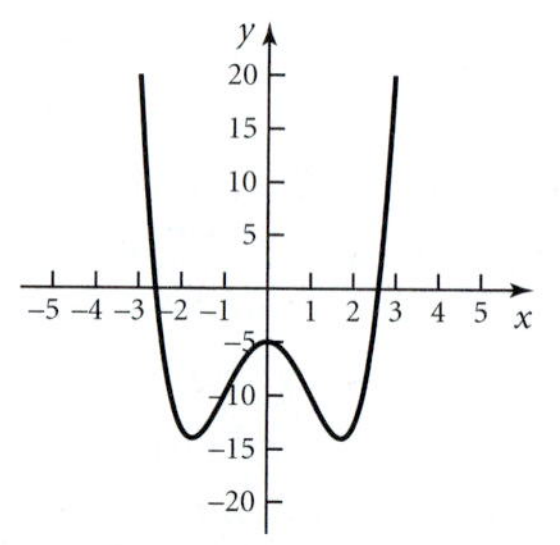

35. intercepts at $(0, -1)$ and $(1, 0)$; increases for $(-\infty, \infty)$; no extreme points; concave up for $(1, \infty)$; concave down for $(-\infty, 1)$; points of inflection at $(1, 0)$

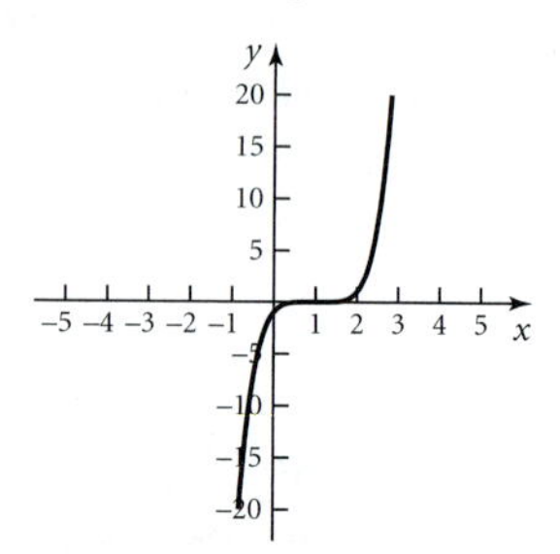

36. intercepts at $(0, 0)$, $(3.6, 0)$, and $(-3.6, 0)$; increases for $(-\infty, -1)$ and $(1, \infty)$; decreases for $(-1, 1)$; relative minimum at $(-1, 4)$; relative maximum at $(1, -4)$; concave up for $(0, \infty)$; concave down for $(-\infty, 0)$; points of inflection at $(0, 0)$

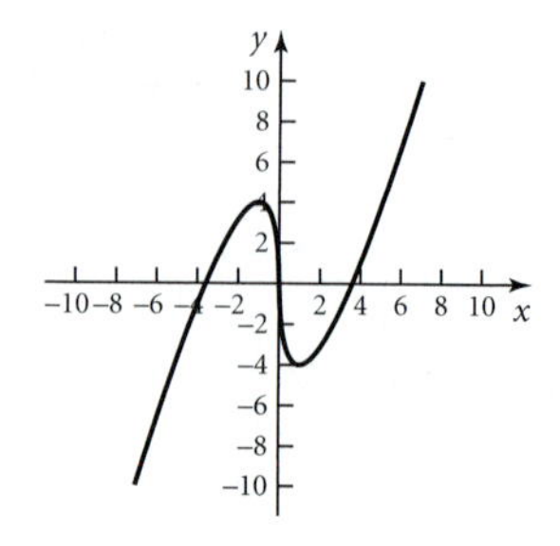

37. intercept at $(0, 0)$; increases for $(-\infty, 1)$; decreases for $(1, \infty)$; absolute maximum at $(1, e^{-1}) \approx (1, 0.37)$;

concave up for $(2, \infty)$; concave down for $(-\infty, 2)$; points of inflection at $(2, 2e^{-2}) \approx (2, 0.27)$

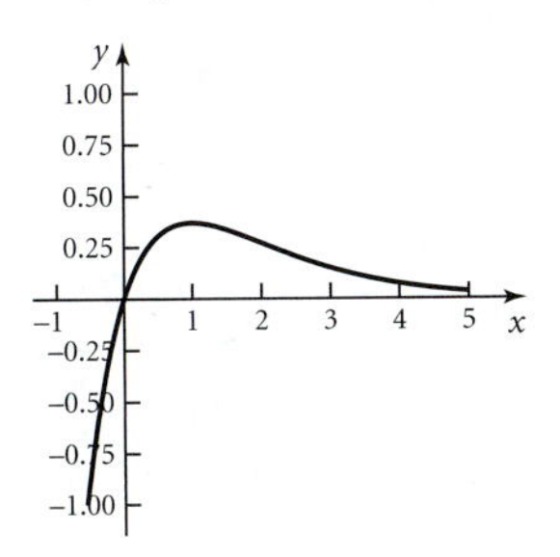

38. intercept at $(0, -\frac{4}{3})$; asymptotes at $x = 3$ and $y = 0$; decreases for $(-\infty, 3)$ and $(3, \infty)$; no extreme points; concave up for $(3, \infty)$; concave down for $(-\infty, 3)$

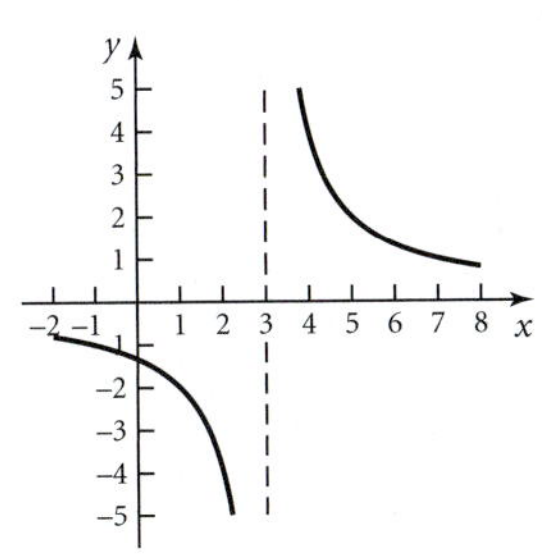

39. intercept at $(0, 0)$; asymptote at $y = 0$; increases for $(-1, 1)$; decreases for $(-\infty, -1)$ and $(1, \infty)$; absolute minimum at $(-1, -2)$; absolute maximum at $(1, 2)$; concave up for $(-\sqrt{3}, 0)$ and $(\sqrt{3}, \infty)$; concave down for $(-\infty, -\sqrt{3})$ and $(0, \sqrt{3})$; points of inflection at $(0, 0)$, $(\sqrt{3}, \sqrt{3})$, and $(-\sqrt{3}, -\sqrt{3})$

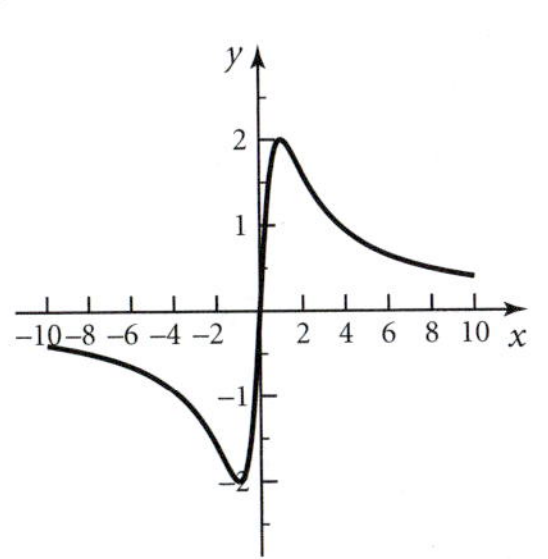

40. intercept at $(0, -\frac{1}{4})$; asymptotes at $x = 2$, $x = -2$, and $y = 0$; increases for $(-\infty, -2)$ and $(-2, 0)$; decreases for $(0, 2)$ and $(2, \infty)$; relative maximum at $\left(0, -\frac{1}{4}\right)$; concave up for $(-\infty, -2)$ and $(2, \infty)$; concave down for $(-2, 2)$

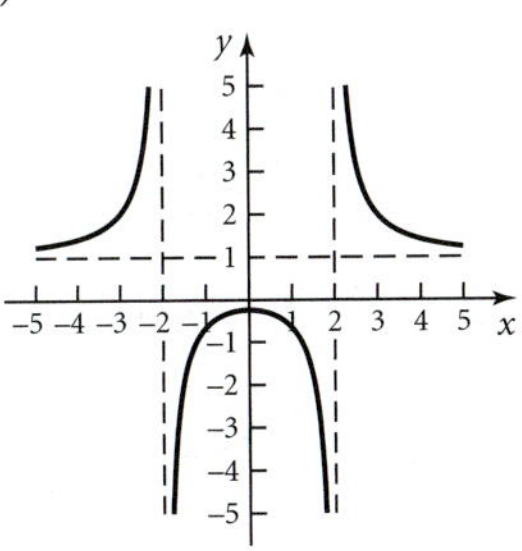

41. after approximately 4 months **42.** After approximately 7 months, the rate of increase in sales begins to taper off. Sales still grow, but at a slower rate. **43.** Absolute minimum is -32 when $x = 4$. Absolute maximum is -5 when $x = 1$. **44.** Absolute minimum is -83 when $x = 2$. Absolute maximum is 302 when $x = -3$. **45.** Absolute minimum is $-2e^6 \approx -806.86$ when $x = -2$. Absolute maximum is $\frac{1}{3}e^{-1} \approx 0.123$ when $x = \frac{1}{3}$. **46.** Absolute minimum is 5.44 when $x = e^{0.5}$. Absolute maximum is 11.54 when $x = 4$. **47.** Absolute minimum is 7 when $x = 2$. There is no absolute maximum. **48.** Absolute minimum is 12 when $x = 6$. There is no absolute maximum. **49.** Between 1950 and 1990, the maximum percent of the U.S. population that is foreign born was 7.92% in 1990. The minimum percent of the U.S. population that is foreign born was 5.316% in 1970. **50.** Between 1995 and 2002, the minimum number of unemployment claims filed was approximately 870 claims in 1998. The maximum number of claims filed was approximately 1250 claims in 2002. **51.** $E = 0.45$; inelastic. Demand is not affected by changes in price. **52.** $E = 3.47$; elastic. Demand is sensitive to changes in price. **53.** $E(14) = 1.2$; elastic. At this price, the demand is affected by price changes. $E(27) = 0.24$; inelastic. At this price, demand is not affected by price changes. **54.** Maximum revenue is $7,746 if the price is $900. Demand at that price is 2499 televisions. **55.** $x = y = 15$ **56.** $x = 10, y = 4$ **57.** 15 and 15 **58.** $\sqrt{30}$ and $\sqrt{30}$ **59. a.** 35 feet by 35 feet **b.** 35 feet by 70 feet **60.** The side with the $10 fence is approximately 45.3 feet; the other two sides are each approximately 28.3 feet. **61.** The box should be 8 inches by 8 inches by 4 inches. **62.** Radius is approximately 1.99 inches, and height is approximately 3.989 inches. **63.** $\left(\sqrt{\frac{7}{2}}, \frac{5}{2}\right)$ and $\left(-\sqrt{\frac{7}{2}}, \frac{5}{2}\right)$ **64.** 58 people at a cost of $174 per person **65.** 42 trees **66. a.** $12.50 **b.** $14.50 **67.** three production runs with 4000 candles per run **68.** About 461 employees and 2993 machines

69. $dy/dx = 3x^2/2$ **70.** $\dfrac{dy}{dx} = \dfrac{8x^3}{9y^2 + 1}$

71. $\dfrac{dy}{dx} = \dfrac{8 - 9xy^2}{6x^2y}$ **72.** $\dfrac{dy}{dx} = \dfrac{x^2 - y}{y^2 + x}$ **73.** $\sqrt{2}/2$

74. $y = -\dfrac{\sqrt{7}}{3}x + \dfrac{16}{3}$ **75.** $dy/dx = -1.385$. To maintain the current output level, if the number of hours worked by full-time employees increases by 1 hour per week, the number of hours worked by part-time employees decreases by approximately 1.385 hours per week. **76.** approximately 923 feet per minute **77.** approximately 1.34 yards per day **78.** approximately 64 mph

Unit 3

TOPIC 20

1. True **3.** False **5.** $4x + C$ **7.** $\frac{b^9}{9} + C$

9. $\frac{x^{12}}{4} + C$ **11.** $\frac{1}{3}x^3 - \frac{3}{2}x^2 + 5x + C$

13. $\frac{1}{3}x^6 + 3x^{-1} + C$ **15.** $-\frac{1}{x^2} + C$ **17.** $-\frac{3}{2x} + C$

19. $4\sqrt{x^3} + C$ **21.** $2x^4 + 2x^2 - \frac{5}{x} + C$

23. $\frac{4}{7}x^{7/4} - x^3 + C$ **25.** $40x^{3/5} + 15x^{7/5} + C$ or $40\sqrt[5]{x^3} + 15\sqrt[5]{x^7} + C$ **27.** $\frac{1}{3}x^3 + 3x^2 + 9x + C$

29. $\frac{1}{2}x^2 - 2x - \frac{3}{x} + C$ **31.** $f(x) = \frac{3}{2}x^2 - 5x - 7$

33. $f(x) = \frac{1}{8}x^4 - x^2 + 3x + \frac{7}{8}$ **35.** $y = 7x - x^4 - 3$

37. $y = 2\sqrt{x^3} - x^2 - 5$ **39.** 7880 sets **41.** 6.4%

43. $4\frac{2}{3}\text{ cm}^2$ **45.** \$20,825 **47.** \$80,250 **49.** 284 feet **51. a.** 120 feet **b.** 26 feet per second **c.** maximum height is about 130.6 feet after about 2.8 seconds. **d.** about 5.669 seconds **53.** $\frac{1}{7}x^7 - \frac{6}{5}x^5 + 4x^3 - 8x + C$

55. $\frac{1}{8}x^8 + \frac{1}{3}x^6 + \frac{1}{4}x^4 + C$

57. $\frac{1}{3}x^3 + 2x^2 + 16x + C$ **59.** about \$5776.85 **61.** Answers vary. **63. a.** $R'(t) = 0.0076t^3 - 0.1644t^2 + 0.8376t + 0.7768$ **b.** $R'(4) = 1.9832$, which is close to the actual value of 2; $R'(8) = 0.8472$, which is higher than the actual value of 0.6. **c.** $R(t) = 0.0019t^4 - 0.0548t^3 + 0.4188t^2 + 0.7768t + 0.1573$ **d.** $R(4) \approx \$6.9445$ million; $R(8) \approx \$12.8997$ million

TOPIC 21

1. true **3.** false **5.** false **7.** false **9.** $\frac{1}{8}e^{8x} + C$

11. $-\frac{3}{4}e^{-4x} + C$ **13.** $40e^{0.3t} + C$ **15.** $4\ln x + C$

17. $2\ln a + \frac{5}{2a^2} + C$ **19.** $\frac{1}{2}e^{2x} - 6e^x + 9x + C$

21. $-\frac{1}{3}e^{-3x} - \frac{1}{4}x^4 + 3\ln x + C$ **23.** $\frac{1}{2}x^2 - 3x + 5\ln x + C$ **25.** $\frac{1}{2}x^2 - 8x + 16\ln x + C$

27. $\frac{7^x}{\ln 7} + C$ **29.** $\frac{3}{\ln 2}\cdot 2^x - \frac{1}{3}x^3 + C$

31. $\frac{6^x}{\ln 6} - \frac{1}{6}e^{6x} + 6\ln x + \frac{x^2}{12} + C$ **33.** $f(x) = \frac{1}{3}e^{3x} + 4x + \frac{5}{3}$ **35.** $f(x) = -2\ln x + \frac{1}{2}x^2 + \frac{5}{2}$

37. $f(x) = x^3 - \frac{5^x}{\ln 5} + x - 4 + \frac{5}{\ln 5}$ or $f(x) = x^3 - \frac{5^x}{\ln 5} + x - 0.893$ **39.** $y = 3\ln x - 4x + 6$

41. $y = -30e^{-0.2x} - \frac{1}{2}x^2 + 3x + 31$ **43.** $N(16) \approx 1015.143$ or about 1,015,143 subscriptions **45.** $M(3) \approx 10{,}968$ mosquitoes **47.** about 172.2 **49.** about 16,127 students **51.** about 2908 computers **53.** The rule in this topic is for e^{kx}. In this integral the power is x^2. **55. a.** $R'(t) = 0.01724(1.808)^t = 0.01724e^{0.592t}$ where t is the number of years after 1990. **b.** about \$378.03 million in 2006 **c.** about \$4038.4 million in 2010

TOPIC 22

1. false **3.** true **5.** false **7.** $dy = (4x^3 - 6x)dx$

9. $du = -6e^{-3x}dx$ **11.** $2.5e^{0.4x} + C$ **13. a.** $\frac{1}{3}x^3 - 3x^2 + 9x + C$ **b.** $\frac{1}{3}(x-3)^3 + C = \frac{1}{3}(x^3 - 9x^2 + 27x - 27) + C = \frac{1}{3}x^3 - 3x^2 + 9x - 9 + C_1 = \frac{1}{3}x^3 - 3x^2 + 9x + C_2$ **15. a.** $\frac{25}{3}x^3 - 20x^2 + 16x + C$

b. $\frac{1}{15}(5x-4)^3 + C = \frac{1}{15}(125x^3 - 300x^2 + 240x - 64) + C = \frac{25}{3}x^3 - 20x^2 + 16x - \frac{64}{15} + C_1 = \frac{25}{3}x^3 - 20x^2 + 16x + C_2$ **17. a.** $\frac{1}{6}x^6 - \frac{3}{2}x^4 + \frac{9}{2}x^2 + C$ **b.** $\frac{1}{6}(x^2-3)^3 + C = \frac{1}{6}(x^6 - 9x^4 + 27x^2 - 27) + C = \frac{1}{6}x^6 - \frac{3}{2}x^4 + \frac{9}{2}x^2 - \frac{9}{2} + C_1 = \frac{1}{6}x^6 - \frac{3}{2}x^4 + \frac{9}{2}x^2 + C_2$

19. $\frac{1}{24}(3x-5)^8 + C$ **21.** $\ln|x+5| + C$

23. $\frac{1}{3}\ln|b^3 - 4| + C$ **25.** $-\frac{5}{6(x^3-4)^2} + C$

27. $\frac{1}{3}\sqrt{(a^2-3)^3} + C$ **29.** $\frac{3}{2}e^{x^2+1} + C$

31. $4\sqrt{p^2+1} + C$ **33.** $3e^{x/3} + C$

35. $\ln|e^x - 3| + C$ **37.** $-\frac{1}{3}e^{-t^3} + C$

39. $\frac{1}{2}(\ln(5v))^2 + C$ **41.** $\frac{1}{28}(x^4+2)^7 + C$

43. $\frac{1}{2}\ln(\ln(x^2)) + C$ **45.** $-\frac{1}{x^3 - 4x + 1} + C$

47. $\frac{1}{2}x^6 - \frac{1}{3(3x-2)^4} + C$ **49.** $-\frac{1}{e^x - 3x} + C$

51. $f(x) = \frac{1}{2}(x^2-3)^5 + 1$ **53.** $f(x) = -\frac{1}{2}e^{-x^2} + \frac{3}{2}$

55. $y = 2\ln|x^2 - 3| - 1$ **57.** $y = \frac{1}{3}\sqrt{(x^2+1)^3} - \frac{1}{2}x^2 + \frac{2}{3}$ **59.** about 76.1 feet **61.** about 9625 grills

63. about \$365.03 million dollars **65.** about 2197 tons **67.** $\frac{1}{6}(x+1)^6 - \frac{2}{5}(x+1)^5 + \frac{1}{4}(x+1)^4 + C$

69. $x + 1 - \ln|x+1| + C_1$ or $x - \ln|x+1| + C_2$
71. Answers vary. **73.** If $u = x^2 - 3$, then $dx = \dfrac{1}{2x}du$, which is not a function of u.

75. a. Tracing gives $P'(2) \approx 1.33$. After two years, the amount of pollution being dumped into the river is increasing at a rate of about 1333 tons per year.

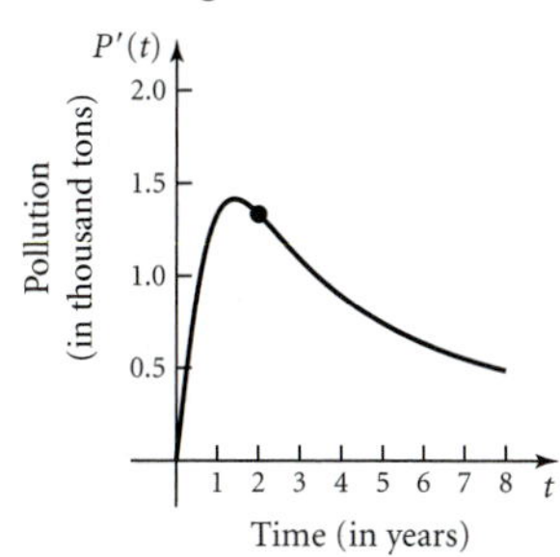

b. $P'(2) \approx 1.33$

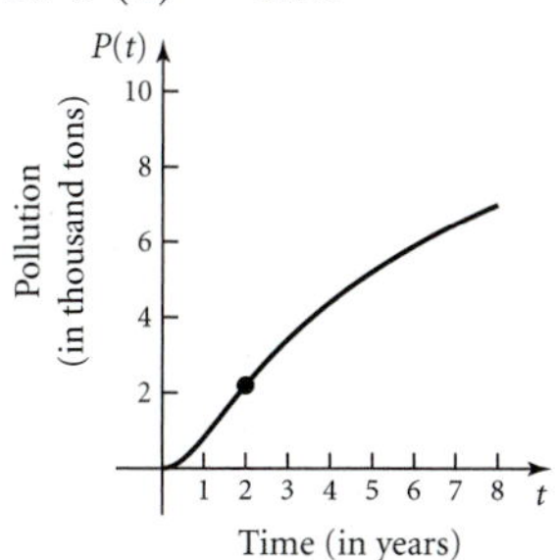

c. Yes, because the graph in part a is the derivative of the graph in part b.

TOPIC 23

1. 30 **3.** 16 **5.** 162 **7.** 3.5 **9.** $\frac{4}{3} - 5\ln 3 \approx -4.160$ **11.** $-\frac{7}{64} \approx -0.1094$ **13.** 90
15. $-2(e^{-4} - 1) \approx 1.963$ **17.** 21 **19.** 57.1
21. $-1 + \frac{3}{2}\ln 3 \approx 0.65$ **23.** $-\frac{1}{8} = -0.125$
25. $8\frac{2}{3} \approx 8.667$ **27.** 3.025 **29.** 0.429
31. 7.460 **33.** $2\frac{1}{3}$ **35.** $\frac{1}{4}(e^4 - 1) \approx 13.4$
37. $\frac{4}{3}\ln\frac{5}{2} \approx 1.22$ **39.** $\frac{2}{5} = 0.4$ **41.** 16.25
43. 7.91 **45.** $-\frac{1}{2}(e^{-4.5} - 1) \approx 0.4944$
47. \$81.72 billion **49.** 476,000 filings **51.** about 416 copies **53.** about 1804 computers **55.** about \$179.7 million **57.** about 104 cars **59.** about \$7000 **61.** about \$31,667 **63.** \$2628.44
65. \$2629.27 **67.** \$34.23 **69.** 6.02 mg/cm^3
71. Answers vary. **73.** Incorrect substitution in step 2. Correct answer is

$$\int_1^3 (3x^2 - 2x)dx = \left(x^3 - x^2\right)_1^3$$
$$= (3^3 - 3^2) - (1^3 - 1^2) = 18$$

75. 34.425 **77.** Answers vary. **79.** Answers vary.
81. a. $R'(t) = 0.0076t^3 - 0.1644t^2 + 0.8376t + 0.7768$ **b.** about \$6.79 million **83. a.** $T(t) = -0.776t^2 + 20.6t - 56.9$ **b.** 70.5°F **c.** 75.4°F
85. 0.8556 **87.** 5.364 **89.** 8.61

TOPIC 24

1. $\int_{-1}^{3}(2x+5)dx = 28$ **3.** $\int_1^3 x^2dx = 8\frac{2}{3}$

5. $\int_{-1}^{1}(2 - x^2)dx = 3\frac{1}{3}$ **7.** $\int_0^2 x^3dx = 4$

9. $-\int_2^4 (2 - x^2)dx = 10\frac{2}{3}$ **11.** $-\int_1^2 (x^3 - 4x)dx + \int_2^3 (x^3 - 4x)dx = 2.25 + 6.25 = 8.5$

13. $\int_0^2 (4 - x^2)dx - \int_2^3 (4 - x^2)dx = 5\frac{1}{3} - (-2\frac{1}{3}) = 7\frac{2}{3}$ **15.** $\int_4^6 (6x - x^2)dx - \int_6^7 (6x - x^2)dx = 9\frac{1}{3} - (-3\frac{1}{3}) = 12\frac{2}{3}$

17. $\int_0^3 e^x dx = e^3 - 1 \approx 19.086$

19. $-\int_{-3}^{-1}\frac{3}{x}dx = -3\ln|x|_{-3}^{-1} = -3(\ln 1 - \ln 3) = 3\ln 3 \approx 3.30$ **21.** $\int_0^4\left(\sqrt{x} + 1\right)dx = 9\frac{1}{3}$

23. $\int_{-1}^{2} e^{-0.5x}dx = -2(e^{-1} - e^{0.5}) \approx 2.562$

25. $\int_0^4 2xe^{-0.5x^2}dx = -2(e^{-8} - 1) \approx 1.9993$

27. $\int_2^5 \frac{2x}{x^2+1}dx = \ln 26 - \ln 5 \approx 1.649$

29. $-\int_0^1 x(x^2-2)^3\,dx = 1\frac{7}{8} = 1.875$

31. $\int_1^3 4x\sqrt{x^2-1}\,dx = \frac{4}{3}(8^{3/2}) \approx 30.17$

33. $\int_{-1}^2 (x^2-(2x-3))dx = 9$

35. $\int_0^3 \left(\sqrt{x}-(-2x)\right)dx \approx 12.464$

37. $\int_{-2}^0 (3-x-x^3)dx = 12$

39. $\int_1^2 \left(e^x - \frac{1}{x}\right)dx = e^2 - \ln 2 - e \approx 3.98$

41. $\int_{-1}^2 (e^{0.5x}-x^2)dx = 2e - 2e^{-0.5} - 3 \approx 1.2235$

43. $\int_0^1 (x^2-x^3)dx = \frac{1}{12} \approx 0.083$

45. $\int_0^2 (2x-x^2)dx = 1\frac{1}{3}$

47. $\int_{-2}^0 (x^3-4x)dx + \int_0^2 (4x-x^3)dx = 4+4=8$

49. $\int_{-2}^1 ((4-x^2)-(x+2))dx = 4.5$

51. $\int_{-\sqrt{2}}^{\sqrt{2}} ((2-x^2)-(x^2-2))dx \approx 7.542$

53. $10\frac{2}{3}$ **55.** ≈ 10.49 **57.** $5+\frac{\pi}{2} \approx 6.571$

59. 9 **61.** Answers will vary but approach 2.54518.

63. Answers will vary but approach 2.051912.

65. $\int_2^6 (20.93t^{0.668} - 22.79t^{0.334})dt \approx 65.9$ billion dollars

67. $\int_4^8 \left(\frac{400}{x} - \frac{450}{x^2}\right)dx = 400\ln 2 - \frac{450}{8} \approx 221$ or \$221,000

69.

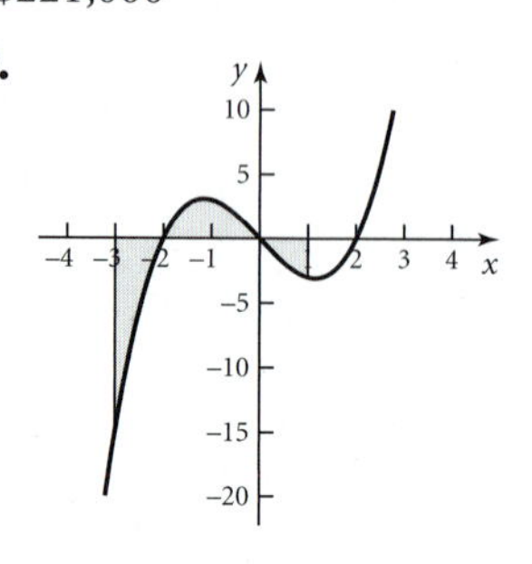

$$A = -\int_{-2}^{-3} (x^3-4x)dx + \int_{-2}^{0} (x^3-4x)dx - \int_0^1 (x^3-4x)dx = 12$$

71.

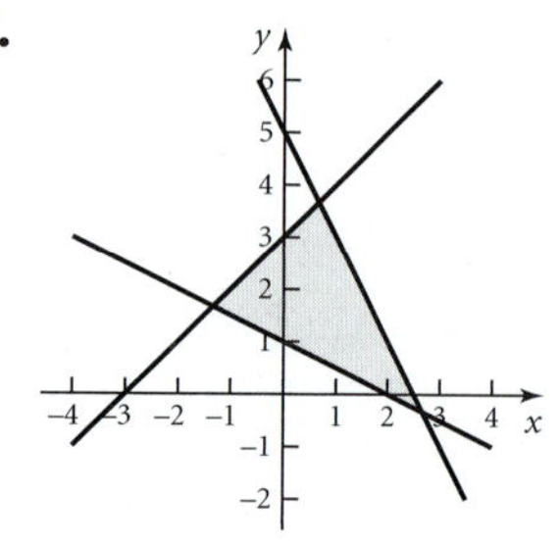

$$A = \int_{-4/3}^{2/3} ((x+3)-(1-0.5x))dx + \int_{2/3}^{8/3} ((5-2x)-(1-0.5x))dx = 6$$

73.

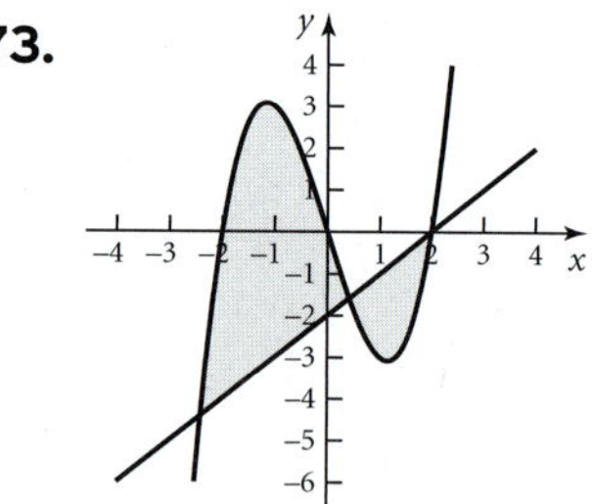

$$A = \int_{-2.414}^{0.414} ((x^3-4x)-(x-2))dx + \int_{0.414}^{2} ((x-2)-(x^3-4x))dx \approx 13.72$$

75.

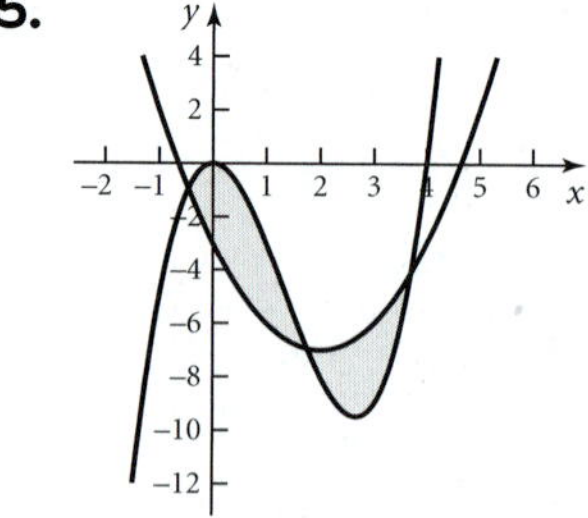

$$A = \int_{-0.4605}^{1.761} ((x^3-4x^2)-(x^2-4x-3))dx + \int_{1.761}^{3.6996} ((x^2-4x-3)-(x^3-4x^2))dx \approx 9.4464$$

77.

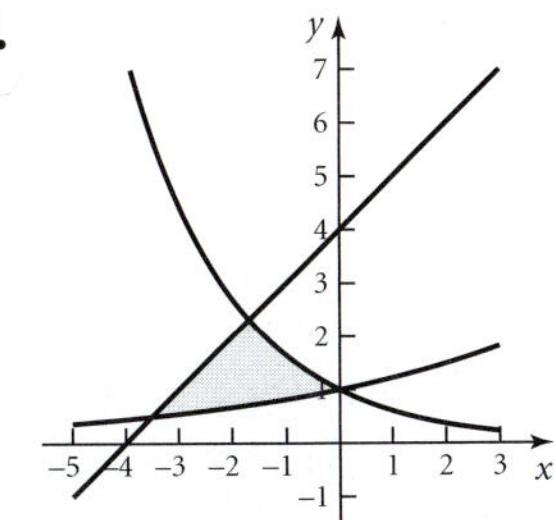

$$A = \int_{-3.504}^{-1.682} (x + 4 - e^{0.2x})dx + \int_{-1.682}^{0} (e^{-0.5x} - e^{0.2x})dx \approx 2.682$$

79. Answers vary. **81.** about \$33.6 thousand

TOPIC 25

1. a. 0.468 or 46.8% **b.** 0.709 or 70.9% **c.** 0.132 or 13.2% **d.** Answers vary.

3. a. $y = 0.72x^{1.757}$ $y = 0.96x^{1.684}$

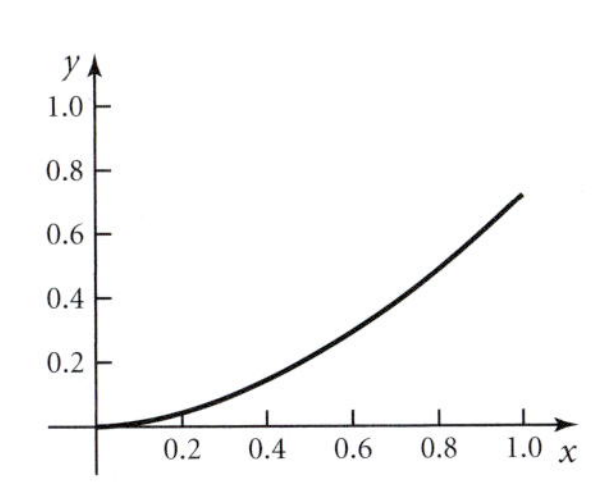

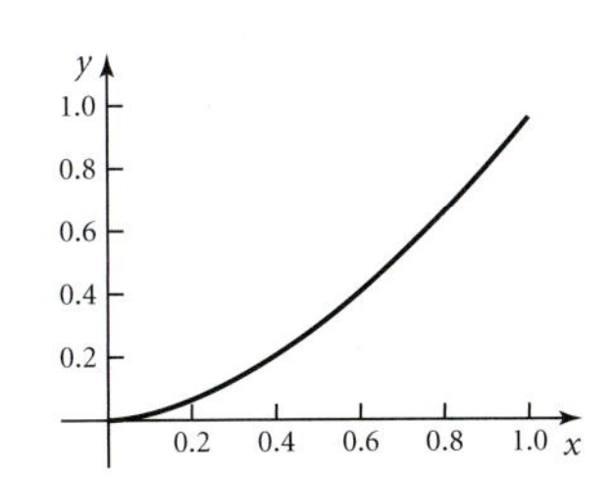

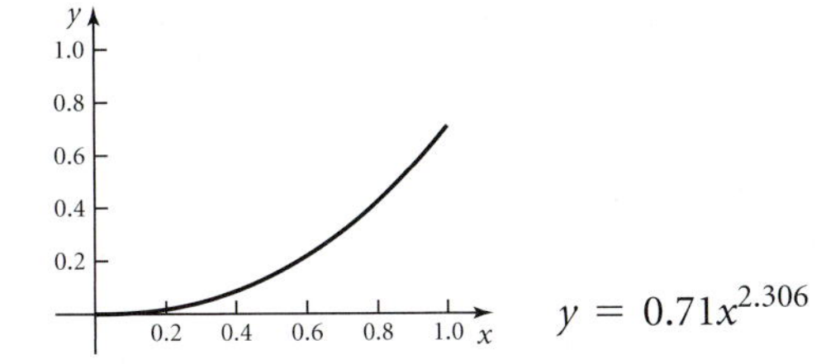

$y = 0.71x^{2.306}$

b. accountants: G = 0.478; lawyers: G = 0.285; doctors: G = 0.570 **c.** lawyers **d.** doctors

5. 0.512 **7.** 0.745 **9.** 0.067 **11.** Bangladesh 0.269; India 0.4; Bangladesh is most equitable.

13. a. before 0.472; after 0.5 **b.** No, the new laws are not more equitable. **15.** \$140 **17.** about \$3482.36 **19.** \$625 **21.** \$18,000 **23.** \$90

25. a. $MS = \int_0^2 ((10.5 - 2q) - (4q - 1.5))dq = \$12$

b. $PS = \int_0^2 (6.50 - (4q - 1.5))dq = \$8;$

$CS = \int_0^2 (10.5 - 2q - 6.5)dq = \$4;$

MS = PS + CS = 8 + 4 = 12, which agrees with the answer to part a

27. a.

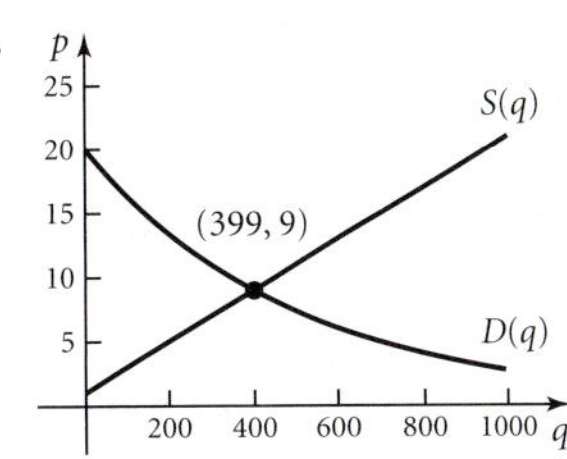

b. $p \approx 9, q \approx 399$ **c.** about \$1906.71 **d.** about \$1594 **e.** about \$3500.72 **29.** 0.238 **31.** 0.549, no **33.** 0.6827 **35.** 0.4987 **37.** 0.9146 **39.** 0.2347 **41.** 0.4934 or 49.34% **43.** 0.8176 or 81.76% **45.** 0.0984 or 9.84% **47.** $P(-1 \le x \le 1) = 0.6827 \approx 68\%$ **49.** $P(-3 \le x \le 3) = 0.9973 \approx 99.7\%$ **51.** .05 or 5% **53.** $P(80 \le x \le 120) = 0.8176$ **a.** 0.0912 **b.** 0.0912 **55.** 0.357 **57.** 0.371 **59.** 0.410 **61.** 0.415 **63.** 1959 **65.** Yes, because as the Gini Index gets larger, the income is more unequal. **67. a.** India $f(x) = 0.857x^{1.484}$; Czechoslovakia $f(x) = 0.893x^{1.365}$; Mexico $f(x) = 0.766x^{1.904}$ **b.** India 0.310; Czechoslovakia 0.245; Mexico 0.472 **c.** Czechoslovakia **d.** Mexico **69. a.** India $f(x) = 3.776x^4 - 7.22x^3 + 5.35x^2 - 1.04x + 0.13$; Czechoslovakia $f(x) = 2.68x^4 - 5.07x^3 + 3.78x^2 - 0.475x + 0.085$; Mexico $f(x) = 6.38x^4 - 12.14x^3 + 8.77x^2 - 2.24x + 0.225$ **b.** India 0.313; Czechoslovakia 0.248; Mexico 0.461 **c.** Czechoslovakia **d.** Mexico **71.** $a \approx 905$ and $a \approx 1147$ (or 900 and 1150) **73.** $f(x) = 0.664x^{2.004}$ **75.** 0.558 **77.** Answers vary.

TOPIC 26

1. $-\frac{x}{2}e^{-2x} - \frac{1}{4}e^{-2x} + C$ **3.** $x\ln x - x + C$

5. $\frac{a^2}{2}\ln(3a) - \frac{1}{4}a^2 + C$ **7.** $\frac{2b}{3}\sqrt{(b-2)^3} - \frac{4}{15}\sqrt{(b-2)^5} + C$ **9.** $\frac{x^4}{4}\ln x - \frac{x^4}{16} + C$

11. $e^x(x^2 - 2x + 2) + C$
13. $\frac{3x}{5}(x-2)^5 - \frac{1}{10}(x-2)^6 + C$
15. $-xe^{-x} - e^{-x} + C$ or $-\frac{x}{e^x} - \frac{1}{e^x} + C$ **17.** 11.287
19. 3.623 **21.** 52.212 **23.** 0.297
25. $\frac{2}{5}\sqrt{(x+8)^5} - \frac{16}{3}\sqrt{(x+8)^3} + C$
27. #10 with a = 2 and b = −3
$\frac{4x+12}{3}\sqrt{2x-3} + C$
29. #8 with a = 3, b = − 2, c = 5, d = 7;
$\frac{1}{31}\left(\frac{7}{5}\ln|5x+7| + \frac{2}{3}\ln|3x-2|\right) + C$
31. #12 with a = 4; $\frac{x}{2}\sqrt{x^2+16} +$
$8\ln|x + \sqrt{x^2+16}| + C$ **33.** #16 with a = 3;
$-\frac{2}{3}\ln\left|\frac{3+\sqrt{9-x^2}}{x}\right| + C$ **35.** 1.3955
37. 0.267 **39.** 116.396
41. $e^{3x}\left(\frac{x^2}{3} - \frac{2}{9}x + \frac{2}{27}\right) + C$
43. $y = -\frac{1}{2}(x-1)e^{-2x} - \frac{1}{4}e^{-2x} + 2 + \frac{1}{4}e^{-2}$ or
$y = -\frac{1}{2}(x-1)e^{-2x} - \frac{1}{4}e^{-2x} + 2.0338$ **45.** about
\$213,883 **47.** 1.434 mg/cm^3 **49.** \$1121.17
51. \$7412.12 **53.** 226.518 **55.** 0.152
57. $x^2\left(\frac{1}{2}(\ln x)^3 - \frac{3}{4}(\ln x)^2 + \frac{3}{4}\ln x - \frac{3}{8}\right) + C$
59. 7.005
61. $(x+2)^{3/2}\left(\frac{2}{9}x^3 - \frac{8}{21}x^2 + \frac{64}{105}x - \frac{256}{315}\right) + C$
63. Answers vary. **65.** Answers vary.
67. Option 1 (\$74,921.91) is better than Option 2 (\$72,154.90).

TOPIC 27

1. $y = \frac{1}{2}x^2 + 4x + C$ **3.** $y = x^3 + \frac{1}{2}x^2 + C$
5. $y = 2\ln|x^2+1| + C$ **7.** $y = -8e^{-x/2} + C$
9. $y = x + 3\ln x + C$ **11.** $y = \frac{2}{5}\sqrt{(x-3)^5} +$
$2\sqrt{(x-3)^3} + C$ **13.** $\frac{dV}{dx} = k\sqrt{x}$ so $V = \frac{2}{3}k\sqrt{x^3}$
15. $\frac{dN}{dt} = \frac{k}{t^2}$ so $N = -\frac{k}{t}$ **17.** $y + 4 = Ce^x$
19. $3x^2 - y^2 = C$ **21.** $2\sqrt{y} = \frac{1}{2}x^2 + C$
23. $2y^2 = 3e^x + C$ **25.** $\frac{y}{x^2+1} = C$
27. $y = \frac{x^2}{8} + 1\frac{7}{8}$ **29.** $y^2 - x^2 = 3$
31. $y = e^{-\frac{x^2}{2}-x}$ **33.** $9y^2 - 4x^2 = 32$
35. a. $P(t) = 8713e^{0.0365t}$ **b.** $P(20) \approx 18{,}080$ people
37. a. $T(t) = 75 + 275e^{-0.061t}$ **b.** about 21 minutes
39. a. $T(t) = 80 - 45e^{-0.027t}$ **b.** about 60°F
41. Answers vary. **43.** $T(t) = T_m + (T_0 - T_m)e^{kt}$
45. $y = Cx$ **47.** $x^2 + 2y^2 = C$

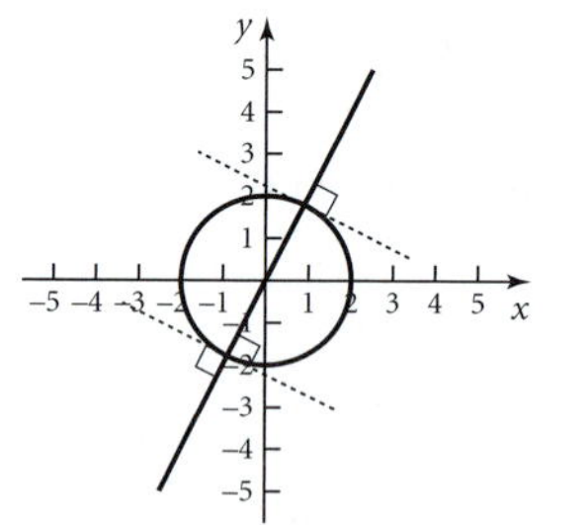
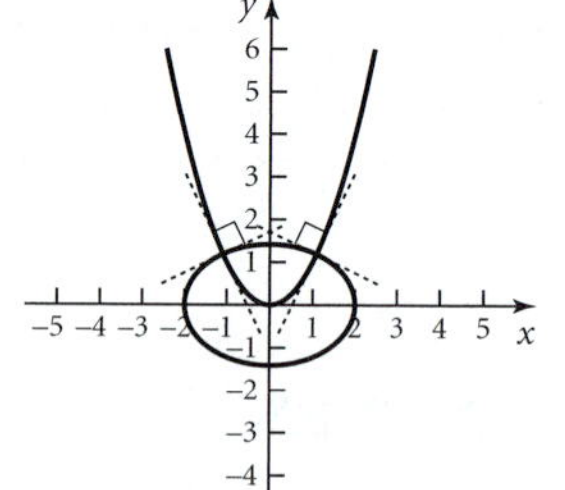

49. $W(t) = 50 - 48e^{-0.5t}$; 50 lb

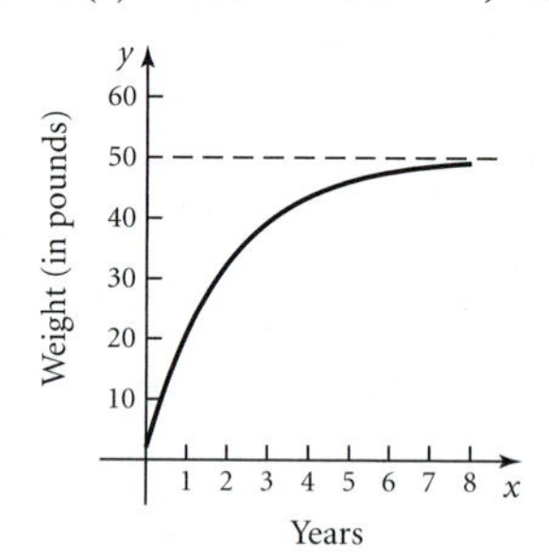

51. $W(t) = 80 - 78e^{-0.9t}$; 80 lb

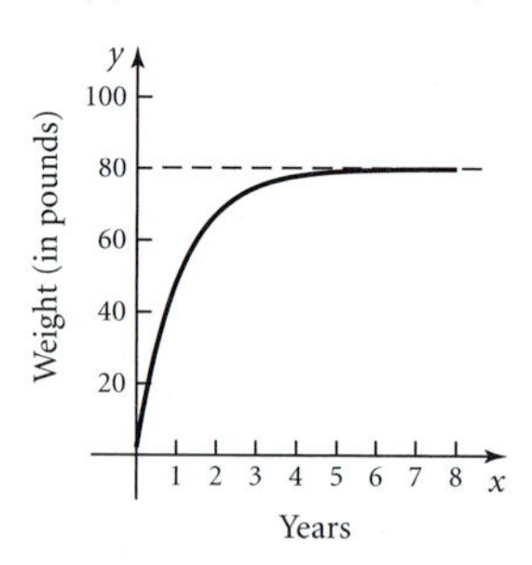

UNIT 3 TEST

1. $\frac{1}{4}x^4 - 2x^2 + 6x + C$ **2.** $-\frac{4}{x^2} + C$
3. $-\frac{1}{3}e^{-3x} + C$ **4.** $\frac{4}{7}x^{7/2} + \frac{1}{3}x^{-3} + C$

5. 104 **6.** $2\ln 2 + 18\frac{2}{3} \approx 20.053$
7. $-4000(e^{-0.25} - 1) \approx 884.8$
8. $6\frac{3}{4} - 3\ln 4 \approx 2.591$ **9.** $f(x) = \frac{7}{2}x^2 + \frac{1}{x^2} - 11\frac{1}{4}$
10. $f(x) = 5e^{0.2x} + 3x - 7$
11. $y = \frac{3}{5}x^{5/3} + x^2 - 2\frac{3}{5}$
12. $y = 3\ln x - 2x^2 + 3$ **13.** $\frac{1}{2}e^{x^2} + C$
14. $-\frac{7}{576} \approx -0.012$ **15.** $\frac{52}{9} \approx 5.778$
16. $\frac{2}{21}(2x^3 - 5)^7 + C$ **17.** $\frac{1}{3}(\ln x)^3 + C$
18. $\frac{1}{5}\ln\frac{242}{31} \approx 0.411$ **19.** $\frac{3}{4}e^4 + \frac{1}{4} \approx 41.199$
20. $\frac{1}{6}(x + 3)^6 - \frac{3}{5}(x + 3)^5 + C$
21. $\frac{x^4}{4}\ln x - \frac{x^4}{16} + C$ **22.** $21\frac{1}{3} \approx 21.33$
23. #10 with a = 2, b = −5; $\frac{x + 5}{3}\sqrt{2x - 5} + C$
24. #6 with a = 2, b = −5; $\frac{x}{2} + \frac{5}{4}\ln|2x - 5| + C$
25. #18 with a = 3; $\frac{1}{6}\ln\left|\frac{3 + x}{3 - x}\right| + C$ **26.** #13 with a = 2; $3\ln\left|x + \sqrt{4 + x^2}\right| + C$ **27.** #20 with a = 3; $\frac{x}{2}\sqrt{x^2 - 9} - \frac{9}{2}\ln\left|x + \sqrt{x^2 - 9}\right| + C$
28. #28 with n = 6; $\frac{x^7}{7}\left(\ln x - \frac{1}{7}\right) + C$
29. #16 with a = 4; $-\frac{3}{2}\ln\left|\frac{4 + \sqrt{16 - x^2}}{x}\right| + C$
30. #31 with n = 1, a = 2, b = 1;
$\frac{2}{5}x\sqrt{(x + 2)^3} - \frac{8}{15}\sqrt{(x + 2)^3} + C$

31. Area $= \int_{-1}^{1}(2 - x^3)dx = 4$ square units

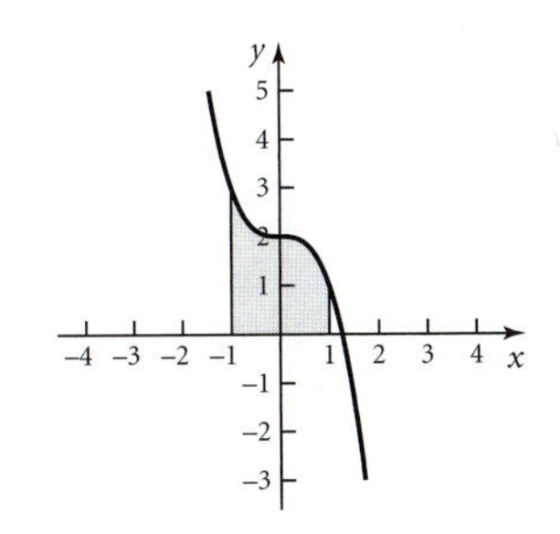

32. Area $= -\int_{0}^{1}(e^x - 3)dx = 4 - e \approx 1.28$

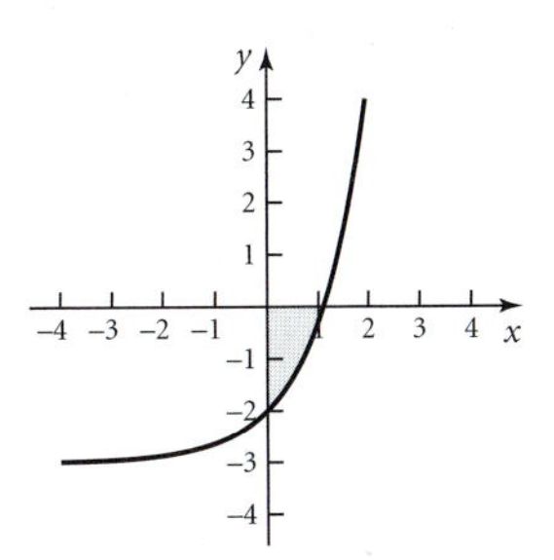

33. Area $= \int_{-2}^{-1}(x^2 - 5x - 6)dx + \int_{-1}^{2}(x^2 - 5x - 6)dx =$ 3.83 + 22.5 = 26.33 square units

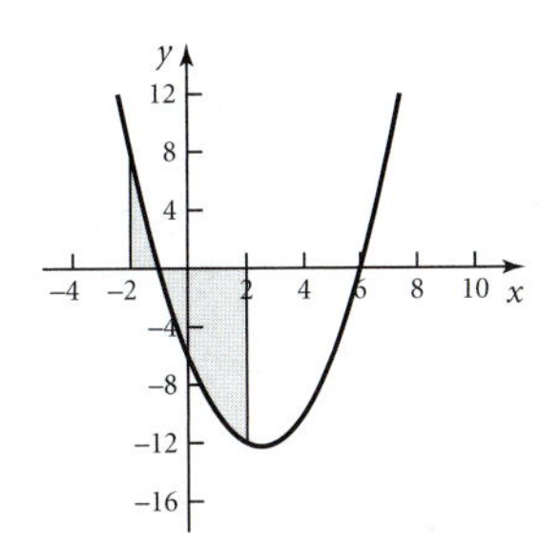

34. Area $= \int_{1}^{4}\left(\frac{2}{x} + x\right)dx = 2\ln 4 + 7.5 \approx 10.273$

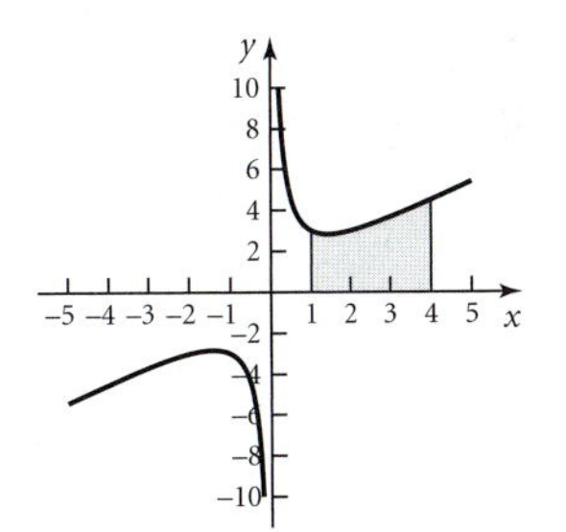

35. Area $= \int_{1}^{3}\left(\frac{2}{x} - (1 - x^2)\right)dx = 8.864$

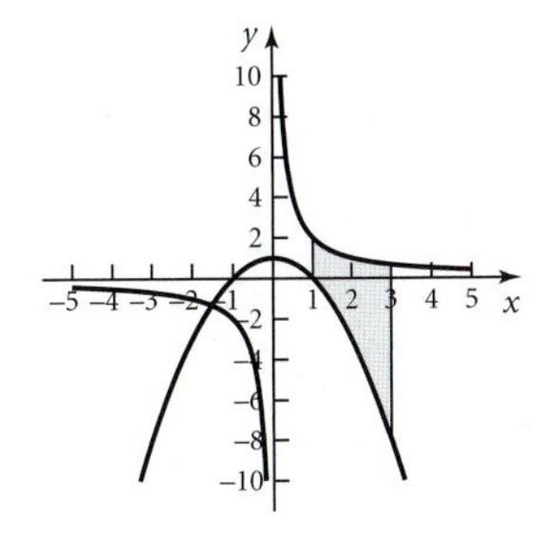

36. Area $= \int_1^5 \Big((x - 11) - (x^2 - 5x - 6)\Big)dx = 10\frac{2}{3}$

Note: $x - 11 = x^2 - 5x - 6$
$x^2 - 6x + 5 = 0 \rightarrow (x - 5)(x - 1) = 0$
points of intersection are $x = 5$ and $x = 1$

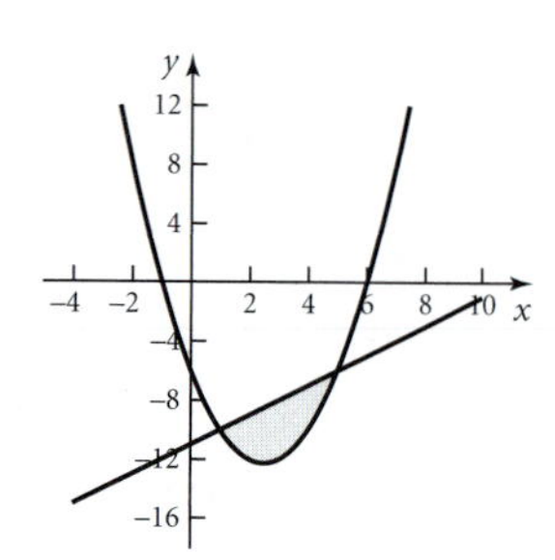

37. $\frac{1}{6 - 2}\int_2^6 (0.076t^3 - 0.528t^2 + 0.28t + 12)dt = 10.048$ or 10,048 students each year

38. $\frac{1}{6}\int_0^6 2te^{-0.3t^2}dt \approx 0.556$ mg/cm³ per hour

39. 1848 cases **40.** 47,357 books **41.** 4214 cases **42.** 17,329 books **43.** \$21.4 million **44.** about \$24,525.71 **45.** 0.487 **46.** 0.250

47. a. accountants 0.458, lawyers 0.460, doctors 0.571 **b.** accountants **c.** doctors **48.** about \$603.57

49. about \$5666.74 **50.** about \$1117.44

51. 0.25 **52.** 0.656

53. $y = 5x - \frac{3}{2}x^2 + C$ **54.** $y = \ln(x^2 + 3) + C$

55. $y = Ce^{4x}$ **56.** $y = Ce^{x^2} - 3$ **57.** $\frac{y}{x} = C$

58. $2y^3 - x^3 - 18x = C$ **59.** $\frac{y}{(x + 1)^2} = \frac{3}{4}$

60. $y = (\ln x)^2 + 2$ **61.** $P'(t) = 0.0345(1.6836)^t$ or $P'(t) = 0.0345e^{0.521t}$ **62.** about 15.25 or 15,250 patients per year **63.** $P(t) = 0.0662e^{0.521t}$

64. 29.260 or about 29,260 patients each year

Unit 4

TOPIC 28

1. 4 **3.** $-3\frac{8}{9} \approx -3.89$ **5.** $\frac{32{,}000\pi}{3} \approx 33{,}510.32$

7. $e^2 - 2 + \ln 3 \approx 6.488$

9. $2000\left(1 + \frac{0.04}{4}\right)^{24} = 2539.47$

11. 75 **12.** $-\frac{141}{32} = -4.40625$

13. all ordered pairs (x,y)

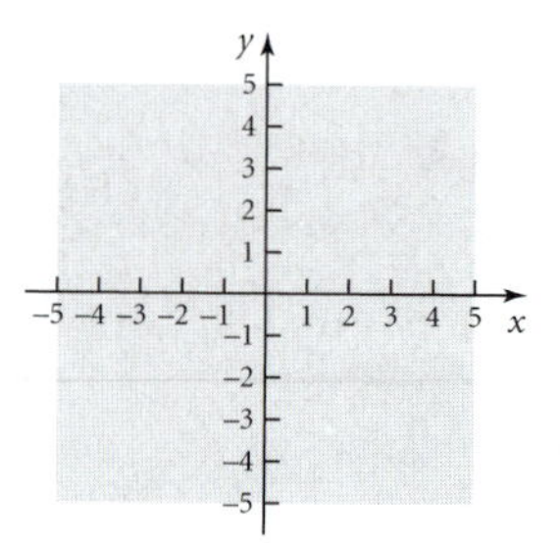

15. all (x, y) such that $y \neq x$

17. all (x, y) such that $y \leq x^2$

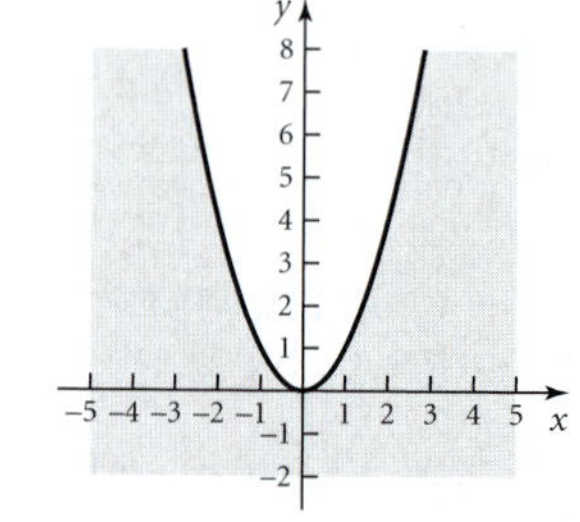

19. all (x, y) such that $y > -2$

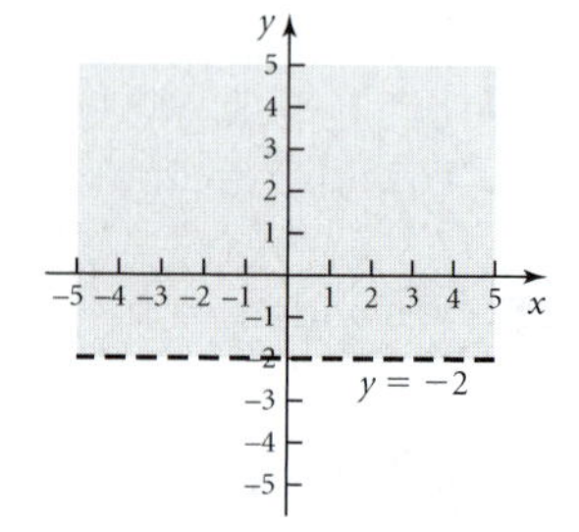

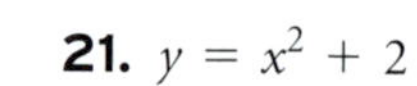
21. $y = x^2 + 2$

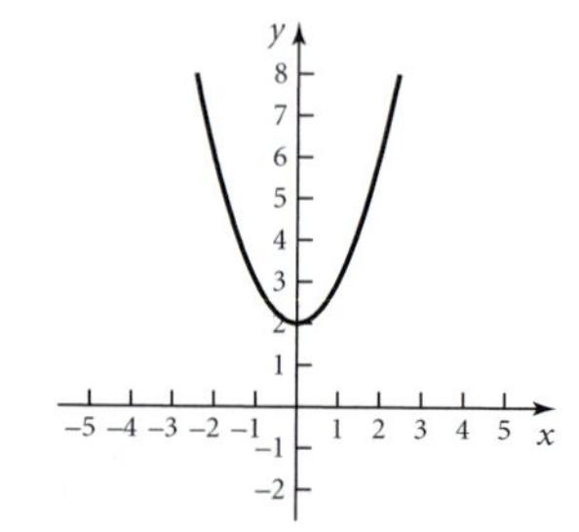

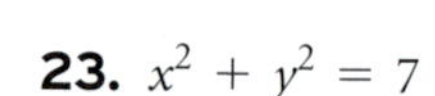
23. $x^2 + y^2 = 7$

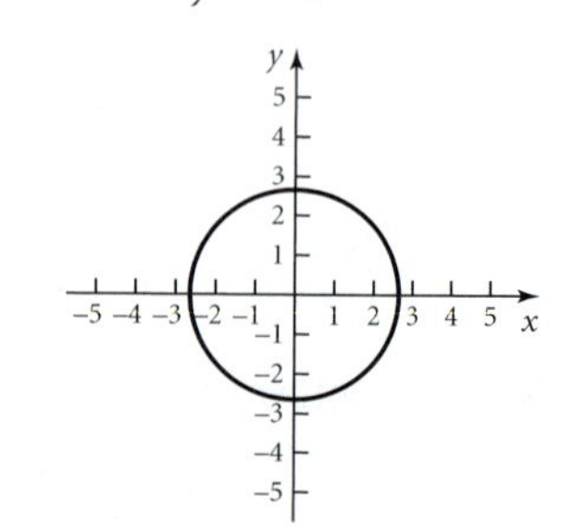

25. Level curves are the family of hyperbolas $y = \frac{k}{x}$.

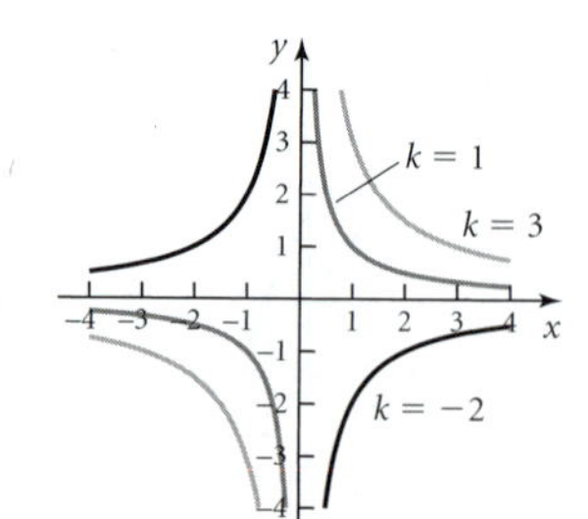

27. Level curves are the family of lines $2x + 3y = k$.

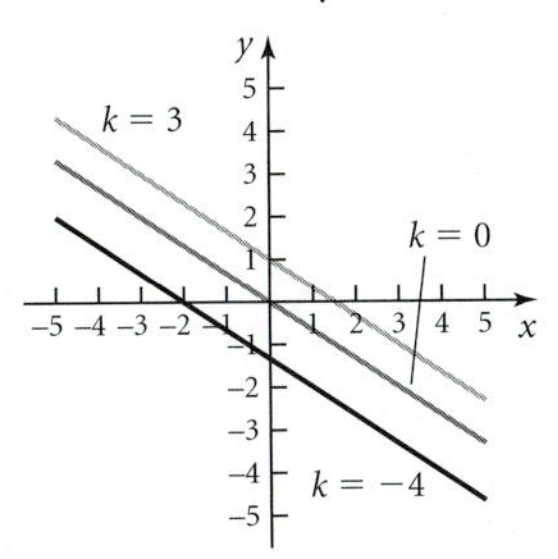

29. d **31.** a **33.** f **35. a.** $C(x, y) = 12x + 19y + 1950$ **b.** \$3858. The cost of grooming 45 small dogs and 72 large dogs is \$3858.
37. a. $R(a, b) = 300a - 5a^2 + 25ab + 125b - 15b^2$ **b.** \$181,500 **39.** $T(90,30) \approx 47$ minutes; $T(80, 40) \approx 36$ minutes **41.** about 55 rackets **43.** about 26,265 items **45. a.** 92.6 **b.** 104.7 **c.** 110.9 **d.** 121.1 **e.** 92.4 **47. a.** about 15°F, if air temperature is 25°F and wind speed is 10 mph, the temperature feels like 15°F. **b.** about −26°F. If air temperature is −5°F and wind speed is 15 mph, the temperature feels like −26°F. **c.** about 12°F **d.** about 27.5 mph **49. a.** $Q(x, y) = 0.23x^{0.13}y^{0.87}$ **b.** about 230 machines **51. a.** \$1538.27 for the 30-year mortgage at 7.5%; \$1946.80 for the 15-year mortgage at 6.75% **b.** \$553,777.20 for the 30-year mortgage; \$350,424 for the 15-year mortgage **c.** \$333,777.20 for the 30-year mortgage; \$130,424 for the 15-year mortgage **d.** The 15-year loan has a higher monthly payment but pays back much less interest over the length of the loan.
53. Graph $y = \dfrac{0.31498}{x^{2/3}}$ (for $k = 3$), $y = \dfrac{3.1748}{x^{2/3}}$ (for $k = 12$), and $y = \dfrac{1}{x^{2/3}}$ (for $k = 6$).

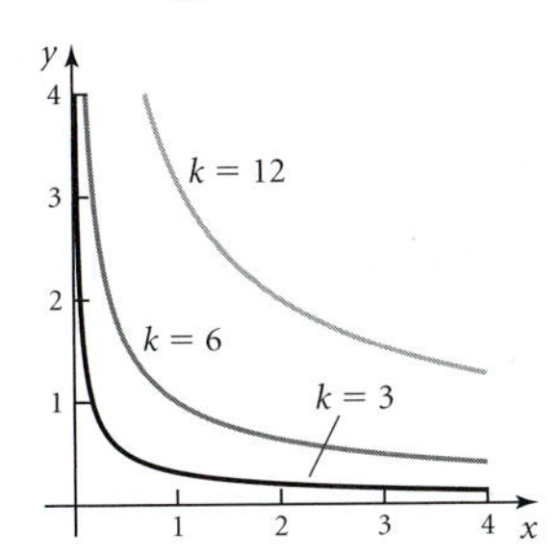

TOPIC 29

1. $f_x = 5$ and $f_y = -2$

3. $f_x = 3x^2 - 2y$ and $f_y = -2x + 8y$

5. $f_x = \dfrac{y}{2\sqrt{x}}$ and $f_y = \sqrt{x}$ **7.** $f_x = ye^{xy}$ and $f_y = xe^{xy}$ **9.** $f_x = y^2e^{xy}$ and $f_y = e^{xy}(1 + xy)$

11. $f_x = \dfrac{x}{\sqrt{x^2 - y^2}}$, $f_y = -\dfrac{y}{\sqrt{x^2 - y^2}}$

13. $f_x = \dfrac{2}{2x - 3y}$ and $f_y = -\dfrac{3}{2x - 3y}$

15. $f_x = \dfrac{4x^3}{3y} + \dfrac{2y^2}{5x^2}$ and $f_y = -\dfrac{x^4}{3y^2} - \dfrac{4y}{5x}$

17. $f_x = -\dfrac{2y^2}{(x - y)^2}$, $f_y = \dfrac{2x^2}{(x - y)^2}$

19. $f_x = \dfrac{4y^3}{4x - 3}$, $f_y = 3y^2 \ln(4x - 3)$

21. $f_x(1, 1, 5) = -2$ and $f_y(1, 1, 5) = -6$

23. $f_x(3, 0, 9) = 6$, $f_y(3, 0, 9) = 27$

25. $x = -2\frac{3}{4}$, $y = -\frac{1}{4}$ **27.** $x = -1, y = -1$

29. $f_{xx} = 4, f_{yy} = 2, f_{xy} = f_{yx} = -5$

31. $f_{xx} = \dfrac{-y^2}{\sqrt{(x^2 - y^2)^3}}$, $f_{yy} = \dfrac{-x^2}{\sqrt{(x^2 - y^2)^3}}$, $f_{xy} = f_{yx} = \dfrac{xy}{\sqrt{(x^2 - y^2)^3}}$

33. $f_{xx} = \dfrac{-2y^2 - 2x^2}{(x^2 - y^2)^2}$, $f_{yy} = \dfrac{-2x^2 - 2y^2}{(x^2 - y^2)^2}$, $f_{xy} = f_{yx} = \dfrac{4xy}{(x^2 - y^2)^2}$ **35.** $f_{xx} = 2e^y, f_{yy} = x^2e^y$, $f_{xy} = f_{yx} = 2xe^y$ **37.** $f_{xx} = f_{yy} = 0$. $f_{xx} + f_{yy} = 0$, so $f(x,y) = 8xy$ satisfies Laplace's equation.
39. $f_{xx} = -ye^x$ and $f_{yy} = xe^y$, so $f_{xx} + f_{yy} \neq 0$ and $f(x, y) = xe^y - ye^x$ does not satisfy Laplace's equation.
41. a. $S_r = 4\pi r + 2\pi h$ and $S_h = 2\pi r$. **b.** $S_r(4, 10) = 36\pi$. If the radius is 4 inches and the height is 10 inches, the surface area is increasing at a rate of 36π square inches per unit increase in the radius. $S_h(4, 10) = 8\pi$. If the radius is 4 inches and the height is 10 inches, the surface area is increasing at a rate of 8π square inches per unit increase in the height. **43. a.** $C_x = 8x$ and $C_y = 5$. **b.** $C_x(20, 35) = 160$. If \$20,000 is spent on advertising and \$35,000 on mailings, the cost increases \$160,000 if advertising costs increase \$1000. $C_y(20, 35) = 5$. If \$20,000 is spent on advertising and \$35,000 on mailings, the cost increases \$5000 if mailing costs increase \$1000. **45.** $T(90, 30) \approx 47$, $T_V(90, 30) \approx 0.52$, and $T_D(90, 30) \approx -0.75$. If the volume of air is 90 and the depth of the dive is 30 feet, the dive lasts

about 47 minutes. The dive time at that depth increases by about 0.5 minutes if the volume of air in the tank increases by one unit. The dive time decreases by 0.75 minutes at that volume if the depth of the dive is increased by one foot. **47. a.** $W(15, 10) = 2.7$. If air temperature is 15°F, and wind velocity is 10 mph, it feels like about 3°F. **b.** $W_T(15, 10) = 1.2$. If wind velocity is 10 mph and air temperature is 15°F, the wind chill increases about 1.2 degrees if the temperature increases 1 degree. **c.** $W_v(15, 10) = -0.7$. If wind velocity is 10 mph and air temperature is 15°F, the wind chill decreases about 1 degree if the wind velocity increases 1 mph. **49. a.** $Q_x = 3x^{-0.75}y^{0.75}$ and $Q_y = 9x^{0.25}y^{-0.25}$. **b.** Marginal productivity of labor is 238. Output increases by 238,000 units if labor increases by 1000 employees and capital is fixed at $750,000. Marginal productivity of capital is 2.09. Output increases by 2090 units if labor is fixed at 2200 employees and capital is increased by $1000. **c.** Output increases more rapidly with $250,000 additional capital. **51. a.** $Q_x = 1.008x^{-0.28}y^{0.28}$ and $Q_y = 0.392x^{0.72}y^{-0.72}$. **b.** Marginal productivity of labor is 2.05. Output increases by 205 units if labor increases by 100 employees and capital is fixed at $820,000; marginal productivity of capital is 0.063. Output increases by six units if labor is fixed at 6500 employees and capital is increased by $1000. **c.** Output increases more rapidly if $100,000 additional capital is used. **53. a.** $R(a, b) = 500a - 0.4a^2 + 1.6ab + 425b - 0.2b^2$ **b.** $R(925, 1475) = \$2{,}495{,}000$. If the freezer-on-top model is $925 and the side-by-side model is $1475, revenue is $2,495,000; $R_a(925, 1475) = \$2120$. If the freezer-on-top model is $925 and the side-by-side model is $1475, revenue is increasing at a rate of $2120 per dollar increase in the price of the freezer-on-top model; $R_b(925, 1475) = \$1315$. If the freezer-on-top model is $925 and the side-by-side model is $1475, revenue is increasing at a rate of $1315 per dollar increase in the price of the side-by-side model. **55. a.** $P(a, b) = 500a - 0.4a^2 + 1.6ab + 425b - 0.2b^2 - \dfrac{a+b}{2} - 8\sqrt{a+b}$ **b.** $P(925, 1475) = \$2{,}493{,}408$. If the freezer-on-top model is $925 and the side-by-side model is $1475, profit is $2,493,408; $P_a(925, 1475) = \$2119.42$. If the freezer-on-top model is $925 and the side-by-side model is $1475, profit is increasing at a rate of $2119.42 per dollar increase in the price of the freezer-on-top model; $P_b(925, 1475) = \$1314.42$. If the freezer-on-top model is $925 and the side-by-side model is $1475, profit is increasing at a rate of $1314.42 per dollar increase in the price of the side-by-side model. **57.** $f_{xx} = e^{xy}(4xy + 2 + x^2y^2), f_{yy} = x^4e^{xy}, f_{xy} = f_{yx} = x^2e^{xy}(xy + 3)$ **59.** $f_{xx} = \dfrac{12y^2}{(2x - 3y)^3}, f_{yy} = \dfrac{12x^2}{(2x - 3y)^3}, f_{xy} = f_{yx} = \dfrac{-12xy}{(2x - 3y)^3}$ **61.** $V_a = 6a^2 + 6ab, V_b = 3a^2 + 4bc, V_c = 2b^2 + 3c^2$ **63.** $f_{xxx} = 6y, f_{yxy} = 8x$ **65. a.** $C_x = 79.62$, $Cy = 153{,}46$. The cost of one additional freezer-on-top model is $79.62. The cost of one additional side-by-side model is $153.46. **b.** The side-by-side model causes the price to rise by $153.46, while the freezer-on-top model causes the price to rise by $79.62. **67. a.** $M_P = \dfrac{i}{1 - (1 + i)^{-360}}$ **b.** $M_P\left(220{,}000, \dfrac{0.07}{12}\right) \approx 0.00665$. For a $220,000 loan at 7%, the monthly payment increases at a rate of about $0.00665 per $1 increase in the amount borrowed. **c.** about $33.25

TOPIC 30

1. critical point is $(0, 0)$; $(0, 0, 4)$ is a relative maximum **3.** critical point is $(0, 0)$; $(0, 0, 0)$ is a saddle point **5.** critical point is $(-2, 1)$; $(-2, 1, 0)$ is a relative minimum **7.** critical point is $\left(1, -\frac{1}{2}\right)$; $\left(1, -\frac{1}{2}, \frac{3}{4}\right)$ is a saddle point **9.** critical points are $(0, 0)$ and $\left(\frac{2}{3}, \frac{2}{3}\right)$; $(0, 0, 0)$ is a saddle point; and $\left(\frac{2}{3}, \frac{2}{3}, -\frac{8}{27}\right)$ is a relative minimum **11.** critical point is $(0, 0)$; $(0, 0, 1)$ is a relative minimum **13.** critical point is $(0, 0)$; $(0, 0, 1)$ is a relative maximum **15.** critical point is $\left(\frac{1}{2}, -2\right)$; $\left(\frac{1}{2}, -2, -12\right)$ **17.** critical points are $(0, 0)$ and $\left(-\frac{1}{3}, -\frac{1}{3}\right)$; $(0, 0, 0)$ is a saddle point; and $\left(-\frac{1}{3}, -\frac{1}{3}, \frac{1}{27}\right)$ is a relative maximum **19.** Maximum is 1600 if $x = 20$ and $y = 20$. **21.** Minimum is $360\sqrt{5} \approx 804.98$ if $x = 9\sqrt{5}$ and $y = 12\sqrt{5}$. **23.** 571 tents from the Atlanta plant and 2429 tents from the Kansas City plant **25.** 250 feet by 250 feet **27.** The base of the carton is 8 feet by 8 feet and the height is 10 feet. **29.** $x = 4, y = \frac{4}{3}$ **31.** 2000 tents from the Atlanta plant and 1000 tents from the Kansas City plant **33. a.** $P(x, y) = 10x + 6y - 0.04x^2 - 0.06y^2 - 505$ **b.** 125 boats at the Tampa plant and 50 boats at the Virginia Beach plant **35.** about $43,636 to development and about $36,364 to marketing **37.** Point P should be located about 2781.8 feet from

Building A. **39. a.** Area is maximized if $a = b = c$. **b.** If the perimeter is fixed at 60 cm, each side is 20 cm and the area is $100\sqrt{3}$ or about 173.205 square cm.

TOPIC 31

1. $xy + \dfrac{y^2}{2} + C(x)$ **3.** $\dfrac{x^3}{3} - xy^2 + C(y)$

5. $\dfrac{x^3e^{2y}}{2} + C(x)$ **7.** $\dfrac{3\ln x}{y} + C(y)$ **9.** 21

11. $\frac{10}{3}$ **13.** $\frac{9}{2}(e^2 - 1) \approx 28.751$

15. $3\ln\frac{5}{4} \approx 0.669$ **17.** $-9 + 24\ln 4 \approx 24.27$

19. $\frac{1}{2}e - 1 + \frac{1}{2}e^{-1} \approx 0.543$ **21.** $-\frac{1}{15}$

23. $e - 1 \approx 1.718$

25. $\displaystyle\int_0^8\int_{-2}^{2-y/2} f(x,y)\,dx\,dy$

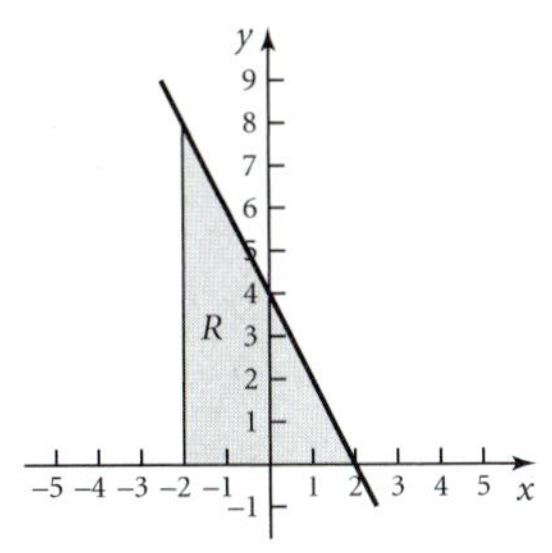

27. $\displaystyle\int_0^4\int_{y/2}^{\sqrt{y}} f(x,y)\,dx\,dy$ **29.** $\displaystyle\int_0^4\int_0^{\sqrt{x}} f(x,y)\,dy\,dx$

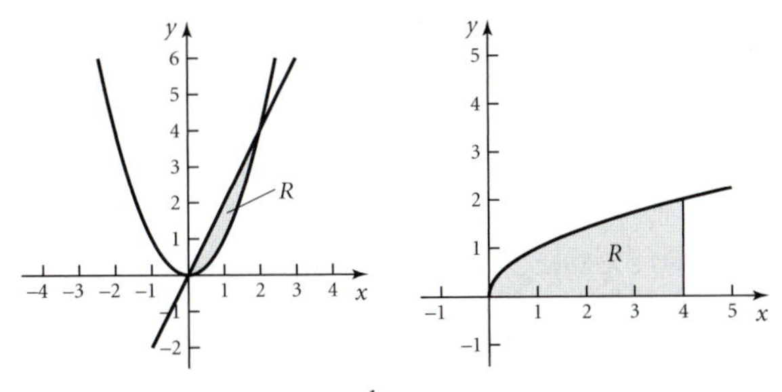

31. 4 cubic units **33.** $17\frac{1}{15}$ cubic units

35. $10\frac{2}{3}$ cubic units **37.** $\bar{f} = \frac{1}{8}\displaystyle\int_0^4\int_0^2 x^2\,dy\,dx = \frac{16}{3}$

39. $\bar{f} = \frac{2}{9}\displaystyle\int_{-1}^2\int_1^{x+2}(x^2 + y^2)\,dy\,dx = 6$

41. about \$411,666.67 **43.** about 5181 boats **45.** $\frac{1}{2}(e - 1) \approx 0.859$ **47.** 120 **49.** 0.1

51. $85\frac{1}{3}$

UNIT 4 TEST

1. -53 **2.** about 7503.69 **3.** $\sqrt{42}$ **4.** 14
5. all (x, y) such that $x^2 + y > 0$ **6.** all (x, y) such that $y \le \dfrac{x}{3}$

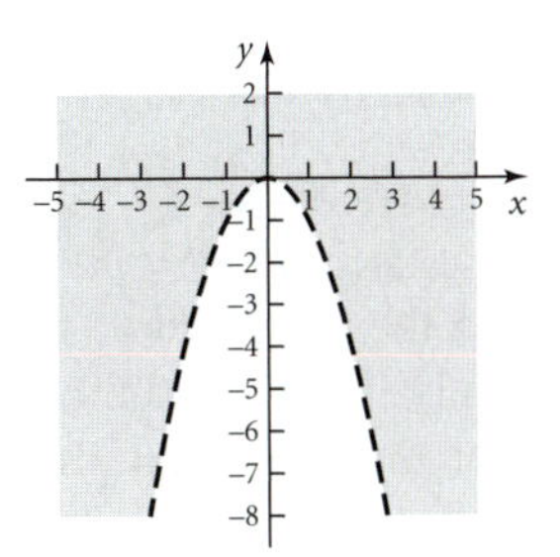

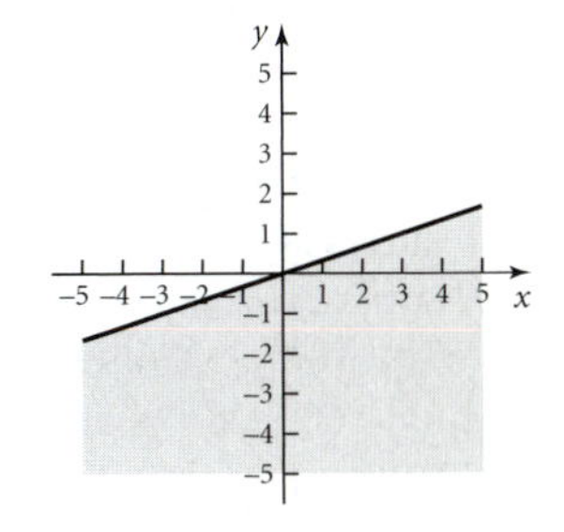

7. $x^2 + y^2 = 4$ **8.** $y = e^{-x} - 1$

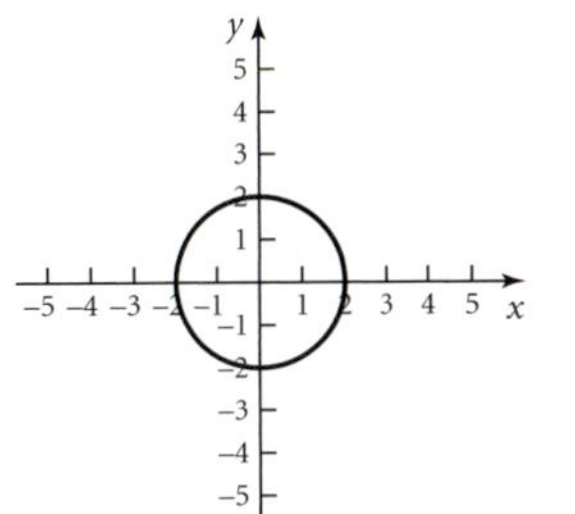

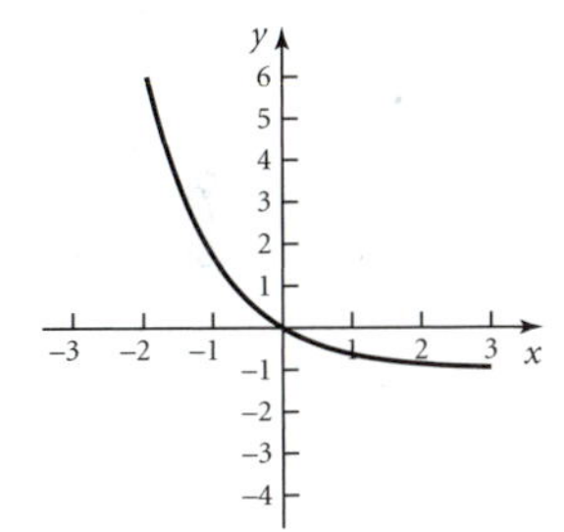

9. The level curves are the family of circles $x^2 + y^2 = C$, where $C = 6 - k^2$. **10.** The level curves are the family of hyperbolas $y = \dfrac{k}{6x}$. **11. a.** $R(a, c) = 5a + 2c$ **b.** $R(80, 64) = 528$. If 80 adult tickets and 64 children's tickets are sold, the revenue is \$528. **c.** $C(a, c) = 3.25a + 1.50c + 275$ **d.** $C(80, 64) = 631$. If 80 adult tickets and 64 children's tickets are sold, the cost is \$631. **e.** $P(a, c) = 1.75a + 0.50c - 275$ **f.** $P(80, 64) = -103$. If 80 adult tickets and 64 children's tickets are sold, there will be a loss of \$103. **g.** 100 adults and 200 children
12. a. $R(x, y) = -3x^2 - 40y^2 + 200x + 250y + 14xy$
b. $R(10, 4) = 2620$ **13.** about 2853 units
14. $f_x = -e^{-x}$ and $f_y = -1$. **15.** $f_x = 2xye^{x^2y}$ and $f_y = x^2e^{x^2y}$. **16.** $f_x = \dfrac{-2}{3y - 2x}$ and $f_y = \dfrac{3}{3y - 2x}$.
17. $f_x = -\dfrac{x}{\sqrt{6 - x^2 - y^2}}$ and $f_y = -\dfrac{y}{\sqrt{6 - x^2 - y^2}}$.
18. $f_x = \dfrac{18y^2}{(x + 3y)^2}$ and $f_y = \dfrac{6x^2}{(x + 3y)^2}$.
19. $f_x = x^2e^{2x/y}\left(\dfrac{2x}{y} + 3\right)$ and $f_y = \dfrac{-2x^4e^{2x/y}}{y^2}$.
20. $f_x(1, 1, 4) = -2$ and $f_y(1, 1, 4) = -2$.
21. $f_x(2, -1, 12) = 18$ and $f_y(2, -1, 12) = 24$.

22. $f_{xx} = e^{-x}, f_{yy} = 0, f_{xy} = f_{yx} = 0$

23. $f_{xx} = \dfrac{-4}{(3y-2x)^2}, f_{yy} = \dfrac{-9}{(3y-2x)^2},$ $f_{xy} = f_{yx} = \dfrac{6}{(3y-2x)^2}$

24. $f_{xx} = y^4 e^{xy^2},$ $f_{yy} = 2xe^{xy^2}(2xy^2 + 1), f_{xy} = f_{yx} = 2ye^{xy^2}(xy^2 + 1)$

25. a. $A_x = 2r$ and $A_r = 2x + 2\pi r$. **b.** $A_x(500, 60) = 120$. If the length is 500 feet and the radius is 60 feet, the area increases at a rate of 120 square feet per 1 foot increase in the length; $A_r(500, 60) \approx 1376.99$. If the length is 500 feet and the radius is 60 feet, the area increases at a rate of about 1377 square feet per 1 foot increase in the radius.

26. a. $Q_x = 54x^{-0.4}y^{0.4}$ and $Q_y = 36x^{0.6}y^{-0.6}$. **b.** Marginal productivity of labor is 34.234; with 50 employees and 16 machines, output increases at a rate of about 34 units per additional one employee; Marginal productivity of capital is 71.32; with 50 employees and 16 machines, output increases at a rate of about 71 units per additional one machine. **c.** At the current level, production will increase more rapidly with five more employees.

27. a. $R(x, y) = -3x^2 - 40y^2 + 200x + 250y + 14xy$ **b.** $R_x = -6x + 200 + 14y$ and $R_y = -80y + 250 + 14x$. **c.** $R_x(10, 4) = 196$. If adult tickets are \$10 and children's tickets are \$4, revenue increases at a rate of \$196 per \$1 increase in adult ticket prices; $R_y(10, 4) = 70$. If adult tickets are \$10 and children's tickets are \$4, revenue increases at a rate of \$70 per \$1 increase in children's ticket prices.

28. a. $P(x, y) = -6x^2 - 41y^2 + 200x + 250y + \dfrac{111xy}{8}$ **b.** $P_x = -12x + 200 + \dfrac{111y}{8}$ and $P_y = -82y + 250 + \dfrac{111x}{8}$. **c.** $P_x(10, 4) = 135.5$. If adult tickets are \$10 and children's tickets are \$4, profit increases at a rate of \$135.50 per \$1 increase in adult ticket prices; $P_y(10, 4) = 60.75$. If adult tickets are \$10 and children's tickets are \$4, profit increases at a rate of \$60.75 per \$1 increase in children's ticket prices.

29. Critical point is $(0, 0)$ and $(0, 0, 6)$ is a relative maximum. **30.** Critical point is $(-3, 2)$ and $(-3, 2, -5)$ is a relative minimum. **31.** Critical points are $(0, 0)$ and $(2, 2)$. $(0, 0, 0)$ is a saddle point, and $(2, 2, -8)$ is a relative minimum. **32.** Critical point is $(-\frac{1}{2}, -2)$ and $(-\frac{1}{2}, -2, -12)$ is a relative maximum. **33.** Critical point is $(0, 0)$ and $(0, 0, 1)$ is a saddle point. **34.** Maximum is 6400 if $x = 20$ and $y = 40$. **35.** Minimum is approximately 1131.35 if $x \approx 14.14$ and $y \approx 22.63$. **36.** Maximum is about 124.766 if $x = 2$ and $y = \frac{4}{5}$. **37.** about 84 boats from the Virginia yard and 36 boats from the California yard. **38.** about \$33,696 for the Windsurfer and \$30,435 for the Pelican. **39.** $\frac{1}{3}x^3e^{3y} + C(y)$ **40.** $-\dfrac{6}{xy} + C(x)$

41. 52.5 **42.** 21.315 **43.** $\frac{11}{3}\ln 4 + 4 \approx 9.083$

44. 4.5 **45.** $\dfrac{8}{35} \approx 0.2286$ **46.** $12 \ln \frac{5}{4} \approx 2.678$

47. $\displaystyle\int_0^{\ln 2}\int_{e^y}^{2} f(x, y)\,dx\,dy$ **48.** $\displaystyle\int_0^{2}\int_{x^2}^{2x} f(x, y)\,dy\,dx$

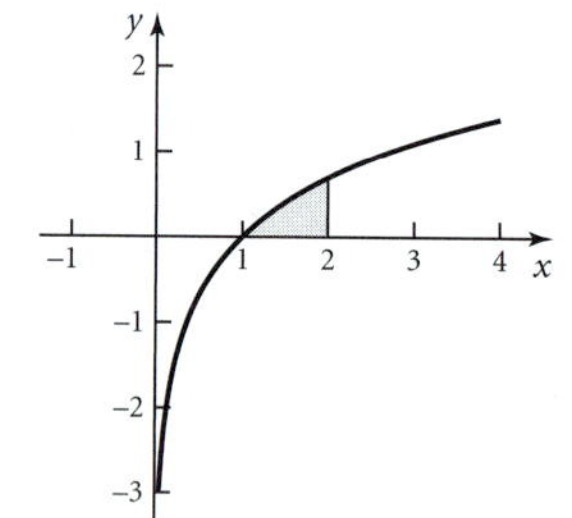

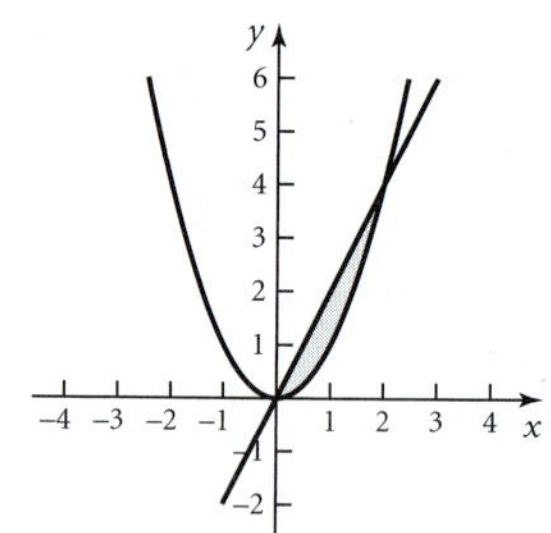

49. $4\frac{2}{3}$ cubic units **50.** $\dfrac{844}{15} \approx 56.267$ cubic units

51. $A = 2$ and $\bar{f} = \dfrac{1}{2}\displaystyle\int_0^2\int_0^1 (4x + y)\,dy\,dx = 4.5.$

52. $A = \dfrac{7}{3}$ and $\bar{f} = \dfrac{3}{7}\displaystyle\int_1^2\int_0^{x^2} (x^2 + y^2)\,dy\,dx = \dfrac{1286}{245} \approx 5.249.$ **53.** $A = 20{,}000$ and

$\overline{Q} = \dfrac{1}{20{,}000}\displaystyle\int_{650}^{850}\int_{500}^{600} 30x^{0.7}y^{0.3}\,dy\,dx \approx 20{,}482$ boats.

Credits

UNIT 0

Page 1, Lighthouse (Unit and Topic Openers throughout); ©Digital Vision/Getty.
Page 1, (Coffee beans); ©Royalty-Free/Corbis.
Page 1, (Starbucks sign); ©Gero Breloer/dpa/Corbis.
Page 2, ©Gero Breloer/dpa/Corbis.
Page 5, ©PhotoDisc Red.
Page 14, ©PhotoDisc Red.
Page 37, ©PhotoDisc (PP).
Page 42, ©Digital Vision.
Page 62, ©PhotoDisc Red.
Page 64, ©Corbis RF.
Page 82, ©Stockbyte Platinum RF.
Page 85, ©Digital Vision.

UNIT 1

Page 97, (Orlando Magic team playing); ©Gary Bogdon/NewSport/Corbis.
Page 97, (Magic player, jersey number 10); ©Leo Dennis/NewSport/Corbis.
Pages 98, 99, 100, (Close-up of Magic player, jersey number 10); ©Gary Bogdon/NewSport/Corbis.
Page 98, (Leibnitz); ©Granger Collection (PAL).
Page 98, (Newton); ©Library of Congress (PAL).
Page 102, ©Fundamental Photos (PAL).
Page 154, ©NBAE/Getty Images.
Page 155, ©Corbis RF.
Page 184, ©Blend Images Getty RF.
Page 200, ©Beth Anderson.
Page 229, ©Corbis RF.
Page 238, ©PhotoDisc Red.
Page 242, ©PhotoDisc.
Page 245, ©Stockbyte Silver GettyRF.
Page 255, ©PhotoDisc.
Page 258, ©StockDisc CD SD174.
Page 279, ©PhotoDisc.

UNIT 2

Page 293, (Comcast van); ©George Wideman/Associated Press.
Page 293, (Televisions); ©Getty Images.
Page 294, ©Getty Images.
Page 308, ©MedioImages RF.
Page 309, ©PhotoDisc Red (PP).
Page 334, ©PhotoDisc.
Page 350, ©Getty Images (Editorial).
Page 370, ©Beth Anderson.
Page 380, ©Dex Image Getty RF.
Page 381, ©Corbis RF.
Page 383, ©Corbis RF.
Page 397, ©Corbis RF.

UNIT 3

Page 409, (Inhaler); ©Stockbyte/Getty Images.
Page 409, (Insulin testing); ©Royalty-Free/Corbis.
Page 418, ©Digital Vision.
Page 430, ©Blend Image/Getty RF.
Page 431, ©Rubberball Productions.
Page 443, ©Stockbyte Silver/Getty RF.
Page 456, ©Digital Vision.
Page 478, ©The Bridgeman Art Library International (PAL).
Page 480, ©The Granger Collection.
Page 516, ©PhotoDisc.

UNIT 4

Page 547, (Achievement test); ©Jen Hulshizer/Star Ledger/Corbis.
Page 547, (Manager in factory); ©Edward Rozzo/Corbis.
Page 547, (Factory); ©Steve Hamblin/Corbis.
Page 548, ©Thomas A. Heinz/Corbis.
Page 573, (Students with calculators); ©PhotoEdit.
Page 573, (Surfboards for sale); ©Rubberball Productions/GettyRF.
Page 591, ©Library of Congress (PAL).
Page 592, ©Beth Anderson.

Index

X

Y